Cultural and Spiritual Values of Biodiversity

United Nations Environment Programme

A Complementary Contribution to the Global Biodiversity Assessment

Intermediate Technology Publications

Cultural and Spiritual Values of Biodiversity

United Nations Environment Programme

Compiled and Edited by

Darrell Addison Posey
Departamento de Ciências Biológicas,
Universidade Federal do Maranhão, São Luís, Maranhão, Brazil
and
Oxford Centre for the Environment, Ethics and Society
Mansfield College, University of Oxford, UK

Chapter Co-ordinators

Chapter 1: Darrell Addison Posey * Chapter 2: Luisa Maffi
Chapter 3: Andrew Gray * Chapter 4: Ranil Senanayake
Chapter 5: L. Jan Slikkerveer * Chapter 6: Gerard Bodeker
Chapter 7: Kristina Plenderleith * Chapter 8: Edwin Bernbaum
Chapter 9: Sarah A. Laird * Chapter 10: Paul Chambers
Chapter 11: Jeff Golliher * Chapter 12: Graham Dutfield

Task Managers

Isabella Masinde and Carmen Tavera

Practical Action Publishing Ltd
27a Albert Street, Rugby, CV21 2SG, Warwickshire, UK
www.practicalactionpublishing.org

First published 1999\Digitised 2013

ISBN 10: 1 85339 397 5
ISBN 13: 9781853393976
ISBN Library Ebook: 9781780445434
Book DOI: http://dx.doi.org/10.3362/9781780445434

A catalogue record for this book is available from the British Library.

Since 1974, Practical Action Publishing has published and disseminated books and information in support of international development work throughout the world. Practical Action Publishing is a trading name of Practical Action Publishing Ltd (Company Reg. No. 1159018), the wholly owned publishing company of Practical Action. Practical Action Publishing trades only in support of its parent charity objectives and any profits are covenanted back to Practical Action (Charity Reg. No. 247257, Group VAT Registration No. 880 9924 76).

PEER REVIEW PANELS

Meeting held at University of Reading, UK, 2–3 October 1996

Chairman: Vernon H. Heywood (University of Reading, UK)

Martyn Bramwell (Guildford, UK), Norman G. Clark (University of Strathclyde, UK), R. J. Daniels (M. S. Swaminathan Research Foundation, Madras, India), Maria L. Gamboa (Rainforest Rescue, Quito, Ecuador), Hanne-Rie Madsen (UNEP, Nairobi, Kenya), Isabella Masinde (UNEP, Nairobi, Kenya), Jerry Moles (Neosynthesis Research Center, Topanga, USA), John Mugabe (African Centre for Technology Studies, Nairobi, Kenya), Christine Padoch (New York Botanical Gardens, New York, USA), Charles Perrings (University of York, UK), Darrell A. Posey (Oxford Centre for the Environment, Ethics and Society, Oxford, UK), Ranil Senanayake (Environmental Liaison Centre International, Nairobi, Kenya), L. Jan Slikkerveer (Institute of Cultural and Social Studies, Leiden University, The Netherlands), Margaret Manuka-Sullivan (Maori Programme, Wellington, New Zealand).

Meeting held at Leiden University, The Netherlands, 22–23 September 1997

Chairman: Vernon H. Heywood (University of Reading, UK)

Kusnaka Adimihardja (Padjadjaran University, Bandung, Indonesia), Martyn Bramwell (Guildford, UK), Joji Cariño (Alliance of the Indigenous–Tribal Peoples of the Tropical Forests, London, UK), R. J. Daniels (M. S. Swaminathan Research Foundation, Madras, India), Tirso Gonzales (University of California, Davis, USA), Reverend Canon Jeff Golliher (The Cathedral of St. John the Divine and the Office of the Anglican Observer at the UN, New York, USA), Maria Clara van der Hammen (Tropenbos, Colombia), Christine Kabuye (National Museums of Kenya, Nairobi, Kenya), Sheila Mwanundu (African Development Bank, Côte d'Ivoire), Isabella Masinde (UNEP, Nairobi, Kenya), Jeffrey McNeely (IUCN, Gland, Switzerland), Darrell A. Posey (Oxford Centre for the Environment, Ethics and Society, Oxford, UK), L. Jan Slikkerveer (Institute of Cultural and Social Studies, Leiden University, The Netherlands), Carmen Tavera (UNEP, Nairobi, Kenya).

CONTENTS

CHAPTER 11

CHAPTER 12

CONCLUSION

APPENDICES

Foreword

Klaus Töpfer, Executive Director, UNEP

AS WE APPROACH THE NEXT MILLENNIUM, 'GLOBALIZATION' has become the dominating tendency. Technology and communication systems are dissolving geographical distance and political boundaries. The positive aspects of such a trend are numerous, including the enhanced access to goods and services propelled by the global economy and cultural change which seems to bring progress and new opportunities to the peoples on earth. However, the trade-offs are less well understood, and among these the impacts of the predominant development model on the global environment should be a major concern for us all. Climate change, loss of biological diversity, depletion of the ozone layer, pollution, exhaustion of water resources, and conflicts over shared resources are some of the most pressing problems faced by humankind.

There is strong evidence that the life support systems on which our economies depend are being overloaded, and unless a shift is made towards sustainable development we might face severe or irreversible damage to our environment. Besides the profound ethical and aesthetic implications, it is clear that the loss of biodiversity has serious economic and social costs. The genes, species, ecosystems and human knowledge that are being lost represent a living library of options available for preventing and/or adapting to local and global change. Biodiversity is a part of our daily lives and livelihoods, and constitutes the resources on which families, communities, nations and future generations depend.

Placing a monetary value on species and ecosystems may be a useful exercise by which to integrate the cost of using and conserving biodiversity into the current global economic system, but it will never be possible to comprehend the true value of life in such a system.

Respect for biological diversity implies respect for human diversity. Indeed, both elements are fundamental to stability and durable peace on earth. The key to creating forms of development that are sustainable and in harmony with the needs and aspirations of each culture implies breaking out of patterns that render invisible the lives and perspectives of those cultures. It is the concern of many people that biodiversity must be appreciated in terms of human diversity, because different cultures and people from different walks of life perceive and apprehend biodiversity in different ways as a consequence of their distinct heritages and experiences.

The separation of spirit from matter seems to be the prevailing philosophical approach in recent times. A re-evaluation of this precept is being shaped by the major religions of the world, in response to the global environmental crisis. This may have a profound repercussion on the way individuals and their societies perceive the environment, leading to more responsible actions.

Most indigenous and/or traditional populations inhabit areas of mega-biodiversity. This illustrates the inextricable link between cultural and biological diversity. The very origins of environmental conservation lie buried in ancient cultures found throughout the world. Modern environmental movements express various ideologies of these original belief systems, yet do not always realize their debt to their forebears, nor towards those who still embody these ideals. Learning and respecting the ways of today's indigenous and traditional peoples, and integrating them into environmental and developmental considerations, will prove indispensable for the survival of diversity.

As reported in Chapter 1 of this volume, 2,500 languages are in immediate danger of extinction and 'an even higher number are losing the "ecological contexts" that keep them as vibrant languages'. Cultural diversity is more than appearance, more than folklore, song and dance. It is the embodiment of values, institutions and patterns of behaviour. It is a composite whole representing a people's historical experience, aspirations and world-view. Deprive a people of their language, culture and spiritual values and they lose all sense of direction and purpose.

The contributions of traditional and indigenous peoples should be made visible. Those who are privileged cannot assume that there is only one world-view. This volume represents UNEP's contribution towards broadening the debate on biodiversity. It is also a call by the United Nations to transform the slogan 'we the peoples' from words into action. This is the only way in which we can become more receptive to the needs of the citizens of the world. This is the only way we can become more flexible in seeking innovative solutions. This is the only way we can capitalize on the abundant human resources and capabilities available within the global society.

This volume on '*Cultural and Spiritual Values of Biodiversity*' presents a sample of the vast array of threads that make up the tapestry of our cultural and spiritual diversity. Some of the articles are philosophical or historical, others are scientific or legal, and some are accounts of personal experiences and beliefs. All are equally enlightening.

Whatever we may think about any particular aspect of this tapestry, it is important for us to contemplate it. If we are to conserve the cultural and natural bounty on earth, we must learn more about it and about the nature of our interactions with that bounty. Taken together, the articles bring out the multidimensional challenges that biodiversity conservation poses, not only to policy-makers and scientists, but to all of us.

This volume has one principal message. We must resolve to weave the life-sustaining customs of all diverse groups on earth into a resilient fabric that will protect the sanctity of all life.

Note From the Executive Editor of the GBA

Vernon Heywood

THIS WORK AROSE OUT OF THE *GLOBAL BIODIVERSITY ASSESSMENT* (GBA), a massive review of our current knowledge in the broad field of biological diversity, commissioned by UNEP. In the Introductory Section of the GBA it is noted that as scientists we now consider that humans form an integral and critically important part of biodiversity, although until recently the tendency was to treat the human species as separate from the rest of nature. Increasing awareness over the past few decades has emphasized the need to improve our understanding of the interactions between human society and biodiversity. As a result, the concept of biodiversity has widened so as to cover these aspects in addition to the normally recognized components of ecological, organismal and genetic diversity. Indeed, during the process of preparing the GBA it became evident at all turns that humans are now the dominant influence on biodiversity and that, moreover, the values, driving forces and human influences, as well as the measures undertaken for the conservation and sustainable use of biodiversity, vary greatly within and between cultures.

Perhaps the most important area where human interactions with biodiversity are intrinsic is agriculture. One of the significant paradigm shifts in the past few years has been the growing appreciation of the concept of agrobiodiversity – the variety and variability amongst the animals, plants and microorganisms that are important to food and agriculture and the ecological complexes of which they form a part – and with it the recognition that biodiversity is essential for agricultural production, just as agriculture should be for biodiversity conservation. A key development was the endorsement, by governments, of the FAO Global Plan of Action (GPA) for the Conservation and Sustainable Use of Plant Genetic Resources for Food and Agriculture, in Leipzig, June 1996. The GPA sets out a global strategy for the conservation and sustainable use of plant genetic resources for food and agriculture that complements the CBD.

Following these two agreements it is interesting to note a developing convergence of interest between the agricultural and conservation sectors. On the one hand, the CBD recognized that agricultural biodiversity is a focal area in view of its social and economic relevance and the prospects offered by sustainable agriculture for reducing the negative impacts on biological diversity, enhancing the value of biological diversity and linking conservation efforts with social and economic benefits (Decision III/11 Conservation and Sustainable Use of Agricultural Biological Diversity of the Conference of the Parties to the Convention on Biological Diversity). On the other hand, the FAO Global Plan of Action covers a number of multidisciplinary areas such as *in situ* conservation of wild plants and crop relatives in natural ecosystems which extend the traditional activities of sustainable agriculture and plant genetic resource conservation, and it is recognized that successful implementation will require the development of new partnerships with a range of intergovernmental and non-governmental organizations as well as with indigenous and local communities.

The problems posed by human and social values are particularly acute in assessing the economic value of biodiversity, and the considerable social and cultural differences in the perception of value are indeed recognized in the main Assessment in the section on 'The Economic Value of Biodiversity'. Also, a major section of the GBA is devoted to 'Human Influences on Biodiversity'. In the latter section, the role of humans as the driving force in causing changes in biological diversity over the ages through agriculture, trade, transportation, urbanization, industrialization and market forces, is explored.

It was felt, however, that a much more detailed review of the diversity of human and social values that have been accorded to biological diversity was warranted. Fundamental are the different ways in which nature and the elements of its diversity – plants, animals, forests, and other elements – are viewed, respected and valued. Accordingly, in September 1994 the Management Group of the GBA recommended to UNEP that the focus on human and social values should be broadened and treated as a separate Section.

Unfortunately, the material was not available in time for inclusion in the GBA, and at the final peer review workshop of the GBA held in Panama in June 1995 it was decided to recommend to UNEP that it should be published, when completed, as a stand-alone volume on 'Human Values of Biodiversity'. A draft was prepared by Dr Jerry Moles and Dr Ranil Senanayake and submitted to a peer review workshop, at Reading, in October 1996. This review recommended a substantial reconceptualization of the project and a general call for additional contributions. It resulted in a volume that has been completely redesigned and reorganized and whose scope has been further widened so as to cover not just human and social but also cultural and spiritual values, with Dr Darrell Posey playing a key role as editor and lead author. A peer review of the new draft was held at Leiden in September 1997 and the present volume is the result.

Because of the nature of the material and the fact that the text was prepared after the completion of the Global Biodiversity Assessment and the disbanding of the organization that led to its production, this volume has not followed the same pattern as that of the GBA. My own role, as Executive Editor of the GBA, in the preparation of the present volume, has been largely supportive and to ensure continuity and complementarity with the main Assessment. In this role I have worked closely with Dr Darrell Posey and am pleased to commend the book as providing a new perspective and a major addition to the growing literature on biological diversity.

1 October 1998

Acknowledgements

Hamdallah Zedan, Chief, Biodiversity Unit, UNEP

THIS BOOK IS THE RESULT OF AN OUTSTANDING RESPONSE to the call by UNEP for contributions to present the complex issues related to the Cultural and Spiritual Values of Biodiversity. The organization is therefore indebted to the many persons and institutions that participated with great enthusiasm in the completion of the task. UNEP extends its gratitude to the Editor, Dr Darrell Addison Posey, for an outstanding job and for his tireless efforts over a period of two years. Without his insight and knowledge this project would not have been possible. The Oxford Centre for the Environment, Ethics and Society at Mansfield College, University of Oxford (U.K.), the Departamento de Ciências Biológicas, Universidade Federal do Maranhão (Brazil) and the Brazilian National Council for Research and Technology provided administrative and institutional support.

The Global Environment Facility (GEF) provided co-financing towards the production of the book, allowing us to complement the *Global Biodiversity Assessment* (GBA) also funded with GEF resources. Leiden University, The Netherlands, hosted the second Peer Review meeting and enthusiastically supported the preparation of the publication.

Special thanks to Prof. Vernon Heywood, Editor of the *Global Biodiversity Assessment*, for his sage editorial, technical, political and practical advice which kept the production process on course. Also to the Task Managers of the project, Isabella Masinde and Carmen Tavera, who contributed their ideas to shape the publication and dedicated long hours to this endeavour in addition to their respective responsibilities within the organization.

The volume would have been impossible to produce with the limited resources and time available had it not been for the generous and dedicated efforts of Chapter Co-ordinators: Dr Edwin Bernbaum (mountains); Dr Gerard Bodecker (traditional health systems); Dr Paul Chambers (aquatic resources); Graham Dutfield (rights, resources and responses); Rev. Dr Jeffrey Golliher (ethical, moral and religious concerns); Dr Andrew Gray (indigenous peoples, their environments and territories); Sarah Laird (forests, culture and conservation); Luisa Maffi (linguistic diversity); Kristina Plenderleith (traditional agriculture and soil management); Dr Darrel A. Posey (introduction and conclusions chapters); Ranil Senanayake (indigenous statements); and Professor L. Jan Slikkerveer (ethnoscience).

Our sincere appreciation is extended to the contributing authors as well as to the traditional and indigenous people who provided freely all the articles and statements contained in the book and whose names appear in the list on page xvi. Their number, 290 contributors, testifies to the importance of the topic and its relevance to the current debate on biological diversity.

We would like to acknowledge the contribution of the Environment Liaison Centre International (ELCI) and its former Executive Director, Dr Senanayake, who provided impetus for the preparation of a special volume to complement the GBA and produced an early manuscript on Indigenous Peoples that served as the basis for the personal depositions found in Chapter 4.

We also thank the peer reviewers for their comments during the Peer Review meetings in both Reading (U.K.) and Leiden (The Netherlands), and likewise to those who provided written comments.

Lastly but not least, our deep thanks to the copy editor, Mr. Martyn Bramwell, for his patience and professional work. Also to the persons who contributed to the various tedious tasks associated with the production of a book of this massive size and complexity: Graham Dutfield, Kristina Plenderleith and Susanne Schmitt who assisted the Editor; Francisco Vásquez who processed the layout; and Neeyati Patel and Hanne-Rie Madsen, Fund Management Officers.

To all those who contributed, UNEP is deeply indebted.

December 1998

Contributors

Darrell Posey * Christine Morris * Paul Chambers * Claus Biegert * Peter Whiteley * Vernon Masayesva * Marie Roué * Douglas Nakashima * Gleb Raygorodetsky * Fleur Ng'weno * Manon Osseweijer * William Wallace Mokahi Steiner * Clark Peteru * Nga Kai O Te Moana * Janet Chernela * Arne Kalland * Níels Einarsson * Beatrice B. Maloba * Jeff Golliher * Chief Oren Lyons * Vandana Shiva * Mary Getui * Rosemary Radford Ruether * John O'Neill * Alan Holland * Mark Sagoff * Timothy Weiskel * Bryan G. Norton * James Nash * Dieter T. Hessel Joan Halifax * Marty Peale * Alastair McIntosh * David Abrams * James Hillman * Fritjof Capra * Tim Jensen Margot McLean * Finn Lynge * Denis Goulet * J. Baird Callicott * William N. Ellis * Margaret M. Ellis * Carter Revard * Graham Dutfield * George Monbiot * Bhaskar Vira * John Wiener * Astrid Scholz * Ignacio Chapela Kathy McAfee * Joseph Vogel * David Stephenson * Carlos Correa * Darshan Shankar * Anil Gupta * Delphine Roch * Tahereh Nadarajah * Gemima Born * Maui Solomon * Christine Morris * Janis Alcorn * Luisa Maffi Jonah Andrianarivo * Tove Skutnabb-Kangas * Mazisi Kunene * Andrew Gray * Joji Cariño * Laurie Anne Whitt David Suzuki * Russel Lawrence Barsh * Andrew J. Chapeskie * Robert H. Winthrop * Elizabeth Reichel D. Mónica del Pilar Uribe Marín * Jaime Rodríguez-Manasse * Glenn Shepard Jr. * Richard B. Petersen Graham E. Clarke * David Bennett * Reimar Schefold * Gerard Persoon * Aroha Te Pareake Mead Davianna Pomaika'i McGregor * Inés Hernández-Ávila * Isabella Masinde * Carmen Tavera * Jerry Moles * Ranil Senanayake * Henrietta Fourmile * Mick * Kathrimila Mulazana * Pera * Kipelelo Waker * Kaichela Dipera Rekamani Mutupi * Daniel Matenho Cabixi * Gamaillie Kilukishak * Meeka Mike * Juan Vargas * Rodolfo Mayorga * Carmen Leiva * Aníbal Morales * Gloria Moyorga * Juanita Sánchez * Eustacia Palacios * Catalina Morales * Cristina Gualinga * Bolivar Beltrán * Piedad Cabascango * Raúl Tapuyo * Nancy Velas * Fredy Pianchiche * Bayardo Lanchimba * Yesenia Gonzales * Melecio Santos * Moi Angulema * Nina Pacari Vega * The Nuaulu People * Patrick Segundad * Eidengab * Johan Mathis Turi * Atentio López * Morea Veratua * Michael Kapo * Adolfo García Laulate * Silvia del Aguila Reina * David Vargas * Anastacio Lázaro Milagro * Susana Dávila * Walter Gabriel Yanesha * Enrique Gonzales Miranda * Walter Reiner Jipa * Glorioso Castro * Mario Sijo Sajami * David Camacho Paqui * Justo Funachi Inuma * Giulianna Montalvo Espejo * Fidel Sánchez * Ciro Cosnilla Olivares * Reyner Castro Martínez * Eliazar Muñoz * Guillermo Gómez. * Elmer Guimaraes * Leoncio García * Abel Alado Tello * Irma Castro de Salazar * Tomas Huaya Panduro * Oswaldo Cairuna Fasaba * Hugo Ochavano Sanaciro * Zoila Canaquiri Bardales * Juan Muñoz Mamani * Marcial Rodríguez Panduro * Teresa Elecputu * Policarpo Sánchez * Emilio Agustín * Josué Faquin * Fidel Sinarahua * Delia Isuiza * Francisco Muñoz Ruth Lilongula * Mudianse Tenekoon * Siosiane Fanua Bloomfield * Korosata To'o * Reverend Asotasi Time Evanson Nyamakao * Ralph Mogati * L. Jan Slikkerveer * Arun Agrawal * Roy Ellen * Holly Harris * Ashish Kothari * Priya Das * Raymond Pierotti * Daniel R. Wildcat * Maj-Lis Foller * Janis Alcorn * Michel P. Pimbert Jules N. Pretty * Tirso Gonzales * Nestor Chambi * Marcela Machaca * Prabha Mahale * Hay Sorée * Kusnaka Adimihardja * Rosemary Hill * Dermot Smyth * Julian Barry * Hew D. V. Prendergast * Stephen D. Davis Michael Way * James Fairhead * Melissa Leach * John Clark * Terence Hay-Edie * Jeffrey A. McNeely * Herbert Girardet * Josep-Antoni Gari * Harold Brookfield * A. V. Balasubramanian * K. Vijayalakshmi * Gerard Bodeker Darshan Shankar * Bhanumathi Natarajan * Jean Goodwill * Lea Bill – Rippling Water Woman * Elois Ann Berlin Brent Berlin * Michael J. Balick * Rosita Arvigo * Gregory Shropshire * Robert Mendelsohn * Francisco Montes Kathleen Harrison * Francoise Barbira-Freedman * Stephen Hugh-Jones * Elizabeth Motte-Florac * Serge Bahuchet * Jacqueline M. C. Thomas * Maurice Mmaduakolam Iwu * Olive Tumwesigye * Francisco Montes Bill Neidjie * S. Davis * A. Fox * Kristina Plenderleith * Miguel A. Altieri * Paul Sillitoe * Wabis Ungutip * Jim Pelia * Kuli Hond * Ben Wabis * Ideng Enris * Sarmiah * Riza V. Tjahjadi * Daniel di Giorgi Toffoli * Rogerio Ribeiro de Oliveira * Richard H. Moore * Deborah H. Stinner * David and Elsie Kline * David Brokensha * Paul Richards * Lori Ann Thrupp * Jean Christie * Pat Mooney * Peter Mann * Kathy Lawrence * Edwin Bernbaum Garma C. C. Chang * Washington Matthews * Stanley F. Stevens * A. N. Purohit * Martin Palmer * Tjalling Halbertsma * Chief John Snow * Evelyn Martin * Thomas Schaaf * Roger Pullin * Sarah A. Laird * Laura Rival Patricia Shanley * Jurandir Galvão * Paul Richards * Julia Falconer * Jeanne Zoundjihekpon * Bernadette Dossou-Glehouenou * Christine Kabuye * Wynet Smith * T. C. Meredith * Michael Dove * Pei Shengji E. Bharucha * Pramod Parajuli * Nandini Sundar * Sargun A. Tont * William Burch Jr. * Alexander Porteous Oliver Rackham * Kenny Young * Patricia Shanley * Jurandir Galvão * Harold Courlander * Orhan Veli Kanik Murat Nemet-Nejat * Melih Cevdet Anday * Talat Sait Halman * Freda Rajotte

Preface

Darrell Addison Posey

IF NOTHING ELSE, THE BIODIVERSITY DEBATE OVER THE last few decades has taught us that life is a mosaic of biological forms, webs of interrelated niches, and myriad ecosystems. Like a mosaic, the intricate pieces that comprise the work, when viewed too closely, may not reveal the overall vision that is the message of the art. Yet, the message of the mosaic is revealed only through its intricacy and diversity.

This volume is a mosaic. It is composed of contributions from people and peoples around the world. The fragments that comprise the text were provided by those who share concerns at the accelerating loss of cultural and language diversity that accompanies the devastation of biological diversity. The papers, stories, poems, case studies, photographs, drawings and song lyrics all contribute to the larger vision of the volume: that of highlighting the central importance of cultural and spiritual values in an appreciation and preservation of all life.

Technical descriptions of biodiversity often give the impression that science and economics are adequate tools with which to characterize the qualities of the intricate web of life. In a world increasingly dominated by mega-modelling, global trading and consumer trends, it is easy to forget that values of plants, animals, landscapes and ecosystems cannot be adequately measured in statistical or monetary terms – and certainly cannot be described using the languages of only a few academic disciplines and markets, no matter how politically favoured and powerful they might be. Values of diversity – biological, cultural and linguistic – are intrinsic to life itself and celebrated by the myriad cultures and societies that have co-evolved with the natural and metaphysical worlds that surround them. Indeed, human beings are an integral part of biodiversity, not merely observers and users of the 'components of biological diversity'.

Indigenous and traditional peoples make this fundamental principle the very core of their societies. For them, nature is an extension of society itself, and the creatures that share life with them are manifestations of past and future generations – of their own flesh and blood. Nature is not, therefore, a commodity to be bought, sold, patented or preserved apart from society, precisely because nature is what defines humanity. The earth is their (our) mother and cannot be compromised, sold or monopolized.

This view is not alien to the teaching of the world's major religions, although ecologists have been disappointed by the lack of decisive guidance by religious leaders in the battle to conserve biodiversity. This has begun to change, and religious organizations are increasingly aware of their role in re-establishing the principles of human stewardship of, and oneness with, nature. The way that each religious group tackles the problem is, again, a reflection of diversity.

Organization of the volume will, hopefully, guide the reader to the overall vision of the mosaic. Each chapter is introduced by its co-ordinator and contains a range of contributions covering different aspects of the general theme.

CHAPTER 1 begins by outlining the inextricable link between nature, society, language and culture. It shows how human beliefs and values have molded the historical 'cultural landscapes' of much of the world and how they continue to affect the conservation of biological and ecological diversity.

CHAPTER 2 pays special tribute to the important and usually overlooked aspects of linguistic diversity and language rights, and illustrates how languages express human observation and understanding of life, natural cycles, and the earth. It explains how cognition and speech help to encode the indigenous and local knowledge systems that are critical to the effective conservation of natural resources.

CHAPTER 3 is an introduction to the complex topic of indigenous peoples and their roles in the conservation of biological diversity. It provides a political and theoretical framework for the evaluation of the diverse peoples who are grouped together as 'indigenous', and outlines some of the principal issues that they confront in expressing and protecting the cultural and spiritual values that lie at the heart of their societies and environments.

CHAPTER 4 is dedicated to the 'Voices of the Earth' – the myriad indigenous, traditional and local peoples who repeatedly – and passionately – warn us of the dangers we all face as biological and cultural diversity disappear from the world. The chapter is a collage of individual depositions from around the world, reinforced by a selection of formal declarations and statements of indigenous and traditional peoples' organizations presented in Appendix 1.

CHAPTER 5 explains the scope, importance and applications of Traditional Ecological Knowledge

(TEK) and illustrates how such knowledge can be utilized to develop more effective and adaptive strategies for conservation. It provides case studies and specific examples of how local communities are already practising sustainability and suggests how traditional knowledge can be utilized in partnership with scientists and policy-makers.

CHAPTER 6 outlines the relevance of traditional knowledge in one of the areas of most concern to human beings – that of medicine and health. It emphasizes the holistic nature of most indigenous and traditional medical systems which view environment and diet as essentials to human health.

CHAPTERS 7 to 10 deal with specific examples of how Traditional Ecological Knowledge is relevant to the use and conservation of traditional agriculture and soil management (CHAPTER 7), mountain and highland ecosystems (CHAPTER 8), trees and forests (CHAPTER 9) and aquatic and marine resources (CHAPTER 10). These chapters provide a general overview and then offer specific examples of how local people – and their beliefs and knowledge – work to conserve biodiversity in an extraordinary range of environments and social conditions.

CHAPTER 11 provides a general overview of the ethical, moral and religious concerns that frame philosophical and political debates on the values of life, nature and culture. It also describes the 'Assisi Process', in which major world religions are reassessing their roles as environmental stewards. Faith statements on religion and ecology are presented in Appendix 2.

CHAPTER 12 emphasizes that people – and peoples – must have rights to their lands and territories, as well as to their knowledge and genetic resources, if successful conservation is to occur. It emphasizes that utilization and effective application of traditional resources are enmeshed in existing international laws and customary practices that defend collective and individual human rights. Too frequently the knowledge and resources of indigenous, traditional and local communities are usurped without the sharing of benefits. Even basic respect for prior informed consent, full disclosure and privacy are ignored. The chapter provides a critique of the inadequacies of existing Intellectual Property Rights laws and the need for the development of a global dialogue to develop new, appropriate and effective systems to enhance control by local communities.

Cultural and Spiritual Values of Biodiversity concludes with some general recommendations based on the contributions contained in the volume. By the end, the reader will hopefully have been presented with compelling evidence that the future of biodiversity must not be left only to technical, scientific and economic experts, but rather depends upon respect for and protection of the myriad views, values and visions that together form the mosaic of life.

Oxford, 1 October 1998

COLOUR PLATES

COLOUR PLATES APPEAR BETWEEN PAGES 284 AND 285

Plate 15. Ploughing a hay field after the bobolinks have finished mating. David Kline farm. (Photograph: Richard Moore: see Chapter 7.)

Plate 16. Black walnut tree, used for high-quality veneer: Leroy Kuhn's farm. (Photograph: Richard Moore: see Chapter 7.)

Plate 17. The main product of the Caiçara agriculture is the manioc (cassava) flour. Since immemorable times, the manioc flour is a substitute of the "European bread", and for this reason it is called the "tropical bread". Other products of the Caiçara agriculture are maize, beans and yam as well as medicinal herbs. (Photograph: Daniel Toffoli: see Chapter 7.)

Plate 18. A young Mende girl winnowing rice. (Photograph: Helen Newing: see 'Casting seeds to the four winds', Chapter 7.)

Plate 19. 'The arrival of "Inlanders" at Fort George Island'. This re-enactment by the Chisasibi Cree, recorded by a film crew for local television, is a highlight of an annual community event called Mamoweedoew Minshtukch – 'the gathering on the island'. The traditional paddle canoes brought out for this occasion (background) are dwarfed by modern canoes equipped with outboard motors (foreground), which are the common means of transport on the river today. (Photograph: Douglas Nakashima: see Chapter 10.)

Plate 20. Under the watchful eye of his son, an Inlander father participates in the re-enactment of setting up the tepee (background). Note the rifle sheaths (left foreground) – ornately decorated to please the animal spirits and bring the hunter good fortune. (Photograph: Douglas Nakashima: see Chapter 10.)

Plate 21. A Tukanoan Wanano basket trap (buhkuyaka) consisting of cylindrical basketry fish traps set in a barrier. (Photograph: Janet Chernela: see Chapter 10.)

Plate 22. Bark-cloth painting by Tukanoan Wanano artist Miguel Gomez showing tree with falling fruit, fish feeding on fruit, and bow-and-arrow fisherman. (Collected by Janet Chernela, Rio Uaupes, Brazil: see Chapter 10.)

Plate 23. Bark-cloth painting by Tukanoan Wanano artist Miguel Gomez showing the detailed coloration on one of the many varieties of bagre, or catfish. (Collected by Janet Chernela, Rio Uaupes, Brazil: see Chapter 10.)

Plate 24. Bark-cloth painting by Tukanoan Wanano artist Miguel Gomez showing two sub-varieties of Cynodon with distinctive features. (Collected by Janet Chernela, Rio Uaupes, Brazil: see Chapter 10.)

Plate 25. "Pyloric Valve", 1992 (artist's collection)

The structure of the leaf corresponds with our own innermost bodily structures –our veins, valves, nodes. The same fragility that exists in an ordinary leaf exists in our human physiology, mysteriously and yet without question. (Artist: Margot McLean: mixed media on linen. See Chapter 11.)

Plate 26. "Moose", 1992 (private collection)
The art of the moose is to be visible and yet invisible. This elusive, lonely, fading, there-but-not-there aspect of the creature contributes to the spirit of the moose –to the ghost of the moose. (Artist: Margot McLean: mixed media on paper. See Chapter 11.)

Plate 27. "Virginia", 1993 (private collection)
Incorporating vegetation from a particular spot in Northern Virginia that is the habitat of a particular kind of land turtle, this painting re-locates the turtle into a landscape of the imagination –allowing, or giving, it another, different life. (Artist: Margot McLean: mixed media on canvas. See Chapter 11.)

Plate 28. "Orinoco", 1994 (artist's collection)
From a trip on the Orinoco River in Venezuela. The "baba" or river crocodile, comes ashore in the evening, lying among the rocks or in one of the tidepools, its presence felt throughout the surrounding area. The image suggests the concealed possibility of life that lurks in the unknown. (Artist: Margot McLean: mixed media on linen. See Chapter 11.)

CHAPTER 1

INTRODUCTION: CULTURE AND NATURE – THE INEXTRICABLE LINK

Darrell Addison Posey

CONTENTS

TEXT BOXES

Introduction: Culture and Nature – The Inextricable Link

Darrell Addison Posey

HUMAN CULTURAL DIVERSITY IS THREATENED ON AN UNprecedented scale. Linguists estimate that between 5,000 and 7,000 languages are spoken today on the five continents (Krauss 1992; Grimes 1996; Harmon 1995; Maffi Chapter 2). Languages are considered one of the major indicators of cultural diversity. Yet an estimated half of the world's languages – the codifications, intellectual heritages, and frameworks for each society's unique understanding of life – will disappear within a century (UNESCO 1993). Nearly 2,500 languages are in immediate danger of extinction; and an even higher number are losing the 'ecological contexts' that keep them as vibrant languages (Mühlhäusler 1996).

Many of the areas of highest biological diversity on the planet are inhabited by indigenous and traditional peoples, providing what the Declaration of Belém calls an 'inextricable link' between biological and cultural diversity (Posey and Overal 1990). In fact, of the nine countries which together account for 60 percent of human languages, six of these 'centres of cultural diversity' are also 'megadiversity' countries with exceptional numbers of unique plant and animal species (Durning 1992).

It is estimated (Gray, Chapter 3) that there are currently at least 300 million people world-wide who are indigenous. There are no reliable figures for how many are 'traditional' societies, but excluding urban populations it could be as high as 85 percent of the world's overall population. These diverse groups occupy a wide geographical range from the polar regions to the deserts, savannahs and forests of tropical zones (IUCN/UNEP/WWF 1991).

Inevitably a volume on cultural and spiritual diversity will dwell upon the beliefs and practices of indigenous and traditional peoples in relation to their use and conservation of biodiversity. But *Cultural and Spiritual Values of Biodiversity* is also dedicated to a survey of issues related to urban populations and industrialized societies – and specifically the ethical, philosophical and religious beliefs and practices that affect them.

Who are indigenous and traditional peoples?

According to UNESCO (1993), 4,000 to 5,000 of the 6,000 languages in the world are spoken by indigenous peoples, implying that indigenous groups still constitute most of the world's cultural diversity (see also Maffi, Maffi and Skutnabb-Kangas, Chapter 2). The definition of 'indigenous' is problematic in many parts of the world. Indigenous peoples are defined by the Special Rapporteur of the UN Economic and Social Council Sub-Commission on Prevention of Discrimination and Protection of Minorities in the following manner:

'Indigenous communities, peoples and nations are those which, having a historical continuity with pre-invasion and pre-colonial societies that have developed on their territories, consider themselves distinct from other sectors of the societies now prevailing in those territories, or parts of them. They form at present non-dominant sectors of society and are determined to preserve, develop and transmit to future generations their ancestral territories, and their ethnic identity, as the basis of their continued existence as peoples, in accordance with their own cultural patterns, social institutions and legal systems. (UN ECOSOC 1986)

This *historical continuity* is characterized by:

- occupation of ancestral lands, or at least of part of them;
- common ancestry with the original occupants of these lands;
- culture in general, or in specific manifestations (such as religion, living under a tribal system, membership of an indigenous community, dress, means of livelihood, life-style, etc.);
- language (whether used as the only language, as mother tongue, as the habitual means of communication at home or in the family, or as the main, preferred, habitual, general or normal language);
- residence in certain parts of the country, or in certain regions of the world, and
- other relevant factors.

The International Labour Organization (ILO) Convention 169 'Concerning Indigenous Peoples in Independent Countries' (1989), identifies Indigenous peoples as:

(a) tribal peoples in countries whose social, cultural and economic conditions distinguish them from other sections of the national community, and whose status is regulated wholly or partially by their own customs or traditions or by special laws or regulations, and

(b) peoples in countries who are regarded by themselves or others as indigenous on account of their descent from the populations that inhabited the country, or a geographical region to which the country belongs, at the time of conquest or colonization or the establishment of present state boundaries and who, irrespective of their legal status, retain, or wish to retain, some or all of their own social, economic, spiritual, cultural and political characteristics and institutions.

A fundamental principle established by ILO 169 is that: ***Self-identification*** *as indigenous or tribal shall be regarded as a fundamental criterion for determining the groups to which the provisions of this convention apply.*

This principle is upheld by all indigenous groups, who, as the Final Statement of the Consultation on Indigenous Peoples' Knowledge and Intellectual Property Rights, Suva, April 1995, says: *We assert our inherent right to define who we are. We do not approve of any other definition.*

Indigenous peoples insist that they be recognized as 'peoples', not 'people'. The 's' distinction is very important, because it symbolizes not just the basic human rights to which all individuals are entitled, but also land, territorial and collective rights, subsumed under the right to self-determination. In contrast, the use of terms like 'people', 'populations' and 'minorities' implicitly denies territorial rights.

This volume does not pretend to provide an absolute definition for 'indigenous'; indeed, such a definition is evolving in international law and customary practice and cannot (and should not, according to the UN Working Group on Indigenous Populations) be precisely defined at this time (Daes 1993). As Willett points out (this volume) the debate over who is 'indigenous' should not side-track the important task of valuing local communities – whether or not they are indigenous. The important task is to rekindle and enhance the spiritual and cultural values that cultures have used effectively to conserve biodiversity.

It is also important not to allow 'traditional' to be used to restrict local innovation and cultural change. In this volume, the term is used as defined by the Four Directions Council (1996) of Canada:

'What is 'traditional' about traditional knowledge is not its antiquity, but the way it is acquired and used. In other words, the social process of learning and sharing knowledge, which is unique to each indigenous culture, lies at the very heart of its 'traditionality'. Much of this knowledge is actually quite new, but it has a social meaning, and legal character, entirely unlike other knowledge'.

Traditional livelihood systems, therefore, are constantly adapting to changing social, economic and environmental conditions. These are dynamic, but – no matter the changes – embrace principles of sustainability (Alcorn, Chapter 5; Kothari and Das, Chapter 5; Pimbert and Pretty, Chapter 5; Bierhorst 1994; Callicott 1989; Johannes and Ruddle 1993; Clarkson *et al.* 1992; Posey and Dutfield 1997). These principles cannot be regarded as universal, but generally emphasize the following values:

- co-operation;
- family bonding and cross-generational communication, including links with ancestors;
- concern for the well-being of future generations;
- local-scale self-sufficiency, and reliance on locally available natural resources;
- rights to lands, territories and resources which tend to be collective and inalienable rather than individual and alienable;
- restraint in resource exploitation and respect for nature, especially for sacred sites.

The sacred balance

Although conservation and management practices are highly pragmatic, indigenous and traditional peoples generally view this knowledge as emanating from a *spiritual* base. All creation is sacred, and the sacred and secular are inseparable. Spirituality is the highest form of consciousness, and spiritual consciousness is the highest form of awareness. In this sense, a dimension of traditional knowledge is not *local* knowledge, but knowledge of the *universal* as expressed in the local. In indigenous and local cultures, experts exist who are peculiarly aware of nature's organizing principles, sometimes described as entities, spirits or natural law. Thus, knowledge of the environment depends not only on the relationship between humans and nature, but also between the visible world and the invisible spirit world.

According to Opoku (1978), the distinctive feature of traditional African religion is that it is: 'A way of life, [with] the purpose of... order[ing] our relationships with our fellow men and with our environment, both spiritual and physical. At the root of it is a quest for harmony between man, the spirit world, nature and society'. Thus, the unseen is as much a part of reality as that which is seen – the spiritual is as much a part of reality as the material. In fact, there is a complementary relationship between the two, with the spiritual being more powerful than the material. The community is of the dead as well as the living. And in nature, behind visible objects lie essences, or powers, which constitute the true nature of those objects.

Indigenous and traditional peoples frequently view themselves as guardians and stewards of nature. Harmony and equilibrium among components of the cosmos are central concepts in most cosmologies. Agriculture, for example, can provide 'balance for well-being' through relationships not only among people, but also nature and deities. In this concept, the blessing of a new field represents not mere spectacle, but an inseparable part of life where the highest value is harmony with the earth (Enris and Sarmiah, Chapter 7). Most traditions recognize linkages between health, diet, properties of different foods and medicinal plants, and horticultural/natural resource management practices – all within a highly articulated cosmological/social context (e.g. Hugh-Jones, Chapter 6).

Chief Oren Lyons (Chapter 11) emphasizes that for his Haudenosaunee people, relationships and obligations are not to some external objects that possess 'life', but rather to kin and relatives. Many indigenous peoples believe they once spoke the language of animals and that their shamans still have this ability (Snow, Chapter 8). Biodiversity, therefore, means the extended family – 'all our relations'. This is the *lokahi* (unity) of the native Hawaiians – the 'nurturing, supportive and harmonious relations' that link land, the gods, humans and the forces of nature (McGregor, Chapter 3). Thus, outsiders (environmentalists, developers, scientists, etc.) may see themselves as working with elements of nature, but indigenous peoples may see their activities as meddling in internal affairs of the 'extended family' (see Barsh, Chapter 3).

Local knowledge embraces information about location, movements and other factors explaining spatial patterns and timing in the ecosystem, including sequences of events, cycles and trends. Direct links with the land are fundamental, and obligations to maintain those connections form the core of individual and group identity. Nowhere is that more apparent than with the Dreaming-places of the Aboriginal peoples of Australia. As James Galarrwuy Yunupingu, Chairperson of the Northern Land Council, explains: 'My land is mine only because I came in spirit from that land, and so did my ancestors of the same land. My land is my foundation' (Yunupingu 1995; see also Bennett, Chapter 3).

As Whitt (Chapter 3) explains, the Cherokee see knowledge itself as being an integral part of the earth. Thus a dam does not just flood the land, but also destroys the medicines and the knowledge of the medicines associated with the land. 'If we are to make our offerings at a new place, the spiritual beings would not know us. We would not know the mountains or the significance of them. We would not know the land and the land would not know us... We would not know the sacred places... If we were to go on top of an unfamiliar mountain we would not know the life forms that dwell there'.

The same is true for the Mazatecs of southern Mexico, whose shamans and *curandeiros* confer with the plant spirits in order to heal. Successful curers must above all else learn to listen to the plants talk (Harrison, Chapter 6). For many groups, these communications come through the transformative powers of altered states or trance (e.g. Shepard's account of Machiguenga shamanism, Chapter 3). Don Hilde, a Pucallpa healer, explains: 'I did not have a teacher to help me learn about plants, but visions have taught me many things. They even instruct me as to which pharmaceutical medicines to use' (Dobkin de Rios 1972; Barbira-Freedman, Chapter 6).

These links between life, land and society are identified by Suzuki (Chapter 3) as the 'Sacred Balance'. Science, with its quantum mechanics methods, says he, can never address the universe as a whole; and it certainly can never adequately describe the holism of indigenous knowledge and belief. But John Clark (Chapter 5) reminds us that Western science itself grew out of reflections upon nature in order to describe 'humanity'; indeed, in past centuries social insects became the focus of explanations of the 'nature' of human society. But science seems to have forgotten its holistic natural history roots.

In fact, science is far behind in the environmental movement. It still sees nature as objects for human use and exploitation ('components' of biodiversity is the term the Convention on Biological Diversity uses). Technology has used the banner of scientific 'objectivity' to mask the moral and ethical issues that emerge from such a functionalist, anthropocentric philosophy. Strathern (1996) makes this clear when discussing the ethical dilemmas raised (or avoided) when embryos are 'decontextualized' as human beings to become 'objects' of scientific research.

Science and technology seldom embrace the values of local knowledge and traditions, and very rarely employ the language of rights and local control over access to resources. Meanwhile, economists would have us believe that markets provide 'level playing fields' that do not moralize globalization and, therefore, work more efficiently to advance the causes of environmental conservation.

With such philosophies and policies there can be little surprise that indigenous, traditional and local communities are hostile to the rhetoric of 'partnerships' and 'sustainable development' – indeed, to the very tenets of 'biodiversity' and 'conservation'. Langton (1996; see also Bennett, Chapter 3) attributes the use of these concepts to an 'anxiety to assert the supremacy of Western resource management regimes over indigenous culture'. Fortunately there are more and more cases of successful co-management of resources, such as with Uluhru in Australia (Bennett, Chapter 3) and the joint forest management programme of India (Sundar, Chapter 9; see also Posey and Dutfield 1997). There are also excellent examples of indigenous peoples themselves implementing effective environmental management plans (Hartshorn 1990; Posey and Dutfield 1996; Ruddle and Johannes 1985; Nietschmann 1985).

The dominant scientific and economic forces assume that traditional communities must change to meet 'modern' standards, but indigenous and traditional peoples feel the opposite must occur: science and industry must begin to respect local diversity and the 'Sacred Balance'. Contributions throughout this volume support this position, and it may be that their views reflect those of the vast majority of people living in industrial, 'modern' societies – leaving us with the possibility that science and industry have lost their legitimate role as responsible global leaders.

As, indeed, may have religions – especially those that encourage 'materialistic, dualistic, anthropocentric and utilitarian concepts of and relations to nature' (Edwards and Palmer 1997). Such philosophies have only served to hasten the destruction of biological and cultural diversity (Jensen, Chapter 11). But as Golliher (Chapter 11) points out, life as an extension of place and family is inherent in many major religions – or, at least once was so before being lost to the dominist theologies that accompanied industrialization. Moore *et al.* (Chapter 7) maintain that Amish beliefs still honour the old holistic traditions. Tont (Chapter 9) shows how love for trees is embedded in Islam because Mohammed compared a good Muslim to a palm tree and 'declared that planting a tree would be accepted as a substitute for alms'; and Graham Clarke (Chapter 3) explains the significance of trees to Buddhists in Nepal. The Alliance of Religion and Conservation (ARC) is undertaking the 'Assisi Process' to establish an alliance of religious leaders to help reaffirm commitments to the environment, 'green' their religious institutions, and rediscover the principles of 'stewardship' for the earth (Palmer, Chapter 11).

Dealing with the problems

It is obvious that not all human impact on the environment is positive. Indeed, the massive, unprecedented destruction of entire environmental systems, extinctions of species and pollution of waters and the atmosphere is due to the greed of human societies running out of control. Population growth, overconsumption, wastefulness and plain arrogance can all be named as reasons for this situation. And although most of this rampant devastation is the result of industrialized societies competing for global markets, an increasing number of indigenous, peasant and local communities are abandoning sustainable traditions in favour of destructive activities.

Some scientists express scepticism that the subsistence and resource management practices of traditional peoples are – or ever have been – guided by a 'conservation ethic' (e.g. Hames 1991; Kalland 1994a). Some argue that mass extinctions of North American megafauna during the Pleistocene era were not caused by climate changes, but by reckless overhunting by humans colonizing the continent from Siberia (Martin 1984). Likewise, the extinction of wildlife on such islands and archipelagos as New Zealand, Madagascar and Australia have been blamed upon the ancestors of today's traditional peoples (Martin and Klein 1984; Diamond 1994, 1997; Flannery 1996). Yet many megafauna species persist in Africa and Eurasia despite human habitation (Flannery 1996; Diamond 1997); and East Africa, where *Homo sapiens* evolved, has retained many of its original megafauna species up to the present day.

Ellen (1986) notes that small-scale and isolated traditional societies are most often regarded as being orientated towards conservation. He argues that it is *because* these societies are small that they impact minimally on the environment. Thus, dependence upon a broad spectrum of useful species is a rational subsistence strategy, but not a conscious means by which to protect species for future generations. Redford and Stearman (1993) are also sceptical that traditional peoples are natural conservationists. They feel it is inappropriate to generalize about traditional communities and make broadly applicable assertions about their environmental values (Stearman 1992). Furthermore, present-day traditional societies are to a large extent part of the global economy and have lost many of their traditional cultural values. It is therefore unfair to expect traditional peoples to continue using traditional, low-impact subsistence technologies and strategies (Redford 1991; Kalland 1994b).

None the less, many anthropologists defend the conservation mentalities of indigenous and traditional peoples (e.g. Bodley 1976; Martin 1978; Clad 1984; Reichel-Dolmatoff 1976). And almost everyone agrees that local communities are more likely to employ environmentally sustainable practices when they enjoy territorial security and local autonomy, (Gadgil and Berkes 1991; Bierhorst 1994; Kalland 1994a; Redford and Stearman 1993; Alcorn 1994; Posey and Dutfield 1997).

It is to no one's benefit to romanticize indigenous and traditional peoples – even those groups that do still live close to the earth and its life-forms – for it is just as easy to find unsustainable practices within their societies as it is to have invented the 'ecologically noble savage' who lives in 'harmony and balance with nature'. All human societies – even the most traditional – are enmeshed in change, and always have been. Human evolution is about adaptation and change, and as cultures and environments adapt to different conditions there will inevitably be practices and customs that become unadaptive and must be modified to fit the new circumstances.

Indigenous peoples themselves reject an 'ecologically noble savage' approach that romanticizes their relationship with nature. Two Native American scientists (Pierotti and Wildcat, Chapter 5) warn that:

> 'Those wanting to embrace the comfortable and romantic image of the Rousseauian 'noble savage' will be disappointed. Living with nature has little to do with the often voiced 'love of nature', 'closeness to nature,' or desire 'to commune with nature' one hears today. Living with nature is very different from 'conservation' of nature. Those who wish to 'conserve' nature still feel that they are in control of nature, and that nature should be conserved only insofar as it benefits humans, either economically or spiritually. It is crucial to realize that nature exists on its own terms, and that non-humans have their own reasons for existence, independent of human interpretation [see below, and Taylor 1992]. Those who desire to dance with wolves must first learn to live with wolves.'

Pierotti and Wildcat (ibid.) also point out that the concepts of biodiversity and conservation are not indigenous and, indeed, are alien to Indigenous peoples. This does not mean they do not respect and foster living things, but rather that nature is an extension of society. Thus, biodiversity is not an object to be conserved. It is an integral part of human existence, in which utilization is part of the celebration of life.

Lynge (Chapter 11, Box 11.2) emphasizes that one of the basic differences between traditional hunters and urban conservationists is that the latter fear, rather than love, nature. As he points out, 'good nature management' depends upon 'the recognition of the interdependent wholeness of humanity', which in turn is based upon a respect for life.

The problem then is not one of whether indigenous and traditional peoples are or are not 'natural conservationists', but rather who (and how) are we to judge them? Different world-views make such judgements tenuous at best. And besides, whose scientific measuring stick is to be used to make the judgements? There are, for example, no universal, nor even standardized, indicators of sustainability, nor universal agreement on how to define, measure or monitor biodiversity. And what are the criteria for judging environmental health? And healthy environments for whom? Never mind the moral question of who are we to judge? Who provoked the biodiversity crisis in the first place?

Two major points remove us from this relativism and judgementalism, leaving scientists, conservationists and indigenous/traditional peoples alike to face the hard facts of observation and experience. One is the unequivocal influence of indigenous and local communities in the formation and maintenance of modern ecosystems. The other is the formidable number of case studies (throughout this volume, and see also WGTRR 1996) showing how traditional ecological knowledge and practices serve to effectively manage and conserve mountain, forest, agricultural, dry-land, highland, aquatic and other ecosystems.

Recognizing indigenous and local communities

Western science may have invented the words 'nature', 'biodiversity' and 'sustainability', but it certainly did not initiate the concepts. Indigenous, traditional and local communities have sustainably utilized and conserved a vast diversity of plants, animals and ecosystems since the dawn of *Homo sapiens*. Furthermore, human beings have moulded environments through their conscious and unconscious activities for millennia – to the extent that it is often impossible to separate nature from culture.

Some recently 'discovered' 'cultural landscapes' include those of aboriginal peoples who, 100,000 years before the term 'sustainable development' was coined, were trading seeds, dividing tubers and propagating domesticated and non-domesticated plant species. Sacred sites act as conservation areas for vital water sources (e.g. Plenderleith, Chapter 7) and also for individual species by restricting access and behaviour. Traditional technologies, including fire use, were part of extremely sophisticated systems that shaped and maintained the balance of vegetation and wildlife. Decline of fire management and loss of sacred sites when aboriginal people were centralized into settlements led to the rapid decline of mammals throughout the arid regions (Sultan, Craig and Ross 1997).

An example of an anthropogenic (human-made) landscape is the 'forest island' (*apete*) created in savannahs by the Kayapó of Brazil (Posey 1997;

Laird, Chapter 9). The Indians use detailed knowledge of soil fertility, micro-climate, and plant varieties to skilfully plant and transplant useful non-domesticated species into wooded concentrations of useful plants. Historically, these *apete* have been considered 'natural' by botanists and ecologists. Likewise, the Kagore Shona people of Zimbabwe have sacred sites, burial grounds, and other sites of special historical significance 'deeply embedded' in the landscape (Matowanyika 1997). Outsiders often cannot recognize these sites during land-use planning exercises, just as Ontario resource managers cannot detect the anthropogenic wild rice (*manomin*) fields of the Ojibway (Chapeskie, Chapter 3).

In societies with no written language or edifices, hills, mountains and valleys become the libraries and cathedrals that reflect cultural achievement (Bernbaum, Chapter 8). For the Dineh (Navajo), the 'Mountain of the South' (Tsodzil) fastens the earth to the sky with lightning, rain and rainbows (Matthews 1897). And Stevens (Chapter 8) describes the sacred mountain groves of Nepal. Martin (Chapter 8) explains how it is difficult for outsiders to understand that not just parts of mountains but entire mountains are sacred to people like the Apache, although it may be certain parts of mountains, such as sacred groves, that are holy for groups like the Khumbu of Nepal.

Sacred groves are one of the most common types of cultural landscape (Laird, Chapter 9). Pei Shengji (Chapter 9) describes how the 'dragon hills' of Yunan Province in China are kept intact because of their sacred nature. Likewise, Falconer (Chapter 9) describes how Ghjanan groves are linked to burial grounds and spirits of the ancestors who protect the forests that surround them. Similar groves are reported in Côte d'Ivoire and Benin (Zoundjihekpon and Dossou-Glehouenou, Chapter 9), as well as Ghana (Iwu, Chapter 6). Sacred groves in India are extensive and well-known in the literature (Bharucha, Chapter 9; Gadgil and Vartak 1981).

Wells and springs are also frequently considered holy, and the areas around them specially protected from disturbance. Whitely and Masayesva (Chapter 10) describe how wellsprings are the 'soul of the Hopi people, representing their very identity'. Oases too can be sacred places for people like the Maasai and Fulani pastoralists, whose lives during severe droughts literally depend upon these protected areas (Chambers, Chapter 10).

Schaaf (Chapter 8) proposes an environmental conservation strategy based on preservation of sacred groves and holy places – an idea that has been taken up by UNESCO and other organizations (Edwards and Palmer 1997; Palmer, Chapter 11; Hay-Edie, Chapter 5). Rival (Chapter 9) warns, however, that there is no guarantee that beliefs about sacred places will continue indefinitely into the future; thus, conservation plans around the concepts of protective spirits and deities should be considered with caution.

A failure to recognize sacred and other cultural (human-modified) landscapes, however, has blinded outsiders to the management practices of Indigenous peoples and local communities (Denevan 1992; Gomez-Pompa and Kaus 1992). Many so-called 'pristine' landscapes are in fact anthropogenic landscapes, either created by humans or modified by human activity (e.g. forest management, cultivation, and the use of fire).

This is more than semantics. 'Wild' and 'wilderness' imply that these landscapes and resources are the result of 'nature' and, as such, have no owners – they are the 'common heritage of all humankind'. This has come to mean that local communities have no tenurial or ownership rights, and thus their lands, territories and resources are 'free' to others just for the taking. This is why Indigenous peoples have come to oppose the use of 'wilderness' and 'wild' to refer to the regions in which they now or once lived.

This is poignantly expressed in an Aboriginal Resolution from the 1995 Ecopolitics IX Conference in Darwin, Australia:

> *The term 'wilderness' as it is popularly used, and related concepts such as 'wild resources', 'wild foods', etc., [are unacceptable]. These terms have connotations of* ***terra nullius*** *[empty or unowned land and resources] and, as such, all concerned people and organisations should look for alternative terminology which does not exclude Indigenous history and meaning. (Northern Land Council 1996)*

Cultural landscapes, and their links to conservation of biological diversity, are now recognized under the 1972 UNESCO Convention Concerning the Protection of the World Cultural and Natural Heritage ('The World Heritage Convention'). A new category of World Heritage Site, the 'Cultural Landscape', recognizes 'the complex interrelationships between man and nature in the construction, formation and evolution of landscapes' (UNESCO/WHC/2/Revised/1995). The first cultural landscape World Heritage Site was Tongariro National Park, a sacred region for the Maori people of New Zealand (Rössler 1993) which was included in the World Heritage List because of its importance in Maori beliefs. Hay-Edie (Chapter 5) reports that UNESCO is also now developing new projects to help local communities conserve and protect sacred places.

The Convention on Biological Diversity (CBD) is one of the major international forces in recognizing the role of indigenous and local communities in *in situ* conservation. The Preamble recognizes the:

'Close and traditional dependence of many indigenous and local communities embodying traditional life-styles on biological resources, and the desirability of sharing equitably benefits arising from the use of traditional knowledge, innovations and practices relevant to the conservation of biological diversity and the sustainable use of its components.'

Article 8(j) of the Convention on Biological Diversity (CBD) spells out a specific obligation of each Contracting Party:

'... subject to its national legislation, [to] respect, preserve and maintain knowledge, innovations and practices of indigenous and local communities embodying traditional life-styles relevant for the conservation and sustainable use of biological diversity and promote the wider application with the approval and involvement of the holders of such knowledge, innovations and practices and encourage the equitable sharing of the benefits arising from the utilization of such knowledge, innovations and practices.'

The CBD also enshrines the importance of customary practice in biodiversity conservation and calls for its protection and for equitable benefit-sharing from the use and application of 'traditional technologies' (Articles 10(c) and 18.4). Glowka and Burhenne-Guilmin (1994) warn that 'traditional' can imply restriction of the CBD only to those embodying traditional life-styles, keeping in mind that the concept can easily be misinterpreted to mean 'frozen in time'. But Pereira and Gupta (1993) claim, 'it is the traditional methods of research and application, not just particular pieces of knowledge, that persist in a 'tradition of invention and innovation'. Technological changes do not simply lead to modernization and loss of traditional practice, but rather provide additional inputs into vibrant, adaptive and adapting holistic systems of management and conservation.

'Traditional knowledge, innovations and practices' are often referred to by scientists as Traditional Ecological Knowledge (TEK). TEK is far more than a simple compilation of facts (Gadgil *et al.* 1993; Johnson 1992). It is the basis for local-level decision-making in areas of contemporary life, including natural resource management, nutrition, food preparation, health, education and community and social organization (Warren *et al.* 1995). TEK is holistic, inherently dynamic, and constantly evolving through experimentation and innovation, fresh insight and external stimuli (Knudtson and Suzuki 1992).

Throughout this volume the reader will find references to TEK. There is no standardization of the term, and a forest of acronyms results – Local Knowledge (LK), Indigenous Knowledge (IK), Traditional Knowledge (TK), Indigenous Knowledge Systems (IKS), Indigenous Resource Management Systems (IRMSs), Local Community Systems (LCSs) and more. For the purpose of this volume, these acronyms all have roughly the same meaning and significance.

TEK is transmitted in many ways. Most is done through repeated practice – apprenticeship with elders and specialists. Walker (Chapter 3) sketches how the 'collective memory' of the Tlingit of the Northwest Coast of North America is embedded in basketry. Oral tradition is critical to this transmission. Schefold and Persoon (Chapter 3) describe how knowledge is transmitted in Sumatra through songs. McIntosh (Chapter 11) documents the 'psycho-spiritual' effect of social upheaval, and shows how poems and music serve to register the resulting devastation of communities and biodiversity. By extension, it seems obvious that poems and music provide important remedies and pathways for environmental restoration. This volume contains many such expressions of ecological knowledge and appreciation for nature and life.

Slikkerveer (Chapter 5) provides many examples of TEK, as well as an extensive treatment of the historical context and principles that underpin the systematic study and evaluation of traditional knowledge systems. This is generally known as ethnoscience, and embraces the now familiar fields of ethnobotany, ethnozoology, ethnoecology, ethnobiology and others. An impressive ethnobotanical summary of useful dry-land plants is provided by Prendergast, *et al.* (Chapter 5). Fairhead and Leach (Chapter 5) provide an eloquent example of how knowledge of termite ecology in West Africa helps local communities improve soils, manage water resources and increase crop production, while Sillitoe (Chapter 7) documents the soil management techniques used by the Wola speakers in the Southern Highlands of Papua New Guinea to conserve the fertility of delicate soils. Adimihardja (Chapter 5) describes how local botanical knowledge in the Mount Halimun area of West Java forms the basis for conserving the diversity of forests and traditional agriculture

Another area where TEK is well understood and exploited is that of agriculture. Many ancient indigenous agricultural and sustainability systems survived until the colonial period. These systems are complex, based on sophisticated ecological knowledge and understanding, highly efficient and productive, and inherently sustainable. Classic examples are the raised bed systems used for millennia by traditional farmers of tropical America, Asia and Africa. Known variously in Meso-America as *chinampas*, *waru waru*, and *tablones*, these were extremely effective for irrigation, drainage, soil fertility maintenance, frost control and plant disease management (Willett 1993). In India, peasants grow over forty different crops on localities that have been cultivated for more than 2,000 years without a drop in yields, yet have remained remarkably free of pests (Willett, ibid.).

Plenderleith (Chapter 7) emphasizes that traditional (folk) crop varieties are vital to modern agriculture for maintaining agro-biodiversity. Indeed, Richards (Chapter 7) suggests that using the ability of farmers such as the Mende in Sierra Leone to combine, select and screen planting materials may be a safer option for maintaining agrobiodiversity than dependence on gene bank collections. Toffoli and Oliveira (Chapter 7) show how traditional fishing communities in Brazil integrate agriculture into their overall natural resource management strategies. This agro-ecological approach is also employed by the Dayak Pasir of East Kalimantan (Enris and Sarmiah, Chapter 7). Traditional agriculture is increasingly shown to be 'productive, sustainable, and ecologically sound', even under extraordinarily difficult conditions such as in the Andean Altiplano (Earls 1989). The *waru-waru* system, for example, employs raised beds, canals and carefully selected plant varieties to overcome poor soils, high altitudes and low temperatures (Alteiri, Chapter 7).

Kabuye emphasizes (Chapter 7) the importance of 'wild foods' that come from managed areas that are not strictly speaking agricultural plots. This underlines the importance of gathered foods which come from a variety of landscapes of cultural significance – and which depend upon the maintenance of biodiversity. Her exposé on Buganda bark cloth (Chapter 9) reminds us that nature still provides not only foods and medicines but also shelter, clothing and, indeed, all of the basic needs of life for many traditional communities. That fact alone is reason enough to make the views and concerns of local peoples considerably different from those of urban dwellers who have lost touch with the link between their basic survival and the survival of healthy ecosystems (Girardet, Chapter 5).

It becomes obvious that local dependence on traditional varieties of crop plants, non-domesticated resources and gathered foods serves to stimulate biodiversity conservation, not destroy or homogenize it as most agro-industrial ('modern') systems do (Thrupp, Chapter 7). Indeed 'modern' agriculture has become one of the major threats to indigenous and local communities, as well as biodiversity, healthy ecosystems – and even food security (Thrupp, ibid.; Mann and Lawrence, Chapter 7).

Large-scale commercial forestry is also a major global threat to traditional knowledge and local communities (Sundar, Chapter 9; Banuri and Marglin 1993; Browder 1986). Participatory and community forestry have helped reduce this threat in many places world-wide (Carter 1996). But one of the major barriers to a fuller appreciation of 'forest-related knowledge' has been a lack of emphasis on non-timber forest products (NTFPs) and non-domesticated resources (NDRs). Methodologies have now been developed to tackle this problem. Smith and Meredith (Chapter 9) describe how assessment of local resource values helped communities in Ngorongoro District, Tanzania, identify their own biodiversity conservation priorities. In Brazil, Shanley and Galvao (Chapter 9) show that Amazon *Cablocos* can earn two to three times more from subsistence use of the forest than from the meagre income from timber sales. Without such economic valuation it is hard for local people to see the economic benefits of maintaining their traditional forest life-styles.

Another important area in which local knowledge plays a major role is in traditional medicines and health systems. Bodeker (Chapter 6) sketches some exemplary systems such as Ayurvedic and traditional Chinese medicines, whose cosmologies define disease as a 'breaking of the interconnectedness of life'. A fundamental concept in traditional health systems is that of balance between mind and body, given that both are linked to community, local environments and the universe (Shankar, Natarajan, Chapter 6). As Rippling Water Woman (Chapter 6) says: 'I have gained an understanding that the relationship a healer has with the environment is a reflection of the depth of understanding achieved of the personal relationship with all creation'. That is a main reason why medicines can not easily be 'extracted' from their 'knowers' or their social contexts (Tumwesigye, Chapter 6, Box 6.1).

Of course, medical concepts can be radically different from Western ones (Berlin and Berlin, Chapter 6), making evaluation of efficacy – not to mention global application – very difficult (Balick *et al.*, Chapter 6). Medicinal plant qualities also vary considerably depending upon when and where they are collected, as any herbalist well knows (Chapter 6). It is also important to remember that the distinctions between medicine, food and health are Western distinctions. For many indigenous and traditional peoples, foods are medicines and *vice versa*; in fact the Western division of the two makes little sense to many traditional peoples (Motte-Florac *et al.*, Chapter 6;

Hugh-Jones, Chapter 6; Hladik *et al.* 1993). And, above all, healthy ecosystems are critical to healthy societies and individuals, because humanity and nature are one, not in opposition to each other (Iwu, Chapter 6).

Despite these difficulties, WHO (1993) estimates that up to 80 percent of the non-industrial world's population still relies on traditional forms of medicines. And many indigenous groups are returning to their ancient medicines and incorporating traditional forms of treatment into their health service programmes (Hall 1986; Goodwill, Chapter 6). Even in industrialized countries, more and more people are turning to alternative health treatments. For example, Americans spend more on complementary approaches than on hospitalization, while Australians pay out more on alternative medicines than pharmaceuticals (Eisenberg *et al.* 1993; McLennan *et al.* 1996). In Britain, the Department of Health reported in 1995 that 40 percent of General Practice (GP) partnerships in England provide access to complementary medicine for their National Health Service (NHS) patients, and 24.6 percent actually make NHS referrals for complementary medicine (Foundation for Integrated Medicine).

Equity and rights

Recognition by the CBD of the contributions of indigenous and traditional peoples to maintaining biological diversity may be a major political advance, but there are major dangers. Once TEK or genetic materials leave the societies in which they are embedded, there is little national protection and virtually no international law to protect community 'knowledge, innovations and practices'. Many countries do not even recognize the basic right of indigenous peoples to exist – let alone grant them self-determination, land ownership, or control over their traditional resources.

'Farmers' Rights', which have been under discussion by the UN Food and Agriculture Organization (FAO) over the last two decades, is one of the few international attempts to recognize the contributions of traditional farmers to global food security. But their legal basis is weak and even meagre guarantees are resisted by some powerful countries. The global fund established to ensure forms of compensation for local farmers remains inoperative. FAO is undertaking a revision of its International Undertaking on Plant Genetic Resources (IUPGR) with the view of strengthening or expanding Farmers' Rights, but the political road to such improvements is rocky and uncertain (GRAIN 1995; UNEP 1997; Plenderleith, Chapter 7).

The International Labour Organisation (ILO) Convention 169 is the only legally binding international instrument specifically intended to protect indigenous and tribal peoples. ILO 169 is clear in its commitment to community ownership and local control of lands and resources. It does not, however, cover the numerous traditional and peasant groups that are also critical to conserving the diversity of agricultural, medicinal and non-domesticated resources. To date the Convention has only 10 States Parties, and provides little more than a base-line for debates on indigenous rights (Posey 1996).

The same bleak news comes from an analysis of Intellectual Property Rights (IPRs) laws. IPRs were established to protect individual inventions and inventors, not the collective ancient folklore and TEK of indigenous and local communities. Even if IPRs were secured for communities, differential access to patents, copyright, know-how and trade secret laws, and to legal aid, would generally price them out of any effective registry, monitoring or litigation using such instruments (Dutfield, Chapter 12; Posey and Dutfield 1996). Box 1.1 summarizes how IPRs are an inadequate and inappropriate way of protecting the collective resources of indigenous and traditional peoples.

The World Trade Organization's General Agreement on Tariff and Trade (WTO/GATT) contains no explicit reference to the knowledge and genetic resources of traditional peoples, although it does provide for states to develop *sui generis* (specially generated) systems for plant protection (TRIPs Article 27(c)). Considerable intellectual energy is now being poured by governments, non-government and peoples' organizations into defining what new, alternative models of protection would include (Leskien and Flitner 1997). There is scepticism, however, over whether this *sui generis* option will be adequate to provide any significant alternatives to existing IPRs (Montecinos 1996).

One glimmer of hope comes from the CBD's decision to implement an 'intersessional process' to evaluate the inadequacies of IPRs and develop guidelines and principles for governments seeking advice on access and transfer legislation to protect traditional communities (CBD Secretariat 1996; UNEP 1997). This provides exciting opportunities for many countries and peoples to engage in an historic debate. Up to now, United Nations agencies have been reluctant to discuss 'integrated systems of rights' that link environment, trade and human rights. However, agreements between the CBD, FAO and WTO now guarantee broad consultations between the World Intellectual Property Organization (WIPO), United Nations Education and Scientific Organization

Box 1.1: Inadequacies of intellectual property rights

Intellectual Property Rights are inadequate and inappropriate for protection of traditional ecological knowledge and community resources because they:

- recognize individual, not collective, rights;
- require a specific act of 'invention';
- simplify ownership regimes;
- stimulate commercialization;
- recognize only market values;
- are subject to economic powers and manipulation;
- are difficult to monitor and enforce, and
- are expensive, complicated, time-consuming.

(UNESCO), United Nations Environment Programme (UNEP), United Nations Development Programme (UNDP), United Nations Commission on Trade and Development (UNCTAD), International Labour Organization (ILO), the Geneva Human Rights Centre, and others. It will take the creative and imaginative input of all these groups – and many more – to solve the complicated challenge of devising new systems of national and international laws that support and enhance cultural and biological diversity.

The principles of *sui generis* systems of rights have in many ways already been established in international conventions like the CBD and ILO 169, as well as major human rights agreements such as the International Covenant on Civil and Political Rights (ICCPR), the International Covenant on Economic, Social and Cultural Rights (ICESCR), and of course the Universal Declaration of Human Rights (UDHR). For indigenous peoples, the Draft Declaration on the Rights of Indigenous Peoples (DDRIP) is the most important statement of basic requirements for adequate rights and protection. Over a period of nearly two decades the DDRIP has been developed, guided by hundreds of indigenous representatives in consultation with the UN Working Group on Indigenous Populations of the Geneva Human Rights Centre (Gray, Chapter 3). It is broad-ranging, thorough, and reflects one of the most transparent and democratic processes yet to be seen in the United Nations. Box 1.2 provides some of the principles affirmed by the DDRIP. The complete text is provided in Appendix 1.18.

The global balance sheet

Although international efforts to recognize indigenous, traditional and local communities are welcome and positive, they are pitted against enormous economic and market forces that propel the globalization of trade. Critiques of globalization are numerous and point to at least two major short-comings: (i) value is imputed to information and resources only when they enter external markets, and (ii) prices do not reflect the actual environmental and social costs of the products. This means that existing non-monetary values recognized by local communities are ignored, despite knowledge that local biodiversity provides essential elements for survival (food, shelter, medicine, etc.). It also means that the knowledge and managed resources of indigenous and traditional peoples are ascribed no value and assumed to be free for the taking. This has been called 'intellectual *terra nullius*' after the concept (empty land) that allowed colonial powers to expropriate 'discovered' land for their empires. Corporations and States still defend this morally vacuous concept because it facilitates the 'biopiracy' of local folk varieties of crops, traditional medicines and useful species.

Even scientists have been accomplices to such raids by publishing data they know will be catapulted into the public domain and gleaned by 'bioprospectors' seeking new products. They have also perpetuated the 'intellectual *terra nullius*' concept by declaring useful local plants as 'wild' and entire ecosystems as 'wildernesses', often despite knowing that these have been moulded, managed and protected by human populations for millennia. It is also common for scientists to declare areas and resources 'wild' through ignorance – or negligence – without even the most basic investigations into archaeological, historical or actual human management practices. The result is to declare the biodiversity of a site as 'natural', thereby transferring it to the public domain. Once it is public, its communities are stripped of all rights to their traditional resources.

It is little wonder then that indigenous groups in the Pacific region have declared a moratorium on all scientific research until protection of traditional knowledge and genetic resources can be guaranteed to local communities by scientists. The 'moratorium movement' began with the 1993 Mataatua Declaration (Appendix 1.5):

> *'A moratorium on any further commercialization of indigenous medicinal plants and human genetic materials must be declared until indigenous communities have developed appropriate protection mechanisms.'*

Box 1.2: Some Principal Rights Affirmed by the Draft Declaration on the Rights of Indigenous Peoples

- Right to self-determination, representation and full participation.
- Recognition of existing treaty arrangements with indigenous peoples.
- Right to determine own citizenry and obligations of citizenship.
- Right to collective, as well as individual, human rights.
- Right to live in freedom, peace and security without military intervention or involvement.
- Right to religious freedom and protection of sacred sites and objects, including ecosystems, plants and animals.
- Right to restitution and redress for cultural, intellectual, religious or spiritual property that is taken or used without authorization.
- Right to free and informed consent (Prior Informed Consent).
- Right to control access and exert ownership over plants, animals and minerals vital to their cultures.
- Right to own, develop, control and use the lands and territories, including the total environment of the land, air, waters, coastal seas, sea-ice, flora and fauna and other resources which they have traditionally owned or otherwise occupied or used.
- Right to special measures to control, develop and protect their sciences, technologies and cultural manifestations, including human and other genetic resources, seeds, medicines, knowledge of the properties of fauna and flora, oral traditions, literatures, designs and visual and performing arts.
- Right to just and fair compensation for any such activities that have adverse environmental, economic, social, cultural or spiritual impact.

The Mataatua Declaration in turn influenced the Final Statement of the 1995 Consultation on Indigenous Peoples' Knowledge and Intellectual Property Rights in Suva, Fiji (Pacific Concerns Resource Centre 1995), which called for a moratorium on bioprospecting in the Pacific region and urged Indigenous peoples not to co-operate in bioprospecting activities until appropriate protection mechanisms are in place. They should:

- demand that bioprospecting as a term be clearly defined to exclude indigenous peoples' customary harvesting practices;
- assert that *in situ* conservation by indigenous peoples is the best method by which to conserve and protect biological diversity and indigenous knowledge, and encourage its implementation by indigenous communities and all relevant bodies, and
- encourage indigenous peoples to maintain and expand our knowledge of local biological resources.

To allay these deep concerns, many scientific and professional organizations are developing their own Codes of Conduct and Standards of Practice to guide research, health, educational and conservation projects with indigenous and local communities (a summary of some of these can be found in Cunningham 1993; Posey 1995; and Posey and Dutfield 1996). One of the most extensive is that of the International Society for Ethnobiology, which undertook a ten-year consultation with indigenous and traditional peoples – as well as its extensive international membership – to establish 'principles for equitable partnerships'. The main objective of the process was to establish terms under which collaboration and joint research between ethnobiologists and communities could proceed based upon trust, transparency and mutual concerns. A list of these principles can be found in Box 1.3.

Similar principles have been elaborated by environmental philosophers, ethicists and eco-theologians (Golliher, Chapter 11). Unfortunately, most of their efforts have been couched in such rarefied discourses that they have had little impact on the practice of science or on public policy.

There are some important exceptions to this, notably 'deep ecology' (Naess 1989; Devall 1988; Fox 1990; Sessions 1995), which has inspired a militant 'Earth First' movement aimed at extinguishing the anthropocentric view that humans have the right to do as they wish with other life-forms. For deep ecologists, 'the *hubris* in asking people "to take responsibility" for the environment is replaced by an invitation to realize the depth of existing ecological relationships' (Golliher, ibid.). Ingold (Ingold *et al.* 1988) has long argued for a discourse that replaces anthropocentrism and ethnocentrism with an 'ontological equality' (see also Clark, Chapter 5). Gari (Chapter 5) believes this will be possible only when the concept of biodiversity is equated with a 'biosphere essence'. To a large extent this requires shifting priorities from instrumental values (in what way is biodiversity useful to humans) to intrinsic values (all life is valuable whether or not it is of use to humans) – not an easy task in a world dominated by economics and global trade.

'Ecofeminism' has also been instrumental in pointing out how unequal gender and power relations have operated to separate 'nature' from 'spirit', thereby catalysing disrespect for biodiversity and the destruction of ecosystems (Ruether 1992 and Chapter 11; Adams 1993; Mies and Shiva 1993; Plaskow and Chirst 1989; Primavesi 1991). This emphasis on 'spirit' provides a much-needed bridge between cultures, since 'cosmovisions' are the organizing spiritual and conceptual models used by indigenous and traditional peoples to integrate their society with the world. These cosmovisions are based on the 'sacred balance' of cosmic forces that unite human beings (males and females equally) with all life (again, equally shared). Reichel (Chapter 3) illustrates this with her theory of Gender-Based Knowledge Systems (GBKS) of Tanimuka and Yukuna Indians of Colombia.

This is similar to the 'web of life' and 'tree of life' concepts that are central to most world religions. Many major religious institutions find it hard to reinterpret their own founding tenets in a modern world torn from ancient traditions and thrust into revolutions in materialism, consumerism, globalized trade and instant global communications. Some are trying to do this through the 'Assisi Process' which encourages soul-searching to find ways of 'greening' major religious institutions and to encourage inter-faith dialogue (Palmer, Chapter 11; Edwards and Palmer 1997; Jensen, Chapter 11; Jensen 1997). Neither process has proved easy, but the latter is especially difficult given the monistic tendencies of each to teach and preach THE truth.

Many people have given up on the fractious world of religions in favour of a more practical 'ethic' of land, biodiversity and environment. This movement takes its inspiration from Aldo Leopold's (1949) ideas of 'land ethic' and 'environmental citizenship'. Callicott (Chapter 11) argues for the need for a global ethic formulated around respect for the diversity of cultures and ecosystems. It may be that the 'need' is not just the artefact of human psychology and moral reflection, but rather that it is spiritually and psychologically grounded. Roszak (1992) believes that the environmental crisis is rooted in the extreme 'disturbance' of the web of life that is a part of human consciousness. Indeed, a basic precept of ecology itself is that disturbance of one element of an environmental system affects all other elements, as well as the whole (Capra, Chapter 11). It may be conjecture as to how *Homo sapiens* is psychologically affected by the overall loss of biological and cultural diversity, but certainly indigenous, traditional and local communities are aware of the negative local effects – and they express their profound concerns in cultural and spiritual terms precisely because they recognize the deep-rootedness of the disturbance.

Box 1.3: Principles for 'Equitable Partnerships' Established by the International Society for Ethnobiology

1. **Principle of Self-Determination.** Recognizes that indigenous peoples have a right to self-determination (or local determination for traditional and local communities) and that researchers shall as appropriate acknowledge and respect such rights. Culture and language are intrinsically connected to land and territory, and cultural and linguistic diversity are inextricably linked to biological diversity; therefore, the principle of self-determination includes: (i) the right to control land and territory; (ii) the right to sacred places; (iii) the right to own, determine the use of, and receive accreditation, protection and compensation for, knowledge; (iv) the right of access to traditional resources; (v) the right to preserve and protect local language, symbols and modes of expression, and (vi) the right to self-definition.
2. **Principle of Inalienability.** Recognizes that the inalienable rights of indigenous peoples and local communities in relation to their traditional lands, territories, forests, fisheries and other natural resources. These rights are both individual and collective, with local peoples determining which ownership regimes are appropriate.
3. **Principle of Minimum Impact.** Recognizes the duty of scientists and researchers to ensure that their research and other activities have minimum impact on local communities.

Box 1.3 (continued)

4. **Principle of Full Disclosure.** Recognizes that it is important for the indigenous and traditional peoples and local communities to have disclosed to them (in a way that they can comprehend), the manner in which the research is to be undertaken, how information is to be gathered, and the ultimate purpose for which such information is to be used and by whom it is to be used.
5. **Principle of Prior Informed Consent and Veto.** Recognizes that the prior informed consent of all peoples and their communities must be obtained before any research is undertaken. Indigenous peoples, traditional societies and local communities have the right to veto any programme, project or study that affects them.
6. **Principle of Confidentiality.** Recognizes that indigenous peoples, traditional societies and local communities, at their sole discretion, have the right to exclude from publication and/or to be kept confidential any information concerning their culture, traditions, mythologies or spiritual beliefs and that such confidentiality will be observed by researchers and other potential users. Indigenous and traditional peoples also have the right to privacy and anonymity.
7. **Principle of Active Participation.** Recognizes the critical importance of communities to be active participants in all phases of the project from inception to completion.
8. **Principle of Respect.** Recognizes the necessity for Western researchers to respect the integrity of the culture, traditions and relationships of indigenous and traditional peoples with their natural world and to avoid the application of ethnocentric conceptions and standards.
9. **Principle of Active Protection.** Recognizes the importance of researchers taking active measures to protect and enhance the relationships of communities with their environments, thereby promoting the maintenance of cultural and biological diversity.
10. **Principle of Good Faith.** Recognizes that researchers and others having access to knowledge of indigenous peoples, traditional societies and local communities will at all times conduct themselves with the utmost good faith.
11. **Principle of Compensation.** Recognizes that communities should be fairly, appropriately and adequately remunerated or compensated for access to and use of their knowledge and information.
12. **Principle of Restitution.** Recognizes that where, as a result of research being undertaken, there are adverse consequences and disruptions to local communities, those responsible will make appropriate restitution and compensation.
13. **Principle of Reciprocity.** Recognizes the inherent value to Western science and humankind in general of gaining access to the knowledge of indigenous peoples, traditional societies and local communities and the desirability of reciprocating that contribution.
14. **Principle of Equitable Sharing.** Recognizes the right of communities to share in the benefits accruing from products or publications developed from access to, and use of, their knowledge, and the duty of scientists and researchers to equitably share these benefits with indigenous peoples.

The worrisome lesson from all of this is that the global environmental crisis cannot be solved by technological tampering ('quick fixes') or superficial political measures. The Native American leader Black Elk puts it thus: 'It is the story of all life that is holy and good to tell, and of us two-leggeds sharing in it with the four-leggeds and the wings of the air and all green things; for these are children of one mother and their father is one spirit.' (Neihardt 1959; Suzuki, Chapter 3).

For industrialized society to reverse the devastating cycle it has imposed on the planet, it will have to invent an 'ecology ' powerful enough to offset deforestation, soil erosion, species extinction and pollution; 'sustainable practices' that can harmonize with growth of trade and increased consumption, and, of course, a 'global environmental ethic' that is not subverted by economically powerful institutions. That may be an impossible task – but there are some viable paths.

One approach is to listen to indigenous and traditional leaders who have become effective leaders in the environment and human rights movements. As Joji Cariño, representing the International Alliance of Indigenous and Tribal Peoples of the Tropical Forests, said in her address to the United Nations General Assembly (June 1997), 'We are part of a powerfully resurgent movement of indigenous peoples world-wide, who today find ourselves at the chalk face of the global crisis of development and environment. Our greatest strength is in the local activities of indigenous peoples who love and care for our territories and environments'. The Declarations and Statements found in Appendix 1 of this volume stand as additional monuments to the eloquence, clarity and forcefulness with which indigenous and traditional peoples' organizations articulate their concerns.

Another path is to relearn the ecological knowledge and sustainable principles that our society has lost. As Bepkororoti Paiakan, a Kayapo chief from Brazil, puts it: 'We are trying to save the knowledge that the forests and this planet are alive – to give it back to you who have lost the understanding'. Learning can come through listening to the 'Voices of the Earth' – the peoples of the planet who still know when birds nest, fish migrate, ants swarm, tadpoles develop legs, soils erode and rare plants seed – and whose cosmovisions manifest the ecologies and ethics of sustainability. Chapter 4 provides a rich collage of depositions from indigenous and traditional peoples to inspire us in this task.

But listening is not enough: we must uphold their basic rights to land, territory, knowledge and traditional resources. And we must discover how economic and utilitarian measures can be countered in our own societies by rediscovering the 'web of life through the cultural and spiritual values of our own traditions.

A postscript on the responsibility of knowing

Homo sapiens was named because of its sentience, which is both a blessing and a curse. To know is a powerful trait, but the responsibility of knowing is a terrible burden. Too often science becomes obsessed with collecting data, but forgets why such information was collected in the first place. Christine Morris (Box 1.4) provides a story inspired by her native Kombumerri/Munaljahlai clans of Australia, which tells of a 'bright boy' who is determined to study a 'Christmas beetle'. He discovers, however, that to do so he must become a beetle and suffer the inflictions that other 'bright boys' have imposed on the non-human world.

Box 1.4: The Christmas Beetle and The Bright Boy

Christine Morris

'Why?', said the cicada to the bright boy, 'Why do you ask?

'Why do I ask? I ask because I want to know!...'

'You want to know! Is that the only reason?'

'Yes! What's wrong with that? I don't need any other reason. If I want to know what it's like to be a Christmas beetle, I don't need any other reason. If I want to know something, I make it my business to find out!'

'Find out? Suppose you could find out what it's like to be a Christmas beetle. And then when you find it, what are you going to do with it?'

'Do with it... Well nothing, I just want to know... and perhaps in the future I'll have a need for it.'

'Perhaps you will have a need for it...mmm...'

'Yes!'

'Then when you've got it you don't always have a place for it? You just think you may have a use in the future... mmm... Tell me, in the meantime, what do you do with it?

'Do with it? Well, I'll just store it in my head!'

'Just store it in your head....mmm'

'I don't like the way you say that. What's wrong with storing it in your head?

'Do you know each new piece of information takes up space in your consciousness?'

'Yes. That makes sense and the brain can hold enormous amounts of information, just like a computer!'

'Ah yes. That's if it has a place to store it!'

'What do you means by a place?'

'Well, the information you have acquired so far in life has a ready framework or set of criteria in which to fit. For example, you may learn the name of every football team on the East Coast, or the scientific names for all the plants in a specific area; however, when you start learning what it is like to be a Christmas beetle, you'll find you don't have any predefined categories because you've never come across such information. I mean, is it an innate Bright Boy desire to be a Christmas beetle, or are you actually asking something that is beyond your physiological knowing? I mean you can slot it into places you think it fits and then you soon find it doesn't fit and so it comes to the conscious again and demands a new place, just like the interfering email sign that interrupts all your computer working. It just overrides everything and demands a response, just like the telephone. So you see, if you do not have the presumed knowledge – that is, a category or known experience in which to fit this knowledge – it will keep resurfacing and resurfacing and if you're not careful it will drive you crazy or distract you from what you're doing and so may cause a nasty accident.'

'An accident?'

'Yes. I'm afraid you just presume that all knowledge fits in somewhere in your Bright Boy capacity for knowing. Then again, maybe you have got that capacity, but you have to know which category or door to open in your mind. Just think what kind of knowing do you need before you can assimilate the experience of being a Christmas beetle? What if you could suddenly fly by your own volition? Would that affect your overall mind? What do you think you would feel if you were suddenly two inches long and a great clumsy Bright Boy was making his way towards you? What if he wanted to catch you and put you in a bottle because you were so pretty? What then Bright Boy? Could you handle the fear? Could you handle the feeling of insignificance? Could you handle feeling your life was in the hands of some clumsy giant that wanted to play with you? What then, Bright Boy? How would your sensibilities stand up to such an onslaught? Let alone your mind?'

Box 1.4 (continued)

'I didn't say I wanted to be a Christmas beetle. I just wanted to know what it was like.'

'Ah my friend, if I told you what it was like you would suddenly wish you were only two inches long and blended into the environment so that you could hide the shame of what the Bright Boys of the world do to those that are other than themselves!'

Bright Boy stared at the Christmas beetle for a while, and then took the beetle from his hand and placed it back in its home.

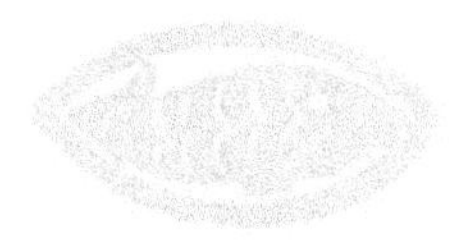

CHAPTER 2

LINGUISTIC DIVERSITY

Luisa Maffi

CONTENTS

TEXT BOXES

FIGURES AND TABLES

Linguistic Diversity

Luisa Maffi

Introduction (Luisa Maffi)

Along with the loss of biodiversity and the erosion of traditional cultures, the world is currently undergoing a third extinction crisis: that of the diversity of human languages. Causes and consequences of all these phenomena reside in the ever more unsustainable exploitation of the earth's natural resources, as well as in the growing marginalization and dispossession of indigenous and minority groups who struggle for survival and self-determination as distinct peoples, with their own land bases and means of subsistence. Increasingly, these peoples see their languages and cultural traditions – and the fight to protect or restore their linguistic rights – as essential elements in this struggle. Language and land are considered by most of them as equally constitutive of their identity as sovereign peoples, and of their right to live as such.

This should come as no surprise. Language plays a key role in all aspects of human life everywhere. It is central to our conceptualization of the world, and for interpreting, understanding and changing it. Initially the language(s) we learn give us the categories to conceive our natural and social world. If an object, process or relationship has been important in the life of our people, it gets named, and by learning that word we also learn what is vital for us to know in our natural and social environment. If we needed to discern every single detail of what we see, our world would be chaotic and our mental energy overtaxed. Language helps us in organizing our world and frees our energy for other tasks. Words for concepts are like pegs on which we hang the meanings that we store in the storehouse of our mind. They are the framework that binds together the details into a totality, a meaningful whole. Verbalizing helps us remember and reproduce meaning and thus make sense of reality. Through the verbalization process we also learn much of our own culture's ethics. Together with the words for objects and phenomena, we learn our culture's connotations, associations, emotions and value judgements. The definition of our ecosocial world, including group identity, status and world view, is again realized through language.

But language plays all these roles in different ways for different language communities. The distinct ecosocial adaptations that each group has elaborated and continues to elaborate in response to changing circumstances are both largely realized by means of language and reflected in it. The particular social and ecological circumstances in which different human groups develop over time – the specific relationships each group establishes among its members and with other people around, as well as with the place in which they live – lead to different and historically changing ways of defining, understanding and interpreting the world via language. The diversity of languages (and cultures) around the world has arisen through these complex and dynamic processes.

Yet this intrinsic and defining role of language in human biocultural diversity is still not well understood in academic, policy-making and advocacy circles alike – while it is salient in the cosmologies, philosophies and traditional narratives of scores of indigenous and minority peoples world-wide. In international debates on biodiversity conservation, it is becoming clear that the link between biological and cultural diversity is an inextricable one, and that it is necessary to think of preserving the world's biocultural diversity as an integrated goal. What has so far largely remained outside the scope of such debates is the role of language, and of the continued presence of a variety of languages on earth, in the maintenance of biocultural diversity (as well as in ensuring equitable and peaceful existence for hundreds of millions of people on earth).

Recently, however, scholars, practitioners and indigenous and minority advocates have begun to address this issue (Nabhan 1996; Maffi 1996, 1997), revealing an emerging convergence of opinion and action concerning the diversity extinction crisis that is affecting the world's languages, cultures and environments. These experts point to the close interdependence of linguistic, cultural and biological diversity, suggesting that the diversity of languages and cultures may share much of the same nature, and serve much of the same functions, as the diversity of natural kinds in ensuring the perpetuation of life on earth. On these bases, they argue that the preservation of the world's linguistic diversity, and of the distinct forms of local knowledge that indigenous and minority languages encode, must be incorporated as an essential goal in bioculturally-oriented diversity conservation programmes.

Fostering this convergence of perspectives requires strengthening the links among the various fields of research and applied work involved in diversity conservation. Experts from many different specialties must overcome disciplinary and other institutional and intellectual barriers that exist among themselves, as well as *vis-à-vis* indigenous and minority peoples, and work together in interdisciplinary and intercultural teams to understand, and devise solutions for, what is perhaps the single most complex issue facing humanity on the verge of the third millennium. It should also become increasingly common for individual specialists to be competent in two or more of the fields involved. Both of these developments will require a profound rethinking of approaches to teaching and training at all levels of education, to promote intercommunication between the biological sciences, the social sciences and the humanities. It will be no less essential to broadcast these views among the general public in order to gain support for research and action in this domain.

Furthermore, it will have to be understood that theoretical and applied issues are two sides of the same coin, as are scientific and ethical issues, and that there must be a genuine commitment to dealing with the two sides together. In particular, it will have to be acknowledged that no effective solutions for biocultural diversity conservation – and for the safety and well-being of hundreds of millions of people on earth – can ultimately be found without taking into account the rights of the world's indigenous and minority peoples: their right to continue to develop as distinct peoples with their own languages and cultural traditions, reversing the culturally and linguistically homogenizing trend that current global socio-economic processes are bringing about along with the drastic reduction of biological diversity.

Nietschmann (1992: 3) has proposed a 'Rule of Indigenous Environments': 'Where there are indigenous peoples with a homeland there are still biologically-rich environments'. If we understand this rule as implying self-determination not only as concerns indigenous peoples' land rights and traditional resource rights, but also their cultural and linguistic rights, we may be close to what it will take to preserve biocultural diversity on earth.

Language and the environment (Luisa Maffi)

Linguistic diversity and language endangerment

As with biodiversity, there are various definitions of linguistic diversity. Most commonly, the number of different languages spoken on earth is used as a proxy for global linguistic diversity. Linguists estimate that there are 5,000 to 7,000 oral languages spoken today on the five continents (Krauss 1992; Grimes 1996). These figures do not take into account sign languages used by deaf people (Supalla 1993), with the addition of which the number of languages in the world would probably go up to over 10,000 (Branson and Miller 1995). *Ethnologue*, the best existing catalogue of the world's languages (Grimes 1996), gives a total of 6,703 languages (mostly oral), of which 32 percent are found in Asia, 30 percent in Africa, 19 percent in the Pacific, 15 percent in the Americas and 3 percent in Europe. Of these languages, statistics indicate that about half are spoken by communities of 10,000 speakers or less; and half of these, in turn, are spoken by communities of 1,000 or fewer speakers (Harmon 1995, based on Grimes 1992). Overall, languages with up to 10,000 speakers total about 8 million people, less than 0.2 percent of an estimated world population of 5.3 billion (ibid.; Figures 2.1 and 2.2). On the other hand, of the remaining half of the world's languages, a small group of less than 300 (such as Chinese, English, Spanish, Arabic, Hindi, and so forth) are spoken by communities of 1 million speakers or more,

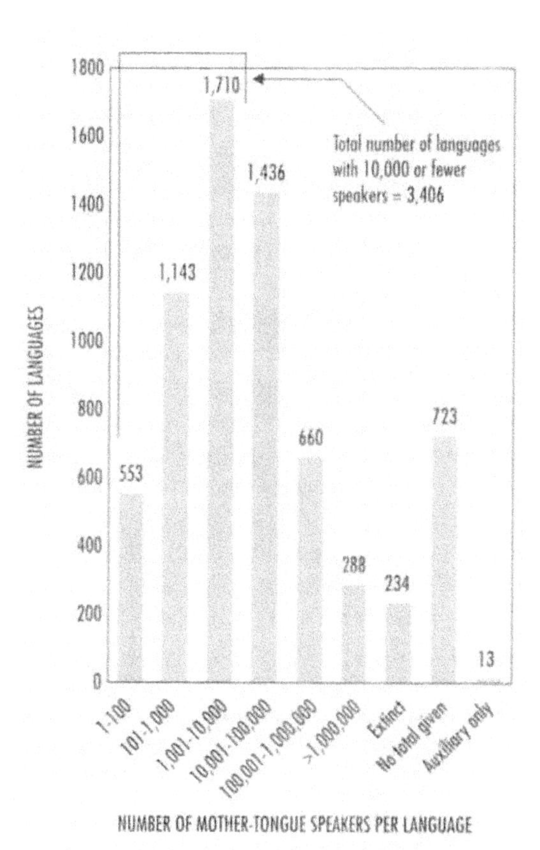

Figure 2.1: Size classification of world's languages by number of mother-tongue speakers (from Harmon 1995).

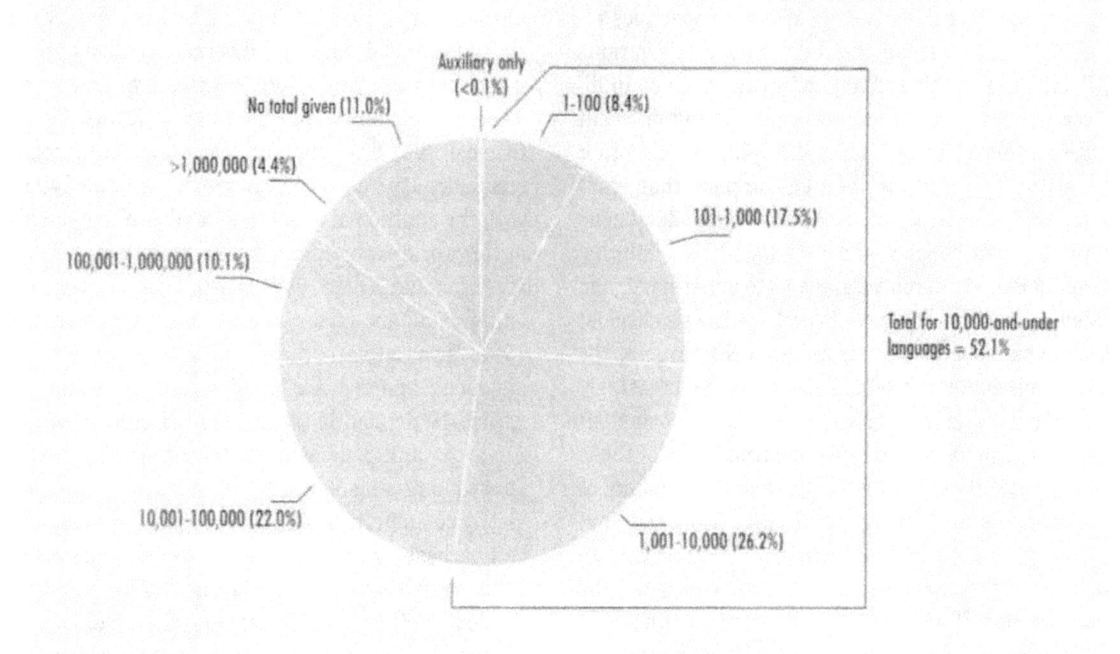

Figure 2.2: Proportion of world's living languages by size category (n = 6,526) (from Harmon 1995).

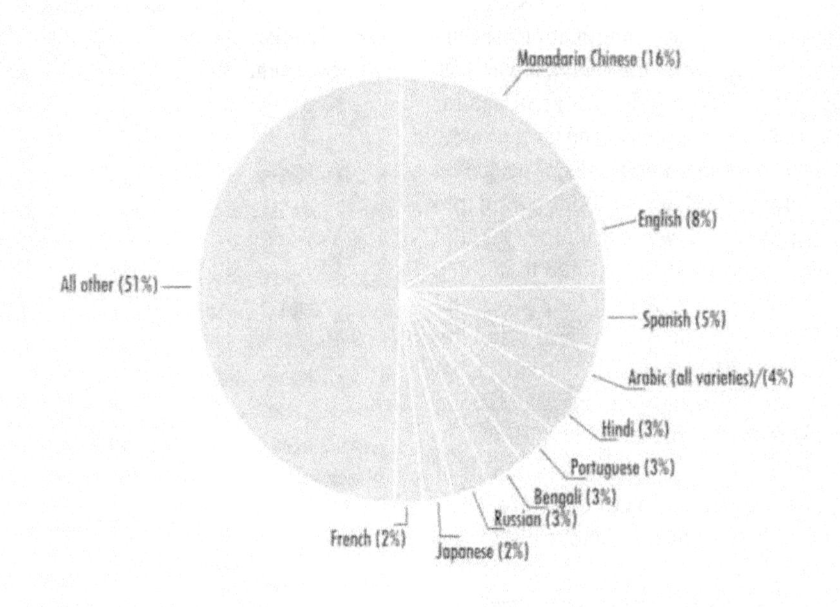

Figure 2.3: Languages with the most mother-tongue speakers: proportion of world population (from Harmon 1995).

accounting for a total of over 5 billion speakers, or close to 95 percent of the world's population. The top ten of these alone actually comprise almost half of this global population (ibid.; Figure 2.3).

Taken together, these figures show that, while more than nine out of ten people in the world are native speakers of one or other of only about 300 languages, most of the world's linguistic diversity is carried by very small communities of indigenous and minority people. The large majority of the world's languages, and particularly the smaller languages, are concentrated in just a few countries, and most are actually 'endemic', i.e. fully comprised within the borders of a single state, and therefore heavily subject to national linguistic policies (Harmon 1995; Table 2.1).

TABLE 2.1: TOP 25 COUNTRIES BY NUMBER OF ENDEMIC LANGUAGES

1. Papua New Guinea (847)
2. Indonesia (655)
3. Nigeria (376)
4. India (309)
5. Australia (261)
6. Mexico (230)
7. Cameroon (201)
8. Brazil (185)
9. Zaire (158)
10. Philippines (153)
11. USA (143)
12. Vanuatu (105)
13. Tanzania (101)
14. Sudan (97)
15. Malaysia (92)
16. Ethiopia (90)
17. China (77)
18. Peru (75)
19. Chad (74)
20. Russia (71)
21. Solomon Islands (69)
22. Nepal (68)
23. Colombia (55)
24. Côte d'Ivoire (51)
25. Canada (47)

Modified from Harmon (1996a). Source of language data: Grimes (1992).

In most recent work on biocultural diversity, the diversity of human languages has been used as the best available indicator of human cultural diversity (e.g. Clay 1993; Durning 1993). Harmon (1996b: 5) argues that, while no individual aspect of human life can be taken as a diagnostic indicator of cultural distinctiveness, language is the carrier of many cultural differences; furthermore, using language as a proxy 'affords us the best chance of making a comprehensible division of the world's peoples into constituent cultural groups'. A proxy is, admittedly, an imperfect tool. Linguistic diversity as a proxy for cultural diversity works better on the global scale than it may work in any individual instances. In many cases, distinctiveness of languages does not correspond to distinctiveness of cultures, or sameness of language to sameness of culture. What matters in this context, however, is the possibility of identifying general trends rather than the ability to satisfactorily account for every single case.

The smaller languages of the world are under threat due to ever-growing assimilation pressures that promote incorporation into 'mainstream' society and the collective abandonment of the native languages in favour of acquired majority languages (a phenomenon known as 'language shift'). Virtually all languages with 1,000 speakers or less are threatened in this sense, although even more widely spoken languages are fully susceptible to the same pressures. Many of these smaller languages are already at risk of disappearing due to a drastic reduction in the number of their speakers, with younger generations decreasingly or no longer learning their language of heritage. Many more have reached a stage of near extinction, with only a few elderly speakers left. Grimes (1996) classifies over 420 languages out of 6,703 (or almost 6.3 percent) in the 'nearly extinct' category, while cautioning that these figures are conservative. Harmon (1995), also conservatively, computed 705 'moribund' (nearly extinct) languages (or 10.8 percent of the 6,526 non-extinct languages reported in Grimes 1992). In some projections, as many as 90 percent of the world's languages may become extinct or moribund in the course of the next century (Krauss 1992; also Robins and Uhlenbeck 1991).

It is a historical fact that languages, like species, have undergone extinction before. Estimates of the extent of the world's linguistic diversity in the undocumented past are bound to remain conjectural and subject to debate. However, informed guesses suggest that the peak of linguistic diversity on earth may have occurred at the beginning of the Neolithic (10,000 years BP), at which time more than twice the current number of languages may have been spoken (Hill 1997). Population movements and political and economic expansion, even well before the era of colonization and empire-building, have long contributed to reducing linguistic diversity everywhere in the world, either by the physical elimination of conquered groups or by their forced assimilation. Awareness of the political implications of linguistic assimilation was perhaps never better expressed than by the 15th century Spanish grammarian Antonio de Nebrija. In 1492, presenting Queen Isabella of Spain with his grammar of Spanish (the first grammar of any modern European language), Nebrija so explained its purposes in his introduction: 'Language has always been the consort of empire' (Illich 1981: 6). Or, as Hale (1992: 1) puts it: '[L]anguage loss in the modern period [...] is part of a much larger process of loss of cultural and intellectual diversity in which politically dominant languages and cultures simply overwhelm indigenous local languages and cultures, placing them in a condition that can only be described as embattled. The process is not unrelated to the simultaneous loss of diversity in the zoological and botanical worlds.'

As with species, what is unprecedented is an extinction crisis of the present magnitude and pace. Already there may be 15 percent fewer languages now that in 1500 AD, when the era of European colonization began (Bernard 1992), losses having been especially marked in the Americas and Australia. And the trend is now accelerating throughout the world, with Australia and the Americas (especially the USA) still in the lead. Of the 420+ 'nearly extinct' languages listed in Grimes (1996), 138 are spoken in Australia and 67 in the USA. Also as with species, for the majority of languages at risk most or all documentation is lacking, so that if they cease to be spoken, this will be a total and irretrievable loss.

Linguistic ecologies

For most of human history, humans have spoken a large number of small languages (Hill 1997), often with high concentrations of different languages coexisting side by side in the same areas. Complex networks of multilingualism in several local languages and pidgins or lingua francas have been a commonplace way of dealing with cross-language communication in situations of contact (Pattanayak 1981; Edwards 1994; Mühlhäusler 1996). Aboriginal peoples of Australia and New Guinea constitute especially well known persisting instances of the extensive multilingualism that once characterized human societies the world over, and that has been a key factor in the maintenance of linguistic diversity historically. Multilingualism *vis-à-vis* neighbouring groups may be explicitly practised as an essential element of social structure, as for example in the Amazonian Vaupés region (Sorensen 1972).

As Mühlhäusler (1996) points out, we are only beginning to realize there may be structure to such linguistic diversity. Mühlhäusler (building on and going beyond Haugen 1972) calls the functional relationships that develop in space and time among linguistic communities that communicate across language barriers 'linguistic ecologies'. He points out that an ecological theory of language has as its focus the diversity of languages *per se* and investigates the functions of such diversity in the history of humanity. He proposes that '[t]he mechanisms that have kept complex linguistic ecologies functioning are the ones a functioning multilingual and multicultural society will require.' (Mühlhäusler 1995b: n.p.). Based on his research in the Pacific region, Mühlhäusler (1996: 47) also suggests that the physical environment is an intrinsic part of traditional linguistic ecologies, in which no separation is felt to exist 'between an external reality or environment on the one hand and the description of this reality or environment on the other'.

These points raise the question of whether human linguistic and cultural diversity and diversification may not share substantive characteristics with biological (including human biological) diversity and diversification – characteristics that ultimately are those of all life on earth. At issue here is the adaptive nature of variation in humans (as well as other species), and the role of language and culture as providers of diversity in humans. Human culture is a powerful adaptation tool, and language at one and the same time enables and conveys much cultural behaviour. While not all knowledge, beliefs and values may be linguistically encoded, language represents the main instrument for humans to elaborate, maintain, develop and transmit such ideas. It becomes then possible to suggest that 'Linguistic diversity... is at least the correlate of (though not the cause of) diversity of adaptational ideas' (Bernard 1992: 82). It is true that diversity characterizes languages (and cultures) not just with respect to one another, but also internally, with patterns of variation by geographical location, age grade, gender, social status and a host of other variables. This internal variation combines with the variation ensuing from historical contact among human populations in propelling language and culture change and all manner of innovation. However, as more and more languages and cultural traditions are overwhelmed by more dominant ones and increasing homogenization ensues, one of the two main motors of change and innovation – the observation of cross-linguistic and cross-cultural difference – breaks down, or is seriously damaged. The end result is a global loss of diversity.

So far, the consequences of loss of linguistic and cultural diversity have been discussed in terms of ethics and social justice, and of maintaining the human heritage from the past (Thieberger 1990). We may now begin to examine them also as posing questions about the future, about the continued viability of humanity on earth: 'any reduction of language diversity diminishes the adaptational strength of our species because it lowers the pool of knowledge from which we can draw' (Bernard 1992: 82; also Fishman 1982; Diamond 1993). From this perspective, issues of linguistic and cultural diversity preservation may be formulated in the same terms as for biodiversity conservation: as a matter of 'keeping options alive' (Reid and Miller 1993) and of preventing 'monocultures of the mind' (Shiva 1993). Mühlhäusler (1995a) argues that convergence toward majority cultural models increases the likelihood that more and more people will encounter the same 'cultural blind spots'

– undetected instances in which the prevailing cultural model fails to provide adequate solutions to societal problems. Instead, '[i]t is by pooling the resources of many understandings that more reliable knowledge can arise'; and 'access to these perspectives is best gained through a diversity of languages' (ibid.: 160). Or, simply stated, 'Ecology shows that a variety of forms is a prerequisite for biological survival. Monocultures are vulnerable and easily destroyed. Plurality in human ecology functions in the same way.' (Pattanayak 1988: 380).

Overlap of linguistic and biological diversity

The relationships between language and environment can be observed at different degrees of resolution. Harmon (1996a) has shown remarkable overlaps between the world's biological and linguistic diversity on a global scale. He compared the IUCN list of 'megadiversity' countries (McNeely *et al.* 1990) to his own list of top 25 countries by number of 'endemic' languages (*cf.* Table 2.1 above). Ten out of 12 of the megadiversity countries (or 83 percent) also figure among the top 25 countries for endemic languages, as seen in Table 2.2.

TABLE 2.2: MEGADIVERSITY COUNTRIES: CONCURRENCE WITH ENDEMIC LANGUAGES. (COUNTRIES IN TOP 25 FOR ENDEMIC LANGUAGES IN BOLD; *CF.* TABLE 2.1; COUNTRIES LISTED ALPHABETICALLY; ENDEMIC LANGUAGE RANK IN PARENTHESES.)

Country
Australia (5)
Brazil (8)
China (17)
Colombia (23)
Ecuador (—)
India (4)
Indonesia (2)
Madagascar (—)
Malaysia (15)
Mexico (6)
Peru (18)
Zaire (9)

Concurrence: 10 of 12 (83 percent)

Notes: Modified from Harmon (1996a). 'Megadiversity countries' have been identified as those likely to contain a large percentage of global species richness. The twelve listed were identified on the basis of species lists for vertebrates, swallowtail butterflies and higher plants. *Source:* McNeely *et al.* 1990.

There are some exceptions, such as Papua New Guinea, the country with the highest linguistic diversity and endemism in the world, with over 800 different languages, but not a site of megadiversity. Harmon (1996a) cautions that no single diagnostic measure will account for phenomena as complex as linguistic and biological diversity. Furthermore, while in both cases assessments of diversity are done mostly on a country-to-country basis, political frontiers are not natural borders for either languages or species. A measure of skewing is to be expected for that reason alone. Yet the overall correlation is striking. Comparison of the top 25 countries by number of endemic languages with the top 25 countries by number of flowering plant species and the top 25 by number of endemic higher vertebrate species yielded additional close correlations (Harmon 1996a, based on Groombridge 1992). The overlap of endemic languages and higher vertebrates is seen in Figure 2.4.

Harmon (ibid.) points to several geographical and environmental factors that may comparably affect both biological and linguistic diversity, and especially endemism: (1) Extensive land masses with a variety of terrains, climates and ecosystems (e.g. Mexico, USA, Brazil, India, China); (2) Island territories, especially those with internal geophysical barriers (such as Indonesia, Australia, Papua New Guinea, Philippines); (3) Tropical climates, fostering higher numbers and densities of species (e.g. Cameroon, Zaire). Nichols (1992) identified similar biophysical factors as promoting high levels of genetic diversity among languages in given geographical regions. In addition, Harmon suggests that biodiversity–linguistic diversity correlations may have been fostered by a process of coevolution of small-scale human groups with their local ecosystems, in which humans interacted closely with the environment, modifying it as they adapted to it, and acquiring intimate knowledge of it. In turn, this knowledge was encoded and transmitted through language. As Mühlhäusler (1995a: 155) puts it: 'Life in a particular human environment is dependent on people's ability to talk about it'. Nabhan (1996) has called attention to the specialized ecological knowledge held by isolated language groups living in areas of high biological endemism by proposing the notion of 'ethnobiological endemism'. Conversely, Smith (1998) notes a correlation between low-diversity cultural systems and low biodiversity, supporting the hypothesis that cultural diversity may require small-scale, localized socio-economic systems, while large-scale systems may lead to a reduction in cultural diversity.

How do these correlations between biodiversity and linguistic and cultural diversity fare when one moves from a global scale to a smaller scale? Individual cases must be examined in light of the spe-

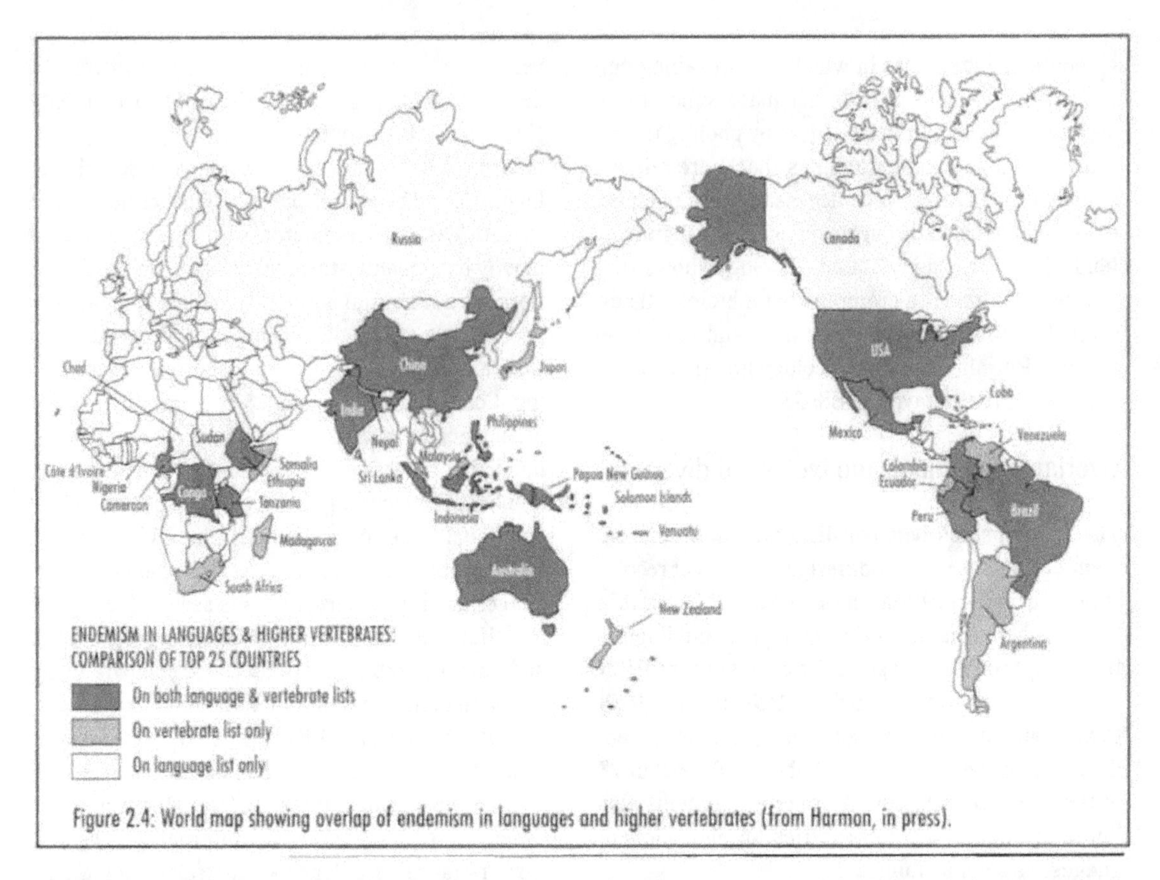

Figure 2.4: World map showing overlap of endemism in languages and higher vertebrates (from Harmon, in press).

cific factors at play, including population movements, culture contact and processes of environmental change. Under local circumstances, linguistic and cultural distinctiveness has often developed even among human groups defined as belonging to the same cultural area or whose languages are considered to be historically related, and who live within the same bioregion. Studies carried out on a bioregional level reveal precisely this kind of local diversity. In comparative research on the ecology of six small indigenous groups in the Amazon Basin, Milton (1996) documented diversity in cultural and ecological practices both between groups coexisting in the same geographical region, but having distinct linguistic affiliations, and between groups speaking related languages and occupying adjacent territories. A strong correlation between biodiversity, linguistic diversity and ethnobiological knowledge was found by de Avila (1996) in the biogeographical distribution of plants used for textile fibres and dyestuffs by indigenous groups of Mexico and Central America. An in-depth study of a single bioregion, the Pacific Northwest Coast of North America (Schoonemaker *et al.* 1997; see also Kellogg 1995), revealed that Northwest Coast Indians are very diverse linguistically, but their languages and cultural traditions share many elements (probably due to a long history of contact and diffusion of linguistic and cultural features among them). On the other hand, apart from its overall common ecological characteristics, this bioregion presents a high variety of local ecosystems and, as a consequence, local subsistence economies may vary considerably from place to place. This diversity is considered 'fundamental to understanding the Northwest Coast as a human environment' (Suttles and Ames 1997: 268).

One may expect this diversity of specialized adaptational knowledge to be reflected in the various local languages, suggesting that the nature and dynamics of the interactions between biodiversity and human languages and cultures may be best observed at the local level. At this higher level of resolution it may become apparent that landscapes are anthropogenic not only in the sense that they are physically modified by human intervention but also because they are symbolically brought into the sphere of human communication by language – by the words, expressions, stories, legends and songs that encode and convey human relationships with the environment and that inscribe the history of those relationships onto the land.

Ethnobiological classification, ecological knowledge and sense of place

The interactions of language and the environment – as mediated by human cognition and social life, and expressed through linguistic features – have been the specific object of investigation of cognitive ethnobiology (the study of the folk categorization and naming of natural kinds). Over the past fifty years, ethnobiologists have demonstrated the systematic and scientific nature of indigenous classifications of the plant and animal world, reflected in local floral and faunal nomenclatures (e.g. Conklin 1954; Bulmer 1974; Berlin *et al.* 1974; Hunn 1977; Brown 1984; Atran 1990; Balée 1994; Martin 1995).

Berlin (1992) provides the most comprehensive treatment of ethnobiological categorization and naming to date. Based on comparative evidence, he proposes that folk classification and naming of natural kinds are governed by a number of general, cross-culturally recurrent principles, based on the perceptual recognition of salient discontinuities and affinities in nature and on observation of the morphological and behavioural characteristics of local plant and animal species. These folk taxonomies are hierarchically organized and reveal a considerable degree of correspondence with Western biosystematic taxonomies. Berlin points out that ethnobiological nomenclature normally reflects this taxonomic organization and is mostly non-arbitrary, in that it is generally descriptive of morphological, ecological or behavioural features of the named taxa.

Berlin's 'intellectualist' approach argues for the prevalence of perceptual factors over practical ones (such as use for food or medicine, or other cultural values) in ethnobiological classification and naming. Other ethnobiologists make the opposite claim, stressing 'utilitarian' criteria (cf. Hays 1982; Hunn 1982). However, perceptual and cultural salience may interact with each other (Hunn 1999). Perceptual criteria appear to be prevalent if one includes among them an organism's ecological salience (as measured by biogeographic distribution of the organism compared to the distribution of the local human population, or by the organism's behavioural habits), and if morphological (phenotypic) salience is understood particularly in terms of the organism's size. On the other hand, exceptions to this general trend mostly correspond to organisms having high local cultural salience, which will be singled out even if they may not be especially salient perceptually. These two dimensions also mutually interact; for example the overall salience of a given set of organisms (mammals, fish, etc.) increases if there is a cultural focus on that set, although recognition of organisms within the set still follows the perceptual criteria.

Recently, ethnobiologists have also begun mapping the biogeographic distribution of traditional environmental knowledge (as represented by linguistically encoded concepts) onto local ecosystems, by means of current mapping technology such as GIS (e.g.Tabor and Hutchison 1994; Bocco and Toledo 1997). These 'cognitive biogeographic' studies are showing remarkable degrees of isomorphism between given ecological niches and humans' cognitive maps of them, each shaped by and shaping the other over the history of local human populations.

Berlin (1999) has undertaken a study of the individual development of folk taxonomic systems rooted in biogeography, stemming from the assumption that each person first learns about the natural kinds found in the localized ecosystem in which he or she grows up (cf. Stross 1973). Berlin finds evidence that these familiar organisms cognitively recognized in early life become the prototypes that guide the classification and naming of unfamiliar organisms in other areas, based on observed perceptual affinities to the prototypes. Thus, over time an individual's classification can go well beyond a localized ecological niche and potentially extend to a whole bioregion.

Another approach to the mapping of ethnoecological knowledge is through the study of indigenous place names, as Hunn (1996) did among the Sahaptin of the Columbia Plateau of North America. Each place name (and the same could be said of plant and animal names) is 'an entry in a mental encyclopedia' (ibid.: 20). Place names describe 'places where things happen' within the local environment, so that 'plotting the distribution of named places [...] is one means to appreciate the ecological niche occupied by local [...] peoples' (ibid.: 18) as well as the 'intensity of cultural focus a region enjoys' (ibid.: 5). Hunn thus stresses both the ecological context in which place naming occurs, and the cultural significance of this activity for indigenous peoples, 'as a framework for cultural transmission and moral instruction, as a symbolic link to their land, and as a ground for their identity' (ibid.: 4). Named landmarks convey and evoke knowledge both about the physical environment and about daily human activities, historical events, social relations, ritual, and moral conduct. As Basso (1996) puts it, 'wisdom sits in places'. Landscapes are networks of such places of knowledge and wisdom and thus, in this sense as well, anthropogenic.

Human relationships with the land can also be expressed metaphorically, such as with the application of concepts of the human body to the land – a tendency to 'embody the landscape' (Hunn 1996: 16). The Maya of Highland Chiapas, Mexico, conceptualize their territory in terms of human physiology (Maffi 1999). Two major ethnoecological categories – labelled 'warm country' and 'cold country' in the local Mayan language – are based on an analogy with the healthy vs. diseased human body, especially in terms of generativity, and refer to the differential fertility and productivity of the land at different altitudes. This metaphorical expression of warmth/fertility vs. cold/barrenness may have arisen through the connection of lunar cycles with both the female reproductive cycle and agricultural and seasonal cycles, as well as via the identification of the earth with a woman's womb. This identification portrays a connection between humans and the land that is at one and the same time as primordially physiological and as powerfully symbolic as an umbilical cord.

Ethnobiologists are thus going beyond folk taxonomies and devoting increasing attention to indigenous concepts of the environment and traditional ecological knowledge (Toledo 1992; Lewis 1993). Pawley (1998) notes that much of this knowledge is not about entities *per se*, such as natural kinds, but about natural processes and relations among entities, such as the relationships among plants and animal species or between humans and the ecosystem. He points out that this knowledge (where 'knowledge' should be understood as including 'perceptions' and 'beliefs') is not simply carried by nomenclature; instead, it pervades the whole linguistic system, all aspects of grammar and language use. It is found in speech formulas, i.e. culturally conventionalized ways of talking about certain subject matters, expressing spatial, temporal and causal relations, and so forth. These formulas range from set phrases to specified ways of building sentences or even whole discourses and narratives, as Pawley (ibid.) shows in the case of the Kalam of Papua New Guinea.

Similarly, Wilkins (1993) found that, among the Mparntwe Arrernte Aborigines of Australia, the interdependence between kinship, land and 'Dreamtime' totemism is reflected in a wide variety of features of their language. These recurrent patterns of linguistic expression show an 'ethno-semantic regularity' that embodies a holistic and mutual link between people, plants, animals, places and the spiritual or supernatural. Wilkins suggests that '...the more important a philosophical theme is to a cultural group, and the longer it has been a theme that has rested in the group, then the more it will have worked itself into the fabric of the language, moving from the lexical to the grammatical, and thus manifesting itself not only at the lexical level but also in aspects of the grammar, and in creative language use' (ibid.: 89–90). If so, Wilkins argues, similar 'ethno-semantic regularities' should be found in other aspects of Mparntwe Arrernte ecological knowledge, as well as in other domains.

Wollock (1997: n.p.) points out: 'It is only when we shift our perspective from language as grammar [...] to language as a pattern of human action, used by human beings with bodies as well as minds and connected with other actions within the social and natural world, that it becomes possible to talk adequately about how linguistic diversity is related to biodiversity'. Lord (1996) synthesizes these concepts in speaking of the Dena'ina Athabaskan language of the Cook Inlet as a 'coevolutionary tongue', a language that confirms the landscape at one and the same time that the landscape shapes it: dictating, for instance, not only an ample vocabulary for talking about salmon and other fish, vegetation, streams and trails, but also extensive development of grammatical devices for expressing 'direction, distances, and relative positionings – very important in a semi-nomadic culture where people needed to be very clear about where they were and where they needed to go for food and other necessities' (ibid.: 46). She adds: 'Languages connected to place help us to respect local knowledge, to ask and answer the tough questions about how the human and nonhuman can live together in a tolerant and dignified way. They can help us extend our sense of community, what we hold ourselves responsible for, what we must do to live right and well.' (ibid.: 69).

The erosion of languages and ecological knowledge

Many indigenous and minority groups (e.g. the Tarahumara, the Basques, the Sámi) have preserved and continued to develop distinct languages as well as identities and cultural traditions despite centuries of encroachment by more powerful outsiders. But in many more cases, language shift is occurring at an accelerating pace. It must be stressed that the demise of local languages encroached upon by majority languages is not due to inability to change and adapt. However, when change is sudden, its pace rapid, and the external pressures strong, the conditions for natural adaptive change may be greatly compromised or altogether removed (Wollock 1997). Language shift then commonly ensues.

Yet, when a language ceases to be spoken, does this necessarily affect the pool of human knowledge? As they shift to majority languages, why should speakers of smaller languages not be able to take with them their traditional cultural knowledge and continue to make use of it? Indeed, there are examples of groups that retained their ethnic identity and some aspects of their cultural traditions after they gave up their native languages (Mohanty 1994a), or even after their languages had become moribund (e.g. Yiddish-speaking immigrants in the USA, the Irish, the Maori, the Hawaiians). Thus, the relationships of language, culture and ethnic identity are complex and not uniformly predictable (Edwards 1985; Haarmann 1986; Fishman 1989).

Nevertheless, as Wurm (1991) points out, changes in the ecology of languages – including the radical alteration or even destruction of (or the removal or alienation from) the natural and sociocultural environments of linguistic communities by external forces – normally provoke the disappearance or drastic modification of the languages. Under these circumstances, a local language '[...] will lose a number of its characteristics which are rooted in the traditional culture of its speakers [...]. It no longer reflects the unique traditional and original world-view and culture of its speakers, which has been lost, but more that of the culturally more aggressive people who have influenced its speakers' (ibid.: 7). Other linguists more radically suggest that 'when a language dies, a culture dies' (Woodbury 1993; also Fishman 1996). Generally, the replacing language does not represent an equivalent vehicle for linguistic expression and cultural maintenance (Woodbury 1993). In particular, local knowledge does not easily 'translate' into the majority language to which minority language speakers switch. Furthermore, along with the dominant language usually comes a dominant cultural framework that begins to take over. Because in most cases indigenous knowledge is only carried by oral tradition, when shifts toward 'modernization' and dominant languages occur and oral tradition in the native languages is not kept up, local knowledge is lost. Due to its place-specific and subsistence-related nature, ecological knowledge is at especially high risk of being lost, as people are removed from their traditional environments or become alienated from traditional ways of life and lose their close links with nature.

The rapid and dramatic erosion of knowledge is heralded by the loss of the linguistic tools with which to express and communicate it. Hill (1995, 1998) has stressed that the processes of language decay are distinct from processes of natural language change. In the latter case, a language may undergo structural reduction at some level, but this is normally compensated by elaboration at some other level (so that overall the language maintains a sort of 'steady state'). Instead, language decay is characterized by a drastic, largely uncompensated erosion of both the structural features of a language (from its phonology to its grammar to its lexicon) and the functional domains of language use (ways of speaking, patterns of conversation, story-telling, and other forms of communication). In other words, language decay consists in a prevailing trend toward progressive loss of structural and functional complexity in a language. In this process, normal intergenerational transmission of a given language is disrupted, so that younger generations learn their language of heritage imperfectly. Furthermore, reduction in the range of contexts available for language use can lead to language decay also in once-fluent adults who are no longer using their language as frequently and in as many different situations as they once did.

Underlying language and knowledge erosion, in both children and adults, is what Nabhan and St. Antoine (1993) have called the 'extinction of experience', the radical loss of direct contact and hands-on interaction with the surrounding environment. They found that, among the O'odham of the Sonoran Desert in the southern USA, school-age children were not fluent in O'odham, although they heard it spoken at home; nor did they go out into the desert as a part of their normal activities. Most of them were not being exposed to traditional story-telling and other forms of oral tradition in their interactions with parents, grandparents or elders. A majority of them indicated that their main sources of information (and of authoritative knowledge) were school and television. The children were able to give the O'odham names for only one-third of the most common plants and animals in their environment – while having no trouble giving the English names for African megafauna seen on television documentaries. Comparable difficulties in naming commonly occurring plants and animals were found by Hill and Zepeda among some O'odham adults (reported in Hill 1995, 1998). Among those elders who had trouble with local names were people who had lived away from the O'odham reservation for long periods of time. They were still fairly fluent in the language but had lost the context of use of the traditional biosystematic lexicon. Furthermore, English did not provide them with an alternative vehicle for maintaining this knowledge, since even the local variant of English makes far fewer distinctions among the local flora and fauna than O'odham does. Hill also points to the foreseeable consequences of this loss in biosystematic lexicon in terms of alienation of indigenous language speakers from traditional forms of knowledge.

The extinction of experience, then, has much to do with processes that promote cultural assimilation and language shift. How radical this language and knowledge loss can be even in a domain as basic as that of subsistence can be illustrated by the case of Scottish Gaelic among the fisherfolk in the East Sutherland region of Scotland (Dorian 1997). Individuals who are formerly fluent speakers or 'semi-speakers' (imperfect learners of their mother tongue who make grammatical mistakes recognized by fluent speakers) and no longer practice fishing or fishing-related activities may not only have trouble with impromptu recall of Gaelic names for kinds of fish or boats, terms for sailing and fish processing, and even the word for 'season'. They may also continue to have trouble remembering the words and using them appropriately in context even after being reminded of what the words are. Dorian suggests that, when socio-economic circumstances drive people away from their traditional mode of subsistence and the local language, not even early intimate familiarity with subsistence lexicon in that language may necessarily guarantee that such lexical items will be easily retrieved from memory at a later time.

In some cases, external factors of language and knowledge erosion may trigger internal processes that enhance the effects of the external forces. Among the Alune of the island of Seram (eastern Indonesia), loss of the local language and environmental knowledge – due to religious conversion, formal schooling and profound political and economic change – is accelerated by the reinterpretation of knowledge by younger people (Florey and Wolff, 1998). The use of incantations traditionally employed in Alune healing practices in association with the administering of medicinal plants has been suppressed through conversion to Christianity. In addition, while some younger Alune still use incantations, they have reinterpreted their function as a means for self-defence and harming others, and have begun reciting them in Malay (the local majority language). The joint effect of these processes is radical language and knowledge loss in a domain as vital as that of health care, where younger Alune no longer rely on local medicinal knowledge but resort to (poor quality) Western medical care.

Where language decline is in process, elders who still speak their native language fluently may get upset at the imperfect learning of the younger generations, at their mixing the native language with the majority language, and particularly at youngsters who use the native language to talk about non-native concepts. Elders may even reconceptualize the language as sacred, or quasi-sacred, to be imparted only to those younger people who seem willing to be instructed in the traditional ways as well. Languages may also be seen as secret for security reasons – as in the case of many Roma/Sinti (gypsies), for whom being able to negotiate between themselves, without being understood by outsiders, was vital during centuries of persecution (Ollikainen 1995). Even if many elders now see restricting access to their language as counterproductive to the survival of both language and culture, there is still pressure for secrecy in numerous instances.

The loss of traditional language and culture may also be hastened by environmental degradation – such as that caused by logging, mining, waste dumping (including nuclear waste dumping), agribusiness, cattle-raising and real-estate development. These processes and the ensuing loss of control over native lands are among the main causes of disintegration of indigenous communities, a phenomenon that has been termed 'ecocide' (e.g. Grinde and Johansen 1995). Land destruction undermines both the physical and the spiritual bases for the maintenance of indigenous groups' identity as distinct peoples with their own languages and cultural traditions. Molina (1998) finds that among his people, the Yoeme (Yaqui) people of the Sonoran Desert, the performance of ritual is hampered by the disappearance of many plant species that were traditionally employed in religious ceremonies. This is due to increasing settlement of the desert by non-Yoeme, unsound harvesting practices, and conversion of land to different uses. Ritual is one of the main contexts for the teaching of the Yoeme Truth, and in particular of the intimate spiritual and physical connection with and respect for nature. As Evers and Molina (1987: 18) put it: 'Yaquis have always believed that a close communication exists among all the inhabitants of the Sonoran desert world in which they live: plants, animals, birds, fishes, even rocks and springs. All of these come together as a part of one living community which Yaquis call the huya ania, the wilderness world. [...] Yaquis regard song [as a part of ritual] as a special language of this community, a kind of 'lingua franca of the intelligent universe' '. K. W. Luckert has described this 'primeval kinship with all creatures of the living world' as 'perhaps the most basic [form of human awareness] in the history of man's religious consciousness – at least the oldest still discernible coherent world view' (cited in Evers and Molina 1987: 201). The Yoeme elders' inability to correctly perform rituals due to environmental degradation thus contributes to precipitating language and knowledge loss and creates a vicious circle that in turn affects the local ecosystem.

Specific patterns and factors of erosion of languages and linguistically-encoded environmental knowledge are beginning to be systematically identified and quantified. Zent (1999) measured ethnobotanical knowledge among the Piaroa Indians of Venezuela, based on individuals' ability to correctly identify local plants by their Piaroa names. He found strong negative correlations with age, bilingualism and schooling – younger, more acculturated Piaroa showing dramatically lower levels of competence than their older, less acculturated counterparts (Figures 2.5 to 2.8). Younger Piaroa who make mistakes in naming plant species are also likely to have problems with the correct identification of the cultural

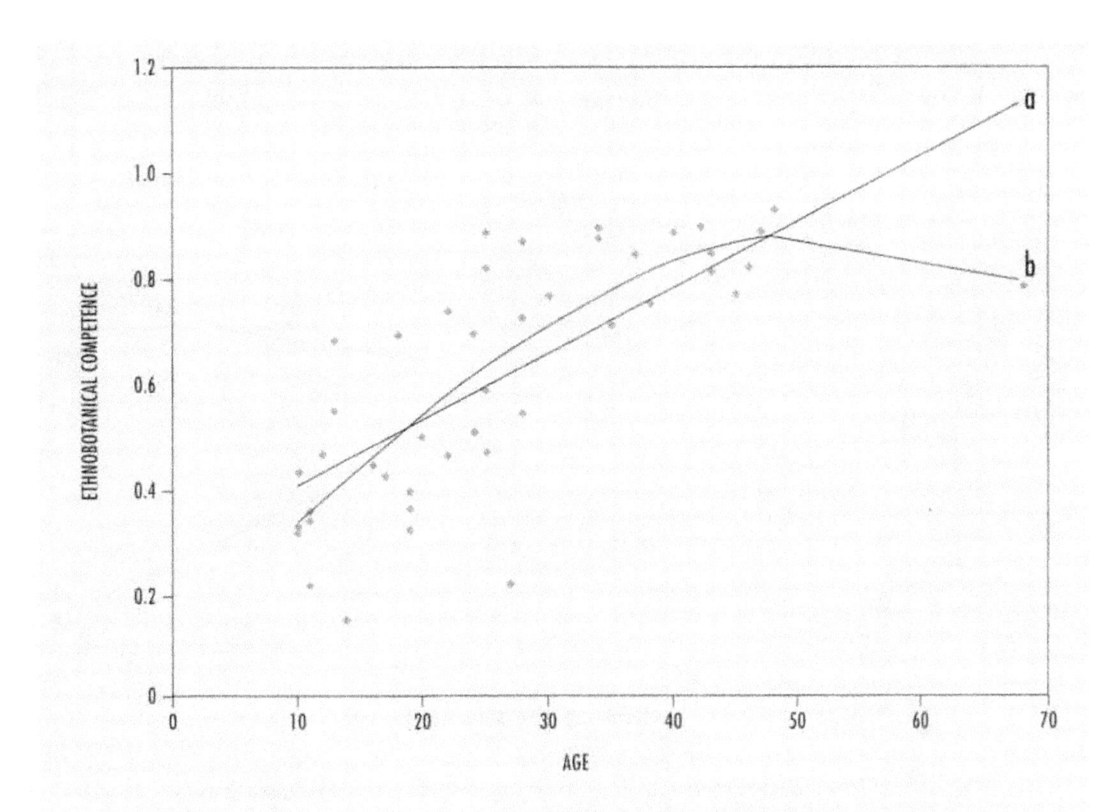

Figure 2.5: Regression of age and ethnobotanical competence among the Piaroa of Venezuela (from Zent, in press).

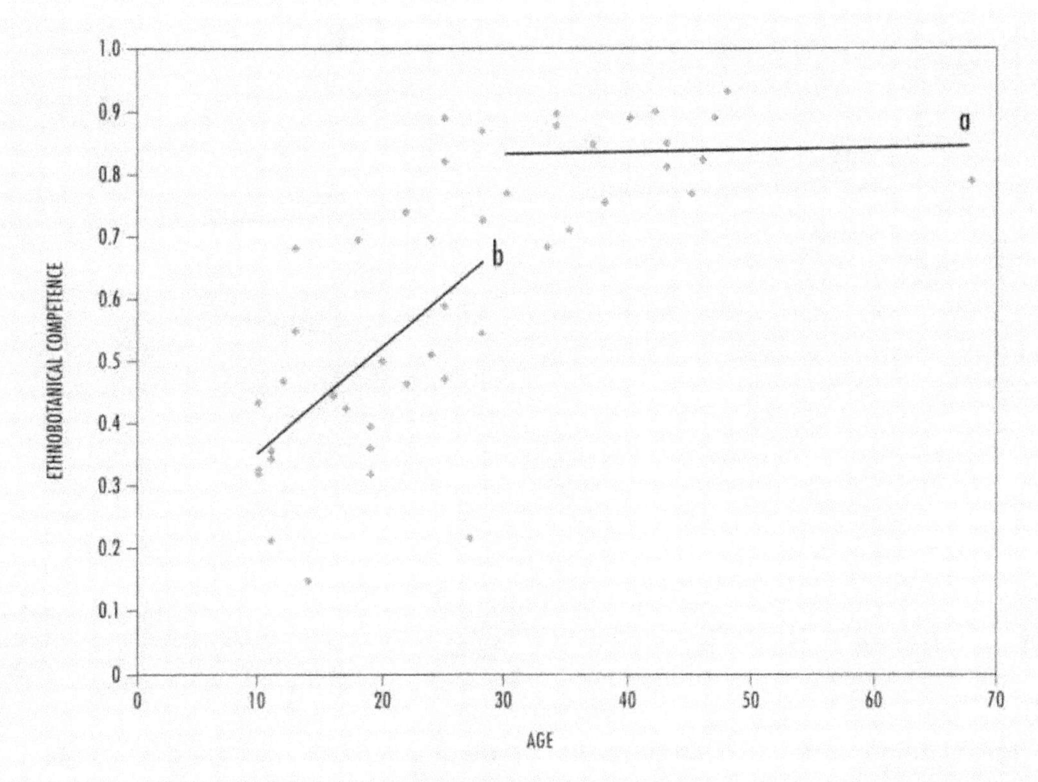

Figure 2.6: Regression of age and ethnobotanical competence according to age subgroups among the Piaroa of Venezuela (from Zent, in press).

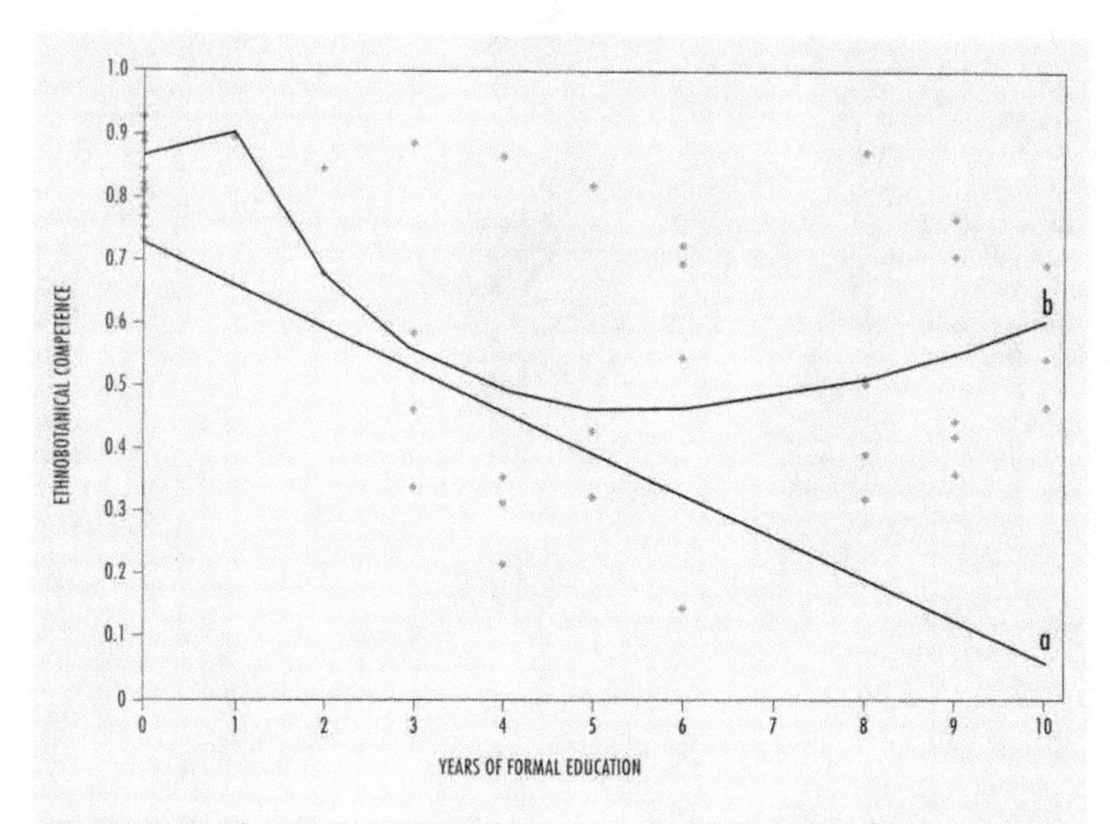

Figure 2.7: Regression of formal education and ethnobotanical competence among the Piaroa of Venezuela (from Zent, in press).

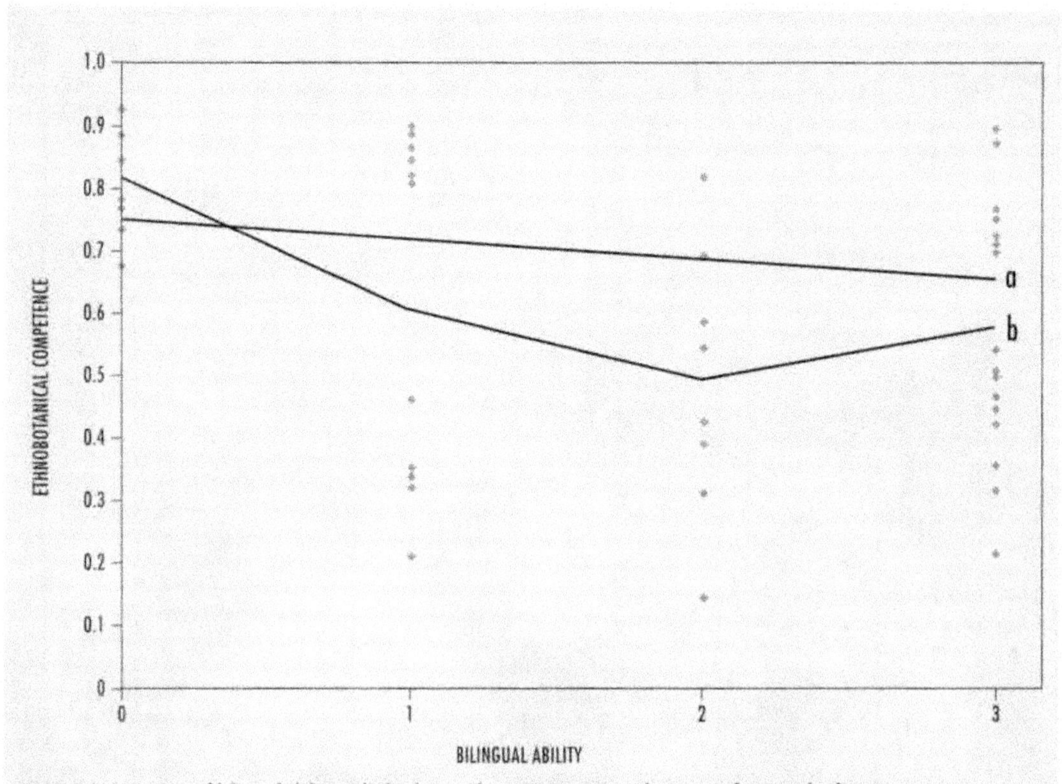

Figure 2.8. Regression of bilingual ability and ethnobotanical competence among the Piaroa of Venezuela (from Zent, in press).

uses of those same species, suggesting that for them ethnobotanical nomenclature is losing its function of anchoring knowledge about the local environment.

For most indigenous and minority groups it may be impossible to assess ecological knowledge loss retrospectively, due to a dearth or total lack of records from the past. Some historical perspective can be gained from a longitudinal study of folk botanical terminology in the English language, for which ample documentation exists (Wolff and Medin 1998). Dictionary data from the 15th century on show that the overall mention of folk botanical terminology is constant or even slowly rising through the eighteenth century, and then takes a precipitous drop in the late nineteenth century and the twentieth century (Figure 2.9). The turning point appears to be the Industrial Revolution, when England's population became increasingly urban and cut off from the natural environment. An investigation of changing attitudes toward nature in England between 1500 and 1800 (Thomas 1983) reveals that many of the same factors that currently operate on indigenous and minority languages were at work in stamping out local knowledge in England. As the use of Latin binomials was being promoted by trained biologists, vernacular names for plants and animals were being stigmatized and fell into disuse or became increasingly the limited province of 'backward' rural populations. Local agricultural and plant medicinal knowledge was also being ridiculed and neglected. The emergence and imposition of more authoritative linguistic and cultural models leading to a reduction of diversity has thus characterized also the internal history of the English language and English society.

Cognitive psychological research suggests that urban populations may also lose the ability for ecological reasoning, that is, reasoning based on the relationships among local plant and animal species. A comparison among city-dwellers in the USA and rural Itzaj Maya Indians from the Petén region of Guatemala (Atran and Medin 1997) showed, on the one hand, that these two groups follow essentially the same taxonomic principles in the classification of flora and fauna. On the other hand, unlike people in the USA, the Itzaj rely upon ecological knowledge over taxonomic knowledge in drawing inferences in reasoning tasks. They 'know too much' about the actual ecological relations among the species in their environment to be willing to apply taxonomic criteria in the abstract (Atran 1996).

The Itzaj's extensive ecological knowledge is matched by their sustainable agroforestry practices (Atran 1993). However, they are a highly endangered group. Linguistically, they would be classified as 'moribund'. There are only about 50 fluent adult speakers left; all children are monolingual in Spanish. Many of the Itzaj's cultural traditions still survive, and they have even established their own small biosphere reserve in which to continue their own way of tending to the environment. However, it is a fragile situation being constantly encroached upon from all sides. Itzaj-speaking adults worry that, as their

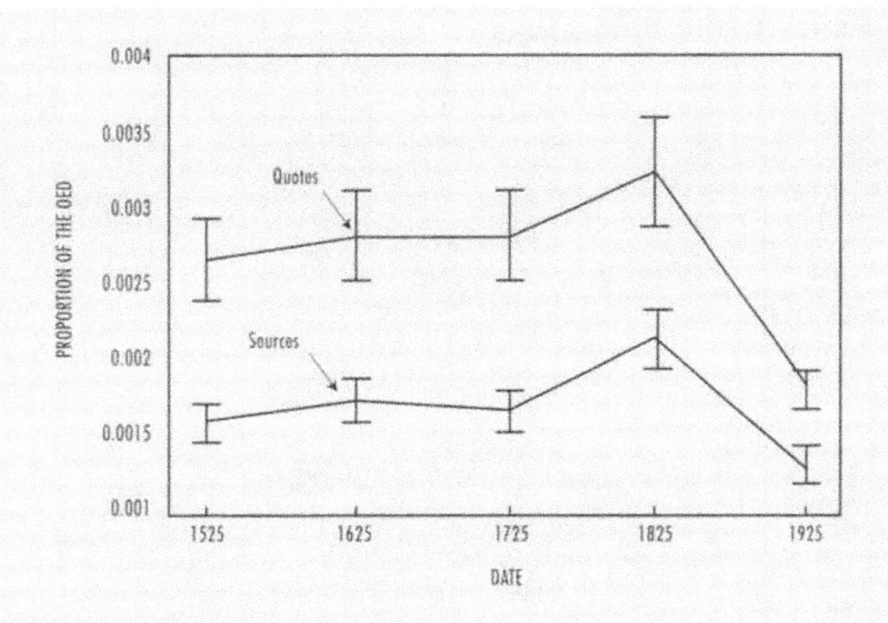

Figure 2.9: Proportion of quotations and sources in the OED referring to trees along with associated 95 percent confidence intervals (from Wolff and Medin 1998).

children are not learning the language and not walking in the forest any more, traditional knowledge about the environment and human interactions with it will be lost.

Yet, as the Itzaj undertake efforts to revive their native tongue, there is some hope arising from the links between language and the environment. Atran (1996) reports that so far the most successful way to awaken Itzaj children's interest in their ancestral language has been through a focus on the plant and animal world. Perhaps deep inside human consciousness lies something like what E. O. Wilson has described as 'biophilia' – an innate human need for contact with a diversity of life-forms (Wilson 1984) – or what traditional peoples around the world conceptualize as an intrinsic spiritual connection between humans and nature. Perhaps this connection is established and evoked, and can be brought back, by language.

Box 2.1: A Sense of Place (Campfire Meditation)

Luisa Maffi

We're hearing so much about
Indigenous knowledge lately
Knowledge about the natural world

We want to know that knowledge
To understand what we've done wrong
To make things better

But knowledge alone won't do that for us

Stories we hear
From indigenous mouths
Are not stories of
Knowledge of place alone
They are stories of
Sense of place

Not what to know about
A place you live in
But how to live in
A place you know

Not just
Humans in nature but
Nature in humans

Stories not of
Knowledge alone
But of wisdom which is
Lessons we draw from knowledge

Information alone is nothing if
There's no lesson to draw

Tucson, April 1996

Promoting the use of native names for species: the case of Madagascar (Jonah Andrianarivo)

The 590,000 square kilometre island of Madagascar is one of the earth's biodiversity hotspots, with high endemism combined with endangered flora and fauna (Mittermeier *et al.* 1994). Despite a multitude of conservation projects that, over the past 15 years, have brought in significant amounts of foreign aid, investments and activities, the country has suffered a worsening of its socio-economic situation and increased pressure on the remainder of its unique non-renewable natural resources. In response to this, and acknowledging that local people will show little interest in conservation unless their welfare is also taken into account, an approach termed an Integrated Conservation and Development Project (ICDP) has been conceived and popularized among conservation organizations operating in the country (Kull 1996). Project leaders are aware that people maintain a strong concern with cultural matters, in spite of economic hardship, and that cultural acceptance is a key factor for the success of conservation. Local artists and religious leaders are being called upon more often to convey the goals of conservation projects. But is their intervention effective?

Use of vernacular names in conservation

One way to increase the likelihood that the conservation of biodiversity will be culturally accepted by local people is to promote the use of native names for animal and plant species (Andrianarivo 1987). Overarching use of scientific and foreign names not only might lead local people to believe that conservation is of exclusive interest to foreign researchers, but also would not help them grasp the concepts of biodiversity and conservation. Furthermore, local knowledge itself cannot be supported, preserved and communicated without using vernacular names. Yet, such use is still shunned in conservation action in Madagascar, largely due to the persisting 'francophonie', the heritage of 64 years of French colonization, during which the use of Malagasy was excluded from all official institutions. In 1972, a student uprising which toppled the Government of Madagascar brought about the restoration of Malagasy in education. Since then, the language has advanced dramatically in the educational domain. However, 'malagasization' was never officially endorsed by later governments, and in administration there is a tendency to avoid the national language in favour of French, the use of which is again being aggressively promoted by France. 'Francophonie' is also implemented in environmental activities, with the tacit agreement of donor nations and institutions. Lack of consistency in local floral and faunal terminologies is often cited as a definite obstacle against using them in conservation projects. However, it is possible to improve the practicality of the use of native names for species, and thus have an efficient tool for exchanging knowledge between field scientists, local people and local leaders, while at the same time supporting the cultural heritage associated with each vernacular name.

Out of Madagascar's estimated 10,000 vascular plants there are 8,500 described species, and no less than 4,500 native names have been noted by field collectors. For the avifauna and the lemurs, the numbers of local names are higher than those of scientifically identified species. More than 400 local names were recorded for the 256 known bird species, and more than 60 for the 50 described lemur species and subspecies. The proposed approach involves first of all conducting an exhaustive inventory of local names used in the investigator's taxa or area of interest (e.g. a plant family, a group of vertebrates, a genus of invertebrates, plant materials used for roofing). Secondly, the terms should be analysed linguistically and matched against the scientific taxonomy. This analysis should result in each scientifically known species being matched with a specific local name or a preferred local name, or being determined to lack any consistently used local name. Thirdly, the local scientist can coin neologisms for species without local names, taking into account the results of the previous linguistic analysis, as well as the biological characteristics of each species. The final result is then a set of Taxonomically Organized Local Names (TOLN) for the taxa (Table 2.3), accompanied by a listing of other culturally relevant knowledge, such as cultural uses of the species and local lore about them, and an etymology of all inventoried names.

The example of lemurs

For the lemur group, 18 of the 50 scientifically recognized species and subspecies have single local names that can be unambiguously attributed to them. A 'preferred' local name can be attributed to 22 more, choosing among alternatives the name that best matches the biological or ecological characteristics of the animal. Neologisms can be coined for 10 of them that have no local name or are referred to inconsistently by more or less common names. Generic names were determined and assigned accordingly as names for taxa at the levels of genus and family.

Some of the local names have a restricted geographical distribution. For instance, *akomba*, derived from the African name for the bush baby, is used only in the northern and north-western part of Madagascar. Local people use it to refer generically to almost all lemur species in the region. In the TOLN, *akomba* is listed as the preferred name for the Black lemur species and subspecies (*Eulemur m. macaco*) because it is the most frequently encountered in that area. The Malagasy neologism *mangamaso* (short for *akomba-mangamaso*) was coined for the critically endangered Blue-eyed black lemur (*E. m. flavifrons*). This neologism tells Malagasy speakers that it is a distinctive subspecies, characterized by the colour of its eyes (*manga* = blue, *maso* = eye). The neologism *akombadredrika*, coined for the Crowned lemur (*Eulemur coronatus*), indicates that it is a lemur from the northern area and morphologically closest to the *dredrika* or Mongoose lemur (*E. mongoz*). By the same token *akombavarika*, coined for Sanford's brown lemur, is the 'northern varika', suggesting that there are some other types of varika that should be known. Both the Crowned lemur and Sanford's brown lemur were indistinctively called *akomba* or *varika* by the local people. The latest described lemur species is the Tatersall's sifaka (Simons 1988). Probably, it had not drawn the attention of biologists before because its reported local names – *akombamalandy* or *simpona* – refer to already known species. The neologism *simpomavoka* is introduced to emphasize that it is a different species, closer to the eastern Propithecus (*simpo*), and recognizable by its yellowish (*mavoka*) pelage. Pelage colour is the most frequently used attribute in ethno-classification of lemurs. It should be underscored that all the newly coined species names are meant to be easily understood and adopted once they are brought back to the local people. Each proposed neologism has been conceived to introduce a basic understanding of biological classification and to sharpen the perception of biodiversity. If any does not, it should be replaced by a better one.

The most mystical living lemur species is certainly the aye-aye. It has at least two vernacular names – *ahiahy* and *fanahy* – which mean 'feared' and 'spirit' respectively. It is interesting that, because of the power of foreign media that use exclusively the westernized name aye-aye, none of the real native names is used any longer, even in Malagasy publications.

Language maintenance and revitalization (Luisa Maffi and Tove Skutnabb-Kangas)

Linguistic diversity and the 'Curse of Babel'

The loss of indigenous and minority languages and of the traditional knowledge they embody is of deep concern to the speakers of those languages, as well as to many scholars. But why should the world at large really worry? There are seemingly excellent social and political reasons why a reduction in the number of different languages spoken on earth should be a good thing. At long last, a widespread attitude has it (e.g. *Science News* 1995; Malik 1996), humanity will be freed of the burden laid on it by the 'Curse of Babel': the multiplicity of languages. With fewer different languages in use, this line of reasoning goes, it will be easier to communicate with people elsewhere in the world; once marginalized populations will be able to develop and prosper; ethnic conflict will decrease; national unity will no longer be threatened; and we will finally be moving toward the globalized cosmopolitan world that is the ultimate destiny of humanity.

However, none of these arguments is supportable. First of all, they are mostly expounded by speakers of languages that are comfortably not at risk of going extinct. Furthermore, the learning of other languages does not have to be subtractive (that is, at the cost of losing one's own language); it can be additive, leading to a situation of stable multilingualism in one's mother tongue and one or more other languages (Skutnabb-Kangas 1984). It is rare for indigenous or minority groups to abandon their languages in favour of a majority one without direct or indirect pressures from governments and other outside forces. Such pressures often include explicit prohibition of and punishment for the use of the local language, or at least systematic disvaluing and stigmatization of the mother tongue by 'mainstream' society, as well as severe restriction of its contexts of use, which leads to exclusion from vital flows of information, such as those on national affairs and matters of subsistence and health (Bodley 1990; Wurm 1991; Batibo 1998; Bobaljik and Pensalfini 1996).

Faced with the challenges of modernity, indigenous and minority language speakers may or may not wish to preserve their own languages and cultural traditions, but should not have to find themselves systematically pressured into the latter choice. Indeed, one may even question whether choice under such pressure can be called choice at all. Moreover, marginalized ethnic groups who opt for or are forced into assimilation into a linguistic and cultural majority often do not overcome their marginalization but end up finding themselves among the dispossessed within 'mainstream' society (Bodley 1990; Mohanty 1994a).

As for the issue of ethnic conflict and national security, specialized studies (e.g. Pattanayak 1988; Stavenhagen 1990, 1996; Mohanty 1994a) show that

TABLE 2.3: TAXONOMICALLY ORGANIZED LOCAL NAMES (TOLN) FOR THE 50 LEMUR SPECIES AND SUBSPECIES, BASED ON MITTERMEIER *ET AL.*'S (1994) CLASSIFICATION, AND FOLLOWED BY AN INDICATION OF THE CATEGORY THEY BELONG TO: SPECIFIC (-S-), PREFERRED (-P-), OR NEOLOGISM (-N-). MALAGASY SELECTED TAXA NAMES FOR FAMILY AND GENUS ARE ALSO GIVEN. ALL MALAGASY NAMES ARE CAPITALIZED AND COMMON ENGLISH NAMES ARE IN PARENTHESES.

	CLASSIFICATION	TOLN	CATEGORY
	Family: Cheirogaleidae	TSIDY	
	Genus: *Microcebus* (Mouse lemurs)	TSIDY	
	Species:		
1	*M. murinus* (Gray mouse lemur)	TSIDY	-p-
2	*M. rufus* (Rufous mouse lemur)	TSIDIMENA	-s-
3	*M. myoxinus* (Pygmy mouse lemur)	KOITSIKY	-p-
	Genus: *Allocebus*	TSIDIALA	
	Species:		
4	*A. trichotis* (Hairy-eared dwarf lemur)	TSIDIALA	-s-
	Genus: *Mirza*	TILITILIVAHA	
	Species:		
5	*M. coquereli* (Coquerel's dwarf lemur)	TILITILIVAHA	-p-
	Genus: *Cheirogaleus*	MATAVIRAMBO	
	Species:		
6	*C. major* (Greater dwarf lemur)	MATAVIRAMBO	-p-
7	*C. medius* (Fat-tailed dwarf lemur)	KELIBEHOHY	-s-
	Genus: *Phaner*	VALOVY	
	Species:		
	P. furcifer (Fork-marked lemur)	VALOVY	
	Subspecies:		
8	*P. f. furcifer* (Eastern fork-marked lemur)	TANTA	-p-
9	*P. f. pallescens* (Pale fork-marked lemur)	VAKIVOHO	-p-
10	*P. f. parienti* (Pariente's fork-marked lemur)	TAKARAKA	-p-
11	*P. f. electromontis* (Amber Mountain fork-marked lemur)	VALOVY	-p-
	Family: Megaladapidae	RAIPAKA	
	Genus: *Lepilemur*	RAIPAKA	
	Species:		
12	*L. doralis* (Gray-backed sportive lemur)	APONGY	-p-
13	*L. ruficaudatus* (Red-tailed sportive lemur)	BOENGY	-s-
14	*L. edwardsi* (Milne-Edward's sportive lemur)	RAIPAKA	-p-
15	*L. leucopus* (White-footed sportive lemur)	SONGIKY	-p-
16	*L. mustelinus* (Weasel sportive lemur)	KOTRIKA	-s-
17	*L. microdon* (Small-toothed sportive lemur)	HATAKA	-p-
18	*L. septentrionalis* (Northern sportive lemur)	MAHIABEALA	-s-
	Family: Lemuridae	VARIKA	
	Genus: *Hapalemur*	VARIBOLO	
	Species: *H. griseus* (Lesser bamboo lemur)	VARIBOLO	
	Subspecies:		
19	*H. g. griseus* (Eastern lesser bamboo lemur)	BOKOMBOLO	-p-
20	*H. g. occidentalis* (Western lesser bamboo l.)	AKOMBAVALIHA	-s-
21	*H. g. alaotrensis* (Lac Alaotra bamboo lemur)	BANDRO	-s-
22	*H. simus* (Greater bamboo lemur)	VARIBOLO	-p-
23	*H. aureus* (Golden bamboo lemur)	VARIBOLOMENA	-p-
	Genus: *Lemur*	MAKY	
	Species:		
24	*L. catta* (Ring-tailed lemur)	MAKY	-s-
	Genus: *Eulemur*	VARIKA	
	Species: *E. macaco* (Black lemur)	AKOMBA	
	Subspecies:		
25	*E. m. macaco* (Black lemur)	AKOMBA	-p-

TABLE 2.3: (CONTINUED)

	CLASSIFICATION	TOLN	CATEGORY
26	*E. m. flavifrons* (Blue-eyed lemur)	MANGAMASO	-N-
	Species: *E. fulvus* (Brown lemur)	VARIKA	
	Subspecies:		
27	*E. f. fulvus* (Common brown lemur)	VARIKOSY	-p-
28	*E. f. rufus* (Red-fronted brown lemur)	VARIKA	-p-
29	*E. f. albifrons* (White-fronted brown lemur)	FOTSIANDRY	-s-
30	*E. f. sanfordi* (Sanford's brown lemur)	AKOMBAVARIKA	-N-
31	*E. f. albocollaris* (White-collared brown l.)	BESOMOTRA	-N-
32	*E. f. collaris* (Collared brown lemur)	MENASOMOTRA	-N-
33	*E. mongoz* (Mongoose lemur)	DREDRIKA	-s-
34	*E. coronatus* (Crowned lemur)	AKOMBADREDRIKA -N-	
35	*E. rubriventer* (Red-bellied lemur)	BARIMASO	-s-
	Genus: *Varecia*	VARIKANDA	
	Species: *V. variegata* (Ruffed lemurs)	VARIKANDA	
	Subspecies:		
36	*V. v. variegata* (Black-and-white ruffed l.)	VARIKANDA	-s-
37	*V. v. rubra* (Red ruffed lemur)	VARIMENA	-s-
	Family: Indriidae	BABAKOTO	
	Genus: *Avahi*	AVAHY	
	Species:		
38	*A. laniger* (Avahi or woolly lemur)	AVAHY	-p-
39	*A. occendentalis* (Western woolly lemur)	FOTSIFE	-p-
	Genus: *Propithecus* (Sifakas)	SIFAKA	
	Species: *P. verreauxi*	SIFAKA	
	Subspecies:		
40	*P. v. coquereli* (Coquerel's sifaka)	AKOMBAMALANDY-p-	
41	*P. v. verreauxi* (Verreaux's sifaka)	SIFAKA	-p-
42	*P. v. coronatus* (Crowned sifaka)	TSIBAHAKA -s-	
43	*P. v. deckeni* (Decken's sifaka)	SIBABE	-N-
	Species: *P. diadema*	SIMPONA	
	Subspecies:		
44	*P. d. diadema* (Diademed sifaka)	SIMPONA	-s-
45	*P. d. candidus* (Silky sifaka)	SIMPOMALANDY	-N-
46	*P. d. edwardsi* (Milne-Edwards' sifaka)	SIMPOVANDA	-N-
47	*P. d. perrieri* (Perrier's sifaka)	SIMPOJOBO	-N-
48	*P. tattersalli* (Golden-crowned sifaka)	SIMPOMAVOKA	-N-
	Genus: *Indri*	BABAKOTO	
	Species:		
49	*I. indri* (Indri)	BABAKOTO	-s-
	Family: Daubentoniidae	AHIAHY	
	Genus: *Daubentonia*	AHIAHY	
	Species:		
50	*D. madagascariensis* (Aye-aye)	AHIAHY	-s-

ethnic differences (whether identified with language, culture, religion or any aspects of social organization) do not normally constitute the source of conflict, although they may be seized upon and attributed special meaning as a basis for mobilization when conflict does arise. In particular, there is no evidence to suggest that the use of different languages by neighbouring populations may constitute *per se* a cause of conflict; nor, for that matter, does monolingualism within or between countries seem to be a guarantee of peace. When populations of speakers of different languages coexisting in adjacent or the same territory do come into conflict, it is more common that the causes of such conflict reside in socio-economic and political inequality, and competition over land and resources, as well as in the denial (rather than the granting) of linguistic and cultural rights.

A particularly telling example is provided by Mohanty (1994a, 1994b). In 1994 there were serious conflicts with physical violence and killings between linguistically (but not culturally) assimilated monolingual Oriya-speaking Kond 'tribals' and Oriya-speaking 'non-tribals' in the Phulbani district of India. The conflicts occurred only in the monolingual areas and, despite similar socio-economic and political conditions, they did not spread to the bilingual areas (where the non-assimilated Konds speak both their own language and Oriya, and where many native Oriya-speakers know a little of the Konds' language). The local explanation was that people in the bilingual area knew each other better and could talk about things instead of fighting.

The idea of Babel as a 'curse' is a widespread interpretation of this element of the Judaeo-Christian religious tradition, yet not necessarily a valid one. It is perhaps more accurate to see the divine intervention that brings about a multiplicity of languages as a way of curbing the arrogance and single-mindedness of monolingual empire-builders. Other religious traditions suggest that a diversity of languages (and cultures) is a good thing. In the Qur'an, 'variations in languages, dialects and modes of expression among the groups and individuals are considered a sign of Allah's omnipotence, beside the creation of the heavens and the earth' (Abdussalam 1998: 57). This seems to be true not only for linguistic but also ethnic/racial diversity: 'Among his signs is the creation of heaven and earth and the variations in your languages [*alsinatikum*] and colours: verily in that are signs for those who know'. (Qur'an 30: 22). As another example, according to the Acoma Pueblo Indians of New Mexico, the mother goddess Iatiku causes people to speak different languages so that it will not be as easy for them to quarrel (Gill and Sullivan 1992).

From *ex-situ* language preservation to *in-situ* integral language restoration and development

Scores of grassroots organizations around the world are at work to ensure the continued viability of smaller languages threatened by encroachment by majority languages and dominant cultures. Heroic efforts are also under way to revitalize ancestral languages that no longer have any speakers. In many cases, language maintenance and revitalization go hand in hand with the reaffirmation of cultural traditions and of traditional relationships with the environment.

For language communities interested in supporting or reviving their languages, linguistic documentation can be a very valuable resource – whether in the form of materials collected by earlier generations of scholars, or of current documentation projects that take pedagogical applications into account (e.g. Moore, D. 1998). In many cases, such documentation may represent the only extant resource about a language that is no longer spoken. Still, these materials alone cannot recreate the natural contexts of language learning and use, i.e. the contexts of verbal interaction in which languages are acquired, used and developed. There is a very close parallel between this form of language preservation and *ex-situ* conservation in biology: while both serve an important function, in both cases the ecological context is ignored. Just as seed banks cannot preserve a plant's biological ecology, *ex-situ* linguistic documentation cannot preserve a language's linguistic ecology: its relationships with people, places and other languages (Mühlhäusler 1996). As Tlingit oral historians Dora Marks Dauenhauer and Richard Dauenhauer put it, 'Preservation [...] is what we do to berries in jam jars and salmon in cans. [...] Books and recordings can preserve languages, but only people and communities can keep them alive.' (quoted in Lord 1996: 68). Terry Supahan, a Karuk Native Californian, concurs: 'I am interested in communication, not preservation!' (quoted in Hinton 1996).

In recent times, and mostly due to indigenous and minority peoples taking up the organization of language support activities on their own, a different trend has emerged, one that places these activities squarely in the hands of linguistic communities and sees them as intrinsically linked to cultural revival and environmental restoration (as well as to land and resource rights and linguistic and other human rights). Again, a close parallel can be drawn with the emergence of a newer trend in conservation biology that promotes integral biocultural conservation models, the *in-situ* interdependent conservation of both biological and cultural resources (Zent, 1999).

The ultimate aim of indigenous and minority groups who reclaim their native languages is the restoration of everyday communication in these languages, normally in a context of stable bi- or multilingualism. In the case of languages that are no longer spoken, this ultimate goal may appear remote or unattainable. Yet languages have been revived that have not been spoken for centuries, as the example of Hebrew will suffice to show. Certainly, Hebrew had abundant documentation (beside benefiting from other uniquely favourable historical circumstances), which is not the case with most of the world's small languages. However, with the help of even limited documentation or a handful of still fluent elders, the desire to keep one's language or have it back can lead to extraordinary advances.

The story of Loren Bommelyn, a Tolowa Native Californian, is remarkable in this respect. In his efforts to learn his ancestral language, and knowing that there were not enough speakers left for him to interact with regularly, he spoke Tolowa to himself – or to others even if they could not understand, before translating into English. Loren is now a fluent speaker (and teacher) of Tolowa (Hinton 1994: 241–242). Even more striking is the case of Klavdia Plotnikova from the village of Ablakovo in the Sajan mountains, the last speaker of Kamass Samoyed, who was separated from her own very small group during the Russian Revolution at the age of 12. All of her people were eventually killed or died, and she was the only surviving speaker of her language. In the 1960s she met an Estonian linguist, Ago Kunnap, who spent several years with her, learned her language and wrote it down. In 1970 she was invited by linguists to the third Finnougrist world congress in Tallin, where she and Kunnap presented a paper on Kamass. When she was asked how she could keep her native tongue alive during the half century when she had nobody to talk to, she replied that she spoke to her God in her own language every day, and God replied in the same language. (Uibopuu 1988: 307; also O. Nuutinen, L. Huss, personal communications 1971, 1997.)

Examples like these, of exceptionally motivated individuals who succeed, against all odds, in maintaining or acquiring their ancestral language, could be multiplied from all over the world. In some cases, such single-minded people have actually achieved more than their individual goals. The Rama language of Nicaragua was on the brink of extinction due to shift to the local English-based creole. Nora Rigby, a fluent speaker of Rama, set out to record her language before she died, helped by linguists who were working in the area. With their support, she later managed to involve the whole Rama community, and a language revitalization project was started, with Rigby as one of the teachers (Craig 1992).

Thus, the work of linguists can be valuable also in terms of how they can aid community-led efforts. Artist and cultural activist L. Frank Manriquez, a Tongva/Ajachmem Native Californian, points out that indigenous peoples do recognize the potential and actual benefits of collaborating with academics and other outsiders. However, she argues, many well-meaning professionals do not devote enough time to listening to what native peoples have to say – to what their views, wishes and needs actually are. Manriquez suggests that Western ears need to open up to the kind of intent listening that alone can bring about genuine mutual understanding and true collaboration in facing the common threats to the world's linguistic, cultural and biological diversity (Manriquez 1998). Like many other indigenous groups the world over, Native Californians are engaging in integrated biocultural conservation efforts (Hinton 1996; Manriquez 1998). The linguistic and cultural revival activities in which they are involved go hand in hand with advocacy for environmental restoration on their lands and the renewed use of native plants for traditional handicrafts, such as basketweaving, and for other purposes.

Attempts to revive indigenous and minority languages, or to stop their decline, may easily meet with resistance from 'mainstream' society, but they are by no means uncontroversial even within language communities themselves (Zepeda and Hill 1991; Hinton 1994). In many cases, under pressure for linguistic assimilation, reinforced by stigmatization of local languages and cultures, and often by actual physical abuse, speakers may have internalized the message that their languages are worthless in the modern world and a hindrance to 'progress' – and that to learn the majority language they have to abandon their own. They may indeed have stopped speaking their language and not have taught it to their children. These people may thus be psychologically ill-disposed toward attempts to bring back the languages they gave up at such great human cost, and much work may be needed to win over their support.

Communities, and especially elders, must also understand that, particularly for languages that have been declining, or that are no longer spoken, the revitalized language will not be exactly the same as it once was – it may, indeed, be largely recreated building on whatever extant materials are available – and that languages change all the time anyway, as they adapt to new conditions their speakers find themselves in (Hinton 1996). People must become aware of how language acquisition occurs and of how language itself works (Dorian 1995). They need to grasp the social nature of language learning, i.e. the dependence of learning on active and sustained exposure to the language, rather than its being a spontaneous phenomenon that does not require input. In turn, lack of adequate understanding of the structure of language may lead to the formulation of language programmes that focus exclusively on the lexicon, i.e. on the learning of words, without providing learners with the tools they need to put together whole sentences and sustain conversations.

Indigenous and minority language support programmes

Once a community has manifested its willingness to work for the maintenance or revitalization of its language, however, a number of activities can be undertaken, depending on specific circumstances and available means (see Fishman 1991 for a useful typology). Minimally, a revitalization programme can aim to foster awareness and appreciation of the ancestral language and culture within the community, in order to create an environment that is conducive to more structured learning. Younger people may then become attracted to the idea of learning their language of heritage, and elders may become willing to impart the teaching, through some form of mentored learning.

One programme that has been very successful in this context is the Master-Apprentice Language Learning Program, created by the grassroots organization Advocates for Indigenous California Language Survival. This programme aims at 'helping deeply committed young adults become proficient in their language of heritage, through apprenticeship to a fluent elder' (Hinton 1994/1995: 14). It stresses acquisition of communicative competence in a way similar to language learning in childhood, i.e. through 'being immersed in an environment where the [ancestral] language is the dominant one being used' (ibid.). Master and apprentice are encouraged to spend as much time together as possible, conducting daily activities entirely in the language of heritage (initially with the help of non-verbal communication and other props). They are also encouraged to carry out traditional activities together – from handicraft making and food gathering or preparation to story-telling and performing ceremonies – as a key for the apprentice to become acquainted with the ancestral cultural traditions and the relevant context for language use.

One major advantage of this kind of programme is that it can be undertaken with limited training and financial resources, and thus is within the reach of most language communities. It is also particularly suited for situations, such as that in Native California, where a fairly large number of small indigenous groups exist, whose ancestral languages are very different from one another, and in which the numbers of remaining speakers and of potential language learners for each native language may be very small and demographically scattered. All this makes native language teaching in classrooms unfeasible or at least inefficient – although in many cases apprentices in this programme have started transmitting what they are learning to their own children and/or to other children in classrooms, at summer language camps, etc: in other words, they are already at work creating the next generation of native language learners.

In the case of larger indigenous and minority groups with higher demographic concentrations and functioning intergenerational communities, native speakers may not be limited to a few isolated elders; there may be more adult speakers, and possibly a number of younger people who have at least some passive knowledge of the language. In such instances, it is possible for a language community to organize native language courses for young adults and children, as well as a variety of communal intergenerational activities (including traditional ones) as outlets for native language use. Under these same circumstances, it may also become possible to introduce native language teaching in formal education for children, either in autonomous programmes, or through modification of existing 'mainstream' public educational structures.

At this level, communities must strive to avoid the pitfalls, and reverse the effects, of the subtractive language learning to which they are normally submitted by assimilationist programmes in the countries in which they are located. Most commonly, subtractive learning takes the form of 'submersion' programmes that force indigenous and minority children to receive instruction entirely through the medium of a majority/official language, thus hampering or totally impeding the learning of the mother tongue, and often not leading to high levels of formal proficiency in the majority language either (Skutnabb-Kangas 1990). 'Transitional' programmes – common, for instance, in many Latin American countries – in which linguistic minority children are initially instructed through the medium of their mother tongue, while they learn the majority language, and are then made to switch to the majority language as soon as they can function in it to some extent, are equally damaging. The mother tongue is attributed no intrinsic value, but is only considered a means to an end, and this normally makes reaching high levels of multilingualism difficult at a group level.

Educational programmes that do work for indigenous and minority language maintenance and revitalization are ones in which the native language is the main medium of instruction at least throughout primary school, and that are based on additive language learning. In these programmes, linguistic minority children have the choice to be taught through the medium of their own mother tongue, in classes with minority children with the same mother tongue only, where they have a bilingual teacher and where they receive good instruction in the majority language as a second/foreign language, also imparted by a bilingual teacher. In higher grades (after Grade 6), a few subjects may be taught through the medium of the majority language. Programmes of this kind are known as 'language shelter' or 'maintenance' programmes (Skutnabb-Kangas 1990), and they can be used for all indigenous and minority children when the native language has been learned by children and is used at home.

Also, majority children may be taught mainly through the medium of a minority language, as in the Canadian 'immersion' programmes, in which English-speaking children voluntarily learn content through the medium of French or some other minority language, in classes taught by fully bilingual teachers. In the latter case, therefore, the language of instruction is a foreign language to all students. Immersion programmes can also be used for revitalization purposes where indigenous or minority children are (re)learning an ancestral language they were no longer learning at home. Here, from a language learning point of view, the children are in the same situation as majority children learning in a foreign language. Immersion programmes are thus totally distinct from the submersion programmes discussed above, in which instruction is entirely in the majority language, with no teaching whatsoever in or of the minority language. A combination of an immersion programme for majorities and a language shelter programme for minorities in the same class is the 'two-way' programme (practised, for example, in over 200 schools in the USA) in which 50 percent minority children (with the same mother tongue) and 50 percent majority children are taught together by a fully bilingual teacher, initially through the medium of the minority language, later through both, and where both languages are taught as subjects to both groups.

Among indigenous peoples, the Maori and the Hawaiians, whose languages had undergone a drastic decline, have pioneered the immersion approach, starting with full native-language-medium pre-school 'language nests' and continuing with immersion primary school. These programmes also aim to involve families in school service and governance, as well as in native language learning through evening classes. Much work is also devoted to teacher training, since the decline that both languages had experienced made it difficult to staff the schools with bilingual teachers. Curriculum development stresses the introduction of native world-views (cultural traditions, philosophy, spirituality) through language. In Hawai'i, these training and curriculum requirements are met through a high-level Hawaiian language programme at University of Hawai'i, which recently also inaugurated a Masters Degree in Hawaiian Studies. A Hawaiian-medium middle and high school has also been started. Since the programme's inception in 1983, the number of native speakers of Hawaiian under 18 years of age has grown from 32 to 1,400 (out of a total Native Hawaiian population of about 200,000 people, of whom less than half were listed as speakers of Hawaiian in the 1990 census). Academic achievement has significantly improved (Kamana and Wilson 1996).

As the Hawaiian and Maori cases suggest, in using education as a support for linguistic and cultural maintenance it is crucial to ensure the congruence between educational form and content. A linguistically positive model, with the indigenous or minority mother tongue as the main medium of education, may not be successful if it has an organization or content that contradicts the positive intentions. It may be difficult for participants to make the distinction between the two and realize that the problem is with the cultural content or organization, not with the language of instruction. In Namibia and South Africa this has been a major obstacle: earlier racist content was taught through the medium of African languages, and many parents are now also rejecting their own languages as school languages because of this link – despite the fact that at least in South Africa government language policy, with 11 official languages, is one of the most enlightened in the world.

Another example is that of the Shuar in Guadalupe, Ecuador (Tynell 1997). In an experimental school, started and supported by the Shuar community, teachers have adopted notions about bilingual intercultural education, through maintenance of the mother tongue and learning of another language. In spite of these stated goals, however, the teachers are reproducing the same Hispanic pedagogical model under which they themselves were formed, i.e. a transitional model. The medium of instruction is Shuar in the first two years and Spanish after that, with Shuar only taught as a subject. Aspects of traditional culture, such as mythology, oral narrative, medicinal knowledge, etc. are incorporated but as separate entities, lessons in Shuar culture. This discrepancy between the aim (maintenance and revitalization of the Shuar language and incorporation of intercultural aspects) and the model chosen (transitional early exit model, which seldom leads to bilingualism) shows how it can be easier to change the goals of education than its practice.

In Australia, several Aboriginal groups have rejected both the assimilationist English-only schooling and an education only given by bilingual Aboriginal teachers. Instead they use an approach whereby they separate 'white knowledge' (literacy, numeracy, etc.), taught by monolingual English-speakers, with White Australian culture and structure, from their own Aboriginal knowledge in school. Their own knowledge is rather 'lived' than taught in schools – which also implies a whole rethinking of what 'school' is, at least for that part of the learning process (Harris 1990).

Among the Hualapai of northern Arizona, a successful bilingual/bicultural programme has been established at the Peach Springs school (Zepeda and Hill 1991). Teaching is imparted through the medium of both Hualapai and English, and the curriculum is community-based, i.e. it aims to develop first of all the children's understanding of Hualapai culture and the local environment. The learning of botany, for instance, starts with the study of the local ecosystem and of Hualapai traditional knowledge about it. Even when additional botanical information is introduced, from other parts of the world, the curriculum continues to emphasize local knowledge through comparison. Several books in Hualapai have been produced on Hualapai ethnobotany, hunting, traditional foods, traditional history and narratives, and so forth.

These examples show that for education to be able to support indigenous and minority language (and culture) maintenance, serious thought must be given to language in its 'cultural envelope', as Smolicz (1990) puts it. Linguistically and culturally transitional models fail to do this. When given the opportunity, indigenous and minority peoples increasingly choose, or strive toward, shelter and immersion models, and see the learning of traditional knowledge (especially through experiential means) as a core element of their education. Crucially, these educational goals are both fostered by and foster the re-establishment of intergenerational linguistic communication and knowledge transmission and help recreate the contexts for language and knowledge use – thus reversing the main processes that lead to language and knowledge erosion and loss.

International language support organizations

In addition to local initiatives, a growing number of non-governmental and other organizations are involved in supporting the preservation of threatened languages, as well as in advocacy in favour of linguistic diversity. Many of these operate on a national or regional level; others are international in nature and/or in the scope of their activities. Only the latter can be mentioned here.

UNESCO (United Nations Educational, Scientific and Cultural Organization) has been devoting attention to language endangerment since the early 1990s. In collaboration with the Permanent International Committee of Linguists (CIPL) and the International Clearinghouse for Endangered Languages (ICHEL), UNESCO's International Council for Philosophy and Humanistic Studies (CIPSH) has developed the *'Red Books on Endangered Languages'*. The 'Red Books' (modelled after the Red Books for endangered species) record and monitor the status of threatened languages by continent or major area. Red Books for Europe, Northeast Asia, Asia and the Pacific, and Africa are currently available in electronic form and can be viewed and searched on line. Through CIPSH, UNESCO also offers Grants for the Study of Endangered Languages.

In addition, several committees and NGOs have been formed in recent years with the specific purpose of increasing awareness of the global language endangerment situation and to provide support for documentation and applied work on endangered languages; among these are the Anthropologists' Fund for Urgent Anthropological Research (USA/UK), the Endangered Language Fund (USA), the Foundation for Endangered Languages (UK), the German Society for Endangered Languages, and the Linguistic Society of America's Committee on Endangered Languages and their Preservation (USA) (see Table 2.4, at the end of this contribution, for details).

Numerous electronic resources also exist that are relevant to matters of linguistic diversity and language endangerment. Of major importance is the on-line searcheable version of Ethnologue (13th ed.; Grimes 1996), to date the most complete catalogue of the world's languages. It provides as reliable information as is available on number of speakers for each language, and pinpoints the especially threatened ('nearly extinct') status of many of the smaller linguistic communities. Also essential are the online versions of UNESCO's *Red Books on Endangered Languages*. These and other resources are listed in Table 2.5, at the end of this contribution.

While these initiatives focus mostly on language endangerment and on supporting threatened languages, the US-based international NGO 'Terralingua: Partnerships for Linguistic and Biological Diversity' is specifically devoted to promoting linguistic diversity and to exploring the links between linguistic and biological diversity (Box 2.2). Such a perspective is in line with the emerging integrated conservationist trend, that increasingly stresses the inherent connections between biological and cultural diversity – and that represents the recognition of integral human-environment relationships traditionally practised, and now being reclaimed, by local peoples around the world. What Terralingua's perspective adds to the new trend is evidence that linguistic diversity is an intrinsic element of the equation, and that it must be protected along with biological and cultural diversity in order to protect, and ensure the continued development of, the diversity of life on earth. Terralingua promotes collaboration and mutual learning across academic disciplines as well as between the concerned scholarly community, the advocacy community, and the world's indigenous and minority peoples struggling to preserve or restore the integrity of their languages, cultures and environments.

Box 2.2: Statement of Purpose

Terralingua: Partnerships for Linguistic and Biological Diversity

Terralingua is an international non-profit organization dedicated to:

- supporting the perpetuation and continued development of the world's linguistic diversity;
- exploring the connections between linguistic, cultural and biological diversity.

STATEMENT OF PURPOSE

A. Terralingua recognizes:

1. that the diversity of languages and their variant forms is a vital part of the world's cultural diversity;
2. that cultural diversity and biological diversity are not only related, but often inseparable; and
3. that, like biological species, many languages and their variant forms around the world are now faced with an extinction crisis whose magnitude may well prove very large.

B. Terralingua declares:

4. that every language, along with its variant forms, is inherently valuable and therefore worthy of being preserved and perpetuated, regardless of its political, demographic or linguistic status;
5. that deciding which language to use, and for what purposes, is a basic human right inhering to members of the community of speakers now using the language or whose ancestors traditionally used it; and
6. that such usage decisions should be freely made in an atmosphere of tolerance and reciprocal respect for cultural distinctiveness—a condition that is a prerequisite for increased mutual understanding among the world's peoples and a recognition of our common humanity.

C. Therefore, Terralingua sets forth the following goals:

7. to help preserve and perpetuate the world's linguistic diversity in all its variant forms (languages, dialects, pidgins, creoles, sign languages, languages used in rituals, etc.) through research, programmes of public education, advocacy, and community support;
8. to learn about languages and the knowledge they embody from the communities of speakers themselves, to encourage partnerships between community-based language/cultural groups and scientific/professional organizations who are interested in preserving cultural and biological diversity, and to support the right of communities of speakers to language self-determination;
9. to illuminate the connections between cultural and biological diversity by establishing working relationships with scientific/professional organizations and individuals who are interested in preserving cultural diversity (such as linguists, educators, anthropologists, ethnologists, cultural workers, native advocates, cultural geographers, sociologists, and so on) and those who are interested in preserving biological diversity (such as biologists, botanists, ecologists, zoologists, physical geographers, ethnobiologists, ethnoecologists, conservationists, environmental advocates, natural resource managers, and so on), thus promoting the joint preservation and perpetuation of cultural and biological diversity; and
10. to work with all appropriate entities in both the public and private sectors, and at all levels from the local to the international, to accomplish the foregoing.

TERRALINGUA: PARTNERSHIPS FOR LINGUISTIC AND BIOLOGICAL DIVERSITY

P. O. Box 122
Hancock, MI 49930-0122, USA
Contact: Dr. Luisa MAFFI, President
E-mail: maffi@nwu.edu
Contact: Dr. Tove SKUTNABB-KANGAS, Vice-President
E-mail: tovesk@babel.ruc.dk
Contact: Mr. David HARMON, Secretary-Treasurer
E-mail: gws@mail.portup.com
Web site: http://cougar.ucdavis.edu/nas/terralin/home.html

Linguistic diversity and language rights (Tove Skutnabb-Kangas)

Language, power and diversity

Language has been seen as an essential, homogenizing element in state-building since the Age of Enlightenment (e.g. Gellner 1983; Hobsbawm 1991). It has played a key role in maintaining colonial structures (Calvet 1974; Mühlhäusler 1990; Phillipson 1992) and in reproducing neo-colonial structures (e.g. Ngugi 1986; Bamgbose 1991). Language has been an important means of control and domination and its importance is rapidly growing (e.g. Bourdieu 1977; Foucault 1980; Habermas 1985; Fairclough 1989). Education through the medium of majority languages or colonial languages has been the most powerful assimilating and homogenizing force for both indigenous children and immigrant/refugee minority children (e.g. Skutnabb-Kangas 1984, 1988, 1990; Jordan 1988; Cummins 1989). There is also continuing controversy over the role of the 'standard' language ('proper' language) as opposed to other variants (dialects, sociolects, genderlects, ethnolects). The enforcement of standard languages has likewise been a powerful method for controlling ordinary people (O'Barr 1982) and for homogenization (Illich 1981). Reducing oral languages to writing is also homogenizing (Mühlhäusler 1996).

But language is also a psychologically, educationally and politically crucial tool for counterhegemony and self-determination; it is one of several mobilizing factors in struggles for national recognition. This is true for indigenous peoples (e.g. Hamel 1994a, 1994b on Latin America; Harris 1990 and Fesl 1993 on Aboriginal peoples in Australia; Magga 1994 on the Sámi in Norway; Karetu 1994 on the Maori in New Zealand; Ekka 1984, Annamalai 1986, Khubchandani 1991, 1994 and Pattanayak 1991 on indigenous peoples in India; Stairs 1988 on the Inuit in Canada). It is equally true for groups reasserting themselves after the disintegration of communist regimes (e.g. Rannut 1994 on Estonia; Druviete 1994 on Latvia). Having control over one's own language and maximizing its official use are also of paramount concern to groups seeking self-determination or more cultural rights before or indeed after decolonization (e.g. Ansre 1979, Mateene 1980, 1985, Rubagumya 1990, Bamgbose 1991, Akinnaso 1994 and Twahirva 1994 on Africa; Hassanpour 1992 on Kurdish; Tickoo 1994 on Kashmiri). Language is also central to the demands of most immigrant groups, regardless of whether or not they are aspiring to become new national minorities (Skutnabb-Kangas and Cummins 1988; Skutnabb-Kangas 1996c).

Thus, the fate of languages is of utmost and growing importance. The struggle over the power and resources of the world is conducted increasingly through ideological means, and ideas are mainly mediated through language. This also partially explains the spread of numerically large languages (English, Spanish, Russian, Mandarin Chinese, etc.) at the cost of the smaller ones. The ideas of the power-holders cannot be spread, nationally or internationally,unless those with less power understand the power-holders' language (e.g. 'international English' world-wide, or standard Italian in Italy). Mass media and formal education are used for this language learning, which is is mostly organized subtractively (the 'major' language is learned at the cost of one's own language, not in addition to it). Language and culture are in the process of replacing 'race' as bases for discrimination. Access to material resources and structural power is increasingly determined not on the basis of skin colour, or 'race' (as biologically argued racism has it), but on the basis of ethnicity and language(s) (mother tongue and competence, or lack thereof, in official and/or 'international' languages). Linguistically argued racism, *linguicism*, and culturally argued racism, *ethnicism* or *culturism*, can be defined as 'ideologies, structures and practices that are used to legitimate, effectuate and reproduce an unequal division of power and (both material and non-material) resources between groups which are defined on the basis of language (linguicism) and culture or ethnicity (ethnicism/culturism)' (Skutnabb-Kangas 1988: 13).

In the post-Cold-War era, not only tolerance of but indeed preservation, promotion and development of linguistic and cultural diversity are vital for world peace and social justice (Galtung 1995; Kalantzis 1995; Huntington 1996). The non-dominant languages and the cultural interpretations and knowledge of the world of which they are bearers are a necessary counterweight to the power-holders' homogenizing and homogenized languages and messages. With the help of (their own) language(s), non-dominant peoples can analyse their own situation and start the decolonization of minds and the reconstruction and revalidation of their own non-material resources (languages, cultures, norms, values, institutions, knowledges, world-views) which have, through stigmatization, been rendered invisible or made to seem handicaps rather than resources.

Linguistic human rights, linguicide and monolingual reductionism

Given the growing importance of language worldwide, state resistance to smaller languages is to be expected. Cobarrubias (1983) provides the following taxonomy of possible state policies *vis-à-vis* indigenous or minority languages: (1) attempting to kill a language; (2) letting a language die; (3) unsupported coexistence; (4) partial support of specific language functions; (5) adoption as an official language. The division of power and resources in the world partially follows linguistic lines along which more accrues to speakers of 'big' languages, because they can use their languages for most official purposes (case 5 or at least 4 above). Speakers of smaller languages are often forced to learn the big languages subtractively, at the expense of their own languages, because the latter are not used for official purposes, including education; i.e. the state has adopted one or other of policies 1–3 above.

In this connection, it is useful to compare the concept of *linguicide* (linguistic genocide) with that of *language death*. The notion of 'language death' does not necessarily imply a causal agent. In the common use of this concept, language death is attributed to circumstances beyond the control of any agents, such as the inevitable processes of social change that come with 'modernization'. It is seen as comparable to the evolution of natural organisms that develop, bloom and wither away. When some liberal economists (e.g. List 1885: 174ff.) considered that nations had to be of a 'sufficient size' to be viable, it followed that smaller nationalities *and languages* were doomed to disappear as collective victims of the 'law of progress'. Their speakers were advised to reconcile themselves with 'the loss of what could not be adapted to the modern age' (Hobsbawm 1991: 29–39). Several Western European liberal ideologists and Soviet language planners in the early part of this century held that linguistically distinct nations were but one phase in the development towards a unified world with a world language, coexisting with national languages which would be 'reduced to the domestic and sentimental role of dialects' (ibid.: 38). This liberal ideology of development is still alive and well. When discussing 'small ethnic groups and languages', we are warned not to 'be idealistic and feel blind pity for everything which *in its natural course* is transformed, becomes outdated or even extinct' (Satava 1992: 80; emphasis added). The concept of language death can be associated with this type of liberal ideology, whether in Eastern Europe, North America (the 'English Only' movement), or in aid policies world-wide, which invariably support dominant languages. Within this paradigm, language death is interpreted as the result of voluntary language shift by each speaker.

Linguicide, by contrast, implies that *there are agents involved in causing the death of languages*. The agents can be *active* ('attempting to kill a language') or *passive* ('letting a language die' or 'unsupported coexistence'). Liberal ideology, however, would not consider all three of these cases as forms of linguicide, but only the first one, since it involves an *active* agent with the *intention* to kill languages; the other two cases would be seen as falling within the domain of 'natural' language death. The *causes* of linguicide and linguicism must be analysed from both structural and ideological angles, covering the struggle for structural power and material resources, on the one hand, and on the other the legitimation, instantiation and reproduction of the unequal division of power and resources between groups based on language. The *agents* of linguicide/linguicism can also be structural (a state, an institution, laws and regulations, budgets, etc.) or ideological (norms and values ascribed to different languages and their speakers). There is thus nothing 'natural' in language death. Language death has causes, which can be identified and analysed.

In preparation for the 1948 United Nations International Convention for the Prevention and Punishment of the Crime of Genocide (E 793, 1948), linguistic and cultural genocide were discussed alongside physical genocide as serious crimes against humanity (Capotorti 1979). In the UN General Assembly, however, Article 3 covering linguistic and cultural genocide was voted down and is thus not part of the final Convention. What remains, though, is a definition of linguistic genocide that was acceptable to most states then in the UN:

> 'Prohibiting the use of the language of the group in daily intercourse or in schools, or the printing and circulation of publications in the language of the group.' (Art. 3.1)

Linguistic genocide in these terms is practised throughout the world. The use of an indigenous or minority language can be prohibited overtly and directly, through laws, imprisonment, torture, killings and threats (e.g. Skutnabb-Kangas and Bucak 1994). It can also be prohibited covertly, more indirectly, via ideological and structural means, such as in educational systems. Every time there are indigenous or minority children in day-care centres and schools with no bilingual teachers authorized to use the languages of the children as the regular teaching and child-care media, this is tantamount to prohibiting the use of minority languages 'in daily intercourse or in schools'. This is the situation for many immigrant and refugee minority children, as well as for indigenous First Nations (e.g. Jordan 1988; Hamel 1994a; Fettes 1998).

Linguicism is a major factor in determining whether speakers of particular languages are allowed to enjoy their linguistic human rights. Lack of these rights, for instance the absence of these languages from school time-tables, makes indigenous and minority languages invisible. Alternatively, minority mother tongues are construed and presented as non-resources, as handicaps which are said to prevent indigenous or minority children from acquiring the majority language – purported as the only valued linguistic resource – so that minority children should get rid of them in their own interest. At the same time, many minorities, especially minority children, are in fact prevented from fully acquiring majority resources, and especially majority languages, through disabling educational structures in which instruction is organized through the medium of the majority languages in ways that contradict most scientific evidence on how education for bilingual education should be structured (e.g. Pattanayak 1981; Skutnabb-Kangas 1984, 1988, 1990, forthcoming; Ramirez *et al.* 1991; Corson 1992; Padilla and Benavides 1992; Cummins 1996).

The linguistic wrongs that are an important causal factor in reducing the linguistic and cultural diversity of the world are partly symptoms of an ideology of *monolingual reductionism*, which consists of several myths. The first four include the beliefs that monolingualism, at both individual and societal levels, is *normal*; that it is *unavoidable* ('it is a pity but you cannot make people cling to small languages which are not useful; they want to shift'); that it is *sufficient* to know a 'big' language, especially English ('everything important is in English, or, if it is important enough and has been written in another language, it will be translated into English'; 'it is the same things that are being said in all languages so why bother?') and *desirable* ('you learn more if you can use all your energies on one language instead of needing to learn many'; 'monolingual countries are richer and more developed'; 'it is cheaper and more efficient to have just one language'). In fact all these myths can be easily refuted – they are fallacies (Skutnabb-Kangas 1996a, 1996b).

The fifth myth says that the granting of linguistic and cultural human rights will inevitably lead to the disintegration of present states. At this level, monolingual reductionism can be characterized as an ideology used to rationalize linguistic genocide (especially in education) by states that claim to be nation-states and purport the existence of (unassimilated) linguistic minorities within their boundaries to be a threat to national unity. The principle of the territorial integrity and political sovereignty of contemporary states is often presented as being in conflict with another fundamental human rights principle, that of self-determination. States fear that minorities that enjoy linguistic human rights will demand first internal self-determination, e.g. cultural and other autonomy, and then independent status, i.e. external self-determination. By denying them linguistic rights and by bringing about homogenization through linguistic and cultural genocide in education and elsewhere, states (or rather their leaders) hope to eliminate the threat of groups that may eventually demand self-determination. To deny minorities those human rights that are most central to their reproducing themselves as distinctive groups – namely, linguistic and cultural human rights, and especially educational language rights – while observing (or appearing to observe) several of the basic human rights for all its citizens, including minorities, is a covert way for a state to make languages disappear at the same time as it retains its legitimacy in the eyes of (most of) its citizens and the international community. Covert linguicide of this type appears to be extremely effective. It is often far more difficult to struggle against covert violence, against the colonization of the mind, where short-term 'benefits' may obscure longer-term losses, than it is to fight physical violence and oppression.

Contra these myths, however, there are strong reasons why states should support, rather than try to eliminate, linguistic and cultural diversity and grant linguistic human rights. Some states might indeed disintegrate in the process, but this should be acceptable if the human right of self-determination is to be upheld (Clark and Williamson 1996). By and large, though, granting linguistic and cultural human rights to minorities reduces rather than creates the potential for 'ethnic' conflict, prevents the disintegration of states, and may avoid anarchy in which even the rights of the elites will be severely curtailed. Linguistic and cultural identity are at the core of the cultures of most ethnic groups (Smolicz 1979). When threatened, these identities can have a very strong potential to mobilize groups: '[...] attempts to artificially suppress minority languages through policies of assimilation, devaluation, reduction to a state of illiteracy, expulsion or genocide are not only degrading of human dignity and morally unacceptable, but they are also an *invitation* to separatism and an *incitement* to fragmentation into mini-states' (Smolicz 1986: 96; emphasis added). Thus, promoting diversity by granting linguistic human rights can actually promote a state's self-interest.

As Asbjørn Eide (1995: 29–30) of the UN Human Rights Commission points out, cultural rights have received little attention both in human rights theory and in practice, despite the fact that today 'ethnic conflict' and 'ethnic tension' are seen as the most important potential causes of unrest, conflict and violence in the world. Absence or denial of linguistic and cultural rights are today effective ways of promoting, not curbing, this 'ethnic' conflict and violence. Linguicide is ineffective as a strategy for preventing the disintegration of present-day states. 'Preservation of the linguistic and cultural heritage of humankind' (one of UNESCO's declared goals) presupposes preventing linguicide. Linguistic diversity at local levels is not only a necessary counterweight to the hegemony of a few 'international' languages, but represents a recognition of the fact that all individuals and groups have basic linguistic human rights and is a necessity for the survival of the planet. The perpetuation of linguistic diversity is a necessary component of any discourse on and strategy for the maintenance of biological and cultural diversity on earth.

Linguistic rights and education

In many international, regional and multilateral human rights instruments language is mentioned in the preamble and general clauses (e.g. United Nations Charter, Art. 13; Universal Declaration of Human Rights (1948), Art. 2; International Covenant on Civil and Political Rights (ICCPR; 1966, in force since 1972, Art. 2.1) as one of the characteristics on the basis of which discrimination is forbidden, together with 'race, colour, sex, religion, political or other opinion, national or social origin, property, birth or other status' (ICCPR Art. 2.1). This suggests that language has been seen as one of the most important characteristics of humans in terms of human rights issues in the key documents that have pioneered the post-1945 UN effort. Yet, the most important linguistic human rights, especially in education, are still absent from human rights instruments. Language gets much poorer treatment in human rights law than other important human attributes, like gender, 'race' or religion.

For the maintenance and development of linguistic and cultural diversity on our planet, educational language rights are not only vital but the most important linguistic human rights. Intergenerational transmission of languages is the most vital factor for their maintenance. If children do not get the opportunity to learn their parents' idiom fully and properly so that they become at least as proficient as the parents, the language cannot survive. As more and more children around the world get access to formal education, much of the language learning that earlier happened in the community must happen in schools. However, beyond the non-binding preambles, in the educational clauses of human rights instruments two phenomena can be observed. One is that mention of language disappears completely, as for instance in the Universal Declaration of Human Rights, where the paragraph on education (26) does not refer to language at all. Similarly, the ICCPR, having mentioned language on a par with race, colour, sex, religion, etc. in Art. 2.2, refers to 'racial, ethnic or religious groups' but omits reference to language or linguistic groups in its educational Article 13. Secondly, if language-related rights are specified, the Articles dealing with these rights are typically so weak as to be *de facto* meaningless. For example, in the UN Declaration on the Rights of Persons Belonging to National or Ethnic, Religious and Linguistic Minorities (adopted in 1992), most of the Articles use the obligating formulation 'shall' and have few opt-out modifications or alternatives – except where linguistic rights in education are concerned. Compare Articles 1.1 and 1.2 with Article 4.3 (emphases added):

> 1.1. States *shall protect* the existence and the national or ethnic, cultural, religious and linguistic identity of minorities within their respective territories, and *shall encourage* conditions for the *promotion* of that identity.
>
> 1.2. States *shall* adopt *appropriate* legislative *and other* measures *to achieve those ends*.
>
> 4.3. States *should* take *appropriate* measures so that, *wherever possible*, persons belonging to minorities have *adequate* opportunities to learn their mother tongue *or* to have instruction in their mother tongue. [Emphasis added.]

Likewise, in the European Charter for Regional or Minority Languages (1992), the formulations include a range of modifications such as, 'as far as possible', 'relevant' 'appropriate', 'where necessary', 'pupils who so wish in a number considered sufficient', 'if the number of users of a regional or minority language justifies it', as well as a number of alternatives, as in 'to allow, encourage *or* provide teaching in *or* of the regional or minority language at all the appropriate stages of education' (emphasis added). Writing binding formulations that are sensitive to local conditions presents unquestionably real problems. However, opt-outs and alternatives permit a reluctant state to meet the requirements in a minimalist way which it can legitimate by claiming that a provision was not 'possible' or 'appropriate', or that numbers were not 'sufficient' or did not 'justify' a provision, or that it 'allowed' the minority to organize teaching of their language as a subject, at their own cost.

A new Council of Europe Framework Convention for the Protection of National Minorities was adopted by the Committee of Ministers of the Council of Europe in 1994. Again the Article covering the medium of education is much more heavily qualified than any other:

> 'In areas inhabited by persons belonging to national minorities traditionally or in substantial numbers, if there is sufficient demand, the parties shall *endeavour* to ensure, *as far as possible* and *within the framework of their education systems*, that persons belonging to those minorities have *adequate* opportunities for being taught in the minority language *or* for receiving instruction in this language.' (Art. 14.2) [emphases added]

Thus the situation is not improving despite new instruments in which language rights are mentioned, and even treated in detail. Even where linguists have participated in drafting new or planned instruments, as in the case of the Draft Universal Declaration of Linguistic Rights, the results are far from perfect.

Draft Universal Declaration of Linguistic Rights

The Draft Universal Declaration of Linguistic Rights, handed over to UNESCO in June 1996, is the first attempt to formulate a universal code of language rights (although it does not mention sign languages). Its 52 Articles, which have already gone through twelve drafts, grant rights to three different entities: *individuals* (= 'everyone'), *language groups*, and *language communities*. When the beneficiary is 'everyone' unconditionally, the rights are *individual* ('inalienable personal rights', Art. 3.1). When the beneficiary is the language group or community, the rights are *collective*. The Declaration considers the following to be inalienable personal rights which may be exercised in any situation:

- the right to the use of one's language both in private and in public;...
- the right to maintain and develop one's own culture;... (Art. 3.1)

These individual rights are the only ones that apply to all unconditionally. Educational language rights are not included.

Collective rights in the Declaration apply to historical language communities – including both traditional national minorities and all numerically small peoples, such as indigenous peoples – or to language groups, that is:

> any group of persons sharing the same language which is established in the territorial space of another language community but which does not possess historical antecedents equivalent to those of that community. Examples of such groups are immigrants, refugees, deported persons and members of diasporas. (Art. 1.5.)

Communities have more rights in the Declaration than either groups or individuals:

> 'All language communities are entitled to have at their disposal whatever means are necessary to ensure the transmission and continuity of their language.' (Art. 8.2)

This could be interpreted to mean that all language communities are entitled to receive from the state the funds needed to organize education in the mother tongue from kindergarten to university. However, no duty-holders are specified for granting the 'means' mentioned in Art. 8.2 – or, for that matter, for granting the 'equal rights' to which language communities are entitled (Art. 10.1) or for taking the 'necessary steps... in order to implement this principle of equality and to render it real and effective' (Art. 10.3). Likewise, no legal obligations are specified when it comes to the Articles dealing with education:

> 'All language communities are entitled to have at their disposal all the human and material resources necessary to ensure that their language is present to the extent they desire at all levels of education within their territory: properly trained teachers, appropriate teaching methods, text books, finance, buildings and equipment, traditional and innovative technology.' (Art. 25)
>
> 'All language communities are entitled to an education that will enable their members to acquire a full command of their own language, including the different abilities relating to all the usual spheres of use, as well as the most extensive possible command of any other language they may wish to know.' (Art. 26)
>
> 'The language and culture of all language communities must be the subject of study and research at university level.' (Art. 30)

Groups have fewer and more vaguely specified rights than communities:

> '... the collective rights of language groups *may* include... the right for their own language and culture to be taught.' (Art. 3.2; emphasis added).

For groups, collective rights to one's own language are thus not seen as inalienable. Again, no duty-holders are specified.

For individuals, only education in the language of the territory where one resides is a positive right (Art. 26.1). There is no mention of bilingual or multilingual territories in the Declaration. This means that for those who speak a language other than the language of the territory, education in their own language is not a positive right. 'Everyone's' right to one's own language is only specified in negative terms:

'This right *does not exclude* the right to acquire oral and written knowledge of any language which may be of use to him/her as an instrument of communication with other language communities.' (Art. 29.2; emphasis added).

Furthermore, Art. 29.2 is formulated so as to suggest that 'everyone's' own language can be learned only if it is a useful instrument when communicating with other language communities. This means that it could in principle be excluded if it is not known by any entity defined as a language community, or if it is not used as a lingua franca between people some of whom represent language communities. The lack of individual rights in the education section makes it likely that all those not defined as members of language communities will be forced to assimilate. According to Art. 23.4, '[...] everyone has the right to learn any language', where 'any language' could also be interpreted as the mother tongue of those who otherwise are not granted positive mother tongue learning rights. However, the article specifies that this right prevails only within the context of principles that support the languages and self-expression of language communities, i.e. not the languages of 'groups' or 'everyone':

1. Education must help to foster the capacity for linguistic and cultural self-expression *of the language community of the territory where it is provided*.

2. Education must help to maintain and develop *the language spoken by the language community of the territory where it is provided*.

3. Education must always be at the service of linguistic and cultural diversity and of harmonious relations *between different language communities throughout the world*. (Art. 23.4; emphases added.)

The Declaration thus gives language communities very extensive rights but leaves 'everyone' with very few rights. This makes the Declaration vulnerable at several levels. Self-determination is not an unconditional right in international law (Clark and Williamson 1996). This means that a Declaration that gives most of the rights to linguistic communities, without specifying firm duty-holders, makes these communities completely dependent on the acceptance of their existence by states. States may claim that they do not have minority language communities, thus confining any such minorities found within their borders to the very limited language rights of 'everyone'. 'Language groups' are even less likely to be recognized as collective entities. And those individuals who are not members of any recognized language communities or groups are in the weakest position. It is for these reasons that establishing firm individual rights is of the utmost importance. Yet the Declaration is most wanting in this respect. It does suggest a monitoring body to be set up by the United Nations, and calls for sanctions against states that interfere with their citizens' rights. At present, however, it remains a draft recommendation that is not legally binding.

In addition, even the language rights for language communities are formulated in such a way as to make them completely unrealistic for anybody except, maybe, a few hundred of the world's language communities, most of them dominant linguistic majorities (cf. Art. 23.4 above). For most African, Asian and Latin American countries, the rights in the Declaration are at present practically, economically and even politically impossible to realize. It therefore seems extremely unlikely that the Declaration will be accepted in its present form. From the point of view of maintaining the planet's linguistic diversity, the immediate fate of the UN Draft Universal Declaration on Rights of Indigenous Peoples may be more important than that of the Draft Universal Declaration of Linguistic Rights. The former has a good coverage of language rights, especially in education (cf. Arts. 14, 15, 16 and 17). If these rights were to be granted in their present form, some 60 to 80 percent or the world's oral languages would have adequate legal support. Despite over a decade of negotiations, several countries, most importantly the United States, are likely to demand substantial changes that will undermine the progress achieved in the Declaration (Daes 1995; Morris 1995). Yet, this Declaration has at least some chance of being accepted and signed, even if in a form that reduces the rights it currently contains. On the other hand, the Draft Universal Declaration of Linguistic Rights remains crucial as the first attempt to formulate language rights at a universal level, opening up the issue to serious international debate.

Recent positive developments

There might also be some hope in two recent developments. The UN International Covenant on Civil and Political Rights, Art. 27, still grants the best legally binding protection to languages:

'In those states in which ethnic, religious or linguistic minorities exist, persons belonging to such minorities shall not be denied the right, in community with other members of their group, to enjoy their own culture, to profess and practise their own religion, or to use their own language.'

In the customary reading of Art. 27, rights were only granted to individuals, not collectivities. And 'persons belonging to ... minorities' only had these rights in states that accepted the existence of such minorities. This has not helped immigrant minorities because they have not been seen as minorities in the legal sense by the states in which they live. More recently (6 April 1994), the UN Human Rights Committee adopted a General Comment on Article 27 which interprets it in a substantially broader and more positive way than earlier (Box 2.3).

It remains to be seen to what extent this General Comment will influence the state parties in relation to linguistic human rights of speakers of smaller languages. It depends on the extent to which the Committee's interpretation ('soft law') will become the general norm observed by the countries where indigenous peoples and migrant, refugee and other minorities live.

The second positive development is the new educational guidelines issued by The Foundation on Inter-Ethnic Relations for the OSCE (Organization for Security and Co-operation in Europe) High Commissioner on National Minorities, Max van der Stoel (The Hague Recommendations Regarding the Education Rights of National Minorities and Explanatory Note, October 1996). These guidelines were elaborated by a small group of experts on human rights and education (including the author of this section). In the section 'The spirit of international instruments', bilingualism is set as a right and responsibility for persons belonging to national minorities (Art. 1), and states are reminded not to interpret their obligations in a restrictive manner (Art. 2) (Box 2.4).

In the section on 'Minority education at primary and secondary levels', mother-tongue medium education is recommended at all levels, including bilingual teachers in the dominant language as a second language (Articles 11–13). Teacher training is made a duty on the state (Art. 14) (Box 2.5).

Finally, the Explanatory Note states that

> '[S]ubmersion-type approaches whereby the curriculum is taught exclusively through the medium of the State language and minority children are entirely integrated into classes with children of the majority are not in line with international standards.' (p. 5).

This means that the children to whom the Recommendations apply might be granted some of the central educational linguistic human rights. The question now is to what extent the 55 OSCE countries will apply the Recommendations and how they will interpret their scope. The Recommendations could in principle apply to all minorities, even the 'everyone' with very few rights in the Draft Universal Declaration on Linguistic Rights. And since indigenous peoples should have at least all the rights that minorities do, these Recommendations might provide them also with a tool, while the Universal Declaration on Rights of Indigenous Peoples is still under discussion.

Box 2.3: Customary Reading and UN Human Rights Committee's General Comment on ICCPR Article 27

In customary reading, the Article was interpreted as:

- excluding (im)migrants (who have not been seen as minorities);
- excluding groups (even if they are citizens) that are not recognized as minorities (or as 'indigenous' – a formulation that has been added to Art. 30 in the 1989 UN Convention on the Rights of the Child, which is otherwise identical to ICCPR Art. 27) by the State (in the same way as the European Charter does);
- conferring only some protection from discrimination ('negative rights') but not a positive right to maintain or even use one's language, and
- not imposing any obligations on the States.

The UN Human Rights Committee sees the Article as:

- protecting all individuals on the State's territory or under its jurisdiction (i.e. also immigrants and refugees), irrespective of whether they belong to the minorities specified in the Article or not;
- stating that the existence of a minority does not depend on a decision by the State but requires to be established by objective criteria;
- recognizing the existence of a 'right' (rather than only a non- discrimination prescription), and
- imposing positive obligations on the States.

Box 2.4: The Hague Recommendations Regarding the Education Rights of National Minorities and Explanatory Note (October 1996) –

Articles 1 and 3

(1) The right of persons belonging to national minorities to maintain their identity can only be fully realized if they acquire a proper knowledge of their mother tongue during the educational process. At the same time, persons belonging to national minorities have a responsibility to integrate into the wider national society through the acquisition of a proper knowledge of the State language.

(3) It should be borne in mind that the relevant international obligations and commitments constitute international minimum standards. It would be contrary to their spirit and intent to interpret these obligations and commitments in a restrictive manner.

What a Universal Convention of Linguistic Human Rights should guarantee at an individual level

In a civilized state, there should be no need to debate the right for indigenous peoples and minorities to exist, to decide about their own affairs (self-determination) and to reproduce themselves as distinct groups, with their own languages and cultures. This includes the right to ownership and guardianship of their own lands and material (natural and other) resources as a prerequisite for the maintenance of non-material resources. It is a self-evident, fundamental collective human right. There should be no need to debate the right to identify with, to maintain and to fully develop one's mother tongue(s) (the language(s) a person has learned first and/or identifies with). It is a self-evident, fundamental individual linguistic human right.

What Linguistic Human Rights (LHRs) should be guaranteed to individuals for linguistic diversity to be maintained? Necessary individual linguistic rights have to do with access to the mother tongue and an official language in a situation of stable bilingualism, and with language-related access to formal primary education (Box 2.6).

Box 2.5: The Hague Recommendations Regarding the Education Rights of National Minorities and Explanatory Note (October 1996) – Articles 11–14.

(11) The first years of education are of pivotal importance in a child's development. Educational research suggests that the medium of teaching at pre-school and kindergarten levels should ideally be the child's language. Wherever possible, States should create conditions enabling parents to avail themselves of this option.

(12) Research also indicates that in primary school the curriculum should ideally be taught in the minority language. The minority language should be taught as a subject on a regular basis. The State language should also be taught as a subject on a regular basis preferably by bilingual teachers who have a good understanding of the children's cultural and linguistic background. Towards the end of this period, a few practical or non-theoretical subjects should be taught through the medium of the State language. Wherever possible, States should create conditions enabling parents to avail themselves of this option.

(13) In secondary school a substantial part of the curriculum should be taught through the medium of the minority language. The minority language should be taught as a subject on a regular basis. The State language should also be taught as a subject on a regular basis preferably by bilingual teachers who have a good understanding of the children's cultural and linguistic background. Throughout this period, the number of subjects taught in the State language should gradually be increased. Research findings suggest that the more gradual the increase, the better for the child.

(14) The maintenance of the primary and secondary levels of minority education depends a great deal on the availability of teachers trained in all disciplines in the mother tongue. Therefore, ensuing from the obligation to provide adequate opportunities for minority language education, States should provide adequate facilities for the appropriate training of teachers and should facilitate access to such training.

Box 2.6: Suggestions on What a Universal Convention of Linguistic Human Rights Should Guarantee at an Individual Level

A universal convention of linguistic human rights should guarantee at an individual level,

in relation to the mother tongue(s), that everybody can:

- identify with their mother tongue(s) and have this identification accepted and respected by others;
- learn the mother tongue(s) fully, orally (when physiologically possible) and in writing (requiring in most cases for indigenous and minority children to be educated through the medium of their mother tongue(s), within the state-financed educational system);
- use the mother tongue in most official situations (including day-care, schools, courts, emergency situations of all kinds, health care, hospitals, and many governmental and other offices).

... and in relation to other languages:

- that everybody whose mother tongue is not an official language in the country where they are resident can become bilingual (or multilingual, if they have more than one mother tongue) in the mother tongue(s) and (one of) the official language(s) (according to his/her own choice);
- that suitably trained bilingual teachers are available;
- that parents know enough about research results when they make their educational choices (e.g. minority parents must know that good mother-tongue medium teaching leads to better proficiency both in the mother tongue and in the dominant language, than dominant-language medium submersion teaching).

... in the relationship between languages:

- that any change of mother tongue is voluntary, not imposed (i.e. it includes knowledge of alternatives and of long-term consequences of choices and is not due to enforced language shift).

... and in the profit from education

- that everybody can profit from education, regardless of what their mother tongue(s) is/are (educational profit being defined in terms of equal outcome, not just of equal opportunity).

A Universal Convention of LHRs must also make states duty-holders, in a firm and detailed way: that is, it must provide enforceable rights. If these rights are not granted and implemented, it seems likely that the present pessimistic prognoses of some 90 percent of the world's oral languages not being spoken any more by the year 2100 may actually turn out to have been optimistic. Languages that are not used as main media of instruction will cease to be passed on to children at the latest when we reach the fourth generation of groups in which everybody goes to school – and many languages may be killed much earlier. There is still much work to be done for education through the medium of the mother tongue to be recognized as a human right. Yet this is what is most urgently needed to ensure that indigenous and minority peoples will be able to maintain and develop their languages and perpetuate linguistic diversity on earth.

Acknowledgements

The authors are grateful to the following people for their insightful comments and suggestions on an earlier version of this chapter: Jonathan Bobaljik, Nancy Dorian, Mark Fettes, David Harmon, Peter Mühlhäusler, Thomas Widlok, David Wilkins and Jeffrey Wollock. Thanks are due to David Harmon and Stanford Zent for permitting the reproduction of graphics from their published and yet to be published work. Maffi and Skutnabb-Kangas also wish to thank each other for mutual feedback on and close scrutiny of several drafts of their respective and common parts of this chapter. Skutnabb-Kangas went far beyond the call of duty as a contributor in her collaboration in this project, and her extensive input is gratefully acknowledged by the co-ordinator. The introduction to the chapter incorporates some of her writing. The section on language rights and parts of the section on language maintenance and revitalization draw extensively on previous writings by Skutnabb-Kangas.

TABLE 2.4: ENDANGERED LANGUAGES SUPPORT ORGANIZATIONS.

ANTHROPOLOGISTS' FUND FOR URGENT ANTHROPOLOGICAL RESEARCH
P. O. Box A
Phillips, ME 04966, USA
Contact: Prof. George N. APPELL, Founding Sponsor

FELLOWSHIPS IN URGENT ANTHROPOLOGY
Royal Anthropological Institute (RAI) and Goldsmith College
University of London, New Cross
London SE14 6NW
Phone: +44 171 919 7800
Fax: +44 171 919 7813

Sponsors research on 'currently threatened indigenous peoples, cultures and languages'. Offers fellowships through RAI.

ENDANGERED LANGUAGE FUND
Dept. of Linguistics, Yale University
New Haven, CT 06520, USA
Contact: Dr. Doug WHALEN, President
E-mail: whalen@lenny.haskins.yale.edu; elf@haskins.yale.edu
Web site: http://sapir.ling.yale.edu/~elf/index.html

Dedicated to the scientific study of endangered languages, support of native communities' language maintenance initiatives, and dissemination of research and applied work results. Offers grants for both kinds of activities.

FOUNDATION FOR ENDANGERED LANGUAGES
172 Bailbrook Lane
Bath BA1 7AA, UK
Phone: +44 1225 85 2865
Fax: +44 1225 85 9258
Contact: Dr. Nicholas D. M. OSTLER, President
E-mail: nostler@chibcha.demon.co.uk

Aims to spread information about language endangerment and to increase scholarly knowledge about the smaller languages of the world. Publishes a newsletter, organizes workshops, and offers funds for language documentation.

GERMAN SOCIETY FOR ENDANGERED LANGUAGES
Institut für Sprachwissenschaft (Institute for Linguistics)
Universität Köln
D-50923 Köln, Germany
Phone: +49 221 470 2323
Contact: Prof. Hans-Jürgen SASSE, President
E-mail: gbs@uni-koeln.de

An independent non-profit society that collaborates with the Committee on Endangered Languages of the German Linguistic Society and whose aim is to promote the use, the preservation, and the documentation of endangered languages and dialects through teaching, research, information and international co-operation.

INTERNATIONAL CLEARINGHOUSE FOR ENDANGERED LANGUAGES (ICHEL)
Dept. of Asian and Pacific Linguistics
Institute of Cross-Cultural Studies, Faculty of Letters
University of Tokyo
Bunkyo-ku, Tokyo 113, Japan
Phone: +81 3 3812 2111 ext. 3797
Fax: +81 3 5803 2784
Contact: Prof. Tasaku TSUNODA, Director
E-mail: tsunoda@tooyoo.l.u-tokyo.ac.jp
Contact: Dr. Kazuto MATSUMURA, Associate Director
E-mail: kmatsum@tooyoo.l.u-tokyo.ac.jp
Web site: http://www.tooyoo.l.u-tokyo.ac.jp

Works with UNESCO on the 'Red Books on Endangered Languages'. Functions as a data bank of materials on endangered languages. Publishes a newsletter.

LINGUISTIC SOCIETY OF AMERICA, Committee on Endangered Languages and their Preservation (CELP)
Survey of the world's endangered languages
1325 eighteenth St. N.W., Suite 211
Washington, D.C. 20036-6501, USA
Phone: +1 202 835 1714
Fax: +1 202 835 1717
E-mail: lsa@lsadc.org
Contact: Dr. Anthony WOODBURY, CELP Chair
E-mail: acw@mail.utexas.edu
Contact: Dr. Akira YAMAMOTO, Survey Coordinator
E-mail: akira@kuhub.cc.ukans.edu

Calls attention to issues about and fosters field research on the world's endangered languages. Conducts a survey of their status and of existing scholarly resources on them.

PERMANENT INTERNATIONAL COMMITTEE OF LINGUISTS
Instituut voor Nederlandse Lexicologie
(Institute for Dutch Lexicology)
University of Leiden
P. O. Box 9515
2300 RA Leiden, The Netherlands
Phone: +31 71 514 1648
Fax: +31 71 527 2115/522 7737
Contact: Dr. P. J. G. VAN STERKENBURG, Secretary General
E-mail: sterkenburg@rulxho.leidenuniv.nl

Works with UNESCO on the Red Books on Endangered Languages.

UNESCO International Council for Philosophy and Humanistic Studies (CIPSH)
Contact: Prof. Stephen A. WURM, Honorary President
Australian National University
Dept. of Linguistics
Canberra, ACT 0200, Australia
Phone: +61 6 249 2369
Fax: +61 6 257 1893
Contact: Prof. Jean BINGEN, Secretary General

Grants for the Study of Endangered Languages

UNESCO CIPHS
1, Rue Miollis
75732 Paris, France
Fax: +33 1 4065 9480

Publishes the Red Books on Endangered Languages. Offers grants for endangered language research.

WORLD FEDERATION OF THE DEAF
13D, Chemin du Levant
01210 Ferney-Voltaire, France
Fax: +33 450 40 01 07

Resource development and advocacy for the world's sign languages.

TABLE 2.5: ELECTRONIC RESOURCES ON LANGUAGES AND LANGUAGE ENDANGERMENT.

Endangered-Languages-L
List address: endangered-languages-l@carmen.murdoch.edu.au
Contact: Dr. Mari RHYDWEN, List owner
Humanities, Murdoch University
Perth, WA 6150, Australia
E-mail: rhydwen@central.murdoch.edu.au

Discussion of language endangerment, linguistic diversity, and indigenous and minority languages.

Ethnologue: Languages of the World, 13th ed. (1996)
Web site: http:/www.sil.org/ethnologue/
Contact: Dr. Barbara GRIMES, Editor
International Linguistics Center
7500 W. Camp Wisdom Road
Dallas, TX 75236, USA
E-mail: editor.ethnologue@sil.org

On-line version of the main catalogue of the world's languages.

GeoNative (electronic database of native place names)
Web site: http://www.geocities.com/Athens/9479
Contact: Luistxo FERNANDEZ, Web site owner
E-mail: txoko@redestb.es

Dedicated to 'putting minority languages on the map'; gathers and displays native place names in the world's smaller languages.

Language Documentation Urgency List (LDUL)
List address: ldul@cis.uni-muenchen.de
Contact: Dr. Dietmar ZAEFFERER
Institut für Deutsche Philologie
Universität München
Schellingstrasse 3
D-80799 München, Germany
Phone: +49 89 2180 2060/2180 3819
Fax: +49 89 2180 3871
E-mail: ue303bh@sunmail.lrz-muenchen.de

Automatic mailbox and database. Provides information on endangerment status of and need for documentation on the world's languages.

List of Language Lists
Web site: http://condor.stcloud.msus.edu:20020/docs/lnglstcurrent.txt
Contact: Dr. Bernard COMRIE, List co-owner
E-mail: comrie@bcf.usc.edu
Department of Linguistics, GFS-301
University of Southern California
Los Angeles, CA 90089-1693, USA
Phone: +1 213 740 2986
Fax: +1 213 740 9306
Contact: Dr. Michael EVERSON, List co-owner
E-mail: everson@indigo.ie

Inventory of electronic bulletin boards devoted to the study of individual languages and groups of languages.

UNESCO Red Book on Endangered Languages: Europe
Compiled by T. SALMINEN
http://www.helsinki.fi/~tasalmin/europe_report.html

Index to the UNESCO Red Book on Endangered languages: Europe.
Compiled by T. SALMINEN
http://www.helsinki.fi/~tasalmin/europe_index.html

UNESCO Red Book on Endangered Languages: Northeast Asia
Compiled by J. JANHUNAN and T. SALMINEN
http://www.helsinki.fi/~tasalmin/nasia_report.html

Databank for endangered Finno-Ugric languages
Compiled by T. SALMINEN
http://www.helsinki.fi/~tasalmin/deful.html

UNESCO Red Book on Endangered Languages: Asia and the Pacific
Compiled by S.A. WURM and S. TSUCHIDA
http://www.tooyoo.l.u-tokyo.ac.jp/asia-pacific-index.html

UNESCO Red Book on Endangered Languages: Africa
Compiled by B. HEINE and M. BRENZINGER
http://www.tooyoo.l.u-tokyo.ac.jp/africa-index.html

On-line versions of UNESCO's Red Books.

Box 2.7: On the Nature of Truth

Mazisi Kunene

'People do not follow the same direction, like water.' (Zulu saying)

Those who claim the monopoly of truth
Blinded by their own discoveries of power,
Curb the thrust of their own fierce vision.
For there is not one eye over the universe
But a seething nest of rays ever dividing and ever linking
The multiple creations do not invite disorder,
Nor are the many languages the enemies of humankind
But the little tyrant must mold things into one body
To control them and give them his single vision.
Yet those who are truly great
On whom time has bequeathed the gift of wisdom
Know all truth must be born of seeing
And all the various dances of humankind are beautiful
They are enriched by the great songs of our planet.

(From Skutnabb-Kangas and Cummins 1988: 176).

CHAPTER 3

INDIGENOUS PEOPLES, THEIR ENVIRONMENTS AND TERRITORIES

Andrew Gray

CONTENTS

TEXT BOXES

FIGURES

Tribute to Dr Gray

This chapter is dedicated to its editor, Dr Andrew Gray, who was lost at sea after a plane crash in the Pacific in May 1999. Dr Gray was a tireless, tenacious and much-beloved defender of human rights, and especially the rights of indigenous peoples around the world. He was also a productive and highly-respected scholar who wrote about the Arakmbut Indians of Peru, with whom he lived and worked for many years. He will be sorely missed by friends and colleagues the world over.

Indigenous Peoples, their Environments and Territories

Andrew Gray

Introduction (Andrew Gray)

General considerations

Over the past twenty years, indigenous peoples have become increasingly vociferous and assertive in environmental negotiations. This is part of a historical process of resistance against colonization which has a history going back thousands of years (Gray 1995: 42-5).

Indigenous peoples strongly resist being defined by others: 'We assert our inherent right to define who we are. We do not approve of any other definition' (IWGIA 1995: 26). This right is recognized by the International Labour Organization's Convention 169 Concerning Indigenous and Tribal Peoples in Independent Countries: 'Self-identification as indigenous or tribal shall be regarded as a fundamental criterion for determining the groups to which the provisions of this Convention apply' (ILO 1989).

As a general orientation to the concept of indigenous, one could say that indigenous peoples are distinct peoples, with their own languages, cultures and territories, who have lived in a country since times prior to the formation of the current nation state. They have become disadvantaged and vulnerable as a result of colonial invasion of their territories either by international colonization or by groups within the countries in which they live.

More than 300 million people in the world are indigenous. Arctic peoples, Native Americans, forest dwellers of the Amazon, mountain peoples of the Andes, African pygmies, pastoralists and Bushmen from Africa, tribal peoples of Asia, Aboriginal and Maori peoples of Australia and New Zealand, and island peoples of the Pacific, all come under the heading 'indigenous'.

Indigenous peoples are culturally distinct and live lives that vary considerably from one locality to another. Of the estimated 6,000 cultures in the world, between 4,000 and 5,000 are indigenous, which means that indigenous peoples make up between 70 and 80 percent of the world's cultural diversity (IUCN 1997: 30). Cultural diversity is one of the fundamental attributes of human beings because 'change, creation and re-creation, interpretation and re-interpretation, are all parts of the fabric of everyday experience' (Carrithers 1992: 9). This is particularly apparent with indigenous peoples whose lives are intimately bound up with their environments, which not only provides for production, but also provides the spiritual inspiration for life.

Indigenous peoples' cultural diversity is grounded in territory and locality, drawing together their social and natural worlds. This creates a difficulty when discussing indigenous peoples. Whereas diversity makes generalizations about indigenous peoples difficult, throughout the world they encounter similar structural threats in relation to access to their resources and defence of their lives and cultures. To emphasize unity at the expense of diversity risks the formation of 'essentialised images' (Brosius 1996: 4), while to talk exclusively of diversity depoliticizes the common indigenous struggle which is based on the assertion of fundamental rights and freedoms.

When looking at the global distribution of indigenous peoples, there is a marked correlation between areas of biological diversity and areas of cultural diversity. This link is emphasized in the opening contribution to this chapter, by Joji Cariño, and is particularly significant for rainforest areas such as the Amazon, Africa and Southeast Asia (IUCN, op.cit. 31).

One reason for this is that the species-diverse environments in which indigenous peoples live are deeply embedded in their production activities and spiritual relations. The diversity present in the natural world adds to the complexity and distinct variety of different interpretations and activities, not only for each people but also for different communities. These interpretations cover food production, curing, mythology, and understanding the world in general. This chapter provides several examples: Shepard from the Machiguenga and Rodríguez-Manasse from the Barí, both show the intense relationship between mythology and the environment. In each case, mythological figures bring crops to human beings and transform people into animals.

However, indigenous peoples are not simply reflections of their environment. To say that biological diversity causes cultural variation would risk falling into a crude environmental determinism. Ultimately, cultural variation is a human matter based on choice and decision. Furthermore, studies increasingly show how indigenous peoples live in anthropogenic landscapes, which they help create (Posey 1997). Indeed, as several contributions to this volume demonstrate, indigenous peoples themselves conserve and create biodiversity. For example, Zoundjihekpon and Dossou-Glehouenou (Chapter 9) show that the forest peoples of Benin and Côte D'Ivoire use religious prohibitions to manage their forests, leaving certain species and certain areas alone, while Chapeskie (this chapter) describes how the Ojibway of Canada plant and manage wild rice in areas thought by outsiders to be 'untouched'. Other examples in this chapter show that for each indigenous people, this nature–culture relationship is defined in specific ways, rooted locally in the territory. This fact is very marked among the Australian Aboriginal peoples described by Bennett. Whereas it is not possible to say that biological diversity determines cultural diversity, or vice versa, both elements of variation rely on each other because both actively contribute to the formation of the other: biodiversity is the inspiration for spirituality and culture for indigenous peoples, who through their production activities and shamanic practices contribute to respect for and the enhancement of biological diversity (Box 3.1).

Biodiversity and nature

The relationship between 'society/culture' and 'nature/biodiversity' lies behind much of this discussion. The distinction has been used (if not over-used) by anthropologists to mark out the gap between human and non-human worlds. This can easily become a taken-for-granted division, reflecting a Western predatory view of nature as separate from society, rather than the intricate and complex relationship between so many indigenous peoples and their environments.

The contribution by Peterson on Zaire (this chapter) draws particular attention to the way in which Western views of the relationship between human beings and nature are 'dualist'– based on a principle of 'either/or'. He contrasts this with a view encountered in many African forest peoples which can be called 'dialectical' or 'holistic' – based on a principle of 'both/and' where human beings and nature are 'creatively intertwined'.

This conclusion has also been encountered in the Amazon where indigenous peoples do not always conceptualize 'nature' as a totalizing non-human phenomenon. Hvalkof (1989: 142) concludes that the Ashéninka of lowland Peru have no real concept of nature 'in which nature is externalized, and thus constitutes an antithesis to culture or 'the humanized'. On the contrary, in the Ashéninka conception what we view as 'nature' to them belongs to the realm of 'social relations'. These findings parallel my own with the Arakmbut (Gray, 1996: 113-4), where there is no unitary notion of 'nature', but a set of relations between humans, spirits and clusters of species.

As with 'nature', indigenous peoples also argue that the concept of biological diversity is alien because it separates the phenomenon of non-human diversity from their knowledge and livelihood. Mead's paper explains how this difference distinguishes scientific from indigenous approaches to knowledge, while Suzuki argues that non-indigenous science should strive to overcome this separation. Biological diversity for most indigenous peoples is taken for granted as part of their everyday existence. The point here is not that 'nature' or 'biodiversity' do not exist, rather that they are part of an approach to the environment which separates what so many indigenous people keep together and simplifies unnecessarily what is so complicated and unique to each people.

Indigenous peoples may respect and enhance biodiversity, but that does not necessarily mean that the 'biodiversity' concept is always appropriate. Indeed, it could have serious repercussions, because emphasizing the notion of biodiversity as separate from human life, its conservation can be promoted by protecting non-human 'wilderness'. This ignores human beings and dispossesses them of their lands and territories in the name of conservation (Colchester 1994).

The relationship between indigenous peoples and biological diversity is not only one of form, but also one of content. Two factors that recur in this context are indigenous knowledge and its link to sustainable development. The contributions in this chapter all draw attention to the intricate 'local knowledge' grounded in locality and biological diversity. Just as there can be no 'nature', there can never be just one 'knowledge' either. The International Alliance article expresses this by saying that one should talk of 'indigenous knowledges', thereby reflecting the range of observations, information and interpretations made by peoples, communities and individuals within groups and communities. The contribution on Nepal shows how indigenous views of the environment are not only conceptualized formally, but can also appear in the vernacular expression of environmental–social metaphors.

However, knowledge is not only a cerebral activity; it is also bound up in practical activity and technological systems reflecting both collective and personal experience and long-term observation. Knowledge is manifest in techniques as well as in perceiving, utilizing and managing resources. Fur-

Box 3.1: Arakmbut Spirituality

Andrew Gray

When an Arakmbut shaman from the Peruvian rainforest performs a curing chant for a sick person, he enters the invisible world to search for the patient's soul (*nokiren*). On the basis of the symptoms observed during the illness, he diagnoses a species of animal, bird or fish which he suspects is responsible and, following its spirit, floats above the forest path calling its name in the chant.

As he passes, he sings out the names of plants, trees, insects or any other species that the spirit causing the illness recognizes. Sometimes he will sing in the spirit's own language. If successful, the shaman finds the soul of the sick person – maybe a group of collared peccary are snuffling around a muddy area surrounding the soul. Then suddenly the shaman will shout 'Look out! There's a jaguar!' The animals flee from the predator and if his timing is good, the shaman can guide the soul back to the patient's body (Gray 1997a).

During this curing ritual, the shaman draws together an encyclopaedic knowledge of species, a practical personal experience of animal behaviour based on a life-time's observation and a visionary insight into the invisible spirit world. His chant, learnt from an elderly relative and developed further from experimentation, will cover as many as fifty different species. This is not a random list, but consist of clusters of rainforest features. For example, the collared peccaries will include – berries the animal likes to eat, areas where they like to shelter, birds that pick insects from their skin, or rough trees good for scratching. Each species is connected to the collared peccary but also relates to the others in the chant.

No species is isolated, each is part of a living collectivity binding human, animal and spirit. According to the Arakmbut, animals originally had a human form, and frequently still appear to each other in that way; spirits can also appear as people or animals, while humans, through shamanic visions and dreams, can communicate with both. In the Arakmbut invisible world, the distinction between animal, human and spirit becomes blurred.

In the act of curing, the Arakmbut demonstrate their relationship with the diversity of species in the rainforest and illustrate the practical application of their knowledge. Often the ritual will involve prohibitions on eating or hunting particular species. Health is thus restored by combining production knowledge and its moral values: for a hunter this means the number of animals that can be killed and how the meat should be distributed. Underlying these practices lie the spiritual dimensions to Arakmbut life, which constitute an imminent presence, constantly manifest and mediating between different species.

The Arakmbut curing ritual draws together several of the themes that will appear in this chapter. The relationship between indigenous peoples and their environment concerns culture, biological diversity, moral values, spirituality and rights. In daily life, these are intimately bound up with each other, and for indigenous peoples such as the Arakmbut, they are expressed in practical activities. A common way of encapsulating the complexity of indigenous peoples' relationship with the natural world is through their mythology and ritual which, as with the Arakmbut curing chant, express the invisible spiritual connection between people and other species.

Sources: Gray 1996, 1997a and 1997b.

thermore, sustainability in indigenous production methods is a consequence of practical knowledge that embodies sustainable principles (IUCN 1997: 36).

Unfortunately, in some parts of the world indigenous peoples' territories have been devastated by un-sustainable economic activities – some carried out by indigenous peoples themselves (Ellen 1993; Posey 1997). These examples almost inevitably arise in areas where indigenous rights have not been respected and the people have been subject to land invasions, leaving them with insufficient resources. At the root of this problem is the commodification of labour and knowledge, the main focus of Whitt's contribution to this chapter. She argues that indigenous peoples have the principles of sustainability embedded in their knowledge, but that these can only be preserved by recognizing rights and respecting cultures.

Indigenous knowledge and technology is passed down through the generations and comes from spiritual revelation and personal innovation. However, as Barsh (this chapter) points out in his contribution, the sharing of information means that indigenous knowledge has important social dimensions: not only is knowledge shared through informal educational methods, but much is also passed around the community in the form of myths, accounts of hunting trips, and descriptions of agricultural activities or curing methods. In this way, knowledge can be owned and shared collectively while each person is also valued for his or her particular innovations and interpretations of the world.

A major feature of indigenous knowledge, as mentioned in Agenda 21 (26.1), is its holistic character. This concept is used in environmental circles to refer to a tendency of nature to form a 'whole' which is larger than the sum of its parts. When holism appears in a socio-cultural context, it refers to those interconnections between different aspects of life where a 'whole' consists of clusters of features (social, cultural, economic, political, spiritual or environmental) which cross-cut non-indigenous defined boundaries. Indigenous peoples, with their distinct cultural principles, will almost invariably provide some 'holistic' insight into their relationship with the environment that is markedly different from that of national society. Particularly important holistic aspects of indigenous knowledge are how the invisible spirit world reflects the environment and connects with production methods. This invisible spirit dimension is too often overlooked by ecologists, although it has a fundamental influence on how indigenous peoples conceptualize and utilize their resources. McGregor's Hawaiian article and Bennett's description of Aboriginal Dreamtime underline the importance of spirituality in looking at indigenous peoples' relations with their environments.

Of particular significance are the 'Voices of the Earth' (Chapter 4) which provide examples, from all over the globe, of traditional and indigenous people's relationships with the earth. Indigenous peoples communicate with the non-human world through invisible spirits. This is particularly apparent in shamanic activities, such as those of the Arakmbut. In these cases, spirituality enters the material world permanently and constitutes the very force of life. For the Desana (Reichel-Dolmatoff 1971), this can operate as a form of 'spiritual conservation'. However, for most peoples, management or use of resources is a combination of several different principles, such as practical knowledge, experience, and taking advantage of specific conditions. Thus, although the spiritual world is ever-present in the environment and largely influences production and curing methods, it would be as misleading to make it a determining factor in biodiversity conservation as it would be to say that cultural diversity is determined by biological diversity.

The form of holism discussed by indigenous peoples is full of connections and links but is not determined by any one factor. Socio-cultural life, spirituality and biological diversity combine to provide the context for indigenous knowledge and sustainable production methods, and the way in which they combine will differ between peoples. Myth and ritual are particularly important ways whereby the forces of the invisible world are drawn into the world in order to ensure growth and fertility. Examples range from the Andean peoples whose rituals draw the power of fertility from the mountains and the rivers to fertilize the earth (Gose 1994); another widespread phenomenon comes from Indonesia where the unidirectional 'flow of life' links together the cosmic, social and material worlds in dynamic movement (Fox 1980).

Holistic clusters of connections link aspects of life in indigenous cosmologies which Western thought separates: this is why indigenous peoples have distinct value systems conceptualizing and acting on the world. Values are given to phenomena according to their perceived context within a system: thus for Western economies value is given to a commodity on the basis of its exchange potential in a system (Appadurai 1988: 3), while moral values are distinct because they are ideas of activities placed within the context of an ethical or religious system. When one talks of spiritual and cultural values in Western contexts these are contrasted with economic values. In addition, social acceptance of these values has a political consequence in that power is embedded in their enforcement over indigenous peoples.

However, whereas one can see a marked variety of the socio-political and cultural aspects of value in all peoples, it is impossible to presume that values will be systematized in the same way. Indeed, indigenous peoples frequently find themselves clashing with this artificial distinction between economic value and moral value. Several contributors have brought this together by referring to the notion of the 'gift'. Gift-giving and reciprocity have long been recognized as fundamental to indigenous economic systems (Mauss 1970; Sahlins 1972). The form of the exchange, the necessity to reciprocate, these are all aspects of the social and personal interactions of daily life in indigenous communities. They take different forms in different places, but they firmly place social, economic and political values within a framework of reciprocity.

Recent work in the Amazon and elsewhere has shown how this reciprocity operates not only on a socio-economic level, but also on a spiritual level (Descola 1992). These examples show how indigenous peoples receive life and sustenance from the invisible world, and in return their souls pass to the species. This is apparent in Whitt's account of the shaman collecting medicinal plants. The notion of a gift is important because not only does it break away from the idea of commodity and price being tied to value, but it also utilizes a concept that transcends the economic to include social, political and spiritual dimensions.

Another important aspect of the invisible spiritual world is that it is the arbiter of legitimacy relating indigenous peoples to their environments. This takes the argument from spirituality to indigenous rights, which emerge at a community level when outside forces, usually in the form of colonizing powers, companies, or the forces of the state, invade indigenous territories and take away their resources, devalue their cultures, and at times threaten their very existence (Box 3.2).

The need to identify indigenous rights has been the main work of the International Labour Organization and the United Nations Commission on Human Rights (Berman 1993). Since 1981, indigenous peoples have participated in the UN Working Group on Indigenous Populations, which in 1993 approved a draft Declaration of the Rights of Indigenous Peo-

Box 3.2: Mobil in the Peruvian Amazon

On 26 March 1996, the Peruvian state oil company Perupetro signed a contract with the international oil company Mobil handing over rights to explore for oil in the rainforest region of the Madre de Dios. In contravention of ILO Convention 169, which Peru had signed several years previously, there was no consultation with the indigenous peoples living in the area. The activities have affected both the Arakmbut people of the River Karene and several voluntarily isolated groups of Mashco Piro and Nahua in the Río de las Piedras (Survival International 1996).

The exploration activities that took place later in the year caused considerable concern locally, particularly for the indigenous organization Federación Nativa del Madre de Dios (FENAMAD) which took a stance to defend the rights of the indigenous peoples of the area. Resisting an oil company is not easy. Some people and organizations felt that the government and the oil companies were too powerful to resist, and as Mobil made offers of compensation to the communities, along with a strong public relations campaign, the communities themselves were uncertain how to react.

Taking advantage of this, Mobil arranged deals with two communities against the advice of the Federación. However, the Arakmbut communities in the main firing line of the exploration activities took a firmer position and succeeded in ensuring that the seismic lines were moved outside community territories. Never the less, the basin of the Karene and Las Piedras rivers, which are the homelands of the Arakmbut and other isolated peoples, are under threat should Mobil decide to exploit oil in the area, and could have serious environmental consequences for the indigenous peoples of the area.

The Arakmbut struggle has not gone unrecognized. FENAMAD, in spite of finding itself opposing a major multinational oil company practically alone, received enormous international and national support from indigenous and non-indigenous sources alike. As a result of massive sponsorship, its efforts were rewarded by receiving, in March 1997, the prestigious Premio Bartolomé de las Casas in Spain. The award will be used to title the lands of the isolated peoples, to secure the recognition of their rights.

Meanwhile, at a local level the discussions in the Arakmbut communities have been particularly relevant to this chapter on indigenous peoples and their environment. The elders in particular were strongly opposed to non-indigenous companies gaining access to their territories and polluting the forest, which provides food, shelter and is the store-house of their knowledge. Particularly vocal were the women, who feared for their children and the survival of the community as a whole.

The reaction of the Arakmbut was that they are not only the owners of their territory, but are also responsible for its welfare. Their rights and the protection of their environment are completely bound together. As the company threatens them, the indigenous Arakmbut assert their rights – the reason they do this is for their survival, which depends on the survival of their environment. The value of the rainforest for the Arakmbut is not only economic: it is also spiritual and cultural. If the forest were purely economic they might have been persuaded to be bought off by the oil company (which certainly tried to convince them). However, more is at stake.

ples. This document has proved more accurate than the ILO Convention in reflecting the perspectives of indigenous peoples. The text of the draft Declaration can be found in Appendix 1.

The main advances for indigenous peoples in the Declaration are: recognition of their territories and their customary institutions, control over all their resources, and consent before any development initiatives take place on their lands. Indigenous peoples' control over their own development, along with self-determination, are meant to provide sufficient protection to ensure the survival of indigenous peoples. In addition to the draft Declaration, indigenous peoples are currently negotiating the establishment of a Permanent Forum within the UN system that will serve as a focal point for discussions on all indigenous issues, including human rights, development and the environment. The draft Declaration on the Rights of Indigenous Peoples is the starting point for the recognition of indigenous rights, and this should be approved and implemented, supported by the Permanent Forum to guarantee the implementation and monitoring of UN agreements on these issues. Rights cannot be separated from the relationship between indigenous peoples and their environment because they are the only peaceful means of protection available. Indigenous peoples have to be respected and the right of their own social and political institutions to determine their destinies recognized. In this way, through acknowledging their religious freedom, the over-arching presence of the invisible spiritual world can be secured, thereby ensuring the sovereign source of legitimation for activities on their territories. All of this can be summed up in the affirmation of self-determination for indigenous peoples, which is the fundamental right recognized in the draft Declaration.

Self-determination for indigenous peoples means 'the right to control over their institutions, territories, resources, social orders, and cultures without external domination or interference, and the right to establish their relationship with the dominant society and the state on the basis of consent' (Berman 1993: 320-1). If there is one conclusion to be drawn from this chapter it is that the rights embraced by the notion of self-determination are sufficiently open-ended to reflect the complexity of indigenous holistic relations with the environment. This point is also made repeatedly in Chapter 4, and in the declarations and statements of indigenous peoples' organizations that constitute Appendix 1.

Recognizing indigenous rights is not a catch-all solution providing automatic answers to the problems of biodiversity conservation. Indigenous peoples need support and resources for their own self-development and in the process of defending their rights may well establish new strategies for conservation of biological diversity. The process of putting self-determination into practice and the recognition of indigenous peoples' rights is not the final answer, but rather establishes conditions – indeed the only conditions – whereby they can carry out their responsibility to their ancestors and descendants to keep their territory and heritage for posterity. Biological conservation will only become possible in any meaningful sense when outsiders recognize indigenous peoples' rights and respect their spiritual and cultural values.

For many years, indigenous peoples have argued that recognition of their rights is the only protection they have against the invasions that threaten their existence. Several contributions provide examples of this: the U'wa in Colombia are threatened by oil companies while Columbia river peoples are defending their territories and fishing stocks from outsiders. The main threats to sustainable biodiversity conservation are those that threaten indigenous peoples. Only when the rest of the world recognizes indigenous peoples' rights and supports and strengthens their organizations, will the indigenous peoples themselves be able to continue their distinct forms of biodiversity conservation as expressed through their spiritual and cultural values and described by the contributors to this chapter.

Indigenous peoples address the UN General Assembly Special Session, 23-27 June, 1997 (Joji Cariño)

In June 1997, Joji Cariño from the International Alliance of Indigenous–Tribal Peoples of the Tropical Forests was invited to address the UN General Assembly on the relationship between indigenous peoples, development and the environment. The following paper is her statement.

Mr. President, Mr. Secretary-General, distinguished Members of the Assembly, and all peoples of the world:

I speak today for the indigenous and tribal peoples of the tropical forests, who have come together in response to the global destruction of our forests. Our network includes the Batwa of Rwanda, the hill tribes of Thailand, the peoples of the Amazon, the Adivasis of India, and many more peoples from 30 tropical forest countries. I am an Ibaloi Igorot from the Cordillera region of the Philippines.

In 1992, leading up to the Rio Conference, we formed the International Alliance of Indigenous and Tribal Peoples of the Tropical Forests, and agreed a Forest Peoples' Charter as a basic statement of our shared principles, goals and demands. The movement arose in response to the destruction of our forests to feed the unsustainable consumption and production patterns of the rest of the world.

Our goal is to secure respect for indigenous rights, territories, institutions and processes, and to promote our indigenous models of development and conservation in tropical forest regions that are more just and more sustainable.

We are part of a powerfully resurgent movement of indigenous peoples world-wide, who today find ourselves at the chalk face of the global crisis of development and environment. Our greatest strength is in the local activities of indigenous peoples who love and care for our territories and environment.

For indigenous peoples, development too often means mining, oil pipelines, logging, dams and biopiracy which devastate our environment, consisting of our lands, territories and spiritual bonds with creation and the earth.

Sixty percent of the Western world's alluvial gold is extracted from tribal lands and waters. The largest single supply of diamonds is mined from a sacred site of aboriginal Australians, while most of the world's remaining tropical forests are on indigenous lands. All over the globe, parks and protected areas overlap with our homelands, encroaching on our rights, while many rivers flowing from our mountains are polluted and dammed.

The earth's biodiversity is intimately linked with indigenous peoples' traditional knowledge. However, our knowledge receives little respect, unless it provides the means to make profits for outsiders. All over the world indigenous peoples are suffering the negative effects of development on the environment.

In East Asia, much of the region's economic growth is paid for by indigenous peoples. For example, the Bakun Dam in Sarawak, Malaysia, will require the clear-cutting of 80,000 hectares of rainforest and the forced displacement of 5,000–8,000 persons from 15 indigenous communities. In the Philippines, 6.7 million hectares (86 percent of the Cordillera region and 23 percent of the whole Philippine land area) has been opened up for mining exploration and development by some of the world's largest mining corporations with notorious relations with indigenous peoples. This has been facilitated by the world-wide liberalization of mining laws and financed by international financial institutions.

Mining and oil exploration affect indigenous peoples in areas as far apart as Papua New Guinea, Indonesia, India, Burma, Suriname, Venezuela, Peru, Colombia, Guyana, Nigeria and many parts of North America. A more recent form of mining is being pursued by biotechnological companies – the 'mining' of our genes, our traditional knowledge and our biological resources.

These problems are not confined to the tropical forest countries. The Inuit and other northern aboriginal peoples are concerned with the contamination of their food by persistent organic pollutants (POPs), most of which come from temperate and tropical lands and are transported to the Arctic.

Not only is our health adversely affected, but indigenous peoples are also negatively and disproportionately represented in all social indicators such as poverty, unemployment, criminal convictions and poor housing. We are being reduced to substandard living conditions due to systematic discrimination and subordination within dominant societies.

Today, the huge imbalance in our human relations is directly feeding the imbalance in our relations with the earth. How can we reconcile indigenous peoples' disproportionate contribution to the world's environmental protection and economic growth with our disproportionate material and social impoverishment, leading even to the extinction of many of our peoples?

When the world is prepared to deal with this paradox, and this social injustice, then I can be convinced that the world is ready for sustainable development.

- Is there not a danger that the economic growth and liberalization and even the environmental protection called for in Agenda 21 will be pursued on the backs of our peoples?
- Are governments prepared to recognize indigenous self-determination and sustainable development as two sides of the same coin?
- Can all governments who have stated their commitment to forests undertake this as a partnership with indigenous peoples?

These problems should be given higher priority by the world's leaders in the coming years.

Often, the campaigns of indigenous communities are misjudged as the ignorance of 'primitives' unschooled in modern economic realities. But make no mistake. We are not peoples of the past – we are your contemporaries and in some ways may be your guides towards more sustainable futures in the twenty-first century. Our heritage, which reaches back prior to the creation of states, is a deeper memory of experience from which to draw upon for the future.

The more than 300 million indigenous people in the world today make up over 4,000 distinct societies representing 95 percent of human cultural diversity, and I would also say 95 percent of humanity's breadth of knowledge for living sustainably on this earth.

The world has woken up to the loss of biodiversity, but not to the disappearance of our cultural wealth. Modern societies are implementing protected areas for wildlife and biodiversity, but now we must look to demarcating and recognizing indigenous lands and territories as guaranteed spaces for sustainable use, rather than treating our lands as expansion areas for failed and unsustainable practices.

Our struggles are very often straightforward conflicts of interest – of securing our local livelihoods and our cultures against the further enrichment of more powerful interests. Globalization has come to mean the incorporation of diverse local economies and societies into the global capitalist system. But indigenous societies suffer only the fall-outs from this inequitable process. Our experience has shown that our incorporation into the world economy often results in our impoverishment – materially, culturally and spiritually – and can even lead to our death as distinct peoples.

The transformation of local sustainable uses of land, forests, mountains, rivers and ice into economic growth and development means more logging, mining, agricultural plantations and toxic dumping in our lands. Globalization and trade liberalization are seriously affecting the ability of indigenous peoples to contribute to global sustainability. We have had 500 years of incorporation. We need to usher in the next 500 years of respect and co-operation. A more sustainable globalization should be a celebration of the many diverse livelihoods, cultures and peoples around the world. Human diversity, together with the earth's biodiversity, are the real riches in this world, and should be enshrined in a revitalized Agenda 21. The current global trends are dismal, but in the few areas where indigenous peoples are granted respect, difficult but impressive progress can be made.

Since Rio there has been increasing policy attention given to indigenous peoples. At the Commission on Sustainable Development, indigenous participation has been limited. However, the Dialogue Sessions with Indigenous Peoples and other Major Groups have been a welcome activity with the potential to evolve into substantive dialogues in the future.

I particularly wish to highlight the International Meeting of Indigenous and Other Forest-Dependent Peoples on the Management, Conservation and Sustainable Development of All Types of Forests, held in Leticia, Colombia, with the sponsorship of the Danish and Colombian governments. The Leticia Declaration, resulting from this meeting, contains detailed proposals for inclusion in the Forest Agenda in the coming years. Any future forest deliberations must have clear mechanisms for full and equal participation by indigenous peoples, women, local communities and NGOs.

Under the Convention on Biological Diversity, an Intersessional Workshop on the Implementation of Article 8j will discuss the knowledge, practices and innovations of indigenous and local communities, other related issues, and the impact of current intellectual property rights agreements on the lives of indigenous peoples.

However, the policy deliberations on issues affecting indigenous peoples have not been entirely enlightened. Many governments come to these negotiations with the objective of limiting the recognition and participation of indigenous peoples. The sum contribution made by some governments is to demand the deletion of the 's' on indigenous peoples – an act of callous bullying that is demeaning to the spirit of openness and co-operation needed for progress in the implementation of Agenda 21.

All governments need to adopt the inclusion of indigenous peoples in national plans for sustainable development and work towards making the United Nations a forum where best practice informs the combined efforts of governments, international organizations and indigenous peoples. However, to date there are no mechanisms for indigenous peoples to share in decision-making in international fora affecting our lands and our lives, such as the CSD, the Intergovernmental Panel on Forests, and the Convention on Biological Diversity.

These bodies must learn from the UN Commission on Human Rights where indigenous peoples sit with governments and international agencies to discuss the UN Draft Declaration on the Rights of Indigenous Peoples. Indigenous peoples are convinced that in the next few years, international standard-setting activities at the United Nations will become extremely crucial for advancing our welfare and our rights.

A declaration on the rights of indigenous peoples can provide some urgently needed standards for the actions of governments, businesses and development agencies in the difficult times ahead of us. These are needed to moderate the actions of mining and oil corporations, logging interests, large infrastructure projects, and pharmaceutical and biotechnological companies operating without regard for indigenous rights, all of which are among the transgressors of sustainable development.

At the national level, where the more difficult negotiations are taking place between governments and indigenous peoples, such standards can encourage the just and equitable settlements that will strengthen the inclusion and participation of the presently marginalized members of society in any national plans.

At a local level, it will give heart to the heroic but highly imbalanced struggles of indigenous peoples to win the respect of dominant societies so that we can live with the integrity of our cultures and continue our special relationship with our lands.

In July this year [1997], indigenous peoples will be celebrating the twentieth anniversary of our entry into substantive dialogue with the UN Commission on Human Rights. The establishment of the UN Working Group on Indigenous Populations with the mandate to develop international standards for the rights of indigenous peoples and to review developments affecting our rights, has provided the space for indigenous peoples to discuss these questions with governments, UN agencies, NGOs and all interested parties. It has proved to be extremely productive.

This dialogue ushered in the UN inauguration of 1993 as the Year of Partnership with Indigenous Peoples, and 1995-2004 as the UN Decade of Indigenous Peoples, turning the partnership into action. Important goals of the UN Decade include:

1. the adoption of the UN Declaration of the Rights of Indigenous Peoples currently being discussed by the Commission on Human Rights, and

2. the establishment of a UN Permanent Forum for Indigenous Peoples, to cover issues such as environment, development, culture and human rights.

This will fill the striking absence of any mechanisms within the United Nations for the co-ordination and regular exchange of information among governments, the UN system and indigenous peoples. In June this year, an International Workshop on the UN Permanent Forum for Indigenous Peoples will take place in Santiago, Chile, to advance concrete proposals for its establishment.

This UN General Assembly Special Session should recommend speedy action by the United Nations bodies on these important political goals to advance the welfare of indigenous peoples. These activities within the United Nations need to inform the post-Rio Agreements in keeping with the principle of the indivisibility of economic development, peace and environmental protection. This will strengthen the affirmation of the welfare of peoples at the heart of sustainable development.

This year, we mark five years after Rio and three years into the UN Decade for Indigenous Peoples with the theme 'Partnership into Action'. This Special Session of the General Assembly should reaffirm two important goals of the UN Decade for Indigenous Peoples. In the next few years, one criterion of Agenda 21 implementation should be the actions taken to secure indigenous peoples' rights and well-being. I look forward to future Agenda 21 reports on the progress being made on the following measures.

1. Securing the demarcation and recognition of indigenous peoples' territories, and the control and management of their ancestral lands.

2. Greater attention given to promoting cultural diversity and indigenous cultural and intellectual rights.

3. The establishment of permanent mechanisms for the shared decision-making of indigenous peoples at the CSD, CBD and the post-IPF Forest Forum.

4. The inclusion in the Forest Agenda of the results of the Leticia Meeting of Indigenous and Other Forest-Dependent Peoples, particularly for the full and equal participation of indigenous peoples, women, local communities and NGOs.

In the past five years, I have sensed a growing acknowledgement of indigenous peoples by other sectors of society involved in the quest for sustainable development. This partnership can only herald a better time for all our children.

I thank you all.

Indigenous peoples, their environments and territories

Metaphor and power in indigenous and Western knowledge systems (Laurie Anne Whitt)

Some years ago, the Cherokee mounted fierce resistance to the construction of the Tellico Dam and flooding of the Little Tennessee Valley. Many of their objections were based on threats to their cultural heritage. Ammoneta Sequoyah, a medicine man, explained that flooding the valley, or digging up the graves, would destroy 'the knowledge and beliefs of [the] people who are in the ground' (Sequoyah v. Tennessee Valley Authority 1990), including knowledge of medicinal plants. Knowledge and land are intimately bound to one another just as the natural world is alive and spiritually replete. This is a significant point of contrast with Western science where knowledge of nature is distinct and separable from nature. This difference is fundamental. It contributes to, and reciprocally sustains, divergent metaphors of knowledge. And these, in turn, have important implications for understanding how and what members of a knowledge community know; how they learn and teach; how they innovate; and how power figures in all of this.

The ideology of the market, and the omnipresence of market forces, have left an indelible mark on the Western conception of knowledge. Just as land and labour became 'fictitious commodities' to accommodate a market economy, so too has knowledge – human intellectual labour – been metaphorically transformed.

Various commentators have noted how the market doctrine obtained political and philosophical hegemony over Western society by pointedly ignoring the distinction between commodities and non-commodities. Commodities, for the economist, have a specific origin and purpose. They are manufactured goods which are produced, sold and consumed. Since human labour – intellectual or manual – is not manufactured, it is not a commodity. This metaphorical transformation of labour and land into 'fictitious commodities' greatly enhanced the power of the market system, ensuring control of virtually all aspects of social behaviour and natural resources.

Law, and most especially intellectual property law, is increasingly central to appreciating the role of power in Western technoscience. It has been, as Alan Hunt argues, a 'primary agency of the advance of new modalities of power and constitutes distinctive features of their mode of operation' (Hunt 1992:21). Intellectual property laws have been a particularly effective strategy for acquiring, commodifying and rendering profitable, intangible indigenous resources, such as artistic expressions and medicinal and spiritual knowledge (Whitt 1995).

Patent law illustrates this. The US Government is currently funding five major industry/university consortia which have planted their bioprospecting stakes throughout the world. The Guajajara people of Brazil, for instance, use a plant called *Philocarpus jaborandi* to treat glaucoma. Brazil earns around $25 million a year from exporting the plant, and the corporations who have patented it reap far greater profits. Yet the Guajajara have been subjected to debt peonage and slavery by the agents of the companies involved in the trade. Moreover, *Philocarpus* populations have been virtually depleted. The concrete repercussions of divergent construals of knowledge – as commodity and as gift – are painfully apparent here. This contrast, its impact, how the law has served to advance new modalities of power and to regulate their mode of operation, arose with the origins and development of intellectual property law.

Copyright law, for example, developed in response to the need for writers to sell their intellectual labour, to turn it into a commodity. As a result of eighteenth century European writers' challenge to the existing power relations, publishers no longer retained the exclusive right to sell and profit from their writings. In the future, writers would have to surrender their copyright before this could take place. But this inclusion of intellectual labourers in the market-place required a significant conceptual transformation. Not only did the nature of writing need to be reconceived, but also the nature of knowledge. Knowledge, or more exactly ideas, became something that need no longer remain in the public domain. It could be transformed into private property, provided that it was 'original' and was fixed in some physical form. It could then be exchanged in the market-place, a commodity to be bought or sold.

There are some obvious points of analogy between the pre-copyright 'inspired craftsmanship' and the process of knowing as construed within many indigenous cultures: these are in marked contrast with the innovation concept of patent law and copyright. The process of knowing is a matter of giving and receiving, and one's behaviour is to reflect that. There are, for example, ceremonies and procedures which a traditional healer must carefully observe in preparing medicines. An Anishinabe healer will be certain to offer tobacco to whatever plants are being collected. The plant will be addressed and thanked for being there, for allowing itself to be used in healing. Only certain plants will be culled, at certain stages of their life-cycles, at certain times of the year and of the day. This reflects the fact that for Indian people, land itself is a 'gift... (so) they assume certain ceremonial duties which must be performed as long as they live on and use the land.... Obligations demanded by the lands upon which people lived were part of their understanding of the world; indeed their view of life was grounded in the knowledge of these responsibilities' (Deloria 1992:262-3).

Something comparable is evident in the agricultural knowledge practices of Andean peoples undertaking the cultivation of their *chacras*. According to Modesto Machaca, 'to open a *chacra* I must ask permission of the Pachamama so that she will allow me to work this soil.... I tell her that I will cultivate this soil with love, without mistreatment, and the fruits she gives me we will all eat' (Rivera 1995: 25). Cultivating a *chacra* is a reciprocal activity, necessarily involving both humans and the land. It is in this sense that Andean agricultural knowledge is to be seen as tied, or tethered to the land. All of the activities that go on in the *chacra* – sowing, weeding, hilling, harvesting and even the storage, transformation and consumption of harvested products – are ritual activities. These rituals express the Andeans' attitude of love, respect and gratitude to the earth for its gifts, including the gifts of knowledge regarding how to cultivate a *chacra*.

The process of knowing of the Andean farmer and the Anishinabe healer is not exclusively or narrowly a cognitive activity: it is also an evaluative one. It is knowledge conditioned by respect and gratitude. This type of example can be found in many indigenous cultures. What they suggest is that when knowledge is construed as a gift, the *process* of knowing, and the relations with others and the natural world that are constitutive of that process, become central. The process of knowing must be undertaken in a way that respects and reflects the fact that each individual, each community, each tribe, nation and species has a responsibility to the workings of the universe, to the generations to come and to those that have passed.

Jake Swamp captures the varied dimensions of this in his account of his training in the gathering of medicinal herbs: 'You don't just go out there and pluck it out by its roots and walk away. You have to prepare. You have to know the words that go with it. What I was taught was that when you see that plant, to first see that it's the one you offer thanksgiving to, that plant is still here with us, still performing its duty and that you wish it to continue. You walk past it and you look for the other one, and that one you can pick. For, if you take that first one, who is to know, maybe that's the last one that exists in the world' (Barreiro 1992: 21).

This constraint, or something like it, lies at the heart of indigenous processes of knowing. Its absence in Western knowledge production marks an important distinction between indigenous and Western knowledge systems. The consequences of that absence, for the natural world and for future generations, are everywhere. The embedding within Western intellectual property law of assumptions about individuality and innovativeness are acutely at odds with the conceptual commitments of many indigenous cultures, and this has directly enabled the continued expropriation of indigenous cultural and genetic resources. Intellectual property law mediates the power of Western technoscience. It provides the mechanism for acquiring, commodifying and rendering profitable, intangible indigenous resources, and serves to consign traditional knowledge to the public domain. It thus provides for facile 'conversion' into private property. In the fifteenth century, Western legal concepts of occupation, conquest and cession provided the justification for the theft of indigenous land and property. Today, intellectual property regimes provide the intellectual cover and 'legal' justification for the expropriation of indigenous knowledge. As Aroha Mead comments, 'I query the concept of 'innovation' as defined by Western intellectual property laws – particularly when no recognition or value is accorded to the customary knowledge which links a species of plant to a particular usage, and details the most appropriate harvest, portion of the plant... and method of preparation' (Mead 1995). Indigenous knowledge and generations of indigenous labour – mental and physical – are minimized and discredited by this 'legal' transformation. All that is credited is the labour of individual corporate and academic scientists who interject 'novelty' into what they have taken.

Western concepts of 'originality' and 'novelty' are thus imposed on the world, as Western law and technoscience conjoin, assimilating the knowledge, resources and labour of generations of indigenous peoples. Meanwhile, indigenous processes of knowing are dismissed as closed, changeless, stultifying and stifling of originality. Such characterizations not only ignore the massive contributions of indigenous peoples – especially medicinal, pharmaceutical, botanical and agricultural – they also egregiously distort indigenous knowledge systems themselves. While 'originality' most definitely is not construed as 'one man's alone', if there were no originality or innovation in indigenous processes of knowing, the incredible richness and diversity of the genes currently being held in gene-plasm banks, to cite but one instance, would be inexplicable.

A very different concept of innovation is to be found in many indigenous knowledge systems. The source of originality is not internalized, as the genius of one individual; the natural world, the community, and the individual are all integrally involved. Individuals are subject to independent forces, and constrained by the need to act with respect for the natural world and for future generations. The community grounds and informs the individual, but since the process of knowing is experientially-based, and what one learns depends on individual development, abilities and preparation, individuals play an essential role in contributing new knowledge to the community. As one young Keres man, Larry Bird, explains: 'You don't ask questions when you grow up. You watch and listen and wait, and the answer will come to you. It's yours then, not like learning in school' (Tafoya 1982).

Such an approach to innovation requires receptivity, reciprocity and responsibility to the natural and human worlds in which one is situated. This is evident in the Andean practice of conversing with the natural world when they strive to increase the diversity of their cultivated plants by 'testing' new varieties. The cultivator does this, 'without obligating the new seed to "get accustomed" by force'. It is accepted for a seed that does not 'accustom' itself to move away... The (cultivator) says simply, 'this seed does not get used to me...' and continues 'testing' others to see if they follow him or her' (Rivera 1995: 33).

The new knowledge that results from such conversing is a gift, and it remains the giver's. When gifts are given, the continuity of social relationships (and, we should add, of relationships with the natural world) has the effect that the gift given always remains the giver's – it is inalienable.

Finding a new story (David Suzuki)

'It's all a question of story. We are in trouble just now because we do not have a good story. We are in between stories. The old story, the account of how we fit into it, is no longer effective. Yet we have not learned the new story.' (Berry 1988)

Over half of all people throughout the world now live in cities, and the largest growth of cities is occurring in the developing countries. The most destructive aspect of cities is the profound schism created between human beings and nature. In a human-made environment, surrounded by animals and plants of our choice, we feel ourselves to have escaped the limits of nature. Weather and climate impinge on our lives with far less immediacy. Food is often highly processed and comes in packages, revealing little of its origins in the soil or tell-tale signs of blemishes, blood, feathers or scales. We forget the source of our water and energy, the destination of our garbage and our sewage. We forget that as biological beings we are as dependent on clean air and water, uncontaminated soil and biodiversity as any other creature. Cut off from the sources of our food and water and the consequences of our way of life, we imagine a world under our control and will risk or sacrifice almost anything to make sure our way of life continues. As cities continue to increase around the world, policy decisions will more and more reflect the illusory bubble we have come to believe is reality.

As we distance ourselves further from the natural world, we are increasingly surrounded by and dependent upon our own inventions. We become enslaved by the constant demands of technology created to serve us. Consider our response to the insistence of a ringing telephone or our behavioural conformity to the commands of computers.

'All domesticated animals depend for their day-to-day survival upon their owners... The human domesticate has become equally dependent, not upon a proprietor, but upon storable, retrievable, transmissible technique. Technology provides us with everything we require... we are further dependent upon the expertise of countless others to provide even the most basic of daily necessities.' (Livingston 1996)

Divorced from the sources of our own existence, from the skills of survival and from the realities of those who still live in rural areas, we have become dulled, impervious, slow.

'If we had a keen vision and feeling of all ordinary human life, it would be like hearing the grass grow and the squirrel's heart beat, and we should die of that roar which lies on the other side of silence. As it is, the quickest of us walk about well wadded with stupidity.' (Eliot 1871)

Through our loss of a world view and our move into the cities and away from nature, we have lost our connection to the rest of the living planet. As Thomas Berry says, we must find a new story, a narrative that includes us in the continuum of earth's time and space, reminding us of the destiny we share with all the planet's life, restoring purpose and meaning to human existence.

How can we restore our connection to the rest of life on earth and live rich, fulfilling lives? Where can we find a new story? We have much to learn from the vast repositories of knowledge that still exist in traditional societies. This was suggested in a report in 1987 by the United Nations Committee on the Environment and Development headed by Norwegian Prime Minister Gro Harlem Brundtland. Entitled *Our Common Future* (World Commission on Environment and Development 1987), it acknowledged the inability of scientists to provide direction in managing natural resources, and called for recognition of and greater respect for the wisdom inherent in traditional knowledge systems:

> *'Their very survival has depended upon their ecological awareness and adaptation... These communities are the repositories of vast accumulations of traditional knowledge and experience that links humanity with its ancient origins. Their disappearance is a loss for the larger society, which could learn a great deal from their traditional skills in sustainably managing very complex ecological systems. It is a terrible irony that as formal development reaches more deeply into rainforests, deserts, and other isolated environments, it tends to destroy the only cultures that have proved able to thrive in these environments.'*

As we approach a new millennium, at the end of a century of explosive growth in science and technology, it is fitting that leading members of the scientific community are starting to understand that science alone cannot fulfil humankind's needs: indeed, it has become a destructive force. We need a new kind of science that approaches the traditional knowledge of indigenous communities; the search for it has already begun.

The inventiveness of our brain has catapulted our species out of the constant need to make a living from our immediate surroundings. Once our world-view embedded each of us within a world in which every part was intricately interconnected. Each of us could be at the centre of this multi-dimensional web of interconnections, 'trapped' in a sense, by our total dependence on all of the strands enfolding and infusing us, yet deriving the ultimate security of place and belonging. We are gifted with a brain which has amplified its reach by science, engineering and technology. With computers and telecommunications we have an unprecedented capacity for information collection and assessment. The challenge is to rediscover those connections to time and space that will reinsert us into the biosphere. Scientists above all understand the wonder, mystery and awe that are all around and within us. The hope is that we may consciously search for a re-enchantment of the world with that sense of its unknowables and its fecundity, generosity and welcome for this errant species we've become.

At a meeting of religious, political and scientific leaders from 83 nations in Moscow, a public statement entitled 'Preserving and Cherishing the Earth: An Appeal for Joint Commitment in Science and Religion' was released (in Knudtson and Suzuki 1992). One remarkable passage in the document stated:

> *'As scientists, many of us have had profound experiences of awe and reverence before the universe. We understand that what is regarded as sacred is more likely to be treated with care and respect. Our planetary home should be so regarded. Efforts to safeguard and cherish the environment need to be infused with a vision of the sacred.'*

Can we incorporate the descriptive knowledge of modern science into a new world-view? More and more the insights we are beginning to acquire hold out the prospect of recreating the story that includes us all. We are creatures of the earth and what we learn about the earth teaches us about ourselves and our indissoluble relationships.

'It is the story of all life that is holy and is good to tell, and of us two-leggeds sharing in it with the four-leggeds and the wings of the air and all green things; for these are children of one mother and their father is one spirit.' (Black Elk, in Neihardt 1959)

Indigenous knowledge and biodiversity (Russel Lawrence Barsh)

Traditional ecological knowledge of indigenous and tribal peoples is scientific in that it is empirical, experimental and systematic. However, it differs in two respects from Western science: first, knowledge is highly localized. Its focus is the complex web of relationships between humans animals, plants, natural forces, spirits and landforms within a particular locality or territory. Therefore, although reluctant to generalize beyond their own field of observations and experience, indigenous peoples can make better predictions about the consequence of physical changes or stresses within a particular ecosystem than scientists who base their forecasts on generalized models and field observations of relatively short duration, often restricted to the university-break season.

Second, local knowledge has important social and legal dimensions. Every ecosystem is conceptualized as a web of social relationships between a specific group of people (family, clan or tribe) and the other species with which they share a particular place. Ecological models often appear in stories of marriages or alliances among species. Hence the structure of an ecosystem is regarded as a negotiated order in which all species are bound together by kinship and solidarity. This social conception of ecology is summarized in five legal corollaries (Box 3.3).

Consistent with these general principles, indigenous peoples possess their own locally specific systems of jurisprudence with respect to the classification of knowledge, proper procedures for acquiring and sharing knowledge, and the nature of the rights and responsibilities that are attached to possessing knowledge. Some categories of knowledge may be attached to individual specialists, and other categories of knowledge to families, clans or the tribe or nation as a whole. In most societies, knowledge is also divided by gender; women are most often the bearers of botanical and medicinal knowledge.

The complexity of local laws governing the distribution of knowledge has important political implications, because no one, and no family or clan, can possess sufficient knowledge to act alone. Decision-making requires the sharing of knowledge, hence a balancing of all interests, including the concerns of non-humans. One clan may speak for its kinfolk the otters, to whom they owe special knowledge and powers, while another clan speaks for the foxes or the trees. This diversity of intellectual property arrangements is a result of culturally different responses to ecological diversity. Indigenous peoples are consequently opposed to the adoption of a

Box 3.3: Five Legal Corollaries of the Social Conception of Ecology

- Every individual human and non-human in the ecosystem bears a personal responsibility for understanding and maintaining their relationships. Knowledge of the ecosystem is moral and legal knowledge, and adepts are not only expected to teach their insights to others, but also to mediate conflicts between humans and other species.
- Since knowledge confers heavy responsibilities, as well as the power to interfere with relationships between humans and non-humans, it must be transmitted personally to an individual apprentice who has been properly prepared to accept the burdens, and to use the power with humility. Teaching is preceded by the moral development of the pupil, and tests of the pupil's courage, maturity and sincerity.
- Knowledge is ordinarily transmitted between kin because it pertains to inherited responsibilities to their own ancestral territory. Since knowledge is localized, it is not necessarily applicable to other ecosystems. Moreover, it could be dangerous for outsiders to obtain information that could be used to meddle with what is regarded as the internal affairs of the local human and non-human 'extended family'.
- Knowledge may sometimes be shared with visitors to the territory so that they can travel safely and subsist from local resources, but knowledge cannot be alienated permanently from the ecosystem to which it pertains. Knowledge can only be lent for a specified time and purpose, generally in exchange for reciprocal loans of knowledge by the borrowers. Lenders retain the right to terminate the loan if the knowledge lent is misused, or if the responsibilities attached to its use are not fulfilled.
- Misuse of knowledge can be catastrophic, not only for the individual who abuses it but also for the people, the territory and (potentially) the world. Misuse of knowledge is tantamount to an act of war on other species, breaking their covenants, and returning the land to a pre-moral and pre-legal vacuum. This is the reason why indigenous peoples tend to take a 'precautionary' approach to the use of ecosystems. Any human activity that goes beyond the bounds of known relationships among species involves a risk of triggering retaliation and chaos.

universal definition of 'traditional knowledge', and would prefer that the international community agree that traditional knowledge must be acquired and used in conformity with the customary laws of the peoples concerned.

Indigenous peoples also feel that their approach to sharing knowledge is incompatible with the 'commodification' of the arts and scientific discoveries in contemporary intellectual property law. Intellectual property law has two objectives: to encourage innovation by providing the innovator with monopoly control of the commercial applications, and to encourage the diffusion of technology by limiting the duration of the innovator's monopoly. Among indigenous peoples, higher priority is given to the proper use of knowledge locally, rather than accelerating growth in the total quantity of knowledge globally.

Indigenous peoples are not indifferent to the importance of innovation. Indigenous cosmologies portray a universe in continuous flux, driven by known forces as well as a diversity of powerful random elements ('tricksters'). Everything is bound eventually to change in ways that cannot be forecast accurately, hence the need for humans to remain vigilant and adaptive.

The use of the term 'traditional' implies the repetition, from generation to generation, of a fixed body of data, or the gradual, unsystematic accumulation of new data. On the contrary, each generation of indigenous people makes observations, compares their personal experiences with what they have been told by their teachers, conducts experiments to test the reliability of their knowledge, and exchanges their findings. All 'tradition', in actuality, is continually undergoing revision.

Individuals must be scientists in order to survive as hunters, fishers, foragers or farmers with minimal mechanical technology. Since every individual is necessarily engaged in a lifelong personal search for ecological understanding, moreover, the standard of 'truth' in indigenous knowledge systems is direct personal experience. Indigenous people are suspicious of second-hand claims (which form the bulk of Western scholars' knowledge), but at the same time they are reluctant to challenge the validity of anyone's own observations.

Thus, what is 'traditional' about traditional knowledge is not its antiquity, but the way in which it is acquired and used, which in turn is unique to each indigenous culture. Much of this knowledge is actually quite new, but it has a social meaning and legal character, entirely unlike the knowledge indigenous peoples acquire from settlers and industrialized societies. This is why indigenous leaders believe that protecting indigenous knowledge effectively requires the recognition of each people's own laws, including their own local processes of discovery and teaching.

There are at least four main reasons why the continued use and management of an ecosystem by indigenous peoples is important for conserving biological diversity.

(1) Indigenous peoples' traditional economic systems have a relatively low impact on biological diversity because they tend to utilize a great diversity of species, harvesting small numbers of each of them. This has minimal effects on intra- or inter-specific niche dynamics. By comparison, settlers and commercial harvesters target far fewer species and collect them or breed them in vast numbers, changing the structure of ecosystems.

(2) Indigenous peoples try to increase the biological diversity of the territories in which they live, as a strategy for increasing the variety of resources at their disposal and, in particular, reducing the risk associated with fluctuations in the abundance of individual species. Since each species is subject to different limitations and its abundance varies in response to somewhat different factors, an increase in the variety of niches and species tends to stabilize the average year-round supply of food and materials. Indigenous peoples' traditional territories are therefore shaped environments, with biodiversity as a primary management goal.

(3) Indigenous peoples customarily leave a large 'margin for error' in their seasonal forecasts of the abundance of plants and animals. By underestimating the harvestable surplus of each target species, they minimize the risk of compromising their food supplies. Unfortunately, this management practice has often led settlers to conclude that indigenous peoples have an abundance of land and resources to spare. Hunting, logging, and clearing by settlers quickly forces indigenous peoples themselves to harvest whatever remains in order to survive (often leading, ironically, to arguments that they are wasteful).

(4) Since indigenous knowledge of ecosystems is learned and updated through direct observations on the land, removing the people from the land breaks the generation-to-generation cycle of empirical study. Deprived of routine direct interaction with the ecosystem, indigenous peoples lose the means of transmitting old models and data, as well as the means of acquiring new knowledge. Basic concepts may survive, but lose their concrete applications. What remains can be as abstract as Western theories of ecological dynamics. Maintaining the full empirical richness and detail of traditional knowledge depends upon continued use of the land as a classroom and laboratory.

Thus, traditional life-styles are sustainable and compatible with the preservation of biodiversity as long as indigenous peoples are not forced to abandon their customary management practices in response to dispossession, intrusions by settlers, or degradation of the ecosystems upon which they depend for their subsistence. The maintenance of sustainable traditional life-styles therefore depends on respect for indigenous peoples' right to the exclusive use and management of their territories.

If empowered to maintain and develop their own knowledge systems, indigenous communities will undoubtedly share a large part of their ecological and medical sciences with other societies. Generosity and reciprocity are core values of indigenous cultures. The determination of what may properly be shared, the means by which it may be shared, and the recompense that may justly be demanded, should nevertheless be left to each indigenous people. The imposition of rules defining or limiting indigenous peoples' claims to their own scientific heritage will be counterproductive. It will result in communities' refusal to reveal what they know, or to deliberately distorting information.

Even if indigenous peoples' right to share their knowledge on their own terms is formally recognized, the extent to which they share equitably in the commercial value of their knowledge will depend on: (1) the degree to which they are fully informed about the potential value of their knowledge, and about the legal consequences of any agreements they may make with outsiders; (2) the extent to which they possess the institutional capacity, at the community level, to engage in effective negotiations, as well as the financial resources to take legal action when necessary to enforce their rights; and (3) the extent to which effective and affordable legal remedies exist at the national and international levels.

In most countries, indigenous communities do not even enjoy legal 'standing' under domestic law. They cannot institute legal proceedings in their collective capacity, nor on behalf of their constituent families and clans. In most national legal systems, moreover, traditional knowledge is not recognized as 'property' that can be defended or recovered through private legal actions. Traditional knowledge also falls outside the definitions of intellectual property used by existing international treaties on patents, copyright and trade law. Even if some governments adopt legislation to give their indigenous peoples special legal protection, other governments will not be bound by international law to respect those laws. Lack of international enforceability is a serious weakness: traditional knowledge is most likely to be exploited commercially by powerful transnational corporations in fields such as biomedicine and biotechnology.

There is a need for the international community to agree that indigenous peoples are the true owners of their ecological knowledge, and that disputes over rights to the acquisition and commercial application of indigenous sciences must be resolved in accordance with indigenous peoples' own laws. Governments, the scientific community and universities should also contribute to ensuring that indigenous peoples themselves possess information, training and institutional structures of their own sufficient to evaluate external research, negotiate collaborative agreements with outside researchers and, if necessary, take private legal action to prevent the licensing or sale of knowledge which was not properly acquired from them. Taking these steps would help reverse the recent erosion of indigenous knowledge systems, and make it possible for indigenous peoples to begin making their own unique scientific contributions to stabilizing and sustainably utilizing the world's forests.

Culture, landscape and diversity (Andrew J. Chapeskie)

In the summer of 1986, I was asked by a young man from the Wabigoon Lake Ojibway Nation to work with him on a 'wild rice' (*manomin*) project. His name was Joe Pitchenese. I agreed, and later that summer, Joe Pitchenese took me on a boat trip to inspect wild rice stands near his community. '...When I was taken to a place called Tobacco Creek on Dinorwic Lake, certain features of the landscape prompted me to ask... how it was that *manomin* in Tobacco Creek was growing in water amongst the trunks of dead trees that were still standing?' 'It was planted there', came the reply, ' All of the 'wild' rice 'natural' resource that I saw on that trip in July of 1986 was in fact the core of an anthropogenic landscape' (Chapeskie 1995: 10-11).

The diversity co-existing with the *manomin* in the river upon which I travelled in 1986 had the look and feel of 'wilderness' – but it was not. When I travelled the *manomin* fields of the Wabigoon River with Joe Pitchenese in 1986 we passed: a moose in the river grazing on plants in the ripening *manomin* fields, scores of moulting ducks of several species racing away from our boat to hide in the 'crop', several bald eagles which feed on the fish growing in the *manomin* fields, muskrat and beaver lodges, red-wing blackbirds, the old fallen-in trapping cabin of Joe's great grandmother (this part of the area is now his trapline), *manomin* harvest landings and campsites, and an 'archaeological' site.

In the time that followed this first visit to the *manomin* fields on the Wabigoon River, I came to learn about and even participate in documenting the widespread extent of these anthropogenic landscapes in part through my work with Ojibway 'harvesters' and in part through community-based participatory 'land use' research projects. Part of this effort involved searching for recollections of customary relationships to land told by Ojibway people themselves. This led to me finding the following recollection of an Ojibway Elder which literally echo the words of Peter Pitseolak about 'farmers' having 'fences':

> *'White man makes a farm to grow hay to feed his animals. He also grows vegetables for food. Indians also feed their animals, only in a different way. Around the middle of April, the Indian trapper looks around to find a bare spot, mostly up on the rocks where the snow goes first, where there is still a lot of snow at the bottom of the hill. They set a match to this bare spot and only burn where it is dry and bare, so there's no danger of a big forest fire because the fire stops when it reaches snow.*
>
> *'Two years later you would find a big patch of blueberries in amongst the bushes. And you would see all the hungry animals of all kinds feeding on those blueberries – fox, wolves, black bear, partridge, squirrels, chipmunks, and all kinds of other birds. No doubt they were happy to find those berries. It was the trapper that got it for them by setting the fire.*
>
> *'This is what I mean when I say Indians feed their animals too. The berries were for our own benefit too. As we would preserve them for our winter use. After a few years, young trees would grow on that burnt place. Then the rabbits would get to feed from those young bushes. In later years, the little trees would get bigger. Then the moose and deer get to feed from it. So, you see, the setting of these small fires can go a long way in feeding many animals.' (Theriault 1992:74-75)*

It is worthwhile to consider the implications of this recollection in light of the Ojibway relationships to the *manomin* fields on the Wabigoon River where they constituted a complex customary Ojibway aquaculture. It involved everything from the initial planting, to *manomin* field maintenance (weeding, cultivating etc.) and elaborate harvest customs, all of which were grounded in co-operative practice (Chapeskie 1990). No one owns these fields. The other striking feature of this *manomin* aquaculture was how the biological diversity of the location where it was practised had both been created and sustained. Then, as now, the Ojibway people from this area feel threatened on the issue of their access to this 'resource'. In fact, by 1986, these people had already lost access to a majority of the *manomin* fields in their ancestral territories.

In accordance with its laws, the province of Ontario both 'owns' and 'manages' (i.e. controls) the 'wild rice resource' I have described above. *Manomin* is governed by the Wild Rice Harvesting Act. Provincial policy with regard to the 'resource' states that: 'Wild Rice occurs *naturally* [emphasis added] in shallow water bodies.' (Ontario Ministry of Natural Resources 1987:2). Why has state law been so blind to the evidence that has always existed to the contrary concerning the true nature of this 'anthropogenic resource'?

This blindness resulted from the very ways that the work of these people has been reflected even in their anthropogenic landscapes. Compared to European agricultural settings these *manomin* fields are positively 'wild' in appearance. I myself have found no word for 'wild' in the Ojibway language that they might have used, for example, to differentiate their Indian corn – *mandaamin* – from a pine tree in the forest. Further in this regard, I believe that it is doubtful that any historical or contemporary geographic boundaries could be found delineating Ojibway anthropogenic landscapes from wilderness. Nevertheless, anthropogenic Ojibway *manomin* landscapes are dramatically different from the rice paddies of the Californian farmer. This is of no small significance when the issue of biodiversity conservation and how indigenous people might contribute to it is considered.

An enclosed 'wild rice' paddy in California is owned to the extent that its owner can exclude virtually all other proprietary interests from it. It is very productive in terms of growing its hybridized crop, but it is monoculture. For the Ojibways from Wabigoon, the notion of 'owning' *manomin* fields is more than just alien; it is offensive in the deepest cultural and spiritual sense. Viewed from the perspective of co-operation and human ecology this is particularly significant. In this sense, even the idea of Ojibway people 'planting' *manomin* is limited. Such 'planting' does lead to an increased abundance the 'use' of which has always been carefully 'regulated' (Chapeskie 1990). But such practices cannot be considered as having the same connotation as they likely do for a paddy 'wild rice' farmer in California. This is because the Ojibway people of Wabigoon do not see themselves 'controlling' *manomin* as a 'resource'; rather, they conceive of themselves in relation to 'it' in a kindred way.

For the Wabigoon Ojibway, the propagation of *manomin* has always customarily been a deeply spiritual, even mystical, experience. Even the harvest of the food is preceded by ritual which celebrates the sacral nature of *manomin* as a living gift and fellow traveller with the Ojibway on their ancestral lands. It is considered spiritually alive as they consider themselves spiritually alive.

What Wabigoon Ojibway think is of value to keep in relation to the plant is not a 'factual historical' memory of who planted *manomin* there. Rather, they value a history of *Manomin* which celebrates the plant as a spiritual gift to them and where *manomin* lives on in relationship to them. A story related to me by a Wabigoon Elder highlights the significance of *manomin* as a spiritual gift (Box 3.4).

Ojibway people must now exist in a 'land use' context where the Government of Ontario has imposed its authority over most of the 'resources' of their ancestral lands. How is the value of their customary relationships to *manomin* in terms of biodiversity conservation to be considered and then addressed?

Ojibway customary relatedness to *manomin* is rooted in values of equity and the practice of co-operation. It is through these values and co-operative practice that they have lived their intimacy with the plant. This has a profound ecological significance when considered from the perspective of biodiversity conservation. This is reflected in one of the most frequently heard Ojibway expressions of how people should work on the land. This expression is: 'take only what you need'. How is it made to work? In the context of customary Ojibway relationships to *manomin* such an expression cannot be separated from the Ojibway practice of reciprocity and the economic security it provided.

Historical Ojibway customs of livelihood equity and reciprocity have supported an affluence without materialism. Such an affluence would not have been possible without a knowledge of landscape that co-operative reciprocal livelihood relationships made possible amongst these people. This awareness of landscape runs deep enough for it to be expressed in terms of kinship. Customary livelihood relations amongst these people in their work on the land have been inextricably bound to an awareness of the intrinsic value of diversity. This is a feature of their culture which should command our respect.

The value of the relationship between customs and biodiversity conservation can be illustrated by reference to the Wabigoon River *manomin* fields. The area is rich in both the diversity and abundance of

Box 3.4: *Manomin* as a Spiritual Gift

Andrew Chapeskie

A great Ojibway leader from Wabigoon named Mis-Koona-Queb was paddling on Rice Lake for some particular purpose. Located on the north end of the lake are sacred rock paintings. In the south-west corner of the lake is a large cliff. From a space in that cliff close to the water came a voice calling Mis-Koona-Queb to come over with his canoe. The figure had long hair, and was dressed in a moose-hide breech cloth. When Mis-Koona-Queb paddled his canoe over, the figure told him to turn the canoe around.

The figure was Nanabozho. There was only one paddle in the canoe. Nanabozho told Mis-Koona-Queb, sitting in the front of the canoe, to paddle and to be careful not to look back. Mis-Koona-Queb was given precise directions as to how he was to work the canoe around the lake. Going through the plants as he was directed to, Mis-Koona-Queb heard sticks working behind him. He didn't dare look back. At the end of the travelling Mis-Koona-Queb was told to turn around and look into the canoe. The canoe was made of birch bark. Mis-koona-Queb asked what was in the canoe. Nanabozho said, 'This is Anishinaabe *manomin*. This is medicine and it will support you and give you nourishment.'

Nanabozho then showed Mis-Koona-Queb how to process the *manomin* that was in the canoe and told him to use a birch bark dish for winnowing the *manomin* in the wind. Mis-Koona-Queb then tasted the *manomin* after this process and it tasted very good. Nanabozho then cooked it in an Anishinaabe cooking pot with water in it over a fire. Mis-Koona-Queb was then told for each and every year from then on when to begin to harvest the Anishinaabe *manomin* and was told that he and his people must first pick some *manomin* and have a feast. The rules of the feast were then given.

During this time, Mis-Koona-Queb was also told how the harvest must be organized and how the Anishinaabe *manomin* must be respected. Nanabozho showed him all of this.

During this time, Nanabozho also taught Mis-Koona-Queb many other things, including how to harvest, preserve and care for blueberries. Finally, after teaching many things Nanabozho made a swing. He swung on this large swing four times in the four directions and after swinging himself in the fourth direction he disappeared from our Anishinaabe people.

(Chapeskie 1990: 130)

its 'natural resources'. As an anthropogenic landscape, the interaction of Ojibway people with it has obviously been intense. Yet, neither the area nor any of its resources is 'owned' by any Ojibway person. Further, everyone in the community has access to this landscape for livelihood purposes. When 'resources' such as *manomin*, whose productivity is cyclical, are abundant here to an extent they are not elsewhere, broad participation in the harvest occurs. Regulation of the harvest does not focus on excluding members from the harvest to limit exploitation of the resource but rather on ensuring that the total amount harvested by the group is sustainable. For other resources of the area which are found more evenly distributed throughout other regions of the ancestral lands of Wabigoon people, some members of the group will access the landscape at different times. All of these livelihood interests in the location must be kept in a consensual balance. Thus it is, for example, that *manomin* must be left to draw ducks for the fall hunt as well as to re-seed itself, given that it is an annual plant, and the *manomin* fields themselves must retain their capacity to support fur-bearers, fish brood stocks and other 'resources' (e.g. 'ethno-botanicals').

It is within this context of equitable access to the total landscape that the diversity it offers is able to yield its greatest value to the Ojibway people living there. Because the value of this diversity is equitably and co-operatively distributed, it is not insignificant that the group as a whole has an interest in conserving it. The practice of equity and co-operation on this landscape opens up its own particular possibilities for the accumulation and application of ecological knowledge. Such Indigenous Knowledge flows synergistically among the 'users' of different 'resources'. Such knowledge of these 'resources' reinforces an intense respect for the 'resources' themselves. The flow of this knowledge has customarily been so important to the affluence of these people that its accumulation and transmission is often accompanied with 'ritual' behaviours associated with the land. Such ritual activities have customarily included 'give-away' ceremonies (historically this meant everything from what Joe Pitchenese has told me), feasting, offering the choice parts of harvested game to Elders and others. At other times, ritual is able to provide a focal point for experience on the land that opens up possibilities for tremendous discovery.

In reflecting on such relationships between Ojibway culture and biodiversity conservation, I do not want to be taken as reducing the existence of 'land use' customs solely to ecological considerations. If the diversity of our world consists of our 'relatives' and not our 'resources', the practice of respect towards them must command our highest attention. This is the capacity for living in companionship with them and the diversity that they constitute.

In customary Ojibway culture, this diversity has been protected and has made possible an affluence that was not dependent upon materialist accumulation but derived through the capacity both to know and to celebrate diversity. Fewer days of work were required to provide the material security for them to spend many more days in social activities and ritual life. Those who might think that such an affluence was possible only in the sense that people simply had to learn to expect little in material terms should be the last to seek knowledge from indigenous Elders about such things as indigenous pharmaceuticals. The fact is that something other than limitations on the capacity for material accumulation has been at work in customary Ojibway society. In fact, my work with Ojibway people, especially the Elders among them who have lived on the land, has led me to conclude that what missionaries and government officials condemned and sometimes still bemoan as wasted time totally missed the point. It was actually time well spent fostering the knowledge and social pre-conditions for an affluence that was not dependent on an environmentally destructive materialism.

For those concerned with biodiversity conservation, customary Ojibway relationships to *manomin* create a paradox. How is control over a 'resource' such as *manomin* to be exercised to promote its best economic 'use' and achieve biodiversity conservation? The provincial regulatory scheme for the 'wild rice resource' outside of the remaining 'community' *manomin* harvest areas of 'Indian Bands' in Northwestern Ontario is focused on the allocation of exclusive proprietary rights over the resource to individuals. This ownership over the resource is seen as providing an incentive to 'licence holders' to increase production for individual economic gain. (Such licence holders have little if any interest in the other resources of the lakes they hold licences to. Animals like ducks or red-wing blackbirds which eat *manomin* become 'pests' which should be 'eradicated' from the rice fields.) The wider economy of Northwestern Ontario is seen as benefiting from such a system. This is the management ideology that currently governs the 'rational exploitation' of the resource.

In such a management policy context, it is understandable that non-aboriginal people, including government officials, have often exhibited a 'general consternation' about Indian 'underproduction of this bounteous food' (Vennum 1986). But Ojibway customary relationships to their landscapes which bear *manomin* provoke us to ask whether the end game of such a state management approach leads to the diminution of the very biodiversity upon which we ultimately all depend. Beyond our immediate efforts to preserve landscapes and protect indigenous rights, the customary relationships of the Wabigoon people to their rice fields challenge us to re-consider the ecological implications of some of our most fundamental economic and social values.

Resource stewardship by Middle Columbia tribes of the American Pacific Northwest (Robert H. Winthrop)

With its tributaries the Columbia River of the American Pacific Northwest constitutes one of the major river systems of North America. Originating in the Canadian province of British Columbia, the Columbia River runs for over 1,900 km, south through the state of Washington and then, joining the Snake River, west to the Pacific Ocean. The concern here is with the ecological and political adaptations of Indian peoples of the middle reach of the Columbia, those groups centred on the Columbia and Snake Rivers from the Cascade Mountains east along the lands of south-eastern Washington and north-eastern Oregon. Today, these peoples compose the Yakama Indian Nation in Washington, the Confederated Tribes of the Warm Springs Reservation, the Confederated Tribes of the Umatilla Indian Reservation in Oregon, and the Nez Perce Tribe in Idaho.

The Columbia, together with its numerous north- or south-trending tributaries, provided a geographic and cultural framework for American Indian lifeways in the region. The Southern Columbia Plateau is an area of major ecological contrasts: river corridors, arid semi-desert lowland zones marked by rocky canyons and buttes; forests and prairies at somewhat higher elevations, and meadows along mountain flanks (Übelacker 1984). Each of these zones contained critical resources used by Indian peoples of the pre-reservation era within a yearly subsistence cycle. Indian winter settlements, however, occurred along rivers. These riverside settlements thus served in a general way to associate communities with territory, although for the most part land and resources in the Columbia Plateau were not under the exclusive control of individual groups (Hunn 1982: 33-34).

Fish have been crucial to the adaptation of native peoples of the region. The anadromous fish, whose life-cycle brings them from the upper reaches of rivers to the Pacific Ocean and back in vast multi-year migrations, have been particularly important. These migrations occur in distinct runs along the Columbia, mainly between May and early October. Significant anadromous fish included chinook salmon (*Oncorhynchus tschawytscha*), coho salmon (*O. kisutch*), sockeye salmon (*O. nerka*), and steelhead trout (*O. mykiss*). Other, non-anadromous species include suckers (*Catostomus* spp.) and lamprey (*Entosphenus tridentatus*). (Linnaean plant and animal terms are taken from Hunn and Selam 1990.)

Edible plants provided an important portion of the diet. Roots were gathered in the spring, generally between February and June, depending on the species and the specific environment. *Lomatium* species such as biscuit root (*Lomatium cous*) and desert parsley (*L. canbyi*) were particularly important in the root diet. Camas (*Camassia quamash*) occurred in abundance in upland prairies, attracting large numbers of Indian people through the summer months. The bulb of the camas was prepared by baking, and served as a staple food. Between June and October, fruits were harvested. In mid summer these were the lowland species such as chokecherries (*Prunus demissa*) and serviceberries (*Amelanchier alnifolia*). Later, usually beginning in August, families travelled to mountain camps to pick black mountain huckleberries (*Vaccinium membranaceum*), a sacred food celebrated in the early-August huckleberry feast. The mountain environment was important for other resources as well: pine nuts from the whitebark pine (*Pinus albicaulis*), cedar roots (*Thuja plicata*) and beargrass (*Xerophyllum tenax*) for making baskets, delicacies such as black lichen (*Bryoria fremontii*), and various medicinal plants.

Game, particularly ungulates such as mule deer and black-tailed deer (*Odocoileus* spp.), also served as an important food source. In the fall, the most significant season of the year for hunting, mountain meadows provided an ideal locale for such game, coinciding with travel to the mountains for the huckleberry harvest.

In the pre-reservation era, the seasonal requirements of fishing, gathering and hunting dictated successive movements to fishing stations and root gathering areas in early spring, to camas fields in middle elevation prairies in the late spring and early summer, to subalpine huckleberry fields in late summer, back to prairies and fishing stations in the fall, and finally returning to permanent villages with the onset of winter.

An anthropological approach to environmental sustainability recognizes that patterns for both the preservation and appropriation of nature are fundamental aspects of any traditional social system. The tribes of the Columbia Plateau possessed sophisticated systems for conserving and allocating the salmon, roots and berries which formed staples of the traditional diet, through the scheduling of harvest via 'first fruits' ceremonies, the distribution of gathering areas and fishing sites to kin groups, and the ritual significance attributed to these foods through their prominence in ceremonial life and gift exchange. Moreover, from the Indian perspective these staples formed only the most prominent elements within an integrated web of resources, physical and spiritual, yielding an ethic of 'holistic conservation' (Stoffle and Evans 1990: 91-92). Not surprisingly, salmon, berries, roots and water remain of key significance for the Columbia Plateau tribes today.

The key symbols of this cultural landscape are both general and regional on the one hand; particular and local on the other. Among native communities of this relatively arid region, pure water can be a spiritual food, a 'medicine'. As such it can have distinctive attributes associated with particular places. Water may be collected for the sick or elderly from particular streams at particular elevations within a mountain landscape, because in such places it has powerful healing properties. At the same time, water has a cultural value generically, and as such is a symbol linking all of the American Indian communities of the region. Traditional meals in mid-Columbia Indian communities still begin with a sip of water, and an exclamation of thanks: _uuš ('water!') (Meninick and Winthrop 1995).

A similar cultural polarity occurs with salmon, a staple food of the region. Salmon is understood generically (*núsux*, in the Sahaptin classification). The first runs of chinook salmon are honoured by spring first-fruits ceremonies (Hunn 1980: 13). At the same time, particular localities may be identified with particular runs of salmon, often interpreted through myths that explain the origins of the fishery. Thus the White Salmon River in southern Washington is named for the late-stage salmon (Sah: *mit'úla*), which has a characteristic white flesh colour. The area was named accordingly *mit'úla-aaš*, place of the white salmon. This association is given cultural emphasis through a creation myth, in which Coyote – creator and culture bearer – made the White Salmon River a fishing place for the Klickitat people (Boxberger and Robbins 1994: 1: 13; Lane and Lane 1981: 46).

The picture of cultural and environmental change from the mid-nineteenth to mid-twentieth centuries suggests a very bleak prognosis for American Indian cultures of the Columbia Plateau and elsewhere. The displacement of Indian camps and villages by Euro-American settlement, the legal confiscation of Indian territories through the reservation system, and the massive environmental change resulting from industry, agriculture, fishing and forestry altered irrevocably the pre-settlement cultural worlds of the Columbia Plateau. In addition, the Columbia system has been heavily utilized to generate electricity: nineteen major dams and five dozen smaller projects have created the world's largest hydroelectric generating system (Lee 1993: 22). Sites appropriate for Indian fisheries tend also to be suitable for dam construction. As a result, 'not only are the migration patterns of the anadromous fish greatly affected by the dams, representing a gauntlet, but the traditional fishery locations themselves were taken away' (Hanes 1995: IV-4).

None the less, from the perspective of the late 1990s the situation appears considerably more positive for the survival of American Indian communities and their traditional resource practices as a result of greater tribal political autonomy, the emergence of political activism, and legal developments allowing more effective tribal intervention in the assessment of proposed environmental change.

Since the 1960s, the emergence of environmental protection as an objective of public policy in the United States has prompted the creation of systematic procedures for evaluating proposed environmental change. This provided American Indian tribes with a new and significant forum in which to fight for the preservation of tribal resources and culturally significant landscapes, and indirectly to preserve the traditional practices (such as dipnet fishing, root gathering or prayer and power questing in remote areas) which such environmental conditions make possible. The National Environmental Policy Act (enacted 1969), which established a system of comprehensive federal environmental review in the United States, is only one of a number of Acts and regulations that provide a role for American Indian tribes in environmental decision-making. The National Historic Preservation Act (enacted 1966, amended 1992) also has proved an important basis for asserting Indian rights. This law now explicitly recognizes an area's 'traditional religious and cultural importance' for an American Indian tribe as a basis for determining it eligible for the National Register of Historic Places, a status conferring certain protections in United States law.

The net result has been to provide a stronger institutional and cultural framework for expressing tribal perspectives on environmental stewardship and advocacy, and the preservation of American Indian cultures. In recent years, tribes have made a concerted effort to reverse the anti-Indian developments of the late nineteenth and early twentieth centuries, seeking to restore the tribal land base; preserve and transmit native languages, cultural knowledge and traditional practices; and regain access to indigenous plant, animal and fisheries resources.

Management of the Colombia River serves to illustrate this process. Over a period of 150 years, urban, industrial, agricultural and recreational uses have transformed the Columbia River system. In the 1990s the environmental consequences of that massive development have become obvious to Indians and non-Indians alike. 'By the late 1970s, salmon runs of 10 to 16 million in the pre-industrial era had dwindled to 2.5 million' (Lee 1993: 23). By the 1990s salmon runs had declined still further. A number of runs of salmon and steelhead in the Pacific North-west have been given 'threatened' status under the federal Endangered Species Act, an ominous sign. Both tangibly and symbolically the salmon stand at the heart of middle Columbia Indian cultures. The disastrous decline of salmon in the Columbia system thus represents the greatest possible challenge to tribal goals of environmental stewardship and the preservation of Indian cultures.

Since the 1960s, the middle Columbia tribes have fought on a number of fronts to reverse the destruction of the Columbia fisheries. Perceiving the tribal allocations of salmon by state fisheries authorities to be a denial of the resource rights retained by treaty, tribal governments throughout the North-west engaged in litigation to obtain greater allocations. At the same time a number of Indian fishermen still residing on the Columbia staged dramatic protests ('fish-ins'), courting arrest for what state authorities considered illegal fishing (Beckham *et al.* 1988: 136-37). These actions in the 1960s and 1970s led to several critical federal court decisions affirming Indian treaty rights regarding fisheries, notably United States v. Oregon and United States v. Washington (Cohen 1986).

Intervention in the legal and scientific complexities of treaty-based fishing rights forced middle

Columbia Indian tribes to establish their own scientific expertise in fisheries biology and related environmental disciplines. As a recent account of this struggle noted, 'When the dams went up, and later as the state [of Oregon] sent a parade of biologists to the stand in U.S. vs. Oregon, the Indians realized they needed their own experts who relied not on traditional Indian wisdom but on facts and figures that would hold up in court' (Baum 1996). One outcome was the creation in 1977 of the Columbia River Inter-Tribal Fish Commission, a consortium for tribal fisheries research and advocacy comprising the Yakama, Warm Springs, Umatilla and Nez Perce tribes. As dams in the Columbia system become subject to re-licensing reviews by the Federal Energy Regulatory Commission (FERC), tribes intervene vigorously to promote technical changes (such as improved fish passage facilities) that will enhance the survival of anadromous stocks, or in some cases simply argue for the removal of the dam in the interest of restoring fish stocks and traditional fishing sites. None the less, the development of tribally-based scientific expertise can only supplement tribal advocacy based on traditional knowledge.

This example shows that the search for local-level innovations that can contribute to the preservation of biodiversity and the sustainable management of resources should not proceed *de novo*. Instead, existing institutional arrangements and culturally salient principles should be sought that can be modified to meet the environmental needs of the twenty-first century. In particular, we need to identify classes of adaptive systems, sets of cultural principles and practices that inherently – by the nature of their functioning – contribute to environmentally appropriate outcomes. This paper has presented an example of one such system, in which the elements of local knowledge, site-specific resource dependence, and political principles balancing autonomy and affiliation contribute to the effective organization of environmental stewardship and advocacy.

Gender-based knowledge systems in the eco-politics of the Yukuna and Tanimuka of Northwest Amazon, Colombia (Elizabeth Reichel D.)

In the Northwest Amazon, gender-based knowledge systems (GBKS) entitle men and women to distinct knowledge legacies which allot a specific eidos and ethos to each person according to gender. The GBKS are named specifically as 'women's knowledge' and 'men's knowledge' and they encompass gender-specific spatio-temporal referents (Hugh-Jones 1979; Reichel 1989, 1993) and supervisory functions for the conservation of particular biodiversity domains and cultural dynamics. As Posey and Dutfield (1996) propose, it is necessary to methodologically understand how the indigenous peoples themselves conceptualize, represent, use and own their cognitive, material and spiritual capital so considerations of 'traditional resource rights' can voice the indigenous concepts of biodiversity, knowledge, rights or natural resources.

In the Colombian Amazon, among the Yukuna and Tanimuka Indians, the GBKS (Reichel 1997) allow women and men an effective management of the rainforest and society and their knowledge bases are encoded in a cosmological trans-explanatory system that empowers each gender as an active and conscious agent in biodiversity conservation. GBKS are knowledge legacies that manage the short-, mid- and long-term existence of society and nature within the context of earth, life and the universe in a cosmological framework where diverse belief systems correlate scientific, rational (Atran 1990; Lévi-Strauss 1962), emotional, artistic, aesthetic, ethical and spiritual principles (Reichel-Dolmatoff 1996) to drive the respect for life and communal solidarity.

Fraternal and sororal *maloca* chiefdoms of the Yukuna and Tanimuka

The Yukuna (Storytellers) and Tanimuka (Ash People) live between longitude 70 and 71 degrees West and latitude 0 to 1 degree south of the Equator. Their environmental politics have allowed them for centuries to live in *maloca* communal roundhouses, have a swidden-foraging subsistence complex, and live in chiefdoms as they abide by their shamanic tradition (von Hildebrand and Reichel 1987). Resource use is made by shamanic negotiations with sentient and supernatural Owners of the rainforest and the universe and with other ethnic groups. Groups of brothers live together in a *maloca* with in-marrying wives, and the agnatic groups occupy traditional terrains which were managed for millennia by their ancestors so men upgrade their local knowledge while women have to manage data from different terrains.

For the Yukuna and Tanimuka, the natural resources in waters, lands, soils and subsoils are considered the property of ethnic groups, lineages and genders, and of individuals, as well as of supernatural and cosmic spiritual Owners. Access to these resources implies knowing their long-term cultural and natural history since this knowledge of ecosystem dynamics and previous interactions between humans and supernatural and spiritual Owners is considered tantamount to legitimate rights over resources. GBKS amass local traditional ecological and social knowledge and record each gender's role in environmental conservation since the well-being of nature is said to be the responsibility of men and women who manage distinct biogeographic regions. The conservation of biodiversity and of society are conceived as the result of a series of negotiations among humans, natural and supernatural beings, who pact life histories and also death histories in a negotiatory universe (Reichel 1987a, 1987b, 1989, 1993, 1997). As men and women, in practical and in enchanted ways, reiterate their ethnomemes (Reichel 1997) or patterns of information, they plan and execute specific forms of bio-social resource management.

Women's 'Thought of Food' and men's 'Travelling in Thought': engendered biodiversity conservation in the swidden–foraging subsistence complex

The Yukuna and Tanimuka collective memory is closely related to a concept of long-term gendered historical agency (Hugh-Jones C. 1979; Strathern 1995) in which group identity is said to result from the complementary recreation of men's and women's knowledge bases. The women's capacity to domesticate plants in *chagra* swidden plots and in the *maloca* house gardens; their ability to tend homes; their capacity to reproduce human life within their bodies, and their ability to connect all life-forms into the life-systems of the soil and subsoil, are said to be achieved through a specialized female knowledge called the 'Thought of Food'. The men's capacity to forage in the rainforest and manage the non-domesticated resources used to hunt, gather and fish; their training to hold authority in patrilineal and patrilocal chiefly positions; their disposition for shamanic activities, and their ability to manage altered states of consciousness, are achieved through a specialized male knowledge called 'Travelling in Thought'.

The imagery of gendered minds and bodies is over-determined to the degree that girls are held to inherit the bones and flesh-blood as well as the knowledge base from their mothers, and boys are held to inherit the bones and knowledge base from their father (his patrilineage) while his flesh and blood are inherited from his mother. During conception, a competition between the father's and mother's knowledge systems and bones and flesh, is said to occur to determine the child's gender (Reichel 1989, 1993). The stronger knowledge base will win if it has been fortified by proper thoughts and ritual practices, and in women's case the knowledge to give birth or cultivate plants well, and in men's case, attendance to rituals such as the Yuruparí male-bonding rites.

The construction of the gender-based identity is achieved in the idiom of male and female complementariness and opposition and each gender is asked to 'defend' its knowledge base against contamination by the other gender. This fear is reiterated as each person is requested to deter the 'theft' of their knowledge base either by other humans (enemy shamans who do Claw and Fang shamanism to sear bodies and disperse knowledge systems) or by supernatural Owners of nature who prey upon people to steal, dispel and disorganize their knowledge systems while seeking compensation for human misuse of nature. Loss or theft of knowledge is said to make people ill or to kill them, and only a Jaguar-Seer shaman can localize, retrieve and reorganize the knowledge base of women and men, and place it back in their bodies, while warning of the consequences of not duly conserving biodiversity and life because the indigenous theory of knowledge is framed in a theory of exchange complexity that includes exchanges beyond humans and nature across the biosphere, this earth and throughout the universe.

A woman's knowledge (Reichel 1987, 1993) is said to accumulate as she acquires plant knowledge and expertise, as well as female secrets from her mother and then from her mother-in-law, and this knowledge system is fortified with each child she has, as well as by her success as a swidden gardener. On the other hand, a man's knowledge is said to accrue by apprenticeship with the father and by acquiring or 'buying' knowledge from experts such as shamans as well as by attending male-bonding rituals. A mother will give her daughter the seeds, tubers, roots and stem cuttings of plants she owns, while imparting the corresponding plant knowledge. Fathers teach sons the natural history of the patrilineally-owned ecosystems and species they use and conserve to hunt, gather and fish, in order to give them rights to manage these. A mother will explain to a daughter how to tend plants and also explain the symbolism of the plants in relation to the way nature is signified and managed. While processing food or making pottery (Reichel 1976), the women also teach girls lessons of resource management, and the significance of the meaning of female-made artefacts in their cosmology (Lévi-Strauss 1984). They relate these to the Ñamatu female ancestress and the female leaders who exist across cosmo-, bio- and ethno-genesis (Reichel 1987, 1993, 1997).

Specialized knowledge is taught to children according to the role they will have as adults: seniors are to be headmen or headwomen, and juniors may be women potters, hammock makers, specialist gardeners, healers, mediators with White people. In turn, the men may be Jaguar-Seer shamans or Curer shamans, tobacco-rollers, chanters, pineapple-brew makers, healers and curers, ritual experts to hold Yurupari rites, or mediators with White society. The gender-based knowledge system is one of the ways corporate secrecy is maintained to uphold specialized resource control and to exert resistance against having other cultures appropriate or displace ethnic identity.

In many ceremonies the children are taught to repeat 'at least ten times' the knowledge that is given to them and the apprenticeship is accompanied by performances where they are said to be given a knowledge that comes not only from the tutor's own knowledge base but from ancestors, parts of nature, and from knowledge that is contained in artefacts of knowledge. These artefacts can be, for example, men's wooden 'thinking stools', hardwood staffs, jaguar-tooth necklaces, feather crowns, or women's pottery or hammocks. That knowledge can accumulate in artefacts of knowledge by its appropriate use and it can also be 'taken out', given or stolen from the artefacts as intersubjective sociality (Strathern 1995) entails the exchange of artefacts and the knowledges they contain.

Women's bodies are said to be protected by the Four (or Three and a sibling) Sisters (also called Aunts, Grandmothers or Mothers) who reside in the nadir of the cosmos, and in the ground, rhizosphere, and wherever women are and biodegradation occurs. Women may become ill because shamans are said to be capable of 'opening or sealing' their bodies or of taking their knowledge systems away and throwing them into the confines of the universe. Men are said to be protected by the Four Brothers (Uncles, Grandfathers or Fathers) who live in the apex of the cosmos, and in the skies and wherever men and spirit-knowledge occur.

Obtaining or using biotic and abiotic resources from the rainforest such as plants, animals or clays or soils, always requires shamanic consultations not only to do resource accountability but also to request 'permission' from nature's Owners and its 'people who work and think' (Reichel 1987a, 1987b). The supernatural Male and Female Owners of the species, ecosystems or seasons are consulted and 'paid' for the 'skins-shirts' of utilized animals and plants. It is the body or the matter which is 'bought' for human consumption while the knowledge base and spirit is requested to stay behind in nature with its people and its Owner in order to not deplete resources and to 'not confuse systems of knowledge'.

When hunting, for example, tapirs, deer or peccary, or when fishing near sacred rapids, or when gathering fruits from sacred groves of ita palms (*Mauritia flexuosa*) or coconut palms (*Cocos nucifera*), the men will 'ask permission' of the supernatural Owner of that species or niche and they will mentally consult the significance of the specific impact on the environment. Cutting the palm is said to be akin to cutting a leg or arm of the Owner or toppling the post of his house (Reichel 1989). Shamans 'pay' the owners with Thoughts and with coca and tobacco, but also with the lives of children or of sick and dying people, while women may offer children to the Female Owners, upon childbirth, since children are said to 'love the earth before they love people'. 'Returning' the knowledge-spirit of plants and animals to their particular Owners and to the Male and Female Four Primal Ancestors is said to 'place every knowledge system in its appropriate site' and hence guarantee the identity of the people in nature and in human societies.

Women plant hundreds of cultivars in their swidden gardens, and each garden is different, revealing the woman's plant knowledge and her expertise in managing her garden ecosystem. The 'buying' of seeds and access to plants requires a negotiation among women either through kin or affinal ties or, in specific cases, is done by payment of other knowledge and plants or with merchandise, although a woman can opt not to disclose certain plant knowledge.

Men hunt, gather and fish in enormous rainforest terrains. The knowledge of Travelling in Thought allows them to be successful foragers and to aptly manage the rainforests over two million hectares. Dozens of species are accessed in distinct ecosystems, and each habitat is known by men by holding the knowledge of that sector of nature. Knowing certain ecosystem's characteristics allows for the sustained use of resources, carefully managing fluctuations in the environmental supply and demand. Many of the animals and plants that men use when foraging are considered to be 'managed' by their human ancestors throughout history by successful pacts with their supernatural rainforest Owners.

Men plant mind-altering plants such as coca and tobacco, which are heavily imbued with religious symbolism, and they also tend the pineapple plants used to make fermented drinks and the gourd plants used to store coca. Men selectively clone and own varieties of coca plants since many rituals involve ritual chewing of coca among men to 'exchange thoughts' and to have 'energy' and stay awake and concentrate while Travelling in Thought. Tobacco and other powerful plants have an extensive plant history. Inheritance of coca stems and of tobacco, as well as of yeasts which are used to make fermented drinks, is transmitted from father to son and men tend to not exchange these with other groups. The knowledge about wild and cultivated species, as well as ecological, meteorological and astronomical correlations, are said to be located as stocks of information in a tree of knowledge and also in the different layers of the cosmo-design.

The Yukuna and Tanimuka cosmo-design locates a flat earth in the middle of multiple male skies and female underworlds which are populated by sentient forces (Hildebrand 1987; Reichel 1987b), surrounded by a cosmic river and populated by natural and supernatural Owners. The supernatural Owners are said to be preying on and negotiating with humans to ensure that humans have environmental and social awareness. The cosmology projects the ethos of a negotiatory universe (Reichel 1997) where wisdom, violence, ruse and conflict as well as mediation occur between and among forces of nature and society to renegotiate access to life itself. The shamans are the experts who communicate with the Owners to tell people how to manage key resources sustainably.

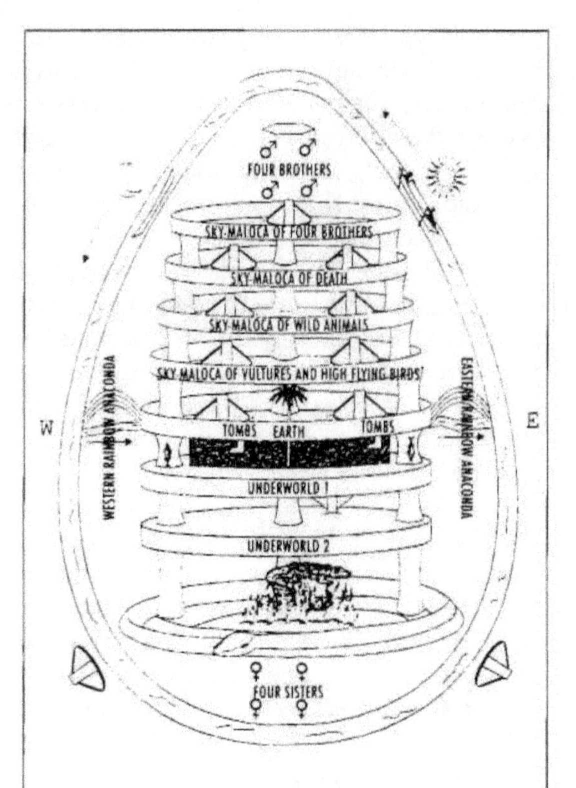

Figure 3.1: The Tanimuka and Yukuna cosmos, by E. Reichel. The sun's canoe is paddled by a howler monkey and a deer; the moon's canoe by mosquitoes. Levels of the cosmos are represented by griddle plates separated by pot-rests. Above the earth are the Malocas of high-flying birds, wild animals, death and The Four Brothers. The shaded area represents the subsoil, containing the tombs of the dead. Below the subsoil lie the underworld levels, and beneath them the cosmic fire, surrounded by a boa. The female symbols of the Four Sisters occupy the nadir of the cosmos.

Shamans as androgynous knowledge-holders: the power of jaguar-seers

Jaguar-seer shamans consult the supernatural Owners of nature when a new season starts or ends, to implement resource use, or to locate and retrieve a stolen or dispersed knowledge system from a sick person. Unlike other people, the shaman does not fall ill if his knowledge base leaves his body, and it is his source of power. The shaman will think from dusk to dawn to explain events within a cosmological framework. He correlates bio-social events within the context of the predicaments of the universe, nature and the ethnic group and does retrospective and prospective long-term thinking to interpret the events. At dawn, the shaman explains to the community or to the sick person what he thought regarding the impact of resource use and gives advice of appropriate future behaviours and thoughts to redress present problems.

After 'Travelling in Thought' throughout society, nature and the several skies and underworlds of the cosmos and this earth (Figure 3.1), the shaman explains the causes of the socio-environmental problem and prescribes corrective measures. His mind is said to have travelled across diverse ecosystems in the guise of powerful predators or far-sighted 'jaguars'. He travels across hydrological systems as an anaconda or sting-ray, in forests as a jaguar or viper, in air as an eagle, and if by night as a bat. His knowledge system is said to be also travelling within his body, since the shaman's body represents mnemonically the topology of the universe and the shamanic geography of the ethnic group.

Significantly, the shaman says that he stores the knowledge bases of men in his right side, and those of women in his left side: he is an androgynous knowledge-holder (Reichel 1997) and knowledge-controller of men's and women's knowledge, hence the most powerful and wise authority of the *maloca*. Unlike shamans, the elder women only amass the knowledge of women to guide, heal and advise younger women.

By shamanic inducements and by sanctions made by the elders, and by conscious commitment, the *maloca* people sustain specific assumptions of their group's future and destiny. This is made with a profound respect for the spiritual properties of all forms of existence throughout the universe. The belief that their gender-based knowledge system and its 'ecology of ideas' (Bateson 1972) allows for their cultural strength maintains a staunch defence (Clastres 1972; Descola 1986; Lévi-Strauss 1962) of the basic patterning of the ideas, images, emotions and behavioural patterns necessary to sustain their mode of life. The action-geared gender groups achieve a cogent agency in the replication of their specific knowledge bases for sustained resource conservation as they self-determine their cultural and ethnic poiesis.

Conclusion

The gender-based knowledge systems, the *maloca* mode of life, and the shamanic world-view, as a form of cultural resistance, persist because of their strength and wisdom, as they deter negative reciprocity, socio-economic stratification, energy overconsumption, and ecocide. They articulate a long-term awareness of the dynamics of bio-social co-evolution which are accounted for in their cosmologies where the interconnections between cosmos, life, humanity and intelligence are interlinked.

The collective imaginary of respect towards the complexity of life and the cultural perspicacity required to reproduce their social and ecological traditions in the rainforest is defended by conventional politics and by the politics of enchantment and of spirituality of their cosmology to ultimately control their resources in the most biodiverse of Amazon rainforests. Such Amerindian cosmologies and knowledge systems impel us to reconsider the value of indigenous peoples and the role of ethnoecology (Posey 1982; Reichel-Dolmatoff 1976) to guarantee sustainable development and a sustainable humanity not only in Amazonia but on earth. The Yukuna and Tanimuka concepts of universe, life, nature, culture and intelligence ultimately reveal a theory of the justice of nature to sustain life in the universe, and a practice of the nature of justice which contributes to critically redefine Western concepts of environmental politics and social and gender equity, and the meaning of knowledge and life itself.

Basketry and biodiversity in the Pacific Northwest (Marilyn Walker)

> *'It happened in those mysterious times when Raven still walked among men, exercising the cunning of his mind in bringing good to his creatures by ways strange and inexplicable to mankind. Already his greatest works had been accomplished. He had stolen the Sun, Moon and Stars from his grandfather, the great Raven-who-lived-above the Nass River, Nass-shah-kee-yalhl, and thus divided the night from the day. He had set the tides in order. He had filled the streams with fresh water and had scattered abroad the eggs of the salmon and trout so that the Tlingit might have food. But not yet had Raven disappeared into the unknown, taking with him the power of the spirit world to mingle with mankind.*
>
> *In those days a certain woman who lived in a cloud village had a beautiful daughter of marriageable age. She was greatly desired by all mortals and many came seeking to mate with her. But their wooing was in vain. At last it chanced that the eyes of the Sun rested with desire upon the maiden, and at the end of his day's travel across the sky he took upon himself the form of a man and sought her for his wife.*
>
> *Long years they lived together in the Sky-land and many children came to them. But these children were of the Earth-world like their mother and not of the Spirit-world of their father, Ga-gahn. One day, as the mother sat watching her children frolicking in the fields of the Sunm-land, her mind filled with anxiety over their future. She plucked some roots and began idly to plait them together in the shape of a basket. Her husband, the Sun, had divined her fears and perplexities. So he took the basket which she had unknowingly made and increased its size until it was large enough to hold the mother and her eight children. In it they were lowered to their homeland, the Earth. Their great basket settled near Yakutat on the Alsek River, and that is the reason that the first baskets in southeastern Alaska were made by the Yakutat women.'*

The origin of basketry and its centrality to life is commemorated and celebrated in this legend, recorded by Frances Paul in 1944 (the name of the story-teller is not recorded). It connects Tlingit origins in an elemental and profound sense with the origin of basketry.

We are told how the earth-world and the spirit-world are connected through basketry. Becoming human transpired through a journey from one world to another in a basket which symbolizes an ancestral connection and a divine one. And we are told how the future of the Yakutat people materialized through the making of a basket. Where the basket touched down became their homeland. The collective memory of a nation is embedded in basketry, one of the oldest and most valued arts of the Tlingit people of the Northwest Coast of North America, and of other Northwest Coast cultures of the United States and Canada from Alaska through British Columbia and into northern California.

Individual women and families became famous for their skills and the beauty and utility of their basketry which permeated all aspects of life. Baskets were woven so tightly and finely they were used as drinking cups or as water storage jars. The traditional way of cooking was in watertight baskets: hot stones from the fire were placed on the food, enough water was added to produce steam, and the basket was covered and set aside until the food was cooked. Baskets were also lightweight travelling trunks. Roots and bark were woven into mats for canoe sails and floor coverings and made into fish nets and cradle swings. Woven hats were made for work and for ceremonies. The high caste wore 'big hats', and special hats were woven for shamans. Shamans used baskets to hold rattles for ceremonies, and charms of goose and eagle down. They drank salt water from baskets with twisted root handles as part of their purification. Some baskets were huge. A 'mother basket', nearly a yard across and as deep, was used as a food dish at Tlingit feasts. Oil storage baskets might hold up to twenty-five gallons of *eulachon* (candle fish) oil. Tiny baskets held tobacco or snuff made from ground and roasted clam shell, dried leaves, and the ashes of the inner bark of yellow cedar, and were hung around the neck on cords to free both hands for berry picking. In the southern part of Tlingit country, and other parts of British Columbia, red cedar bark was the usual basket-weaving material. Above the northern limit of red cedar trees, baskets were woven from yellow cedar and from spruce roots.

Today, clear-cut logging has devastated much of Southeast Alaska and British Columbia and endangers the biodiversity of the Northwest Coast rainforest. Basketry has much to say about how cultural and spiritual values embedded in traditional knowledge are intertwined with biodiversity. Knowledge about materials – when to gather them, where to find them, how to work with them – requires 'knowing' about the environment in a profound way, through direct experience built up over time rather than through abstract or intellectual knowledge that characterizes Western science's involvement with a place or a resource. Traditional harvesting practices ensured the sustainability of the resources on which basket-makers relied. Scars on old but still vital trees are reminders that a tree has given – for clothing, utensils, or shelter. The inner bark of cedar was used for fishing lines, twine and rope, netting, and even hand towels for use after eating. Mats, and of course baskets, were woven from it.

Just one strip was taken, usually from a tree on the steep side of a mountain, and with no branches on one side because these are reaching towards the light, away from the hillside. A horizontal cut is made near the base of the tree and the bark is pulled with two hands. This way a long, tapering strip of bark can be peeled up the length of the tree, leaving the tree to heal and to continue to grow. Even whole planks were harvested this way. Some scars are 150 to 200 years old on trees that were much older when they were harvested. 'The Tree of Life is what we call the cedar', Theresa Thorne, a Cowichan elder and basket-maker told me. 'It gives us everything we need'. Roots were taken also without killing the tree.

To a basket-maker, the place her materials come from is key, not simply the materials themselves. The long-term sustainability of the resources needed by the basket-maker is part of a complex traditional land and resource management system that spans many generations – past and future. 'Taking care' encompasses not only the materials themselves but also the places in which these materials are found, and the meaning these places have in First Nations/Native American history and consciousness.

Stewardship of the resources needed for basketry protects biodiversity in the broadest sense. Understanding the role of basketry in traditional culture, and the role of the basket-maker, requires us to expand out understanding of biodiversity to include not just the physical elements of the biosphere but the metaphysical also. A basket is a physical object, but like the land it is derived from it is also a container of meaning, of memory and identity, of myths, teachings and dreams. Western science divides the world into the organic and inorganic, the animate and inanimate. In one of these baskets, however, body, mind and spirit are inseparable. Looking at such a basket or holding one in our hands if we are lucky enough, we are forced to reconsider such distinctions as well as our answer to the seemingly simple question, 'What is alive?'

The spirit of a basket connects a maker to her past, present and future, to her family and her community. Basketry is a link with the ancestors, with the children who receive the knowledge passed on in the baskets she's making, and with the land in which her history is etched. Northwest Coast basketry acknowledges the land and the objects derived from it as sacred places to be attended to with respect and humility, and reminds us that planning and sustainability mean more than our lifetimes. Reading the Origin of Basketry story once more, we are reminded that the earth is the source of all creativity. We are reminded also that biodiversity is about the physical world and the metaphysical. It is about body and spirit and the interconnectedness of all life.

Basket-makers talk about understanding their craft and the materials they use from 'within', 'from the heart', 'from the centre of the body...', 'the soul', 'the core' – not from the head, I was told, as Western science is seen to do. I was told how Western science misunderstands this profound difference – how we are used to working with the head. It is even difficult to talk about basketry in this way, to try to convey such feelings to a Westerner, people told me, because these are things that must be experienced or felt, not simply spoken about.

First Nations have shown me how our relationship with the materials we use is one of reciprocity. Gathering plants means preparing yourself, putting yourself in the 'right' frame of mind, 'thinking good thoughts' and 'having a pure heart'. This may mean fasting, entering a sweat lodge, or 'smudging' by burning herbs for cleansing and purification. When Theresa Thorne, Bob Sam, Larry Louie, Judy Good Sky and others showed me how they gather plants, they told me how they ask the plant's permission: 'You ask the plant to help you', I was told. 'Perhaps someone in your family is ill or needs help in some way. You tell the plant how it will be used, and ask it to help you.' First Nations/Native Americans talk about this as showing respect for a plant and the gift it offers us. A gift must be made in return. You give a little tobacco, some tea or sugar; perhaps a handful of corn meal or rice is scattered at the base of the plant. A hair from your head or even spit will do if you are poor or have nothing else to give. Afterwards you must say thank you.

To someone who knows about baskets and their meaning, to a basket-maker herself, this respect and energy may be retained in baskets themselves. A basket-maker may feel the memory of the maker and the materials she used in the delicate strength and power of old baskets in museum collections. She might also feel sad that they have been removed from their spiritual context and that they are judged now by criteria very different from when they were made and used. But for a basket-maker without older women who know how to make baskets, the baskets themselves are mentors, as are the plants from which they are made. If we know how to listen.

Amongst First Nations and Native Americans who speak about these things, the making of baskets is intuitive and multi-sensory. Basketry is about the *making* of the basket: it is not just about the finished piece. The process of creation connects you to other basket-makers – past and future – to your materials, and to the place that offers such a gift. Meaning and spirit are in the materials themselves – in the bark and in the strong but flexible roots of cedar and spruce which connect us with the earth and retain its energy and strength.

Footnote: In Canada, Indigenous People refer to themselves as First Nations; in the United States, Native Americans is the term of self-identity.

Acknowledgement
The author thanks the many people who have taught her about plants over the years, a few of whom are named here.

Oxy in U'wa territory - the announcement of a possible death (Mónica del Pilar Uribe Marín, translated by Peter Bunyard)

'Our law is to take no more than is necessary. We are like the earth, which feeds itself from all living beings but never takes too much, because if it did, all would come to an end. We must care for, not maltreat, because for us it is forbidden to kill with knives, machetes or bullets. Our weapons are thought, the word; our power is wisdom. We prefer death before seeing our sacred ancestors profaned.' (The U'wa people.)

The U'wa – whose name means 'intelligent people that know how to speak' – belong to the Chibcha linguistic macro-family. They are one of the few peoples that have managed to survive in Colombia while maintaining their ancestral culture in a living form. They live for the most part among the lofty heights of the Sierra Nevada del Cocuy-Guicán, from where they have seen their sacred lands disappear over the years. Their lands once embraced the Eastern Cordillera from the Sierra Nevada del Cocuy to the Sierra Nevada de Mérida in Venezuela.

Two visits to Washington in 1997 by the President of the Association of Traditional U'wa Authorities, Roberto Afanador Cobaría, represented yet another intense effort to save the indigenous U'wa from the loss of what for them was a last chance of an independent existence and a way of living that they have never wholly lost despite five hundred years of white domination. The U'wa have a culture that is deeply rooted in tradition: today that culture could be on the verge of extinction in the face of the imminent intrusion into their territory of a multinational oil company.

Cobaría was in Washington to explain to the Inter-American Commission on Human Rights the circumstances that are affecting his community since the Colombian Government approved the environmental licence it had granted to Occidental of Colombia for petroleum exploration and exploitation in the U'wa area. At the same time, Cobaría had sought – since the beginning of the year – a closed-door session with Stephen Newton, president of Occidental's operations in Colombia, and two vice-presidents of the parent company. He informed the Occidental chiefs why his community rejected any intrusion into their territory by 'Oxy' or anyone else. But the reality is that neither Oxy nor the Colombian Government appear to be listening to the indigenous peoples, even though, in September 1997, the Organization of American States recommended Oxy to withdraw immediately from U'wa territory.

In fact Oxy, as described in its proposal 'Seismic exploration of the Samoré Block', has plans to explore for oil in a region that impinges on the departments of Arauca, Boyacá and the north of Santander, a total land area of 208,934 hectares, of which one-quarter is in U'wa territory.

As of now, the international tribunal has yet to decide on the legitimacy or otherwise of the licence. Occidental, too, has still to show itself open to reason in respecting the cultural, physical and socio-economic integrity of the U'wa. Their land is everything to them, and the U'wa are hanging on to their threat to commit collective suicide should all attempts fail to keep Occidental and others away.

According to their eco-philosophical thinking and their conception of the world, the exploitation of petroleum promises the U'wa a false future and a development in which they have no place. For that reason they have announced an ancient strategy of resistance: 'Faced with inevitable death, with the loss of our lands, with the extermination of our history, we prefer a dignified end, worthy of our ancestors who challenged the dominion of the conquistadors and missionaries. Our U'wa communities therefore prefer collective suicide.' In effect, legend has it that a precedent for collective suicide occurred in 1726 when a Western Highland clan of the U'wa threw themselves off a cliff, now known as the 'Rock of the Dead', to avoid Spanish domination and the loss of their lands.

The U'wa have many reasons for keeping Oxy at bay, as well as any others who may wish to exploit the natural resources in their lands. The social fabric of life for the U'wa is woven into ancestral knowledge, into respect for the natural environment as imbued with spirit, into their customs, their medicines and healing, and not least into the way that property and land belongs to all collectively.

For the U'wa, petroleum (*Ruiría*) contains blood that gives them strength and life, as it does to all living matter, whether plants or animals. The oil in the earth is, 'the mother of all the sacred lakes' and the U'wa believe it to be 'working' in the same way that emeralds, gold and coal are 'living and working' as active agents in Mother Earth and therefore they should be 'left alone, not touched'.

Their social and cultural life is therefore organized according to their model of the creation. Land, laws and customs are therefore all part of a system of bartering with nature that the U'wa need to employ to fulfil the order that derives from the cosmos. For them the Law is sacred, 'because it is the design of the eternal father and mother. The Law is therefore inviolable and cannot be changed, as one can change the laws of the *Riowa* (Whites). The Law therefore determines for the U'wa very special and specific forms of behaviour that have allowed a balanced development in the environment they have occupied... The sacred is traditional U'wa Law and any deviation from this foundation brings with it grave dangers for the world and for Man, because the Law has not been written by us; on the contrary, we narrate it, we sing it, we practise it. Such laws, our traditional codes, are the pillars of our culture and the posts that sustain the world.'

It is this knowledge that has allowed the U'wa to elaborate an advanced eco-philosophical thinking and to develop their particular approach in fragile ecosystems. By means of ritual practices, uses and customs, the U'wa know perfectly how to manage and make use of the different altitudinal levels so as to gain access to different resources, and how to achieve a rich and diversified production of foodstuffs. However, it is now difficult, they say, 'because the *Riowa* have taken over the best lands and some indigenous members of the U'wa have modified their traditional cultivation practices, assimilating those of the Whites which are more harmful because they exploit the land excessively, both through burning and through creating cattle pastures'. Those who are more traditional opt for a method of clearing in which they do not cut down large trees nor those that bear fruit. Instead they care for the soil so that it will always continue to reward them with its produce. Moreover, through traditional celebrations they bring about a favourable climate for plants and animals, such that the U'wa can live in harmony without sicknesses. 'However, the 'Whites' (*Riowa*) don't let us continue, they continually importune us with their projects; they cut the path we wish to follow; they deny us our autonomy'.

It is therefore remarkable how, despite years of intervention, exploitation and genocide, the U'wa have kept their customs intact, and that they have not renounced their songs and rituals, of which the most important are the sung myths of *Reowa*, which correspond to the ritual of blowing and is in essence for processes of purification and *Aya*, celebrated after *Reowa*, the purpose of which is to seek the ordering of the universe and of those beings that exist in it. Thanks to these songs and rituals the U'wa remain united to the distinct levels of the universe and achieve the equilibrium of a society that uniquely yields primacy of place to the ancestral knowledge that rules the *Werjayá* (the spiritual leaders). Moreover, through their cultural practices they help maintain harmony between the forest, gardens, humans and spirits. Nevertheless, the U'wa people recognize that 'in the frontier with the *Riowa*, a zone of transition lies between the traditional and the Western, where indigenous peoples predominate that do not practise a grand part of the collective ritual acts and who live with *mestizos* and humble peasants'.

For the U'wa one of the inevitable consequences of Oxy exploitation of oil will be colonization, with all the havoc and destruction it brings in its wake. Colonization results in a terrible attrition of the natural world, with the wiping out of species; it leads to a loss of identity among indigenous people, to a collapse of traditional culture as well as the bringing in of incurable diseases that can ravage a population. Colonization also implies the taking over of land that is vital for the survival and well-being of the indigenous population and it may well destroy the spiritual life of the U'wa. As it happens, the loss of territory over recent years has taken the U'wa population close to famine, thereby affecting their health. Any environmental impact study of Oxy's incursion into their lands must take account, they insist, of all future consequences, including demographic changes, social violence and disease.

The U'wa have an oral tradition which binds and roots them to the ultimate limits of their ancestral territories, marked for instance by ridges, lakes and rivers. The U'wa have a clear memory of ancient landmarks, and the present-day *resguardo* and reserve do not begin to encompass the traditional lands. 'If we speak of territory we must go back to the time before our lands were invaded by colonists. We have to understand the deep relationship that exists between our concept of what are our lands and between our cosmology and behaviour. When Yagshowa was organizing the world, neither the Gringo, the American nor the Spaniard were yet here – just indigenous peoples, the *wejaya*. As soon as the creator Yagshowa had finished his work, then the U'wa appeared. The father eternal gave petroleum (*Ruiría*) for all the world, but he laid down limits: he knew precisely where the Spaniards and others would come and for that reason he made this territory untouchable. They (Oxy) cannot touch here. Perhaps they'll be able to get authorization in another part, but not here.'

The U'wa find it incomprehensible why the government and Oxy should play games with their laws of life. 'Why don't they respect our right to live and be different from the *Riowa*? We want nothing other than to live in our world, and not to abandon the joy that always accompanies us in our own rhythm of life. We do not want to live in the contaminated land of the Whites.'

Given the current environmental crisis in Colombia and elsewhere in the world, with all the concerns over levels of consumption and exploitation of natural resources, the U'wa struggle to protect themselves and their land has profound relevance for all of us in seeking alternative modes of development. U'wa beliefs and cosmology go way beyond their localization: they should be part of a desperately necessary reconsideration of the relationship between society and the natural world that has been imposed upon all of us by the West.

Barí: knowledge and biodiversity on the Colombia-Venezuela border (Jaime Rodríguez-Manasse, translated by David Simon)

The Barí are a Chibcha-speaking indigenous people whose territory lies in a rainforest region on the borders of Colombia and Venezuela. Barí territory in Colombia is recognized as 'Indigenous Protected Land', while in Venezuela it is an 'Indian Reservation'. Both areas are classified as National Forest Lands. This area of rainforest forms a part of the Catatumbo ecosystem (which in Barí means 'the land of thunder and lightning', referring to a natural phenomenon known as the 'Lights of Catatumbo' which illuminates the night sky). The Catatumbo includes Lake Maracaibo, the Juan Manuel swamps, the Catatumbo river basin valleys, rainforest, the Perijá mountains and cloud forest area of the Motilones.

Whereas science frames the structure of Western knowledge, for the Barí, myths provide a form for ancestral knowledge that enables someone to understand *'Bakiarúna'* (Barí mythic knowledge). Mythical structure is linked closely to language and provides a logical model for resolving socio-cultural contradictions. Until now, ancestral myths have constantly been assimilated into the dominant ideology of the West. This has the effect of breaking up thought. The following extracts from Barí mythology show how their ancestral knowledge is bound up in biodiversity as an integrated whole, and not as a form of knowledge that breaks up thought. The following myths, *'Bikogdó risó Boborayi'*, tell of biodiversity.

No light, no water, no sun, no stars existed... everything was dark. Everything was mountainous without desert plains. Sabaseba came from the place where the sun hides... he could see clearly... at night as if it were day. When he appeared, everything was submerged in darkness. Sabaseba came smelling good. He came from the mountain where one can find the fruit tai chirokba. He never set foot on land, but rather moved softly through the air.

Sabaseba worked all the time, from sunrise until sunset. The world was completely mountainous.

Figure 3.2: Bari territory in 1900.

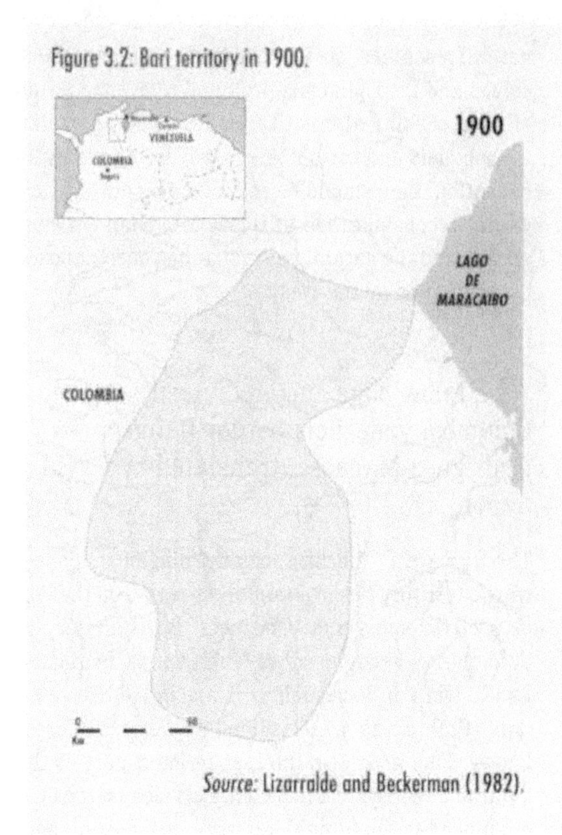

Source: Lizarralde and Beckerman (1982).

Figure 3.3: Bari territory in 1983: note territorial contraction.

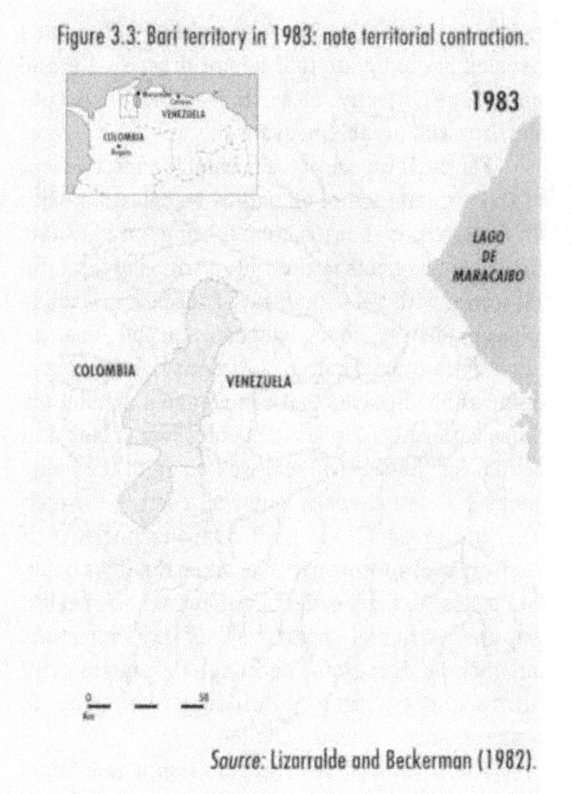

Source: Lizarralde and Beckerman (1982).

Sabaseba laboured all hours to make the world flat, to do away with the overwhelming presence of the mountains. Then, tired of so much hard work, he pulled up a pineapple to eat and broke it in half. In its heart there was a man, a woman and a baby Barí.

Sabaseba opened a great many pineapples, and from within them came men, women and children. Those that came from the purple pineapple were called Ichigbari and those from the yellow were called Barí. Sabaseba told them, 'You will be Barí and will not look at the faces of the Ichigbari. The Barí will walk on the ground while the Ichigbari will fly'. Then Sabaseba told the Barí to unite and build houses.

Sabaseba means both 'soft wind' and 'knowledge'. Pineapple in Barí is called 'Nangandú'. It is one of the oldest plants in existence. The Ichigbari have their farm land on top of the trees between the land and the sky. The Taibabioyi were born out of rotten pineapples. They live in the water where they have their farm lands. Sabaseba made a flint axe, from the haft of which he made téchi bows and arrows.

Bashunchinba (Spirit God) protects the Barí from the Taibabióya with the Bikogdó: a rainbow. A Bashunchinba is a Barí who has died physically but not spiritually. When they die, the Barí begin to have a great headache, and after death they go upwards to Barún where they sit on the ground and wait. Bashunchinba, who have died previously, begin to come to pick up the dead body. They place the dead person in a small boat and play the drums loudly: 'Bam, bam, bam...' As the people play... the spirit of the dead Barí sees them, climbs down from the small rowboat and goes running towards the people. When in the sky, an elder brother who has died first welcomes his brother...

The new Spirit-Brother told elder brother Bashunchinba, 'I'm hungry', and was given a huge plantain. It was enormous! Bashunchinba told him, 'If you eat this you will have to live here'. When he saw some huge sugarcane he said, 'If you really want to eat them you should cut them and come and live with me in the sky'. Then he cut the huge plantain and began to eat it.

But one day the younger brother decided to cut the dry stem and tie up the roots to take them to the earth where his living brothers resided. As he was tying them up, Bashunchinba told him, 'That is mine'. The younger brother replied, 'I am going to take this to my house from where you brought me'. The young Bashunchinba walked but did not fall. Everything was completely silent. But he kept on walking and the sun darkened while he walked on. Another sun rose and yet another. Then he began to perceive the earth heard the footsteps of the Barí.

But he was not happy. The elder Bashunchinba had already infected him with the smell of death. People said to him, 'Leave us in peace'. His sister whispered to him, 'Your laughter will make you sleep, you will not be able to stay'. Then, he realized that Bashunchinba had already contaminated him and his

Figure 3.4: Crop rings around a Bari house: the traditional method in the Atshirindakaira, Ichidirrankaira, Bari community.

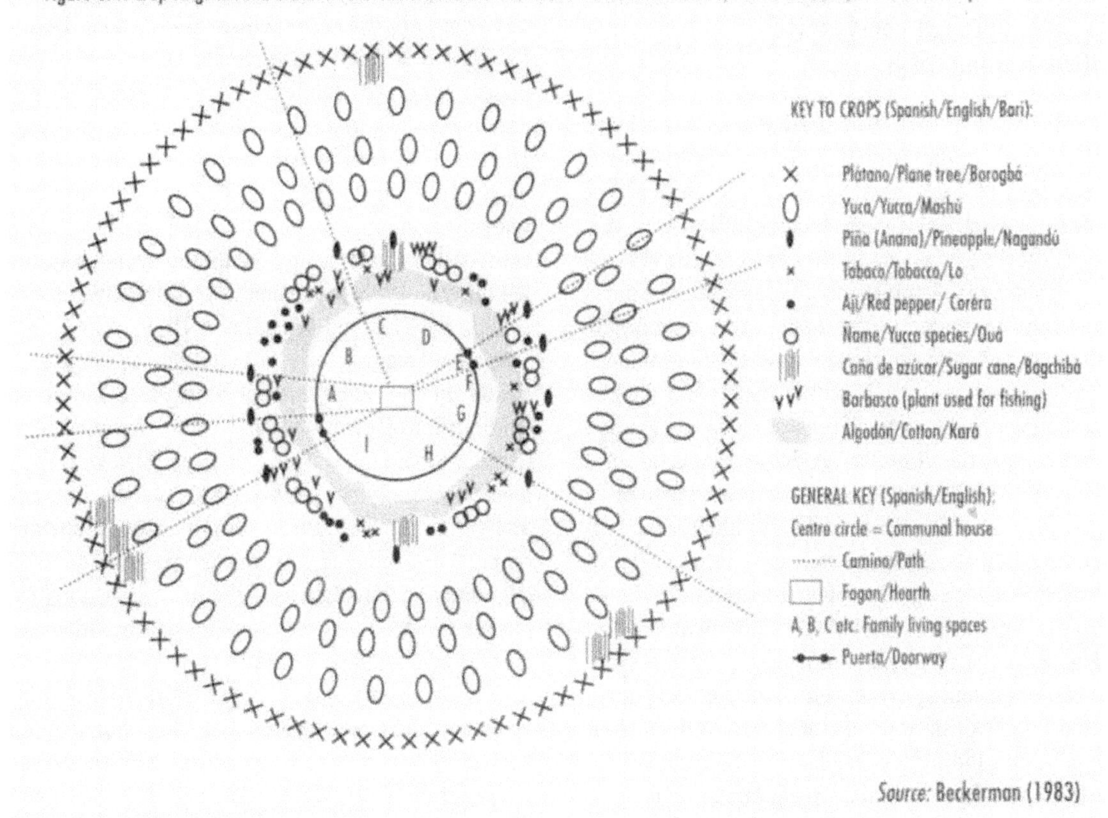

Source: Beckerman (1983)

skin had dried out; he was forced to return to the place where he had come from.

When he came to earth, he had brought the few seeds of spicy pepper, pineapple, plantain and sweet potato, and told his living brother to plant the roots, which he did. From the seed that Bashunchinba brought we cultivated farm lands and now we have equipment to fish and hunt with. From the seed that Bashunchinba gave to the Barí, they were able to cultivate all the lands.

Dababosá told the Barí, 'You will learn ways to cure yourselves'. They were surprised, but never forgot what Dababosá taught them. Thanks to him we are still curing the sick. From that time on, we have been able to protect ourselves so as not to die from illness. Dababosá could take on the guise of a jaguar. Yet, although he taught us much, we are now suffering more than before. Still, we are saving the lives of our children with his teachings.

Dababosá took young people and made them grow before returning them to their parents. One day he took a little girl and brought her under the earth to his cave where the sun rises. Then he left, turned into a jaguar and spoke to the girl's father. But he could not understand Dababosá and shot him with an arrow. Then Dababosá cured himself and returned to the father, speaking in Barí, and taught him the secret of how to cure people. He did this at night.

Then the little girl returned home. She brought a turtle for her parents and she brought them all kinds of meat: monkey, mountain pig and turtle. Even though the door of the farmhouse was shut, Dababosá and the little girl entered without having to open it. Dababosá showed the father the secret of curing wounds. If Dababosá stayed with the Barí, women would not suffer while giving birth.

Sabaseba had told Dababosá, 'Teach them the secrets of tobacco that can cure wounds'. When he had done so, Dababosá told the parents of the little girl, 'I will not return again. I am going to stay with Sabaseba'.

This is true, because this is what wise men have taught us. This is biodiversity – *'bobórayi'* in the Barí language.

Acknowledgement

The author is grateful to David Simon M.A., Latino-American Studies, for the translation of this contribution.

Shamanism and diversity: a Machiguenga perspective (Glenn Shepard Jr.)

For the Machiguenga (Matsigenka) of southern Peru, the exuberant diversity of life in the upper Amazon is a fact of daily existence. At the core of the Machiguenga world-view, biological diversity is a prima facie expression of the prowess and virtuosity of shamans. A transformative, shamanistic act is thought to underlie the observable taxonomic disjunctions between related species of organisms. The process we call evolution is driven, for the Machiguenga, by past and present shamans. Primordial beings, the *Tasorintsi* or 'blowing spirits', first used the transformative powers of tobacco to breathe diversity into the animal and plant kingdoms. Legendary human shamans ascended into the spirit world in hallucinogenic trances to obtain cultivated plants and other technological innovations for Machiguenga society. Modern shamans are responsible for obtaining new varieties of crops and medicines, for calling wary game animals out of their enchanted hiding places, and for fighting off illness and other demonic forces. Without the shaman's transformative power there can be no adaptation, innovation or evolution – cultural or biological.

To this day, shamans protect the Machiguenga people from evil forces thought to inhabit the rugged cloud forests of the Andean foothills. Dangerous animals such as spectacled bears, jaguars and aggressive troops of giant monkeys that take refuge there are a threat to careless travellers. Demons living in caves and unpopulated forests may rape and kill people on long treks. Evil sulphurous smells and hellish flames are said to emerge from certain mountain cracks, sending illness up into the atmosphere. Powerful shamans living in isolated headwater settlements scout these regions on their magical flights, keeping the evil forces at bay. The Machiguenga consider lightning to be the weapon of shamans, who seek out and explode demons where they lurk in wait for human victims. Fear of the remote headwater regions keeps the Machiguenga out of some of the most fragile and threatened cloud forest ecosystems. Today, Machiguenga worry that recent oil-drilling activities in their territory may threaten the health and safety of isolated villages where powerful shamans still reside, while unleashing evil vapours and illness into the atmosphere. Their fears are probably not unjustified.

Shamans receive their transformative powers from diverse psychoactive substances, including *ayahuasca* (*Banisteriopsis*), *floripondio* (*Brugmansia*), *chirisanango* (*Brunfelsia*), the toxic eggs of *Bufo* toads, and especially tobacco (Wilbert 1987; Baer 1992). During hallucinogenic trance states, powerful shamans are said to be able to convert themselves into various species of animals. Research into the psychological effects of narcotics and hallucinogens shows that they potentiate analogic and intuitive reasoning and give free reign to the chains of association typically found in dreams and hypnotic states (Berlin *et al.* 1955; Joralemon 1983; Winkelman 1986; Dobkin de Rios and Winkelman 1989). The transformative cognitive effects of hallucinogens are invoked by the Machiguenga to explain the same conundrum of evolution that Linnaeus once pondered: how did the world come to be filled with such a great diversity of plants and animals, all different and yet apparently more or less related to one another? For Linnaeus, the answer was the Mind of God. For the Machiguenga, the answer is, to paraphrase an often-parodied public service announcement, God's Mind on Drugs.

Shamans maintain their transformative powers by cultivating a relationship with the *Sangariite*, the 'invisible ones' or 'pure ones' of the forest (Rosengren 1987). When a shaman dies, he or she does not meet the fate of ordinary human souls, which is to be converted into a tapir and hunted down and eaten by the moon spirit Kashiri. Rather, shamans' bodies are said to disappear mysteriously from their place of burial. The shaman avoids the pitfall traps set by the moon for human souls, and walks off into the forest to join the invisible and immortal *Sangariite* spirits.

The *Sangariite* are said to live in small, natural clearings in the forest. These clearings are created by the symbiotic relationship between the shrub *matyagiroki* (*Cordia nodosa*) and mutualistic ants (*Myrmalachista*). The shrub provides stiff, protective corridors of hairs and spacious nodes or domatia for the ants to live in, while the ants clear the understorey around the plant, allowing the *Cordia* colony to grow (Morawetz, Henzl and Wallnfer 1992; Davidson and McKey 1993). To an ordinary person, the clearings are simply small colonies of *Cordia* plants in the forest. To the shaman, partaking of hallucinogenic plants, these small clearings open up into vast villages, humming with the voices and songs of the joyous *Sangariite*, and surrounded by spectacular gardens. The *Sangariite* raise as their pets all the game animals eaten by the Machiguenga (Baer 1984). The currasows are the chickens of the *Sangariite*, the jaguar is their watchdog. They release these pets from their invisible villages for people to eat.

The shaman goes into trance by switching places with a spirit twin from the invisible world of the *Sangariite*. The spirit twin comes to sing and heal among the humans gathered at the ceremony, while the shaman visits the invisible village of the *Sangariite*. There, the shaman may make deals with the *Sangariite*, so that they release more of their pets for the Machiguenga to hunt and eat. The *Sangariite* may provide the shaman with new varieties of crops or new medicines to heal the sick.

The relationship between shamans and the *Sangariite* is an active part of modern Machiguenga beliefs. Even in those communities where shamanism has been supposedly lost due to the influence of missionaries and changing society, people try to maintain trading relationships with the more remote villages where shamans still have access to the Sangariite's enchanted banks of new genetic material. This situation creates a set of power relationships between the upriver and down-river communities, between the traditional and the modern, between the 'savage' and the 'civilized' (Taussig 1987). Furthermore, such customs are probably necessary to maintain the genetic diversity and vigour of their crops.

Genetic diversity is a ubiquitous feature of indigenous agriculture, lending natural resilience in the face of pests and diseases (Posey 1983). This is especially important for such clonally propagated crops as manioc and plantain bananas, the two Machiguenga staples. Vegetative propagation has the advantage that any desired trait can be reproduced faithfully in successive generations of clones. The disadvantage, however, is that genetic diversity cannot be maintained through sexual recombination and selective breeding, as in seed crops like corn, but only through the random and slow process of mutation. Pests and diseases can quickly overtake such genetically stagnant clonal lines. The Machiguenga are well aware of this problem, noting how some cultivars tend to lose their vigour quickly through time. Shamanistic innovation provides a metaphor for understanding the genetic vigour of new cultivars. New varieties, brought from remote villages where powerful shamans are at work, breathe new life and diversity into ailing crops. For this reason, it is necessary for the Machiguenga to maintain constant contact with an outside source of novel cultivars. For the Machiguenga, the ultimate source of such innovation are the *Sangariite* and their invisible villages. Diversity of cultivars is not only good agriculture, it is a manifestation of divine providence.

The most important plant varieties the shamans bring from the gardens of the *Sangariite* are manioc and the *piri-piri* sedge (*Cyperus*), both propagated vegetatively. While manioc is the principal food staple of the Machiguenga – its name *sekatsi* synonymous with 'food' – *ivenkiki* (sedge) is synonymous with medicine. The Machiguenga cultivate astounding numbers of manioc and sedge varieties. Manioc varieties are individually named and distinguished by leaf shape and colour, stem colour, maturation period, resistance to inundation, tuber size and texture, and so on. I have counted some thirty named varieties to date, though the total number is certainly greater (cf. Boster 1984). Likewise, each sedge has a distinctive medicinal use: for headache, for fertility control, for difficult births, for insanity, for hunting skill, for gardening skill, for skill in spinning and weaving, for bathing infants... the list goes on, certainly in excess of three dozen varieties (cf. Brown 1978).

When manioc goes to seed, it typically produces an inferior inedible variety called *kanirigompireki*. The sedge is host to a parasitic fungus of the Claviceps family (like ergot, the natural source of lysergic acid) which destroys the flower and fruit, making it incapable of reproduction without human manipulation. The ergot alkaloids produced by this fungus are likely the compounds that produce the medicinal activity of these plants (Plowman *et al.* 1990). In both cases, the Machiguenga depend on shamans to bring new varieties and renew the genetic vigour of their gardens.

In some cases, the owner of a cultivar can actually name the shaman who brought it down from the sky. One shaman, Perempere, who died no more than two decades ago, was the proud recipient of eight or more manioc varieties from the *Sangariite*, and all are still cultivated in the upper Manu. One former shaman, still living, admitted to having brought down a manioc variety himself. How did he accomplish this? He took the hallucinogen *Brugmansia*, wandered through a swamp of *aguaje* palms (*Mauritia*) in a state of trance, and returned home with a manioc cutting in his hand. He named the manioc variety, *koshishiku*, 'from the *aguaje* swamp,' and many of his family members now cultivate it.

Such feats are hard to rationalize from a sceptical, scientific viewpoint. Yet as Machiguenga farmers observe, manioc goes to seed in old gardens and the seeds can remain dormant for years as secondary forest growth returns. When this forest is felled and burned again, wild manioc with its small, inedible tubers grows abundantly. By sheer chance, useful varieties are sometimes produced (Sodero 1996). Are manioc seeds in old garden growth the secret to the shaman's power of agricultural innovation? I suggested the hypothesis to a Machiguenga companion and he laughed at the preposterous suggestion. 'Of course not! Wild manioc is useless, and grows everywhere. Shamans get manioc cuttings by climbing into the sky.'

Missionary activity among the Machiguenga over the past three decades has disparaged the practice of shamanism and traditional medicine. The arrival of new diseases, social as well as physical, has threatened the existence of all Machiguenga, shamans and mortals alike. The diversity of traditional cultivars is being supplanted by commercial monocrops, making indigenous populations more dependent upon expensive pesticides and fertilizers and putting them at risk of crop failure and famine (Rhoades 1991). The Machiguenga believe that when their last shaman dies, so dies humanity's access to food, medicine, cultural innovation and protection from the multitude of demonic forces. The record of ecological and social havoc wreaked over past decades by invading colonists, gold miners, petrochemical and lumber companies, terrorists and drug traffickers in indigenous territories would only confirm the Machiguenga's worst fears.

An appreciation of Machiguenga belief confirms the principle behind the Declaration of Belém: the 'inextricable link between cultural and biological diversity'. In the origins of biodiversity in the mythical past, and in the maintenance of crop diversity to this day, the Machiguenga rely upon the transformative powers of shamans. Conservation of biodiversity involves not only protecting endangered species and habitats, but also protecting the myths, the religions and the cultures that know those species and habitats best, and that depend upon them for their survival. We hope that conservation efforts in Peru and elsewhere in the world will take indigenous cultures and values seriously, and thus help the Machiguenga, and humanity in general, maintain contact with the invisible spirits of immortal shamans, and with the richness and vitality of the forces of nature.

Central African voices on the human-environment relationship (Richard B. Peterson)

From a Central African perspective, human beings and nature are related in a 'both/and' dialectical manner rather than in a manner characterized by an 'either/or dualism'. Such dialectical perspectives also characterize Central African social thought, particularly in regard to the relationship between the individual and society. (Throughout this article I use the term 'dialectical' not in its formal Hegelian sense but in a more informal sense of connoting 'both/and' rather than 'either/or' thinking. 'Both/and' thinking consists of delving into the creative tension inherent in synthesizing what are seemingly opposite characteristics, propositions or processes.) These two sets of relations are themselves inextricably linked: that is, the relationship between the individual and the community holds very real implications for the human/environment relationship, and for the environment itself. This article addresses both of these key relational dilemmas. I illustrate how Central African thought can help to correct prevalent perspectives in the West regarding the relationship between the individual and the community, and that between humans and the environment.

'Both/and' relations between the individual and the community

In his recent history of the Central African forest region, Jan Vansina (1990) reveals how the human communities that inhabited the rainforests of Equatorial Africa were geniuses at maintaining a balance between their needs for autonomy and for security. Although some groups did tend toward centralization and experienced rapid growth, many more creatively intertwined both decentralization and co-operation. Historically, myriad groups were involved in a repeated dynamic of decentralizing in order to maintain their autonomy and sense of group identity, while simultaneously working to promote good relations with outsiders in order to reap the benefits of security and co-operation in the face of large-scale threats.

The same balancing act played out between one community and another continues to be played out, to varying degrees, within the community between the individual and the group. Any African environmental ethic rests on the same base that supports all African traditions – that of communalism. The African community is not composed of a group of individuals 'clinging together to eke out an existence' (Omo-Fadaka 1990: 178). Nor is it, as one African writer described community in the West, '...a conglomeration of individuals who are so self-centred and isolated that there is a kind of suspicion of the other, simply because there isn't enough knowledge of the other to remove that suspicion' (Malidoma Somé, quoted in van Gelder 1993: 33). Rather, in Africa the community is imbued with a certain bondedness. Bondedness entails respect, which in turn entails taking responsibility for one's fellow human being, not as an atomized individual but as a member of the common fabric of life. Since life's fabric is of one piece, connections within the fabric have to be maintained. If there is social or personal disharmony or illness, something has become disconnected and needs restoration.

Therefore, for the good of the whole, the responsibility to restore this broken connection falls on everyone. In such a manner African communalism provides a strong source for individual morality (Onyewuenyi 1991).

Yet, although African communalism may have much to contribute to social ethics, its influence does not necessarily mean that the individual is smothered or ignored as some Western writers have been wont to believe. Again, as in the situation with autonomy and co-operation, individualism and communalism do not exist in a dualistic and oppositional relation but in a dialectic whereby each reinforces the other (Gyekye 1987).

Pre-Westernized systems of land tenure in Central African forest environments illustrate well certain aspects of an African both/and way of thinking with regard to individuals, communities and land. Unlike our Western emphasis on individual ownership and on seeing land as a commodity, under Central African tenure systems, the goals, aspirations and property of the individual and those of the community exist hand in hand within a total system in which the two ideals are held in some degree of balance. Land is neither private property nor is it communally owned and worked in the socialist sense. Rather, land in most cases is held in communal trust; it belongs to the group, to all members of the community, extending usually at least to the level of the clan. However, within that common property ownership, each individual at the same time has their own piece of land that truly 'belongs' to them, and for which they and they alone (including family and extended family) are responsible, and to which they and they alone have usufruct.

This dialectical manner with which Central African societies treat both communal and individual drives regarding land allows the two to play themselves out in tandem. Under such methods, the community does not forgo the benefits of individual responsibility, effort and motivation that come through individual 'ownership' (but an ownership very different from our Western sense of private property). At the same time, the community keeps individualism from getting out of hand by preserving a communal sense and communal systems whereby the land belongs to everyone. With individual usufruct comes communal responsibilities and various social levelling mechanisms that keep individuals mindful of their obligations to others.

Other examples of this dialectic can be found in Central African systems of labour. Common among many societies are various communal institutions created to allow for a group sense of co-operation, helping each other out, and making sure the whole village survives. One particularly important organization of such type is what is known in Lingala and Swahili as *likilemba* – shared communal labour groups whose labour rotates from one individual's project (in this case usually garden cutting) to another's. With most *likilemba* no payment is involved; only the obligation to feed the group of workers. It is unlikely that a *likilemba* ever includes the whole village, and in villages of mixed ethnicities and families it is common to find them remaining within the extended families or among neighbours. One finds variations of the *likilemba* in urban centres – mutual aid groups (*mutualités* in French) whose members all contribute to a common pot and then rotate the use of the pot as specific needs arise.

Similar to the case of land tenure, we see that the *likilemba* and *mutualité* are means to provide space for both the individual and the communal to exist and be lived out simultaneously. What belongs to the individual is preserved – each person has their own garden or project – but at the same time, a sense of community, mutual help and co-operation is fostered by people coming together to cut each other's gardens or contribute to a common pot.

Application of Central African individual/community dialectics to actual development initiatives

How might these ideas translate into the very real development dilemmas facing organizations working on environmentally-sound development in Central Africa? Perhaps rather than choosing a primarily individual-based strategy or a primarily communal strategy, organizations need to find a way to combine both in a single system, a middle way that allows the individual and the communal to co-exist.

For example, rather than trying to build communal projects that are owned by everyone yet owned by no one, it would work better to encourage and support individuals in their personal projects (fishponds, vegetable gardens, fruit tree orchards, reforestation plots, agro-forestry gardens, animal husbandry, etc.), yet also encourage a communal system of labour to build such projects. Modelled after the *likilemba*, such a system could help provide the people-power often needed to get individual projects off the ground. Under such a both/and system, each person would also have the certainty that they would benefit directly from development, thus providing the necessary motivation and responsibility to make projects successful. But, at the same time, communal or co-operative labour that rotated between individual initiatives could keep development from becoming a completely individualistic money-making enterprise and enhance the communal spirit many grassroots development organizations strive for.

Efforts to improve livelihoods are difficult if not impossible to instil from the top down through a series of different committees and animators organized hierarchically. Rather, such efforts have a greater degree of success if they begin with real live individuals who truly desire to undertake certain development initiatives. Perhaps grass-roots development projects need to start with such individuals – encourage them, teach them and learn from them, and provide the seeds for them to realize their own individual projects. But they also need to promote projects that can be achieved only through individuals coming together to help each other and work co-operatively. Out of that co-operative labour, people might then begin to learn from each other, and to meet together, doing so not because it is required by the committee or centre above them in the hierarchy, but because they really have a reason to meet, they really have a desire to share ideas generated by their individual projects.

Both/and relations between humans and the environment

The same sort of dialectical relationship between the individual and the community in Central Africa can also be observed in Central Africans' relationships to the natural world itself. Again, Central Africa has a lot to teach us in regard to our Western perceptions of humans' place in nature.

Social ecologist Murray Bookchin (1990: 19-30) has remarked on the tendency of Western environmental thinking to fall prone to either of two extreme and fallacious views on the relationship between human society and nature: on the one hand the view that society and nature are totally separate realms (the hallmark within both capitalist and certain conservationist schools), and at the opposite extreme, the view that dissolves all differences between nature and society such that nature absorbs society (prevalent among sociobiologists and extreme biocentrists).

Such dualistic and reductionist views would be quite foreign to Central Africans whose understanding of the relationship between society and the natural world is more complex, holistic and dialectical. For example, a mutual and complex interaction between Central African forest-dwellers and their surroundings has allowed them to develop a rich knowledge of the environment that goes beyond the purely utilitarian (Vansina 1990: 255). Widespread knowledge provides them with the room and directions in which to innovate in the face of change. If something no longer works, if a natural disaster wipes out a certain resource, they know what else to try. If one year the forest gives only a little of the preferred *asali* honey, they know where to look for the less sweet but also good *apiso*. If hunting proves poor in one locale, switching camps to a new area is not difficult. In other words, nature and humans interrelate with some degree of flexibility and slack. Unlike Western biocentrists who tend to view humans as victims under the heavy hand of nature as taskmaster, Central Africans see nature as offering them some freedom of choice rather than forcing their fate upon them. The experience of Central Africans again provides us with a lesson. It affirms that although we cannot do with nature whatever we please, neither does nature leave us freedomless. Instead there exists the opportunity and (dare we say) the responsibility for us to play a creative role in shaping the future of the natural and social evolutionary process. We are co-creators, not simply victims of natural deterministic forces.

Malawian theologian Harvey Sindima, in describing the African concept of creation writes, 'The African understanding of the world is life-centred. For the African, life is the primary category for self-understanding and provides the basic framework for any interpretation of the world, persons, nature and divinity' (Sindima 1990: 142). Elsewhere he speaks of African cosmology as stressing the 'bondedness, the interconnectedness, of all living beings' (1990: 137). Although their denotative meanings are the same, the African idea of life-centredness contrasts and corrects the Western meaning that has been given to biocentrism. The latter has not escaped the trap of dualism such that it has often come to imply a certain misanthropic and oppositional understanding of the relationship between humans and nature, or conversely a relationship offering no distinctions between the two. Environmental ethicists such as Paul Taylor uphold a biocentrism in which the human species has no special status *vis-a-vis* other species, and *Homo sapiens*, like all other species, must be judged only on a morally individualistic basis (Taylor 1986).

Instead of focusing on the either/or debate between anthropocentrism and biocentrism, life-centredness focuses on the bondedness of all forms of life. Rather than analysing the place and standing of different human and non-human life-forms on the basis of their comparative rights, African life-centredness focuses on life itself, in a holistic rather than analytic fashion. It is not a matter of seeing what is most important, or of deciding if one thing is more important than another, but of believing and acting on the basis that all of life is important; even more, that all of life is sacred. Further, life-centredness is oriented less toward individual entities (rocks vs. trees vs. animals vs. people) and more toward the relations between them. More attention is paid to processes and the flow of forces between entities than to the entities themselves. Emphasis falls on relating rather than existing since it is the nature and quality of relationship that determines whether the whole will sink or swim. The relationship between any two living entities affects all the rest of life since all of life is bonded. Therefore, deciding whether humans or life is central is, in an African understanding, a non-question.

Application of Central African human/environment dialectics to actual conservation initiatives

Central African perceptions of the human/environment dialectic (that humans are part of, rather than apart from, nature) hold certain lessons for conservationists and all of us in the West who are concerned about the disappearance of the world's biodiversity. Such perceptions affirm what some Western ecologists have also come to realize: we are misguided to manage for a 'pristine' nature because nature does not exist 'pristinely'. We only place our desire for pristineness (i.e. no human influence or presence) on it. African dialectical thought reminds us that we are nature, we cannot get ourselves out of it. Such a fact also makes us aware that neither can we view nature outside of ourselves – we will always be looking at it through some degree of subjectivity. We would do well to examine and know what our own subjectivities are and how they influence what we see. When we try to manage according to the subjectivity of 'natural pristineness', we often end up moving more against nature's grain than with it.

Thus, an 'African ecology' if you will, can contribute a corrective to our Western dualistic ecology, and help to amend our disconnection with nature by emphasizing that we must manage for a whole system, humans included. Africans certainly recognize the important differences between humans and animals (as one major non-human part of nature), but they have less of a tendency than we in the West to set the two up in dualistic opposition. The two are parts of one whole, or as one forest farmer put it to me, 'God made us and animals together. If people leave from this forest, the animals will also disappear'.

An African ecology can also do much to clarify that the real problem, the real destructive glitch, is not human beings *per se*, but distinct human-created socio-economic institutions that foster unsustainable uses of the environment. Modern capitalist markets, one example of such institutions, interacting with a complex of other forces including technology and human (African as well as Western) greed, have been a key factor in destroying the relative balance that had existed between humans and the natural forest environment that supported them. One farmer shared with me a poignant example of how these 'new ways' penetrated and changed the relatively balanced systems of land use that had existed in the time of his father.

> *'This problem of poison in the waters: it came really only with this civilization of the Europeans. They have this poison to put in the soil next to the crops in order to kill pests, but crafty people have taken it and put it in the rivers and streams to kill fish. People took it for a good thing, but it is only ruining our waters, some is even killing people. These ways, they began to change ... well some of it is due to the whites, those who came to us; it was their knowledge that began to change our knowledge. We saw how much easier it was to get things with these bad ways, we see the ease and we jump into it and even though the rivers may be ruined, I get my fish and I sell it and I get wealthy.'* (Elanga 1995)

In destroying traditional resource use patterns, this commercialization of nature also succeeded and continues to succeed in destroying the natural ecosystems on which all of life, human and non-human, depends. Western conservation initiatives in Central Africa have tried to solve the human/non-human needs equation not by directly restraining these market forces as much as by establishing State (and in the minds of many of the villagers I talked with, 'American') control over vast areas of forest seen by local people as God's gift to them from which they can live. Given the history and philosophy of state-administered conservation in many Central African countries, such control ends up weighing the needs of the non-human ecosystem over the needs of people, not to mention the undeniable overture for state exploitation of the local population such control has always facilitated.

In short, commercialized use more than indigenous people's use of the forest lies at the root of Central Africa's environmental problems. We would do better to try and control the market forces that lead to over-exploitation of the environment rather than unjustly restrict the subsistence practices of people who have lived in these forests much longer than ourselves.

Thinking through nature in the Nepal Himalaya (Graham E. Clarke)

This paper is an account of populist ways of thinking through images of nature in the highland Himalaya of Nepal, in particular of the Yolmo region. It focuses on the indigenous understanding of plants and the use of land in agriculture as models for politics and society. The discussion uses examples, abstractions of images and metaphors from routine, daily comments and discussions of the local people, and indicates how ideas of the natural environment are extended outwards for use as general models of life, and as reasons for success and failure.

Structural anthropological theory has taken the conceptual distinction between 'nature' and 'culture' as a general feature of human modes of knowing. However, the social world of the villages of highland Nepal has ceremonies that specifically link the worlds of nature and culture. For example, in a highland Buddhist village of northern central Nepal, dance ceremonies always led on smoothly from the recitation of textual religious precepts. Often, one of the first dances started with the following origin myth recited in song:

If it were not for the sky and mountains there could be no clouds,
If it were not for the clouds there could be no rain,
If it were not for the rain there could be no streams,
If it were not for the streams there could be no grass (crops),
If it were not for the crops there could be no livestock (produce),
If it were not for the produce there could be no village,
If it were not for the village then we, people, could not be here.

In this example, the dependency of the world of man on nature is clear. This thought is stylized and prescribed, whereas our main concern here is on the use of the informal ideas of nature through which the mass of the population habitually thinks. Primarily, this is an account of the ways of explaining and arguing of people who are Buddhist or Hindu, rather than the cultural representations of these great religions in themselves.

For reasons of presentation these ideas are given in an idealized form; that is, a system of logic that illustrates a unitary whole. In practice, of course, the empirical order is more fragmented and neither natural metaphors nor the literate ideologies of the great religions are the sole mode of thought available to individuals in these regions. There is also, of course, the ideology of economic modernity, with all the attendant rhetoric of development and progress. Modernity implies cultural contact and change, and one of its key features is that it embodies doubt and half-truths rather than certainty and absolutism. Choice and individualism do not reinforce any such absolute perception, and the contrast between the traditional notion of change as distortion from the ideal (such as heresy), and the modern notion of change as progress or development, is a striking change in perspective.

The dominant world-view presented here is not the collective world-view of economic progress and modernity, nor that of the traditional great religions of Asia. Rather it is a routine, vernacular world, in which individuals think through their own daily experience using metaphor and allusions from popular culture.

Conscious, popular thought and reflection in Nepal concern explanations of daily experience and activity as well as the variation in success and failure of various enterprises, individuals and groups. On a day-to-day basis, the principal model available to farming peoples is that of biological, and in particular agricultural, processes, which are applied to individuals, politics and development as well as to many other domains.

One common example is the way in which people's characteristics may be likened to various types of animals, such as bears, foxes and monkeys. Another equally common idiom is the extension of biological terms beyond their literal sense, such as the use of 'to plant' as a common metaphor for to sow the seeds of a course of action. An illustration would be the farmer who, when asked who he was going to vote for in an election, replied as follows, 'When we plant seeds we may be sure that we will get some fruit; but will that fruit be sour or sweet? We have to wait and see'.

Such 'folk wisdom' is deep in a number of ways. It embodies a modern virtue of empiricism that is an open-mindedness, even a scepticism. Such natural metaphors are commonly applied to economic, bureaucratic and political processes. In the above example the context was an election, and in electoral campaigns the cultural norm is for a promise of material assistance against promises of votes; the speaker was himself signalling openness to such an offer.

The context of such poetics is implicit, rarely signalled overtly in everyday speech, and would be used by a speaker without him possessing any special skill in oral debate, or even literacy. Natural metaphors can be used in a highly sophisticated and creative manner by farmers. For example, one elaborate biological analogy for political activity came from a local political representative. He was non-literate, and not himself a candidate in that election, and was chatting over how local political activists came together to discuss things in secret at election time:

'If a tree can change its leaves every year then why shouldn't we change a candidate once every five years? [The difference is that] leaves come down one after another, and fall at their right time from on high without making any noise; but when they [the old politicians] come down, there is a noise, because each is cursing the one following them, the large one's (of yesterday) say to the small buds (of today): 'today it is my turn, so I fall down, saying, "tomorrow it will be your turn": that is why there is all that whispering during an election.'

This natural language has an implicit model of process, and provides a clear model of growth and decline. As such it may direct habitual thought, and predispose certain associations, as indicated in the following example.

'Those two have been exploiting and pressing-down on the people: it is all very well for those who have taken [literally eaten] thousands and thousands [of rupees], but how about those who have not even received tuppence?'

The term 'eaten' (*kannu*) has a wide extension in Nepalese and can be used for the consumption of almost any material other than alcohol. Material success here is seen in terms of feeding and growth to the extent that the term can sometimes be best translated as 'to consume' or 'to use up' rather than to eat; the extension to other domains is so normal that it has no necessary association to foodstuffs. Yet the underlying model is still biological and cyclical: as a baby suckles milk at the breast of its mother, so to grow in political stature a candidate has to eat.

'Milking the system' is one such expression that we still use in the West, but in highland Nepal the notion of milking of all processes is commonplace and has no negative moral connotation: that which in the West connotes corruption and immorality would here be seen as normal and natural. Moreover, this metaphor of milking and imbibing as a natural path for growth is extended to mechanical areas where, in an industrial society, it may appear not so much immoral as technically inappropriate and dangerous. For example, a helicopter stopped regularly in one village. One day the pilot allowed the members of the household where he was taking tea to take kerosene from the fuel tank. This was by pressing the fuel-dumping stopcock on the 'belly' of the helicopter, and the villagers later referred to this as 'milking the bird'. When next the helicopter came, the villagers gave the pilot tea, and many households sent a member down to 'milk the bird'. The idea that a 'bird' might need to 'eat' all its own 'milk' to fly home did not enter such a pattern of thought, and the course of action nearly led to the emptying of the fuel tank, and so possibly to disaster.

Such biological images of production carry over to success in secular activity. A region that has received assistance for economic development may be referred to in pastoral images of abundance such as an area 'flowing with milk and honey'. To us the idiom appears biblical, but to the farmer the idea of a reciprocal or direct return of the produce to the cultivator is almost a reflex thought. Overall, politicians redirect and redistribute the wealth, or 'fruit', to those who are their supporters. The dominant model of state-craft here is not the disinterested technocrat, allocating goods and factors of production according to utilitarian procedures to where they will produce the greatest good for the greatest number; rather it is a natural model of growth through feeding, with redistribution of fruit through processes like 'milking', conducted on the basis of personal relations.

The model also provides an explanation for failure and the lack of material fortune. As has been commented a number of times in the general anthropological literature, notions of impersonal or abstract forces or statistical chance are rarely seen as a sufficient explanation for individual events, and those close by, in particular outsiders who are present or those from one's own community, are likely to be seen as causal agents (Evans-Pritchard 1937). An explanation of good or bad fortune may take an abstract or supernatural form, as in magic and witchcraft; here it is viewed more prosaically as an immoral capturing or redirection of natural processes, as a personal agency responsible for the absence of the conditions for germination and growth.

For example, the following are comments from discussions on why no development assistance had come to a village:

'And why are we oppressed in this way today? We are pressed down, aren't we? And what is it that has pressed us down? Is it the earth? Is it the sky? Or is it people who have pressed us down? Let us have a look at this too!'

'Some say, "Just wait, and I will do something for you". Yes, some of our friends and brothers are like that; but later they say, "Oh, brother, I can't do it now". And in the future, this same person may take a knife to us, throw rocks at us, or knock our house down. The types who speak like that are our friends and brothers too!'

The terms 'press-down' and 'oppression' appear to us as a colourful and extreme idiom. Though an etymological link between the English concepts 'to press down' and 'oppression' is obvious enough, this association is not at the forefront of modern English usage where the normal idiom would be one of having been 'held-back'. Again, the Nepalese expression derives from the habitual associations of the idiom of thought, based on agricultural conditions of growth in which a plant that has been planted too deep (pressed down too far) will not germinate, will not rise-up and grow, will not flower or bear fruit.

There is a contrast and complementarity here, and explanations of success and failure circle around stories of who has 'risen up' and of who has 'pressed one down' (*dabaunu* or *thicnu*). 'To rise-up' (*utnu*) is the opposite of 'to force down' (*thicnu*), and has a complex of senses as in 'to wake-up', 'to be alert', 'to germinate', 'to come alive', and 'to rise in position', and 'to stand-up' both literally and 'to stand' for office in an election. The term is also used in the sense of to 'wake-up' as in the US political slogan 'Wake-up America'; to say that someone has 'woken-up' is to liken them to a seed that has germinated and is growing; by contrast, a person who is 'pressed down' is the opposite, like a seed that is dormant and does not germinate.

This particular pair of 'rise-up' and 'press down' illustrates the way in which vernacular speech follows such complementary ideas. Such pairing and complementarity is normal, and other relevant pairs here are 'up and down', 'big and small', 'give and take', 'to eat and dry-out'. In most cases, the one term corresponds to high rank and status and forms the positive, dominant term of a contrast, and there is an inferior complementary meaning coming from the contrast between the two.

In these ideas, it is natural for a politician on the way up in his career to be fed, that is to receive material gain and the rewards that allow him to acquire more power and grow further, to 'give and take' (*linu-dinu*) and become a 'big man'. This is a cyclical and finite process. One is fed, one swells up, rises and grows and bears fruit; but in this virtuous cycle, ultimately, as in the earlier example of the leaf above, one declines.

There is a converse pathway to this virtuous, natural, cycle of growth that comes from being fed and eating, and that is of being 'dried out' (*sukeko*), of decay and desiccation, rather than bearing fruit. Here the terms to 'dry-out' and 'pressed-down' are associated. For example, crops that have not been watered are termed *sukeko*, and a child who is stunted in growth is termed *sukeko*. There is also a close association between *sukeko* and the term for non-irrigated or 'dry' land, *pakho*. *Pakho* has the multiple senses in Nepal of land left at the side, land on a hill-side, and land that is barren or marginal. *Pakho* contrasts to the rich, irrigated flatter land at the valley-floor (*khet* or *bari*) which bears abundant crops. Most significantly, the term *pakho* is also used for the people who live on such dry land on the side: hence the sense of simple and rough, and backward and undeveloped, that attaches to people from poor, dried-out, hill-land, who can be known simply as *pakhey*, which has the connotation of a 'hill-billy'.

It is not only that those who have germinated and who are growing can be contrasted to those who have failed to grow and have dried out. As the language of kin, neighbour and community suggests, the functional link between the two also can be understood in terms of natural processes. This is especially the case if the two are physically close to each other. A large tree will have an extensive root system that draws off all the water from below, and a full canopy that prevents sunlight reaching down from above, thereby keeping other similar plants dry and cold so that they do not grow. In the same way, the rise into power of one relative, neighbour or adjacent community is thought to explain why others close by do not develop.

To this way of thinking, the proximity of one's own kind can be a precondition for damage. Hence the commonplace in rural Nepal of the assertion that those close by, such as brothers or neighbours, have kept one 'pressed down', have taken that which one needs. As it is with individuals and households, so it is with communities, and a community's failure to receive material benefits from the State will often be seen as the result of the success of a community close by.

For a person or community to be so dried out and 'exploited' (*sosak samanthe*) is the complement and opposite to another enjoying himself or themselves with material pleasures, 'imbibing goods' (*moj garnu*). There is an opposition, a complementarity, in the linkage between neighbours in this model of biological development, which forms an implicit, concrete model both of structure and of the processes of a progressive social differentiation. Complaints may be couched in this manner, namely that one group has 'eaten everything', from the moral standpoint of 'equality' and fairness. There is an emphasis on the importance of equality and redistribution in social relations that cannot easily be stated in images of nature; that is, of a cultural corrective, a natural tendency to inequality.

As a whole, the biological model may be seen in contrast to a religious model which presents an 'other-worldly' alternative. Other-worldliness is linked to ascetic ideas which, though obviously not part of the daily routine of a householder, are readily accessible in Nepalese and South Asian culture as an institutional whole. In the popular view such power comes from a denial or a renunciation of consumption, with images of self-mortification, pilgrimage, and Buddha fasting to enlightenment under a tree. However, this stands outside the life-cycle of rise and decline.

The power of religious purity, like that of the forest monk, comes from a divestment of the material possessions of this world, of formal office or the holding of goods. Such use engenders a contradiction, as 'this-worldly' power comes from an 'other-worldly' attitude; once the worldly path is returned to, there is a necessary decline, as with 'contamination' with this world, decay, and hence reduction of purity and power to act, must necessarily follow. To become involved is to enter a cycle that ultimately results in mortal decline. Overall, the notion of purity and sacred power contrasts with that of secular or worldly power, the one as transcendent and timeless, the other as a mortal cycle of life and death.

Other models occasionally come to the fore in Buddhist communities that imply that respect for life is a public value. The popular Buddhist doctrine of salvation, in which killing is proscribed, is used to back up the decision not to use pesticides on apple trees; but this should not be thought to exclude a 'rational choice'/ material self-interest explanation at the same time. The same Buddhist people are not averse to eating meat, only to killing animals, and it is routine to maintain a milk herd of yak crossbreeds by keeping young male calves away from the mother, so that they die 'naturally'. As with ourselves, there are many legalistic ways through moral dilemmas, and one must be careful not to take single-dimensional, ideological statements as sufficient explanations to cultural practices.

At the same time, the 'line' of decision shifts culturally. For example, I recall men who customarily left leeches on their feet, to have their fill of blood, before dropping-off 'naturally'. Our cultural classifications (such as extending the terms 'pest', 'vermin' and 'weed' to plants and animals we do not like) objectify, as does the extension of more abstract concepts such as 'labour' to our humanity itself. The cultural classifications and soteriology of popular Buddhism tend to extend sentience outwards into the animal world, and people can have the same moral dilemmas over insects and leeches as do we over foxes and deer.

The main account given here has been one of populist perceptions, the everyday logic among farming communities, the ways of thinking and arguing of people who are Buddhist or Hindu, in their everyday lives. Here a natural, developmental cycle of growth through feeding, of the production of fruit, that is reproduction, and of natural ageing and decline, that is decay, has been outlined. In describing these perceptions as 'natural' no simple dichotomy with 'culture' has been intended. If anything, the populist model of nature on which we have focused is the reverse, a form in which nature itself is taken as the model for culture, which is viewed as an aspect of nature. Yet people habitually perceive and argue from what they know on a daily basis, such as the natural processes that link land, animals and people, to more abstract ideas of society, including the state and economic activity.

This position is as distinct from that of Western phenomenology, or a High Buddhist theology in which all is an aspect of 'mind' or 'nothingness', as it is from a medium-level Buddhist theology or Western Judeo-Christian thought, both of which separate the worlds of the animate and the inanimate. Darwinist evolutionary theory extended the world of nature to incorporate the world of man; this popular Nepalese Himalayan model similarly views the human and social worlds as an extension of the processes and cycles of the natural worlds.

Stepping from the diagram: Australian Aboriginal cultural and spiritual values relating to biodiversity (David Bennett)

Australia is one of the twelve megadiverse countries, which cumulatively contain 60-70 percent of the world's biodiversity, and it is the only developed megadiverse country. The Australian continent's 7,682,300 square kilometres contain '80 biogeographic regions representing major environmental units' (SoE 1996: 4-23). These range from the wet tropics of the Cape York Peninsula in the north-east to the arid and semi-arid 'red centre'. There are alpine forests, moist open forests, dry open forests, open woodlands, native grasslands, wetlands, salt marshes, mangroves and deserts among the 80 types of bioregions.

Aboriginal people have inhabited Australia for at least 60,000 years. In that time the approximately 250 language groups developed intimate relationships with their particular bioregions or segments of the landscape. Australian Aboriginal peoples 'consequently had an affection for, and a feeling of oneness with, nature that few of the present-day generation of white Australians can even comprehend, let alone feel in their own hearts' (Strehlow 1950: 17). The conservation of nature and the landscape is for them not an issue separate from their own survival as individuals, as societies, or as cultures.

The Royal Commission into Aboriginal Deaths in Custody states in its discussion on the importance of land, 'Aboriginal attachment to land is non-transferable, and relates to the particular land itself, in a way which is connected with their religious beliefs and with the renewal of social life' (RCIADIC 1991: 473). Aboriginal attachment to land is non-transferable because for each group there are specific ceremonies, songs, legends and other cultural and spiritual ties that link a specific group to a specific 'country'.

While there are some strong similarities in cultural and spiritual practices among all the groups across the continent, of the 250 Aboriginal groups, each has its own specific cultural and spiritual ties to a particular country. For example, many groups have remarkably similar cosmological beliefs about the origin of their particular ancestral beings; nevertheless, and most emphatically, these resemblances can never be conflated into a single story of a single people.

To understand Aboriginal people and their relationships to their 'country', it is imperative to understand the complex interrelationships among living people, their country, their totems, and their ancestral beings as embodied in The Dreamtime or The Dreaming. The Dreamtime is both an active creation period, when the world was given shape, and the norms, values and ideals of Aboriginal people were fixed like the landscape, and a continuous passive time now 'existing alongside secular time but not identical with it' that reaffirms the norms, values and ideals set during the creation period (Berndt 1974: 8). In the creative phase the moral universe was established and fixed once-and-for-all-time. The Dreaming 'determines not only what life is but also what it can be' (Stanner 1979:29).

To generalize some common elements in different stories of Dreamtime genesis, the world was without form, but it was not void. Unlike the Judeo/Christian biblical Genesis, creation in The Dreaming is not *ex nihilo*, but a transformation and culmination of a formed but featureless world already existing. No explanation is given, nor thought needed for existence prior to The Dreaming. The Dreaming is not a theory of mystical explanation of moral origins, but an account of what Stanner calls 'the principle of assent to the disclosed terms of life', which governs the ritual behaviour and descriptive statements about the moral relations among people, their 'country' and other species. In other words, Aboriginal groups are not attempting to analyse why things are they way they are, but are accepting Dreamtime stories as demonstrations of the way things are.

The agents of change or transformation were 'ancestral beings'. Other names for ancestral beings are cultural heroes, deities, mythical beings, gods, spirit beings, or spirit ancestors. Ancestral beings (to call them but one name for convenience) are usually area specific. Their impact and significance varied regionally. Very few, if any, ancestral beings were known throughout the continent. Although a deed may be repeated at another location, each deed is always fixed to a specific geographical location. These ancestral beings had limited, or perhaps more accurately, specific powers. They were more like personified forces or superhumans to be co-operated with and to be used, than gods to be worshipped and to be entreated for their aid.

With the exception of some superhuman or supernatural powers the ancestral beings 'were like men and women of today in that they had similar thoughts, strivings and feelings, could be hurt and suffer pain, could age and in a certain physical sense die, although a part of them, a second soul, could not die. Otherwise they were free of the limitations, restrictions and inhibitions that affect today's men and women' (Stanner 1976: 5-6). Two of their superhuman powers were the creative capacity to transform and give definition to the world, and the ability to change their own shape. Their creative capacities extended over the physical and moral realms.

The complementary relationship of the landscape and the moral system provides a physical manifestation of the truth of the moral system, if such proof was needed, and it also gives a sense of the solidity, resilience and resistance to change of the moral system founded in The Dreamtime. The Dreamtime provides an aetiology of landscapes while the landscapes are empirical, irrefutable proof of the existence of The Dreamtime. The argument is circular, but stories of The Dreamtime are accepted as axiomatic and do not depend on empirical proof. The empirical 'proof' functions as a reassurance, not a guarantee.

In giving shape and definition to the landscape, the ancestral beings created the meaning of the country. Each Aboriginal group has many stories about the creation of their country or land by their ancestral beings. 'The actions of the spirit ancestors are recorded in stories which provide the moral and ethical framework by which Aboriginal people live on the land and relate to one another, in other words, their "Law"' (Rose 1995: 7). These stories connecting people to their 'country' may be conceived as 'highly complex systems of understanding, with the power to generate and determine social behaviour' (Schama 1995: 209).

From an Aboriginal point of view, land and the ancestral beings particular to that land are inseparable. Aboriginal peoples hold that there is a direct connection between themselves and their ancestral beings, and because they hold that their country and their ancestral beings are inseparable, they hold that there is a direct connection between themselves and their country. 'The place from which a person's spirit comes is his or her Dreaming-place, and the person is an incarnation of the ancestor who made the place. A person's Dreaming provides the basic source of his or her identity.' (Myers 1986: 50) Chairperson of the Northern Land Council, James Galarrwuy Yunupingu states, 'My land is mine only because I came in spirit from that land, and so did my ancestors of the same land. My land is my foundation' (Galarrwuy Yunupingu 1995).

Within every country there are locations, often called Dreaming Sites, where ancestral beings have left a mark of their passage, 'evidence' of some event or act, or in the case of 'increase sites' the point at which they 'went-down'. When ancestral beings finished their wanderings, which gave definition to the face of the earth, they would 'go-down' marking the end of the creative phase of their existence. The spot at which they 'go-down' is called an 'increase site'. It is the place where the ancestral beings have continued their spiritual existence since the active, creation period of The Dreaming. Within these sites, Aboriginal people believe, there are life-forces, or the potential for all life – human, animal and plant. But each site is specific to a given ancestral being, so the life-force is specific to a given totem. For example, if in Aranda country there is a waterhole that is the site of Karora, the bandicoot, then that is the source for bandicoots and humans of the bandicoot totem. It is not a site, for example, of green parrots or red flying-foxes.

It is easy to understand why country is so important and is non-transferable. Events took place at specific locations. People are connected to these places through these events. They have very strong spiritual and cultural connections to their 'country'. Each person with a connection to a site of significance has a responsibility for maintaining that site. The effect is to parcel out the 'country'. No one owns the land in a fee simple sense, but the people of that land have a collective right to use the land in a way similar to the 'commons' of Britain. And those who use the land have a collective, but distributed, responsibility to protect, sustainably manage and maintain their 'country'.

Native title

The High Court of Australia in its decision of 2 June 1992 on Mabo vs. Queensland (No 2) (1992) 175 CLR 1 rewrote history by overthrowing the doctrine of *terra nullius*. *Terra nullius*, meaning land belonging to no one, was the legal fiction that in 1788 *Terra Australis* (Australia) was a wilderness, neither owned nor affected by humans, and the people were mere incidentals on the landscape. Or as Noel Pearson, former Director of the Cape York Land Council, has stated so succinctly, 'legally invisible vermin'. The most salient point about native title is that the identity of Aboriginal and Torres Strait Islander people comes from their association with country. The importance of the demise of *terra nullius* is the recognition of the existence of Aboriginal and Torres Strait Islander people in addition to the recognition that the land was owned. Their very identity is bound up with their land. Without attachment to their land, they are not merely dispossessed, they are without identity.

With land comes meaning. With land comes land management. Land management for Aboriginal people is a complex of traditional ideas and attachment to land, an understanding of contemporary environmental problems, a desire for self-management and an awareness that all of these are factors are interrelated. The result of these activities for Aborigines is 'country; land cared for, known, named and managed on a sustainable basis by its owners' (Head 1992: 52).

Native title clearly has environmental ramifications. Subsequent to the Mabo decision, the Federal Government passed the Native Title Act 1993 to clarify the meaning of native title. Section 223(1) of the Act states:

> 'The expression 'native title' or 'native title rights and interests' means the communal, group or individual rights and interests of Aboriginal peoples or Torres Strait Islanders in relation to land or waters, where:
> (a) the rights and interests are possessed under the traditional laws acknowledged, and the traditional customs observed, by the Aboriginal peoples or Torres Strait Islanders; and
> (b) the Aboriginal peoples or Torres Strait Islanders, by those laws and customs, have a connection with the land or waters; and
> (c) the rights and interests are recognized by the common law of Australia.'

Section 223(2) clarifies section 223(1) by specifically mentioning that 'Without limiting subsection (1), 'rights and interests' in that subsection includes hunting, gathering, or fishing, rights and interests'. For instance, the Kowanyama and Malanbarra communities of Cape York Peninsula, who manage their traditional lands (although not acquired under native title), see land management and self-management as a complex of rights, including:

1. the right to share in natural resources as an assurance to a social and economic future;

2. the right to responsibility for management of those resources;
3. the right to a chosen life-style, and
4. the right to a clean and healthy environment. (Daphney and Royee 1992: 43)

The way these rights are practised has caused considerable controversy. Some contend that traditional or customary practice includes the use of traditional implements and excludes the use of motor vehicles, while others contend that traditional or customary practice means following those practices but does not dictate the implements used or excluded. Sharing natural resources, managing those resources, choosing a life-style, and maintaining and promoting a clean and healthy environment, does not demand that the land management techniques are pickled in time. The point is that whatever technique is used, it must be used according to customary or traditional practice, not that land management is limited strictly to traditional implements.

As Sweeney comments: 'While the existence of Aboriginal rights is to be ascertained as at the date of the acquisition of sovereignty, the means of exercising those rights are not limited to the means utilized at that time. In particular, the use of present-day tools in the harvesting of plants, modern transport and firearms in hunting animals, boats and nets made of present-day materials in fishing, still comprise the exercise of a traditional right, albeit in a modern way. To hold otherwise would be to commit Aboriginal peoples to a living archaeological museum' (Sweeney 1993: 115-116).

'The implements used for hunting, fishing and gathering in accordance with traditional rights are not frozen in time, and indigenous people may use modern implements to carry out their traditional practices' (Bennett 1996: 7). For example, the Mowanjum community near Derby, West Australia, utilize offshore islands, reaching them with dinghies and outboard motors. At least some of the reasons for using the islands relate 'to renewing ties with traditional country' and 'the way that the islands are used today is probably exactly the same way they were used before the advent of aluminium boats' (Kohen 1995: 21). It would be foolish or in Mick Dodson's term 'essentialist' to believe that no Aboriginal person would cross the line from land management to land exploitation. Yet, Aboriginal people have more to lose from 'long-term degradation to the detriment of the environment', because their attachment to the land is spiritual as well as economic, social and environmental and it 'provides the basic source of his or her identity'.

'Country is sustained as a lawful place, through the conscious participation of living things' (Rose 1996: 44). The geography and special features of the land are marks of ancestral beings' activities. 'For almost all of the landscape (hills, water holes, and so on), the country... takes its name from the Dreaming, either from the event or from the associated rituals and songs' (Myers 1986: 50). The country is not self-sustaining. 'Many Aboriginal people believe that the landscape and its resources are conserved largely through ritual means rather than by managing the countryside in a practical sense' (White and Meehan 1993: 38).

Since the Mabo decision in 1992, the importance of the spiritual and cultural values of the environment and biodiversity have become better accepted and higher profile. 'The cultural value of biodiversity conservation for present and future generations is an important reason for conserving it today. Human cultures co-evolve with their environment, and the conservation of biological diversity can be important for cultural identity throughout Australia.' (Biodiversity Unit 1993: 21) At least some Australian national policies and federal legislation recognize that Aboriginal spiritual values and traditional ecological knowledge have a role to play in maintaining the environment.

Nature in songs, songs in nature: texts from Siberut, West Sumatra, Indonesia (Reimar Schefold and Gerard Persoon)

The Mentawai archipelago forms part of a chain of non-volcanic islands running parallel to Sumatra about 100km off the west coast. There are four large inhabited islands in the group: Siberut, Sipora, North Pagai and South Pagai. Until recently they were covered with dense tropical rainforest. Siberut is the largest and most traditional of the islands, with a population of about 23,000 native Mentawaians and 2,000 ethnic 'strangers', mainly of Minangkabau origin.

The island of Siberut is an elongated rectangle measuring about 110km by 50km. The rainforest landscape is quite hilly – the highest elevation being 384 metres. These hills are interrupted by wide valleys, and the eastern slopes facing Sumatra drop gently to the sea. Along the coast, white coral beaches alternate with mangrove swamps and muddy rivers. The island's two main administrative centres are here: Muara Sikabaluan in the north and Muara Siberut in the south. The western shore forms the extreme edge of the Asian continental shelf – dropping steeply beneath a giant surf which rolls across from Africa.

Until quite recently, the people who live here retained many ancient and complex indigenous traditions. Nowadays most Mentawaians are Christians: however, communities that practise the old religion do still exist on Siberut: they live in the interior of the island and stay away from the government-controlled villages along the coast.

The Mentawaians are descended from some of the earliest groups of peoples to have migrated to Indonesia several thousand years ago. They have preserved Neolithic Austronesian traditions with some Bronze Age elements which have remained virtually unaffected by subsequent Hindu, Buddhist and Islamic cultural influences. Nowhere else in Indonesia has this archaic heritage been preserved so well.

The Mentawai people have neither chiefs nor slaves. Traditional society is organized in clan groups consisting of about ten families living in large communal houses, called *uma*, built at irregular intervals along the rivers. The inhabitants live off the land, cultivating sago, yams, tubers and bananas. Their form of agriculture is different from many other forest-dwelling people as they do not use fire after cutting down the trees. Gradually, and without erosion or loss of soil fertility, they replace the original forest by one dominated by fruit trees. They also raise chickens and pigs and the men go out hunting and the women do the fishing. There is no economic specialization: every man is a canoe builder, a pig farmer, a bow and arrow maker and a vegetable gardener, and every woman is equally capable of performing all the tasks assigned to the women.

Because of its isolation, Siberut stayed out of the mainstream of political events for a very long time. Only at the beginning of this century did external influences begin to step in. When they did so, however, their impact was significant. Under the direction of European missionaries and colonial administrators, head-hunting and traditional religious ceremonies were forbidden.

The pace of change speeded up after Indonesian independence, when the indigenous religion was banned altogether and everyone had to embrace either Islam or Christianity. Punishment for the practice of the old religion was severe, and ritual objects were burned. The government further insisted that communal houses be constructed around a village church and school. Men were forbidden from wearing long hair, glass bead decorations and loin cloths, and teeth filing and tattooing were also prohibited.

The majority of the Mentawai people became Christians at this time. They learned German Protestant or Italian Catholic church songs. They built simple single-family dwellings, dressed in the modern style, and complemented their subsistence economy with products for trade like tobacco and iron-ware. But under this veneer of conformity the old beliefs lived on, for example in healing ceremonies, such as the *pangurei* 'wedding celebration', in which a sick man is dressed up as bridegroom to attract his soul back to him.

Large-scale logging operations have greatly changed the amount and quality of the rainforest over the past few decades. Since 1993, however, the logging concessions have been cancelled and about half of the island has now been turned into a National Park. For a long time already nature conservationists have taken a great interest in Siberut because of its biodiversity. As a relatively small island, it has a high degree of animal and plant endemism. The four endemic primate species are of particular importance (Whitten 1980; WWF 1980).

Traditional Mentawai beliefs hold that creation is a harmonious whole. There are spirits everywhere, and all elements in the natural environment – plants, animals and natural forces – are thought to have souls. Because they are invisible, man's activities can unwittingly disturb them. For example, a tree that is cut might fall on a forest spirit's dwelling. Before beginning any major activity, therefore, ceremonies are held to please and placate the spirits.

Everything – man and beast, plant and rock – has not just a soul, but also a personality. The few implements that the Mentawaians possess are made with special care to match the external form with its inner essence. Compatibility is important: a man may possess an excellent hunting bow, but unless it 'gets on' with him, he may frequently miss his target. Religious practices are directed at restoring harmony between man and the natural and man-made environments in which he lives.

In the traditional longhouses, or *uma*, each family has its own section. In addition, the families have field houses some distance away where they raise pigs and chickens, cultivate tubers and bananas, gather sago and go fishing. Today, only a few of the large communal dwellings remain: most were abandoned as a result of the policy of bringing islanders into the mainstream of modern Indonesian life. In some areas in the south, however, people have built *uma* in new villages, and these convey some idea of their former state.

An *uma* is a village under one roof. Floor plan and usage reflect a strong community spirit. A notched trunk serves as a stairway to the front verandah. Skulls of pigs and game (monkeys, deer and wild boars), some of them carved and decorated, hang from the rafters. This verandah is the living room during the day, but at night it becomes a sleeping place for men and guests. The first inner room has a communal hearth used for ceremonial events: here dances are held. In a second room one can find the main fetish of the community, a bundle of holy plants. It is here that the women and small children sleep.

Music and songs

The music culture of the people of Siberut consists of a number of elements. Apart from the songs, some of which are accompanied by drums, there are also the slit drums, the main purpose of which is to send messages about important events like a successful hunt or the passing away of an *uma*-member. These slit drums (*tuddukat*) consist of sets of three or four drums of various lengths and, accordingly, pitch, and the messages can be heard over large distances. The drums are also used to invite people to come to the *uma*. Messages of successful hunts are used to impress neighbouring *uma* and the beating of the drums and the blowing of the triton shell can sometimes go on for hours (Schefold 1973).

In the forest fields people also use sets of three or four bamboos or pieces of solid wood of varying size as a kind of xylophone (*boboman*). These are beaten with two sticks just for entertainment. Cylindrical drums (*kejeuba*), made of stems of a particular palm tree and covered with the skin of a deer, snake or monitor lizard, are used to accompany the dancing. The only true melody instrument is a bamboo flute (*pipiau*) which is rarely heard. It is used mainly during periods of mourning.

Songs (*urai*) are by far the most important element in the music culture of the people of Siberut. There are various types of song texts. First there are the *urai silainge*, literally 'youths songs': songs about daily matters, about social relations, about love and love affairs. Frequently, there are special events which inspire people to make a song. These can include almost anything: the arrival of a new trader or teacher in the village, the sight of a beautiful animal, the activities of a logging company, or the misbehaviour of one of the *uma*-members. In principle anybody can make a song, but once the song is there, it can be sung by anyone, irrespective of age, gender and status. Songs are learned through careful listening and repetition, and only spread through oral transmission. If the theme of a song as well as its words and melody are appealing, they can spread rapidly across the island. In this way songs often accompany rumours and gossip. All of these songs basically tell a story or give a point of view as a kind of reflection on a particular event. In some cases different song texts are used with the same melody.

The lyrics and the melody are made by an individual person. They are usually sung during evening times while smoking, chatting or just 'sitting in the wind' on the verandah of the communal house. But they may also be sung while canoeing on the river or working in the forest fields or while minding the pigs. Other people attracted by a song might pick up the text and add it to their own repertoire.

There are no professional singers on the island. Just as with all other skills and capacities, everybody learns to sing, but some men and women are certainly better than others and they might take a greater pleasure in singing. Nor, necessarily, do good singers also make good songs. Some of the people with good voices acknowledge that they have not made a single song. Their repertoire consists of songs made by others.

Another category of songs, the *urai turu* ('dance songs') are sung while dancing. One of the dancers sings along while the drums are played, and he and the other dancers move around on the dance floor in a circular movement, rhythmically stamping on the boards and thereby joining in with the beat of the drums. The songs are often about animals and animal behaviour in which the primates and birds take a prominent place. Dancing is done by both men and women. Together with the beating of the drums, the stamping of the feet makes the dancing a very inciting event which sometimes leads into a trance. Someone takes the lead in the dancing movement while the others follow making identical movements. Depending on the theme of the dance there can also be two parties, for instance monkeys teasing another animal.

A third category of songs consists of the *urai kerei*, the songs of the medicine men, or *kerei*. They are complex because they are phrased in a special language of which many words are not known to the general audience. These songs are basically a medium for the medicine men to communicate with the spirits and not, as with the other songs, to tell a story or to communicate a feeling of joy, love or fear.

Some of the *kerei* songs are used while the medicine men collect plants with magical powers in the forest or while they prepare the medicines for rituals. These plants form a special case in the Mentawaians' involvement with their natural surroundings. Their applicability to a specific curative or ritual goal is always derived from certain morphological characteristics. One of hundreds of examples is a shrub with very hard wood. Its name, correspondingly, is 'Hard Interior' (*kela baga*). Parts from this shrub can be applied in ceremonies to 'harden' participants against illness and evil influences. In the accompanying songs the spirits of the plants are addressed to perform their particular healing or protecting task. When the medicine men go out to collect such plants, they document a comprehensive knowledge of the natural world which draws meaning from even the apparently most insignificant herbs: they move through the forest as if through an agglomeration of helpful spirits.

Many *kerei*-songs are part of the communal rituals of the group. They are aimed at expelling evil spirits from the house, at attracting good forces, or at the enhancement of the offering of sacrificial animals. In healing ceremonies and in ritual, the singing is often accompanied by the ringing of little bells. In performing their tasks, the medicine men are dressed up with glass beads, flowers and other adornments to please and impress the spirits.

Learning these songs by heart is done through imitation and repetition. Young and old medicine men as well as novices get together in order to practise their singing and to learn new songs. They usually do so during evening time. Many songs do not have clear rhythmical patterns: it is up to the singer to decide how long he will hold the different tones of a melody. When singing in groups, someone, with the best voice or the one who knows the lyrics of the song best, takes the lead while the others join in, following the melodic lines ('cueing'). Frequently, runs in the highest falsetto are inserted into a melody, designed to create a magical mood (Yampolski 1995).

Nature in its widest sense is very important in these songs. In all three categories nature is frequently used in a metaphorical sense, or elements of the landscape (mountains, islands, rivers, sun, moon, stars, rain or rainbow) are used as markers for time and place. Physical characteristics of animals or animal behaviour are often referred to. In a song directed to her mother a girl describes her beloved one: 'gently moving and with a waist like a *joja* (Mentawaian langur), while look at me, I am just like a *bulukbuk* (big basket), my waist is as big as my chest'. Also interesting is the interaction between different categories of animals, and between animals and humans. From all these texts it becomes clear that people are impressed and fascinated by the natural world in general and the animal world in particular. The texts reveal a great deal of knowledge about this natural world. Singing is not limited to human beings. Animals can sing too, and the Mentawaians believe that there is also a special meaning to this animal singing – just as with the singing of humans: for example the singing of the gibbon (*bilou*) in the early morning is said to be calling for the sun to come out.

Song texts

Here we give some examples of song texts that illustrate the way in which various types of songs reflect themes derived from nature. They are just a few examples of the hundreds of songs that are sung on the island.

Gibbon (*urai turu*)

After a hot day, the evenings in Mentawai bring refreshing coolness. This lasts until the early hours, when the chants of the gibbon (*bilou*) – which accompany the grey dawn – are slowly dying away in the forest and the misty veils from the valley are rising along the hillsides slowly to dissolve in the blue sky. There is a dance song reflecting this mood: it reports on a 'song in nature'.

> The *Bilou* from the mountain ridge, with sorrow in his heart,
> when the sun declines,
> climbs up into the *Eilagat* tree.
> In the fogs of the forest I can hear him,
> the *Bilou* from the mountain ridge, the *Bilou* who is lonely,
> far up in the tree top,
> *ko-a-ii*.
>
> He soars up into the *Eilagat*-top,
> and waits for the sun, the *Bilou* from the mountain ridge.
> Far above him the child of the eagle is sailing along,
> and he calls for the sun which rises on the bright sky,
> the *Bilou* from the mountain ridge is calling his sun,
> he is calling for the rays of sunshine,
> *ko-a-ii*.

Sea gull (*urai silainge*)

The mouths of the main rivers on Siberut are rich fishing grounds, not only for people but also for birds.

> The little sea gull at the mouth of the river
> is waiting for a floating tree,
> on which he will be fighting,
> with other sea gulls.
> A tree, to be used as a canoe.
> Looking for food on top of the waves,
> little sea gull, enjoying yourself
> in the middle of the beach.
> As soon as the people come to look for fish, you take off.
> You fly away and shout: *kri-kri*!
> You will be fighting again on the coral reef,
> near the mouth of the river.

Eagle (*urai turu*)

Along the coast, sea eagles hunt for fish. Once in a while they rest in the mangrove trees. People are impressed by these mighty birds though they are also scared by them while they are fishing in their little canoes. This song was made after somebody was drowned at sea.

> You, eagle, you go to a place to rest.
> You, eagle, out at sea, you rest on the waves,
> the waves of the sea.
> But my father rests in the sea.
> He was taken by a wave.
>
> You scare people who are fishing out at sea.
> You are also out while it is raining.
> You fly high and you dive in the clear sea.
> You look for food along the sea shore.
> You can also overlook the forest, you eagle.

Forest hut (*urai silainge*)

The only place for real privacy and tranquillity for a young couple is a hut in the forest fields at some distance from the *uma* settlement. But inside the forest there are also other living beings which are not co-operative or which disturb the couple.

> I am walking along a forest path.
> I am looking for a red chicken, my love.
> I call him, but he does not come.
> What can I do, my love?
> Tonight you will be sleeping on my field.
> In my field hut, my love,
> which I have already cleaned.
> It is made of rattan and leafs, my love.
> Let's look for *durians*,
> on top of the mountain.
> I search for fruits but it is a pity, my love.
> Squirrels have eaten all the fruits.
> So I do not find them my love.
> I only see beetles, please don't tease us!
> What can I do about it, what can I do?
> Let us go home, my love, let us go home!

Sailing (*urai silainge*)

The trip from Siberut to the island of Sipora in the south, a distance of some 30km through open sea in a dugout canoe, is a big adventure for people of the island who usually stay close to the shore. The trip is seldom made because of the big waves and strong currents between the islands. No wonder the trip inspired one of the men to compose a song which is sung with a great deal of tension in the voice.

> Come, my friend the wind, come.
> We put up our sail.
> Come a little stronger from the east,
> and blow into our sail.
> It is getting dark already.
> The sea is turning black.
> But do not come too strong.
> We will be scared.
> We roll on the waves,
> and we row with our paddles.
> Come slowly friend, come.
> We return home like a shell,
> drifting on the waves.
>
> The sun is down,
> and there is no wind,
> which will bring us to the other side.
> Behind us is Pananggalat.
> We are now heading for Simaritu.
> There is only little wind.
> Come, friend, come, but come slowly.
>
> Take care, take care!
> We have to go straight.
> Take care that the waves don't eat us.
> Take care, but don't be afraid!
> Don't be sad.
> Shortly we will arrive at a beautiful island.
> We will not be hungry any more.
>
> Look, the wind is coming!
> It is getting stronger.
> It is blowing our sail.
> We ride on the back of the waves
> (in between two big waves).
> I don't know where we are going.
> I can see no island.
> Let us follow the birds,
> assuming they return home.
>
> But we don't know where they are going.
> Do not be sad.
> Shortly we will be eating and drinking.
> Let us ask for a gentle wind,
> so that we can move safely.
> You in front and you at the back,
> Take good care!

Sacrificial pig (*urai kerei*)

During the religious feasts of the *uma*, pigs are sacrificed to the ancestors on various occasions. The most important of these is the *paeru*, literally, 'making good', which is designed to ritually prepare the pig (*eruket*, 'the one who is to make good') for the offering.

The medicine men (*kerei*) assemble in the house in a semi-circle in front of the pig (which is bound to a carrying pole) and address it in a rhythmical chant, ringing their bells in accompaniment. The song, starting with the cry *'pánorá'*, and with an evocation of the adornment of a medicine man, presents an associative stream of images which call up the living space of the people, from the sea shore right up to the longhouse far inland. Besides a few interjections, it has no direct relation to the sacrificial pig. At the end of the chant, the *kerei* compare the bent climbers they sing of with the arched leaves they themselves hold in their hands. The medicine men, too, are 'arched' as if from very old age, i.e. sage and experienced.

> *Kerei, kerei, pánorá,* hey *kerei!*
>
> A ripe-coloured glass bead,
> one of the equally formed, the equally formed beads
> with the equally formed, the equally formed naval openings
> on the necklace string made from the stripes of the rattan liana
> which grows where the forest is swampy,
> with its clinging thorn claws
> with its round arched leaves.
>
> To the string belong also the threads of the barkbast tree.
> The women of my *uma* twine you.
> If its leaf falls, it turns fluttering on the mountain slope.
> They twine and they sing close by on the verandah.
>
> To our adornment belongs also the tortoise shell which never shatters,
> of the four-paddled tortoise
> whose brothers paddle in the sea with the eightfold breakers.
> They break over the pebbles with the hard surface,
> the tones then ringing over the long beach
> on whose back the snails lie
> with their spiral countenances.
>
> On the bank stand the *loigigi* plants with eight sharp thorns
> and on the spit mangroves of equally high foliage.
> Who is there in their tops? Only I, the maiden, the seagull of the island
> with yellow-round eyes, with shrieking voice,
> over the waves trickling away.
>
> To them belong also the stilt-root mangroves
> on the eight punting stakes,
> punting through the fog shreds, the fog shreds of the sky,
> to you, my *eruket* pig.
> To them belongs also the *ogaga* tree with jackfruit like fruits,
> the goal of our young brother, the creeping cat,
> at night its voice wails under the silent sky
> to you, my *eruket* pig.
>
> Let us enter into the stream with the many branches, with the many obstacles.
> Quickly I arrive at the notched log-stairway
> which is cut off level at the top, after the eight indented steps.
> Who is there on the verandah?
> Only I, the boy, the pointed-tooth boar
> bound to the carrying pole, the carrying pole of wood.
>
> Let us go further to the coconut trunk,
> it grows next to the wide landing platform,
> trunk with the thief-proofing thorns.
> Sky high sits the owl with the clinging claws.
> To it belongs also the *simakainau* bush with the widely-branched shoots.
> With the rainwater-cool blossoms,
> cool and slippery,
> slippery as my hand which squeezes them
> over you, my *eruket* pig.
>
> *Kerei, kerei, pánorá,* hey, *kerei!*
>
> To that belongs also the *mairup* tree
> with its hibiscus-like blossoms.
> They are the goal of our young brother the humming-bird
> who turns his head from side to side.
> As climbers it has the yam with the many husked tubers
> whose leaf buds extend and clamber up the *lakoba* tree
> with boughs like cross-beams.
>
> The *potse* liana embraces it, the forest *potse*,
> with the reddish skin and blossoms like leaf veins,
> forest leaf veins, thrice wound, thrice bent.
> Who is there on the top?

Only I, the boy, the eagle with yellow-round eyes.
He likes it, travelling there like a dugout,
to travel there like a small dugout with hollowed out insides,
his arms like paddles, with double bended oar blades, with double sides.
He bathes in the sky in the rain while the sun shines
for you, my *eruket* pig.
Pánorá!

Kerei, kerei, pánorá, hey, *kerei!*

To that belongs also the *toktuk* tree
with a thick bark, with a thick core,
for you, my *eruket* pig.
Pánorá!

Kerei, kerei, pánorá, hey, *kerei!*

To that belongs also the *bairabbi* tree
with the red shining fruit, goal of our young brother
the flying fox. Come hither,
he flutters to and fro like his image
on top of the feast pole made from bamboo,
the top of which bends down
adorned with *pelekag* flowers,
the sweet smelling ones.
Sweet smells the deep-hollow longhouse
for you, my *eruket* pig.
Pánorá!

Kerei, kerei, pánorá, hey, *kerei!*

To that belongs also the *toilat* tree.
Who is on its top?
Only I, the woman, the hawk with the curved-in face
paddling near to the sky
to you, my *eruket* pig.

To that belongs also the *lemurat* tree with the double-tip fruit,
they punch through the early mist of the sky
for you, my *eruket* pig.
Pánorá!

Kerei, kerei, pánorá, hey, *kerei!*

Let us climb up to the coolplant ground,
cool and slippery, slippery like my hand.
To that belong also the *tobe* plants
they grow criss-cross through one another
with firm-hooking creepers, with hibiscus-like blossoms.
Their companion is the *toimiang* bamboo,
bent like the thorny *laibi* climbers,
climbers like the leaves we now hold, we the *kerei*!
Ready-ho!

Conclusion

These song texts show that nature in its broadest sense is an important idiom or medium for the people of Siberut to express themselves in. Nature is not just a collection of resources ready to be utilized. The whole environment – animals, plants, trees, rocks and mountains as well as natural forces like wind and thunder – provides an inexhaustible reservoir for depicting human emotions and human relations.

However, this is only part of the Mentawaians' interpretation of their environment. Although it is not within the scope of this paper, it should at least be mentioned at this point that seen from another angle nature is in the first instance a religious category. The Mentawai world is full of spiritual powers, powers with which humans have to establish and maintain good relations. These powers are a source of support and protection and can be addressed for magical means. They also bring about dangers which threaten the well-being of the group and its individual members.

This other view of the environment, not as a source of metaphors for this world, but as a spiritual world of its own, forms the second aspect of the Mentawaians' conception of nature. In this view the seeming wilderness of the forest appears as a mask that conceals what in reality is another cultural domain, a 'culture of the beyond'. It is a domain of spirits and ancestors who are related to the living and who lead a similar existence, albeit some of them a more beautiful one and of course always one evoking a life without death. There are many *kerei*-songs referring to this other reality. However, since the purpose of these songs is to paraphrase 'culture' rather than 'nature', we have not included them in our collection.

To give just one brief example: The *Bilou*-ape which in our first song was depicted as calling up a mood of human loneliness and nostalgia, appears in ritual chants as a dangerous companion of evil spirits, ghostly creatures which inhabit the otherworldly domain and which just like enemies in the human world try to harm people entering their realm. In a curing song to heal a hunter who upon return from the forest was struck with disease, the *Bilou* is addressed by the *kerei*, among several other threatening spiritual beings of the jungle, with the stanza:

> Come here, all of you with the scratching hands, we cool down your radiation.
> You, *Bilou*, come, come, we are on the search for the cause of the illness of your grandchild, *ko-a-ii.*
> You, who is mocking us from above on our search for the cause of the illness of your grandchild, come,
> *ko-a-ii.*

Songs and song texts do not necessarily reflect in a straightforward manner the 'real' world of actions and daily life. People on Siberut have gone through rapid processes of change including missionary work, large-scale logging, incorporation into a market economy, and, during the last few decades, ecotourism. To some extent these forces have affected their world-view, life-style and range of material possessions, including also their hunting devices. This has certainly widened the gap between views and beliefs as expressed in rituals or texts of myths and songs on the one hand and patterns of activities related to the environment on the other. In other words, there may be symbolic continuity in spite of ecological or economic change.

The changes have also given rise to new songs focusing on these modern elements. But the past has also taught that many of these changes come and go without lasting effects. The material wealth, for instance, which came after the *gaharu* boom (see below) in the late eighties, disappeared within a few years. Basically people fell back on familiar ways and possessions. Traditional culture was repressed during some periods in the past but regained some of its strength once outside pressure was released. To a certain extent these changes can be temporary, but they help to redefine, in their own manner, the relations between humans and between humans and nature. In that respect it is important to note that the people of Siberut themselves hang on to the symbolic expressions of biodiversity and its relevance for the world of humans as well as for the spiritual world contained in its 'natural' manifestation.

[*Gaharu* is a highly valued non-timber forest product. It is produced in the core wood of the *Aquilaria* tree as a result of fungus infection. *Gaharu* is used in the pharmaceutical industry in the Middle East. In the late eighties traders offered large sums of money or products in kind to Mentawaians for this infected wood. Within a few years the supply was exhausted and the economic boom was over.]

Sacred balance (Aroha Te Pareake Mead)

'Putauaki (Mt. Edgecumb) is my sacred mountain, Rangitaiki is my sacred river. Ngati Awa is my tribe, Ngati Pahipoto my sub-tribe, and Kokohinau my meeting place. Mataatua is the canoe whose genealogy binds me to my tribe, the nine tribes of my region and the other tribes of the North whose genealogy stems from the same ancestors as mine. These are what give identity and standing in the world and enable me to acknowledge you.'

Customary Maori greeting

According to Maori, the creation of all living things on earth is a great epic love story, one in which Ranginui (Sky Father) and Papatuanuku (Earth Mother) sacrificed their deep love for each other and separated, because in their earthly embrace they were suffocating their children. Ranginui was pushed into the sky in order that their children could have air, light and space and be able to move freely and grow between them. Their separation was a heart-wrenching and violent one and it is recorded that the children had to cut off the arms of Ranginui because he would not let go of Papatuanuku. The form of Papatuanuku is female with her head lying face down to the east, her legs to the west. When Ranginui was finally positioned in the sky, where he remains today, the blood from his arms dripped down on to Papatuanuku, and hence the *horu* [red oxide of iron, used to paint houses, canoes and the face], and the *pukepoto* [blue phosphate of iron] that his descendants use in painting.

Initially, Papatuanuku faced upwards, towards her husband, but such was the impact of the sorrow of Papatuanku and Ranginui, that his tears throughout the day and night caused heavy rain and snow storms, and her tears caused clouds and mists. Combined, they both muffled the light and stifled the air. Their children still could not see and so Papatuanuku's head was turned to face the ground away from the vision of Ranginui. This is why the inner earth still trembles with earthquakes.

The seventy children of Ranginui and Papatuanuku were all male and are the ancestors that we still acknowledge today, such as Tanemahuta (god of the forests) and Tangaroa (god of the oceans). Every living being, in its smallest microbial form, is descended from Ranginui and Paptuanuku. Far from this being a vague notion, Maori genealogy (*whakapapa*) is very detailed. Every species of marine life, every plant and animal endemic to this region, can be traced back to Ranginui and Papatuanuku. Maori traditional knowledge of both the genealogy and properties of species and natural conditions is equivalently detailed.

My non-Maori colleagues often assume that as an educated indigenous professional I must consign this particular aspect of my cultural heritage to a lesser status than the accepted European Darwinian cultural explanation. Well, for the record, I do not. To my mind, the Maori explanation of creation and evolution teaches me all I need to know to understand my role in life and attitude towards nature. My heritage teaches me about concepts such as the integrity and inter-dependency of living things. It makes me quite comfortable with the notion that as a human being I am but one part of a whole and that my generation is also simply one strand in the rope of humanity. It pre-determines that the relationship I have with nature is based on kinship and respect and that in order for me to survive in a culturally rich way, I depend on the survival of others, not just other humans, but also plants and animals, in the sea as well as on the land. It clarifies that both male and female elements are necessary to create and sustain life, be it human, plant or animal. It provides me with a proven framework from which I can analyse and identify risks and benefits to the well-being of all those areas that form my cultural heritage and which encourages me to accept responsibility that in my lifetime I will not contribute to, or allow others to cause, any diminishment to the cultural heritage of my ancestors and descendants, including those yet unborn.

'Biological diversity', which is not translatable to the Maori world-view, is in every sense a post-Darwin reductionist construct which separates culture and spirituality from nature. It removes the inter-relationship between humankind and other living things. Like science, it is not neutral, objective, nor a universal value.

Biotechnology, and gene technology in particular, are also borne of the reductionist ideology. Biotechnology is not the technological complement to nature or biodiversity, nor is human gene technology necessarily the complement to human health and well-being: rather both are a reinforcement that the external world is the principal force by which natural objects must be made 'to fit'. In today's context the external world is the global economy. The criteria for determining whether something 'fits' or not, is profit. This may seem an unduly harsh viewpoint, but research for the sake of knowledge or for solely humanitarian reasons is unfortunately a thing of the past: a casualty of commercial imperatives. Instead of encouraging diverse systems of production for local consumption as a response to a global food shortage, clean water and poverty crisis, priority is given to the establishment of monocultures of 'super-crops' for trade and export.

The news is filled with stories about cloned sheep, stags than can be altered to mate all year round, cows that can be bred to mature quicker, and the list goes on. I can walk into a 'Big Fresh' supermarket and buy *kuku* (mussels), and *pipi* (cockles) all year round. Two questions. Who determined that this was necessarily a good thing? And what are the risks? I do not need to be a marine biologist to know that when a seasonal marine resource is turned into a year-round product, it has been genetically modified, or bred under artificial conditions (i.e. farmed for the trade market rather than harvested as one of many inter-dependent species within the coastal marine environment), isolated from the natural ecosystem that depends on the presence and interaction of a diverse range of species. The cost of making available year-round seasonal resources is that the natural cycle and food chain is adversely affected, and the traditions and knowledge that form the *whakapapa* (genealogy) of that resource are lost to those peoples (Posey and Dutfield 1996: 33).

As a Maori consumer I am not convinced that year-round accessibility of seasonal foods is a benefit. I treasure the memories and experiences of joining with my family to gather food. My heritage includes the precept that at certain months of the year we will have special foods and at other times we will not. I associate all of these resources as indicative of a season, and within my cultural framework there are lessons to be learnt about the life-cycle, sustainable management and protocols for gathering, that are superior in value to the taste of those foods in my stomach. Respect for the reproduction of life as a continuation of genealogy strikes me as a paramount cultural concern.

Biotechnology in itself will not serve the needs of many if the driving force is the profit of a few; thus, gene technology presents more of a risk than a benefit. There is sufficient reason to take a precautionary approach to gene technology. I am concerned at the rapid pace at which we are embarking into this uncharted area when it is very clear from my education in Maori *tikanga* that there are significant risks associated with the objectives of this field.

When the philosophical leap has been taken to commoditize natural living things and to alter their genetic composition, the effects transcend the mere resultant product: belief systems are eroded, traditional knowledge of how to care for plants and animals is lost, customary use of properties of plants are alienated, and local communities are prevented from earning a living and providing for their families on their own lands.

In 1992, the Maori Congress wrote in its report to the UN Conference on Environment and Development that 'economic utilization of the environment must not compromise traditional values, the needs of future generations, or the earth's spiritual integrity'. A sacred balance of sustainable utilization and protection of the earth's resources is necessary to ensure the survival of all, rather than the profits of a few.

Whatungarongaro he tangata; toitu he whenua.

Man will always perish, but the land will remain for ever.

Hawaiian subsistence, culture and spirituality, and natural biodiversity (Davianna Pomaika'i McGregor)

Na Hawai'i, the Hawaiian people, were the original inhabitants of the island archipelago of Hawai'i. Hawaiian oral traditions passed on through chants, legends, myths and *mo'oku'auhau* or genealogies, trace the origins of the Hawaiian people to early Polynesian ancestors, and beyond them to the life forces of nature itself. The Hawaiian people are believed to be the living descendants of Papa, the Earth Mother and Wakea, the Sky Father. Many Hawaiians trace their origins to them through Kane of the living waters found in streams and springs; Lono of the winter rains and the life force for agricultural crops; Kanaloa of the deep foundation of the earth, the ocean and its currents and winds; Ku of the thunder, war, fishing and planting; Pele of the volcano; and thousands of deities of the forest, the ocean, the winds, the rains and the various other elements of nature.

Many Hawaiian people believe themselves to be a part of nature and nature to be a part of them. This unity of humans, nature and the gods formed the core of the Hawaiian people's philosophy, world view and spiritual belief system. In Hawaiian, the term that expresses this harmonious fundamental relationship is *lokahi* – 'unity'. Related terms expressing this fundamental relationship include *aloha 'aina* – 'love the land' and *malama 'aina* – 'care for and protect the land'. *Aloha 'aina* – 'love the land', *aloha in na akua* – 'love the gods', and *aloha kekahi i kekahi* – 'love one another', express the three precepts that form the core of the Hawaiian people's philosophy, world view and belief system. It is important for Hawaiians to sustain supportive, nurturing and harmonious relations with the land, the gods and each other, particularly their *'ohana* or extended family.

Moreover, the Hawaiian, the land and the gods are also spiritually, culturally and biologically united as one – *lokahi* – by lineal descent. In their *mo'oku'auhau* – family genealogy chants, Hawaiians trace their lineal ancestry to historical figures and ultimately, through them, to various deities and gods of the land, ocean, forest and nature. The land and all of nature is the source of existence for the Hawaiians – not only as the origin of humanity but also as the source of natural resources for day-to-day subsistence. The Hawaiians honour and worship the life forces of nature as gods. They do not possess or own the land or its abundant resources. Instead, they maintain stewardship over it – planting and fishing according to the moon phases and the changes from rainy to dry seasons. The traditional Hawaiian land system evolved to provide Hawaiians with access to the resources they would need for subsistence and to allow for stewardship over the land to the lineal descendants associated with particular ancestral *'aumakua* – deities and *akua* – gods.

Throughout the islands of Hawai'i are isolated rural Hawaiian communities surrounded by pristine, diverse and abundant natural resources in the ocean, the streams and the forests. These rural Hawaiian communities were bypassed by the mainstream of economic, political and social development. The Hawaiians living in these communities continued, as their ancestors before them, to practise subsistence cultivation, gathering, fishing and hunting for survival in accordance with the *'ohana* (extended family) cultural and spiritual values and responsibilities taught to them by their ancestors. These included the practices of *aloha 'aina/kai* (cherish the land and ocean) and *malama 'aina/care* (care for the land and ocean). Thus, we find that the subsistence life-style sustained the biodiversity of the natural resources and, in turn, the diverse natural resources sustained the subsistence life-style.

Rural communities where Hawaiians have maintained a close relationship to the land through their subsistence livelihoods have played a crucial role in the survival of Hawaiian culture and Hawai'i's natural resources. An analogy that conveys a sense of the significance of these areas can be found in the natural phenomena in the volcanic rainforest. Botanists who study the volcanic rainforest have observed that eruptions, which destroy large areas of forest land, leave oases of native trees and plants called *kipuka*. From these natural *kipuka* come the seeds and spores for the eventual regeneration of the native flora upon the fresh lava. Rural Hawaiian communities are cultural *kipuka* from which Native Hawaiian culture can be regenerated and revitalized in the contemporary setting. Protection of the natural resources and the integrity of the life-style and livelihoods of the Hawaiians in these rural districts is essential to the perpetuation of Hawaiian culture.

Only a handful of rural Hawaiian communities have survived the onslaught of post-statehood (1959) development. These include the islands of Moloka'i and Ni'ihau; the districts of Hana and Kahakuloa on Maui; Kahana, Hau'ula, La'ie and sections of the Wai'anae coast on O'ahu; the districts of Ka'u and Puna and small communities in Kona, excluding Kailua, on the Big Island; and Kekaha and Anahola on Kaua'i.

Cultural *kipuka* were traditional centres of spiritual power. In traditional Hawaiian chants and mythology, major *akua* (gods) and Hawaiian deities were associated with the areas. The districts were isolated and difficult to access over land and by sea. Due to the lack of good anchorage and harbours, early traders often bypassed these districts in favour of more accessible areas. The missionaries entered these areas and established permanent stations during a later period than in other parts of Hawai'i. Thus, traditional Hawaiian spiritual beliefs and practices persisted there, without competition, for a longer period of time. As Christian influences entered these areas, they co-existed with traditional beliefs and practices.

The geography of these districts discouraged the widespread or long-term development of sugar plantations. In the arid areas, the lack of water resources made development of sugar plantations unfeasible. In the areas with sufficient rainfall, the terrain was too steep or rugged for plantation agriculture. Where plantation agriculture failed, such as in Moloka'i and the Hana district, ranches were able to succeed. The ranches employed Hawaiian men as cowboys and allowed them to live with their families in these isolated districts and pursue traditional fishing, gathering and hunting activities to supplement their wages.

Where neither plantations nor ranches were established, traditional subsistence activities continued to be pursued, undisturbed by modern economic development. In the wetland areas, taro continued to be farmed, often in conjunction with rice. In the arid areas, sweet potatoes, dry-land taro, and other traditional and introduced crops suited to the dry soil and climate were cultivated.

The diverse undeveloped natural resources in these areas provided an abundance of foods for the Native Hawaiians who lived in there. Forested lands provided Hawaiians with fruits to eat; vines, plants and woods for making household implements and tools; and herbs to heal themselves. They provided a natural habitat for animals that were hunted for meat. Marine life flourished in the streams. The ocean provided an abundance of food. Subsistence activities continued to be the primary source of sustenance for the Native Hawaiians. Production in these districts was primarily oriented around home consumption. We find in these areas that the natural resources sustained a subsistence life-style, and a subsistence life-style, in return, sustained the natural resources.

'Ohana cultural and spiritual values

The quality and abundance of the natural resources of these rural Hawaiian communities can be attributed to the persistence of *'ohana* values and practices in the conduct of subsistence activities. An inherent aspect of these *'ohana* values is the practice of conservation to ensure availability of natural resources for present and future generations. Ancestral knowledge about the land and its resource has been reinforced through continued subsistence practices. While travelling to the various *'ili* (sections) of the traditional cultural practices region, through dirt roads and trails, along spring-fed streams and the shoreline, practitioners continuously renew their cultural knowledge and understanding of the landscape, the place names, names of the winds and the rains, traditional legends, *wahi pana* (sacred places), historical cultural sites, and the location of various native plants and animals. The practitioners stay alert to the condition of the landscape and the resources and their changes due to seasonal and life-cycle transformations. This orientation is critical to the preservation of the natural and cultural landscape.

To Hawaiians, the land and nature are the foundation of cultural and religious custom, belief and practice. The land and all of nature is alive, respected, treasured, praised, and even worshipped. The land is one *hanau*, sands of birth, and *kula 'iwi*, resting place of ancestral bones. The land lives as do the *'uhane*, or spirits of family ancestors who nurtured both physical and spiritual relationships with the land. The land has provided for generations of Hawaiians, and will provide for those yet to come.

When Hawaiians live on, and work, the land they become knowledgeable about the life of the land. In daily activities they develop a partnership with the land so as to know when to plant, fish or heal the mind and body according to the ever-changing weather, seasons and moons. Hawaiians acknowledge the *'aumakua* and *akua*, the ancestral spirits and gods of special areas. They even make offerings to them. They learn the many personalities of the land, its form, character and resources and name its features as they do their own children.

Hawaiian subsistence practitioners speak of their cultural and spiritual relation to the lands of their region and their commitment to take care of it and protect it for future generations. The land is not viewed as a commodity: it is the foundation of their cultural and spiritual identity as Hawaiians. They trace their lineage to the lands in the region as being originally settled by their ancestors. The land is a part of their *'ohana* and they care for it as they do the other living members of their families.

Hawaiian stewardship and use of natural and cultural resources

There are certain basic principles of Hawaiian stewardship and use of natural resources which, in practice, have protected the biodiversity of important natural resource areas in the Hawaiian Islands.

(1) The *ahupua'a* is the basic unit of Hawaiian natural and cultural resource management. An *ahupua'a* runs from the sea to the mountains and contains a sea fishery and beach, a stretch of *kula* or open cultivable land, and higher up, the forest. The court of the Hawaiian Kingdom described the *ahupua'a* principle of land use in the case In Re Boundaries of Pulehunui, 4 Hawa. 239, 241 (1879) as follows: 'A principle very largely obtaining in these divisions of territory [*ahupua'a*] was that a land should run from the sea to the mountains, thus affording to the chief and his people a fishery residence at the warm seaside, together with products of the high lands, such as fuel, canoe timber, mountain birds, and the right of way to the same, and all the varied products of the intermediate land as might be suitable to the soil and climate of the different altitudes from sea soil to mountainside or top'.

(2) The natural elements – land, air, water, ocean – are interconnected and interdependent. The atmosphere affects the lands which, in turn, affects running streams, the water table and the beaches and ocean. Cultural land use management must take all aspects of the natural environment into account. Hawaiians consider the land and ocean to be integrally united and that these land sections also include the shoreline as well as inshore and offshore ocean areas such as fishponds, reefs, channels and deep sea fishing grounds. Coastal shrines called fishing *ko'a* were constructed and maintained as markers for the offshore fishing grounds that were part of that *ahupua'a*.

(3) Of all the natural elements, fresh water is the most important for life and needs to be considered in every aspect of land use and planning. The Hawaiian word for water is *wai* and the Hawaiian word for wealth is *waiwai*, indicating that water is the source of well-being and wealth.

(4) Hawaiian ancestors studied the land and the natural elements and became very familiar with their features and assets. Ancestral knowledge of the land was recorded and passed down through place names, chants and legends which name the winds, rains and features of a particular district. Hawaiians applied their expert knowledge of the natural environment in constructing their homes, temples, cultivation complexes and irrigation networks. Hawaiian place names, chants and legends inform Hawaiians and others who know the traditions of the cultural and natural resources of a particular district. Insights about the natural and cultural resources inform those who use the land about how to locate and construct structures and infrastructure so as to have the least negative impact upon the land.

(5) An inherent aspect of Hawaiian stewardship and use of cultural and natural resources is the practice of *malama 'aina* or conservation to ensure the sustainability of natural resources for present and future generations. These rules of behaviour are tied to cultural beliefs and values regarding respect of the *'aina*, the virtue of sharing and not taking too much, and a holistic perspective of organisms and ecosystems that emphasizes balance and coexistence. The Hawaiian outlook which shapes these customs and practices is *lokahi* or maintaining spiritual, cultural and natural balance with the elemental life forces of nature.

Hawaiian families who rely upon subsistence for a primary part of their diet respect and care for their surrounding natural resources. They only use and take what is needed in order to allow the natural resources to reproduce. They share what is gathered with family and neighbours. Through understanding the life-cycle of the various natural resources, how changes in the moon phase and the wet and dry seasons affect the abundance and distribution of the resources, the subsistence practitioners are able to plan and adjust their activities and keep the resources healthy. Such knowledge has been passed down from generation to generation through working side-by-side with their *kupuna* or elders.

All of the ethnically diverse people living in Hawai'i today enjoy the rich natural resources which make Hawai'i a unique and special place to live. The persistence of Hawaiian subsistence, cultural and spiritual values and practices has contributed significantly to the survival of the rich and diverse natural resources in the islands. As society enters the 21st century, it will become increasingly important for everyone in the islands to adopt the Hawaiian way of loving and caring for the land. It is the responsibility of everyone who lives in and visits the islands to protect the precious natural resources of the islands.

Box 3.5: 'Grampa's Song for Little Bear'

Inés Hernández-Ávila

For the spirit of the bear cub stoned to death by Boy Scouts in Yosemite National Forest in the summer of 1996.

Little one, I must sing for you
if not my throat will close in heartbreak
Each time I see your startled body receive the first blow
sadness wants to overtake me and leave me
almost surrendering to petrifying grief

As I imagine a moment to moment replay of the beating
that caused you to cross over into spirit
it is as if time has not passed
I hear your cries and I am helpless
My will cannot enter that moment to stand between you and the callously
irreverent and unconscious arms of death
My screams now cannot halt the stones that have already been thrown

As your life is wrenched from you,
your mother in the shadows of the trees
stands frozen in space beside her other little one
she cannot believe her eyes and goes deep inside herself
For several moments she holds back from shock
even when the pain of her vision reaches all the cells of her body
she still remains with your sister in fear
of the godless ones who have done this to you
Your sister cannot begin to fathom the vehement assault

Just a short time ago you were a strong Little Bear
walking with your mother and your sister
touching the earth touching you back with so much love
now your life on earth is gone
and you are spirit

Little Bear
I sing Grampa's song for you
this song for the Ancient Ones,
the Oldest Bear Couple from the Beginning of Time
They are sleeping when I begin and as I sing they begin to wake
As the song calls them forward in time they are not as drowsy as before
Soon they are dancing to their own music and feeling pretty good
Suddenly, Little One, they see you in the distance
standing so forlorn and alone their hearts leap forward to you
It comes to them why they have been awakened
Grampa shows up, too, and he takes the lead in moving towards you
dancing a gentle grampa bear's dance so you will not be afraid
Your little head is hanging down in sorrow
the hurt of your wounded spirit creating a path to you

Box 3.5 (continued)

As Grampa reaches you he kneels and holds out his scruffy oldtimer arms
and you accept the embrace of his love
he croons to you his songs and gives you the doctoring your spirit needs
He calls you *precious one* over and over
and you sigh huge shuddering sighs of accepted comfort
When he is done, he stands and offers his hand to you
You take his hand and begin walking with him towards the Ancient Ones
who have received your sighs washed clean with their tears
Your steps are faltering at first, but as you continue,
they begin to be assured once more and your spirit begins to be in peace
Your mother and sister have been dreaming Grampa's song
traveling through dreamtime they see how you have been helped
and their spirits begin to be comforted
for the song is for them, too.

Inés Hernández-Ávila is Nez Perce Indian of Chief Joseph's band on her mother's side and Mexican Indian on her father's side. She is Associate Professor and Chair of the Department of Native American Studies at the University of California, Davis. She is a poet, a scholar and a cultural worker.

CHAPTER 4

VOICES OF THE EARTH

Ranil Senanayake

CONTENTS

Voices of the Earth

Ranil Senanayake

Introduction (Isabella Masinde and Carmen Tavera)

This chapter is perhaps the core of the entire volume as it contains without further interpretation a number of statements freely given by traditional and indigenous people from many different parts of the world. Translation into English has to some extent impoverished these accounts, but they show with great clarity how biodiversity is valued from the perspective of the infinite combinations of social and cultural background and individual experience of which language is the most prominent expression.

The issues addressed in these statements represent a mosaic of the key experiences and concerns that form the very foundation of the existence of a number of indigenous peoples, and underline the inextricable link between cultural and biological diversity. The ties to the land; the priceless value of all forms of life; the wisdom of their ancestors and elders and the collective memory; the links between themselves, nature and the spiritual world; and the impact of change, are some of the issues they chose to share with the global community.

Western science has taken almost 2,000 years to understand and value biological diversity. Only at the close of this century has the dominant global community coined an expression to reflect the complexity and variability of life on earth and accepted the dependence of human societies on the maintenance of life support systems and the myriad species, genes and ecosystems that have evolved on our planet. For cultures other than the Western culture, the word biodiversity has no direct translation as it encompasses much more than the variability of life on earth. For Richard Aiken (Kuarareg) the word biodiversity is just a word: 'The meaning of biodiversity does not fully describe the spirituality of the people, the land and the sea that we experience'. Biodiversity, more than a concept, is a practice and a central aspect of the material and spiritual life. It is the tapestry of life.

Traditional beliefs and practices – which make up spirituality – are dynamic, resilient and living. Great emphasis is placed on the preservation of linguistic diversity by traditional communities and indigenous peoples because it is the expression of their culture and the vehicle for the maintenance and transmittal of their knowledge. Respect, understanding and acceptance of this knowledge are central to the survival of these communities and to that of all human kind.

We do hope that this modest array of views will contribute to the understanding of the implications of forcing change among these cultures and of the loss of knowledge and wisdom that would come along with these changes.

The ties to the land

The strong and indivisible relationships between communities, culture, spirituality, nature and territory are invariably present in indigenous peoples' statements. While political boundaries are considered artificial in terms of the Mother Earth and its creatures, the familiar environments in which countless generations have been born, and in which specific cultures have evolved, are the centrepieces of the complex puzzle of adaptation and innovation. The words of Mukalahari Kaichela (Botswana) are an eloquent example of the impact of uprooting any community from its land: 'The experience of moving away is so painful when you think of it because they are moving from a place where they have been living for a long time. They know what the plants are for; they know the sources of water and food. When people are moved to a new place they are cut off entirely from their culture and are moved to a place where they must start a new culture'.

In the statements provided by NGOs and indigenous peoples for the CSD 1995, it is stated that 'indigenous peoples live in territories'. This means that a people and its communities are responsible for the control and use of the total environment – soil, sub-soil, trees and other plants, animals and birds. All the resources of an area are included in this generic sense of territory, including land, shores, lakes, rivers, islands and sea areas. They also state that, 'indigenous territories are considered to be inalienable'. This means that they are owned by a people as a whole and are passed from ancestors to descendants as a part of that people's heritage. This is emphasized in the statement provided by Patrick Segundad from Malaysia, whose community believes that, 'land cannot be owned because it belongs to the countless who are dead, the few who are living, and those yet to be born'. That territories are part of a holistic vision of the universe which includes political and social control over resource use, and spiritual reverence for the invisible religious aspects of forest life, is shown in the statement provided by Moises O. Pindog, Omis Balin Hawang and Baliag Bugtong from the Philippines.

The value of biodiversity

No matter what the respondents' cultures or backgrounds, all have agreed that all attempts to price biodiversity must fall short of its real value. Life cannot be measured in monetary terms, and loss of biodiversity would never be fully compensated. 'To the indigenous population, the vital elements of nature do not have a price: they are invaluable. To others, nature does have a price. When the oil companies arrived in Ecuador, they destroyed much of nature. Now they need to repair what they destroyed, and this has a price for them. Despite what they pay, they cannot bring back what was destroyed.' (Bolivar Beltrán, Ecuador).

'To some people, these valuable things may not matter, but to an Ife they do matter. The place where his ancestors were buried is important. Sometimes the government wants to build something on a burial ground, and our people resist. While the government may offer other land, the people will disagree. They want to preserve that particular place despite the offer by the government. A burial ground is a very important aspect of our lives which an Ife cannot actually quantify in terms of money.' (Topko-Ayikue, Ife).

Coping with change

Innovation, adaptation and resilience are attributes of all cultures. However, imposing sudden, forceful and unwanted change has proved dramatic for most indigenous peoples: reduction of their territories, imposition of Western law over customary law, destruction of key ecosystems and resources, and general loss of control over their destinies are some of the main concerns expressed in the statements. The impact of inappropriate land-use practices and new consumption patterns in Brazil are voiced in the statement of Daniel Matenho Cabixi: 'There are other changes which also threaten our way of life. For example, three years ago clouds of grasshoppers invaded the plantations and the white farmers blamed the indigenous communities. We have become the scapegoats for an accident that was provoked by the landowners themselves. Today, there are greater numbers of grasshoppers because their natural predators have been reduced or eliminated. The hawks, armadillos, snakes and other birds are now gone because of the clearing of the native flora and the use of pesticides in the fields. Until the coming of the outsiders, there was no problem of recycling our waste. All of the indigenous products went back to the land. Waste is a problem with industrial products.'

Displacement from traditional territories to give way to 'development' is a major change from which communities may never recover. In the statement provided by Evanson Nyamakao (Korekore) he records that: 'We were removed from a town there because of Lake Kariba which was built in 1957. This was a bit of a change for us because where we used to stay there were plenty of animals. Here there are no wild animals. To carry on our tradition, you need certain animals to be the totem. When you need certain plants that are no longer available, it changes or influences traditions. The difference in soils between here and near Nakuti caused a great change in our agriculture. Here we couldn't hunt for meat, and if you can't get enough bush food, the health of the tribe goes down'.

Spirituality and nature

Traditional/indigenous communities' relationships with the environment are expressed in all the statements provided. These relations are based on beliefs rooted in the myriad myths of origin, featuring celestial and terrestrial phenomena: the sun, the moon and shooting stars, the Sky Father and Earth Mother, as shown in Aroha Te Pareake Mead's statement (New Zealand). In this statement, the Maori worldview of the origin of life is a love story between Ranginui, the Sky Father, and Papatuanuku, the Earth Mother. Theirs was a love of devotion and sacrifice. They were once joined, but as their children grew, they separated in order to enable their children to move freely between them so the children had space to play, light to grow and air to breathe. Ranginui and Papatuanuku sacrificed their love for their children and this is a value which every Maori individual inherits. The statement goes on to say that Ranginui and Papatuanuku grieved so deeply and overwhelmingly for each other that even with the space, light and air their separation created, their children still could not prosper. Ranginui's tears for his beloved wife are what the Maori know as rain, snow and hailstones. Papatuanuku's grief and longing for her husband manifested itself as the moods of the earth – mist and fog. Such was the grief and longing between Ranginui and Papatuanuku, that their children still could not breathe, the tears from the Ranginui and the mist and fog of Papatuanuku were suffocating. Eventually, Papatuanuku in an act of supreme sacrifice as a wife and mother, turned herself around so that she faced the underworld rather than the husband she deeply loved. That, her moods and her longing for her husband are now expressed as eruptions of volcanoes and the tremors of earthquakes as well as mists and fog.

Talking bushes; sacred rocks, hills and groves; prophetic birds and animals, all symbolize something. Henrietta Fourmile (Australia) talks about the beliefs attached to animals and birds such as the cullier bird greeting her tribe, the bush turkey's value as a representative of her ancestors, and her own experience with the bush turkey before and after having her baby boy. She believes that the bush turkey represented her grandfather and wanted the baby to be named after him. The naturally derived aesthetics that these communities saw and continue to see in biological diversity, as anecdotes of their spirituality, are their sustenance, their survival and the only meaningful pointers to their destiny. The role of traditional beliefs and taboos associated with the way the communities handle sacred trees, religious sites and groves are present in most of the statements.

Man's special relationship with plants and animals

Man holds a special and inherent attachment to biodiversity. In both Chapter 5: Ethnoscience, 'TEK' and its application to conservation and Chapter 6: Valuing biodiversity for Human Health and Well-Being: Traditional Health Systems, good examples have been given of how the traditional communities and indigenous peoples have utilized the resources available to them over the years. Forests have remained intimately linked with ancestry and cultural heritage. In Chapter 8: Forests, Culture and Conservation, it is clear that forests and culture have influenced each other's development. Forest landscapes are formed and are strongly characterized by cultural belief and management systems, and cultures are materially and spiritually built upon the physical world of the forests. Up to 2,000 years ago, many traditional communities derived virtually all their material resources, including food and shelter, from wild vegetation and other natural resources. They slept on mats and made baskets from grasses and reeds, and pots from clay for storing food. Pastoral communities supplemented these with resources from animals, with animal skins providing clothing, bedding and containers for food and water. In a statement provided by Cristina Gualinga (Ecuador), she states that 'my grandfather never required money or clothes. All that he needed could be found in the jungle and he was so strong that he could transform himself into a tiger and travel quickly wherever he wanted to go'. She further states that, 'our medicine is from the drugstore that is the jungle. Everything is available to us, this is the richness of the Amazon. The forest also gives us perfumes and plants that give fertility, infertility and endurance for the long walks'.

Traditional communities hold strong beliefs in the healing power of plants and animals or parts of them; in their ability to provide people with strength, both physical and spiritual; and in their ability to be in tune with the universe. Specific forest sites such as groves and individual trees or animals are considered sacred, and are valued for particular cultural occasions and as historic symbols by different communities.

Conservation of resources, beliefs and customary law

Customary practices have contributed to the conservation of resources by several inadvertent or indirect local controls and some intentional traditional management practices. Taboos, and seasonal and social restrictions on hunting and the gathering of plants, have helped to limit harvesting of these resources to some degree. In a number of cases, the communities live in villages and each village has a hunting reserve where different species of animals and plants are allowed to grow. Group hunting is done once a year during the dry season. Tradition prohibits frequent hunting, night hunting, fishing and gathering, and allows selective hunting of animals. Sacrifices are performed in the hunting shrines before the hunt. All hunters partake in the catch. This instills the spirit of oneness for which the society stands.

To the Ikalahan people of the Philippines, biodiversity means all the things of a forest. When they open a virgin forest they test the area by placing about ten cooked sweet potatoes in a given spot. If they are touched or removed, it is a sign that they must stop clearing and leave the patch to remain as a forest. This means that spirits are present in that spot. Similarly, if they intend to cut a big tree, they light a fire by the tree. If the fire goes out, it is a sign that the tree is protected by a spirit and they should not touch it. They also offer prayers to the spirits before they embark on mining in their forests. In Chapter 3: Indigenous Peoples, their Environment and Territories, Elizabeth Reichel D. points out that, 'obtaining or using biotic and abiotic resources from the rainforest such as plants, animals or clays or soils, always requires shamanic consultations, not only to do resource accountability but also to request 'permission' from nature's Owners and its 'people who work and think'. She further reports that 'the Supernatural Male and Female Owners of the species, ecosystems or seasons are consulted and 'paid' for human consumption while the knowledge base and spirit is requested to stay behind in nature with its people and its Owner in order to not deplete resources and 'not confuse systems of knowledge'.

The effect is that these communities see a connection between territory, culture and their identity. Their rights to use resources are based on customary legal systems. A statement by Michael Kapo from Papua New Guinea expresses the importance of their traditional legal systems. For example, people are not allowed to shout and stamp their feet in the village in the middle of the night, because this is when the spirits look for food. To go against this rule would risk disturbing and angering them, and they could make you sick. The villagers believe that noises in the night disturb the biodiversity of the village. They also have rules governing the harvesting of certain crops, especially large trees. A whole village has to come to a consensus before a large tree is cut down. In this case, their dignity protects biodiversity.

Taboos preventing women and young people from cutting certain trees, hunting, or fishing during certain seasons, have provided unshakeable conservation of many species. For example, in Botswana they believe that as soon as they start ploughing and planting, the kudus should no longer be killed. They are left alone until after the sorghum has been harvested. They believe that the kudu is a sorghum specialist, and if it is killed the crop will be affected. In certain communities where roots are used for treating illnesses, only lateral roots are harvested. This avoids ripping the plant out by its roots and the soil is usually returned after harvesting. If this is not done, it is believed that the sickness will return.

In order to limit the number of users of both plant and animal products for medicinal purposes, and as a way of protecting their knowledge, traditional practitioners often pound the products into powder or burn them to ash. Many go further and associate the medicinal potency of plant and animal material with special rituals and spiritualism, thereby limiting the number of practitioners.

The collection of the indigenous peoples' statements (Jerry Moles and Ranil Senanayake, ELCI)

The statements that follow are a representative selection of the depositions provided by representatives of indigenous and traditional communities from America, Africa, Asia, Europe and the Pacific in response to a request from UNEP for information on how these communities value biodiversity. UNEP commissioned the Environment Liaison Centre International (ELCI) to carry out this task, and the statements were collected under our supervision.

A number of different methods were employed in collecting these depositions. The Pacific Concerns Centre in Fiji sent an interviewer to neighbouring Island Nations to request that indigenous peoples speak their mind concerning biodiversity and its values. Confederacion de Nacionalidades Indigenas del Ecuador (CONAIE), the Confederation of Indigenous Nations of Ecuador, invited representatives of some of the indigenous nations of Ecuador to participate in a workshop on biodiversity. On the first day, the workshop focused upon what Western science terms biodiversity. After a day of discussion, the participants constructed three-dimensional table-top models of landscapes and watercourses and explained how they envisioned biodiversity, interpreted with the use of the models. In Peru, Dr Thomas Moore organized a workshop in the Amazon region and the representatives of indigenous peoples and small farmers' (*campesinos*) organizations came together and arranged themselves in groups to discuss and come to some agreements upon the meaning of biodiversity to them and its value to their people. Mr Dennis Martinez invited indigenous and traditional peoples from North America to assemble in Tucson, Arizona, USA, and discuss biodiversity and its values in a two-day meeting. Dr Claudia Menezes, President of EarthKind–Brazil, organized a three-day meeting at the University of Rio de Janeiro. Indigenous people from around the country met for one day discussing the meaning of biodiversity from the perspective of the biological sciences, and for two subsequent days presented their response from the perspectives of their communities.

While all of these contributions have been edited, and redundancy reduced when necessary, the structures of the depositions remain unchanged. Individual statements were collected from people in a number of settings, ranging from their village homes to major international meetings concerning indigenous peoples and the protection of biodiversity. The interviews were tape-recorded, transcribed, edited and wherever possible returned to the people who contributed them.

In some circumstances, interviews were conducted through interpreters, while in others information was collected in national or Western languages and translated into English. Editing was done with the intent of sharing these indigenous statements with representatives who are responsible for charting our ways across the next century.

There are no complete translations when ideas and terms represent meanings that are relevant to the experiences of a particular culture or people within their native and traditional landscapes. If there are errors from the best expression of the meanings of the collected statements and reports, it is in the direction of making the information accessible to those who are neither indigenous nor traditional. It is hoped that the traditional and indigenous peoples find their Truth presented fairly in the statements that follow.

As you read each statement and report, remain mindful of how the information was collected. As the various Indigenous and Traditional Peoples presented their information in ways that were appropriate to them, diversity of perspectives was added. Each statement speaks of the earth and the presence of our species here. That different means were used to present their viewpoints allows us to appreciate the diversity of understanding and opinion that we share as members of the species *Homo sapiens*.

Voices of the earth

Australia – Henrietta Fourmile – Polidingi

I'm Henrietta Fourmile and, traditionally, I am from the area of Cairns. I'm a member of the Polidingi Tribe and fall in the line of my fathers which is part of the Kimo Clan or Wallobalo Clan of the Cains Edmonton Region. I was born in Yaraba which is my mother's country of the Pulanji people, and I was brought up there until I moved out to attend the schooling system in Western society.

My tribal group, the tribal group of my fathers, has been here from the time of colonization. The same is true of my mother's tribe in their present location. We are the rainforest people of the Cains region. The entire environment around us is pretty much the way of life of our own people. Unfortunately, at the present time, we have development and this is something that gets in the way of our people maintaining our traditional practices.

When I discuss biodiversity with my people, they want to know what it is. When I start explaining the concept in terms of the species, animals, plants, the whole of existence, they start to realize what I am describing. It's part of their land, of their very existence as Iwingi people in this entire area. Not only is it the land and soil that forms our connections with the earth but also our entire life-cycle touches much of our surroundings. The fact that our people hunt and gather these particular species on the land means emphasis is placed on maintaining their presence in the future. At the same time, we want to maintain our practices of eating some of the flora and fauna. What is sometimes called 'wildlife' in Australia isn't wild; rather it's something that we have always maintained and will continue gathering. In addition to the foods that the environment provides, there are other symbols, totems, and other things that we as a group of people place value on. Many of the species are our totems.

Our existence is part of the land, and that can be seen by the way we name in our language. I was given a language name, Mupopopo, a part of a very thin vine. It's associated with the land and country in which my grandfather was born. My grandfather, at the time of colonization, was named as Douglas King of this area of land so my language name connects me with my territory. The practice of giving a language name by our grandparents that is a species on our land continues. Obviously, as you mature, you see the species and know the ones that live on your land very well. You feel a sense of commitment to the one that shares your language name.

The connection I have to my mother's tribe is the bird we call the cullier, which is the sea hawk. It has a very strong significance to our birth, and especially to our family. Our connection with it is very strong spiritually. When we get out and see the cullier, it greets us and we know that something else is strengthened in terms of our connection. My father was given the name of the bush turkey from his grandfather's name and was part of the bush turkey family. Now my son has the name of bush turkey and, just to show the connection he has with that particular turkey, let me share with you a story. When he was born, we were living in a suburban area in Brisbane and, for the first time, a bush turkey actually came into our land and stayed in the garden until the day I went into labour. The bush turkey continued coming up into the tree, looking through the window, and then going down again. This was worthwhile. After I had been at home with my son for two weeks, the bush turkey paid us its last visit and disappeared. We never saw it again. The connection is very strong.

The value of the bush turkey is a whole inner feeling that my grandfather is there, my great grandfather is there, and the whole essence of how it feels. You don't understand it unless you are a part and it is you. This kind of value has nothing to do with money.

We feel biodiversity because it is our total existence whether we live in modern society or not. We regard all the different species as part of our existence, irrespective of whether we have them or don't have them now. They are still part of us. They are our identity and we must fight to be able to manage the species, fight to be able to preserve them. We have to become part of the Western culture to be able to push for some kind of joint management and control over the species to protect their habitats. It is not a matter of just coming together to preserve biodiversity: rather it is a matter of knowing that we have the species there for our next generation and the next generation. Not only do we know the species but we also know how they can be used.

At the moment in the North Queensland area, for my tribe and the Fungaji people, the Filangi people in the north, the Liganji people, the Mamu people and Tapu people, in all our travel group we eat the sea turtle species as a part of our main diet. Since we still hunt them, and in order to ensure that they do not go extinct, we really need to get into some kind of management practice. One way of doing that is to possibly have some kind of farming of turtles. At the same time, there is a need to maintain the cultural aspect of eating the turtles and bringing them into ceremonial activities so they are protected. The whole issue of protecting biodiversity is the same as protecting what belongs to us as well.

It is very important that indigenous peoples participate in the deliberations that lead to the setting of government policies on biodiversity because we value it differently. Our way is very different from the way the economic system sees that value. This can be shown by the fact that for years we've gathered, we've lived off the land as much as we could, but today we rely heavily on the shops. The truth is that we still live off the land because everything we eat comes from the land. We must go in our current direction for a certain period of time; however, probably not for a lifetime. There are movements about again coming back to the way we formerly used our resources. When you go to the reserves or what were called reservations which were set up by the government and churches, most of those lands still have a variety of species remaining. The reason that they have not been destroyed is because the habitat is still there which was protected by the way the people valued the species as they lived off the land. Where we see ourselves again today is that we face the problem with the Western society where we live in a contemporary historical urban tradition. None the less, the aboriginal peoples who have moved to this area from other areas, irrespective of their purpose for moving, continue to maintain some traditional practices whether it is turtle eating, hunting, gathering, or using the raw materials of the rainforest. The people feel themselves becoming very much assimilated into Western society. What we are trying to do at the moment is to hold on to what we can of our cultures and, at the same time, have the benefit of Western society. This can be destructive because you never know which culture you are living in. It can become very mixed up and a threat to your very own existence if you want to maintain your identity with a particular tribal group. This is very confusing for a lot of people and, unfortunately, something which we cannot control. We cannot control the government, the movement of developers in the region, or the kind of management being practised at the moment. We are pushing for joint management, to be involved at all levels of government in terms of ensuring that the management staff recognize our abilities to know how to manage species in a particular area. We are also looking at the use of Western technology with traditional technology to see how we can pull them together.

We are also working at establishing our rights as indigenous peoples to continue our traditional practices. This is a very difficult process because it requires that we be involved with our own traditions as well as in a cash economy society. We just have to continue pushing our line of traditional rights, maintain our practices and, at the same time, work in the society of today for laws to ensure that the rights of management and protection are still maintained.

Australia– Mick – Near Cairns

From my father's side I'm native and I'm an Australian now as well. So I'm a mixture as an individual. I have been in the army for six years and travelled around a little bit but this is my home area. We were moved here long ago. My grandparents had moved and changed from their old travel areas. We have become part of the business community in Yaraba. Yaraba is a combination of different tribes that have been moved from Queensland. We are now in someone else's travel area but, over the years, we have inter-married and become one community. At one time, it used to be a closed reserve and we had to get a pass to go to other areas, but now this is pretty much like a local government area. We can come and go as we please.

There has been a new government decision about land titles and there is a process under way to decide who or what tribes in Yaraba have rights to the land. This decision gives peoples the right to their native soil. When the Whites first came to Australia they declared the place vacant and did not recognize any Aboriginal rights. To them, Aborigines were just animals. But now there is recognition that there were people here, and therefore tribal bands have a right to land. Many of us who don't come from here are waiting to see what will happen. Whether we can stay or not depends on the tribes of people who were here originally, and what they think. In the extreme, we may have to move out. But if they say we can stay provided we recognize them as tribal people, that's all right, that's what we are hoping for.

It's their land and I'm hoping that they can also recognize us. If they say we pay a bond, we will pay a little rent to them. But if they start talking about moving us out because we are historical peoples or disbanding us all over the town areas, that will be a bit sad. Most of us, like me, have been in the area for generations. If it weren't for many historical peoples there wouldn't be a Yaraba to claim. Much of the work that established Yaraba was done by historical peoples. My father was one of them.

What makes all of this difficult is the fact that people have been moved, against their will, away from their home areas, so we don't have access to our native biodiversity. In terms of the sea rights and the Great Reef Park, the authority set up policies and we have to go change them and try to dismantle what could be a disadvantage to our people. Barriers were put up on the reef with long nets on our traditional hunting grounds. They put up all those multimillion dollar platforms out there for tourists, and that is where we have to go and look for our food. And on the mainland too, a lot of the forest foods are gone because of clearing for farms, and even the animals and the birds are finding it hard because they have no place to settle down.

People are destroying all the time. You see, there has been one law for the rich and one for the poor. People are quiet up here. You can build a road right through the place, smashing down the forest without asking anybody. This is nonsense for everybody. If those were black people doing it, we would all be in jail.

All of our traditional food resources are natural things that keep us going. But now many of our Aboriginal people have got diabetes because we all eat refined foods. In the past, all we had to eat was from the bush – bush tucker – so then we didn't have diabetes. And the use of fertilizer has changed the botany by running into creeks and destroying the ecosystem. All that fertilizer running out into the ocean affects the shoals, especially in the flood time when the rivers are fast, and this changes biodiversity too.

We are returning to bush medicines again but it's being revived very slowly. This depends upon biodiversity too. It seems that more and more of an area is being taken for a handful of tourists and farming. And you just can't go take what you need like in the past. When you fish, you need bigger boats, faster boats, and to get approval to look for your tucker. Bukiburn, white pigeons, ginger, plums, *Kuban*, they are shellfish, and *Kuban Bino* is shellfish too. All of this was our food. We've got certain trees that you use for your peas and your white potato and you use different kinds of sap from the trees for sauce. Some people go and collect leaves for washing. We have a rich biodiversity. Camping out is still our thing.

People and kings. It isn't the people and kings of yesterday. This is psychological. We are happy knowing that the mangroves are there. They are green and part of the ecosystem. But they are thinking of putting in a subdivision where the mangroves are, and this is against the wishes of even the white people. They say, 'Look, we don't want a subdivision there. We want the mangroves. We may not go and walk around the mangroves, but just glancing over there and seeing the trees and knowing it's there is psychologically good for us'. And this is coming from white people. It's the same with us psychologically. As long as we know we have a shrub there; as long as we know we can govern, camp, and hunt; we know we've got a place to go.

Thinking about biodiversity reminds me of a pop song in the sixties called 'I said you don't know what you've got until its gone'. And that is what it is, you don't really know what you want until it is gone. I think a lot of people would really wake up in these days of awareness. Some things should remain and not be touched. The traditional people like the usual. It has been that way and they want it to stay that way. It's a part of home. But now it's like someone moving into your home and removing all your things – all your food and all your religious stuff – from your house. Taking all of that, imagine doing that to your house? You would feel a little bit upset too.

And this is what it's like living in our environment. When the trees, plants and animals are gone, there is something missing. It is the same as having someone remove things from your house. You have a certain feeling when you are going along and all of a sudden see another branch missing and another net keeping you away from your hunting ground. You know that your hunting ground is getting smaller and smaller.

There is no regulation against this. I think that if you've got the big dollars you can get anywhere. You can get into college or anywhere you want. There is everything around for multimillionaires. Look at the tourist areas. Supposed to be booming, right? You walk around and see how many of the premises are vacant because with all the tourists coming in, the money is all channelled, channelled into specific groups. So it is benefiting a few rich people and even some of the white people don't benefit either. The white people miss it in one way. We miss it in a different way. That's our environment, our real home.

Botswana - Kathrimila Mulazana - Bowankez

My name is Kathrimila Mulazana. I belong to the Bowankez tribe which is situated in the southern region of Botswana. Like the majority of northern tribes in Botswana, we are pastoralists. Historically, we wouldn't say we are not hunters and gatherers for this has been another aspect of our lives.

From a cultural point of view we would take biodiversity to be the trees and plants around us, the land on which we do our ploughing, and the grass known for grazing. Even now, I think of the materials used for building our house, and to me, the meaning of biodiversity would include the protection and sustainable reclamation of all these resources.

There are traditional ways in which biodiversity is protected. We have certain trees that should not be cut or used for certain duties like, for instance, firewood. On the other hand, we identify certain other trees for special uses as, for example, centre pillars of houses, wood carving, and so on. There is a timing that is given to the cutting of such trees, or traditionally there was a timing where you could only cut these tree in a specific season of the year. Unfortunately, a lot of this type of information is escaping, is getting lost. I think it's because of development. Now you find that because of each culture, no integration is possible. People are losing their interest in taking resources from their surroundings. At the same time, no one carries the responsibility of resource management that was normal within the traditional community. This sole responsibility now rests with our government. At the same time, the people no longer see resources as having value to themselves, but as objects that can be used by people to sell and gain opportunity around some obstacle. Because the responsibility has been taken away from the communities, the people don't see any direct link any more between them taking care of it and the benefits they gain from it.

Traditional people did respect the taboos, and also a system of tribal or village management. In a village session, we have the chief as the person who would say, 'Now is the time for people to go and cut specific trees' or during the harvesting or even ploughing seasons, it is the chief who gives instructions that people could go out and start ploughing. Now it has become the government's responsibility to say, 'Start ploughing from this time of the year' – say November, December and January – and what is taken away is the decision to start ploughing. It is no longer a community initiative but rather a part of a government subsidy programme. People are not really looking at the land as a responsibility now that the government must pay them for ploughing.

We are now going back to the whole discussion on biodiversity. Traditionally, when the field was cleared for ploughing, certain trees were left standing. Now I clear my field specifically to satisfy the government official. Now my field has to be 'clean', and therefore I will take out every tree that is there.

Unfortunately, some of the taboos are not really well explained. Some people think that taboos are just to say that you've got to pay in respect, and if you didn't, this or that would happen. The whole essence of the taboos is not being put across to the young generation.

My generation is still fighting to dig out the traditional information about the meaning behind certain practices and beliefs. We now know that some of the trees can be good as a control mechanism for pests. Indigenous peoples know our competitors and how to control some of them. Other trees were planted only for the sake of providing shade during the harvesting season. Trees are even planted in the field itself if they are known to be good as sources of food. There are all these linkages that used to be understood but are not really spelt out other than in daily practice.

And now, because of the fact that it comes from the government, it is no longer a community thing. You either have the indigenous knowledge or it is getting lost. Our traditions are dying. They were recorded but now people cannot know and take them as part of their lives.

The value of biodiversity? During the good old days people really saw the environment around them as a source of everything that we need now, like food, shelter and even grazing for the cattle. Now there is unfortunate competition for land in the sense that there are different land uses which affect people on a day-to-day basis. This being the case, you find that people's lives are also changing now quite abruptly and they fail to maintain the environment around them.

There are other problems that we are facing. One is drought, and as pastoralists we are not looking at the quality of the cattle we are herding but only at their numbers. There is too much declaration of what the land is to be, and we are looking at the need not to have ploughed fields. We just go out and clear as much land as we need for a purpose, and if this is not done with care, destruction of the environment around us can occur. It is this kind of competition for resource use that puts us in a predicament as to whether we need to preserve the land as it is or whether we need to go out and plough it. At other times, we need to put a lot of cattle on the land while the government says that we must reserve land for wildlife purposes. These and other kinds of conflicts really put people in uncertain situations.

Let's look, for instance, at the administration of the Land Boards here in Botswana. The Boards are responsible for allocating boreholes for residential plots, and for ploughing fields and the like. Therefore, the people in any given area are not really responsible for the management of the land. The allocation of the land is up to the Boards, and at the end of the day the people are just looking at the Land Board as an institution that is responsible for taking care of the resource. It is much the same as the management of wildlife, which is on the shoulders of the Department of Wildlife and its helpers. The communities know, by default, that no one is taking care of things. What you find is all of this sectoral business, with different parts of the environment having different management plans carried out by different people. This places the communities in a position not to have any interest.

Life has changed in quite serious ways, and if you pick the example of the people in North Botswana, for instance, they are known traditionally to be hunters and gatherers. But they were forced to move out of this way of life because of development thinking. People say that we develop because of the need to be integrated into mainstream society. The people didn't want the main society from the beginning. The other thing is that given all the management responsibility for resources like wildlife, they knew that they could go out and hunt any time. They had their own regional conservation systems that allowed them to know that certain animals should not be killed during specific periods of the year. Even if they were on a hunting mission and tracking animals, they could detect whether a particular animal was male or female and, on this basis, know very well which animal to kill. Now that they have run short of access to hunting any time they feel like it, there is a control mechanism. They are now given a hunting permit which stipulates what they can take. With the permit system, the people wouldn't really take on the conservation aspect as their responsibility. They are just out now to kill. Part of the reason is because of the changing land systems. Now there are boundaries and they can only hunt in a given or controlled hunting area in a hunting or game reserve. They cannot hunt in a National Park and all of these restrictions are denying them the opportunity of going into those areas that they know where they will find a certain plant or animal. Again, the whole thing has to do with a policy which was adapted in 1976, 1978 and 1979. The people have become confined to small areas and their life patterns have changed quiet dramatically.

In my tribe, we are pastoralists and agriculturists. In our situation, it was a question of population increase and that there were no places in a given area where we knew we could graze our cattle. People didn't really raise livestock for getting money. Rather it was sort of a cultural identification that you needed to have cattle. Now, with the whole thing of money coming in, people are moving towards the north where livestock is a business and you find people's lives changing as the management system changes. It is the economic element that is pulling people out from where they are but I think those people who have social attachment to the environment wouldn't really go for the economic opportunity. In a lot of instances we have seen people staying at a place because they have a cultural attachment to it, despite the lack of certain amenities. When asked to move, people say no, our ancestors lived here and we are going to die here. I think in the larger aspect, a lot of people in my tribe will not really want to see themselves moving from the area. It's a combination of things. People would say I like this place because it has a big forest or I like this hilly part of the area because it provides me with one, two, three, things. It is really a combination of those kinds of reasons.

Botswana - Pera - Bakalaharil

My name is Pera, as is my father's name. I belong to the Bakalaharil tribe and we are from the western part of Botswana.

Biodiversity to me means two things, or perhaps better said, two groups of things. There are the things that are found under the soil in term of roots, tubers, and sometimes small animals such as ground squirrels and the like. Then if you look on top of the soil, you will see some grasses, plants, trees and wildlife. We depend upon the things on top of the soil for our livelihood. Some of these things provide food while others are used to form shelter, provide medicine, and for carving utensils. We do eat some foods from beneath the soil, like tubers and roots. Also, some of our food is from the wild – like fruits and some of our meat. We are having some small problems with the issue of meat because of the conservation rules that have been established. We are happy to conserve, but some conservationists come and say that preservation means that we cannot use the animals at all.

To us, preservation means to use, but with care so that you can use again tomorrow and the following year. We have our own beliefs about the environment and there are a lot of things in our world that support these beliefs. These beliefs protect our resources. For example, as soon as we start ploughing and planting, the kudus are no longer killed. Instead, they are left alone until after the harvest of our sorghum. The kudu is a sorghum specialist and if it is killed, the sorghum will be affected. We protect certain trees in a similar fashion, like the tall one standing there. At this time of the year we will not harm that tree even if we need a piece of it for firewood, for carving, or for making stools or chairs. You do not touch the tree until after the harvest is passed. In our traditional worshipping, when we carry out our rituals, we go under a tree in the area like that one.

It would be very difficult for us to move away from this place because I don't know how we would survive. Even if we lost our trees, it would take a long time for replacements to grow. Our medicine, like some of the tubers, are here and we would not know where to look in another place. What we need is around us.

All of the things that are here, the tubers, plants, trees and wildlife, whatever, are God-given. To put a value on them is difficult. The money teller would be very little interested. What is equivalent to the biodiversity here, to the things that surround us, is my life. If you took these things away, it would be like taking part of my life, and then my survival would be questionable.

Botswana – Kipelelo Waker

I am Kipelelo Waker. I was born in the small rural village of Shoshoa and by association with my father I belong to the Longato tribe, but my mother comes from the Pakalaka tribe and I was largely brought up in my early years amongst the Baroro people. In this respect, I find it really difficult to identify myself specifically with any tribe. I'm currently an adopted Mukhata by residence and I studied and lived with the Barolwa when I was doing my Masters degree. I feel somehow strongly that I would like to associate myself also with the Barolwan tribes in terms of resource.

From my learning point of view, biological biodiversity is the future of our life-style and how completely can we live on traditional resources and sustain them intact. To sustain the resources is not just for us, but to be able to show our children what they are and what they have meant to us.

In spite of the pressures that are currently being placed on the resource from what seems to me to be a simple and modern cultural point of view, I was brought up with the natural resource as something you use and have available in the future. My education as a young person living in a cattle post amongst the ordinary Bathara community was like a nursery education. In many ways I think of Dolly Parton's song, which says, 'I missed the young when in my youth while so many things could have been explained'. It is like a nursery rhyme. To me the nursery was the bush itself from which I learnt to harvest, to feed the livestock that sustained our livelihood, and a point of view from which the modern African child is largely divorced. I feel a very strong conflict, or what I consider my dilemma as an enlightened person, over the education my child is receiving, which presents a very international version of the resource areas, divorced from the traditional understanding of natural resources in their day-to-day utilization.

For somebody brought up the way I was and to practice my current work the way I do, it is difficult to decide what we should be teaching the children. The privilege of growing up as ordinary children in a rural background dependent solely on the resources as an extension of the household is very much a part of our lives. The aspect of rural life in learning to conserve what you don't know or don't know the value of was an important part of my upbringing.

I don't know to what degree other people have lived with the resource. If you live in a bushman community, the resource is much closer than it is to other people. You are brought up to believe in your contribution at your various ages in relationship to the different biological components. When you grow up as a young child, you spend your time collecting the different fruits; then, as you grow older, you begin to look after the livestock. You collect things to carry home and that's a different stage from going around with your mothers who are responsible for collecting a certain aspect of the biological biodiversity. As you move on to look at the livestock, your focus mentally becomes the grass and the various profuse trees. So at each stage, you move to different levels of learning about the resource. You are not introduced to or swamped with all the detail at the same time, even though what you have in the end is a deep understanding of the resources.

The placing of value on biodiversity or resources is the issue that people fail to compromise over or understand fully as far as I am concerned. What value can you put on the resource? It's always very difficult for me to do. All I know is that it is immensely important. If you ask me, I will probably inflate the value of the resource beyond what any economist would do because the price must reflect the value over a period of time. The economic world deals with today and what can be used in order to achieve the next development project. An economist and I are two incomparable people in terms of the perception of the long term. People may look at our differences as a consequence of ignorance from a development perspective. Yet we must look in the long term. If we depend on the natural resource, its sustainability is not related to the income you can get now. The income that is available at present is exactly what causes the traditional or rural society not to be as conservative of the resources. As perceived from our side, the pressure of acquiring money now is great, and we have a second way of looking at the resource. Because of this, we recognize the very important and serious concern of a document that I'm about to produce which looks at the Wangatu tribe and the current methods of harvesting resources for sale and the encroachment of other people from the outside. The people from the outside are not as conservative as the Wangatu people who live with the resource and see it as part and parcel of their lives now and tomorrow. For people who are involved for their entire lives in the natural resources around them, the loss of the resources ruptures the whole structure of their communities and really effects their tradition and culture. You can't put a money value on that.

There are critical issues for both education and resources. The value that I always find and try to prove is in the context of the lives of the people. It's very difficult for us to maintain the life-style with dependence on the resource because we are constantly bombarded with information. What is important to know is that we are going to stay the way we were at the beginning. At the same time we are willing to incorporate what comes to us by grafting it on to what we already have, but the pace at which it comes makes it difficult to retain what we have on the ground.

Our participation is always individual and is affected by outside pressures which we cannot control. But if given the opportunity, and despite how much I value the number of cattle I have, it's nice to be seen in a Mercedes. I'm caught between the different values that society forces on me.

If there was a single thing that represents what biodiversity means, it would have to be the forest which has everything that is really worthwhile. I'm often irritated by attitudes toward wildlife because, to a lot of people, wildlife is what natural resources are all about. Yet the forest is the resource upon which the wildlife depends. The trees hold even the grass and maintain it by holding the water. The role of trees in maintaining the structure of the soil is often ignored. We have a word, *dukwa*, which is quite important. It not only means the forest itself but represents a whole concept of what the forest protects. This is why we traditionally had a *naghayasiuva*, the piece of land where your best livestock or wildlife is being raised and is the protected area. We don't hunt in that area to allow for the recovery of particular species. You cannot go into one of these areas without the permission of the chiefs who maintain control and act as the overseers of the resource. If you are caught with an axe in that place, you don't have an argument. You know you are charged without trial.

Botswana - Kaichela Dipera - Mukalahari

My name is Kaichela Dipera. I am a Mukalahari in Botswana.

Biodiversity means the plants, like grass and trees, and the animals including the birds. There are many things of value here to highlight, and it goes beyond just the source of food and shelter. We are sitting under these trees which provide me and my family with shade and medicine.

What is more important to me is that my husband and my household are going to be sure that the animals and plants do not become extinct. We have all of these traditions which conserve the things around us. For example, the pieroria is one of the fruits that we have always used. We only harvested after all the fruits are ripe and then the fruits are dried. We eat the fruits in two ways. Part of the fruit we eat without grinding but we make sure that each seed is thrown back into the world for another generation. We grind another part and then add water so it becomes juicy and we use the juice as food. Again the seed is taken back to the world so that it can grow. We also have some taboos which protect things. For example, after the harvest of the fruit, nobody is allowed to go and cut a branch from that tree to use as a walking stick. If you do cut a branch, it is as if you have gone against your way of life. While we do use some branches to form a hedge or use as firewood, you would not just cut any branch. You have to do this in a very selective manner. We only select thick branches for use as hedge and, when they are dry, we use the branches as firewood.

I cannot think of any specific value that I can attach to all the things that surround me. I can say biodiversity means my whole life in relation to the bits and bits of biodiversity I mentioned earlier.

As an example of what the environment means, the government is asking the Bushmen people to move away from the centre of Kalahari Game Reserve from a small place called Gards. The government wants to give an opportunity for the area to be developed. The experience of moving away is so painful when you think of it because [the people] are moving from a place where they have been living for a long time. They know what the plants are for, they know the sources of water and food. When people are moved to a new place, they are cut off entirely from their culture and moved to a place where they must start a new culture. This doesn't always work out because most of them die. In the new place, if one gets ill they don't know where to get *chuba* as medicine. Even in terms of their rituals, moving them to a new place means they have to start from scratch in developing a new culture. This doesn't work out well. Most die in the process, and I wouldn't like an experience like that to come here.

Botswana - Rekamani Mutupi - Mukalahari

My name is Rekamani Mutupi. I am from the Mukalahari tribe and we are found in western Kweni in western Botswana. I am a church leader and a member of the farmer community.

Biodiversity to me means a lot more than just my day-to-day livelihood. Some of these things are linked to our culture. For example, we have some groups which started with buffalo while others started with other wild beasts and these are respected as parts of our culture. We would not kill these animals, and if someone else did kill one of them, we would not eat the animal. In general, there are a lot of smaller animals and plants that are very much close to our hearts and we do a lot of things to protect all of these as parts of our lives. This is our place and we don't want to be moved from here. If you took us to another place, we would not be pleased. We always want to live our lives where great-grandfather stayed.

Brazil – Daniel Matenho Cabixi

Biodiversity is a Western concept that has no correspondence in the language of Paraci people. The daily practices of my community and the ecological niche that we occupy, however, demonstrate that we have a specific awareness of biodiversity. We have our mythological hero who is called Wasari.

If we relate his teachings to the notion of biodiversity, we realize that Wasari was not only a great ecologist but also had a profound knowledge of the place where the Paraci live. Wasari allocated territory to the different Paraci groups in the Chapada (the name of the Paraci region) and taught them the technologies of hunting and preparing and consuming the natural resources. Wasari further established political, economic and political principles revealing how to deal with other human beings, animals and nature.

Wasari, during his pilgrimage among the Paraci, established a set of principles that the people should follow in the management of their resources.

In relationship to the fauna, Wasari indicated the species that should be consumed in the daily diet. Spiritual symbols and the relationship to the supernatural have determined, until our time, which game cannot be consumed under any pretext. The game is not the property of the individual hunter but rather belongs to the entire group. The same is true of the flora of the Chapadas. Similarly, Wasari determined that crops associated with the practice of agriculture, for example corn, sweet potatoes, bananas and peanuts, also should be consumed collectively. Wasari named the headwaters that feed the big rivers of the Amazon Basin and the Rio Plata as the source of life. Important geographic accidents that exist in the Chapada were also recognized by Wasari and the Paraci people still preserve his insights as an active part of their cultural knowledge. In my opinion, we can compare Wasari to the important environmentalists of the modern age when we consider his profound knowledge of our biodiversity. Without Wasari and the possession of his knowledge, the people would not have any possibility of surviving in the ecosystem of the Paraci Chapadas.

The balance established by the teachings of Wasari was interrupted when we were conquered by a different civilization bringing problems to us and other indigenous peoples in this part of the world. Our traditional territory of 12 million hectares where we had maintained a cycle of utilization of natural resources by organizing lands for hunting, agriculture and collecting has been reduced to 1,200 hectares today. Now the Paraci have to face a number of serious limitations in order to survive. The political system established by Wasari is now disorganized. Many important areas for gathering and hunting were taken away, including the headwaters of the rivers, the source of all life according to the Paraci understanding of the world. Our former lands are now in the hands of the white landowners, and this new form of management has destabilized the natural resources of our territory.

This imbalance is also the consequence of mining activities (for gold, diamonds and other precious stones); exploitation of latex; the construction of highways; and, at present, the creation of plantations of monocrops of soya and sugar cane and the introduction of cattle. In turn, these changes have led to a substantial loss of Paraci knowledge of our surroundings, our ethnoscience.

In recent times, a major concern of the Paraci people has focused upon the larger national economy because this is important for our political survival. Many want to copy the model of development that the outsiders implemented in the Chapadas of the Paraci and are involved in extending monocultures that have nothing to do with our traditional knowledge. So the vision that the Paraci people have of their future doesn't give them a dignified or secure position. About six percent of the traditional territory is occupied by extended plantations of soya, corn and beans.

We are very worried because we notice big changes in our environment. The rain cycle is not following the traditional pattern that we knew 40 years ago. In the middle of the wet season, we sometimes get 20 days of heat and no rainfall. The raising of temperatures is another phenomenon that we can talk about. In the past, we always had mild weather but now the heat has become unbearable. There are other changes which also threaten our way of life. For example, three years ago clouds of grasshoppers invaded the plantations and the white farmers blamed the indigenous communities. We have become the scapegoats for an accident that was provoked by the landowners themselves. The grasshoppers have been known since the first decade of the century when Rondom (the Marshal who set up the Indian Bureau of Brazil) made several expedition to the Paraci. In his writings, he mentioned the grasshoppers in the Chapadas. Today, there are greater numbers of grasshoppers because their natural predators have been reduced or eliminated. The hawks, armadillos, snakes and birds are now gone because of the clearing of the native flora and use of pesticides in the fields.

Until the coming of the outsiders there was no problem with the recycling of our waste. All of the indigenous products went back to the land. Waste is a problem with industrial products. It is important to stress that Indians live in villages and not in concentrated populations. We had open skies in backyards and waste was never a problem because the villages were not permanent. The Paraci never stay more than a decade in one place. With time, nature absorbed all the trash.

We think about the ecosystem of the Chapada and about when Wasari walked the territory pointing out to the people the medicinal herbs and edible plants and roots, and teaching how to prepare them. He taught similar things about the animals and the fish. We remember that Wasari taught religious rules and how to cook and consume the animals. For example, game should be consumed after religious rituals. Until now the Paraci have kept all those beliefs and ceremonies. In order to have good a game harvest and good production of wild plants and roots, the rules taught by Wasari are a foundation of spirituality and guarantee the physical and spiritual survival of my people.

The Paraci also believe in the existence of a supreme being called Inore. He takes care of both the spiritual dimension and the material life.

Wasari and his brothers were the mythical heroes of the people. The strongest cultural reference for the Paraci are the sacred flutes. Our religiosity is maintained through the sacred flutes in our ceremonies. They are mementoes of the past saga of old people that we call *Kinohiriti*. These elders or *Kinohiriti* are respected with seriousness and have a great charisma in the flutes. The beliefs are transmitted from generation to generation and there are several rules and dogmas in terms of the use of the flutes which are used in important moments of our lives. For example, they are played in the initiation and nomination ceremonies and also when there is an abundance of food. The sacred flutes are the fundamental links of our congregation and allow the Paraci to overcome internal problems our their society.

One hundred or two hundred years ago, religious feelings permeated every aspect of our lives. In the daily activities of hunting and collecting, bathing in the rivers, manufacturing our goods, and in the relationship between parents and children, the religiosity was always present. Now it is different.

We have the example of the Saluman Indians who are our neighbours living on the Juina River. They have maintained their religiosity untouched. Every minute is filled by religious feeling. The Paraci have lost a lot of this. We have such feelings but only in special circumstances. We are very worried about our young people because they are not as interested as they should be. We are noticing a divorce between our old traditions and the present reality as a consequence of the interjection of new values from the outside. This can be seen as a rupture between the young generation and the traditional patterns. The new generation is more open to the values available in Western society, including those found in the religions. This is a crucial problem for us. That is why this year we are starting a project called *Tucum* that aims at the training and capacity-building of teachers who could help to revitalize our culture. The project is not only for the Paraci people but also for neighbouring groups. Through the schools and a special pedagogy we are giving back the past and reconnecting with our traditional values. In our treatment of illnesses, every day the Paraci are more into taking Western medicines. At the same time, I am personally stressing the use of traditional medicine. In large part, the people trust more the allopathic medicines than the traditional medicines. Only a few herbs are still used by the shamans, Pagi.

Maybe the only way to restore the use of traditional medicines would be to conduct field research on all the herbs that can be used and define, through technical test, the use and function of these herbs. Once the evaluation is complete, a report should be given to the Indians proving that the medicines are important for health treatment. I see that my relatives are losing credibility when the use of traditional medicinal knowledge is suggested. Very reduced numbers of roots and plants are still used as tea and vapours. This trend must be reversed. The scientific institutions could have an important role in the process of returning the knowledge to the Indians.

Wasari's presence among the Paraci is still very strong. I believe that if you excite the new generation to practice the orientation passed on by Wasari – the respect for nature and all living beings – the Paraci will overcome their social problems and the current political and economic crisis. I believe that all native peoples should return to their traditional ethical and religious practices as terms of reference. For the indigenous peoples whose lives are disrupted, this is a turning point: not in the sense that we will live exactly as we did in the past but in the sense of integrating our traditional knowledge with Western knowledge. Within this conjunction, we must establish principles and new criteria that will guide the behaviour of native peoples into the future. I see that this is the path that we must follow to survive as ethnic peoples. Every indigenous group should design its own rehabilitation programme for its youth, allowing them to be with their mythology, their traditional history and values for human relationships and religiosity. At the same time, they must be given information produced by Western society and be allowed to analyse and criticize both indigenous and Western ways. A critical analysis of these values must be made, determining at which point there are correspondences for better human, religious and spiritual understandings. The youth must be given the freedom to make their own choices. This is the only way to stimulate the new generation to find out about traditional values. Both the indigenous and Western ways must be placed at the same level of importance for the comparison.

Canada – Gamaillie Kilukishak – Inuit (Meeka Mike – translator)

I am Meeka Mike. I am the translator for Gamaillie Kilukishak who is our elder and representative of the Baffin Region Inuit Association.

What is biodiversity? Gamaillie's response is something that he tells researchers who want to study in the north where he is from. You need to have the knowledge of the animals and the land that they live on. You must be in constant contact with the land and the animals and the plants.

There are different things that live around us: things for food and things that are not for food. There is value in things that are used for food. There are other things like birds in the sky, a little green stuff that comes up in summer, whatever, but we are very dependent upon the living animals, mostly wildlife, and plant life is less important. The Arctic is surrounded with snow and ice most of the years. Plants are not always available.

When Gamaillie was growing up he was taught to respect animals in such a way as to survive from them. At the same time, he was taught to treat them as kindly as you would another fellow person. While we don't have ceremonies to what the Europeans call nature, we were taught certain values and laws to live by. For example, if you came across an abundance of a certain kind of animal, you would take only what you needed. We do not over-hunt for the sake of the animals.

There is a yearly cycle but it isn't talked about directly in our culture. Rather we are aware of the cycles by living and experiencing them. For instance, we hunt the caribou in August because then it has new fur. This new fur lasts longer when used as clothing. In the early spring, which is March, the seals start to have their pups and that's when we start hunting for them. Their skin is new and good for clothing as well. In the summer time, the women start picking berries which they save for winter or harvest as needed during the summer. There are other small plants that you eat as you walk along when hunting or travelling to another camp.

We live in a very harsh environment. With any animal, some years are good and some years are bad. Life is tough. You try to get any food you can – the birds, ducks, and so on. These are birds that start coming up in the spring and are gone again by August or September. All these things are the resources you can see, and the seasons that they come in determine when you use them.

One more thing that we believe is that if there was any cruelty to animals, if they were made to suffer or were shot just for the sake of killing them, then some time in the future the animals will attack back or take revenge on the person, his dependants, his grandchildren, and his great grandchildren for such misbehaviour.

Costa Rica – Juan Vargas, Rodolfo Mayorga, Carmen Leiva, Aníbal Morales, Gloria Moyorga, Juanita Sánchez, Eustacia Palacios, Catalina Morales

In our indigenous communities, each generation teaches our history to the next generation. That is how we, today, have inherited our knowledge from the first generation of indigenous people, created by Sibö. In those times, Sibö walked among our first ancestors here in Talamanca, and he told them many things. He told them, for example, how he made the earth and the sea, how he brought the first animals to the earth, and how he planted the first trees. He also gave certain laws to us and told us how we will be punished if we fail to obey them. We also know that all things that exist in this world have an origin in another world. What we see here as one thing, Sibö sees as another, because he made all things and knows their origins. He made the first indigenous people out of corn seeds, for example, so for him we are corn seeds.

All of the things we know here in this world are like shadows or reflections of their origins in another plane of reality. For example, the venomous snakes are Shula'kma's arrows; the tapir is Sibö's cousin; the wind is Serke, Owner of the Animals; the sea is MnuLtmi, a woman who Sibö converted into the sea; the rainbow is ChbëköL, a snake that ate people during the time of our earliest ancestors; the stones that the */awapa/* use are Siá, Sibö's sister. We are very respectful toward these manifestations of celestial beings, and sometimes fearful, because we know they have meanings and powers that are invisible to us.

Sibö had a relative named PlékeköL, which means King of the Leaf-cutter Ants. PlékeköL worked with Sibö and the other immortal beings in the other world. When Sibö decided to create this world, the earth, the first people he made were the indigenous people. He brought corn seeds from the other world and planted them in the soil of the earth, and they grew into the first indigenous people. We are called */dtsö/*, which means corn seeds.

When Sibö was instructing the first indigenous people, he showed them the image of PlékeköL and said to them, 'This is PlékeköL. He is my relative, and he will be the origin of another race of people, the */sîkwa/*'. White people have their origin in an immortal being, a relative of Sibö, but indigenous people have their origin in Sibö's corn seeds. We belong to Sibö: he is our owner; we are his things, not his relatives.

Sibö made white people in the day. That is why white people are more scientific, because Sibö gave them the intelligence to do many things that the indigenous people can't do. White people make cars, planes, boats, money; they make many things. But Sibö only taught the indigenous people to plant, to raise animals and to hunt and fish.

There is a big difference between white people and indigenous people The origin of white people is the King of the Leaf-cutter Ants. Just look at the leaf-cutter ants, how they all work together cleaning and clearing all the land around their nests. Where the leaf-cutter ants live, all the vegetation is gone because they cut every last leaf and take it back to their big nests. That's how the white man is. He works very hard, but he destroys nature. He chops down all the trees to make his big cities, and where he lives all the vegetation is all gone. There is nothing there. The white man cuts down everything that is green, and where he lives there are no trees, no rivers, no animals. He destroys everything in his path.

On the other hand, the indigenous people don't work so hard. We plant corn, raise animals and live in the forest. We like to see the plants and the animals, the birds and the rivers around us. We don't like to destroy nature, we like to live in nature.

Sibö is the Owner of indigenous people. He takes care of us like an owner takes care of his possessions. Animals and plants also have Owners that take care of them, just as Sibö takes care of human beings. The Owners of plants and animals are supernatural beings, and they are very powerful. They don't like to see us mistreat their possessions, and in fact they punish us if we abuse their animals and plants. That's why we have to be sure to obey Sibö's laws. He taught us how we should live with all the things on earth.

Sibö gave this law to the indigenous people: we are not to misuse or abuse the animals. When we go out to hunt, it is a sin to leave an animal wounded. We have to kill it quickly so it won't suffer. And if you hunt an animal knowing you're not going to be able to eat it all, eventually you will be punished. You will go out to hunt and you won't find a single animal. You may hunt and hunt and hunt, but you won't get anything because Sibö is hiding the animals from your sight. I have known of cases like this.

If we sell the meat of wild animals, Sibö punishes us. When indigenous people kill a wild animal, it is to eat the meat, not to sell it. We have pigs and chickens and cows to raise and sell, but if we sell the meat of wild animals, we will die sad.

Right now there are squatters occupying parts of our Reserve. They say, 'Why do you Indians want so much land? You don't even cultivate it!' We have explained to them that we are taking care of the forests and the animals, but they only want to destroy everything. They don't understand that we want to protect the forest so the animals can live, and that we need the forest because it gives us many things that we need.

We are poor. We don't have money to buy zinc, so we need to protect the plants whose leaves we use to make thatch roofs. The forest gives us lianas that we use to make baskets and fences and houses. The forest gives us medicinal plants. And we like to have the forest around us. To us, a big pasture, say ten or twenty hectares, is not beautiful. The forest is beautiful because Sibö created it. Yes, we clear small plots to plant, to live, but we would never cut down the whole forest because it is valuable to us just as it is. We need it to live, too.

An indigenous person without the forest is a sad person. We know that without the forest, we wouldn't have water. We only have a few sources of water in our Reserve, and already the squatters have destroyed several of them. One squatter family took some land high on a hill at the source of a river. They cut all the trees along the riverside to make a big pasture, and the river is almost dry now. What water it carries is contaminated from their cows, so the people below can't drink it. It's hard to move these people off the land. Sometimes we feel sorry for them and we tell them they can stay but they can't cut any more trees. But really, they are endangering all of us.

In about 1956 or 1957, the government made a law that said that everyone has to respect private property. That means that you can't hunt on private property without getting the permission of the owner. But we didn't enforce that law on our farms here. White people started coming in to hunt, and we didn't do anything, we didn't say anything to them. They would walk right through our yards, by our houses, with their rifles and their dogs, and we never said anything to them. Why? Because we were so timid, we were afraid to say anything to them. At that time this was not an Indigenous Reserve, and we didn't know our rights.

The white people don't respect the borders of the Reserve. They come right through my yard to hunt. One day I went out to report them to the authorities, and they got furious with me. They said this place doesn't belong to the indigenous people, they said they would beat me, they said they would kill me, and I don't know what else. They keep coming in to hunt. They live from that. I hear their dogs about three times a week. For the animals to live, we have to preserve the forests. And that is what we want. We want the animals to return and reproduce. We have the faith and the hope that the animals will come back again.

Selected narrations from *Taking Care of Sibö's Gifts* by Paula Palmer, Juanita Sánchez and Gloria Mayorga.

Ecuador – Cristina Gualinga – Quichua

My name is Cristina Gualinga. I am a descendent of Pando, from the Quichua group, and I belong to a big Sarayac community.

My grandfather never required money or clothes. All that he needed could be found in the jungle and he was so strong that he could transform himself into a tiger and travel quickly wherever he wanted to go. My grandfather's story is our past, our history. My uncles are always with our grandfather when they take *ayahuasca* (sometimes called the visionary vine because it transports people into other perceptions of reality) because, for us, the spirit doesn't die. Grandfather talks and we can hear his voice. When someone takes *ayahuasca* and I decide that I want to talk with my father's spirit, the shaman calls him and we talk together. My father gives us advice and he tells us things we want to know. We believe – and it is true – that there are lakes and big hills in the jungle where our souls and the souls of our ancestors live. We communicate and live together with everything that is alive because everything has a spirit which strengthens us. In the river there are many things that are alive and that are a part of our lives. These are our friends. We live with them together, dreaming and helping each other in whatever is necessary for our maintenance. To communicate with the spirits of the men and women from the lakes and the big hills in the jungle we have to have certain discipline to gain their friendship. This discipline consists of alimentation, sexual relations, harmony, a good heart for others, and the like. The same corporal discipline is necessary for friendship with the animals.

In my family there have been two shaman women, my aunts Valvina and Shimona. The possibility of shaman women has to do with our ancestor Pando. One shaman woman was Pando's granddaughter and the other was my mother's sister. To be a shaman is difficult: You have to isolate yourself from the world and do nothing but good things. As shamans, we don't have to provide our body with everything it wants. Instead, we just live like spirits having the strength survive and continue forward.

There were some shaman women in the Quichua group, but no longer, because of colonization and the influence of outside cultures. The life we now live isn't like it was in the time of my grandfather. People have grown closer to the Western culture than to their own traditional culture. This causes a lot of damage to our way of life and there are just a few groups of indigenous peoples who want to keep alive the traditions that are beneficial for them.

My community has a big river and the jungle is without any exploitation, all virgin rainforest. Along the side of the river we have our plantations, the traditional *chacras* (agricultural fields) that we use for our own sustenance. There are several different potatoes, bananas, yucca, fruits, and so on. We use the forest far less frequently than in the past for hunting and collecting fruits because we have realized that if we continued exploiting the jungle, it will be empty. This cannot be if we want to survive.

Our medicine is from the drugstore that is the jungle. Everything is available to us. This is the richness of the Amazon. The forest also gives us perfumes, and plants that give fertility, infertility and endurance for the long walks.

Nature, what you call biodiversity, is the primary thing that is in the jungle, in the river, everywhere. It is part of human life. Nature helps us to be free but if we trouble it, nature becomes angry. All living things are equal parts of nature and we have to care for each other.

For us, the value of nature is very great. In the jungle there are also men and women who are our friends, and with them and all else that is alive we join together and build a unit. In our thoughts we can talk and listen with all the love that we have. Love to nature because everything in nature is alive. Monkeys are a group of persons and, at the end of the destruction that man is causing, they will come together like the indigenous peoples and attack those who are destroying our environment. They will know who their friends and enemies are. This is our point of view and it doesn't matter that they can't speak because we can speak for them and defend them.

When we have to kill or hunt an animal, we must ask permission from the owner because they are not free. The animals form groups like we human do and each group has an owner. We have talked a lot about this with our ancestors. The tapirs, for example, have their superior who guides them through the forest. The collared peccaries also have a chief who directs them to places where their relationship and their communication is good. But men cannot see all of this, and that's why they continue the destruction. The oil companies don't see it so we have to defend the jungle against the oil companies that destroy nature.

The drinking of *guayuza* in the mornings is a sort of seminar for planning, to advise the children, to plan the work for the day, and to see how young people should be to live well in the future. While drinking *guayuza* we also comment about our dreams, about our experiences, and who we have seen at the lagoon. *Guayuza* is a plant that makes us stronger.

In my community there is still respect for the shamans and people believe in them. There are some of the young people who don't want to believe, and who just want to kill the shamans like in the Western world, but this is very bad. The shamen are very powerful people. They are the scientists of nature. Without having been to the university, they know everything. They are very important.

Ecuador – Report on the Conference/Workshop on Biodiversity organized by the Confederación de Nacionalidades Indígenas del Ecuador (CONAIE), the Confederation of Indigenous Nations of Ecuador

Participants: Bolivar Beltrán, Confederación de Nacionalidades Indígenas del Ecuador; Piedad Cabascango, Federación Pichincha Riccharimui; Rafael Tapuyo, Federación de Centros Chachi de la Provincia de Esmeraldas; Nancy Velas, San Antonio de Machachi; Fredy Pianchiche, Federación de Centros Chachi; Bayardo Lanchimba, Federación Pichincha Riccharimui; Yesenia Gonzales, Costa de la Península de Santa Elena; Melecio Santos, Comuna San Pedro.

The Confederation of Indigenous Nations of Ecuador (Confederación de Nacionalidades Indígenas del Ecuador – CONAIE) assumed the responsibility for collecting statements about biodiversity from a number of indigenous communities. After consultations, it was decided to hold a conference/workshop in Quito in which the concept of biodiversity would be explained and the participants would then be asked to give their interpretation of biodiversity from the perspectives of their communities.

In order to facilitate the collection of information, the leadership of CONAIE decided to have the participants create visual models of biodiversity which would serve as a focus for the discussions. The indigenous leader of the conference/workshop, Mr Bolivar Beltrán, served both as an educator in presenting information on biodiversity and as a facilitator leading the subsequent discussions. The information that follows was taken from the workshop/conference report entitled, 'Memoria, Curso – Taller, Tema: Biodiversidad,' (Record, Course – Workshop, Theme: Biodiversity).

Bolivar Beltrán: One definition of biodiversity describes the concept as all of the variety of life in the oceans, on the earth, and in the rivers. We know that a plant is alive and even the rocks go through a transformation process meaning that they too are alive. They said, if we talk about protecting biodiversity we are not only talking about the territories that have been declared by the government as reservations but the whole country and even the whole continent because the political boundaries are only imaginary. Protecting biodiversity is the responsibility of mankind around the world.

The Shuar (an indigenous group in Ecuador), at the beginning of the creation of their community, had not discovered fire and used to eat everything raw. There was a big tree where there was always a fire burning. Every animal in the jungle wanted to own that fire in order to be protected during the storms, but none of them knew how to take it. Then a jacobin bird, called *Jempe* by the Shuar, came. He was fleeing from the storm and casually found the big tree with the fire. When he got closer, he realized that his tail was on fire. The jacobin shook his tail and flew off, finally resting on a pile of withered leaves and starting a big fire.

What we learn from this story is that even that small bird helped the people of the community keep the balance between the living nature, earth, water and the fire that the jacobin bird brought. Every plant and animal has its place and helps maintain the balance of life.

To the indigenous population, the vital elements of nature do not have a price: they are invaluable. To others, nature does have a price. When the oil companies arrived in Ecuador, they destroyed much of nature. Now they need to repair what they have destroyed and this has a price for them. Despite what they pay, they cannot bring back what was destroyed. Nature is invaluable.

What would happen to us if the animals that help us by announcing deaths, visits, etc. all disappeared? All these beings are essential in nature, and the scientists ignore this. If our yoke of draught animals disappears, this will affect the whole community. But nature is more than just the plants and animals, the birds and fish. Dreams are a part of the spirituality of a nation and of nature itself. Every nation has its origin in relation with nature. What would happen if nature is lost?

In the Amazon on a rainy afternoon I could hear the noise of big trees being cut. The old man who owned the house where I was resting asked me to listen carefully how the tree screamed because it did not want to be cut. He was talking about the terrible noise of the tree branches while it was falling. The tree was trying to grasp the other trees and screaming for help. If we think about this image, we can see that a tree is alive. The old man asked me to tell this story and to talk about the terrible feeling I had so that other people could understand the tree's pain.

I also saw two little birds who were flirting and asked the tree permission to rest on one of its branches and build their nest there. When a tree is cut, a whole life chain is affected – including the flirting birds.

Humans are the only animals that have developed their intelligence to the point of language and complex thought and, at the same time, the only species that is destroying nature. As a consequence we are destroying ourselves. Humans know what they are doing but cannot stop. We are breaking the balance and will pay the costs in the end.

There are some things in nature which do not have a price for us or any other indigenous nation. Things, like air, water and soil have very deep meaning. Biodiversity cannot be measured but it does have a value, a spiritual value that is like life itself. The spiritual element that represents the yachag, a wise man who cures the others, shows that there is not only a material aspect but also a spiritual aspect which is very important to biodiversity.

At one time, we believed that biodiversity was the Amazon and that we could only talk about biodiversity there. Now we know that biodiversity exists even in the cities. Latin America and the remainder of the world is becoming increasingly urban. Everybody is moving to urban communities and this causes many problems, like the creation of marginal zones around the cities. Houses are built where forests once stood, creating a serious problem for biodiversity. The rural areas are becoming deserts because the only people remaining are the very old who cannot work any more. As a consequence, there is no one to care for the land. This is a way of destroying nature.

The Amazon is threatened with deforestation to provide pasture for cattle and land for other agricultural activities. To protect the jungle, the two-hectare Omaere Ethnobotanical Park was created in Puyo-Pastaza in 1993. The purpose of this park, managed by indigenous groups, is to protect the plants and trees that live close to the Puyo River. Many of the species are used as medicines by several groups like the Shuar, Achuar and Huaorani. One of the objectives of the park is to help indigenous people maintain their knowledge about the Amazon jungle.

If we talk about biodiversity, we have to embrace the whole country. Biodiversity concerns every living being. While the government has designated 38 areas as forests in the Amazon, we must go beyond these forests in our effort to protect nature and include the entire territory of our country and the continent as a whole. Biodiversity is a world human community responsibility. It is very important that we have our objectives very clear.

We, as members of the CONAIE want to present our own interpretation of biodiversity. We invite you to build a model of what biodiversity means for each one of us for the people who will come in the future. We want to share what we think about nature and to make a proposal for environmental conservation by the indigenous communities and by the larger society. The questions to be answered are:

1) Who are we, and where do we come from?
2) What is nature and biodiversity for us?
3) What kind of value does biodiversity have for us?

Piedad Cabascango: My name is Piedad Cabascango and I speak on behalf of the Pichincha-Richarimui Federation. I am from the Pedro Moncayo Canton of the Tupigache Patronage in the Province of Pichincha and I live in the Loma Grande community. My nationality is my language, once called Runashimi, then Incashimi and now Quichua.

In my region we preserve our culture and some traditions. Culture and biodiversity are intimately related. If we consider the birds and animals in my community, the people say that when a *cuzungo* cries next to one's home, it means that a thief is around and some people are going to die. When two sparrow-hawks fly around a house, it means that somebody in that house is going to get married. When the *cuyes* (guinea pigs) are screaming, they say it's going to rain or that strange and unknown people are going to arrive. When a turtle dove cries on top of the house, it means that people are going to die and that everyone must be prepared to cry.

There are also several beliefs about dreams that reflect our biodiversity. When one dreams about avocados or about cutting a *zapallo*, it means that married people are going to die. When one dreams about corn and barley, one hopes to receive money the next day. If one dreams about wheat, its because somebody is going to arrive.

Our elders are important because of their understanding of nature and biodiversity. They know the dates on which we must seed our fields. For example, corn is always sown in November when the offerings are made. Potatoes are planted in late December with another planting around Easter and on the Eve of Saint John. In September, we always organize parties with typical music and we gather in Tabacundo with a lot of people. We always bring a godmother and the people bring all the fruits that have been gathered during the year.

The elders say that these parties are held to give thanks to nature, the sun, the moon, the soil, and the mountains for providing fruits during the year for nourishment and to sell in the market. Then, the elders say that we must begin to plant the seed again and prepare the soil; and that we must do it well.

At the same time, things are changing in our region. Because of the establishment of the flower growing enterprises, people are leaving their traditional activities and going to work for these companies. Today, many youngsters don't even finish primary school. They are too busy making money, working in the factories, and they don't even care about the community activities or working their pieces of land. This is sad because it's important to value nature. Unfortunately, in some areas our young friends are spending all their time working and have almost totally lost their culture, even to the point of forgetting our language and music. These outside forces pose a serious threat to indigenous communities because we are the ones that must maintain and generate our culture, our customs, and cultivate the soil.

Nature and biodiversity are represented in all the plants, the animals and human beings themselves. We are connected to all the elements, the mountains, the sun, and everything around us. Our rivers are also important places where life is found. In considering biodiversity, we have the testimonies of our ancestors who lived with nature. We consider spirituality to be fundamental, because its something integral. To me, biodiversity is all the beings that are related in nature: man, animals and plants, even vegetables, rivers, seas, animals in the jungle, and all the beliefs we have kept from our ancestors and from our dreams. Wisdom itself is also a part of biodiversity. I think that everything is important, because it is related to all of us, to all living beings on this planet.

Mari Yagual: There are some traditional beliefs in my community too. If my right hand itches, it means I will get money. I must scratch my hand and keep it in every pocket I have because I'm going to receive money the next day. Another belief is that if a spoon falls down while you are cooking, you are going to have visitors at lunch or dinner time. We have lost our traditional robes and our language, but some beliefs remain respected. When we go to our fields, we dig deep and often find ceramic pieces that are from our ancestors. We used to belong to an indigenous community of which we don't know much, but we do know we were part of the community.

Bayardo Lanchimba: We have some customs and beliefs. We have always worked with a yoke of draught animals, as well as our traditional work groups called *mingas*. We still celebrate Saint Peter's

Feast which is a thanksgiving to the earth. It goes from May to June during harvest time. The people from the community get organized and prepare *chicha* (typical drink) and dance the whole day.

Planting time is not clearly defined any more. It used to be in October and November but now, with all the weather changes, we plant and harvest whenever we want. In earlier times, the rains used to come immediately after we planted and we harvested in the beginning of summer, but now everything is changing and we are suffering. Our land used to be green: it was forested. There used to be animals – birds, rabbits, deer, and a lot of sheep.

We used to cultivate in ways that maintained the soil in good condition. We only used organic fertilizers and we had good results, but today there are only chemicals which give us smaller harvests and poor quality products. There used to be enough water because of the rich vegetation. Our ways of cultivating the fields to obtain food and following our habits and beliefs are important parts of our biodiversity.

To me, nature means the recovery of what we have lost and that we have been losing without being aware of the changes.

Bolivar Beltrán: A moment ago somebody said that we all are Mother Nature because of the naturalness that we have. Without being technicians, we have talked about biodiversity and what it means to us. This afternoon we will become architects and engineers and build a small flat model of what we think about biodiversity. We want to make things to demonstrate the meaning of the term because this is sometimes easier than talking.

In Chile there is a community called Mapuche and they have a cinnamon tree which is sacred. Every feast is organized around this tree: the tree is their first guest. We wondered why this tree was so important and the answer is that it provides them with food, gives them a shadow on sunny days, and protection from the rain. It is the most important part of their community and if their holy tree were destroyed, their history would also be lost.

In the communities we come from, we also have sacred plants and animals. We have, for example, our curative plants and forests that are essential to our existence. We all know the symbol of peace, the dove; the symbol of hell, fire; the symbol of the region of Imbabura, its lake; the symbol of the region of Pichincha, the mountain 'Pichincha'; and the symbol of the Chachi community, the forest itself. We all identify with symbols. Our history tells us that our flag is a national symbol; yellow from gold and richness (which is not well distributed); blue that represents the ocean which has been exploited by a few men; and red symbolizing the blood of those who died not only during the battle of 24 May 24 but of all indigenous people who died through our long history of suffering. How can we rescue a symbol like our thanksgiving ceremonies to nature for all the products it shares throughout the years.

You have been working in three groups to answer the following questions:

1) How was nature in the past?
2) How is nature at the present time?
3) How do we want nature to be in the future?

Please use the models you have constructed to illustrate what you have concluded.

Group 1: Subject – How nature used to be

Consuelo Méndez: In our community we used to work with ploughs like some communities do even today. We planted and harvested crops, we cared for cattle, and all of our fertilizers were organic. Today our fertilizers are chemicals. There used to be more vegetation and richer forests than in the present.

For this discussion we have divided our community into two regions, the ocean region and the river region. In the ocean region, as you can see, there are palm trees, big trees, and people fishing in their small boats. In the past, there was a lot of fish and marine life. The houses were tall and there were not many people. In the river region, there are small hills where people used to plant corn, watermelons, and even tomatoes, while other people raised animals such as cattle, birds, fish and shrimp.

Segundo Bautista: For me, it is like my friend Consuelo explained. There used to be more forests and lots of water and trees. Our houses were simple and we used to have more animals and other things.

Group 2: Subject – How nature is at the present time

Raul Tapuyo: We can see here the marks of the cut trees. With this I want to show that our forest is being destroyed in different ways. When you cut the trees, we get less rain, and the rivers will go dry. We can see here a river that is almost dry. You notice that there are only stones in the river and here is a skinny fish because there is no food for him. Not only this but the whole river is polluted.

Here is a little house and we understand that the construction materials come from the forest. If we destroy the forest we won't build good houses made of wood.

There is a skinny boy because there isn't enough food. We see that the animals die with the destruction of the forest. Man is in a difficult situation because it isn't easy to find food in the forest any more and this child is really sick.

There you observe birds that are flying because they cannot find trees to rest in and food to eat. With a little more time they will disappear. With the trouble with Pe, there was fighting in the forests. Every war, every bomb that explodes, destroys not only the forest but the surface of the earth and the animals, and pollutes the environment. This is what we see in our days.

Piedad Cabascango: Some time ago we used to cultivate our lands to produce food, but now in Pedro Moncayo and in Cayambe there are a lot of different flower industries. On the one side we have the flower industry, on the other side are a few trees – but these are being sold to the flower industries for wood.

Because of the pollution and the use of fungicides in the area, plantations have been destroyed and children born with genetic malformations. We have observed increasing amounts of rubbish and plantation waste. This causes problems for local people also.

Fredy Pianchiche: The situation is very difficult within the communities because of the exploitation of the forests in our region. Here is a bird on a fallen tree because he has nowhere to go. Here are the schemes about the cutting of trees, we have a little horse, another cut tree, and all this represents the situation in the Amazon. This young man suffers from malnutrition because of the destruction of the forest. I painted a river but it isn't dry yet.

Group 3: Subject – How we would like nature to be

Melecio Santos: We must have an unpolluted atmosphere so that it can rain. Water is necessary for our products like corn, bananas and our different animals. Here is a duck and other animals. We want these big industrial boats to disappear in the future. These boats are full of fish and prawn and these are sold to other countries. A long time ago just canoes went fishing and there was always enough fish. We also want a certain control over the quantity of fish taken because their survival in the sea must continue. There is never enough money and it is very important to us that there is an exchange of different products between the Sierra (mountains) and the coast. We could exchange fish for potatoes, salt, etc. as it was done in the past. In this way our lives would be calmer and our children could live in peace. In our model, we haven't included the shrimp or oil companies who have come and made false statements to us. Because we don't oppose these companies, there will not be any changes.

Maribel Guacapia: I don't think the trees have to be cut: we must take care of them. We must cultivate more potatoes, corn and other things because we and our animals need food.

José Aapa: In the coastal region we plant bananas, corn and coconut and we have canoes. We fish with a spear, we live in houses and we have an animal called monkey in our forest.

Bolivar Beltrán: Before we continue, I would like to think about what this work means to us, what it means to rebuild our country. We have to understand better what biodiversity means. At noon they asked me for more information about biodiversity. I told them that after the afternoon activity we could explain it better. We have tried to express through the graphics in your plans what we couldn't express with our words.

What is missing in our present understanding of the word biodiversity? Piedad said earlier that, in her opinion, nature did not have any value. Nature is composed by everything; the animals, plants, and human beings. We already have all the elements, we just have to put them in order and make a concept. It is all life that exists on the earth. It includes the rivers, seas, land, atmosphere, birds and stones. Do you think that we can give the water an economic value? No, we can't quantify it. Can we put a price on the ground? No, but it has a very important spiritual value. We cannot quantify water or the ground, but in our country they put a price on it. For example, the ranch Yuracruz which has 600 hectares costs 3,000,000,000 sucres (monetary unit in Ecuador). There are people who want to privatize water and sell it. There are things in nature and indigenous villages that we can't put a price on because they have a very important value. We can't put a price on biodiversity but it has a value, a spiritual one. Biodiversity has value like our own lives have value to us. My friends, the Chachi (indigenous people in Ecuador) say that in their region a fifty year old tree costs about five to fifteen thousand sucres. In the city you can find a single board that costs 2,000 sucres. I wonder what these people who continue cutting the trees earn. We have seen the people taking trees but we do not know exactly what is happening. We just know that they don't sell trees one at a time but rather by the hectare.

Melecio Santos: In my region, the lignum-vitae tree is the best wood and it is sold at price of 5,000 sucres and a board made of it costs between 35,000 and 40,000 sucres. This is too expensive if you compare it with the price they pay for just one tree.

Bolivar Beltrán: I asked you about the cutting of trees because we talked about the direct relationship between nature and man and we know that we are a part of nature. I don't know how many hectares of forest are cut in Ecuador but I do know that Ecuador is high on the list of countries in America in terms of the number of trees cut. A good number of trees have been cut and when we talk about the relationship between man, nature and animals, it means that everything has to be in order. It means that we don't have to cut more trees than is absolutely necessary. If we continue on our present course, there won't be any forests, any zones where animals can reproduce, and there won't be enough food. We are practically killing ourselves if we continue this way. There is biodiversity, but it is deteriorating.

What are we doing as indigenous people? Everybody says that before the exploitation of nature, indigenous people used to collect fruits from the trees and that's how they lived. Unfortunately, all this development of society has brought industry, which destroys. We have expressed our wish about how nature should be, but this is not the responsibility of those who come and destroy. It is our responsibility to understand how things are and to do something because an indigenous village or a bird is as important as our own lives.

There are thousands of different ways to understand nature and everyone has his or her own way. That's why today we are going to express our understanding through these models in which we used the elements of nature.

For the indigenous people is it very hard to express what biodiversity is, but the building of these models is our expression. I want to invite you to interpret these models that represent biodiversity.

Model 1

Piedad Cabascango: Our group had four persons and we represent both the mountain and the coastal regions. Each of us has our own concept of what nature and biodiversity mean. We have tried to represent all the plants, animals and humans because of our relationship with all these elements including the sun. We have painted a river in which an important part of life concentrates. We have a boat, palm leaves and other plants. Within biodiversity we have the testimonies of our ancestors who lived with nature. We have considered that the spirit is something very important.

Raul Tapuyo: We have represented the clouds because they are the breath of the jungle. Clouds come down as rain and form rivers. These rivers carry the most important element for human life and plants – water.

We have painted the sun because it is also an element of life. It helps the plants in their photosynthetic processes to produce energy for man and animals. We have also painted this mountain with snow because that is the origin of the rivers. We consider all these elements important because they are part of biodiversity and we have to take care of them.

Model 2

Yessenia Gonzales: In this part of our model we see the region of Esmeraldas, the city, in which we can't find any type of forest. We have painted our houses, hills, mountains, rivers, animals and a child planting a tree. We have tried to show handicrafts in the elaboration of a hat. We have painted an iguana which is part of the Galapagos Islands.

Melecio Santos: I want to start with this zone which has been exploited by the oil companies. Here we find just rubbish. From our trees, we take the materials for brooms which is an important help. We do not have to buy manufactured brooms. The people in the mountains collect materials for the roofs of houses because they are so poor that they cannot afford an alternative. There is the river that serves us as a medium of transportation and communication. The sun is a very important element. Boys, girls, men and women are all part of nature. We are all accustomed to live in this environment.

Model 3

Fredy Pianchiche: We are representing these coconut and banana palms in the model. Within the river we have a lot of rocks, a few fishes, a duck, and a canoe. All of this defines biodiversity and this is all I can say about our work.

Bayardo Lanchimba: We have represented this couple as practising typical indigenous customs. I think it is part of biodiversity too, as are the different arts of cultivating the ground and the obtaining of our food.

Segundo Bautista: We have represented this plant called totora (a reed used in the manufacture of a number of items) which grows on the border of lakes. You can find it in abundance in the province of Imbabura.

Bolivar Beltrán: We have tried to show a lot of the elements that are parts of biodiversity. We do not say that any one of these elements has more value than any of the other elements. They are all important. We talked about man, animals, mountains, the sun, and also about the cities and the modern buildings. We have represented our country in one model without the division of mountains, coastal and jungle regions. We have made an indigenous house that represents all the country. The spiritual element is the yachag, an indigenous wise man, that is curing. This means that not only the physical part is biodiversity but also the spiritual part, and human society must be included.

Summary of the models

I. What is biodiversity?

- All living and non-living elements that are in our universe. The elements are human beings, plants, animals, water, air, rivers, seas, earth, mountains, clouds, and so on.
- All of the elements that exist on the earth and in the seas, rivers and air.
- Everything that exists on this planet: earth, air, water, beliefs and customs.

II. What will happen to nature if we remove one of these elements?

- Nature would be destroyed because all the elements are the complements of nature.
- The balance among all of the elements of nature will no longer be in equilibrium.

III. What happens when we bring new elements into nature, for example chemicals?

- The balance of nature will be disrupted and we begin to lose the species that exist.
- We need all the species. Because they help us in our lives, they cannot be allowed to disappear.
- In our representations, everything is in equilibrium. Who destroys this balance?
- Human beings. We have actually destroyed much of nature but we can also do something to make things better.

Ecuador – Moi Angulema – Huaorani

My name is Moi Angulema, one of the Huaoranis. Huaorani is also the name of our language. The word Huaorani means a group, several persons, or a person who lives in a group. Foreigners name us 'savages' or *'aucas'* but to us, we have always been the Huaorani. I represent parts of Ecuador and Peru which was my territory but our land was split in half by the boundary between the two countries. We have settled in the provinces of Napo and Pastaza in Ecuador and have a jungle where I live in a community of the Amazon.

My group is but one of several Huaorani groups. It's a shame but in the Amazon every nationality speaks their own language. We have defended parts of Ecuador and Peru where we have lived for centuries. We still live there as Huaoranis but today we don't have contact among the Huaoranis of the various provinces. A lot of people know me. I have contacts to protect and I go around talking about the environment, the forest, protecting the jungle, and how to maintain respectful relations with the government. We want to look at other indigenous peoples for social and cultural exchanges so that we can get ideas in order not to disappear and to be able to live together.

We think that the Huaoranis have a right to live in peace and have enough jungle to make this possible. We have a right to have unpolluted rivers and to live without the destruction of our forest. Every group is fighting for its own interests. They have to look around on our planet to see how they can defend their own environment and help other indigenous communities do the same. We want contact with other indigenous peoples of the world to know how they defend themselves, how we can fight together, and how we can join to defend the lives of the Huaoranis.

The plants and animals are different types of life and also different from human life. The medicinal plants, tigers, eagles, birds and the forest all have shaman spirits. We must protect and defend the environment so that no one can cause damage.

There are seven types of beliefs: how the earth was born; how and where the forest and the water grew; how we live together; about the clay of the pots; where we come from, and why everyone on the entire planet has just one mother. Foreigners, Quichuas, everyone has a legend, a history. My grandfather showed the world that there were these types of beliefs.

There is a Huaorani legend in the zone of Pegonca. The Niemen was born woman as a group of rafts. The Peconga was man. The Peconga is a plant. The legend is like the story of Adam and Eve in the way that we believed in the Peconga. The Peconga came first and screamed and then all the evil came out. The sons of the Morete palm, the mother of all, then grew. From this place grew tall people about two meters in height. Until the present, the Huaoranis are still growing because we were small. These tall people didn't think about having food, they didn't know how to build houses, but then suddenly the anaconda 'boa' came from the sky to the earth. The tigers and the birds saw the boa and cut off its tail. It bled on the birds and that's how they got their different colours. The reason we have blood is because of the cutting off of the tail of the boa. The boa then acted as a god and punished everyone and that's why there is suffering and damage to life until our world is finished.

In the beginning there were just four women and five men and they were the origin of all people on the entire planet. They developed into too many people. As a consequence, we began to cultivate yucca, banana and many other things. Since that time, we have eaten this food. Because of what we learned from the legend, we think a great deal about the boa, the earth, the water, and the forest because they gave us life.

Now the bad civilization is cutting down the jungle, the river is polluted, and this is why the boa is crying. It was our mother, our grandfather and our father. The hearts of the Huaoranis are strong because the father is really strong. The boa developed all the birds, the rivers, the planet. Without the boa there wouldn't be a Huaorani shaman. He is most dangerous. When he gets angry, he calls the boa and it could come. When the boa dies and a Huaorani is buried, the blood goes inside the earth and that's what they call petroleum.

The Huaoranis are strong in everything they do and the land of Huaoranis has many forces. Life is Huaorani. The Huaorani have to be like they were in the past, united, not moving to the cities. Without earth we can't live, but with earth we live.

The *bigay*, which is similar to *ayahuasca* which we use to cure all kinds of sickness, is disappearing slowly. This root helps only the Huaoranis. Without this plant Huaoranis wouldn't exist. It is the most important thing we have. All the shaman say that without it they can't cure. Every plant has its legend. In saying this, it also says that the Huaorani has a lot of history because the Huaorani has parts of plants, the forest and the rivers.

Ecuador – Dr Nina Pacari Vega – Leader of Earth, Territories and Environment, Confederation of the Indigenous Nations of Ecuador (Confederación de Nacionalidades Indígenas del Ecuador)

To talk about the indigenous woman, the environment and biodiversity, we must start from the philosophy of indigenous peoples in *Abya-Yala* (an indigenous name for the Americas). It is only in this way that we can understand that the relationship between the environment and biodiversity is not casual and is not a modern topic. Further, this relationship does not simply follow purely ecological and economic reasoning. Rather, it comes from a profound rationality of the indigenous peoples, their philosophy, and their cosmic vision.

The indigenous communities of this continent established a deep respect for their collective individualities. This means that every indigenous group (Maya-Quiche, Quichua-Inca, Aymara, Apache, etc.) developed their own language, agriculture, architecture and science. At the same time, however, they developed a common conception about life, the world, humans, nature, etc. because the central axes in architecture, nature, etc. are the same. Humans integrate themselves to nature at the moment of their birth, and will never separate. They are a part of this greater whole. Nature is the big society – the society of life – in which man and woman have no distinguished place beyond that of other living things. In the bond of humans with nature, the priority is respect and admiration for the other living beings.

This is why plants and animals are sacred: indigenous peoples see linkages between themselves, nature and the spiritual world in many ways. Woman and nature come together and find their beginning in fertility. Corn, potatoes, coca and *ayahuasca* are the 'plants of gods'. Among the animals, the jaguar, the eagle, the python and the condor symbolize force, power and death. These are the foundations of an agricultural-ecological-environmental civilization which decoded the measurement of time in the sky and related it to the temporality of nature. In turn, this understanding serves in procuring food, maintaining health, and in the integral development of the people. They perfected methods of agriculture; moulded clay, gold and silver; built temples, pyramids and villages; and created our customs through myths and legends. This integral conception is very common in all the indigenous cultures. That's why there is no doubt that in the process of civilization that we all share, we have developed our collective individualities, our similarity as demonstrated in the conceptions and thoughts that we have about the world and life.

All indigenous peoples share the capacity of being abstract and concrete at the same time because of their complete comprehension of the phenomenon of life and knowing that all the different aspects exist as parts of everything else. We indigenous peoples do not separate the earth from the territory, the ground from the underground, the human from nature, etc. From our point of view, it is another perspective all together to conceive of the world, life and what they call environment and biodiversity today.

The creation of humankind

According to the indigenous nations' vision of the cosmos, the creation of humankind comes from the idea of imperfection. In the case of the Mayas, men and women were first made of dirt. The *Popol Vuh* (ancient Maya history of creation) says that human's flesh was first made of soil and mud but that they realized that it was not consistent, not strong, and could not move. The first man and woman created from soil and mud could talk but had no understanding or reason. They quickly fell and became wet in the water and, as a consequence, could no longer stand. The second form of humankind was made of wood. Wooden men and women existed and reproduced. They had sons and daughters but did not have a heart or intelligence, could not remember their creators. As a consequence, they died. The third trial was made of corn. Men and women were made of white and yellow corn and through this transition finally discovered how to be made of flesh and blood.

In every indigenous nation, the beliefs about humans, community, and universal creation is closely related with the idea of imperfection, with nature, with agriculture, with environment, with the cosmos. That is why one of the philosophical principles of every indigenous nation is the idea of balance and respect between human beings and nature.

Unilinear conception vs. Integral conception

According to our vision of the cosmos, every aspect exists as long as it belongs to a whole. But when the Western culture talks about its collective memory, it follows a unilinear direction which excludes, separates, classifies and breaks things apart. Let us remember that Hegel, one of the most important Western philosophers, said, 'If you want to know the nut, break it'. Western culture is used to destroy, observe and rebuild in a different way. For example, in the eighteenth century Francisco Jimenez discovered the *Popol Vuh* and decided to translate it into Spanish. He made changes he considered necessary without realizing that he was changing the text into a Western world conception. Later, in the nineteenth century, French intellectuals studied the changed text and decided to divide it into chapters because they could not accept that the *Popol Vuh* was such a lengthy text. Further, they tried to adapt it using their logic and reason – a strange and different way of thinking – which they supposed to be an improvement.

The numeric conception

Talking about the numeric system, we see that according to the Western conception, zero represents nothingness, emptiness. According to the indigenous conception, it is the end and also the beginning of something. These different conceptions are reflected in counting, calendars, and astronomy which was applied in agriculture and engineering and very important to indigenous methods of cultivating. In the case of some Mesoamerican nations, it focuses attention on the corn in terms of the cycles of planting and harvesting and other ratios, for example, concerning amounts of corn required to plant certain quadrants of land plus all that corn represents in their ideas about the creation of humankind.

Cultivation and other elements – the basis of the integrated conception

The associated cultivation is not recent and is not just any cultivating system. It is much more than that. This process serves as the basis of the integral conception that we, the indigenous nations, have about humankind, nature, production, and the use and conservation of the soil and other natural resources. In the department of Loja in Ecuador, for example, the indigenous group has developed associated, complex and integrated cultivation of seven crops, taking care of the soil with a varied range of foodstuffs for good human nutrition.

Many kinds of cereals, legumes, tubers, roots, vegetables, spices and fruits were part of our ancestors' diet. The quinoa, for example, is one of the richer cereals of the Andes. Among the legumes, there are many kinds of beans. The Bolivian and Peruvian Andes contributed potatoes and other kinds of tubers like *melloco* and *mashua* to the world food supply. In the Amazon, the yucca was an important contribution. There are also rich and nutritious drinks like the *chicha* made of fermented corn or yucca which are an important part of the ritual lives of the indigenous communities. Our knowledge of the tropical forests provide us with fruits, resins, nuts and soft and hard woods that have many uses.

Indigenous physicians recognized the therapeutic qualities of plants and animals in their environment based upon thousands of years of community experience and passed along through hundreds of generations. The medical contributions are enormous because much of what is actually known in pharmacology comes from our biological resources. It has been determined that 75 percent of the commercial medicines we use came from indigenous regions. But what have we received? What kind of guarantee do we have that our resources and knowledge will be recognized by those who gain financially from our contributions?

What modern people call the 'wild' or non-domesticated creatures were also included in our integrated understanding and care of the environment. The animals were divided depending on their spiritual influence over the nations. The jaguar, the snake, the eagle and others have important places in our religious worlds and we often represent them in our feasts and handicrafts. They are included in our mythology and related to our communities and nations.

It must be added that indigenous groups have found many ways to protect the soil without sacrificing productivity. They created a culture of agriculture based on diversity and applied their knowledge of astronomy to prevent disasters. All this represents the balance that indigenous nations have which maintains the harmony between humans and their environments. We cannot forget that with the excuse of a modernizing agriculture, monocultures and chemical fertilizers were imposed as well a whole knowledge of agriculture that did not belong to us and that is destroying our environment and depreciating indigenous knowledge.

Windbreak screen, micro-climate and integral management of the resources

Indigenous communities developed their own integrated management of the water, the soil, the plants, prior to contact with modern technology. That is why the existence of a windbreak screen made of native trees (*capuli*, *arrayan*, *uvilla*, *chimbalito*, *chilca*, etc.) allowed the creation of micro-climates for agricultural production. The very same trees that provided the windbreaks also provided products and space for medicinal plants and other valuable crops. In turn, this reinforced the economic development and consolidation of indigenous cultures. This integrated management represents a way of thinking, a culture, and the way of living of these nations.

All the cultivation and management systems that I have discussed were weakened during the colonization process and during the Republican period. During colonization, the land took on a new destiny as did everything built upon it. The integrated management of resources was destroyed. We were occupied and did not even own our lands any longer. In the face of this new reality, we had to start taking advantage of our forests without the care and respect we had demonstrated in the past. Instead of protecting our forests, other artificial forests were imposed like eucalyptus and pine that have contributed to the destruction of our lands.

The government responsible for the introduction of the new trees never taught the people how to manage these exotic forest and this aggravated the situation. Besides, we actually face other serious problems like demographic pressure, soil erosion, extreme poverty, and the necessity of regaining the integrated management system if we are to have economic alternatives and a better life.

Indonesia - The Nuaulu People

About we Nuaulu people. Our own government here in Indonesia allowed large lumber companies to come in here looking for lumber, such as *onia* (Malay 'kayu meranti', *Shorea* spp.). So they levelled the tops of mountains, digging them all up. At the heads of rivers they cut down punara (*Octomeles sumatrana*) trees, they cut down *onia* along the edges of rivers, and vehicles levelled and filled in the heads of rivers. While they lived here it was still good. We got around well because they were working.

Vehicles went up and down the roads so they were clear. Or if we went hunting we rode on their vehicles with them. But when they went home, our roads were covered up, trees started to grow on them and then we couldn't travel about well because when it rained landslides covered the roads. Game animals moved far away as did cuscus. Land slid into the rivers because they had cut down the big trees along the edges of the rivers.

Therefore we are really suffering because we have to go around the roads. Before they came here we knew when it was the rainy season and when it was the dry season. But when they levelled our lands and rivers here in our forest it wasn't the same when it rained and when it was sunny. It was sunny all the time so land slid into all the rivers. Therefore we do not feed well because it is no longer like before.

Before, the rivers flowed well and the sun shone well so they looked good to us. But now that the vehicles levelled them, so much of the fish in the rivers and the game animals in the forest have moved far away. They electrocuted all the fish in the rivers so there are no more fish. So where can we look for our food? Even if we look for our food in rivers that are far away we do not find any fish. We do not find any game animals. The deer have moved far away.

Therefore we want to ask for money to cover the price of our forest but the government in Masohi and Amahai forbid us from doing so. So we are quiet and are obeying them. But because of our village and forest we are suffering. We suffer when it is so difficult to go to our forest and look for our food because they levelled all the rivers. They levelled all the mountains. The rivers do not flow well. It is difficult to find game animals. Therefore, we do not feel well about this.

They destroyed the lands and rivers. They took away all the big trees. They sold them and made a profit but they did not give any of it to us. Therefore the Nuaulu elders do not want anything to do with them because they did not think of us. We let them take the wood because they said that they would plant new trees to replace those they cut down. But when will those trees grow? They will never grow like the trees before. How will they grow like those big trees? And when will they plant the trees to replace them? It will be a long time before those little trees are big.

Therefore the elders do not want anything to do with them because the lumber companies came here making things difficult for us with our forest. Our lands and rivers are no good at all. They have been gone a long time, like the Filipinos. When we go to the river Lata Nuaulu or Lata Tamilou we have to cut the thorns that have grown with our machetes until we are almost dead because they block the path. If it rains just a little there are landslides cutting off the path and then we have to go far around them before we can find a straight path. Therefore, we are suffering a lot just because of this.

So, if there is any help or any word that can be given here in Indonesia that would help the officials here in Indonesia, help quickly so that they will not agree that all our lands, rivers and trees be taken. So the heads of the rivers would not be levelled so we cannot eat well or find food well.

We people find our food in the forest. There are a lot of Nuaulu people who do not fish well so they look for food in the forest. This is just us Nuaulu people. Other people look for food and have a lot of people who fish but there are only a few of us who fish. Therefore these people look for food but do not find it. We are all dead from hunger. Before the lumber companies came we got around well. We found food well because the deer slept nearby, pigs lived nearby, and cassowaries lived nearby. But when they levelled and destroyed these animals' places and caves they ran away. So it is very hard for the Nuaulu people to find food because they chased away all the pigs and deer so that they are now far away.

Therefore, if someone can find a little help and wants to talk to the officials here in Indonesia, I ask that he help us a little so that they do not come here and work again. We do not want them to because we are already suffering a lot.

That is all.

Malaysia – Patrick Segundad – Kadazan

My name is Patrick Segundad and I am from the Kadazan community. We are the main indigenous people in the State of Sabah in Malaysia, which is located in the northern part of Borneo Island.

The term 'biodiversity' does not exist in my people's understanding or language. If I were to translate the term into our language, I would say it is everything in this world, down in the sea, and things that we can touch. At the same time, it is more than this – more than just things that can be touched or things that are alive. The air, the water, and the sun also must be included. If one being or part of biodiversity is disturbed or not kept in the perfect manner, an imbalance is created which will affect all other things.

Also, there is a spiritual aspect to what is also part of biodiversity. Although our peoples embrace Christianity, Islam, or whatever, we believe in the existence of spirit. The spirit is more like a guide, something that you must respect or be conscious of. It could be the spirit of the land, the spirit of things that live on trees or rocks, or even your ancestor spirits. In our language there is something called *'adat'*, an unwritten understanding of common things that everybody should know.

Adat is not only important in how we deal with our resources but also in how we live. It isn't like the concept of managing but rather that two things happen in the same time. While you might manage something, what you manage is also managing you. A person is a part of a greater single action, a larger balance or harmony.

Adat is often described as a traditional legal system but, to the indigenous peoples, it is much more, encompassing a set of beliefs and values that effect all aspects of life. Further, *adat* is a set of unwritten rules and principles that extends to everything and to relationships within both the physical world and the spiritual world.

Everything is inhabited by some kind of spirit and there is a proper way to conduct relationships with them. All things are in balance and any disturbances in the spiritual world may affect other members of the earthly family or community. Indigenous communities recognize the creator, spirits of the dead, and demons. *Adat* is closely linked to agricultural practices and management of the ecosystems.

Normally, each indigenous community has a number of elders – men and women well versed in *adat* and its rituals – and others, who command great respect within the community. Most village leaders are members of the higher social strata, although this is not a stipulation for the position. These elders form a council which takes collective decisions on important matters and also presides over village courts in which all community disputes are settled.

The Kadazandusun village or community, which is the basic unit of the traditional society, usually has a headman named *mohoingon/molohing* (old person), who is skilled in *adat*. This position has been given official recognition under both the British administrations and local governments since the formation of Malaysia. In the past, Kadazandusun communities in each area were sometimes headed by warriors. These men of wisdom and bravery were generally known amongst the Kadazandusun as *Pangazau* (head-hunter). One outstanding man would be respected throughout several areas among the various *Pangazau*. This paramount chief, generally known as *Huguan Tosiouo/Huguan Siou* (*'Huguan'* meaning 'tough leader'; *'tosiouo'* from *'osiouo'* meaning 'brave, with supernatural prowess'), is considered a 'leader', 'supreme head' or 'the one who shows the way'. These non-hereditary leadership positions involved heavy responsibilities rather than privileged status.

The concept of *adat* is also embedded in the agricultural system. There is a wealth of ritual and ceremony involved, especially with the swidden system, which aims to redress the balance of nature that agriculture temporarily interrupts. Spirit worship is practised through these ceremonies rather than in specific places, temples or at regular intervals. The whole process of work thus brings individuals into contact with the spirit world, and if this should cause conflict between the spirits, the consequences may be felt by the whole community. This, in turn, undoubtedly encourages communal work and the sharing of responsibility for any activities that may adversely affect the spirit world.

Women form the vast majority of those who exercise the priestly functions. The status of women in Kadazandusun society was high in the distant past and has changed over time due to influences from outside. The predominance of women may be due to the psychology of females who, in matters of religious belief, take a longer time to be convinced, but once they acquire conviction become more committed and faithful in their observance. Whereas man's emotion is to meet a challenge, woman's emotion is to create an atmosphere. And because of the maternal instinct in women, her nature is to transmit not only natural life but also spiritual life. Another factor is that over the centuries women have been agriculturists and men have been hunters. Because of this tradition women tend to be more consistent than men.

The function of the Priestess is to endeavour to control or alter events that are considered to be causing problems in life. This is done by appeasing the spirits or forces responsible for the crisis. The Priestess asks the spirit (in this case the devil) upset by human action to accept the sacrifice and at the same time calls on *Kinoingan id Sawat* (God in the highest) and *Id Suang Tanah* (God below the Earth) as judges and witnesses. The offering is not made as an act of worship or adoration of the devil but rather to pacify his anger at human negligence. The ceremonies carried out by the Priestess can usually be classified into three types: (1) those connected with agriculture; (2) communal ceremonies for the benefit of the whole village; and (3) personal ceremonies for the benefit of a single individual or household, for example to cure sickness, bad dreams, and so on.

As with other indigenous societies within the region, land is not owned. The concept of land ownership is alien to indigenous peoples where they believe that land belongs to the countless number who are dead, the few who are living, and those yet to be born. They see themselves as passing their lands on unharmed to the generations that follow, and consciously manage their resources to ensure sustained yield. Individual families have well-defined farming sites where they enjoy exclusive use and are, in effect, temporary residents with protected rights. The forest, however, is almost always communal property, although individual trees may be claimed by a single family.

Communal property is not free for all to use but rather is organized within a management system where rules are developed, group size is known and enforced, incentives exist for co-owners to follow the accepted institutional arrangements, and sanctions are enforced to ensure compliance. Most areas of forest will be claimed by a community, but boundaries – particularly for hunting grounds – are often vaguely defined.

Humans are merely a transient part of this world and land belongs to God as the Creator. Indigenous peoples have strong ties with the land. The land gives the people life, it gives life to the trees which, in turn, give life to various micro-organisms and a resting place for the dead. The community's rights centre on three sources of life – the land, air and water – which refer to rivers, beaches, trees, wild plants and wildlife among other things.

When we talk about land we do not distinguish between forest and other lands. Both are the same, whether used by humans to plant or where plants grow by themselves as an act of nature. Indigenous peoples know where useful forest trees are located, where the best rattans can be found, and the whereabouts of deer and other valuable game. Hunting and gathering are important not only for economic existence but also for religious and cosmological reasons.

The indigenous peoples of Sabah utilize as least a quarter of Borneo's floral species for food while most of the world relies on only 20 major crops for staple food. Plants are used in concocting traditional herbal medicines to treat a wide range of ailments from simple coughs, diarrhoea, consumption, eye infections, skin problems, sores, cuts, wounds, and so on to physiological diseases like hypertension and even malignant cancerous tumours. The method of preparation, however, also depends on taboos and religious beliefs.

Apart from precious stones, bones and other animal by-products such as feathers, beaks and shells, Sabah's indigenous peoples also use a variety of plants for their cultural and social needs. Different parts of plants are used to make shelters, boats, hunting equipment and handicrafts, to carry out ceremonies, and to prepare dart poisons.

Adat influences the right to collect forest products and to hunt, and is often expressed as religions restrictions on over-exploitation of trees and animals. These *adat* controls are very strict, with systems of taboos which require communities to fulfil a host of activities before, during and after collection. When the individual or community violates these taboos, the community must make amends.

Guidelines expressed though *adat* for opening land for agriculture are usually very simple and practical. Farmers clear small areas of forest and burn the debris. Most plots are secondary forest that has previously been cleared for agriculture because it takes less energy to clear than primary forest. Burning the debris releases potash and phosphates immediately into the soil, prior to planting crops which will need them. Burning is done at the end of the dry season. The normal average size of a swidden plot for a family varies from 0.5 to 2.5 hectares. Only small areas are cultivated leaving most of the land in fallow. Clearing small areas is a major factor contributing to the reduction of soil erosion. The other major factor that reduces soil erosion is the variety of crops grown on any one site. Different crops are planted throughout the year, providing the farmer with a steady supply of food. Subsistence crops such as rice and corn are grown on freshly cleared sites and cash crops are grown only after the land has been cultivated for sometime.

It cannot be denied that the swidden agriculture system requires relatively large tracts of land but it is labour-intensive and requires few tools. It is more appropriate to measure the system's efficiency not by output per unit labour but by yield per unit area. It should be noted here that traditional swidden agriculture within forested areas evolved to meet the needs of the local economy, not to provide raw materials for export. To this end, the swidden agriculture system is remarkably efficient.

Indigenous farmers usually have an extraordinary wealth of scientifically sound knowledge about plant species and soil qualities. This is indicative of people who are far advanced technologically. They have such a highly defined and reliable knowledge about their environment because their very survival is dependent upon the validity of their information. Their traditional practices are well adapted to local environmental conditions because these practices are the product of an intensive process of natural selection over many generations. Unlike scientific researchers in government-funded laboratories and experimental stations, farmers do not receive a monthly pay check regardless of the success or failure of their 'experiments'. Failure means hardship, even death. Consequently, selection strongly favours accurate and reliable knowledge.

Modern political, economic and cultural forces have changed indigenous peoples. Introduced religions, such as Christianity and Islam, have often failed to incorporate old beliefs which were important to the sustained use of lands and forests. Traditional agriculture is a manifestation of the indigenous peoples' concepts of a world of balance and renewal, which is rapidly eroding under modern conditions and circumstances.

In the end, traditional values are about balance and renewal and have little to do with what is called supply and demand. Biodiversity is a part of life and related to land. Without biodiversity, life would be meaningless. Without something like *adat*, it is difficult to have balance and renewal. It is similar to having a pen but no paper, or paper but no pen. Both things must be in place before we can write – just as *adat* must be in place before we can control over-use and have balance. For example, in my community we are only allowed to fish in the river for certain periods of time. So, if somebody goes and tries to catch fish before the agreed time, and is discovered, he will be asked to compensate or *sogit* which could burden him afterwards. This must be done to please and compensate both the spiritual world and the community. Because of this person's behaviour, the whole community is affected by the actions of the spiritual world.

This type of environmental control has been in operation for hundreds of years. When *adat* is not respected, you can see and feel the consequences. Today, many people don't respect *adat*. You can see this lack of respect in our forests and villages. Ten to fifty years ago when people still believed in and respected *adat*, there were forests and people who lived in the forest who appreciated and respected their surroundings. They saw no reason to change their way of life. But now they are within the influence of the modern world and there are things in the outside world that they want. Those things can be purchased with money, and the question becomes how does one get money? The majority of the younger generation are frequently influenced by money and are the ones who go and cut and over-use the forests. Because these younger people do not respect *adat*, the forests are diminished. When *adat* is in control , e.g. 'You can't cut this tree', 'You're not allowed to take more than this', 'When you cut one rattan, you plant two', 'That is for you and for your son or daughter in the future', then the forests will be preserved. By following the guidelines decided by *adat*, they are helping the forest to exist in the future for future generations.

It is like the durian tree. When you plant the tree, you're thinking about 20-30 years from now when it will bear fruit. Some people would say, 'What's the point of planting this tree? Probably your son will be old by the time of first harvest and have his own son and perhaps a grandson'. But as we eat durian today, we will remember that these are the fruits given to us by past generations. They had their own visions and their own thoughts. While they didn't go to school, they were educated in other ways. They followed what their forefathers did because they understood that this knowledge came through the actions of their forefathers. Without this knowledge they would not have enjoyed what they enjoyed in their time. It is interesting that this cultural information is like genetic information, and the evolving social and cultural forms have their analogies in biological evolution Our ideas have evolved and served us well.

It is bothersome that in some of the writings about indigenous peoples' knowledge, it is either presented as unchanging or as having undergone changes that leave it in a new and disorganized state. As for all peoples, change for us is all the time too. To meet new conditions as the world changes around us, *adat* must be compensated. *Adat* is permanent and ongoing, and imbalances have been created. There is no compromise with *adat:* it is not the law set up by humans but is rather something that has been negotiated between humans who are still living and the spirits.

Some people might be confused and think that *adat* is like the Ten Commandments. But the Ten Commandments are different: they are written and, when things are written, people can interpret them in different ways. But *adat* was never written. It cannot change so there is no interpretation as it is remembered with spiritual values.

Adat has been here for hundreds of years so it's already agreed that this is a 'must do' for you so that your grandchildren or the people who are going to live in the future are given the same mandate. *Adat* is always, it's a fixed one, you cannot change this thing. *Adat* is the constant that guides the way you live, the way you are going to take care of your surroundings, the animals, the air, the water, the spirits... everything, and you see the spiritual, your family, your community, and other things like that.

Nauru – Eidengab, from Buada District

I have earned my living by making handicrafts from local materials for the past 40 years. Biodiversity is all we have to live on. The plants, trees, birds and shrubs have gone: only the fish remain. At the present time we must go very far to look for plants to make our handicrafts. The place where we used to plant our crops has now been dug up, and the phosphate company is still digging on our island.

There used to be thousands of frigatebirds, but now they are not very many because their nesting places have been disturbed by people mining phosphate. People now breed fish in pools and we can no longer fish freely where we please. The rain is very rare now but in the past we had frequent rainfall. Biodiversity is our life here in Nauru but it is almost all gone. The land is destroyed, the money gone, and Nauru is left behind with nothing.

Our forest and water, both sea and fresh water, is life to us. It is our food, our medicine, and our culture. We are sad today that all of these plants and animals are gone. We now have to go to the store all the time to buy food, and go to the hospital for all our sicknesses. We use synthetic materials for our handicrafts. Whatever is left we want to preserve. We try to tame the frigatebirds now but it is hard because they do not have their natural habitat. This really makes me sad.

New zealand – Aroha Te Pareake Mead – Maori

Ko Putaunaki te maunga
Ko Rangiteiki te awa
Ko Mataatua te waka
Ko Ngati Pahipoto te hapu
Ko Ngati Awe te Iwi
No reira, tena koutou, tena koutou,
kia ora koutou katoa.

I am Maori from Ngati Awa, Te Arawa and Ngati Tuwharetoa on my father's side and Ngati Porou on my mother's side. My tribes are along the eastern coast and central region of the North Island of New Zealand. The Maori name for the North Island is *'Te Ika a Maui'*, the fish of Maui. *Te Ika a Maui* reminds us of the creation of this island – the interaction between humans and nature. It tells a story of human greed, jealousy and mischievousness, elements that still remain at the core of the relationship between peoples and nature.

The Maori word for New Zealand is *Aotearoa*, 'Land of the long white cloud'. When Maori introduce themselves, as I did at the beginning of this interview, we begin by naming the sacred cultural landscapes of our particular tribal region. Telling you my name is the end of the introduction, not the beginning. First you learn that *Putauaki* is the name of Ngati Awa's mountain, and *Rangitaiki* the name of the river. You find out that the name of the canoe that brought my ancestors from Rarotonga and beyond, and who started the Mataatua tribes, is *Mataatua*. If you were familiar with Maori topography, you would be able to identify the community and even my family lines from my introduction, before I ever told you what my name was.

This is not a custom that is relegated to the past. Whether you live in the city or in a rural area, if you're young or old, employed or unemployed, male or female, all Maori still introduce themselves to each other and to outsiders in exactly this way.

I may not make specific mention of 'biological diversity' very often, but that is not because this contribution is separate and distinct from the issue, rather it is because the use of the term is limited to non-indigenous cultures. Biological diversity does not easily translate into indigenous peoples' world-views.

I have stood in front of my own peoples for over seven years now and told them about this 'thing' called biological diversity. I have explained its meaning and provided them with copious amounts of documents, reports and opinions. I can say unequivocally that the term biological diversity is not easily translatable into an indigenous culture, such as Maori. It comes with a whole range of cultural baggage that simply confirms that protection for future generations and trade liberalization are opposing paradigms.

What does biological diversity mean for me? Ultimately, it means that anyone who commodifies biological resources; separates them from cultural heritage; attempts to exert exclusive individual ownership, is consigning the diversity of life to solitary confinement in a prison that condemns all those who regard nature and peoples as being more important than trade.

I think that how a people and then a nation names its environmental landscapes and biological diversity speaks volumes about who they are, what their values are, and how they define their responsibilities to future generations.

For instance, the difference in the names '*Te Ika a Maui*' and 'North Island' can not be explained as purely linguistic. How can an island in the Southern Hemisphere be named as the 'North'? Clearly the early European explorers James Cook and Abel Tasman did not yet comprehend the significance of the concepts of a 'hemisphere' and following on from that, the notion of a Northern and a Southern hemisphere.

Many indigenous peoples of the Pacific wonder why Tasman and Cook are regarded as 'discoverers' and experts who knew what they were doing, when our own ancestors, who systematically populated the entire South Pacific by voyaging in canoes and navigating by the stars, are still portrayed as being primitive, and their methods unscientific and accidental?

My observation of the Western European system of naming is that it is steeped in the notion that individuals 'discover' what has existed for generations (a classic example is the Halle-Bop comet whose creation spans millennia). How can one discover what already exists? For me this is a question that I place at the same level of those who ask, 'Is their life beyond earth?' I too experience absolute excitement at sighting the first photographs of Mars. I can't wait to see what science uncovers from this latest voyage. I'll be very disappointed however, if specific sites on Mars are mapped and named in recognition of scientists who arrogantly assert that their 'knowledge' has enabled the world to acknowledge the existence of natural phenomenon that pre-dates humanity itself.

To name on this basis places an individual above all others, and one generation as being superior and more advanced than previous generations. This notion is the antithesis of all that I have been taught. Names are very important in my culture. The way in which we assigned names to geographical places of cultural significance was part of how we transmitted history and environmental knowledge. The same is true for the naming of native birds and plants and other types of indigenous flora and fauna. For every species there is a story that tracks its origin. Like the naming of *Te Ika a Maui*, it is often a story about families, inter-relationships and the joys and difficulties of trying to stay together or learn to live apart.

The Maori world-view of the origin of life is a love story between Ranginui, the Sky Father, and Papatuanuku, the Earth Mother. Theirs was a love of devotion and sacrifice. They were once joined but as their children grew they separated in order to enable their children to move freely between them so the children had space to play, light to grow and air to breathe. Ranginui and Papatuanuku sacrificed their love for their children, and this is a value which every Maori individual inherits.

Ranginui and Papatuanku grieved so deeply and overwhelmingly for each other that even in the space, light and air their separation created their children still could not prosper. Ranginui's tears for his beloved wife are what we know as rain, snow and hailstones. Papatuanuku's grief and longing for her husband manifested itself as the moods of the earth: mist and fog. Such was the grief and the longing between Ranginui and Papatuanuku that their children still could not breathe, the tears from the Ranginui and the mist and fog of Papatuanuku were suffocating. Eventually, Papatuanuku in an act of supreme sacrifice as a wife and mother, turned herself around so that she faced the underworld rather than the husband she deeply loved. Her moods and longing for her husband are now expressed as eruptions of volcanoes and the tremors of earthquakes as well as mists and fog.

I am a 'modern' Maori. I live in the city, have studied as an academic, and work now as a government employee. I am really proud to be Maori. I marvel at the wisdom of my ancestors. Ranginui and Papatuanku are not a myth to me: they are my past and my vision for the future. Their story teaches me what I must do and why. Their love teaches me to love, to regard all living things as my own: every species of life, including humans, as descendant from Ranginui and Papatuanuku.

I know that there are many in the world who reject indigenous knowledge and explanations of the evolution of life such as the understanding of the Maori, but I don't believe that sceptics have any valid basis for their views, other than, perhaps, racism. I have been taught, in both indigenous and non-indigenous educational systems, that science is a discipline that transcends all cultures. A scientific approach is one that takes a problem or a proposition and through a systematic process comes up with a result. I've learned that the interpretation of any given result is highly prone to cultural perceptions, be they indigenous or non-indigenous, Western or Eastern, male or female.

I can only reiterate that biological diversity is not an easily transferable notion across cultures. My heritage places the environment in a pivotal role, but biological diversity carries with it a story of Western reductionism and commodification. I am firmly convinced that my traditions are much more suitable than [those of] my Western counterparts.

Norway – Johan Mathis Turi – Saami

My name is Johan Mathis Turi and my *si'ida* (work group) is the Unihurt. Reindeer herding is my main occupation, but this is not all that I do. When we have an occupation, we use our qualities as human beings, including our intelligence. Herding is an intelligent occupation and a very good platform even if you think of changing your career. Reindeer herders are strong in their beliefs because they utilize nature and learn from it. Herding is a good school.

I was born into this life of herding and saw reindeer as part of nature. I was five years old when I first saw a Norwegian. I'm not an old man, nor am I from the Western world. Instead, I'm from the Stone Age and have gone to the Space Age in one generation. This is not a negative thing because we have learned so much through contact with this development. I don't feel I have lost anything of my Saami identify as a consequence. In fact, it is stronger than ever.

I speak on behalf of reindeer herders. This is a very exciting time for my generation. Of course we have gone through many problems, but they would have come anyway as it is a part of human life to face problems. We are surrounded by other nations, by majority peoples, and while this is a problem, it has not changed us. There is a Saami nation.

To explain more fully our situation, let me explain something about my childhood. We did not have money coming in but we had a very rich life. When the new time came, we moved more into the commercial world. We couldn't have got through this time if we had not been so strong because of our life-style. The reindeer is the centre of nature as a whole and I feel I hunt whatever nature gives. Our lives have remained around the reindeer and this is how we have managed the new times so well. It is difficult for me to pick out specific details or particular incidences as explanations for what has happened because my daily life, my nature, is so comprehensive. It includes everything. We say *'lotwantua'*, which means everything is included.

When I think of biodiversity it is the same: everything is included. I could not be a reindeer herder without it. It is a necessity. Biodiversity is both art and necessity. The other species, the animals and plants, are important and we use them. The predators are also a part of biodiversity. We do not want any animal extinct. Each is a part of the whole.

Thinking of nature is difficult because, again, everything is included. I think the problem is with how we think when it comes to nature. Environment is that which has been discussed with authorities. They never talk about the human being so we can never agree, for instance, on the protection of nature because for nature to be protected, they must consider the human. The most serious aspect of preserving nature is the preservation of the human being.

When you ask if nature and culture are linked, it is difficult for me to answer. I have lived my life-style and I cannot think of things without thinking of that life-style because it includes me. Whatever I say and think, of course, links nature and culture. All peoples living like the Saami, linked to nature as reindeer herders, have spiritual contact. We talk about holy places and whatever they are, they are. At the same time, these holy places must have come about because of a need. There is a necessity to please nature so that it gives to you. You have to be kind to nature so that nature will give you more. You have to be friends with the environment, otherwise it won't work in co-operation with you on whatever you call the spiritual.

I think like my grandfather. When you produce something, you make it on purpose. Then if it turns up in the common belief, you believe it. If you make yourself a *starlo* (a Norwegian word), then you end up believing it. This is like an aspect of religious experience. We live adoring nature and then it becomes sacred. Then if you have a symbol or make a symbol for something, a goddess or god for example, and you start to name it or whatever, then it becomes a belief. This is how we ask each concept whether it existed previously. But first you made them yourself and then you started believing in them.

We have a word, waiyulat, the richness or abundance of nature, which means its comprehensiveness. Waiyulat means that when you need something, it is always there. When I need grass for my cow, for instance, it is there. Nature was here when I was born and I am still using it after all this time. This must be abundance.

It is difficult for me to answer what value I would place on biodiversity or nature because I have never thought in these terms. I cannot imagine being without nature. It is priceless. For the sake of discussion, let's say that some peoples can maintain their livelihood artificially. They keen their animals in stables, and don't have the hassles of herding them or protecting them from predators. They don't have to follow them from place to place, and in the end they have their reindeer and their income. They have their livelihood, but they don't have nature to put it in. But then the reindeer would not be a reindeer if you fenced it in. That wouldn't be allowed. I could not accept it. The reindeer is my equal. How can I take my equal and arrest him? It would not work. The reindeer might as well be a cow. I wouldn't do that because the reindeer is a proud animal. The reindeer have to have their freedom. This is why we follow them. We don't fence them in.

Even if reindeer production is now industrialized, with modern slaughter-houses, it still depends on nature. Reindeer herding hasn't changed into farming. Even now, with our selling to urban people, herding is still a profession. Traditional reindeer husbandry is such an advanced school to a human being. This is what gives you the pleasure and the confidence which you cannot get elsewhere. If you fenced the reindeer in, you wouldn't understand the reindeer. As he walks north and you follow and watch, from there you learn. Without this opportunity you would not learn about nature and the Saami traditions, and it wouldn't be a tradition if you fenced them in and started to feed them. The reindeer might as well be chickens. This way we would earn much more money.

Now, in newer terms, it is easier to explain things because we have contact with other minority and indigenous peoples and can see that those people don't have the platform that we have and are much weaker that the Saami. The reindeer herding traditional way gives you so much spiritual strength. I can face the Saami. I can face the problems in a proper way because I have this self-confidence. Because of my reindeer herding profession I can see and feel I am equal to anyone and can discuss issues with them on equal terms. This is what makes me a human being or equal to everyone. The revenue income is known to me and I find that it is not good. Reindeer herding doesn't earn much. It's a low-income profession compared to sea herding. Even factory workers earn more. But there's so much more in reindeer herding that adds up making it a richer profession. You can't count these riches in money.

When I look back from today, I see the effort the Saami have made to preserve nature without even knowing it. And now the awareness comes to everyone and they understand that protection of the environment is a necessity. It is important to include the indigenous minorities living in nature in deciding how to protect nature. Environmental protection is easier to discuss now and that's why there is a great future for everyone, and especially for the reindeer herders. I am not fearful because we can discuss things. I believe that the active minorities offer the best possibilities for a secure future.

Panama – Atentio López - Kuna

Recently I made a trip through Peru and Chile, visiting indigenous communities. This visit arose from a concern by indigenous peoples facing neo-liberalism and globalization which are reaching the gates of our communities. It is necessary for us to define a common strategy to face these new winds of change which sooner or later will shake the foundations of the millenarian cultures of our territory Abya Yala.

In Peru, a group of indigenous specialists from all parts of the world discussed the topic 'Development and Indigenous Peoples' in the Sacred Valley of the Incas whose fertile lands have produced so many varieties of maize. In the setting we looked at how the traditional knowledge of indigenous peoples continues to be consumed by Western technology and modernity which floods our countries. Even though our knowledge is stolen from us, no one recognizes the wealth and collective property rights of indigenous peoples.

The conclusion of the discussion was that indigenous knowledge and its methods should be conserved, protected, promoted and improved for the benefit of our communities. Indigenous technologies should be strengthened and consolidated, and indigenous values revived. The cultural dynamics and the collective well-being of indigenous peoples must be respected along with equitable distribution and sustainability of resources, so that future generations can benefit. Decisions should continue to be taken collectively and the wisdom of experienced elders should be listened to. We indigenous peoples should be proud of our identity and our cultures.

The second part of the journey took me to Chile, to a preparatory meeting on the establishment of a Permanent Forum for Indigenous Peoples at the United Nations. The United Nations was discussed and how the different agencies relate to indigenous peoples. A Declaration of Temuco supporting the creation of the Forum was drafted. In Chile, in spite of the 'economic miracle', we could see the serious problems facing indigenous peoples. For example, the conflict over the Bio Bio dam, which will flood lands, cemeteries and houses of the Mapuche-Pehuenche. The government and the electrical company ENDESA are initiating five dams with financial support from the World Bank. Another threat to the Mapuche has been the construction of the Temuco by-pass road which will cross the Mapuche territory of Truf Truf.

Misery and harassment face our brothers and sisters as gradually they are invaded by non-indigenous peoples. Land-owners and local politicians are problems everywhere – whether in Chile, Peru, Panama or Canada. However, all the continent is embraced by the arms of peoples of soil and maize, who one day will make their revolution and demand the space and respect which they had prior to the arrival of the colonizers.

Papua New Guinea – Morea Veratua

Right now they call me a diplomat because I'm on an assignment to Jakarta with our mission, but ultimately I am a villager from Paka in Guinea about 40 miles or 65 kilometres from the capital city. I consider myself a villager because I belong to a village family and I live in the village when I am in my country. I also belong to a clan and a tribal group. I know my own folk law, customs and culture because I speak my tribal language. I exercise all my social, cultural and traditional obligations at the village level as well. I use traditional knowledge in using and analysing natural resources, particularly land, rivers, trees, tools, implements and so forth. So, in totality, I am an educated person but I am an educated villager.

In our culture, nature, biodiversity, or the environment within which man lives, is his true world. Anything beyond this is secondary and forms another reality. His daily life and mere existence is totally dependent on what surrounds him. He has, over time, conditioned himself or has been conditioned by others to live in harmony with nature, relying on what is necessary and refraining from excessive abuse of both flora and fauna for his survival and that of his line. He considers his existence as total affinity with nature and what he understands to be his. The relationship with and attitude towards the natural resource bases of land, rivers, coast, the beach, reefs, hills, mountains and so on are very much built into his cultural village systems in both physical and social-spiritual worlds. The environment is, to him, the very basis of his existence and he respects that.

The value of biodiversity or nature? To me, indigenous knowledge systems of nature, environment, and the various ecosystems within it, represent the most valuable survival kit in rural areas. Such knowledge is much treasured and guarded with pride by those who have it. Its application is certainly not practised for increased economic security but rather to attain a place in the social-cultural community and status within the society. Traditional or indigenous knowledge is related to biodiversity and its utilization, protection and conservation, and has been the only effective basis upon which our people have been able to survive comfortably for generations. Therefore, I believe that modern knowledge of biodiversity and its utilization, protection and conservation would be better served by acknowledging and using traditional knowledge as well.

The value placed on what surrounds my parents and relatives is what they use. This is important to the point where they feel disheartened when outside forces like the government or developers come in and destroy their gardens and tree crops to build a huge road or do excavation and forestry development work. They know that they cannot replace the things that they have attached themselves to over generation. For instance, the clan's historic sites or the trees that have been there for generations and have certain attachments to the clan's history cannot be replaced. All of these things are very valuable. These things are valued very highly but it is in the modern day rush for economic development that these things are disregarded and becomes less valuable when compared to money and what money can bring in.

I am talking about spiritual attachment to land and to the biodiversity that is found within that land or near the river. That we associate ourselves with life and land is a problem. In fact, our people have a long history of fighting and actually dying for their land. We have folk laws and stories that actually bid you to believe that you come from the land. This is your right. This is your land. You don't sell it. You've been given the land or you acquired it from generations before. Generations are tied to our traditional lands and we expect future generations to act with the same attitude. The value of our land includes its value to our ancestors. Attachment to such resources is spiritual and very valuable.

Traditionally, we would register our land in different ways. For instance, you could plant certain types of trees. Another way is related to the fact that we never had cemeteries. If you feel very strongly about a particular land that should remain in the family line, the old man of the clan may actually nominate himself, when he dies, to be buried in that particular land in order to avoid any future disputes over it. We have folk law belief systems that acknowledge his bones being in that piece of ground as a permanent register of your own or the clan's ownership. If there is a dispute, all you have to do is direct people to this particular burial site and, if the bones are found there, this basically ends the dispute with a neighbouring clan who may have had an interest. The neighbours absolutely respect such claims.

There are other ways to recognize land or property. For example, your mango tree or your best tree or any fruit that can be recognized as being of value by custom registers your possession. The respect for property recognized by custom is total. In addition, the people traditionally had packets of lands registered in a much wider area. It would be proper to feel obliged to have traditional activities or ritual practices in these packets of land. These aren't just single huge tracts of land. Instead, the tribe would have several, maybe 50 to 100 pieces of land, scattered within a radius of say 50 to 100 kilometres. Maybe ten tracts or pieces of land would have some evidence of your ancestors' ownership, and the knowledge of these lands would have been handed down from generation to generation. Even now, even though we don't access these pieces regularly, we still have affinity with this land because our ancestors have occupied it a long time ago. We go back and say, 'Yes, that's our great grandfather's tree' or, 'That's where he was buried', and this no one can dispute. Value to land or to nature is set by your ancestors. We make conscious decisions to place ourselves and thereby give land to the future. As I said, the ancestor's knowledge of the environment and the biodiversity of that particular locality is very great.

An important knowledge that was treasured and guarded would only be passed down to certain members of the family. So, for example, if you have four brothers, maybe one of them would be fully versed with the traditional knowledge of your land situation. The remaining three may know a little of this, a little of that, but the one who has all that knowledge is respected for his knowledge. What has been given to him is an honour for his good deeds and for what he does. This does not necessarily mean he is the oldest of the brothers: he could be the last born or the middle born. Rather, it depends very much on the old man who feels he is capable of having this knowledge.

In our tradition, we have never had mono-cropping. It has always been a mixed cropping subsistence system because that was again part of our traditional indigenous knowledge. Our system was defined over time and demonstrated that mixed cropping was and still is the most dynamic and most comfortable for us and most adequate in supporting our families. When I look at mixed cropping systems, I see more beauty than with mono-cropping. There are a lot of fascinating aspects of mixed cropping which a traditional farmer or subsistence gardener would be able to tell you. For example, they can explain why you can only mix a certain crop with one crop and not with others. The farmer has his own reasons: he sees knowledge in the sun and the slope of the hill. His knowledge of the soil tells him which land is suitable for certain crops and not suitable for others. The crop mixes, the crop rotation, and the diversity he puts in that garden comes with a lot of traditional reasoning and a lot of spiritual reasoning as well. This knowledge is attached to even a simple crop like bananas. In my area, our people traditionally wrap the bananas in a certain way. There are so many varieties of bananas and each variety has a different way of wrapping. The people have a reason for wrapping each one because they serve a purpose like when we pay bride price. The banana plays an important role in showing your status as a farmer who can produce the best quality. We still continue these practices. Our traditional yam cultivation system is *taro*, every crop has its own unique way of propagation and cultivation and all are attached to a social purpose. So biodiversity in our land is tied to the social fabric and the way we conduct ourselves.

The monetary gain in the modern era now is very individualistic. You may earn a lot of money but that money may only help your family to get a new car and material goods. Although the money a farmer gets does help the individual family to become wealthy, at the same time it alienates the family from the rest of the community very quickly. In the traditional regional situation, what was produced from the farmer's garden or farm automatically benefited the village or the clan. At the same time, the farmer gained respect and earned credibility for such an achievement in a different way in a different context. I feel the money economy as it is now is basically destroying the core of our village cohesiveness.

Papua New Guinea – Michael Kapo

My name is Michael Kapo and at the moment I am working as a social justice activist and environmental campaigner with the Individual Community Life Forum in Port Moresby, Papua New Guinea. I am a trained journalist and I come from Madang Province in Papua New Guinea. The sub-district is Pogea and my village is Arawa. I come from a chiefly clan and my duties, obligations and tasks are those of a modern-day chief or someone who has come out from an indigenous village set up to work in town. Moving from the village to the town is very demanding.

My activities back in the village relate to and rotate around the village court of conduct. I'm concerned with the village laws and the tribal laws of the land which relate to the duties of members of clans that make up a village which belongs to a tribe. One way to describe a tribe is as a very large clan. Our tribal customs or laws have been in existence for thousands of years. Basically, when we look at a village set-up we have both people and our spirits, our dead fathers, our dead mothers, and our ancestors. It is because of our ancestors that we have our land, which is very sacred to us. In addition to the land, we have our environment, and when we look beyond the village we also know the other villages who live next to us. The other villages may have a different language and a different way of doing things.

When we encounter aid, it is the discovery of modernization. With modernization, we once again see that there is a different way of doing things. We find that the people from other places want us to do things in their way rather than respecting our way. These people from other places go around saying that they can look at how we are dressed and know something about us. They also want to make sure that there is peace all the time in the village. This leads me to a number of issues, but I would like to touch on one important issue that is affecting Planet Earth at this point in time – the issue of biodiversity.

Biodiversity is a very big word, and different people from different backgrounds, training, influences, and experience can explain what biodiversity means to them personally. At the same time, a group or people pretending to be organizations might say that biodiversity is this or that. Personally, when I look at the word bio-diversity, there is a very spiritual connotation to the word. Biodiversity has a more spiritual sense than in the sense that 'bio' means life and 'diversity' means having different kinds of things in life or different kinds of things representing life. For example, if I look at the Guam River which runs from Musorua, one of the villages upstream, and goes through my village of Arawa downstream, I know that the biodiversity of Guam River consists of fish, crabs, hills, birds, people, little creeks, water plants, houses, villages, the sun, the moon, the rain, and the wind, so almost anything that lives around the Guam River makes up its biodiversity. If we speak of something other than a river, say a mountain or a village, we can also speak about the biodiversity found there.

Biodiversity, to an extent, can only be explained by a person who happens to live in a particular area and then only when he or she thinks that what they are seeing is represented by the word biodiversity. Almost everywhere on Planet Earth and apart from Planet Earth, there is biodiversity. Biodiversity is different forms of life existing in one village or different forms of life existing in different villages.

Biodiversity is described by science in terms of the metals, rocks, sand, water, and so on, the things that the scientists themselves can explain. On the other hand, an indigenous person's explanation of the biodiversity of a river, sea, coral reef, island or waterfall has a fullness in that [the] person also includes the biodiversity of a village, both in the world where humans live and in the world where spirits live. All of this makes up just one unit. For us to know more than others about biodiversity is not surprising. We've told people for thousands of years that you don't shout and stamp your feet in the middle of the village in the middle of the night because this is when the spirits look for food. If you do this sort of thing, you are running against the path the spirits might take and they could make you sick. So the stamping of the feet, the shouting very loudly and the laughing and calling disturbs, if you like, the biodiversity of the village. This is because if we don't know to respect the indigenous laws, the relation of the culture to the environment, man, and spirit which must be maintained at all times will be upset. If the relation is not maintained harmoniously, then things go wrong with biodiversity. If we need, at this point in time today, to explain what biodiversity is all about and what has gone wrong with Planet Earth, we only need to look back for thousands of years at the indigenous communities. They were able to sustain themselves because our ancestors were advanced, were building, breaking, and learning to correct their mistakes. At this time in history, our ancestors were still learning and the biodiversity of the human person was still developing. In indigenous society, this still continues today as it has for more than fifty thousand years.

In the village, we listen to each other. People tell us not to drink from that vine because if you drink your body will start to stretch. We know that what we are told is true. We also know that if you don't believe such things, that if we break that law, then we are going to suffer. When I look at biodiversity, it is much more than the scientific explanation. Those who promote biodiversity must look into what I see as respecting our traditional values in the villages, in our tribe, in our clan, and in our houses. If I go back to a house, let's say in Madang where I come from, at six or seven in the evening, when we have our dinner, we will probably have fresh fish from the river along with bananas, cassava, taro and some greens. We sit around the fire and maybe roast a fish or two; then we look around this place that we know, and we see our buildings, our houses, which are made up of everything that we find in our environment. The explanations that we know come from the minds of men and for indigenous societies, minds taught by thousands of years of experience in the past. We know that man has come to extract resources from the land like trees, like wood, like palm leaves to make houses, like black palms to make floors of the houses, and so on. In the first place all of these things are not to be chopped down for nothing. Men, from where I come from, will have to go and say that they want to cut that big tree so already a message is being passed. And when the tree is cut, they know exactly when to cut it and how to cut it. This respect for biodiversity was not outwardly shown but was at the back of the mind of the indigenous person when he thought about cutting the tree and when he cut the tree.

My question now is how can scientists come along and say that this is what biodiversity is all about. The indigenous ways of understanding biodiversity or problems like the hole in the ozone layer can be best explained by people who are doctors, prophets, and scientists in their own right in the indigenous societies. Somebody from the continent of the great Turtle Island, today called America, said that I don't need to go to university to learn about biodiversity. I believe what he said is the truth. Biodiversity comes down to us. Voice should be given to individual people in indigenous societies because their lives involve all of biodiversity. Indigenous peoples recognize concrete things like land which, at the same time, is non-concrete in a sense that it is spiritual.

When we look at non-concrete things, when we go back into the spiritual world, science can't explain why you've seen something like a ghost or why a shadow moved. Indigenous societies have explanations for such things that move. As an indigenous person, land is much more of a spiritual symbol and is much stronger than its physical representation or projection.

When I look back into what biodiversity is to we indigenous peoples presently working with NGOs (Non-Governmental Organizations) and government organizations, the first question is, 'What does biodiversity mean to me?' If we can link biodiversity to the spiritual aspect, things will change. I'd like to see biodiversity and the spirit together. I'd like to test it in ways that we could all make a judgement on and say, 'Yes, this made a difference' or, 'No, this did not make a difference'. For example, if someone dies from bomb blast in Mururoa, the dead person is there for all to see. Then I can say, 'All right, fair enough, now someone is dead, now we know that weapon is bad'. If spirit and biodiversity are together, then we can say biodiversity is better now or is worse now. We can say that the people in the village are better now or worse now.

Scientists are jumping up and down saying, 'What's going wrong with Planet Earth?' Indigenous peoples know a great deal about what is going wrong. About five or six years back, I saw a documentary about the indigenous peoples who live in modern-day Colombia in South America and they said that there is a good explanation as to what has gone wrong. But to understand, it is necessary for people to come and listen to them for thousands of years. Science never listens to indigenous communities that want to explain what biodiversity is all about. The scientists say that what we have to say is trash. Instead, the scientists come with a rod and go to a river and test it and say this or that is so bad. We don't need a rod to know that a chemical is changing things. When we plant root taros near swampy areas and find that our taro is not coming up very good, we know things are bad – but nobody dares listen to us.

So, as I've said, if you wish to explain biodiversity, you have to look far beyond the scientist's style. It's not that I am snobbish about science: instead I am not grateful for the fact that science does not consider the indigenous way of explaining what biodiversity is all about. Scientists are flying in and flying out of Fiji. Divers fly in and fly out of Tonga. Everywhere in the Pacific, we see different kinds of people saying, 'We are going to explain biodiversity'. They tell us, 'We are going to teach you how to catch your own fish in a sustainable way'. But what they don't seem to recognize is that we have been doing this for thousands of years. Nobody in his right state of mind can come and tell me, 'Now Michael, we will teach you the sustainable way of catching a fish'. We've actually been catching fish for thousands of years in a very sustainable way. We know it is sustainable because we survive. So you know if indigenous societies all across the world have been able to live for thousands of years and we were able to explain and maintain ourselves in the Pacific, we come to the conclusion that science, at this point in time, consists of a record of what has gone wrong with the world. The scientists explain what's gone wrong by saying that a lot of people are using CFCs, air-conditioners, whatever all those different kinds of things are, and burning gases and all this and that. In the meantime, we've been making gardens, doing what we call shifting cultivation, and the difference between us and the scientists talking about CFCs is that for fifty thousand years we've been doing our gardens. There was nothing wrong with biodiversity as we cultivated our gardens, shifting from place to place on the land.

It comes back to the idea that biodiversity is pretty much a big word and I think it can only be explained by, if you like, an indigenous person. But if science has a much better way of explaining and containing what is going wrong with Planet Earth, and if it deals truthfully with what it sees, then it will have to respect the fact that indigenous societies can explain biodiversity one hundred plus one times better than science itself can explain biodiversity.

Indigenous Peoples' Perspectives on Biodiversity. Pucallpa, Peru, February 16-17, 1996

Twenty-four people from the Peruvian Amazon Region met in Pucallpa on February 16 and 17, 1996, in response to an invitation to contribute statements on the value of biodiversity to their communities. Nineteen represented indigenous organizations and the remaining five represented small farmer or peasant (*campesino*) organizations. Rather than give individual statements, those attending decided to organize themselves into five groups and, within each group, produce consensus statements. In a plenary session, an overall format for the presentation of the final report was adopted. Each group selected a presiding officer or president and a recording secretary responsible for the final report.

Group 1

Members: Adolfo García Laulate (President) EGA; Silvia del Aguila Reina (Secretary) HIFCO; David Vargas; Anastacio Lázaro Milagro; Susana Dávila; Walter Gabriel Yanesha.

I. What does nature or biodiversity mean to us? What importance does it hold?

Biodiversity represents all that exists and all that surrounds us as we go about our daily lives. Biodiversity exists without the intervention of the hands of human beings. The total of biodiversity represents all life forms on the planet, all of which are interrelated.

Importance:

- It is the main source of life, without which nothing can exist.
- It is the guarantee of life for our people and our planet.
- It provides us with protection, food, health, homes, clothes, hunting, fish, music, etc.

II. What value is given to the elements that compose biodiversity?

1. Economic: The elements that we extract from nature serve us in order to meet our needs for clothing, education, health, food, and so forth.

2. Religious: We have many beliefs and trust in the elements of nature such as the trees, animals, tobacco, coca, and the like.

3. Cultural: There exist various cultures and we should value them. It is our style of life to live in harmony within our ethnic group.

4. Medicinal: In nature, there exist a great variety of medicinal plants, for example, *uña de Gato, Oje, Gredo*, seven roots and *catahua*.

III. Suggestions, alternatives and recommendations

Suggestions:

1. We suggest that public, private, national and international institutions certify medicinal plants as to their value rather than simply conduct research.
2. These same institutions should support us in the conservation of biodiversity.

Alternatives that must be followed:

1. We must manage our resources well.
2. We must use the land in accordance with its capacity

Recommendations: To strengthen grassroots organizations, in order to defend us from those who would despoil our natural capital.

Group 2

Members: Enrique Gonzales Miranda, CARE; Walter Reiner Jipa, FENAMAD; Glorioso Castro FUCSHICO; Mario Sijo Sajami, Asoc. FUSEVI; David Camacho Paqui, CECOC; Justo Funachi Inuma, CEPRODA; Giulianna Montalvo Espejo, Asoc. FUSEVI.

I. What does nature or biodiversity mean to us? What importance does it hold?

Biodiversity defines the ecosystem where we live. It is the conjunction of all the resources that exist on Planet Earth, including the animals, plants and life forms in the soils. Biodiversity is important because it gives us life. It comes from nature, it is the creation of diverse life-forms without the intervention of human beings and, at the same time, is the joining together of the resources which surround us, the light, air, sun, water, and so on. It is composed of all the animals and plants with which we co-exist. The importance of biodiversity is that we can only exist with it and, as a consequence, we must conserve it. Nature and biodiversity are all that is natural: water, plants, animals, soils, and the like. Nature and biodiversity represent the joining together of biotic and abiotic entities that exist in relation to humans and from which we draw benefits.

Importance:

It is from nature that we live and if we don't plan carefully, we could suffer. An example of poor planning is the indiscriminate cutting of the forest which destroys our way of life.

- The soil is important because we walk upon it and we are fed from it.
- Water is important because we can not live without it.
- Human beings are important because they create the intelligent capacity for development.
- Flora and fauna are important because they provide us with what we require to live.
- Solar energy is important because all life is dependent upon the sun to exist.
- Air is important because it is an essential element of life.

II. What is the value of the elements of biodiversity?

1. Economic

- Forests are a store of value.
- Humans search for ways to live in order to develop.
- Biodiversity, the plants and animals, are a natural factory upon which humans depend for food, clothing and shelter.
- We should value biodiversity at a just price and market it in order to maintain our families.

2. Religious

- Some plants have religious value.
- We should respect religious values (beliefs) from generation to generation.
- We should respect mystical powers as well as traditional powers.

3. Cultural

- We make musical instruments and other artefacts from biodiversity.
- We should value old customs in respect to nature and biodiversity.
- We should respect the mother tongue, identity, customs, music, crafts, dances, theatre and foods.
- We should have a balance between ourselves and nature or biodiversity taking enough to sustain ourselves but not too much which would threaten biodiversity.

4. Medical

- Because the plants are used to cure illness, biodiversity has value.
- We must value natural medicine and continue to use it.
- We value the variety of plants that have healed us for a long time and the knowledge that indigenous people have to use these plants. For example, we use *piri-piri* for snake bites.

III. Suggestions, alternatives and recommendations

- We must plan for the future so that we and the resources are not negatively affected.
- Studies are required on how to get profit or benefit from resources while at the same time conserving them for future generations.
- Laws and norms should be established to conserve and protect the resources in accordance with indigenous needs.
- We must search for management methods for the natural forests.
- National and international organizations should defend indigenous rights.
- The government and international organizations must give the necessary support to grassroots organizations for the management of their resources.
- People must be made aware through the various mediums about the current destruction of nature.
- Peasant and indigenous communities should make their voices known at the national and international level about the proper management of natural resources.
- Institutionalize COICAP (Coordinatora Indigena y Campasina del Peru – Co-ordinators of Indigenous and Peasant (*campasino*) Peoples of Peru) so that indigenous and peasant peoples can be defended and given the capacity to properly manage the natural resources.
- In regard to environmental pollution, financial resources must be provided to grassroots organizations to better development and sustainably manage their resources and for organizational activities.
- To restore devastated forest, efficient development programmes must be promoted.

Group 3

Members: Fidel Sánchez, AMETRA; Ciro Cosnilla Olivares, ECOSA; Reyner Castro Martínez, FUCSHICO; Eliazar Muñoz, MAROTISHOBO; Guillermo Gómez, FECONAU; Elmer Guimaraes, FORMABIAP; Leoncio García, ORDESH

I. What does nature or biodiversity mean to us? What importance does it hold?

Nature is a living process in which all living beings interact in the realization of life. Nature is the process in which the diverse mechanisms that support life and human beings are emergent. Biodiversity is the existing life within nature that functions by maintaining constant interrelations among its various components. Nature or biodiversity is all that which surrounds us and includes culture, our ways of life, traditional medicine, and so on. Biodiversity is the joining together through creation of the animate and the inanimate.

Importance:

Nature to us is important because within her we maintain our ways of life which enable us to survive and create the mechanisms that allow us to develop as a people and society.

II. What value do the elements that compose biodiversity have?

1. Economic

The elements of nature have much value, for instance, in terms of animals skins, vegetables, traditional medicines, cultural artefacts, and other things that allow us to exist.

2. Religious

Previously we believed in an Inca religion but currently we practice Western religions. The value of biodiversity is respected through these mediums.

3. Cultural

Biodiversity allows our practical culture to continue to function.

4. Medicinal

Biodiversity provides cures for different illnesses.

III. Suggestions, alternatives and recommendations.

- That the governments recognize and respect the existence of each people and their function.
- We require a process of forest regeneration.
- International and national organizations should intervene to assist in saving what remains of our biodiversity.

Group 4

Members: Abel Alado Tello; Irma Castro de Salazar; Tomás Huaya Panduro; Oswaldo Cairuna Fasaba; Hugo Ochavano Sanaciro; Zoila Canaquiri Bardales; Juan Muñoz Mamani; Marcial Rodríguez Panduro

I. What does nature or biodiversity mean to us? What importance does it hold?

Nature is the joining of elements that form the ecosystems of which we humans are a part. Biodiversity is all that lives and joins together the biotic and abiotic, including, the sun, air, water, animals, plants and humans.

Importance:

Because there is an interrelationship among people, animals, and plants, in order for all to survive, the integrity of the system must be self-sustaining.

II. What value do the elements that compose biodiversity have?

1. Economic

Biodiversity serves to elevate the socio-economic level and to raise the standard of living.

2. Religious

Each ethnic group has its own particular religious values of biodiversity.

3. Cultural

We must reconsider the indigenous philosophy of each ethnic group in relationship to biodiversity.

4. Medicinal

Thanks to plants, we can cure many illness and also we can transform the plants into creams, powders, soaps, and the like which make our lives more pleasant.

III. Suggestions, alternatives and recommendations

Suggestions:

- When we use the medicinal plants, it is necessary to have days of dieting in co-ordination with the preparation to achieve the wanted effects.
- We must take precautions in the preparation and dosage of our medicines.
- We must consider the use of medicinal plants in family planning.

Alternatives:

- We must transform the raw materials to give added value and maintain strict quality control.
- We must reach a point where medicinal plants are less expensive than modern manufactured pharmaceutical products.
- We must organize ways to distribute our own products.

Recommendations:

- Each group should organize a credit bank to meet urgent needs with respect to the production of medicine and to have medicinal plants available at all times.
- Each ethic group should restore its culture.
- We should share the experiences of our teachers and publish their discoveries.

Group 5

Members: Teresa Elecputu (President) AMETRA; Policarpo Sánchez (Secretary) SHINANYA JONI; Emilio Agustín, IDIMA; Josué Faquin, FECONADIP; Fidel Sinarahua, PALMICULTURA; Delia Isuiza, AMACAU; Francisco Muñoz MARUTI SHOBO.

I. What does nature or biodiversity mean to us? What importance does it hold?

Nature or biodiversity is the conjunction of beings and resources in terms of flora and fauna and serves to protect and allow the development of human life. The elements that make up biodiversity possible include the forests, land, animals, water, and the like.

Importance:

- Biodiversity is important because it gives life.
- If we are not careful, we can break the balance of nature and threaten our own existence.
- Biodiversity makes possible a healthy life.
- Biodiversity and the forests are the lungs of the world in that they maintain the air that we breathe.
- Each element of biodiversity has its own unique value as part of the ecosystem.

II. What value do the elements that compose biodiversity have?

1. Economic

- The extraction and commercialization of the elements of biodiversity provide economic income for our families.
- We use the biological resources for economic income in order to improve our health, education, housing, and so on.
- At the same time, biodiversity is priceless and allows constant use for our daily food.

2. Religious

- Our resources have allowed the continuation of traditional religious practices.
- The traditional celebrations provide the opportunity to show gratitude for the benefits of nature.

3. Cultural

- Encourage people to identify with their ethnic groups and receive their traditional knowledge (values and customs) in terms of the uses of biodiversity.

4. Medicinal

- In nature, the resources required for primary attention for health care can be found.

5. Political

- The rights of indigenous people and peasants must be recognized in terms of the uses of biodiversity.
- Remove the threat of law enforcement when dealing with the problems and rights of indigenous people and peasants.

Philippines – Moises O. Pindog, Omis Balin Hawang and Baliag Bugtong – Ikalahan

We are the Ikalahan people. Our name means 'the people from the oak forests'. Our grandparents came here from Tinoc in the north. We moved to these hills many years ago because of the tribal wars and head-hunting.

We distinguish ourselves by the type of forest that we live in. The animals of this forest are very important to us because we are hunters. We also practise slash and burn (swidden) agriculture. We are very conscious that there are other powers in a forest in addition to its plants and animals.

What you call biodiversity means to us all of the things of a forest. When we open a virgin forest, we test the area by placing about ten cooked sweet potatoes in a given spot. If any are touched or removed, it is a sign that we must stop clearing and leave that patch to remain as forest. This tells us that *Bibyo* (spirits) are present in that spot. Similarly, if we intend to cut a big tree, we light a fire by the tree. If the fire goes out it is a sign that the tree is protected by a spirit and we will not touch it. We also offer prayers to these spirits before we embark on mining in our forests. These spirits are ages old. Some came here with us, some were here before. In the forest there are spirits who are settled in some place and there are spirits who roam. Usually they are malevolent to those who would disturb them. We feel the presence of these spirits of the forest as we ourselves feel different when we encounter their territories. Sometimes, when a large tree is cut, we can hear the crying of voices that tell us of its spirits. At other times they appear in our dreams if we seek to disturb their areas. We appease the spirits if they are disturbed by our swidden clearing with the sacrifice of a pig or a white chicken. If the forest is cut and cleared the spirits will go away. Thus the life of the forest (biodiversity) around us contains the rivers, rocks, plants, animals and its spirits.

All of our medicines are taken from the forest. This lore was taught to us by our grandparents. In addition we collect rattan, honey and a multitude of other products. Certain plants are essential to our ritual and culture. For instance, the leaves of a certain species of ground orchid are rubbed on cooked rice to be placed on the graves of dead people for their last meal. A carved head from an old tree fern substitutes for a human head in rituals around victory or other ceremonies. It is also mounted on a spear made from forest bamboo. We use specific tree species to make our coffins. Fibres from the *tabac* tree are used for the production of string, and spoons and bowls are made from the *Alstonia* tree. Thus specific trees provide us with our cultural needs.

Today we appreciate the need to protect the forest as a natural system due to educational input. The functional uses of some species are losing their value due to the inroads of modern technology. But we would like to maintain our forest and all of its attributes as it gives us protection in so many different ways. It is, in the end, a representation of who we are. It is our life.

Solomon Islands – Ruth Lilongula

I am Ruth Lilongula and I come from the Solomon Islands. I was born on the Island of Chocho, into a tribe known as Kukumagiwa which is part of a larger tribe known as Guruvat. At the moment I am working for the Ministry of Agriculture and Land as Director of Agriculture Research.

Biodiversity is a very broad issue which, to us, is environment, plants, and everything else including you and me. Let me start with my identity to describe how we think of our place in nature and biodiversity. I am Ruth and I identify with my father and mother, my sisters and brothers, my extended family, my clan, and my tribe. On top of this, I identify with the land that is given to me, with the trees that grow on it, the animals that live on it, the medicinal plants that grow on it, the streams and rivers that run through it, the birds and the snakes that live on it, the spirits that are in the trees, and the land, and the rivers. I also identify with the sea and the fishes and the creatures that live there. My tribe identifies with the eagle and the bird. At sea, we identify with the shark and the crocodile, and on land, the lizard. The rivers mark the boundaries where we live. The spirits, both good and bad, live in the trees and on the land and take a part in our everyday existence.

How do we value the biodiversity that surrounds us? Biodiversity is the very core of our existence within our communities. You cannot say how many dollars this is worth because it is our culture and our survival. In this context, biodiversity is invaluable.

We don't see our land and everything else in terms of money. Rather, we value our surroundings as our identity, as who we are and our inheritance that is given to us. It is like giving birth to a child. The child is yours and is of value to you. This cannot be put in monetary terms. You look after the child and the child looks after you. This relationship is the value. If you separate the mother and the child, then the value is lost.

There are problems within our society when a lot of monetary value is placed on the riches of the land and sea. You are not supposed to do this. You must respect your environment so that you may live and survive. This is the way that we live by the rivers. In my tribe, we have special values about the river that belongs to us. There are special spiritual and supernatural powers that connect the river to us. Now that the logging companies are cutting down trees and polluting the rivers, we get a lot of earthquakes. While we do not associate the pollution with the cutting of the trees, we do relate it to the fact that we have made the spirits and the Gods within our environments cross.

We are conscious of the sustainable use of natural resources and biodiversity. We live on the richness of biodiversity. In the past, there was never any need for us to do plant and animal breeding work in order to pass them along to clients. For example, in our system, it is my tradition and upbringing that if you are walking along the reef and you see two clam shells together, by tradition you are not allowed to take the two. You leave one behind, not for your children, but for the future children, for your grandchildren. Your children will be eating what you collect today, but the one that is left is for the future. And when you take out the meat from the shell, you don't leave the shell open in the sea. That is traditionally taboo because the sea will then be no good. So we turn the shell upside down so that it can be home for other creatures that are living in the sea. When you make a garden, you do not harvest every mango or every crop that you plant. You always leave a few for the mother and, in this way, we continue to have what we need.

Our genealogies are connected with food and our land. We know how tribal identity is related to land ownership and boundaries and which people have the right to use which lands. We know exactly where our boundaries are, and if you step across those boundaries where you do not belong, you are dealt with by the spirit that lives on that land. The genealogies are also related to what can be hunted. For example, there are leatherback turtles that go into a lake and you can go there and wait to capture one. You cannot get one unless you are an immediate descendant of my grandfather. Once my father caught one for our church and the next time my brother caught one for the church. These were the only occasions during my lifetime that the leatherback turtles were taken. Other people have come to look for the turtles but they can never take them because they are not of that blood. They don't belong because they are not connected with our genealogy and have no harvesting rights. You just can't hunt if you are from a different tribe. Sorry. So you see that our genealogies protect biodiversity. Even those people who are distantly related to us and of the same tribe cannot take any turtles because they don't know when the turtles are coming ashore. We can tell whether the leatherback turtle is coming because we can see the special lightening just over the horizon of the sea and we know it's coming tonight. We believe that this is controlled by our ancestors. We don't have the urge to take it. Usually we go admire the leatherback turtle, play zig-zag with it, and then it goes back to the sea. Only when we feel right about taking a turtle, do we take it. We believe it is not just right for you to take it just because you are going to eat it. It can only be taken for a public event or some event that is ceremonial. You are not allowed to take it at any other time: it is like a pet.

In our tradition, management is a part of the value system. A lot of demeaning things have been said about the shifting cultivation that we do. While I haven't done any formal research on our system of cultivation, I am a scientist and I know that such criticisms are not valid. We have a management system that keeps us going, that respects the earth, and that is sustainable. Critics say that slash and burn cultivation brings about erosion, but if done in the traditional way, this is not true. The clearing and burning allows us to make a one-time crop and we have protection from insects for a short period of time. But as the plants grow, so do the insects, and after harvest they start to have an impact. When we work on the hillside we start at the bottom of the hill and don't clear everything. We cut the trees across the slopes as opposed to up the slopes. The plants grow back quickly. Until the plants grow back along the base of the hill, you cannot clear the land again. This, to me, is part of erosion control.

We place value on everything to do with the land, and even those trees that we cut when we clear. The logs go right across the hillside and under these logs is where you will find the tubers of sweet potatoes that haven't been infected with any sweet potato fever. Everything that is done in the garden is for a purpose. There are trees that grow up straight away after you plant and they are not removed. Further, certain trees are always left to grow in the garden and you don't remove them.

Because of the economic drive of people, monocultures are coming in and changing the way agriculture is done. The risk for us as farmers is far greater. In our traditional farming systems, the risks are spread out so that if one thing fails, we can rely on other crops. Thus, when we have disasters, we still have food. We have a storage system for the times of famine. When the usual foods are not available, you harvest other crops. Even when harvesting at sea, we are mindful of how it relates it to the ploughing and cutting of trees in the bush. If you take too much, if you take more than you need, nature loses its ability to replenish itself and this is true both at sea and on land. Traditionally, there was always some control over what you were doing. For example, for the hunting of tortoise you have to go through certain rituals before you are entitled to hunt. At the moment, the people who use all this modern technology have destroyed the system that we already have for sustainable use of our resources.

There are spiritual values in nature also. In the past we saw spirits in every tree. But now, with Christianity coming in, things are changing. It was easy for we Solomon Islanders to accept Christianity because we believe that God lives everywhere. It was just that we had different names for such things in the past. God lives on the trees and there are times, if you are close to something like that, you can specifically call a shouting tree or a hunting tree. You go in and give Him everything in your praise. Whenever you are there, you feel good about things, and as you come home, you stop by a small stream or waterfall and wash. You go out in your disappointment and in your happiness, you have a special place within the environment where you can shout and dance and whatever else you want to do. Our environment is many things, a classroom, a pharmacy, and a supermarket.

These values about the environment are passed down through the woman. So there is an important role, the role of women, in sustainable development – at least in my country. Unfortunately, this is not clearly addressed. If you want sustainability and you want things looked after day to day, you have to put some effort into understanding the role of women. It is between women and children that all this knowledge is passed along. But this isn't done in the house. You do it in the bush as you walk along. So, when you go into the garden, you learn all there is to know about sustainable farming. When you go hunting for food and other things in the bush, you learn about which trees to take and which trees not to take for medicines. We believe that you have to go and ask the tree first. We ask, 'Do you allow me to take you, to use you for medicine?' and we believe the trees answer. If they say yes, you take the medicine, but if they say no, then you do not. They could direct you to another tree which is okay to use. You don't take any others, and this is part of my upbringing. I am quite well versed in this because we are a country of healers.

Not until you are aged fifteen do you get from under your mother's arm. She is the educator. And every time your father gets cross about something, he doesn't have to explain. It is mother's responsibility to explain why daddy is cross. Today, with the education system we have, the educated teachers are trying to act, or the organizations are trying to act, as mothers and fathers to these children. This is where we are getting really broken up culturally and not learning very much. The biodiversity around us is the teaching tool that a mother uses to educate the children, and keeps the culture alive.

Mothers teach the children how to stay away from danger. We normally live along the coast. The mothers warn if there are shark infested places and show the children what they must do in case a shark is coming, and how to chase it away. We are given all this educational background. Even so, before we go to the sea our mothers remind us again. A mother will say, 'That shark is dangerous, its aggressive, don't go into the sea'. There are certain times that by just reading the shadows of the trees and such that the mother will tell you not to go into the sea or to cross the rivers.

Biodiversity means our identity and culture survives, and it's also the very core of our existence and life itself. There is no parameter of biodiversity that is divorced from our identity, heritage and pride. There is no separation between our people and biodiversity. We are all part and parcel of one system. Biodiversity is a God-given heritage and inheritance with me, as the human being, as a part of the whole system.

Although I am a scientist, when I go home I don't tell people how to farm because they know better than me how to do it. I've just been to a conference of tribal communities on my islands and I told them that the knowledge that they have is much better than what I have. I need that knowledge as a scientist to build upon in order to come up with the answers. I can't operate on my own, and a lot of work that I have done in agriculture is built upon the indigenous knowledge that is already there. This is why I am quite livid when people say they have made discoveries about our agriculture and that they know more than us because they have studied our knowledge and written it down. I am very concerned about outsiders exploiting indigenous peoples' knowledge, using it and calling it their own when, really, it's not theirs and is without any basis in our culture. We live in a society that shares for our own benefit. With the type of biodiversity activity we are seeing in the UN and other places, this aspect of our culture will be lost because there is no sharing. People, in effect, are saying, 'I take this information and plant material and I put a patent on it and bye to you. I don't recognize you and the good service that you have done'. For us, it is give and take between tribes and peoples.

Sustainability exists throughout our culture. When you go to fish, you get certain fish and when that's enough for the family, you come home. Only when you are fishing for feasting are you allowed to take more. And if you fish today, and you haven't finished eating it by tomorrow, it isn't right to go out again because you have your food. You must respect the sea. I have a story which shows what I mean. My step-children are English. I took them fishing and they fish every day. My mother got cross and she came to me and said, 'Daughter, where is the upbringing that I gave you? You fish every day? You know this is food to us and not something to play around with just to satisfy your step-children's need for pleasure. You should prevent all this'. I had to tell that to my husband, who does not understand.

Sri Lanka – Mudianse Tenekoon

My name is Malvila Sribrahmana Vanninnayake Tenekoon Mudianselage Mudianse Tenekoon. My name reflects my history. *Malvila* means 'field of flowers' and was an ancient name for this island of Sri Lanka. *Vanninayake* means 'the leaders of the Vanni'. The Vanni is an area within what today comprises the North Central and Eastern provinces of Sri Lanka . This region is characterized by forest, punctuated with great rocky outcroppings and an intricate system of lakes, ponds and rivers. The area is rich in biodiversity and even boasts herds of elephants.

The Vanni has a very ancient history. We practice a ritual called *Dena Pideniya* in which we recognize our continuity from the line of Queen Kuveni. Queen Kuveni is known to have lived in Sri Lanka prior to the arrival of Buddhism 2,300 years ago. With the coming of the Buddhists, written historical records were initiated and have been maintained to the present day. Queen Kuveni was from the Yaksha who were the indigenous people of this island. The Yakshas practised integrated, irrigated agriculture for thousands of years before the arrival of Prince Vijaya from India who became the husband of Queen Kuveni. Prince Vijaya is often associated with the modern Sinhalese, but not with us: we are of the people indigenous to this land.

Our agricultural ways maintained the wildlife and biodiversity that the British recorded when they first came here. We recognize that the first people, animals or plants of a land have within them the ability to resist local catastrophes and infections. Consider malaria. We have been living here for ages without huge infections while visitors and invaders were perishing in large numbers. Today, conditions have changed: our children succumb in large numbers too because of the changes in the mosquitoes and the dilution of the blood.

Biodiversity to me is made up of the things and conditions that maintain the balance that we have lived with for centuries. It included the animals, plants, rocks, rivers and spirits. For instance, to us, trees are not only living beings that shelter animals, insects and other small plants. They also shelter spirits. We value the identity and services of spirits differently. Different deities require different types of offerings and times spent in their worship. On the landscape, this resolved into different villages having different identities and responsibilities relative to these deities. The villages are also identified by different service functions. For example, a *devealagama* or temple village would grow crops and offerings particular to some deity while a *gabadagama* or granary village focused on the king and grew and stored grain. This diversity of life-styles and patterns of land-use made biodiversity a vibrant and living thing for us. The value of this thing to us was its ability to sustain our children and our culture.

The removal of the living environment and the cultural environment means an inability to sustain ourselves. It is clear from the destruction of my land today that both the natural and cultural environment has lost this value. Regaining this value as a culture will sustain us. Regaining this value is what is most important for any culture to sustain itself. If biodiversity is to be valued it has to be appreciated in this way. To give all credibility for making decisions to a Western consumerist system, judged by so-called 'economic values', is not only folly: it is doom.

Tonga – Siosiane Fanua Bloomfield – Promoter of Traditional Health Care, Fatai Village

Biodiversity is my home. Everything is familiar to me and I am familiar to everything around me too. It is where I feel rooted and nurtured. We've just had a whole lot of rain which everyone enjoyed in its coolness and cleansing effect. Now I look out and see our hibiscus trees bloom red and proud in the sun. I love Tonga's environment, its predictability, its gentle safeness and its natural bounty. Being a promoter of traditional health care, however, I sometimes feel sad because some of our traditional medicinal plants are hard to find. At the same time, I know many people who are making efforts to rejuvenate these plants.

I do not like pesticides, fertilizers, and some of the modern farming equipment, but nobody seems to be able to keep that tidal wave away from our shores. They say there are introduced unseen pests which spoil cash crops and so on, claiming pesticides are needed. It is true that the land has given to many small farmers through cash cropping a lifestyle undreamed of ten or fifteen years ago. Still, I am sad to see that where once there were trees, shrubs and small wild-life the land now looks vacant and tired, covered only by some kind of cash crop. I am still thankful to God for arranging to have me born in Tonga.

Western Samoa – Korosata To'o, Director O le Siosiomaga Society Inc.

Biodiversity is the basis of our survival. It represents the total processes that are interlinked to give us the full realization of nature. The richness of Samoan biodiversity should never be understated. For years, the people have depended upon the forest for building their *'fales'* (houses), for firewood, for building canoes, etc. The knowledge and the availability of the plants used for medicinal purposes play an important part in this context. Biodiversity is culture itself. It restores the elements that strengthen cultural values because the degradation of biodiversity is also associated with cultural deprivation. It is therefore important to keep our mangrove forests because they are breeding grounds for fish and numerous other marine organisms. It is part of the life cycle. Biodiversity not only effects the air we breathe but also strengthens the brighter opportunities for our children and future generations. This is our value and no money in the world can pay for this.

Western Samoa – Reverend Asotasi Time – Uafato Conservation Project

Biodiversity is the sum total of all species – plants, insects, humans, fish, birds, trees, and so on. It tells us the abundance of nature and how much is the wonder of God and His creation. It is within this realm that explains the responsibility of humankind and how to utilize resources sustainably. The forest is like a deep freezer: anything you need is kept there. All resources that are needed by the community are kept there. The community should do all it can to sustain these resources. The Samoans call it *'pule faafaeetia'*. They should control and manage resources that are productive and should never forget that we are the trustees for future generations. As humans, we should not abuse nature.

Zimbabwe – Evanson Nyamakao - Korekore

My name is Evanson Nyamakao. I am of a Korekore tribe. At the present time we are living at Fledge, which is Ward 15 here in Hurongwe, but we used to stay near Makuti which is on the Zambezi Escarpment. We were removed from a town there because of Lake Kariba which was built in 1957. This was a bit of change for us because where we used to stay there were plenty of animals. Here, there are no wild animals. To carry on our tradition, you need certain animals to be the totem. When you need certain plants that are no longer available, it changes or influence traditions. Our traditions changed because when we stayed near Makuti, we ate a lot of meat which was hunted by our parents in the bush. Since the time we came here, there has been no meat. So when we came here, the government used to give us beans and *kapenda*. To balance things out and because the people were new in this area and could not get meat anywhere, the government provided other foods. Certain people began going to the Sanyati River to fish.

The difference in soils between here and near Makuti caused a great change in our agriculture. There was a rich soil in our old place. This place has poor soils. You can plough it for just two years. Then the soil comes to expire and needs more fertilizer. In our old place, we did not use fertilizer even if you ploughed for the whole of your life.

There were other changes that resulted from our move. Here, we can't hunt for meat and, if you can't have enough bush food, the health of the tribe goes down.

When we have decisions to make, we call the names of the forefathers. When we are moving to another area, if we want to carry our forefathers along, we have to clap hands to them and take a little bit of soil from their grave. We wrap the soil in a black cloth, and we take it to the place where we are going. When we arrive and are going to put them on a certain place, we have to say, 'We have put you here. So you are now here with us so you are to guide us'. The ancestors have participated in our lives in many ways. For example, say somebody killed you. In the past we used to brew beer to hand you over to the forefathers. You are already passed. You are called by your name and handed over to the forefathers who have called on your children to guide them. If someone in your group was murdered, you should fight back with the people who have killed him. When there is a marriage, we tell the forefathers that your daughter and son-in-law are married now and the forefathers are to guide his house. It is the same with the birth of a new baby. By giving him or her a name, its just the same, the ancestors are involved.

If our plants are guided by the ancestors, the things that eat the plants will not enter the fields. We have certain days, Mondays and Thursdays, when we are not allowed to go to the fields. These are the days for the ancestors. If you plough at this time, it's against the ancestors. So they will have to send the animals now to eat the crops in your field.

Zimbabwe – Ralph Mogati – Korekore

My name is Ralph Mogati. I was born here and I've been married. I also have a traditional name, which is Raphomore. I am from the Korekore tribe. We are from the Rongoiya area in Zimbabwe.

The environment in this beautiful valley is so useful to us. For example, the Korekore long ago started using a particular tree for their medicine. Because they couldn't go to hospital, they would just treat each other with this tree. Animal life is very rich here. We've got everything, for example, the buffalo. Each year, we sell some animals from each and every area. We have an animal service that counts the animals so we know exactly how many live in our area. In this way, we protect our land from over-grazing.

Before we make any decisions, we consult our ancestors first . We remind ourselves about our central spirits, we talk to them, and then they give us a go-ahead. Before we touch the land, we go to the spirits. The spirits are linked with certain animals or trees. For example, we have totems here. Some people don't eat elephants because they are the totem. We've got trees that we respect very much, like the tree I spoke of earlier. Not long ago, our people could ask for food from those trees and they could obtain the food. But now, alas, they tell me, it's no longer happening so easily for us.

We don't want to cut trees down. For firewood, we gather dead wood from the forest. The reason we don't cut is not for the spirits but only for the environment. We are teaching people every day not to cut the trees, and if they still want to cut, then we can sue that person. I'm their leader. I've got to tell them.

We are also concerned that some people in the area burn for gold and this threatens our environment. Of course, there is some good to the burning for gold because there is a lot of gold there. At the same time, you see, if we search for gold we will find ways to destroy the environment so we don't allow it. Gold doesn't have value to us as to other people, who replace the owners. The people who are burning for gold are doing it illegally.

CHAPTER 5

ETHNOSCIENCE, 'TEK' AND ITS APPLICATION TO CONSERVATION

L. Jan Slikkerveer

CONTENTS

TEXT BOXES

FIGURES AND TABLES

Ethnoscience, 'TEK' and its Application to Conservation

L. Jan Slikkerveer

Introduction (L. Jan Slikkerveer)

Background

'The practice of science, including belief and magic, forms a fundamental characteristic of all human societies.' (Bronowski 1981)

Almost parallel to Western 'scientific', 'cosmopolitan' or 'global' disciplinarity, 'indigenous', 'traditional' or 'local' knowledge has developed *in situ* to encompass the holistic, inter-disciplinary wisdom, practices and experiences of local communities and ethnic groups. Despite efforts to list theoretical binary oppositions between indigenous and global knowledge (such as 'qualitative' versus 'quantitative', 'intuitive' versus 'rational', 'holistic' versus 'reductionist', and 'spiritual' versus 'mechanistic'), such distinctions cannot easily be made. Many components of indigenous knowledge systems are presently being studied, documented and made available to the world community, to link up eventually with the *ex situ* global knowledge systems that are easily accessible in print and in data bases.

Indigenous and global knowledge systems are alternative pathways in the human/scientific quest to come to terms with the universe, and are the result of the same intellectual process of creating order out of disorder. Indeed, as Clark (this chapter) illustrates, the very roots of Western science and philosophy are based on the observation and political interpretation of local observations of nature. Thus the 'scientific' methods and techniques developed and used for the study of global knowledge systems can be – and have been – adapted, refined, tested and used to study and understand indigenous knowledge resources.

These different pathways reflect an arduous process of assessment, recognition and emancipation of indigenous peoples and their knowledge systems. This progress has passed from the interest of early ethnologists in the 'exotic', through the colonial and post-colonial concerns over 'primitive' peoples, and on to the currently emerging 'new' ethnoscience focused on indigenous knowledge systems in the context of development and change. Finally, attempts have been made to ensure the protection of traditional and indigenous peoples and their knowledge systems (*cf.* Posey and Dutfield 1996).

As Arun Agrawal highlights in his contribution to this chapter, the debate about the supposed 'critical difference' between indigenous and scientific knowledge (recently initiated in the *Indigenous Knowledge and Development Monitor*, 1995–1996) has largely ignored the prevailing power structures in which these knowledge systems are embedded. In his view, the value and usefulness of indigenous knowledge for sustainable development lies ultimately in the empowerment of local communities, many of which are now under serious threat.

With the new and expanding study of indigenous knowledge systems, and increasing interest from research foundations and supporting agencies in allocating modest funds for small-scale projects in health care, agriculture and natural resource management, the position of the indigenous peoples themselves is finally attracting the attention it has long been denied.

Thus the debate has shifted from *what* is the value and usefulness of indigenous knowledge systems for sustainable development to *how* can such knowledge systems be used to ensure equitable benefit sharing of the resources with the contributing communities. Or, as Posey (personal communication 1997) aptly puts it, 'How can humanity benefit without further undermining the health and well-being of traditional and indigenous peoples?'

A valuable aspect of this process is the way in which indigenous knowledge will be understood, respected and synthesized with global knowledge in a balanced, humane way. This international 'knowledge debate' is largely framed by the Convention on Biological Diversity (CBD, Article 8.j, UNEP 1992) which calls on each state to:

> *'Subject to its national legislation, respect, preserve and maintain the knowledge, innovations and practices of indigenous and local communities embodying traditional lifestyles relevant for the conservation and sustainable use of biological diversity and promote their wider application with the approval and involvement of the holders of such knowledge, innovations and practices and encourage the equitable sharing of the benefits arising from the utilization of such knowledge, innovations and practices.'*

Posey (this volume) points to the fact that respect and equity cannot be achieved without recognition of the basic rights of indigenous peoples, traditional societies and local communities, including full disclosure about, and Prior Informed Consent for, all activities that affect them. Basic rights include: the right to self-determination for indigenous peoples, right of local-determination for traditional and local communities, guarantees of territorial and land rights, right to development, equitable benefit-sharing, collective rights for communities, and religious and customary rights. It is these rights that underlie the shifts in power necessary to enhance and support local, *in situ* biodiversity conservation.

Changing attitudes towards indigenous peoples will be reflected in a 'new' dynamic ethnoscientific view which combines both formal and empirical observations. The operationalization and implementation of a participatory 'ethnosystems' research methodology is urgently needed for the assessment, documentation and use of alternative solutions to the world's present environmental predicament. 'Community-Controlled Research' (CCR) and 'Collaborative Research' (CR) are beginning to replace the paternalistic, authoritarian investigation and development methods of the past. The role of 'catalyst' or 'consultant' is unfamiliar (and uncomfortable) to most scientists and policy-makers, but convincing and impressive examples of collaborative activities and co-management with indigenous and local communities are now abundant. Some indigenous and local communities have established their own research and development guidelines and have developed their own policies on all activities that affect their lands and territories. For example, communities have developed 'community biodiversity registries' and data banks, sometimes with the help of outside scientists and consultants, but in their own languages and controlled by their own experts (Posey, this volume).

Cognitive anthropology, ethnoscience and TEK

For a long time, indigenous knowledge was ridiculed as 'primitive', 'native', 'old-fashioned' or even 'magic' – and as such was ignored, marginalized and in some cases even wiped out (*cf.* Jiggins 1989; Slikkerveer 1989; Warren 1989). However, during the 1950s and 1960s a growing academic interest emerged in the study of local ideas, taxonomies and classifications from the participants' own point of view. Conklin (1957), Goodenough (1956) and Kay (1966) developed the sub-field of 'cognitive anthropology' following earlier ethnologists such as Franz Boas and Bronislaw Malinowski who had stressed the importance of the 'native point of view'. The new ethnoscience approach sought to replace the researcher's own perceptual categories with those describing the organizing principles that underlie culture and behaviour.

Participants' categories were elicited using the local language, often with the help of interpreters, and this method was particularly successful in gaining a deeper insight into the structures of language and culture. Today, cognitive anthropologists focus their attention on the study of universal cognitive processes of creation and use of language which, according to Noam Chomsky (1966), are uniquely human and genetically transmitted through neuropsychological structures in the brain. Through comparative studies in different cultures, researchers using this approach attempt to discover how the human mind functions in order to infer general clues to behaviour.

In contrast to anthropology as an empirical science based on observable data, cognitive anthropology has become a formal science involving a 'mentalist approach' to the study of the principles and ideas underlying behaviour, rather than the behaviour itself (Applebaum 1987). Assumptions about the culture under investigation are made on the basis of the generalizations perceived by the informants/participants.

In the present volume, a particularly interesting sub-set of ethnoscience deserves special attention, and that is ethnoecology – defined by Hardesty (1977) as, 'the study of systems of knowledge developed by a given culture to classify the objects, activities and events of its universe'. Ethnoecology focuses primarily on the ideas, perceptions and classifications of the environmental relationships of members of a particular community or culture. Thus, an *emic* view on representation of the environment from within – as opposed to the *etic* view from outside – is constructed as it is perceived by the members of the community themselves, and inevitably involves cosmology as it regulates the people's interactions with their environment.

Traditional Ecological Knowledge (TEK) – also referred to as Indigenous Ecological Knowledge (IEK) – primarily encompasses local knowledge of plants, animals, soils etc. and the associated experience and wisdom of human interactions with the environment (*cf.* Johannes 1989; Inglis 1993). Berkes (1993: 3) defines TEK as, '*... a cumulative body of knowledge and beliefs, handed down through generations by cultural transmission, about the relationship of living beings (including humans) with one another and with their environment.*'

Initially TEK remained rather formal in its orientation, but gradually it developed to encompass the use and management of agricultural and other natural resources, eventually becoming an essential element in local decision-making and planning activities. TEK also includes the 'invisible' socio-cultural context of knowledge systems. Berkes (1993: 5) observes the following three dimensions: a) symbolic meaning through oral history, place names and spiritual relationships; b) distinct cosmologies or world-views as conceptualizations of the environment; and c) relations based on reciprocity and obligations towards both community members and other beings, and resource management institutions based on shared knowledge and meaning.

This wider perspective has recently attracted renewed attention, particularly in agriculture where farmers' experimentation and accumulation of experience is guided by their holistic cosmology, or 'cosmovision' (Van den Berg 1991; Grillo 1991). Cosmovision refers specifically to the way in which the members of a particular culture perceive their world, cosmos or universe. It represents a view of the world as a living being, its totality including not only natural elements such as plants, animals and humans, but also spiritual elements such as spirits, ancestors and future generations. In this view, nature does not belong to humans, but humans to nature. As the concept of cosmovision includes the relationships between humans, nature and the spiritual world, it describes the principles, roles and processes of the forces of nature, often intertwined with local belief systems. As Haverkort (1995: 456) notes: '...it makes explicit the philosophical and scientific premises on the basis of which intervention in nature (as is the case in agriculture and health care) takes place'. Although not always immediately identifiable, the prevailing cosmovision of the members of many indigenous communities guides and regulates a complex of socio-cultural phenomena such as the organization of the culture and the way of daily life, and determines to a large extend the way in which goals are achieved. The study of cosmovisions requires a special research methodology to deal with 'extra-scientific' factors and variables that are often 'invisible' to outsiders, including variations along gender lines IUCN/UNEP/WWF (1991).

The dynamic context of knowledge systems interactions

In the practical setting of the age-long processes of interaction and acculturation among the different peoples of the world, the Western-oriented Transfer of Technology (TOT) model used in 'development co-operation' has dominated. Often, this process evolved in a unilinear mode from research institutes, laboratories and universities in 'developed' countries and was subsequently transferred to the 'under-developed' countries of the Third World (*cf.* Beal, Dissanayake and Konoshima 1986).

This strategy initially fostered high hopes for socio-economic development and growth, but an increasing number of disappointing development programmes and projects introduced by outsiders who tended to ignore local experience and wisdom provoked an international call for alternative solutions. The pioneering work *Indigenous Knowledge Systems and Development* (Brokensha, Warren and Werner 1980) heralded a 'new' ethnoscience in which indigenous knowledge systems were seen in a more dynamic perspective within the development process. Since the publication of this work, an increasing number of applied studies have highlighted the crucial role that 'alternative' knowledge systems play in rural extension projects and programmes. Likewise, the 1978 World Health Organization conference in Alma Ata highlighted the potential of indigenous systems for more participatory and sustainable health systems. This led the way to a new, more dynamic field of ethnoscience research in other sectors such as agroecology – the management and conservation of agricultural and natural resources. Since then, studies in disciplines such as anthropology, agriculture, fisheries, forestry, ecology, biology, botany and medicine have begun to document the adaptability and viability of local systems for the international development process (*cf.* Posey 1985; Richards 1992; Chambers, Pacey and Thrupp 1989; Warren 1989; Mazur and Titilola 1992; Fairhead 1992; Slikkerveer 1991; Leakey and Slikkerveer 1991; Mathias Mundy 1992; Warren, Slikkerveer and Brokensha 1995; Posey and Dutfield 1996).

This new approach has also had implications for the dominant 'top-down' development paradigm. In international development strategies, the newly-developing field of 'Indigenous Knowledge Systems and Development' paved the way for the current shift toward sustainable development which explores alternative modes of knowledge generation and exchange. The strategies of 'sustainability' and 'grassroots participation', promoted by the 'Brundtland Report' (1987), have pointed to a more realistic 'Farmers First' paradigm (Chambers, Pacey and Thrupp 1989) in which indigenous practices are included in 'bottom-up' agricultural decision-making and planning processes.

Indigenous knowledge extends beyond the *'in situ'* label because it involves interactions and transformations among indigenous knowledge systems in conjunction with global systems of knowledge and technology 'imported' from the West. Thus the 'new' ethnoscience seeks to understand the complex interface between indigenous and global knowledge systems in relation to the intertwined processes of development and change. Differences, contrasts and oppositions will further give way to multidisciplinary R&D activities. According to Harrison's (1990) assessment of the current environmental crisis in Sub-Saharan Africa, for example, interfaces between local and global ecological knowledge systems are well documented in agriculture, fisheries, forestry, ecology, soil science and botany.

The *political* context is also important. In his study of agricultural development in Ecuador, Bebbington (1993: 275) notes that indigenous knowledge provides, 'a dynamic response to a changing context constructed through farmers' practices', indicating the political importance of indigenous knowledge as a local point of identity. Adams (1996: 146) points out that 'Indigenous Knowledge' (IK) not only has to be observed in a wider context of social and economic change, but also that it needs an understanding of the critical political importance of 'the definition and appropriation of knowledge and meanings'; thus, 'IK becomes, in this sense, a central issue in empowerment'.

This approach places the concept of 'interaction' above that of 'intervention' and it is clear that protection of both integrity and resource and intellectual property rights links up with the political context of indigenous knowledge (*cf.* Posey and Dutfield 1996). This important position reflects Arun Agrawal's plea (this chapter) for our continuing attention to the way in which power structures knowledge as a prerequisite to achieve the aim of working in the interests of indigenous peoples and their knowledge systems.

Ethnomethodology and the 'ethnosystems approach'

For a better understanding and explanation of the indigenous perceptions, practices, beliefs, values and philosophies associated with biological and cultural diversity, answers have to be found for complex methodological questions. Could appropriate parameters be developed in conjunction with local communities in order to validate such indigenous concepts in comparison with similar global phenomena? How could 'invisible' phenomena such as indigenous cosmologies, belief systems and attitudes be recorded and analysed in a 'scientific' mode? In what way could individual (subjective) variables of perceptions and ideas be transformed into systems (objective) variables for value-free measurement and comparison? Could indigenous knowledge systems in general be compared with Western reflective knowledge systems? The answers would eventually be found in a methodology that goes beyond mere qualitative interpretations to include comparable quantitative data that could be encapsulated in multivariate models.

'Ethnomethodologies' were introduced in the 1960s by Garfinkel (1964) and Cicourel (1964). Their methods were used by people in everyday life to order, understand and make sense of behaviour and the social world. Thus 'the people's methodology' reflected the practical common-sense reasoning in everyday situations that is mostly ignored by conventional sociologists who regard the accounts of ordinary people as insufficient and replace them with their own versions. Two major techniques were used: disruption of the process of everyday routine in order to reveal the bases of the social order; and the use of conversation analysis in order to disclose the skills that people use to make 'sense of reality'. Critics stress that this presents an over-ordered notion of everyday life, without an explanation of the social structures and constraints. Although ethnomethodology proved useful for the development of interaction research, it has since lost much of its initial support.

Current efforts to establish a comprehensive body of knowledge on 'indigenous knowledge and development' obviously need more than the mere accumulation of data from case studies and narratives on the subject. The 'ethnosystems' methodology has shown to further increase the understanding and clarification of the interaction process between indigenous and global knowledge in several study areas of East Africa, Indonesia and the Mediterranean (*cf.* Leakey and Slikkerveer 1991; Adimihardja 1995; Adams and Slikkerveer 1996). Numerous key elements of local knowledge systems have been recorded, analysed and successfully integrated into pluralistic forms of agroecology, comprehensive health care and integrated wildlife management. Moreover, this ethnosystems methodology has facilitated the design and testing of a computer model of integrated agricultural behaviour and its components (Figure 5.1; *cf.* Slikkerveer 1996a).

'Ethnosystem methodologies' facilitate the assessment of the cognitive and behavioural components of particular groups or communities as 'systems' in a holistic mode, and facilitate the elaboration of the concept of culture as the result of historical processes of acculturation and transculturation in a more dynamic way. They also employ the well-known techniques of 'participant observation', 'semi-structured interviews', 'triad ranking', the construction of 'transects' and the use of 'community cartography' in close co-operation with members of the community. However, as the 'ethnosystems methodology' seeks to pertain to a better understanding of indigenous knowledge systems, and at the same time enhance a non-normative, more realistic comparison between indigenous and global systems of knowledge and technology, it adds a combination of three more methodological principles: 'Participant's View' (PV), 'Field of Ethnological Study' (FES), and 'Historical Perspective' (HP).

Participant's View (PV)

This view includes the assessment of symbolic representations, world-views (cosmologies or cosmovisions), culture-bound philosophies of nature and the environment, perceptions, attitudes, opinions, etc. as part of the underlying structure of values, norms and belief systems, which characterize specific cultures. It links up with the formal ethnoscience concept of an *emic* view of cultures (from within). This technique pays special attention to the *in situ* aspects of indigenous knowledge.

Field of Ethnological Study (FES)

This approach is rooted in the 'Leiden Tradition' in structural anthropology, which evolved during fieldwork in Indonesia during the thirties (*cf.* Van Wouden 1935; De Josselin de Jong 1980). It is related to the concept of 'culture area' in which cultural features can be compared among different ethnic groups within the same region. Regional comparative research has been shown to be more realistic – less normative – as it seeks to include the practical setting in the analysis.

Historical Perspective (HP)

This perspective facilitates the (pre)historical analysis of complex contemporary configurations, be they in agriculture, medicine or resource management. Particularly in transcultural research settings, strict contemporary-oriented approaches have failed to unravel the dynamics of the origins and development of processes that have led to present-day complexes. Anthropologists and (ethno-)historians are collaborating closely on this historically-oriented methodology to complement the method of *ethnographic analogy* (*cf.* Wigboldus and Slikkerveer 1991).

Thus, the ethnosystems approach seeks to contribute to the establishment of a common ground for realistic comparisons of different indigenous and global knowledge systems in order to 'bridge the gap' through a broader, non-normative (*emic*) framework for regional comparative and (pre)historical analysis. For TEK, the implementation of this methodology refers to the further operationalization of its adaptive potential to complement and synthesize with global knowledge systems in complex ecological settings that include such areas as practicality, world-view and belief systems.

Similarly, the 'inextricable link' between biological and cultural diversity, as highlighted by Posey (this volume) also transpires through the analysis of the complementarity of cultural and biodiversity manifest in the project on 'Indigenous Agricultural Knowledge Systems' (INDAKS) carried out in Kenya and Indonesia in collaboration with LEAD at Leiden University in The Netherlands and MAICH at Chania, Crete (1993-1997).

Using the 'ethnosystems' methodology it has been possible to quantify the *individual* participants' view – perceptions, beliefs, cosmologies, attitudes and opinions – as *systemic* 'socio-demographic' variables and insert them into the 'predisposing variables' as 'background characteristics'. In this way, it became possible to compare different 'blocks' of predisposing, enabling and system variables in a newly-designed computer model of Integrated Agricultural Behaviour and its components, and identify specific determinants of sustainable use of agroecological resources (*cf.* Slikkerveer 1990, 1996a; Figure 5.1).

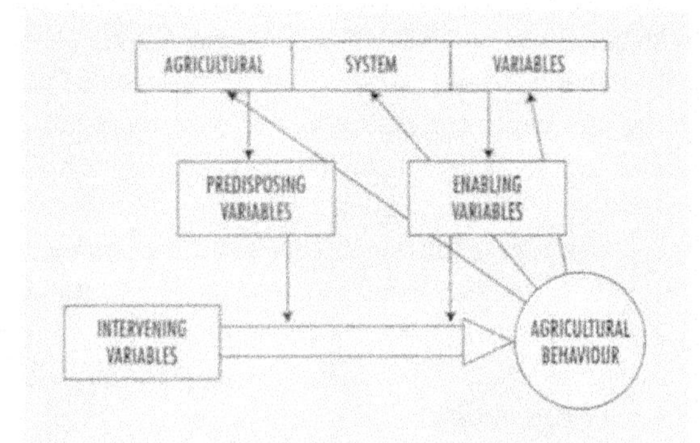

Figure 5.1: The INDAKS computer model of integrated agricultural behaviour and its components (Slikkerveer 1996a).

In Kenya, for example, growing concern about the loss of biodiversity and cultural diversity, particularly with regard to the use of indigenous plants and crops for food and medicine in the context of agricultural and primary health care development, has also contributed to the recent interest in the 'new' ethnoscientific approach of TEK which takes into account the interdisciplinary and historical dimensions of the complex agricultural and medical configurations in the country. In addition to the spread of agricultural monocultures for cash cropping, such as coffee, tea, sugar cane, cotton, sisal, and tree crops such as mango, cashew nut, orange, lime, lemon, banana and coconut, a number of small-scale farming systems in rural areas are also growing a wide range of local varieties of cereals, pulses and horticultural crops, still applying the conservational indigenous management practices handed down over many generations.

Towards conservation of bio-cultural diversity

In the quest for a global solution to the conservation and management of biodiversity, one of the key issues is related to the age-long enigma of, 'Humans' place within – and not above – Nature' (Leakey and Slikkerveer 1993). While Western ecological theories have tended towards 'stewardship of the earth', and control over and exploitation of natural resources, most indigenous ecological principles are largely concerned with experience, sustainment and prediction in relation to human subsistence and survival. These principles include values, norms and beliefs regarding the maintenance of the 'balance of nature' which have evolved over generations, and which encapsulate specific conservation methods and practices.

As Posey (Chapter 1) rightly notes, cultural diversity is shown to be closely linked to biodiversity in several respects. The argument is particularly strong in the comparison of cultural diversity in terms of different systems of local, regional and global knowledge *vis-a-vis* biodiversity in terms of different genes, species and ecosystems (*cf.* Brush 1989; Warren 1995; Slikkerveer 1996b). Indigenous knowledge is currently at risk of extinction just as is biodiversity.

In addition to the growing international concern over the loss of *biodiversity*, particularly since the political debates at FAO and the UN Conference on Environment and Development (UNCED, Rio de Janeiro, 1992), a similar apprehension emerged at the beginning of the 1990s among anthropologists, ecologist, geographers and ethnoscientists. Their concern is over the approaching loss of the related *cultural diversity* in terms of rapidly disappearing local and regional systems of knowledge and practice of the predominantly traditional and indigenous peoples around the globe. In this context, the concept of *bio-cultural* diversity has been elaborated in order to stress their crucial complementarity for achieving an alternative, less exploitative philosophy of nature and the environment for improved sustainable natural resource management and conservation (*cf.* Slikkerveer 1992).

The concept has evolved largely from 'new' ethnoscientific research which shows that several indigenous systems of knowledge and technology – despite their manifestly sustainable orientation – are being marginalized or put at risk of extinction. Indeed, the loss of cultural diversity in many countries is largely the result of unbalanced historical processes of acculturation and transculturation in which traditional and indigenous societies have been dominated by knowledge and technology from the West. The subsequent unidirectional 'Transfer-of-Technology' approach continues to encompass the rigidly advancing, commercial resource management schemes from Western countries. This process is particularly rampant in international development co-operation activities in human and animal health, agriculture, forestry and fisheries.

Parallel to the growing awareness during the 1980s of the urgent need to conserve indigenous plant, crop and animal species from an ecological point of view, interest in indigenous peoples' conservation and management practices has emerged from studies of local farmers' knowledge and skills in how to select, breed and utilize their floral and faunal resources. Cultural systems have been described as directly related to biological systems, as 'folk knowledge, 'farmers' knowledge', 'local knowledge', or 'rural people's science' (Barker, Oguntoyinbo and Richards 1992; Brokensha, Warren and Werner 1980; Chambers, Pacey and Thrupp 1989).

This kind of TEK has been only partially explored and recorded, despite the threats that face the rich faunal and floral resources of many regions of the tropics. Among the pioneering projects studying indigenous perceptions and practices of biodiversity management and conservation in West and East Africa, Southeast Asia, the Pacific and the Amazon region, is the interdisciplinary 'People, Land Management and Environmental Change' (PLEC) programme, currently being implemented by the United Nations University (UNU). In Box 5.1, Harold Brookfield provides a synopsis of the organization and execution of this important work on small farmers and their management practices in different settings.

Several contributions to this chapter similarly provide illustrations of the progress that advanced ethnoscience research in the form of TEK can provide at different levels to the theory and practice of the conservation of bio-cultural diversity around the globe. In this regard, the unique significance of TEK is highlighted in its specific localized context in the historical overview of Roy Ellen and Holly Harris. Rosemary Hill and Dermot Smyth show in a case study among the Kuku-Yalanji people in Australia that collaborative research and planning between Aboriginals and planners can be attained only if the indigenous leaders are fully involved and the

Box 5.1: People, Land Management and Environmental Change (PLEC)

Harold Brookfield

PLEC is a project of the United Nations University (UNU), approved in 1997 by the Council of the Global Environment Facility (GEF) for implementation under the GEF work-plan, with the United Nations Environment Programme (UNEP) as 'implementing agency', and executed by UNU. The project, initiated in a small way in 1992, has had a long preparatory phase during which a great deal of targeted research has been undertaken on small farmers and their management of the environment. PLEC brings together the guided work of five 'clusters' of scientists for common aims in the testing and demonstration of good farmers' practices in conservationist and sustainable resource management. There is particular emphasis on conservation of biodiversity, including crop biodiversity. Diversity in resource management practices, crop biodiversity, and the dynamism of change in farmers' adaptations, together define the 'agrodiversity' which is central to PLEC work (Brookfield and Padoch 1994).

The 'clusters' are in West Africa (Ghana and Guinea), East Africa (Kenya, Uganda and Tanzania), China (Yunnan province), Papua New Guinea and Brazilian Amazonia. There are associated groups, outside the ambit of the GEF project but within UNU/PLEC, in Thailand, Mexico, Jamaica and Peru. Each contains experienced as well as more junior scientists, and students. All groups are multidisciplinary, with both natural and social scientists working together. The core of PLEC work lies in its 'demonstration sites', where scientists learn from farmers, and facilitate the demonstration to others of successful techniques developed by the best farmers. The demonstration sites belong to the rural people, and the work done in them is the farmers' own. The scientists' role is to measure and evaluate what they do, help select what is best and most likely to be sustainable, and get behind the farmers' own experimentation. Demonstration sites have already been set up in Ghana, China and Amazonia, and are being formed in other areas. In those already in operation, co-operating farmers have been identified, and plots have been set up for measurement and experiment.

PLEC places its primary emphasis on what farmers do. The local knowledge that underlies their practices is unequally distributed among farmers of both sexes and is not always shared. Moreover, it cannot be understood outside the social and cultural context in which knowledge is handed on, and in which new information and experience are interpreted. Scientists' interpretations of information they gather on local knowledge have to be drawn up in relation to the farmers' own perceptions, best first approached through close observation of the practices that are followed. PLEC views 'local knowledge' not as a fixed or traditional pool, but as constantly changing, being renewed by information and experience, and tested by ongoing experiment. This in turn is the basis of dynamism in peoples' farming systems on which PLEC relies for success in extracting best elements from peoples' own resource management practices.

Each cluster has its special strengths. The analysis of biodiversity in an agricultural context is strongly developed in Yunnan (China), West Africa (Ghana, Guinea) and Amazonia (Brazil). The Papua New Guinea group has gone further than others in the quantitative analysis and mapping of agricultural systems using diagnostic characteristics. The East African cluster has particular strength in the dynamics of soil and water conservation in diverse and sensitive environments. Promotion of South–South interchange across the network is one important strategy by which to share expertise. North–South networking in support of clusters' attainment of their objectives relies on the three Scientific Co-ordinators who keep in close touch with clusters, and on Scientific Advisors appointed on limited-term contracts to provide particular expertise in areas such as, for example, biodiversity measurement, soil-degradation assessment, GIS applications, or experimental method. The Co-ordinators visit clusters from time to time, and Scientific Advisors will be specifically charged with missions to particular clusters. Networking has also been achieved by correspondence, and by regional and general meetings held in the period 1993–1997 in West and East Africa, Southeast Asia and Amazonia. The project periodical *PLEC News and Views*, a newsletter with substantive articles, is of major importance, being made available to all PLEC members as well as being a vehicle for wider information to interested parties. Eight issues have been published on an approximately half-year schedule, and *PLEC News and Views* now has distribution of 400 copies.

The method used in a demonstration site varies in detail from country to country. Often, it begins with a joint inventory of resources and their uses, and goes on through participatory appraisals (using both

Box 5.1 (continued)

formal and informal methods) of participatory re-structuring of land use, which includes conservation of threatened natural or quasi-natural areas. Along the way are biodiversity inventories and soil surveys, and determinations of the problems faced by farmers which constrain their freedom of decision. Operationally, the joint team of farmers and scientists then experiments with tried crops and methods, and introduces innovations. Scientific observation, monitoring and measurement of outcomes are continuous; technical help is provided on demand. Later, the participatory approach is popularized among other communities, and demonstrated to officials and policy-makers. Finally, GIS and field methods are used to find areas in which similar solutions can be adopted.

A guiding principle in the choice of methodology is the use of short-cut methods that will achieve reliable results without serious loss of scientific rigour. A simple method of rapid diversity measurement for small areas within a landscape framework is one such technique (Zarin 1995). Another is the use of 'participatory environmental monitoring' in which particular indicators, selected by both observation and enquiry, are used to build up a reliable general view of trends that can be used as it stands or form the basis for more detailed inquiry into local knowledge about the environment and its processes. Not all PLEC work can be done by rapid methods. Monitoring in the demonstration sites has to be illuminated by the sort of understanding that can only be gained through time. Long-resident work in village communities is mainly carried out by students who are simultaneously being trained, or by assistants employed locally. It is an essential complement to the reconnaissance methods, and briefer visits by the scientists. Moreover, experimental work measuring the utility of farmers' methods requires a long period. It can be initiated within the life of PLEC, and indicative results can be obtained, but it is so structured in collaboration with the farmers and the agricultural authorities that it can be sustained beyond project life.

Farmers have to cope with many problems, in economic, political and environmental fields, and also respond to new opportunities. Experimentation is constant, leading to adoption or rejection of new crops and practices, and to restructuring of the use of 'farming space' in response to changing needs and opportunities. Where they are faced with declining rewards for their efforts, farmers often experiment with ways to manage its consequences and reverse the trend. PLEC is a practical programme in support of farmers' own goals, but it also aims to inform policy. Farmers' adaptations to environmental and socio-economic change can often be successful if the conditions of wider society permit freedom and provide encouragement, but the real conditions of farmers' decision-making are often poorly understood. Recent work in several regions of the world has shown that prevailing conclusions about environmental change and its causes rest on incorrect assumptions, even about the direction of environmental change itself. Policy interventions are very likely to miss the mark if trends and causation are not firmly established.

resource, cultural and intellectual property rights are secured. Janice Alcorn also emphasizes the significance of such collaboration for improved Indigenous Resource Management Systems (IRMS), based on a case study on collaboration among various castes in Kanataka (India).

The crucial role of cosmology and world-view in the use and management of resources is further demonstrated in different settings. So, Raymond Pierotti and Daniel Wildcat compare and analyse local beliefs about the close relationship with nature among the indigenous peoples in North America as opposed to the concepts of the environment of Western European immigrants. Terence Hay-Edie demonstrates that local landscape perception and sensory emplacement are important factors to be included in decision-making for environmental strategies. The role of local belief systems in the management of resources is an important issue, and Tirso Gonzales, Nestor Chambi and Marcela Machaca describe in this context the central role of the seed in the cosmovision in contemporary Andean agriculture, while Kusnaka Adimihardja provides an account of the Sundanese cosmology in traditional agriculture in West Java (Indonesia). Cosmovisions prevalent in the 'Great Tradition' and 'Little Tradition' of India are documented by Mahale and Soree, who conclude that traditional and modern knowledge systems can be integrated for sustainable management and conservation.

Michel Pimbert and Jules Pretty pay special attention to the great potential of community-based TEK for conservation, arguing that decentralized local systems are critical to successful conservation policies. Ashish Kothari and Priya Das show that Western science and resource use have eroded most local knowledge systems in India, as in the rest of the world. They describe a number of efforts that are being made to revive these systems for enhanced biodiversity conservation in India. Julian Barry describes in a case study of a joint forest management project in collaboration with the Anangu Aboriginal people how the involvement of their unique TEK can significantly contribute to the successful conservation and management of Australia's natural and cultural heritage. Prendergast, Davis and Way highlight the value of dryland plants for the local peoples that depend on them in different regions. Josep-Antoni Gari addresses the controversy between political economy and political ecology of biodiversity. Given the strong economic interests associated with biodiversity, he pleads for a new system of 'natural values' that also takes indigenous communities' value systems into account.

That not only rural but also urban environments include a great diversity of habitats that challenge biodiversity conservation is highlighted by a comparative contribution on urbanization and biodiversity by McNeely. In his analysis of the loss of biodiversity in the urbanization process, particularly the loss of habitat, he elaborates on various approaches to improve the management of urban areas in order to conserve biodiversity. Herbert Girardet shows that cities have become centres for consumption where decisions with a destructive impact are often being made. He pleads that, instead, cities should become centres for respect for the environment. A general assessment by Follér of the current threat to indigenous peoples and their knowledge systems leads on to Arun Agrawal's conclusion that empowerment is among the major pre-requisites of integration of both knowledge systems, pertaining to the concept of power-sharing with indigenous and traditional peoples.

In order to encapsulate the strong and direct relationship between biodiversity and cultural diversity, it seems useful to further employ and operationalize the concept of bio-cultural diversity. This requires recognizing the diversity of genes, species and ecosystems as well as the various sustainable and conservation-oriented principles and strategies of indigenous peoples with regard to their plants, crops, animals, soils and other natural resources and components of their environment. Such a concept links up with the coincidence of areas of high 'cultural diversity' with areas of high 'biological diversity, or – analogously to Vavilov's assessment of the world's diversity centres (1951) – with the coherence of 'Centres of Biodiversity' with 'Centres of Cultural Diversity'. Such 'Centres of Bio-Cultural Diversity' would not only serve as most valuable research areas for multidisciplinary teams, but could also be designated as areas where indigenous and global knowledge systems could integrate on the basis of shared power and self-determination.

In this way, the role of anthropology and the related 'new' perspectives in TEK could assist in multi-disciplinary research of not only the indigenous knowledge systems, but also of the adaptive processes and behavioural patterns which pertain to the sustainable management and conservation of biodiversity. In this context, Juma (1989: 93) notes, 'The social sciences can mediate between indigenous resource use patterns and other institutions by examining indigenous knowledge to understand the problem of biodiversity loss and the methods for biodiversity conservation'.

The contributions that follow show that in the current global strategy of sustainable management and conservation of biodiversity as operationalized by TEK, the concept of bio-cultural diversity has to be further explored and elaborated in order to enable different disciplines and multi-disciplinary teams to co-operate with local communities in the study and documentation of the human values, perceptions, spiritualities and belief systems involved in the conservation of culture and the environment around the globe.

Ethnoscience, 'TEK' and conservation

On power and indigenous knowledge (Arun Agrawal)

In discussions about how development can best be brought about so that it really is, finally, in the interests of the poor and the powerless, indigenous knowledge has come to occupy a privileged position. This welcome shift in the fortunes of indigenous knowledge has taken place as a result of the early work of scholars such as Conklin (1957), and more recent contributions from such pioneers as Stephen Brush, Robert Chambers, David Brokensha and Michael Warren among others. The efforts of these scholars have brought home the need to pay greater attention to indigenous knowledge, and cautioned against easy dismissals of the worth and utility of such knowledge.

Although the attention to indigenous knowledge is to be welcomed, existing conversations on the subject continue to be vulnerable to a danger: that of losing sight of the initial objectives that led scholars to focus on indigenous knowledge. Most scholars have now come to accept that there are no simple or universal criteria that can be deployed to separate indigenous knowledge from Western or scientific knowledge. Attempts to draw a line between scientific and indigenous knowledges on the basis of method, epistemology, context-dependence, or content are intellectually barren and have produced little that is persuasive. But even as advocates of indigenous knowledge recognize the futility of postulating a strict separation between indigenous and other forms of knowledge, current conversations on the subject possess a dynamic that continues to feed upon this separation. Thus, scholars remain highly preoccupied with using 'scientific' criteria to show the validity of indigenous knowledge, with collecting, systematizing, and storing indigenous knowledge, with proving the utility of indigenous knowledge to development initiatives. Many wish to use the international patents system to protect the interests of the poor and the marginal. The danger is that such activities on behalf of indigenous knowledge can become ends in themselves, and the interests of the marginal can become sidelined. Such dangers are especially acute if the question of power remains in the background.

Even if we keep our field of vision confined to the role of indigenous knowledge in development – i.e. the development of those who are supposed to possess indigenous knowledge – the question of power, how it is exercised, and the effects it produces must remain central. The very defences that scholars of indigeneity and indigenous knowledge offer in favour of their enterprise show that those who possess indigenous knowledge have not possessed much power to influence the course of history. Indigenous peoples have remained, for the most part, in positions of local resistance to the effects of domination produced by those who possess and apply scientific knowledge. And this, it can be argued, is true no matter which analogue of indigenous or scientific knowledge one invokes. That is, it is on the basis of the relation to power (and perhaps only on that basis) that one can define the difference between local/traditional/practical knowledges on the one hand, and global/Western/theoretical knowledge on the other. One comes to distinguish between different forms of knowledge on the basis of particular practices and institutional inscriptions, themselves products of differentiated relations of power and its exercise.

This contribution probes the relationship between power and practice, admittedly in a sketchy and preliminary fashion, to highlight some of the risks that accompany efforts to keep in the foreground knowledge rather than people or their social and political context. It is important, if indigenous knowledges and our investigations of it are to serve the interests of the poor and the marginalized, to bring to the fore the institutions and practices sustained by different forms of knowledge. A satisfactory reading of how power operates in conversations on indigenous knowledge, I suggest, needs not just to consider the points at which different forms of knowledge connect with nodes of power, but also the ways in which we come to see indigenous knowledge as necessary for development. In this context, Foucault's Nietzschean insight is critical, that 'knowledge is not gained prior to and independently of' the uses to which it will be put to gain power (Hoy 1986: 129). In the case of indigenous knowledge, we need to think how its postulated relationship with development leads its advocates to inscribe a series of practices that convert indigenous knowledge into an instrument of scientific progress, development, and the institutions that claim to control both development and the knowledge needed to develop.

Consider the most popular strategy for preservation and protection advocated by those who believe in indigenous knowledge: *ex situ* conservation. Systematic studies so that particular instances of indigenous knowledge can be documented and collected are seen as the prime means to safeguard such knowledge. As more such studies become available and as more instances of the relevance of indigenous knowledge are found and archived in national and international centres, development practitioners, it is believed, will naturally become persuaded of the importance of indigenous knowledge. The greater appreciation of the benefits of indigenous knowledge will lead, in turn, to greater efforts to further the interests of those who possess such knowledge. The pursuit of knowledge, in the above rendering by the advocates of indigenous knowledge, becomes politics by other means.

Before considering the gaps in this logic, it may be useful to point out the ways in which the instrumental logic of development transforms the character of efforts on behalf of indigenous knowledge, and indeed, indigenous knowledge itself. The first demand of this logic is that useful indigenous knowledge be separated from among the various other kinds of indigenous knowledge in combination with which it exists. Only those forms of indigenous knowledge that are seen as potentially relevant to development, then, need attention and protection. Other forms of such knowledge, precisely because they are irrelevant to the needs of development, can be allowed to pass away. One can call the identification and separation of useful knowledge the process of particularization.

But particularization can take place only accompanied by other processes. The second demand of the instrumental logic of development is that particularized knowledges be tested and validated using the criteria deemed appropriate by science.

Indigenous knowledge must first be recast in the image of science before being utilized for development. Independently, such knowledge has no validity, only possibilities. The use of scientific criteria to test and examine, and the documentation of these tests, can be referred to as validation.

Validation has a corollary: abstraction. Not all elements of useful indigenous practices are necessary for development. Only the strictly useful elements need be abstracted for maximum effect. Particular rituals, or words, or movements that are the concomitant of the administration of a herbal medicine or drug can be divested and discarded as not being part of the crux of the useful practice of administering a herbal medicine. They can form no part of interest from the point of view of development. Only those elements of indigenous practices need be retained that can more easily be transplanted into other contexts. The stripping away of non-essentials also facilitates the next stage of the process through which indigenous knowledge is made ready for development.

Once knowledge is particularized and validated (abstracted), it needs to be catalogued, and archived, and then circulated before it can be used more widely. This can be termed the process of generalization. Only insofar as a particular element of indigenous knowledge is capable of being generalized is it really useful for development. If suitable only for an individual and particular context, indigenous knowledge need not be studied at all – not at least by those interested in development.

I use the neologism *scientization* to refer to the three stages of particularization, validation and generalization All efforts to make indigenous knowledge useful to development must run the gamut of practices from particularization to generalization. The application of these criteria and practices in relation to indigenous knowledge follows a particular logical relationship between power, utility and truth. It helps instantiate a division within indigenous knowledges so that only useful indigenous knowledge becomes worth protection. But the valid doubt that should assail one at this point is whether there is anything particularly indigenous about knowledge that has undergone the sanitization implicit in the movement from particularization to generalization. In the very moment that indigenous knowledge is proved useful to development through the application of science, it is, ironically, stripped of those very characteristics that could even potentially mark it as indigenous.

The identification of valid scientific elements in the host of practices that are termed indigenous knowledge is no more nor less than any other scientific pursuit. It is scientific not because there is anything self-obviously true about it, but because it conforms to the procedures whereby science is reproduced and new knowledge classified as scientific. This becomes the preoccupation of those who seek to change the fortunes of the powerless when they focus on knowledge rather than interests and politics.

A second issue becomes pertinent at this point. Even if a scientific logic can be identified within the indigenous, even if particular indigenous practices can test true on the criteria of science, there is no reason such evidence should help those from whom indigenous knowledge is abstracted, then archived. The instrumental logic of converting the indigenous into the scientific can certainly further the perception that indigenous knowledge is worth saving. But the prevalence of such perceptions might do little to modify prevailing relations of power among different social groups, especially since it is these same relations of power that lead in the first place to social changes that disadvantage indigenous groups. There are evident gaps in the argument that once the value of indigenous knowledge becomes obvious, efforts to channel greater resources and power to indigenous populations will begin to take place. The reverse might also hold equally true.

By their own efforts, conforming to an instrumental logic of development, advocates of indigenous knowledge make clear that there is no necessary unity between indigenous peoples and their knowledges. By scientizing (particularizing, validating, abstracting and generalizing), they realize (in the sense, 'make real') the possibility of separating 'useful' from 'useless' indigenous knowledge. If the usefulness of knowledge possessed by indigenous peoples is the justification for pursuing their knowledge, the strategies that demarcate useful and useless knowledge bear the unfortunate implication of condemning those knowledges that are not useful. Once the knowledges of indigenous peoples are separated from them and saved, there is little reason to pay much attention to indigenous peoples themselves.

Efforts to document and scientize indigenous knowledges can thus be doubly unfortunate. One, they channel resources away from the more vital political task of transforming the relations of power. Two, they provide a means to more powerful social actors to appropriate useful indigenous knowledges. In the absence of efforts to change the relations of power that define interactions between different social groups, weaker groups seen to possess valuable knowledge can simply be coerced or persuaded to part with it. Inferior in asymmetric relations of power, they would be ill-equipped to resist such appropriation. The history of colonialism, replete with examples of such unequal exchanges, should warn against any easy consolation that the strong, when coming in contact with weaker groups who have valuable possessions, will attempt to bolster the interests of the weak.

The irony involved in the *scientization* of indigenous knowledge is driven by a particular relationship between development, science and power. Development is itself founded on the belief (some might say, 'conceit') that scientific knowledge can help transform social processes. Because the current attention to indigenous knowledge finds justification in the claim that it is useful for development, advocates of indigenous knowledge are compelled to submit such knowledge to the touchstone of scientific practice. The same practices of power and truth that create science, then, also constitute the indigenous into the scientific.

To point to the above is not to defend the possibility of a pure state in which the indigenous can, even less, should exist. It is, rather, to indicate the impossibility of escaping a particular instrumental logic of science and development once we begin to make claims about the significance of indigenous knowledge for development.

Nor do I wish to suggest that weaker groups are never able to protect their interests, or that the structures of power relations always determine outcomes. I point, instead, to the Utopian nature of particular kinds of attempts to strengthen the position of indigenous peoples *vis-a-vis* others. I insist, simultaneously, on the need to keep in the foreground the ways in which power works. Without explicit and continuing attention to how power structures knowledge, it will remain impossible to achieve the aim of working in the interests of indigenous or other marginal peoples.

Attention to power relations would necessitate that advocates of indigenous knowledge use the strategy of documenting and archiving knowledge as only one of the elements in their arsenal of weapons. They must, simultaneously, follow other courses of action – among them lobbying governments, questioning science, channelling resources towards more independent processes of decision-making among indigenous peoples, and mobilizing and organizing indigenous peoples.

Acknowledgements
I would like to thank Ajay Skaria and Shiney Varghese for their thoughtful comments on an earlier draft of this short paper.

Embeddedness of indigenous environmental knowledge (Roy Ellen and Holly Harris)

Indigenous knowledge at home in the West

The West often assumes that it has no Indigenous Knowledge (IK) – in terms of local environmental knowledge of plants, animals, soils and other natural components – that is relevant, in the sense of 'folk' knowledge; that it once existed but has now disappeared, and that somehow science and technology have become its indigenous knowledge. Certainly there is plenty of evidence that the existence of, for example, codified pharmacopoeias such as the *De Materia Medica* of Dioscorides displaced local knowledge and oral tradition extensively in Europe and the Mediterranean, but uncodified knowledge persisted and gradually filtered into organized texts as the number of modern remedies of European folk origin manifestly attest (Cotton 1996: 10-11).

But Western folk knowledge – non-professional, experimental, uncodified, *ad hoc*, often orally-transmitted – is arguably just as important as it ever has been; just different, informed by science where appropriate, and located in different contexts – domestic horticulture, dog-breeding, bee-keeping etc. (see Clark, this chapter). The folk are no less creative. Moreover, in parts of Europe, urban folk actively seek out the authoritative knowledge still regarded as being present in their own rural traditions, as in truffle-hunting, geese-rearing or the preservation of rare breeds of sheep. This is splendidly illustrated in the work of Raymond Pujol (e.g. 1975) in France. Peasant or rural knowledge becomes in this context Europe's own inner indigenous other. Interestingly, and paralleling a development we will examine later for Asian indigenous knowledge, such European folk traditions have in the last 40 years or so been reified, reinvented, celebrated and commoditized, as demonstrated in the contemporary cultural significance of living folk museums, craft fairs, etc. One of the ironies of this is that such 'folk' traditions have become highly codified. As we shall see, the double irony is, however, that the process of codifying folk knowledge into organized scholarly knowledge has ever been thus.

During mediaeval and early modern Europe, proto-scientific knowledge of plants and animals superseded folk knowledge by classification, analysis, comparison, dissemination – usually through books and formal learning – and thus generalization. The process was not sudden: for a long time common experience, oral tradition, personal experience and learned authority contributed to the 'aphoristic' knowledge or 'received wisdom' upon which organized, specialized knowledge, particularly medical knowledge, depended (Wear 1995: 158-159). And it is not at all easy to know where unorganized folk knowledge, professionally restricted organized knowledge, and proper scientific knowledge began and ended. In such proto-scientific technological practices, it is significant that elements of discrete knowledge do not usually disclose how they were arrived at. In other words, their 'epistemic origins' are hidden. Sometimes they are of European folk origin, but from the sixteenth century onwards, they incorporated medicines of Asian and American origin.

It was this anonymity that helped to define an emergent scientific practice in opposition to folk knowledge. Even after scientific discourse and practice had become distinct, methodologically self-conscious and discriminating, it continued to draw on practical folk experience. Darwin, for example, depended extensively on the knowledge of pigeon fanciers in working out the details of natural selection (Secord 1981, 1985; Desmond and Moore 1991: 425-30). Indeed, we can see that modern natural history arose through a combination of such indigenous scholarship and field studies (Zimmermann 1995: 312), and field studies themselves often in turn drew heavily on the knowledge of local experts. Some have argued that the phylogenetic taxonomies of contemporary post-Linnaean biology are based on a European folk template (Knight 1981; Ellen 1979; Atran 1990). Arguing a rather different tack, others have gone further by claiming that the European folk scheme and that of modern biology are no more than variants on a single cognitive arrangement to which all humans are predisposed through natural selection (Atran 1990; Boster 1996).

What we now recognize as scientific knowledge of the natural world was constituted during the eighteenth and nineteenth centuries in a way that absorbed such pre-existing local folk knowledge as was absorbable and, ultimately, confined what was not to oblivion; being at best of some antiquarian interest, at worst denied any existence as a meaningful and credible set of practices, precisely because of the inability of the new paradigm to absorb it. Part of this residue re-emerged as recognized folk knowledge in the late twentieth century and has been subjected to the kind of cultural revival we have already referred to.

The rest, unlabelled and unloved, continues as that vast body of tacit knowledge that is necessary to operationalize book and theoretical knowledge, and which continues to inform the practical engagement of ordinary skilled people: the informal un-codified knowledge of house workers, of Durrenberger and Palsson's (1986) Icelandic fishing skippers, or of any number of skilled professionals who take their cue from real-life situations unmediated by books. Unfortunately, the current economic pressures of publishing and the demand for useful information are leading us to the further codification of the hitherto uncodified, of the '1001 handy household hints' and 'tips from the greenhouse' variety, thus giving the appearance of removing even more from the realm of IK.

The impact of Asian folk knowledge on the development of Western scientific traditions

As we have seen, much Western science and technology emanates from indigenous European folk knowledge (e.g. herbal cures), but from the earliest times ideas and practices were also flowing into Europe from Asia, and vice versa. By the later Middle Ages, however, and the beginnings of modern European global expansion, a self-consciousness emerged about the desirability of obtaining new knowledge. We can see this process at work by examining some recent scholarship relating to European scientific interests in India and Indonesia.

As early as the sixteenth century, travellers were being advised to observe indigenous practices and to collect material with a view to extending the European *materia medica*. For example, Garcia da Orta, a Portuguese physician living in Goa, provides us with a description of plants of the East which formed the basis of medicines available in Europe and the Portuguese colonies, and from which they could be extracted. Da Orta relied on personal medical experience, fieldwork and indigenous knowledge, and initially depended on Arabic sources, thus reflecting the centre of gravity of the international trade in materia medica. Da Orta's *Colóquios dos Simples e Drogas de Coisas Medicinais da India* (published in Goa in 1563) was translated into Latin in 1567 by Charles d'Ecluse (Clusius) who went on to establish the *Hortus Medicus* in Vienna and, in 1593, the *Hortus Botanicus* in Leiden (*cf.* Markham 1913). In turn, Jacob Bondt relied heavily on the *Coloquios* for his pioneering book on tropical medicine, *De Medicina Indorum*, published in Leiden in 1642 (*cf.* Noteman 1769; Grove 1996: 125, 129, 131, 133).

The decline of Portuguese power in Goa and the establishment of the Dutch in Cochin was marked in botanical terms by the project of the *Hortus Indicus Malabaricus*, initiated by Hendrik van Rheede tot Drakenstein (1636–1691) in response to the medicinal needs of the Dutch East India Company (*cf. Continans Regioni Malabarici* 1678-1694; Grove 1996: 126, 134). We can thus observe a remarkable chain linking Indian medical ethnobotany, compilations of Middle Eastern and South Asian knowledge organized on essentially non-European precepts, Portuguese and Dutch political interests, and the formative period of modern scientific botany and pharmacology.

The perfection of European printing, the establishment of botanical gardens, global networks of information and materia medica transfer, together with the increasing professionalization of natural history, have facilitated the diffusion and dominance of an Ayurvedic and tribal medico-botanical knowledge, and imposed an indigenous technical logic on subsequent 'European' texts of South Asian botany. These retain the essentially indigenous structure of the *Coloquios* and the *Hortus*, thereby transforming European botanical science through contact with South Asian methodologies of classification, rather than the other way round. But given the long history of mutual knowledge transfer going back to ancient times, any division between European and Asian botanical systems might be construed as arbitrary (Grove 1996: 127-8). We can see a similar although less complex process in the work and influence of George Rumphius. He was a German naturalist employed by the Dutch East India Company, who between 1653 and 1702 lived in Amboina where he systematically recorded the natural history of not only the islands immediately around Amboina, but also the islands of Southeast Asia at large. Here, he relied heavily on local assistants and their knowledge. His most important work on plants was published posthumously as the *Herbarium Amboinense*. Remarkable about this work is not only its importance in listing many species hitherto unknown in European botanical descriptions, but also the heavy reliance both on indigenous descriptions of plant ecology, growth patterns and habits, and the extent to which Rumphius relied upon Malay and other local folk classifications and terms to provide a meaningful and comprehensive account (*cf.* Peeters 1979; Beekman 1981).

Compared with Van Rheede's work on Western India, however, rather than finding interference from classical Javanese and other politically dominant schemes, we find instead a reliance on Malay – which was essentially a new language at that time in the Moluccas – as a linguistic filter for indigenous ideas and knowledge. In turn, Linnaeus, in particular in 1740, fully adopted the Cochinese classification and affinities in establishing 240 entirely new species, and to a lesser extent relied on the Ambonese and Malay classifications and descriptions provided by Rumphius.

The influence of the *Herbarium Amboinense* and the *Hortus Malabaricus* immediately established Holland as the centre of tropical botany. French, English and Dutch naturalists employed by respective East India companies, following Dutch methods, were instructed to collect as much indigenous knowledge as possible. Their influence on the canonical Linnaean texts meant that subsequent authorities came to depend on essentially Asian organizing frameworks: Roxburgh, Buchanan-Hamilton and Hooker in India, and Burman, Blume, Henschel and Radermacher in Indonesia. Even more so, as Grove (1996: 3, 139, 140-141) notes, 'the seeds of modern conservationism developed as an integral part of the European encounter with the tropics and with local classifications and interpretations of the natural world and its symbolism'.

During the nineteenth and twentieth centuries, local knowledge was increasingly systematically tapped and codified. Such 'routinization' resulted in the publication of scientific accounts of new species and revisions of classifications which, ironically, depended upon a set of diagnostic and classificatory practices which though represented as Western science, had been derived from earlier codifications of indigenous knowledge. Numerous encyclopaedic inventories began to appear, such as George Watt's *Dictionary of the Economic Products of India* and Burkill's (1935) similar *Encyclopaedia on Malaya*, inspired by the work of Watt, which had all the hallmarks of the scholarly arm of imperialism. Thus, the European relationship with local Asian knowledge was, paradoxically, to acknowledge it through scholarly and technical appropriation, and yet somehow to deny it by reordering it in cultural schemes which link it to an explanatory system that is proclaimed as Western. While on a personal level, scientists may have acknowledged the contributions of their local informants, at the professional level the cultural influences which those same informants represented were mute.

Indigenous knowledge marginalized

If, in the context of late European colonial scientific fieldwork in Asia, traditional knowledge was evident but mute, with the inexorable rise of modernity it became a kind of ignorance (Hunn 1993). Tradition was something to be overcome rather than encouraged, and several generations of 'top-down' development experts and organizations engaged in resource extraction and management in the underdeveloped world have either deliberately avoided it on the grounds that their own models were superior, or simply never realized that it might be a resource to be tapped. The dominant model of development has for some fifty years or more been based on useful knowledge generated in laboratories, research stations and universities, and only then transferred to ignorant farmers (Chambers and Richards 1995: xiii). Such attitudes are now recorded by Paul Richards (e.g. 1985) and others.

But not only has IK been grossly undervalued by Western-trained 'scientific' managers in terms of its potential practical applications: when it was, at last, absorbed into 'scientific' solutions it was, curiously, insufficiently 'real' to merit any certain legal status or protection from the battery of patents and copyrights which give value and ownership to Western scholarly knowledge and expertise. Even when the knowledge was clearly being utilized, it was often redescribed in ways that eliminated any credit to those who had brought it to the attention of science in the first place (see Dutfield, Chapter 12). The point is made very effectively by Harris (1996: 11) in her discussion of *Azadirachta indica*, or the 'neem tree'. 'Whether or not the chemical properties assumed to provide the active substance for cure have been identified by the communities does not appear relevant. Rather, it appears that the method used by these Western firms, and their ability to synthetically reproduce the compounds, was perceived as the true science, and consequently, deserving of patent.'

The inherent ethnocentrism and elitism of late twentieth century Western science, therefore, has made it difficult for scientists themselves to accept that the folk have any knowledge of worth (Johannes 1987: ix). This view is reinforced by perceptions that traditional peoples often adopted wasteful, even delinquent, patterns of resource extraction, as classically exemplified in the literature on shifting cultivation (*cf.* Dove 1983); and that when subsistence practices were evidently damaging, it was a matter of preference rather than an outcome of poverty.

The rediscovery and reinvention of Indigenous Knowledge.

Since about the mid-1960s, the process of marginalizing IK as outlined above has been put into reverse, and is indeed accelerating to a remarkable – some would say, alarming – degree. This is due to both romantic and practical reasons.

The romantic reasons have their immediate political renaissance in the counter-culture of the 1960s (Ellen 1986), with the notion that traditional, indigenous or 'primitive' peoples are in some kind of idyllic harmony with nature. Such a view was initially prompted by a crisis in the perception of science and technology, in terms of the increasing remoteness and arcane character of science, its perceived arrogance and negative technological outcomes, and its inability to explain much about the world that ordinary people sought explanations for. This, as has been suggested on various occasions (Budiansky 1995: 3, n7 p. 251, quoting Kaufman, quoting Chesson) amounted to 'good poetry' but 'bad science'. Others have gone even further and questioned the poetry. It often involved the selective remodelling of Asian and other exotic traditions to suit the needs of Western environmentalist rhetoric drawn from an intellectual pedigree that favoured idealized native images. Conklin and Graham (1995: 697) put it this way: 'Contemporary visions of transcultural eco-solidarity differ in that native peoples are treated not as peripheral members whose inclusion requires shedding their own traditions but as paradigmatic exemplars of the community's core values'.

In this new vision, indigenous peoples are given central focus because of, rather than in spite of, their cultural differences. But, as Conklin and Graham (1995: 696) point out, this perception and consequent alliance between indigenous peoples and science is a fragile one, based upon an assumed ideal of (indigenous) realities which contrasts with the realities for the local people themselves. Such assumptions are in danger of leading to 'cross-cultural misperceptions and strategic misrepresentations'. The reconstitution in an Asian context has often involved both the Great Tradition and the Little Tradition – the Scholarly and the Tribal – often failing to distinguish between the two and confusing ideal symbolic representations with hard-headed empirical practice. It is not altogether surprising, therefore, that this muddle has confirmed some scientists in their worst prejudices, and led to the inevitable backlash summed up in phrases such as 'the environmentalist myth' (*cf.* Diamond 1987 *versus* Johannes 1987). The difficulties of separating rational, empirical knowledge from religious, moral or symbolic knowledge are well illustrated in an analysis by Lemmonier (1993) of Ankave-Anga eel-trapping in Papua New Guinea.

Nevertheless, with the discarding of the more fanciful portrayals of the wisdom of traditional peoples, a more practical approach has emerged. This has been encouraged by anthropologists and development professionals eager to make IK palatable to technocratic consumers, and by technocrats themselves who were already predisposed to see a role for IK. Its dissemination has been part of a rhetoric praising the virtues of 'participation', 'empowerment', 'bottom-up', and 'farmer-first'. Some measure of the institutionalization of this version of IK – and in this version abbreviations of the IK and TEK variety are rhetorically essential – is the number of networking organizations and research units (*cf.* Warren, Slikkerveer and Brokensha 1995: xv-xviii,

426-516). One of the difficulties with this approach, however, has been that categorization of IK effectively becomes a direct consequence of the limited parameters of Western science and development theories which rely upon an ordered conceptual framework from which and in which to work. As Hobart (1993: 16) observes, 'There is an unbridgeable, but largely unappreciated gap between the neat rationality of development agencies' representations which imagine the world as ordered and manageable, and the actualities of situated social practices'.

As a consequence, we seem to end up with a theory that misrepresents the context in which certain knowledge occurs and is experienced. Hobart (1993) has pointed to the limitations of 'scientific' development knowledge in that it ignores or undervalues contexts. By uncritically placing local knowledge systems under the umbrella concept 'indigenous knowledge', decontextualization is necessarily implied and the unique and important knowledge of specific groups becomes subject to the same limitations and criticisms that we make on behalf of Western science and development theories. Moreover, the tendency to define indigenous knowledge in relation to Western knowledge is problematic in that it raises Western science to a level of reference, ignoring the fact that all systems are culture-bound and thereby excluding Western knowledge itself from the analysis. This limits the analysis of indigenous systems by narrowing the parameters of understanding through the imposition of Western categories. Fairhead and Leach (1994: 75) draw attention to this problem, particularly regarding the tendency to isolate bits of knowledge which are fitted into a *'mirror set of ethno-disciplines'* for the purposes of analysis and documentation.

By examining local knowledge in relation to scientific disciplinary distinctions, Fairhead and Leach point to how this can lead to the construction of certain aspects of local knowledge as important, while excluding or ignoring other areas and possibilities of knowledge which do not fall within the selective criteria of Western scientific parameters. They argue, moreover, that this risks overlooking broadly held understandings of agroecological knowledge and social relations. So, for example, research and extension agents who examine Kouranko farmers' tree management practices in Guinea, fail to take into account the farmers' tree-related knowledge which involves knowledge and management of crops, water, and vegetation succession as well as the ecological and socio-economic conditions that influence them. By failing to include the broader constitutive processes surrounding Kouranko tree management, extension workers risk obscuring and decontextualizing local knowledge, and jeopardize the potential it may have for development on specific and general levels (Fairhead and Leach 1994: 75).

Thus, in this depleted vision, IK becomes a major concept within the development discourse, and a convenient abstraction which consists of bite-sized chunks of information that can be slotted into Western paradigms, fragmented and decontextualized; a kind of 'quick fix 'if not a panacea. Such approaches are in danger of repeating the same problems of simplification and over-generalization that Richards (1985) and Hobart (1993) identify as the major limitations in development theory, and in science that is applied to development 'ignoring specific and local experience in favour of a generalizable and universal solution' (Harris 1996: 14).

Furthermore, in the hands of NGOs – which in the last few decades have become significant 'knowledge-making institutions' – and within the 'universalizing discourse' of environmentalism – IK has become further reified. Because environmentalists and indigenous NGOs are now influential moral and social forces that stimulate public awareness, act as whistle-blowers and watch-dogs, and are moving from the role of critics to that of providers of policy proposals to governments, often using the rhetoric of science, they are gaining enormous authority (*cf.* Yearly 1996: 134). This process has evidently yielded results in terms of projecting a more positive image of IK.

In some important respects, the way in which development professionals have contextualized and formalized IK by codifying it, while rejecting the cultural context, has simply repeated what has happened in previous scientific encounters with traditional knowledge, as discussed above. And, similarly, once IK is drawn within the boundaries of science, it is difficult to know where to draw the boundaries between IK and 'science' (see Agrawal, this chapter). As we have already seen, changing the boundaries is often sufficient to redefine something as 'science', as what defines it is to a considerable extent determined by who practices it, and in what institutional context the practices are taking place. However, the danger of turning local knowledge into global knowledge is, as Hunn (1993: 13-15) notes, that 'at the empirical level, all IK is relative and parochial, no two societies perceive or act upon the environment in the same ways. Science, by comparison, is a system of knowledge in rapid flux that seeks universal rather than local understanding'. It is precisely the local embeddedness of IK that has made it successful.

Local community knowledge and practices in India (Ashish Kothari and Priya Das)

The diversity of life-forms, estimated at anywhere between 5 and 50 million species of plants, animals and micro-organisms, is the cumulative product of a still ongoing evolutionary process. Humanity, perhaps the most complex and evolved product of this process, has right from its emergence been dependent on biological diversity. Our continual interaction with and manipulation of nature has had variable impacts on this diversity, ranging from enhancement, regeneration and conservation to depletion, destruction and extinction of its elements. The impact of human activities on biodiversity in particular, and on nature in general, has been conditioned by (a) the way human society perceives nature, and (b) the way we have organized our societies. Thus, the enormous range of organizational forms, socio-political systems and socio-economic relations plays a decisive role in the way humans relate to and use resources, and the subsequent impact on biodiversity.

The social evolution of humanity has been paralleled by the emergence of several modes of resource use, and each mode adopted by any society has been characterized by a definitive set of ecological, ideological and technological features. Broadly, the various modes of resource use, and the societies named after the most predominantly employed modes, are hunting, gathering, pastoral nomadism, shifting cultivation, and the present mode of industrialization. While industrialization has been the hallmark of modern and 'developed' societies, traditional societies have employed the other modes with varying degrees of importance. In sharp contrast to the traditional modes, the industrial mode of resource use, within the short span of its existence, has had a devastating impact on nature and on biological diversity. Taking cognisance of this in conjunction with the fact that the world's high biodiversity areas overlap the habitats of indigenous and local communities, it has become increasing clear that the conservation of biodiversity is closely tied to the protection and continued use of traditional and local community knowledge – including ongoing innovations from outside – that are related to natural resources.

Being closely associated with and dependent on nature, traditional societies have largely (though by no means always) shared a symbiotic and sustainable relationship with nature. They relied heavily on the diversity of biological resources in order to meet their requirements: dependence on diversity helped lessen the pressures on any single component. Thus, through trial and error, aided by natural selection, these communities have continuously enlarged their knowledge systems on natural resources and have found several new uses for it. Traditional ecological knowledge related to natural resources and their use has thus been defined as representing a collective understanding attained over a long period of time, in particular places, of the relationship of a community and the earth, encompassing spiritual, cultural and social aspects as well as substantive and procedural ecological knowledge (Doubleday 1993). Under the pervasive influence of modernization, there has been significant usurpation of other knowledge bases, without due acknowledgement, or the latter have been marginalized. Only very recently has the modern world recognized the inherent worth of, and the need to respect, local community knowledge.

Characteristics of Local Community Systems

Local Community Systems (LCSs) which refer to resource-use systems developed over generations in specific communities, have both their advantages and disadvantages, and have both a positive and a negative impact on biodiversity.

Advantages/positive points

1. Located within the socio-cultural milieu: LCSs are deeply integrated within the social, cultural and political milieu of the community, deriving their legitimacy and strength from this milieu. Thus, rules regarding restraints on resource use are embedded in cultural and religious systems which give them a legitimacy that goes beyond scientific/ecological prudence. In most cases, therefore, resource management systems are an integral part of their tradition and culture.

2. 'Landscape' integration: LCSs do not typically have hard and fast divisions between the various kinds of land/water use, but rather form one continuum. Thus, from the point of view of usage and function, forests merge into agricultural fields, which merge into wetlands, and so on. This gives rise to highly-integrated resource-use systems in which land-use, for instance, does not militate against a wetland, or vice versa. Local communities can practise and manage a range of resources simultaneously.

3. Conscious/unconscious restraints on use: LCSs have several rules (usually unwritten, but codified in the oral tradition) regarding the use of resources. These include seasonal, functional, geographical or other restraints on the use of biological resources; rotational or restricted use of habitats, etc. While there is a limited technological

capacity to exploit resources, as also limited demand on resources from typically small populations, conscious restraints on exploitation marks most LCS. Gadgil and Guha (1992) have classified, under various categories of restrained resource-use practices, the judicious measures of resource use employed by traditional communities. These include community-imposed restrictions on the amount harvested, subject to the density of the resource available; incidental conservation by according religious protection to species or patches of landscape; prohibiting hunting methods that are exhaustive or have a debilitating effect; protecting certain life stages critical to 'population replenishment'; and disallowing certain groups on the basis of age, sex and social standing; certain methods of harvest; types of harvest, and harvest from particular areas.

4. Resource-use and conservation integrated: No distinction is usually made between resources/habitats for conservation, and those for use: there is usually no concept of 'wilderness'. The only exception would be sacred landscapes/habitats/species which are off limits for use. In all other cases, including agricultural fields, forests, wetlands and pastures, both conservation and usage are integrated.

5. Use of high level of biodiversity: A high diversity of biological resources and resource use systems marks every part of the life-cycle of local communities. Every species is used in many ways, several different species are used, and within species, genetic diversity is maximized.

6. Relative self-sufficiency: Most (though often by no means all) essential needs of the community come from local resources. This includes food, shelter, clothing, household and agricultural implements, products for ritual use, etc.

7. Dynamic/innovative (gradual): There is considerable dynamism and innovation in LCS, especially in the forms of resource-use. This is best seen, for instance, in agriculture, where farmers' ingenuity in the use of habitats and species is remarkable. Typically, though, change is gradual in LCS, making them appear to be static.

8. Egalitarian: Many local communities, especially tribal communities, are marked by a high degree of egalitarianism in resource access and use, with everyone being assured of at least their basic needs. However, this does not necessarily hold in all communities, especially non-tribal ones (see below).

Weaknesses/negative points

1. Inflexibility to sudden or large-scale changes: Though there is dynamism within LCS, there is generally an inability to cope with sudden or large-scale changes, for instance, those introduced by the sudden entry of the outside market, or of government take-over of common lands. Local institutional structures, rules of resource-use restraint, and so on, tend to break down in the face of such changes.

2. Fragility due to complex web of linkages: Since all parts of the LCS are intricately linked, much like a rainforest, a change in one part can have a chain effect on others. The introduction of market mechanisms, for instance, or those of government-controlled institutions, will effect local traditional institutions, which in turn will effect the way resources are managed/used.

3. Ignorance of certain elements of biodiversity: While a large range of biodiversity is used, there are also gaps in local knowledge, relating especially to species that are not in use or not in some way impinging on the lives of the villagers, such as small fauna or micro-organisms. Apart from agricultural pests that needed to be countered, and insects of significant value (honey-bee etc.), there is little or no knowledge of the plethora of insects and micro-organisms that are an equally critical part of biodiversity.

4. Tendency, at times, to over-use: Resource-use restraints are not always honoured, or do not exist at all, for certain elements of biodiversity. This could be true, for instance, of items used for ritual purposes. In north-east India, for instance, there is an overhunt of several species of hornbills as many tribal communities value the 'horned' beak of these species for their alleged medicinal properties as well as for use as part of their traditional headgear.

5. Socio-economic deprivation: In many local communities, especially non-tribal caste-based ones, considerable inequalities exist in access to resources, and in the ability to make decisions regarding their management and use. Lower castes or economic classes, or women, may face deprivation in resource access and use.

Despite the above negative characteristics, overall, LCS display a high degree of sustainability and equity with regard to the use and management of biological resources.

Context of LCS resource-use

LCS resource uses occur within three larger contextual levels:

Belief systems

Two kinds of belief system form the context of resource uses: localized folk or tribal religions (e.g. animism), and widespread 'classical' religions (in sociological literature, the 'Little' and 'Great' Traditions). Complex ritual and cultural practices are codified within these belief systems.

Religions, dominant or tribal, have traditionally envisaged man as a part or a subordinate of nature and in equal standing with all of nature's other living beings. Most world religions through various writings, exhortations and preachings have provided a system of moral guidelines towards environmental preservation and conservation. The Hindu religious scriptures professed the ethics of inter-connectedness, emphasizing the total community of life in nature and the oneness of nature with the human race in order to preserve the biosphere and to enhance the evolution of all species and societies (Dwivedi 1994). Hindu theology very prominently alludes to conservation of species by espousing a belief in the incarnation of God in various elements in nature. The ethic of non-violence propagated by Buddhism and Jainism entailed compassion towards all living creatures and a ban on killing animals, as well as protecting trees. Islamic tenets strongly conform to the belief that 'Allah is Unity', the implications of which are held to reflect the man-nature unity in its several parameters. Serving human beings is only the part function of natural elements and thus not the only function. Sikhism too, in the Guru Sahib, proclaims the glory of God in nature and environment.

Spirits, the most predominant aspects of tribal religion, have been defined as, 'systems of religious beliefs and practices which regulate the relations of social organization with those of the habitat and environment' (Vidyarthi 1963). These spirits, which reside in the 'sacred geography', are an integral part of the 'sacred complex'. The inexplicable acts of nature were inordinately attributed to the supernatural. Thus, in their cosmovision, nature is seen as an integral part of the sacred realm. These precepts and injunctions of most religions thus provided a strong foundation for nature worship. Animals, plants and natural landscapes were accorded divinity either as abodes of or as incarnates of spirits. Once they were attributed sacredness they were neither harmed nor killed out of reverence or out of fear of incurring the wrath of the deities or spirits. These religious beliefs provided a framework within which protection accorded to either patches of landscape or single species was legitimized.

Sacred groves, sacred ponds, patches of grassland, animals and others are examples of traditions of conservation backed by religious sanctions. Sacred groves are essentially tracts of relatively untouched forest, preserved since time immemorial around sacred or temple structures, or set aside as abodes of local deities or spirits. In terms of conservation, sacred spaces are analogues to the protected areas of today, except that they represent a community-based system of conservation, where in the absence of formal laws, social fencing was effected by religious codes and sanctions. Violations were held in check by religious threats and were dealt with through community-evolved modes of punishment. Found in various parts of India, these are variously known as *sarna* (Bihar), *oraon* (Rajasthan), *deorais* (Maharastra), *kavus* (Kerala), and other names. These groves are important today because they are the best of the forests that might have flourished in the region, housing rare and endangered plant species, many of which may have disappeared from the region outside the groves. In the Uttara Kannada region of southern India, the only remaining natural stands of *Dipterocarpus* and a large patch of *Myristica indica* persist in a sacred grove of the Goddess Karikannama (Gadgil 1987a). They also serve as a community's medicine chest (Gadgil and Vartak 1976), and perform crucial ecological functions such as the *oraons* as 'green oases' in the Rajasthan desert, and the *kuthuvals* of Tamil Nadu as wind-breaks (Mitra and Pal 1994).

Apart from sacred groves, certain fallow lands are also set aside for protection. These are exemplified by the *Aands* found in some parts of Rajasthan. Various water bodies (village tanks, ponds, rivers and others) are attributed sacred qualities and are protected against over-fishing or over-extraction of any other resource available. Some of them exist within the bounds of sacred groves. Preserved in their pristine condition, they allow underwater forms of life, even at the micro-level, to flourish undisturbed. The only surviving population of *Trionyx nigricans*, the freshwater turtle, is found in Bangladesh, in a sacred pond dedicated to a Muslim saint (Gadgil 1985). Certain species of plants and animals have been preserved on account of sacred qualities attributed to them by mythical tales or by identifying them with gods of the Hindu Pantheon. The most common example is that of *Ficus religiosa*. Considered sacred in most parts of the country, these trees in the Uttara Kannada region serve as a keystone species that supports a whole range of insects, birds, primates and other organisms (Gadgil 1987b). Identified with Lord Hanuman, the rhesus macaques abound in the Indian landscape. The religion of the Bishnoi community in Punjab and Rajasthan protects the blackbuck, an endemic species which is considered to be an incarnation of Lord Shiva and thus symbolizes prosperity.

Plants and animals are worshipped as totems that symbolize the kinship ties of humans and nature. The practice of certain taboos with regard to resource extraction could be interpreted as conservation practices cast in religious idioms, as most of them allow for the regeneration and perpetuation of species. Among the Onges of the Andaman Islands in the Bay of Bengal, the religious regard for certain wild edible roots prevents the people from uprooting the tubers, so that when they collect these roots from the scrub jungle they ensure that they replant the top of the root left connected to the vine (Cipriani 1966). Certain taboos ensure protection to certain 'critical life history stages'. The Phasephardis, a endogamous group found in the semi-arid regions of Western Ghats, never harm the pregnant does, or fawns, of antelopes or deer. The females of most species are less hunted as they are the progenitors. Similarly, taboos against hunting and fishing during particular seasons more often than not coincide with the breeding and spawning seasons. Egrets, storks, herons, pelicans, ibises and cormorants, considered fair game in non-breeding seasons, are almost never targeted at their colonial nesting grounds.

In contrast to these beliefs, there have been notions that have equally effectively degraded nature. Settled cultivators used religious idioms to justify the agrarian take-over of the forests for agriculture. Large patches of forests were burnt as offerings to the divine forces. This practice is still prevalent among some indigenous communities in Gujarat and Rajasthan. An instance of such a kind is mentioned in the *Mahabharata*. *Arjun* and Lord *Krishna*, symbolizing the custodians of the predominant agricultural communities, are known to have set fire to the entire Khandava forest at the behest of the Fire God, who came in the guise of a Brahman. The *Adivasis* (tribals) living in the forests were termed *Rakshas* (demons), and vanquishing them was symbolic of the victory of the good over the evil (Gadgil and Guha 1992). Animal sacrifice has been a dominant ritualistic practice among several communities.

Socio-political economic systems

In addition to the belief systems that condition relations between humans and biodiversity, the social, political and economic relations between humans themselves influence or control resource use. Thus, local community and wider institutional structures form the second context of resource use. These include structures for the management of common property resources, customary tenure rights, customary laws/rules regarding resources, localized economies (such as the village market or *haat*), and so on. These structures exist at various levels, from a group of users of a particular resource, or the village as a whole, to a cluster of villages or an entire region, and more widely to the nation and the globe. These structures enforce the actual rules for resource use, through social sanctions against misuse, reward for compliance, etc.

Traditional community-based management of common property resources typically imposes restrictions on the indiscriminate use of resources and ensures some form of distribution of benefits and livelihood opportunities. The self-managed village commons (water bodies, forests, pastures etc.) are often equitably managed, though by no means always. In community ventures for fishing among some of the fisher folk of coastal Andhra Pradesh, irrespective of an individual's catch the total harvest is divided in accordance with the needs of each family. Among the Bhotiyas of Himachal Pradesh, the community-owned pasture lands are managed in such a fashion that all families have equal opportunities to graze their yak herds on both good and bad pasture lands (Haimendorf 1985).

The traditional management system entails 'resource partitioning', and diversification of resource use either in common areas or by territorial isolation. This checks over-use and inter-group competition. The three nomadic groups that occupy the semi-arid regions of Maharastra – the Phasephardis, the Nandiwallas and the Vaidu – indulge in hunting, but each group has its own specialized techniques, and specific targets, and the extent to which they are dependent on hunting also differs. Phasephardis snare deer, blackbuck, antelope and game birds. The Vaidus trap small carnivores and other small mammals. And the third group, the Nandiwallas, hunt monitor lizards, wild pigs and porcupines with the help of dogs. Only the Phasephardis are specialist hunters, while the other two combine hunting with other occupations such as sooth-saying and trade (Malhotra, Khomne and Gadgil 1983).

In terms of both property ownership and social relations, the inter- and intra-societal relations directly or indirectly bolster conservation practices. In communities that are cohesive and have strong kinship leanings, the group interest takes precedence over individual interests. Thus, according to Alcorn (1994), 'Traditional tenure is a partnership between individuals and the broader community to maintain the community's resource base'. It can be hypothesized that the joint family system dominant in Hindu agricultural societies has been maintained to prevent the fragmentation of holdings into unviable units. Communal labour and co-operative ventures are an integral part of the traditional societies. Gadgil and Guha (1992) have asserted that the traditional hereditary occupational division in the caste system affects resource use by judicious partitioning of resources. However, it remains a contentious issue whether the ecological gains offset the social inequalities generated by this system, or whether in the long run the inequalities themselves could undermine sustainability.

The extraction and use of non-timber forest produce (NTFP) is measured and regulated in accordance with the traditional tenure system, both within and across defined territories. For example, the Cholanaikans of Karnataka have well-defined principles allowing them to collect NTFP for self consumption from within their respective territories, otherwise known as *chenam*. However, there are rigid norms regarding such collection beyond one's own territory (Misra 1980). The political processes involved in the social enforcement of these measures and practices of conservation remain a critical aspect of this system. The headmen of the *Jamas* of the Jenukuruba tribe of Karnataka ensure that there is no trespassing beyond territorial bounds; religious heads mediate cases involving the transgression of sacredness and the chiefs in the shifting cultivator's community regulate the appropriate allotment of land. There exists a structure that monitors the community-evolved and community-sanctioned code of conduct, the violation of which merits penalties, including expulsion from the community.

Knowledge systems

Corresponding to the above distinction between local and larger-scale belief systems, there are two kinds of knowledge systems which form a third context of resource use: localized practical knowledge (e.g. tribal medicine), and widespread 'classical' knowledge (e.g. *Ayurveda*). In both there is a high degree of knowledge of biodiversity. These knowledge systems are of profound significance to conservation as both these systems result in practices such as the conservation of sacred spaces and sacred species, and the organization of resource uses (including agriculture) which are given ritual meaning. These knowledge systems may be codified or non-codified, but both are in a state of dynamic interaction and are mutually influential.

Implications for biodiversity

The impacts of local community resource uses on biodiversity are mixed in any given situation; they could help to maintain or conserve it, enhance it, or reduce it. An analysis of the overall impact is complicated by the fact that the same resource use might have different impacts on different elements or levels of biodiversity. For instance, agricultural practices might enhance diversity at the genetic level (e.g. by developing several landraces of a particular species), reduce it at the species level (e.g. by clearing forests for making fields), but enhance it at the ecosystem level (e.g. by creating a mosaic of micro-ecosystems). Hence we will now broadly analyse the implications that local community practices in forestry, agriculture, animal husbandry, ethnomedicine and fisheries have had for biodiversity.

Forestry

The diverse types of forest in India are perhaps the country's richest assemblage of biodiversity, and the dependence of traditional communities on this forest resource is enormous – the forests contributing directly to the survival of millions of tribal people and members of other rural communities. The forests are of such critical socio-economic importance that these communities have tended to exercise self restraint in their use of forest resources, thus conserving both the constituent biodiversity of the forests and the diversity of forest types as a whole. Besides, these communities are also vast repositories of forest-related knowledge of biodiversity. For instance, hunting and gathering societies are said to have extensive knowledge of the immediate resources as well as a profound understanding of the local ecological interrelations. Consequently, their practices of prudence (as delineated by Gadgil and Guha 1992) involving a qualitative and quantitative control on over-exploitation, seem geared towards maintaining an ecological equilibrium. A simple instance of the mode of hunting of the South Indian tribe Irula clearly exemplifies this. 'Irulas have no formal methods of assessing the sustainability of their uses of wild species, but their sensitivity to changes in habitat, changes in season and knowledge of the biology of these species allows them to be effective exploiters. They will not hunt depleted areas for the simple fact that it is not 'energy effective'.' (Whitaker and Andrews 1994).

Even among the settled cultivators, whose fields were removed from the forests, customary regulation of people's access to land and forest produce checked the indiscriminate use of these resources. These primarily included a quantitative restriction on the amount harvested from the community-owned village woodlots. Besides this, the extractive methods followed the same stipulation of prudent use, as delineated earlier. It is interesting to note that some communities restricted the lopping of branches during the rainy season as it inhibited growth (Gadgil 1992).

But not all practices have been conducive to forest conservation. The development of agriculture and the spread of the agrarian society in the Deccan Peninsula between 6000 and 1000 BC led to gradual deforestation (Gadgil and Guha 1992). Shifting cultivation in many places has destroyed the primary composition of natural forests, although practised traditionally it allowed for regeneration of secondary forest cover. The nomadic pastoralists in places have contributed to the gradual expansion of arid regions by allowing overgrazing of pastures, although the assumption that all grazing is detrimental is invalid. The areas grazed by *Gaddis* in the alpine meadows of the Himalayan landscape are known to have a high plant species diversity (Saberwal 1996). The adverse effect of certain religious practices has already been mentioned.

Agriculture

The Indian subcontinent is known as the 'Hindustan Centre of Origin of Crops and Plant Diversity' (so termed by Vavilov, 1955). At least 166 species of crops (6.7 per cent of the total crop species in the world) and 320 species of wild relatives of crops are known to have originated here. Crop diversity of such immense magnitude is accredited to the ingenuity and skills of the traditional farmers. Operating within limited possibilities, these farmers have evolved sophisticated and complex agricultural systems and practices. These include the use of diversity over both time and space within the farm, with practices like multiple cropping and inter-cropping of a mix of species or intra-species variety, crop rotation, maintaining fallow periods, incorporating wild and weedy relatives of crops, experimental and deliberate selection for a variety of traits, and interspersing of trees and other non-crop species. The *barahnaja*, an inter-cropping pattern practised by farmers of the Tehri Garhwal region of the Uttar Pradesh Himalayan foothills, involves the use of about 12 types of crops grown in a single field, each with a different growing cycle and nutrient requirement, and all combining into a highly productive, sustainable system (Jardhari and Kothari 1996). Documenting the varietal diversity of rice, scientists at the IARI (International Agricultural Research Institutes) are known to have collected several thousands of cultivars from the region between 1800 and 2700 metres in Meghalaya and Arunachal Pradesh (IIPA 1994). Shifting cultivation is known to use up to 35 crop species in a single cultivation cycle (Ramakrishnan 1992).

So far as technological inputs were concerned, the farmers were more or less self-reliant, with most of the inputs being drawn from their own farmlands and surrounds, entailing a recycling of the crop plant and preservation of natural nutrient cycles (Pereira 1993). *'Vrkshayurveda'*, the ancient plant science dealing with all aspects of agriculture, embodies the basic tenets of this form of self-reliant and integrated system (Vijayalakshmi 1994). Apart from directly affecting crop diversity, traditional agricultural farms also supported an array of life-forms: 'weeds' and other plant species (either in the hedges or in the fields), innumerable species of micro-organisms, fishes, insects, birds and some mammals. Settled farming systems, constituting a part of the micro-ecological region, entailed 'a simultaneous preservation of diverse ecological regions' such as pastures and wastelands and other organisms therein, which may inadvertently affect agriculture (Pereira 1993).

Animal husbandry

Pre-dating settled cultivation, the domestication of animals was an indispensable aspect of the local community economy and formed a crucial part of the technological inventory. In the absence of any external inputs, traditional communities selected and domesticated breeds that were best adapted to their micro-habitats, bearing an ability to resist the extreme and difficult environmental conditions. The varietal diversity is clearly attributed to such indigenous methods of selection and breeding. As a consequence, we find diverse breeds existing under different environmental conditions. The double-humped camel, a distinct species, is found in the cold deserts of Ladakh in trans-Himalayan India, while some breeds of the single-humped camel, like the Bikaneri breeds and Jaisalmeri breeds, are found in the hot deserts of Rajasthan and Gujarat. The Deoni cow of Maharastra, the Punganur cow of Chittor, the Osmanabadi goat and the Deccani sheep, have all originated as best responses to their 'agri-ecozone', effectively meeting the production goals generated by 'social needs' (Ghotage and Ramdas 1997).

In India, livestock genetic diversity includes 26 breeds of cattle, 8 breeds of buffalo, 40 breeds of sheep, 20 breeds of goat, 8 breeds of camel, 6 breeds of horse and about 18 breeds of poultry (Sahai 1993). In addition, the pastoralist tribes and semi-pastoralist tribes of the northern Himalayas and the north-east have domesticated breeds of yaks and mithuns.

Ethnomedicine

India is known to have one of the richest ethnomedical traditions in the world, with the largest use of local biodiversity for medicine (Shankar 1994). The indigenous and traditional health care system that serves more than 70 per cent of the population is based on the rich diversity of medicinal plants and associated knowledge. Most of the *Materia Medica* of traditional communities dates to the *Vedas*. A *Rig* vedic hymn of several centuries ago is known to mention 107 plant-based cures for diseases. According to the classical medical treatise *Charak Samhita*, all plants have a potential medicinal value. This is amply reflected in a myth, according to which Brahma, the God of Gods, proclaimed sage *Jivaka* a great physician when, after 11 years of wandering, *Jivaka* expressed his inability to find a plant without medicinal qualities (Vijayalakshmi and Shyam Sunder 1993).

Over 7,500 species of plants have been used in Indian medical traditions. In the classical medical systems of *Ayurveda*, *Unani* and *Siddha*, over 2,700 documented species of plants are used (Shankar 1994). In terms of numbers, something like 1,800 plants are documented in various Ayurvedic tenets. There are approximately 400 in *Unani* and about 500 in the *Siddha* system. Similarly, there exist equally impressive localized knowledge and practices of plant-based ethnomedicine. Practically every tribe or forest-based community that is documented is known to possess knowledge of a number of medicinal plants and their uses. Among the Gaddis of Western Himachal Pradesh, at least 50 species of plants are used for curing different ailments, and are well known to the local medicine men, the *Vedus* (Lal, Vats, Singh and Gupta 1994). The Madav Koli tribals of Western Ghats use 202 plant species for medicine and 109 for veterinary purposes (Shankar 1994).

The traditional systems are not only known for the diversity of plants used, but also for the diverse ways in which these plants are administered. Each community within its cultural context makes an independent appraisal of its local resources and develops its own classification. For example, the plant *Centella asiatica*, known to both the classical and folk traditions, is put to 33 different uses across different communities in southern India (Shankar 1994).

Fisheries

India has approximately 8,000 kilometres of coastline, with a high diversity of marine biological resources. About 2,500 species of fish are found in Indian waters. These and other marine biological resources supply about 13 per cent of annual protein to the Indian population. This coast has been the 'lifeline' of the large number of traditional fishing communities, who at present also control more than 50 per cent of production of fish in the country. Fishery, both marine and freshwater, is an age-old tradition. The practices and technologies of these communities were geared towards a sustainable harvest and consequently its regeneration and conservation. According to Kocherry (1994), prior to the dawn of industrial fisheries development in the country in the middle of the sixties, the fishing communities in India with their traditional knowledge of the sea and its environment harvested the resources on a moderate scale. In this process, the fisher-folk were their own masters. The craft and gear deployed were the most appropriate to suit the environment and these were developed by the fisher-folk over centuries of experience and with skills passed from parent to child. The catamarans, the small canoes, big canoes and different gears were all results of traditional innovations to meet the dynamics of tropical water, fish behaviour and changes in seasons. The fishermen almost never overfished the resources which they considered their common property. Every fisherwoman and fisherman sees the sea as something very fundamental, as 'Mother Sea'. These technologies, unlike modern trawlers and purser-seiners, did not destroy marine ecology, nor harvest seedlings by raking up the seabed.

Raychaudhuri (1980) describes how the fishermen of *Jambudip* (India) co-ordinate the complex variables of seabed topography, seawater conditions and sequences of tide with fish behaviour, to ensure both successful catches and their safety at sea. In their selection of the appropriate seabed over which to conduct their activities, these fishermen are like the agriculturists who tend to classify the soil according to its relative fertility and the types of crops grown. The 'soil' of the seabed is classified by its capacity to support the net poles and by its fertility regarding the types and quantity of fish in the waters above it. Such practices have thus helped to conserve a considerable amount of marine diversity.

Erosion of Local Community Systems

In India, as in the rest of the world, Local Community Systems (LCSs) have been severely eroded by a variety of factors.

1. Displacement and devaluation by modern systems, such as the replacement of traditional medical practices by the modern allopathic system, and of community customs of conservation by state-sponsored practices of conservation. In more recent times, local knowledge has even been appropriated by the state and private sectors in the form of intellectual property rights (such as patents) on products and processes derived from LCSs.

2. Institutional take-over of resources by state and private sector: common property resources such as forests and wetlands have been taken over by the state or by private corporations, resulting in the alienation of local people and the breakdown of traditional management systems.

3. Other factors of erosion include the over-exploitation of resources by the state or private sector: the diversion of resources from the rural to the urban sector; the physical displacement of communities by development projects (estimated by some people to be in the order of about 25 million in the

last few decades in India alone); reduction in access to resources due to top-down conservation policies, and factors internal to the community such as the rise in population (of humans and livestock), inequities, and changes in life-style and aspirations.

4. The erosion of LCSs leads to negative consequences for biodiversity, as well-tried systems of management, use and conservation break down. This can be seen especially in the case of common property resources: instead of a tightly regulated system of use by the community, the system tends towards an open property one. Forests, wetlands, pastures, have all suffered serious consequences. Another factor is the lack of secure tenure: without the security of realizing sustainable revenues or returns from commercialization and industrialization of agriculture, forestry and fisheries, the result is unprecedented biotic impoverishment. For instance, where large-scale chemical-based intensive farming has replaced small-scale organic farming as part of India's 'Green Revolution' there has been an enormous loss of agro-biodiversity. Finally, the breakdown of religious and spiritual sanctions has resulted in several sacred groves being cut down or neglected.

Reviving Local Community Systems

In the face of the above erosion, a number of efforts are being made to revive LCSs, either in the traditional form or by regenerating and protecting natural forests through local initiatives. In eastern India more than 10,000 communities are now regenerating and protecting forest areas (Poffenberger and McGean 1996). In the Alwar district of Rajasthan, villagers with the help of NGOs have regenerated forests and converted a water-deficient area into a water-surplus one, using predominantly traditional knowledge of hydrology and water-harvesting structures. At least one forest (of 1200 ha) has been declared a 'public sanctuary' for wildlife (Singh 1997).

The *Beej Bachao Andolan (BBA)*, or 'Save the Seed Movement', initiated by the workers of the Chipko Movement in the Tehri Garhwal region of the Himalayas, has begun restoring indigenous crop diversity by growing several hundred varieties of rice, beans, millets and other crops. Similar work is being done by farmers and groups associated with the network *Navdanya*, and farmer-level gene banks along with revival of rice diversity is being promoted by the Academy of Development Science in the Karjat area of Maharashtra (Jardhari and Kothari 1996; Ramprasad 1994; Richaria and Govindswami 1990).

Herbal medicinal practices have been re-instituted in several tribal areas of Udaipur district and by the Foundation for the Revitalization of Local Health Traditions in several areas of southern India. Revivals of 'sacred groves' has taken place in Maharastra, Bihar, Rajasthan and the North-east (Mitra and Pal 1994). In some parts of India, efforts are being made at participatory documentation of the biodiversity that is used by local communities in the form of Community Biodiversity Registers (CBRs). Initiated by a network of groups working on environment, health, agriculture and traditional technology, these CBRs aim to revitalize traditional knowledge and assess its economic and socio-political value (including protection against imposition of intellectual property rights by outsiders by providing proof of prior use).

Traditional knowledge, culturally-based world-views and Western science (Raymond Pierotti and Daniel R. Wildcat)

In recent years there has been considerable discussion of the differences between the world-views and knowledge bases of indigenous peoples, and those of the 'dominant' or 'Western' culture (e.g. Suzuki and Knudtson 1992). One major difference is that Western Europeans look backward and forward in time to obtain a sense of their place in history, while indigenous peoples look around them. Native Americans think spatially, not temporally (Deloria 1992).

The idea of a human history existing independent of place and nature is foreign to Native Americans whose history cannot be separated from the geography, biology and environments to which they belong. As Deloria (1990: 17) notes, 'In the traditional [way of knowing], there is no such thing as isolation from the rest of creation'. Understanding the notion of thinking spatially allows us to understand the significance of the native tradition of invoking and praying to the four directions, and the sky and the earth. Persons making such prayers are acknowledging the space in which they live, and their understanding that the creative forces that shape their lives exist in all of these directions. In essence, this native world-view requires one to be native to a place (*cf.* also Jackson 1994) and to live with nature, in contrast to the dominant Western world-view which assumes humans live above, separated, or in opposition to nature.

Those wanting to embrace the comfortable and romantic image of the Rousseauian 'noble savage', however, will be disappointed. Living with nature has little to do with the often voiced 'love of nature', 'closeness to nature,' or desire 'to commune with nature' one hears today. Living with nature is very different from 'conservation' of nature. Those who wish to 'conserve' nature still feel that they are in control of nature, and that nature should be conserved only insofar as it benefits humans, either economically or spiritually. It is crucial to realize that nature exists on its own terms, and that non-humans have their own reasons for existence, independent of human interpretation (*cf.* below, and Taylor 1992). Those who desire to dance with wolves must first learn to live with wolves. Being native is expressed in the ability to experience a sense of place which casts off the modern view that 'space' exists to be conquered. The cultural diversity of Native Americans reflects their intimate ties to the land and the biology of the places they call home. Not surprisingly, native peoples have considerable insights into the workings of nature, and in many ways modern Western science is only just beginning to catch up with ancient Native American wisdom.

The ideas expressed in this contribution are the intellectual property of many native peoples with whom the authors have studied and lived during the last ten years at Haskell Indian Nations University in Lawrence, Kansas. They are synthesized for an indigenous theory and practice of politics and ethics of the natural world that is more applicable to current issues than the dominant model of the so-called Western Civilization.

Box 5.2: Principles and Elements of Revitalization of LCS

Reviving community rights to resources:

Given that one major reason for the loss of LCS is the alienation of people from common property resources they previously managed, the revival of community controls over these resources would help to revive LCS. Serious policy and legal changes are needed to make this possible, along with appropriate checks to ensure that communities do not misuse the powers they regain from the state.

Adapting community control institutions:

Traditional community institutions such as Village Panchayats, and variations on them such as Van Panchayats, need to be built upon for conservation purposes. The potential of the 73rd Constitutional Amendment on Panchayats, and its extension to tribal areas, needs to be fully explored.

Enhancing and building on local knowledge:

A considerable amount of traditional and local knowledge continues to be relevant from the conservation and sustainable use point of view. Building upon this, rather than replacing it by modern systems, ensures compatibility with other elements of LCS, suitability to local resources, easier acceptability within the community, and greater social sustainability.

Ensuring traditional intellectual rights:

In the face of the increasing power of private intellectual property rights (IPRs), communities should be empowered with community intellectual rights, which safeguard their knowledge and ensure that they are fairly benefited by the use of this knowledge by the outside world.

Integrating LCSs and larger/modern systems:

LCS knowledge and practical systems can be complemented by introducing ecologically and socially appropriate and sensitive new technologies, at all times being careful not to displace essential elements of the LCSs which are important for conservation and sustainable use. Research that takes place in laboratories or governmental fields far removed from local communities should shift to where these communities are based; and farmers, forest dwellers and fisherfolk themselves should be equal partners in R&D programmes, using their own knowledge and skills and learning from new ones introduced by the formal sector. In turn, formal sector students should be constantly exposed to LCSs.

Using the market creatively:

The market can become an ally in conservation of biodiversity and sustenance of LCSs, provided that local communities retain some control over it, that local demand and needs are given priority over outside ones, and that ecological sustainability is maintained.

Positive social and economic incentives:

Instead of the reverse discrimination faced by LCSs today (e.g. organic farming versus 'Green Revolution' farming, with the latter getting subsidies and support), there should be a system of incentives for practices that are oriented towards sustainable use and conservation. This could include financial and developmental inputs of various kinds, provision of property and intellectual rights, and others.

Native peoples and ecology

One feature separating native peoples from other American minorities, and in fact from the dominant culture, is that they lack an immigrant experience within their memories (Deloria 1995a, 1995b). This has yielded a very different view of land-use and resource management from that which occurs in other cultures. The world-views and cultures of native peoples evolved in the environments of the continent we now call North America. These peoples depended upon the animals and plants of these environments for food, clothing, shelter and companionship, and as a result developed strong ties to the fish, the land animals, the forests, the grasslands, and to the very earth of North America. As these places and beings existed and changed along with them for tens of thousands of years, Native Americans developed the sense of place that led them to think spatially.

As Native American peoples developed through observation of their fellow beings, they noted that each species had characteristics that set them apart from other species, and enhanced their chances of survival (Marshall 1995). Humans had free hands, understanding and intelligence, but lacked the speed, horns, teeth, claws and strength of other species. The way for humans to survive and prosper was to pay careful attention, and learn as much as possible about the strengths and weaknesses of all the other organisms, so that they could take them as food and avoid being taken by them as food. The body of knowledge acquired through this careful observation was passed on to others through detailed stories, which had to be repeated constantly so that the knowledge would be passed on to future generations. Several themes emerged through these stories. First, it was realized that all things are connected. This is not simply an empty phrase, but a realization that it was impossible for any single organism to exist without the connections that it had to many other organisms. Our native ancestors had a rich knowledge of nature; they observed other organisms killing and eating: wolves, for example, were good hunters from whom much could be learned. Predators were recognized for their power, and humans recognized a kinship with them since, like the predators, humans also depended upon the taking of life for their food. As Deloria (1990: 16–17) explains, 'Little emphasized, but equally as important for the formation of (Native) personality, was the group of other forms of life which had come down over the centuries as part of the larger family'.

Taking the lives of other organisms and consuming their tissues in order to sustain one's own body tissues establishes the connectedness of living things. By eating parts of other organisms you demonstrate that they are made of the same materials as yourself. This connectedness is a fundamental tenet of ecology, called cycling. Thus the idea of a cycle, or circle, of life is not a mystical, ethereal concept based upon great mysteries, but a practical recognition of the fact that all living things are connected. In fact, cycling goes beyond animals and plants, since nutrients, such as phosphorus, nitrogen, calcium, and sodium also cycle through water, soil, rocks and the atmosphere. Thus, the native idea that humans, and other living things, are also connected to earth, water, air and sun also reveals an understanding of the geochemical aspects of ecology.

Native Americans do not think of nature as 'wilderness', but as home, the place where they became human beings (Standing Bear 1993). Natives do not leave their 'house' to 'go into nature', but instead feel that when they leave their shelter and wander about encountering other non-humans and natural physical features they are simply moving into other parts of their home. As Reichel-Dolmatoff (1996: 8-9) further notes: 'What we call nature is conceived by Native peoples as an extension of biological man, and therefore a (Native) never feels "surrounded by nature"... a (Native) walking in the forest, or paddling a canoe is not in nature, but he is entirely surrounded by cultural meanings his tradition has given to his external surroundings'.

For native peoples this insight has profound implications for their conceptions of politics and ethics. Although some current Western philosophers are exploring the area of environmental ethics (Taylor 1992), this is ancient terrain for the natives. Unlike dominant Western political and ethical paradigms which find knowledge of how human beings ought to act imbedded in the life of one's social relationships, our ancestors found knowledge of how to behave from membership in a community consisting of many non-human persons – the four-leggeds, winged-ones, plants, etc.

Western thought followed the lead of Aristotle, and defined politics and ethics as exclusively human realms. Aristotle proposed that human values are learned from our fellow community members. From a native perspective Aristotle's basic reasoning was right, but his notion of community and its membership was wrong. Native cultural practices defined politics and ethics as existing in the realm of ecosystems, and would argue that it makes no sense to limit the notion of politics and ethics only to human beings. How we human beings live will indeed reflect the communities to which we belong. By limiting the definition of 'persons' to human beings, however, Aristotle created a false and far too narrow sense of community and the corresponding spheres of political and moral life.

The inclusion of other living beings and natural objects into the category of 'persons' requires political and ethical constructs that include these other community members. The best illustration of how native peoples include many other natural objects and living beings as members of their community is found in native clan names and totems. 'There seem to have been a series of covenants between certain human families and specific birds, grazing animals, predatory animals, and reptiles' (Deloria 1990: 17). These animals are connected to families over prolonged periods of time, and offer their assistance and guidance during each generation of humans.

It is frustrating and uncomfortable to constantly hear non-native peoples speaking romantically of the Indians' 'closeness to nature' or 'love of nature.' The relationship is more profound than this expression connotes. To be Wildcat, Bear, Deer or even Wasp clan means that you are kin to these other persons – they are your relations, your relatives. Ecological connectedness is culturally and ceremonially acknowledged through clan names, totems and ceremony. In nearly all native creation stories, animal-persons and plant-persons existed before human persons. These kin exist as our elders and, much as human elders, function as our teachers and as respected members of our community. Acknowledging non-humans as teachers and elders requires that we pay careful attention to their lives, and recognize that these lives have meaning on their own terms (*cf.* also Taylor 1992).

Perhaps the best way to think of this traditional knowledge borne of experience is that native people lived their lives as though the lives of other organisms mattered. Natives experienced other creatures in their role as parents, as offspring, and ultimately as persons within a shared community. They realized their own lives were intimately intertwined with those of these other organisms. Most importantly they recognized that the human being, 'man', is not the measure of all things, but exists as but one small part of a very complex ecosystem, unlike the Western view that places human beings above the rest of nature.

Recognizing connectedness and the meaningfulness of other lives did not mean, however, that animals or plants should not be taken or used for food or clothing. Native people were dependent upon them for these very reasons. Instead, each taking was accompanied by recognition of the fact that the take represented loss of life to a fellow being whose life had meaning on its own terms. Such a perspective leads inescapably to these conclusions: 1) lives of other organisms should not be taken frivolously, and 2) other life-forms exist on their own terms and were not put here solely to be used and exploited by humans.

People of European heritage often develop a love of nature because they love a particular cat or dog, or perhaps animals in general. They often imbue these animals with human emotions and thoughts, i.e. anthropomorphize them, which leads them to oppose the taking of other animals through hunting, and in some cases to refuse to eat meat. It is often assumed that such attitudes are similar to those of native peoples, since it is assumed that opposing hunting, or eating a non-meat diet, brings people closer to 'harmony with nature'. Such people are often shocked when they realize that native peoples regard hunting, fishing and meat-eating as part of a strong cultural tradition, even when the animal being hunted is endangered or subject to federal regulation, such as bowhead or beluga whales.

This conflict of views results from a failure to realize that native people do not identify with and anthropomorphize animals, but instead recognize that the lives of animals and plants exist on their own terms, and have value independent of any we human beings place on them (Taylor 1992). Being taken as food is a common fate of species within their natural environment, and eating the flesh of these animals establishes the connectedness that is such a profound aspect of spirituality. The lives of human beings and their families often depended upon taking the life of the animal, and the act of giving up its life so that humans could survive was recognized as a profound sacrifice for the animal. In such situations, the native people understood themselves as predators, part of the world of the prey, and connected to the prey in a profound experiential sense. By contrast, Europeans often identify with the prey in an extremely anthropocentric and psychological sense, and react as if their loved ones were being taken for food.

This attitude may manifest itself in the form of hostility towards all predators, leading to a belief that wolves, cougars and bears are creatures of evil, capable of 'slaughter of helpless prey' (Lopez 1978; McIntyre 1995). In early times, Europeans even captured and hanged wolves that took livestock as if the predators were human criminals who had committed murder. To this day, Western culture demands the killing of any individual predator that attacks a human, even including sharks and crocodiles. In contrast, native peoples identified with the predators (Buller 1983; Welch 1986; Marshall-Thomas 1994; Marshall 1995), for as hunters who had to rely upon knowledge of prey they recognized the connection between themselves and the wolf, the bear or the lion (Marshall-Thomas 1994). They also recognized themselves as potential prey for other large carnivores. This knowledge of connectedness and of ecological similarity allows the native to respect the predator, since they know how difficult it is to take lives, and that the predator feels most connected to the prey when it has taken its life (*cf.* Marshall-Thomas 1994).

All predators were respected for their strength and their weapons, but among some native peoples one predator spoke particularly to humans. This was the wolf, who lived in family groups and was not strong enough or swift enough to kill large prey alone. 'From the dawn of our spiritual and psychological being, our closest relative in the wild has been *Makuyi* – in English, wolf.' (Jack Gladstone, Blackfeet, in McIntyre 1995). Wolves working as a group, however, could bring down even large plant-eaters. As hunters, wolves did not have the greatest strength or sharpest claws and teeth, and their weapons, formidable as they are, are of little use without endurance, patience and perseverance: qualities the first peoples could develop in themselves (Marshall 1995). More important, however, was that if people were to emulate the wolf, they also had to exist to serve the environment, and to accept the mutuality (connectedness) of life. 'Understanding this reality made them truly of the earth, because every life ultimately gives itself back to the earth.' (Marshall 1995: 6-7)

Native people recognized that wolves, and their relative the coyote, had relationships to many other creatures, and that these relationships were part of the connectedness of all things. In some cases these relationships were described as friendships, which implied that both participants received benefits from the interaction. Only recently has the empirical basis of some of these stories been examined by Western science, which revealed relationships either previously unknown or misunderstood.

For example, a story often told by natives was that badger and coyote were 'friends' and hunted together (Ramsey 1977). Western thought, driven by the idea that competition among species drives community dynamics, categorized the relationship between coyote and badger as competition between these two predators. Recent study, however, revealed that coyotes and badgers hunting ground squirrels each had significantly higher rates of capture when hunting together (Minta *et al.* 1992). Coyotes and badgers wander around together, and when they see a squirrel, coyote gives chase. If the squirrel goes into a burrow, badger will dig up the burrow, or both will dig together. If the squirrel stays in the burrow, badger will get it and have a meal. If the squirrel leaves by another burrow, coyote often gets it and has a meal. Both coyote and badger catch more squirrels (and probably woodchucks and prairie dogs as well) when they hunt together than when they hunt alone.

Similarly, stories of friendship between wolves and ravens (e.g. Welch 1986) have been supported by discoveries that these species may forage co-operatively, and even play games with each other (Lawrence 1976; Heinrich 1988; Pierotti 1991). Ravens fly through the forest and look everywhere for animals that have died. In winter, if a raven spots a carcass that is frozen and too hard for the raven to peck open, it will call loudly, which attracts wolves, or coyotes, which then chew open the carcass, making food available for the ravens as well as the wolves. Some Western scientists studying wolves have realized that native people have far greater knowledge of the behaviour and ecology of wolves than Western science, and have turned to native people to help them in their study of these animals (Stephenson 1982).

Until recently within the Western tradition the only individuals who understood connectedness between predators and prey were ecologists, since a fundamental tenet of ecological theory is that predators cannot exist without prey, and that prey populations are often regulated by predators. In his essay, *Thinking like a Mountain*, Aldo Leopold (1948) argues that if deer fear the wolf then a mountain must also fear its deer, since without the wolf to limit the number of deer, the deer could kill the vegetation, and even risk the mountain itself through erosion. Thus, the connectedness of all things can be understood not only as a native belief, but as the fundamental concept of ecology.

Traditional knowledge and evolution

Another major teaching of native peoples is that all things are related, e.g. Lakota end all prayers with the statement *Mitakuye oyasin* ('All my relations'). This teaching emerged from the concept of connectedness, since it is obvious that if you can take another organism into your body and have it become part of you, you and it must be made of the same material and are thus related. The relatedness of all things is one of the central tenets of the Western science of evolution, since we now know that all organisms on Mother Earth contain DNA and RNA, which contain similar instructions that allow individual organisms to develop in a way to resemble their parents, but not be identical to them.

The idea that each organism is an individual, with unique qualities, is both a basic understanding of native perception (Deloria 1992: 88–89), and a fundamental tenet of Darwin's theory of evolution through natural selection. To examine this point in more detail, let us look at a statement of *Okute* (Shooter), a Teton Lakota (McLuhan 1971).

> 'Animals and plants are taught by *Wakan Tanka* (the Lakota creator) what they are to do. *Wakan Tanka* teaches the birds to make nests, yet the nests of all birds are not alike. *Wakan Tanka* gives them merely the outline. Some make better nests than others... Some animals also take better care of their young than others...All birds, even those of the same species, are not alike, and it is the same with animals, or human beings. The reason *Wakan Tanka* does not make two birds, or animals, or human beings exactly alike is because each is placed here to be an independent individual and to rely upon itself... From my boyhood I have observed leaves, trees and grass, and I have never found two alike. They may have a general likeness, but on examination I have found that they differ slightly. It is the same with animals (and) with human beings. An animal depends upon the natural conditions around it. If the buffalo were here today, I think they would be very different from the buffalo of the old days because all the natural conditions have changed... We see the same change in our ponies... It is the same with the Indians...'

In his book *The Origin of Species*, Charles Darwin (1859) laid out the conditions necessary for natural selection to occur and lead to evolutionary change. These included variation among individuals, which has an effect upon the ability of individuals to survive and reproduce. *Okute* describes similar phenomena by emphasizing variation among individuals (the major theme of his statement), and this variation leads some individuals to build better nests or take better care of their young, i.e. to reproduce more successfully as a result of this variation. Thus *Okute* has described the process of natural selection in the context of traditional Lakota knowledge.

It is obvious from these statements that *Okute* understands that individuals vary within species, and that the conditions under which an animal or plant exist shape its appearance and behaviour. This is as clear a statement of natural selection as exists in Darwin. Perhaps more significantly, *Okute* realizes that evolutionary change can take place over very short periods of time, something that has only recently been recognized by Western science, as recent numerous studies have shown that evolutionary change can take place within only a few generations (Grant and Grant 1991; Weiner 1993). Finally, it is important to note that *Okute* includes Indians within this statement, which indicates that he is aware that humans evolve in the same manner as do non-humans.

Okute's statements were originally taken to be statements about his 'holy beliefs' and a 'mysterious power whose greatest manifestation was nature' (McLuhan 1972). To natives, spirituality and holiness are manifested in nature, and because of the connectedness between human beings and nature, spirituality suffused every moment of a native person's being. It is difficult for people who follow Western traditions and religions to understand that to native people, spirituality is constant, thus one must show respect for non-humans at all times. Natives sometimes contend that they have no such concept as religion, which they interpret as rituals that are practised part of the time, but not rigorously adhered to. Instead, native people live every moment in a state of respect and awareness of the power of creation as manifested by the place where they live, and the fellow beings with whom they share that place.

Native people were also aware of inheritance between generations. When selecting seeds for planting corn, we deliberately select seeds from as wide a variety of ears as possible, for we believe that corn is a gift from the creator, and to discriminate against small ears would show a lack of appreciation for the gift (Jackson 1994). This applies evolutionary principles by preserving genetic diversity, which means that regardless of environmental conditions during the next planting season, some corn is likely to grow and prosper. As with *Okute*, this approach recognizes that individuals vary, that diversity among individuals is a gift from the creative spirit of nature, and that plants (and animals) depend a great deal on the natural conditions around them. Second, by preserving genetic diversity we acknowledge that traits are inheritable and can be passed from one generation to the next. In contrast, Western agricultural practices in recent years have artificially selected for fewer and fewer highly productive lines, with a resulting loss of genetic diversity. Thus, if environmental conditions change, Western agriculture may lose the genetic diversity in crops necessary to respond to new environments. In contrast, traditional practices will allow native peoples to continue to grow crops, which is one reason why we have survived and are still here.

The recognition that changes in the environment cause changes in the form of beings, along with their non-human centred world-views, can be seen in the creation myths of native peoples. A major difference between native and Western world-views is that in nearly all Western belief systems creators tend to be human, or human in form. In contrast, in native myths creators are typically animals, e.g. raven (Haida, Koyukon), wolf (Comanche), bear (Menominee). The key question is, 'How would it change your world-view if the entity that created your world or culture was not a human, or even human-like?'

One consequence of viewing your creator as non-human is that you would not be troubled by the idea that humans like yourself came from organisms that would not be recognized as human. In fact, the clan systems of many natives are a means of acknowledging the relatedness of humans and non-humans. Viewing animals as creators also implies that the animals existed before the humans did, since to be a creator it is necessary to exist before your creation does. '*Sungmanitu Tanka Oyate*, the Big Dog from the Wilderness People (wolves), were a nation long before human beings realized and declared themselves a nation.' (Manuel Iron Cloud, Oglala Lakota, in McIntyre 1995: 260.)

The meaning of the term 'creation' has led to confusion between native and Western peoples. Western religious practice based upon literal interpretation of the Bible uses the concept of creation to argue that God created all organisms within a short period of time, and that these organisms have remained unchanged since the time of creation. For this reason, Western creationists argue that evolution, as described by Darwin and other Western scientists, cannot have occurred since it assumes change in organisms over time (contrast this with *Okute*'s statements). This biblical view of nature is antithetical to views held by native peoples, since in its essence it argues that all things are not related, in fact humans were specially created in the image of the creator to have dominion over non-humans. What many fundamentalist Christians oppose most strongly is the idea that humans are related to animals. They cannot accept that other creatures have their own reasons for living, and do not exist solely for the benefit of humans, since the Bible tells them that they have dominion over these other organisms.

Native peoples also have creation stories, which tell how specific peoples came to recognize themselves as distinct cultures, separate from other groups. These stories typically recognize the importance of non-humans in the development of a distinct culture; for example the Comanche or Lakota describe how they learned to hunt from Wolf (Buller 1983; Marshall 1995), or the Koyukon tell how they live in the world that Raven made (Nelson 1983). A tendency has arisen among biblically oriented religious instructors to co-opt these stories and imbue them with Christian religious symbolism, e.g. Black Elk Speaks (Neihardt 1932). This practice has been employed to take native peoples away from traditional beliefs, and to turn them from their connection with nature and their recognition of the connectedness and relatedness of all things. Native creation stories do not deal with the exact time when these events happened, since they happened so long ago that they exist 'on the other side of memory' (Marshall 1995: 207). Instead, these stories emphasize that the natural world and its creatures helped shape humans into what they have become, which is an evolutionary view of the origins of humans and their cultures.

It is possible to reinterpret the Western creation story of Adam and Eve as a story about life for human beings prior to the 'discovery of knowledge', which led humans to invent agriculture. It has also been argued that the story of Cain and Abel is a metaphor for the destruction of people with a biocentric world-view by agriculturalists (Quinn 1992). Following this logic, the 'fall of mankind' resulted from losing connection with nature, i.e. leaving the Garden of Eden, and embarking upon the domestication of animals and plants only for human use. The tendency to identify the beginning of human civilization with this biblical 'fall of man' (the beginning of agriculture?) results in the fundamental temporal orientation of Western civilization. The experiences of Adam, or Moses, are presumed to have meaning today because, in this largely abstract non-geographical or spatial sense of history, their history is now our history. As a result, such traditions derive instructions for living in the modern world, regardless of place, from written words which may be thousands of years old, and totally unrelated to the places where we actually live.

In contrast, the spatial orientation of native peoples leads them to find their spiritual experiences in their encounters with their fellow beings in the places where they live. There are always new experiences and knowledge in the world, and verbal traditions can be adjusted to respond to changing conditions so that instructions for living are fit to the current ecological and historical context. Native knowledge and spirituality derive from the physical and biological environment that is part of daily life (Deloria 1992). Thus, native attitudes and beliefs evolve as environmental conditions change, and as the knowledge and experience gained through daily interaction with that environment change. This is why it is important for native people to state, 'We have survived', because despite major changes in our world resulting from encounters with Europeans and Western culture, we are still here. If the environment continues to change to such a degree that the dominant culture cannot survive, Native American people will continue to survive. We survived the Ice Age, which lasted longer, and had at least as great an ecological impact as the arrival of Europeans. However, to survive we must not allow the Western way of thought, disconnected from nature as it is, to become the dominant way of thought among our people.

An example of how Western ideas can affect contemporary native thought is illustrated in attitudes towards wolves. As discussed above, native peoples of North America typically considered wolves to be friendly, if cautious animals, that would never attack a human without provocation. This observation has been supported since there has never been a reported fatal attack upon a human by a wild wolf in the history of North America (Tubbs, in McIntyre 1995: 353–357). In contrast, the relationship between wolves and humans in Europe is based on stories of wolves acting as predators upon humans or human analogues, e.g. Little Red Riding Hood, etc. (Lopez 1978).

The origins of these stories are unclear. What is clear, however, is that many Europeans hate and fear wolves, and brought this attitude with them to America (Lopez 1978; McIntyre 1995). It is distressing that this prejudice against wolves is being passed on to young native people. A recent study among native people indicated that elders (over 60 years of age) did not fear wolves and viewed them as brothers; middle-aged people (30–60) did not like wolves, but also did not really fear them; however, young people (15–30) feared wolves and stated that they would shoot them if they saw them. This is a classic case of native people being poisoned by the attitudes of the dominant culture, and revealing fear based upon ignorance, for of the people questioned, only the elders had actually seen wild wolves in their lifetimes.

We close with a description of both a dream and a nightmare. The dream is that all peoples can be educated to be native to the places where they live and to live their lives as though the lives of all beings were important and had meaning on their own terms. If this can be achieved, humans will recover their place as one species among many, no longer the measure of all things, but at home in places where they belong, and part of a large family of related beings. The nightmare is that not only will the dream not happen, but that our own children will lose this knowledge, and when they lose this knowledge, it may not be possible to regain it. What is at risk is that our descendants will not follow our views, but those of the dominant culture. If this happens, a rich tradition of knowledge based on long experience may be lost, and with it may go the hope of humans to coexist with our non-human kin in a spiritually fulfilling way, as well as hope for the future of our planet. As the Onandoga scholar and university professor Oren Lyons has stated, 'What happens to wolf will happen to us (native peoples), and if we disappear, you (Western culture) will follow'.

Are indigenous groups and local knowledge threatened in today's world society? (Maj-Lis Follér)

According to a prevailing view, indigenous peoples are seen as threatened by the 'modern' world. This threat is debated in today's anthropological literature and in organizations supporting indigenous peoples' rights, such as IWGIA (International Work Group for Indigenous Affairs). The notion 'threatened' is often used with different meanings, therefore an elucidation of it might be useful. The first use is related to extinction of whole groups of indigenous peoples, amounting to genocide.

Genocide on indigenous groups mostly happens to hunter-gatherers, slash-and-burn agriculturalists and horticulturalists and is often committed – directly or indirectly – by the government of the nation state. The reason might be that the state wants to exploit the territory in a more profitable way. In the Americas it started some five hundred years ago in the name of God and the lust for gold and minerals (e.g. Hemming 1978; Stannard 1992). Outrageous examples from our own century include those against the Yanomami of northern Brazil and southern Venezuela, the Guayaki indians in Paraguay in 1974, and the tribes in the Chittagong Hill Tracts in Bangladesh during the 1980s (Chalk and Jonassohn 1990; Jackson 1984; Rifkin 1994; Van and Lee 1996). This is, of course, the worst violation of groups of people and individuals.

Another concern is related to indigenous groups' rights to live in accordance with their own traditions, their rights of access to land which their ancestors used, and the right to use their own language. This is a threat to the peoples' cosmologies and cultures.

A third threat is also from the dominant Western society and is directed against nature, and the natural environment used by indigenous groups. This threat is often defined as 'merely' loss of biodiversity. This might happen through the activities of large-scale capital and national corporations in the traditional territories of indigenous peoples with a high potential economic value for governments and for national and international stakeholders. These activities include oil exploitation, mining, dam building, logging, mono-agriculture of cash crops, cattle ranches, the establishing of national parks, nature reservations and tourism. Such exploitative interactions between one party with power and one without happen all the time, all over the world (1996).

This contribution deals with the consequences – for the livelihood of indigenous groups – of these threats and the current globalization process, with a focus on the loss of local knowledge. Increased information flows and the dependency between different groups with unequal power positions in world society are some of the manifestations of this process. There are also clear differences between what is accepted as knowledge and what are accepted as sources of knowledge. Some examples of sources of knowledge are: experience, tradition, authority, revelation and intuition (Bärmark 1993).

The framework within which this will be analysed is the interdisciplinary field of science called anthropology of knowledge. One of the goals of anthropology of knowledge is to analyse the scientific way of thinking in relation to ways of thinking in other cultures. There are different scientific approaches – the anthropology of knowledge (e.g. Bärmark 1988; Elkana 1981) and the social study of science (e.g. Woolgar 1988) – which strengthen the idea that the social environment influences the development of understanding and is the context within which knowledge is created.

Ecological anthropologists and human ecologists have also shown the importance of nature for the development of knowledge. This approach is called historical ecology (Balée 1995), and is aimed at understanding the long-standing man-environment interaction based on 'Traditional Ecological Knowledge' (TEK). Anthropology of knowledge can be seen as a way to 'widen the scientific context' and to critically reflect upon our own preconceptions, most of which are taken for granted as universals (Follér 1993). The intention is to start a dialogue to facilitate our understanding of other ways of understanding the world. A central issue is that each form of knowledge has to be understood within its cultural and natural context. One point of departure in the anthropology of knowledge is to avoid seeing the indigenous groups with a nostalgic retrospect of times long past – making them stereotype. We need to find a balance between the two extremes of ethnocentrism and xenocentrism if we are aiming for a sustainable and pluralistic future.

Today's international concern about indigenous knowledge has arisen through various factors internal to our Western society. One of these factors is what is referred to as the 'ecological crisis' – the implication that human activities are heading for a collision with nature. The United Nations conferences on the environment – in Stockholm in 1972 and Rio de Janeiro in 1992 – reflected this awareness. The results, including documents such as *Agenda 21* and the Biodiversity Convention, have served to focus attention on ecological issues. In the same spirit are various initiatives to support indigenous groups, especially those living in tropical forests, and an interest in biodiversity, conservation and the protection of nature. Indigenous peoples also face an intensifying challenge to the integrity of their societies by an increasing pressure from 'outsiders' to document, utilize and commercialize indigenous knowledge.

Some of this exploitation is justified in the name of conservation and the documentation of indigenous culture. It is also being justified in the wider interest of humankind, and in various resolutions and declarations on indigenous peoples' rights. But, at the same time, indigenous individuals and communities are part of the globalization process and thereby seeking access to the market economy. They are becoming 'modern' and thereby also a part of, for example, the ecological problem (Conklin and Graham 1995). My own research has mainly been in the Amazon and there are many signs there of the intensified contact between global society and indigenous groups with consequences for the people, their cosmology and the ecosystem (Follér 1990, 1997; Follér and Garrett 1996). The perceived risks are social, cultural and environmental (Colchester 1989; Sponsel 1995).

Anthropology of knowledge and thinking through cultures

This short essay is about us. It is an attempt to interpret today's threats to indigenous groups and their local knowledge, and therefore it becomes the study of us. The ethnoscience, modernization and globalization theories undoubtedly tell us more about the nature of social scientists, anthropologists and ethnobotanists than they do about the actual situation of the 'natives' we are studying. We have cultural lenses through which we interpret the indigenous peoples and their local knowledge, and these tinge our way of thinking and acting in front of them.

One may 'think through' other cultures by means of the other (viewing the other as an expert in some realm of human experience); by getting the other straight (rational reconstruction of the beliefs and practices of the other); by deconstructing and going right through and beyond the other (revealing what the other has suppressed and kept out of sight); and witnessing in the context of engagement with the other (revealing one's own perspective on things by dint of a self-reflexive turn of mind) (Shweder 1991:2). The indigenous peoples can be regarded as a distant mirror in which we can see the strength but also the shortcomings of our own way of living. The 'thinking through' culture is a means for critical examination of our culture's basic assumptions on epistemology and ontology. We cannot deny that the very structure of our thinking about goals, the framework of our mind, is provided by our own culture. All knowledge has a social character, and the selection of knowledge is socially determined.

Another characteristic of Western culture in general, and its science in particular, is ethnocentrism – an attitude of being superior to other cultures. We perceive people in Asia, Africa and Latin America as 'The people without history'. Entire generations in these parts of the world have been taught to believe that they had no technical past or history to speak of prior to colonialization. What has been forgotten from the perspective of the Westerner is that these societies have survived for thousands of years – long before any contact with Western societies – many of them with well-developed knowledge systems including technologies such as irrigation systems, dams and terrace-building for agriculture, climate-appropriate house constructions, and systems for locating food and organizing their social groups. This is the 'hidden history' of non-Western people.

Knowledge is borne by human beings and it is thereby a part of their culture. Clifford Geertz (1973: 89) has formulated the conception of culture as, 'a historically transmitted pattern of meanings embodied in symbols; a system of inherited conceptions expressed in symbolic form by means of which men communicate, perpetuate and develop their knowledge about and attitudes towards life'. All knowledge is in some sense local knowledge, but here it refers to indigenous peoples' knowledge. It becomes visible when it is transformed into activities primarily related to agriculture, hunting, fishing and ethnomedical practices, but is also seen as part of the cosmology in myths and narratives. It is characterized as a form of knowledge that has evolved over the centuries by learning from experience and living close to nature.

An appealing description of local knowledge can be found in Turnbull's (1988) *The Forest People*. It deals with his visits among the pygmies of the inner regions of the Congo. It is written with an empathy for the lives and feelings of a people; about their love for their world and their trust in it. Turnbull emphasizes the 'we' and 'them' when he describes what is knowledge, what different human beings perceive, hear and see. Talking about the silence in the forest he notes that, 'these are the feelings of outsiders, of those who do not belong to the forest' (op. cit. 1988: 17). 'Even the silence felt by others is a myth' he stresses (*op. cit.* 1988: 17). The anthropologist is here a spokesman for the locals, describing what we Westerners miss, and what they – the pygmies – experience.

He conveys the meaning of local knowledge and its range, and its function as a prerequisite for these peoples' survival. 'The people of the forest say it is the chameleon, telling them that there is honey nearby. Scientists will tell you that chameleons are unable to make any such sound.' (op. cit. 1988: 18).This is what it is about: who knows what, and which knowledge is counted?'

The pygmies have lived in the forest for many thousands of years. Turnbull (1988) tells us with respectful voice that, 'They do not have to cut the forest down to build plantations, for they know how to hunt the game of the forest and gather the wild fruits that grow in abundance there, though hidden to outsiders' (op. cit. 1988:19). The cosmology of these people is revealed in front of us. The narrative we are told outlines the close relationship between people, forest, belief and their universe. A threat to the forest and the biodiversity that existed there is a threat to the people and their culture, and this is today's reality. The forest people probably don't exist today in the way that Turnbull experienced them during the 1950s. All societies change, in some sense, but have these societies changed according to their own prerequisites?

To Hunt In The Morning by Siskind (1975) is another of these brilliant anthropological narratives telling us about the life of the Sharanahua Indians in a remote region of the eastern part of the Peruvian tropical rainforest. She assigns the link between the local knowledge, language and belief system: the importance of the myths when constructing *malocas* (typical houses), during hunting, and when the shaman is healing illnesses or stabilizing the social order. But, with all its characteristics, it is a society in transition and people are becoming 'modern' in the sense that, according to Siskind (1975), they are releasing themselves from the collective traditions.

In the Amazon, indigenous ways of perceiving, utilizing and managing resources involve six different forms of what is called folk knowledge: gathered products, game, aquaculture, agriculture, resource units and cosmology (Posey *et al.* 1994). Likewise the value of TEK (Posey 1996) is shown for the Kayapó Indians in the Brazilian Amazon, whose agricultural knowledge and practices enhance biodiversity. The co-existence over a long time span of humans and natural resources creates what is called the 'anthropogenic landscape'. According to the anthropology of knowledge, and the contextualization of knowledge, this is an example of a development of a technology, knowledge and competence that has been suitable for the survival of the culture. Today there are abrupt breaks, for example among the *Kayapó* , caused by invading loggers, ranchers, and landless settlers (Posey 1996). In such circumstances the group with less power has to move further away from the invaders or adapt its knowledge and cosmology to new conditions in order to survive.

Globalization of local knowledge

Today a new element has slipped into the text – that TEK can be of great value for preserving the biodiversity of the tropical forest in the future, and that indigenous people are the equals of the conservationists. They are hereby becoming the 'noble ecological savage', saving the planet for our common future. This is the position the 'green movement' advocates. But this is not the only view in the present situation: the prevailing image in world society is the one found in the globalization tendencies that the world is one and that the market economy will solve the problems by giving wealth to everybody. The globalization image means giving an instrumental value to nature with consequences such as deforestation, mining and hydro-electric dam enterprises with ecologically disastrous effects for indefinite time. But also within the 'greening' discourse there seem to be two extreme ideas of which future we are aiming for: on the one hand the idea of throwing the whole developmental paradigm away, thus just seeing the negative impacts of modernization, and on the other hand, seeing the local ecological knowledge paradigm as a magical solution for the future of the indigenous people and for sustainable development. The two extremes have to be avoided. Escobar (1995: 170) writes: '...to embrace them uncritically as alternatives; or to dismiss them as romantic expositions by activists or intellectuals, who see in the realities they observe only what they want to see, refusing to acknowledge the crude realities of the world, such as capitalist hegemony and the like'.

The 'monocultures of the mind'

Vandana Shiva (1993) has paid attention to the phenomenon of the disappearance of traditional knowledge systems in her home country India, and she speaks of it as the 'monoculture of the mind'. She gives many examples of the loss of traditional knowledge in the encounter between Indian forms of local knowledge and the dominating Western knowledge, and she also shows different levels within society where it emerges. The most evident reason why local knowledge disappears is by it simply not being seen – its 'invisibility'. Its very existence is negated (Shiva 1993). She hereby stresses the power aspect of scientific knowledge as a part of Western dominance and the colonial heritage. One example Shiva (1993) gives of the delegitimizing of local knowledge concerns the seeds from the neem tree which had been used by Indian women as a medicine for hundreds of years. Western experts 'discovered' the pesticide capacity of the seed, and patented it. By valuing indigenous knowledge as 'unscientific' and thereby worthless, the scientists could commercialize an already available and recognized knowledge system. According to Shiva the fragmented linearity of the dominant knowledge disrupts the integration between systems. The two paradigms are cognitively and ecologically incommensurable (Shiva 1993: 20). The question is: is it as Shiva says, or is it possible to integrate the practical knowledge of traditional medicine into the discourse of scientific knowledge of biomedicine without any of the extreme biases – of throwing away the scientific knowledge and technology; of romanticizing the traditional knowledge and its technology; or of decontextualizing it by picking out pieces or fragments and using them as mere technical instruments.

Examples of the latter are the adoption of acupuncture as an instrument for treatment while throwing away its underlying ontological assumptions, and the use of medicinal plants as isolated drugs without putting them into their cultural context. It is obvious that we need a meta-theoretical platform from the anthropology of knowledge in order to analyse problems of the kind just mentioned. The central point is to stress the culturally specific in each case and to begin a dialogue: to try to initiate a form of interchange of experience and to find the type of knowledge that will best fit in the culture in question with its social and economic conditions and natural prerequisites.

Conclusions

Genocide, indigenous groups' human rights and other legal rights are, of course, the overwhelming threat. The focus of this contribution is the threat from the global capitalistic market economy, and indigenous peoples' loss of control of their knowledge and their not having access to sufficient land for their livelihood due to the increased pressure from deforestation by national and international companies, and mega-constructions of dams, mines, oil-drilling, hydro-electric plants etc. expropriating land and resources. These impacts represent an impairment and debasement of knowledge by violation of indigenous peoples' right to live in accordance with their own traditions. Today's large-scale economic activities have consequences for the loss of biological diversity by destructive use of natural resources.

Biotic impoverishment is a reality today, and biodiversity is a foundation of natural ecosystems. Globalization is a politically structured ideology, produced in world society within the prevailing economic growth theory. Indigenous groups all around the globe are in many places forced on to marginalized land. They thereby destroy the rainforest with their extensive slash-and-burn agriculture, and cultivation on steep slopes in mountainous regions causes erosion. But these are consequences of the globalization process --politics on an international or a national level. People are displaced, which is a violation of their right to live in accordance with the wishes of the group.

The monoculture of the mind, as Shiva (1993) calls the ethnocentric view that local knowledge is biased and exploited by science, has to be revealed. The myth of economic growth has to be questioned and the economic incentives to destroy ecosystems must be eliminated.

The threat of extinction exists. The narrow way of looking at science as the only form of knowledge must also be questioned. Anthropology of knowledge pleads for other world visions and other forms of knowledge and sources of knowledge to be considered. An awareness from world society that the global threats to our environment may represent an interest of indigenous groups and their local knowledge, who through millennium have survived, can be seen as a good sign. As scientists we must re-evaluate the idea of scientific knowledge as the only rational choice for a sustainable future. A widening of science to cover other forms of knowledge, together with less ethnocentrism and an aspiration for 'thinking through' other cultures to better understand our own way of thinking, might be a good starting-point.

Acknowledgement

This study is based on ongoing work within a project called 'Knowledge, Culture and Medicine' sponsored by the Swedish Council for Research in the Humanities and Social Sciences.

Indigenous Resource Management Systems (IRMS) (Janis Alcorn)

The term Indigenous Resource Management System (IRMS) includes local strategies, institutions and technologies of farming, herding, hunting, fishing and gathering. Local people often have a rich and detailed knowledge of local plants, animals and ecological relationships: sometimes called Traditional Ecological Knowledge (TEK) or Indigenous Knowledge Systems (IKS). People who have derived resource management systems appropriate to their local ecological and social situations are sometimes called 'ecosystem people', as opposed to 'biosphere people' (such as the urbanized citizens of industrial societies) who depend on resources imported from distant places (Dasmann 1984). The specific knowledge of ecosystem peoples is but one aspect of their resource management systems.

In most traditional societies, the earth is understood to be the source of all that is good. Local folklore warns of the misfortunes that befall those who fail to respect the earth, water, wildlife and trees. These values and beliefs are learned from relatives and neighbours as part of childhood experience. They are embedded in the local language, including songs and stories, and reflected in art. The value given to nature is evident in decision-making in all spheres of life. 'Making a living' and 'taking care of things' are not separated from 'conservation' as is the case in urban societies.

The successful evolution and functioning of an IRMS depend on shared cultural values, social rules, and systems for conflict management that have local legitimacy. In other words, indigenous resource management systems cannot be separated from other aspects of life in those areas where people depend on their immediate environment for their livelihood. Appropriate social behaviour includes appropriate behaviour toward nature, for example, correct ways to hunt animals, showing respect for the prey and its family, etc.

Agro-ecosystems

The land and waters of a group (or coexisting groups) form the agro-ecosystem within which their IRMS operate. Usually, IRMS maintain wild species and their habitats in some parts of the group's territory, while altering habitats in other areas to favour the growth of crops and livestock. For example, wild species such as edible grubs, caterpillars and termites are often managed, in so far as their food plants and/or other habitat requirements are maintained or encouraged within the agro-ecosystem. In some cases, farmers increase the food plants of insects or game by planting or protecting them. In other cases, their habitats are normal by-products of the farming system, such as secondary growth in fallows of swidden systems.

When they make farming decisions, farmers weigh the benefits of maintaining the habitats of useful insects and game. Agroforestry systems (the inclusion of trees in agricultural systems) are a common, complex type of IRMS, in the humid tropics; such systems may maintain trees among crops, create successional situations where trees follow crops in a given field, or maintain forest patches separate from fields (particularly where watershed management is a concern).

Through their IRMS, communities and individuals, 'embroider on a canvas of nature' (De Schlippe quote on the Zande) to create mosaics over the existing ecological diversity. Some of the 'embroidery' is very subtle. The mosaics are made of many more pieces than 'residential areas' and 'agricultural fields' (those which outsiders usually recognize as areas under management). Areas important for fishing, hunting game and gathering fuelwood, medicine, artisan's materials and other wild products are often very important for local livelihoods and regulated by subtle mechanisms such as the rules governing inheritance of tenurial rights to use particular areas or resources. Often it is these zones that are 'cut out' of village territories during demarcation of protected areas. Such alienation undermines the traditional dispute-management regime (including the authorities who traditionally allocated rights to resources in those areas) and undermines existing curbs on land/resource use.

Among ecosystem people, local feedback leads to recognition of resource over-exploitation. The response may be to substitute another species, if one is available, or the feedback may lead to taboos on the use of a species or its exploitation. In some cases, despite feedback, over-exploitation may lead to local extinction of a species, particularly if a substitute is not available. But if damage to the ecosystem becomes clearly visible, a shift in livelihood strategy is also likely to occur.

Institutions and tenure

In response to feedback and tensions among individuals seeking access to resources, institutions have arisen to ensure continued community access to resources and to restrict their use by outsiders. These institutions result from a political process of trade-offs among members of a community who must work together because of their interdependence. These are often referred to as 'communal property' management systems, since access to the resource in question is regulated by the local group of individuals, as opposed to public property claimed by the state. Communal property regulation is integrated into a broader community system that defines and allocates individual and group rights to particular resources within the lands held by the community.

Tenure refers to a bundle of rights and responsibilities in regard to specified resources – who can, and who can't, do what with which resource. The effectiveness of tenure systems depends on their widespread acceptance and adherence to rules governing access; on the strength of local institutions and organizations that administer local justice; and on the guidance of local leaders committed to the values of the system. Some have described traditional tenurial systems as a form of 'institutional capital', because compliance is sustained with a low investment in enforcement (Field 1984). Within a given community, some rights to resources may be close to individual ownership (for example they may include the right of inheritance). At the same time, rights to other resources may be shared within the community: a form of communal property. Often, while farmers have individual tenure over crop land, the community recognizes the rights of all community members to gather from non-cultivated lands; or private individual rights (usually given to a kin group or family) may hold for part of a year, while community members have the right to use the space and its resources at other times.

Farmers may have rights to their crop yields, but may also recognize the rights of migratory pastoralists over wild forage and crop residues after harvest, as well as the rights of anyone to collect medicinal plants from the field at any time of the year. In addition, rules often regulate the harvest of particularly valuable wild resources. There may be also communal labour obligations for maintaining wild resources. Enforcement is often by social pressure, but may include stiff penalties determined by a group of elders or other traditional authoritative body. In societies that rely primarily on gathering, different groups may have rights to particular geographic areas, but still allow neighbouring groups to use their territories when local weather conditions reduce productivity during particular years. Communities often recognize reciprocal rights to share the resources of other communities during times of famine or social unrest.

While communal property is a type of tenure, in traditional societies the community to which the resources belong includes the ancestors, the spirits, and the unborn, as well as the living people of a community. These resources are part of a unit that includes living things, air, water, land, forest, reefs and the subsurface space. Rituals often mark the boundaries of the lands and waters belonging to the community. An individual's rights to community membership and hence to community resources are usually determined through kinship. Disputes over 'who' has rights to 'what resources' for 'what purposes' are resolved locally through dispute-resolution mechanisms that evolve as the community changes with time.

Management rules

Local rules that restrict who uses how much of a biological resource require effective local social institutions, accepted rights and obligations, and a shared vision for interpretation and action. Traditional conservation ethics support local tenurial institutions using social pressures to influence an individual's decisions and encourage compliance. This involves not only people vested with authority within particular local organizations designed to regulate resource access, but also includes local curers and diviners who use shared ethics to identify and apply social pressure against those who break the rules. The effectiveness of the tenure system depends on widespread acceptance of and adherence to rules governing access, strong local institutions to administer local justice, and guidance by local leaders committed to the values of the system itself.

Farmers, fishers and pastoralists generally value the diversity of available ecological zones and allocate resource use in ways that are both: (a) conscious of the spatial, distributional and ecological consequences on the broader landscape-wide mosaic; and (b) conscious of the social impacts of resource distribution on individuals and on the community at large. An IRMS can include rules for allocation of resources within a community and/or between communities. Less obvious rules, such as those involving marriage, may reinforce the desired resource allocation. For example, Tukanoan fishing communities in the rich waters of the upper Amazon are responsible for distributing fish to other Tukanoan communities with few fishery resources (Chernela 1993). Marriage rules require out-marriage between resource-rich and resource-poor villages and support reciprocity.

Maintaining watershed forests and fishing waters often requires co-operation within and among villages that share access to the resource. Annual rituals are often used to reaffirm villagers' respect for nature and the spirits that will punish them if they damage nature. Offerings are made to the forest and water spirits. These annual rites reaffirm villagers' commitment to each other and to everyone's right to enough of the resource for subsistence needs. The rituals may initiate formal meetings where people discuss substantive issues and disputes that have occurred during the previous year. They may also provide opportunities to amend regulations about resource distribution, maintenance of infrastructure (such as barriers and canals), conflict management and watershed forest preservation.

IRMS manage game-hunting in a variety of ways similar to those used for regulating extraction of wild plants, e.g. through regulating the number of hunters (as in the lineage husbandry described by Marks (1994)), restricting access to areas that can be hunted or fished, or establishing the seasons when hunting is allowed. Hunting, like gathering, depends on the availability of natural habitat as part of the land-use mosaic. Hence, management of agricultural areas also affects game management. IRMS regulation of hunting is not well-documented, but some surprisingly complex systems have been found. Traditional use of fire technology for game management,

Box 5.3: Caste and Resource Management

In India, IRMS often require co-operation among castes. These are endogamous groups who are bound to each other by kinship, reciprocal obligations and customs, and have a particular profession that relies on a particular set of resources. Access to the resources necessary for a given profession is restricted to a particular caste, which in turn functions as an essential element of the larger society. For example, in Uttara Kannada District of Karnataka state, there are 19 castes (Gadgil & Iyer 1989). People who fish from boats are divided into three castes, which use three different areas for fishing: river, estuary and coast. The sub-clans of each of these castes use particular territories within their caste's larger territory.

Three castes are agriculturalists who also collect shellfish, and hunt mammals and birds. Each agriculturalist caste has other special differences; two weave mats but from different species of plant. In addition, there is one horticulturist caste, two entertainer castes (one of which taps toddy palms), barbers, washermen, artisans (potters, goldsmiths, carpenters, blacksmiths, lime-makers, stone-workers, tanners and basket-weavers), and traders. Bamboo is reserved for the use of hide tanners, and deer can only be hunted by one of these 19 castes.

All the castes work together in close-knit villages. Prior to colonization, the castes managed their common forest and fishery resources successfully through their local IRMS.

for example, relies on ecological knowledge in order to use fire in the right season and right places in order to maintain the desired mosaic of micro-habitats. In northern Alberta, Canada, trappers maintain 'fire yards' (meadows, prairies and small forest openings) and 'fire corridors' (banks of streams, lake shores, sloughs and trails) to create micro-habitats as a 'fire mosaic' in the larger forested landscape (Lewis 1989). Aborigines in northern Australia use fire to create a mixture of different ages of successional situations within a given habitat type, and to protect other sites from fire, so enhancing the available range of habitats for game (Lewis 1989).

Sacred forests may be strategically placed to cover different ecological zones where they provide a haven for animal reproduction and other wild resources, especially medicinal plants that are not used in large quantities. Societies often delimit sacred forests in areas where most natural forest has been cut down for agriculture, degraded by overgrazing, or threatened by other land-use changes. Sometimes these forests serve as burial grounds for high-status individuals, or as the grounds of temples or homes of spirits. IRMS include a variety of means for creating and maintaining crop and livestock genetic diversity through social mechanisms such as seed trading networks, lineage ownership, etc. as well as through specific local techniques for propagation and experimentation.

Conservation as culture

While the techniques and tools of resource management are easily seen, and some aspects of traditional knowledge are easily documented, direct discussion of 'resource management' is not usually a productive way to understand local IRMS. Local people often do not view nature as a bundle of resources; there may be no translation of the term 'resource' in their language. IRMS themselves are rarely visible and labelled in local languages. Their patterns, however, can be identified by studying the landscape, and by exercises in which the outsider attempts to make the choices necessary to carry out livelihood activities as if he or she were a naive new member of the community.

Social factors are the most fragile components of IRMS; they are most susceptible to damage or loss due to changes in apparently unrelated spheres of local life. Formal schooling and loss of local language are among the most radical agents of change. Cultural values that support IRMS are shared and passed on to younger people through songs, stories, ritual texts and other verbal communications in the local language. When language is changed, the new values of the new language are adopted. These new values often do not support the old ways. Second, increasing influence of the market economy has a profound indirect influence on IRMS through transformation of non-monetary values into monetary values. It introduces the idea that land, labour and nature are commodities, instead of a sacred heritage that binds the members of the community to one another. The labour requirements of IRMS (building communal fish traps, patrolling forest areas, serving as game bosses, etc.) often involve reciprocal exchanges. If people choose instead to take jobs for pay, then IRMS may fail for lack of contributed labour.

The loss of authority of elders' councils and other traditional decision-makers is the third critical threat to IRMS. When the central government imposes a new local government and fails to recognize the tenurial rights of communities as mediated by traditional governing bodies, then the traditional rules regulating resource access lose their legitimacy. When a community's legitimacy as an authority has been usurped by the state, community property becomes no one's property. Hardin's famous (1968) *Tragedy of the Commons* model describes the problems of open access that can occur in this way.

Where social and economic conditions are in flux, migrants, contract labour or people resettled by states often enter areas claimed by other communities. They may not recognize the subsistence values of many species, and may harvest known valuable wild products using a 'deplete and switch' strategy. This type of activity often occurs along roads opened by logging or mining companies. The immigrants may be able to run IRMS appropriate to their old resources, but they often lack the knowledge and institutions necessary to manage the new sets of resources. Their activities often come into conflict with the IRMS of the original residents.

Diversity and sustainability in community-based conservation (Michel P. Pimbert and Jules N. Pretty)

Top down, imposed conservation all too often entails huge social and ecological costs in areas where rural people are directly dependent on natural resources for their livelihoods. A growing body of empirical evidence now indicates that the transfer of 'Western' conservation approaches to the developing countries has indeed had adverse effects on the food security and livelihoods of people living in and around protected areas and wildlife management schemes (Ghimire and Pimbert 1997; Ghimire 1994; Kothari *et al.* 1989; IIED 1995; Wells and Brandon 1992; West and Brechin 1991). On several occasions, local communities have been expelled from their settlements without adequate provision for alternative means of work and income. In other cases, local people have faced restrictions on their use of common property resources for food gathering, harvesting of medicinal plants, grazing, fishing, hunting, and the collection of wood and other wild products from forests, wetlands and pastoral lands. National parks established on indigenous lands have denied local rights to resources, turning local people practically overnight from hunters and cultivators to 'poachers' and 'squatters' (Colchester 1994).

Resettlement schemes for indigenous peoples removed from areas earmarked for conservation have had devastating consequences. So have the coercive wildlife conservation programmes that were implemented by the former pro-apartheid governments of Rhodesia (Zimbabwe) and South Africa (McIvor 1997; Koch 1997). Denying resource-use to local people severely reduces their incentive to conserve it. Moreover, the current styles of protected area and wildlife management usually result in high management costs for governments, with the majority of benefits accruing to national and international external interests. All these trends may ultimately threaten the long-term viability of conservation schemes as local populations enter into direct conflict with park authorities and game wardens.

This deep conservation crisis has led to the search for alternative approaches that re-involve local communities in the management of wildlife and protected areas. 'Community-based conservation' and 'peoples' participation' have indeed become part of the conventional rhetoric and more attention is being paid to this approach on the ground by international and national conservation organizations. There are now several examples of projects that involve local communities and seek to use economic incentives for the conservation and sustainable use of wildlife and protected areas (Kiss 1990; McNeely 1988; Sayer 1991; Stone 1991; Wells and Brandon 1992). However, the practice of community-based conservation remains problematic because of its high dependence on centralized bureaucratic organizations for planning and implementation. Some of these initiatives are nothing more than 'official accommodation responses' to the growing opposition to parks and local resource alienation in forests, wetlands, grasslands, mountains, coasts and other biodiversity-rich sites. None the less, a few of them are clearly challenging the dominant conservation approaches and seem to be based on more equitable power- and benefit-sharing arrangements (for recent reviews see IIED 1995; Borrini-Feyerabend 1996). But these more progressive initiatives are limited in number and scope. They are still relatively isolated examples in mainstream conservation practice.

Community-based conservation: from blueprints to process

There are few examples of community-based conservation based on indigenous knowledge and rule-making institutions. A recent survey of forest conservation implemented by WWF International in Africa, Latin America, Asia and Europe has shown that there are relatively few community-based programmes in which there is significant devolution of power to local people (Jeanrenaud, pers. comm. 1997, and 1998). The situation is similar in wetlands, mountain areas and other ecological contexts where rural people live.

The way conservation bureaucracies and external institutions are organized, and the way they work, currently inhibit this devolution of power to local communities. The methods and means deployed to preserve areas of pristine wilderness largely originated in the affluent West, where money and trained personnel ensure that technologies work and that laws are enforced to secure conservation objectives. During and after the colonial period, these conservation technologies, and the values associated with them, were extended from the North to the South – often in a classical top-down manner. Positivist conservation science and the 'wilderness preservation' ethic hang together with this top-down Transfer-of-Technology model of conservation. They are mutually constitutive elements of the blueprint paradigm which still informs much of today's design and management of protected areas and wildlife schemes in developing countries (Pimbert and Pretty 1995).

The main actors in this approach are normal professionals who are concerned not just with research, but also with action. Normal professionals are found in research institutes and universities as well as in international and national organizations where most of them work in specialized departments of the government (forestry, fisheries, agriculture, health, wildlife conservation, administration). The thinking, values, methods and behaviour dominant in their profession or discipline tends to be stable and conservative. Lastly, normal professionalism generally, 'values and rewards 'first' biases which are urban, industrial, high technology, male, quantifying, and concerned with things and with the needs and interests of the rich' (Chambers 1993).

Conservation usually reflects the priorities of regional, national and above all international interests over local subsistence needs. The design, management and infrastructure of protected areas and wildlife schemes all too often reinforce the interests of global conservation and those of the international leisure industry and other commercial groups. Local people often express their sense of deep frustration with these externally imposed priorities by saying that, 'people should be considered before animals' (Hackel 1993), and they often view wildlife conservation as, 'alien, hypocritical, and ... favouring foreigners' (Munthali 1993).

Declaring biodiversity-rich areas 'internationally important' conservation sites is meaningless for local resource-users if the issues that emerge from such declarations have not been discussed and resolved to the satisfaction of local communities. Farmers and forest dwellers who have lost land and/or traditional rights over resources cannot appreciate the value of vague 'long-term' conservation benefits for society or humanity. In their view, conservation benefits should be immediate and quantifiable, with local people getting a fair share of the benefits accruing from the successful management of the protected area and wildlife schemes.

A radical shift is required, from imposed conservation which aims to retain external control of the management and end uses of biological resources to an approach that devolves more responsibility and decision-making power to local communities. Community-based conservation is likely to be sustainable ecologically, economically and socially only if the overall management scheme can be made sufficiently attractive to local people for them to adopt it as a long-term livelihood strategy (Pimbert and Pretty 1995). In that context, dialogue, negotiation, bargaining and conflict resolution are all integral parts of a long-term participatory process which continues well after the initial appraisal and planning phases.

Existing conservation institutions and professionals need to shift from being project implementors to new roles which facilitate local people's analysis, planning and action. The whole process should lead to local institution-building or strengthening, so enhancing the capacity of people to take action on their own. This implies the adoption of a learning process approach in conservation and a new professionalism with new concepts, values, participatory methodologies and behaviour.

Reversals for community-based conservation

To spread and sustain community-based conservation, considerable attention will have to be given to the following needs, social processes and policies.

1. Debunk the 'wilderness' myth and reaffirm the value of historical analysis

Most parts of the world have been modified, managed and in some instances improved by people for centuries. The very biodiversity which conservationists seek to protect may be of anthropogenic origin since there is often a close link between moderate intensities of human disturbance and biodiversity. Much of what has been considered as 'natural' in the Amazon is, in fact, modified by Amerindian populations (Posey 1993). Indigenous use and management of tropical forests is best viewed as a continuum between plants that are domesticated and those that are semi-domesticated, manipulated or 'wild', with no clear-cut demarcation between natural and managed forest. Species richness and the abundance of wildlife in indigenous peoples' agriculture in the Sonoran Desert (USA) is greater than that in adjacent or analogous habitats that are not cultivated (Reichardt *et al.* 1994), and in agricultural landscapes it is mainly local people who create and manage biological diversity (Haverkort and Millar 1994; Salick and Merrick 1990).

Many of the areas richest in biological diversity are inhabited by indigenous peoples who manage, maintain and defend them against destruction (Alcorn 1994; Colchester 1994). Ethnoecological studies are increasingly discovering that what many had thought were wild resources and areas are actually the products of co-evolutionary relationships between humans and nature (Gomez Pompa and Kaus 1992; Pimbert and Toledo 1994; Posey 1994). UNESCO (1994) introduced the term 'cultural landscapes'. Designating landscapes and the species they contain as 'cultural' has a number of important implications for community-based conservation and the concept of rights over biological resources. Local communities may therefore claim special rights of access, decision, control and property over them. This historical reality should be the starting point of community-based conservation wherever local people have shaped local ecologies over generations. To transcend the 'wilderness myth', community-based conservation must begin with the notion that biodiversity-rich areas are social spaces, where culture and nature are renewed with, by and for local people (Ghimire and Pimbert 1997).

2. Strengthen local rights, security and territory

Colonial powers, international conservation organizations and national governments have a long history of denying the rights of indigenous peoples and rural communities over their ancestral lands and their resources. For example, most of the very large area earmarked for conservation in Costa Rica is under a strictly protected regime that excludes local communities, unlike in Germany and France where protected area regimes represent more of 'social compromise' (Bruggermann 1997; Finger and

Ghimire 1997). This negation of the prior rights of indigenous and other local communities has been one of the most enduring sources of conflict and violence, both in the developing world and in industrialized nations such as Canada where aboriginal people seek greater self-determination by regaining control over territories now enclosed in the country's protected area network (Morrison 1997). Denying resource-use to local people severely reduces their incentive to support conservation and undermines local livelihoods. Policies for community-based conservation clearly need to reaffirm and protect local rights of ownership and use over biological resources for ethical as well as practical reasons. Two immediate priorities in many developing countries would be to:

- reform protected area categories and land-use schemes to embody the concepts of local rights and territory in everyday management practice, and
- strengthen local control over the access to and end uses of biological resources, knowledge and informal innovations.

3. The need for genuine peoples' participation and professional reorientation in conservation bureaucracies

Despite repeated calls for peoples' participation in conservation over the last twenty years, the term 'participation' is generally interpreted in ways which cede no control to local people (Forster 1973; III World Parks Congress 1982; McNeely 1993). It is rare for professionals (foresters, protected area managers, wildlife biologists) to relinquish control over key decisions on the design, management and evaluation of community-based conservation. Participation is still largely seen as a means to achieve externally-desirable goals. This means that, whilst recognizing the need for peoples' participation, many conservation professionals place clear limits on the form and degree of participation that they tolerate in protected area and wildlife management.

4. Build on local priorities, the diversity of livelihoods and local definitions of well-being

From the outset, the definition of what is to be conserved, how it should be managed, and for whom, should be based on interactive dialogue to understand how local livelihoods are constructed and people's own definitions of well-being. Participatory, community-based conservation starts not with analysis by powerful and dominant outsiders, but with enabling local people, especially the poor, to conduct their own analyses and define their own priorities. Whilst the above examples of professional biases are also rampant in the wider community of development planners, economists and agricultural scientists, the problem is compounded in public and private conservation organizations because they have few, if any, sociologists or anthropologists working in the field or at headquarters. Clearly this must change (Chambers 1993).

5. Build on local institutions and social organization

Local organizations are crucial for the conservation and sustainable use of biodiversity. Local groups enforce rules, incentives and penalties for eliciting behaviour conducive to rational and effective resource conservation and use. For as long as people have engaged in livelihoods pursuits, they have worked together on resource management, labour-sharing, marketing and many other activities that would be too costly, or impossible, if done alone. Local groups and indigenous institutions have always been important in facilitating collective action and co-ordinated natural resource management providing striking evidence of active conservation. These institutions include rules about use of biological resources and acceptable distribution of benefits; definitions of rights and responsibilities; means by which tenure is determined; conflict resolution mechanisms, and methods of enforcing rules, cultural sanctions and beliefs (Alcorn 1994). Similarly, the literature on common property resources highlights the importance and resilience of local management systems for biodiversity conservation and local livelihoods (Arnold and Stewart 1991; BOSTID 1986; Bromley and Cernea 1989; Ostrom 1990; Jodha 1990; Niamir 1990).

In most societies, however, the composition of institutions is likely to reflect and reinforce imbalances of power, with the weaker and underprivileged social groups being least represented in decision-making structures. Most communities show internal inequities and differences, based on ethnic origin, class, caste, economic endowments, religion, social status, gender and age. These inequities can create profound differences in interest, capacity and willingness to invest for the management of natural resources. Under these circumstances, what benefits one group and meets some conservation goals may harm others and the natural resources on which local livelihoods depend. Ensuring fair and equitable representation of different community interest groups in local institutions will often require considerable negotiation, bargaining and conflict resolution both within and between local communities. At times, it may be necessary to support the formation of new, more equitable, local institutions which some community groups (e.g. women, marginalized people) may want.

6. Locally available resources and technologies to meet fundamental human needs

Community-based conservation that seeks to provide benefits for local and national economies should give preference to informal innovation systems, reliance on local resources, and local satisfiers of human needs. Preference should be given to local technologies by emphasizing the opportunities for intensification in the use of available resources. Sustainable and cheaper solutions can often be found when groups or communities are involved in identification of needs, design and testing of technologies, their adaptation to local conditions and, finally, their extension to others. The potential for intensification of internal resource use without reliance on external inputs is enormous. Greater self-reliance and reduced dependency on outside supplies of pesticides, fertilizers, water and seeds can be achieved within and around protected areas by complicating and diversifying farming systems with locally available resources.

Similarly, health, housing, sanitation and revenue-generating activities (e.g. tourism) based on the use of local resources and innovations are likely to be more sustainable and effective than those imposed by outside professionals. There are, nevertheless, opportunities to combine the strengths of modern science and local traditions of knowledge in some contexts. The advantages and skills of professionals (at the micro- and macro-levels) can be effectively combined with the strengths of indigenous knowledge and experimentation by empowering people through a modification of conventional roles and activities. This participatory research and development would permit the generation of diverse, locally-controlled technologies which may be more sustainable in the long term than the classical Transfer-of-Technology approach.

7. Economic incentives and policies for the equitable sharing of conservation benefits

Many of the schemes designed to provide local economic incentives for community-based conservation need to pay greater attention to equity and human rights issues. Community wildlife management and participatory protected area management have little chance of success where benefits are not distributed equitably among various members of the community. 'Equity' should entail the sharing of benefits in a way that is commensurate with the varying sacrifices and contributions made by, or damages incurred in, the community (e.g. through lost access to resources, damage to crops, and through the physical danger presented by many wild animals). The distribution of benefits within the community should also be administered by a local institution that carries out its activities in a transparent way and is accountable to the community.

8. Build on local systems of knowledge and management

Local management systems are generally tuned to the needs of local people and often enhance their capacity to adapt to dynamic social and ecological circumstances. Although many of these systems have been abandoned after long periods of success, there remains a great diversity of local systems of knowledge and management which actively maintain biological diversity in areas earmarked for conservation (Kemf 1993; West and Brechin 1991). Such systems are sometimes rooted in religion and the sacred. Sacred groves, for example, are clusters of forest vegetation that are preserved for religious reasons. They may honour a deity, provide a sanctuary for spirits, or protect a sanctified place from exploitation; some derive their sacred character from the springs of water they protect, from the medicinal and ritual properties of their plants, or from the wild animals they support (Chandrakanth and Romm 1991). Sacred groves are common throughout Southern and South-eastern Asia, Africa, the Pacific Islands and Latin America (Shengji 1991; Ntiamoa-Baidu *et al.* 1992). The network of sacred groves in countries such as India has since time immemorial been the locus and symbol of a way of life in which the highest biological diversity occurs where humans interact with nature. As Apffel Marglin and Mishra (1993) note, sacred groves are preserved by villagers: 'not because it represents the antithesis of their productive activities but because it safeguards their livelihoods and their continued existence. When the commons of local communities are still protected by the Goddess, nature's diversity is preserved'. Clearly these pockets of biological diversity could be the focus for the conservation and regeneration of forest cover, so perhaps forming the basis of more 'culturally appropriate' protected areas.

Some indigenous peoples and rural communities have established protected areas that resemble the parks and reserves codified in the CNPPA's system and in national protected area policies. In Ecuador, for example, the Awa have spontaneously decided to establish conservation areas. They have secured rights over a traditional area, which has been designated the Awa Ethnic Forest Reserve (Poole 1993). Sacred places such as the Loita Maasai's 'forest of the lost child' in Kenya (Loita Naimana Enkiyio Conservation Trust 1994) are also widespread forms of vernacular conservation. This form of conservation is based on site-specific traditions and economies; it refers to ways of life and resource-utilization that have evolved in place and, like vernacular architecture, it is a direct expression of the relationship between communities and their habitats (Poole 1993). The similarities between vernacular and scientific models of conservation, however, obscure the fact that motivations for setting up such areas are quite distinct from those leading to national parks and wildlife management schemes, even though the ultimate contribution to biodiversity conservation may be identical.

The crucial distinction is that such areas are established to protect land *for* use rather than *from* use; more specifically for local use rather than appropriation and exploitation by outside interests. Indigenous ways of knowing, valuing and organizing the world must not be brushed aside by so called 'modern' technical knowledge which claims superior cognitive powers. Despite the pressures that increasingly undermine local systems of knowledge and management, community-based conservation should start with what people know and do well already, so as to secure their livelihoods and sustain the diversity of natural resources on which they depend.

Conclusions

Sustainable and effective conservation calls for an emphasis on community-based natural resource management and enabling policy frameworks. These are not the easy options. Contemporary patterns of economic growth, of modernization and nation-building all have strong anti-participatory traits. The integration of rural communities and local institutions into larger, more complex, urban-centred and global systems often stifles whatever capacity for decision-making the local community might have had and renders its traditional institutions obsolete. This contribution has nevertheless tried to identify some of the key social issues and processes that could be acted upon to decentralize control and responsibility for conservation and natural resource management. It should be emphasized here that the devolution of conservation to local communities does not mean that state agencies and other external institutions have no role. A central challenge will be to find ways of allocating limited government resources so as to obtain widespread replication of community initiatives. Understanding the dynamic complexity of local ecologies, honouring local intellectual property rights, promoting wider access to biological information and funds, designing technologies, markets and other systems on the basis of local knowledge, needs and aspirations call for new partnerships between the state, rural people and the organizations representing them.

Building appropriate partnerships between states and rural communities requires new legislation, policies, institutional linkages and processes. Community-based conservation is likely to be more cost-effective and sustainable when national regulatory frameworks are left flexible enough to accommodate local peculiarities. It requires the creation of communication networks and participatory research linkages between the public sector, NGOs and rural people involved in protected area and wildlife management. Legal frameworks should focus on the granting of rights, access and security of tenure to farmers, fishermen, pastoralists and forest dwellers. This is essential for the poor to take the long-term view. Similarly, the application of appropriate regulations to prevent pollution and resource-degrading activities is essential to control the activities of the rich and powerful, e.g. timber, bio-prospecting and mining companies. Economic policies should include the removal of distorting subsidies that encourage the waste of resources; targeting of subsidies to the poor instead of the wealthy, who are much better at capturing them; and encourage resource-enhancing rather than resource-degrading activities through appropriate pricing policies.

Such changes will not come about simply through the increased awareness of policy-makers and professionals. They will require shifts in the balance of social forces, power relations and economic organization. Indeed, the implementation of community-based conservation invariably raises deeper political questions about our relationship with nature and how we organize society – towards more centralization, control, uniformity and coercion, or towards more decentralization, democracy, diversity and informed freedom.

Agriculture and cosmovision in the contemporary Andes: the nurturing of the seeds (Tirso Gonzales, Nestor Chambi and Marcela Machaca)

In the same way as there is not just one way of doing agriculture, the seed does not mean the same in every language and, by extension, cosmovision. The seed does not mean the same, nor does it have the same role, in Western contemporary societies as it does in 'indigenous' agricultures. Capitalist agriculture, commercial seeds (hybrid or improved) and scientific agricultural knowledge are not what Quechua and Aymara agricultures are about. This contribution is an invitation to approach both worlds in their own terms, and this means acknowledging that the seed is part of, and is related to, different ways of being (ontologies), different ways of knowing (epistemologies), and different ways of being related to the world (cosmologies).

For these reasons, it is necessary to acknowledge that terms such as 'traditional farmers/agriculture', 'modern farmers', '*in situ* conservation' of 'plant genetic resources', 'germplasm' and 'agrobiodiversity', among others, are not neutral, nor are they universal. They are part of a set of Western concepts related to theories of modernization and development, rural and agricultural development blue prints, or conservation strategies. These terms are coherent with the contemporary Western ways of being, ways of knowing, and ways of being related to the world. However, the Western cosmovision informing the three ways noted do not necessarily inform the ways of knowing, being, and being related to the world of non-Western 'indigenous peoples'. This crucial recognition should highlight the problems embedded in Western blueprints of development proposed for, or forced upon, indigenous peoples' lives, territories or environments throughout this century.

Pueblos Originarios (Original Peoples) such as the Quechuas and Aymaras in the Andes, acknowledging their particular cultural and institutional diversity, share, in general, a rich and unique cosmovision that is far removed from the contemporary Western one. Today the West is starting to articulate and search for alternatives such as 'alternative agriculture', 'sustainable agriculture', 'organic agriculture' as a way to make a transition or move out of conventional, non-sustainable, anti-environment capitalist agriculture. The principles of 'sustainable agriculture' are found, at least in part, in the agricultures of the *Pueblos Originarios* which, as in the case of the Andes, are 8,000 years old. The appropriate Western institutions should make a wider and more open recognition of the paradigms embedded in non-Western peoples such as the Quechuas and Aymaras, thus favouring the strengthening of the search for sustainable ways of doing agriculture, such as those present all over the Americas (North, Meso and South) and the world over.

There are about 6,000 languages in the world, of which about 5,000 are indigenous. Of the 5.5 billion people living upon mother earth, between 200 and 600 million are indigenous peoples (Durning 1992). The current indigenous population of the Americas is around 42 million, with a total of 900 languages.

In Peru, there are about 57 indigenous ethnic groups, with a total population of more than 9 million people. In the Peruvian Andes the organizational unit is the *ayllu* and in the case of the Aymara people from Conima, Puno, currently, the *ayllus* have been divided into *Comunidades Campesinas* (Peasant Communities) and *Parcialidades*. Each *ayllu* divides into three or four *Parcialidades*. Previously denominated *Comunidades Indígenas*, in 1992 there were 4,976 *Comunidades Campesinas* officially recognized by the Peruvian State. Despite destructive colonial policies and contemporary development policies applied by the nation-state, the Peasant Communities possess their own institutions, rituals, religions, languages, cultures and customary laws, as well as their own 'philosophical' terms: ways of being, knowing, and being related to the world.

Cosmovision and the culture of *crianza* (nurturing) in the Andes

In the Andes, Quechuas and Aymaras have always talked of the nurturing of life. In our world everything is alive; nothing is excluded. We all are relatives. Our *Pachamama* (Mother Earth) is sacred and alive. That is why we may talk about the living Andean world. Life in the Andes is a culture of the nurturing of harmony. Harmony is not given; we have to procure it. Contrary to what happens in the Western contemporary world, the agricultures in the Andes are important for the continuance of life. In our localities we all accompany each other so that in this way life may continue to flow and regenerate. The living world that is the Andes has three principal components: the community of *sallqa* (nature), the community of *runas* or *jaques* (humans), and the community of *wacas* (deities). The *chacra* (or plot of land for Andean cultivation) is the place, par excellence, where the nurturing is rendered in the most complex and intense way, among the three communities. This is possible because in the Andes the members of the *sallqa*, the *runas* or *jaques*, and the *wacas* (*Pachamama, Achachilas, Mallkus, Kuntur Mamani, Serenos*, local and universal deities) are persons, have the attributes of living beings, and 'find themselves intimately related' (Chambi y Chambi 1995). *Sallqa, runas* and *jaques*, and *wacas* don't live independently: on the contrary, each one of these forms is complementary; they depend upon and nurture one another.

In the Andes, nurturing is reciprocal. When we practice reciprocity, we make *ayni*. One must know how to nurture to be deserving of the nurturing of others. Those who nurture (mountains, water, clouds, *runas*, *jaques*, plants) are at once those being nurtured: *uywaypaq uywanchik*, 'nurturing we nurture' (Machaca 1996). This takes place in this manner because what is dealt with here is a world of equivalents, where everyone is a person, and each one has dignity, and treats every one with respect and esteem. Additionally, it is important to remember that:

Figure 5.2: The Andean cosmovision.

'any person (whether they be man, tree, or rock) may present themselves in any of the three forms, when it is convenient as the physiology of the animal world is rendered in the nurturing of the harmony which sustains it. ... In this way, each one of these forms nurtures the other two, and is nurtured by them in turn. This is so because the Andean world is not a world of things, of objects, of institutions, of cause and effect relationships, but rather we are in the presence of a world of renderings, recreations, of renovations' (Grillo 1993: 5).

Conversation takes place in a reciprocal manner among the members of the three communities which compose and nurture the local *pacha* (earth). 'Conversation' is not a metaphor. Conversation denotes that the beings that communicate with one another do so in this way because they are able to understand one another. The term 'conversation' includes every form of expression, whether it be feelings, emotions, diverse manifestations – not necessarily conveyed in speech, as occurs with dialogue. Conversation is possible as it is rendered between equivalent and incomplete beings. Life in its entirety, that is to say, the three basic communities of the *Pacha* are regenerating themselves in every instance.

As Van Kessel y Condori Cruz (1993: 17) noted: 'In this affectionate and respectful conversation, [the *sallqa*, the *wakas*, and the *runa/jaque*] fill with life and flourish. This nurturing is symbiotic: at the same time as the *chacra*, cattle, water is nurtured, these nurture the [*runa/jaque*] giving [them] life and making [them] flourish. A similar mutuality develops between the community of *wakas* and the community of humans: while the first, headed by *Pachamama*, feeds the human life, the second feeds the divinities by means of its *wilanchas* and its *mesas*. It is the 'offering to the earth', according to the principle of reciprocity. Good labour in the *chacra*, responsible and dedicated, is another way to feed the earth, 'nurturing *chacra*' and producing the fertility of the earth.'

The *ayllu* is a kinship group, but it is not restricted to human lineage: it includes each member of the local *pacha* (local landscape) (*cf.* Grillo 1993, Figure 1).The *ayllu* in the Aymara *pueblo* (village) of Conima, Puno, in its spatial aspect, is divided into *ayllus* for a better conversation and reciprocity in the nurturing of life (Chambi y Chambi 1995: 12). A similar situation occurs in the Quechua communities, for example in the Community of Quispillacta, Ayacucho. The *ayllu* is found in the local *pacha* where the three components that comprise the natural collectivity live. The *pacha* is characterized by being animated, sacred, variable, harmonious, diverse, immanent and consubstantial.

In the Andean communities, time is not linear just as the community is not based on writing. The Aymara and Quechua cultures are not cultures that have utilized writing extensively: they are fundamentally oral cultures. On the other hand, time is intimately linked to the pulse of life, to its cosmic and telluric pulsing, as in for example, the rhythms and cycles of the moon, of the sun, of the climate, of the agricultural cycle (plant, harvest, plant). Because of this, it is said that time in the Andean world is cyclical. This does not mean that the cycles are the same every year. Quite the contrary. The cycles are rather varied (rains, wind, hailstorm, temperature changes, etc.) for which reason the activities corresponding to the different *crianzas* (nurturings) – the respective rituals and festivities – also vary in date. The agricultural activities are not determined by a calendar, but are carried out according to the rhythm of the cycles of nature. It is also noteworthy that for the Quechua and the Aymara, the 'present', the 'past', and the 'future' do not have the same meanings as they do in the contemporary Western world.

In this context, Grillo (1993: 8) notes: 'The 'present' in the Andean world, which is alive, renews itself, recreates itself, by digesting the 'past', that is, by including the past. ... In the Andes there exists the notion of sequence, the notion of before and after, but these do not oppose one another as past and future do in the modern West, but rather they find themselves included in the 'present', in 'the ever always', always being renewed, always re-created.'

This brief presentation describes the way of living and seeing in the Andean world, that is to say, the Andean cosmovision, which intimately links the Quechuas and Aymaras with their terrain, mountains, waters, crops, animals and rivers. For example, the Quechuas of the Community of Quispillacta would call this their particular way of life: '... our customs, which differentiate us from other realities and cultures. Our custom is born from nature, from the soil, from the mountains, from the rivers, that is to say from the *sallqa* (nature itself), and from the *Pachamama*. The *runa* is part of nature and lives harmoniously with each one of the components in a reciprocal and equitable relationship' (Machaca 1992: 8).

The *chacra* and the seed in the Andes

The *chacra* (a plot of land for cultivation, usually between one and two hectares) is part of the *pacha*, and is not in opposition to it. The *chacra* is not simply a plot of land for cultivation. To make *chacra* is a ritual and a festivity. The agricultural calendar illustrates it: festivities and rituals all have their proper times. All festivities, be they at a communal or familial level, are directed at thanking the deities for the fruits obtained through agriculture. To make *chacra* for the *runa* or *jaque* ('man' in Quechua and Aymara respectively) is to contribute to enriching and regenerating the local *pacha*. In this way, it would be possible to say that the *runa* and the *jaque* are agriculturalists, cultivators of life, or they may be shepherds who care for the animals of the deities. The *Pachamama* (Mother Earth) is the one who generates or engenders life, nourishing and regenerating in conversation, in nurturing (*crianza*), in reciprocity and in complementarity the *sallqa*, *wacas* and *runas* and *jaques*. The members of these three communities make and have their own *chacras*. To make *chacras* is not the privilege of *runas* or *jaques*. Notwithstanding, for the present contribution we will attempt to present and highlight certain aspects of the *chacra* and the seed.

Life in the Andes does not revolve around humans. The *runa* or the *jaque* knows that she or he is just one more member of the natural collectivity. The contribution that humans make to the regeneration and the festivity of life is participating and intimately experiencing the rituals, making *chacra*. To make *chacra* is to accompany one another, among the members of the natural collectivity, with the participation of the *runa* or the *jaque*, of the *sallqa*, and under the protection of the *wacas*. To make *chacra* implies nurturing the diversity of persons. It requires a great sensibility in order to attune to one and all, among one another. Each *chacra*, like every seed, is unique in its way of being and in its personality. This demands that the *ayllu*, the families, their members, know how to converse with each *chacra*. The seed is a living being (*muju o kawsay* in Quechua, *jatha* in Aymara). As such, it is part of the *ayllu* and a member of the *sallqa*. Like every other person from each of the three communities of the *pacha*, the seed is sensitive and has its own culture. In Quispillacta, Ayacucho, as in Conima, Puno, and in many other Quechua and Aymara communities, culture is not a particular and exclusive attribute of humans, 'because all [beings] are persons and know how to live in their own way'.

As Machaca (1996: 104) notes: 'The seed, in this sense, has its own culture; it lives with you and nurtures you, but it also leaves when it is not appreciated or is mistreated. Just as, for the Andean population, food security was and is still today the most primordial need, understanding and practising the culture of the *chacra*, that is agriculture, has constituted and still constitutes the central activity as it proportions enduring health to nature and the human communities. To nurture seeds means above all to nurture the *chacra*, to strengthen the processes of circulation or the ambling of seeds throughout different paths. As a result, the recuperation of variability isn't only a question of seeds, but above all is an understanding of Andean culture in its real magnitude. The habituating of new seeds, and the fluidity of circulation, will not be rendered while the forms and knowledge (*saberes*) of Andean life and the *chacra* itself are not invigorated.'

We will now touch upon some aspects of how the nurturing of the seed takes place in Quispillacta. The Quechua from Quispillacta and the Aymara from Conima, in accordance with her/his abilities and knowledge, 'attunes' herself/himself to the set of living beings or persons, for the nurturing of the *chacra*. Aymaras and Quechuas converse permanently in a very fine-tuned conversation with the elements of the natural collectivity. In Conima, during the nurturing of the *chacra*, the community dwellers (*comuneros*) observe a set of signs or *lomasas* ('indicators') on a daily basis so that in this manner they may conveniently converse with the climate of that year (*mara* in Aymara, *wata* in Quechua) throughout the agricultural or livestock season. In Conima, Puno: [the Aymara] 'converses with 80 to 100 or more signs for agricultural activity, for the simple fact that in the Andes, because there is quite a varied and dense climate, one must converse constantly and firmly with the signs or *lomasas*, and what is more, each one of these signs must be confirmed and conjugated with a series of other signs, and on different dates or opportunities throughout the year' (Chambi y Chambi 1995: 81).

The *chacra* is the place where each being remembers and commits itself to nurturing new plants, terrains, waters, etc. for the benefit of each one of them. The *chacra* will be the scenario for a greater relationship among the living communities, in which the living beings interact, converse, reciprocate and develop a mutual caring. (ABA 1993: 5) In the vision of the members of the Quispillacta Community, one nurtures other living beings during some moments so that later on one is nurtured by other living beings, and vice versa. For this reason, the fruits of these plantings are called *kawsay*, whose equivalent would be 'life' or 'giving life'. *Kawsay* is then sometimes a 'living being' who accompanies the human community, and at other times is a source of life. In Quechua, the word is *uywa*, once nurtured, and *uywaq*, that which nurtures (ABA 1993:7).

The concept of *crianza* (nurturing) in this case and particular context refers to the cultivation of plants and the nurturing of animals, soils, seeds, mountains, waters, etc. This concept also defines Andean agriculture as the culture of nurturing *(crianza)*. The crops are nurtured in *chacras* located in different ecological floors. It is worthy of mention that for the great diversity of crops, raising animals is a complement to the activity of nurturing. (ABA 1993: 6) Moreover, no distinction is made between wild and domesticated plants; all are nurtured. The following are the 'steps' through which the seed incorporates itself into the *chacra*.

Incorporation of the seed into the family of crops in Quispillacta: 'visible' and 'less visible' forms

The incorporation of the seed follows a number of processes: 'visible' and 'invisible' rites. Through rituals, the seed is incorporated into the family as a new member. The ritual facilitates that the community dweller (*comunero*) and the seed approximate one another. Both of them will become a part of a 'trial' or a 'more intimate knowing'. The object of incorporating a new variety of seed into the family *chacra* is the diversification of crops. To increase the varieties as much as is possible, because the climate, the soils and the waters ask for this, is part of the nurturing (*crianza*). When there is a desire to incorporate a new variety, one looks, 'one shows (how)', such that finally, the new variety becomes accustomed to that particular *chacra*. (ABA 1993: 7). However, nothing guarantees that the new seed will remain forever.

The complexity of the world of nurturing (*crianza*) requires that the *comunero* of Quispillacta (follows) 'a set of steps and processes in order that the approximation (of the *runa*) to the new seed be the most intimate possible (something 'magical'), and this is achieved through rituals that are little visible, in the sense that the *runa* will maintain an intimate relationship with the seed, avoiding being noticed in 'his intentions'. It could easily be expressed as 'deceiving' the new being, 'in order to invite a sharing' of all of the life experience of the interested *runa*. If we could compare this to the human community (its courtships for example), we would say that we are treating [dealing] with exactly the same process of how a young man or woman becomes attracted, or captivated. This means that the *runa* interested in an ecotype or crop must inspire much affection from this being. In this way, the probability that the 'deceived' seed will remain in the particular *chacra* may be higher.' (ABA 1993: 8).

'Visible' forms of initial approximation of the Quispillactan toward the seed

Approximating the seed in Quispillacta involves two phases: obtaining or acquiring the new seed, and the 'trial'. The approximation takes place during the agricultural cycle, during harvest time and after, through various modalities (*cf.* Table 5.2 for Quispillacta and Table 5.3 for Conima). Agricultural activity is carried out with the participation of various families. Almost always, any type of work is realized in collective groups, where there is the participation of more than two *ayllus*. Each modality implies that the participants have different roles and specific names.

The time of harvest presents a singular opportunity for the *runa* to approximate the seed and appreciate its way of being – its colour, the number of tubers produced, a never before seen appearance, culinary quality, etc. (ABA 1993: 8; Machaca 1993: 164). The collective work (*ayni, minka*) also permits the recognition of the peculiarities of the *chacra* – the type of *chacra*, where it is located, the climate – just as it permits the recognition of which *ayllus* are the best nurturers of plants and animals, and which are not (Machaca 1993: 164). Once the Quispillactan has obtained a seed, the next step is the 'trial' (Table 5.2). The trial ends up being the process of accommodation of the members of the collectivity (the three communities: *sallqa, runa, wacas)* to its own and even more ample collectivity or within a different collectivity. The trial consists of planting in special or designated *chacras*, for example in family plots near the family dwelling, with the sole purpose of living fully with the seed, which in technical terms seems to correspond to evaluations of the phenotypical characteristics of the new plant. The plots

TABLE 5.1: THE MOST IMPORTANT 'VISIBLE' FORMS OR MODALITIES OF APPROACHING THE SEED.

DURING THE HARVEST

Modality: *Hurquchakuykuy* (to separate silently without the owner's awareness).

Actors: *Allaq* (3 to 5 men, heads of family, open the furrows and uncover the tubers). Search for *wanllas* (the biggest potatoes).
Pallaq (in general women and children pick up the tubers).

AFTER THE HARVEST

Modality: *Maskapa* (re-search of tubers)
Pallapa (re-search of grains)

Actors: *Pallaq*

OTHER MODALITIES

- Beyond the family circle or groups of collective work within the *ayllu*: communal assemblies.
- *Ruykay* (barter/*trueque*)
- *Haymay* (to help later)
- *Yanapakuy* (to co-operate)
- *Llankin* ('gift')

Source: **Adapted from Machaca (1993).**

TABLE 5.2: INCORPORATION OF THE NEW SEED INTO THE NEW QUISPILLACTAN FAMILY.

YEARS OF BECOMING ACCUSTOMED				
1ST YEAR	**2ND YEAR**	**3RD YEAR**	**4TH YEAR**	**5TH YEAR**
Approaching	*1st Agricultural Season*	*2nd Agricultural Season*	*3rd Agricultural Season*	*4th Agricultural Season*
	GARDEN	SOIL TRIAL	WEATHER TRIAL	INCORPORATION
	Special plot	In the same ecological niche	In different ecological niches	In different plots
	PHASES OF THE 'TRIAL'			

YEARS OF UNDERSTANDING, CARING FOR, TEACHING, AND PROTECTING EACH OTHER.

Source: **Adapted from Machaca (1993).**

TABLE 5.3: MODALITIES OF STRENGTHENING THE DIVERSITY OF SEEDS IN CONIMA, PUNO.

MODALITIES OF RECIPROCITY AND EXCHANGE

Jathacha: To give seeds.

Waki: Lending terrain allows for gathering half of the harvest.

Mayt'a: Loan of a mixes of seeds.

Turka: Barter of seeds.

Chhala: Exchange seeds for other products.

Partiira: To work the *chacra al partir*.

Paylla: Payment with a product for having helped with harvesting or for having lent pack animals

Chiki: To asign a portion of the crop to the person that helps in the nurturing of the chacra; this person will increase the variability of his/her seeds as s/he receives these crops.

Laurunasiña: To take under concealment, a few seeds from a *chacra*, with the intention of returning them in the next harvest without any notice.

Apjhatas: When somebody gets married; the young couple is provided help with seeds.

Jaljhata: When somebody passes the charge of authority, the *comunero* families help this person (1) with products, taking into account that this authority did not have time to nurture his/her *chacras*, (2) organizing seed fairs, (3) through purchases, (4) by taking seeds from undomesticated plants, among other activities.

Seed diversity is enriched through: the use of mixed native varieties, the enhancement of the circuits of spatial rotation of the seeds, and the strengthening of the modalities of reciprocity and exchange.

Source: **Nestor Chambi, personal communication (1997).**

constitute 'farmer research centres', as much regarding the adaptation to the soil, climate and other elements of Andean agriculture, as the *ratay*, or becoming accustomed or 'getting along'.

Nothing guarantees that the new seed will stay to coexist and incorporate itself into the new family *chacra*. During the first year it may go, or it may stay. If the seed stays, its slow incorporation will take various cycles and agricultural seasons. Even when, at the fifth year, the new plant-person demonstrates that it is accustomed to living in its new *ayllu*, nothing guarantees that it will stay for ever. If at the end of several cycles, the person-plant retires, the *comunera/o* will in a very careful manner evaluate what she/he did or did not do with the new seed, so that she/he will not repeat, during another opportunity, what motivated the seed to leave.

Even then, the seed, like all living beings, tires and deserves a just rest after having contributed to the nurturing of the natural collectivity. This does not mean the death or disappearance of the person, but rather the step from one form of being to another. The Aymaras of Conima, for example, in the festival for the *Virgen de la Candelaria* (on 2 February) – known as *Ispallanakan Phistapa*, and clearly a celebration of the *chacra* – celebrate the *Ispallas* (deity of the producers)awaawa. 'Not only are the new *Ispallas* found, but they must meet with the *Ispallas*' 'mothers' or 'grandmothers', so that they may embrace, as if [the *Ispalla* mother or grandmother] were showing them that she would be entrusting to the 'new *Ispalla*' ... that they should make them (*jaques*) eat, or that these 'new *Ispallas*' must now nurture persons, with the type of passing or with the act of blessing from the 'old mother or grandmother *Ispalla*'. This act, in Aymara, is described: *Machaq ispallampi merq'e ispallampi qhomantasiyaÒawa*... When the grandmother *Ispalla* is giving this blessing, it is said that she leaves them with the following charge: "Just as we have nurtured these people, now it is time for them to nurture."' (Chambi y Chambi 1995: 59–60).

Conclusion

Any contemporary Western effort aiming to protect and/or enhance 'biodiversity' or 'agrobiodiversity' located within indigenous peoples' territories should have a clear understanding, in their policies and practices, that such 'biodiversity' and 'agrobiodiversity' is inextricably linked to indigenous cultural diversity. Despite 500 years of policies and practices prejudicial to the environment and indigenous communities, agrobiodiversity is to a great extent the outcome of non-Western indigenous practices in which underlying cosmological principles guide the nurture and regeneration of life. In the short term, such recognition may contribute to wiser and more appropriate decisions on the allocation of scarce budgetary resources, and to greater and more decisive participation of indigenous peoples according to their agendas and communal decisions. An unavoidable key factor postponed by most states all over the world is the resolution of the 'Indigenous Question', that is, the struggle of indigenous peoples for self-determination, and control over their territories and resources.

Acknowledgement

The authors wish to thank Maria E. Gonzalez for the translation of this contribution.

Cosmovisions and agriculture in India (Prabha Mahale and Hay Sorée)

India has two main forms of agriculture: irrigated and rainfed (divided into rainfed dryland with 400 mm rain and 75 rainy days, and rainfed wetland with 1500 mm rain and 200 rainy days). With agricultural land totalling 170 million hectares, 27 per cent is irrigated and 73 per cent rainfed. The peak potential for irrigated agriculture is almost attained, while there is still considerable room for developing rainfed agriculture.

Indian rainfed agriculture is an extensive mixed cropping (including forestry) and mixed farming system of production that forms the traditional way of life, not merely an economic activity. With producers being largely autonomous and relatively free from external political and economic controls, production was highly dependent on indigenous knowledge. Colonization by the British saw transformations like the enclosure of forests and commercialization of agriculture. Later came the Green Revolution, emphasizing high external inputs.

Indian cosmovisions

There are two major traditions of cosmovision in India: the 'Great Tradition', representing the Sanskrit or classical tradition, and the 'Folk Tradition', representing popular Hin du tradition and the tradition of the tribals. The rituals and practices of the Hindu tradition, both classical and folk, is a matter of continued history representing a living cult which is deeply connected with social, religious and cultural traditions, both in the orthodox and popular sense. They mostly converge but occasionally diverge from each other. Then there is the indigenous system of the original inhabitants, the tribals, who have another knowledge base. Vedas, originally an oral tradition, are a collection of hymns, mantras and prayers in Sanskrit, which communicate the sacred knowledge of cosmic order visioned by the seers. They cover diverse branches of learning such as astrology, medicine, law, economics, agriculture and government. The Vedic tradition is the root of the cosmology and knowledge system of the vast majority of Indians, the Hindus and Jains. In India's traditional thought there is no distinction between the sacred and the profane: everything is sacred. The essence of this tradition is to live in partnership rather than exploitation of nature.

The most complete holistic perspective of the universe, evolved by the Vedic culture about 6,000–8,000 years ago, has been sustained by the Indian civilization through millennia. The Vedas have played a major role in bringing together man and his faith in nature and have guided man through *Rta*, the cosmic morality. The cosmology, the total world-view, had the man–nature relationship at its core. All life is believed to be interrelated and interwoven. Indian science and philosophy are thus based on the perennial postulate of the perpetual cyclic degeneration and regeneration of movement of creation. According to Hindu mythology, Brahma is the creator, Vishnu the conserver and Shiva the destroyer of the universe.

The basic theory of cosmovision is known as *Siddhanta*. The *Sarva Tantra Siddhantas* cut across all areas of traditional Indian science. The following elements of these *Sarvatantra Siddhantas* are important for agriculture:

- understanding the composition of all material, animate as well as inanimate, in terms of the five primordial elements (*pancha mahabhutas*): air, water, earth, fire and ether/sky/space;
- understanding the properties and actions of human beings, animals and plants, in terms of three biological factors, *Vaata*, *Pitta* and *Kapha*. (*Vaata*: slender, light and averse to sunlight; *Pitta*: medium sized, abundant and fond of sunlight; *Kapha*: stout and bulky, abundant flowers and fruits, house many creepers);
- understanding the property and composition of *Dravya* (matter), and the basic units of *Guna* and *Karma* (quality and action) (Nambi *et al.* 1996).

The universe consists of living (*bhutas*) and nonliving elements (*mahabhutas*), the five basic elements; *prithvi* (earth), *aap* (water), *tejas* (fire or light), *vayu* (air) and *akas* (space). All living beings are born and evolve out of the five *mahabhutas*. In death they go back to them. The *mahabhutas* are the primary natural resources essential for all agriculture. Through myths and rituals man is ever reminded of his duty (*dharma*) to sustain these elements.

The cosmovision of *Krsi Sastra*, the science of agriculture, can best be summarized with the help of a quotation from *Chakra Saamitha*, the Ayurvedic text. 'The basic aim of the concepts and fundamental principles of all the sciences is to establish happiness in all living beings. But a correct and thorough knowledge of the basic principles of the universe and the (human) body leads to the correct path to happiness, while deceptive knowledge leads to the wrong path.' (Mukundan 1988). The cosmic forces were personified in the form of various gods and goddesses, whose influence or failure to maintain *Rta* was considered the main cause of any health imbalance. They thus played a role in healing and hence it was the responsibility of every individual to observe the prescribed rules (Vatsyayan 1992). Most of the Vedic rituals are institutionalized in Hindu *dharma* and are a part of day-to-day life of the people. Varuna is the God of waters and all the rivers Ganga, Yamuna, Saraswati, Kaveri are deities of the vast water cosmogony. No ceremony for birth, death or marriage is complete without the ritual purification of water.

The vegetative and animal life-forms like lotus, coconut, mango, snake, tiger and cow are central in Hindu myths. Cows (the symbolic representation of the earth) have been traditionally objects of great worship and reverence and the killing of cow is listed as one of the major sins in Hinduism: 'All that kill, eat and permit the slaughter of cow would rot in hell for as many years as there were hairs on the body of the cow' (Artha Shastra of Kautilya). The teachings in the ancient scripts like the *Upanishads* emphasize the importance of trees. Reverence for trees is expressed in various tree worships of *Ficus* species. Trees have also been linked with penance, education and religious activities. *Prithvi*, the Mother Earth, is the divine mother who sustains plant and animal life. She is perceived as a powerful goddess and the world as a whole. The cosmos itself is perceived as a great being, a cosmic organism. Different parts of the world are identified as parts of her body. The earth is called her loins, the oceans her bowels, the mountains her bones, the rivers her veins, the trees her body hair, the sun and moon her eyes, and nether worlds are her hips, legs and feet. *Vayu* (air) in Vedic Pantheon is associated with *Indra*, the God of the firmament, the personified atmosphere. He is the pure breath of life (*prana*). Finally the sun, the great ball of fire, is the energizer, the life giver.

The goddesses too, illustrate important ideas of Hindu philosophy. For example, *Prakriti* denotes physical reality. It is nature in all its complexity, orderliness and intensity. The Goddess *Sri*, or *Lakshmi*, is today one of the most popular and widely venerated deities. In early Vedic literature she is invoked to bring prosperity and to give abundance. In the *Sri-Sukta* (an appendix to the *Rig Veda* of pre-Buddhist date) she is described as moist in cow dung. Clearly, *Sri* is associated with growth and the fecundity of moist rich soil. Villagers, particularly women, worship *Sri* in the form of cow dung on certain occasions. *Lakshmi* is associated with lotus (symbolizing vegetative growth) and elephant (whose power brings fertilizing rains). Together they represent the blossoming of life. *Durga* is one of the most formidable goddesses of the Hindu Pantheon. Her primary mythological function is to combat demons who threaten the stability of the cosmos.

Cosmovision of the rural people

In the context of villages, the above-mentioned goddesses are worshipped by the upper caste Hindus. Taboos on ploughing are observed on *Dashahara*, *Ambubachi* and *Naga Panchmi* among the villagers of Pachara. *Shiva* is often pictured in folk tales in Bengal as a harassed farmer. On two of his festivals, *Shivaratri* and *Shivar Gajan*, farmers refrain from ploughing (Chakrabarti 1988). However, these 'Great' gods and goddesses, though acknowledged to be in charge of distant, cosmic rhythms, are only of limited interest to most other villagers.

Every village has its own village deities. They often share the names or epithets of deities from the Sanskrit Pantheon. But they do not necessarily have similarity to the 'Great Tradition' goddesses in question. Unlike the 'great' gods whose worship is often restricted to certain castes, these deities are goddesses of the whole village. All over southern India, these village deities are almost exclusively female. They are usually not represented by anthropomorphic images but are represented by uncarved natural stones, trees or small shrines.

The village and its immediate surroundings thus represent for the villagers a more or less complete cosmos. The central divine power impinging on, or underlying, this cosmos is the village goddess. The extent to which order and fertility dominate the village cosmos is bound by the relationship between the goddess and the villagers. Their relationship is localized and aims not so much at individual welfare as the welfare of the whole village. In return for the worship of the villagers, the goddess ensures good crops, timely rain, fertility and protection from diseases, spirits and untimely death. The entire ritual complex built around agricultural operations has protective, prohibitive and promotional values. For example, the villagers in Pachara (West Bengal) propitiate *Laksmi* and *Manasa* a number of times every year. While *Manasa* is worshipped generally during the cultivation season, *Laksmi* is worshipped during pre-harvest or post-harvest periods. Many rituals performed for a living human being are extended to the Mother Earth.

The *Adi Perukku* agricultural festival is celebrated in Tamil Nadu. On the eighteenth day of the Tamil month of *Adi* (mid July to mid August) this festival hails the arrival of the monsoon. Reverence is paid to the River Goddess. Farmers are encouraged to sow the seeds. An important aspect of this festival is the sowing of nine varieties of seed – wheat, paddy, toordal, greengram, groundnut, bean, sesame, blackgram and horsegram (*navadanya*) in a pot. It is called *mulaipari* and is a forerunner of the present germination test (Nambi *et al.* 1995). Apart from the festivals of the village goddess, a number of ritual performances directly or indirectly related to various stages of management of agricultural production and consumption are observed by individual families and by particular caste communities. These rituals vary from region to region and from community to community but the ultimate goal is the same: worship of deities, implements, bullocks, spirits in the fields, etc. to ensure a good harvest.

Cosmovision of the tribals

Among the tribals in India, the link between man and earth is deeply ingrained in their perception, social life and rituals. The Oraons, the fourth largest tribal group in India (dwelling in the Jharkhand areas of Bihar, Madya Pradesh and Orissa) believe in the existence of a supreme spiritual being (*Dharmes*, the Creator or Sun King), ancestors (*pachabalar*) and spirits, placed in a hierarchical order. They are appeased by offerings and sacrifices lest they destroy crops, forests, vegetation and natural resources.

Traditionally the ritual music and dance of the Oraons are directly related to the nature cycle: the blossoming of plants and trees, the position of the moon, and the seasons. They celebrate *Sarhul/Khaddi*, their most important festival, when the *sal* trees are in full blossom. In this festival a symbolic marriage of *dharmesh*, the sun, with *khekhel*, the earth, is enacted through worship. With conversion to Christianity, the shift in faith led them to view their role in relation to nature as masters of the earth and of the creatures living therein. However, in recent years a process of indigenization of Christian rules and rituals has taken place. The Christian Oraons celebrate the traditional Oraon festivals in almost the same way and manner as the non-converts (Xaxa 1992).

Traditional Indian agricultural science

In the Vedas, particularly *Rig Veda* and *Atharva Veda*, a great deal of attention is paid to agriculture, implements, cattle and other animals, rains, harvests etc. Ancient texts related to agriculture are the *vrkshayurveda* (*ayurveda* of plants),the *krshisastra* (science of agriculture) and the *mrgayurveda* (animal science). They provide a wealth of knowledge on a variety of subjects, such as collection and selection of seeds, germination, seed treatment, soil testing and preparation, methods of cultivation, pest control and crop protection, rearing of cow and care of draught cattle, etc. Outbreaks of disease, and pest attacks on plants, are viewed on the basis of the same principles as epidemics in humans. The basic understanding is that epidemics occur due to imbalances in the ecosystem. One of the major causes of such vitiation of the environment is human error leading to wrong intervention in natural processes. Such errors are termed *prajanaparadha*. Other causes could even be the prevalence of *adharma* or wrong ways of living.

The main protection against epidemics is to have a thorough knowledge of nature so as to avoid causing serious imbalances in the ecosystem. *Vrkshayurveda* – the science of treating plant health – was accorded a prime position in the history of agriculture in India. The three major ancient texts that provide a strong basis to *vrkshayurveda* are compiled by Varahamihira, Chavandarya and Sarangadhara. These texts provide indications of an integrated approach to the control of crop pests and diseases through soil, seed, plant and environmental treatment. The common characteristics of these different methods are, according to Chandrakanth and Basaveradhya (1995):

- a multi-pronged attack on the pest/disease;
- improving plant health, thereby increasing the natural resistance capability;
- enriching soil with nutrients and increasing useful microbial activity; and
- broad spectrum effect on pests/disease.

The link between the traditional texts and farming practice was through proverbs, verses and idioms in all Indian languages which expressed the folk knowledge and contained the same information as the classical texts. In the Tamil Nadu study (Box 5.4) folk knowledge, available in a large number of proverbs, is compared with the classical knowledge. However, at some later stage this link between the classical and folk knowledge seems to have disappeared. At present we have two traditions: the Sanskrit classical knowledge contained in the manuscript form, and the oral tradition residing with the farmers, rural and tribal. Hence the proverbs are considered as much a collection of scientific information as the texts themselves.

Folk knowledge

The perception and understanding that farmers have about ecology, crops, land, labour, livestock, agricultural implements, etc. has a profound bearing on the strategy they adopt in their day-to-day agricultural operations. Their ideas about climate, crops, optimum climatic conditions required for cultivation, and beliefs related to crops and fruits, come from the knowledge they have received from their forefathers and their own long experiences in the natural laboratory of the field. The farmers are able to identify various types of seeds and seedlings, often based on morphological characteristics. Looking at the flowers of a plant, an estimation of the yield is made. Chakrabarti, in his book *Around the Plough* (1981), presents detailed information on the cultivators' notion regarding the onset of rainfall, the optimum climatic condition for cultivation.

There are hundreds of farmers' innovations already documented. Gupta, Capoor and Shah (1990) compiled an *Inventory of 1200 Farmer Innovations for Sustainable Development* from all over the world. Publications like *ILEIA Newsletter* and the *Honey Bee* try to diffuse these innovations. The case studies from Karjat, Tamil Nadu and Andhra Pradesh are full of them. They are mainly from the dry regions, in which the poorer part of the population lives and where the influence of conventional agriculture has been limited. It definitely proves that though these farmers may lack economic resources, they are rich in knowledge resources. The technically useful items of indigenous agricultural practices are often documented without reference to the symbolic or ritual matrix in which they exist. It is a moot question whether, by looking at these practices from a mere scientific rationale, are they not devaluing them. Despite the fact that the farmers have been subjected to external influences, they are still experimenting and making innovations – sometimes adapting the external knowledge to the indigenous knowledge, and sometimes revitalizing their own knowledge.

There are indigenous institutions that regulate community administration, decision-making and elements of farming, and the rites and rituals related to cosmovision. In the villages, the religious functionaries like the Brahmin priest, and traditional functionaries in the tribal communities like the *Naiks* and *Disari*, play an important role. The functioning and strength of the institutions that have kept the environment protected depends on how successfully future citizens are introduced to the heritage that generates respect for these institutions. There is no

Box 5.4: 'Folk' Practitioners as Innovators of Traditional Science

A. V. Balasubramanian and K. Vijayalakshmi

The Centre for Indian Knowledge Systems (CIKS) has compared the practices and knowledge of farmers in the North Arcot district in Tamil Nadu with traditional Indian knowledge as described in classical texts of the Vedas. The cosmovision of the farmers is described from the celebration of festivals as they show the awareness of the relationship of the people to various forces of nature. For farmers there is a special relationship with water which is the very basis of agriculture. Before the commencement of any agricultural practice, they worship their implements and also the field. Worship is done by offering sacred water, incense, lighting of the lamp and worship of the Mother Earth: the Mother Earth being the bestower of all wealth, who is the mother of all creatures in this world, and who brings happiness and prosperity. After offering prayers and doing *pradakshina* to the field, keeping the field always to one's right side, one should also worship the Sun God, Wind God and other gods.

A very important feature of our traditional Indian science and technology is that its knowledge, theories and principles are not meant to be reposed in a small number of experts, institutions or texts, but are also found with the day-to-day practitioners of the arts and sciences. The 'folk' practitioners are also equally the innovators in the frontiers of their discipline, and the theories and technical categories belong to their domain as well. If we consider, for example, a highly developed branch of Indian science such as medicine, the basic theories at its foundation – such as the *Panchabhuta* theory of matter and the *Tridosha* theory of causation of disease and its treatment – are part of the common knowledge of our people, and a number of technical terms, such as *vata, pitta, kapha, agni, rasa, ushnna, sheeta, veerya* etc. are all part of the vocabulary of our households. The expert, or specialist, seems to play a very different kind of role here, namely that of systematizing the corpus of knowledge.

The major change due to the interaction of indigenous practices with outside sources of knowledge has been attributed to the Green Revolution, leading to a dramatic decrease in biodiversity. High-yielding varieties are central to this approach. Yet, biodiversity and sustainability debates have led to a rethinking of the value of traditional agriculture. It has been shown that several traditional varieties are being preserved by farmers which fare well on marginal lands, assuring farmers at least food security for one year. Several initiatives can be envisaged to assist farmers in maintaining their identity and strengthening their traditional scientific knowledge:

- use of natural products for pest control and crop protection (provision of information, workshops, assessment of the efficacy of use and survey of farmers' knowledge and innovations);
- set up of rural gene banks for traditional paddy varieties (collections, making a selection of seeds available, maintaining specific varieties in small plots, and providing natural pesticides to farmers);
- the publication of a *panchangam* ('almanac') for farmers containing auspicious and inauspicious dates for various agricultural activities; indications of rains and methods of forecasting; omens, proverbs, day-to-day entries.

Farmers reported that they consult the *Panchangam*, at least for critical operations. They are read by the local astrologer, while the impetus to use them comes from the women. They offered a number of suggestions to improve the *Panchangam*, such as: including addresses of farmers practising organic agriculture, and listing available literature; methods for detecting ground water; traditional varieties of cattle; fairs and festivals in the region; and advice for agricultural operations, depending on the birth star of the farmer.

Source: CIKS, No. 2

way by which knowledge systems can grow if the traditional cultural anchors are not properly located (Gupta 1992). Culture provides a 'grammar', while technology provides new words. The meaning of life can only be discovered by the blending of the two.

Indigenous and Western science

Writers such as Pereira and Willett have divided the world into two: the West (reductionist and materialistic) and the indigenous (holistic). The cosmovisions of the two obviously differ dramatically. The authors describe how influential and overriding the Western scientific approach has been and still is. It has pushed traditional knowledge systems to the back yard and has become the overriding paradigm (Pereira 1993; Willett 1993). Willett (1993) pleads for an active research programme, geared at the 'purity and pureness' of indigenous knowledge. These writers represent a 'revivalistic' school of thought that aims at discarding the Western knowledge system and reviving the indigenous knowledge systems. This approach in our view is not sustainable, because external knowledge has influenced and is influencing the agricultural system and it may not be possible to reverse that. A transition from the Western to the indigenous system necessarily should have a transition period, where a cross fertilization between the two could be envisaged.

Vandana Shiva, and the organization *Navdanya* which she leads, take up indigenous knowledge from the concept of a people's movement for biodiversity conservation. They are carrying out research on the erosion of biodiversity and on the social and ecological impact of monocultures in forestry and agriculture. They found that Indian farmers in diverse ecosystems are still the custodians of a tremendous richness of diversity, and they continue to be the main suppliers of seed, in spite of the emergence of the seed industry. Their main objective is the strengthening of farmers' rights and seed supply systems. The name of the organization, *Navdanya*, literally means, 'nine seeds'. Metaphorically it represents the balance based on diversity at every level, from the cosmic domain to the community, from the ecology of the earth to the ecology of the body .

While accepting the significance of indigenous knowledge, we feel that there is a need for cross-fertilization of indigenous and conventional agricultural knowledge systems. There has recently been a great resurgence of interest by scientists in the study of traditional methods of agriculture. Diverse areas of plant science are discussed – such as agronomic methods, pest control and plant protection techniques, climatology, biodiversity, water harvesting and traditional animal health care practices. Accordingly, there are efforts to see how indigenous knowledge can be assimilated into the mainstream of scientific or modern technical knowledge generation. There is the need to reorganize agricultural research and extension to facilitate lateral flow of information between indigenous systems and formal institutions.

Options and dilemmas for tomorrow

There is a close-knit association between cosmovision, or how the relationship between man and nature in its widest sense is viewed, and agricultural practices. Folklore, proverbs and songs are the vehicles of this process. This, however, does not necessarily mean that farmers always have a thorough insight into what they are doing, and why. Traditional knowledge often implies that rites, practices and customs are continued out of sheer habit, or out of an undefined fear of bad influences if they are abolished. This could mean a slide towards superstition, which means that there is a kind of mysterious belief. Practices remain the same and knowledge development comes to a standstill. Constant review of any knowledge system is essential for further development. Still, it is true that in India the sacred and the profane are strongly interrelated. We have to make sure that this relationship is further strengthened in order to avoid beliefs and customs becoming mere superstitions.

The advantage of modern science is that it is analytical and tries to arrive at general truths by discovering the parts of the whole. The advantage of indigenous knowledge is that it is locale- specific, holistic, and relates to diversity. The disadvantage of modern agricultural science is that it deals with parts only and is obsessed with the general. As has been described earlier, knowledge systems are not static but dynamic. They change. Farmers and researchers experiment. This again may be much too simple a remark, at least in the short run. The Western scientific approach did not integrate indigenous knowledge into its concept. Likewise, we could not find real instances where the Western concept has been integrated into the 'holistic cosmovision'. However there are indications that this is changing.

Synthesis of local and external knowledge

Is a synthesis of local and external knowledge, or of traditional and conventional knowledge, possible? In our opinion a synthesis is not possible if traditional knowledge is viewed as an 'alternative knowledge' to the external knowledge. Both have their base of existence and both have their limitations. Influencing each other would mean a change of fixed paradigms and the creation of new ones, which no doubt will bring about major improvements and changes.

That means having an openness to question one's own system and the courage to enter into a dialogue with the other.

Cosmology and biodiversity of the *Kasepuhan* community in the Mount Halimun area of West Java, Indonesia (Kusnaka Adimihardja)

'Ema, Bapa, abdi neda widi titip Nyi Sri, ulah aya nu ngaganggu ngaguna sika, berkah doa salametanna kalawan rahayu sadayana'.

('Mother, Father, I entrust *Nyi Sri* , the Goddess of Rice, to your authority to watch for the disturber and destroyer and to provide us with blessing and welfare.' – Prayer of the *Sesepuh Girang* .)

Surrounding Mount Halimun there are small communities that still follow a traditional way of life. They are called the *kasepuhan* and they belong to the Sundanese ethnic group of Java. They live in small groups in the southern highland areas of Banten, Bogor and Sukabumi on the slopes of Mount Halimun, a complex of mountain ranges in West Java. The size of their region is about 122,000 hectares, consisting of a conservation forest of 82,000 hectares and a natural preserve 'National Park' of 40,000 hectares – the largest in the area, established by the Ministry of Forests in 1979. The *kasepuhan* people earn their living predominantly by collecting forest products, gardening, and rice planting in both wet rice fields (*sawah*) and in dry swidden fields (*huma* or *ladang*). Their swidden fields constitute the major agricultural pattern of the *kasepuhan* people which is rooted in the local vision of the origin of plants that are properly cultivated in the swidden fields.

These indigenous land-use systems form the oldest pattern of agriculture known among the Sundanese people: for centuries the *kasepuhan* people have lived and farmed on these hillsides and in the valleys of the forested region. Each generation has passed on the agricultural knowledge, beliefs and practices which have evolved from living and working so closely with nature. Their age-old experience is based on their unique cosmology which as an indigenous philosophy of nature and the environment seeks to understand and explain the essence of humans' relation with the universe. In this way, their cosmology is directly related to the unique way in which they have been using and managing their agricultural and natural resources for many generations.

Kasepuhan: cosmovision

For the *kasepuhan* people, the universe will continue to exist as long as its laws of regularity and equilibrium – controlled by its cosmic centre – can keep the universe's elements in a harmonious balance. This local belief continues to influence all aspects of the lives of the *kasepuhan* people, and their agricultural system is rooted in their cosmology and their belief in the ancestors, called *tatali paranti karuhun*, 'the ancestors' way of life'. The people also believe that the physical elements and social systems in the universe are intimately connected. Such cosmovision is reflected in their original 'animism-dynamism' belief system which has later been influenced by Hinduism and Islam. Rambo (1981: 46) who theorized about the ways of understanding the relationship between social systems and ecosystems, suggested that within an agroecosystem, the social systems and ecosystems interact with each other through the exchange of energy, material and information. Similarly, through their political, economic, ideological, belief and knowledge systems, the *kasepuhan* affect and are affected by their surrounding ecosystems.

The *kasepuhan* strongly believe that disturbing the regularity of various physical and non-physical components – which are perceived as 'alive' in the universe – can cause disaster for human life. Therefore, the main duty of the people is seen as being to maintain the balanced relationships that exist between the various components. These principles are clearly manifested in various ceremonies, and in the way the health of local ecosystems is taken into account in day-to-day activities. According to the *kasepuhan*, these beliefs originated from a source that is called *pancer pangawinan*, or 'heredity': *'kami mah turunan pancer pangawinan'*, 'we are coming from the hereditary line of *pancer pangawinan*.'. The Sundanese word *pancer* means *lulugu*, which refers to *asal usul* or *sumber* in Bahasa Indonesia, a 'source' or 'origin' in English. Rigg (1862: 210) explains the word *pangawinan* from *pangawin*, which means 'to carry spears in procession'. This is reflected in the folklore from South Banten, Sukabumi and Bogor, and the *kasepuhan* people have a tradition of carrying a spear in various victory ceremonies as an expression of the continuation of their ancestral tradition that acknowledges their direct descent from *pancer pangawin*..

Several examples of sustaining ceremonies, such as the circumcision ceremony, *sunatan* or *nyalametkeun*, and the wedding ceremony, *kawin*, which also means to unite day-to-day life with spiritual life, to unite *Dewi Sri*, 'the Goddess of Rice', with the soil, and to unite the earth with the heaven, have been extensively studied and described (*cf.* Adimihardja 1995). These ceremonies seek to unite the microcosm with the macrocosm in order to reach a status of *rasa manunggal*, 'the one unity of life'. In this context, a doctrine is known as *mawas diri*, 'self introspection', and also incorporates *koreksi diri*, 'self correction', which forms a guide for the *kasepuhan* people in an effort to help people to overcome bad attitudes and to avoid transgressions of the ancestral order. To reach a state of harmony, order, safety and peacefulness in life, the *ucap jeung lampah* , 'statement and action', must be in balance, not in opposition to each other. This significant notion is also established in the belief that each member of the *kasepuhan* people must consult the *karuhu*, the 'supernatural', to avoid various dangers. Therefore, every social activity must be preceded by a prayer called *doa amit*, 'prayer for permission', to *karuhun*, the 'ancestors', the deities, and 'God the Almighty'. The *sesepuh girang* or *sesepuh kampung*, the local 'village leader', usually recites *doa amit* at the beginning of every social activity which involves all members of the *kasepuhan*. Once a year, a devotional visit to the graves of the ancestors in South Bogor and South Banten is made, where, at the gravesides, the *sesepuh girang* recites a prayer to ask for forgiveness from the ruler of heaven and earth and for blessing and welfare.

The *doa amit* is also recited by the *sesepuh girang* before the rice planting and rice harvesting in the wet paddy fields and dry lands. After praying, an important ceremony is held in the house of the *sesepuh girang*, attended by all the *sesepuh kampung*, the formal and informal leaders of the village, where the times for planting and harvesting the dry lands and paddy fields are determined for several villages of the region. This, however, does not imply that the *kasepuhan* do not deviate from their traditional customs. As an ideal, the moral concepts help people to learn and to maintain the customary social attitudes that they call *tatali paranti karuhun*, 'ancestor's manners', but they believe that every violation of the rule or law of *tatali paranti karuhun* will cause *kabendon*, 'disaster', which will happen not only to those who violate the law but also to the entire community. When the balance is found, the people will be able to achieve the feeling of peacefulness and harmony in their heart and thoughts, and in their social relationships with other people in this universe, as a provision for life hereafter. This state of harmony, as we will see below, is also an important principle in maintaining the balance with nature, its plants, trees and animals.

Within the *kasepuhan* circle there are guiding principles of life that are in accordance with the belief of *ngaji diri*, 'self introspection', which is crucial to achieving *rasa manunggal*, 'unity feeling', a guide in life. If this is followed well, punishment or *kabendon* for deviating from the 'ancestors' manners' can be avoided. The guiding principles are also expressed in speech, attitude, manners and friendship (*cf.* Adimihardja 1995). As Warnaen (1986) notes, such principles are commonly exercised among the Sundanese people, and they are still practised by the members of the *kasepuhan* community. These pattern of attitudes, acts, rules and norms have originated from those rules and norms that were practised in the past by the Sundanese people, prior to the influence of the Islamic religion in West Java.

Kasepuhan perception of the forests

According to the *kasepuhan* belief system, the surrounding forests are the principal source of life. They recognize three types of forests, which form the base of their way of forest management and conservation:

- the ancient forest, called *leuweung kolot* or *leuweung geledegan*, a dense or *geledegan* forest of large and small trees that are the home to various kinds of animals;
- the exploited forest, called *leuweung sampalan*, where people establish their farms, attend to their cattle, and collect firewood;
- the sacred forest, which they call *leuweung titipan* , a forest traditionally recognized by all the members of the *kasepuhan* as sacred. It is not to be exploited without the approval of the foremost leader of the community, the *sesepuh girang*. The cultivation of this forest is possible only if a message of permission is received from the ancestors through the *sesepuh girang*. In this way, this sacred forest is continuously protected and maintained.

The *leuweung titipan* around southern Sukabumi is located on Mount Ciawitali and Mount Girang Cibareno. Nowadays, the management of entrusted or holy forests in the Mount Halimun region, and ancient forests that have become nature preserves, fall under the responsibility of the Directorate General of Forest Protection and Natural Preserve, the *Perlindungan Hutan dan Pelestarian Alam* (PHPA), whereas the exploited forest *leuweung sampalan* is the management responsibility of the National State Forest Corporation, *Perusahaan Umum Perhutanan Indonesia*, *Perhutani*, Unit III West Java.

Sampalan forest: traditional land use and conservation

The opening-up process of land utilization around the *sampalan* forests by the *kasepuhan* people of Sirnarasa Village in South Sukabumi generally develops by cultivating the dry land as *huma* or *ladang*. The main species planted in this multicropping system are rice (*pare gede* or *cere* and *angsana* variety) mixed with millet (*kunyit: Panicum viride*), pigeon pea (*Cajanus cajan*) and sesame (*wijen: Sesamum orientale*), while Job's tears (*hanjeli: Coix lacryma-jobi L.*) forms the boundary around the *huma* or *ladang*. In order to enrich their food, they also cultivate various kinds of wild plants in combination with cultivation plants.

The practice of planting trees is a sign that the plot is cultivated by someone. Even if someone else re-opens the land for *huma*, the products still remain the property of the previous farmer. Mixed gardens are formed by the integration of wild and cultivated plants. A *jami* area not far from the village which they cultivate intensively with dominant plants for food, medicine and sale is usually called a *kebon* or *kebun*, 'garden'. The *sawah* or wet-rice fields can be developed when there is an adequate supply of water flowing to the fields. The *kasepuhan* people also cultivate fish in the wet-rice fields during the three months' interval after the harvest. There are also wild plants in the *sawah* that can be eaten as vegetables, such as *genjer* (*Limnocharis flava*), *eceng* (*Monochoria vaginalis*) and *gelang* (*Portulaca oleracea*). The farmers may also plant such species as *arbila* (*Dolichos lablab*), *kacang panjang* or string bean (*Vigna sinensis*), and *mentimun* or cucumber (*Cucumis sativus*) on the edge of the *sawah* . In the recent INDAKS study (cf. Adimihardja 1995), an inventory of not less than 114 crop varieties was completed in the swidden agriculture, garden and wet-rice fields, also indicating their planting area and their general uses. When people come and stay in the area of the *jami*, and establish a new settlement, it is called *ngababakan*, a 'pioneer settlement'. Sometimes, these new settlements may develop into a new village or *kampung*.

Kasepuhan: a sustainable polyculture strategy

The *kasepuhan* people perform various ceremonies directly related to the different aspects of cultivation in swidden fields, especially with regard to rice, which is considered a holy plant. Rice and the other plants that are usually planted in swidden fields are believed to have originated from the cemetery of *Dewi Sri*, the 'Goddess of Rice'. According to a Sundanese legend, *Dewi Sri* was born from an egg that belonged to the *Dewa* or God *Anta*, the holder of the power on earth. Later, *Dewi Sri* was killed after eating the apple from the Garden of Eden, and after her burial, plants and some other trees began to grow on her grave: several varieties of coconut trees, arrenga trees, rice, bamboo, spreading plants and grasses. *Dewa Batara Guru* later presented all these plants and trees to *Prabu Siliwangi*, a wise King of the Padjadjaran Kingdom, so that *Prabu Siliwangi* could plant and cultivate them throughout Java. According to the traditional Sundanese belief, all the plants that are now growing throughout the world have originated from *Dewi Sri's* graveside, thus believed to be the 'centre' of the first varieties of plants which are today available to support all human life. As the cemetery, according to the tradition, is also believed to be the 'centre' of the macrocosm and microcosm, the *kasepuhan* people still believe that the centre of the *huma*, the 'dry field', constitutes the manifestation of the cemetery of *Dewi Sri*.. The location of a 'centre' in the *huma* is also a symbol of the belief that the beginning and ending of life are manifested within the 'centre' of the macro- and microcosm. Following this belief, the local farmers still start their planting and harvesting in the centre of the *huma*. Before planting and harvesting, farmers offer a prayer called *doa amit*, which is followed by brief ceremonies to honour *Dewi Sri*. The centre of the *huma* among the *kasepuhan* people is called *pupuhunan* (*puhu (n)* means 'centre').

The legend of the killing of *Dewi Sri* is to be interpreted as a symbol of the importance of keeping 'order' within the life of the community. A 'taboo' would have been broken if *Batara Guru* had fallen in love with his adopted daughter *Dewi Sri*. It is believed that the death of *Dewi Sri* was a 'sacrifice' to help increase the people's prosperity, symbolized by the growth of many species of plants and trees in the cemetery. In the Islamic tradition, there is similarly a 'sacrifice' in the form of the killing of a goat as a symbol of the Prophet Ibrahim's faithfulness to God the Almighty who ordered him to kill his son, the Prophet Ismail. For the *kasepuhan* people, 'death' is 'destiny'; they believe in a concept called *pasrah*, namely to offer oneself to 'destiny' and accept whatever comes. Learning to accept 'destiny' is part of the 'self evaluation', a task they believe every human in the world must undergo.

All plant varieties, especially those grown in the *huma*, are treated in the indigenous farming practices as if they possess human qualities. So, rice is believed to be the manifestation of *Dewi Sri*., and in consequence, before planting and harvesting the *kasepuhan* undertake various ceremonies to pay respect to *Dewi Sri*, as well as to all their ancestors. The wide variety of plants and crops that are growing in the cemetery of *Dewi Sri* serves as a symbol of the importance of maintaining diversity for the *kasepuhan* farmers in their farming activities. This has happened because the traditional Sundanese communities lived in a forest ecosystem which was still uncultivated, and today, the traditional farming pattern of cultivation of multiple crops on dry land, *huma*, and its related ceremonies, are still practised by the *kasepuhan* communities. By following their traditions, they hope that their ancestors will protect them from failing harvests.

Today, the *kasepuhan* do not live in isolation from the outside world: their traditional values and management and conservation practices have been challenged by many outside influences, such as imported agricultural technology introduced by extension programmes, 'modern' education, Islamic teachings, new products introduced by salespeople from outside the villages, and radio and television programmes. Although these influences have had a certain impact on the local socio-economic and cultural systems as well as on the surrounding forest ecosystems, the basic traditional values of the *kasepuhan* are still being followed. Traditional ceremonies, agricultural practices and generation-to-generation teachings of the local customs, values and wisdom are continuing in the Mount Halimun area, still forming a strong guidance for their work and living.

Local management strategies and change

It is clear that several groups and institutions, such as the *Perhutani*, the PHPA, and the local *kasepuhan* and non-*kasepuhan* people will play a role in the future of the Mount Halimun ecosystem. The conservation of the Mount Halimun forest ecosystem is primarily dependent on the mutual understanding between the government forestry officials and the *kasepuhan* people, particularly with regard to the forestry management and conservation practices. Recently, forestry officials have begun to change the forestry management system, which has existed since colonial times. This system is based on the notion that 'the forest is only for the lives of flora and fauna'. This perception is, of course, not shared by the *kasepuhan* who live around Mount Halimun, and in the past, differences in opinion between the forestry officials and the local people have led to serious conflicts.

As it has recently become evident in West Java – as elsewhere in the tropics – that indigenous communities can participate in more sustainable forest management schemes on the basis of their local knowledge, practices and cosmology, forestry officials have begun to try to understand the belief systems, customs and management practices of the *kasepuhan* people. Both groups are starting to work co-operatively to maintain the integrity of the forests through the development of special programmes. So, degraded forest can be repaired by using a local practice called *tumpangsari* or 'inter-cropping'. The forestry officials are now even trying to mobilize the *kasepuhan* people to intensify their farming practices in accordance with the socio-cultural, economic and environmental values of their community. In this way, 'inter-cropping' will allow for the local practice of swidden agriculture without exploiting the forest resources. Equally, the mixing of subsistence-rice varieties with certain agricultural commodity production could extend the basis of the farmers' productivity (Evers 1988; Adimihardja and Iskandar 1993). Recently, the government forest officials have offered support to improve the quality of local handicrafts, and to supply plant seedlings for 'Regreening Programmes' with the support of a local NGO.

The maintenance of the local swidden system is very important for the *kasepuhan* people, as the related farming activities are embedded in their cosmology: all their traditional ceremonies are based on the growing of rice in these systems. The ceremonies associated with *huma* are the traditional way to maintain a balanced relationship between humans, nature and God the Almighty, and in order to continue to live and survive, the *kasepuhan* recognize that the forest and its inhabitants must be sustained and preserved. For them, the practice of swidden agriculture and the forest form a unity, in which humans, animals, plants and trees all have their place. The *kasepuhan* interpret the word *huma* or *uma(h)* as *imah* or 'home'. According to this perception, this 'home' is not only for sleeping in but also serves as the main source for their spiritual and physical life. The destruction of the forest means virtually the destruction of their homes, and as such the destruction of the *kasepuhan* as a socio-cultural group. For the *kasepuhan*, swidden agriculture in the present day is not just a continuation of their *tatali paranti karuhun*, 'ancestral traditions'; it is also a form of survival of the *kasepuhan* in their ecosystem, in which a particular local institution has been identified as the basis for continuation of the community.

Co-operative activities between government forestry officials and the *kasepuhan* people may help to develop a shared view about the management of land areas and the conservation of biodiversity in the area around Mount Halimun. These activities, however, must be regarded as an effort to support the *kasepuhan's* role in maintaining the integrity of their local ecosystems. The dynamic agricultural and natural resource management practices of the local people can be regarded as a form of social energy that has a potential to further develop rational creative actions through the process of change of the management systems as a social learning process (Soedjatmoko 1986; Adimihardja 1993). In this regard, the interaction between humans, culture and environment can be perceived as a progressive contextualization process in which local knowledge, cosmology and practice can play a major role in the conservation of biodiversity in the region (Vayda 1969; Poerbo 1986; Adimihardja 1993).

In order to maintain their livelihoods without destroying the environment, the *kasepuhan* people continue to protect and utilize various local species and varieties of plants as part of their polyculture strategy in accordance with their ancestors' orders. They cultivate these various plants and trees in order to support their daily needs, supplemented with several seasonal plants and vegetables that have a high economic value, such as clove, coffee, bananas and red chilis. These activities not only sustain the livelihood of the community: they also bring some cash into the community, and help to conserve the forest by modestly increasing the returns from their agricultural land. Their efforts to enhance agrodiversity have brought them both economic and social benefits, and at the same time have contributed to the protection of the forest ecosystem. The participation of the *kasepuhan* communities in the newly-introduced 'Regreening Programme' of the government for the surrounding upland forest area develops since it also seeks to link up with the ancestral teachings and customs which are based on the local cosmology on conservation and management.

Conclusion

The cosmology of the indigenous forest peoples of Mount Halimun in terms of their cultural, natural and spiritual perceptions are most relevant to the conditions of their ecosystem and its biodiversity as they have been regulating the use, management and conservation of the agricultural and natural resources in a sustainable way over many generations. These perceptions and their associated practices and methods have recently been shown to be rather adaptive and highly selective to processes of development and change in the area, introduced by outsiders such as representatives of commercial organizations, experts of NGOs, and government officials.

Over the past few years, joint research efforts have assessed, documented and analysed the 'invisible' yet strong influence of the local people's cosmology on their agricultural and natural resources management practices, and have also revealed the prominent position of the *kasepuhan* principles and practices of protecting and preserving the biodiversity of both wild and non-wild food and medicinal plants in the area. By implementing one of the major principles of the 'ethnosystems methodology' – the 'Participant's View' (PV) – a substantial contribution has been made to the further understanding and explanation of the underlying local beliefs and perceptions and the complex relationships among humans, nature and culture in the all-embracing cosmos of the *kasepuhan* communities of Mount Halimun (*cf.* Leakey and Slikkerveer 1991).

Only those development programmes and projects from outside that seek to link up with the indigenous cosmology and its related principles of use, management and conservation of resources are expected to be successful, particularly if they take the local perceptions and practices of the people seriously into consideration to jointly achieve the conservation and protection of Indonesia's rich and unique biodiversity for future generations.

Collaborative environmental research with Kuku-Yalanji people in the Wet Tropics of Queensland World Heritage Area, Australia (Rosemary Hill and Dermot Smyth)

The Wet Tropics of Queensland World Heritage Area (WTWHA) is a forested region of some 900,000 ha, located in the north-east of the State of Queensland, Australia. It was inscribed on the World Heritage List as a natural property in 1988 and contains the principal or only habitat for numerous species of threatened plants and rare or endangered mammals. It includes the most diverse assemblage of primitive angiosperm plant families in the world, and elements of its biota represent eight major stages of earth's evolutionary history (WTMA 1992). Norman Myers, in a speech to the International World Heritage Tropical Forests Conference in Cairns in 1996, described it as, 'the hottest of all hot spots' for biodiversity.

It is also an area of outstanding cultural diversity. Aborigines in the region have a distinctive rainforest culture, with unique features in their material technology, languages and social organization (Horsfall and Hall 1990). There were sixteen major language groups, each with a number of different dialects in the Wet Tropics region before European contact (Dixon 1976). Today, after more than a century of European occupation, centres of survival of Aboriginal rainforest culture and language are scattered throughout the region (Horsfall and Fuary 1988). Since the Wet Tropics region was added to the World Heritage List, Aboriginal people have made submissions to the government seeking a higher level of involvement in its management, through joint management, and have recommended the re-nomination of the Wet Tropics as a joint natural and cultural world heritage site (Fourmile *et al.* 1995). In response, the Queensland Government provided for the intent of developing joint management arrangements with Aboriginal people when enacting legislation to protect the site (WTWHA Protection and Management Act 1993). A Review of Aboriginal involvement in the WTWHA is currently in progress; and Aboriginal people seek the development of a framework regional agreement as an outcome of this process (Bama Wabu 1996). Australian Federal legislation requires one of the five members of the board that oversees management to be an Aboriginal person from the region. Noel Pearson, Chairperson of the influential Cape York Land Council held the position from July 1994 until September 1997; it is currently vacant. A decision by the relevant Australian and Queensland Government Ministers to remove mechanisms for the recognition of native title rights from the Draft Wet Tropics Plan is currently the subject of litigation by the Cape York Land Council.

The northern part of the WTWHA is the traditional land of some 3,000 Kuku-Yalanji people, who are now concentrated in two small towns, Wujal Wujal where land is held 'in trust' by the Aboriginal Community Council, and Mossman. The terrestrial common law native title rights of the Kuku-Yalanji extend over several hundred thousand hectares, most of which is inside the WTWHA. They have recently registered a Native Title Claim under the Native Title Act 1993. In addition to their use of the rainforest, Kuku-Yalanji, like other Aboriginal groups in the Wet Tropics, utilize a number of sclerophyll vegetation areas, which are scattered throughout their traditional estates (Anderson 1984). Some of these are being changed through invasion by rainforest, which Kuku-Yalanji attribute to the prohibition of their burning practices by the government. This loss of biodiversity at the community scale is greatly regretted by Kuku-Yalanji, who wish to light 'corrective fires' (Lewis 1994) to return the open forest in some sites. Important cultural resources associated with open forest, including particular grasses used for making dilly bags sold in local cultural heritage tourist ventures, are no longer available without considerable journey by motor vehicle.

A collaborative research project

The Kuku-Yalanji concern over the loss of open forest is shared by a number of scientists, albeit for different reasons. Harrington and Sanderson (1994) showed that half the area of wet sclerophyll forest in the region has been lost to rainforest invasion over the last fifty years, and related the observed changes to disruption of traditional Aboriginal fire regimes. The survival of several important species may be threatened if this loss of open forest habitat continues.

We approached Kuku-Yalanji people in 1995 with a proposal to develop a collaborative research project focused on the role of their fire practices in creating and maintaining biodiversity at the community level in the Wet Tropics region. There is no formal mechanism to guide collaborative research with Kuku-Yalanji people who have not yet had their traditional land returned to them with an Australian statutory title, or developed institutional co-management arrangements with the government managers of the WTWHA. The collaborative process was shaped by Kuku-Yalanji interests and concerns raised in initial meetings, and has continued to evolve throughout the research process. The *'Guidelines on Research Ethics Regarding Aboriginal and Torres Strait Islander Cultural, Social, Intellectual and Spiritual Property'* recently adopted by the James Cook University (Centre for Aboriginal and Torres Strait Islander Participation, Research and Development 1996) were very useful. There are four major foci in the process developed to date: (a) ensuring outcomes of relevance to Kuku-Yalanji; (b) protecting intellectual, cultural and spiritual property rights; (c) ensuring their participation in the research; and (d) establishing a process to reach agreement about the publication of the research data.

Outcomes of relevance to Kuku-Yalanji

The very first requirement for any successful collaboration is to ensure that the actual content of the research is meaningful and useful to the indigenous society. Kuku-Yalanji people's main interest in this research is as a means of regaining control of fire and related management practices in their traditional lands. The Cape York Land Council, which represents Kuku-Yalanji people in relation to their native title claim, supported the project, noting that it would provide information useful to ensuring a primary place for Kuku-Yalanji ecological knowledge and resource management practices in the Wet Tropics Plan (WTMA 1995). In addition, cultural and ecological data that show evidence of long-term occupation of the land by Kuku-Yalanji continuing to the present may be useful should the native title determination proceed to an adversarial situation in the courts. The need for indigenous people to gain management control as well as legal title to protected areas has been reinforced by other Australian experiences. Woenne-Green *et al.* (1994) argued that in the initial years following the handover of Kakadu World Heritage Area, while Aboriginal people gained ownership of their lands, the Australian National Parks and Wildlife Service gained control of the Park.

The two local Kuku-Yalanji Aboriginal organizations, the 'Wujal Wujal Community Council' and 'Bamanga Bubu Ngadimunku Incorporated' also supported the research because of its potential relevance to enhanced management control. Both these organizations employ Community Rangers to undertake natural and cultural resource management (Chittenden 1992), and to interact with government agencies in order to bring a greater indigenous perspective to their management activities.

For the Senior Custodians who are the prime holders of the traditional environmental knowledge, more immediate concerns take precedence in assessing the relevance of a research project. They welcomed the research because it was seen to strengthen the processes of cultural renewal and maintenance. Senior Custodians almost invariably are very concerned to ensure that younger Kuku-Yalanji people are properly educated in cultural terms, and are very aware that acquisition of traditional environmental knowledge requires considerable and consistent time, and that there are many competing demands on the time of the younger Kuku-Yalanji. They frequently discussed ways of using the project to enhance education, and requested we compile a plant collection which could be later used to produce curriculum material for Kuku-Yalanji school children. We were also asked to promote in our report the idea of a field study centre as a co-operative project between Kuku-Yalanji and the Wet Tropics Management Authority, and to write Kuku-Yalanji language and Aboriginal English versions of our reports.

The Wet Tropics Management Authority provided funding support for the project, which ensured that any Kuku-Yalanji people involved were paid for their time at the relevant university rate of pay for Senior Custodians and Senior Research Assistants. This payment, although relatively small for each person, was very important; people appreciated the payment not only because of their economic needs, but also because of the implied acknowledgement of their professional status.

Protecting intellectual property rights

Even the most useful research cannot be truly supported by indigenous people unless control of intellectual property is assured. Kuku-Yalanji people's desire to protect their intellectual property rights is reflected in the 'Julayinbul Statement on Indigenous Intellectual Property' (Rainforest Aboriginal Network 1993), to which they contributed at a conference in North Queensland in 1993. In this statement, indigenous people declare their willingness to share their intellectual property provided that their fundamental rights to define and control the property are recognized. Kuku-Yalanji customary law, like that of other Australian Aboriginal societies, controls access to information on criteria including membership of a group with custodial rights, age and gender. Rose (1996) Fourmile (1996) and Posey (1996) have argued the need for legal mechanisms to protect indigenous knowledge from exploitation by pharmaceutical companies and to ensure indigenous societies to receive an appropriate share in profits from commercial drugs and other products. Indigenous knowledge also needs to be protected from commercial exploitation in tourism ventures and publications, and from appropriation by government land managers. While the desirability of re-establishing traditional Aboriginal land management practices in relation to fire is recognized by the Queensland Department of Environment (Stanton 1994), there is no associated formal recognition of the need to involve Aboriginal people in the process.

The first step in ensuring protection of intellectual property in this research was a written statement from the principal researcher to Kuku-Yalanji organizations and people that no cultural data would be published without their consent. Australian law appears unable to protect communally-held intellectual property such as traditional environmental knowledge, although the Native Title Act of 1993 may offer protection (Bennett 1996). Legal protection for intellectual property was subsequently provided by the contract agreed with the Wet Tropics Management Authority when funding the project, which stipulated that ownership of cultural knowledge and information provided with the consent of the Aboriginal Corporations shall remain the property of the corporations.

A Reference Group was established to oversee the whole project. The group includes Senior Custodians from each of the main clan groups in the areas under detailed investigation, and representatives of two Aboriginal Corporations and three other organizations. Although envisaged initially as the prime decision-making body, the Reference Group was not empowered by Kuku-Yalanji people to take that role. Rather, it has become part of the decision-making process, operating more as a forum in which decisions reached separately with the various corporations and individuals can be endorsed.

Two repositories were established for the primary raw data from the project, one at each of the two local Aboriginal Corporations. These collections consist of photographs, transcripts of taped interviews, plant specimens, and data sheets and tapes. Also included are copies of material about Kuku-Yalanji people we collected from urban libraries when undertaking the literature review; this material had not previously been seen by most Kuku-Yalanji people in the community. As the intellectual property associated with traditional environmental knowledge is communally held, there is a need to take account of the interests of Kuku-Yalanji people who are not directly involved in providing data or making decisions. A set of guidelines which describes categories of information available for release has been developed and disseminated to take account of this broader interest; information about medicinal or specialist food plants that others may be able to capitalize on is prohibited from publication, as are site-specific cultural data.

Ensuring Kuku-Yalanji participation in the research
In addition to the Senior Research Assistant responsible for providing information directly to the project, we also employed other Research Assistants to accompany the principal researcher in all work, and to enhance capacity-building of the community. These assistants provided the overall service of ensuring culturally appropriate conduct of the project, and also received training in ecological survey and other research techniques. A joint presentation about Kuku-Yalanji fire practices was made to the International World Heritage Tropical Forests Conference held in Cairns in September 1996. This conference was notable for its lack of involvement of indigenous people, and many participants regarded indigenous people more as a threat to biodiversity conservation than as knowledgeable custodians whose management has previously ensured a high level of protection of biodiversity.

Information about the project is disseminated through publication of a three-monthly newsletter, and through regular meetings with the corporations and other organizations, one of which is open to all Kuku-Yalanji people.

Negotiating about publication of research findings
We gave a commitment at the beginning to reach agreement about the publication of the research, and have developed a process of ongoing meetings and reviews to achieve this. One set of data has been approved for public release from the project so far. The data were initially presented orally to a meeting of the Reference Group, and subsequently incorporated into a written paper. The paper was read individually to all Senior Custodians whose information was used, and written agreements were obtained permitting publication. The paper was sent to the two Aboriginal Corporations who were legally bound as copyright holders. Meetings were held with the corporations to discuss the contents of the paper, and written agreement obtained permitting the publication of the material. Meetings also occurred with other Aboriginal organizations to seek feedback on the paper even though official approval for release of the data was not required.

A similar process has been initiated for release of the subsequent data from the project. An initial round of meetings has occurred with individuals and organizations explaining what is proposed for inclusion in the final report on the project. Drafts of material will be sent to all individuals who have supplied information, and to the relevant organizations. These drafts will be read to non-readers and feedback sought regarding the appropriateness of the proposed contents at a second set of meetings with individuals and organizations. A final meeting of the project Reference Group will be brought together to confirm that agreement has been reached for publication of the report.

Research findings
Most of the detailed research findings are not yet available for publication. A Fire Protocol has been developed as an interim means of negotiating the customary law obligations of Kuku-Yalanji, and the Australian law obligations of the WTWHA managers; in the long term, recognition of Kuku-Yalanji native title rights is essential to ensuring good fire management. Kuku-Yalanji use fire both to promote fire-prone open forest, and to protect fire-sensitive rainforest. The data are generally giving support to the view that rather than being a natural area representing millennia of evolutionary development free from human intervention, the Wet Tropics of Queensland is a cultural landscape, whose vegetation patterns have been altered by tens of thousands of years of human occupation and management. The need to provide carbohydrate resources is emerging as a critical aspect of their management, as it has for other rainforest hunter-gatherer peoples (Bailey *et al.* 1989).

The data also reflect Kuku-Yalanji people's emphasis that the most profound aspect of heritage for them is the spiritual meaning and significance imbued in the landscape (Pearson 1995). The research has focused attention on the different outcomes Kuku-Yalanji and the government organizations seek from their management strategies. That sought by Kuku-Yalanji has been characterized as maintaining the integrity of the cultural landscape, while that sought by the government managers is described as maintaining the integrity of the natural landscape. Renomination of the WTWHA as a joint natural and cultural World Heritage site would be useful in mediating the different management strategies that are associated with these different desired outcomes.

Kuku-Yalanji responses: improving the collaboration
The collaborative process that developed has undoubtedly given Kuku-Yalanji people more control than has been available for any previous research

that has focused on them. It has provided Kuku-Yalanji people with experience in mobilizing a research effort to support their own strategic interests and goals, which is important to strengthen the social vitality and political efficacy of a community, and thereby its resilience to change (Lane and Rickson 1997). Senior Custodians who have contributed information about the project have been the most enthusiastic participants, emphasizing the usefulness of this work in cultural revival and renewal. They are concerned about the loss of environmental knowledge that is occurring because of the cultural disruption that followed colonization of Australia (Rainforest Aboriginal Network 1993).

However, there were instances where other Kuku-Yalanji people became concerned about the use and future ownership of the cultural data being collected, and were unaware of the arrangements in place, despite the efforts being made to disseminate information. This resulted in the principal researcher being asked by one of the Aboriginal groups in the region to stop work for a period until the concerns were resolved. It may be that the communal nature of traditional environmental knowledge is such that this is inevitable – any Kuku-Yalanji person has the right, under customary law, to question the use being made of the material and the authority under which it has been released to the public. The concept of nominating leaders to act on behalf of the whole community in regard to intellectual property does not fit well with Kuku-Yalanji social structure (Posey 1996).

A community strategic planning process which involves Kuku-Yalanji people widely, and identifies research priorities and collaborative processes, is probably the best means of overcoming this potential problem. The Kuna Yala in central America have successfully developed such a process (Posey 1996). However, this is clearly much more possible when indigenous people have land and co-management institutions in place. The lack of such mechanisms should not stop research proceeding – the initial research may be extremely useful in ensuring that indigenous people regain the control of their lands. The documentation of cultural knowledge by anthropologists has been critical in enabling the return of land to Aboriginal control in Australia (Burke 1995).

Contractual arrangements between the University, the Co-operative Research Centre, and the Kuku-Yalanji Corporations would be very useful in the absence of a formal co-management structure. Such arrangements could look more broadly at training, the access to university programmes and a strategic research plan. These matters cannot be appropriately dealt with in the context of one, necessarily limited, research project. Mechanisms for dispute resolution should be built into such contracts.

Conclusion: management implications

The biodiversity of the WTWHA is clearly linked to the cultural diversity and associated management practices of the rainforest Aboriginal people. Collaborative research such as this project can help elucidate this link and enhance the management of the area. However, this will only occur fully when Aboriginal peoples' aspirations for recognition of their native title rights, adoption of a framework regional agreement, and renomination of the area as a joint natural and cultural property have been achieved (Bama Wabu 1996; Fourmile *et al.* 1995) Protected area managers also need to recognize that Aboriginal environmental knowledge can only be shared in a process that involves them – it is forbidden under customary law for them to make certain information available unless they maintain control of that information and its subsequent use.

Balkanu (1997), an Aboriginal Corporation based in North Queensland, has recently released a *Draft Statement of Principles Regarding Bio-physical Research* in the Aboriginal lands, islands and waters of Cape York Peninsula. These guidelines aim to ensure that all research occurs in a process that respects Aboriginal land and cultural rights, and ensures their equitable participation in all stages of the research. As common law native title rights of the rainforest Aboriginal people may extend over more than 80 per cent of the WTWHA, all regional bio-physical research in the future may be collaborative research with Aboriginal people (Yarrow 1996).

Collaborative research is about balancing the need to protect indigenous intellectual property with the need to ensure that it is functioning fully in the indigenous society, which can only occur through direct application in the management of their traditional lands. It is inevitable that the processes of cultural maintenance have been affected in all indigenous societies by the colonization process: the challenge for researchers is to empower indigenous societies to utilize the resources available in research institutions to enhance their own strategic interests and goals, as well as the mutual benefit of cross-cultural understanding. On the ground, research is fundamentally about highly-trained people from one culture interacting with highly-trained people from another. Provided that mutual respect is maintained and the process is not dehumanized through exploitation, there is a great potential for mutual benefits to emerge from the interaction. This collaborative process was developed after many years of personal association of the researchers with Kuku-Yalanji people. A significant level of trust and cross-cultural understanding had developed before the project began. Researchers aspiring to undertake collaborative research should expect to invest time and effort in learning to know and be known by the cultural group with whom collaboration is sought.

Enhancing protected area management through indigenous involvement: the Uluru model (Julian Barry)

Involving indigenous people in protected area management has considerable benefits that have not been adequately recognized within Australia. With a few notable exceptions, Australian conservation and other agencies have not allocated the resources necessary to adequately involve indigenous people in the management of Australia's natural and cultural heritage. Uluru – Kata Tjuta National Park (U-KTNP) is one such exception. By focusing on joint management of the park, this contribution demonstrates that very significant benefits result from the involvement of indigenous people in protected area management.

U-KTNP has been inalienable Aboriginal freehold title land since it was handed back to the park's *Anangu* (Aboriginal) traditional owners on 26 October 1985. The park has been jointly managed by *Anangu* and the Australian Nature Conservation Agency from the time of the hand-back. Handing back the park to its traditional *Anangu* owners has been one of the most significant acts of reconciliation between Australia's indigenous and non-indigenous inhabitants. It has also given rise to considerable debate within the Australian community. Some sectors of Australian society predicted the hand-back would limit tourist access to the park and result in disastrous consequences for the tourism industry. Ten years later and despite these dire predictions, joint management at Uluru has produced significant benefits to tourists, the tourism industry and the scientific community. Furthermore, the hand-back has enhanced the image and prestige of the park nationally and internationally.

Uluru (previously known as Ayers Rock) has for some decades been a national icon and one of Australia's primary tourist destinations. It is a place of great spiritual significance to *Anangu* and one of the world's most significant semi-arid ecosystems. The hand-back has provided a range of benefits to its *Anangu* traditional owners and enhanced park management in a number of ways. Since hand-back, *Anangu* have been able to participate in park activities from a position of strength. As owners and lessors of the land, they have a majority on the park's Board of Management and a lease agreement that ensures that they are involved in all major management decisions. Since 1985, *Anangu* have demonstrated a willingness and enthusiasm to share a great deal of their knowledge with non-*Anangu* people. As a result, *Anangu* have made an enormous contribution to park interpretation, scientific research, land management, visitor management and infrastructure planning at U-KTNP.

Interpretation

Central Australia has a number of significant national parks, although Aboriginal interpretations generally play a limited role within them. The Uluru model shows that when Aboriginal people are in a position of strength they are enthusiastic about sharing much of their cultural knowledge with non-Aboriginal people. At Uluru, tourists have exceptional access to Aboriginal cultural information and interpretation material. A number of factors account for the extensive Aboriginal interpretations at U-KTNP. Since hand-back, park management has been required to seek *Anangu* people's views in respect of how the park should be interpreted. *Anangu* have insisted that the park's, 'interpretative materials promote *Anangu* perceptions as the primary interpretation of the park' (Uluru Plan of Management 1991: 19). In addition, *Anangu* have monitored the park's interpretation programme and ensured that all park staff provide accurate information about Aboriginal culture. Although park guests can purchase excellent publications on subjects such as park geology, Western interpretations of the park are not accorded primary status.

At U-KTNP there are currently three *Anangu* guided walks – the *Kuniya* (woma python), *Liru* (poisonous snake), and Cultural Centre tours. The park's fourth guided walk, the *Mala* (rufous hare wallaby) walk, is not always led by *Anangu* people. This walk, delivered by rangers, focuses on the activities of the *Mala* creation ancestors and the park's system of joint management. All walks in the park are based on *Tjukurpa*, which is a word referring to *Anangu* people's law and the period of creation. *Anangu* want visitors to the park to understand the central role that *Tjukurpa* plays in their lives. In addition to the guided walks at Uluru, written information and self-guided walks provide tourists with detailed information about *Anangu* culture. Since late October 1995, tourists have also had the benefit of the new Uluru - Kata Tjuta Cultural Centre. This huge building, which *Anangu* and non-*Anangu* designers created in the shape of two creation ancestors, houses a range of visual, audio-visual and interactive information about *Tjukurpa*, *Anangu* culture, *Anangu* interpretations of the environment, and how *Anangu* and non-*Anangu* work together to manage the park.

Scientific research and land management

Anangu knowledge of the environment, based on over 20,000 years of occupation of the Uluru region, is intimate and extensive. This knowledge has provided invaluable assistance to scientists who have worked within the park. Subsequently, *Anangu* people have built a strong relationship with senior figures in the scientific community. Between 1987 and 1990, Australia's most comprehensive semi-arid ecosystem fauna survey was completed as a joint exercise between *Anangu* and Australian government scientists. Since 1985, there has also been extensive co-operative research into ecosystem delineation and maintenance, rare and endangered fauna, refugia, flora and invertebrates. *Anangu* have given scientists the benefit of their extensive knowledge about introduced animals and their effects on the environment.

At the direction of senior *Anangu*, the park's system of mosaic or patch burning has been given major priority. For thousands of years *Anangu* burnt their country in small patches. This traditional system creates fire breaks, scarifies seeds, promotes vegetation regrowth and in a variety of ways protects and maintains the arid-land ecosystems. In 1990, two large fires threatened the park. In twelve hours, one fire travelled thirty kilometres to the park's border where it was broken up by the patch burns and brought under control by rangers.

Infrastructure development

At U-KTNP, all infrastructure development is undertaken co-operatively with *Anangu* to avoid compromising the cultural values of the park. *Anangu* bring to the planning process valuable skills, knowledge and advice about the effects of rain and water movement, shade and wind. Infrastructure development at Uluru is complicated by the fact that the park is invariably dry but experiences occasional flooding. Also, the fragile ecosystem faces high visitor densities at many park sites. *Anangu* have an advanced appreciation of these factors and are very good at predicting where tourists are likely to walk, sit, photograph, and even get lost! *Anangu* are experts at determining whether infrastructure will directly or indirectly create erosion or dust, and the effect of such problems.

Conclusion

In view of the achievements of the last 10 years, it is no surprise that since hand-back Uluru has enhanced its status as one of Australia's premier tourist destinations. Since 1985, tourist visitation to the park has almost tripled, in addition to enhancing the management of the park, benefiting its traditional owners, and contributing to the reconciliation process. *Anangu* ownership and joint management of U-KTNP has resulted in a range of benefits to the national and international community.

The joint management of U-KTNP is not without its challenges and difficulties. However, the park's Aboriginal traditional owners and the Australian Nature Conservation Agency remain firmly committed to joint management and are proud of its achievements. I hope that the Uluru model will inspire non-indigenous people to become more enthusiastic and committed to the prospect of empowering Aboriginal people and involving them in the natural and cultural resource management of protected areas.

Dryland plants and their uses (Hew D. V. Prendergast, Stephen D. Davis and Michael Way)

Some dryland plant species occur widely in different habitats and continents and are used by a diversity of cultures in ways that can be either remarkably uniform or very variable at a local level. *Dodonaea viscosa* (*Sapindaceae*) is a good example. In Peru its leaf is chewed like coca *Erythroxylum coca* (*Erythroxylaceae*), and in Australia it is still known as native hop because of its former use as a substitute for hop *Humulus lupulus* (*Cannabaceae*) in beer making. In Botswana it is an important bee plant. In many places its qualities as a good firewood, igniting readily, have been recognized, as has its potential for carving and as a hedging material. From Gabon and Nigeria to Sudan, India and the Philippines, *D. viscosa* is a well-known febrifuge, whilst an infusion of the roots has been claimed as one of South Africa's oldest remedies for the common cold.

Away from mainly local and non-commercial usage, there are other species that have assumed a national or regional importance as trading commodities. Two examples show the relative novelty of such trade and how it is being threatened. In the Sonoran Desert which straddles the frontier of Mexico and the USA, an important 'keystone' tree is the ironwood *Olneya tesota* (*Leguminosae*; Nabhan and Carr 1994). Ecologically it acts as a nurse plant for other plants, including cacti, by creating favourable micro-habitats for germination and seedling establishment, and it is important too for the survival of many animals. Local people have traditionally carved ironwood to make implements needed for their hunter-gatherer life-style. But in the 1960s, carving entered a new phase altogether when figurines of animals were produced for visiting tourists. Now, up to 3,000 craftsmen are involved, using an estimated maximum of 5,000 t/year of ironwood. The whole industry could collapse, however, if land clearance and charcoal production continue to deplete ironwood populations and to push up the price of the raw material.

In southern Africa the palm *Hyphaene petersiana* is an important species for basket-making (based on the leaves), for its edible fruits, and for building and fencing. In north-central Namibia, baskets are destined almost exclusively for local or regional markets (Konstant *et al.* 1995). In the early 1990s the annual offtake of leaves, some 10 per cent, was well within the sustainable limits of the 30 per cent recommended from studies elsewhere, but severe browsing by livestock could now be a threat. In Botswana a subsistence exploitation of the same species for basket-making was turned, in the 1970s, into a commercial venture for the tourist and export market (Cunningham and Milton 1987). The effects of a sevenfold increase in the value of the industry from 1976 to 1982 were to reduce both the supplies of adequate raw material, especially near settlements, and the possibility for the species to reproduce through seed.

Any destructive harvesting of dryland plants has to be closely reviewed given their low natural productivity and long regeneration time. In Western Australia, the highest quality incense oils from sandalwood (*Santalum spicatum*, *Santalaceae*) are produced from trees in semi-arid areas (Loneragan 1990). Whole trees (including their root systems) are harvested, but it takes 50–90 years in these areas for them to reach a commercial size. The sustainability of the trade depends on controlled exploitation and, eventually perhaps, cultivation.

Of the quarter of a million or so species of higher plants, relatively few have become so valued that they are traded globally. Among dryland plants the one most fabled, traded, used and written about is the frankincense (*Boswellia sacra*, *Burseraceae*; Miller and Morris 1988). The Ancient Egyptians were using its incense (the tree's resin) by the Fifth Dynasty (c. 2,800 BC) and believed it to be the sweat of the gods; Greek and Roman physicians, as well as those of India, applied it to an impressive array of illnesses; and today it is still burned in religious ceremonies and used medicinally.

Another internationally traded dryland tree

Box 5.5: Uses of Dryland Animals

International Panel of Experts on Desertification (IPED)

In addition to their uses for food and hides, animals have many other traditional uses. In dryland areas in Nigeria, Adeola (1992) found that wildlife products had important roles in cultural ceremonies (e.g. funerals and installation of rulers), the performance of traditional rites (e.g. invoking and appeasing traditional gods and witches), and as constituents in traditional medicines or for aphrodisiac, fertility or potency purposes. Animals used included large game, rodents, reptiles, birds and a mollusc, the African giant snail.

Like plants, (*cf.* Box 5.6) animals have an important role as indicators of the seasons. On the Logone River in Cameroon, the floodplain fishing season starts after the waters have risen in the annual flood. The timing of the ritual *moutwak* opening ceremony depends on the arrival of a particular fish species called *hiya* (*Alestes nurse*) (Drijver *et al.* 1995).

In drylands the arrival of rain at the end of the dry season is a long-awaited event. Many species are used to predict the arrival of coming rain (Niamir 1990). In western Kenya people use frogs, birds and white ants as indicators. In north-east Tanzania, changes in the patterns of behaviour of birds, insects and mammals are important signals. The Ariaal of Kenya cite several indicators of rain including the appearance and singing of several birds, the movement of safari ants, frogs calling, and changes in the texture of termite mounds (Clarfield 1996). The Turkana of Kenya say that several birds (ground hornbill, green wood hoopoe, spotted eagle owl and nightjar) and frogs are prophets of rain.

Attitudes to wildlife are often far from positive; for agriculturists many animals represent a threat to their precious crops, while to herders other species represent a threat to their stock. Many 'biosphere' people (Gadgil 1993a) are in direct competition with wild animals for food. Clarfield (1996) reports that among the Ariaal of Kenya, animals were predominantly classified according to their threat to livestock (predators) or their danger to people (mainly lions, snakes and buffalo).

[*Biological Diversity in the Dry Lands of the World.* 1994. Secretariat, UN Convention to Combat Desertification in countries experiencing serious drought and/or desertification, particularly in Africa.]

resin, but of far higher commercial value, is gum arabic, a polysaccharide produced by *Acacia senegal* (*Leguminosae*). Its high water solubility and low viscosity confer much prized emulsifying, stabilizing, thickening and suspending properties which are exploited in many food products, such as confectionery; in the pharmaceutical industry for the manufacture of tablets; in the printing industry for treating lithographic plates, and in ceramics for strengthening clay. The source is still 'wild' although exploitation is now also taking place of the plantings of *A. senegal* established for desertification control in the Sahel. To collect the gum a strip of bark is levered up and off the wood, causing minimal damage, and the tears of gum are subsequently picked by hand.

The above are only a few examples of the uses of dryland plants. By mid-1998, the *Survey of Economic Plants of Arid and Semi-Arid Lands* (SEPASAL) at The Royal Botanic Gardens, Kew, contained information on over 6,200 tropical and subtropical dryland species.

Termites, society and ecology: perspectives from West Africa (James Fairhead and Melissa Leach)

'If an old teacher comes upon a termite mound (the original says 'ant-hill', the older name in English for a termite mound) during a walk in the bush, this gives him an opportunity for dispensing various kinds of knowledge according to the kind of listeners he has at hand. Either he will speak of the creature itself, the laws governing its life and the class of being it belongs to, or he will give children a lesson in morality by showing them how community life depends on solidarity and forgetfulness of self, or again he may go on to higher things if he feels that his audience can attain to them. Thus any incident in life, any trivial happening, can always be developed in many ways, can lead to telling a myth, a tale, a legend. Every phenomenon one encounters can be traced back to the forces from which it issued and suggest the mysteries of the unity of life, which is entirely animated by Se, the primordial sacred Force, itself an aspect of God the creator.' (Ba 1981: 179, speaking of Bambara wisdom, West Africa).

Introduction

This paper addresses the place of termites and their soil machinations in the ecological and social thought and practice of peoples living in a large region of West Africa, and reflects on certain similarities in ecological representations and techniques which seem to exist over much of West Africa. It shows how farmers work with their understanding of termites in choosing or generating moist and fertile farm sites. Termites are, however, important not only in farming but also in general considerations of humidity and fertility which are equally important in human health and the structuring of power relations.

Our interest in termites was stimulated by several incidents during our social anthropological fieldwork among Kuranko (Mande) and Kissia (Mel/ West Atlantic) farmers in 1992-94. The first was during a visit to Bamba, a village where between 1906 and 1908 the French colonizing army installed a military post in a savannah, now long overgrown by forest. Elders there recounted how the villagers had called in specialists to evict the French – whose success in colonial war was attributed to their letter-based communication system and their fortresses. This they did by introducing termites which destroyed the fortress buildings and ate the papers. We thought little of this anecdote until later finding mention in the national archives of the extreme difficulty that termites caused the French in Bamba. Subsequently, the French took care to site new camps in places 'naturally' free of termites, failing to see, perhaps, how social 'natural' phenomena might be. This incident indicated that the Kissia claim to, and perhaps do, influence termite activity, and that this is a specialist activity.

In a second incident, in the Kissi village of Toly, when we were noting field boundaries, our close friend Dawvo mentioned that 'when we the Kissia find a termite mound near a field boundary, we place the boundary over it to leave half in one field, and half in the other, to prevent jealousy'. It transpired that Kissi farmers prefer fields with more termite mounds, desiring at least four in the field which they clear from fallow annually.

'There are trees that we don't cut, like *kondo* (*Erythrophleum guineense*) as the leaves make the rice yield very well. If you find *puusa* termite mounds as well, the rice does well. Everywhere that there is one, you should cultivate – masses of rice!'

Without exception, farmers suggested that termite mounds are associated with humid soil conditions. They say that they pipe water to the soil surface: '*Puusa* mounds have canals (*yoyowan*) and water comes from them'. Yoyowan (canals) are likened to bone marrow. 'When you have termite mounds, the area down from it tends to be relatively well watered. If you don't have a forest fallow, it is best to choose a savannah site where there are several termite mounds, as there is sure to be sufficient water.'

In our other fieldwork site, Kuranko-speaking farmers identified the mounds of *togbo* and *tuei* varieties of termites as important for bringing soil to the surface (*ka du la yele*). In the dry season, soil around them remains damp (*sumasuma*). Such mounds are 'good in fields' (*a ka nyi senindo*), and are singled out as sites for intensive gardening of peppers, tobacco, squash and other crops. These mounds are contrasted with *birE* (*Cubitermes*) mounds which themselves have dry soils, but which nevertheless indicate the presence of mature fallows.

Termite ecology and farming

At the time, we did not realize that farmers throughout much of Africa deliberately seek out land with many and large termite mounds, whether for reasons of fertility or soil water.

This is the case, for example, in Tanzanian *chitemene* farming where farmers choose sites with termite mounds (Mielke and Mielke 1982). In parts of Malawi, mounds are owned and traded by farmers as they are valued for gardening (L. Shaxon, personal communication). In the West African region, Iroko (1982, citing Quénum 1980: 65) notes how on the Sakete and Pobe plateaux of Benin, an abundance of termite mounds is taken to indicate good soils for cereal farming, and in Atacora (also Benin) an abundance of large mounds is a prerequisite for high-fertility-demanding yam cultivation (Iroko 1982: 58). Many farmers prefer soils where termite mounds are found to have high clay content, although such preferences certainly depend upon specific conditions (cf. *Spore* 1995, 64: 4). Mounds that one can feel have more clay also indicate better soils. Citing Mercier (1968), Iroko goes on to note how this does not just concern agricultural aspects of fertility, but also ritual ones. The agro-pastoralist Fulbe of Benin choose their encampments in areas of many termite mounds, 'signalling the presence of the goddess of fertility, of fecundity and of abundance, a major preoccupation of these farming and herding people' (Iroko 1982: 54).

We were equally unaware of the dispersed but none the less considerable research showing the potential fertility and soil water benefits which termites can bring (e.g. Lóbry and Conacher 1990; Lal 1987). In particular, several studies of termite activity on soils in West Africa show its importance in soil enrichment. As Hauser (1978) notes in Burkina Faso, for example, termites carry clay material rich in minerals from deep below the ground (10-15m; up to 70m), and enrich the soil surface with it, which compensates considerably for erosion. Termites tend to homogenize soil horizons, preventing the formation of sharp boundaries in the soil, which can form natural boundaries to plant rooting, and favouring water infiltration. Notably, their activity dismantles iron pan enabling tree roots to penetrate more deeply. Termite galleries (tunnels) improve water penetration, and plant roots can follow them. The tunnels can similarly improve air circulation during the wet seasons. Certain mounds have concretions of lime (calcium carbonates) where none appears present in the upper soils. The only negative aspect of termite activity, as Hauser notes, is in losing 20 percent of the land to inhabited mounds. Others in this region draw similar conclusions (e.g. Lepage *et. al.* 1989).

Concerning lime, and perhaps other mineral concentrations, it has been noted in certain parts of Africa that farmers break up the mounds and distribute their soil over their fields, noting the fertility boost that results (Iroko 1982: 64 in Benin) and also mix the enriched soil with cattle feed. Agronomists suggest that this fertility boost derives from the lime concretions, which reduce soil acidity, or from the enrichment of termite soils in other minerals and nitrates. The latter are concentrated in the mound when termites collect and store plant matter and/or because nitrogen-fixing bacteria are sometimes active (cf. Lapperre 1971). The lime is also used in the manufacture of mortar (Milne 1947).

Early in our research, we also did not know the extent to which African land-improving practices encourage and manipulate termite activity. In Zai farming in Burkina Faso, for example, fallow soil pits are filled with organic matter less to fertilize the area, as one might suspect, but expressly to attract termites to improve soil physical properties, especially water infiltration. This makes a millet crop possible and subsequently enables denser vegetation to colonize (Mando *et al.* 1993). The farmers consider that termite tunnels allow rainwater to accumulate in the holes and percolate into the soil. Farmers in western Sudan pile wood branches on poor land. The action of the termites burrowing through the soil around this wood, possibly aided by an increase in fertility from its breakdown, was reported to improve the land in about four months (Tothill 1948). In several cases, farmers put organic matter around incipient mounds to encourage their development, recognizing the incipient mounds and their improved fertility, and identifying these in areas where tobacco, for example, might be planted (Helena Black, personal communication).

In conversations with Susu farmers in Sierra Leone we learned how they upgrade their fallows from savannah to forest vegetation partly by intensively grazing the land to reduce burnable material, but partly also by working the land to incorporate organic matter into the soil and – certain farmers said – to stimulate termite activity. (We are grateful to Kate Longly for making these conversations possible during her fieldwork in Kukuna, Sierra Leone, in 1993.) Farmers in much of Africa thus seem to be working with termites in their soil management. Such termite management is often indirect (through manipulating the ecological conditions in which termites thrive), but it can also be direct, as when Kuranko farmers speak of certain trees and fruits that 'seed' termite mounds (*Dichrostachys glomerata*, *Acacia sylvicola* and *'Tinkumansore'* (unidentified) whose red fruit is said to seed mounds) or when evicting the French, although this latter case was not for farming and involved specialist powers.

Both Mondjannagni (1975) and Iroko (1982) note how termite mounds are seeded in West Africa, stressing how this is a specialist and generally secret endeavour:

> '... it is this land chief who established the first relations with the land divinity and who installed on this land the divinities of his people. This pact takes the form, among the Aïzo for example, of burying a piece of termite mound containing or not the termite queen mother. At the place chosen, one pours some millet or maize flour, mixed with palm oil. If, at the end of a few months, the termite mound reconstitutes itself, it signifies that the pact has been favourably registered by the land divinity. In such a case it is the definitive settlement under the authority of the chief of the people who proceeded with the rites of the pact, and who is by this fact the veritable chief of the land, that is to say, the land tenure priest, intermediary not only with the land divinity but also between all the other divinities and the member of the new community. It is he who is charged with distributing to members of the community, periodically, the new land to clear following precise directions, as a function of the space already occupied by chiefs of surrounding land' (Mondjannagni 1975: 163 my translation; cf. Iroko 1982: 55).

Why people might want to seed termite mounds should become apparent later in this paper.

Termites and vegetation: settlement and society

Our research on vegetation in West Africa was focused on how farmers manage vegetation in a part of the forest-savannah transition zone. In this ecological zone, when water, soil and fire conditions are more favourable, dense humid forest can become established, but where they are absent there are starkly contrasting grassy savannahs, stabilized by regular fires during the particularly marked dry season. In these savannahs, trees tend to be grouped on and immediately around inhabited or abandoned termite mounds, presumably because of the higher fertility, improved water access, or reduced fire intensity (as grasses fare worse on these mounds). Tree clumps come to form islands of dense vegetation – almost forest – surrounded by savannah.

Some ecologists have argued that in forest-savannah transition areas, this effect of termite mounds in altering soil conditions in favour of forest vegetation can assist in the extension of forest cover into the savannah lands. 'Where favourable conditions permit, these islands of woodland increase in size until they merge with each other and produce a closed canopy' (Harris 1971: 73-4). This association of termite mounds with woody clumps, especially mounds partially or totally abandoned by termites, is noted in most ecological zones in these regions, from much farther north, in the Sudanian zone of Burkina Faso (Hauser 1978) to the forest margin zone (cf. Begue 1937; Abbadie *et. al.* 1992) but not in the Sahelian zone, where the converse is the case (see Guinko 1984). In a similar ecological zone in South America, farmers are known to initiate new forest island sites in savannahs by transplanting termites to new locations, creating the conditions for their establishment and eventual mound-building (Anderson and Posey 1989). (In this, farmers use special practices which require the co-transplanting of particular ant species alongside particular termite species. Islands of woody vegetation (*apêtê*) in savannah are formed through the active transfer of litter, termite nests and ant nests to selected sites. This substrate then serves as a planting medium for desired species, and also facilitates the natural succession, which is further enhanced through active protection of *apêtê* when the savannahs are burnt. Compost mounds are prepared from existing islands where decomposing material is beaten with sticks. Macerated mulch is carried to a selected site (often a small depression) and piled on the ground. Organic matter is added from crushed *Nasutitermes* sp. (termite) and *Azteca* spp. (ant) nests. Live termites and ants are included in this mixture. According to the informants, when introduced simultaneously the termites and ants fight among themselves and consequently do not attack newly established plantings. Ants of the genus *Azteca* are also recognized for their capacity to repel leaf cutter ants (*Atta* spp.). Seeds, seedlings and cuttings are then planted. Mounds are formed at the end of the dry season (Anderson and Posey 1989)).

Farmers are familiar with this association of trees with termite mounds, and identify particular tree species normally associated with mounds (cf. Guinko 1984). These are often fruiting species which certain mammals, such as palm rats, bring to abandoned mounds, where they make their digs. These animals, as well as those that live around termite mounds, also fertilize them. And just as trees tend to grow around termite mounds for all of these reasons, so termite mounds tend to form around certain trees where the termites find food.

The way termite settlements influence vegetation patterns offers Kissia and Kuranko ways to consider their own influence on vegetation. Their villages almost invariably lie in open clearings in the middle of dense, semi-deciduous forest patches, which form 'islands' of forest in the surrounding savannah. These peri-village forests provide protection against fire, convenient sources of forest products, shelter for tree crops, concealment for initiation activities, and, in the past, fortification. The forest islands are formed largely through everyday activities. Targeted grazing and thatch collection reduce flammable grasses on the village edge. Gardening on village fringes 'ripens' and 'matures' the soils, animal and human manure fertilizes the soil, and trees establish themselves quickly either when transplanted or when their seeds are distributed by domestic and wild animals (Fairhead and Leach 1996). In short, in a way analogous to termite 'settlements', people's settled social life in savannahs tends to promote forest.

Farmers also associate forest islands with abandoned village and farm hamlet sites (in Kuranko *tombondu*; in Kissie, *ce pomdo*). The forest vegetation they carry exists today as the legacy of the everyday lives of past inhabitants. The spatial patterning of people's settlements and forest islands is remarkably similar to the patterning of termite settlements and islands; similarities on which villagers reflect when interpreting their own landscapes.

In their effects both on soils and on forest island formation, termites thus seem to provide Kissia and Kuranko with 'hints and clues' about living and working with ecology (cf. Richards 1992). Analogies between people's and termites' manipulation of ecology gain further plausibility in local thought from the analogies in their social organization. Kissia and Kuranko recognize in termite organization a social world parallel to their own: one of male and female chiefs, and of different categories of worker, all living within a village. Kissia describe termite society as led by two *Kolatio* (leaders) which lie on an east-west axis; a male on the east, and a female on the west. They inhabit the heart of the mound (*telekotin* in Kissie). They and the majority of termites are protected by *Kangua*; the soldiers.

The social metaphor of termites was noted by early Portuguese visitors to West Africa.

Let us conclude by discussing the *baga-baga*, a kind of ant. Its king is one of the same kind, but bigger, that is, longer and thicker. The society of this animal is a well organized natural republic. The royal palace is a mound of earth like a pyramid, filled with cells inside. The female subjects serve their leader by surrounding him in the centre of the tower. Among them is to be found the heir and successor to the monarchy, an ant that has a body similar to that of the king, but which has the capacity to grow larger. This superior ant is served with all respect and all the signs of natural love. It never leaves the ant-hill; the others bring it the delicacies of mother nature. These ants make war on other, smaller ants (Alvares 1615 II, 1, p.15).

The metaphorical significance of termites to society and authority is alluded to by Iroko in Benin, where royalty eat products incorporating royal termites as an ingredient: 'A sovereign who subjected to such treatment is supposed to be always and loyally obeyed by his subjects, in the manner in which royal termites is unquestionably the object of tender care on the part of the latter' and numerous sovereigns have derived the power of their speech and domination from this practice' (Iroko 1982: 67).

That termites work in large, highly co-ordinated groups, and in a disciplined manner, embodies values strongly upheld within people's initiation societies and labour group organization. Termites seemingly 'know what to do' and respect authority, just as initiated people should; and reveal – as the opening quotation notes – how community life depends on solidarity and forgetfulness of self. In Bambara, the word *ton* (termite mound) is synonymous with 'rule' and 'work group'. In Kissia initiation societies, the sign of future leadership, carried by the senior-most new initiate in the coming out ceremony, is the chameleon – an animal which, as we shall see, has a role in termite control.

Iroko (1982) shows how the course of people's migration history, and decisions of where to settle, not just where to farm, can be strongly influenced by interpretations of signs found in termite mounds, made by those initiated into the knowledge of how to interpret such signs. He goes on to show how choice of settlement location is, throughout Benin, dependent at least at an ideological level upon the presence of termite mounds. (He notes this for 'most localities of Borgou and Atacora, *Aïzo* localities such as Adjan (Allada), *nago* localities such as Sakete, and the land of *Ganmis* (Ifangny)' (9182: 54)). For example:

> 'The impressive number of termite mounds scattered across the landscape of the majority of the zone inhabited today by the Tchangana of Borgou seems to have been, in this region, one of the reasons for settlement of the land. Un-

der the reign of Adandozan of Abomey (1797-1818), an old migrant called Dandji, followed by 31 children, left Ato-Agokpou in Togo and settled definitively on a site where the abundance of termite mounds was for him a foreteller of prosperity. Klouekanmey was thus founded. He subsequently discovered shards of pottery, old wood pipes and stone tools, and concluded that others before him had lived there, attracted by the termite mounds. (Iroko 1982: 54). (However, it is possible that this conclusion is the wrong way round, and that the proliferation of termite mounds was due to the fact that others had lived and worked there.)

Termites also give people hints and clues about the timing of their agricultural activities, a key element to considerations of fertility in the region. The flights of termite alates (winged reproductives) of the different species which occur with absolute regularity each year provide one of several natural clocks by which people synchronize their farming activity.

People do not always welcome termites. Certain species destroy buildings and others (in Kissi, those inhabiting the *telin* type of mound) prevent rice growth and farmers attempt to evict them, despite the edible fungi they furnish. Everyday activities in the village can sometimes serve to deter termites from infesting habitations, and the incessant pounding of rice in Kuranko and Kissi villages, for example, is said to scare them away. In Mid-Western State, Nigeria, people invite drummers to beat drums in their houses in order to evict termites (Malaka 1972). Kissia and Kuranko have told us of a range of ways in which to evict termites from their mounds. They channel surface water flows into them, or dig out the queen. *Telin* mounds in fields can be destroyed by inserting into them the scales of a pangolin – an animal that eats termites – or a chameleon. One can also ask a young, presumably uninitiated, girl to urinate on the mound. (Some understanding of the logic of this can be derived from comparing it with Cros 1990). Once unwanted termites are evicted, their mounds are often colonized by beneficial species.

In some cases, control is exerted over termites in fields in order to improve rice production. Kissia elders responsible for sowing rice mix it with *wangaa*, a substance that enhances fertility. A main ingredient of this is crushed chameleon bones, which, one might surmise according to Kissi logic, limits termite damage to crops, just as chameleons evict termites from their mounds. A second ingredient could be a product that attracts termites (see later – for example *mangana* seed), thus creating a substance that simultaneously harnesses termite's local beneficial effects to a growing seed (especially in moisture control), while preventing their negative ones (by limiting termite grazing). As such knowledge lies on the verge of secrecy and sorcery, it is difficult to obtain detailed information.

These aspects of termite ecology and society and their manipulation provide a context for considering some further significances of termites in regional traditions. In this we should state at the outset that we are going to gloss over the fact that, in certain cases, knowledge of such traditions is linked to social and political institutions, not least to the many initiation associations (secret societies) in the region, and access to the education they offer. Proper consideration of this knowledge would involve examining the political use of termite symbolism and secrecy, rather than the symbolism per se (cf. Bledsoe 1984; Bellman 1984). Equally, a longer and more deeply researched analysis would examine alternatives and challenges to explanatory possibilities offered here, perhaps in the expression of different initiation associations, social groups, or institutions of state education, Islam, etc. Such an analysis would be linked to modern political and economic struggles and competition for authority. Neither of these issues falls within the scope of this short paper which seeks only to highlight the potential importance of termites within certain symbolic and political orders, and to show how their role in these is consistent with some aspects of everyday experience in both agro-ecology and human ecology.

Ecology, society and water relations

In the myth and symbolism of the region, considerations of humidity are central to the way people consider the origins and hence the nature of many things. Myths of creation frequently allude to a state of dry barrenness which is given life through receiving humid breath and water. And as might be expected from the way farmers interact with termites in everyday farming, termites and their influence on humidity are central to such narratives.

Perhaps the most elaborately documented West African mythology is that of the Dogon. A Dogon view of life is that 'the more a being is alive, the more it needs water, the faster it dies of thirst' (Calme-Griaule 1965: 247). In their eyes, according to this author at least, certain animals and plants seem to manage easily without water, particularly the Pale Fox which astonishes the Dogon with its ability to live in the dry season and its hydrophobic tendencies. It shares this status with the tree *Acacia albida* which perversely flourishes during the dry season when it comes into leaf, but drops its leaves in the wet season. It is partly for this reason that the Pale Fox is the enemy of *Nommo*, the Dogon saviour. Dogon contrast these hydrophobic enemies with beings that conserve humidity during the dry season, notably the tree *Urena altissima*; and more central to our story here, the termite. As Dogon note, the places where they live are always humid. These are 'friends of water' (Calme-Griaule 1965: 247).

This association is well portrayed in a Dogon funeral libation: 'People of the termite mound of Enguele, people who have trampled clay, people of the water of termites drawn by god (*Amma*), that day is had, goodnight' (Dieterlen 1987: 88). In Dogon origin stories, when God, imaged as male, created the universe he did it with two assistants, sometimes called his 'wives': the termite and the ant. In this sense, the termite exists prior to creation, extant within the 'egg of creation' and as Griaule notes, 'the only witness to the creative thought of God (Griaule and Dieterlen 1991: 205). Termites assisted God, for example, when he was having problems with the Pale Fox, who had stolen the seeds of his creation and taken them to earth. God sent termites and ants to watch over the fox. The ant was ordered to retake the stolen grains, while the termite was ordered to divert all the humidity of the initial earth (a primordial placenta at the time) away from the hose where the seeds had been sown to prevent them from germinating, and to eat any grain that did germinate (Griaule and Dieterlen 1991: 208). (Note how this is the opposite of the termites' normal role – to bring humidity, and not to eat seeds – a role seemingly enhanced by the *wangaa* mentioned earlier.) While in this case the termite dries the soil, the importance alluded to is the termites' ability to channel humidity within it. In this same part of the myth, the termite (*tu*) actually gains a second name, 'the water drawer of God'. It was the termite that drew to the surface the water that enabled the Pale Fox to drink (Griaule and Dieterlen 1991: 205). As we have seen, ecologists accord with this old wisdom, finding that termites do 'bring water to the surface'.

This importance of termites to water flow finds numerous echoes in West African mythology. Where termite mounds are found at water sources or beside swamps or rivers, they are often shrines to the perennial nature of the water flow, and the sites for offering prayers and sacrifices to ensure it. The mythical association of humidity and water flow with termite mounds seems to be fairly general (Zahan 1969: 25-26).

Shrines to water-related termite mounds become more comprehensible when one considers a second element in this hydro-ecological tradition, *ninkinanka*; a water motivating force often incarnated as a rainbow in the sky and as a python on land. In Kuranko villages, when a rainbow is visible between the viewer and an approaching cloud it is usual for the cloud to alter its course, a phenomenon which villagers pointed out to us time and again, and which seemed (to us) to hold. Within regional mythology, the rainbow, *ninkinanka*, is taken to exert control over rain and the weather, and rituals to it have been noted across the region (Griaule and Dieterlen 1991: 156; Paques 1953: 70).

In both our study villages, *ninkinanka* is said to emerge from particular termite mounds: in Kuranko, of the *togbo* sort; in Kissi of the *telen* sort. Should one hit the earth of these mounds, the ground echoes, suggesting a cavernous space beneath. The soil around them is hard, and one knows not to cultivate. One Kissi informant suggested that within these termite mounds, *ninkinankalen* rainbows derive from the open mouth of an animal, *koka bebendou*. (This animal, with a long tail, we cannot identify. Perhaps it is mythical; perhaps it is the Great Pangolin, which lives in such holes, and which in local opinion feeds on termites, ants and palm fruits). Others suggest that the rainbow emerges from the python, *piowvo*, represented as its breath or spittle (Millimouno 1991: 25-26). In Kuranko, within the earth, *ninkinanka* takes the form of a snake, which lives in these mounds, and the rainbow seems to be a metamorphosis of it.

Wherever we have worked, farmers have explained how the rainbow emerges from one termite mound, often in damp conditions or beside a pool, and arcs over to another, from which these damp places receive their water. As the python, *ninkinanka* moves underground. Some Kuranko explain how the underground movement of the python and the airborne arc of the rainbow enjoin in a circle, and find in this a way to image the path of underground water flow. The underground half of the circle describes the paths of water flow which render water sources and particular field sites humid, often more humid that similar places elsewhere. *Ninkinanka* thus replenishes the rain clouds and upland water sources which enable these underground flows to keep going. It explains the existence of sources which seem never to dry up. Indeed, certain of these sources are associated with ancestors who are said to have built the particularly fruitful relationship with the *ninkinanka* on which a community can thrive.

Elsewhere, the rainbow-snake is sometimes said to circulate between large hills. In the Kissi village of Lengo-Bengo, for example, the rainbow-snake circulates between a drier male part of the mountain (*lengo piandu*) and a more humid female one (*lengo laandou*), the latter being a vast hillock which thus remains moist and good for cultivation. The rainbow-snake's movement assists the inhabitants who make offerings to it, and its movement is accompanied by rattles, an instrument played by women. (This information is taken from a work whose author is unknown to us. The reference is: *La portée philosophique des contes, légendes et proverbes en milieu Kissi - Centre d'application: Préfecture de Guékedou*. Memoire de diplome de fin d'études superieures, Université de Kankan (19 ème promotion). Faculté de sciences sociales, Dept. Philo-Histoire. pp 57-59.) In Bambara regions of Mali, Paques (1953) describes several examples where the rainbow-snake moves between hills, showing again their movement between male and female hills, where the latter are humid. Indeed the latter is sometimes not a hill but a pool. (The role of the rainbow-python in relation to water transfer is generic to West African mythology: see Rouch 19xx; Ellis 1965).

The python's underground pathway between the points described by the rainbow is sometimes said to be along channels dug by termites. This role of termites as assistants to such snake spirits was noted by the earliest European visitors to West Africa: 'It is said that these great [snakes] are found in swarms in some parts of the country, where there are also enormous quantities of white ants [termites], which by instinct make houses for these snakes with earth which they carry in their mouths. Of these houses they make a hundred or a hundred and fifty in one spot, like fine towns' (Cadamosto, fifteenth century, in Crone 1937: 44). The idea of termites as 'servitors of spirit snakes' persists in modern myths. In a Koniagui tale, for example, the snake-spirit asserts to his woman captive in a mound that 'the termites are mine: feed them with your fat, and then I will eat them' (Houis 1958). (Termites are, of course, not only god's assistants in creating humidity, but also in disposing of corpses, and blood spilt on land.) In the Kuranko village where we lived, areas inhabited by pythons in amongst many termite mounds were indeed termed *nyina* 'villages'. The close relationship between pythons and termites is elaborated throughout West Africa (for example, Hambly 1931: 11 for Benin; Calme-Griaule 1965 for Dogon).

Conclusion

In this brief essay we have placed termites centrally in examining certain aspects of farming and ecological knowledge. In the region, termite mounds are considered to bring benefits to crops, improving soil fertility and water relations, and in the long term influencing vegetation succession and patterns, and the dynamics of fallows.

The ways in which termites are thought to play this role are linked to broader understandings of influences on water flows and fertility, elaborated in the cosmogony of the region in which termites feature. Given that people derive power from linking in to, and altering, these broader influences on water flows, and fertility, and given that termites and their mounds provide a potential way of exerting influence over the basic forces of fecundity, termite mounds seem to provide a focal point for the diverse political institutions in the region's authority structures. Thus we find the power and position of termites in ritual procedures to ensure humidity and fertility; and in facilitating 'tenurial' authority over particular fields and swamps. In a longer paper we would have elaborated on the importance of termites to the region's village power associations which deal with fecundity - especially twinning - and humidity at a wider level.

Ecologists and agronomists are beginning to appreciate the importance of termites in soil formation, and are seeing possibilities of using termites in soil 'rehabilitation'. This moves away from considering termites as a 'pest' merely to be eradicated. In this shift, the agronomists are moving closer to the ways that many African farmers have long manipulated termites in their struggle to produce. Agronomists and ecologists interested in this aspect of indigenous science could usefully, we suggest, study local knowledge of termites in the narrower ecological sense (e.g. in understanding the effects of particular tree species on termite activity). Like the region's farmers, they would gain a more profound understanding through attention to the broader political traditions within which specific ecological knowledge is located.

We hope that this review of 'traditions' relating to termites exemplifies how social scientists might profit from closer consideration of peoples' experience and representation of ecological phenomena. Anthropologists themselves have been tempted to treat termite mounds as banal, seeing little in their use as shrines other than the fact that they are there – as strange features punctuating the landscape just as do certain hills, rocks, trees and streams. There are also many theories, now fashionable, that permit social analysts to overlook the content of specialist knowledge, as if it were more the fiction of colonial enthnographers than the lore of West African society, and irrelevant anyway to everyday things. Yet the phenomena in which termites are implicated comprise a roll call of the very phenomena that have long been central in the study of West African society. Termites depict the cardinal points in the construction of their mounds, and some say adapt the air vents of their constructions in keeping with the stellar calendar. The annual flight of the winged reproductives (alates) of each termite species occurs predictably (and rather miraculously) at a certain hour on a certain night each year, providing a precise seasonal timekeeper for agriculturalists and ritualists alike. When they emerge, the alates head directly to a light in the night sky; to the moon, the stars (or the fire that people wave over the mound to collect them as food). Termite mounds control and relocate moisture. They create fertile soils, enriching the topsoil not only from the subterranean earth but also from the recycling of all flora and fauna, and from the blood, sweat, tears and corpses of people. Termite mounds provide homes (and food) for pythons, pangolins, chameleons, palm rats, and Pale Foxes. They regulate the growth of certain flora, most notably the germination of baobabs (*Adansonia digitata*) and the growth of fungi. Their mounds are also associated not only with the growth of the specific *Ficus* and other species mentioned here, but also provide the favoured locations for the growth of *Khaya senegalensis*, *Milicia excelsa* (iroko, *syn. Chlorophora excelsa)* and *Antiaris africana*. Dug out, termite mounds serve as prototypical blast furnaces for iron smelting. And termites provide an enviable example for social altruism (albeit rather monarchical). When a granary, eaten out by termites, collapses on those shading themselves beside it (Evans-Pritchard 1937) perhaps the injury to those involved might be attributable to witchcraft for more reasons than an inopportune coincidence of events.

Mirrors to humanity? Historical reflections on culture and social insects (J. F. M. Clark)

'I have... some observations to make: come, follow me; you will find them interesting.' So instructed the exiled Napoleon to his physician and companion, Francesco Antommarchi, when rain interrupted both their efforts to excavate a large water basin and their discussion of Napoleon's military campaigns. Upon investigation, Antommarchi discovered that ants were the objects of Napoleon's observations. With appropriate military metaphors, Napoleon described the tiny insects' responses to his various manoeuvres with a sugar bowl:

> 'This is not instinct,' said he, 'it is much more – it is sagacity, intelligence, the ideal of civil association...'. 'You see it is not instinct alone that guides them.... However, be the principle which directs them what it may, they offer to man an example worthy of observation and reflection.... Had we possessed such unanimity of views...' (*cf.* Antommarchi 1825)

Using his observations upon ants as a springboard for a polemic about the legitimacy of his leadership, Napoleon argued that patriarchal theories of government were, 'ridiculous pretensions' when compared to 'a man of the people' such as himself. Antommarchi was no stranger to the possible conflation of nature and society, or, indeed, of science and sedition. In London, en route to St. Helena, he believed himself under suspicion, because he possessed anatomical plates for a projected book: 'In the present age everything conspires; and muscles and tendons might compass the death of kings or communicate with usurpation!' Through his narrative of Napoleon, he orchestrated the confluence of theories of government and observations upon the behaviour of social insects.

Perceived similarities and differences between humans and other animals have long provided a focus for discussions about the natural world. Definitions of humanity often hinge upon comparisons between 'man and animals'. Whether for the purposes of self-definition or legitimation, humans tend to project their own social and cultural arrangements onto their interpretations of nature (*cf.* Thomas 1983). Anthropocentric agenda usually determine the level of interest in particular facets of nature. In the late eighteenth century, for instance, naturalists acknowledged a pervasive 'imperfect acquaintance with the various tribes of insects'. Because of their estrangement from the rest of the 'larger animals', insects provided insufficient grounds for anthropomorphic analogies. What, naturalists complained, were they to make of antennae? (Paley 1802).

The collective organization of social insects, however, provided analogies laden with social and cultural significance. Although endowed with an alien morphology, ants, bees, wasps and termites became subsumed within discussions about the relative intelligence and moral utility of animals. Throughout the nineteenth century, they were discussed at length in professional scientific journals, in various periodicals of the new higher journalism, in popular lectures to working men and women, in travel narratives, and in didactic children's primers. As social insects, bees and ants in particular provided scientists and social commentators with biological analogues to social and cultural concerns, such as the governing principles of society, gender roles, the division of labour, cleanliness, funeral rites, altruism, slavery, evolutionism and 'character'.

Traditionally, bees were nature's penultimate proof of the legitimacy of monarchical government. When the spectre of Catholicism threatened the stability of an already fragile monarchy, Moses Rusden came to its defence with his *Further Discovery of Bees* (1679). Knowledge of the hive, he argued, affirmed

Figure 5.3: John Gedde's box-hive, as presented in Moses Rusden, *A Further Discovery of Bees* (1697).

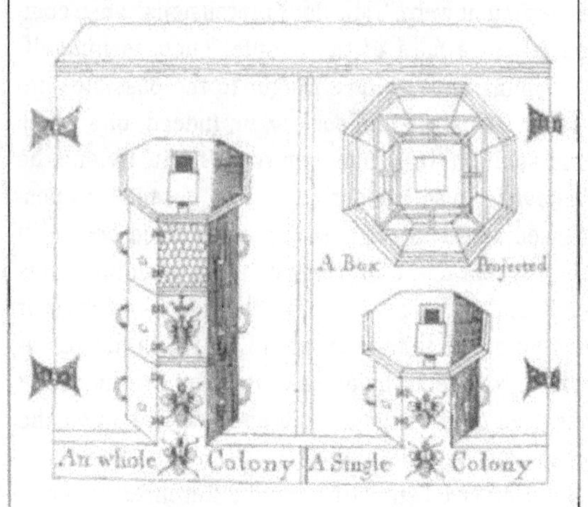

Source: Photograph taken from D. J. Bryden, 'John Gedde's bee-house and the Royal Society', *Notes and Records of the Royal Society of London,* **48** (1994), 192–213 (Figure 4, page 202).

Figure 5.4: The bee-hive as a model of social organicism: George Cruikshank's 'The British Bee Hive'. Process engraving. Designed in 1840, altered and etched in February 1867, and published in March 1867.

Source: Photograph taken from William Feaver, *George Cruikshank*, ed. Hugh Shaw (Arts Council of Great Britain, London, 1974). Copyright William Feaver and Arts Council of Great Britain.

Figure 5.5: 'Prince Albert's Bee-Hives.'

Source: Punch, or the London Charivari, **7** (1844), pp 90–91.

Figure 5.6: *Punch*'s 'Fancy Portrait' of 'Banking Busy Bee' Sir John Lubbock, author of *Ants, Bees, and Wasps* (1882), and numerous other publications on social insects.

SIR JOHN LUBBOCK, M.P., F.R.S.
How Doth The Banking Busy Bee
Improve His Shining Hours
By Studying On Bank Holidays
Strange Insects And Wild Flowers!

Source: Punch, or the London Charivari, 19 August 1882. Photograph of a reproduction done by the Wellcome Institute Library, London.

the natural roots of a patriarchal monarch, and refuted 'such as look with malicious Eye upon Kingly Government, as being the effect of necessity and force, and not of natural inclination, and of choice ...'. Appearing under the heading, 'Monarchy founded in Nature, and proved, by the History of Bees, etc.', the body of the text informed the reader that king bees have 'a certain spot in their forehead, in the shape of a Crown'. And although Charles Butler believed that the 'first swarm went forth at the will of the Commons', 'Bees never swarm but by order and consent of the King', for they 'naturally abhor Rebellion and Treason'. Ostensibly, Rusden (1679) arrived at these conclusions after observing bees through the transparent box-hives of John Gedde. The King was impressed enough to order the erection of similar hives at Whitehall, Windsor and at Falkland Palace. Apothecary and royal Bee-Master, Rusden effectively combined experimentation and observation with polemic.

Although the bee-hive as a hierarchical, organic model of society proved long-lived and resilient, a number of factors contributed to its eclipse in the late eighteenth century. With its emphasis upon observation and experimentation, the seventeenth-century scientific movement ushered in a natural philosophy, which was committed to a mechanistic world-view. Nevertheless, the ascendance of natural theology demonstrated that scientific inquiry could continue to draw moral guidance from nature. The search for evidence of the benevolent hand of God – the Architect, Craftsman, Contriver – in nature produced a happy marriage of science and religion. And one of the most popular arguments from design was the construction of the cells of bees. The bees' ability to construct geometrically precise hexagonal cells with rhomboidal bottoms was proffered as evidence of a supernatural template implanted in animals as part of the beneficent design of the Creator. The bees' maximum economy of wax and space provided a cogent example of teleological utilitarianism; mathematicians manipulated figures to confirm the perfection of God's contrivance (Merchant 1980). Aware of its natural theological cachet, Charles Darwin acknowledged that 'so wonderful an instinct as that of the hive-bee making its cells will probably have occurred to many readers as a difficulty to overthrow my whole theory' (Darwin 1859).

From the martial skills of Napoleon's ants to the architectural talents of Darwin's bees, a common concern underpinned discussions about the social Hymenoptera. In the latter half of the eighteenth century, bees and ants became more than just the traditional preserve of hierarchical, organic utopias: they supplied important evidence for Enlightenment values, which challenged the hierarchy in nature and society. As natural history metamorphosed into a history of nature, the behaviour of insects threatened to blur or extinguish the distinction between man and beast (Foucault 1970). Discussions pertaining to instinct, reason, intelligence and habits became an ideological battleground. Drawing upon Enlightenment rationalism and its theories of knowledge, social reformists and early evolutionists sought to rupture the divide between man and beast by negating a mechanistic brute instinct. In direct contradiction to the Cartesian belief that animals were hard-wired machines, naturalists and other intellectuals maintained that they were capable of experience and learning. Furthermore, they saw the mind as an active agent in species' adaptation or transformation. The experience of different environments, they argued, produced different habits in animals, which, through constant repetition, created new anatomical variations. This depiction of an atomistic, self-developing nature had important social and political implications. If all ideas were simply manifestations of sensations, and if rational behaviour was simply the association of ideas, all sentient beings were capable of perfectibility. Discoverable natural laws of behaviour could sanction social and educational reform from the bottom up – from insects to elephants, from radical dissenters to reactionary conservatives. Consequently, natural philosophers argued in earnest whether the complex behaviour of bees and ants was the result of instinct or reason (Richards 1992).

At the same time, social theory began to displace abstract constitutional and political arguments. The emphasis, therefore, shifted from discussions of political authority, obligation and the right of rebellion, to concerns about moral restraints and sanctions, and the social hierarchy (Hole 1989). In the 'age of revolution' these changes were reflected in descriptions of social insects. From the early seventeenth century, naturalists had grappled with the realization that the 'king' bee was perhaps a 'queen'. Admission that a patriarch did not rule the hive threatened the order and stability provided by this natural model of a monarchical government. Moreover, the suggestion that the 'queen' might take multiple mates vitiated the dignity of the monarch. Consequently, natural philosophers equivocated about the gender and sexuality of the 'ruling' bee for more than a century. In Revolutionary France, however, naturalists displayed a decided discomfort with a 'king' or 'queen' bee. They concluded that there was no monarch at all in nature, and redefined the queen as the 'mother' bee, whose principal role was that of egg layer. The real power in the hive belonged to the workers. Between 1789 and 1793, the old aristocracy and the monarchy were murderously displaced in Revolutionary France. Under the banner of scientific objectivity and exactitude, French naturalists redefined the queen bee as the 'mother' or 'female' bee, and attributed harmony in the hive to a universal dedication to the common laws (Drouin 1992).

In the early decades of the nineteenth century, the question of gender became increasingly complex. Detailed anatomical investigations of the worker or neuter bees and ants affirmed that they were infertile or 'imperfect' females. This fact became especially significant when the commentators of genteel Victorian society confronted a perceived increase in the number of unmarried women. Census statistics revealed that women far outnumbered men; surplus, or 'redundant', women faced no hope of marriage. In the 1850s and 1860s, a plethora of articles and pamphlets addressed the 'spinster problem' (Martineau 1859), 'and discussions of insect sex and gender became fraught with metaphorical ambiguities' (Clark, 1997).

Whereas bees traditionally represented the ideal monarchy, ants were usually considered the archetype of communal organization: 'for the good of their democratical state, each ... [ant] mutually employs his pains by turn' (Muffet 1658). As the spectre of socialism spread throughout Europe in the latter half of the nineteenth century, these distinctions between ants and bees were often forgotten or blurred. In fact, critics of socialism frequently turned the hierarchy of the bee-hive back upon itself. T. H. Huxley warned secularist William Platt Ball: 'Have you considered that State Socialism (for which I have little enough love) may be the product of Natural Selection? The Societies of Bees and Ants exhibit socialism in excelsis.'(Huxley 1900).

Interestingly, Ball's response focused almost exclusively on bees. First, he identified the uniqueness of bee and ant socialism. It was limited to a single family, with 'only one mother to the whole community'. This essential preliminary to the evolution of insect socialism 'is not easy of realisation among mankind'. In addition, he argued, the socialism of bees was hardly a desirable model: drones are massacred; jealous mothers kill their fertile daughters; industrious working bees work themselves to a rapid death; sufferings of fellow bees are dealt with callously; and 'socialistic hospitality' towards strangers amounts to a murderous, xenophobic stinging frenzy (Huxley 1890).

In contrast to Ball, Russian émigré Peter Kropotkin drew upon social Hymenoptera for his systematic analysis of co-operation or 'mutual aid'. In opposition to conservative social Darwinists, the 'anarchist prince' argued that co-operation, rather than competition and conflict, was the universal basis of ethical principles, and the engine of evolution. He turned to nature for evidence: 'And if the ant ... stands at the very top of the whole class of insects ... is it not due to the fact that mutual aid has entirely taken the place of mutual struggle in the communities of ants?' (Kropotkin 1914). The same, he continued, held true for bees. Although Kropotkin used it to affirm his own anarchist politics, the 'co-operative' behaviour of social insects was a staple of socialist and collectivist arguments in the late nineteenth century.

Writing in 1904, H. G. Wells produced a short story in which divisive human individualism faced the apocalyptic advance of myrmecological collectivism *in excelsis*. Entitled *'The Empire of the Ants'*, it related the progress of a gunboat, which had been sent to provide relief to a remote Amazonian village that was besieged by carnivorous ants. Ostensibly, the boat and its crew set off on a positivist pilgrimage to assert control over nature and man. Ironically, the divisive multinational crew proved no match for the newly evolved species of intelligent army ants. They were left to a futile display of their impotent technology; they fired several rounds of their cannon, and then retreated with the knowledge that the fearsome ants would reach Europe by the 1950s or 1960s. These intimidating insects walked in an upright position and wore military clothing: 'they were', bemoaned the protagonist, 'intelligent ants. Just think what that means!' In effect, anthropomorphized ants portended the end of humanity. Trained under T. H. Huxley at the Normal School of Science, Wells deployed his knowledge of biology to produce a sceptical *fin de siècle* degenerationist vision (Bowler 1989). Evolution was not a guarantee of the progress of human civilization. Drawing upon H. W. Bates's description of army ants in his travel narrative *'The Naturalist on the River Amazon'* Wells created the *doppelganger* of confident British imperialism. Intelligent nature from the 'periphery' had begun its slow, inexorable, triumphal march towards Europe.

Fostered by an interest in travel narratives of distant lands, the eighteenth- and nineteenth-century passion for the exotic could be turned towards the 'alien empire' of the insects. After noting that travellers provide a steady stream of tales and observations, Latreille (1798) had pleaded: 'Would I also not have the right to publish an account of my voyages; to relate what I have seen or what I believe that I have seen?' The land that he explored was the 'empire of the ants' beneath every French person's feet. Like European travel and exploration writing, descriptions of insect societies provided self-affirmation and self-definition through explicit and implicit comparisons. Both travel narratives and literature on insect microcosms introduced readers to realms beyond familiar experiences and perceptions. The insects and non-European peoples that populated these unfamiliar realms served as objectified and culturally distant 'others'. Throughout the nineteenth century, however, social evolutionists most often denigrated non-European native peoples under the rubric of *'savage'*, and elevated the ant as a *'small people but exceedingly wise'*.

As travel narratives carried readers to distant lands, glass hives and colonies transported observers to unseen subterranean and hidden realms. They offered transcendent and synchronic views of nature: Lilliputian worlds which could be endued with social and cultural meanings (Stewart 1984). Thus, in the early 1840s, Prince Albert had glass bee-hives installed at Windsor Castle; and the public was treated to a trenchant satirization of the event in one of Richard Doyle's first cartoons for *Punch* (Houfe 1978). In a similar vein, Henry Cole promulgated the nineteenth-century 'gospel of work' with his announcement that the Great Exhibition commemorated 'the working bees of the world's hives' (Briggs 1982). And in the ultimate act of anthropomorphic irony, the Crystal Palace – that glass temple to human industriousness and ingenuity – later featured a 'Royal Exhibition of Working Ants'. As people paid their threepence to see 'glass nests of living ants at work', Sherlock Holmes was composing his *Practical Handbook of Bee Culture, with Some Observations upon the Segregation of the Queen*, by watching 'the little working gangs [of bees] as once .. [he] watched the criminal world of London'. 'Social insects could be mirrors to humanity.' (Conan Doyle 1981).

Although glass bee-hives and ants' nests shared much in common with the nineteenth-century passion for placing nature under glass, there were important distinctions. William Kirby, the doyen of nineteenth-century British entomology, exhorted 'every patriot Zoologist' to collect and classify, because once 'an animal subject is named and described, it becomes a ... possession for ever, and the value of every individual specimen of it, even in a mercantile view, is enhanced' (Kirby 1825). John Lubbock (1856), who shared his large country home with thirty to forty glass ants' nests, reacted against this myopic bias on 'collecting'. Drawing on a common analogy, he argued that insects in a collection should be like books in a library; they must be studied to be of value. More important, 'collecting for the sake of collecting' represented an unconscionable lack of appreciation for the variety and diversity of habits of the living insect world. Edmond Wells, the enigmatic myrmecologist in Werber's (1996) recent novel *'Empire of the Ants'* draws similar insights from his observations on ants and the human tendency to anthropomorphize them: 'Nature, with all due respect to Mr. Darwin, does not evolve in the direction of the supremacy of the best (according to which criteria, anyway?). Nature draws its strength from diversity. It needs all kinds of people, good, bad, mad, desperate, sporty, bed-ridden, hunchbacked, hare-lipped, happy, sad, intelligent, stupid, selfish, generous, small, tall, black, yellow, red and white. It needs all religions, philosophies, fanaticism, and wisdom. The only danger is that any one species may be eliminated by another.'

Perhaps Wells was more prescient than he probably intended. Unreflective anthropomorphism may represent an apocalyptic vision. The true value of nature does not necessarily lie in possessing it or transforming it into human terms. Motivated by the same sentiments, Tim Ingold (1993) has campaigned to free discussions of human beings and non-human animals from anthropocentrism and ethnocentrism through a recognition of ontological equality. We must, he argues, acknowledge humans and other animals as co-participants, 'as centres of perception and action, in a continuous life process'. In the words of Mitya, the Russian firefly, 'It all depends ... on what side you look at things from. Stop. Just who is looking, anyway? And at whom? ' (Pelevin 1996)

Landscape perception and sensory emplacement (Terence Hay-Edie)

In the phenomenological description of place such as the 'set apart' or numinous presence of a sacred site, it is important to pay sufficient attention to the auditory and olfactory *patina* which may not appear in lexical descriptions and categorizations of the world. Sound and smell, by their 'invisible' nature, have too often been readily subsumed by Western oculo-centric observers into the symbolic realm without exploring the encompassing material reality created by sonic and olfactory features of a natural landscape. Only recently, ethnobotanists have begun to study the fragrance of flowers and plants. Whilst the task of describing existential memory inscription focuses on discrete environmental features such as specific mountain peaks, tree species, bird calls or rocks and other landmarks transformed through memory into spirit-places or the abodes of ancestors (features of 'memory magnetism') the encompassing superstructure of a more daily 'synesthesia' needs adequate attention.

The enveloping and comforting familiarity of lingering smells and sounds in a particular place, such as the fragrance of incense and the voices of monks chanting in a Buddhist temple, provide the constant backdrop of a stage where human actors move around the physical setting. In the case of natural sacred sites the cognitive differentiation of mundane sounds pervasive in the environment, such as the rustle of wind through porous textures of tree leaves, swaying grass or vegetation; across variably smooth flat surfaces such as standing or turbulent water; or the flowing of water down streams and mountainsides (sometimes cataclysmic in high altitudes when glacial lakes burst) present a continuous background 'texture' on the eardrum, only consciously 'foregrounded' at will.

Dense forest environments such as those inhabited by numerous 'indigenous' peoples in Papua New Guinea, and parts of Africa, Southeast Asia and lowland South America, present acoustic, visual and olfactory spaces quite unlike the open fractured landscapes inhabited by the majority of the population of the planet. Forest-dwelling peoples live in enveloping environments where sounds are trope-like as they travel through vegetation of varying density and rebound through space along multiple reflective surfaces of height and depth. For these peoples, cultural and symbolic signifiers, whether lexical or mythological, will wither and die without the soil and water of a physical landscape of these pervasive elemental essentials.

Steven Feld (1996) has recently coined the term 'acoustemology' to describe an inter-sensuality or 'synesthesia' as a 'perfume of hearing' which recognizes the diffuseness of the senses as a 'motional sensorium' where sound, sight and smell are incorporated into the same being-in-the-world. In the Bosavi region of Papua New Guinea, Feld (1996) points out that the Kaluli verb *dabuna* signifies absorption of sensory information through both ear and nose: multi-sensory 'sago-places' emit a smelling 'presence' of the aroma of fresh or rotting sago pith accompanied by the social history and mythological memory inscribed in the place by the activities of the ancestors. The composer Murray Schafer also describes the use of church bells in Northern Italy to signal different occasions; daily onshore-offshore wind cycles which carry a 'complete circumference of distant sounds' around a Breton fishing village; and Australian Aborigines' use of their eardrums to detect vibrations at ground level, suggesting research into 'soundscapes' for a 'post-literate' phase in the West, where the ear would return as a primary sensing instrument.

To 'capture the spirit' of a sacred place, a complete phenomenological description using all the senses must be invoked to achieve a complete picture of the 'essential elemental envelope' which enfolds a diverse array of different ecological indicators such as the timing of rain, solar intensity or wind direction. However, to essentialize landscape in this way may invite criticism for failing to differentiate between contested and appropriated perceptions of the land and its surface and for overlooking the role played by a cultural system in framing the perception of place. Yet in terms of a material phenomenology, once a landscape/soundscape envelope is removed from a cultural system, the symbolic syntagms lose all meaning on a 'bare stage' devoid of props (*cf.* Pimbert and Pretty 1995). In the case of forest-dwelling people this means the destruction of the auditory-olfactory enveloping cosmos of vegetation in a forest world enfolded back into a human internal cognitive map.

In a culturally and ecologically situated being-in-the-world the capacity to differentiate and categorize the diverse sounds of water flowing through a particular habitat; the taxonomic categorization and behaviour of birds and animals; and the patterns of sonic frequencies and solar intensities produced by different weather conditions, each provide indicators of the sensory and ecological engagement in an environmental envelope. Feld (1996: 105) describes how in Bosavi the land is nearly always wet from rain, allowing for 'foot indicators' to act as 'sensory ratios' relative to bodily emplacement in this life-world: '... the presence of wetness in the air and the slick, slippery feel of different thicknesses of mud on the feet are central to orienting oneself in visually dense places. Additionally, one simultaneously hears what kinds of water presences are above, below, ahead, behind, or to the sides and whether these waterways are diminishing or augmenting in and out of presence. This sensuality of locating and placing, along with its kinesthetic-sonesthetic bodily basis of knowing, is critical to a Kaluli acoustemology, a sonic epistemology of emplacement.'

For coastal, mountain or open plains environments the elemental envelope represents a less three-dimensional totality compared to the 'interiority' of peoples living within forests (Molyneaux 1995). Some societies recognize 'sacred caves' as an underground spirit world with an atmosphere comparable to churches and cathedrals. In the open space of plains, environmental perception relies less on the encompassing totality of vegetation on all sides and the location of sacred sites may become linked to a complex ecological knowledge at the micro-level to categorize the productivity of grass for forage, the quality and type of soil, or a highly developed sense of where to find water in arid areas. Mountains, peaks and vantage points allow for a semi-aerial perspective providing views and vistas over a landscape unavailable to groups living in forests and flat plains. However, virtually no human group can actually live on the top of a mountain – a practical nuisance if not an outright impossibility as mountainous situations are precarious and fragile environments where the elements can change very fast making human occupation a difficult and risky business. Until the Romantic period, mountains in Europe were feared as evil and dangerous (Thomas 1983). Beliefs and perceptions of sacred mountains are therefore, perhaps inevitably, 'bottom-up' where everyone has a common vantage point looking up towards a single peak.

A crucial component of any cultural categorization of symbolic features of the natural world will also depend, however, on the *stability* of the ecological system. Where significant change occurs on a regular basis all categorization systems including vision, sound and smell will through necessity attempt to address these ecological patterns. Environments are dynamic and change considerably over time as forests are cut and planted, soil is leached and erodes, long-term weather cycles alter rain patterns. Historical ecology which analyses the ongoing dialectical relations between human acts and acts of nature that are made manifest in the landscape may thus be a fruitful avenue to analyse *structural inversals* in associations of the 'sacred' with features of the landscape such as forests whose geographic spread and morphology can be severely altered by logging and clearing for agriculture.

On the coast of Kenya a number of Giriama 'sacred groves', or *kayas*, are protected by committees of elders from each village, responsible for restricting access into the clearings at the heart of the groves. They are recognized as 'National Monuments'. As Parkin (1991) shows, Giriama rites and beliefs have been reworked to accommodate locally variable influences of Islam and Christianity, and new spatial economic and ecological relations between communities. Parkin (1991) demonstrates that the Giriama concept of *'kaya'* refers to the 'homestead' clearing inside the grove, a 'sacred void'. The site of occasional large-scale ceremonies, it acts as a 'moral core of Giriama society, and a defence against total domination and assimilation'.

In ecological terms the 'sacred void' of the homestead clearing reflects a relatively recent conversion of former continuous forest by the agricultural ancestors of the Giriama, who migrated down from Ethiopia (Spear 1978). However, with pressure to clear more land for cultivation near the *kayas*, the concept of the 'sacred' has been increasingly identified by conservation organizations with the 'fortress' of vegetation *surrounding* the space in the middle. A recent study (1996) has shown that the Giriama themselves have also re-fashioned the interior space within the *kayas* as 'sacred' over the former domestic space of their ancestors who are sometimes associated with particular trees within the groves where they have been buried. An interesting counterpoint case of the *kaya* inversal is found in the forest clearings of the Huoarani in lowland Amazonia. Reminiscent of Melanesian material cargo cults, Rival (1993) shows how Huoarani have restructured their forest clearings to make room for oil companies' helicopters by adapting their 'traditional' housing settlement patterns at the perimeter of the clearing. In this instance the ecological situation has remained one of a domestic space within continuous forest, but the space therein has been re-arranged according to new social relations ('interior inversal'). In this way the 'memory work' of experiential (sensory) events in a dwelt-in landscape creates new networks of place-names and places of power. However, a question to be addressed is whether these 'networks of exchange' are relative to ancestral, social or contemporary inter-relations?

In mountainous bio-climatic conditions the forest resources have hundreds of other useful values besides timber, fuel-wood and fodder extraction, such as for the collection of wild fruits and vegetables and the preparation of herbal medicines and remedies. In Nepal, nine tenths of the population remain rural, depending on local forests to provide 90 per cent of household energy needs and 85 per cent of animal-fodder requirements; and an established culture of landscape management exists where particular forested areas serve a host of ecological functions to guard against soil erosion, maintain micro-climates, and balance watershed and hydrological run-off. Ethnographies of the world-view of the many different ethnic groups in Nepal also reflect widespread spiritual associations with forests, mountains and other 'natural sacred places' such as caves and specific rocks (Aris 1990; Ramble 1995).

The Sherpa population living in the Solu-Khumbu region of high altitude Nepal, bordering Tibet, have been well documented by anthropologists such as Fürer-Haimendorf who describes the *'shinggi nawa'* forest guards, and a strong tradition of landscape veneration in the form of animistic spirits dwelling in features of the landscape and propitiated to keep villagers safe from their divine wrath and evil misfortune. Beliefs in *Lu* spirits as half-human, half-serpent, female spirits inhabiting springs, trees, shrines and some houses are widely held among different Sherpa groups in Northeast Nepal. In a comprehensive volume on the cultural (or 'historical') ecology of the Sherpas, Stevens (1993) uses oral historical perspectives to analyse Sherpas' environmental history and adaptation to high-altitude conditions, and shows that as a part of their agro-pastoral and forest management system, Sherpas recognize a variety of different types of 'sacred forests' in Solu-Khumbu.

A determining factor remains the demand for timber for use in the construction of monasteries and for fuel-wood in the cold mountain climate. 'Lama's forests' *(lami nating)* established by the personal intervention of revered local religious leaders within the network of Sherpa monasteries; smaller 'temple forests' allied to specific monasteries; *'rani ban'*, the secular preserves created by the monarchy between 1912 and 1915 to prohibit unauthorized logging; along with a few private forests, ensure a lasting supply of timber for beams and joists in the construction and upkeep of monasteries and bridges. (*cf.* Stevens 1993: 196) However, a number of pre-monastic Buddhist Sherpa myths also surround particular 'sacred groves' – the home of *lu* type spirits in the upper reaches of Khumbu associated with the migration of the ancestors of the Sherpa over the mountains from Kham in Tibet some four centuries ago. These highly visible organic 'cultural markers' in the landscape provide a strong sense of historical depth and continuity nurturing the identity of Sherpas in the whole area of Solu-Khumbu. 'The sacred trees and forests that are such striking elements of Sherpa village landscapes represent substantial gestures of faith in a land where trees are so useful and so scarce.' (Stevens 1993: 196).

The earliest 'nodal' sacred grove is situated near the temple at Pangboche and is considered to be 'sacred' by the Sherpa not only because it is inhabited by *lu*, so that to cut any of the trees would bring bad luck, but also because the trees are believed to have been implanted four centuries ago by the earliest Sherpa religious hero, *Lama Sanga Dorje*, when he dispersed a handful of his hair to the wind – the hairs falling back to earth and rooting themselves as the juniper forest. This myth may be related to a version of the pan-Himalayan perception of forests along the mountain range representing the hair on Shiva's head, introducing wider questions relating to the symbolic significance, resonance and pilgrimage to the mountains from across South Asia and beyond (Einarsen 1995).

Conclusion

A host of scientific and cultural questions are raised by the dynamic changes taking place in and around natural sacred sites. Critically, a distinction remains to be made between idiosyncratic and collective memory inscription in place; constructed monuments such as temples, churches and other religious edifices; inorganic enduring features of landscape such as boulders, caves and mountain peaks 'made animate' through symbolism and world-view; and organic 'renewable' features of a landscape considered sacred such as forests, groves and water springs.

In an overview of many sacred sites from all over the world, Colin Wilson (1996) notices that many of the earliest religious monumental structures such as the pyramids at Giza and in Mexico were meant to resemble mountains, thus reproducing the permanence of a human-made structure as substitute 'natural' mountains. In this way all features of the natural and cultural world are affected by a degree of impermanence and transience – a theme central to all the major religions of the world, as well as the domain of historical ecology. These scientific and spiritual questions of linking environmental stability and cultural change may be fruitfully examined through the exploration of the shifts in spiritual identity and attachment invested in the varying degrees of associative and ecological change of sacred sites.

In a rough scale of their fragility and vulnerability to disappearance through loss of respect and traditional knowledge, organic natural sacred sites display most vividly the overlap between the continuation of the reverence for nature and the preservation of a healthy ecosystem. Nowhere is this more evident than in the case of those 'environmental envelopes' such as tropical forests where indigenous populations are quite literally re-enfolded 'within' the acoustic, visual and olfactory landscape. However, within the ongoing dialectic relationships of fragmentation and ecological change that take place in formerly extensive forests that are reduced to vestigial groves, a series of crucial 'inversals' may develop in relation to the 'sacred' and its phenomenological emplacement. Large forest areas may thus become disconnected from their former inhabitants and 'spiritually empty', while remnant forests may become spiritually 'alive' yet scientifically too small to regenerate and 'standing dead'. Future research will need to focus on this type of historical and mythological depth to reveal the interweaving of the spiritual and the scientific in and around natural sacred sites.

Cities, nature and protected areas: a general introduction (Jeffrey A. McNeely)

The process of urbanization

That the world is becoming urbanized is scarcely news. It is, nonetheless, worth considering present and possible future trends, especially for their implications for protected areas, people and biodiversity. The broad patterns of urbanization are similar to those for population as a whole: low numbers and slow growth for the past few thousand years; increasing numbers and rates of growth since about 1750; and an explosion since about 1950.

Throughout the past two millennia and more, towns have been an important part of the human condition. Several pre-industrial civilizations supported very large cities, such as Classical Rome with an estimated half a million, or Ming Dynasty Peking with a million people (Chandler and Fox 1974). But for most of human history the vast majority of populations lived in sparsely-settled rural areas, often quite isolated; estimates are that no more than 5 or 6 percent of the world's population lived in urban areas until around 1750. The use of fossil fuels made urban living possible, and even necessary, as it increased human productivity, fostered industrialization and trade, and reduced transportation costs. Cities as we know them today are made possible by the energy provided by oil, which enables them to draw food and other raw materials from anywhere in the world. The result has been rapid urbanization, beginning in the so-called 'first wave' countries of industrialization, in Northern and Western Europe and North America. By the middle of this century these industrial countries were about 70 percent urbanized (UN 1995).

The developing countries were still over 80 percent rural as late as 1950, and most Asian and African nations still have mostly rural populations (70.1 percent and 67.3 percent, respectively). In contrast, in developed countries and in Latin America about 75 percent of the population is urban. But as the developing countries become more industrialized, their cities are growing and their countries are becoming more urbanized. The urban population of the developing countries, which stood at 296 million in 1950, is expected to rise to 2 billion by the end of this decade, and to reach 4 billion in another quarter century, making the world dominantly urbanized. By 2000 nearly half of the world's population will be living in cities (WRI 1994).

Urbanization has led to the concentration of political power in cities, which has tended to foster policies that favour urban over rural areas. Various forms of subsidies make food and other basic goods cheaper in the city, discourage agricultural investment, and attract rural people into urban areas. As Brown and Jacobson (1987) point out, 'Many cities have been enriched only by impoverishing their hinterlands'. Such inequity is only a short-term, unsustainable relationship, however; sooner or later a more appropriate distribution of costs and benefits will evolve.

Most developing country governments have encouraged the growth of their large cities as a means of linking their domestic economy with that of the rest of the globe. Thus for many countries, the capital city controls the trade between urban areas and both rural and international markets, so cities like Manila, Mexico City and Bangkok may have more in common with Tokyo, London and Washington than with their rural hinterlands. Again, this cultural disparity between urban-dwellers and the rural population must be considered temporary – in the long run, a sustainable relationship between urban and rural will evolve. The rapid growth of cities is unprecedented in human history, making it difficult for governments to provide an adequate physical infrastructure for the burgeoning populations. Thus some experts predict squalor, pollution, poverty and misery for the world's megalopolises, though they continue to attract people from the countryside where opportunities presumably are less. What does all of this mean for nature, culture, people and protected areas?

Urban habitats

The urban environment includes a great diversity of habitats – refuse tips, sewage farms, waste ground, derelict ground, abandoned railway yards and railway rights-of-way, golf courses, buildings (both inside and outside), pavements, parks, etc. – and the literature documents their value for wildlife (e.g. Ahern and Boughton 1994; Goudie 1993; Shepherd 1994; Trojan 1994; Wee and Corlett 1986, Whiteley 1994). In some cases these habitats play a significant role in the conservation of rare species (Ahern and Boughton 1994; Whiteley 1994; Shepherd 1994) and the modern urban planning ethic is placing more emphasis on the maintenance of biodiversity in the urban environment (Nicholson-Lord 1987; various papers in Platt *et al.* 1994; Smith and Hellmund 1993). Thus, urban areas should be seen as management challenges for those concerned with biodiversity, not simply as sterile biological wastelands of interest only to humans.

The urban environment is a mosaic of man-made, natural and semi-natural habitats with climatic and hydrological conditions that distinguish it from adjacent rural areas (Berry 1990; McPherson 1994). While only 18 percent of cities are open space (Nicholson-Lord 1987), as much as 40–70 percent may be green and photosynthesizing (Nicholson-Lord 1987; Ignatieva 1994; Loucks 1994). Cities create their own climates. Sukopp and Werner (1982) have summarized the considerable literature on climate changes caused by urbanization, concluding that ultraviolet radiation is less, especially during winter;

Box 5.6: The Delhi Ridge Forest

The final few kilometres of the Aravalli Hill Range of Northwest India runs through India's capital city, and is covered by dense dry deciduous forest and thorn scrub. This 7,770ha sliver of vegetation harbours a large amount of biodiversity, including nearly 200 species of birds. It also serves as the city's green lungs, cleaning the air, acting as a giant air conditioner, and trapping tons of dust coming in from the Western Indian Desert. It also provides Delhi's citizens a place for relaxation, enjoyment and education; several thousand people use it daily for such purposes, and schools use it for outdoor biology classes.

As the city has grown, however, the Ridge Forest has faced a series of familiar threats, including construction, road-widening, garbage dumping and quarrying. Perhaps nearly half the Ridge has been destroyed or badly degraded in the past three to four decades.

Sensitive government officials were able to save certain pockets over the last few decades. But the real demand for the conservation of the Ridge has come from citizens' groups, school children and local residents, as the Ridge became one of the symbols of the fight to make the growing metropolis more liveable.

the duration of sunshine is 5 to 15 percent less than in the rural surroundings; annual mean temperature is 1°C–2°C more; the annual mean wind speed is 10 to 20 percent less; relative humidity is 2 percent less in winter and 8-10 percent less in summer; precipitation is 5 to 30 percent more; cloud cover is 5 to 10 percent more and fog is as much as 100 percent more; and condensation nuclei are 10 times more. These climatic conditions often support a rich diversity of species, of a mix that is quite different from that of the countryside.

Middleton (1994) points out that in Canada the greatest diversity of organisms in cities is found in the ravines, stream valleys, abandoned industrial sites, rail lines, waterfronts and other derelict or undeveloped areas that form a significant proportion of the land area of most Canadian cities. Stanley Park in Vancouver and Mount Royal in Montreal contain substantial remnants of the original ecosystems, and the Lesley Street Spit in Toronto harbour – a four-kilometre artificial peninsula made from construction debris – was home to 150 species of plants and was visited by 150 species of birds within 20 years of the start of construction, all within a few kilometres of the centre of Canada's largest city. Gemmell (1980) found that industrial habitats are of significant botanical importance in Northwest England. They include several types of calcareous waste, notable for their orchid-rich calcicolous floras and the presence of uncommon species. Pulverized fuel ash and habitats created by industrial excavations such as clay pits are also important for some species. He concludes that industrial habitats are valuable for the conservation of uncommon species, research and education, but new habitats need to be created from industrial workings, using techniques to encourage colonization and habitat diversity.

In short, urban habitats are highly diverse and help contribute to the world's biodiversity. They should be considered as part of national strategies and action plans for conserving biodiversity, as called for in the Convention on Biological Diversity (which was signed by 157 countries at the Earth Summit in 1992, entered into force at the end of 1993, and has now been ratified by nearly 120 countries).

Biodiversity in cities: which species survive?

Urban metropolises are burgeoning, crowded, overstressed habitats with little room for nature. Or so it seems at first glance. Hidden within the maze of concrete, steel, plastic and thatch, however, a surprising amount of wildlife still survives. Pigeons, swifts and kestrels fond of rocky crags are at home in a forest of man-made crags, as are black rats, house mice, and shrews in man-made burrows, where they feast on the detritus of human society. One of the reasons that cities can support relatively high numbers of species is that they provide a mosaic of different types of habitats, ranging from skyscrapers to gardens to forests and fields. In a few cases, urban populations of certain wildlife species can be a significant factor in their survival. For example, a small lake in the centre of Hanoi in Vietnam harbours the only surviving population of the giant freshwater turtle (*Pelochelys bibronii*) (Quy 1995).

In Canada, the corridor stretching from Quebec City to Windsor includes both Toronto and Montreal, forming a large urban system of around 34,000 square kilometres. This region coincides with Canada's deciduous forest zone, one of the most species-rich in Canada; it is now home to almost half of Canada's threatened or endangered species. One of only four remaining Canadian populations of the Massassauga rattlesnake (*Sistrurus catenatus*) is found within the city of Windsor, Ontario; and rapidly-declining Great Lakes populations of the common tern (*Sterna hirundo*) are almost totally dependent on breeding colonies in the harbours of Toronto, Hamilton and Port Colborne (Middleton 1994). While the specific land-use changes associated with urbanization (Leidy and Fiedler 1985) lead to a decline in biodiversity, economic development in urban and peri-urban habitats does not invariably lead to the loss of all of the original biodiversity and, in addition, new habitats such as urban parks, urban forests, urban wetlands, domestic gardens and roadside plantings, often support a surprisingly rich fauna and flora.

In some parts of the world, species typical of non-urban habitats have successfully adapted to urban habitats. Some birds typical of mature closed forests, such as Goodie's lorikeet (*Trichoglossus goldiae*) from Papua New Guinea, are able to colonize urban environments. The species was originally a rare resident in primary forests, but in the 1970s began feeding on casuarina seeds in highland towns of PNG, where it is now often the most abundant bird. Similarly, the white-collared kingfisher (*Halcyon chloris*) was originally an exclusively coastal bird, but in a matter of decades it has extended its niche to gardens and city parks in urban Singapore. Such species show behavioural plasticity and ecological flexibility which may benefit the survival of populations in urban situations (Diamond 1986).

Wildlife is very adaptive. For example, in New Guinea the greater wood-swallow, which traditionally hunts from tall ridge-top trees over the rainforest, in towns perches on radio towers and telephone poles from which it sallies forth to seek its prey. Shining starlings transferred their nests from holes in trees to vents of air conditioners and have become a pest, while swiftlets that typically nest in caves now use tunnels excavated by the Japanese army during World War II. A nocturnal ground-feeding predator, the Papuan frog-mouth, has learned to hunt rats under street lights (Diamond 1986). In Europe, many species of birds have become commensal with human towns. Starlings, pigeons and sparrows thrive in concrete habitats, and the sparrow is now so closely associated with humans that its original niche is unknown.

Birds and rodents are only part of the vertebrate fauna of urban ecosystems, which also have their fair share of predators. For example, foxes are widespread in cities throughout England and Wales at least (MacDonald and Newdick 1980). Urban foxes have a relatively small range in cities, averaging about 86ha. It is interesting to note that in most parts of England, foxes are more inclined to inhabit larger towns than smaller ones, perhaps because the larger towns have relatively more suburban habitat which makes them attractive to foxes. The breeding dens of foxes are often found in close proximity to houses, beneath outbuildings or warehouses. In North America, racoons and opossums are also commonly found within city limits, at least where sufficient habitat is found; and Los Angeles, a huge conurbation, even has a thriving population of coyotes.

Cities also support numerous species of plants, often including threatened species. Metropolitan Manila, with a population of 7 million or so, has at least 2,389 plant species, and Berlin has nearly 1,400. In Europe, at least, larger cities tend to have greater numbers of plant species, and surprisingly, cities with denser populations also tend to have greater numbers of species (Klotz 1990). The small German city of Wurzburg has 554 species of wild plants compared with Warsaw's 1,416, even though Warsaw has 2.4 million people. Some cities are so rich that they hold more species than the countryside immediately surrounding them. For example, 2,100 species of flowering plants and ferns are growing wild in London, according to David Bevan, conservation officer of the London Natural History Society. The native wild flowers, grasses and ferns of the whole of Britain amount to no more than 1,500 species. While cities and their surroundings often have more species than surrounding agricultural lands or forests, the species diversity tends to decline from the outskirts to the inner city, at least in central Europe (Kowarik 1990). The flora of modern cities is greatly increased by introduced species. For example, Berlin has 839 native species of plants and 593 species introduced through human activities (167 before AD1500 and 426 after AD1500). Many of the introduced species are far more successful than the native species; Berlin's *Red Data Book* lists 58 percent of the native species but only 12 percent of the species introduced after 1500 (Kowarik 1990).

The conclusion is that despite a few exceptions, the species that adapt best to urban settings are those that have the relatively flexible behaviour that is often typical of species found in secondary habitats.

Urbanization and biodiversity: what are the impacts?

Urbanization affects biodiversity in four main ways:

- Geographic expansion of settlements and infrastructure destroys wildlife habitat, displacing the existing vegetation through land conversion which alienates land as wildlife habitat, fragments the remaining habitat, and isolates remnant natural ecosystems.

- Cities require concentrations of food, water and fuel on a scale not found in nature. Urban demands for biomass require fuelwood, industrial wood, sawnwood and other products such as fruits and flowers from surrounding areas, often leading to deforestation in the surrounding regions. Around cities, plantations of genetically similar trees are displacing the local vegetation to meet the urban demands for biomass.
- Just as nature cannot concentrate the resources needed to support urban life, neither can it disperse the waste produced in cities (Brown and Jacobson 1987). Urban activities thus produce pollutants which have a significant impact on hydrological and atmospheric systems at both the local and global level, degrading soil, air and water.
- Native species are replaced by introduced ones. Urban dwellers plant many species of plants around homes, along avenues and in parks. These are largely ornamental and often exotics which displace the native vegetation while adding to overall diversity. In some cities, by far the majority of the plant species are introduced. For example, of the 217 tree species in St. Petersburg, 35 are native and 182 are introduced (Ignatieva 1994).

The effects of urbanization on biodiversity can be considered from two perspectives: the direct effects urbanization has on biodiversity (the loss of habitat; the fragmentation of habitat; the creation of new man-made habitats such as cemeteries, derelict lands, rubbish tips, etc.); and the indirect effects it has through its impact on hydrological systems and the atmosphere.

Indirect effects

Urbanization covers the urban landscape with impervious surfaces (one study of an average US urban area estimated this at 12–37 percent of total urban area – Loucks 1994) and these have a dramatic effect on run-off, an effect that is exacerbated by extensive sewers and drain systems (Binford and Buchenau 1993; Goudie 1993). A comparison of contaminant profiles of urban run-off and raw domestic sewage indicates that run-off contributes more suspended solids, pesticides, chlorides and heavy metals while the sewage is the main source of nitrogen and phosphorus (Goudie 1993). The less soil available, the more concentrated are these chemicals when they arrive in the rivers.

Most African and Asian urban centres lack sewage systems (UNEP 1992) and human sewage is the most important pollutant of the freshwater and coastal zones in developing countries (Markham 1994). Sewage, both treated and untreated, contains high nitrate and phosphate loads and, together with the nutrients and contaminants from urban run-off, produces an assault on the aquatic environment that has resulted, at best, in eutrophication, and at worst in the almost total destruction of wetlands and associated loss of biodiversity. A common phenomenon near and within urban areas is channelization of natural rivers and streams, carried out as a means of flood control or to facilitate navigation. This drastically alters the physical characteristics of a stream, increasing water velocity and reducing habitat diversity and riparian vegetation (and thus nutrient input to the stream from these sources). As a result, some species are eliminated and species composition is altered (Allan and Flecker 1993; Binford and Buchenau 1993; Goudie 1993). Other impacts include reinforcing of the banks, filling in of surrounding wetlands for the development of industry, construction of wharves along the banks, and so forth. In addition, urban water courses are typically highly polluted as a result of influx of sewage and runoff. When undammed, running waters often demonstrate increased water volume, a rise in the incidence of flooding, and a higher flow rate, thereby leading to increased bank erosion.

City-produced contaminants (such as CO_2, SO_2, nitrous oxides, ozone, etc.) have effects within the city, close to the city and globally (Berry 1990; Hawksworth 1990; Westman 1990). Lichens have proved to be excellent monitors of sulphur dioxide pollution which has caused the absence of any lichen species in the central areas of some major cities and their severe reduction in cover and diversity (over 80 percent reduction in the worst cases) in polluted urban areas with potential secondary effects on a range of vertebrates and invertebrates associated with them (Goudie 1993; Hawksworth 1990; Loucks 1994). Loucks (1994) suggests that such pollution effects, although not well documented, occur in other groups such as plants, birds and other biota. Urban air pollution also affects adjacent habitats; Westman (1990) recorded changes in the structure of the Californian coastal sage scrub ecosystem with replacement of native shrubs by exotic annual plants as a result of urban pollution, and Goudie (1993) reports on smog damage to Ponderosa pines 129km away from the pollution source – Los Angeles. Finally, at the global level urban air pollution is implicated in the greenhouse effect, global warming and sea level change, with all the associated effects on biodiversity (Berry 1990).

As the 'urban shadow' of pollution spreads concentrically around a city, expensive adaptations are required so that water supplies can remain safe.

Shanghai, China, for example, has had to move its water supply intake 40km upstream at a cost of $300 million because of the degradation of river water quality around the city. In some urban areas, environmental degradation also results from household attempts to compensate for inadequacies in formal water supply services. In Bangkok, Jakarta, and Mexico City, for example, excessive pumping has also led to land subsidence, causing damage to property, housing and infrastructure. In Bangkok, excessive ground water pumping has caused some land to subside by up to a metre, resulting in cracked pavements, broken water and sewage pipes, intrusion of seawater into aquifers, and increased flooding in low-lying areas.

Direct effects

Obviously urbanization contributes to one of the key factors in loss of biodiversity – the loss of habitat – and since urban centres tend to be near rivers and coastal areas, wetlands and coastal ecosystems have been especially severely affected (Schmid 1994; Walker 1990). For example, human settlement is seriously implicated in the current threats to important Asian wetlands (Scott and Poole 1989); the loss of natural flood plains in North America and Europe (Allan and Flecker 1993); and the loss of much of California's coastal sage scrub habitat (Westman 1990). However, in terms of the extent of habitat lost the effects at a global scale are not yet very serious. For example, the densely-populated city-state of Singapore (45,869 people per square kilometre) still has considerable vegetation. And the United Kingdom is highly urbanized and densely populated yet the built-up area of 1.3 Mha (Wynne *et al.* 1995) constitutes only 5.3 percent of total land area. It should be noted, however, that urban development requires vast amounts of materials – rock, sand, gravel – which destroys natural habitat and alters the rural landscape (Douglas 1990); and of course the urban population is a serious drain on resources in the surrounding regions. Perhaps a more serious effect of urbanization is that of habitat fragmentation and its effect on ecosystem structure and function. This has been the subject of much debate for over two decades (for a review of this issue see Saunders *et al.* 1991).

The process of fragmentation reduces the size of the habitat and causes changes in micro-climate which affect the biota as do the size, shape and location of fragments in the landscape (Saunders *et al.* 1991). In addition, species respond differently to fragmentation depending on their size, life-history strategy, and role in ecological interactions (predator-prey; pollinator-plant, etc.) (Cody 1986; Renman and Mortberg 1994; Soulé and Simberloff 1986; Wilcove *et al.* 1986). A useful general conclusion is that the smaller the remnant, the greater the chances of extinctions and disruptions of ecological processes altering species composition; and the greater the 'edge' effect which may alter species composition and may enhance invasion of alien species (Saunders *et al.* 1981; Wilcove *et al.* 1986). While the effects of fragmentation on natural and semi-natural urban areas are poorly documented, the effects mentioned above have been recorded in several cases (for example, Renman and Mortberg 1994; Turner *et al.* 1994).

Man-made habitats generally have lower biodiversity than their natural counterparts (e.g. the landscaped park versus forest) (Corlett 1992; Hails 1992). Despite some exceptions, as human populations increase, species richness declines. For example, 199 resident and migratory bird species were recorded for the Jakarta area in 1938, but fewer than 100 species were noted during a 1978–79 survey (Indrawan and Wirakusumah 1995). On the other hand, biomass (living weight) of wildlife may increase and higher densities of certain species may be recorded (Gliwitz *et al.* 1994; Goudie 1993; Luniak 1994). In Helsinki, Finland, for example, some 21 species of birds survive, well under half the number found in uninhabited forests; but these birds are over three times more numerous and have ten times the biomass of the forest bird populations. The avifauna in urban woods in Tokyo has fallen by half over the past two decades. Even common birds in urban habitats are declining in numbers, including the tree sparrow, crow, great tit and domestic pigeon (Numata 1980). On the other hand, about 200 species of beetles are found in urban woods in Tokyo, about one-third of the total species in Japan.

In short, the impacts of urbanization on biodiversity are profound but complex, with some species gaining in population size, some losing population, some disappearing and some invading. Ecosystems undergo profound change, as species change relative abundance, with unpredictable results.

Approaches to managing urban habitats to conserve biodiversity.

City and nature are not necessarily contradictory. Certainly biotic communities are transformed by urbanization, but – as indicated above – many plants and animals are capable of living in close proximity with people, especially if people create the conditions that enable biodiversity to prosper. Those interested in conservation need to find – indeed are finding – ways of ensuring that a reasonable level of biological diversity is maintained in urban settings. This requires the establishment of an appropriate range of biotopes in settled areas, designed in such a way as to be attractive to a range of species. This need not necessarily be intensive management, as much vegetation will develop spontaneously if appropriate conditions are provided and the plants selected for habitat rehabilitation are appropriate to the climate and soil conditions prevailing. Local nature reserves, especially in cities, provide opportunities for community-based projects which in turn engender local pride; provide contact points for local authorities to meet residents; and stimulate public interest in wider issues of nature conservation. English Nature has suggested targets of providing a local urban nature reserve of at least one hectare per thousand population and a natural green space of at least 2ha within 0.5km of every home, at least one 20ha site within 2km, at least one 100ha site within 5km, and at least one 500ha site within 10km.

One clear message is that connecting the biologically rich areas of a city – ravines, stream beds, abandoned land, parks – to each other and to the larger expanses of wild habitat in adjacent farmland, will enable more species to survive and thrive in the city. Landscape management techniques that may be appropriate include selective placement of different materials, varying the topography, exposing different substrates, using variable lime and fertilizer treatments, controlling soil moisture, and introducing indigenous plant species which are otherwise unable to colonize because of the geographical isolation of their sources of emigrants. Many cities, especially in Europe, have significant green areas that make numerous contributions to human welfare:

- improving climatic conditions through shading, effects of cooling, and regulation of air interchange;
- reducing air pollution as green surfaces provide a medium on which dirt particles can settle and eventually enter the soil;
- acting as filters to reduce wind velocity and thus reduce the capacity of the wind to hold particles of pollutants;
- reducing noise from traffic and other sources;
- using urban wetlands as sewage farms, as for example in Calcutta or Riyadh;
- providing living space for animals, which can then also occupy micro-habitats throughout the urban area;
- helping to protect river banks against floods and other temporary disturbances, and
- providing a range of psychological benefits to urban dwellers, including providing a link with the past; providing images of temporal changes in natural systems with the changing of the seasons; and so forth.

Conclusions and recommendations

Cities extend far beyond their municipal boundaries, drawing on resources from the surrounding countryside and eventually serving as a market for resources from the entire world. Therefore, forests, wetlands, the coastal zone and farmlands are essential to the survival of cities, even though they typically are not considered sufficiently important to attract the attention of urban planners. And because cities have so much influence, they must play an increasingly important role in maintaining the protected areas which in turn support the resource systems upon which urban welfare depends.

For cities to become more sustainable, they need to start developing a stronger awareness of the ways in which they affect the rest of the world. They must create their own feedback systems, continually monitoring their global and local environmental impacts and responding effectively to the messages received. They need to reorganize their energy, food, sewage and transport systems to ensure that they have maximum efficiency and minimal environmental impact.

Since cities are especially responsible for the conversion of fossil fuels into energy in forms that makes urban living possible, thereby producing more carbon dioxide, they must take more responsibility for action to enhance the vegetation cover of our planet which will re-convert the CO_2 to oxygen.

It clearly is in the enlightened self-interest of cities to become as green as possible within the metropolitan area, and to nurture their relationship with their hinterlands, contributing to the management of protected areas which can ensure water supplies, sources of inspiration, destinations for tourism, and various resources desired by urban-dwellers.

Acknowledgements

My thanks to George Barker, of English Nature, for getting me started with the literature; to Jeremy Harrison of the World Conservation Monitoring Centre for his advice; Cécile Thiery of IUCN for her assistance with research; Adrian Phillips, Gayl Ness and Madhav Gadgil for technical advice; and Sue Rallo for her usual brilliant secretarial support.

Toward urban sustainability (Herbert Girardet)

Urban growth is changing the face of the earth and the condition of humanity. In one century, global urban populations have expanded from 15 to 50 percent of the total (Kirdar 1997), which itself has gone up from 1.5 billion to nearly 6 billion. By 2000, half of humanity will live in cities, with much of the other half depending on urban markets for their economic survival. Urban agglomerations and their resource uses are becoming the dominant feature of the human presence on earth, profoundly changing humanity's relationship to its host planet and its ecosystems. The concern here is with the physical impacts of urbanized people, as well as the changes that occur in our minds as we urbanize.

In a world dominated by cities, the international community is starting to address the issue of urban sustainability. The process began in Rio with Agenda 21 and continued at the 1996 UN City Summit in Istanbul. The 100-page Habitat Agenda, signed in Istanbul by 180 nations, states, 'Human settlements shall be planned, developed and improved in a manner that takes full account of sustainable development principles and all their components, as set out in Agenda 21'.

We need to respect the carrying capacity of ecosystems and the preservation of opportunities for future generations. 'Science and technology have a crucial role in shaping sustainable human settlements and sustaining the ecosystems they depend upon.' (Habitat Agenda, UNDP 1996). Mega-cities depend on mega-structures. Large-scale urbanization greatly increases per capita use of fossil fuels, metals, timber, meat and manufactured products, with major external environmental implications. Unlike most traditional cultural systems, modern urban systems crucially depend on a vast system of external supply lines facilitated by global transport and communications infrastructures. This is not culture, or even civilization in the traditional sense, but mobilization – of people, resources and financial capital.

City people often have very limited understanding of their use of resources. Energy is a case in point. When we think heat and light we don't think firewood, but switch on electric or gas appliances – yet we are hardly aware of the power station, refinery or gas field that supplies us. And we hardly reflect the impacts of our energy use on the environment because they are rarely experienced directly, except when we inhale exhaust fumes on a busy street.

Demand for energy defines modern cities more than any other factor. Most rail, road and aeroplane traffic occurs between cities – for business, social contact or pleasure. Most urban activities depend on fossil fuels – to warm, cool or illuminate us, or to supply us with goods and services. Without routine use of fossil fuels, mega-cities of ten million people and more would not have occurred. As far as I am aware, there has never been a city of more than one million people not running on coal, oil or gas. But there is a price to pay: not only is air pollution a continuing menace in our cities, but most of the increase of carbon dioxide in the atmosphere is attributable to combustion in or on behalf of the world's cities. Yet, most city people find it hard to make the connection with something that is not happening here, but 'out there'. Urban food supplies are another case in point. The direct experience of growing food is largely absent in urban life: we harvest at the super market and most people have come to expect food to be served up packaged and branded for enhanced recognition. As city people we are hardly aware of the impacts of our food consumption on the fertility of the farmland supplying us, often in some distant place.

As we put cling-wrapped meat or fruit in a supermarket trolley, we are blissfully unaware that humanity now uses nearly half the world's primary production from photosynthesis and that most of this is utilized by urban people. Our knowledge system fails to inform us that the human species is changing the very way in which the 'the web of life' on earth itself functions: from the geographically scattered interaction of a myriad of living species, to which local cultures are intimately connected, into an assembly of concentrated urban centres into which the one species, humanity, funnels resources from all over the world; cities today take up only 2 percent of the world's land surface, yet they use over 75 percent of the world's resources (Girardet 1996).

Arising from the work of Rees and Wackernagel (1996), I have examined the ecological footprint of London – i.e. the land surfaces required to feed it, to supply it with wood products and to re-absorb its C02 output. In total, these extend to 125 times London's own territory of 159,000ha, or nearly 20 million hectares (Jopling and Girardet 1996). With only 12 percent of Britain's population, London requires the equivalent of Britain's entire productive land. In reality, these land surfaces, of course, stretch to far-flung places such as the wheat prairies of Kansas, the soya bean fields of Mato Grosso, the forests of Canada, Scandinavia and Amazonia, or the tea gardens of Assam or Mount Kenya. But this global dependence of Londoners has never been a big issue. Food is there to be enjoyed – the environmental impact of food supplies, including the energy used to produce and supply them, is rarely discussed.

The same applies to the city's metabolism. Like other organisms, cities have a definable metabolism (Girardet 1992, 1996). That of traditional towns and cities was characterized by interactions between dense concentrations of people and their local hinterland, with transport and production systems centred on muscle power. Beyond their perimeters, traditional settlements were usually surrounded by concentric rings of market gardens, forests, orchards, farm and grazing land for use by town people. Today, urban farming is still alive and well in cities in many countries. In Chinese cities, for instance, people still practise returning night-soil to local farmland to assure sustained yields of crops. With their unique systems of governance, Chinese cities administer vast adjacent areas of farmland and aim to be self-sufficient in food from this land (Sit 1988). Is this model of urban-rural linkages relevant to cities elsewhere in the world? The metabolism of many traditional cities was circular, whereas that of most 'modern' cities is linear. Resources are funnelled through the urban system without much concern about their origin, or about the destination of wastes; inputs and outputs are treated as largely unrelated. Contemporary urban sewage systems are a case in point. They have the function of separating people from their wastes. Sewage, treated or not treated, is discharged into rivers and coastal waters downstream from population centres, and its inherent fertility is lost to the world's farmland. Today coastal waters everywhere are polluted by sewage and toxic effluents, as well as the run-off of fertilizers applied to farmland feeding cities. This open loop is utterly unsustainable.

The linear metabolic system of most cities is profoundly different from nature's own metabolism which could be likened to a large circle: every output by an organism is also an input which renews and sustains the whole living environment. Urban planners and educators should make a point of studying the ecology of natural systems. On a predominantly urban planet, cities need to adopt circular metabolic systems to assure their own sustainability and the long-term viability of the environments on which they depend. Urban outputs will need to be regarded as potential inputs into urban production systems, with routine recycling of paper, metals, plastic and glass, and composting of organic materials for re-use on local farmland. The local effects of urban resource use also need to be better understood. Cities accumulate materials within them. The 1.6 million inhabitants of Vienna, every day, increase the city's actual weight by some 25,000 tonnes (Brunner, pers. comm.). Much of this is relatively inert material, such as concrete and tarmac. Other substances, such as heavy metals, have toxic effects as they leach into the local environment. Nitrates, phosphates or chlorinated hydrocarbons accumulate in local watercourses and soils, with as yet uncertain consequences for future inhabitants. These issues need to be addressed by national and urban policy, to establish new ways in which to engineer and plumb our cities. They also need to be addressed at a subtler level. The value system of city people needs to ascertain that this is not taken for granted indefinitely. Our separation from natural systems and our lack of direct experience of the natural world is a dangerous reality as it reduces our understanding of our impacts and of the ways in which we might reduce them.

Can we maintain living standards in our cities whilst curbing their local and global environmental impacts? To answer this question, it helps to draw up balance sheets comparing urban resource flows (Jopling and Girardet 1996). It is apparent that similar-sized cities are supplying the needs of their people with a greatly varying throughput of resources. Many cities could massively reduce their throughput of resources, maintaining a good standard of living whilst creating much-needed local jobs in the process. Cities in the North often have a much less impressive track record than those in the South, though poverty there is a significant driving force for the high levels of waste recycling. Towards sustainable urban development, how can we improve our understanding of the impacts of our current lifestyles? Can large modern cities adopt more local, more frugal, more self-regulating production and disposal systems?

An answer to these questions may be critical to the future well-being of the biosphere, as well as of humanity itself. Maintaining stable linkages between cities and their hinterland – local or global – is a new task for most city politicians, administrators, business people and people at large, requiring new approaches to urban management. Many of the world's major environmental problems will only be solved by city people conceptualizing new ways of running their cities. Some cities have already made circularity and resource efficiency a top priority. In Europe, many cities are installing waste recycling and composting equipment. Throughout the developing world, too, city administrations have made it their business to encourage the reuse of wastes.

Given that the physiology of modern cities is currently characterized by the routine use of fossil fuels, a major issue is whether people will see the potential of new, clean and efficient energy technologies for powering their cities, such as combined heat-and-power systems, heat pumps, fuel cells and photovoltaic modules. In the coming decades enormous reductions in fossil fuel use can be achieved by incorporating photovoltaic modules in urban buildings. Some writers have argued that cities could be beneficial for the global environment, given the reality of a vast human population. (Gilbert *et al.* 1996). The very density of human life in cities could make for energy efficiency in home heating as well as in transport. Public transport and waste recycling systems could be more easily organized in densely inhabited areas. And urban agriculture, too, if well developed, could make a significant contribution to feeding cities and providing people with livelihoods (UNDP on Urban Agriculture 1996).

With the whole world now copying Western development patterns we need to formulate new cultural priorities and this should centre on formulating value systems for urban living, giving cities the chance to realize their full potential as centres of creativity, education and communication. Cities are nothing if not centres of knowledge and today this also means knowledge of the world and our impact upon it. Reducing urban impacts is as much an issue of the better uses of technology as of education and of information dissemination. We also need to revive the vision of cities as places of conviviality and above all else of sedentary living. As I have suggested above, currently cities are not centres of civilization but mobilization of people and goods. A calmer, serener vision of cities is needed to help them fulfil their true potential as places not just of the body but also of the spirit. The greatest energy of cities should be directed towards creating masterpieces of human creativity.

The future of cities crucially depends on the utilization of the rich knowledge of their people, and that includes environmental knowledge. Urban communications systems have a particularly important role to play in helping city people to understand their impacts and to bring about the necessary changes in the way we run our cities. In future, cities need to develop communication strategies that help people to confront the global impacts of their economic power and consumer habits. City people need new communication channels to help them improve their decision-making, particularly regarding the impacts of their life-styles. Here we can learn a great deal from the cultural feedback methodologies practised by traditional cultures which use regular village meetings to reflect their impacts on the local environment.

Today we can establish similar feed-back processes in modern cities using advanced communication technologies. The global economic and environmental reach of cities needs to be matched with smart early warning systems that will enable city people to ring alarm bells when new, unacceptable developments occur – monitoring and ameliorating their impact on the biosphere. The smart city of the new millennium will find means to provide its citizens with regular insights into the consequences of their actions, thus changing those actions. Today new communication technologies should also be utilized to enhance the way cities function by improving communications within them, leading to better decision-making. Urban intranets, now in place in a growing number of cities, should improve the communication flow between various sectors of urban society. If such changes occur, using the best of modern communication systems, we may yet learn to run our cities in smarter ways, improving their metabolism and reducing their ecological footprints. Large cities are not going to go away for the time being, but the way they work certainly need not be as damaging and wasteful as it is at the present time. Cities for a new millennium will be energy- and resource-efficient, people-friendly, and culturally rich, with active democracies assuring the best uses of human energies. Prudent investment in infrastructure will enhance employment, improving public health and living conditions. But none of this will happen unless we create a new balance between the material and the spiritual and cultural realms. Eco-friendly urban development could well become the greatest challenge of the twenty-first century, not only for human self-interest, but also to create a sustainable relationship between cities and the biosphere. Ultimately that cannot be done without changing the value systems underpinning our cities. Adopting circular resource flows and helping cities reduce their ecological footprint is ultimately a cultural issue. In the end, it is only a profound change of attitudes, a spiritual and ethical change, that can ensure that cities become truly sustainable.

Towards a political ecology of biodiversity (Josep-Antoni Gari)

The meaning of biodiversity remains an open question, despite the commonly accepted definition of biodiversity as 'diversity of genes, species and ecosystems' (CBD, Art. 2 1992). Current controversies around biodiversity on issues such as bio-prospecting, biotechnology, property rights regimes over biodiversity, and conservation of natural areas, emerge from conflicts between different socio-ecological constructs of biodiversity. These conflicts can be regarded as a main controversy of biodiversity as a natural resource versus biodiversity as an ecological praxis in nature and inside human societies. It is what we may call the controversy between the political economy and political ecology of biodiversity. In facing biodiversity, the capitalist system constructs biodiversity as a natural resource, in the sense of mere biological raw materials (e.g. genes and enzymes).

That is the dominant approach, and it has largely spread in science, research, industry and government dealings with biodiversity. In this way, Rick Cannell, a natural products scientist of the pharmaceutical company Glaxo Wellcome, defines the rainforest as, 'a rich bank of genetic resource' in which 'we can move from tree to tree, across the canopy, collecting samples as we go, with the freedom of fishermen who have just discovered boats' (*cf.* Cannell 1997). In general, biodiversity is regarded as a source of information for agricultural and pharmaceutical research (Swanson 1996) or as a necessary precondition for biological resources (Wood 1997). In capitalist societies, biodiversity arises as a resource because that is the view of nature from the urban-industrial system (Escobar 1996a). Biodiversity seen as a product of nature is hence free for commodification and exploitation. That is a political economy view over biodiversity. Then, in understanding and constructing biodiversity, most of the ecological depth is missing due mainly to the hegemony of the commodification process.

If, however, biodiversity is taken as complex, dynamic and openly-evolving, it erodes the capitalist version of nature, which is based on a rather static and commodifiable perspective. Nevertheless, we can identify divergent views, especially from inside both the ecological sciences and many local and indigenous communities.

On the one hand, the significance of biodiversity has been emphasized as a source of resilience for ecosystems (Perrings *et al.* 1995), or in a broader view as a cornerstone in the dynamic functioning of ecosystems (R. Costanza, pers. comm.). On the other hand, it is a characteristic of many indigenous and local communities to have inserted biodiversity deep into their cultural values, their social processes and their economic system. Their attitudes towards biodiversity are based on the view of biodiversity as a dynamic quality of nature and as an integral part of themselves. In fact, traditional communities embody spiritual and aesthetic values of biodiversity that tie human societies to the earth (Posey 1996). Several international documents have already recognized this socio-ecological integrated perspective of indigenous and local communities, and hence their intrinsic positive contribution to biodiversity conservation (CBD Art. 8j 1992; Rio Declaration on E&D, Preamble and Principle 4, 1992; Agenda 21, Ch. 26-1 1992; ILO Convention 169, Preamble and Art. 13-1 1989).

In this controversy over biodiversity, political ecology emerges as a very valuable criticism of the capitalist view of biodiversity, standing for alternative hybridizations (*cf.* Escobar 1996b; Peet and Watts 1996). Political ecology is not a mere additional nature-society perspective, but one that can really cope with the biodiversity conflicts. Thus, for instance, by valuing the constructs of indigenous and local communities, but not by imposing them, political ecology stands for a radical social change, involving a new way of constructing and seeing nature–society.

Political ecology involves new forms of democracy, decentralization, pluralism and social movements. They arise to face the current biodiversity crisis because they can better integrate the modern and the non-modern, the local and the global, Western science and traditional ecological and health knowledge. Political ecology thus goes beyond closing and self-legitimating doctrines, and opens the society–nature construct to pluralism and co-evolution. In addition, political ecology, in considering that reasoning is the basis for political praxis, becomes a platform for liberation (Peet and Watts 1996: 263). Consequently, the new political ecology perspectives on the socio-natural construct do not only cope with current conflicts over biodiversity, but they may also arise as liberation discursive processes.

CHAPTER 6

VALUING BIODIVERSITY FOR HUMAN HEALTH AND WELL-BEING: TRADITIONAL HEALTH SYSTEMS

Gerard Bodeker

CONTENTS

TEXT BOXES

FIGURES

Valuing Biodiversity for Human Health and Well-being: Traditional Health Systems

Gerard Bodeker

Traditional health systems (Gerard Bodeker)

The understanding and use of biodiversity as a means of sustaining human health and well-being in non-Western societies reflects a cultural, and often spiritual, appreciation of both the biological environment and the deeper forces perceived to influence it. Traditional health systems extend to an appreciation of both the material and non-material properties of plants, animals and minerals. Here, the term 'systems' is used to reflect the organized pattern of thought and practice – diagnostic, clinical and pharmacological – that shapes and maintains most bodies of traditional health knowledge. These are systems of knowledge that include concepts of both the sacred and the empirical; frameworks for understanding health and healing; assumptions of cosmos and causality, and taxonomies which address a perceived order in nature. They range from the cosmological to the particular in addressing the physiological make-up of individuals, their communities, and the specific categories of materia medica – plants, animals and minerals – used for therapeutic purposes to enhance health and well-being. Food and medicine may not be separated into discrete categories.

From early palaeontological records it is possible to see that plants and animals were used in the past for both medicinal and ritual purposes. Indeed, it is likely that healing practices synthesized both spiritual and medicinal modalities. In the late twentieth century, the World Health Organization estimates that around 80 percent of the population of most non-industrial countries still relies on traditional forms of medicine for everyday health care (WHO 1993).

A fundamental concept in traditional health systems is that of balance: between mind and body, between different dimensions of individual bodily functioning and need, between individual and community, individual/community and environment, and individual and the universe. The breaking of this interconnectedness of life is a fundamental source of *dis-ease*, which can progress to stages of illness and epidemic. Treatments, therefore, are designed not only to address the locus of the disease but also to restore a state of systemic balance to the individual and his or her inner and outer environment (Bodeker 1996).

Speaking in Caracas in 1995, at the South and Central American Regional Meeting of the Global Initiative For Traditional Systems (GIFTS) of Health, Dr F. Pocaterra, a Western-trained medical doctor from the Wuayuu people, put it this way: 'Traditional Wuayuu medicine, as that of other indigenous peoples of the continent, and other great cultures like India and China, is based on universal laws that can be divided into three categories: the divine or supernatural, the human, and the earthly. The Wuayuu people are animist. We attribute life to all things – to the flora, the fauna, to natural phenomena, to water and to mountains. We mystify our relationship with nature and the superior entities, as it is a material culture, with spiritual symbolism, in which magical–religious rituals are transcendental. For the Wuayuu, everything has a soul, everything has a life, a spirit. Everything is eternal.' (Pocaterra 1995).

Native American communities incorporate traditional forms of treatment into the US Indian Health Service (IHS) alcohol rehabilitation programmes. In a study of 190 IHS contract programmes it was found that 50 percent of these offered a traditional sweat lodge or encouraged its use. Treatment outcomes were found to be better for alcoholic patients when a sweat lodge was available. In addition, the presence of medicine-men or healers, when used in combination with the sweat lodge, greatly improved the outcome (Hall 1986).

In the Vedic tradition of India, the source from which the Ayurvedic medical system derives (*Ayur* = life, *Veda* = knowledge), consciousness is considered the basis of all material existence. 'The infinite consciousness alone is the reality, ever awake and enlightened...'; '... wind comes into being, though that wind is nothing but pure consciousness.' 'Within the atomic space of consciousness there exist all the experiences, even as within a drop of honey there are the subtle essences of flowers, leaves and fruits.'; '... even what is inert is pure consciousness...'; 'It is pure consciousness alone that appears as this earth.' (Venkatesananda (trans.) 1984.)

Within this framework, consciousness is of primary significance and matter is deemed secondary. Accordingly, pure forms of Ayurvedic medical treatment will first address the spiritual and mental state of the individual – through meditation, intellectual understanding of the problem, behavioural and lifestyle advice, etc. – and then address the physical problem by means of diet, medicine and other therapeutic modalities (Sharma 1996).

The seminal text of Traditional Chinese Medicine, the Yellow Emperor's Classic, dating *circa* 300BC, also details a cosmology in which matter is secondary to ethereal dimensions of existence. The cosmos is described as composed of ethers of heaven and earth, which are *yang*, with the attributes of bright, light and male – and *yin*, with the attributes of dark, heavy and female. The universe contains phenomena created by the dynamic action upon *yin* and *yang* of the Five Agents (*wu-hsing*): the elements water, fire, metal, earth and wood, which mutually create and destroy each other. The concept of *qi* refers to subtle energy or life force and is further described in categories that govern nourishment, defence systems, flow of energy, physiological functioning of organs, respiration and circulation. The relationship among these phenomena in their natural state is one of balance and harmony.

The cosmologies of traditional health systems, then, ascribe life, spiritual value and interconnectedness among all life-forms to the aspects of the natural world used in the process of promoting human health and well-being.

While having a theoretical foundation in the non-material realms of existence, indigenous medical traditions may be systematic in their view of nature, drawing on plant and animal taxonomies to classify and select medicines. Berlin (1973) proposes that consistent categories exist within indigenous systems of classification – ethnotaxonomies – that serve as a catalyst for traditional naming systems. The result has been new insights into the effects and adaptiveness of indigenous applications of plants. Andean shamans of the Sibundoy Valley, for example, exhibit a greater degree of discrimination and a more diverse range of information about plants than Western botanists. The gardens of the shamans reflect their knowledge of ecosystem management as well as serving as symbolic reconstructions of the web of relationships between plants, humanity and the cosmos (Pinzon and Garay 1990).

In northern Brazil, Yanomami Indian herbalists, faced with a severe epidemic of malaria, developed an empirical approach to identifying plants with potential anti-malarial effects. One major criterion was bitterness (Milliken 1997). Bitterness is also the taxonomic category used in Ayurveda and Traditional Chinese Medicine for selecting plants that have antipyretic effects. Indeed, Ayurveda uses six tastes as the basis for classifying the principles of food and medicines. In Ayurveda, individual constitutions are classified according to these three basic organizing principles or *doshas* – known as *Vata*, *Pitta* and *Kapha*. These principles apply across all forms of life, and a species' medicinal value is determined by the category to which it belongs.

The choice of plants and other medicinal materials is based on their capacity to influence the *doshas*. The classic Ayurvedic text, *Caraka Samhita*, notes that while folk knowledge exists regarding the form and effects of medicinal plants, a higher form of knowledge is that pertaining to the principles governing correct application of the plants to human health. 'One who knows the principles governing their correct application in consonance with the place, time and individual variation, should be regarded as the best physician' (*Caraka Samhita*, Vol. 1, p. 59). Natarajan points out in her contribution that in India certain plants, e.g. *tulasi*, are objects of veneration and worship based on tradition concerning their origins and knowledge of their medicinal properties, thus contributing to simultaneous use and protection of the species.

Food and medicine may not be separated into discrete categories in non-Western societies. The Hausa of Nigeria use certain plants as both food and medicine, including plants identified as having anti-malarial effects (Etkin 1997). Stephen Hugh-Jones points out in his contribution to this chapter that *coca* use in the Amazon reflects an implicit categorization of substances into contrasting but analogous and complementary systems.

The Maasai of East Africa cook the bark of *Acacia goetzei* (*Leguminosae*) and *Albizia anthelmintica* (*Leguminosae*) with their traditional diet of boiled meat, milk and blood. Research has shown that combining the bark with the other foods results in the lowering of cholesterol levels to one-third that of the average American. Unique saponins in these plants are considered to be implicated in producing the cholesterol-lowering effects (Johns, in Balick and Cox 1996). Conservation, cultivation, classification, documentation, knowledge recovery, biodiversity prospecting, integrated health care service delivery, parallel health care service delivery – all of these themes are being incorporated into programmes addressing the interface between traditional foods, medicinal plants and their use in human health and welfare (Quansah *et al.* 1996; Tumwesigye 1996 – see Box 6.1).

As described in the contribution by Motte-Florac and colleagues, the Pygmy groups of Central Africa have separate languages but a common vocabulary related to the classification of rainforest products. As these traditions disappear, the value of plant

Box 6.1: Bumetha Rukararwe: Integrating Modern and Traditional Health Care in South-West Uganda

Olive Tumwesigye, Pharmaceutical Technologist, Rukararwe Bumetha, Bushenyi, Uganda

Bushenyi Meditrad Healers' Association (Bumetha) of Rukararwe, in Bushenyi District in South-West Uganda, was established in 1988 through the joint efforts of traditional healers and Western-trained health-care workers. The goal of Bumetha is to have good health for all members of the community. The services of Bumetha started in the small community of Rukararwe and later expanded to include neighbouring communities.

The herbal medicines used traditionally have nutritive value, especially for patients who have no appetite and are showing signs of malnutrition or chronic anaemia, or who are suffering from side-effects of Western medicine. It is for this reason that Bumetha does not encourage research that focuses on extraction of active ingredients with a view to synthesizing these and throwing away the whole plant. Most medicinal plants have multiple value and are capable of naturally manufacturing drugs and foods that are needed for good health. Healers and farmers are encouraged to plant medicinal trees using integrated agroforestry methods. Bumetha and other associate members have special gardens for herbs, and appeal to the public to conserve and protect trees and medicinal herbs for the sake of life and health.

The combination of Bumetha healers at Rukararwe – i.e. traditional and Western healers, spiritual healers and pharmacists, and their medicinal plants – provide holistic healing to the patient. The community and, increasingly, the nation is getting inexpensive and effective medical services from Bumetha. Knowledge and awareness of preserving and conserving the environment – especially medicinal plants – has been promoted. The public generally has changed its attitude positively towards traditional healers.

(Adapted from 'Bumetha Rukararwe: Integrating modern and traditional health care in South-West Uganda', *J. Altern. and Complem. Med.*, 2,3, 373-377, 1996. Mary Ann Liebert Publishers, Inc., New York.)

biodiversity and its benefit to human health is less well understood and is valued less as a result (Motte-Florac *et al.* 1993).

The ancient Ayurvedic text *Caraka Samhita* states that 'The best of the mountains, the Himalayas, are the excellent habitat of medicinal plants. Therefore, fruits of *haritaki* (*Terminalia chebula*) and *amalaki* (*Emblica officinalis*) which grow on the mountain range, should be collected in the proper season, when they are matured and rich with manifested *rasas* and potency by following the prescribed procedure. These fruits should have been mellowed by the sun ray, wind shade and water and un-nibbled at by birds, unspoiled and unafflicted with cuts and diseases.' (*Caraka Samhita,* p. 15, Vol. III.)

Today, the Himalayas are in danger of losing their medicinal plant biodiversity. India, which reportedly harvests 90 percent of its medicinal plants from uncultivated sources, exports tens of thousands of medicinal plants a year and is one of the main suppliers of the European herbal market. China, which uses/exports four times the volume of medicinal plants used by India (Lange 1996), gathers 80 percent of its medicinal plants from uncultivated sources.

Herbalists world-wide are concerned about the growing scarcity of medicinal plants. Michael Balick and colleagues draw attention to this in their contribution. Balick, with Paul Cox, has reported that in 1940 the Belize healer Don Elijio Panti had to walk only ten minutes to collect medicinal plants. By 1988, Don Elijio had to walk seventy-five minutes to reach adequate sites (Balick and Cox 1996). Balick *et al.* note in this chapter that it seems likely that the value of tropical forest for the harvest of non-timber forest products will increase relative to other land uses over time, as these forests become more scarce. The Chiang Mai Declaration of 1988 (Box 6.2), called a global alert to the critical situation with regard to loss of medicinal plants through over-harvesting and habitat destruction (Akerele *et al.* 1991).

In Africa, which has the highest rate of urbanization in the world, the larger the urban settlement, the larger the traditional medicine markets tend to be. To meet the demand from urban communities, traditional medical practitioners must resort to

Box 6.2: The Chiang Mai Declaration

Saving lives by saving plants

We, the health professionals and plant conservation specialists who have come together for the first time at the WHO/IUCN/WWF International Consultation on Conservation of Medicinal Plants, held in Chiang Mai, 21-26 March 1988, do hereby reaffirm our commitment to the collective goal of "Health for All by the Year 2000" through the primary health care approach, and to the principles of conservation and sustainable development outlined in the World Conservation Strategy.

We:

Recognize that medicinal plants are essential in primary health care, both in self-medication and in national health services.

Are alarmed at the consequences of loss of plant biodiversity around the world.

View with grave concern the fact that many of the plants that provide traditional and modern drugs are threatened.

Draw the attention of the United Nations, its agencies and Member States, other international agencies and their members, and non-governmental organizations to:

- the vital importance of medicinal plants in health care;
- the increasing and unacceptable loss of these medicinal plants due to habitat destruction and unsustainable harvesting practices;
- the fact that plant resources in one country are often of critical importance to other countries;
- the significant economic value of the medicinal plants used today and the great potential of the plant kingdom to provide new drugs;
- the continuing disruption and loss of indigenous cultures, which often hold the key to finding new medicinal plants that may benefit the global community;
- the urgent need for international co-operation and co-ordination to establish programmes for conservation of medicinal plants to ensure that adequate quantities are available for future generations.

We, the members of the Chiang Mai International Consultation, hereby call on all people to commit themselves to Save The Plants That Save Lives.

Chiang Mai, Thailand, 26 March 1988

Source: Akerele, O., Heywood, V. and Synge, H. (eds) (1991) *The Conservation of Medicinal Plants.* Cambridge University Press.

commercial sellers, who are at the end of a chain which starts with the large-scale gathering of medicinal plants from wild sources. This harvesting occurs on a scale that is far greater than the sustainable levels that were the case when local practitioners gathered herbs in their area for use in local health care. Consequently, 'the supply of plants for traditional medicine is failing to satisfy demand' (Cunningham, UNESCO, 1993).

As Shankar highlights in his contribution, intelligent use of medicinal plants and a generational perspective on sustainability are needed to ensure their role in sustaining the health of the nation. Such thinking is reflected in the Arusha Declaration of 1991, issued from a WHO-sponsored meeting held in Arusha, Tanzania. The Arusha Declaration called on national and international organizations to seriously address the role of medicinal plants and traditional medicine services in providing important health care to the majority of the population in Africa (Box 6.3).

Spirituality may be inseparable from empirical validity, as reflected in the ancient Vedic text the *Shrimad Bhagavatam*: 'The sun is the soul of the deities, men, beasts, reptiles, creepers, and seeds...' (Rishi Sukadeva, XXII, p. 508-9; Shrimad Bhagavatam 1988). Maurice Iwu's contribution, for example, draws attention to the symbolic and spiritual values of medicinal plants in traditional African medicine.

Box 6.3: The Arusha Declaration

'The use of traditional medicinal plants has been the basis of the practice of traditional medicine in the South. Most of the medicinal plants of the world are, by and large, located in the tropical areas of the South, which contains about two-thirds of the plant species of the world. A number of these plants appear to be on the verge of extinction because of man's irresponsible destruction of their natural ecosystems, which makes such plants even more valuable.

The countries of the South should vigorously strengthen their co-operation in the field of drawing up inventories, nationally and collectively, of their medicinal plant resources, as well as in the cultivation, processing, marketing and in general widening the use of herbal medicine to meet the health needs of their peoples, in accordance with the objectives of self-reliance, respect of cultural heritage and the integrity of the natural ecosystem.'

(Arusha Declaration, in Mshegeni *et al.* 1991).

Indigenous perspectives offer new directions in planning for medicinal plant biodiversity conservation. Where a spiritual approach to healing guides the therapeutic use of medicinal plants and other forest biodiversity, high value is placed on the life of the forest (see contributions by Montes and Harrison, and by Barbira-Freedman). Some traditions, such as those of the Native North American indigenous peoples (see contributions by Goodwill and by Bill) consider themselves being called to serve as guides to the spiritual meaning of balance and the capacity of the environment to heal or harm human life according to its condition.

Honouring and using, revering and understanding, harvesting and conserving – these may not represent the polarities that they appear to be. Their co-existence in indigenous relationships with medicinal plants may be more than the tolerance of inconsistency or the capacity to live with paradox. Rather, it may express an appreciation for deeper levels of unification in life, where paradox is seen more as the reflection of a superficial or fragmented perception than as an accurate rendition of a meaningless universe. It is this dimension of traditional health knowledge that is recognized when cultural and spiritual values are called on in the assessment of biodiversity.

The spiritual dimensions of medicinal plants in the Vedic tradition of India (Darshan Shankar)

Ayurveda (*Ayur* = life; *Veda* = knowledge) has evolved from a profound cultural and philosophical perspective that is also shared by other Indian *Sastras*, or branches of Vedic literature. In this perspective the world is conceived in terms of matter and spirit which exist in conjunction. The spiritual dimension therefore is intrinsic in every material form, and matter cannot exist without spirit. According to the Samkhya philosophical school there are two aspects of creation. The manifest (*Prakrti*) which is expressed in the form of a matter-spirit combination and the unmanifest (*Purusa*) which is formless. It is the grosser expression of matter (*Prakrti*) that is usually appreciated through sensory perception but the spiritual expression in matter lies beyond normal sensory perception. Consciousness is the name given to the spiritual expression in nature. In addition to the material world known to physicists, chemists, botanists and geologists, there is this spiritual world. The two worlds are in fact not seen to be separate but, rather, interconnected. The spiritual world is the subtler dimension of what appears to the normal senses to be purely a material world. The manifest world is phenomenal and ever-changing. When one orients one's mind and senses inwards the subtle states of existence can be seen and experienced. *Padarath Vijnan*, 'knowledge of that which exists', deals with this spiritual–material world-view.

Thus spiritual dimensions of medicinal plants are implicit in their existence. The Cakrapani commentary of the *Carakasamhita*, an Ayurvedic treatise of 500BC, refers to consciousness in plants. This consciousness is reflected in the morphology and biological properties of the plant, but the properties of the plant can be significantly enhanced if one knows how to access its spiritual state. To the traditional mind the subtler spiritual properties are as real and certainly more powerful than the gross properties.

In local health traditions, collectors of plants worship plants and pray to them before collecting them. The real intention here is to recognize and acknowledge the spiritual properties of the plant. A plant when processed both spiritually and materially becomes more potent. Preparation and administration of herbal medicines entails use of mantras to enhance the efficacy of the drug.

Traditional conservation practices

In India, 40–70 percent of the plant diversity in any local ecosystem is used for human and veterinary medicine. Today, biodiversity-dependent rural communities are facing a serious resource threat because of the rapid erosion and loss of natural habitats. The biodiversity loss is not only a threat to the ecology of the planet but also a more immediate threat to the livelihood and security of rural communities. Biodiversity conservation is therefore of immediate relevance to these communities which are dependent upon, interact with, and are a fundamental dimension of, local ecosystems. The fact that communities 'know' and put to use a large proportion of local biodiversity is the foremost evidence that they have been conserving plants.

The intimate knowledge of local communities about their bio-resources is clearly seen in the tremendous diversity of local names for, and uses of, the same plant and animal species as one moves across what may referred to as 'ethnobiogeographic regions'. The etymology of local names and the content of local knowledge also reveals the understanding communities have of the properties, morphology, phenology, reproductive biology and habitats of plants.

Local traditions regarding the 'collection times' of plants also illustrate the conservation ethic. Collection times are related to seasons: for example, tender leaves are to be collected in the monsoon; barks of trees and sap are to be collected only in autumn; tubers in early winter; roots in late winter; flowers in spring and in the flowering seasons. The times are related to the stages of the plant life-cycle (e.g. leaves of certain plants should be collected before the flowering of the plant). There are practices that involve non-destructive collection (e.g. only the north-facing roots of a plant should be collected), or in tribal areas it is preferred to use only fresh plants, so that the plant is collected only when needed, and then in limited quantity. There are guidelines related to the time of the day when a plant should be collected (e.g. plants are usually not collected after sunset or at midday). In certain cases, plants are collected only at certain times of the year when particular constellations occur (e.g. the *Pushya Nakshatra* or on a full moon or a new moon day). All these regulations, which are part of 'culture', can be seen as indicative of a conservation ethic. More obvious forms of conservation like 'sacred groves' and gardens for the 'snake gods' (*Sarpa Kavu*) are only highlights of a more generalized all-pervasive conservation ethos.

Today, for various sociological reasons, many of these traditional conservation practices related to medicinal plants are being eroded. In terms of the sustainable use of medicinal plants today the situation is that a few hundred species of economically valuable medicinal plants are being over-exploited by commercial enterprises who use the 'ecosystem people' as their raw material collectors to gather plants from the wild. Several hundred other non-commercial species that are used by local communities for their own self-help health care are under threat because of the rapid pace of degradation, disturbance and outright loss of natural habitats all over the country.

'Market cultures' are unlikely to promote a comprehensive approach to medicinal plant conservation since the market uses only a fraction of the biological resources known to communities. Of 7,500 species known to Indian tribals, fewer than 500 are used in the market place, with real benefits going to traders rather than primary collectors. Out of over 300 medicinal species known to the Kanis only one found its way to the market. Thus the argument that market incentives can serve to promote conservation appears rather hollow. This is not to argue the one should not attempt to commercialize indigenous knowledge and medicinal plants. But commercialization – through herbal medicine development or bioprospecting – is not enough to ensure wholesome and substantial community benefits.

Traditional knowledge, culture and resource rights: the case of *Tulasi* (Bhanumathi Natarajan)

Tulasi is an aromatic woody herb that occupies a unique position in India due to its spiritual and medicinal properties. The word *tulasi* comes from the Tamil language, but the plant is also known by many other names in various other Indian languages. Its Latin name is *Ocimum sanctum* (*O. tenuiflorum*).

Ocimum sanctum belongs to the family Lamiaceae, which includes about 200 genera and 3,200 species. The etymology of the specific name *O. sanctum*, means 'used for religious purposes'. The origin of *O. sanctum* in Indian spiritual and medicinal traditions is still obscure (Vartak and Upadhye, n.d.), but it is mentioned in the *Puranas*, and is still used in the classical and folk traditional systems of health. Hence, the case of *tulasi* well exemplifies how closely the spiritual and medicinal domains are connected in traditional societies.

Tulasi in myths and legends

Many stories about *tulasi* are found in the ancient Vedic texts, the *Puranas*, which contain the most important stories about the *devas* (gods) and other sacred beings. In the *Brahmavaivarta Purana*, *tulasi* is anthropomorphized as the wife of the demon Sankhacuda or Jalandhara (Ocean's son). Sankhacuda performed austere penances to Lord Vishnu and received a boon of being invincible as long as his wife was faithful to him. Because of this boon, Jalandhara became arrogant and started to perform atrocities on men. When this went beyond endurance, Vishnu, who himself granted him the boon, decided to do something. Vishnu disguised himself as Jalandhra and enticed Tulasi into being unfaithful to her husband. Jalandhara was then killed easily. After having learnt that she was tricked by Vishnu, Tulasi asked for an explanation. Vishnu explained patiently that to get rid of evil it is sometimes necessary to kill. In order to pacify Tulasi, Lord Vishnu gave an assurance that she would be worshipped by all women for her faithfulness to her husband and that she would be immortal. Tulasi then committed *sati* and out of the ashes came a plant which was called *tulasi*. Even today this plant is worshipped by all Hindu women in India. (Dange 1990: 1482; Gupta 1971: 75). This story is also mentioned in the *Padma Purana*.

In another story connected with this plant, Lord Vishnu was very attracted to the woman Vrinda. In order to divert his enchantment, the gods appealed for help to the wives of Lords Vishnu, Shiva and Brahma. Each of the wives gave them a seed to be planted, and out of the seeds arose three different plants, of which one was *tulasi*. The three plants appeared before Lord Vishnu in the form of three beautiful women, and thus Vishnu forgot his infatuation for Vrinda. The three women later assumed the form of plants (Dange 1990: 1482; Gupta 1971: 76). This story is also found in the *Padma Purana*, Chapter 105, Part VIII, under 'The Greatness of Dhatri and Thulasi' (Deshpande 1991: 2689). Some tribal myths also contain stories about Vrinda. Krishna, an avatar of Vishnu, outraged Vrinda's modesty by disguising himself as her husband. Vrinda cursed Krishna and said, 'I shall be born in the form of the sacred *tulasi* plant and you will have to bear my leaves on your head for the wrong you have done to me'. Krishna repented and granted her desire, and so nothing was more dear and acceptable to him than *tulasi* (Gupta 1971: 77–78).

Tulasi is believed by Brahmins to be the wife of Lord Vishnu. Some others, like the Nairs of Kerala, connect the plant with Lord Shiva. They say if you worship *tulasi* it is like worshipping Lord Shiva, and one does not need to seek other places to worship Shiva. In the past, temples had their own herbal gardens where medicinal plants were cultivated (Sampath 1984: 13), including *tulasi*. The water in which *tulasi* leaves have been soaked is supposed to have medicinal properties (Gupta 1971: 81). Even today in temples in India a spoonful of water in which *tulasi* leaves have been soaked is given to the worshippers.

It is little wonder then that this plant is considered holy in India. It is cultivated in almost all Hindu houses and temples throughout India by all indigenous, rural or urban peoples. Every morning, women worship this plant and after worshipping eat two or three leaves as the leaves are believed to have medicinal properties. The seeds of *tulasi* are given to widows as the seeds have the property of removing sexual desires. Hence their chastity may be preserved. Women worship *tulasi* for the safe return of their sons and husbands if they undertake a journey. Virgin women also worship *tulasi* to be successful in matrimony (Gupta 1971: 81). *Tulasi*, then, represents the feminine. She is the representative of a woman who is good, virtuous and faithful. From her different names we can also deduce some of the various medicinal characteristics, as well as the different manifestations of goddesses.

Medicinal properties

Tulasi has been used in the classical health systems of India, namely Ayurveda and Siddha. According to Ayurveda, *tulasi*'s characteristics and uses are that it is aromatic, carminative, antipyretic, diaphoretic and expectorant. This plant is used for heart and blood diseases, against leucoderma and strangury, asthma, bronchitis, vomiting, foul smells, lumbago, pains, hiccough, eye pain, and purulent discharge of the ear. It is also used against malaria and poisonous afflictions. It is widely used as a specific for all kinds of fevers (Sivarajan and Balachandran 1994: 485). *Tulasi* is also important in oral and folk systems. This oral knowledge is held by millions of housewives, thousands of medical birth attendants or *Dais*, tribal healers, wandering monks and indigenous communities.

Scientific literature on *O. sanctum* shows the presence of several chemical compounds primarily of the terpenoid class (Chopra, Nayar and Chopra 1956: 179; Gulati and Sinha 1989; Dey and Choudhuri 1984; Knobloch and Hermann-Wolff 1985; Parekh, Gupta and Maheswari 1982). These compounds possess antibacterial, anti-viral and anti-fungal activities. Some other compounds are used as analgesics, expectorants, for their antiseptic qualities, and as local anaesthetics (Harborne and Baxter 1993). Biological activity tests on *tulasi* leaves displayed anti-bacterial and anti-fungal properties (Dey and Choudhuri 1984; Gulati and Sinha 1989), and insecticidal activity (Chokechijaroenporn, Bunyapraphatsara and

Kongchuensin 1994). Biological activity tests performed with an extract of the plant have indicated that it can enhance immune functioning by increasing antibody production (Mendiratta *et al.* 1988).

Given the role of *tulasi* in myths and legends, and in various medicinal traditions, biodiversity and the knowledge associated with it can be seen to be very much alive. Their survival can be attributed to a combination of enduring cultural, spiritual, religious and traditional practices.

Manipulation of genetic resources and knowledge

Traditional knowledge is embedded in the social, cultural and moral aspects of people (Banuri and Marglin 1993). Genetic and cultural/spiritual information has been produced over millennia by people: yet all this information goes unrecognized. The modern system of knowledge compartmentalizes this knowledge as economic, technical, etc. (Banuri and Marglin 1993). Value is given to the knowledge only when the resources are manipulated by using modern scientific processes such as biotechnology and genetic engineering, where scientists and corporate owners are responsible (Kloppenburg 1991). Increasingly, genetic resources and the knowledge associated with them are being appropriated and patented by industrial corporations in the North, a situation that conflicts with the ethical, epistemological and ecological organization of traditional knowledge systems. Furthermore, the values of biodiversity in various cultural contexts, such as sacred species, sacred groves, seeds etc. shows that the cultural means of treating biodiversity as inviolable often go unrecognized (Shiva 1996; Shiva n.d.).

Health care systems in India have long been based on the use of locally available plants, animals and minerals. The traditional systems of medicine have developed over thousands of years, side by side with a widespread folk culture concerned with primary health care. These highly decentralized systems have survived until now due to their autonomous and self-reliant nature and an oral means of transmission. Despite two hundred years of colonization, where attempts were made to suppress traditional health care systems, these practices survive (TWN and CAP 1988).

Strengthening the foundations of biodiversity

Tulasi is largely taken care of by women. It is the women who do the planting and worship the plant every morning. This plant is even kept by women

Figure 6.1: A woman worshipping *tulasi* early in the morning in the city of Madras, Tamil Nadu, India. (Photograph: Bhanumathi Natarajan)

who live in high-rise buildings in cities. Can the survival of the knowledge associated with the plant be attributed to the fact that it belongs to the women's domain? Throughout history, biological diversity has been conserved by local communities, especially by their women, who have used, developed and nurtured this diversity. Therefore, in order to strengthen the foundations of biological diversity, women's knowledge and their rights and control of biological resources need to be prioritized (Shiva 1996). For such knowledge to survive generations, it must be maintained and passed on through such practices as those already mentioned. Simultaneously, governments must take action to encourage these practices, as they are required to do according to the Convention on Biological Diversity.

Traditional health care in native Canadian communities (Jean Goodwill)

As indigenous peoples, we do have our own way of interpreting our health, our values and our traditions in North and South America. I am from the Cree tribe, the largest tribal group in Canada. We have a lot in common with one another across the North and South, such as the use of herbal and other plant remedies, and the use of tobacco and sage in our ceremonies. At the Global Initiative for Traditional Systems (GIFTS) of Health conference in Venezuela, someone said, 'We need to think of survival. We need to try to convey our traditional knowledge to the younger generation'. In Canada this was hindered in the past by religious groups. We have to rescue what is left. I would like to outline briefly what has transpired in our area where we are now attempting to convey traditional knowledge to the younger generation.

Our history in North America is similar to that in other places. We were colonized, and our lives were controlled for many years. European settlers took over the land, and tried to assimilate our people into their culture. As a result, our people lost their language, their traditions, their cultural values and ceremonies. Our political leaders made many attempts to change, in order to meet the educational needs of our people. They recognized that education was important and must recognize Indian people in the context of their values and their unique legal, political and economic circumstances.

In 1976 our political leadership – The Federation of Saskatchewan Indian Nations – initiated the Saskatchewan Indian Federated College, in co-operation with the University of Regina. From 7 students in 1976, the enrolment now exceeds 1,200. The college programmes sponsor development and success of students through a holistic, indigenous cultural approach to courses and programmes, incorporating spiritual, emotional and physical as well as intellectual capabilities. Counselling and guidance is given by Elders, who are the main core of campus life. They join with faculty, staff and students in prayer and in all cultural ceremonies.

One of the Indian Health Studies courses is on traditional Indian health concepts. Its main purpose is to provide students with a background to and knowledge of our traditional systems and values. And we tell our students we are not there to teach them to become medicine people: we are only there to provide them with the knowledge and understanding of our many tribes in North America and their commonalities, just as there are commonalities between North and South American indigenous peoples.

We also emphasize the importance of ceremonies; the use of herbal and other plant remedies; rituals, pipe ceremonies and vision quests. We do not teach these things in class. We just tell students that these things do exist, they are valuable, and above all they must respect all traditional values. If they want to expand their knowledge, and expand any other practices, it is up to them to go and find that knowledge through other people, through other Elders that they may find, and we give them guidance on how they should approach these people.

Within the current situation of Indian health professionals in Canada there are approximately 60 Indian physicians and 300–400 Indian nurses. They hold annual meetings in different parts of the country. A recent theme was respecting Mother Earth and the environment, and the role of nurses with regard to the environment. Our main speaker was a nurse who is also a traditional healer in her own right. Many of our Indian physicians and nurses have been trained in Western medicine, but some of them are becoming traditional healers and are beginning to learn some of the values of their own people. It has taken a long time to come about, but I think the combination of these two factors is very important.

In the area of traditional health, spiritual leaders perform ceremonies in many communities. Some are herbalists, some are spiritual leaders, some a combination of both. A lot of people think that most of these healers are shamans or Elders, but that is not necessarily so. Some of them are young adults. It depends on who has received a calling, or spiritual guidance, to become healers. We don't know who they are, but they themselves eventually know when they are ready to accept their responsibility.

The healers that we have today, and in particular over the past two to three years, have been telling us that the message they are getting from the spiritual world is that we should tell the people of the world that we have made a mess of the environment. It is time we changed: it is time that something was done; we have to go out and teach others rather than hold back and hide from so-called exploiters. We have to express these things more publicly than we have done in the past. Many of our healers are very sceptical, because our ceremonies in the past were banned by religious groups and governments. As a result they are not ready to express these things publicly. But I think that their scepticism will eventually go by the wayside. We are looking to the next generation of young people to support this move as they become more knowledgeable and respectful of their traditions and values. We are also very concerned about the ethical issues related to research among indigenous people. Aboriginal and indigenous organizations have begun to address these issues from their own perspectives.

In Canada we have many medicinal plants that have been used for centuries by indigenous people and many are still in use today. We have found that Western and traditional medicine and practices can work well together. Some herbalists are slowly revealing their sources of herbal remedies. We just have to be patient and when they are ready many people will benefit from the large store of knowledge that exists among our people – provided that everyone co-operates and respects each other's cultural diversity and value systems. The greatest challenges will be convincing Western medical professionals and trying to reduce exploitation by big corporations and other entrepreneurs.

Learning to connect spirit, mind, body and heart to the environment: a healer's perspective (Lea Bill – Rippling Water Woman)

This contribution attempts to present the depth of relationship involved with the environment when traditional learning is applied to healing and becoming a traditional healer. I base this perspective on my own experience. My first lesson involved communication with the spirit of the self and the essence of all creation. Self understanding was acquired through experiences in visioning, fasting and utilization of all creation. The more centred with spirit I became, the more my ability to understand and apply the information received from the natural world flowed with ease. Understanding myself as a spirit essence and connecting at this level with the natural systems opened a world of understanding and knowledge that cannot be received through books or lectures. I have gained an understanding that the relationship a healer has with the environment is a reflection of the depth of understanding achieved of the personal relationship with all creation.

Communication with the environment requires a willingness to be open to the subtleties of natural communication. Prayer and the use of prayer chants were part of the initial learning about communication with the spirit. The intonations of the voice are a language in their own right, although the intellect may not register recognition of the meaning of the language. One of the greatest challenges of this process is to overcome the imprinting received through childhood learning. I was a very shy child who had experienced ridicule and discrimination both in school and in out-of-school situations. I acquired basic spiritual and healing knowledge through my grandparents, who were both traditional healers. I had to learn how to overcome fear and being self-conscious, along with the imprinting of 'you're not of any worth'. I learned to ask for guidance and help through prayer. Prayer became a way of listening to my voice and how it affected the environment. The most profound experiences have been in the mountains where I became more connected to the environment which continues to be a gentle and non-discriminatory supporter of my learning. There were times when the world would become silent to listen to my songs, to my prayers and to my weeping as I became connected to my spirit. I listened to the land and its patron, and I observed the movement and presence of all elements, water, wind, sky, plants and animals. These were my messengers, who brought affirmation to my prayers and applauded my songs and comforted my being as it underwent its healing. I have included here a prayer from my first language, Cree. I have translated the words of the Cree language using the most appropriate words to capture the essence of the prayer. All prayers will vary from one medicine person to another and will be particular to the ceremony and/or healing that is being conducted. This prayer is a teaching prayer. It is a prayer that reflects and sets the intent of my journey as a healer.

> Sacred Father of all Creation, thank you for the Beauty and Wonder of life. Thank you for the Beauty and Wonder of this day. I sit before you as your humble servant. Guide me in all I proceed with. Open my heart, open my mind, and clear all negative imprinting held in my body in order that my spirit may be a conduit for divine knowledge, truth and purpose.
>
> Sacred Mother of all Creation, with utmost love and respect, I express gratitude for the gifts you provide to humanity. Thank you for the nourishment that comes from your being. Thank you for the medicines that enhance my capacity. Thank you for the unyielding strength you reflect for me to learn perseverance. Thank you for the Beauty of your being, the synchronicity of all action, creation, cycles and evolution.
>
> Sacred Mother of all Creation, continue to be my teacher and guide as I walk upon your being. Open my heart to the beauty of your spirit. Open my mind to the abundant wisdom dwelling in all Creation. Guide my body with the principles of Love and Order. Guide my spirit in divine truth and purpose.

The second principle I learned in healing is that all living matter has spirit, and therefore must be approached as a living spirit entity. The natural world has a supreme ability to contain its spirit essence according to natural law and thereby holds the key to spirit reclamation. Humanity seeks his/her spirit in natural surroundings as they reflect the nature of his/her spirit in a physical manner through landscape, interactions of and with wildlife, and in sudden shifts of environmental elements. Vision-seeking in the natural environment is a means of reuniting with the spirit essence of life. The natural world reflects the individual aspects of spirit that become overwhelmed by day-to-day confusions and life processes. An example of the power of natural spirit is water, which is my namesake. Humanity may attempt to procure the power of water, but its spirit essence remains undaunted by manipulation or abuse. Quietly, it continues to influence life with its interactions. When its physical essence is overburdened, the spirit flows forth with unwavering and mighty strength, clearing all that is before its path, resurrecting elements that have long been submerged by the hand of man. Principles such as this one, reflected by nature, become my teaching and healing tools. It is from this perspective that I, as a traditional healer, view the world and the elements that enhance the gift of healing. During healing my assistants are the natural elements around me because I have developed a strong relationship with this world. I never cease to be filled with wonder when I have taken a client to the mountains or by a stream for healing and all becomes quiet with expectation. As I prepare and set the intent by outlining the needs of my client through prayer, all that is around me adjusts to begin with the healing. A healing ceremony has several stages.

The first phase of healing involves a prayer ceremony that encompasses all the elements for healing. The second phase involves the use of songs and rattles. This phase focuses on healing the physical body with vibration and by releasing negative imprinting and accumulated stress in the body. When the healing is performed in a natural setting there will often be complete calm: the birds will stop singing, and the winds will become gentle. There are times when the winds will pick up and swirl around my client or a bird will sing a loud or a gentle song, aiding in the healing. Once this phase is complete, a post conference is held with the client to teach and provide guidance. Teaching and counselling can sometimes take several hours and several sessions following the process outlined above.

I acknowledge and understand that there are parallels in the systems between humanity and the natural world. The systems within the world of plants, animals, trees, rocks and the atmosphere all contribute to the overall understanding of life process and the principle of cause and effect. The imbalance in one system ultimately affects the other. The natural world celebrates our healing as it means greater opportunity for increased balance and biodiversity. Both systems strive for balance in their own manner, whether it is of mind, body, spirit or emotion. Humanity and the environment are stewards of each other.

General overview of Maya ethnomedicine (Elois Ann Berlin and Brent Berlin)

In the traditional medicine of the highland Maya, as with many other ethnomedical systems, the maintenance or re-establishment of a state of health is dependent on events and interactions in two separate realities: the natural, usually visible, reality that follows predictable physical norms, and a frequently non-visible reality that relates to extranatural phenomena. We adopt Foster's (Foster and Anderson 1978) dual division of medical systems into *naturalistic* and *personalistic* to characterize these two cognitive frameworks. In the naturalistic system, a health condition is empirically determined and is based primarily on immediately apparent *signs* and *symptoms*. For example, the naturalistic gastrointestinal condition *ch'ich' tza'nel* ('bloody diarrhoea': literally 'blood' + 'faeces') is unambiguously recognized by the presence of blood in the stool and severe pain in the lower abdomen. For a naturalistic condition such as bloody diarrhoea, it is the norm that one treats oneself with medicinal plants or, lacking this knowledge, consults with individuals who are themselves knowledgeable about such plants.

In contrast, diagnosis of personalistic conditions is based on *retrospective presumption of etiologic agent*. For example, the personalistic condition *jme'tik jtatik*, lit. 'our ancestral mothers and fathers', may result from an inadvertent encounter with these ancestral spirits. Any given case of *jme'tik jtatik* could involve gastrointestinal symptoms. The initial diagnosis could possibly be personalistic; most frequently, however, such cases are first treated with plant medicinals, and later classed as personalistic in cases that are unresponsive to herbal remedies or that are either prolonged or progressively worsen. These patterns of diagnosis have been extensively described by virtually everyone who has studied the subject. Diagnosis and treatment frequently involve the intervention of healers with special powers, such as a pulser or diviner. While personalistic conditions may at times also be treated with herbal medications, Maya curers normally employ remedies that require ceremonial healing rituals and special prayers.

Almost all earlier work dealing with highland Maya ethnomedicine focused on its personalistic bases, and the general outlines of this aspect of Maya ethnomedicine are well known (for some of the more important examples see: Fabrega, Metzger and Williams 1970; Fabrega 1970; Fabrega and Silver 1970, 1973; Guiteras-Holmes 1961; Harman 1974; Holland 1963; Holland and Tharp 1964; Metzger and

Williams 1963; Silver 1966a, 1966b; Vogt 1966, 1970, 1976). This preoccupation with the cosmological aspects of Maya healing has tended to de-emphasize the wealth of medical ethnobiological knowledge possessed by the Tzeltal and Tzotzil, leading generally to the conclusion that highland Maya medicine incorporates a poor understanding of human anatomy, that it has but a weak relationship to physiological processes, and that healing primarily satisfies psycho-social needs through magical principles (Arias 1991: 40-43; Fabrega and Silver 1973: 86, 211, 218; Holland 1963: 155, 170-1; Vogt 1969: 611). Since magical principles have little to do with science, a reading of these works supports a view that Maya ethnomedicine is anything but scientific and that the Maya themselves lack a scientific understanding of health and disease.

Our data on the naturalistic aspects of Maya ethnomedicine lead to quite opposite conclusions. The extensive medical ethnobiological materials we have collected demonstrate that the highland Maya have a remarkably complex ethnomedical understanding of anatomy, physiology and the symptomatology of particular health conditions. Our findings show that almost all of the medicinal plant species comprising the highland Maya pharmacopoeia specifically target individual health conditions. In this sense, highland Maya traditional medicine is an ethnoscientific system of traditional knowledge based on astute and accurate observation that could only have been elaborated on the basis of many years of explicit empirical experimentation with the effects of herbal remedies on bodily function.

Many conditions are considered health-related, for which clear folk definitions can be elicited and for which preventive and protective measures are taken, but which are not defined as illnesses. From the moment of birth, ritual and physical preventive measures are taken to ensure health and strength and to prevent sickness and misfortune. For the Maya, like most lay persons, there is no clear-cut definition of health. A state of health and well-being exists when one is not suffering infirmity or disease, perhaps including misfortune. *Health is the absence of illness, debility or dysfunction.*

John Harris, a physician who studied health conditions among the Tenejapa Tzeltal during the late 1960s, characterized the definition of illness as the inability to work (Harris, n.d.). Our ethnomedical studies lend strong support to Harris's observations since descriptions of treatments for a large proportion of health conditions include injunctions to care for oneself and to refrain from work. However, one modification seems in order. It seems more appropriate to consider the inability to work, or to carry out normal activities, as the point at which one assumes the patient role. *A patient is one who is limited in the performance of normal daily activities due to a state of ill health.*

The term *chamel* ('sickness') is derived from the term for death, *cham* ('to die'). When one becomes a *jchamel* ('patient'), one has entered into, as it were, the (not irreversible) process toward death. In contrast to the biomedical concept of infection and disease as 'threats to health', the Tzeltal and Tzotzil concept is more accurately expressed as 'threats to life'. Most health problems are considered 'cold' conditions. This is apparent inferentially in the causes of disease where cold agents (water, wind, weather) are frequently attributed aetiologies, as well as in prescribed treatments and dietary proscriptions. Most proscriptions involve the prohibition of ingestion of cold things. Those conditions that are not cold typically present with signs of excess heat, such as fever, localized infection, inflammation or skin eruption (boils, rashes). *Death is the ultimate cold state, and the sick patient has entered the cold path to death.* It follows, then, that a primary healing quality of treatments is the ability to affect the thermal state/condition of the patient. An extensive set of additional qualities also exits, directed at treatment of aspects other than the thermal state.

In ethnobotanical surveys among the Maya, of the 27,333 responses to the question, 'Why does plant x have the power to cure?', 13,370 (49 per-cent) included the quality of warmth. If we were dealing with a dichotomous system, these results would be unremarkable. However, using only those medicinal virtues listed in our ethnobotanical databases, we find a total of 36 unique terms, 10 of which can not be analysed at the time of writing. It is clear from an examination of the English glosses for the Maya names of medicinal virtues that these curative properties are, for the most part, organoleptic qualities (i.e. detected by sensory responses such as taste and smell). Respondents vary in their evaluation of the medicinal virtue of a plant, as to both the plant's perceptual and attributed qualities. None the less, clear-cut patterns in the frequency of informants' responses allow one to make statistically valid assessments of the medicinal virtue of the most significant plant species in the highland Maya pharmacopoeia.

Although we lack systematic data from a large number of people, our Maya collaborators have suggested the kinds of effects produced by the various medicinal virtues. Taking as examples the terms that occur most frequently in our data, a bitter virtue alleviates pain, a caustic virtue is good for burning away the problem (such as an itchy skin rash), an astringent virtue promotes superficial healing, and a sour virtue counteracts nausea. The taste of the prepared medication is sometimes disagreeable (as in the case of very bitter treatments). It is likely that the good, sweet scent and sweet properties are used for palatability. This is said to be especially important for children's medicines. As our collaborators explained, if two plants have a similar ability to effect a cure, the more pleasantly flavoured will be selected for children. Alternatively, good-tasting herbs will be added to a concoction primarily to improve flavour.

More detailed examination of the pattern of use of medicinal virtues lends further support to the specificity of function. The primary use of warm virtue is in the treatment of gastrointestinal conditions. However, in keeping with the general healing quality, warm preparations are employed across a broad spectrum of condition types. The therapeutic goal of the virtue of warmth is general healing of the body or curing of the condition.

The virtue of coldness has a more restricted application. While the overall goal of cold virtue therapy is temperature reduction, there appear to be two distinct kinds of pharmacological effects produced: systemic reduction of body temperature and alleviation of a localized inflammatory process (including increased temperature of tissue). The dermatological conditions treated are furuncles (41 percent), burns (31 percent) and rashes (28 percent). Furuncles or boils are characterized by circumscribed swelling, inflammation and pain, and they have a central necrotic core. There is also localized hyperthermia, which would be the rationale for application of cold plant medicinals as pastes or poultices. The cold-treated rashes are the eruptive types, such as measles and chickenpox. Treatment of localized inflammatory response would apply to ophthalmic problems (e.g. conjunctivitis), wounds and bites, dental problems (caries, periodontal disease) and sprains. The condition glossed as 'struck' may indicate a physical (*Chamula*) or metaphysical (*Tenejapa*) blow. Presumably localized swelling would be the object of cold virtue therapy. Thirty-one percent of the dermatological conditions treated are burns. Some of the plants used to treat burns have physical qualities that indicate their cold virtue – e.g. gelatinous inner material which feels cool.

It is evident from the frequency of treatment of health conditions with a bitter virtue that the pain type treated is specific to the spasmolytic properties that have been so strongly associated with these species in our pharmacological studies. This would affect not only gastrointestinal motility, but also probably alleviate respiratory symptoms through relaxation of constricted respiratory passages and reduction of paroxysmal cough response. The caustic virtue is used primarily to treat rashes (such as scabies), warts, pimples and some cutaneous fungal infections.

The complexity of the Maya system we present here, along with the gradual extension of the numbers and kinds of healing qualities reported in the literature, lead us to conclude that there is much more to be done in reaching a complete understanding of the ethnopharmaceutical principles employed by indigenous healers.

[Adapted from: Berlin, E. A. and Berlin, B. (1996). *Medical Ethnobiology of the Highland Maya of Chiapas, Mexico: The Gastrointestinal Diseases.* Princeton: Princeton University Press.]

Ethnopharmacological studies and biological conservation in Belize: valuation studies (Michael J. Balick, Rosita Arvigo, Gregory Shropshire and Robert Mendelsohn)

A great deal of attention has been given recently to the value of non-timber forest products in the tropical forest. One method of ascertaining this value is to inventory a clearly defined area and estimate the economic value of the species found there. Peters, Gentry and Mendelsohn (1989) were the first to elucidate the commercial value of non-timber forest products found within a hectare of forest in the Peruvian Amazon. This study did not include medicinal plants in the inventory, and at the suggestion of the authors, this aspect was evaluated in Belize. From two separate plots, in thirty- and fifty-year-old forest respectively, a total biomass of 308.6 and 1,433.6 kilograms (dry weight) of medicines whose value could be judged by local market forces was collected. Local herbal pharmacists and healers purchase and process medicinal plants from herb gatherers and small farmers at an average price of US$2.80/kilogram. Multiplying the quantity of medicine found per hectare above by this price suggests that harvesting the medicinal plants from a hectare would yield the collector between $864 and $4,014 of gross revenue. Subtracting the costs required to harvest, process and ship the plants, the net revenue from clearing a hectare was calculated to be $564 and $3,054 on the two plots. Details of the study can be found in the original article (Balick and Mendelsohn 1992).

Not enough information is available to understand the life-cycles and regeneration times needed for each species, and therefore we cannot comment on the frequency and extent of collection involved in sustainable harvest. However, assuming the current age of the forest in each plot as a rotation length, we calculated an estimate of the present value of harvesting plants sustainably into the future by using the standard Faustman formula: $V = R/(1\text{-}e\text{-}rt)$, where R is the net revenue from a single harvest and r is the real estate rate; t is the length of the rotation in years. Given a thirty-year rotation in plot 1, this suggests that the present value of medicine is $726 per hectare. A similar calculation for plot 2, with a fifty-year rotation, yielded a present value of $3,327 per hectare. These calculations assume a five percent interest rate.

These estimates of the value of using tropical forests for the harvest of medicinal plants compared favourably with alternative land uses in the region such as *milpa* (corn, bean and squash cultivation) in Guatemalan rainforest, which yielded $288 per hectare. We also identified commercial products such as allspice, copal, chicle and construction materials in the plots that could be harvested and added to their total value. Thus, this study suggested that protection of at least some areas of rainforest as extractive reserves for medicinal plants appears to be economically justified. It seems that a periodic harvest strategy is a realistic and sustainable method of utilizing the forest. On the basis of our evaluation of the forest similar to the second plot analysed, it would appear that one could harvest and clear one

Box 6.4: Spirit, Story and Medicine: *Sachamama* – an Example of The Ancient Beings of the Amazon Rainforest

Francisco Montes, translated and edited by Kathleen Harrison

When we speak of *la Sachamama*, we usually refer to a gigantic boa, a serpent of the forest. This boa is forty metres long, with a diameter of two metres. The head has a form similar to that of the reptile called an iguana. *Sachamama* possesses magnetic power that can pull to her whatever type of animal or person that passes in front of her head. This boa is incapable of moving to either side, due to its gigantic size, and also because it is covered with trees, herbs, bushes and lianas. I am telling you that upon this giant boa grow trees.

We know very well that all the plants, trees, bushes and vines that grow on top of *la Sachamama* are healing plants, with strong curative powers. These plants that grow on *la Sachamama* have special qualities, their own unique curative powers, different than the plants that grow elsewhere. The following are some examples.

- *Puma sanango*, used to cure witchcraft and the effects of sorcery (*brujeria* or *hechizos*). For this one uses the cooked root. The genie or spirit (*el genio*) of the *puma sanango* is an enormous fierce jaguar.
- *Puma huasca*, a liana with properties to cure *brujeria* (sorcery) or *hechizos* (spells). One takes the cooked stem of the vine. This liana teaches one to be a very good *curandero* (curer). Its *genio* is the *yana puma*, or black jaguar, a very fierce animal. All these animals have magnetic powers like *la Sachamama*. This is because they grow right out of *la Sachamama*; it is she that gives them the power to pull or attract.
- *Zorrapilla* or *shapumbilla*, a small herbaceous plant which we have introduced into our garden, El Jardín Sachamama. The curing property of this herb is that it heals cuts or whatever kind of injury.
- *Boa huasca* is a liana that also grows upon *la Sachamama*. The resin of this plant has healing qualities.
- *Lluasca huasca* is a vine that has a resin resembling phlegm. With this vine the *indigenas* (natives) cure blemishes on the face.

Sometimes there also grows there a gigantic tree, very difficult to find in the forest. The name of this tree is *trueno caspi* or *rayo caspi*, also called *lluvia caspi*. This tree is known by all three names: 'thunder tree', 'lightning tree' or 'rain tree'. The *curandero* takes in his diet the cooked bark of this tree to learn; then this man is capable of causing thunder, lightning and rain, but is also able to stop the rain or silence the thunder. The man that eats of this tree is very dreaded, and nobody messes with him. The spirits of the *trueno caspi* are dedicated to controlling the rain, thunder and lightning, because together these are the forces of *la Sachamama*. You may see by all this that *la Sachamama* has many relationships with plants and people. Because in reality the spirit of *la Sachamama* is connected to the spirits of the plants, you yourself may see and believe the effects of these plants that I describe.

hectare per year indefinitely, assuming that all the species found in each plot would regenerate at similar rates. More than likely, however, some species, such as *Bursera simaruba*, would become more dominant in the ecosystem while others, such as *Dioscorea*, could become rare.

The analysis used in this study is based on current market data. The estimates of the worth of the forest could change based on local market forces. For example, if knowledge about tropical herbal medicines becomes even more widespread and their collection increases, prices for specific medicines would fall. Similarly, if more consumers become aware of the potential of some of these medicines, or if the cost of commercially produced pharmaceuticals becomes too great, demand for herbal medicines could increase, substantially driving up prices. Finally, destruction of the tropical forest habitats of many of these important plants would increase their scarcity, driving up local prices. This scenario has already been observed in Belize with some species. It seems that the value of tropical forest for the harvest of non-timber forest products will increase relative to other land uses over time, as these forests become more scarce.

[Excerpted from: *Medicinal Resources of the Tropical Forest: Biodiversity and its Importance to Human Health*. (1996) Michael J. Balick, Elaine Elisabetsky and Sarah A. Laird (eds). Columbia University Press, New York.]

'Vegetalismo' and the perception of biodiversity: shamanic values in the Peruvian upper Amazon (Francoise Barbira-Freedman)

The sub-montane rainforest east of the Andes includes a wide range of altitudes, micro-climates and soils, and is one of the most biodiversity-rich areas of Amazonia. It has been colonized by mixed Hispanic and Andean settlers since the Conquest and the major part of its landscape has been modified. In spite of ethnocide with each wave of colonization, particularly that induced by the rubber boom, a relatively high number of indigenous minorities have adapted to the dominant society. An interface between towns and the surrounding forest developed as the forest–gardens–human settlement continuum emerged.

Paradoxically, the values attached to this continuum were extended to the urban–forest interface through the practice of shamanism which provides a popular religion and medicine in the towns and the hinterland of western Amazonia (Barbira-Freedman 1997). The three main towns are Iquitos, Pucallpa and Tarapoto. The indigenous forest people to whom I refer are the Jakwash Lamista who now are scattered along the middle Huallaga River, San Martin, Peru. The shamanic knowledge and practice system known in Peru as *vegetalismo* developed during the colonial period from the syncretic merging of Amerindian shamanism and an Andean/Hispanic herbalist tradition. It is the main component of the informal health sector, even in areas well endowed with health services. In their practice, *vegetalistas* refer to a widely shared folk cosmology modified locally by Ashaninka and the Shipibo Indian influences.

Vegetalismo is characterized by the use of hallucinogenic plants to induce ecstatic experiences and cognition, a pharmacological intervention involving the use of a wide range of medicinal plants and the use of magical songs, spells and ceremonial objects.

Far from being the vestiges of shamanic practices remaining among urban dwellers (Dobkin de Rios 1972; Luna 1986), *vegetalismo* is integral to an ongoing process of indigenous exchange. *Vegetalismo* involves constant interaction between town and forest. Shamanic training must take place in a forest environment, while urban practitioners rely on suppliers of medicinal psychoactive plants which they cannot cultivate in their town gardens. *Vegetalistas* are both *medicos vegetalistas* (doctors in their own right) and *brujos* (infamous sorcerers). Their control of the exchange between knowledge systems determines their power and status as shamans.

In *vegetalismo*, there are layers of relationships between and among models of the forest as cosmos. Interpretation of shamanic perceptions of biodiversity – and its possible implications for conservation – must take account of these layers, all of which may be activated at once in a ritual healing performed by an urban *vegetalista*. Psychoactive plants dissolve boundaries between self and outside world in the shaman's ecstasy. They enable the constant redefinition of cosmic landscapes in which social relations with kin and neighbours are re-negotiated on patients' behalf in terms of relations between humans and the forest. Shamanic knowledge is shared to a variable extent by the wider community. As such, it constitutes a collective store of knowledge. Different modes of preparation of plants reflect the level of association sought with their 'mother-spirits'. Although knowledge of potencies and effects reflects an understanding of the influence of location, harvesting conditions (waxing or waning moons, dawn or evening), as well as insights from personal experimentation, the knowledge conferred by psychoactive plants is characterized as the main source of practice. What is taught in visions or dreams by the plant teachers amounts to ecology – a system of dynamic interdependences.

In visions or in dreams induced by the ingestion of a 'plant-teacher', its 'mother-spirit' 'presents itself' to the novice and instructs him/her directly on the qualities of the plant, including aspects of its habitat, habitus, spiritual qualities and effects on the human organism. The relationships between plants, animals and humans, and their place in the cosmic domain, are also shown. This knowledge is found in urban *vegetalistas* as well. Don Hilde, a Pucallpa healer who did not undergo shamanic training, explained: 'I did not have a teacher to help me learn about plants, but visions have taught me many things. They even instruct me as to which pharmaceutical medicines to use.' (Dobkin de Rios, 1992: 146)

In urban areas, the market is abundant in forest foods and materials often imported from distant areas. Medicinal plant stalls are staffed by herbalists, the majority of whom are women. There is constant interaction between the herbalists and the *vegetalistas* as the latter often rely on the local herbalists to provide them with specific plants, or send patients to them with prescriptions.

Jakwash Lamista shamans and other native forest people express the notion of balance and reciprocity between humans and the forest environment. In the urban practice of *vegetalismo* this relationship is expressed through the counselling that shamans give to their patients. In an echo of the Jakwash Lamista shamans' reliance on the guiding influence of the plant-teachers, contemporary 'ethnobotanists' extol the spiritual values of shamanic plant knowledge: '... shamanism's legacy can act as a steadying force to redirect our awareness toward the collective fate of the biosphere... plants are the missing link in the search to understand the human mind and its place in nature' (T. MacKenna 1992: 13, 265).

The transmission of the sophisticated knowledge of biological relationships which hunters require may be lost in deforested and urbanized areas of western Amazonia. *Vegetalistas*, however, continue to hold this knowledge in their visions and to impart the values with which it is associated to their patients and communities.

Luna (1991) and McKenna *et al.* (1995) have stressed the scientific potential held in *vegetalismo* as an 'uninvestigated folk pharmacopoeia'. While conservationists often appeal to the world-view and practices of native Amazonians, the shamanic values transmitted by *vegetalistas* among the population of the Upper Amazon reflect historical inter-connections bridging natives and non-natives, forest and town, and the forest as a 'natural resource' and a 'healing forest'.

'Food' and 'drugs' in north-west Amazonia (Stephen Hugh-Jones)

Debates about the 'drugs problem' are frequently characterized by the explicit or implicit assumption that demand for drugs is psychological or physiological in origin. There is also a tendency to reify the issue by focusing attention on substances rather than on people. Attention is thereby diverted away from the social forces that lie behind the consumption or prohibition of psychoactive substances – what people do with drugs – onto the apparent power of the substances themselves – what drugs do to people.

I want to use a specific case to suggest that whilst the category 'drug' often seems to divert attention away from 'food' and 'drink', an anthropology of 'drugs' might usefully begin by thinking more about the consumption of psychoactive substances in relation to the consumption of more ordinary fare.

With reference to the Barasana Indians of north-western Amazonia, coca powder, prepared from the leaves of the coca bush (*Erythroxlum coca*) accompanies verbal exchanges and is used as a vehicle for social interaction, and patterns of coca consumption are related to social divisions based on gender, age and kinship. Coca use is also related to the use of tobacco in the form of cigars and snuff, alcohol in the form of beer, and a hallucinogenic drink called *yagé*. Although some or all of these might be referred to as 'drugs', rather than imposing an alien category, the overall pattern of consumption of these substances in relation to that of other food and drinks reveals an implicit categorization of substances into two contrasting but analogous and complementary systems.

Though indigenous use of coca is usually associated with the Andean zone, it is in fact also quite widespread in western Amazonia. Coca is a semi-tropical plant and the coca consumed by the highland Quecha and Aymara is actually grown in the warmer mountain region on the eastern slope of the Andes towards the Amazon forests; in the lowlands further to the north and east, a different Amazonian variety of coca is found. This variety, grown from cuttings rather than seeds, and with a much lower alkaloid content, is used mainly by the Tukanoan and Witotoan-speaking groups in south-east Colombia and north-east Peru. Unlike in the Andes, where coca leaves are chewed whole mixed with lime, here they are first dried by stirring them rapidly into a pot over a fire, then pounded to a fine green powder to which ash from burned *Cecropia* or *Pourouma* leaves is added. The powder, stored as a wetted lump in the cheek, is slowly swallowed – hence the widely used phrase 'coca chewing' is not really appropriate in this context.

The Barasana are one of some twenty Tukanoan-speaking Indian groups living in the Vaupés region of Colombia near the frontier with Brazil. These intermarrying groups are exogamous, patrilineal units and, in theory at least, each speaks a different language. Relations between these groups are characterized by reciprocal exchanges of food and material goods, of feasts and of spouses in marriage. The external equality of status between groups stands in marked contrast to the ranked hierarchy within. Each group is subdivided into a series of clans related as 'brothers' and ranked according to the birth-order of their founding ancestors. In addition to this linear ranking of clans, there is some evidence to suggest that Tukanoan Indian society as a whole was once also divided into three ranked strata of 'chiefs', 'commoners' and 'servants'. Present-day Barasana state that, as in the pre-conquest Andes, coca consumption was once the prerogative of the higher ranking groups. Ideally at least, the clan is a single, co-residential unit. Each such unit, living in a communal, multifamily longhouse or *maloca*, is made up of a group of brothers or parallel cousins together with their in-married wives and children. Relative age and birth order are not only important determinants of status between clans but also between brothers within the clan or longhouse. This age-based hierarchy is reflected in the details of how coca is prepared and consumed.

The Barasana treat coca with a respect that gives the plant and all the products and activities associated with it a semi-sacred status. The planting of coca bushes, the picking and processing of coca leaves, and the consumption of coca powder are all ritualized activities surrounded by elaborate etiquette. Coca is eaten by all adult men and by a few adult women. It thus takes on the role of a sign, both of adult (principally male) status and of certain powers and attributes that go with this status. In particular, it signals the capacity to engage in communication with other human and spiritual beings in the outside world. Though it is sometimes eaten alone, coca consumption is essentially a social activity, accompanied by conventionalized speech and formalized behaviour which focuses not so much on the ingestion of the powder or on its effects but on the act of exchanging or sharing.

Coca is used in three different but overlapping contexts: at work during the day; in the men's circle at might; and during periodic ritual dances involving visiting groups from other *malocas*. At work, coca is usually accompanied by a shared cigar and occasionally by snuff; at night, cigars and snuff are almost always taken and *manikuera* (boiled manioc juice) is usually served. At dances, *yagé* is added to the list of coca, cigars and snuff and *manikuera* is replaced by manioc beer. In addition to these differences in kinds of substances consumed, each different context involves a different degree of formalization of behaviour and is associated with differences in both the style and content of speech.

Though people do not often comment directly upon the physiological or psychological effects of coca, when they do, the effects they perceive or choose to emphasize also depend on context. In relation to work, people would emphasize that coca gives them energy, stamina and concentration and that it staves off hunger; in relation to the men's circle, they say it elevates their mood, makes them more convivial and able to talk, helps them to think and to meditate, and keeps sleep at bay; in relation to ritual dances, they say that both coca and *yagé* help them to learn and concentrate on the complex verbal and bodily routines involved in dancing and chanting and that coca helps them to stay awake and do without food for twenty-four hours or more. Resistance to sleep and hunger are two of the virtues of adult men. They would also emphasize the role of coca in facilitating communication with their fellow men and with their ancestors.

When they are engaged in solitary work, men will sometimes eat coca on their own. Consumed in this way, coca comes close to having a purely instrumental use and is on a par with 'snacks' which are eaten to satisfy hunger. When work is done in company, eating coca leaves takes on a more social aspect. Such work is usually done on behalf of a particular man who is expected to supply his fellow workers with coca and cigars and who determines the periods of rest when they are consumed. The coca is now referred to as a 'treat' or 'reward' (*bose*) and, like those for tea, coffee or cigarettes elsewhere, coca breaks allow people to rest and give rhythm and structure to the work in hand. After the break, the men return to their tasks with a wad of coca in their cheeks; when the wad is finished, it is time for another break.

If coca-breaks give structure and rhythm to work, more generally, the production and consumption of coca is intimately related to time. Like the women's harvesting and processing of bitter manioc, the men's picking and processing of coca is a regular, daily, time-consuming process. Men usually hunt, fish or do other work in the morning and then begin to pick and process coca in the afternoon, the process being timed to end around dusk. Men say that, without coca, their day would have no structure, and a myth about the origin of the day and night describes just such a time when men had neither night nor coca and were forced to sit around aimlessly, not knowing what to do when.

Coca is strongly associated with its owner, the man who plants it, who organizes its picking and processing, and who gives it out to others. This association between plant and person carries over to the level of the group as a whole. Each group owns one or more specific varieties of coca, planted from cuttings and coming from a common clone. The coca plants of each group are part of their ancestral inheritance, passed across the generations and maintained by an unbroken line of vegetative reproduction which is used as one of the principal images when speaking of the lineal continuity of the group through time. Like the group itself, their coca plants come from a common source, a continuous line of growth from an ancestral stock obtained by the ancestor of the group at the beginning of time. Just as the rows of coca in the gardens are compared to the individual men who own them, so the original plant is identified with the body of the ancestor (see below). Coca serves as an intermediary between people and to give coca is to give out part of oneself, an aspect of one's identity. Each group also speaks and owns a distinctive language as another aspect of its patrimony and identity. This too is given out in the speech that goes with coca.

Men eat coca preferentially at night and they eat it continuously throughout both day and night during rituals. By contrast, the men will normally eat food only during the day. At the end of each ritual dance, food in the form of a large meal, preceded by shamanic spells to render the food safe, marks the men's return to normal life. Men prepare coca towards the front of the house and eat it on their own in the middle whilst women prepare food at the rear of the house, eat together with their menfolk at the sides of the house by day, and sometimes continue eating after dark, sitting by themselves towards the rear of the house.

Drinks are likewise never consumed with food or meals. They are either taken just after eating or else consumed at other times when no food is served. Drinks and coca are thus both peripheral to meals but they are often taken together. This combination is most marked at dances when large amounts of beer and coca are served in a seemingly endless supply, but it also happens each evening when *manikuera* (boiled manioc juice) is served and also in the day when men drink *mingau* (manioc starch boiled with water) or *farinha* (manioc granules) mixed with water as a refreshment after bouts of work. Finally, coca is routinely accompanied by some form of tobacco, either cigars or snuff or both, and there is a strong expectation that tobacco and coca should always go together.

From these and other observable practices we can provisionally isolate two opposed complexes which I shall refer to as 'food' and 'non-food'. 'Food' comprises fish, meat and cassava which are usually eaten together at meals, and also various fruits, tubers and gathered foods which are more usually eaten as snacks. 'Non-food' is made up of coca, hallucinogenic *yagé* and tobacco in the form of cigars and snuff, together with drinks of *farinha, manikuera, mingau* and manioc beer. 'Food' or *bare* is a named category for the Barasana but there is no equivalent name for 'non-foods'. However, the category is evident in their behaviour and also in their statements that coca, tobacco and *yagé* are the foods of the spirits and ancestors. The division between these two complexes thus corresponds very roughly to that between the mundane and the spiritual, the everyday and the ritual.

[Adapted from the original article of the same title in: Hladik, C. M., Hladik, A., Linares, O. F., Pagezy, H., Semple, A. and Hadley, M. (1994) *Tropical Forests, People and Food: Biocultural Interactions and Adaptations to Development*. UNESCO, Paris.]

The role of food in the therapeutics of the Aka pygmies of the Central African Republic (Elisabeth Motte-Florac, Serge Bahuchet and Jacqueline M. C. Thomas)

The Aka pygmies, from the frontier between the Central African Republic and the Popular Republic of Congo, have a permanent and easily accessible reserve of food at their disposal. However, although they can draw their food from the great diversity of the rainforest resources throughout the year, the value given to the products differs according to the ways in which they are acquired (harvesting, gathering, hunting) which entail differing degrees of difficulty, uncertainty and consequently different effort/profit ratios.

Without denying its biological aspect or minimizing its possible seriousness, disease is interpreted as the outward sign of disrupted equilibrium. This upset balance may affect the relationships that one maintains with oneself and with the 'worlds' one is part of; in other words with all the components one perceives in nature, society, the cosmos and even the invisible. Therefore the term 'therapeutics' will not be restricted to medicine only but will refer to every action allowing the preservation or restoration of health. So, all that is commonly known as 'magical practices', that all-embracing term which reveals the Western world's ignorance and limitations, will be taken into account with the same significance as medicine.

For the Aka, 'health' is indissociable from the concept of 'life'. Both are translated by the same word *mò.siki* and both can only find their perfect expression in conditions of equilibrium and harmony. This search for balance is present everywhere, even in food, where it expresses itself in their liking of moderation; they enjoy a sauce without excess – judiciously salted, correctly seasoned with oleaginous seeds (but without reaching a glutinous state) and not too highly spiced (Thomas and Bahuchet, 1981-1992).

This need for equilibrium excludes both lack and excess from the diet. The latter (*Ø.ngòlè*) is perceived, beyond banal indigestion, as dangerous, indeed carrying a connotation of death. That is why, as stoutness is a sign of (and synonymous with) fitness, rounded forms express more satiety than obesity and above all are opposed to thinness, an evident mark of disrupted equilibrium, sickness and disease. According to the Aka, the feeling of repletion required for well-being can only be obtained from an essentially meat plus honey diet, eaten to satiety (Bahuchet 1985; Thomas 1987).

Being in possession of all their faculties implies for the Aka pygmies not only perfect fitness but also, in total interdependence, an inner equilibrium, a feeling of well-being which can only be derived from harmonious relations with oneself, others, nature and the supernatural, as may be inferred from the different individual or group rituals and practices that precede a hunt.

The aim of these gestures and practices is to increase the individual's abilities, above all those of the body as they conceive it: keenness of eye, skill (hands and arms), strength provided by 'vital energy' (belly), 'intuition' or more exactly 'ability to see beyond material appearances' which is generally translated by the occidental notion of luck (between the eyebrows). It is interesting to note that the localization of these last two 'centres' taps a knowledge shared by traditional medicines in all the continents. The traditional medicine of the Aka pygmies is holistic and does not neglect any of the human being's dimensions. Plants occupy an important place in the somato-therapeutic aspect of curative treatments (Motte 1979). The Aka use plant species to cure the majority of the most common illnesses and diseases. Several plants are known and used to treat the same disease: because they grow in different types of forest, they allow the Pygmies to cure themselves when travelling, wherever they find themselves.

It sometimes happens that plants are used as a support for therapeutic practices from other fields of knowledge. These procedures use psychological, social and symbolic aspects of treatment, in ways specific to the culture. Thus, some treasures of what could be called 'the Aka art of preventive therapy' can be encountered in 'magical practices'. Their *savoir-faire* in defusing all sources of individual malaise and breakdown in group equilibrium pre-dates (probably by several millennia) the most effective psycho-verbal and psycho-corporal therapies that are flourishing at present in Western countries. In the terms of these therapies one would say that emotions, needs, problems etc. are recognized, revealed, and verbally, physically and symbolically expressed as well. Loosening stresses in order to release energies permits them this emotional mobility which avoids pathological knots and finds expression in a way of being that has always roused the admiration of those who have approached them.

Their mastership goes still further, reaching levels that are not just curative but also preventive, and in which the whole community takes charge of an individual's problems. The participation of the whole group in rituals allows uneasiness to be eliminated in a community framework which brings security. It also avoids any increase of individual suffering due to a feeling of exclusion or guilt with respect to the group, which is also in the interest of the group. Thus, although specific individuals are recognized as 'sooth-sayer–healers', each person is at once his own healer and contributes to the health of the group, which in turn supports the health of the individual. To carry matters to extremes, we could say that the Aka pygmies' individual and community life conception is in itself the best prevention against the different forms of stress and the numerous diseases that are linked to it.

Food and curative therapeutics: the treatment of disorders of alimentary origin

Rub-down

Leaves are very often used, just as they are, as a rub-down. The person who has broken one of his food prohibitions rubs his whole body with the leaves of *Combretum* sp. and, so purified, will recover all his dancing abilities, which are required to come into contact with the spirits (Motte 1980).

Unctions

Skin salves are often used among the Aka pygmies; generally a paste prepared from the wood of the tree *Pterocarpus soyauxii* (*Fabaceae*). Its use always has a ritual aspect because of its colour, which is a sign of sacredness (Thomas 1991), and various animal parts will be incorporated into the paste, such as fragments of bone, or their ash, which are added to greasy compounds such as cabbage palm butter (*Laccosperma* spp.) or palm oil.

Scarification

Incisions, which allow active compounds to pass into the circulatory system, are made in places varying according to the patient and the disease. Pomades are introduced into these incisions, for example a mixture of cabbage palm butter and the powdered, charred, scales and bones of the pangolin, or the powdered, charred and pulverized bones of the eaten animal such as the Gabonese grey parrot or green parrot.

Plasters

A leaf of *Whitfieldia elongata* (*Acanthaceae*), heated over cinders, is applied to the ears to prevent deafness from becoming irreversible.

Draughts

Medicines given to drink are either cold macerations or decoctions.

Fumigation

Smoke, passing from a perceptible to an unseen state, links the world of men and the invisible world. Accordingly, this procedure is carried out behind the huts, and on the camp boundaries, which are considered as a junction area between the two worlds.

Snuff

In some cases of prohibition transgression, the sick person must take a pinch of snuff (e.g. a mixture of charred mongoose skin and bones, bark scrapings and powder of redwood; cf. above).

Oral medicines

In Aka therapeutics, the number of oral medicines is quite limited: they represent only one-third of all the medicines studies (350). Most are taken as drinks. Non-liquid oral medicines are rare (8 percent). They are generally parts of a plant which are eaten raw (roots as aphrodisiacs, young leaves against coughs, seeds against stomach aches) or sometimes mixtures.

The Akas' eating habits (acquisition, preparation, sharing, consumption) show that over the centuries they have established a remarkable system of management of individual health (physical, mental, psychic, social) as well as of group vitality and perenniality. Over the past decades this system has shown, through the progressive disappearance of rituals, some signs of weakening. These changes do not show that the procedures are becoming less effective but rather that there is a deep transformation of the way of life. Hunts in particular are becoming smaller and less frequent. And if game becomes scarce, hunting will decrease and the life/health equilibrium will be disrupted. The resources of biomedicine may sometimes be used to restore this disrupted equilibrium. Will these upheavals result in the Aka forming new strategies or in them losing their equilibrium, their harmony, their health?

[Adapted from the original article of the same title by Elisabeth Motte-Florac, Serge Bahuchet and Jacqueline M. C. Thomas, in: Hladik, C. M., Hladik, A., Linares, O. F., Pagezy, H., Semple, A. and Hadley, M. (1994) *Tropical Forests, People and Food: Biocultural Interactions and Adaptations to Development.* UNESCO, Paris.]

Symbols and selectivity in traditional African medicine (Maurice Mmaduakolam Iwu)

Any attempt to understand the fundamental conceptions of health and disease among the Africans must deal with the broader attitudes that they have to life itself. To most African communities, every object or form reflects in its existence the dualism of life: a recognition and translation of reality into its composite parts, the physical and psychic, not as two separate entities but as sections of an indivisible whole. Living is a religious act: healing, therefore, includes rituals, incantations and medicines that will remedy both the diseased body and the sick spirit.

Perhaps the overriding belief in Africa is that everything animate or inanimate has a sort of spirit or power within it – a life force which can only be properly harnessed and utilized by the knowledgeable and the initiated. This belief has erroneously been classified as animism, fetish worship or voodooism. That would be a gross distortion, since the objects and animals are not 'worshipped' but consulted or used as instruments or means of achieving certain objectives, including healing. The object or animal is essentially inert and the force inhabited by nature in it is restricted by the physical manifestation of its being. Does this belief not parallel the present knowledge of the powers trapped with the 'innocuous' atom?; which though ordinarily inert can be source of great force when properly harnessed. This force of nature is actually the quintessence of the thing itself: man, beast, tree or stone. Thunder, lightening and other natural phenomena usually explained away ignorantly by Western oriented minds, are personified and given a cosmic place. Each has its own personality, specific, limited in definite way, and very much like the object it inhabits. It is common to see a traditional healer talking to a tree or a stone – challenging the material obstacle of the object to appeal to the life force within, to control and use it for the benefit of his patient.

If animism is simply a belief in spiritual beings then all religious people are animists, but if the word is taken to mean a belief that animals have indwelling spirits (souls) like man, then animism is not a common pattern of African culture. The concept of the Holy Trinity was therefore easy for the Africans to accept from the Christians since they already believed in the possibility of the same spirit to be manifest in countless forms. Africans, however, found particular attraction with the Holy Ghost, and since then they have created many Christian religious sects that embody the fundamental Christian belief with the traditional African concepts of life, disease and death.

It is this belief in the indwelling spirits that accommodates what Freud referred to as the ambivalence of the primitive over the dead. The soul of the deceased is treated with elaborate ceremonies and rituals, but the body is handled with the caution it rightly deserves. The ancient Africans were aware of the contagious nature of certain diseases and had to impose taboos to protect the community against the possibility of epidemic disease from corpses. The spirit of the deceased, on the other hand, has to be assisted with rituals and sacrifices to make the journey to the land of the ancestors. It is the duty of the traditional medicine man to determine the nature of death and the manner of disposal of the corpse.

To the African, the struggle for existence between all creatures, including plants, and even stones, is a very real struggle, each individual holding its place by some innate strength of personality. On the other realm is the delicate co-existence of the living forms, inanimate forces, spirits, gods, the apparently dead and the living dead. Life itself is seen as inseparable from religion. The preservation, restoration and enhancement of health necessarily involves the whole human community, the living dead, spirits and gods, the natural environment and God.

In the African cognitive experience there are two levels of the intellect – the 'normative' or practical, or the so-called 'active intellect', and the 'residual' or habitual. It is the active intellect that controls cleverness, tactfulness, slyness and learning; the residual intellect is concerned with what Jahn calls 'habitual intelligence' – wisdom, active knowledge. It is therefore possible for one to be endowed with an active intellect, excel in book intelligence, and yet remain empty in the residual intellect.

The rural African, though unable to read and write, still considers his urban brethren and their Western neighbours as 'really disarmingly naive'; and as having no intelligence. Jahn (1961) recounted that when a Rwandese woman was challenged with the question, 'How can you say something so stupid? Have you been able, like them, to invent so many marvels that exceed our imagination?', she was credited with offering the following reply, with a pitying smile: 'Listen, my child! They have learned all that, but they understand nothing!' The 'uneducated' woman was merely emphasizing the basic ignorance of Western oriented minds as to the fundamental nature of the world and laws governing the relationship between things.

A healer's power is not determined by the number of efficacious herbs he knows but the magnitude of his understanding of the natural laws, and his ability to utilize them for the benefit of his patient and the whole community. He should be knowledgeable of the taboos and totems, without which the entire community will disintegrate.

The African healer therefore questions strongly any form of treatment that focuses only on the organic diseased state and ignores the spiritual side of the illness. The sterile use, *per se*, of different leaves, roots, fruits, barks, grasses and various objects like minerals, dead insects, bones, feathers, shells, eggs, powders, and the smoke from different burning objects for the cure and prevention of diseases is only an aspect of traditional African medicine. All creatures and objects are believed to possess some psychic quality in them. If a sick person is given a leaf infusion to drink, he drinks it believing not only in the organic properties of the plant but also in the magical or spiritual force imbibed by nature in all living things and also the role of his ancestors, spirits and gods in the healing processes. The patient also believes in the powers of the incantation recited by the healer and assists him in the designation of the ingredients of the remedies given, from mere objects into healing tools. The art of healing is part of African religion; there is a peculiar unity of religion and life that is characteristically African. Healing in Africa is concerned with the restoration or preservation of human vitality, wholeness and continuity: it is a religious act, perhaps superseded only by the rituals of birth, puberty, marriage and death.

A fundamental belief that is central to all African philosophy is the belief in the existence of one supreme being. The One Great God is viewed as an integral member of society in contradistinction from the Western and Christian idea of God staying aloof in Heaven in the community of good spirits, looking down on the evil ones in Hell, and yet seeking to govern a mixture of sinners and the righteous on earth. He is not even personified; neither can he be confined in shrines and fetishes nor excluded from everyday happenings on earth. It is considered foolish to have shrines erected to God as he is part of all creation. Man is viewed just as an immanent ray from the supreme being and therefore the question of death, birth and living does not arise. 'We have always been here', is a common answer given by the ancestors to explain their existence on earth. Death is not seen as an irretrievable separation from life but a mere retraction of the soul to the reservoir of power, the supreme being. The deceased is not considered 'dead and gone' in the strict sense of the phrase. He is still recognized as an important member of the community, with real presence, affected by the actions of the living, and one who participates in their lives by his wisdom and advice. This belief that the dead are an integral part of the community is essentially responsible for the elaborate funeral rites observed by various African tribes.

[Excerpted from an award lecture delivered at the University of Nigeria, Nsukka, in 1989.]

Box 6.5: Australian Poems

Bill Neidjie, with S. Davis and A. Fox

We want goose, we want fish.
Other men want money.
Him can make million dollars,

but only last one year.
Next year him want another million.
Forever and ever him make million dollars ...
Him die.

Million no good for us.
We need this earth to live because ...
we'll be dead,
we'll become earth.

This ground this earth ...
like mother and brother.

Trees and eagle ...
You know eagle?
He can listen.
Eagle our brother,
like dingo our brother.

We like this earth to stay,
because he was staying for ever and ever.

We don't want to lose him.
We say 'sacred, leave him'.

Aboriginal song from Oenpelli region

Come with me to the point and we'll look at the country,
We'll look across the rocks,
Look, rain is coming!
It falls on my sweetheart.

Cited in Broome, R. (1982) *Aboriginal Australians*, George Allen and Unwin, Sydney.

Colour Plates

1. 'Los Amazonicos reclaman sus derechos' (Amazonians protesting their rights). Tigua paintings reflect indigenous culture as well as contemporary politics. This painting depicts a major march of indigenous people from the Amazon on their way to Quito, the capital, to ensure their rights are included in a new national constitution, drafted in March 1998. (From Tigua, Ecuador. Artist: Julio Toaquiza. Photo/caption courtesy of Jean Colvin.)

2. 'Shaman curando' (Shaman healing). Shamans are still an important part of Quichua culture. In remote communities, Western medical assistance is difficult and expensive to access. Most people first consult the local shaman. (From Tigua, Ecuador. Artist: Alfredo Toaquiza. Photo/caption courtesy of Jean Colvin.)

4. 'Anuncio de la Muerte' (Announcement of death). The legendary condor is a common figure in many Tigua paintings. Flying high over the *paramo* (highlands), the condor is the bearer of important messages. Despite the altitude and harsh conditions, the people of Tigua cultivate small plots of potatoes and grains on the steep slopes of the Andes – land granted to them after agrarian reform in the mid 1960s. (From Tigua, Ecuador. Artist: Gustavo Toaquiza. Photo/caption courtesy of Jean Colvin.)

5. 'Contaminacion de la Amazonia' (Pollution of the Amazon). Indigenous federations in Ecuador's Amazon basin have been struggling for years to limit oil exploration and development in their traditional lands. Texaco's activities in the last decade caused serious environmental pollution which affected both human and animal populations and dramatically disrupted the harmony and care that characterize the relationship between indigenous people and nature. (From Tigua, Ecuador. Artist: Gustavo Toaquiza. Photo/caption courtesy of Jean Colvin.)

3. 'Adoracion del Inca al Pachacama' (The Incas worship Pachacama). Although the Incas occupied Ecuador for less than a century, many indigeous people today claim this heritage as an expression and validation of indigenous culture. Mountains and clouds are often represented in human form in Tigua paintings, to show they are living elements of nature. (From Tigua, Ecuador. Artist: Alfredo Toaquiza. Photo/caption courtesy of Jean Colvin.)

6. View of a Karañakaék community, Barí territorial recuperation. (Photograph: Jaime Rodriguez-Manasse: see Chapter 3.)

7. *Argania spinosa*, the argan, is an economically, ecologically and culturally important tree of dry, south-west Morocco, especially in the Souss Valley around Agadir. The seeds of its olive-like fruits contain a highly nutritious cooking oil which is rich in essential fatty acid. The leaves and shoots are heavily browsed by goats. Where it grows – even in desert areas with as little as 50mm of annual rainfall – the argan is often the dominant tree or shrub. It is the subject and inspiration of poetry, and the focus of the lives of many people. (Photograph: Hew Prendergast: see Chapter 5.)

8. Quechua children from the indigenous community of Pampallacta, Cuzco, Peru. (Photograph: Tirso Gonzales: see Chapter 5.)

9. Seed diversity as an expression of cultural diversity. Corn varieties presented at a Seed Fair in Chivay, Arequipa, Peru. (Photograph: Tirso Gonzales: see Chapter 5.)

10. Indigenous farmers from Yauri, Cuzco, in Peru, tending a potato "chacra" with a unique tool designed by them. (Photograph: Ramiro Ortega: see Chapter 5.)

11. Quechua Indians from a community near Cuzco, Peru, doing communal work. (Photograph: Tirso Gonzales: see Chapter 5.)

12. An indigenous farmer from Chivay, Arequipa, in Peru, participating at a Seed Fair. Every seed is a "person" and has a name. (Photograph: Tirso Gonzales: see Chapter 5.)

13. A proud indigenous woman from Chivay, Arequipa, in Peru, exhibiting at a local Seed Fair the outcome of nurturing various types of seeds. (Photograph: Tirso Gonzales: see Chapter 5.)

14. A bobolink in flight over a hay field before ploughing. The bird's nest is in the field. David Kline farm. (Photograph: Richard Moore: see Chapter 7.)

15. Ploughing a hay field after the bobolinks have finished mating. David Kline farm. (Photograph: Richard Moore: see Chapter 7.)

16. Black walnut tree, used for high-quality veneer: Leroy Kuhn's farm. (Photograph: Richard Moore: see Chapter 7.)

17. The main product of the Caiçara agriculture is the manioc (cassava) flour. Since immemorable times, the manioc flour is a substitute of the "European bread", and for this reason it is called the "tropical bread". Other products of the Caiçara agriculture are maize, beans and yam as well as medicinal herbs. (Photograph: Daniel Toffoli: see Chapter 7.)

18. A young Mende girl winnowing rice. (Photograph: Helen Newing: see 'Casting seeds to the four winds', Chapter 7.)

19. 'The arrival of "Inlanders" at Fort George Island'. This re-enactment by the Chisasibi Cree, recorded by a film crew for local television, is a highlight of an annual community event called *Mamoweedoew Minshtukch* – 'the gathering on the island'. The traditional paddle canoes brought out for this occasion (background) are dwarfed by modern canoes equipped with outboard motors (foreground), which are the common means of transport on the river today. (Photograph: Douglas Nakashima: see Chapter 10.)

20. Under the watchful eye of his son, an Inlander father participates in the re-enactment of setting up the tepee (background). Note the rifle sheaths (left foreground) – ornately decorated to please the animal spirits and bring the hunter good fortune. (Photograph: Douglas Nakashima: see Chapter 10.)

21. A Tukanoan Wanano basket trap (*buhkuyaka*) consisting of cylindrical basketry fish traps set in a barrier. (Photograph: Janet Chernela: see Chapter 10.)

22. Bark-cloth painting by Tukanoan Wanano artist Miguel Gomez showing tree with falling fruit, fish feeding on fruit, and bow-and-arrow fisherman. (Collected by Janet Chernela, Rio Uaupes, Brazil: see Chapter 10.)

23. Bark-cloth painting by Tukanoan Wanano artist Miguel Gomez showing the detailed coloration on one of the many varieties of bagre, or catfish. (Collected by Janet Chernela, Rio Uaupes, Brazil: see Chapter 10.)

24. Bark-cloth painting by Tukanoan Wanano artist Miguel Gomez showing two sub-varieties of Cynodon with distinctive features. (Collected by Janet Chernela, Rio Uaupes, Brazil: see Chapter 10.)

25. "Pyloric Valve", 1992 (artist's collection)
The structure of the leaf corresponds with our own innermost bodily structures –our veins, valves, nodes. The same fragility that exists in an ordinary leaf exists in our human physiology, mysteriously and yet without question. (Artist: Margot McLean: mixed media on linen. See Chapter 11.)

26. "Moose", 1992 (private collection)
The art of the moose is to be visible and yet invisible. This elusive, lonely, fading, there-but-not-there aspect of the creature contributes to the spirit of the moose –to the ghost of the moose. (Artist: Margot McLean: mixed media on paper. See Chapter 11.)

27. "Virginia", 1993 (private collection)
Incorporating vegetation from a particular spot in Northern Virginia that is the habitat of a particular kind of land turtle, this painting re-locates the turtle into a landscape of the imagination –allowing, or giving, it another, different life. (Artist: Margot McLean: mixed media on canvas. See Chapter 11.)

28. "Orinoco", 1994 (artist's collection)
From a trip on the Orinoco River in Venezuela. The "baba" or river crocodile, comes ashore in the evening, lying among the rocks or in one of the tidepools, its presence felt throughout the surrounding area. The image suggests the concealed possibility of life that lurks in the unknown. (Artist: Margot McLean: mixed media on linen. See Chapter 11.)

CHAPTER 7

TRADITIONAL AGRICULTURE AND SOIL MANAGEMENT

Kristina Plenderleith

CONTENTS

FIGURES AND TABLES

Traditional Agriculture and Soil Management

Kristina Plenderleith

The role of traditional farmers in creating and conserving agrobiodiversity (Kristina Plenderleith)

Agrobiodiversity, one of the greatest expressions of collaboration between people and nature, is under threat. The diversity of farm animal breeds and crop plant varieties that were present in the world at the beginning of the twentieth century existed as a result of thousands of years of selection and breeding by local farmers. However, during this century that creative survival process has been overwhelmed by the commercialization of farming. Markets have been opened up in the shift from a predominantly rural to a predominantly urban world population, and there has been a growth of monopoly controls to enable breeders and farm suppliers to protect their commercial investment in seeds and domesticated animal breeds, fertilizers, pest control chemicals and veterinary medicines. As a result, local diversity, knowledge and skills have been lost – with the extent of that loss only now being assessed. The prospect of world food shortages resulting from population growth and urbanization gave rise to the Green Revolution, whose high-yielding crop varieties (HYVs) – developed by commercial breeders from farmers' varieties – were acclaimed as the solution to food insecurity. But that revolution not only failed to live up to its promise, it also destroyed much of the skills and knowledge of traditional farmers that resulted in the agrobiodiversity upon which commercial plant and animal breeding is founded, and which continues to be endangered by twentieth-century erosion of traditional culture and life-styles.

Indigenous and traditional communities, for the most part, still retain their holistic, adaptive and innovative approach to the land. They plant, breed, experiment and conserve. They grow food and cash crops, herbs and medicinal plants, and rear livestock. They make use of off-farm resources by gathering, either regularly (firewood), seasonally (fruits, edible plants, medicinals), or occasionally (for famine foods when crops are meagre or fail, and for construction materials such as rattan), and they boost their protein intake by hunting, trapping and fishing. As agroforesters they plant and tend trees to provide fruit, building materials, fodder and oils. These trees enrich the soil and improve its structure, regulate water retention, and provide shade for crops, people and livestock. Unlike the monocultures of modern industrial agriculture, traditional farmers' mixed systems maximize the land's productivity whilst minimizing the effects of pests and diseases. All these activities are the product of knowledge and experience passed down through the generations and shared in the communities.

These traditional farmers are individuals (frequently women) or family groups, who work either small settled landholdings, or swidden plots, and still remain largely independent of the wider market for their livelihood. Their understanding of the local environment has enabled them to take full advantage of available resources. For example, in the deserts of North Africa, cultivators exploit the high water table surrounding oases by using layered plantings to create a humid microclimate which mimics the natural forest structure. Tall palms shade smaller fruit trees, beneath which in the cool moist air further layers of crops are planted. Such efficient use of a limited resource in a harsh climate made possible the development of the town of Nefta in southern Tunisia, where the inhabitants cultivate the oasis and live in houses built into the sandstone cliffs encircling it. Miguel Altieri (this chapter) describes another sophisticated form of land management in a harsh environment – the centuries-old *waru-waru* farming systems created by the inhabitants of the Andes.

Farmers would not have been able to retain the sustainability of their land without understanding the functions of the soil. The importance of soil fertility and soil structure to crop yield is fundamental, and understanding the complexity of soil management is an essential feature of local farming systems that recurs in several contributions to this chapter. The continuing fertility of the mounds cultivated by the Wola in Papua New Guinea (Sillitoe and Ungutip), and the swidden cultivation of the Caiçara in Brazil (Toffoli and de Oliveira), are dependent on this knowledge. Also, their sophisticated use of fire to recycle nutrients is a totally different activity from the wide-scale forest burning carried out by plantation managers and shifted cultivators that is decimating tropical forests today (see also Enris and Sarmiah, this chapter).

Traditional taxonomic systems demonstrate local communities' understanding of their environment, yet these systems have received inadequate attention from scientists in spite of their practicality and deep knowledge base. Taxonomic systems are essential for identifying and recording the elements of biological diversity. Where indigenous and local community taxonomic systems extend to non-biological components of the physical environment – such as soils, seasons, meteorological phenomena etc. – they could usefully be included in formal assessments of diversity made at the national, regional and sub-regional levels (UNEP 1997c), particularly in developing indicators for monitoring and assessing diversity. A system of soil classification developed by farmers in the Northern Province of Zambia, for example, uses features in the topsoil to direct how they will use the land (Sikana 1994). Their categories are dynamic, based on colour in the top layer, texture, consistency and organic matter content, or according to a characteristic environmental feature such as the grass growing on a soil or an insect found in it. Consequently, a farmer may assign a single category to soils which a scientist would classify as being of different types or vice versa. In the Peruvian highlands, the Quechua's system is linked to rights administered by the community, such as rights to land (Tapia and Rosas 1993). Their classification system is based on experience and observation of irrigation, humidity, texture and soil compaction, and land is allocated to community members according to social status, marital status and age.

Indigenous and traditional communities have an impact on their environment that extends beyond the lands they cultivate, resulting from the intimate proximity of cultivated and non-cultivated land in traditional farming systems, which blurs the divisions between them, and through their use of Non-Domesticated Resources (NDRs) (Posey 1996). This chapter offers examples of traditional societies' management of their food resources, ranging from the gathering of fruits from uncultivated parts of their farms by the Amish (Moore and Stinner) to Christine Kabuye's description of traditional foods used in Africa, where NDRs provide essential famine foods in regions with recurrent food shortages. Christine Kabuye also describes how different foods are eaten by different age groups. Furthermore, medicines and foods for health cannot necessarily be separated from food for daily consumption, as illustrated by Stephen Hughes-Jones' portrayal of the place of coca in the lives of the Barasana Indians of north-west Amazonia (Chapter 6).

A study carried out by Melnyk (1995) in southern Venezuela compared the extent to which reduced access to forest resources resulting from deforestation had affected use of NDRs by two Piaroa communities. The study found that forest foods remained an integral part of household resources in spite of the considerably greater time needed for gathering them. The importance of NDRs to people's livelihoods, and their need to protect them, means that communities have always actively sought to conserve resources *in situ*. In doing so they have protected landscapes that may appear totally 'natural' but which are to a greater or lesser extent managed by the local inhabitants.

Cultivation and animal husbandry in traditional and indigenous societies are a part of the cosmovisions of those societies that have provided a system of checks and balances for conserving essential elements of the landscape. Where there has been external intervention, ancient belief systems may be supplanted, but frequently they are incorporated within the new context. In the Andes, the superimposition of Christianity on indigenous beliefs has resulted in the co-existence of indigenous and Christian gods in the agricultural calendar (Revilla and Paucar 1996). Fiestas that are a mixture of pantheism and Christianity take place throughout the year, celebrating planting, the harvest, the animals, the dead and the living. Celebrations such as harvest thanksgiving also have a reinforcing social function, for they bring people together where they can exchange ideas and experiences, tell traditional stories to the young, and strengthen community links. The extensive research into the life-style of the Amish in North America, carried out by Moore and Stinner (this chapter), provides an insight into the interweaving of religion and folklore in Amish lives in marking the seasons and cultivating the land. The Amish have retained a respect for the experience of their forebears and living older generation, and younger farmers still turn to parents and grandparents for advice. This dual dependence is demonstrated by farmers who may have adopted high-yielding varieties of seeds and the chemicals recommended for them, but who still revert to traditional remedies as a form of insurance. For example, the Pesticides Action Network in Indonesia has found that farmers regularly consult the village shaman when they fail to control pests or diseases on their land.

In places where indigenous beliefs remain dominant they demonstrate understanding of the available resources. A resource that is limited, or infrequent in its use, may be supported by customary laws that maintain prohibitions – such as those placed on hunting, fishing, tree cutting and other extractive activities – and help regulate use to prevent over-extraction. Bennagen (1996) describes how in the Philippines indigenous peoples still believe that the land is God's gift, and that deities and environmental and ancestral spirits are the owners of the land and its resources. Individuals, households, groups and communities may use the land to nourish life, but in the long term the land is held in trust by the village. Users of the land have to consult the spirits and deities through rituals, and areas are protected by being declared 'sacred places' where particular spirits dwell. Among the traditional Hanunoo, the ridges which bound their territory are considered sacred, and are protected by climax forest that provides a natural boundary for the community and a refuge for wildlife. Traditional custom and spiritual beliefs direct the rice cultivation of the Dayak Pasir Adang of Indonesia (Enris and Sarmiah, this chapter). They still use their own calendar, and cultivate according to calculations made from the positions of the moon and stars. However, their culture is under threat from encroachment on their lands by logging companies, threatening their long-established sustainable relationship with the forests in which they live, with deforestation, loss of biodiversity and soil erosion.

Community hierarchies, family and kinship structures also help to regulate land usage and ownership of resources. In Burkina Faso the *saaka* system is based on kinship (Kakonge 1995) with an emphasis on individual qualities such as solidarity, responsibility, accountability, collaboration and initiative: qualities that secure and perpetuate community life. However, because women generally have remained outside the power base of local politics and land ownership, their contribution to traditional agricultural systems has been widely ignored. Bayush Tsegaye (Fisher 1994), however, reports that women are the key element in the Ethiopian strategy for farmer-based genetic resource conservation because they have traditionally been involved in seed selection, storage and utilization. But, in spite of women's specialized knowledge of local crops and micro-ecosystems (Zweifel 1997), they are frequently restricted by their society and remain outside decision-making processes. For example, in sub-Saharan Africa, cultivation is often women's work, but control of the community is the men's responsibility. Millar (1996) describes the role of Boosi women living in the savannah grasslands of the Upper East Region of Ghana. Land control in Boosi society is divided, with ritual control vested in the land priest and legal control in the chief. There are no written rules or organized bodies regulating land use beyond the belief that land is a sanctuary for the gods. Boosi women are subject to many restrictions in spite of their essential role in food provision. They may not own land and so cannot perform sacrifices on their own. Married women may not go into the granary without their husbands' permission. They may not go into a farm during their menstrual period, and they may not search for firewood in the fetish groves or adjacent areas except at specified times. Women grow groundnuts on their own plots which they use as food and as a cash crop. Some cultivate cereals, generally millet, and grow vegetables on borders or bare patches in their husbands' plots. Women also have small livestock. They keep poultry for sale or sacrifice, and a few goats or pigs.

Beyond local community rules, the wider influence of government and mainstream religions may also limit women's access to the land, sometimes in rather unexpected ways. For instance, in a patriarchal Muslim society in the Gezira region of the Sudan, individuals traditionally owned land, and families worked their land together (Barnett 1977). However, government intervention destroyed this harmonious structure when the creation of the Gezira Scheme for extensive irrigated cotton cultivation fragmented landholdings. The new system exposed women to possible contact with men who were not family members, which was unacceptable for devout Muslims. Thus an essential part of the labour force was lost when the women could no longer join their men working in the fields.

All contributors to this chapter emphasize that the most effective traditional agriculture and soil systems tend to be supported by strong cultural and spiritual traditions. Usage rights may be extremely complex and based on intimate knowledge of the carrying capacity of the immediate environment. This may not be obvious where rights take the form of cultural or spiritual taboos, or family or clan rights. Rights and taboos can, however, be seen to be pragmatic, relating to maintaining sustainability and social cohesion. In fragile environments where populations have increased rapidly, due either to inmigration or family growth, there is often a simultaneous collapse of culture and traditional sustainable practices. Government policies, and/or aid, exacerbate this where there is either ignorance of the rational nature of local culture and practice, or where it is not in the larger political interest to take notice of them. Implementation of intensive irrigated agriculture in Sahelian countries of West Africa, for example, has displaced pastoral nomads from their seasonal use of river flood plains without providing access to substitute lands. Transmigration policies implemented by the government of Indonesia similarly encroached on societies living sustainably in the forests (Denslow and Padoch 1988). More than

4 million peasant farmers were moved from densely populated Java, Madura and Bali into unfamiliar locations, mixing people from different cultures and disrupting cultural patterns and land use. Where communities have been able to maintain their cultural and spiritual identity their land-use practices have every chance of remaining vigorous and effective.

Traditional knowledge and cultivation techniques promote long-term sustainability. Yields may be lower than from HYVs, but overall the practices are more predictable, do not harm people or their environment, and farmers can retain their independence and their cultural identity as they are the innovators as well as the practitioners. One of the major obstacles to maintaining traditional farming systems has been the persistent misunderstanding of them by Europeans and the scientific community. David Brokensha (this chapter) provides an illustration of this in his account of scientists' attitudes to African agriculture. It is impossible to estimate the damage done to traditional systems – and the biodiversity lost – by outside intervention in traditional agriculture, but dynamic systems throughout the world have been damaged in the move to mass production and industrial agriculture. Paul Richards suggests (this chapter) that breeding programmes be returned to farmers like the Mende in West Africa, who demonstrate skill and imagination in combining, selecting and screening genetic material, pinpoints the contribution traditional farmers can make towards ensuring food security by maintaining genetic diversity if government and industry are prepared to work with them.

The difficulty of returning to traditional systems for the supply of germplasm, however, is that much of that supply base has been lost. Farmers' self-reliance was weakened when they were encouraged to adopt Green Revolution farming techniques. This happened in Brazil, where they were offered incentives to use commercial hybrid maize seeds in place of local varieties (Cordeiro 1993) in a strong propaganda campaign against 'outdated' traditional maize varieties. However, when yields from the hybrid varieties dropped because the farmers could not afford to buy essential fertilizers and pesticides, they no longer saved their own seeds or knew how to cultivate local varieties. This wholesale adoption of new technologies and the abandonment of traditional practices meant that not only were traditional, well-adapted varieties of maize lost, but also knowledge of traditional low-input cultivation techniques. Communities need help to revive traditional systems in such circumstances, and sharing knowledge is one effective way of ensuring that old skills are not lost. In India, the *Honey Bee* magazine, founded and edited by Anil Gupta, is published for the purpose of farmers sharing and promoting their traditional techniques, remedies and innovations. It underlines its focus on farmers by publishing in local languages; in Tamil, Gujarati, Bhutanese, Telegu, Hindi and Kannada, as well as in English.

The failure of the Green Revolution to assure food security has made it necessary for breeders to return to traditional farming systems in search of suitable genetic material and techniques for sustainable land use. But breeders have resisted acknowledging that traditional farmers have a right to be rewarded for sharing their knowledge and materials with industry. States have failed to agree on a workable and acceptable form of 'Farmers' Rights'. Under pressure from grass roots organizations, the UN Food and Agriculture Organisation (FAO) has been addressing the problem of Farmers' Rights since the 1980s, and is currently renegotiating its International Undertaking on Plant Genetic Resources (IUPGR) to find a means of effecting some form of Farmers' Rights. These must also harmonize with the Convention on Biological Diversity (CBD) through the Subsidiary Body on Scientific, Technical and Technological Advice (SBSTTA). After three years of meetings and drafting sessions little progress has been made, though the Fourth Extraordinary Session of the Commission on Genetic Resources for Food and Agriculture (CGRFA: formerly the IUPGR, renamed in October 1995, FAO Conference Resolution 3/95) in December 1997 appeared to bring greater agreement between countries than previous sessions. The renegotiation should provide the Conference of the Parties to the CBD with a workable form of Farmers' Rights. SBSTTA is compiling, *inter alia*, a core set of indicators of biological diversity (UNEP 1997a: 11) as guidelines for enacting bodies to the CBD, which states that agrobiodiversity is unique because, 'Unlike most natural resources, agricultural genetic resources require continuous, active human management', thus offering a link between agricultural diversity and farmers. In stating that, 'Wild species and obsolete crop cultivars and races of livestock provide a source of genetic resources – possessing desired genetic traits – that may be used in crop or livestock breeding programmes', SBSTTA also confirms breeders' inability to maintain genetic diversity without recourse to the gene pool conserved by traditional farmers to invigorate their programmes. One of SBSTTA's areas of consideration for maintaining agrobiodiversity (UNEP 1997a: Decision III 11) is the, 'Mobilization of indigenous and local farming communities for the development, maintenance and use of their knowledge and practices with specific reference to gender roles', through capacity building, participation and/or empowerment; building on traditional knowledge, practices and innovations; technology transfer; access to genetic resources; and partnerships and fora between stakeholders. SBSTTA is not setting out a new agenda. It is already there, manifested in the campaigns and work of NGOs and farmers' groups throughout the world. But it is giving recognition to the dependence of crop plants on continuing management, and to the part that traditional farming communities may play. Examining what the next steps for linking biodiversity and agriculture should be, Lori Ann Thrupp (this chapter) suggests policy changes that would attack the root causes of agrobiodiversity loss and ensure people's rights.

If the underlying reason for rural societies' loss of autonomy is their powerlessness, their lack of land rights, and their interpretation of 'ownership', then ways must be found to protect them from the clash between their cultural values and the drive to patenting and monopoly control by governments and industry. Jean Christie and Pat Mooney of Rural Advancement Foundation International (RAFI) (this chapter) present four common threads that run through the vast cultural diversity of farming communities and separate them from industrial societies. These centre around ideas of stewardship or custodianship which are antithetical to intellectual property rights. They argue that without protection and support from national and international measures, not only will the wealth of knowledge of those communities be lost to the world but also the peoples who have developed and maintained that knowledge.

Loss of that knowledge is manifested in the ruthless way in which industrial agriculture has altered the rural landscape. By disempowering and impoverishing self-sufficient communities, food distribution is now even more unequal, and hunger today is a reality for more than 800 million people, whilst the developed world has an oversupply of food. Mann and Lawrence close the chapter by seeking the reasons for this, and examining the inequity of a food supply system that emphasizes the powerlessness of the people and the greed of a global food system where hunger is often, and needlessly, most prevalent in rural areas. Food security is bound to be elusive because harvests are dependent on a good growing season, but the common feature between all the traditional farmers mentioned in this chapter is that through diversity they minimize risk. They obtain their essential foods from diverse sources, including NDRs. Their intercropping systems discourage pest and disease infestations whilst promoting soil fertility. Their cultures provide mutual support for the common good. In 10,000 years their strategies have enabled experimentation to take place hand in hand with ensuring security, and the revolutions of the twentieth century that have transformed agriculture have subsumed the understanding of the need to maintain diversity beneath the goal of maximizing production regardless of risk. While the CBD and FAO processes may be indicators of changes of focus at the top, they are yet to adequately acknowledge the inextricable link that exists between maintaining agrobiodiversity and protecting indigenous and traditional cultures and peoples. This inextricable link must be widely proclaimed, supported and strengthened.

The agroecological dimensions of biodiversity in traditional farming systems (Miguel A. Altieri)

Despite the increasing industrialization of agriculture, the great majority of the farmers in the developing world are peasants, or small producers, who still farm the valleys and slopes of rural landscapes with traditional and subsistence methods. In many areas traditional farmers have developed and/or inherited complex farming systems, adapted to the local conditions, that have helped them to sustainably manage harsh environments and to meet their subsistence needs, without depending on mechanization, chemical fertilizers, pesticides or other technologies of modern agricultural science (Denevan 1995).

A salient feature of traditional farming systems is their degree of plant diversity, generally in the form of polycultures and/or agroforestry patterns (Clawson 1985). This peasant strategy of minimizing risk by planting several species and varieties of crops stabilizes yields over the long term, promotes dietary diversity, and maximizes returns under low levels of technology and limited resources (Richards 1985).

Traditional multiple cropping systems provide as much as 20 percent of the world food supply (Francis 1986). Polycultures constitute at least 80 percent of the cultivated area of West Africa, while much of the production of staple crops in the Latin American tropics occurs in polycultures. Tropical agroecosystems composed of agricultural and fallow fields, complex home gardens, and agroforestry plots, commonly contain well over 100 plant species per field and provide construction materials, firewood, tools, medicines, livestock feed and human food. Home gardens in Mexico, Indonesia and the Amazon display highly efficient forms of land use, incorporating a variety of crops with different growth habits. The result is a structure similar to a tropical forest, with diverse species and a layered configuration. Small areas around peasant households commonly average

80–125 useful plant species, many for food and medicinal use (Toledo *et al.* 1985; Altieri 1995).

Many traditional agroecosystems are located in centres of crop diversity, thus containing populations of variable and adapted landraces as well as wild and weedy relatives of crops. It is estimated that throughout the Third World more than 3,000 native grains, roots, fruits and other food plants can still be found. Thus traditional agroecosystems essentially constitute *in situ* repositories of genetic diversity (Altieri *et al.* 1987). Descriptions abound regarding systems in which tropical farmers plant multiple varieties of each crop, providing both intraspecific and interspecific diversity, thus enhancing harvest security. For example, in the Andes farmers cultivate as many as 50 potato varieties in their fields (Brush *et al.* 1981). Similarly, in Thailand and Indonesia farmers maintain a diversity of rice varieties in their paddies: these are adapted to a wide range of environmental conditions, and regularly exchange seeds with neighbours (Grigg 1974). The resulting genetic diversity heightens resistance to diseases that attack particular strains of the crop, and enables farmers to exploit different microclimates to meet their nutritional needs and gain other use benefits. By safeguarding native crop diversity, peasants have provided a major ecological service to humankind, for which they have not been appropriately recognized or compensated.

Biodiversity is not only maintained within the area cultivated. Many peasants maintain natural vegetation adjacent to their fields, thus obtaining a significant portion of their subsistence requirements from habitats that surround their agricultural plots, through gathering, fishing and hunting. Such activities afford a meaningful addition to the peasant subsistence economy, providing not only dietary diversity but also firewood, medicines and other resources that support non-agricultural activities in the households. For the P'urhepecha Indians who live around Lake Patzcuaro in Mexico, gathering is part of a complex subsistence pattern based on multiple uses of their natural resources. These people use more than 224 species of native and naturalized vascular plants for dietary, medicinal, household and fuel needs (Toledo *et al.* 1985).

Depending on the level of biodiversity of closely adjacent ecosystems, farmers also accrue a variety of ecological services from surrounding natural vegetation. For example, in Western Guatemala the indigenous flora of the higher forest provides valuable native plants which serve as a source of organic matter to fertilize marginal soils as leaf litter is carried from nearby forests and spread each year over intensively cropped vegetable plots to improve tilth and water retention. Some farmers may apply as much as 40 tonnes of litter per hectare each year, and rough calculations indicate that a hectare of cropped land requires the litter production from 10 hectares of regularly harvested forest (Wilken 1987).

The nature of traditional farming knowledge

The development of traditional agroecosystems has not been a random process; on the contrary, intercropping, agroforestry, shifting cultivation and other traditional farming methods are all based on a thorough understanding of the elements and the interactions between vegetation and soils, animals and climate. Perhaps the most complex classification systems are those used by indigenous people to group together plants and animals. Usually, the traditional name of a plant or animal reveals the organism's taxonomic status, and in general there is a good correlation between folk taxa, especially as applied to plants. Indeed, the ethnobotanical knowledge of many traditional farmers is prodigious: the Tzeltals of Mexico, for example, can recognize more than 1,200 species of plants, while the P'urepechas recognize more than 900 species, and the Yucatan's Mayans some 500. Such knowledge enables peasants to assign specific crops to the area where they will grow best (Toledo *et al.* 1985).

The strength of rural peoples' knowledge is that it is based not only on acute observation but also on experimental learning. The experimental approach is very apparent in the selection of seed varieties for specific environments, but it is also implicit in the testing of new cultivation methods to overcome particular biological or socio–economic constraints. In fact, it could be argued that farmers often achieve a richness of observation and a degree of discrimination that would be accessible to Western scientists only through long and meticulous research (Chambers 1983).

Indigenous knowledge about the physical environment is often very detailed. Many farmers have developed traditional calendars to control the scheduling of agricultural activities believing, for example, that phases of the moon are linked to periods of rain. Other farmers cope with climatic seasonality by using weather indicators based on the phenologies of local vegetation (Richards 1985). Soil types, degrees of soil fertility, and land-use categories are also discriminated in detail by farmers. Soil types are usually distinguished by colour, texture, and even taste. Shifting cultivators usually classify their soil based on vegetation cover. Aztec descendants recognize more than two dozen soil types identified by source of origin, colour, texture, smell, consistency and organic content. These soils are also ranked according to agricultural potential, and each soil class specifies characteristics adequate to particular crops (Williams and Ortiz Solorio 1981).

Using their traditional knowledge, indigenous agriculturalists have met the environmental requirements of their food-producing systems by concentrating on key ecological principles resulting in a myriad complex agricultural systems. In such systems, the prevalence of diversified crop assemblages is of key importance to peasants as interactions between crops, animals and trees result in beneficial synergisms that allow agroecosystems to sponsor their own soil fertility, pest control and productivity (Altieri 1995).

By interplanting, farmers achieve several production and conservation objectives simultaneously. With crop mixtures farmers can take advantage of the ability of cropping systems to reuse their own stored nutrients and the tendency of certain crops to enrich the soil with organic matter (Francis 1986). In 'forest-like' agricultural systems, cycles are tight and closed. In many tropical agroforestry systems, such as the traditional coffee under shade trees (*Inga* spp., *Erythrina* spp. etc.) total nitrogen inputs from shade tree leaves, litter and symbiotic fixation can be well over ten times higher than the net nitrogen output by harvest, which usually averages 20 kg/ha/yr. In other words, the system amply compensates the nitrogen loss by harvest with a subsidy from the shade trees. In highly coevolved systems, researchers have found evidence of synchrony between the peaks of nitrogen transfer to the soil by decomposing litter and the periods of high nitrogen demand by flowering and fruiting coffee plants (Harwood 1979).

Crops grown simultaneously enhance the abundance of predators and parasites, which in turn prevent the build-up of pests, thus minimizing the need to use expensive and dangerous chemical insecticides. For example, in the tropical lowlands corn/bean/squash polycultures suffer less frequent attack by caterpillars, leafhoppers, thrips etc. than corresponding monocultures because such systems harbour greater numbers of parasitic wasps. The plant diversity also provides alternative habitat, and food sources such as pollen, nectar and alternative hosts to predators and parasites. In Tabasco, Mexico, it was found that eggs and larvae of the lepidopteran pest *Diaphania hyalinata* exhibited a 60 percent parasitization rate in the polycultures as opposed to only a 29 percent rate in monocultures. Similarly, in the Cauca Valley of Colombia, larvae of *Spodoptera frugiperda* suffered greater parasitization and predation in the corn/bean mixtures by a series of Hymenopteran wasps and predacious beetles than in corn monocultures (Altieri 1994).

This mixing of crop species can also delay the onset of diseases by reducing the spread of disease-carrying spores, and by modifying environmental conditions so that they are less favourable to the spread of certain pathogens. In general, the peasant farms of traditional agriculture are less vulnerable to catastrophic loss because they grow a wide variety of cultivars. Many of these plants are landraces grown from seed passed down from generation to generation, and selected over the years to produce desired production characteristics. Landraces are genetically more heterogeneous than modern cultivars and can offer a variety of defences against vulnerability. By contrast, a pest or a pathogen has a much less difficult barrier to breech when it encounters a genetically uniform modern cultivar grown under continuous monoculture over wide areas. Consequently, today entire crops promoted by the Green Revolution are at times attacked and seriously damaged or even destroyed.

Many intercropping systems prevent competition from weeds, chiefly because the large leaf areas of their complex canopies prevent sufficient sunlight from reaching sensitive weed species, or because certain associated crops inhibit weed germination or growth by releasing toxic substances into the environment (Reinjtes *et al.* 1992).

Many of these positive interactions and synergisms that emerge as traditional farmers assemble crop mixtures result in polycultures that outyield monocultures. In Mexico, 1.73ha of land have to be planted with maize to produce as much food as one hectare planted with a maize/squash/beans mixture. In addition, a maize/squash/bean polyculture can produce up to 4 tonnes/ha dry matter for ploughing into the soil, compared with 2-3 tonnes/ha in a maize monoculture (Francis 1986).

Integration of animals (cattle, swine, poultry) into farming systems provides milk, meat and draft power whilst adding another trophic level to the system, making it even more complex. Animals are fed crop residues and weeds with little negative impact on crop productivity, turning otherwise unusable biomass into animal protein. Animals recycle the nutrient content of plants, transforming them into manure. The need for animal feed also broadens the crop base to include plant species useful for conserving soil and water. Legumes are often planted to provide quality forage, but also serve to improve soil nitrogen content (Beets 1990).

Andean farming systems as a case study

The terraces throughout the Andean slopes, and the *waru-waru* (raised fields) and *qochas* in the *altiplano*, are sophisticated expressions of landscape modification that have historically provided more than a million hectares of land for agricultural purposes (Rengifo and Regalado 1991). The past and present existence of these and other forms of intensive agricultural system document a successful adaptation to difficult environments by indigenous farmers. In fact, applied research conducted on these systems reveals that many traditional farming practices, once regarded as primitive and misguided, are now being recognized as sophisticated and appropriate. Agroecological and ethnoecological evidence increasingly indicate that these systems are productive, sustainable, ecologically sound and tuned to the social, economic and cultural features of the heterogeneous Andean landscape (Earls 1989).

The types of cultural adaptation that farmers have developed in the Andes include (Araujo *et al.*, 1989): domestication of a diversity of plants and animals and maintenance of a wide genetic resource base; establishment of diverse production zones along altitudinal and vertical gradients; development of a series of traditional technologies and land-use practices to deal with altitude, slope, extreme climates, etc.; and different levels and types of social control over production zones, including sectorial fallows.

A highly sophisticated indigenous farming system known as *waru–waru* evolved over 3,000 years ago in the high plains of the Andes, at an altitude of almost 4,000m. This system enabled farmers to produce food in the face of floods, droughts and severe frosts by growing crops such as potatoes, quinoa, oca and amaranthus in raised fields consisting of platforms of soil surrounded by ditches filled with water. This combination of raised beds and canals has proved to have remarkably sophisticated environmental effects. During droughts, moisture from the canals slowly ascends plant roots by capillary action, and during floods furrows drain away excess runoff. *Waru-warus* also reduce the impact of temperature extremes. Water in the canal absorbs the sun's heat by day and radiates it back at night, thereby helping protect crops against frost. On the raised beds, night-time temperatures may be several degrees higher than in the surrounding region. The system also maintains its own soil fertility. In the canals, silt, sediment, algae and organic residue decay into a nutrient-rich muck which can be dug out seasonally and added to the raised beds (Erickson and Chandler 1989).

At the end of the twentieth century, *waru-warus* are again dotting the landscape of the *altiplano* as a result of a joint project between development workers and farmers (*Proyecto Interinstitucional de Rehabilitación de Waru-waru en el Altiplano*) aimed at reconstructing this indigenous farming system. This ancient technology is proving so productive and inexpensive that it is now actively being promoted throughout the *altiplano* through a project where technicians initially assisted local farmers in reconstructing some 10ha of the ancient farms. The encouraging results lead to a substantial expansion of the area under *waru-warus*. As no modern tools or fertilizers are required, the main expense is for labour to dig canals and build up the platforms. Potato yields from *waru-warus* can outyield those from chemically fertilized fields. Recent measurements indicate that *waru-warus* produce 10 tonnes potatoes per hectare compared with the regional average of 1–4 tonnes/ha (Sanchez 1989).

Another example of a sophisticated traditional production strategy is the practice of dispersing agricultural fields over the landscape (Golan 1993). In many areas Andean farmers utilize a sectorial fallowing system which encompasses a 7-year rotation (4 years cropping and 3 years fallow) using fields located between 2,700 and 3,800 metres. Field scattering is a risk-minimization strategy. Research shows that when a harvest pools the produce of several fields it reduces the yield variance relative to yields experienced from year to year if households relied on production from a single field. Actual yield data from a family that planted eight fields shows that the minimum yield experienced in a single field was 958 kg/ha and the maximum was 11,818 kg/ha. If the family has the same amount of land as their eight-field total, but planted in only one of the eight locales, then there is a 1/8 chance that their potato yield would have been a disastrous 958 kg/ha. If the family had planted in only two of the eight fields, the average minimum rises to 1,421 kg/ha and the maximum drops to 9,026 kg/ha. Although planting in more plots reduces the maximum expected yield, what is more important is that it raises the minimum yield farther above the disaster level. (Figure 7.1)

There is no doubt that Andean traditional agroecosystems exhibit important elements of sustainability. They are well adapted to their particular environment; they rely on local resources; they minimize risk; they are small in scale and decentralized, maintain biodiversity and conserve the natural resource base (Rengifo and Regalado 1991). Realistically, however, research for sustainable agriculture models for the Andes will have to combine elements of both traditional and modern agroecology. Agroecologists have argued that traditional patterns and practices encompass mechanisms to stabilize production in a risk-prone environment without

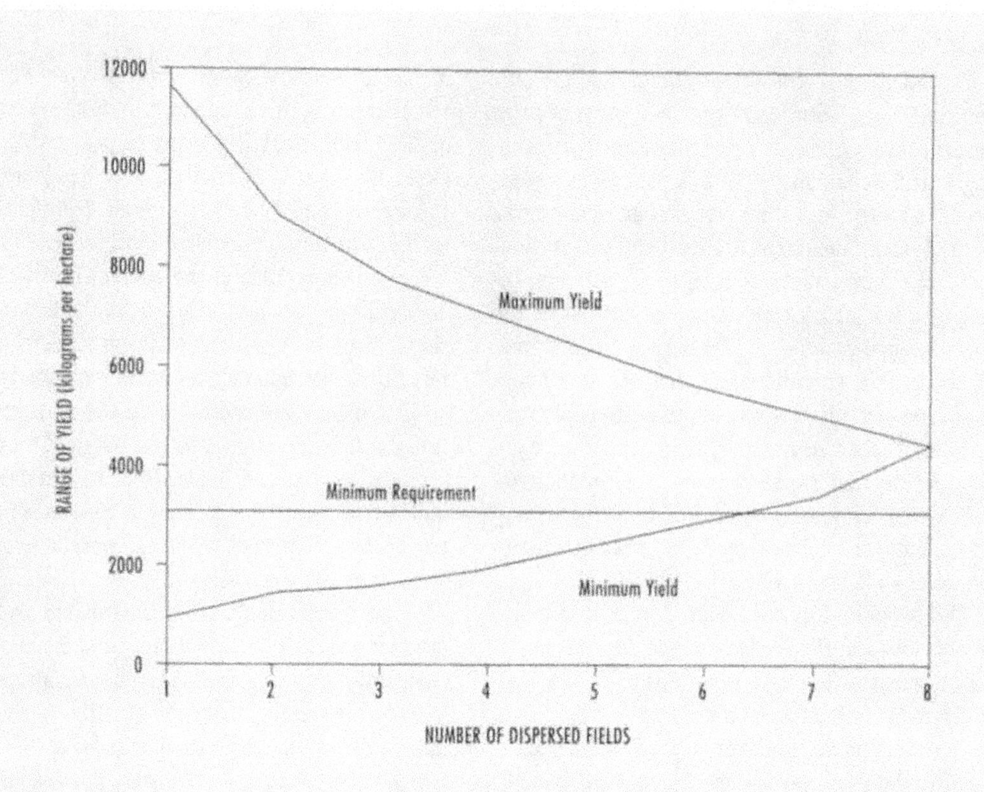

Figure 7.1: Yield variance and risk reduction for an Andean household according to the number of farmed fields.
Source: After Golan (1993).

external subsidies and to limit environmental degradation. Such stabilizing qualities of traditional agriculture must be supported and complemented by agroecological practices that enhance the soil, water and germplasm conservation potential of traditional technologies, and that also provide diversification guidelines on how to assemble functional biodiversity so that peasant systems can sponsor their own soil fertility, plant health and productivity (Chavez *et al.* 1989). Since most traditional systems are part of larger landscapes, an agroecological approach to watershed management is absolutely essential for understanding and managing the processes that affect the natural resource base in areas broader than just individual or communal plots (Figure 7.2).

Traditional farming and agrobiodiversity: the future

The persistence of more than three million hectares under traditional agriculture in the form of raised fields, terraces, polycultures, agroforestry systems, etc. documents a successful indigenous agricultural strategy and embodies a tribute to the 'creativity' of peasants throughout the developing world. It also demonstrates that the peasant strategy towards complexity has a deep ecological rationale as the kinds of agriculture with the best chance of enduring are those that deviate least from the diversity of the natural plant communities within which they exist. Many lessons on how to preserve and manage agrobiodiversity are intrinsic to traditional forms of agricultural production.

These microcosms of traditional agriculture offer promising models for other areas as they promote biodiversity, thrive without agrochemicals, and sustain year-round yields. It is particularly evident from the examples provided here, and in the extensive literature on the subject, that ancient agricultural systems and technologies can aid in the rescue of today's peasants from the vicious cycle of rural poverty and environmental degradation. For agroecologists, what have been especially useful are the ecological principles that underline the sustainability of traditional farming systems, which – once extracted and systematized – can be combined into alternative production systems for peasants, such as those promoted by hundreds of NGOs throughout Latin America, Africa and Asia (Pretty 1995). Most agroecologists recognize that traditional systems and indigenous knowledge will not yield panaceas for agricultural problems. However, traditional ways of farming refined over many generations

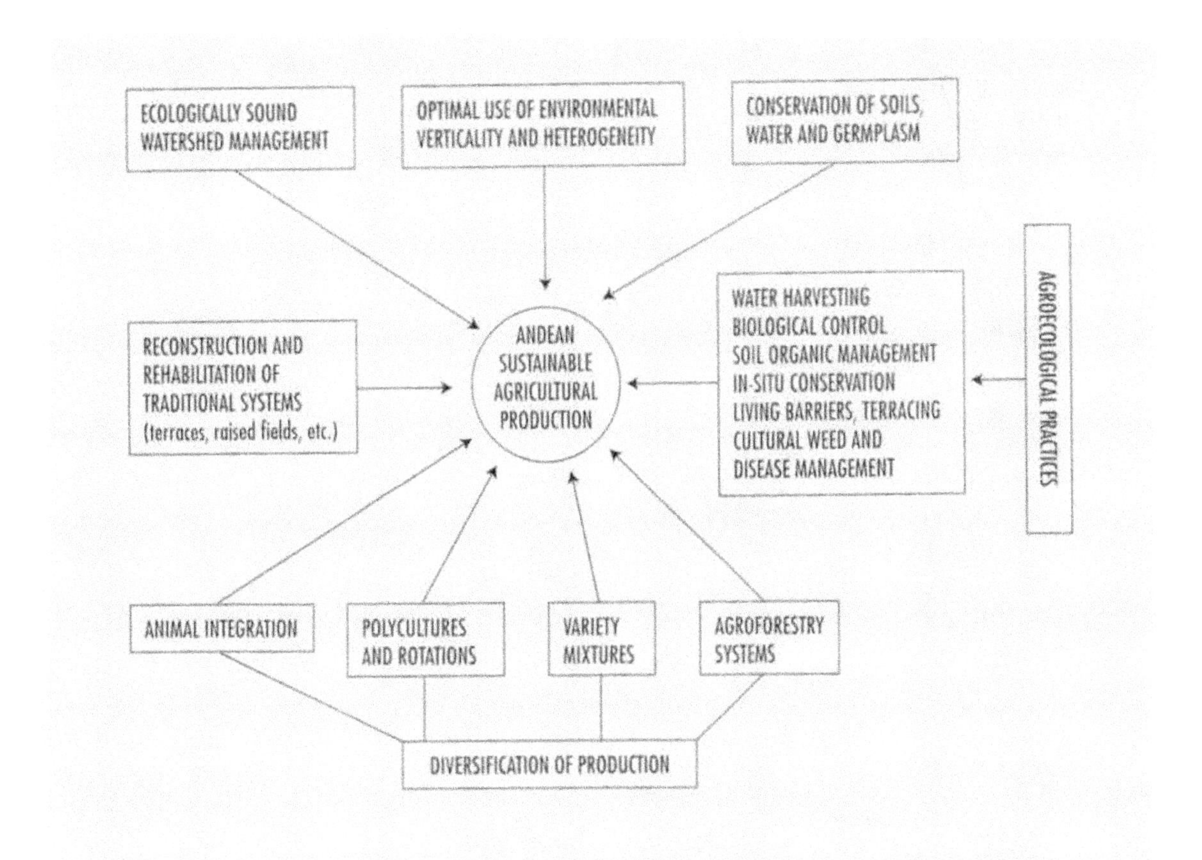

Figure 7.2: The integration of traditional and agroecological technologies to achieve sustainable production in the Andes.

by intelligent land users provide insights into managing soils, water, crops, animals and pests. Research can assess the benefits of aspects of traditional systems – their economic structure, genetic diversity, species composition and function as agroecosystems – as well as their social and economic characteristics and potential for wider application. The research process can have additional benefits by fostering collaborative relationships between researchers and indigenous people, and providing the groundwork for successful local development projects.

While it generally lacks the potential for producing a meaningful marketable surplus, subsistence farming does ensure food security. Many scientists wrongly believe that traditional systems do not produce more because hand tools and draft animals place a ceiling on productivity. Productivity may be low, but the cause appears to be social, not technical. When the subsistence farmer succeeds in providing sufficient food there is no pressure to innovate or to enhance yields. Nevertheless, agroecological research shows that traditional crop and animal combinations can often be adapted to increase productivity when the biological structuring of the farm is improved and labour and local resources are used efficiently (Altieri 1995). Such an approach contrasts strongly with many modern agricultural development projects, characterized by broad-scale technological recommendations, which have ignored the rationale and heterogeneity of traditional agriculture, resulting in an inevitable mismatch between agricultural development and the needs and potential of local people and localities.

What is alarming is that economic change, fuelled by capital and market penetration, is leading to an ecological breakdown that is starting to destroy the sustainability of traditional agriculture. After creating resource-conserving systems for centuries, traditional cultures in areas such as Mesoamerica and the Andes are now being undermined by external political and economic forces. Biodiversity is decreasing on farms, soil degradation is accelerating, community and social organization is breaking down, genetic resources are being eroded, and traditions are being lost. Under this scenario, and given commercial pressures and urban demands, many developers argue that the performance of subsistence agriculture is unsatisfactory, and that intensification of production is essential for the transition from subsistence to commercial production. Actually, the challenge is how to guide such transition in a way that yields and income are increased without raising the debt of peasants and

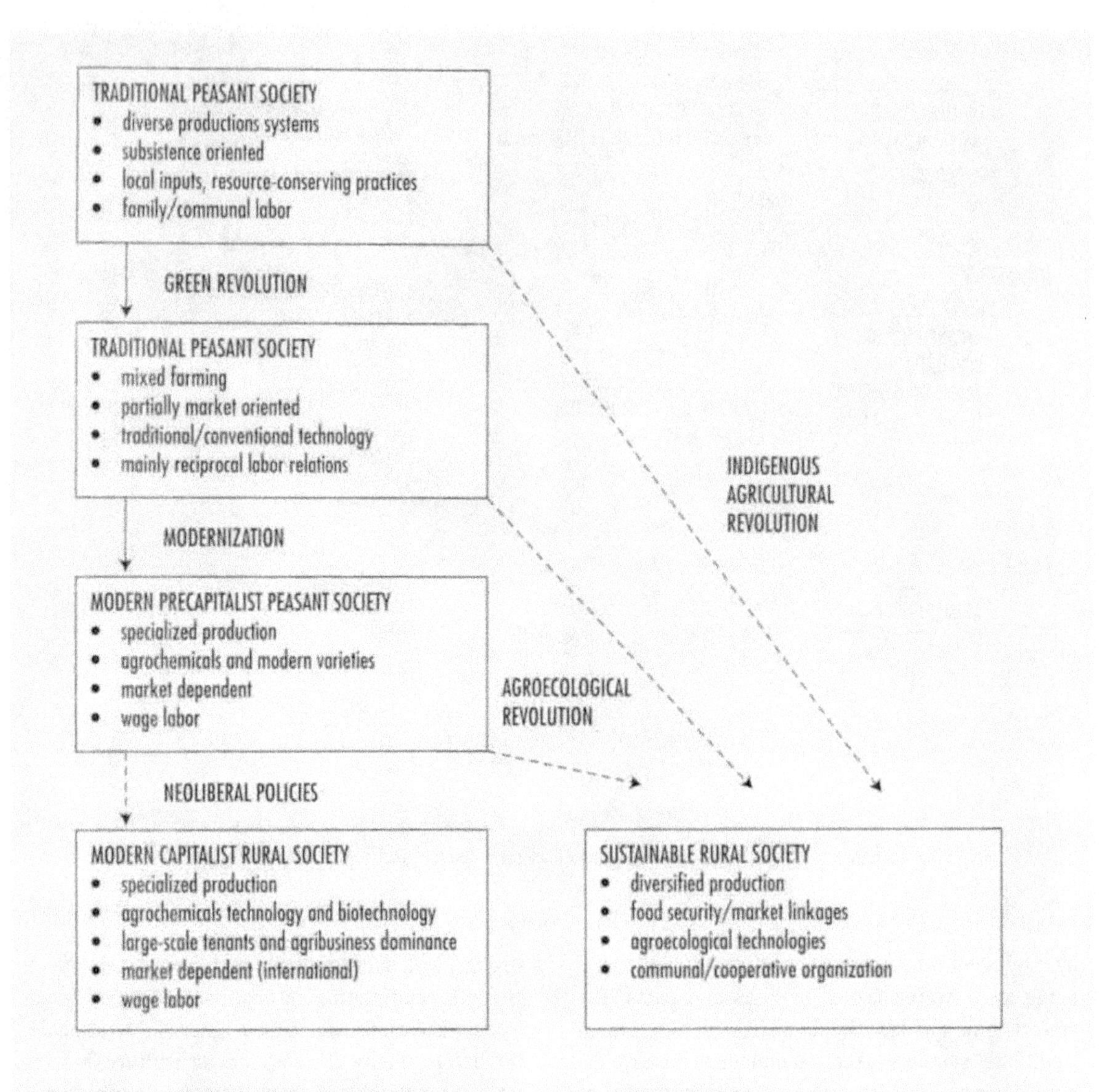

Figure 7.3: Alternative paths for the conversion of peasant societies toward sustainable development.

further exacerbating environmental degradation. Many agroecologists contend that this can be done by generating and promoting resource-conserving technologies, a source of which are the very traditional systems that modernity is destroying. Although it may be impossible to return traditional agriculture to its original state of equilibrium, what is possible is to reverse the current process of agricultural 'involution' spearheaded by short-sighted development, guiding the transition of the various phases of 'modified' peasant agriculture to a more sustainable rural society (Figure 7.3).

As the inability of the Green Revolution to improve food security, production and farm incomes for the very poor becomes apparent, a quest has begun in the developing world for affordable, productive and ecologically sound small-scale agricultural alternatives. The emergence of agroecology has stimulated nongovernmental organizations (NGOs) and other institutions to search actively for new kinds of agricultural development and resource management strategies, based on local participation, skills and resources, that have enhanced small-farm productivity while conserving resources in many parts of the world (Thrupp 1996). What are urgently needed are the right policies and partnerships to scale up such local successes.

The challenge of stationary shifting cultivation: farming in the Papua New Guinea highlands (Paul Sillitoe, with Wabis Ungutip, Jim Pelia, Kuli Hond and Ben Wabis)

The idea of stationary cultivation under a shifting regime appears a contradiction in terms, but this antilogy characterizes farming systems in the Papua New Guinea Highlands. Although people here may move garden sites and clear new cultivations when crop yields fall below certain tolerable levels, in some locales they are able to keep gardens under cultivation for decades, even generations. Their ability to maintain some of their long-term cultivation within the broad context of a shifting cultivation strategy challenges widely-held assumptions about the nature of traditional agricultural practices in the tropics.

The Wola speakers of the Southern Highlands practise this type of agriculture. They live in small houses scattered along the sides of five valleys, in extensive cane grassland below heavily-forested watersheds. Dotted across the landscape are neat gardens. The Wola depend almost exclusively on horticulture to meet their subsistence needs, living on a predominantly vegetable diet in which sweet potato is the staple. They keep pig herds of considerable size. These creatures are handed around with other items of wealth, such as seashells, cosmetic oil, and today cash, in an unending series of ceremonial exchanges that mark all important social events. These transactions are a notable force for order in their fiercely egalitarian acephalous society. Their supernatural conceptions centre on a belief in the ability of their ancestors' spirits to cause sickness and death, in various other forest spirit forces, and in others' powers of sorcery and 'poison'. Today the majority profess to be Christians.

The Wola cultivate plots for varying periods of time. At one end of the scale, some of their garden sites are cleared for just one, or possibly two, plantings of mixed crops, and then fallowed under natural regrowth for many years. At the other extreme, plots are kept under more-or-less permanent cultivation for decades, with occasional relatively brief periods of grassy fallow between some cultivations. When they clear an area of natural vegetation and break the ground for a new garden, the farmers themselves often cannot say for how long it will remain under cultivation. They plant it and see how their crops fare. As long as the yields are respectable, and the location of the garden convenient, they may continue to replant it indefinitely. (These are not annual replantings but usually extend over one to two years, depending on the plants intercropped.)

A remarkable feature of this semi-permanent farming system is that, other than planting material, nothing comes into the sites from the outside during their entire productive life-cycle. In the initial clearance all vegetation of suitable size is burned, except for that used for fence stakes or log barricades. Nutrients locked up in the vegetation are rapidly returned to the soil in a readily-available form via the ash, except for that fraction lost as gas to the atmosphere. When a plot is recultivated and planted, weedy regrowth and crop residues are composted into earth mounds, called locally *mond* by the Wola, on which the sweet potatoes are grown. Similarly, if a site is left fallow for a longer period, the coarse grasses and herbs are uprooted and incorporated, either as compost or ash (Floyd *et al.* 1987). According to the literature, such a regime should lead to a decline in productivity (Nye and Greenland 1960; Sanchez 1976; Proctor 1989) due to nutrient losses, weed proliferation, disease build-up, or soil depletion through erosion.

An investigation of soil fertility under cultivation suggests that the chemical properties of the Wola region's soils that limit crop nutrition and production are: i) low levels of phosphorus availability and high rates of phosphate fixation; ii) acid conditions, which interfere with the supply of some nutrients and reduce total base saturation; iii) depressed cation exchange capacities and lowered availability of exchangeable cations, which is particularly problematic with potassium; and iv) the high organic matter content which makes low available nitrogen levels likely, as indicated by the high C:N ratios. By contrast, the physical properties of the soils, with their high organic matter content, low bulk densities, fair topsoil aeration and good drainage, are generally favourable to crop production.

Controlled burning of a site gives a critical, though short-lived, boost to the availability of several elements, notably increasing pH and recycling elements contained in the cleared vegetation. This increase in availability is particularly significant for phosphorus, but is also important for potassium and nitrogen. It is not that crops massively mine the nutrients made available and held in the virgin soil, for although these decline to new equilibrium points they remain relatively constant, even after years under cultivation. It is that conditions promoting the availability of critically limiting nutrients are encouraged. The availability of limiting nutrients is sufficient to permit a wide variety of crops, several of them annuals, to flourish in newly cleared gardens, including a range of beans, green leafy vegetables, aroids and cucurbits. This crop diversity is short-lived, paralleling the ephemeral nature of improved nutrient availabilities following clearance. Sites cultivated over again, maybe repeatedly, largely support sweet potato only.

Regarding crop production, the supply of all three major plant nutrients potentially limits yield. After one or two cultivations of a site, nitrogen, potassium and phosphate levels all fall below the nutritional requirements for healthy growth of many crops. Sweet potato is one of the few crops that has the capacity to continue producing tolerable yields under these particular nutritionally-depleted conditions, and its relatively low phosphorus requirement particularly suits it to these soils. Sweet potato can yield adequately under low phosphate conditions if the fall in nitrogen during cultivation outpaces the fall in potassium, because high nitrogen levels encourage prolific vegetative growth over tuber development when other nutrients are not limiting. The evidence suggests that phosphate availability and potassium supply are probably the principal nutritional limitations in the majority of gardens. While sweet potato may yield adequately on soils relatively low in extractable phosphorus (and even potassium if its ratio with nitrogen remains favourable), minimal levels of these nutrients must none the less be maintained for tolerable tuber production to continue.

The traditional method of soil management, featuring the building of soil mounds, enables nutrients stored in plant residues to be incorporated into the soil during cultivation, where it soon rots down into a soft compost. Organic matter plays a central role in maintaining soil fertility, and adding crop residues or straw is successful on a range of soils under shifting cultivation, giving yield responses as good as, or higher than, fertilizer or manure applications. Compost is especially effective as sweet potato manure because it provides balanced nutrition in a slowly available form and improves soil moisture availability and soil aeration. The mechanism of nutrient uptake that the *monds* afford is especially effective at overcoming phosphate fixation and poor base saturation. Grasses are high in potassium and the boost in available potassium from grasses in the compost is probably central to the success of composted mounds in sustaining the region's semi-permanent sweet potato cultivation.

The sweet potato occupies a central place in this farming system (Yen 1974). It makes up by far the largest area under crops, and comprises something like 75 percent of all food consumed by weight (Waddell 1972; Sillitoe 1983). The agricultural regime, notably the semi-permanent garden plot, depends on this crop's ability to continue yielding adequately, regardless of changes in soil fertility status with time under cultivation. The changes that occur do not necessarily reduce sweet potato yields; contrary to expectations, farmers maintain that the soil on some sites improves with use, becoming better with time for sweet potato cultivation. Far from experiencing a decline in staple crop yields, as the accepted model of low-input subsistence agriculture predicts, some sites experience the reverse.

The cultivation of sweet potato on mounds of soil and compost is a characteristic feature of subsistence agriculture throughout the Central Highlands of Papua New Guinea (Waddell 1972), and is central to the feasibility of near-continuous sweet potato production on some of the region's soils. The plano-convex mounds vary in size between two and three metres across. The plant residues incorporated into the centre of them vary from garden to garden. If women, the builders of mounds, rework gardens while they still support some crops, the residues will include herbaceous and grassy weeds together with uprooted crop remains, notably sweet potato vines. If gardens are left fallow for many months and they grass over, the principal vegetation incorporated will be coarse grasses.

The process of mound building begins with women pulling up the vegetation on a site, sometimes with men's assistance. They use small one-metre long digging sticks, or small spades, to loosen the soil, and pull up the plants with roots attached. Sometimes they clear heavy growth using bush knives, chopping against log cutting blocks. The vegetation is strewn across the surface for several days to dry out, which incidentally protects the loosened topsoil from rainfall erosion. When they are ready to mound a garden, women work systematically, preparing small areas at a time and heaping up the crumbled soil. They prepare square depressions bounded by four-sided soil ridges into which they place the weeds and crop residues. Occasionally dry material is burned. To start the mound they scoop soil from the surrounding ridge over the compost, and continue building it up by digging up soil around its perimeter. Finally, they plant the mound by pushing several sweet potato slips into its surface.

Mounding ensures that the soil is friable, while the compost provides a soft centre into which tubers can readily swell into long and straight roots. Topsoil depth is also increased. The incorporation of plant material into the mounds probably increases the total microbial population and decomposition rates (Sanchez 1976), and burying surface-germinating weed residues deep in the heart of the mounds gives sweet potato a head start in the competition for light. Large amounts of organic compost also improve a mound's water-holding capacity, increase its internal temperature and encourage vigorous root growth. But care is required when using compost not to increase the chances of disease. For example, if using sweet potato vines as composting material, caution is needed where severe weevil infestation exists (Leng 1982). Local people say that mounding reduces disease and the rotting of tubers. In Enga Province in the Central Highlands, black rot (caused by *Ceratocystis fimbriata*) was four to five times lower in tubers from composted mounds than from uncomposted ones (Preston 1990). Another benefit found in trials in Aiyura in the Eastern Highlands Province, was that planting into level heavy soil was inferior to traditional mounds which also gave higher yields than ridges of corresponding size (Kimber 1970, 1971).

There is little chance, however, that local farmers can significantly increase their rates of composting. The major factors restricting composting rates are: i) the limited availability of suitable grassy and herbaceous composting material, and ii) the time and hard work required to collect it. Under the traditional agricultural regime, only the vegetation uprooted from a site is incorporated into the mounds. If people collected more material from elsewhere this would in all probability amount to depriving other fallowed garden sites of compost material, undermining the long-term maintenance of their fertility. Even if compostable vegetation were available from elsewhere, requiring women, the soil tillers and mound builders, to collect it and incorporate it into mounds would place an intolerable burden on them. None the less, traditional methods of organic manuring and soil management deserve close attention since we have a long way to go before we understand the variety and dynamics of such tropical agricultural systems.

Rice cultivation by the Dayak Pasir Adang community, East Kalimantan (Ideng Enris and Sarmiah, edited by Riza V. Tjahjadi)

Rice cultivation by the Dayak Pasir community in Sepan, East Kalimantan, is a long-established sustainable way of life now threatened by invasive logging and agricultural development programmes that disrespect the people's cultures, traditions and rights.

Dayak Pasir agriculture is regulated by the Dayak Pasir calendar and begins at the start of the agricultural cycle in June. The first month of the Dayak year begins when the seventh star, *turu*, appears, and is set by calculations made on the counting board of time of the Pasir people (*papan pembilang*), a pocket-sized piece of blackwood tree carved on both sides. *Papan pembilang*, or *papan waktu*, is a mathematical system for understanding and forecasting nature by drawing on indicators from a matrix of seven calculations. It even includes a correction variable, used if one or two measurements do not point towards a clear conclusion. It is officially used by the chieftain to make decisions for his community. The auspicious time to begin clearing the land will be chosen by observing the moon and the stars. When *turu* (7th) and *tolu* (3rd) stars appear, the eyes of the year also appear; for it is not only humans and animals that possess eyes, the moon and stars also have eyes. The Dayak Pasir people believe everything in the world has eyes.

The first task for a farmer is to select land (about 1.5 hectares) for clearance in a suitable location. Rights to use the land must be verified and permission obtained from the owner's family or descendants. Those who ignore the custom will be fined in accordance with the rules of the Dayak Pasir people. If the owner refuses to give the land, or to lend it, an alternative location should be looked for instead. Some lands may not be converted for agriculture: these include ground with fruit-bearing trees inherited from an ancestor, and land where there is an ancestral burial ground. Converting these lands into agricultural lands will bring misfortune or calamity upon those breaking the custom.

The Pasir begin by purging the area of the evils that dwell in the woods, forests and land in order to protect them from sickness while they work in that area of forest, and to make whoever else lives there accept them. One uncooked egg, *bore* powder, and water with flowers are used to wash the sky and the land to be cleared, and they pray that their intentions for the year will be blessed by the Almighty. If the agricultural plot is large it generally takes one month to clear and clean it. The land will be cleared for three consecutive days, unless there are signs indicating an obstacle or a constraint. For example, breaking a machete (*mandau*) signals there is a taboo, and it is advisable to stop working and find a replacement. Violating the taboo will bring disaster: one of those involved, or a close family member, will die within the year.

Preparatory work before planting is often done communally by calling a *sempolo*. *Sempolo* is a system of mutual work done in rotation in a community. For example, if one hamlet has ten households, they will clear land together farm by farm until the basic work is accomplished for all the farms. *Sempolo* is mostly used for dryland farms, while shallow, swampy or wet land is usually worked by individual households. The person or the family who owns the *sempolo* is responsible for providing food for the workers.

When the area is cleared it is left for 3–7 days whilst trees are chopped with an axe. First, the tree is cut until it is as big as an arm. Then it is cut until it is as big as the *lanjung* (a cylindrical basket made from rattan, carried on two shoulders with a headband). The people clearing the land have to decide whether to cut the trees themselves, or whether it would be better and faster to organize a *sempolo*. After the trees are chopped, the branches and twigs are cut into small pieces so that they burn easily.

Burning the plot is only done after the branches and twigs are completely burned. Many men, both young and old, are needed for a tree-cutting *sempolo*. Women who have the strength for the work may also participate.

The plots are burnt when *tolu*, the 3rd star, is in the middle of the sky. Care must be taken to prevent the fire from spreading to the forest, particularly if the forest bears fruits, so first a fire boundary is created. Whilst making the fire boundary the people also look for a dry bamboo to burn the plot. The tip of the bamboo is broken, set alight, then thrust into the plot to be burned. Burning is carried out by men and women, including young persons who are family members of the owner of the agricultural land. Young and old, men and women – all are included. Fruit trees on agricultural lands that are owned by another family, or by other descendants, should be left alone. If the fruit trees are accidentally burned, a fine should be volunteered to the tree's owner. Custom demands it. They make a buffer zone or fire boundary around the fruit trees and place bundles of fresh leaves there to absorb the heat of the fire. It is taboo for a neighbouring cultivation area to be burned. If this happens due to poor control of the fire, those involved must stop preparing for cultivation and move to other land areas, or wait for the next planting season.

According to the Pasir people, land that is to be used for agricultural crops should be cleaned first to protect it from rats, squirrels, lizards and birds, all of which like to eat newly-sown seeds. The Pasir people therefore burn their agricultural land meticulously. If there are still unburned twigs left they will gather them together and burn again. If there are still uncut trunks they will be chopped again. When the land is burned clean, seeds of the local spinach and a local bitter brassica are planted, then they plant corn seeds mixed with cucumber seeds. Cucumber seeds should also be mixed with rice seeds to make it smooth. They plant vegetables while waiting for the right time to plant rice.

Rice is the priority crop of the Dayak Pasir. The Pasir Adang believe it is taboo to put rice seeds in a pail or basin: they should be placed in the *lanjung*, the special place for rice seeds. The first rice seeds in the field are placed in holes, each with its own name. The name of the father is *Haji Harab* (head of the rice seeds/head of the group); the name of the mother rice seed is *Putri Kse* (mother of the rice group); followed then by *Juru mudi* (pilot/captain of the ship); *Jaga ruang timba* (to guard the water inside the ship; to pump water in the ship); *Jaga papan laut tiang* (to guard the ship's mast; to watch the direction of the wind); *Jaga luan* (to guard the front).

The story of planting rice, according to the Pasir Adang people, is the return of the rice bringing many members and friends as promised during the first rice planting. The rice seeds descend to *penian*. *Penian* is the harbour for the rice seeds, before they reach their final destination, which is to feed humans. Before planting rice, the rice harbour should be prepared for the seed *lanjung*, to prevent it from sticking to the ground. This is made from wood cut into four two-inch long pieces for the base of the *lanjung* pare. Then they will find wood for the traditional ingredients of Pasir which will be embedded in the agricultural land as a medicine, or they will plant it around their *lanjung* seed. A small space should be left to serve as the door of the fence, or as a space for reaching the rice seeds inside the *lanjung*. The Pasir Adang people believe the door of the rice harbour should face the direction where the sun rises, unlike other Pasir communities, such as the Pasir Pematang and Pasir Telake people, whose rice harbour doors always face the setting sun.

Rice will grow faster if a communal *sempolo* is organized. This *sempolo* may involve 50–100 people, including men, women, young people, even children. It is an activity that especially delights young men and women. The mature seeds should be planted before *sempolo* begins. If it is done during *sempolo*, it must be done in the morning before the people arrive. The suitable location for planting the first mature seeds is decided by finding the trace of the *toda luwing*. If this cannot be found, they should look for the root that encircles their agricultural land, which indicates that rice seeds planted here will produce many rice yields. The Pasir people know they can get many rice yields if they follow the requirements and times indicated by the counting board of time. The men generally make the hole for the seed using a 1.5–2.0 metre long dibber made from *ulin* wood or other kinds of wood. Then, either the women or the men will bring the seeds using a small *lanjung* tied on the waist or tied over the shoulder.

When the rice is half a month old it is said the 'navel' of the rice will start to fall. This is called *lempung puser* by the Pasir people, meaning the tip of the rice stalk, which is already quite yellow, will fall off. When the rice is one month old, the grass is weeded for the first time to prevent it growing too thick. If necessary, a second weeding is carried out when the rice is 1.5–2 months old and a grass gathering *sempolo* is called. Grass gathering is often carried out by women and their daughters, who may be reluctant helpers. The rice is protected and tended until at about four months there is a sign that a grain is starting to form, and the rice becomes pregnant indicating that the grain of rice will emerge. To hasten the production of grain, a bamboo *bamban* is cut into four and a folded leaf is inserted in the *bamban*. The *bamban* is placed in the stump of a felled tree. After this, its flower will bloom. This is the time to guard rice fields against various kinds of pests, especially fields that have yielded grains already as they are susceptible to a certain rice-seed bug and to grasshoppers. The method of controlling this rice-seed bug is by hanging a dead crab until it becomes rotten so the rice pests will concentrate on it. If the rice pests persist they will use a mantra and black ants. They also watch out for other pests, such as worms, that could destroy their rice fields. If there is a worm in the four-month-old rice a mantra for worms is used. They continuously observe if their rice seeds already have three parts, indicating that two are already filled and one is still empty, and watch the condition of their rice fields for the right time to harvest.

In harvesting the rice, the rice stalks are cut, then the good and clear grains are selected and put in a *lanjung* or a sack which they carry with them. After three weeks harvesting the rice fields privately, a *sempolo* should be organized to finish harvesting quickly to avoid attracting pigs, rats and squirrels. A harvesting *sempolo* is carried out by men and women, young and old, except for children who might lose many of the grains. Harvesting is often accompanied by songs and traditional Pasir poems to entertain the coming of their rice yields and welcome them from a foreign land. Usually young men and women sing alternately, especially if they are in love, but old men and women, married or single, can also perform solo so as not to feel the heat of the sun that burns their skin while they work.

In a *sempolo*, 4–5 people are in charge of carrying the rice to the house. These could be either men or women, whoever is strong enough to carry the rice in a big *lanjung*. When all is harvested the unhusked rice grains that will be used as seeds are cleaned first. The rice is trampled and then dried under the sun for 2–3 days. When it is dry, it is winnowed in a winnowing basket. The winnowed rice is measured using a particular oil can, and stored in a tightly sealed sack. Provided it is not opened it will keep for several months. The rice is stored in a *selipik* made of *nipah* leaves or sometimes of tree bark.

Caiçara agroforestry management (Daniel di Giorgi Toffoli and Rogerio Ribeiro de Oliveira)

As the Portuguese colonized what was later to become Brazil, the Indians were gradually exterminated or expelled from the coastal region, leaving a heritage that still survives in some places today. The Caiçaras are a living example of the Indian-colonist miscegenation that took place around the oldest towns which the Portuguese founded in the sixteenth and seventeenth centuries, such as Iguape, Cananéia, Ilha Bela, Paraty and Angra dos Reis. The word *caa-içara* comes from Tupi-Guarani, the Brazilian Indians' main linguistic branch, and refers to a trap made of stakes, still used for catching fish today. But the word has also come to mean the fishing communities that dot the coast from southern Rio de Janeiro (RJ) to São Paulo (SP) and northern Paraná (PR) (Figure 7.4), and the economy of these mixed-blood groups is characteristically artisanal fishing and subsistence agriculture. For almost a century, the Caiçara communities had very little contact with large urban centres, and outside influences were minimal, consisting mainly of purchases of clothing, fuel (diesel oil and kerosene), ammunition, salt and tools. Money needed to acquire these items came from selling goods such as fish, bananas, cassava flour, etc.

Caiçara lands are home to many different kinds of ecosystem. Five major habitats associated with the Caiçara culture are: lower montane forests, sandy coastal plains, mangroves, lagoons/estuaries, and rocky shores. These habitats are included under the broad 'Atlantic forest' category, which is one of the world's most threatened biomes as only 5 percent of its original area remains. Some of these areas are still quite well preserved, such as the Praia do Sul Biological Reserve (Ilha Grande, RJ), and have been incorporated into the Brazilian system of Protected Areas. Most Caiçara communities are found in such environmental protection areas, which were created by governmental decrees that essentially ignored their presence. Consequently, prohibitions placed on repeated use of natural resources have limited them within their own environment and driven them to resort to other activities in order to survive. Recent 'other activities' include a significant increase in tourists to the beaches on long weekends (Toffoli 1996). As Caiçara lands are coveted for their real-estate value there has been a decrease in areas planted to crops.

Fishing is still a major Caiçara activity and they are knowledgeable fishermen. Today, many adult males are employed by sardine boats fishing offshore. They return home to fish inshore when sardine fishing is prohibited and artisanal fishing remains a viable alternative. They make canoes from hollowed-out tree trunks of *'cedro'* (*Cedrela glaziovii*), *'guapuruvú'* (*Schizolobium parahybum*) and *'ingá'* (*Inga* spp.) and use spears, snare nets, or fixed or floating traps to earn their subsistence from the coves and bays. Since the end of the 1970s, medium-sized to large boats made of boards have become common (Toffoli 1996).

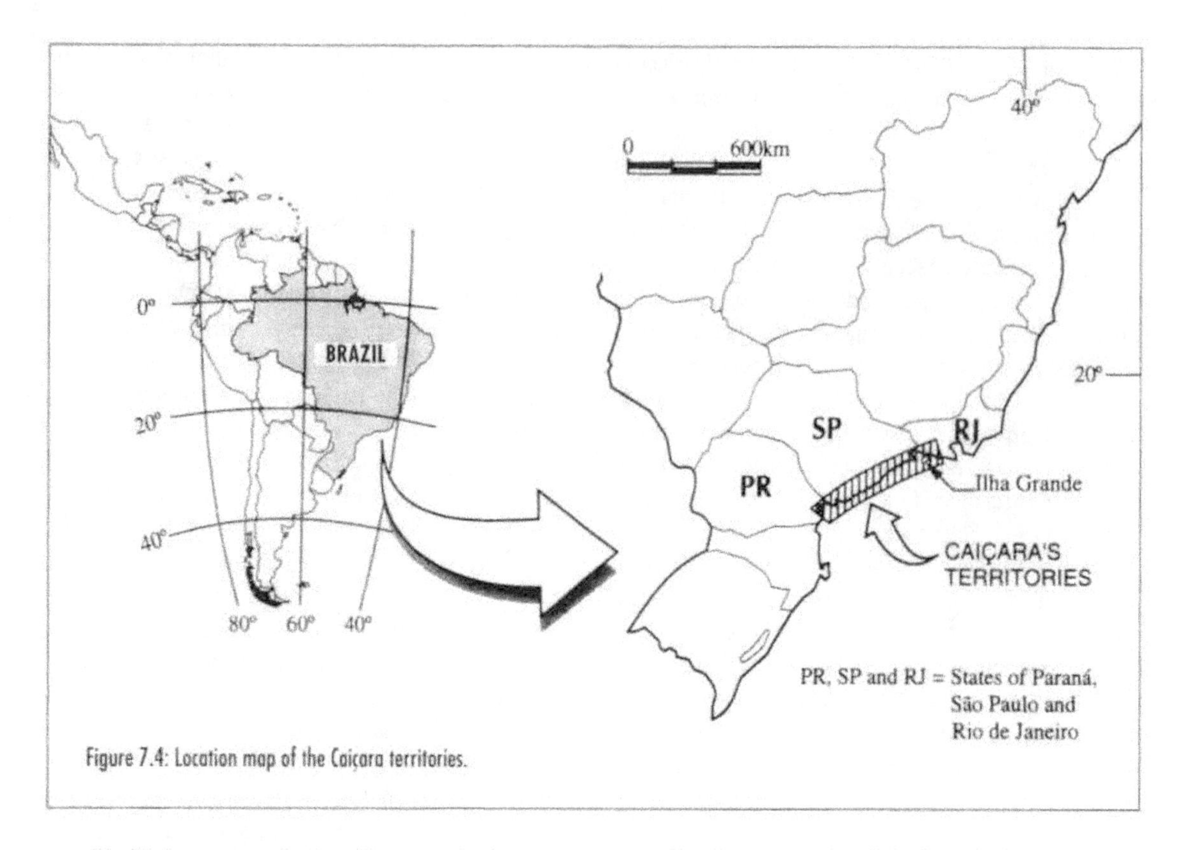

Figure 7.4: Location map of the Caiçara territories.

The Caiçaras use slash-and-burn agriculture to supplement their fish-based diet. Their diversified subsistence farming is based on Atlantic-forest regeneration traits, where nutrients contained in the biomass of the secondary forest are temporarily put to use. Elements of the indigenous culture are observed in both the management of the environment and in the produce that is processed. Their main product is cassava flour (eaten at most meals), which has been used as a substitute for the bread of the Europeans since time immemorial. For this reason it is sometimes called the 'Bread of the Tropics' (Mussolini 1980). A wide range of other crops and medicinal herbs are also a part of Caiçara agriculture.

Caiçara agroforestry management techniques are very similar to those of slash-and-burn agriculture in the tropical world. The soil is prepared for planting by first cutting down the forest or regrowth vegetation. A few weeks later, the brush and stumps are burned, but first a wide swath is cleared around the perimeter. The fire is set early in the morning when the soil is still cold. Since the brush is scattered more or less uniformly over the entire area, the fire sweeps rapidly over the ground, so the soil structure and biota are not greatly affected. Moisture loss in the upper 5cm of soil is less than 20 percent which greatly reduces impact on the soil fauna (Oliveira *et al.* 1994). Tree stumps are left standing, which helps to reduce erosion, improves the infiltration capacity of the soil and contributes to regeneration of the native plant cover when the plot is abandoned.

The plots are used mainly for subsistence crops. Crops are grown in rotation, but it is common to see many different crops planted at the same time, including cassava (*Manihot* spp.), beans (*Phaseolus vulgaris*), pigeon peas (*Cajanus cajan*), taro (*Dioscorea* and *Colocasia* spp.) squash (*Cucurbita* spp.), corn (*Zea mays*), rice (*Oryza sativa*), papaya (*Carica papaya*), watermelon (*Citrullus vulgaris*), sweet potato (*Ipomoea batatas*), etc. Vegetables such as kale (*Brassica oleracea*), lettuce (*Lactuca sativa*) and onions (*Allium cepa*) are also planted in small beds spaced between rows of the main crop. Many of these crops have several different varieties, and more than ten different varieties of cassava are known at Ilha Grande. Tuberous species are preferred because these do not require storage; the edible part of the plant (root) can be left in the soil for a long time without rotting. Some secondary foods like yams (*Dioscorea bulbifera*) are also grown in the plots, as well as edible weedy species like the green amaranth (*Amaranthus viridus*) and elephant's ear taro (*Colocasia esculenta var. antiquorum*). Medicinal herbs are also tended in the plots (e.g. *'carqueja'*, *Baccharis* spp. and *'melão-de-São Caetano'*, *Mormodica charantia*). Chaos apparently reigns in these plots because everything is planted haphazardly. But this system reproduces, on a smaller scale, the diversity and spatial distribution of the forest species, which helps to control pests and reduces local competition for nutrients (Meggers 1977; Posey 1983). The *'saúva'* ant (*Ata* spp.) is a serious problem and so the Caiçaras block entrances to ant hills with beach sand.

As productivity in the plots declines due to reduced soil fertility and the export of nutrients through harvesting, they are abandoned to lie fallow for 4–5 years. During this period fertility is restored almost to former levels when the regrowth vegetation can be felled and burned for another planting. The fallow period is an integral part of this system, though it is important to note that the plots are not left totally to random regrowth since management is part of the Caiçara culture. To some extent useful species are selected to meet the peoples' needs for firewood, medicinal herbs, dyes or spices, or to recuperate the soil. They select multiple-use species such as the leguminous tree *Anadenanthera colubrina*. People like to have this tree near their homes because the bark is used to dye fish nets. It is also one of the species that swiftly colonizes cut-over areas, since the stumps are not removed and quickly resprout at the beginning of the fallow period. As a legume, *Anadenanthera colubrina* plays an important role in nitrogen fixation and consequently is very important in the initial phases of ecological succession. This species has high dominance and frequency values in succession stages up to 25 years.

Caiçara management favours diversity and soil micro-organism density, a very important factor for the recovery and fertility of the soil. A study done at Ilha Grande (N. G. Rumanek, personal communication) found a total of 2.5×10^4 individuals/gram of soil of *Rhizobium* (nitrogen fixing bacteria) in a one-month old plot, and 5.1×10^4 individuals in a one-year-old plot. The number of bacteria fell to 5,200 individuals per gram in five-year-old regrowth vegetation and to only 12 individuals in a 25-year-old secondary forest. In a parallel study done in the same plots, E. R. Silva (personal communication) found a total of 555 spores/gram of soil of arbuscular mycorrhizal fungi (two species) in a one-month-old plot, and 638 belonging to three species in a one-year-old plot. In the five-year-old regrowth, 1,470 spores from two species were found, while in the 25-year-old regrowth there were 401 spores from three species. Similar results were found for microarthropods (Toffoli 1996); that is, a greater abundance of these organisms in recent plots and fewer in the older regrowths. The Caiçaras have managed to increase genetic variation of cultivated crops through ecosystem management. In the case of cassava, the existence of a sexual cycle leads to intra- and inter-specific breeding among the varieties. This improved genetic variation provides a strategic reserve for the farmer if his plots are attacked by a new disease. Plot rotation also promotes the germination of the seed bank (Toffoli 1996).

The regrowth vegetation that surrounds the plots also represents an important source of pioneer species and early secondary species. In a study of regrowth plots of different ages at Ilha Grande, Oliveira *et al.* (1994) found that after a three-year fallow period, there were 19 species of trees and shrubs in the regrowth vegetation. This phase is characterized by species from the initial stages of succession which show rapid growth and have low wood density. The dominant species are *'embaúba'* (*Cecropia lyratiloba*), *'maria mole'* (*Aegiphila sellowiana*), *'cobi'* (*Anadenanthera colubrina*), *'crindiúva'* (*Trema micrantha*) and *'capororoca'* (*Rapanea schwackeana*). These species spring up spontaneously at the site; seeds are either carried in by the wind or in animal faeces, or come from the soil seed bank. Most species, however, resprout from the stumps that are left after felling when the plot is first prepared. As time goes by, new species gradually arrive and become established in the plot. After 10 years, the regrowth vegetation consists of 29 tree and shrub species. After 25 years of succession, this number increases to 72 species. Since most of the pioneer and early secondary species disappear after 50 years, the number of tree and shrub species falls to 57, most of which are characteristic of advanced succession stages (Delamonica 1997). Caiçara management is therefore responsible for the creation of a certain type of regrowth vegetation characterized by rapid growth, and the presence of species that tolerate cutting and fire and are very efficient at taking in nutrients from the atmosphere. As time goes by, the regrowth vegetation assumes the structure of a mature forest and develops efficient nutrient cycling through the reuse of nutrients and the incorporation of new inputs. It is important to remember that initial stages of regrowth take place under conditions of edaphic oligotrophism, so, as succession progresses, nutrient cycling probably becomes more efficient.

As with most tropical ecosystems where soils are poor in nutrients, atmospheric input (rain, dust, sea spray) is an important nutrient source. One of the mechanisms used by regrowth vegetation to capture nutrients from the rain is a network of fine roots near the soil surface that penetrate the litter layer (leaves and branches that fall to the soil) absorbing nutrients and incorporating them into the system (Stark and Jordan 1978). Furthermore, the root-fungi association (mycorrhiza) forms a dense absorption zone from which very few nutrients escape. Therefore, the development of this cycling structure (roots + litter) is basic for the incorporation of nutrients from outside the soil-plant system. Oliveira *et al.* (1994)

report on the various cultural practices of the Caiçaras that lead to vegetation selection and the greater efficiency of regrowth vegetation due to the development of this 'filter', which leads to system self-sustainability. The litter from the three-year-old regrowth vegetation is found to be similar to that of the 150-year-old forest in both quantity and chemical composition because management practices have created nutrient cycling structures in three-year-old regrowth forest similar to those of 150-year-old forest. Therefore, although the managed regrowth vegetation is low in species richness, in terms of function it resembles a mature ecosystem. According to Oliveira *et al.* (1994), Toffoli (1996) and Oliveira and Coelho Neto (1996), Caiçara management techniques are characterized by the maximization of nutrient use contained in the regrowth biomass, which considerably increases fertility, infiltration capacity and abundance of soil organisms. The self-sustainability of this agricultural system is linked to low productivity per plot, to the need for large areas for fallow periods and consequently to a low population density.

The fisher-farmer Caiçaras have learned by experience to adjust to the biogeochemical cycles of the forest. The composition and structure of the forest has also become intimately intertwined with this type of agriculture. This interaction between the people and the forest has left its mark on most forest formations of Brazil's Atlantic coast in lands occupied by the Caiçaras. Their group values, associated with a profound identity awareness and a sense of place, are also evident in agroforestry management. However, if social and economic structures continue to change to the point of destroying the Caiçara culture, this agricultural system may become extinct, leading to an increased dependency on processed foods and loss of the managed ecosystem. Therefore, the recovery and diffusion of the ancient knowledge of the environment held by these minority peoples must play a fundamental role in any self-sustaining management practice.

Acknowledgement: The authors wish to thank Dorothy Sunn Araujo for the English translation, and Luiz Carlos Toffoli for preparing the map.

Honouring creation and tending the garden: Amish views of biodiversity (Richard H. Moore, Deborah H. Stinner, and David and Elsie Kline)

The Amish people of North America came from Switzerland and Alsace in 1693 as an off-shoot of the Anabaptist Protestant Reformation (Smith 1957; Hostetler 1993). The North American Amish, located primarily in the states of Ohio, Pennsylvania, Iowa and Indiana, preserve much of their Germanic heritage in language (an oral folk German dialect) and culture, and have evolved a highly interconnected social system of co-operative, mutualistic and unifying interactions which sustains them as a separate subculture.

The primary unit of Amish society is a patriarchal family (Huntington 1956; Hostetler 1993). Groups of families are tightly connected as parts of an Amish church community or *Gemeinde* – a redemptive community (Cronk 1981) – who meet for worship in each others' homes and barns (Hostetler 1993). Each church community has its own lay church leaders and a set of socio-religious rules called the *Ordnung*, which create boundaries between them and the world, and limit the scale of many aspects of their culture, including technology and farm operation (Hostetler 1993). For example, to encourage a slower pace of life and more local connectiveness, *Ordnungs* do not allow church members to own automobiles (Hostetler 1993). Horses are the main source of power in agricultural field work, so farm size is limited to the area a family with a team of draft horses can manage (20–60 hectares) (Kuhns 1989; Kline 1990).

The agrarian life-style is the core of traditional Amish society. Tilling the soil has religious significance for the Amish based on biblical interpretations, and as a result the Amish have a strong sense of land stewardship (Hostetler 1993). A bishop from Pennsylvania visiting Ohio was quoted as saying, 'We should conduct our lives as if Jesus would return today, but take care of the land as if He would not be coming for a thousand years' (Kline 1995a). In the view of many Amish, farming offers, 'a quieter life and one feels closer to God' (Stinner *et al.* 1989). For example, much ploughing is done with walk-behind ploughs giving direct feedback on soil conditions, and time to observe and enjoy the clouds, birds and beautiful pastoral landscape created by Amish farm communities. The rhythm of the northern temperate zone seasons and associated farming and religious rituals provides a deep sense of order to life for Amish farm families. Their self-sufficient farming systems reinforce their cultural separateness and are an integral part of the culture's persistence and growth (Stinner *et al.* 1989; Hostetler 1993).

Amish farm sustainability is maintained by single-heir succession of the family farm which prevents land fragmentation and enhances conservation and biodiversity through long-term stewardship of the farm land and its respective environmental zones. While the Amish practise bilateral descent with a patrilineal bias, succession of the farm may be by any one child and spouse, male or female. The decision of who will be the successor rests with the parents, and payments for the farm will become the retired grandparents' income. Thus, from the Amish view, it is essential for the farm value to be something that can reasonably be achieved by the successor and his/her spouse. A rising population, tourism, and a chronic shortage of land have contributed to rising real-estate values, yet many people wish to farm. As a result there has been a rise in farms partitioning an acre or two for one or more of their non-successor children.

Influences of Amish land management on biodiversity

'We farm the way we do because we believe in nurturing and supporting all our community – and that includes people as well as land and wildlife.' (Kline 1990). The farming practices used by today's Amish farm families have developed over 300 years and have sustained the Amish as one of the most persistent and successful subcultures in North America (Stinner *et al.* 1989, 1992). Their farming systems generally are much more diversified than non-Amish farms. Use of 'solar powered' draught horses (Belgians and Percherons in Holmes County) rather than fossil-fuel powered machines (although small amounts of fossil fuel are used) also contributes animal manure, which is valued highly, both for building and for maintaining soil fertility (low levels of chemical fertilizers are used by most Amish farmers).

Although a growing number of Amish families are shifting into market vegetable production, dairy and diversified livestock farms still dominate. Natural breeding, rather than artificial insemination, is the primary breeding method for dairy cows, hogs and horses. This helps to increase genetic diversity within livestock species. Holsteins are the dominant dairy animal, although a few Amish farmers use smaller breeds such as Jerseys (Guernseys are preferred but existing gene lines do not meet with Amish farmer approval). Horses and cows all have names – such as Tom, Barney, Maggie, Tony, and Linda – and this helps both to identify and create a close bond with individual animals.

In addition to crop and animal production, many Amish farm families manage woodlots for hardwood lumber, wood stains, maple syrup, nuts, soil and fuel. All Amish families (even the growing number who are not farm based) have vegetable gardens, which the women typically manage. Meat for the family is raised from chickens, hogs and bull calves. On the mixed livestock and dairy farms, a four- to five- year rotation is used: hay – a mixture of medium-red clover (*Trifolium pratense*), alsike (*T. hybridum*), alfalfa (*Medicago sativa*) and timothy (*Phleum pratense*); corn (*Zea mays*); corn silage; oats (*Avena sativa*); wheat or spelts (*Triticum æstivum*); emmer wheat (*Triticum dicoccum*); and barley (*Hordeum vulgare*). In addition to crops, Amish livestock farms have permanent meadows/pastures. Increasingly, some form of rotational grazing (Voisin 1960) is being used on pastures, hayfields and crop fields after harvest.

The diversity of crops, and the rich sources of organic matter from legumes and manure, create several ecological and economic benefits for Amish farm families which contribute to their sustainability. Most insect and disease cycles have been broken, therefore little is spent on insecticides or fungicides, and there are healthy communities of beneficial below- and above-ground organisms. Soil quality is good and fertility levels are sufficient for high levels of production with low purchased fertility costs. Indications are that farms are operating on extremely efficient nutrient cycles for nitrogen. Finally, Amish farming practices create a diversity of ecological zones which serve as habitats for rich biodiversity on their farms and in their communities, which is an important quality-of-life value for many Amish families.

The Amish still tend to combine observation of the signs of nature with calendar dates as a guide to planting times in the spring. For example, clover is best sown when the ground reaches a 'honeycombed' state brought about by the freezing and thawing of the soil. When the serviceberry (*Amelanchier arborea*) blooms, it is an indication that the soil temperature is warm enough to sow oats. This usually occurs around mid-April. However, the flowering of serviceberry can vary by up to three weeks between early April and May. Several weeks after the serviceberry has bloomed, when the dogwoods (*Cornus florida*) burst forth in bloom and the young leaves on the white oaks (*Quercus alba*) are the size of a squirrel's foot, the soil is ready for the corn. When the first monarch butterfly (*Danaus plexippus*) comes winging across the hayfield it is time to think about cutting the hay. Watching barometric pressure, reading weather forecasts, cloud watching (Kline 1997) and 'gut' feeling are used to decide precisely when to mow hay, which needs three days of sunshine to cure properly for storage for the winter months.

The Amish still prefer the traditional mixture of clover for their hayfields. Not only does this diversity assure a good crop of hay, but it also makes a much more attractive and interesting field for humans, domestic animals and wildlife. A mixed hayfield in early to mid-June is a colourful and delightful place to be. Bobolinks (*Dolinchonyx oryzivorus*) abound in its varied habitat, as do many other species of birds and mammals (Kline 1990, 1995b), and butterflies congregate in the hayfields to feed on the sweet nectar of the clover blossoms. Mixed hayfields tend to mature later than alfalfa/orchard grass (*Medicago sativa/Dactylis glomerata*) hayfields now so common in industrial farming. Thus the late-nesting bobolinks have much better nesting success in the clover/mixed hay fields (Kline 1990), and indeed, bobolinks are observed much more frequently on Amish farms than on non-Amish farms.

Some Amish still grow heirloom varieties of open-pollinated field corn (usually raised for millers who grind it into corn meal for human use), but the majority of Amish farmers plant hybrid varieties of field and sweet corn. However, Amish farmers are helping to preserve the genetic diversity of maize because, even though the seed corn is purchased, farmers prefer to buy it from small family-owned seed corn companies instead of from the giant seed corporations

Amish farmers watch the weeds on their land as indicators of soil deficiencies. Sorrel (*Rumex acetosella* or *R. acetosa*) indicates a soil with low pH. Often sorrel is seen along field edges where no lime was spread recently. Quack grass (*Agropyron repens*), a persistent pesky alien grass, thrives on soils deficient in calcium. Even pest insects such as alfalfa weevil (*Hypera postica*) are more of a problem when the soil nutrients are out of balance. As the Amish say, *'das unzunt Tier hut die Lice'* (the unthrifty animal has the lice).

The majority of Amish enjoy wild fruits such as black and red raspberries (*Rubus* spp.) and elderberries (*Sambucus canadensis*), and many odd corners and some fence-rows on their farms are left to grow wild and free. These brambles then provide food and shelter for a host of wild things (Kline 1990). When the children pick blackberries in late July the fencerows still ring with birdsong. Gray catbirds (*Dumetella carolinensis*) scold, common yellowthroats (*Geothlypis trichas*) and house wrens (*Troglodytes aedon*) sing; so do the song and field sparrows (*Melospiza melodia* and *Spizella pusilla*). In the heat of the 'Dog Days' of August a myriad insects buzz, chirp and rasp, adding to the joy of the time and season.

Many Amish families have a healing salve recipe that includes wild plants. Herbal medicine is a tradition among the Amish that has a long history back into Europe (Seguy 1973) and local *brachers* (healers) do exist. There is interest in encouraging Amish students to begin learning local medicinal plants while they are still in school, in hopes that one will grow into a healer for the community. A few Amish herbalists collect and use wild plants medicinally; however, most use pre-prepared herbs. Herbs such as pennyroyal (*Hedeoma pulegioides*) from the woods, peppermint (*Mentha x piperita*) from the permanent pasture, and sage (*Salvia lyrata*) and spearmint (*Mentha x spicata*) from the garden are mixed together. Dandelion (*Taraxacum officinale*) wine is used to cure 'threshing sickness' from the dust breathed in the barn during threshing.

Clear traces of some northern European indigenous traditions can be seen today in Amish farming practices and agricultural life (albeit unconsciously). The popularity of the Amish almanac (Raber 1996) suggests that some Amish still plant seeds according to certain astrological signs: for example, sowing spring clover in the sign of Leo in March; or planting potatoes on or the day after the May full moon, the radishes in a waning moon, and aboveground crops in a waxing moon (Kline 1995a). Corn shocking in the full moon was mentioned as a particularly enjoyable time to do this activity. Autumn Thursdays are still the preferred wedding days for many Amish. This gives the bride's family opportunity to use the harvest bounty for the wedding supper, and Thursday (*Donnerstag*) is the day of Donar, the old South German god of marriage, agriculture, livestock and fertile growth (Schreiber 1962; Kline 1995a). These attitudes and practices contrast sharply with those of industrialized agriculture and suggest that contemporary Amish still maintain their ancestors' connection with the natural world which affects their views of biodiversity.

Social systems that promote sustainability must have either a low level of system perturbations or some type of regulating mechanism (Odum 1962) to provide for flexibility (Moore 1996). It is also expected that such systems provide means for characterizing (Conklin 1954, 1957) and accessing local knowledge (Geertz 1993; Brush and Stabinsky 1996), including strategies to maintain biodiversity (Moran 1996; Orlove 1996) and to respond appropriately to abnormal system fluctuations. The role of a flexible farm management system is to maintain natural cycles on the farm through an understanding of these cycles. Unexpected events such as a hard frost, a dearth of honeybees, or a leaf virus, may set in motion a series of events that push the system out of equilibrium.

Rotations can be varied somewhat. For example, the winter of 1995–96 damaged the spelts crop, so some farmers ploughed under the spelts in the spring and grew a second year of corn there instead as corn was in short supply from the previous year. It is possible to grow three years of hay if the combination of legumes seems appropriate. Sometimes an extra year of hay is grown. Manure applications in the late winter/early spring are applied according to subplots by easing the lever on the manure spreader. Several corn varieties are always planted because it is important to have both silage corn and feed corn. The faster-maturing varieties, such as 90-day corn, are usually planted around the outside of a corn plot so that it will mature in September before the slower varieties, enabling household labour to be more evenly distributed. Some long-eared varieties of corn are hand picked and the corn is stacked in shocks.

Families normally cut and shock the spelts/wheat and oats together. After about ten days of drying in the fields, the shocks are gathered and threshed by threshing rings, usually consisting of 4–6 neighbours sharing a threshing machine. Work begins on the farm whose grains mature first. The silo filling rings consist of as many as 8–15 different farmers helping put corn silage into the silo. The rings are slightly overlapping, and membership of the rings of adjacent farms varies slightly, so co-ordinating the silo filling schedule is quite challenging. Membership in threshing and silo rings also changes over time according to the needs of individual farms. Tasks in production, as well as the communal meals afterwards, are age and gender specific. Unmarried daughters can help with the small grain threshing (the married women cook the threshing meal) and hay baling, but not silo filling because there is heavy lifting involved.

Labour and equipment exchanges happen on an almost weekly basis. This is particularly true for families with young children where the wife is busy taking care of the children. Youths in their teens or twenties work on other farms, depending on the family and church relationship. Church members sometimes help their church leaders harvest their corn or provide firewood.

Local knowledge of biodiversity

Local knowledge – based on local experience, local social relations and local environment – maintains a viable and flexible farm management system and a rich quality of life. Knowledge gained from experience is passed on orally through everyday life examples. Both elderly and young alike pass on critical information. Amish are sometimes described as 'traditional', but local knowledge is a very dynamic process of adapting local experience to the current situation. For example, children may see on their own farms that fireflies are most abundant over the oat fields, but this information is also reinforced by parents, grandparents and neighbours who tell stories about fireflies.

The elderly hold a cache of knowledge concerning past experience which comes in handy when unusual weather or pest cycles present themselves. In the spring of 1996, steady rains prevented a timely planting for most farmers, but for the elderly it was comparable to the spring rains of 1947. Their advice was critical in helping farmers utilize the high number of subplots on the farms, enabling the draught horses to plough fields that had drained. Ploughing techniques were modified for ploughing in wetter conditions, and muddy or poorly drained spots were temporarily skipped. This contrasted with the conventional corn belt farmers who were not able to plant until the end of June – a full month and a half after normal scheduled planting.

While most Amish enjoy nature and prefer to live in rural areas with their pastoral landscapes, detailed knowledge of local flora and fauna is widely variable. Science is not taught in Amish classrooms (Hostetler 1993), but natural history is, particularly in schools where the teachers come from backgrounds that value such knowledge. Other important sources of information on local biodiversity are field guides, both commercially available ones and some published specifically for Amish children and families. Amish magazines give an indication of their appreciation of the natural world and knowledge of local biodiversity with poems and stories on nature, and quizzes and crossword puzzles testing readers' knowledge of biodiversity. A few of the articles express the idea of humans learning important spiritual lessons from other creatures. For example, in *Inspired by Nature – Taught by a Titmouse*, the writer, identified as 'Renewed in Spirit' (Anon. 1996a), tells of being depressed and watching a titmouse at the bird feeder taking seed after seed to a nearby branch and hammering them open for the food within: 'Suddenly a thought entered my mind. As the titmouse keeps flying back for food, I must keep on returning to the Throne of Grace for patience and help to overcome my faults. And if God gives me a hard trial, I must, with the Holy Spirit's help, break it until at last in the end it discloses the 'food' I need.... And as my little feathered friend sings praise, so may I too bring songs of praise to our Creator. Dear little birds, how many lessons you teach me as you brighten my days.'

Conclusions

While widespread surveys of biodiversity in Amish farming communities have not been conducted, it is clear that the small scale and diversity of cropping patterns in time and space create landscape patterns very different from those of conventional large-scale industrialized farming communities (Stinner *et al.* 1989). The numerous ecological edges and corridors in Amish landscapes created by fences, woodlots, meadows, riparian strips and diverse cropping fields are most certainly habitats for a great diversity of flora and fauna.

Amish farm stewardship and knowledge of local biodiversity begins with religious values and a desired rural quality of life that resonates from field to watershed. Examples of Amish farmers intensifying production and losing this ethic and knowledge are becoming more common due to the growing population problem, lack of farm land, increased wage labour, and commercialization. Nevertheless, increased biodiversity on farms has positive ecological, economic and social impacts on their quality of life. Nature's 'free services' (Odum 1997) are reaped in many forms, ranging from 'free' fertilizer from cow and horse manure to 'free' pest removal by the multitude of birds constantly flying overhead.

For many Amish families, a very important part of their quality of life is living and working close to the land, surrounded by a diverse array of God's creations. Their workplace and homeplace is one and the same, which seems to cultivate a deep connection to their place on earth and the flora and fauna that share it with them. Their small-scale animal-powered farming practices create diverse habitats which maintain a rich biodiversity on their farms and in their communities, and which sustain not only their spirits but also their economic and ecological viability.

The slower pace of life and non-industrial technology of the Amish seem to offer them opportunities to observe and learn from nature. 'We often joke that where tractors can plough a six-acre field in two hours, I figure two days – but my time includes listening to vesper sparrows and meadowlarks and watching clouds scud across the sky' (Kline 1997). Since the Amish choose not to have radio and television, they naturally watch the signs of nature – the song of the cardinal and robin, the leaves on the maples, the wind and the clouds.

The theology for living espoused by Kline (1995a) summarizes at least one Amish person's view of the spiritual value of biodiversity: 'Am I wrong in believing that if one's livelihood comes from out of the earth, from the land, from creation, on a sensible scale, where we are a part of the unfolding of the seasons, experience the blessings of drought-ending rains, and see God's hand in all creation, a theology for living should be as natural as the rainbow following a summer's storm? And then we can pray, *'Und lasz uns deine Creaturen und Geschöpf nicht verderben...* ' ('And help us to walk gently on the earth and to love and nurture your creation and handiwork....').

What African farmers know (David Brokensha)

African farming systems were complex and flexible, demonstrating impressive knowledge of the local ecology, and – in terms of the population and technology available – were generally satisfactory. The farmers had adapted well to survival in what very often were harsh environments, with poor soils, low and uncertain rainfall, and problems of pests and plant diseases. The farming systems were site-specific, although they shared many common features – such as being labour intensive. African farmers – and it must be stressed that this usage includes significant, sometimes superior, numbers of women farmers – emphasized survival and the avoidance of risk, and were accustomed to making rational decisions. Decision-making in African agriculture has been the subject of many studies, which have shown that farmers daily face many choices, and that their decisions have significant results on their lives. Some farmers are more experienced. Sometimes this is acknowledged, and an accepted 'good farmer' will be emulated by others who recognize his/her skills and knowledge.

The past tense was used above because these once-viable, adaptive, flexible and successful African agricultural systems have been modified and distorted, in some cases out of all recognition. Many factors have caused these changes. Perhaps the most influential has been the rapid and dramatic population increase – Kenya, for instance, has an annual population growth rate of 4 percent. This means that in many cases there is no longer sufficient land to support people under the old ways, especially, as is often the case, where the land is marginal. Meanwhile, 'Green Revolution' type innovations have led to impressive increases in yields of particular crops, but the net result has been to increase rural inequality, so that the rich get richer and the poor get poorer. Commercialization of crops and the spread of market economies have further changed old patterns. New cash crops, often grown for export, have changed both landscapes and societies, e.g. *mange tout* in Zimbabwe, or carnations and strawberries in Kenya, both air-freighted to European markets. Migrant labour has taken many of the younger men away from the farms, as a result of both 'push' (the need to earn money to pay taxes; the desire for new things) and 'pull' factors – the latter including the attraction of the cities, and new social pressures to acquire both goods and status by migrating.

Governments, both colonial and independent, have imposed a myriad laws and regulations, some pertaining directly to agriculture (e.g. requirements to build soil conservation terraces; grow specific new crops; use new technologies; comply with compulsory planting of famine crops) and many others indirectly affecting farmers – such as the taxes mentioned above. Governments have drastically altered land rights, either by claiming that *all* land (and all forests) is 'crown' land, belonging to the government, or promoting individual land title. This happened under the well-known Kenyan 'Land Reform', and had far-reaching results on Kenyan farmers, especially on rural inequality. In addition, African governments, everywhere, and of all types, have promoted centralization, which struck at the heart of traditional local authorities, further weakening the farming system. Traditional leaders, in many cases, had exercised some control and guidance over agricultural activities, in such matters as allocating land, determining the rights of strangers, and even of announcing times of planting.

Also, over the past three decades a series of bloody conflicts and wars has led to the creation of at least ten million refugees in Africa. Their numbers have been augmented by others who have been resettled by government action, usually because of the construction of large hydroelectric dams. These people have had to move somewhere, and in almost all cases have put further pressures on the farming systems of their (often-reluctant) hosts.

Given all these disturbing factors, it is not surprising that the original systems have changed, sometimes drastically. Nevertheless, in many parts of the continent they are remarkably intact. And, whatever the shape of 'traditional agriculture' it is imperative to understand what farmers believe and do before introducing any changes. But most scientific 'experts' resist this imperative as they strongly believe in the power of science, which is hardly surprising as so many of them are highly qualified scientists. The trust in science often goes with a set of stereotypes about African farmers, who are seen as lazy, conservative, reluctant to accept change, backward, superstitious and ignorant. De Schlippe (1956) wrote perceptively of the mutual incomprehension between agricultural officials (in this case Belgian agricultural officers) and the local Zande farmers in what was then the Belgian Congo. He described Zandeland as a classic example of the African intermittently wet tropics, presenting a bewildering variety of types of soil and vegetation, with hardly a few square metres of uniformity. He characterized it as, 'hidden order in seeming chaos'. 'When one enters a Zande homestead for the first time no fields can be seen. The thickets of plants surrounding the homestead seem as patchy and purposeless as wild vegetation. It is impossible to distinguish a crop from a weed. It seems altogether incredible that human intelligence should be responsible for this tangle....'

Such passages encapsulate so much of the thinking that still persists today among many experts. There is still a tendency among donor officials and most African government officials to be authoritarian in approach, 'top-down' in their interpretation of development, dogmatic, and in favour of a strong centralized government.

The question of intercropping versus 'pure stands' is a good example. Belshaw (1980) conducted pioneering studies in Uganda in the 1960s and 1970s (see also Innis 1997), and documented the following advantages of indigenous intercropping techniques:

- different plants' varied rooting systems mean that they do not compete for plant nutrients and soil moisture;
- one crop may provide a favourable microclimate for another, e.g. bananas providing shade for young coffee bushes;
- there may be a complementary effect when nitrogen-fixing plants are grown with non-nitrogen-fixing plants (e.g. beans grown with maize);
- a scatter of seeds among another species means that the minority species may escape diseases and pests that could ruin them in a pure stand;
- mixed cropping can lead to lower labour requirements by producing a quick vegetation cover that will smother weeds;
- soil and water resources are protected under the plant cover;
- mixing food crops provides a wider variety of foods over an extended harvesting period, and
- mixed cropping lowers the risk: there is a good chance that in adverse conditions at least part of a mixture of crops would survive.

These are all clearly compelling reasons for intercropping, but twenty-odd years since Belshaw published his findings pure stands are still encouraged by modern agriculture.

The experts often see Western techniques and methods as superior to indigenous ones, and sometimes extend this attitude to the crops and trees. So, for example, exotic crops have been encouraged (e.g. hybrid maize supplanting millet, although the latter is both more nutritious and more drought resistant). Also, exotic trees have been introduced, sometimes too enthusiastically and without considering the local society and ecology. Eucalyptus and *Pinus* spp., though popular with foresters, are not always the best solution. However, some exotics, e.g. jacaranda, and *Grevillea robusta*, proved most successful in the Mbeere area of Kenya.

Another area of misunderstanding concerns the supposed 'communal nature' of African societies. On the one hand, there are those, like Garrett Hardin (1968), who blame many of the ills of peasant farming on 'The Tragedy of the Commons' whereby any form of common rights brings ruin to all and is antithetical to progress. According to Hardin, everyone has the right to use the commons, yet no one has an incentive to control their own exploitation, resulting in catastrophic over-use of the resource. This can only by averted by delegating responsibility for the commons to individuals (privatization) or to bureaucrats (socialism). Peter Castro (1995), in a reasoned critique of Hardin's thesis, presents a detailed account of Kirinyaga (Kenya) where in pre-colonial times there were explicit, sophisticated and effective institutional arrangements for managing common property resources and maintaining trees on the land, based on conditions of social predictability rooted in kinship, neighbourhood relations and religious customs. It is precisely these sorts of social/cultural factors that are so often omitted from any discussion of development in Africa, with predictably distorting effects.

Others have taken an entirely opposing view of the nature of communal efforts in Africa, making African societies almost into model 'communitarian groups'. One of the most extreme examples was the doomed *Ujamaa* villagization project in Tanzania, in the immediate post-independence years. This was based on the (false) premise that Africans were naturally co-operative, and would happily work together for the common good. This may happen, as Castro demonstrated for pre-colonial Kikuyu society, but only in special circumstances. It is by no means a universal characteristic. Less grandiose attempts to base projects on the assumed communal nature of African society include numerous village woodlots and village co-operative societies, nearly all of which have been dismal failures because they were based on invalid premises.

Another view of African farming systems is that they may have worked in the past, with fewer people and without massive outside interference, but now they are completely disrupted, so the best thing is to make a fresh start. Those holding such a view are usually appalled by physical manifestations – especially by soil erosion or 'desertification' – and earnestly seek a drastic remedy. However, they often use over-simple analysis, and omit the social content of the ecological scene.

But, despite the resistance, considerable data have been generated from extension studies that provide an overview of the subject. These range from Leakey's (1934) study of the Kikuyu, and Audrey Richards' (1939) study of the use of fire by the Bemba in the then Northern Rhodesia, to the work of Chambers and his colleagues (Chambers *et al.* 1989; Scoones and Thompson 1994). (For more information, see the series on Indigenous Knowledge, published by Intermediate Technology Publications, London.) This body of literature documents the knowledge and innovative nature of African farmers developed over centuries. To summarize and repeat the main arguments, there is ample evidence that:

1. African farmers developed an extensive and deep body of knowledge about those resources (soils, vegetation, grazing) on which their lives depended.

2. This knowledge had evolved over the years, and was based on rational observation and experimentation.

3. The knowledge varied between groups and localities, and also between individuals.

4. We now have a wide and accessible range of comprehensive studies describing and analysing indigenous agricultural knowledge in Africa.

5. It is necessary to understand the capacity of existing resource-use systems and the rationale of present management practices (which have evolved in terms of local needs and the limits of the environment).

6. This holds even where the original farming systems have been grossly distorted. It is still necessary to find out first what farmers do, and what their perceptions and beliefs are.

7. Local people should be involved in discussions and decisions regarding their production systems; this concerns the area of 'people's participation'. (For further discussions on this important topic, see Chambers 1983; Chambers, Pacey and Thrupp 1989.)

If these points are accepted, then there is a need to counter the prevailing emphasis on high technology agriculture that is dependent on mechanization, agrochemicals, irrigation and, increasingly, on large-scale farming. Not only do these practices promote an increase in inequality, which must eventually be politically threatening, but they also deny millions of small-scale farmers a chance to achieve, at least, sustainability. Chambers, Pacey and Thrupp (1989 pp. xviii-xix) argue that resource-poor farm families have benefited least from industrial and Green Revolution agriculture because simple and high-input packages do not fit well with the small scale, complexity and diversity of their farming systems, nor with their poor access and risk-prone environments. However, these farmers have again and again been found to be rational and right in behaviour which at first seemed irrational and wrong to outside professional observers.

No one approach will be appropriate for all situations, as much depends on specific socio-economic, biophysical and cultural-historical circumstances; on traditions of collective action; and on the quality of local leadership, as well as on the role of government. In most cases, effective natural resource management will neither be a resuscitated traditional form, nor will it be imposed from outside. Rather, it will be a blend of old and new institutions and will represent a collaborative venture between government and people.

The cultural values of some traditional food plants in Africa (Christine Kabuye)

The inherent culture of survival of our ancestors led them to experiment with plants in their search for food. For thousands of years, humans thus subsisted on fruits, leaves, roots and tubers, grains and nuts. At first these foods were gathered, but later for convenience a few were grown in various garden systems while the rest continued to be collected from the wild. Food production systems in the traditional setting revolved – and in many African countries still revolve – around a few cultivated species and many others that are tended or managed in the wild.

Seasonal variations govern the growing and gathering of food, and many stories are told of lean periods when people were forced to migrate to new lands in search of food and pasture for their animals. The nomadic life-styles of some communities demonstrate such migrations. Elsewhere, precautions had to be taken to ensure the availability of food. These involved selecting food plants to be grown, obtaining seed, locating suitable land for cultivation, and finding those plants that could be collected from the wild. Accumulated knowledge about factors related to all these activities became so closely knit with the daily life of people that it formed part of their traditional cultures.

Traditional food plants have many values attached to them. Traditions are, however, not static as people keep on adopting elements from neighbouring cultures or from new immigrants. For example, the most important traditional food in Buganda, the banana (*matoke*), which is of Malaysian origin, is said to have been introduced to Africa in prehistoric times. Yet it is so entrenched in Buganda traditions that one would think it is native. It is a traditional belief that the first banana was brought to Buganda by Kintu, the first *Kabaka* (King) of Buganda when he arrived from the north-east. He is also said to have come with the cow, while his wife Nambi brought with her the fowl and the finger millet. Kintu came looking for land with favourable conditions to settle, which he did – and even founded a kingdom. As for Nambi, halfway through that journey she realized she had forgotten the millet, which she had to go back for. This unfortunately meant that her brother Walumbe (death), who they had left behind, caught up with them!

The story of Kintu and Nambi shows the elements of the culture of survival. First, the search for favourable lands and then ensuring hold of them, while on the other hand seeds and planting material were brought to ensure food security. The new arrivals must have depended on those they found already living there to locate other sources of food, while the earlier settlers were happy to incorporate the new sources of food. But the fact remains that it is inevitable that the culture of survival is always accompanied by the threat of Walumbe (death).

The well-being of a people starts with the family as the smallest unit within a community. Good health is supported by the kind of foods people take. The values attached to such foods go a long way in traditional cultures. Thus there are foods that a mother must have, there are those a child has to be given, and there are general foods for the whole family. It is recognized that food plants give health through their nutritional values. Sometimes certain foods are prescribed to get rid of some bodily discomfort, or to avoid ill health, showing that the line between food and medicine is very thin. At the family level, therefore, one has to know the sources of good food: what to plant and what to obtain elsewhere. The mother, who is invariably responsible for preparing food for the family, is equipped with specific knowledge as she grows up; and when she is starting a new home she takes with her some seeds for her new home garden.

Ensuring the availability of food within a community is a preoccupation of many societies. In African traditions, activities undertaken encourage communal responsibility for food security. Times of planting, weeding and harvesting could be important communal occasions. In some communities, sharing work in building storage facilities and granaries meant a common responsibility for ensuring food security. The same went for the sharing of seed with relatives, in-laws or neighbours, which ensured maximum variety. In addition, certain foods were a must at ceremonies concerned with birth, initiation into adulthood, marriages and funerals, and sharing them meant that everyone in the community was assured of having them in times of want.

As not all food plants are grown, traditions recognize the importance of *wild* sources of food. Some fruit trees are owned and inherited in the *wild*, while others are left standing when a clearing is made for cultivation fields. Leaving pieces of wild land or wild patches are some of the strategies used to ensure that vegetables, fruits or roots not available in cultivated fields can be collected to add to cultivated foods or during times of want.

TABLE 7.1: SOME IMPORTANT TRADITIONAL FOOD PLANTS IN AFRICA.

(I) GRAINS

Digitaria exilis, Hungry rice, *Fonio*	Staple crop, grown on poor soils in West Africa. Grains are ground into flour for porridge, or mixed with other cereals. It is also used for making beer. (Purseglove 1975)
Eleusine coracana, Finger millet, *Bulo* (Uganda) *Wimbi* (Kenya)	Important cereal in several African countries. Grains are ground into a flour for porridge, or a stiff meal is made with boiling water, and eaten with a sauce. In Ankole, western Uganda, a sowing ceremony was undertaken to ensure good harvest (Williams 1936). In Kenya, nursing mothers eat porridge made from fermented grain to increase milk production. The porridge is a good weaning food for children and digestible for the elderly. Often served at social functions and festivities. The flour is used to make beer.
Eragrostis tef, Teff	Over half the grain in Ethiopia comes from teff, the only teff-producing country. Used for a staple food, *ingera*, similar to a pancake, made from the fermented flour and served with various sauces.

(II) LEGUMES

Lablab purpureus, Hyacinth bean, *Njahi* (Kikuyu – Kenya)	A bean rich in iron grown in the Kenya highlands. Served at social gatherings. The Kikuyu of Kenya give it to nursing mothers. Young pods and leaves can be used as vegetables.
Vigna subterranea, Bambara groundnut, *Mpande* (Luganda - Uganda)	Cultivated in western Kenya, eastern Uganda, and West and Central Africa. Rich in protein (16–21 percent) and carbohydrate (50–60 percent), the hard seeds are usually soaked, boiled or roasted and made into flour. The roast *bambara* with *simsim* are kept in small pots by the Luhya of Kenya to be offered to special visitors. As a sign of binding two friends together a nut is shared between them.
Vigna unguiculata, Cow pea, *Mpindi* (Luganda - Uganda)	Pulse with seeds rich in protein (23.4 percent) and carbohydrates (56.8 percent). The young leaves (28.5 percent protein) are cooked as a vegetable. In Uganda leaves are cooked, dried and pounded into powder, stored in small packets of banana fibre. It can keep for a long time. A sauce is made by adding water or adding it to groundnut or *simsim* sauce. *Mpindi* is one of the clan totems in Buganda and people of the clan do not eat any of its parts.

(III) PLANTS USED FOR OIL

Sesamum indicum, Simsim, *Entungo* (Luganda – Uganda)	The plant, an erect annual with lilac flowers, is usually grown as a source of sesame oil. Fruits are oblong capsules producing many seeds, containing 45–55 percent oil, 19–25 percent protein and 5 percent water. Seeds eaten whole, roasted with groundnuts, or made into balls by adding sugar, or roasted, pounded and made into a paste, stored in small packets of banana fibre and eaten whole, or made into a sauce added to vegetables. At weddings, the seeds are thrown at the bride and bridegroom as a sign of fertility, to wish them many children and plenty of food. Various parts of the plant are used in medicines.
Vitellaria paradoxa, Shea butter tree, *Yao* (Luo - Uganda)	A butter is prepared by roasting the seeds then pounding them into a paste. This is boiled and skimmed to obtain the butter which contains 45–60 percent oil by weight and 9 percent protein. The tree is very important in the diet of people in northern Uganda. It is common where it occurs and is conserved *in situ*. The Acholi of northern Uganda apply the oil to their bodies during ceremonies. The tree is also important in West Africa with a number of varieties, and it is protected and inherited in northern Nigeria.

TABLE 7.1 (CONTINUED)

(IV) LEAFY VEGETABLES	
Cleome gynandra, Spider herb	Grows in home gardens or wild and is widely distributed in wastelands in Africa. Leaves and young shoots are used as a vegetable, boiled and left to simmer with or without fat. Sometimes milk is added before eating. During some Kisii ceremonies in western Kenya it is eaten without salt. The leaves are rich in protein (35.8 percent by weight). It is recommended for mothers after birth to increase breast milk. Leaves are rubbed on the head to relieve headaches.
Solanum nigrum, Black nightshade	Found in wasteland or as a weed in cultivation. Often saved when weeding, and the leaves harvested as a vegetable, or grown in home gardens. The leaves contain 29.3 percent protein and are said to be good for pregnant mothers.
Vernonia amygdalina, Bitter leaf	Shrub often planted as a hedge in home gardens in West Africa. The leaves, although bitter, are a favourite vegetable. Young leaves are soaked in several changes of water and used in soup as a digestive tonic. The leaves are rubbed on the skin to stop itching or to treat other skin diseases.
(V) FRUITS	
Adansonia digitata, Baobab	The fruit pulp is eaten raw or made into a drink. The juice is sometimes added to porridge or milk. Seeds are roasted like groundnuts. The tuber-like root tips are cooked and eaten in times of famine. Young leaves are used as a vegetable and can be dried and ground into powder to be used when needed for sauce. Used for medicine (fruit pulp can also be used in treating fever), and as a source of fibre for making ropes and baskets.
Sclerocarya birrea	A shrub common in wooded grasslands of East, Central and southern Africa. The seeds are oily and are eaten raw. The greenish fruit pulp is eaten raw or made into a drink. In southern Africa the tree is well known for *marula*, a drink made from the fruits.
(VI) BEVERAGES	
Coffea spp., Coffee	Both *Coffea arabica* and *C. canephora* are widely cultivated in various parts of Africa. *C. arabica* originated in Ethiopia where the coffee drink still has a unique social role in the care taken to prepare it for visitors. In Uganda *C. canephora* is the common species. The Baganda cook the fruits with *Ocimum basilicum* and keep them tied in small packets of banana fibre ready to be offered to visitors.
Musa spp., Banana	Beer-making is important in African customs, and is made from various plants. In Buganda, the most important is the banana, two kinds being used: the small sweet banana, *kisubi*, and the big one, *mbidde*, especially cultivated for beer. The juice is extracted by pressing and squeezing the bananas with blades of a grass, *Imperata cylindrica*, in a wooden canoe-like trough or any suitable container. Ground sorghum or finger millet is added, left to ferment for one or more days, and strained. In some parts of East Africa beer is prepared from a mixture of warmed honey and water to which halved fruits of *Kigelia africana*, the sausage tree, are added.
(VII) MUSHROOMS	
	Mushrooms are collected from the wild in season, dried, and stored to be used when wanted. The small white-capped *Termitomyces microcarpus*, which grows in swarms in the first few weeks of the rains, is important in Baganda customs. Grows in waste nest material cleared by termites out of mounds and scattered. Traditionally, if a woman sights these mushrooms, *obutiko obubaala* in Luganda, she must call others in the village to harvest with her. This ensures each family has a supply for the occasions they are needed. The mushrooms are cooked fresh, or mostly dried and stored in banana fibre packets. They are soaked and cooked in a sauce prepared for a girl by her mother the night before she gets married (Bennett *et al.* 1965). The girl also takes some to prepare for her husband. They may be added to groundnut or simsim sauce. *Obutiko* is a clan totem and people of this clan do not eat them.
(VIII) ROOTS	
Dioscorea spp., Yams	Cultivated for their root tubers, they are a very important food in West Africa, and to a small extent in East and Central Africa. In West Africa they are used to make *Fufu*, prepared by pounding boiled yams (Purseglove 1975). Yams are also boiled, roasted or fried, and eaten alone or with a sauce. *Dioscorea dumetorum*, found wild in forested areas in tropical Africa, was used for food in famine years.

There are many food plants that play a role in traditional life-styles. Whether these are cultivated, semi-cultivated, or tended in the wild, they all contribute to food security in terms of quantity and quality. The undomesticated food resources are usually not appreciated by developers, and need attention. Some of these are relatives of cultivated plants and could provide genetic material for their improvement. Others provide essential vitamins and proteins which may be lacking or insufficient in the crops that are grown. Keeping and protecting wildlands where they occur will ensure their continued availability as a supplement to cultivated food sources. Many have been conserved in a regulatory way because of their cultural values.

Casting seed to the four winds: a modest proposal for plant genetic diversity management (Paul Richards)

Management of genetic resources is an important aspect of biodiversity conservation. It is an activity that draws upon the science-based theory of genetics, whose rapid advances often mask indigenous knowledge of patterns and processes of variation among animals and plants. But indigenous knowledge also contributes to the conservation of genetic resources. Valued human activities are regularly protected by theorization. Many times this theorization takes the form of an explicit 'story', designed to make sense of the phenomena and to facilitate transmission of those valued practices. Anthropologists and folklorists have long included this theorization under the category 'myth'. Midgley (1992) reminds us that modern science cannot proceed without creating its own myths. This is no bad thing, she concludes, so long as scientific storytellers do not become too carried away by their literary prowess.

Mythic materials tend to stabilize practices that attempt to 'code'. At times the effect is to entrench vested interests and harmful practices. At other times, myth might usefully help to reinforce socially or ecologically desirable action. Anthropologists show that it is sometimes possible to work back from myth to an underlying social or technical practice. For example, the Mende-speaking peoples of southern and eastern Sierra Leone inherit an ancient rice-farming tradition based on the local domestication of a distinct African species of rice, *Oryza glaberrima*. Mende cultivators inhabit a region where three distinct rice-farming traditions – dryland (rainfed), coastal wetland and riverine wetland – converge. This ancient rice-farming technology, with its distinct sets of adapted planting materials, was from the fifteenth century onwards usefully diversified by the introduction of Asian rice varieties and production techniques.

Rice is mainly a self-pollinated crop, so that when farmers select the types they want for seed at harvest they tend, progressively, to arrive at genetically purer planting materials. But rice cross-pollinates to some limited extent. Genetically varied materials tend to be found more frequently among seed reaped from the edge of the farm. A colonial observer in the 1940s reported Mende farmers carefully selecting seed from the centre of pure stands (Squire, quoted in Richards 1985). Mende farmers recognize many different types of rice adapted to specific soil moisture conditions, and are often very anxious to obtain seed that reproduces true to type. But in some specific circumstances – to adapt to misfortune or a perceived change in climatic or economic conditions – Mende rice farmers seek to experiment with new planting materials. Some have come to rely on agricultural projects to offer up new material, but others – especially in areas not covered by projects – actively explore material they know, or suspect, to be genetically varied as a result of cross-pollination.

One very important class of material of this kind apparently results from cross-pollination of Asian and African rice. For some years scientific breeders experimented in a desultory way with hybridization of the two species, but an interspecies sterility barrier hampered progress. Only recently have breeders succeeded in developing hybrid lines. But in this they appear to be following in the wake of local farmers who, over a number of decades, have selected and diffused a number of 'natural' hybrids combining the hardy growth characteristics of *O. glaberrima* with the higher yield characteristics of *O. sativa*.

Mende farmers see themselves as managing, or riding upon, a strong current of rice genetic variation. One farmer reported that he was always on the look-out for any strange material appearing in a field ready for harvest. He would collect any such off-type material and plant it in the following year in a trial plot close to his farm hut, to 'see if there was anything there for him'. He was alluding to a proverb that says, 'Water intended for you will not run past you'. His basic idea was that plant variation flows (seed flow, to be precise), and within a broad stream of continuous variation there are crosscurrents and eddies suited to the circumstances of individuals. People vary and the environment changes – and thankfully, across the years, the 'flow' of rice provides options for everyone. The standard improved variety recommended by the extension service year-on-year was not the only option as far as this man was concerned.

For the Mende, plant variation has three components. First there is God. The material has within it the capacity to vary. Nothing stays the same forever. God made it so. Then there are natural processes. Regularly, farmers report that natural processes or living creatures give them access to new planting material. Seeds are carried about by floods, or by birds. One man found a new rice variety when tapping a palm tree. The seed had germinated on the crown of the palm where it had been dropped by a passing bird. Lastly, there are the activities of humans. Poor farmers sometimes depend on borrowing planting material from friends or merchants. Trade seed, perhaps originally intended to be sold for food, is notoriously mixed, and may even contain some sweepings from the merchant's store. The merchant handles seed from far and wide. Neighbours sometimes presume to judge a man's financial standing by assuming that a farmer with very mixed rice must have received a loan, though it could not have been on favourable terms. But the poor sometimes hit back at their misfortune by selecting novelties in their 'lucky dip' and putting them to work. Other farms may contain small experimental patches where seeds, selected and sent as gifts from friends or kin, are being assessed. Material may be handed on under the name of the first farmer who selected and stabilized an exotic find or spontaneous local cross.

These three processes are buffered by distinctive myths and rituals that together comprise the local theory of seed flow. Prayers will be offered to God as the ultimate author of natural beneficence. The activities of natural and human agents of seed flow are framed within linked sets of ideas about natural and ancestral spirit forces.

One particularly important set of such ideas concerns elephants as precursors of human settlement in the forests now occupied and converted by Mende farmers. Elephants are 'bulldozer herbivores'. They tend to open or enlarge swampland vents in the forest cover as dry-season wallows. Puddled moist soils in these sunlit vents are a favourable environment for wild rice. Elephants roam far and wide, and graze unguarded rice fields, depositing undigested seeds in their dung. Mende farmers consider the elephant an important source of exotic planting materials. Rice varieties acquired in this way are known *hele kpoi* ('elephant dung').

But not all elephants are wild animals. Some are human 'shape shifters'. In some Mende villages, members of longer-established lineages, perhaps fallen on hard times, might point to their family's pioneering role by claiming elephant shape-shifting powers. One old man I knew was convinced that he would not die a normal death. When his human span was filled he expected that, like his older brothers before him, he would shuffle off into the forest to rejoin his natal elephant herd.

Thus the world of elephants merges with that of the ancestors, from whom so many blessings flow. Useful variation extracted from familiar seeds is considered a form of ancestral blessing. The fact that there is something latent in the material, there to be extracted and enjoyed today, reflects the care exercised by the ancestors in handing on the heritage in good working order. Our responsibility is to keep the process in being.

Modern scientific theory is ultimately reductive. Genes can be snipped out and shifted from plant to plant. The breeder acquires rights of ownership over such concrete manipulations. (Whether the biotechnologist will be equally keen to 'own' problems that might arise from the re-insertion of such transgenic creations into a natural environment is less clear.) The Mende farmer lives in a more stormy, dynamic world of long-term seed flow. Ancestors are revered for having marked and shaped elements in the world which the present generation inherits. But cultivators no more own the stream of seed flow than they can own the breeze. The proper attitude is one of respect, for God, nature and the ancestors, coupled with a hope that our own actions might one day also be perceived as having favoured the onward transmission of ancestral blessing. For this we might merit a little remembrance, before we fade into that elephant-haunted twilight world of forgotten pioneers.

And so to a suggestion. It is sometimes assumed that poor agrarian districts of the globe are rich in biodiversity and poor in human resources. Scientific breeding, undertaken by a remote elite on behalf of the poor, is one way round this assumed poverty of human resources. Breeders make the clever choices, and all farmers have to do is plant what they provide. But what if the problem is wrongly conceived? The lesson of the Mende is that the human capacity to combine, select and screen planting materials is locally present in hyper-abundance. Maybe it makes more sense to concentrate on enriching the gene pool, leaving local talent to do the rest. Forget the Green Revolution. Treat local myths seriously. Charter a plane and scatter duplicates of the international rice gene bank collections to the four winds.

Linking biodiversity and agriculture: challenges and opportunities for sustainable food security (Lori Ann Thrupp)

Biodiversity, and detailed knowledge about it, have allowed farming systems to evolve since agriculture began some 12,000 years ago (GRAIN 1994).

Although modern agriculture is sometimes seen as an enemy of biodiversity, traditional agriculture is actually *based* on richly diverse biological resources and agroecosystems that benefit from resources in natural habitats. Agrobiodiversity is a fundamental feature of farming systems around the world. It encompasses the many ways in which farmers can exploit biological diversity to produce and manage crops, land, water, insects and biota (Brookfield and Padoch 1994). The concept also includes habitats and species outside of farming systems that benefit agriculture and enhance ecosystem functions.

Traditional agrobiodiversity depends on folk varieties or landraces, which are geographically or ecologically distinctive populations of plants and animals that are conspicuously diverse in their genetic composition (Brown 1978). Landraces are the result of the development by farmers of complex techniques for selecting, storing and propagating the seeds of landraces.

The numerous practices for enhancing biodiversity are also tied to the rich cultural diversity and local knowledge that are valuable elements of community livelihood. Rural women are particularly knowledgeable about diverse plants and tree species and about their uses for health care, fuel and fodder, as well as food (Abramowitz and Nichols 1993; Thrupp 1984). Many principles from traditional systems, as well as intuitive knowledge, are applied today in both large- and small-scale production. In fact, 'traditional multiple cropping systems still provide as much as 20 percent of the world food supply'. (UNDP 1995: 7)

Figure 7.5: Vavilov Centres of Plant Genetic Diversity: Areas of High Crop Diversity and Origins of Food Crops, according to N. Vavilov

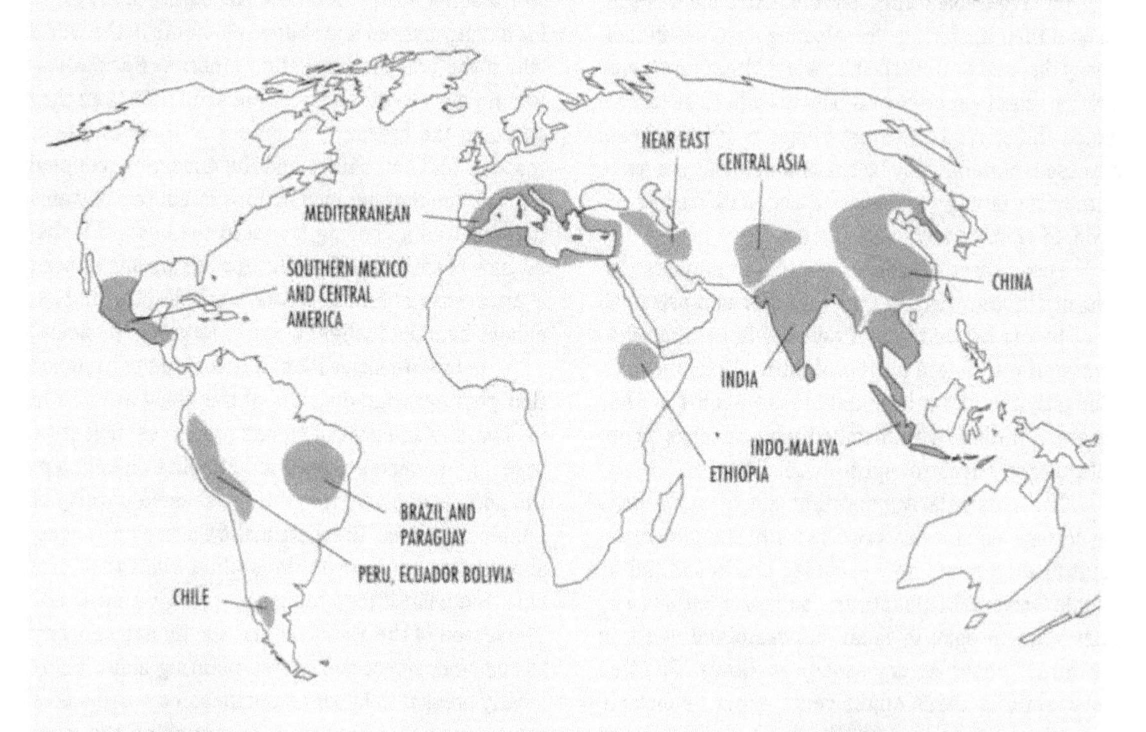

1. Ethiopia
barley, coffee, sorghum

2. Mediterranean
oats, olives, wheat

3. Asia Minor
barley, lentil, oats, wheat

4. Central Asia
apple, chickpeas, lentil

5. Indo-Burma
eggplant, rice, yam

6. Indo-Malaya
banana, coconut, sugar cane

7. China
sorghum, millet, soybean

8. Central America
bean, corn, tomato

9. Peru-Ecuador-Bolivia
bean, potato, squash

10. Southern Chile
potato

11. Brazil-Paraguay
peanut

12. North America
sunflower

13. West Africa
millet, sorghum

14. Northern Europe
oats, rye

Source: N. Vavilov, 1949, *Chronica Botanica* Vol 13. Waltham, Massachusetts, adapted by Reid, Walter and Kenton Miller, 1989. *Keeping Options Alive: The Scientific Basis for Conserving Biodiversity.* World Resources Institute, Washington DC.

These components of agrobiodiversity yield an array of benefits. They reduce risk and contribute to resilience, food security and income generation. They also improve the health of soils and benefit nutrition and productivity. Experience and research (UNDP 1992; Altieri 1987; Brookfield and Padoch 1994) have shown that agrobiodiversity can:

- increase productivity, food security and economic returns;
- reduce the pressure from agriculture on fragile areas, forests and endangered species;
- make farming systems more stable, robust and sustainable;
- contribute to sound insect, pest and disease management;
- conserve soil and increase natural soil fertility and health;
- contribute to sustainable intensification;
- diversify products and income opportunities;
- reduce or spread risks to individuals and nations;
- help maximize effective use of resources and the environment;
- reduce dependency on external inputs;
- improve human nutrition and provide sources of medicines and vitamins; and
- conserve ecosystem structure and stability of species diversity.

Agrobiodiversity loss: conflicts and effects

Although the predominant patterns of agricultural development in the last several decades have increased yields they have also reduced genetic diversity of crop and livestock varieties and agroecosystems, and have led to other kinds of biodiversity loss. People consume approximately 7,000 species of plants, but only 150 species are commercially important, and about 103 species account for 90 percent of the world's food crops. Just three crops – rice, wheat and maize – account for about 60 percent of the calories and 56 percent of the protein people derive from plants. Along with this trend towards uniform monocropping, the dependence on high levels of inputs such as irrigation, fertilizers and pesticides has increased world-wide. The reduction in diversity often increases vulnerability to climate and other stresses, raises risks for individual farmers, and can undermine the stability of agriculture.

Homogenization also occurs in high-value export crops and increases vulnerability to insect pests and diseases that can devastate a uniform crop, especially on large plantations. In addition, there has been a serious decline in soil organisms and soil nutrients. These losses, along with fewer types of agroecosystems, also increase risks and can reduce productivity. Furthermore, many insects and fungi commonly seen as enemies of food production are actually valuable. Some insects benefit farming: for pollination, contributions to biomass, natural nutrient production and cycling, and as natural enemies to insect pests and crop diseases. Mycorrhizae, the fungi that live in symbiosis with plant roots, are essential for nutrient and water uptake.

The global proliferation of modern agricultural systems has eroded the diversity of insects and fungi, a trend that lowers productivity. Agrochemicals generally kill natural enemies and beneficial insects as well as the 'target' pest. This disruption in the agroecosystem balance can lead to perpetual resurgence of pests and outbreaks of new pests – as well as provoke resistance to pesticides. This disturbing cycle often leads farmers to apply increasing amounts of pesticides, or to change products – a strategy that is not only ineffective but that also further disrupts the ecosystem and elevates costs. This 'pesticide treadmill' has occurred in countless locations (Murray and Swezey 1990; Daxl 1988). Reliance on monocultural species and the decline of natural habitat around farms also cuts beneficial insects out of the agricultural ecosystem.

Agricultural expansion has also reduced the diversity of natural habitats, including tropical forests, grasslands and wetland areas. Projections of food needs in the coming decades indicate probable

TABLE 7.2: THE EXTENT OF GENETIC UNIFORMITY IN SELECTED CROPS

CROP	COUNTRY	NUMBER OF VARIETIES
Rice	Sri Lanka	From 2,000 varieties in 1959 to less than 100 today. 75% descend from a common stock
Rice	Bangladesh	62% of varieties descend from a common stock
Rice	Indonesia	74% of varieties descend from a common stock
Wheat	USA	50% of crop in 9 varieties
Potato	USA	75% of crop in 4 varieties
Soybeans	USA	50% of crop in 6 varieties

Source: World Conservation Monitoring Centre (1992). *Global Biodiversity: Status of the Earth's Living Resources* (Brian Groombridge, ed.). Chapman & Hall, London.

TABLE 7.3: REDUCTION OF DIVERSITY IN FRUITS AND VEGETABLES, 1903 TO 1983 (VARIETIES IN NSSL COLLECTION)

VEGETABLE	TAXONOMIC NAME	NUMBER IN 1903	NUMBER IN 1983	LOSS (PERCENT)
Asparagus	*Asparagus officinalis*	46	1	97.8
Bean	*Phaseolus vulgaris*	578	32	94.5
Beet	*Beta vulgaris*	288	17	94.1
Carrot	*Daucus carota*	287	21	92.7
Leek	*Allium ampeloprasum*	39	5	87.2
Lettuce	*Lactuca sativa*	497	36	92.8
Onion	*Allium cepa*	357	21	94.1
Parsnip	*Pastinaca sativa*	75	5	93.3
Pea	*Pisum sativum*	408	25	93.9
Radish	*Raphanus sativus*	463	27	94.2
Spinach	*Spinacia oleracea*	109	7	93.6
Squash	*Cucurbita spp.*	341	40	88.3
Turnip	*Brassica rapa*	237	24	89.9

Source: Cary Fowler and Pat Mooney (1990). ***The Threatened Gene – Food, Politics and the Loss of Genetic Diversity.*** The Lutworth Press, Cambridge.

TABLE 7.4: PAST CROP FAILURES DUE TO GENETIC UNIFORMITY.

DATE	LOCATION	CROP	EFFECTS
1846	Ireland	Potato	Potato famine
1800s	Sri Lanka	Coffee	Farms destroyed
1940s	USA	U.S. crops	Crop loss to insects doubled
1943	India	Rice	Great famine
1960s	USA	Wheat	Rust epidemic
1970	USA	Maize	$1 billion loss
1970	Philippines, Indonesia	Rice	Tungo virus epidemic
1974	Indonesia	Rice	3 million tons destroyed
1984	USA (Florida)	Citrus	18 million trees destroyed

Source: World Conservation Monitoring Centre (1992). ***Global Biodiversity: Status of the Earth's Living Resources*** (Brian Groombridge, ed.). Chapman & Hall, London.

further expansion of cropland, which could add to this degradation.

Policy and institutional changes

Effective reforms and dramatic agrobiodiversity conservation policies that benefit the public, especially the poor, are urgently needed. Policy changes that attack the roots of problems and ensure peoples' rights include:

- ensuring public participation in the development of agricultural and resource-use policies;
- eliminating subsidies and credit policies for HYVs, fertilizers, and pesticides to encourage the use of more diverse seed types and farming methods;
- policy support and incentives for effective agroecological methods that make sustainable intensification possible;
- reform of tenure and property systems that affect the use of biological resources to ensure that local people have rights and access to necessary resources;
- regulations and incentives to make seed and agrochemical industries socially responsible;
- development of markets and business opportunities for diverse organic agricultural products, and
- changing consumer demand to favour diverse varieties instead of uniform products.

Building complementarity between agriculture and biodiversity will also require changes in agricultural research and development, land use and breeding approaches. The types of practices and policies constitute potential solutions and promising opportunities. Such changes are urgently needed to overcome threats from the ongoing erosion of genetic resources and biodiversity. Experience shows that enhancing agrobiodiversity economically benefits both small-scale and large-scale farmers, while at the same time serving the broader social interests of food security and conservation. Implementing the changes and policies suggested in this report will support agrobiodiversity and lead to wide-ranging socio-economic and ecological gains.

Rural societies and the logic of generosity (Jean Christie and Pat Mooney)

Rural societies differ greatly from one another in their views of knowledge-sharing and their approaches to innovation. Concepts of property, land and nature also vary. Many communities look upon most property as communal. Others confer personal or family custodianship over land and living resources. It is not unusual for agricultural communities to permit *de facto* ownership over crops and livestock, including the succeeding generations of domesticated species. It is unheard of, however, for non-industrial farming communities to grant unlimited rights to land and resources, or to permit ownership of the processes of life. Concepts such as stewardship or custodianship come much closer to rural realities than those such as exclusive monopoly, private property or intellectual property.

In most rural societies, knowledge and innovation are not seen as commodities but as community creations handed on from past to future generations. The earth and nature are used and managed but are not exclusively owned. In contrast, European-based intellectual property rights are founded on the belief that innovative ideas and products of human genius can be legally protected as private property. Plant breeders' rights and recent applications of patent law further assert that a vast array of living things are also products of human genius, subject to private monopoly controls.

There is logic to rural systems of generosity. Farmers understand that they must experiment, and that new genetic combinations must be introduced into their fields in order to compete with diseases and pests. The freer the exchange, the greater the potential benefit. This simple truth has been lost in the industrialized countries where intellectual property has created secrecy and reduced scientific exchange.

Every society is complex, and farming communities manage extraordinarily complex ecosystems. It is not surprising that knowledge specialization and apprenticeship systems are common, and that reward mechanisms ensure that knowledge is preserved, shared and enhanced. Patent lawyers like to compare these customary practices to the medieval European guild system that gave rise to intellectual property laws in the North. Although the comparison is fair, neither today's farming communities nor yesterday's guild members would recognize the intellectual property regimes now being imposed by the World Trade Organization.

All peoples have laws, customs and well-defined practices to regulate land ownership, land and resource use, and the acquisition of different types of knowledge. Yet within the vast cultural diversity of farming communities there are striking common threads which unite them and distinguish their view of nature and innovation from the values and worldview that are enshrined as law in industrial societies. Some of these are especially relevant to the intellectual property debate:

Knowledge and innovation cannot be isolated from land and culture

'[When we talk about biodiversity] we are really talking about our whole world view, our cultures, our lands, our spirituality ... These are all linked.' (Stella Tamang, Federation of Nationalities, Nepal)

For indigenous and farming communities, and for all rural peoples, their relationship to land is an important part of their identity. The lands and waters they live with underpin who they are and are the foundation of their very survival. Over and over again, when reflecting on biodiversity or indigenous knowledge, rural people insist that living things cannot be understood separately from the land that nurtures them. Peoples' myriad uses of natural resources cannot be separated from their culture; their culture cannot be separated from the land. For them, this oneness of land, people, knowledge and culture is the only basis for meaningful consideration of biodiversity. But the intellectual enclosure reflected in European intellectual property regimes is dissecting knowledge and fragmenting flora and fauna into unrecognizable genetic bits and pieces. At stake is the intellectual integrity of rural communities.

Farming communities nurture biodiversity and respect the land

Ninety per cent of the earth's most biologically-diverse lands have no government protection, and are cared for exclusively by farming communities and other traditional resource users. Almost all of the earth's most biologically diverse 'hot spots' are home to or bordered by the South's farming communities.

Stewardship not exploitation is the preferred relationship with natural resources

Non-industrial agrarian and indigenous peoples use land, manage natural resources, and pass on knowledge about them to future generations. Their relationship with nature is multidimensional and complex. In many rural societies, the earth itself and life are sacred. Monopoly control over the use and exploitation of living things, including food crops, is an entirely alien concept to many farming societies. The notion of intellectual property over living things is often a sacrilege.

Knowledge and innovation are collective creations
Innovation and adaptation to change have been a part of rural societies for millennia, and knowledge has been passed on from generation to generation. While specialized knowledge about plants and crops is often entrusted to particular social groups or to honoured individuals, it is not their private property.

These qualities of knowledge systems and innovation in rural and indigenous societies stand in stark contrast to the requirements of the World Trade Organization's TRIPs agreement (on Intellectual Property) and the 1992 Biodiversity Convention. Between them these two legally-binding accords are imposing a system of private property and greed on thousands of community-based systems of generosity – the WTO by obliging signatory states to introduce intellectual property over living organisms, and the Biodiversity Convention by requiring that prevailing intellectual property regimes be respected. Unless measures are introduced soon to protect the knowledge and innovation systems of indigenous and rural peoples within the cultures that sustain them, both the wealth of knowledge they have produced over millennia, and the very peoples who have developed and maintained it, will be lost.

It is an urgent task, but an achievable one.

Rebuilding our food system: the ethical and spiritual challenge (Peter Mann and Kathy Lawrence)

The deepening crisis in our food system is not only economic, social and political in nature, it is also an ethical and spiritual crisis that can only be overcome through different ways of thinking, new strategies for action, and changed patterns of living. The attention being paid to our food system today is one aspect of a wider shift in human consciousness that has arisen since the world was first seen from outer space: that the earth is an interconnected whole. The basic elements of the food system – the food, the meal, the local and global community, the seed, the soil, the farm and the land – must be seen in terms of such values as diversity, sustainability, democracy and justice. Regional food systems need to provide high-quality food and sustainable livelihoods while enhancing the natural resources and biodiversity on which sustainable agriculture systems are based. Translating this vision into action will challenge the food system as it is now constituted to build on the many alternatives emerging in communities around the world.

Hunger and poverty are a global reality. More than 800 million people in the world are persistently

TABLE 7.5: THE ROLE OF COMMUNITY KNOWLEDGE IN GLOBAL DEVELOPMENT

HEALTH AND MEDICINE	FOOD AND AGRICULTURE	ENVIRONMENT AND DIVERSITY
Local: 80 percent of the South's medical needs are met by community healers using local medicine systems.	Almost 90 percent of the South's food requirements are met through local production. Two-thirds are based on community farming systems.	Almost 100 percent of the biodiversity 'hot spots' are in areas nurtured by indigenous communities and/or bordering the South's farming communities.
Global: 25 percent (and growing) of Western patented medicines are derived from medicinal plants and indigenous preparations.	90 percent of the world's food crops are derived from the South's farming communities and continue to depend on farmers' varieties in breeding programmes.	The wild relatives of almost every cultivated crop are found in biologically-diverse regions of the South and are nurtured by indigenous communities.
Market: The current value of the South's medicinal plants to the North is estimated conservatively at US$32 billion annually.	The direct commercial value derived from farmers' seeds and livestock breeds is considerably more than US$6 billion a year.	90 percent of the world's most biologically diverse lands and waters have no government protection and are nurtured exclusively by rural communities.
Expertise: Well over 90 percent of all health practitioners are community healers.	99 percent of all plant breeders and other agricultural researchers are based in rural communities.	99 percent of all practised biodiversity expertise resides in indigenous and other rural communities.
Risk: Almost all local knowledge of medicinal plants and systems as well as the plants themselves could disappear within one generation.	Crop diversity is eroding at 1–2 percent per annum. Endangered livestock breeds are vanishing at rates of about 5 percent a year. Almost all farmers' knowledge of plants and research systems could become extinct within one or two generations.	Rainforests are coming down at a rate of 0.9 percent per annum and the pace is picking up. Much of the earth's remaining diversity could be gone within one or two generations.

Source: RAFI 1997, Table 6.

hungry and 1.3 billion people live in poverty. Last year alone, hunger and disease related to malnutrition – direct consequences of poverty – silently claimed the lives of 10–12 million children under the age of five. Yet the world produces an abundance of food.

Paradoxically, hunger is most widespread where food is produced – in rural areas, the home of the vast majority of the world's poor. Often the abundance displayed at the global supermarket in the cities of the North is built upon the dispossession of the rural poor in the South who are forced to grow cash crops for export. And hundreds of millions of peasant farmers and small farmers throughout the world are threatened with the loss of their land as corporate agribusiness strengthens its grip on the world food system.

When anti-hunger groups around the world look at this scandal of hunger in the midst of abundance, and ask what is wrong with this food system and how it can be fixed, they come up with essentially the same answer. They see poverty as the root cause of hunger, and poverty is rooted in the powerlessness of the poor. 'The poor are hungry because they do not have the assets to produce enough food or do not earn enough to buy the food they need. And this points to the solution: to fight hunger, we have to enable the poor to produce food or empower them to buy food. This not only ends hunger; it also enables the poor to contribute to their local and national economies, if empowered to do so. And as many now understand, if we succeed in reducing rural poverty, we can also curb urban poverty.' (Vision Statement: Conference on Hunger and Poverty 1995.) From this anti-poverty perspective, groups fighting hunger in the North and South are impelled by a common vision of justice, of solidarity with the poor, and of social change. They see hunger as part and parcel of a global economic system that is driving inequity, exploitation, waste and poverty.

The first task of the anti-hunger movement is to change consciousness, starting with our own, and overcome the tenacious grasp of the hunger myths on our minds. For hunger is not caused by too many people or too little food, by nature, or laziness, or fate, or God. Hunger is caused by lack of power, lack of access to resources, lack of democracy. Therefore we must learn to see poor and hungry people as our partners, not dependants; as citizens, not clients. They are the agents of their own liberation, and our task is to help them achieve this.

In recent years, the anti-hunger movement and social justice groups concerned with sustainable agriculture – farmers and farm workers, educators, consumer advocates, environmentalists, chefs, urban gardeners, and other community groups – have begun to join forces to address food system issues. A starting point for collaboration has been to increase the accessibility of high-quality, nutritious food to people of all income levels. But as sustainable agriculture and anti-hunger advocates begin talking, they often realize that 'access' does not mean the same thing to everyone. While anti-hunger groups focus mainly on access to large quantities of commercially-produced food, sustainable-food-system advocates look more critically at the food system itself. They are worried about a shocking decline in the number of farms and farmers, the growing consolidation of agribusiness, and a globalized food system that heedlessly destroys good soils, clean water, genetic diversity, and knowledge of local ecosystems, crops and climates.

Both hunger and the decline in small-scale farming are part of a global economic system that is driving inequity and poverty in tandem with the exploitation, waste and destruction of precious natural and human resources. The global food system itself is creating hunger because the industrialized food system is replacing people and livelihoods with chemicals, centralized genetic manipulation and monopolies, in order to maximize short-term corporate profit and consolidate control over food. With no jobs, no access to land, and no control over food prices, is it any wonder that we have become separated from our food system – from the sacred and life-sustaining wonder of real food, lovingly produced and consumed?

Al Gore (1992) has argued that the deepest root of the crisis is a pervasive loss of connection, a separation. He argues that all three crises facing us today – the personal crisis affecting individuals and families, the social crisis breaking up communities, and the environmental crisis affecting the earth itself – are interconnected. In his view, we are separated from ourselves, from our deepest feelings, from our soul, outer from inner, and this is our *personal crisis*. We see this separation daily: anomie, desensitization, feelings of personal powerlessness, and isolation. We are separated from each other, from solidarity and real community, and this is the root of our *social crisis*. Signs of this social crisis are everywhere: poverty, growing inequality, joblessness, social disintegration, and economic and social apartheid. We are separated from the earth, from nature and the cosmos, and this creates the *ecological crisis*. Yet the heart of this crisis is ethical and spiritual.

Ethical and spiritual principles underlie the strategies to rebuild our local, regional and global food systems. Vandana Shiva (1991b) has reminded us that in traditional farming systems peasants, as custodians of the planet's genetic wealth, treated seeds as sacred, as the critical element in the great chain of being. Seed was not bought or sold, it was exchanged as a free gift of nature. Corporate agribusiness, by contrast, treats seeds as private property, protected by patents and intellectual property rights. Shiva contrasts the corporate centralization of seed research in the North with the decentralization of local cultivators and plant breeders, and the diversity of thousands of locally-cultivated plants in the agricultural system, built up over generations on the basis of knowledge generated over centuries.

There is a growing movement in the South and the North to recover the sacred meaning of seeds, restore biodiversity and revolutionize the way we think about food. Fred Kirschenmann (1997) sees sustainable agriculture as, 'more than an alternative farming system'. It is part of the ecological revolution which is believed to be essential to our survival, and a truly sustainable agriculture must 'encompass this larger vision'.

One of the ways we can bring about this revolution is to become 'lovers of the soil' in Kirschenmann's words, which means learning to 'see' the soil – 'see its life and beauty, smell its rich aroma, hear its voice'. And one of the quickest ways to learn to see is to grow and prepare our own food. Wendell Berry is on a similar path in suggesting that we locate the farm economy within 'the Great Economy' which connects us to human virtues and human scale, and to an order which, like the Tao or the Kingdom of God, is both visible and invisible, comprehensible and mysterious (Berry 1987). We are being called to recover the sacred traditions that see the land as a gift of the whole community, to be treasured and its produce to be shared.

Perhaps this Chinook blessing litany is an appropriate start for reaffirming our respect for the dignity of all beings, and reconnecting, through a renewed and revitalized food system, with our spiritual and ecological selves and a world free of hunger and despair:

> ... We call upon the land which grows our food, the nurturing soil, the fertile fields, the abundant gardens and orchards, and we ask that they: Teach us and show us the way.
>
> ... We call upon the creatures of the fields and forests and the seas, our brothers and sisters the wolves and deer, the eagle and dove, the great whales and the dolphin, the beautiful orca and salmon who share our Northwest home, and we ask them to: Teach us and show us the way.
>
> ... And lastly, we call upon all that we hold most sacred, the presence and power of the Great Spirit of love and truth which flows through all the universe... to be with us to: Teach us and show us the way.
>
> (*Earth Prayers* 1991).

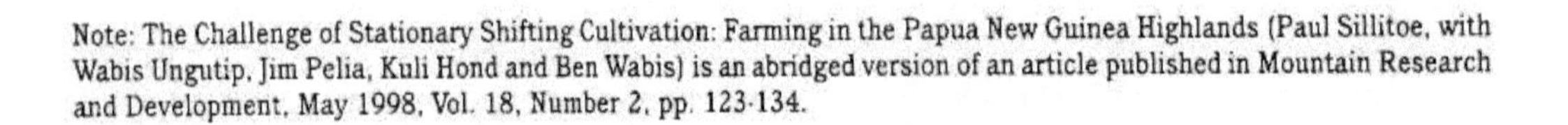

Note: The Challenge of Stationary Shifting Cultivation: Farming in the Papua New Guinea Highlands (Paul Sillitoe, with Wabis Ungutip, Jim Pelia, Kuli Hond and Ben Wabis) is an abridged version of an article published in Mountain Research and Development, May 1998, Vol. 18, Number 2, pp. 123-134.

CHAPTER 8

MOUNTAINS: THE HEIGHTS OF BIODIVERSITY

Edwin Bernbaum

CONTENTS

TEXT BOX

Mountains: The Heights of Biodiversity

Edwin Bernbaum

Introduction (Edwin Bernbaum)

For many people around the world, mountains have a peculiarly evocative nature. The sight of a peak reaching toward the clean blue heights of the sky lifts the mind and spirit, conjuring up visions of a higher, more perfect realm of existence. The eerie clouds and mist that frequently veil a mountain, cutting it off from the world below, enshroud it in an aura of mystery that seems to conceal some deeper, more meaningful reality. The storms that swirl around them, punctuated with flashes of lightning and crashes of thunder, make mountains awesome places of power, both natural and supernatural. The harsh environment of high peaks – forbidding cliffs, tangled forests and treacherous glaciers – speaks of an alien realm of existence, the abode of forces outside human control, to be feared and avoided. The rain clouds that form around their summits and the rivers that well out of their slopes lead many to look up to mountains as divine sources of water and life. The massive form of a hill or mountain can also serve as a natural fortress and point of reference, giving people a sense of security and orientation, both physical and spiritual.

The evocative power of mountains highlights basic cultural and spiritual values and beliefs that deeply influence how people view and treat the world. These values and beliefs determine to a great extent which resources people are willing to exploit and which features of the environment they feel motivated to protect. For any assurance of long-term success, measures directed toward environmental preservation and sustainable use need to take such spiritual and cultural factors into account; otherwise they will not win the support of local communities and others with the greatest stake in what is going on. Moreover, the reverence that mountains tend to awaken in cultures around the world provides a powerful source of motivation for preserving biodiversity.

The physical characteristics of mountains have made them important strongholds of biological and cultural diversity. Their remoteness, difficulty of access, and rugged terrain have combined to make them among the last places on earth to be exploited and developed. It is no coincidence that a large number of wilderness and protected areas around the world lie in mountainous regions where environments and ecosystems have remained relatively intact. However, with the advent of helicopters and modern technology, even remote mountain areas have become vulnerable to exploitation. Clear-cut logging is wreaking havoc on the forests of previously inaccessible ranges of British Columbia, while a mine is chewing an enormous hole in the upper slopes of the highest range in New Guinea, deep in the jungle-filled interior of the island (Messerli and Ives 1997).

The inaccessibility and remoteness of mountains have also helped to preserve cultural diversity. The isolation of mountain communities has kept them out of the way of modernization and allowed them to retain traditional ways of life abandoned in the more accessible lowlands. At the same time, dissenting groups with cultures opposed to mainstream societies, or persecuted by them, have traditionally sought refuge in mountains in order to preserve their cultural integrity. Heretical European sects such as the Albigensians and the Vaudois fled to the safety of the Alps and Pyrenees while myths of hidden valleys free from religious and political oppression have drawn Tibetans to settle the Himalayan border regions of Tibet, most notably the principality of Sikkim (Bernbaum 1990; Ramakrishnan 1996). The advent of modern communications, with satellite television and VCRs, however, has opened even the most isolated communities to the corrosive cultural influences of the outside world.

The vertical topography of mountains means that they harbour a great multiplicity of microclimates. Variations in altitude, steepness of terrain, the direction a slope faces, and the composition of the soil all make for widely differing climatic conditions and habitats. Within a few thousand vertical feet, we find plant and animal species that are separated at sea level by thousands of miles. In order to survive, people need to know and cultivate many indigenous crops to fit the different ecological niches of their fields and land, thereby reinforcing the natural biodiversity of the mountains (Messerli and Ives 1997).

In fact, of all the different kinds of natural landscapes that possess spiritual or cultural value, mountains contain the most diverse and complete environments and ecosystems. If we consider them in their entirety, from their bases to their summits, they can include jungles, forests, shrub lands, meadows, deserts, rivers, lakes, tundras, glaciers – the full range of habitats from the tropical to the Arctic. A mountain that has spiritual or cultural value can, therefore, give us a valuable picture of what features of nature in general a particular society values and feels motivated to protect. In addition, mountains whose natural environments have been protected provide rich gene pools for replenishing plant and animal species lost in the outside world.

Spiritual and cultural values can preserve the biodiversity of mountains in a number of ways. People may avoid desecrating the environment of a mountain or mountain site because they consider the place sacred or value it for other reasons. The Dai people of south-western China, for example, have set aside Holy Hills as gardens of the gods, to be left untouched by farming and hunting (Pei 1993). National parks and local communities frequently protect mountain environments for aesthetic rather than scientific reasons, barring commercial development and exploitation in order to preserve the unspoiled beauty of a primeval forest or a spectacular view.

The conservation of biodiversity may be the by-product, rather than the intention, of beliefs and practices associated with a mountain site. The prohibition of female animals on Mount Athos, based on the belief that the patron saint of the Holy Mountain, the Virgin Mary, allows no other females, has meant that monks have not permitted goats to graze on the peninsula. This has had the unintended consequence of preserving flora that has been chewed up and lost elsewhere in Greece and the Balkans (Oikanomou et *al.* 1993). Beliefs and practices can also protect the biodiversity of mountains by imbuing them with a negative or forbidding value. People may keep away from an awe-inspiring peak because they fear it as the abode of evil spirits or of powerful forces that forbid human presence. Or, as in past centuries in Europe, they may avoid mountains as undesirable, repugnant places – warts, pimples and blisters on the face of the earth.

On the other hand, people may protect the biodiversity of mountain environments because they value the particular flora and fauna found there. Many societies around the world, from Asia to North America, regard mountains as natural pharmacies stocked with medicinal plants. In a well-known episode of the *Ramayana*, a major epic of ancient Hindu culture with power in India today, the monkey deity Hanuman flies to the Himalayas for a magic herb that will restore his comrades to life. Navajo or Dineh singers go to four sacred mountains in the south-western United States to obtain medicinal plants which they wrap in bundles and use to cure people in healing ceremonies. When they pick them, they do so in rituals that show respect for their indwelling spirits and minimize damage to the ecosystem (Bernbaum 1990; Matthews 1897).

Beliefs and practices associated with mountains frequently imbue particular plants and animals with special value and grant them divine protection. Because wildlife such as vicuñas belong to the *apus* or mountain deities of the Peruvian Andes, many indigenous people of the region refrain from hunting them. Hindus have special regard for *bhojpatra* or *Betula utilis*, a species of Himalayan birch revered for the use of its bark in ancient times as paper for sacred scriptures. A *swami* I met at Badrinath, the major pilgrimage place in the Indian Himalayas, was encouraging everybody who stayed at his *ashram* to plant the trees. The Japanese take special care of mountain groves that provide certain kinds of wood needed for temple construction and rituals.

Many cultures revere mountain peaks, above all, as places of natural and supernatural power, to be approached with caution and respect. Maori warriors crossing the plateau beneath Mount Tongariro in New Zealand would avert their eyes from its summit for fear of provoking divine wrath in the form of a blinding snowstorm (Orbell 1985; Yoon 1986). This kind of fear can provide a powerful reason for leaving the ecosystem of a mountain alone. On the other hand, people may value the natural fertility and biodiversity of mountains as expressions of their life-giving powers. Native Hawaiians see the lush foliage growing out of old lava flows on Kilauea as a living manifestation of the creative power of Pele, the goddess of the volcano (Taswell 1986; Bernbaum 1990).

The sacred power associated with a mountain generally takes the form of a deity. Traditional stories and paintings of such deities frequently depict them in the company of certain plants and animals. The Hindu god Shiva often appears in a flowering paradise high on Mount Kailas with his mount, a sacred bull, beside him; and Hindu texts will identify his hair with Himalayan forests (Purohit and Bernbaum, this chapter). As the story in 'The Huntsman and the Deer' shows, holy men such as Milarepa can also embody this kind of power and protect the surrounding foliage and wildlife (Chang, this chapter). Frequently depicted in temple murals, this story about Tibet's most famous and beloved yogi has inspired Tibetans over the centuries to value wild animals and refrain from killing them. Mountains associated with such deities and holy men will tend to be set aside as places where hunting and gathering of the particular species under their ownership or protection are banned.

Many cultures view mountains as temples and places of worship. Mountain shrines built to house deities frequently provide the spiritual basis for sacred groves that have preserved biological diversity lost in surrounding regions. For example, forests that have historically received the best protection in Khumbu, the area near Mount Everest, are ones associated with Buddhist temples and monasteries (Stevens, this chapter). Mountains may also take the form of a place of worship, imagined as a kind of natural church or cathedral. Here a deity does not necessarily reside at the peak, but rather the mountain provides a special, evocative setting for making contact with the sacred through prayer and contemplation. This view can provide a strong rationale in monotheistic traditions for maintaining the ecological integrity of mountains in order to enhance their spiritual value as places of worship. John Muir, a major figure in the genesis of the modern environmental movement, used this kind of argument to rally public support for protecting valleys in the Sierra Nevada such as Yosemite and Hetch Hetchy – although in the case of the latter he was not successful (Muir 1988).

Mountains are commonly pictured as gardens and earthly paradises, created by a supreme deity or inhabited by the gods. The monks of Mount Athos refer to their sacred peninsula as the 'Garden of the Mother of God'. Some of them see their roles as gardeners or stewards entrusted with the task of caring for the natural environment of the Holy Mountain (Oikanomou *et al.* 1993). Notions of primordial harmony connected with ideas of paradise encourage many people to see wildlife, even predators, as part of the sacred order that reigns there. The view of mountains as gardens and paradises can, in fact, inspire both modern and traditional societies to protect wilderness areas as sacred places of beauty and spiritual value – instead of regarding them as wastelands to avoid or subjugate.

Another widespread view of mountains is as cemeteries and places of the dead. Many societies bury their dead on a mountain or facing a prominent peak, such as Kilimanjaro in Africa. Others believe that the spirits of those who have died have gone to sacred mountains: traditional Chinese belief holds that they go to the foot of Tai Shan (Bernbaum 1990; Chavannes 1910). In either case, through rituals centred around such mountain sites, communities maintain a continuing relationship with their ancestors, who may provide them with blessings, such as children, crops and water. The respect paid to the departed frequently transfers to the flora and fauna of mountains where they lie buried or live in another form. The fear that ghosts inspire also prompts people to stay away from places associated with the dead, which can result in the inadvertent conservation of biodiversity on a sacred mountain.

People throughout the world look up to mountains as sources of blessings such as water, life, healing, health and well-being. Beliefs and practices intended to maintain the spiritual as well as physical purity of these sources can work to preserve biodiversity. Some societies, such as the Sherpa near Mount Everest, avoid polluting mountain springs for fear of provoking the ire of disease-causing spirits inhabiting such places (Stevens, this chapter). Others set aside meadows and forests surrounding special water sources as areas that must be kept pristine to preserve the healing power drawn from them in rituals. Sacred mountains are also a major source of medicinal plants. Traditional attention paid to the provenance of these plants and the careful ways they should be harvested have provided a sound basis for protecting the environments in which they grow. Even contemporary secular societies feel motivated to preserve the pristine environment of mountains as natural sources of spiritual and physical well-being.

Finally, mountains are revered around the world as places of religious revelation, spiritual transformation and artistic inspiration. The highest revelations in the Old and New Testaments of the Bible take place on sacred peaks: Moses receives the Torah on Mount Sinai and Jesus is transfigured and revealed as the Son of God on Mount Tabor. Mountains like Hua Shan in China and Mount Athos in Greece are a favourite haunt for hermits, monks and holy men intent on practising meditation and attaining spiritual realization. Pilgrims go to sacred peaks like Mount Kailas in Tibet in the hope of touching the divine and transforming their lives. Evocative views of mountains ranging from the Alps to Mount Fuji have provided inspiring subject matter for poets and artists, who have depicted them in divine terms. Kuo Hsi, one of China's greatest landscape painters, wrote: 'The din of the dusty world and the locked-in-ness of human habitations are what human nature habitually abhors: while, on the contrary, haze, mist, and the haunting spirits of the mountains are what human nature seeks, and yet can rarely find... Having no access to the landscapes, the lover of forest and stream, the friend of mist and haze, enjoys them only in his dreams. How delightful then to have a landscape painted by a skilled hand! Without leaving the room, at once, he finds himself among the streams and ravines; the cries of the birds and monkeys are faintly audible to his senses; light on the hills and reflection on the water, glittering, dazzle his eyes. Does not such a scene satisfy his mind and captivate his heart? This is why the world values the true significance of the painting of mountains.' (Kuo Hsi 1935: 30-31, 33)

Since the environment itself contributes to the efficacy of a mountain as a place of revelation, transformation or inspiration, people will have reason to do what they can to maintain its biological integrity. Such motivation underlies much of the rationale and impetus for the modern-day environmental movement, as well as many of the measures traditional societies have taken to set aside and protect their sacred mountains.

Although mountains may seem unchanging and eternal, their environments are, in fact, quite fragile. Natural processes of erosion work there at their highest pitch, constantly shaping and re-shaping the landscape. Earthquakes, volcanic eruptions, landslides, rock falls, avalanches and floods can easily sweep away forests, inundate meadows or scour out valleys. A misplaced road or rice terrace can inadvertently cause extensive damage, undermining entire slopes and causing villages to slip away. The short growing season, harsh climate, and thin soil at higher elevations means that vegetation regenerates very slowly and the plants often have shallow roots, making them especially vulnerable to grazing or hiking. Because they extend up to the limits of possible life, mountains are among the first places affected by global warming and climate change. Moreover, flooding and drought caused by mountain deforestation can have devastating effects on heavily populated regions downstream, impinging on the lives of hundreds of millions of people living in lowland areas like the Gangetic plains of northern India (Messerli and Ives 1997).

Because of their physical fragility, mountains require special care and attention. The fact that they preserve sanctuaries of biological and cultural diversity lost in much of the outside world makes their protection all the more urgent. Under the impacts of modernization, the influx of outside forces and population growth, traditional controls and practices that have successfully protected many mountain environments for centuries are now being overwhelmed. Local communities need to find ways to strengthen and adapt these controls and practices to new circumstances.

As the papers in this chapter suggest, the basis for doing so lies in acknowledging and renewing the underlying cultural and spiritual values that have worked to conserve biodiversity in mountain areas. The first priority in such efforts is to include in any proposed conservation or sustainable use programme the groups and individuals for whom the particular mountain or mountain site under consideration has special significance. Stakeholders and other concerned parties should be involved from the beginning as full participants in the process. Their views, needs and preferences need to play a leading role in determining whatever programmes are ultimately implemented. As Evelyn Martin's contribution implies, if efforts had been made to include the Apaches as full participants in the initial stages of site selection and planning, much of the controversy over putting an observatory on Mt. Graham in Arizona might have been avoided or mitigated. Since she wrote the article in 1992, two of three planned telescopes have opened on the peak. The Apaches and environmentalists have lost their legal appeals to stop the project, which they hold, respectively, violates sacred ground and threatens the survival of the endangered Mt. Graham red squirrel (the population of the squirrels has remained stable, but perilously low).

Outside organizations and public agencies can help by recognizing and reinforcing the expertise and authority of local people and institutions traditionally entrusted with protecting features of the environment. The case studies discussed in Stanley Stevens' contribution to this chapter suggest, for example, that wardens of the Sagarmatha National Park around Everest should work closely with lamas to enhance their standing in the Sherpa community and broaden the protection they afford to sacred forests in Khumbu. Thomas Schaaf proposes in his paper that 'environmental education be targeted specifically at younger people in informing them that the beliefs of the elders can be very beneficial for the preservation of the environment for future generations'. Educational materials can highlight myths, such as one told in the contribution to this chapter, 'The Creation of the Sacred Mountains', that sanctify various species of plants and animals. The recitation of this creation myth by Dineh or Navajo singers plays a major role in healing ceremonies intended to restore the sick to a state of harmony with themselves, their society, and the world – the basis for restoring the health and well-being of nature as well. Educational programmes can also demonstrate the scientific value and practical uses of medicinal and ecological knowledge possessed by elders, healers and other traditional figures.

Many societies draw vitality and cohesion from their relationship to mountains and sacred sites. In New Zealand, for example, members of each Maori tribe and sub-tribe identify themselves with a hill or peak that establishes their descent from a legendary ancestor and embodies the *mana* or power that sustains them as a group. In the latter part of the nineteenth century, British colonists were buying up parcels of land on Tongariro, the mountain of the

Ngati Tuwharetoa. In order to preserve the integrity of the sacred volcano and keep the *mana* of the tribe intact, the Paramount Chief gifted Tongariro to the British Crown as a national park for everyone – the first one in New Zealand and the fourth in the world (Lucas 1993). Environmental organizations and government agencies need to realize that weakening what imbues such a mountain with value may undermine a culture, resulting in adverse social, economic and environmental impacts as the society falls apart and traditional controls are lost.

Spiritual and cultural values promoting biodiversity are often highlighted in the views that people have of sacred mountains and their significance. These views usually take the form of metaphors, such as the mountain viewed as a temple or as a place of the dead. Many of these metaphors link mountains and their environments to features of places where people live and work, such as houses, gardens, temples and cemeteries. Emphasizing such links with vital aspects of daily life can help to revitalize traditional controls and practices that promote biodiversity in mountain areas. Elders and religious leaders can remind people that they should, for example, treat the forests of mountains that hold the spirits of the dead as they would groves of trees in cemeteries where their parents and grandparents lie buried.

The metaphors underlying views of sacred mountains also provide bridges of mutual understanding between cultures that can help to promote biodiversity. In particular, they can make traditional beliefs and practices directed toward the natural environment more comprehensible and worthy of respect and empathy by people from modern Western societies familiar with monotheistic religions. In the following passage from his contribution to the chapter, Chief John Snow makes very effective use of metaphors for this purpose. 'These mountains are our temples, our sanctuaries, and our resting places. They are a place of hope, a place of vision, a very special and holy place where the Great Spirit speaks with us.' With the help of such metaphors, he shows how the Stoney people have traditionally established spiritual relationships with animals and nature through vision quests high in the Canadian Rockies.

Before proceeding we should introduce a word of caution. Some cultures, such as the Zuni in North America, place great value on keeping knowledge of sacred places, beliefs and practices secret. They may not want ceremonial sites, for example, to be identified or marked, even for their protection. In such cases, the secrecy of sacred places, beliefs and practices should be respected: outside organizations and individuals should not research or document them. Any measures of environmental protection should be left entirely to the discretion of traditional leaders in charge of such matters.

Many mountain societies have been able to preserve spiritual and cultural values protecting the environment that have been lost elsewhere. Traditional practices and ways of life based on these values can serve as models for those who live in the lowlands. Stories of hermits and saints living in mountains, such as the story of the yogi Milarepa protecting the deer from the hunter, have inspired ordinary people to follow their lead and care for plants and animals as a spiritual or religious practice. As the contribution on Badrinath in the Indian Himalayas indicates, mountain shrines that draw pilgrims from other regions provide ideal places to disseminate ideas and practices that promote biodiversity and environmental awareness grounded in religious and cultural traditions. The actual experience of participating in conservation measures at a pilgrimage site, such as planting trees at Badrinath, can reinforce these ideas and practices and motivate people to implement them back in their home communities.

Up to this point, we have been discussing the spiritual and cultural value of biodiversity, but as a modern scientific concept biodiversity may have little or no meaning in traditional societies. Many of them have no equivalent terms in their languages. Moreover, most traditional ways of knowledge do not distinguish among species in the same way science does. Their systems of classification or taxonomy usually have different purposes that have more to do with how people related to flora and fauna in their daily lives. Rather than single out biological differences, many of them focus, for example, on anthropomorphic similarities between humans and animals – as well as among animals themselves – in order to establish hunting, clan, totemic or other relationships. Such relationships, based on what other species, including plants, have in common with people, may actually provide a more meaningful and sustainable basis for biodiversity conservation than the abstract rationales given by science.

Schaaf suggests in his contribution that sacred sites, particularly those in mountains, can provide innovative alternatives to legally designated protected areas based on Western scientific ideas. The Holy Hills of the Dai people present an example of such an alternative: they form 'green islands' preserved by traditional beliefs and practices that function as 'stepping stones' for the flow of genetic material, helping to maintain natural reserves (Pei 1993). The contribution on Badrinath reports on a project to restore an ancient sacred forest at the major Hindu pilgrimage place in the Indian Himalayas. It tells how Indian scientists from the G. B. Pant Institute of Himalayan Environment and Development – formerly under the direction of Prof. A. N. Purohit, co-author of the article – have worked with priests to motivate pilgrims and local stakeholders to replant trees for religious reasons in an innovative programme that provides a promising model for engaging people in environmental measures based on their own spiritual and cultural values. Purohit remarked to me that official programmes such as those designating Biosphere Reserves mean little to most villagers in India. On the other hand, sacred groves scattered throughout the Himalayas have preserved many different species and could serve as kernels around which larger protected areas could be established that would make more sense to the local populace. I saw an example of just such an approach used with success at Binsar Sanctuary near Almora, where a core forest and shrine sacred to the Seven Sages and a locally important protector deity became the basis for rallying the support of villagers in creating a surrounding preserve.

The various contributions to this chapter provide support for the proposition that policy-makers and researchers should broaden scientific conceptions of biological diversity to include spiritual and cultural dimensions as well. Programmes to preserve biodiversity need to take into account the many different facets of what plants and animals – as well as features of the natural landscape, such as mountains – mean to people of different cultures and religious traditions. If they do not, they will fail to receive the popular support needed to make them sustainable. As Martin Palmer and Tjalling Halbertsma point out in their contribution, developing environmental programmes with religious communities who have been around for centuries is 'a better long-term strategy than working with transient political structures'. If their projects are to have lasting impacts, researchers, policy-makers and organizations dedicated to protecting species and their mountain habitats cannot afford to ignore the spiritual and cultural values of biodiversity.

The huntsman and the deer (Adapted from *The Hundred Thousand Songs of Milarepa*, translated by Garma C. C. Chang, 1970)

The yogi Milarepa went to a secluded place at Nyishang Gurda Mountain on the border between Nepal and Tibet. To the right of the mountain towered a precipitous hill where one could always hear the cries of wild animals and watch vultures hovering above. To its left stood a hill clothed with soft, rich meadows, where deer and antelopes played. Below, there was a luxuriant forest with all kinds of trees and flowers, within which lived many monkeys, peacocks and other beautiful birds. The monkeys amused themselves by swinging and leaping among the trees, the birds darted here and there with a great display of wing, while warblers chirped and sang. In front of the hermitage flowed a stream fed by melting snow and filled with rocks and boulders. A fresh, clear, bubbling sound could always be heard as one passed by. It was a very quiet and delightful place with every favourable condition for devotees.

One day, Milarepa heard a dog barking. He thought, 'Hitherto, this place has been very favourable for meditation. Is some disturbance on the way?' Before long, a black, many-spotted deer ran up, badly frightened. Seeing this, an unbearable compassion arose within the yogi. He thought, 'It is because of the evil *karmas* this deer has acquired in the past that he was born in such a pitiable form. Though he has not committed any sinful deeds in this life, he must still undergo great suffering. What a pity! I shall preach to him the *Dharma* [the Buddhist teachings] and lead him to eternal bliss.' Thinking thus, he sang to the deer a song of compassionate teaching.

Affected by the yogi's compassion, the deer was relieved from its painful fear of capture. With tears streaming from its eyes it came near to Milarepa, licked his clothes, and then lay down at his side. He thought, 'This deer must be hunted by a ferocious dog'.

A red bitch with a black tail and a collar round her neck ran toward him. She was a hunting dog – a savage and fearful creature. Milarepa thought, 'She is indeed ferocious. Full of anger she regards whatever she sees as her enemy. It would be good if I could calm her and quench her hatred'. Great pity for the bitch rose in him and he sang her a song of *Dharma*.

Hearing this song, the bitch was greatly moved and her fury subsided. She then made signs to the yogi by whining, wagging her tail and licking his clothes. Then she put her muzzle under her two front paws and prostrated herself before him. Tears fell from her eyes, and she lay down peacefully with the deer.

Before long a man appeared, looking very proud and violent. In one hand he had a bow and arrow, and in the other a long lasso for catching game. When he saw Milarepa with the bitch and deer lying beside him, he thought, 'Are the bitch and deer both bewitched by this yogi?' The hunter drew his long bow, aimed at Milarepa, and shot. But the arrow went high and missed.

Milarepa said, 'You need not hurry to shoot me, as you will have plenty of time to do so later. Take your time, and listen to my song'.

While the yogi was singing, the hunter thought, 'Usually a deer is very frightened, and my bitch very wild and savage. Today, however, they lie peacefully together. Hitherto I have never missed a shot, but today I could not hit him. He must be a black magician, or a very great and unusual Lama'.

Thinking thus, the hunter entered the cave, where he found nothing but some inedible herbs. Seeing such evidence of austerity, a great faith suddenly arose within him. He said, 'Revered Lama, who is your companion, and what do you own? If I am acceptable to you, I should like to be your servant; also I will offer you the life of this deer'.

In reply, Milarepa sang to the hunter:

The snow, the rocks, and the clay mountains –
These three are where Milarepa meditates;
If you they satisfy,
You may come with me.

The deer, the argali, and the antelope –
These three are Mila's cattle;
If you they satisfy
You may come with me.

The lynx, the wild dog, and the wolf –
These three are Mila's watchdogs;
If you they satisfy
You may come with me.

The grouse, the vulture, and the singing Jolmo –
These three are Mila's poultry;
If you they satisfy
You may come with me.

The sun, the moon, and the stars –
These three are Mila's pictures;
If you they satisfy
You may come with me.

The hyena, the ape, and the monkey –
These three are Mila's playmates;
If you they satisfy
You may come with me.

Porridge, roots, and nettles –
These three are Mila's food;
If you they satisfy
You may come with me.

Water from snow, and spring, and brook –
These three are Mila's drink;
If you they satisfy
You may come with me.

The hunter thought, 'His words, thoughts, and actions are truly consistent'. The uttermost faith thus arose within him. He shed many tears and bowed down at Mila's feet, crying, 'I pray you to free my bitch, thus delivering her to the higher realms and to bring this black deer to the Path of Great Happiness. I pray you grant me, the huntsman Chirawa Gwunbo Dorje, the teaching of the *Dharma* and lead me to the Path of Liberation'.

Chirawa Gwunbo Dorje was completely converted to the *Dharma*. He then remained with Milarepa and did not return home. Through practising the teachings given him, the huntsman eventually became one of the heart-son disciples of Milarepa and was known as Chira Repa – the Cotton-clad Huntsman. The deer and the bitch were also forever removed from the Paths of Sorrow. It is said that the bow and arrows which the huntsman offered to Milarepa are still in that cave.

[Adapted from Chang, C. C. (1970) *The Hundred Thousand Songs of Milarepa*. Harper Colophon, New York.]

Creation of the sacred mountains of the Dineh (Navajo) (Adapted from Washington Matthews, 1897)

After emerging into this, the fifth world, First Man and First Woman, along with Black Body and Blue Body, set out to build the seven sacred mountains of the present Dineh land. They made them all of earth, which they had brought up from similar mountains in the fourth world. The mountains they made were *Sisnaajini* or Blanca Peak in the east, *Tsoodzil* or Mount Taylor in the south, *Dook'o'ooslid* or the San Francisco Peaks in the west, *Dibe nitsaa* or Mount Hesperus in the north, with *Dzil na'oodilii*, *Ch'ool'i'i*, and *Ak'i dah nast'ani* in the middle of the land.

Through *Sisnaajini* in the east, they ran a bolt of lightning to fasten it to the earth. They decorated it with white shells, white lightning, white corn, dark clouds and harsh male rain. They set a big bowl on its summit, and in it they put two eggs of the pigeon to make feathers for the mountain. The eggs they covered with a sacred buckskin to make them hatch, which is why there are many wild pigeons on the mountain today. All these things they covered with a sheet of daylight, and they put the Rock Crystal Boy and the Rock Crystal Girl into the mountain to dwell.

Tsoodzil, the mountain of the south, they fastened to the earth with a great stone knife, thrust through from top to bottom. They adorned it with turquoise, with dark mist, soft female rain, and all different kinds of wild animals. On its summit they placed a dish of turquoise; and in this they put two eggs of the bluebird, which they covered with sacred buckskin. This is why there are many bluebirds on *Tsoodzil* today. Over all they spread a covering of blue sky. The Boy who Carries One Turquoise and the Girl who Carries One Grain of Corn were put into the mountain to dwell.

Dook'o'ooslid, the mountain of the west, they fastened to the earth with a sunbeam. They adorned it with haliotis shell, with black clouds, harsh male rain, yellow corn, and all sorts of wild animals. They placed a dish of haliotis shell on the top and laid in this two eggs of the yellow warbler, covering them with sacred buckskins. There are many yellow warblers now on *Dook'o'ooslid*. Over all they spread a yellow cloud, and they sent White Corn Boy and Yellow Corn Girl to dwell there.

Dibe nitsaa, the mountain of the north, they fastened with a rainbow. They adorned it with black beads, with dark mist, with different kinds of plants, and many kinds of wild animals. On its top they put a dish of black beads; and in this they placed two eggs of the blackbird, over which they laid a sacred buckskin. Over all they spread a covering of darkness. Lastly they put the Pollen Boy and Grasshopper Girl in the mountain to dwell there.

Dzil na'oodilii was fastened with a sunbeam. They decorated it with goods of all kinds, with the dark cloud, and the male rain. They put nothing on top of it. They left its summit free, in order that warriors might fight there; but they put Boy Who Produces Goods and Girl Who Produces Goods there to live.

The mountain of *Ch'ool'i'i* they fastened to the earth with a cord of falling rain. They decorated it with pollen, dark mist and female rain. They placed on top of it a live bird, the bullock oriole – such birds abound there now – and they put in the mountain to dwell Boy Who Produces Jewels and Girl Who Produces Jewels.

The mountain of *Ak'i dah nast'ani* they fastened to the earth with a sacred stone called mirage stone. They decorated it with black clouds, male rain and all sorts of plants. They placed a live grasshopper on its summit, and they put Mirage Stone Boy and Carnelian Girl there to dwell.

[Adapted from Matthews, W. (1897) Navaho *Legends*. American Folklore Society, 5. Boston.]

Sherpa protection of the Mount Everest region of Nepal (Adapted from Stanley F. Stevens, 1993)

Sherpas, a people of Tibetan origin, have their own geography of their Khumbu homeland in the Mount Everest area of the Nepal Himalaya, one much richer than that of outsiders in its intimate familiarity with every corner of the region. The Sherpa conceptual map of Khumbu, which is today both a national park (Sagarmatha, Mount Everest, National Park) and a World Heritage Site, is far more complex than even the finest Western maps. Their geography recognizes another level of features altogether, naming a multitude of field sites, pastures, forest areas, tributary valleys, mountainsides and settlements that do not appear on any foreign maps. The awareness of local geography reflected by place names, moreover, is only the surface of Sherpa geographical knowledge of Khumbu, a working vocabulary by which to organize their enormous understanding of distinct, local microenvironments. For Sherpas, the landscapes of Khumbu are also a continual evocation of their individual and collective past, the history which took place in these places. Beyond that, for Sherpas Khumbu is a world alive with supernatural forces – gods, spirits and ghosts who are also associated with certain places. The mountain Khumbila, for example, is the main residence of the Khumbu Yul Lha, the great god of the region, and there are other peaks and places that are the homes of clan gods. Springs, boulders, caves, forests and individual trees are the domain of particular spirits and demons. This religious geography is not separate from ordinary life, for it is believed to have great bearing on the luck and well-being of people and communities and is an important consideration in the choice of settlement and house sites, use of water and attitudes towards its pollution, and forest use and protection.

Sherpa religious values and respect for sacred places have played an important role in shaping local land-use and natural resource management, and remain today a conservation cornerstone for Sagarmatha (Mount Everest) National Park. Sherpas believe that all wildlife and plants have consciousness and spirits, and many take the non-violent traditions of Buddhism so seriously that they refuse to harm even insects. For centuries they have shunned hunting for either subsistence purposes or sport, in effect creating a wildlife sanctuary throughout the region long before a national park was proposed. Sherpa respect for wildlife has undoubtedly contributed to the abundance in parts of the region of musk deer, Himalayan tahr, Impeyan pheasants (Nepal's national bird) and other species that are rare or endangered in many other high mountain areas of Nepal.

Sacred trees and forests are an integral part of the landscape of Khumbu and an expression of the historical depth of Sherpa Buddhist faith. Villages and fields are dotted with trees believed to be the homes of *lu*, spirits worshipped by particular families who pass down the caretaking of the tree, spirit, and shrine through generations. Temples are surrounded by sacred groves. Other forests, called *lama nating* (lama's forest) are set apart because certain lamas sanctified them as places where no tree must be cut and into which no cutting implement might be taken. These sacred forests and groves were the earliest, most strictly regulated, and most enduring of the protected forests. They encompassed substantial areas of trees in what were often prime sites for harvesting timber and fuel-wood adjacent to villages.

Sherpa sacred forests may have evolved from Sherpa and Tibetan beliefs about the spirits known as *lu*. Sherpas believe that several types of these half-human, half-serpent, female spirits live in Khumbu, inhabiting springs, boulders, trees, shrines and houses. The *lu* of springs control the flow of water and can, if offended, withhold it. Most houses have a *lu* and a special shrine is built inside the home for it. These are generally small stone shrines tucked in unfrequented corners of the house, usually in a lower-storey corner. Such *lu* can influence family health and luck for good or ill and hence must be very carefully respected and given regular offerings. Tree *lu* sometimes live in trees near springs and sometimes in forests. It is said that these forest spirits sometimes follow people home and take up residence in trees near their house. *Lu* trees within villages are a distinctive phenomenon. These may be few or many and can be of any species. Each belongs to one of the families whose house is near by and the women of that family have responsibility for carrying out rites at it. Often a small shrine is built at the foot of the tree.

Beliefs in *lu* are widely shared by many different Sherpa groups in north-east Nepal. Khumbu, however, seems to be particularly densely populated by *lu*, for nowhere else in the Sherpa world are *lu*-inhabited trees so common. The trees are usually old juniper (*Juniperus recurva*), but occasionally rhododendron, willow or fir which so distinctively dot Khumbu villages and fields. Many have shrines at their base, but even when *lu* trees are not marked by shrines they are still trees that are well-known to all villagers and which they treat carefully. To offend a *lu* is considered dangerous, for serious illness and other family misfortune can follow.

The beliefs and bans surrounding what is reputedly the earliest sacred forest, the grove at Pangboche, greatly resemble those associated with *lu*-inhabited trees. It is believed that to cut trees in the grove brings bad luck and illness. The juniper trees near the Pangboche temple are said, in fact, to be inhabited by *lu* as well as to be sacred because they were created four centuries ago by the great early Sherpa religious hero, Lama Sanga Dorje, who here scattered a handful of hair to the wind which, on falling back to earth, took root as forest.

The largest sacred forests in Khumbu are those that were established by the personal intervention of revered local religious leaders. There are several examples of such lama's forests, including those in the vicinity of Yarin, Phurtse, Mong and Mingbo. The Yarin forest, said to have been declared a protected forest by Lama Sanga Dorje, is the largest of these lama's forests, extending more than a kilometre along the upper Imja Khola. The smallest, a tiny grove near Mingbo, was protected only in the late nineteenth century by Lama Zamde Kusho, who then had a hermitage in the area.

The most famous of the lama's forests among the Sherpas is at Phurtse, where all remaining forest within a particular area was declared protected forest by a lama well over a century ago. According to the legend, a lama who lived in a hermitage just above the village warned villagers during an especially powerful ceremony that grave misfortune would attend the family of anyone who took any sharp instrument into the forest, be it an axe, a *kukri* (the Nepali machete) or a knife. This prohibition apparently also originally included the forest on the other side of the Dudh Kosi below Mong. Much of the Phurtse lama's forest has been very stringently protected, and there at 3,800 metres there are more than ten hectares of birch, some probably a century-and-a-half old, whose gnarled and mossy limbs show no sign of ever having been lopped.

Trees growing around temples are respected, and small groves are associated with the temples of Khumjung and Kerok and several small monasteries and nunneries, most notably at Thami Og and at Tengboche. Temple and monastery groves are much smaller than lama's forests. The largest, Kerok, encompasses little more than a hectare of old birch and rhododendron. There is a conspicuous lack of oral traditions regarding these groves, suggesting that these trees acquired sanctity from the founding of the temples rather than the temples themselves having been sited in already sacred groves. There are also no oral traditions that these Khumbu temple groves were planted (unlike those in some other Sherpa areas). It thus seems possible that some temple groves are relics of once-more-extensive forests or woodlands. Felling trees in temple groves is considered an inauspicious act, and no special forest guards are considered necessary. Few trees in such groves show signs of having had limbs lopped.

Sherpas did not extend the same degree of conservation consciousness to the region's non-sacred forests. Some of these other forests were carefully managed to provide continuing sources of highly-valued products such as house beams, but most timber and fuel-wood was obtained from unprotected, unmanaged local forests and woodlands. Oral traditions of historical degradation and clearing of these heavily-used areas contrast sharply with the extraordinary respect with which sacred forests were protected. The first Khumbu protected forests, and those that have remained the best conserved, were created not out of concern with the supply of forest products or avoiding immediate environmental calamities but out of a community dedication to honouring sacred places and trees.

The sacred trees and forests that are such striking elements of Sherpa village landscapes represent substantial gestures of faith in a land where trees are so useful and so scarce. Throughout the region, Sherpas largely continue to protect them as carefully today as they have for generations, despite population pressure, tourism, forest nationalization in the 1960s, national park forest regulation since 1976, and changing local life-styles and values. Some sacred groves and forests have been affected by traditionally-authorized grazing, the collection of leaves and conifer needles for compost from the forest floor, and in some cases by the collection of dead wood. Sherpa protection of sacred forests has overall, however, superbly safeguarded forests which are often among the region's oldest and least altered, and in some cases are the last remaining forests in their localities. The largest of these sacred forests provide important habitat for musk deer and other wildlife. And sacred forests, along with the practice of protecting individual *lu* trees, have been critical to the survival of many of the oldest and largest individual juniper, fir, birch and rhododendron trees in the region. Today, as for centuries, Sherpa protection of sacred trees, groves and forests continues to be a crucial contribution to the conservation of biological diversity in what is now one of the world's most renowned mountain national parks and World Heritage Sites.

[Adapted from Stevens, S. F. (1993) *Claiming the High Ground: Sherpas, Subsistence and Environmental Change in the Highest Himalaya*. University of California Press, Berkeley and Los Angeles.]

Badrinath: pilgrimage and conservation in the Himalayas (A. N. Purohit and Edwin Bernbaum)

In 1993, scientists from the G. B. Pant Institute of Himalayan Environment and Development (GBPIHED) visited Badrinath, the major Hindu pilgrimage place in the Indian Himalayas, and noticed how the surrounding slopes and valley had been stripped bare of forest. More than 400,000 pilgrims a year come to Badrinath from all over India, arriving on roads built in the early 1960s. Their influx has had a substantial impact on the local environment. At the suggestion of the scientists, the Chief Priest agreed to use his religious authority to encourage pilgrims to plant seedlings for the restoration of the site.

A tree planting ceremony was organized on 16 September 1993, with the enthusiastic participation of pilgrims, local priests, holy men, mendicants, villagers, local residents, military, government officials and other interested parties. A number of outside dignitaries, including the Secretary of the Ministry of Environment and Forests of the Government of India, attended and spoke at the event. The Chief Priest blessed the seedlings and gave an inspiring talk highlighting religious beliefs and myths about the spiritual importance of trees.

One of these myths with strong environmental implications for the entire Himalayan region relates the origins of the Ganges. A sage prayed to the goddess of the sacred river to come down from Heaven. As an excuse to stay put, she claimed that the force of her descent would shatter the earth. Shiva, the form of the supreme deity most closely associated with the Himalayas, stepped in to break her fall with the hair on his head. The Chief Priest noted that Hindu texts identify the trees of the Himalayas with Shiva's hair. In the summer the Ganges does indeed fall from Heaven in the form of monsoon rains, and when the Himalayan forests are stripped away, the earth literally shatters under the water's impact with landslides and floods. He concluded by encouraging the participants to take the seedlings and plant them as an act of devotion to Lord Shiva, regenerating his sacred hair and protecting the environment. They responded with enthusiasm and planted an estimated 20,000 trees.

Because of heavy snowfall and a lack of adequate selection, care and acclimatization of the trees, most of the seedlings planted at the ceremony in 1993 did not survive the next two winters. The weight of snow and the shearing effect of sliding crust broke off the tips of many of the plants, and those that survived showed little or no growth. In May 1995, GBPIHED started a tree nursery down the valley from Badrinath at Hanumanchatti, at an altitude of 2,500 metres, to raise, harden and acclimatize seedlings for plantation in the Badrinath area.

On 5 June, a second ceremony was organized at Badrinath Temple to plant broad-leaved trees just behind and above the shrine. According to rough estimates, about 250 people participated in the ceremony and a number carried plants up to the higher plantation sites.

The original intention was to plant trees to help control erosion on slopes threatening the shrine, but a controversy over earlier use of this land by the government forced most of the planting to take place on private land owned by *pandas* or local priest-guides. A few seedlings were planted on the edges of the area under dispute and on steep avalanche slopes nearby. Metal guards placed around the latter were flattened and twisted by avalanches the following winter.

During the summer of 1995, trees selected from the most promising species and acclimatized at the Hanumanchatti nursery were planted in various sites around the valley, including a major *ashram* or pilgrim guest-house in Badrinath. More than 90 percent of them survived up to the end of the fall before the closure of the area for the winter. GBPIHED scientists developed special tree covers and guards and installed them on plants in November. As a result, more than 80 percent of the broad-leaved trees planted survived the winter snows and were alive when people returned in the spring of 1996. Everyone but the military and a few holy men abandons the area during the winter.

In 1996, more trees were planted at the Garhwal Scouts' Camp outside Mana and plantation began at the nearby Sikh Regiment Camp. In order to promote the theme of restoring Badrivan, the ancient sacred forest of Badrinath, each plantation site in the valley is named for the local community or group and identified with Badrivan.

In September of that year, another tree-planting ceremony took place, at Hanumanchatti, a shrine dedicated to the monkey god Hanuman and guarding the approach to Badrinath. After speeches by various dignitaries including a *swami*, several scientists and a brigadier, a well-known holy man, revered throughout India for his asceticism and spiritual powers, blessed and dispensed seedlings to an eager crowd of about 500 pilgrims, villagers, priests, holy men, government officials and military personnel. Villagers of the local village had donated land on a small sacred mountain next to Hanumanchatti, and people proceeded there to plant 1,500 trees in pits that scientists and villagers had prepared in advance.

Although the number of trees actually planted has been relatively small, and implementation of certain intended measures has yet to be achieved, the programme at Badrinath has been extraordinarily successful in changing local opinions about reforestation. The failure of attempts to reforest the area before 1993 had convinced the vast majority of the population that it could not be done. The ceremonial planting of seedlings blessed by the Chief Priest in 1993 generated a great deal of enthusiasm. After seeing how trees have survived with special covers to protect them from snow, nearly all the population is now in favour of tree planting and thinks the programme to re-establish the sacred forest of Badrivan will succeed.

The next step will be to strengthen the programme at Badrinath and to extend it to other sites and conservation measures. Some specific measures that have been suggested by various stakeholders include cleaning up litter and disseminating seeds for pilgrims to plant back in their home communities. Intended as a collaborative effort including all the concerned parties, particularly the local people, the expanded programme will be used to develop and test guidelines for replicating the approach at other sites in the Himalayas and, where appropriate, elsewhere in the world.

Sacred mountains of China (Martin Palmer and Tjalling Halbertsma)

Until this century, the great Taoist sacred mountains of China were immune to most changes down on the plains. Legendary homes of gods, goddesses, immortals and sages, they were protected both by the care and attention of the Taoist monks on the mountains and by the reverent respect afforded these vast mountain ranges as sacred sites by the ordinary people of traditional China. This was important not just for the spiritual well-being of the people, but also environmentally, for the sacred mountain ranges – some over 320 kilometres (200 miles) long – are crucial watersheds and diverse habitats for flora and fauna.

This century has seen an unprecedented attack on these mountains and the faith that both sustains them and is sustained by them. The collapse of Imperial China in 1911; the coming of Westernizing forces in the Republic who saw Taoism as old-fashioned; the warlord period of *c.* 1920–1936; the Civil War between Communist and Nationalist forces; the Japanese invasion, and finally half a century under frequent attack from the Communist Party and rulers of modern China, has broken and crushed much of traditional Taoism. During the excesses of the Cultural Revolution (1966–76) almost all the temples, shrines and hermitages were destroyed or looted. Ancient sites of natural beauty were polluted or damaged, and for many years on many mountains no Taoists were allowed to live or worship. Meanwhile, changes in forestry, agriculture, urban development and latterly internal tourism have begun to seriously intrude upon the formerly sacrosanct sacred mountains.

By 1993 the Taoists had been allowed to return to many of their sacred mountains. They have regained many of their former temples and monasteries but they have not been given back control of the land on the mountains. With little access to the bodies controlling most mountains, and virtually no standing or links overseas beyond the Chinese Diaspora, they were unable to have more than a very marginal impact on the way the mountains were being developed. This was perhaps most tellingly shown in the Chinese Government's papers submitted to UNESCO when seeking World Heritage status for the most holy of Taoist mountains, Mount Tai. Taoists were not consulted, nor did UNESCO inquire after them.

This invisible presence was the status quo until the Alliance of Religions and Conservation (ARC) joined forces with the China Taoist Association (CTA) in 1995 to help protect the sacred mountains. There were many who said, in the first stage of this project, that it would be impossible to get such sensitive issues as religion and ecology – both of which are problematic for the Chinese Communist Party – linked together with full government approval. However, ARC's ability to bring gentle international pressure to bear and to ensure full involvement of the key religion made this possible.

But it is Taoism itself that provides an effective approach for the preservation of sacred mountains. In its declaration, the CTA stipulated the importance of biodiversity from a religious point of view: Taoism has a unique sense of value in that it judges affluence by the number of different species. It is this notion of diversity that for the Taoists has provided the structure for the religious dimension of the sacred mountains project.

This respect for diversity and the natural environment in general is represented in a diverse pantheon of Taoist deities such as earth gods, or those protecting clean waterways, mountain sites, etc. and is embedded in principals such as *wuwei*, or active non-interference with natural processes and ways, for Taoism follows the *Tao*, literally 'The Path' or 'The Way'.

While this approach is effective for nature conservation on sites where local communities value the environment as sacred through religious and spiritual standards, this diversity of species is also valued by Taoists as crucial on an even wider scale, for as the CTA put it, if all things in the universe grow well, then a society is a community of affluence: if not, this kingdom is on the decline.

There is one final point worth making, and one which we have made forcefully to UNESCO, one of the organizations working on sacred mountains through the World Heritage Programme. Making strategic alliances with governments to preserve environmental or cultural features is short term. However, for thousands of years the Taoists have been on these mountains. They will be there in thousands of years' time. Developing an environmental programme with such religious communities is a better long-term strategy than working with transient political structures. In China, with the Taoist Sacred Mountains, this is already bearing fruit.

[The China Sacred Mountains Project is jointly run by CTA and ARC. Project papers, the Taoist faith declaration and further information are available from: Alliance of Religions and Conservation, 9A Didsbury Park, Manchester, M20 5LH, UK]

The Rocky Mountains: sacred places of the Stoney People (Adapted from Chief John Snow, 1977)

Long ago my ancestors used to go to the mountain tops to pray. They were a deeply religious, sincere, tradition-oriented people, who followed, observed and upheld the teachings, customs and beliefs of our forefathers, respected the creations of the Great Spirit, and lived in harmony with nature. They were Stoneys – members of the Great Sioux Nation who spoke a dialect of the Nakota branch of the Siouan language family. Today we, their descendants, speak the same tongue.

In order to understand the vital importance the mountains had – and still have – it is necessary to know something of our way of life before the coming of the white man. It is not enough to say the mountains were the Stoneys' traditional place of prayer because our life was not a fragmented one with a compartment for religion. Rather, our life was one in which religion (and reverence for nature, which revealed religious truth) was woven throughout all parts of the social structure and observed in conjunction with every activity.

In the days prior to the coming of the white man, we lived a nomadic way of life, hunting, fishing and gathering from the abundance of this good land. There were literally millions of buffalo roaming on the western prairies, along the foothills, and even into the Rocky Mountains themselves. There were game animals of all kinds – moose, elk, deer, wild sheep and goats – readily available for us to hunt and to enjoy. The land was vast, beautiful and rich in abundant resources.

We talked to the rocks, the streams, the trees, the plants, the herbs and all nature's creation. We called the animals our brothers. They understood our language; we, too, understood theirs. Sometimes they talked to us in dreams and visions. At times they revealed important events or visited us on our vision quests to the mountain tops. Truly, we were part of and related to the universe, and these animals were a very special part of the Great Spirit's creation.

Some of the most exciting and unusual legends and stories in the history of the Stoneys concern Iktumni, a great medicine man, a wise teacher, a prophet. Iktumni was wise because he was able to talk and commune with both the natural world and the spiritual world. He was able to interpret the animal world to the tribal members in human terms and understanding. The animals and birds were his brothers and he lived with them, so he was able to talk with them in their own language. The rocks and trees were also his brothers, so he communed with them and learned the secrets and ways of nature. He communicated and talked with the spirits of the mountains, who revealed ancient truth, philosophies and prophecies to him. In turn, he taught these things to the Stoney people.

Iktumni was perhaps the most famous of all the Stoneys in finding truth in nature, but he was by no means the only such seeker or even the only successful one. Indeed all the Stoney people were continually searching for the truth by observing the universe around them.

The most sacred search was a special religious journey into the rugged mountains, seeking wisdom and divine guidance. This was known as the 'vision quest', a tradition handed down through the centuries and practised by us as a means of approaching the Great Spirit. If the seekers were favoured, the Great Spirit would deliver a revelation and thus give direction and guidance to our tribe.

Sacred ceremonies and rituals were observed by these seekers of truth before they journeyed into the rugged mountainous country. In this preparation they were guided and aided by many members of the tribe who spent much time fasting and praying in the sweat lodge.

After the purification ceremonies were prayerfully observed in the sacred lodge, the seeker of truth and insight into religious thought would prepare to set off on the vision quest. There in the mountainous wilderness he would be alone; he would live close to nature and perchance he would receive a special revelation. It might come through a dream or a vision, through the voice of nature, or by an unusual sign. It might be that wild animals or birds would convey the message of his calling to him.

Many a brave has sojourned to these sacred mountains of ours in search of his calling – the purpose for which he was born. He searched in hope that the Great Spirit would make known to him his future task so that he could take his place in the tribal society, and help his people.

Century after century the rugged Rocky Mountains sat here in majesty and nature seemed to say, 'Your thoughts must be as firm as these mountains, if you are to walk the straight path. Your patience and kindness must be as solid as these mountains, if you are to acquire understanding and wisdom'.

Upon these lofty heights, the Great Spirit revealed many things to us. Some of my people received powers to heal. They could heal the physical body with herbs, roots, leaves, plants and mineral spring waters. They could also heal the broken and weary soul with unseen spiritual powers. Others received powers over the weather. These gifted religious men and women could call for a great storm or calm the weather; they could call on the winds, rain, hail, snow or sleet, and they would come. From these mountain-top experiences my fellow tribesmen and women were given unique tasks to perform to help the tribe prepare for things to come.

Therefore the Rocky Mountains are precious and sacred to us. We knew every trail and mountain pass in this area. We had special ceremonial and religious areas in the mountains. In the olden days some of the neighbouring tribes called us the 'People of the Shining Mountains'. These mountains are our temples, our sanctuaries, and our resting places. They are a place of hope, a place of vision, a place of refuge, a very special and holy place where the Great Spirit speaks with us. Therefore, these mountains are our sacred places.

[Adapted from Snow, Chief John (1977) *These Mountains are our Sacred Places: The Story of the Stoney People*. Samuel Stevens, Toronto and Sarasota.]

Of telescopes, squirrels and prayers: the Mt. Graham controversy (Adapted from Evelyn Martin, 1993)

If you want to travel from Mexico to Canada, but have only half a day to spend, try visiting Mt. Graham, 75 miles (120km) north-east of Tucson. This 'sky island' hosts the visitor to five of America's seven biological zones. Ascending from the Lower Sonoran desert to the Hudsonian spruce-fir forest, the mountain is home to 18 plant and animal species found nowhere else. But as you near the 10,720-foot (3,267-metre) summit, you will be in a battle zone between heaven and earth.

Fanning the flames of controversy is the University of Arizona's plan to build the Mt. Graham International Observatory. The kind of research to be carried out on the mountain is the stuff of which scientific dreams are made: studies of the early universe and galaxy formation from up to 10 billion years ago, star and planet formation in the Milky Way and other galaxies, and the search for other planetary systems. These studies would be conducted in collaboration with the Vatican Observatory and other international partners.

The university has telescopes on mountains such as Kitt Peak near Tucson. But explosive population growth has created 'light pollution', and the new wave of advanced telescopes requires higher altitudes than are found at existing sites. To house its own new telescopes, the university originally surveyed 280 potential sites in the United States. Of three astronomically-preferred locations, Mt. Graham alone was not in a designated wilderness or national monument.

However, an analysis by two scientists at the National Optical Astronomical Observatories indicated that Mt. Graham's 'merit ranking' was only 38 out of 57 sites reviewed. But Peter Strittmatter, Director of the university's Steward Observatory, praises the site's excellence: 'Mt. Graham is expected to at least match the best image sharpness demonstrated anywhere in the continental U.S. People around here are all dreaming of the day it will open'. But others are dreaming of the day the telescopes will be torn down.

The Mt. Graham red squirrel (*Tamiasciurus hudsonicus grahamensis*) is an endangered subspecies whose only earthly habitat is the summit area of Mt. Graham. Genetically isolated within its sky island, it faces a precarious future as a result of the previous logging of some of its habitat. The mammal is one of 25 currently-recognized subspecies of red squirrel in North America. Listed as an endangered species in 1987, the average estimate of its population declined to a low of 123 in 1989, but rose steadily to a high of 377 in spring 1992, before dropping to 332 in the fall.

What has so vehemently fuelled environmentalists' concerns is the way that Congress went about approving the observatory. Prior to the issuance of the Forest Service's final environmental impact statement, and based on a biological opinion of the U.S. Fish and Wildlife Service, Congress attached a rider to the 1988 Arizona-Idaho Conservation Act (AICA). The rider stated that provisions of the Endangered Species Act and the National Environmental Policy Act shall be deemed satisfied with respect to the observatory.

Mark Hughes, staff attorney for the Sierra Club Legal Defense Fund, comments, 'AICA was completely irresponsible, a perversion of the system'. In 1990, observatory opponents forced Congress to hold oversight hearings. The two Fish and Wildlife Service biologists who had written the biological opinion testified that they had been ordered, against their professional judgement, to conclude that the observatory would not harm the squirrel. And their regional director was caught explaining to Congress that he had allowed non-biological information – that is, the prestige and importance of the observatory – to affect the finding, contrary to provisions of the Endangered Species Act.

For three years, the Sierra Club Legal Defense Fund has lost most of its court challenges because of judicial support of AICA. Two remaining counts concern what Hughes calls the university's scientifically-indefensible monitoring of the red squirrel. According to Steve Emerine, Associate Director of the university's public information office, the university has, during the first three years, spent double the required minimum of $100,000 per year on monitoring. 'The most recent squirrel numbers show that the observatory is not having a negative effect', notes Emerine. 'Although the fall 1992 numbers were down somewhat from the earlier numbers, the recent estimate is still about double that of 1989.' But Hughes claims that every party *except* the university agrees that clearing part of the habitat harms the squirrels.

Adding fuel to the controversy over Mt. Graham is the fact that the mountain is sacred to the San Carlos Apache Indians. *Dzil nchaa si an*, or Big-Seated Mountain as traditional Apaches call it, is home to the *ga'an* spirit dancers. The Apaches want their prayers to flow unimpeded from the summit.

Beginning at least in the 1600s, several Apache tribes lived and travelled in the area around Mt. Graham. During the so-called Indian Wars in the mid-to-late 1800s, the mountain variously provided sanctuary to the Apaches and served as a summer recuperation area for ailing U.S. soldiers and their families. Reflecting Apache territorial patterns, the original San Carlos Apache Reservation encompassed *Dzil nchaa si an*. But, as happened so often, the U.S. subsequently decided it wanted the land – the mountain for logging, and the adjacent Gila River valley for mining and agriculture. In 1873, the Apaches lost their mountain.

Ola Cassadore Davis, a San Carlos Apache, has matched the astronomers' figurative dreams of scientific inquiry with her own literal dream of saving the sacred mountain. Thus was formed the Apache Survival Coalition, with Davis as chairperson.

Dzil nchaa si an is 'of vital importance for maintaining the integrity' of cultural traditions, according to a 1990 resolution passed unanimously by the San Carlos Tribal Council. Elders, as well as medicine men and women, use the mountain for religious activities. The *ga'an* spirit dancers teach the medicine people how to heal the sick through song and prayer, and how to apply special herbs and plants found only on *Dzil nchaa si an*. In addition, the mountain contains ancient religious shrines and burial grounds.

What is difficult for outsiders to appreciate, though, is that the entire mountain, not just a given site, is considered sacred. 'It's the mountain that gives life, sustains life', emphasizes coalition vice-chairman Ernest Victor, Jr. 'If God created what is on the earth, holy places, it is wrong for man to desecrate them.' The Apaches accept for now the mountain's other uses. What makes the observatory different, claims Victor, is that 'nature will reclaim the other uses. The communications towers have only a small amount of cement, and the tree roots will go through the pavement. But the concrete of the telescopes will be there for centuries – it will be scarred for centuries.'

Coming under special attack by the coalition is the Vatican, for two reasons. The first is the irony by which the Vatican is building its telescope, when Pope John Paul II visited the south-west in 1986 and urged American Indians to be bold and fight for their land rights. The second is the affront to traditional religious interests that the Apaches perceive in the Vatican's official statement. Reverend George Coyne, Director of the Vatican Observatory, writes, 'We are not convinced ... that Mt. Graham possesses a sacred character which precludes responsible and legitimate use of the land.... there is to the best of our knowledge no religious or cultural significance to the specific observatory site'.

Exacerbating the Apaches' uphill fight is the fact that they themselves did not take their plight public until 1989, in part because Native Americans are extremely reluctant to openly discuss sacred matters. Although Forest Service records show numerous notifications, starting in 1985, about the environmental reviews, San Carlos tribal records do not indicate that the notifications were received. Yet, according to the Forest Service's Abbott, 'It's hard for me to imagine that anyone was not aware of the proposed observatory'. In 1987, a group called the Friends of Mt. Graham did send the Forest Service a letter indicating that traditional Apaches used the mountain for ceremonial purposes.

The coalition's lawsuit against the Forest Service cites provisions of the American Indian Religious Freedom Act (AIRFA), the National Environmental Policy Act, the National Historic Preservation Act, the National Forest Management Act, and the First and Third Amendments of the U.S. Constitution. The gist of the lawsuit concerns the broad element of religious freedom, and the technical element of the requisite cultural surveys. However, the toothlessness of AIRFA has already resulted in the U.S. Supreme Court placing strict interpretations on religious freedom, including access to sacred sites. An archaeological survey was conducted for the Forest Service by the university, but critics emphasize that this survey is insufficient and misses the point, especially in the case of Apaches, who do not leave physical evidence of their shrines.

University spokesman Emerine hopes that some access solution can be found. To the coalition, the issue is not access, which already exists, but respect for the mountain. For, as Apache Burnette Rope stresses, 'If the spirits leave, we don't know where they'll go'. Adds Victor, 'It's very hard to say what will happen'.

At the heart of the controversy is the failure of America to come to grips with broad issues of land ethics. Aldo Leopold urged us to become plain members and citizens of the land community, rather than conquerors of it. The Vatican Observatory was established in part to acknowledge the centuries-old criticism of the Holy See forcing Galileo to deny that the sun, not the earth, was the centre of the solar system. Now, the debate is increasingly about whether humans are at the centre of the earth community, or are one among many at the centre.

The Mt. Graham controversy presently pits astronomers who revere the skies through technology against a small mammal with a cloudy future, and against Apaches who engage the heavens with only their eyes and hearts. Mainstream society locks away its equivalent of sacred lands in wilderness designations. But until our land ethic also embraces whole ecosystems, sacred Indian lands, and other areas that are the foundation of both our physical and spiritual existences, controversies like Mt. Graham will continue to be addressed in less than comprehensive fashion.

Environmental conservation based on sacred sites (Thomas Schaaf)

When discussing nature protection, one generally thinks of legally protected areas (national parks, strict nature reserves, forest reserves, etc.). Although these legally designated areas are important and have many merits, they are often met with incomprehension and even opposition by local people. Land that could be used for direct economic usage is suddenly protected by law. This creates dangerous land-use conflicts between economic interests and nature protection interests.

Land-use conflicts between nature protection and economic utilization exist throughout most industrialized countries as well as in the developing world. In the latter, however, the problem is all the more profound since the concept of legally designated and protected areas is by and large a Western one which is not always understood by non-Western societies. With the designation of 'Yellowstone National Park', the world's first legally designated national park came into being 125 years ago. Before the arrival of the white settlers, this area had only been very sparsely populated and the life-styles of the Indian population had only marginal impacts on the environment. However, the designation of Yellowstone as a national park had one important conceptual consequence: legally protected areas generally exclude human populations.

How, then, can environmental conservation best be practised in non-Western societies without reverting to strictly legal and largely Western-oriented concepts? It appears that sacred sites – many of which are sacred mountains – can perhaps provide an alternative and innovative approach to environmental conservation. Is it possible to base environmental conservation on traditional values and religious beliefs which are of direct cultural relevance to people in developing countries?

In many parts of Africa, Asia and the Pacific, as well as in Latin America, sacred sites exist with inherent spiritual or religious significance. They can be abodes of gods or burial grounds of ancestors. Access to such sacred sites is generally taboo and/or restricted to a small circle of persons (priests, pilgrims). Due to access restrictions to the sites, sacred places have often preserved a very rich biological diversity. Often very visible in the landscape, vegetation-rich sacred sites can differ dramatically from adjoining degraded environments that are not sacred. Many sacred mountains are plant or wildlife *sanctuaries* (the English term expresses this concept very well) as the picking of plants or the hunting of animals is generally not permitted on holy ground. They are veritable genetic reservoirs.

For ecological research and nature protection, such sacred sites are of great value since they reflect the state of largely unimpacted and 'intact' ecosystems. They can serve as reference areas for assessing the potential natural vegetation of a whole region. One can generally assume that the plant species occurring in the sacred sites are native species (eventually even endemic species) which are well adapted to the local climatic and edaphic conditions.

As already evidenced through an earlier UNESCO project on sacred groves in the savannah of Ghana, sacred sites can also be used for the rehabilitation of degraded environments. Afforestation schemes that included the establishment of fodder banks for livestock and the planting of cash crops at the periphery of the sacred groves have helped to enlarge the sacred groves through an additional buffer zone around the holy site.

Of even higher importance is that an approach that is based on cultural and religious values will probably result in greater sustainability. Many very well-meant conservation and development projects fall into pieces once external donor funds for environmental conservation have ceased to flow. Any type of intervention activity will be all the more sustainable the more it takes into account local traditions and socio-cultural conditions. Legal decrees or financial means can be subject to temporal limitations; a religious belief is timeless.... Religious beliefs as a determining factor in human behaviour have often been ignored by Western conservation scientists or development experts. Cultural identities need to be taken into account in the field of development aid and environmental conservation.

Basing environmental conservation on traditional religious beliefs will also lead to a recognition and better appreciation of non-Western cultural identities. In this respect it is essential that environmental and conservation scientists collaborate closely with anthropologists. Moreover, environmental education should also be considered since younger generations often show signs of disrespect for the values held sacred by older generations. This is particularly true for societies that are in a transformation process (e.g. through 'Westernization'). It would be good if environmental education were targeted specifically at younger people in informing them that the beliefs of the elders can be very beneficial for the preservation of the environment for future generations.

UNESCO is currently undertaking a study on sacred places (mountains, but also rivers and lakes) with a view to assessing their potential for enhanced environmental conservation. This trans-disciplinary project will include ecologists, anthropologists and environmental conservation experts. Several sites from Africa, Asia and Latin America will be analysed through a comparative methodology. Some sites will be taken from the World Network of Biosphere Reserves and from the World Heritage sites, which are two networks serviced by UNESCO. Many of these sites are sacred mountains (e.g. Machu Picchu World Heritage site in Peru, Changbaishan Biosphere Reserve in China). However, we will also consider many sacred sites that have no legal protection but which are solely preserved through their religious significance. We hope that we will then be able to formulate guidelines for the enhancing of environmental conservation through religious beliefs and cultural values.

Box 8.1: 'The Land is Yours'

Roger Pullin

The land is yours, but the morning is mine
Walking your mountainside in the golden sunshine.
The land is yours but I don't envy you.
You'll see no fence round me, my possessions are few.

The land is yours, but you don't respect it.
It's just another figure in a book.
It will turn on you, if you neglect it.
It will pay you back for everything you took.

The land is yours, but it's our children's dream
To journey round the world and then come home again.
Before the madmen came, to ask where they'd been,
They could trespass on the mountains of a thousand different men.

The land is yours, but you don't respect it.
It's just another figure in a book.
It will turn on you, if you neglect it.
It will pay you back for everything you took.

The land is yours but the time marches on
And the things that you leave will be here when you're gone.
The land is yours but the morning is mine.
Walking your mountainside in the golden sunshine.

CHAPTER 9

FORESTS, CULTURE AND CONSERVATION

Sarah A. Laird

CONTENTS

TEXT BOXES

FIGURES AND TABLES

Forests, Culture and Conservation

Sarah A. Laird

FORESTS AND CULTURE HAVE INTERTWINED THROUGHOUT human history. Forest landscapes are formed and are strongly characterized by cultural belief and management systems, and cultures are materially and spiritually built upon the physical world of the forest. Just as people have acted upon and altered forests throughout history, so too have forests profoundly influenced human consciousness and culture. As a result of these long histories of co-development between forest ecosystems and local cultures, cultural and biological diversity are intimately connected and cannot effectively be separated from each other. Programmes designed to achieve the conservation of biodiversity, therefore, must be undertaken with a full grasp of the cultural context within which forests exist.

'Culture', as it broadly relates to forests, includes what Balée (1994) refers to as a 'mental economy'. This mental economy is composed of the elements forest-dwelling communities employ to strategically manipulate the natural environment. It includes not only traditional ecological knowledge, such as that on relations between species of flora and fauna, edible versus inedible foods, and forest management systems, but also shared notions of kinship, marriage, taboos, cosmology and ritual.

Traditional forest management relies on a complex of ecological processes, management of 'eco-zones', and a variety of habitats to maximize the range of products and services that forests can provide. These systems are highly site-specific, diverse and ever-changing, and generally promote and require the maintenance of forest structure and species diversity (Gomez Pompa and Burley 1991; Alcorn 1990; Posey and Balée 1989; Anderson 1990; Balée 1994; Posey 1997; Padoch and de Jong 1992; Redford and Stearman 1993). In fact, Balée (1994) defines 'management' in reference to the Ka'apor in Brazil, as '... the human manipulation of inorganic and organic components of the environment that brings about a net environmental diversity greater than that of so-called pristine conditions, with no human presence'.

Traditional systems of management have shaped forests throughout the world, and many so-called 'natural' forests in fact represent arrested successional forest once managed by people. For example, Posey (this chapter) describes a system of resource use and management, and the cultural landscapes, of the Amazonian Kayapó Indians, and cites examples throughout the region whereby presumed 'natural' ecological systems are products of human manipulation. The Kayapó initiate and simulate the formation of islands of forest, or *apêtê*, in the *campo-cerrado* (savannah), through the careful manipulation of micro-environmental factors, knowledge of soil and plant characteristics, and the concentration of useful species into limited plots. The integration of forest and agricultural management systems depends upon the manipulation and use of transitional forests, and agriculture is but one phase in a long-term forest management process.

The creation of mosaics of 'eco-units' by the Mexican Huastec and Peruvian Bora farmers means that at any given time lands are open for agriculture, secondary successional species are reproducing elsewhere in the mosaic, and in another unit mature forest species are reproducing. In this way, the elements necessary to regenerate forests are retained in the system (Alcorn 1990). Ka'apor management in the Brazilian Amazon also involves managed vegetation zones in different phases of recovery, from house gardens to young swidden, old swidden and fallow. Within these diverse vegetational zones exist environments in which unusually high densities of useful plants and animals exist (Balée 1994).

In the Jharkhandi region of India, the landscape is managed on a continuum that does not recognize breaks between farm, forest and settlements (Parajuli, this chapter). In parts of Ghana, it is similarly difficult to distinguish between 'forested' and 'cultivated' land, and generally there exists a gradient of vegetation types passing from recently cleared farm land to secondary forest in farm fallows, to forest stands that have not been cleared for hundreds of years (Falconer, this chapter).

Traditional management and knowledge systems also include regimes to sustainably harvest and process materials from individual species. For example, *Ficus natalensis* and *F. thonningii* bark harvesting in Buganda involves elaborate protocols, including the tying of banana leaves to the de-barked tree, followed by the application of cow-dung. Subsequent beating of the bark to make cloth is an extremely complex process, and also the result of long experience of traditional use (Kabuye, this chapter).

Integral to traditional forest management is the use of elaborate taboos, myths, folklore and other culturally-controlled systems which bring coherence and shared community values to resource use and management. Today, under pressure from urbanization, cash economies and other socio-economic, political and cultural changes, many of these systems are breaking down, to the immediate detriment of forests and valuable species. But in many areas these forms of management and control remain strong.

In the eastern Amazon, for example, hunters who are greedy, or do not heed the wishes of the giant cobra, giant sloth and the *curupira*, will become lost, or otherwise suffer some punishment. These creatures require respect for the forest, and what might be called 'sustainable' approaches to the harvesting of game and plants (Shanley and Galvão, this chapter). Interdicts, totems and sacred areas based on religious beliefs, and implemented by religious chiefs, are widely used throughout West Africa to control the use and management of forests and resources. For example, in Benin the panther is venerated by the Houegbonou clan, who can neither kill nor eat it. The Houedas people adore, protect and breed pythons. Fon and Torris people identify monkeys with twins, which leads to the protection of some monkey habitat. *Chlorophora (Milicia) excelsa* – iroko – is a sacred tree protected and revered throughout West-Central Africa (Zoundjihekpon and Dossou-Glehouenou, this chapter).

On the western edge of the Gola Forest Reserve, the largest element in a complex of four surviving areas of high rainforest on the Sierra Leonean side of the border with Liberia, villagers have a taboo against bringing the wood of *Musanga cecropioides* into the village for use as fuelwood. Richards (this chapter) describes this as a good example of 'the true understanding of biodynamic processes sometimes encoded in mythic knowledge'. In the Yoruba story *Trespassing on the Devil's Land*, the 'devil's land' is the forest, a dark and fearful place in which the bush spirits reside and where – if you dare to clear forest for farming – 'they will make trouble for you'. 'No one farms in the bush', the young character Kigbo is warned, 'and ill befalls him when he goes against this warning and clears the forest to plant his farm'.

Traditional systems of forest management can also yield valuable insights to foresters. The processes that foresters utilize as part of silvicultural systems can gain from management techniques employed by forest peoples for millennia. These include: repairing patches created from agroforestry systems and tree fall gaps; timing the creation of these gaps to promote desirable species; managing succession, including the selection and protection of desirable species and the weeding out of the undesirable; managing regeneration to promote fruit trees and species to attract wildlife; manipulating the scale and timing of agricultural and fallow phases to minimize pests and maintain soil fertility; planting forest groves for low-intensity product extraction on steep slopes and along streams to prevent soil erosion and protect watershed; the 'use' of birds, bats and other animals, to 'plant' forest groves; and enrichment of forests by transplanting propagules and seedlings of rare or non-native seedlings. To date, however, there are few examples of traditional management activities – developed over millennia by the inhabitants of extremely unique and complex ecosystems – being employed by the forester (See discussion in Laird 1995; Salick 1992a; Hartshorn 1990).

Traditional management and knowledge often conserve, and in some cases enhance, biodiversity. Given the pressing need to re-think and re-construct a new resource management science that is better adapted to serve the needs of ecological sustainability, Gadgil and Berkes (1991) and others have suggested that traditional management systems form the basis of sustainable ecosystem and biodiversity conservation. The unique complex of knowledge–practice–belief (Gadgil, Berkes and Folke 1993) found in traditional management systems (and demonstrated in many of the contributions to this chapter), can act as an effective basis for forest and biodiversity conservation.

It must also be recognized, however, that – for example in the case of Bastar district in Madhya Pradesh, India – the bulk of threats to forest land come in the form of commercial interests or large-scale development projects such as tourism, bauxite mining or proposed power plants and paper mills. In these areas of activity there is very little scope for the application of local TEK and management systems, and no real commitment is given to local communities' active participation or viewpoint (Sundar, this chapter). It is clear that, in addition to promoting sustainable livelihoods based on traditional ecological knowledge, there must exist a parallel effort to mitigate the negative impacts of powerful commercial interests on forest lands and resources.

The economic importance of traditional knowledge and forest management systems

Traditional knowledge of species use, and the management of forest products and ecosystems, provide numerous economic benefits for local communities, as well as having wider applications in regional and international economies. One powerful way in which traditional ecological knowledge and forest biodiversity contribute to local socio-economic conditions is in the provision of affordable health care. Traditional, largely plant-based medical systems continue to provide primary health care to more than 75 percent of the world's population. Of varying scale, age and formality, these systems produce effective, affordable and culturally-appropriate medicines. In many cases throughout the world, the erosion of traditional knowledge systems has led to an impoverishment of local medicine and a dependence upon expensive pharmaceuticals, one of the major cash expenditures of forest communities. A number of programmes have been established to shore up and build upon local knowledge of medicinal plants and existing medical systems in order to develop more effective health care systems within the context of local health, economic and social conditions (See Chapter 6).

A number of industry programmes are under way to research and record traditional uses of plants as a starting point for commercial product development by companies. These companies might come from the pharmaceutical, agricultural, personal care, cosmetic, fragrance, phytomedical, horticulture or other industries, each of which depends in distinctly different ways upon traditional knowledge and forest biodiversity. Some companies go directly in search of traditional knowledge and forest biodiversity to inform their research strategies, but the bulk consult the ethnobotanical literature, which is vast and growing, or use intermediaries to collect material. In general, large-scale commercial interest in forest biodiversity is based narrowly on species use, rather than management, beliefs or other manifestations of forest culture (ten Kate and Laird 1997).

The most important economic value for forests is far and away that held by local communities. Local forest communities make wide and varied use of forest resources and products, which are sometimes called NTFPs (non-timber forest products) or NDRs (non-domesticated resources – see Posey 1997). NDRs might include food, medicine, fodder, fuel and construction materials. Forest species are almost always of multiple use to local communities. In Southwest Province Cameroon, for example, a single tree species or shrub might be used for half a dozen or more of the following: timber, building and carving, food (e.g. fruit, edible oil or leaves), spice, dye, living fences or boundary markers, fish poison, attractants for game, insecticides, ornamentals, personal care (e.g. soaps, lotions, cosmetics) and for ceremonial use or personal protection (e.g. from witches, evil spirits, jealousy) (Laird and Sunderland, 1996).

Because NDRs are often part of subsistence or localized economies, they tend to be under-valued in comparison to large-scale, often unsustainable options such as timber, and large-scale clearing for agriculture. But when quantified, the story is quite different. A growing body of research and literature is demonstrating the greater value of NTFPs when compared with alternative land uses (e.g. Perez and Arnold 1996; Clay 1996; Balick *et al* 1996; Falconer 1992; Cunningham 1990; Abbiw 1990; De Beer and McDermott 1989; de Jong and Mendelsohn 1992; Nepstad and Schwartzman 1992; Peters *et al* 1989; Plotkin and Famolare 1992). In the Capim area of the eastern Amazon, for example, Shanley and Galvão (this chapter) found that non-timber forest resources are critical to the daily livelihoods of local communities, providing a significant number of products for subsistence use. Results of their research demonstrate that while entire trees are commonly sold as timber for $2.00 each, ten fruit of the same tree can earn $2.00 in the marketplace. Additionally, preliminary results indicate that the majority of NTFPs used over the course of one year derive from mature forest: 87 percent of the fruit, 85 percent of the fibres and 79 percent of the game.

Traditional ecological knowledge has a range of potential economic applications, from the identification of valuable medicinal plants, fragrances and foods, to silvicultural systems for complex tropical environments. This knowledge base could provide the foundations for sustainable forest management, alternative economic activities, and leads for the development of new commercial products (which might in turn create an economic incentive for the conservation of biodiversity and forests). Some efforts are being made to realize the economic and management potential of TEK, although it still remains to be seen if scientists can aptly apply the TEK they acquire from indigenous peoples to develop alternative models for sustainable forest management and use (Posey 1997; Convention on Biological Diversity, e.g. Articles 8, 11 and 15; Balick *et al* 1996; Nepstad and Schwartzman 1992; Gadgil and Berkes 1991).

Trees, forests and sacred groves – myths, lore and legends

In addition to the 'practical' and 'tangible' benefits resulting from the interface of culture and forests, there exist strong cultural and spiritual values for forests which do not take material form. These are, however, integral parts of the 'mental economy' of societies, and in many cultures cannot be clearly

Box 9.1: Forest Products as an Economic Alternative

International Markets

Clay (1996) has published lessons gleaned from the Cultural Survival Enterprises work to market forest products regionally and internationally in order to generate alternative, sustainable income for local communities from the forest. In order for the commercialization of products to contribute to sustainable development and conservation, he suggests the following *Twenty Lessons from the Field.*

- Land and resource rights are essential to both income generation and conservation.
- Undertake a community resource inventory.
- All things being equal, start with products that are already being produced and have markets.
- Capture the value that is added as the product travels through the market system.
- Improve harvesting techniques of existing products.
- Reduce post-harvest losses.
- Increase the competitiveness of a community's exiting products in the market.
- Keep a strategy as simple as possible.
- Diversify production and reduce dependence upon a single product.
- Diversify markets for raw and processed forest products.
- Add value locally.
- Identify and use appropriate production and processing technology.
- What goes up must come down: use the business and marketing structures that are being established in a community to buy manufactured products in bulk as well as sell products.
- Know what you are selling.
- There is strength in numbers.
- Make a decent profit, not a killing.
- Don't create or reinforce patron/client relations within the community or between the community and outsiders.
- Solutions must be equal to the problems.
- Require community/producer investments and where outside finance is needed, use loans not gifts for income-generating projects.
- International markets are for the protection of ecosystems, not for the people who live in them.

Local markets and subsistence use

To realistically assess local peoples' motivations to conserve forests for NTFPs consumed as part of subsistence or local economies (which represent the bulk of NTFP/NDR consumption world-wide), a five year project is under way in the eastern Amazon. Shanley and Galvão (this chapter) briefly describe this project, and research conducted to acquire a basic knowledge of the ecology, use and markets for locally valued resources. The project includes floristic inventories of over 60ha; the creation of 40km of trails linking economic species; the weighing of all game, fibres, medicinal plants and fruit utilized by 30 families over the course of one year; and production/yield studies in which fruit production of over 200 trees throughout a 6,000ha area were monitored over a three-year period. Market research was conducted in the principal port city, Belem, in order to follow the collection, transport, chain of sale and final revenue of leading Amazonian forest fruit and medicinal plants over a three-year period.

In order to realistically incorporate subsistence and non-timber values into conservation and development programmes, this sort of complex, time-consuming, multi-disciplinary approach is required. However, truly multi-disciplinary, multi-year field research programmes remain rare in this area, so our understanding of the relationship between forest products and conservation remains, in most cases, ambiguous.

separated from the physical forest. In fact, to add up the material benefits of the forest – such as the protection of game, wildlife and watersheds; the provision of fertile land; timber and non-timber products; medicinal herbs, seeds, roots and stem bark; snails and mushrooms; and fuelwood – is to assess 'only a fraction of the total potential of the forest' (Abbiw 1990). Religious, spiritual and other cultural beliefs are integral to the value of forests for local people.

Forests and sacred groves

Every culture has narratives or beliefs which answer in different ways the fundamental questions about how we came to be, and articulate how and where people originated, collective transformations undergone by the community, and how people should behave towards one another and their environment (Elder and Wong 1994). Forests are the subject of a great deal of myth, legend and lore. Societies most closely entwined with forests tend to regard them with a healthy respect, an awe at their splendour and majesty, sometimes dread and fear of the powerful spirits that lurk within them. Ancestors often find their resting places in forests, many wandering in various states of unease and spitefulness.

In European culture, the word 'savage' was derived from *silva* meaning a wood, and the progress of mankind was considered to be from the forest to the field. Schama (1995) describes how from Ireland to Bohemia, penitents fled from the temptations of the world into forests, where in 'solitude they would deliver themselves to mystic transports or prevail over the ordeals that might come their way from the demonic powers lurking in the darkness'. The 'indeterminate, boundless forest', then, was a place where the faith of the true believer was put to a severe test. The forests in European culture were also considered to be a more positive site of miracles, the source of great spiritual awakenings; and the forest itself was held to be a form of primitive church or temple. The first temples in Europe were forest groves, and although progressively replaced with temples made of wood, and subsequently by churches made of stone, places of worship – particularly those of Gothic architecture – continue to evoke the forest with their design and proportions (Rival, this chapter; Schama 1995; Burch, this chapter).

Schama (1995) quotes a poem by Bryant called *Forest Hymn* which expressed the American , or New World, version of the forest as a form of primitive church or temple:

> The groves were God's first temples. Ere man learned
> To hew the shaft and lay the architectrave
> And spread the roof above them – ere he framed
> The lofty vault, to gather and roll back
> The sound of anthems; in darkling wood
> Amidst the cool and silence, he knelt down,
> And offered to the Mightiest solemn thanks
> And supplication.

In other regions of the world there also exists a relationship to forests that combines fearful respect and awe at the beauty and mystical source of life held within forests. Budda would sit alone in the depths of the forest, lost in meditation, and it was in the midst of a beautiful forest that he was shown the four great truths (Porteous 1928). In Ghana, beliefs about forests include the belief that they are the home of dwarfs, and the domain of the mythical *Sasabonsam* – a legendary figure responsible for all the woes of mankind and to which mishaps and everything evil are attributable (Abbiw 1990). The Dai people of Yunnan Province, China, believe that the forest is the cradle of human life, and that forests are at one with the supernatural realm. They believe that the interrelationship of human beings with their physical environment consists of five major elements: forest, water, land, food and humanity (Pei Shengji, this chapter).

In sacred groves are manifested a range of traditions and cultural values of forests. Although they occur throughout the world, sacred groves share many similar features, which are summarized in part by Pei Shengji (this chapter) in his reference to the four hundred 'dragon hills' (*lung shan*) in the Yunan Province of China: '... a kind of natural conservation area... a forested hill where the gods reside. All the plants and animals that inhabit the Holy Hills are either companions of the gods or sacred living things in the gods' gardens. In addition.... the spirits of great and revered chieftains go to the Holy Hills to live following their departure from the world of the living'.

Sacred groves are specific forest areas imbued with powers beyond those of humans; they are home to mighty spirits that can take or give life; they originate from a range of roots, and include: sites linked to specific events; sites surrounding temples; burial grounds or cemeteries housing the spirits of ancestors; the homes of protective spirits; the homes of deities from which priests derive their healing powers; homes to a powerful animal or plant species; forest areas that surround natural sacred features such as rivers, rocks, caves and 'bottomless' water holes; and sites of initiation or ritual (Falconer, Pei,

Bharucha, Zoundjihekpon and Dossou-Glehouenou, Pramod Parajuli – all in this chapter;Vartak and Gadgil 1981).

Access to most sacred forests is restricted by taboos, codes and custom to particular activities and members of a community. Gathering, hunting, wood-chopping and cultivation are strictly prohibited in the Holy Hills of China. The Dai people believe that these activities would make the gods angry and bring misfortune and disaster upon the community. A Dai text warns: 'The Trees on the Nong mountains (Holy Hills) cannot be cut. In these forests you cannot cut down trees and construct houses. You cannot build houses on the Nong mountains, you must not antagonize the spirits, the gods or Buddha' (Pei Shengji, this chapter). In Maharashtra, India, regulations and religious customs are set down by priests (known as *pujaris* or *bhagats)* with a knowledge of the forest deities, their ties to the surrounding landscape, and their influence on the daily lives of the community. Ancient folklore and stories are told which include fairly specific detail on the supernatural penalties that will result should the groves be desecrated, for example by felling trees. However, control over extractive activities in sacred groves varies by village, and in many places a complete ban is not in place, and limited collection of fallen wood, fruit from the forest floor, medicinal plant collection, honey collection, tapping of *Caryota urens* to make an alcoholic beverage, and other activities are permitted, if strictly controlled (Bharucha, this chapter).

Sacred groves have survived for many hundreds of years and today act as reservoirs of much local biodiversity. The 40 contiguous groves studied by Bharucha (this chapter) account, as a whole, for most of the plant species present in the Maharashtra region. The forest structure is also unique, representing the least disturbed islands of old growth. The Holy Hills in China also make a significant contribution to biodiversity conservation on a number of levels: they contribute to the conservation of threatened forest ecosystems; they protect a large number of endemic or relic plant species; and the large number of Holy Hills distributed throughout the region form 'green islands' or 'stepping stones' between larger nature reserves (Pei Shengji, this chapter).

As a result of the high conservation and biodiversity values held in sacred groves, increasing attention is being paid to their potential as a tool and model for biodiversity conservation. For example, in its 1996 *Sacred Sites – Cultural Integrity, Biological Diversity* (1996) UNESCO found that:

'Sacred groves have served as important reservoirs of biodiversity, preserving unique species of plants, insects, and animals. Sacred and taboo associations attached to particular species of trees, forest groves, mountains, rivers, caves, and temple sites should therefore continue to play an important role in the protection of particular ecosystems by local people. Particular plant species are often used by traditional healers and priests who have a strong interest in the preservation of such sites and ecosystems. In some regions of the world, beliefs that spirits inhabit relict areas have served to quickly regenerate abandoned swidden plots into mature forest. In other areas, sacred places play a major part in safeguarding critical sites in the hydrological cycle of watershed areas. Furthermore, in a number of instances sacred sites have also been instrumental in preserving the ecological integrity of entire landscapes. For these reasons, sacred sites can help in assessing the potential natural vegetation of degraded ecosystems or ecosystems modified by humans.'

Sacred groves have survived in many regions despite tremendous economic pressure on forest resources. In some parts of India, for example, sacred groves have retained high levels of biodiversity and remain largely intact, while government-controlled forest reserves are often in poor condition. Local-level control has been vital to the protection of these areas, but economic pressures are mounting, and changing land-use patterns have contributed to a serious depletion of resources and a phenomenal rise in the price of land. This in turn has provided an irresistible incentive for some local people to sell the groves, irrespective of the sentiment that at one time was sufficient to preserve them (Bharucha, Pramod Parajuli, this chapter).

Even in cases where local communities are determined to retain sacred groves, they are often as vulnerable to outside political and economic forces as other forest areas. In East Kalimantan, for example, oil palm plantation and logging operations are clearing ancestral *(adat)* forest. The *adat* covers four types of forest: *Sipung Bengkut* (perennial tree gardens which have been developing since 1912), *Sipung Bua* (fruit tree gardens), *Sipung Payo* (swampy areas) and *Sipung Uwe* (rattan gardens). The companies promise in return to encourage 'community participation', 'the development of sustainable forest management', and 'income generating schemes', which are considered 'empty and pointless' offers for a priceless ancestral forest that cannot be equated with monetary and material conditions (Enris and Sarmiah, Chapter 7).

Although sacred groves undoubtedly contribute to the conservation of biodiversity, it is questionable whether the complex history and traditions that have created and maintain these areas can be operationalized as a tool or model for further conservation efforts. Conservation is often a side-effect of customs that associate or dedicate forest resources to the deities. In the Western Ghats of India, rather than managing resources for future use, communities are instead attempting to benefit from the protection and good-will afforded by the deity in return for not disturbing the sanctity of the sacred grove (Bharucha, this chapter). This would be a difficult dynamic to reproduce in a conservation programme. In Southern Ghana, Falconer (this chapter) argues that sacred groves exist as part of a system so complex and variable that a much clearer understanding of the spiritual, mystical and political functions and beliefs of sacred groves is needed before they can be incorporated into conservation programmes.

In South India, sacred groves are populated by dead spirits prevented from transforming and hence remaining ghosts forever. Their life force engenders trees to grow wild, and gives rise to highly fertile but extremely dangerous sacred groves, which are frightening and highly ambivalent (Rival, this chapter). Rival warns against environmentalists' views of sacred groves and trees as sanctuaries of biodiversity, home to benign and protecting deities. The suggestion that the belief systems that have protected these groves should be promoted to encourage the conservation of larger forest areas ignores the fact that – while both environmentalists and local peoples view trees as vital and holding regenerative power – trees in traditional India are not benign protectors: they are frightful and the power of their life force is extremely dangerous. While important reservoirs of biodiversity, it is unlikely that with the exception of a few areas, the cultural beliefs and management systems that have led to the conservation of sacred groves could easily be incorporated into the Western cultural conservation ethos.

Trees

Trees are universally powerful symbols, a physical expression of life, growth and vigour to urban, rural and forest dwellers alike. They can symbolize historical continuity and human society. They are often of frightening magnitude, linking earth and heavens, arbiters of life and death, incorporating both male and female aspects, and home to both good and bad spirits, including the souls of ancestors. Trees provide protection from harm, cure disease and increase fertility. Trees preside over marriages, are planted at the birth of a child and at burial sites. In some origin myths, the first men and women were made of wood.

The Tree of Life in Mesopotamia and India brings fertility by linking death with life. The birds visiting its branches are the souls of the dead. The cross on which Jesus died grew into a tree on Mount Golgotha. The fig tree opened for Mary to seek shelter for the infant Jesus from the soldiers of Herod. The date palm was the staff of St Christopher which helped him to carry the weak and small across a raging river. The birch in Scandinavia, larch in Siberia, redwood in California, fig in India, and iroko in West Africa are widely revered and respected.

The Cosmic Tree, Tree of Life, or *Axis Mundi*, features in many of the world's religions. In Amazonia, the World Tree is often ceiba/kapok, *Ceiba pentandra*, or a yuchan, *Chorisia insignis*. The trunks of these tall emergent trees are characteristically bulbous, hollow and spongy, and the wood is rather soft. The ceiba has a life-span of up to 200 years and is arguably the tallest of Amazonian forest trees. It reaches maturity and starts to flower some time between its fortieth and sixtieth year, thus beginning its reproductive cycle at the oldest age people live to in the region. It lives a life corresponding roughly to four Huaorani generations. In Huaorani culture, the Amazon basin was born from the fallen giant ceiba tree (Rival, this chapter).

Ties to nature manifest themselves most notably in Turkish culture through attitudes to plants in general, and trees in particular. After conversion to Islam, the importance of trees grew in local culture because Mohammed compared a good Muslim to a palm tree and declared that planting a tree would be accepted as a substitute for alms. Trees are planted after children are born, when a son is drafted into military service, after a wedding, and as a memorial to the dead (Tont, this chapter). In one of the oldest collections of Turkish tales which make up the *Book of Dede Korkut*, the unknown poet agonizes over his failure to find a more exalted name for his beloved plant:

> Tree, tree, do not be embarrassed because I call you by that
> [after all] The doors of Mekka and Medine are made of wood
> The staff of Moses is also made of wood
> The bridges spanning over big rivers are also made from you
> The ships which roam the black seas are also of wood.

The oak tree was worshipped by Romans, Druids, Greeks and Celts as the home of deities. In Europe, fairies were said to make their homes in old oak trees, departing through holes where branches had fallen; it was considered healing to touch the fairy doors with diseased parts (T. Shanley 1997, personal communication). In Scotland in the last century, mistletoe growing on the famous Oak of Errol was bound up with the fortunes of the family Hay, acting as a 'sure charm against all glamour and witchery' (Porteous, this chapter). Cowley and Evelyn in seventeenth-century England wrote about the oak (as in Schama 1995):

> Our British Druids not with vain intent
> Or without Providence did the Oak frequent,
> That Albion did that Tree so much advance
> Nor superstition was, nor ignorance
> Those priest divining even then bespoke
> The Mighty Triumph of the Royal Oak

The cultural context of conservation

'Forest ecosystems are vital for biological diversity as they provide the most diverse sets of habitats for plants, animals and micro-organisms, holding the vast majority of the world's terrestrial species. This diversity is the fruit of evolution, but also reflects the combined influence of the physical environment and people. Forests and forest biological diversity play important economic, social and cultural roles in the lives of many indigenous communities.' (The Clearing House Mechanism, The Convention on Biological Diversity.)

Incorporating the perspectives and voices of forest people

The bulk of the world's biodiversity and forests exist outside protected areas, and outside the active management authority of conservation projects and forestry departments. Local communities are, therefore, and often *de facto*, the primary stewards of forests. Lack of secure land tenure, rights to resources, self-determination and other factors obstruct and interfere with what in many cases are long histories of traditional management of forests and resources on the part of indigenous peoples and local communities. Genuine participatory mechanisms need to be developed to allow indigenous peoples and local communities a decisive voice in evaluations of deforestation processes and the evolution of an appropriate policy response. For example, the Leticia Declaration, December 1996 (Appendix 1.16) states:

> ***Underlying causes of deforestation and forest degradation***
>
> *Deforestation and forest degradation is exacerbated by a lack of understanding of the holistic world views and ways of life of indigenous peoples and other forest dependent peoples. The underlying causes of deforestation and forest degradation which urgently need to be addressed include the following:*
>
> *the failure of governments and other institutions to recognize and respect the rights of indigenous peoples and other forest dependent Peoples to their territorial lands, forests and other resources;*
>
> *the increasing problem of landlessness among impoverished peasants who are denied access to land outside forest areas due to inequitable land ownership patterns, and who also have no alternative economic opportunities;*
>
> *government policies and those of private sector industry [that] are geared to exploit forest and mineral resources to the fullest extent for purely economic gain. These policies are often incompatible with other existing forest conservation policies. Such policies include substituting forests with industrial tree plantations for the pulp industry; oil and gas exploration by trans-national corporations; uncontrolled mining operations; and establishing nuclear waste storage sites on indigenous territories.*

Given this identification of the underlying causes of deforestation, suggested solutions and prerequisites for sustainable management include: secure land tenure and long-term guarantees to land and territories; recognition that forests are fundamental to the survival of cultural diversity, and the welfare of indigenous peoples, and form the basis of their livelihoods, cultures and spirituality; recognition that traditional forest-related knowledge is essential to sustainable forest management, and is intimately bound up with ownership and control of lands and territories and the continued use, management and conservation of all types of forest; traditional knowledge must remain alive, cultures must continue to develop, and indigenous contemporary knowledge and technologies must be respected; and it must be recognized that nearly all forests are inhabited.

On a local level, however, most forest peoples' voices are not incorporated into the identification of species and ecosystem conservation priorities. In North-Central Tanzania, for example, there exists a great deal of international interest and involvement in conservation, which has meant that local peoples' voices can often be drowned out, or ignored, or their

Box 9.2: Forest Policy: Out of the UNCED

Forests occupied a prominent position in the international deliberations at the United Nations Conference on Environment and Development (UNCED) in 1992. Two products were dedicated entirely to forests: Chapter 11 of *Agenda 21*, 'Combating Deforestation', and the 'Non-legally-binding authoritative statement of principles for a global consensus on the management, conservation and sustainable development of all types of forests' (also known as the 'Forest Principles'). During the preparations for the Third Session of the Commission on Sustainable Development in 1995, it was decided to establish the 'open-ended *ad hoc* Intergovernmental Panel on Forests' (IPF). The IPF was given a two-year mandate in order to co-ordinate international policy on forests, and build consensus on eleven priority issues, including traditional forest-related knowledge. Its successor is the *ad hoc*, open-ended Intergovernmental Forum on Forests (IFF), also established under the aegis of the CSD (CSD 1995; CBD Clearing House 1997).

The broad categories addressed by the IPF included: implementation of UNCED decisions relating to forests; international co-operation in financial assistance and technology transfer; scientific research, forest assessment and development of criteria and indicators for sustainable forest management; trade and environment relating to forest products and services; and international organizations and multilateral institutions and instruments (CSD 1995).

Subsequent to the establishment of the IPF, and in order to ensure coherent support from the UN System, an Inter-Agency Task Force on Forests (ITFF) was created to contribute to the work of the IPF. The collaborating organizations are: FAO, UNDP, UNEP, ITTO, World Bank, UNDPCSD, CIFOR and the Secretariat of the CBD. In June 1997, the ITFF issued an Implementation Plan as a first response to the proposals for action adopted by the IPF, and endorsed by the CSD at its Fifth Session. The ITFF was called upon to '... undertake further co-ordination and explore means for collaboration and coherent action at the international, regional and country levels, in support of any continuing intergovernmental dialogue on forests' (ITFF 1997).

Cultural issues are addressed in various forms throughout these forest policy processes. The Conference of the Parties to the CBD in November 1996, for example, introduced the programme of work for forest biological diversity (Decision III/12) by 'Recognizing that issues related to forests must be dealt with in a comprehensive and holistic manner, including environmental, economic and social values and issues'. Included in 'Key Words' for programme elements from the Third Session of the CSD and the First Session of the IPF are: 'underlying causes of deforestation', 'traditional forest related knowledge' and 'valuing multiple benefits'. The IPF's 'expected outputs' also included 'new approaches to forest resources assessment including social and economic values of forest resources'.

In May 1997 the Executive Secretary of the CBD convened a Liaison Group on Forest Biological Diversity to develop a programme of work. The results of the meeting were used at the third SBSTTA in September 1997 to formulate recommendations for the consideration of the Fourth Meeting of the Conference of the Parties in May 1998. These include: '(1) holistic and intersectoral ecosystem approaches that integrate the conservation and sustainable use of biological diversity as well as socio-economic considerations; (2) scientific analysis of the ways in which human activities, in particular forest management practices, influence biological diversity...' (Juma, 1997).

The Convention on Biological Diversity Clearing-House Mechanism introduces its section on Forest Biological Diversity with the following: 'Forest ecosystems are vital for biological diversity as they provide the most diverse sets of habitats for plants, animals and micro-organisms, holding the vast majority of the world's terrestrial species. This diversity is the fruit of evolution, but also reflects the combined influence of the physical environment and people. Forests and forest biological diversity play important economic, social and cultural roles in the lives of many indigenous communities.'

Sources:

Secretariat to the Commission on Sustainable Development. *Intergovernmental Panel on Forests: its mandate and how it works.* United Nations, New York.

Secretariat to the Convention on Biological Diversity. *Forest Biological Diversity*. Clearing House Mechanism, Montreal, Canada.

Secretariat to the Convention on Biological Diversity. *The Biodiversity Agenda: Decisions from the Third Meeting of the Conference of the Parties to the Convention on Biological Diversity,* November 1996, Buenos Aires.

Box 9.2 (continued)

Department for Policy Co-ordination and Sustainable Development (DPCSD). *Interagency Partnership on Forests: Implementation of IPF Proposals for Action by the ITFF,* June 1997. United Nations.

Juma, Calestous. *Statement by the Executive Secretary of the Secretariat of the Convention on Biological Diversity, to the second Committee of the United Nations General Assembly*, 6 November 1997.

Documents on forests available through the CBD Clearing-House Mechanism (http:\\www.biodiv.org)

Forests and Biological Diversity (UNEP/CBD/SBSTTA/COP/3/16)

Traditional Forest-Related Knowledge and the CBD (UNEP/CBD/SBSTTA/3/Inf.5)

Draft Programme of Work for Forest Biological Diversity (UNEP/CBD/SBSTTA/3/5)

Report of the Meeting of the Liaison Group on Forest Biological Diversity (UNEP/CBD/SBSTTA/3/Inf.5)

Forests and Biological Diversity (UNEP/CBD/SBSTTA/3/Inf. 22)

Indicators on Forest Biological Diversity (UNEP/CBD/SBSTTA/3/Inf.23)

knowledge and perspectives inadequately represented in conservation programmes (Smith and Meredith, this chapter). However, local values for ecosystems and resources will often not correspond to externally-established values. Working with the Batemi, Smith and Meredith have begun to explore procedures to assign species conservation priorities based on the relationship between three key factors: the importance to local people and traditional use; the relative scarcity of the species in the landscape; and the intensity of pressure that may serve to reduce species availability, including the availability of substitutes. They acknowledge that tools such as these will ultimately fall short because they 'entail quantification and presume that value is objective and static rather than subjective and dynamic'. 'However, if such tools strengthen the position of local groups in protecting their culturally-defined resources, they will contribute to shifting the power balance in biodiversity conservation decision making.'

However, the process of incorporating indigenous perspectives, and the application of TEK, to conservation and development problems is a complex and difficult one, and can at times be problematic. Based on experience with the Joint Forest Management Programme in India, Sundar (this chapter) is wary of external efforts to extract indigenous perspectives and knowledge for conservation and development purposes, and doubts the sincerity of some programmes attempting to do so. For example, in Bastar, the most forested district in Madhya Pradesh, India, while larger political and economic processes conspire to marginalize certain types of rural people, both through physical displacement and by the denigration of their life-styles and systems of knowledge, the government is also seeking to institutionalize their participation and valorize their knowledge in particular schemes.

Valorizing indigenous knowledge and local community perspectives *per se* is not necessarily a boon to local people. If the incorporation of local communities' perspectives, and the application of traditional ecological knowledge to local environmental concerns requires a quantifiable, static, cultural and knowledge base, then it will not succeed: 'Indigenous or local knowledge is not a frozen, inert, or timeless entity, but is dependent on the material conditions of those whose knowledge it is, the changing environment in which they find themselves and the uses to which it is put.' (Sundar, this chapter.) In many cases environmentalists underestimate the multiplicity of interests which different groups within a village can have, and 'projects end up ventriloquizing villager's needs... because artful and risk-adverse villagers ask for what they think they will get. It is also because various development agencies are able to project their own various institutional needs onto rural communities' (Mosse 1995).

The culture of forest conservation

The way in which the field of conservation identifies problems and crafts solutions is a culturally-framed activity, and one which to date has largely ignored the importance of culture as an element in conservation. Cultural values, which are highly variable and difficult to quantify, often contrast markedly with values illuminated through modern scientific approaches, which are the foundation for most conservation and forest management programmes today (Dove, Parajuli, both this chapter; Burnham 1993; Richards 1996; Sharpe 1997).

Culturally-driven debates about 'untouched' and 'wild' spaces are currently under way in Europe and

North America, including that on 'wildwood' in Britain (Box 9.3). 'The first law' of forest conservation might be said to be that 'any human society is unable to appreciate the value of woodlands until a substantial proportion of them have been converted to other uses. There is therefore a direct correlation between rate of woodland decline and rising rates of concern' (Burch, this chapter). Or, as Daniel Abbiw (1990) said of Ghana, 'the paradox of the forest is that where it occurs man's tendency is to destroy it, and where it does not occur the tendency is to create it'.

Regardless of the society, forest practices and responses to trees – including approaches to conservation – mirror our predominant social values, class systems, attitudes to personal liberty, relations between young and old, men and women, and the sacred and the profane (Burch; Sundar, both this chapter; Peluso 1992; Schama 1995). Additionally, each profession involved in conservation brings with it a code of behaviour and operating mythologies. For example, Burch (this chapter) suggests that some myths associated with professional forestry in North America – a system that has been heavily influenced by German practices of silviculture (in turn based on forest scarcity felt in fourteenth century Germany) – now include 'greatest good for the greatest number', ' multiple use' and 'sustained yield'.

Traditional cultures have a great deal to teach the fields of conservation and development, but not always in the ways commonly imagined. Dove (this chapter) describes the Kantu' system of augury in West Kalimantan, which is based on the belief that the major deities of the spirit world have foreknowledge of events in the human world and that, out of benevolence, they endeavour to communicate this knowledge to the Kantu' primarily through seven species of forest birds. Rather than build upon empirical links between the environment and human decision-making, however, the system of augury systematically severs these links. Any systematic ecological information that the birds might convey (e.g. fixed and predictable habitats, feeding patterns and mating seasons) are obscured by the investment of meaning in non-systematic phenomena such as the number, type and direction of the bird calls. In effect, the system recognizes the impossibility of correctly predicting agro-ecologi-

Box 9.3: 'Wildwoods' – Culture and Conservation in Britain

Oliver Rackham

[In Britain, a culturally-driven debate rages about 'untouched wildwood': one which Oliver Rackham addresses in this excerpt from his letter to the Editor of *Tree News* (Spring 1996).]

Coppicing and conservation

... The fact is that in Britain, of all countries, human influence is pervasive and not always negative; it has destroyed species and habitats, but has also created new species and habitats. The cultural landscape, formed by the interaction between mankind and the natural world, is all the landscape that we fully know about. Many aspects of it are of value in their own right as biological habitats, such as meadows, chalk-pits and medieval churches. Coppice woods do not claim to be wildwood, and are not just an inferior substitute for it. Biological conservation cannot be isolated from the conservation of artefacts and cultural values.

Trees in this country have a continuous history of some 12,000 years since the end of the last Ice Age. Human activity has been an obvious major influence for a little over half that period. British wildwood disappeared so long ago that we do not know in detail what it looked like. Even 12,000 years ago, Palaeolithic hunters had already taken their toll of the great Pleistocene beasts, which may explain why the ensuing wildwood was different from that of corresponding periods in earlier interglacials.

Wholly natural wildwood is an elusive concept. In Britain it lasted for a relatively brief period during the Mesolithic. Who knows how it would have changed over the last 6,500 years had it been spared? In North America and Australia, human activities have determined whole continents throughout the post-glacial period. By Columbus' time there were not many forests on earth, other than on uninhabited islands, that had never been altered by peoples' activities. The obsession of some conservationists with untouched wildwood has been steadily undermined by archaeological research.... Wildwood cannot be re-created either by doing nothing or by management. An abandoned coppice wood will not turn into either wildwood as it was in 4500BC, or the wildwood that might have developed by 2000AD had it never been encoppiced...

Oliver Rackham, Corpus Christi College, Cambridge.

cal conditions, and the consequent need to devise pluralistic rather than deterministic agricultural strategies. Therefore, while the content of the Kantu' system of forest augury is not environmentally meaningful, it is culturally meaningful.

Dove (this chapter) argues that – guided by a single deterministic vision, which is a reflection of modern scientific approaches – in the face of the uncertainty that characterizes tropical forest ecosystems, conservation and development programmes try to eliminate that uncertainty. The Kantu' augural paradigm, on the other hand, embraces ignorance and, paradoxically, by trying to make sense of the limits of knowledge, transcends them. It contributes to a pluralistic strategy that can cope with a wide range of conditions and minimizes extremes of either success or failure (a phenomenon common in conservation and development programmes): 'This traditional system of adaptation provides a needed lesson to development planners in the virtues and ills of, respectively, indeterminate versus determinate strategies for resource use in the tropical forest.'

The links between culture, knowledge and the conservation of forests are complex and varied. Critical to the success of any conservation undertaking, however, is a sound understanding of the local cultural context. Traditional practices, beliefs and myths provide the foundation upon which practical solutions to specific conservation problems must be based. Additionally, individuals and professions working in the field of forest conservation will be at their most effective when fully aware of the limitations, as well as strengths, of their own very powerful forest cultures.

Conclusion

We have seen that culture, as broadly understood, is intertwined with forests in many and diverse ways. Traditional ecological knowledge and systems of management, which have co-evolved over millennia alongside forests, have transformed forests to such an extent that it is only in isolated pockets that one can find a forest unaltered by people. Likewise, the forest has acted upon human culture. It is rare, even today, to find a society untouched by forests, and most have a belief system relating to forests and trees. In forest areas, complex and site-specific cultures incorporate elaborate taboos, myths and beliefs relating to forests and forest species. Cultural diversity is integral to the conservation of forests and biological diversity.

Forests are home to ancestors, are the origins of life, are threatening and powerful, but are living-giving and sacred. We see in the contributions to this volume an enormous range in the ways in which forests are used and valued. This provides ample warning to conservationists who would over-simplify and frame conservation solutions based upon limited understandings of local cultural beliefs for trees and forests. However, the tradition in forest conservation of late has been one which more likely discounts culture, and marginalizes values and beliefs that are not 'linear' or 'deterministic'. In many parts of the world, this has meant that conservation programmes have failed to incorporate large parts of the 'mental economy' of local communities, with likely inopportune consequences for the success of programmes.

In forest culture we find the common threads of human experience. Whether the 'dangerous and highly fertile' sacred groves of India, the oak tree in Britain, the graveyard forests in Côte d'Ivoire, or in the fall of the great *Ceiba pentandra* which created the Amazon Basin: throughout the world we see a shared focus on the origin, force and power of life expresssed in trees and in the forest.

Trees and the symbolism of life in indigenous cosmologies (Laura Rival)

The Amazon 'World Tree'

The Cosmic Tree (the Tree of Life or *Axis Mundi*) is the subject of a vast body of philosophical, mythological and religious writings (Brosse 1989). In parts of Amazonia, the World Tree is botanically identified as a *ceiba* (*Ceiba pentandra*), or a *yuchan* (*Chorisia insignis*), both members of the silk-cotton tree or bombax family (*Bombaceae*), or with the largest species of the closely related *Tiliaceae* family. The trunks of these tall, emergent trees are characteristically bulbous, hollow and spongy, and the wood rather soft. These trees are central to the social life and mythology of many Amazonian societies, and the Huaorani Indians of Amazonian Ecuador are no exception. The Huaorani myth of origin centres on the giant *ceibo* tree, *bobehuè* (*Ceiba pentranda*), which, it is said, contains all forms of life. With a life-span of up to 200 years, the *ceibo*, arguably the tallest tree of the Amazon forest (it is not uncommon to find sixty-metre-high specimens, with a three-metre basal diameter and a forty-metre-wide crown – Villajero 1952), reaches maturity, and starts flowering, some time between its fortieth and sixtieth year. In other words, it starts reproducing at the oldest age people can live to, and has a life span roughly corresponding to four Huaorani generations. Box 9.5 contains an abridged version of the myth.

Can this myth provide us with clues to Huaorani cultural understandings of nature and society? There

Box 9.4: Spirit of the Forest

Song lyrics by Kenny Young

Up from the canopy I can see
A thousand chainsaws comin' for me
For millions of years this has been my home
Turn around, turn around, it soon may all be gone

They'll never break the spirit of the forest
They'll never cut the heart from the tree of life
They'll never break the spirit of the forest

I hear the cry of the Yanomami
The kookaburra and the golden monkey
Thousands of acres up in smoke every day
Millions of species fadin' away

They'll never break the spirit of the forest
They'll never cut the heart from the tree of life
They'll never break the spirit of the forest

Jamais quebrarao o espirito das matas
Jamais cortarao o coracao na arvore da vida
Matipu, Makuxi, Waimiri, Pataxo
Bororo, Tikuna, Kanela
Tukano, Xavante, Juruma, Guarani, Kayapo

Up from the canopy I can see
Flames of extinction coming for me
Hanging in the balance there lies our fate
Turn around, turn around, before it's too late

They'll never break the spirit of the forest
They'll never cut the heart from the tree of life
They'll never break the spirit of the forest

is ground to argue that the myth expresses the fundamental characteristics of the Amazon ecosystem, such as, for instance, the fact that its fertility is not located in the soil, which is arid and devoid of nutrients, but in the intricate and luxurious vegetation for which the soil acts as a mere support. There would be no life on earth without the existence of trees (one at least is needed) for they provide shade, food and shelter, and prompt the formation of rain. The primordial tree is a small ecosystem in itself, and the world expands when this perfectly self-contained microcosm collapses. The organic expansion, however, cannot occur without the simultaneous creation of hills, forests and rivers, which all originate with the fall of the great canopy tree. The new ecosystem, the outgrown successor of the primordial tree, is as integrated and self-generative as the original one. If sky and earth are no longer connected by a vine, they are still interacting through cycles of rain, flood, evaporation and growth. At the beginning of the story, the soil is scorched by the heat of the sun, and the earth water runs beneath the surface, as if in hiding. The fall of the giant tree suddenly reverses the situation, and the once overheated earth is now flooded. The myth thus makes clear that neither of these two situations are propitious to vegetal growth and life. The tropical vegetation engendered by this extraordinary energy requires a balance between heat and humidity, shade and light. It is only with the twin creation of the Amazon tributaries and forested hills that an equilibrium is finally achieved.

Roe (1982) would give a very different interpretation. In *The Cosmic Zygote*, his impressive comparative survey of Amazonian mythologies, he argues that the Amazon world tree is the centre of a 'bio-logic perpetual-motion machine' (Roe 1982: 264), a kind of cosmic womb which gets fertilized and fecundated by the sun. The tree's outside is hard, male and eternal; its inside soft, moist and female. According to Roe (1982: 143), the world tree is 'a symbolic continuum incorporating both male and female aspects, life and death, in a single concrete object'. It symbolizes the sexual nature of cosmic creation, which unites female - periodic and cyclical – and male – immortal and continuous – energies. Roe supports his thesis that the world tree is a 'phallic-uterine' symbol with a wealth of examples taken from a wide range of Amazonian myths, which all seem to share common traits with ancient myths found in other parts of the world. Could all religions be founded on the belief that a sacred tree, the central axis around which the universe was ordered, preceded the existence of humans on earth (Brosse 1989)?

In many cultures we find the belief that the Cosmic Tree is attached to heaven, its roots reaching the centre of the earth. The roots plunge into the primordial, underground river, the river of life and death. This river, manifested as a spring or well, sprouts out of earth from a stone. In many European beliefs, childless women who bathe in these springs, wells or rivers will conceive. The Tree of Life in Mesopotamia and India brings fertility by linking death with life. The birds visiting its branches are the souls of the dead. The fig tree under which Buddha meditates can also be classified as a cosmic tree, and so can the tree of Genesis (and the cross on which Christ died: it grew into a tree on the Mount Golgotha). In China, we find another type of Cosmic Tree, the *Kine-mou*, the tree of the absolute beginning. The Cosmic Tree, which is often represented as inhabited by two antagonistic forces fighting each other, although almighty, is not indestructible. It can be uprooted by tempests, fire or flood. All living beings die, except for a surviving couple, the ancestors of humanity. In some Indo-European versions, the first man and woman were made of wood. In Europe, the first temples were sacred groves (i.e. live trees); these were progressively replaced with temples made of wood. It is only subsequently that churches were made of stone. These stone buildings not only reproduced the wood structures, but also, at least for some, such as the Gothic churches, evocated forests.

If Amazonian Cosmic Tree myths contain similar themes, some – characteristically European and Asian – are entirely absent. For example, in Germanic beliefs, the Cosmic Tree is linked to a god or king who sacrifices himself to enhance his power and knowledge. He hangs himself from the tree, dies and is reborn. Human sacrifices to trees replace this original self-sacrifice. In Ancient Crete, Minos, the king-priest, whose name means 'lord, protector and saviour', is the master of time and fertility for the limited period during which he officiates (then his strength weakens, his vitality is too low to remain in office). Anthropologists, unlike mythologists, have looked beyond religious parallels, and tried to anchor cosmological systems within specific cultural matrixes. Practical knowledge and symbolic significance are intimately associated. For instance, there is only one Huaorani word, *ahuè*, to translate 'wood' and 'tree'. The root of this word */hu /*, found in many words and expressions translating as 'to live', 'to feel', 'to be alive' or 'to have emotions', means 'life'. Plant life is not only identified with life and growth, but also with beauty, vigour and energy. Humans are associated with leaves (new leaves are shiny as newborn babies, falling leaves are losing colour and vitality as the elderly), longhouses are associated with emergent trees, and endogamous groups with palm groves (Rival 1993).

Box 9.5: The Huaorani Myth of Origin

In the beginnings of time, the earth was flat, there were no forests, no hills. The earth was like a dried, barren and endless beach, stranded at the foot of a giant *ceibo* tree. This tree, attached to heaven by a strong vine, was the only source of shade against the strong sun. Only seedlings under its direct protective shade could escape from the merciless heat of the sun; this is why there were no hills and no forests. There was no moon, and no night either. All that was alive dwelled in the giant tree. In those times of beginning, most living beings, neither animals nor humans, formed one group. Only birds were different and lived apart: the doves which were the only game available to hunters, and two dangerous individuals, Eagle (eater of raw meat) and Condor (eater of rotten flesh), who preyed on people and doves alike. Life in those times would have been good to live, if it had not been for these two preying birds. Every time someone would leave home, Eagle or Condor would descend on their victims, kill them, and take them back to their nest on the highest branch of the giant tree. Eagle would be heard eating, undisturbed by the grief of the victims' relatives watching the bones of their dear kin fall to the ground, one after the other.

Fortunately, Squirrel and Spider decided to take action and put an end to this dreadful business. One day, they climbed up to the very top of the tree. Eagle had fallen into a deep sleep following a copious meal. Spider wove a tight and intricate web all around his body. Eagle did not feel the web until his attempt, the following day, to swoop down upon new prey. Caught in his impetus, he was soon to be seen hanging upside down, his head swinging in the air, and his feet still attached to the branch. Squirrel's plan was to detach him, and let him fall heavily to the ground, like a dead log. But the manoeuvre went wrong. Instead of cutting the web off the branch, Squirrel's teeth incised the vine that linked up the tree to the sky. While the vine sprang up, with Squirrel still biting on its end, the giant tree crashed onto the ground, westward. To this day Squirrel's tail can be seen in the sky, especially on the bright nights that precede heavy rains, where it glows like a fluffy trail of golden dust.

The Amazon Basin was born from the fallen giant tree, and the many species of fish from its leaves. The roots became the Amazon headwaters. Before the dramatic fall, there was hardly any water, except for a scarce rainwater preciously collected in clay pots. But in his fall, the giant tree exposed the peg which blocked the underground waters. Incapable of resisting the pressure without the support of the roots, the peg loosened and the whole country was soon submerged by the enormous flooding which killed almost all that was living in its wake. A few people survived by taking refuge in a hollow branch, in which they painstakingly made their way up river. They died of exhaustion, except for a brother and a sister, who became husband and wife. The world was turned into an immense muddy flood plain, until Woodpecker, the cultural hero, managed to lift the hills up out of the mud. The hills were soon covered with the forests in which the first Huaorani, our ancestors, found refuge and dwelled. It is in these forests that they multiplied and grew numerous again.

The *ceibo* tree, like other plants and trees, is a physical expression of life, growth and vigour. It is primarily praised for its kapok (a fluff used as wadding and feather), an essential part of the hunting gear. One of the three 'seasons' is named after this material: the 'season of wild cotton' (*bohuèca tèrè*). From August to September, *ceibo* kapok, blown by the winds, floats all over the forest and becomes available in great quantities. The *ceibo* tree, with its impressive height and solitary character, is also used as a landmark. Trekkers climb to its top to orientate themselves when lost, or whenever they wish to embrace in one panoramic view the forested landscape – their homeland. The theoretical challenge remains, however, to understand adequately the relationship between a real *ceibo* tree as observed and used in the forest, and its ideal representation, as found in the myth of origin.

Trees and humans in South India

In South India, the vitality of trees is seen as emanating from the supernatural life force of ghosts, the wandering spirits of the recently dead (Uchiyamada, in press). Middle caste Nayars bury the cremated remains of their dead relatives in funeral urns at the foot of jack-fruit trees. A year or so later, the urns are excavated and the exhumed bones brought to the coast, where they are left to drift in the Arabian Sea. After this ceremony, the soul attains liberation and the deceased finally becomes an ancestor. The jack-fruit tree, which represents the life-span of the family (including the ascendants) and its ancestral house, is sacred. By contrast, the souls of dead Untouchables, considered inherently impure and polluted, never attain liberation. They remain attached to the soil, and become wandering lineage ghosts rather than proper ancestors. Milk trees (*paala*), which are extremely fertile but bring misfortune to the landed castes, grow on sites where Untouchables are said to have succumbed to violent treatment and died. Sacred groves often form and develop around milk trees. South Indian funeral rites, therefore, offer an excellent illustration of the complex ways in which tree symbols may be used in a society deeply structured by hierarchical inequality and cross-cut by diverse cultural and religious traditions. The spatial organization of lineage temples further demonstrates the real contrast in fate between high caste and Untouchable spirits. High caste deities are seated in hierarchical order around the temple's rectangular, paved, and neatly swept yard, at the centre of which stands a single sacred tree (a pipal and a mango growing intertwined) whose name means 'soul transcending death'. At the back of the temple lies the sacred grove, peopled with evil deities of Untouchable origin. Whereas high castes view the sacred grove as a source of wild, dangerous and capricious powers, Untouchables represent them as their own lineage temples. What do we learn, then, from these funeral rites? We learn that the life force, when properly controlled by the living, is transitory and transforming, and that the trees associated with it (the coconut planted on the cremation pit, the jack-fruit tree under which the funerary urn is buried, or the pipal-mango tree in the temple yard) are sacred and harmless. But when dead spirits are prevented from transforming, hence remaining ghosts for ever (a common fate for Untouchables), their life force engenders trees that grow wild, and give rise to highly fertile, but extremely dangerous, sacred groves. The life force contained in trees, therefore, is frightening and highly ambivalent, for its regenerative power comes from the dead.

Do such religious ideas still prevail in modern India? Uchiyamada assures us that the answer is 'yes', but my answer would be less affirmative than his. Transformations in the landscape greatly undermine the transmission of indigenous ways of seeing the world, as these are deeply embedded in direct experience and implicit knowledge. As transmission now relies more on story-telling than on phenomenological engagement with the forest ecosystem, tree symbolism gets easily displaced by rigid categorizations of trees and other natural elements that pervade education and politics, and regulate social life. When not discarded, indigenous beliefs are subject to processes of fixation, normalization and incorporation not dissimilar to those affecting native tree species. New policies on sacred groves, such as those recommended by Mitra and Pahl (1994) are a case in point. These two authors argue that sacred groves, real sanctuaries of biodiversity, are unique examples of ecological understanding and management, and that 'social fencing' religious beliefs which have encouraged people to protect vast tracts of virgin forests for centuries should be encouraged or reintroduced through government legislation. Does their recommendation square with what we have learnt from Uchiyamada about people's fear of sacred groves and of their resident gods' implacable wrath? Not quite. Mitra's and Pahl's modern environmentalist view of sacred groves reinterprets traditional Hindu beliefs in the light of present-day anxiety about pollution and deforestation. Their 'deifying conservation', like traditional Hindu beliefs, contains the same notion of vitality and regenerative power, and of trees as pollution and sin cleansers. Popular green writings are full of such parallels between traditional religious beliefs and ecological wisdom (Guha 1989). There is, though, a major difference: trees in traditional India are not benign protectors; they are frightful, and the power of their life force is extremely dangerous.

Trees – symbols of life and regeneration

'Memory', says Boyer (1996) 'is what touches us even more than a tree's extreme age, its species or its beauty. [...] As soon as a tree lives beyond [its] age, [it] becomes a parable for the passing of time, an image of the living and yet petrified memory'. Trees, adds Schama (1995), have become the living and enduring memorials of historical heroes and great historical events; they unquestionably symbolize historical continuity in the West. They provide, as such, a visible symbol of human society and give Euroamericans a sense of identity. After having studied tree symbolism for three years, I have come to the conclusion that in non-Western cultures trees are not so much symbols of time and history, but, rather of life, health and potency. They provide excellent models based in sentient experience and practical knowledge used by people to conceptualize the growth, maturation and durability of individual beings and communities. In many parts of the world, close analogies are drawn between tree growth and the development of the human body. It is not surprising, therefore, that these formidable life-giving plants also figure prominently in numerous life-cycle rituals (Rival 1998). Almost everywhere, the human imagination has identified trees with humans, and glorified their great proportions, old age, potency and self-regenerating energy.

Cultural landscapes, chronological ecotones and Kayapó resource management (Darrell A. Posey)

Traditionally, Amazonian Indians have been thought of as merely exploiters of their environments – not as conservers, manipulators and managers of natural resources (e.g. Meggers 1996). Researchers are finding, however, that presumed 'natural' ecological systems may, in fact, be products of human manipulation (Alcorn 1981, 1989; Anderson and Posey 1985; Balée 1989a, 1997; Balée and Gély 1989; Clement 1989; Denevan and Padoch 1988; Frickel 1959; Roosevelt 1994; Sponsel 1995; Sponsel, Headland and Bailey 1996; and others). Likewise, old agricultural fallows reflect genetic selection and human enhanced species diversity (Anderson 1990; Balée 1989b; Denevan and Padoch 1988; Irvine 1989; Redford and Padoch 1992).

The Kayapó Indians of the Middle Xingu Valley, Brazil, provide a good example of how scientific assumptions of 'natural' landscapes have hidden the complexity and potential of local management practices to modify ecosystems. The modern Kayapó population is still under 5,000, but pre-contact populations were many times larger and presumably had even greater impacts on the vast region they exploited (Posey 1994). They live in an ecologically diverse region that comprises nearly 4 million hectares of *reserva indigena* in the states of Para and Mato Grosso. Ethnohistorical research with the Kayapó Indians shows that contact with European diseases came via trade routes and preceded face-to-face contact with colonizers. Epidemics led to intra-group fighting, fission and dispersal of sub-groups which carried with them seeds and cuttings to propagate their foods, medicines and other resources (Posey 1987).

A form of 'nomadic agriculture' developed, based on the exploitation of non-domesticated resources (NDRs) intentionally concentrated in human-modified environments near trail sides, abandoned villages and at camp sites (Posey 1985). Agricultural practices also spread, along with techniques for the management of old fields to enhance the availability of wildlife and useful plants. During times of warfare, the Kayapó could abandon their agricultural plots and survive on non-domesticated species concentrated at trail sides, former village sites, forest openings and ancient fields.

Agricultural plots were engineered to develop into productive agroforestry reserves dominated by NDR species, thereby allowing the Kayapó to oscillate between (or blend together) agriculture and gathering. Such patterns appear to have been widespread in the lowland tropics and defy the traditional dichotomies of wild vs. domesticated species, hunter–gatherers vs. agriculturalists, and agriculture vs. agroforestry. Even today, over 76 percent of the useful plant species collected to date are *not* 'domesticated', nor can they be considered 'wild' (Posey 1997; Roosevelt 1994). (I suspect that as a more complete floral inventory is completed, this percentage will approach 98 percent.)

Nowhere is this more evident than in the formation of 'islands' of forest, or *apêtê*, in the *campo-cerrado* (savannah). The Kayapó initiate and simulate the formation of forest patches through the careful manipulation of micro-environmental factors, knowledge of soil and plant characteristics, and intentional concentration of useful species into limited plots. Although most *apêtê* are small (under 10ha), elders reported plant varieties in a 1ha plot as having been introduced by villagers from an area the size of Western Europe (Anderson and Posey 1989).

The principle elements of Kayapó management have been previously described in some detail (Posey 1983, 1985, 1987, 1995, 1997) and include:

- overlapping and interrelated ecological categories that form continua;
- modification of 'natural ecosystems' to create ecotones;
- emphasis on long-term ecotone utilization (chronological ecotones);
- concentration on non-domesticated resources;
- transfer of useful plant varieties between similar ecological zones, and
- integration of agricultural cycles with forest management cycles

Several options are possible for representing indigenous resource management models. I believe the most inclusive and descriptive representation of the Kayapó system places savannah or grasslands (*kapôt*) at one end of a continuum as the 'focal type' (example that most typifies the category) and forests (*bà*) at the other end (opposite focal type). *Kapôt* types with more forest elements would be represented to the right of the diagram, while *bà* types that are more open and with grassy elements would lie on the continuum diagram to the left, or toward the savannah pole.

This would put *apêtê* at the conceptual centre of the continuum, since forest elements are introduced into the savannah to produce these anthropogenic zones. Agricultural plots (*puru*) also lie conceptually near the centre of the continuum, because sun-tolerant vegetation is introduced into managed forest openings. *Apêtê* can be thought of as the conceptual inverse of *puru*: the former concentrates resources in the forest using sun-tolerant species, while the other does the same in the savannah using forest species.

Even though ecological types like high forest (*bà tyk*) or transitional forest (*bà kamrek*) are securely located at the forest pole, they are not uniform in their composition. All forests have edges (*kà*), margins (*kôt*), and openings caused by fallen – or felled – trees (*bà krê-ti*) which provide zones of transition between different conceptual zones. Thus, a plant that likes the margins of a high forest might also grow well at the margin of a field (*puru-ka* or *puru-kôt*) or in an *apêtê*. A plant that likes light gaps provided by forest openings might also like forest edges (*bà-kà* or *bà-kôt*) or old fields (*puru-tum* or *ibê-tum*). Plants from open forest types or forest edges can predictably proliferate along edges of trails or thicker zones of *apêtê*. Using this logic, the Kayapó can transfer biogenetic materials between matching micro-zones so that ecological types are interrelated by their similarities rather than isolated by their differences. These interfaces can be considered ecotones, which become the uniting elements of the overall system.

There is another interesting dimension to the model that appears when looking diachronically (temporally or historically) across the system. Agricultural clearings are initially planted with rapidly growing domesticates, but almost immediately thereafter are managed for secondary forest and NDR species. This management depends upon planting and transplanting, removal of some varieties, allowing others to grow, encouraging some with fertilizer and ash, and preparing and working the soils to favour useful species.

Management aims to provide long-term supplies of building materials, ceremonial objects, medicinals and other useful products, as well as food for humans and animals. The old fields (*puru tum*) are at least as useful to the Kayapó as agricultural plots or mature forest. A high percentage (an initial estimate is 85 percent) of plants in this transition have single or multiple uses. When the secondary forest grows too high to provide undergrowth as food for animals (and hunting also becomes difficult), then the large trees are felled to create more hospitable conditions for management and/or reinitiation of the agricultural cycle. Likewise, *apêtê* are managed to maximize useful species in all stages of the forest succession. When their centres become dark and unproductive, openings (*irã*) are created which allow light to once again penetrate the forest and initiate a new cycle.

The Kayapó resource management system is, therefore, based on the conservation and use of transitional forests in which agriculture is only a useful (albeit critical) phase in the long-term process. *Apêtê* exhibit parallel transitional sequences in the *campo-cerrado* and depend almost exclusively on non-domesticated resources. The degree to which genetic materials are transferred between similar micro-zones of different ecological types points to how the Kayapó exploit ecotones that host the highest diversity of plants. Management over time can be thought of as management of *chronological ecotones*, since management cycles aim to maintain the maximum amount of diversity and the greatest number of ecotones.

The Kayapó model illustrates how previously assumed 'natural' ecosystems in Amazonia have been consciously modified by indigenous residents through time. The degree to which this has taken place has yet to be quantified, but Kayapó 'forest islands' data show concentrations of plant varieties from a vast geographic area. This case underlines the necessity for historical studies to understand the long-term effects of management of cultural and anthropogenic landscapes. Above all, it exposes the inadequacies of our scientific, educational and political institutions which separate agriculture from forestry and ignore the importance of non-domesticated resources.

[Adapted from: 'Indigenous Knowledge, Biodiversity, and International Rights: Learning about Forests from the Kayapó Indians of the Brazilian Amazon'. *Commonwealth Forestry Review*, 76(1): 53-60. Special Issue: *The Contribution of the Social Sciences to Forestry*.]

Invisible income: the ecology and economics of non-timber forest resources in Amazonian forests (Patricia Shanley and Jurandir Galvão)

Non-timber forest resources (NTFPs) are intrinsic to the daily livelihoods of rural dwellers and yet continue to be ignored and undervalued in global, national and household accounting systems. Products with the potential to reach the medicine chests or dinner tables of First World households are preferred candidates for study over the thousands of plant species utilized on a daily basis by rural communities. While underscoring the economic promise of non-timber forest products for the world's future, such a preference obscures the critical yet 'invisible' subsistence value of extractive products within rural households today.

To more accurately appraise the subsistence value of non-timber forest products to rural families we measured the use of fibres, fruit and game by 30 families in a 3,000-hectare community forest over the course of one year, and by one family on one forested hectare over the course of three years. Preliminary results demonstrate that most of the recorded non-timber resources utilized over the course of one year – 87 percent of the fibres and 79 percent of the game – were extracted from mature forests. Use of game, fruit and fibre offered the equivalent of over 25 percent of the average family income, while capable hunters earned the equivalent of over 50 percent of their annual income from game.

Other components of the forest-based research undertaken include floristic inventories of over 60 hectares, the creation of over 40 kilometres of trails linking economic species, and production/yield studies in which fruit production of over 200 trees throughout a 6,000ha area were monitored over a three-year period. Market research in Belém, the principal port city of eastern Amazonia, studied the collection, transport, chain of sales and final revenue of leading Amazonian forest fruit and medicinal plants over a three-year period.

Economic valuation of forest as used for subsistence, as compared with its timber value, revealed the economic potential of one hectare to be worth two to three times its value as sold for timber. The value of a tree sold for wood – $2 – is the equivalent of the market value of 10 fruit of the same tree. Depending on species, and discounting half a tree's production to account for predation and home consumption, one tree's annual fruit production can offer $20–$60. Residents noted that game capture, and the nutritional and health benefits offered by forests, influence their decisions to conserve instead of selling or deforesting their land.

The timber industry has aggressively penetrated rural areas in eastern Amazonia since the 1970s, and although offering relatively trivial sums, cash-poor farmers commonly accept grossly disadvantageous land and timber deals. Therefore, the raw material most often extracted from select fruit and medicinal trees in the eastern Amazon is not medicinal barks, roots, fruits or oils, but wood. Three decades ago, a handful of species were extracted as timber in eastern Amazonia: currently more than one hundred species are targeted. Today, all of the most significant NTFPs occurring in the region's forests are logged. Whereas timber extraction previously existed in a symbiotic relationship with other uses of the forest by rural communities, offering cash, trails and jobs, the growing overlap of timber with locally useful NTFP species now threatens the livelihoods of forest residents. If timber sales and swidden agriculture continue at the present rate, the most highly valued non-timber forest resources will be threatened with extinction in the coming decade.

In spite of the exceptional value that standing forests offer, cash-poor families sell logging rights or land for meagre sums, with severe consequences: decreased food sources, increased distances to resources, and decreased populations of game. As wild food resources decline, forest residents must substitute native animal and plant products with market goods. However, for rural families whose annual

Box 9.6: Lore, Legend and Forest Conservation in the Eastern Amazon

Patricia Shanley and Jurandir Galvão

Plants are important not only for our stomachs but also for our spirits. Unrecognizable to many modern city dwellers, people living close to the earth have perceived that forests offer not only food and material resources but spiritual benefits as well. Many indigenous tribes have customs to thank nature through prayer and offerings. But the customs, legends and myths that traditionally protect forests are eroding. Today, many people think that lore is mere superstition; however, myths and legends serve important practical ecological, socio-economic and spiritual functions. Emotional ties with nature, and legends, are two of the most powerful incentives to conserve and respect forests.

In the Brazilian Amazon, both scientists and indigenous groups believe in the legends of the grand cobra, the giant sloth and the curupira. The grand cobra swims the depths of the rivers and curbs fisherman from over-fishing. The giant sloth roams deep forests sending men screaming in fear, and the curupira tricks disrespecting woodsmen by making them lose their way in the woods. The curupira has feet pointed backward, curly black hair and sleeps in the huge buttresses of the tall *tauri* tree. When you are lost in the woods, walking in circles, always returning to the same place, it is the curupira that is following you, because you did something greedy or foolish to make him angry.

To find your way out of the forest, you need to leave tobacco for the curupira. If you have no tobacco, you must instead find a vine and make a complicated knot, hiding the point as well as possible in the middle. Next, toss the knot over your shoulder for the curupira to discover. While the curupira is looking for the point, you must run quickly from the woods. If you enter the forest and do not respect her – the curupira will get you. She makes sure that you do not hunt excessively and that you treat the plants and animals as you care to be treated.

incomes approximate $1,000, regular purchases of meat, fruits, fibres and pharmaceuticals are out of economic reach. If forests are to remain standing and continue to support livelihoods, both the subsistence and market benefits of non-timber resources must be recognized and included in local and global assessments of forest worth.

Musanga cecropioides: biodynamic knowledge encoded in mythic knowledge (Paul Richards)

Musanga cecropioides, sometimes known in English as the 'corkwood tree' because of its light wood, or as the 'umbrella tree' from its fine spread of leaves, is a fast-growing tree of the West African rainforest, found in gaps created by natural tree-falls and in secondary successions. Its unusual characteristic is that it is found in dense abundance in such clearings but is squeezed out by other slower-growing trees after about thirty years, and it is never found in regular successions associated with repeated cycles of forest shifting cultivation. In other words, it is a once-and-once-only marker of transition from or back to high forest. In the language of social anthropologists it is 'liminal' – a marker of an important transitional event in the life-cycle of a living entity.

Prior to the civil war that has ravaged large parts of interior Sierra Leone since 1991, the village of Lalehun was a small community of farmers and foresters on the western edge of the Gola North reserve, the largest element in the complex of four surviving areas of high rainforest on the Sierra Leonean side of the border with Liberia. Among the traditions of these Mende-speaking people is the idea that *Musanga* should never be brought into the village and used as firewood. The tradition states that this would 'spoil' the spirit of the town, represented by a small shrine driven into the ground in an open space at the northern end of Lalehun's main 'street'. Without this taboo on use of *Musanga* the co-operation among families essential to the village's survival would be undermined.

The rationalist cries 'nonsense' – why shouldn't people make use of this readily available tree, found all along the edge of the forest reserve where farming has nibbled into the forest margin? It is of no interest as a rare species to be protected. As a quick-growing light wood it is easily felled and transported. Indeed villagers know it makes quite a decent fuelwood, since there is no prohibition on its use for making cooking fires in the farm huts where most Lalehun villagers spend the daylight hours during the farming season.

But villagers' sincere attempts to maintain the taboo is in fact a fine example of the true understanding of biodynamic processes sometimes encoded in mythic knowledge. The taboo speaks to the separateness of town and forest. The life of the town is at the expense of the life of the forest. *Musanga* has come to symbolize that transition. It is only seen where actively there is a process of forest becoming settled land, or cleared forest returning to high forest. The tree occurs in no other context. The moral force behind the taboo is the recognition of the need to grasp the difference between these two spheres. I doubt it could be argued that a ban on *Musanga* in the village supports what outsiders might consider a 'conservation ethic'. What it speaks to, however, is the truth and reality of significant transitional states. Humans, indeed, live life on a tightrope. Wise management of nature requires first that we develop an understanding of life as process. That this understanding is already encoded in a minor ritualized aspect of local resource use indicates the possibility of Mende people on the edge of the Gola Forest appreciating the need for wise management of forest-edge biological processes in the longer-term interest. This kind of knowledge should not be despised in the rush by science to colonize the field of biodiversity management.

Non-timber forest products in southern Ghana: traditional and cultural forest values (Julia Falconer)

Sacred and protected forest groves are common features in southern Ghana. They are attracting an increasing amount of interest as possible sites of biological diversity and as models for forest conservation. Sacred groves have different origins and serve different functions in some communities. The origin of some are linked to specific historical events, while others are believed to have existed since creation. Some sacred areas are burial grounds housing the spirits of ancestors, while others house protective spirits. Some sacred groves are renowned for the healing powers of their deity, and priests and healers of these groves derive their powers from the spirits. In other cases, rivers and other features are sacred and the forest vegetation serves to protect them.

The prominence and protection of these groves vary considerably between and within communities, as can the degree of 'sacredness'. The younger generation and immigrants appear most likely to disregard the sacred taboos. In comparing 'study villages', the most significant factors that affect people's attitudes to the groves seem to be the strength of the traditional political and spiritual leaders, the influx of immigrants, and land pressure.

Box 9.7: Trespassing on the Devil's Land: African and West Indian Variants

Harold Courlander

This story is found in a number of versions in West Africa, and has found its way to the West Indies as well as to Surinam. The following is a Yoruba version, from Nigeria.

In a certain village there was a young man named Kigbo. He had a character all his own. He was an obstinate person. If silence was pleasing to other people, he would play a drum. If someone said 'Tomorrow we should repair the storage houses,' Kigbo said 'No, tomorrow we should sharpen our hoes'. If his father said 'Kigbo, the yams are ripe. Let us bring them in,' Kigbo said 'On the contrary, the yams are not ready'. If someone said 'This is the way a thing should be done,' Kigbo said 'No, it is clear that the thing should be done the other way around'.

Kigbo married a girl of the village. Her name was Dolapo. He built a house of his own. His first child was a boy named Ojo. Once, when the time came for preparing the fields, Kigbo's father said to him, 'Let us go out tomorrow and clear new ground'.

Kigbo said: 'The fields around the village are too small. Let us go into the bush instead'.
His father said: 'No one farms in the bush'.
Kigbo said: 'Why does no one farm in the bush?'
His father said: 'Men must have their fields near their houses'.
Kigbo said: 'I want to have my fields far from my house'.
His father said: 'It is dangerous to farm in the bush'.
Kigbo replied: 'The bush suits my taste'.

Kigbo's father did not know what else to say. He called Kigbo's mother, saying 'He wants to farm in the bush. Reason with him'.
Kigbo's mother said 'Do not go. The bush spirits will make trouble for you'.
Kigbo said 'Ho! They will not trouble me. My name is Kigbo'.

His father called for an elder of the village. The village elder said, 'Our ancestors taught us to avoid the bush spirits'.
Kigbo said 'Nevertheless, I am going'.

He went to his house. His wife Dolapo stood at the door holding Ojo in her arms. Kigbo said, 'Prepare things for me. Tomorrow I am going into the bush'. In the morning he took his bush knife and his knapsack and walked a great distance. He found a place and said 'I will make my farm here'.

He began to cut down the brush. At the sound of his chopping, many bush spirits came out of the trees, saying 'Who is cutting here?'
Kigbo said ' It is I, Kigbo'.
They said 'This land belongs to the bush spirits'.
Kigbo said 'I do not care' and went on cutting.

The bush spirits said, 'This is bush spirit land. Therefore, we also will cut'. They joined him in clearing the land. There were hundreds of them, and the cutting was soon done.
Kigbo said 'Now I will burn,' and he began to gather the brush and burn it.
The bush spirits said 'This is our land. Whatever you do, we will do it, too'. They gathered and burned brush. Soon it was done.

Kigbo returned to his village. He put corn seed in his knapsack. His father said 'Since you have returned, stay here. Do not go back to the bush'.
His mother said 'Stay and work in the village. The bush is not for men'.
Kigbo said, 'In the bush no one gives me advice. The bush spirits help me'. To his wife Dolapo he said 'Wait here in the village. I will plant. When the field is ready to be harvested, I will come for you'.

He departed. And when he arrived at his farm in the bush, he began to plant. The bush spirits came out of the trees, and said: 'Who is there?'
He replied ' It is I, Kigbo. I am planting corn'.
They said 'This land belongs to the bush spirits. Therefore, we also will plant. Whatever you do, we will do'. They took corn seed from Kigbo's knapsack. They planted. Soon it was finished.

Box 9.7 (continued)

Kigbo went to a village where he had friends. He rested there, waiting for the corn to be grown. In his own village, his wife Dolapo and his son Ojo also waited. Time passed. There was no message from Kigbo. At last, Dolapo could wait no longer. She went into the bush to find her husband, carrying Ojo on her hip. They came to Kigbo's farm. The corn stalks were grown, but the corn was not yet ripe.

Ojo said 'I want some corn'.
His mother said: 'The corn is not yet ripe'.
Ojo said 'I am hungry'.

Dolapo broke off a stem of corn and gave it to him. The bush spirits came out of the trees, saying 'Who is there and what are you doing?'
She replied 'It is I, wife of Kigbo. I broke off a stem of corn to give the little one'.
They said 'Whatever you do, we will do,' and they swarmed through the field breaking off the corn stalks. Soon it was done, and all the broken stalks lay on the ground.

At this moment Kigbo arrived. He saw Dolapo and Ojo, and he saw all the corn lying on the ground. He said 'The corn is ruined!'
Dolapo said 'The bush spirits did it. I broke off only one stalk. It was Ojo's fault. He demanded a stalk to eat. I gave him a stalk, then the bush spirits did the rest'. She gave Ojo a slap.

The bush spirits came out of the trees, saying 'What are you doing?'
Dolapo said 'I slapped the boy to punish him'.
They said 'Whatever you do, we will do'. They gathered around Ojo and began to slap him.

Kigbo shouted at his wife 'See what you have done!' In anger, he slapped her.
The bush spirits said 'What are you doing?'
He said 'Slapping my wife for giving me so much trouble'.
They said 'We will do it too'. They stopped slapping the boy and began slapping Dolapo.

Kigbo called out for them to stop, but they wouldn't stop. He cried out 'Everything is lost' and he struck his head with his fist.
The bush spirits said ' What are you doing?'
He said 'All is lost. Therefore, I hit myself'.
They said 'We will do it too' and they gathered around Kigbo, striking him on the head.

He called out 'Let us go – quickly!' Kigbo, Dolapo and their son returned to the village, leaving the farm behind. He saw his father. Because of shame, Kigbo did not speak.

His father said 'Kigbo, let us go out with the men tomorrow and work in the fields'.
Kigbo said 'Yes Father, let us do so'.

[From: Courlander, H. (1996), *Afro-American Folklore*. Marlowe and Company, New York.]

Attitudes towards, and beliefs in, traditional deities and groves also vary according to the spirit and grove in question. In Nanhini, for example, no villagers enter the grove of the goddess Numafoa or ignore her taboos. In the same village, a second deity has less influence and so the taboos associated with its grove are not so strictly followed. Each grove has particular governing rules. In some cases, entry to a sacred grove is strictly limited, but in others the area maybe exploited or restricted for certain forest resources and not others. In Nanhini, for example, one grove cannot be farmed or hunted in, nor can snails be collected, but the palms can be tapped for wine, and goods and medicines gathered. In all the cases we encountered, clearing sacred areas for farming is strictly prohibited. Generally, the sacred groves are important sources of medicines used by healers, especially by those whose healing powers are derived from the spirits.

Most sacred groves and rivers have sacred days (*daboni*) associated with them, at which times farming activities are prohibited. The days are often in remembrance of particular historical events such as battles or settlements of disputes in which the particular deity played a role. For example, in Essuowin the sacred day is Thursday, and no farming activities are permitted. In some communities there are several *dabonis* and adherence to one or the other depends on personal circumstances. For example, in Koniyao several rivers and streams are considered sacred. Each has a *daboni* which prohibits people from crossing it on that day; thus, depending on the location of individual farms in relation to different sacred rivers, different people adhere to different non-farming days.

In Koniyao, there is no sacred grove since the shrine associated with the river Kyrirade was destroyed during the bush fires in 1983. There are, however, several important Asante shrines north of Koniyao near Kokofu where people go for serious health and spiritual problems. In Banso, the protected area is where the Gwira Royal Family is buried, and entry into it by strangers is strictly prohibited. Most of the people in Banso believe in the protective powers of the Anko god of the Ankobra river. Every child born here is dedicated to this god in a baptismal ceremony, and it is believed that these children will never drown in the river, and that the god will protect them from evil forces.

The sacred groves in Nanhini and Essuowin provide interesting case studies. In Nanhini there are two protected areas of note: the Kobri Kwaye and Numafoa. The Numafoa is especially revered; even the most ardent Christians believe in the goddess's protective powers. Her grove has been protected by the Pekyi chief for many generations. The grove was even spared in the 1983 bush fires although much of the surrounding area was burnt and as a result its boundaries sharply define the surrounding landscape. Many people recount testimonies of her recent visits to them and of her healing prowess. Numafoa is believed to be actively participating in people's daily lives, and stories about her abound. In this sense, the Numafoa goddess and grove differ from the others in the Nanhini area. The second god, Kobri Kwaye, is less revered than Numafoa. People tend to take less important problems to this god and while they respect some taboos, they do collect forest products and hunt in the Kobri Kwaye grove whereas they would not think of doing the same in the nearby Numafoa grove.

The Kobri Kwaye in Essuowin has several fetishes associated with it as well as a keeper who is appointed by the chief. The Kobri is a healing god and his fetish priestess is renowned for her healing skills. He is the most revered of all forest and river deities in Essuowin, especially by the older generation. The boundaries of the Kobri sacred grove are well respected despite increasing land pressure, although some damage occurred in 1983 during the bush fires. At the outskirts of the village there is a shrine where the stools of all the past fetish priests of the grove are housed. Libations are poured here by the chief and the elders on behalf of the village as a whole. While most people will seek permission of the chief before entering the grove, some of the younger generation enter it regularly without permission for small-scale mining of gold. In time the area will probably be farmed as some of the younger generation, especially the immigrant farmers, have little faith in the deity and grove.

In Essuowin and Nanhini the sacred groves have custodians, and traditions associated with the deities are strictly adhered to. The gods are believed to participate in people's daily lives and the traditions have protected the groves over generations. In places such as Essamang and Kwapanin, by contrast, where there are no 'active' sacred groves, there is less interest and belief in particular deities. Where the gods are not believed to be actively involved in people's lives there is less reverence towards them and many people ignore the associated taboos of using sacred groves. This was most aptly illustrated when the chief of Essamang offered for sale some figures dug from their ancestral burial ground. By comparison, in Banso, strangers are strictly prohibited from entering the burial ground. In Nkwanta, the Mintiminim deity is held in high esteem and both immigrant and native Wassa strictly adhere to its traditions and customs.

A variety of sacred plants and animals are associated with particular spirits in sacred groves and rivers, with fetish priests, or with clans of people. Several forest plants are believed to be sacred throughout the entire region: for example, the forest emergent *odii* (*Okoubaka aubrevillei)*, the 'odum' tree (*Chlorophora (Milicia) excelsa*) and the liana, *ahomakyem* (*Spiropetalum heterophyllum*) are sacred in all the study communities. In many cases there are rituals associated with the use of spiritual or sacred resources: for example, an egg must be given to the *ahomakyem* climber before a piece is used. To use any part of the *odii* tree, libations must be poured. It cannot be approached at midday and according to some people it must be approached naked. If exploited, these species are not felled or uprooted, and they are used only in small quantities.

Some sacred plants may be used to protect an entire community. For example, when disease strikes a village, the village is swept with the *sumee* plant (*Costus afer*) and the debris is left piled on the village outskirts. It is believed that the plant drives evil spirits from the village. In all the Ashanti villages studied there are some plants with magic–religious powers.

Some animal species also have sacred or fetish value and are used in spiritual healing and for protection. The forest is believed to be home to these animals, even if they are not caught in the forest itself. Common animal products used in healing come from chameleons, duikers, leopards, elephants, alligators, tortoises and pythons. Part of a chameleon, for example, is tied to the wrist of a new-born baby to ward off evil spirits and disease. The belief is that the chameleon imparts its camouflage to the child and evil forces are unable to impart disease to the child who becomes invisible.

All the deities have particular animal taboos associated with them, and rearing such animals in a deity's domain is strictly prohibited. For example, for Numafoa in Nanhini dogs and goats are taboo and no one in the village owns either. In Nkwanta the goat, dog and pig are taboo on account of the Mintiminim god. Some animal species are protected by particular gods. For example, the Bongo (a large antelope) is protected by the Mintiminim god in Nkwanta and no one may kill it in his domain.

The prevalence of sacred groves throughout southern Ghana indicates the high value placed on forest vegetation and its spiritual associations. However, from our brief and limited study it is clear that the establishment and protection of sacred forest areas is complex and varies immensely from area to area and even within communities. It is difficult to imagine how such areas can be 'type cast' and used as models for conservation, or how their protection can be assured as 'biodiversity' sites. It is evident that much clearer understanding of the spiritual, mystical and political functions and beliefs of these groves is needed *before* research is conducted on their physical condition or conservation value. In addition, a study of the traditional rules associated with the use of such areas and an assessment of the younger generation's views on them would provide valuable insight into the evolving values of protected forest vegetation.

[Excerpted from: *Non-Timber Forest Products in Southern Ghana* (1992). Main report to ODA and the Forestry Department, Ghana. It is part of a wider study on forest use and value in Southern Ghana carried out in 1989.]

Cultural and spiritual values of biodiversity in West Africa: the case of Benin and Côte d'Ivoire (Jeanne Zoundjihekpon and Bernadette Dossou-Glehouenou)

In West Africa, for thousands of years, peoples have established and preserved social behaviour patterns in order to control the relationship between nature and society and to promote the sustainable use of natural resources. Social practices over the years gradually confer important cultural and spiritual values on biodiversity, expressed in beliefs about divinities identified with diverse elements of the universe, and the veneration of ancestors.

Soothsayers are the exclusive intermediaries for these divinities or spirits, which people have to respect for the preservation of social and natural order because 'in these cultures...whether traditional or exotic, people are related to Nature through invisible links which lead each person to preserve or affect the order of things' (Toffin 1987). Thus, the health of individuals and the social order are conditioned by the quality of the relationship between society and nature. Relationships with natural resources are controlled by prohibitions on food, plant species used for fire, or areas such as sacred woods.

In ancient African civilizations, nature was perceived as the residence of ancestors who 'control the behaviour of living people and are therefore permanently present among them' (Coulibaly 1978). This justifies the veneration of nature and the existence of taboos protecting natural resources such as sacred areas.

Prohibitions derive from laws instituted by religious chiefs who ensure their effective application through systems of control and punishment. According to Dossou (1992), these laws are based on prohibitions requiring the peoples of a given geographical area not to consume or use all their resources in order to avoid their exhaustion. These prohibitions favour a distribution of resources, and are supported by spiritual and cultural thinking.

For example, access to forests is regulated, and forbidden to non-initiates. In particular, these restrictions apply to fetish forests, graveyard forests, initiation forests and protected water areas. According to Dossou (1992), fetish forests are often places for gathering, exchanging and testing religious chiefs and medicine men, while graveyard forests are places where they bury the bodies of those who have died tragically through accidents, drowning, burning or infectious diseases such as smallpox. Initiation forests are kept by secret groups (*Oro, Kouvito, Zangbéto*) who are in charge of educating young villagers and maintaining social order according to the rules and discipline of the locality.

These are not ordinary woods. According to Loucou (1984), 'the sacred wood is situated generally near the village. Of a quite small area, 2 to 4 hectares and often round, this 'wood' is a portion of forest in the middle of a savannah region. It is a sacred area which divinities and ancestors are supposed to visit permanently; a sanctuary where an altar is erected for sacrifices and where the paraphernalia of rites is stocked, where education is given to initiates and where certain village ceremonies take place. In order to preserve the inviolability of this sacred area, local people often build false access paths'.

Sacred woods, created by human beings, often serve for initiation rites and also conserve biodiversity. According to Coulibaly (1978), 'sacred woods are natural islands spared by human beings; they are relics, witnesses of the ancient flora which occupied the area before human habitation; even during the dry season, apart from some species such as the silk-cotton tree and the baobab tree, the whole sacred wood is still green and dark. The survival of these woods can be explained by their sacred characteristics. They are, in fact, intangible, inviolable areas; no one dares go there for game hunting'.

For the Aoua people in Côte d'Ivoire, Fairhead and Leach (1994) indicate that the goddess Assié prescribes certain environmental and agricultural activities. In so doing, the regulations designate those who will be in charge of clearing and cultivating which part of the land outside the village, they determine the breaks with culture, and forbid some animals or the cultivation of certain plants, including rice. They consider that the cultivation of rice will lead Assié to withdraw her control of fertility (rain and fire) which will result in the destruction of the environment and to the collapse of human society.

In West Africa, socio-economic activities in relation to nature in general and biodiversity in particular, such as agriculture, hunting and fishing, are ruled by prohibitions, totems or sacred areas which occupy an important place in the spiritual life of traditional African peoples. These activities, when based on religious beliefs, help in the preservation of people's health and harmony while allowing for the traditional management of flora and fauna.

For coastal peoples, the great fishing period (May to October) is initiated by an opening rite over the 'Aby' lagoon, sometimes carried out simultaneously in the different areas (Ibo 1996). The priest of the spirit called Assohon opens the fishing in May and closes it in October (Perrot 1989). Sacred catfish of Sapia are sheltered in the Dransi River which is formally forbidden to fishermen. Together with sacred crocodiles from Gbanhui, all the aquatic species are covered by food prohibitions to the villagers. During the day dedicated to sacred and venerated crocodiles, it is forbidden to go to the Yonyongo River.

In traditional societies, the lives of men, women and children are partly subject to prohibitions covering animals and plants, which are all aspects of biodiversity. The respect or non-respect of prohibitions can be a matter of life or death. Outside of the role played by religion in prohibitions, the ones mentioned here constitute forms of rational planning and management of natural resources. It is important to stress that urbanization and religion are factors that can distort or transgress traditional laws. Even if some prohibitions are losing ground today with the extension of cities and modern life, certain practices are still in force and inviolable, and contribute to the preservation of biodiversity.

Bark cloth in Buganda (Christine Kabuye)

Bark cloth is very old in Bugandan culture, and making the cloth is an art that goes back to early history. As an item of clothing it was first used by people in the Kabaka's court. It was only during the eighteenth century that it was allowed for ordinary citizens and adopted as national dress (Thompson 1934) for both men and women.

Bark cloth was also used for bedding, for screens in homes and for wrapping bodies for burial. It was an item of wealth and the Kabaka expected his subjects to bring back cloth to his court as tax. There are, however, royal bark cloth makers who are from the Otter Clan. Customarily, even up to now, the mother of a girl to be married must be presented with bark cloth. At one time there was so much demand for bark cloth that the most important occupation for men was bark cloth making. As demand decreased with the introduction of cotton cloth, the art was left to a few specialists. It is still used by some people at the Kabaka's court and adorns the Kabakas' tombs. It is still in great demand for burials and related ceremonies as well as other works of art.

Bark cloth is made from fig trees, the most important being *Ficus natalensis* followed by *F. thonningii*. Another species *F. amadiensis* and a related genus *Antiaris* produce slightly inferior bark cloth. *Ficus ovata* was also rarely used. Both *F. natalensis* and *F. thonningii* are forest species, mostly starting as epiphytes, but widely cultivated in Buganda mainly as shade in banana and coffee plantations, although they are sometimes found in open cultivated areas. Over 50 varieties were recognized at one time (Eggeling 1951). The harvest of bark is made sustainable in that after removal, new layers of the bark develop, and through re-planting a stock of trees from which to harvest is retained. Bark can be harvested from a tree every year, the first bark harvested producing more inferior bark cloth than the subsequent ones.

Bark is removed by two circular cuts and one vertical cut so that the bark is peeled off the trunk as one sheet. The outer surface is scraped as well as the inner surface to remove the sap. It is tied in banana leaves and left for one night. To protect the tree, the exposed trunk is wrapped with fresh banana leaves and left for three to four days to dry.

The leaves are then removed and the trunk is plastered with wet cow dung and tied with dry banana leaves to protect it from direct sunlight. The cow-dung is left until it drops off. The bark cloth is produced by several beatings and sun treatments. Before beating, the bark can be steamed to produce a superior quality of bark cloth known as *kimote*.

The bark is laid on a long log (*mukomago*) made from *Sapium ellipticum*, *Trilepisium madagascariense*, *Spondianthus preussii* or *Ficus ovata*. The beating is done by using three different wooden mallets or hammers (*ensaamu*) in succession. These are made from *Teclea nobilis* and have a handle and a round head which has circular grooves of varying width (Lanning 1959). The bark is first worked upon with wide-grooved mallets, beating twice on the inner side and twice on the outer side. It is folded first in half, then in four layers, and beaten each time. The second mallet with finer grooves is then used when the bark is folded in eight layers. It is then unfolded, and gently beaten in reverse order. The cloth is spread in the sun for five to fifteen minutes. The third mallet with yet finer grooves is now used to beat once on each side of the sheet. The bark is then put in the sun to dry completely and gain a deep colour. The sheet is made damp by leaving it outside at night for about two hours. It is then folded into a long strip and made soft by kneading with hands and fingers for two hours, by which time it is ready.

The best bark cloth comes from Buddu county. It is always steamed before beating, which takes a week. The cloth here has a distinctive deep red-brown colour. The importance of bark cloth in Buganda is demonstrated by the fact that on accession to the throne, the new Kabaka is presented with two ceremonial ivory mallets by the royal bark cloth makers. A wooden mallet is also handed to male heirs, equipping them for bark cloth making.

Identifying biodiversity conservation priorities based on local values and abundance data (Wynet Smith and T. C. Meredith)

Ngorongoro District in north-central Tanzania is comprised of a number of world-renowned biodiversity conservation sites, such as the Ngorongoro Conservation Area and the Serengeti National Park. It is also home to a sizeable population of Maasai pastoralists and Batemi agro-pastoralists, who rely on the bio-physical environment for the basis of their existence. It is therefore important that conservation strategies for the region take into account the cultural and social context and needs of the human population. While some conservation priorities can be established by the international scientific community, locally important species and ecosystem components can be identified only by the local population (Oldfield and Alcorn 1991; Boom 1989; Johannes and Hatcher 1986).

The approach described in this paper is based on three assumptions. The first is that biodiversity conservation cannot be left exclusively to protected areas but must take place where real communities live and work. The second is that local populations are experts in matters regarding their local environment and that their expertise is required to make biodiversity conservation effective. The third is that, recognizing the first two assumptions, it is none the less essential that new methods be found to define and communicate local conservation priorities. This is required because local populations do not live in isolation and, particularly in internationally significant ecological areas such as north-central Tanzania, important decisions that affect land use and conservation will never be made independent of regional, national and perhaps international interests. To be effective in multi-stakeholder negotiations, the local 'voice' must be as clear, consistent and comprehensible as possible.

Through work with a local conservation NGO in the Batemi Valley in north-central Tanzania, it became evident that local communities needed much stronger tools to participate effectively in controlling their own environmental future. We explored a procedure for assigning conservation priorities for woody species based on traditional use, the availability of substitutes and the relative scarcity of the species in the landscape.

Sale and Loliondo Divisions in Ngorongoro District are experiencing increasing land-use pressures as populations and the demand for land grow. An overall conservation strategy is being developed for the area by Korongoro Integrated Peoples Oriented to Conservation (KIPOC), a regional, indigenous non-government organization (NGO). The strategy is based on local needs and existing use patterns of the Maasai pastoralists and Batemi agropastoralists who inhabit the area. This requires an understanding of the human use and traditional management of ecological resources, and of the species that play a particularly important role in the land-use practices of the local populations.

The Batemi occupy a relatively small area within Loliondo and Sale Divisions. The main valley inhabited by the Batemi formed the focus of this study. Located to the north-east of the Serengeti National Park and north of the Ngorongoro Crater, the area consists of a valley, bordered by an escarpment to the north, the Rift Valley to the east and the

Sale Plain to the south. Elevations range from about 1200 metres above sea level on the valley floor to 2000 metres in the hills to the north of the valley.

As an agro-pastoral group, the Batemi are quite reliant on their environment. Their subsistence strategy consists of the gathering of local resources, herding of goats, and both irrigated and dryland agriculture. They have created a system of irrigation that relies on springs and seasonal streams. Recent studies have explored the intricacies of their irrigation system (Adams, Potanski and Sutton 1994), their medicinal use of plant species (Johns *et al. 1994*) and general uses of woody plant species (Smith, Meredith and Johns 1996). The Batemi use a wide range of tree and shrub species (90 species in total), which accounts for 79 percent of species found in the area (Smith, Meredith and Johns 1996). They also have a number of traditional conservation practices regarding wooded areas surrounding the springs and streams and the vegetation along the irrigation channels (Smith, Meredith and Johns 1996).

Ecological value can be based on many things ranging from aesthetic and spiritual value through to cash commercial value. It is not necessary to distinguish these at the local level – people value what they value. But because local values will not necessarily correspond to externally established values, there does need to be some way of expressing priorities. An assessment of the literature on valuing biodiversity shows that there are really no effective tools that allow for comparison between market and non-market values or between utilitarian and non-utilitarian values. (Prance *et al.* 1987; Pinendo-Vasquez *et al.* 1990; Phillips and Gentry 1993a, 1993b; and Phillips *et al.* 1994 have done innovative work in this area.) This problem was explored by looking at woody species (selected because they are so influential in determining habitat characteristics) in the Batemi Valley. This study focused on consumptive use because of the increased pressure on biodiversity resources related to consumption and because of the human consequences of loss of the resource although we believe the procedures can apply equally to non-consumptive uses. A procedure for determining local conservation priorities by determining three conditions for each species was developed: its importance to local people, its relative abundance in the local landscape, and the intensity of pressure that may serve to reduce the availability of the species. The intent is nothing more than to provide a basis for local communities to define conservation priorities and to argue their position in a larger context (i.e. with government or international NGOs).

Field data from the Batemi Valley are used to generate qualitative indices that should be useful in conservation work. The work is exploratory and is intended to show how these factors could be used to identify conservation priorities. Ultimately, even tools like this necessarily fall short because they entail quantification of value and presume that value is objective and static rather than subjective and dynamic. However, if such tools strengthen the position of local groups in protecting their culturally defined resources, they will contribute to shifting the power balance in biodiversity conservation decision-making.

The assessment of locally important biodiversity resources is an essential step in the identification of conservation priorities for biodiversity. If no attempt is made to ordinate or index relative values, the default is that all species are viewed as equally important conservation priorities. This implicitly overlooks both differences in local pressure on resources and differences in human consequences of resource depletion. The logic of our approach is this: species that are highly valued, intensively exploited and/or locally rare are likely to be high conservation priorities; species that are not valued, not exploited and are locally abundant will be low priorities.

Factors influencing the local value of an ecological resource include both the importance of the societal need that is being met (whether utilitarian or not) and the range of options or substitutes available for meeting that need. So value is directly proportional to the importance of the need being met but inversely proportional to the number of substitutes. For example, a plant that is used for an essential ceremony and that cannot be substituted will have a high value. A plant that is used only for fuel and can be replaced by many other species will have a low value.

The vulnerability or endangerment of a renewable resource is related to its availability relative to the rate of loss or consumption. Establishing conservation priorities based on local cultural needs involves relating value to endangerment. There is an advantage to quantifying these values simply because it allows long species lists and complex arrays of uses to be simplified and presented visually for discussion. For example, value (called V) can be set to be equal to the total of the importance (I) of all uses for that species divided by the number of substitutes (S) for each use. The importance scores and the array of substitutes are determined by asking people. Likewise, the degree of endangerment (called E) can be set in relation to the availability (A) of the species and the rate of consumption (C) or other causes. The availability score can be determined by ecological mapping, and consumption rates can be estimated through interviews or from records of landscape change (which may even, as in our study, be based on archived air photos).

These values can then be used to estimate conservation priority ranking. Species considered as high priorities will be those with high values (V) and high endangerment (E); low value species with low vulnerability would not be. This can all be calculated mathematically and plotted on graphs, and while it must not be taken as definitive, it provides useful points for discussion and could be used by local groups to discover vulnerability that may not have been evident, to reach agreement on local conservation priorities, or to present a consistent and locally supported position to other groups in land-use negotiations.

For reasons of length, this section does not show any of the detail of computation, but rather shows how the results can be used. Calculated use values for selected species, plotted against abundance, reveal obvious differences in the abundance of species with similar values. There are many valued species that do not appear to be abundant in the area. *Acacia mellifera*, for example, although one of the most prized species for construction, was not very abundant. *Haplocoelum foliosum*, *Mimusops kummel* and *Dichrostachys cinerea* are other examples of species with high values and low abundance. Other species that may warrant attention include *Ozoroa mucronata*, *Cordia africana* and possibly another *Cordia* sp. (*mgombeha*) and *Zanthoxylum chalybeum*.

Other species have relatively high abundances in relation to use value, such as *Ficus sycomorus*. On the other hand, there are species – *Croton dictygamus*, *Acacia tortilis*, *Grewia bicolor* and *Vangueria apiculata* – that are quite abundant but appear to have relatively low value. These species might therefore be low on the list of conservation priorities. Still other species, such as *Euclea racemosa* have both low abundance and low use values, and thus presumably lower conservation priority.

When a volume-of-use coefficient is added to the equation (V*C), the resultant values are slightly different. The most highly valued and used species are basically the same, with *Acacia mellifera* having the highest value, followed by *Haplocoelum foliosum*. *Dichrostachys cinerea* is slightly higher than *Acacia nilotica* using this approach. A few other species also change positions relative to each other, including *Strychnos henningsii* and *Commiphora africana*, and *Commiphora pelleifolia* .

Differences in abundance and importance of species allow for informed discussion of conservation strategies. Is this conceptualization useful? Beyond formalizing ideas about conservation priorities, it may also help to highlight concerns or structure management plans where sufficient data exist. Quantifying the parameters in practice may be difficult because of subjective value assessments that may vary between individuals, between interest groups and certainly between culture groups. The number of substitutes also changes with economic,

TABLE 9.1: NUMBERS USED TO CALCULATE VALUES FOR THE VARIOUS TYPES OF USE.
(Construction and implement uses are subdivided into their respective categories).

USE	CITATIONS	IMPORTANCE	CONSUMPTION
Construction			
Poles	53	60	3
Bee	39	10	1
Gate	3	2	1
Trap	5	2	1
Door	5	5	1
Rope	3	5	1
Furniture	2	5	1
Fences	10	20	3
Implements			
Walking stick	13	5	1
Rungata	1	1	1
Digging stick	4	1	1
Handle	2	2	1
Bow and arrow	7	2	1
Toothbrush	5	1	1
Other	2	1	1
Firewood	32	75	4
Services	7	15	1

TABLE 9.2: LIST OF VALUES FOR VARIOUS SPECIES BASED ON THE TWO METHODS OF ASSIGNING VALUE.

ABBR.	GENUS AND SPECIES	VERNACULAR	B. CIT.	B. VALUE	BUILD V*C	IMPL.	FIRE	SERV.	TOTAL.
Abp	*Abrus precatorius*	*mnyete*				0.5			0.5
Ab	*Acacia brevispica*	*mhereki*					9		9
As	*Acacia senegal*	*mhuti*	1	1.13	3.4				3.4
Ag	*Acacia goetzei*	*msigisigi*	2	1.4	3.65		9		12.65
Am	*Acacia mellifera*	*mng'orora*	10	10.3	27.8	0.77	28		56.57
An	*Acacia nilotica*	*kijemi*	6	5.7	14.3		18		32.3
At	*Acacia tortilis*	*mkomahe*	5	5.3	13.9		37		50.9
Ax	*Acacia xanthophloea*	*mrera*					18		18
Cm.a	*Commiphora africana*	*kidirigheta*	2	4	12				12
Ce	*Commiphora ellembekii*	*mwaraheta*	6	2.41	4.68				4.68
Cm.p	*Commiphora pelleifolia*	*muluba*	6	4.2	10.4				10.4
Caf	*Cordia africana*	*mringaringa*	7	2.7	4.93				4.93
Cd.s	*Cordia sp.*	*mgombeha*	3	3.4	10.18		9		19.18
Cd.g	*Cordia gharaf*	*muhabusu*	1	1.13	3.4				3.4
Cd	*Croton dictygamus*	*mgilalugi*	1	1.13	3.4	0.92	9		13.32
Cr.m	*Croton megalocarpus*	*ekitalambu*	1	1.13	3.4				3.4
DC	*Dichrostachys cinerea*	*kiholi*	4	5.4	16.2				16.2
Du	*Dombeya umbraculifolia*	*gwaretu*	2	2.3	6.79	0.57			7.36
Er	*Euclea racemosa*	*mraganetu*	3	0.77	0.77	0.29			1.06
Ec	*Euphorbia candelabrum*	*kiroha*					9		9
Et	*Euphorbia tirucalli*	*kidigho*					18	2.14	20.14
Fs	*Ficus sycomorus*	*mkoyo*	7	2.4	2.37		9		11.37
Gb	*Grewia bicolor*	*ebusheni*	4	4.5	13.58	4.2			17.78
Gt	*Grewia trichocarpa*	*esere*				0.77	9		9.77
Hf	*Haplocoelum foliosum*	*egirigirya*	6	6.8	20.4		18		38.4
M	*Maerua sp.*	*kasingiso*						4.28	4.28
Mk	*Mimusops kummel*	*ghanana*	7	5.3	14.35			2.14	16.49
Sq	*Sterculia quinquelobia*	*mugurumetu*	6	4.5	6.73				6.73
Sh	*Strychnos henningsii*	*kibunja*	3	3.4	10.19	1.8	9		20.99
Va	*Vangueria apiculata*	*mgholoma*	1	2	6				6
Vam	*Vernonia amygdalina*	*mtembereghu*	1	1.5	1.5				1.5
Zc	*Zanthoxylum chalybeum*	*mulongo*	3	3.4	9.65	0.2	9		18.85

technological and environmental circumstances. None the less, the difficulty of determining absolute values should not be seen as a major barrier because the results are, at least initially, intended only as aids to discussion and planning.

The assigning of use-values to various species illustrates the difficulties involved in establishing objective or at least consistent conservation priorities. By combining biological data on the abundance and distribution of the various woody species with use data, key socio-economic species that may represent conservation priorities can be identified. Certain species, including *Acacia mellifera* and *Haplocoelum foliosum* were identified as being of potential concern, or priorities for further work. The quantitative values presented serve as a way of integrating information, and they do appear to be effective in identifying species that should at least receive closer scrutiny as conservation strategies are developed.

Biodiversity conservation is an activity that must recognize the legitimacy of human actors in the landscape and must acknowledge the absolute importance of local involvement in and support of conservation strategies. The tools described in this paper provide a way in which local priorities can be defined, communicated and defended. They are therefore potentially tools of empowerment that will allow local groups to participate more fully and fairly in local conservation. This is the first step in incorporating the full array of biodiversity values into decision-making, and wresting control over decision-making from technocrats.

Figure 9.1: The Kantu' territory in Kalimantan, Indonesia.

Forest augury in Borneo: indigenous environmental knowledge – about the limits to knowledge of the environment (Michael R. Dove)

Current interest in 'indigenous environmental knowledge' is based on the premise that indigenous communities possess knowledge of their physical environment that is privileged by virtue of great time-depth, intimate daily association, and unique cultural and ritual perspectives. It is argued that scientific study of this knowledge can yield lessons that the wider world can use to improve its own environmental relations. This is an overly narrow view and an inadequate appreciation of indigenous environmental knowledge. One of the lessons of such knowledge involves not extending but circumscribing the bounds of knowledge, to tell us not what we can know about the environment, but what we cannot know. This counter-intuitive view is in keeping with the current turn in science toward less deterministic theorizing, as exemplified in chaos theory and the study of complex systems. This analysis is based on a case study of the forest augury of the Kantu', one Dayak group of West Kalimantan (Figure 9.1). The Kantu' are a tribal people who meet subsistence food needs through the cultivation of upland rice and a wide variety of non-rice cultigens in forest swiddens; they meet market and trade needs through the cultivation in forest gardens of Parà rubber (*Hevea brasiliensis*) and, to a lesser extent, black pepper (*Piper nigrum*).

Kantu' augury is based on the belief that the major deities of the spirit world foresee events in the human world and that, out of benevolence, they endeavour to communicate this knowledge to the Kantu'. By reading the intended meaning of these communications correctly, they believe that they too can possess this foreknowledge. The most common media through which the deities express themselves are believed to be seven species of forest birds, which are thought to be the sons-in-law of the major deity *Singalang Burong*. The birds vary in age and hence in authority. In ascending order they are the *Nenak* (white-rumped shama, *Copsychus malabaricus*), *Ketupong* (rufous piculet,

Sasia abnormis), *Beragai* (scarlet-rumped trogon, *Harpactes duvauceli*), *Papau* (Diard's trogon, *Harpactes diardii*), *Memuas* (banded kingfisher, *Lacedo pulchella*), *Kutok* (maroon woodpecker, *Platylophus galericulatus*) and *Bejampong* (crested jay, *Blythipicus rubiginosus*). In practice, the Kantu' take most of their omens from the first three birds, or from three variant ritual practices called beburong besi (taking omens from iron), betenong kempang (taking omens from the kempang stick) and beburong pegela' (taking omens from an offering).

The Kantu' deem omens from these birds to be relevant to many facets of life, including travel, litigation and, especially, swidden cultivation. Omens are observed through most of the stages of the swidden cycle and typically are honoured by proscription of swidden work on the day received. The most important omens, however, are those received during the first stage of the cycle, selection of the proposed swidden site. This stage of the swidden cycle, called *beburong* (to take birds or omens), consists of traversing a section of forest proposed for a swidden and seeking favourable bird omens. The character of the omens received at this time – *burong badas* (good birds) versus *burong jai'* (bad birds) – is believed to be a major determinant of the character of the eventual swidden harvest. Accordingly, if a sufficiently ill omen is received, the site should be rejected for farming that year.

The key to interpreting site-rejection in particular, and the system of augury in general, is the indeterminacy of the physical environment of Borneo, the impossibility of correctly predicting critical agro-ecological conditions, and the consequent need to devise pluralistic rather than deterministic agricultural strategies. For example, during my fieldwork the members of one household observed that rice-destroying floods had not occurred for several years. They reasoned on this basis that the likelihood of such a flood occurring during the coming year was relatively high, and so they decided to locate all of their coming year's swiddens on high ground. This reasoning is apparently flawed: there is no evidence of cyclical patterns in rainfall or flooding in Kalimantan, and thus anticipation of a flood (or a drought) will be wrong as often as right. The best strategy would be to prepare neither for a year with a flood nor a year without a flood, but for an average year – that is, a year with some percentage likelihood of a flood. The system of augury promotes such 'averaging' strategies by making agricultural decision-making less deterministic – by systematically severing empirical linkages between the environment and human decision-making (see Dove 1993).

Augury and the tropical forest environment

The key to my interpretation of Kantu' augury is the fact that I could find no empirical linkage between the behaviour of the omen birds on the one hand, and on the other hand the success or failure of the Kantu's swidden harvests. Whereas there is an empirical basis to the birds' behaviour, in the sense that they have fixed and predictable habitats, feeding patterns and mating seasons, there is no temporal or spatial pattern in the birds' behaviour that correlates with temporal or spatial variables critical to swidden success. With regard to temporal patterns, the principal birds from which omens are taken are not known for seasonal variation in behaviour. There is some diurnal variation: the Kantu' say that most birds (such as the Rufous Piculet) do not call during the heat of the day, nor (with the exception of the White-rumped Shama) when it is raining. But there is no indication that this variation is relevant to swidden success or failure. The evidence is somewhat more suggestive as regards spatial patterns. For example, Freeman (1960) found that the omen birds vary in the extent to which they approach human habitations; for example the Rufous Piculet is commonly seen and heard in the immediate vicinity of longhouses. There also is related variation in the extent to which the omen birds frequent primary forest; for example the Kantu' say that the Rufous Piculet, along with most other birds, does not frequent primary forest. Since proximity to the longhouse and the primary/secondary forest distinction are important variables in the Kantu' swidden system, such spatial variation suggests a possible empirical linkage between specific bird omens and specific swidden outcomes.

In any case, the rules of augural interpretation thoroughly scramble any possible empirical linkage between bird behaviour and swidden success or failure. A linkage is ruled out not by ecology, therefore, but by culture. The rules for interpreting omens are ecologically arbitrary, which effectively disassociates interpretation from the behaviour of the birds. According to Kantu' augural lore, for example, some omen birds have more than one type of call, and the meaning of an omen varies according to which call is heard. Thus, the normal call of the Rufous Piculet is auspicious but its variant trill is inauspicious (Freeman 1960). There appears to be no ecological significance to this variation. Similarly, there is great augural, but no agricultural, significance attached to whether one or more calls of the Rufous Piculet (or other omen bird) are heard. Equally important to interpretation (and equally irrelevant from an agricultural point of view) is whether the call is heard (or the bird is seen) to the observer's right or left. Augural interpretation is subject to an extravagant number and variety of additional rules and caveats, all of which appear arbitrary in an agro-ecological sense.

The agro-ecological arbitrariness of augury is most clearly illustrated in the performance of *betenong kempang* (to divine from the kempang tree), a variant type of augury that the Kantu' sometimes practice instead of seeking omens at the prospective swidden site. It consists in cutting a pole from the kempang tree (probably *Artocarpus elasticus*) and measuring and marking one's *depa'* (arms-breadth) on it. The augurer then proceeds to cut some of the underbrush on the site, after which he or she re-measures his or her arms-breadth against the kempang pole. If this measurement exceeds the initial one (indicating that the pole has 'shrunk'), this augurs ill for the proposed site, but if the measurement falls short of the initial mark (if the pole has 'grown'), this augurs well. Although this procedure is susceptible to unconscious influence on the part of the augurer, it none the less represents a cultural statement regarding the fundamental randomness of the augural system.

The absence of an empirical causal association between omens and agricultural ecology is in fact a prerequisite to Kantu' belief in the efficacy of omens. If a particular omen is obviously associated with some spatial or temporal variable relevant to swidden success, it undercuts its own supernatural character. In most systems of divination, it is precisely the impossibility of any such empirical connection that confers supernatural authority on the system (Aubert 1959). This explains why any systematic ecological information that the birds might convey to the Kantu' must be obscured by the investment of meaning in such non-systematic phenomena as the number, type and direction of their calls.

The content of the Kantu' system of forest augury, although not environmentally meaningful, is culturally meaningful. For example, the importance of the oppositions of left/right and hot/cold in interpreting omens relates to a wider cultural tradition of symbolic opposition that is found throughout Southeast Asia. However, these meaningful relations within the body of augural lore do not correlate with meaningful relations between the forest environment and swidden agriculture. The absence of any such correlation means that the content of the augural system is arbitrary and could have been taken from many possible phenomena besides the behaviour of forest birds. The birds came into use by virtue of their projective value. One of the most salient sensory inputs in the tropical rainforest is sound, especially that from (usually unseen) birds. This suggestion is supported by the fact that the seven omen birds are remarkable less for their appearance (Freeman 1960) than for the arresting character of their calls. These calls are notable for their anthropomorphic character, evoking qualities of excitability, taciturnity and so on (Richards 1972; Freeman 1960).

Another distinctive feature of the rules governing augural interpretation is the proscription of inter-household sharing of omens and promotion of idiosyncratic interpretation of them. Augury is performed by each household on its own, usually by the eldest male. 'Omens cannot be shared', the Kantu' say. If an auspicious omen became known to a neighbouring household, the latter would want to join in taking it. Such sharing might abrogate the auspiciousness of the omen or, minimally, make it difficult for the original recipient household to obtain that omen again in future years. The Kantu' minimize sharing of omens by the simple expedient of keeping their own household's omens secret from other households. Sharing is also minimized by augural rules that tie omen interpretation to the varying composition and fortunes of the individual household. For example, the meaning of certain omens (e.g. the *bacar* call of the Rufous Piculet) is said to vary depending upon whether elders live in the household. Of more importance, many omens have no meaning other than to signify a reversal of the household's prior swidden fortunes, regardless of whether these were good or ill (Sandin 1980). For example, if a household hears the *bacar* call of the Rufous Piculet when selecting a swidden site, they must abandon that site unless they (viz. their particular household) have never obtained a good harvest of rice. This arbitrary reversal of the meaning of omens makes it difficult to share them, increases inter-household diversity in responses to omens, and generally enhances the randomizing effect of augury.

Sharing of omens and systematization of augury's impact is also mitigated by the belief that augural interpretation is personal and idiosyncratic. The Kantu' say, *'Utai to'ngau bidik kitai'* ('This thing is a matter of our own fortune'). As Metcalf (1976) claims for the Berawan (of Sarawak, Malaysian Borneo), each person builds up over his lifetime a personal and distinctive relationship with each of the omen birds. It is quite possible, as a result, for two augurers to assign completely opposing meanings to the same omen. This personal relationship, coupled with the fact that there is considerable inter-household variation in the knowledge and intensity of observance of augural rules, ensures that there is considerable inter-household variation in both the seeking of omens and the interpretation of the omens obtained. If omens conveyed to the Kantu' empirically valid information about the environment, we would expect inter-household agreement on what information is conveyed by what omen, and we might also expect inter-household sharing of this information, but this is not the case.

Development implications

The principles that underpin the Kantu' system of augury, like the pluralism just discussed, are very different from those of modern science. Analysis of these differences can help to defamiliarize and thus make more accessible to critical review the concepts being used by contemporary scientists and planners to try to understand and manage tropical forests. The aspect of modern science that stands out in greatest relief by comparison to Kantu' augury, and that may benefit most from such a review, is its linear, deterministic, and monistic character. Holling, Taylor and Thompson (1991) critique this character in an intriguing essay, which they end with a poem by William Blake, the last, memorable line of which reads, 'May God us keep, From Single vision and Newton's sleep!' Such critiques notwithstanding, this paradigm of single vision continues to be embraced by scientists, international donors/lenders and national planners working in the field of development.

The difference between the deterministic development paradigm, and the paradigm that lies behind Kantu' augury, can be seen in their capacity for distinguishing between system and rules. Almost all development strategies for exploiting the tropical forest are written at a level that is analogous to that of augural rules as opposed to the augural system. It is as if the development strategies were intended to inform people about the meaning of particular omen birds, but not the meaning of the overall system. Development strategies tend to promote whatever is associated with the most recent success as *the* 'bird' for the future. Since no single solution is ever successful for long, development strategies become serially committed to one 'bird' after another. This process is flawed, however. My analysis of Kantu' augury shows that the Kantu' are protected from bad decision-making by their search for the right bird; they become vulnerable when they start searching for the right system. The wisdom of Kantu' forest augury, therefore, lies in the way that it focuses such efforts on the birds as opposed to the system itself. This dimension is missing from the development paradigm, which typically lacks an over-arching framework within which the limitations of individual strategies can be seen. The lesson for development planners, therefore, is to look not just for new 'omen birds', but for a system of 'omen-taking' within which the individual birds make sense, to write at the level not just of rules but also of systems.

Another important distinction between the augural and development paradigms is the way that they cope with the uncertainty that characterizes the tropical forest ecosystem of the Kantu'. Both paradigms represent responses to uncertainty, but while the development paradigm tries to eliminate it, the augural paradigm embraces it. The latter approach has received increasing support from scientists. Converging studies in a variety of fields (one of the most explicit of which is chaos theory) suggest that we need to come to better terms with the limits of our ability to know, in a deterministic way, the unknown. Common to these studies is the belief that embracing our ignorance is, paradoxically, the best way to overcome it. Thus, Ludwig, Hilborn and Walters (1993) write: 'Confront uncertainty. Once we free ourselves from the illusion that science or technology (if lavishly funded) can provide a solution to resource or conservation problems, appropriate action becomes possible'. Thus, Kantu' augury, by not just emphasizing but indeed celebrating uncertainty, does in fact reduce the uncertainty of agro-ecological futures in the tropical forest. Augury, by trying to make sense of the limits to knowledge, transcends them; while development, by assuming that there are no limits, circumscribes itself.

The critical difference between the two types of process may be the presence or absence of feedback. In the development paradigm, any success tends to be rewarded with ever-greater commitments of resources to the particular strategy that produced this success: there is positive feedback to success. The augural rules of the Kantu', in contrast, minimize any alteration in resource allocations based upon the past success (or failure) of a particular omen or swidden strategy: there is no feedback, or neutral feedback, to success. The purpose of the positive feedback in the development paradigm is to find the *right* solution; the purpose of the neutral feedback in the augural paradigm is to avoid finding *wrong* solutions (viz. solutions based on apparent but invalid ecosystemic patterns). The positive feedback paradigm exacerbates the volatility of relations between society and environment, while the neutral feedback paradigm dampens it. The former generates successive, ever more deterministic systems for managing environmental relations, which are ever more productive if successful but ever more disastrous if unsuccessful; while the latter contributes to a pluralistic strategy, which can cope with a wide range of conditions and minimizes extremes of either success or failure.

The available evidence suggests that there is no systematic relationship between augury and favourable conditions for swidden cultivation, and that the lack of any such relationship is culturally enhanced by the rules of the augural system itself. This metaphoric throw of the dice at this critical point in the swidden cycle is a statement about the indeterminacy of the environment, the imperfection of our knowledge of it, and the inappropriateness of systematic management strategies. The selection of swidden sites is problematic for the Kantu' because of the large number of environmental variables that differentiate sites and because the particular variables associated with swidden success change unpredictably from year to year. The augural system of the Kantu' helps them cope with this unpredictability by minimizing a tendency towards systematization in swidden behaviour. This promotes intra-household and inter-household diversity in swidden strategies, which helps to ensure a successful adaptation to a complex and uncertain environment. This traditional system of adaptation provides a needed lesson to development planners in the virtues and ills of, respectively, indeterminate versus determinate strategies for resource use in the tropical forest.

The Kantu' expression for the search for omens at a potential swidden site is *ninga tanah* (listen to the earth). It should now be clear that the augural system does indeed allow the Kantu' to listen – at a number of different level – to the earth. Interest in such listening is not limited to Bornean tribesmen. With increasing appreciation of the complexity of our environment and the need to be receptive to its patterns (or lack thereof), scholars are coming to recognize that a variety of different phenomena, in many different cultures, represent attempts to 'listen to the earth'. For example, in an inspired comparison of the rainforest and baroque music, Diamond (1990) writes:

> 'When, late in life, Bach wrote his Lord's Prayer, I suspect that he was trying to express the view he had reached of nothing less than life itself, and of his own struggles to hear God's voice despite the obstacles that life poses... It's as if Bach were praying: yes, yes, by all means forgive us our trespasses, and all those things – but above all, God, give us the will and ability to hear Thy voice through this world's confusion. With this metaphor, Bach also unwittingly captured better than any other metaphor I know, the sense of what it's like to come to know the rainforest. This conclusion is neither blasphemous nor trivializing, because to biologists the rainforest is life's most complex and wonderful creation. It overwhelms us by its detail. Underneath that detail lie nature's laws, but they don't cry out for attention. Instead, only by listening long and carefully can we hope to grasp them.'

If Bach's music is a metaphor for creation, then, so is augury a (constructed) metaphor for the rainforest and the principles that govern it.

The continuities among Bach, the rainforest, and Kantu' augury are based on a common, underlying element: the challenge of trying to understand our environment. Interpretations of the challenge vary: Diamond and Bach think that it comes from the wealth of detail in the world and, by inference, the human penchant for missing the patterns for the detail; while Kantu' augury suggests that the challenge comes from a human penchant for just the opposite – seeing false patterns where there is only detail. The differences among these interpretations are not without interest, but of more importance here is the fact that they all focus on the same issue: human sentience and culture appear to both facilitate and frustrate the 'knowing' of nature. Bateson (1972) asked, 'How can consciousness be used to comprehend phenomena that transcend it?' The Kantu' system of augury, I suggest, evolved in response to the related question, 'How do we make sense of nature in cultural terms?'

Acknowledgements

I first carried out research on the Kantu' of West Kalimantan from 1974 to 1976 with support from the National Science Foundation (Grant #GS-42605) and with sponsorship from the Indonesian Academy of Science (LIPI). I gathered additional data on local systems of adaptation to the tropical forest during six years of subsequent work in Java between 1979 and 1985, making periodic field trips to Kalimantan, with support from the Rockefeller and Ford foundations and the East-West Center and sponsorship from Gadjah Mada University. A recent series of field trips to Kalimantan, beginning in 1992, has been supported by the Ford Foundation, the United Nations Development Programme, and the John D. and Catherine T. MacArthur Foundation, with sponsorship from BAPPENAS and Padjadjaran University. Earlier versions of this analysis were published in Dove 1993 and 1996 and are drawn on here with the generous permission of the original publishers. The author alone is responsible for the analysis presented here.

The holy hills of the Dai (Pei Shengji)

The Dai (T'ai), an indigenous ethnic group in south-west China, inhabit the Xishuangbanna region in Yunnan Province. They have a long tradition of biodiversity and habitat conservation characterized by the management of Holy Hills through formal and informal norms, ethical rules and religious beliefs. The Dai were first recorded in Chinese historical texts as far back as the early years of the Han Dynasty (*c.* 2000BCE). For centuries they have depended upon the natural world for survival, which has involved, in addition to settled agriculture, hunting, fishing and the collection of wild plants.

The Dai originally followed a polytheistic religion that was heavily bound to the natural world and embraced a forest-oriented philosophy. The Dai perception of the interrelationship of human beings with their physical environment is that it consists of five major elements: forest, water, land, food and humanity. They believe that the forest is a human's cradle. Water comes from the forests, land is fed by the water, and food comes from the land that is fed by the water and the rivers. Human life is supported by the forests, and the forests are one with the supernatural realm. One of the Dai folk songs states, 'Elephants walk with the forests, the climate with bamboo'. Another folk song of Xishuangbanna says, 'If you cut down all the trees, you have only the bark to eat; if you destroy the forests, you destroy your road to the future'.

Today the Dai practice a predominantly Buddhist religion. In their traditional concepts a Holy Hill or *Nong* is a forested hill where the gods reside. All the plants and animals that inhabit the Holy Hills are either companions of the gods or sacred living things in the gods' garden. In addition, the Dai believe that the spirits of great and revered chieftains go to the Holy Hills to live, following their departure from the world of the living.

Holy Hills can be found wherever one encounters a hill of virgin forest near a Dai village and are a major component of the traditional Dai land management ecosystem. In Xishuangbanna approximately 400 of these hills occupy a total area of 30,000 to 50,000 hectares, or 1.5 to 2.5 percent of the total area of the prefecture. There appear to be two types of Holy Hill. The first, *Nong Man* (or *Nong Ban*), refers to a naturally forested hill, usually 10 to 100 hectares in area, that is worshipped by the inhabitants of a nearby village. Where several villages form a single larger community, another type called *Nong Meng* is frequently found. Forested hills of this second type occupy a much large area, often hundreds of hectares, and they belong to all the villages in the community. In respecting the Holy Hills, the Dai villagers not only keep the sanctity of the Hills, but they also present regular offerings in the hope that the gods will be pleased and protect their health and peace.

Traditionally, the Holy Hills constitute a kind of natural conservation area with great biological diversity, founded with the help of the gods; and all animals, land and sources of water within the area are inviolable. Gathering, hunting, wood-chopping and cultivation are strictly prohibited. Although they are intimately associated with their beliefs and rituals, the Dai people do not use these hills as cemeteries; areas of burial are confined to separate hills called *Ba hao* in the Dai language. The Dai people believe that such activities on the hills would make the gods angry and bring misfortune and disaster as punishment. A Dai text warns, 'The Trees on the Nong mountains (Holy Hills) cannot be cut. In these forests you cannot cut down trees and construct houses. You cannot build houses on the Nong mountains, you must not antagonize the spirits, the gods, or the Buddha'.

Xishuangbanna lies in a transitional region between tropical and subtropical zones. Different types of forest vegetation are found in this mountainous region, ranging from rainforest (below 800 metres),to seasonal rainforest (between 800 and 900 metres) and evergreen mountain forest (above 900 metres). Almost all Holy Hills are located in the seasonal rainforest areas. This might be explained by the fact that Dai villages are settled along the distributive line of seasonal rainforest vegetation in this region. A number of studies indicate that the vegetation on these hills closely resembles the patterns of vegetation in large tracts of pristine, regional forests in terms of character, structure, function and species composition.

Near the village of Man-yuang-kwang, for example, the Holy Hill covers 53 hectares at an altitude of 670 metres above sea level. The hill's forests contain 311 different plant species belonging to 108 families and 236 genera. The structure of the forest community can be divided into three layers of trees, of which 20 to 30 percent are deciduous or semi-deciduous, one shrub layer, and one layer each of herbs and seedlings. The forest's ecological characteristics, including energy flow, material flow and meteorological functions, after eight years of continuing observation, indicate strong similarities between the patterns of the Holy Hill and those of the tropical seasonal rainforest nearby.

Despite modern development interventions that have covered some of these hills with cash crops, the Holy Hill concept has made a significant contribution to the conservation of biological diversity in Xishuangbanna. First, it has contributed to ecosystem conservation; as a consequence there are hundreds of well-preserved seasonal rainforest areas, which are characterized by species of *Antiaris*, *Pouteria*, *Canarium* and others. Second, a large number of endemic or relic species of the local flora have been protected, including about 100 species of medicinal plants and more than 150 species of economically useful plants. Third, the large number of forested Holy Hills distributed throughout the region form hundreds of 'green islands'. This pattern could help the natural reserves, which were established by the state government in recent years, by exchanging genes and playing the role of 'stepping stones' for the flow of genetic materials. The natural reserves are separated into five large sections and seven locations totalling 334,576 hectares and are usually surrounded by larger and smaller Holy Hills.

Human culture is built upon and developed on the basis of the physical world, and they are interdependent on each other. Hence cultural diversity largely depends on the biological diversity which provides tangible materials in enormous varieties from which humans can establish societies and lifestyles. As the Xishuangbanna region is characterized by a wide biological diversity and ethnic cultural traditions, the Dai people have developed their own traditional culture based on the available resources in their surrounding environment.

The principle of the co-existence of biological diversity and cultural diversity has resulted in distinctive physical phenomena in the landscape of the Dai people in Xishuangbanna. The Holy Hills with their natural forest vegetation, and other traditional practices such as the preservation of plants in temple yards, have become a part of Dai life. The interdependence of cultural diversity and biological diversity has strongly demonstrated that the principle of co-existence of both diversities has been established through the process of human history. Thus, it suggests that the conservation of biodiversity and cultural diversity should be considered as interlinked in the process of development today.

[Adapted from Pei Shengji (1993) 'Managing for Biological Diversity Conservation in Temple Yards and Holy Hills: The Traditional Practices of the Xishuangbanna Dai Community, Southwest China. In: Hamilton, L. S. (ed.) *Ethics, Religion and Biodiversity*. The White Horse Press, Knapwell, Cambridge.]

Cultural and spiritual values related to the conservation of biodiversity in the sacred groves of the Western Ghats in Maharashtra (E. Bharucha)

Several areas of India, particularly the forest belts, are inhabited by tribal people known to have patches of vegetation associated with religious and cultural sentiments that have been preserved through local customs. Most authorities have suggested that their pro-conservation behavioural patterns are related to an urge to preserve nature.

The present paper deals with the socio-cultural and socio-religious aspects of the lives of forest 'ecosystem people' of the Mawal and Mulshi Talukas of Maharashtra in the Western Ghats. It describes the lives of local people and the linkage with the veneration they have for local deities around which 'Sacred Groves' are located. This work explores the local control systems that have led to the preservation of plant species in this 'hot spot' of biological diversity.

The Western Ghats of India are an acknowledged 'hot spot' of plant and animal diversity. People have lived in close association with these forests for thousands of years. These communities once had a relatively low impact on the forest, preserving clearly-demarcated patches as sacred forests. Contemporary thought stresses these issues, however it frequently fails to point out how such mechanisms work. It also fails to explore the way in which current changes in land and resource use patterns affect these traditional values, including retention of sacred groves.

Sacred groves in the Western Ghats are surrounded by very heterogeneous populations. Among the people who live in this region are communities highly dependent on forest resources for food, fuelwood, fodder, medicines, fibre and timber, and materials for creating artefacts and a variety of household goods. There are also key specialist groups who know where to gather medicinal herbs, and the leaves, fruit and roots of medicinal plants, and how to process them for specific ailments. There are specialists within the community who know where to fish in the hill streams and how to hunt and capture wildlife. Least understood are the people who claim a knowledge of the forest deities and how these are linked to the surrounding landscape, and the relationship of the deity to the daily lives of the community. These priests, known as *pujaris* or *bhagats*, have been chiefly responsible for the implementation of localized rules and regulations, based on religious customs, that have led to the conservation of regional floral diversity.

Of the 40 groves in this area, 30 are less than a hectare in size. Four are between one and two hectares and four are between two and three hectares. One covers four hectares and the largest is eight hectares in size. Thus, most groves are extremely small. The forest is mainly semi-evergreen in nature: however, groves to the west are evergreen. The region's species diversity as a whole is remarkable. In the adjacent Khandala hills, Fr. Santapau recorded over 800 plant species, of which 162 are trees and 82 are shrubs. Reddy reports 300 plant species from the Ambavne hills in this region. In the 15 groves studied, the total number of species of trees and shrubs inventoried is 223. The species richness of trees and shrubs in a grove varied from 10 to as high as 86. Contrary to contemporary thought, each grove by itself cannot be said to be of great species richness. Taken together, however, they include most of the plant species that are present in this region of the Ghats. The forest structure of the groves, however, is remarkable, as the groves represent the least disturbed islands of old growth in the region.

This inventory does not account for the large number of ground flora and annuals that appear in the monsoon. The small size of individual groves is an obvious factor that determines the relatively low species richness of individual groves. However, the size of each grove does not correlate well with their species richness, indicating that the number of species is more closely related to the level of local protection rather than with size. Most groves have a canopy cover of 80 percent; however groves that have been disturbed usually have a 40 percent canopy.

During the British period the hill forests of the Ghats were notified as Reserved Forests or as Protected Forests. The former were kept for timber extraction by the government. In the latter, local people were permitted to extract resources, regulated by the Forest Department. Though restrictions on use of resources from Reserved Forests continued after Independence, those relating to Protected Forests were relaxed considerably, leading to a complete disappearance of some forests and the partial degradation of others. Groves within Reserved Forest landscapes thus form less clearly definable entities than those in Protected Forests or in village Revenue lands in which the surrounding forest has been totally degraded. In 12 of the 40 groves studied there has been a recent change in surrounding land-use, to develop tourism or expensive housing for urban people. Five are surrounded by agricultural land. Seven are within Reserved Forests. Others have a mixed landscape that forms a mosaic around them.

Two forms of land-use can be said to form the matrix of landscape elements in which the groves are situated. One type of grove forms an island of green, in an area where hill slope cultivation over hundreds of years has led to highly lopped stunted trees and shrubs with patches of cultivated and fallow fields. The branches and leaves of the trees are used year after year for 'wood-ash cultivation' in which biomass collected from the trees is burnt in crop nurseries where rice or a hill-slope grain *nachni* (*Eleusine coracana*) is grown. The crops are rotated and the land is then abandoned until it recovers. Shortening of the length of fallow periods over the years due to increasingly intensive use of the land has led to the formation of a degraded ecosystem. The other type of matrix consists of Reserved Forests, notified since British times, with varying degrees of forest preservation. The fact that the Sacred Groves stand out as much older trees, with an intact canopy, in contrast to these government-controlled forests, points to the greater level of protection afforded by local sentiments compared to that afforded by policing and enforcement in this area.

A total of 156 respondents from 30 villages were interviewed from the Mawal and Mulshi Talukas in the Western Ghats. These included local farmers, the *pujaris* or priests of the deities, people who collect resources, and medicinal plant collectors. The area in which these agropastoral communities live has over 40 sacred groves dedicated to a variety of local deities. The settlements vary in size from 200 people to over 1,000 people. Each family usually has around three to five head of livestock which are maintained for cattle dung, used mainly as a fertilizer and as a fuel.

Each village may have several temples, some of which are Brahminical and presided over by the local Brahmin priest. These are situated within the village. There are also local forest deities to which a patch of forest is dedicated and traditionally preserved. These are mostly situated at a remote place. The simple temple in the depth of the forest patch is located high up on an adjacent hill, or at the origin of a stream. The situation thus promotes the view that sanctifying the area around the water source has led to forest conservation, which in turn maintains the spring's perennial nature. Local people who look after the shrine itself are either Mahadeo Kolis or Marathas. Rituals include animal sacrifice, rites to control resource use and rituals for predicting future events.

Several of the groves are associated with folklore that strengthens local sentiments for the deity and has led to pro-conservation behaviour. These stories are of ancient origin, or are fairly specific about when supernatural events in connection to the deity are supposed to have occurred. Some are claimed to have a fairly recent origin. Some folk tales refer to supernatural events that have occurred following the desecration of a grove's sanctity. Others are related to the origin of the grove.

About half the local respondents reported a sense of reverence for the deity, and people frequently expressed the sentiment that the deity is a personal protector of the individual and the community. Approximately 40 percent of the respondents showed a high degree of veneration for the local deity, and indirectly for the grove. Most respondents estimated that the groves are between 200 and 250 years old. This could be related to the fact that people such as the Mahadeo Kolis migrated into this region around that time. They could thus have established these groves after taking over control of the region, or could well have used sites that were venerated before their arrival. Other respondents claimed that the groves are ancient and have been present from the beginning of human history: 'they were there from the beginning'.

There are clearly discernible institutional mechanisms related to 'ownership' of the site that are based on local traditions which have been responsible for preserving the groves. Among the 40 groves, 15 are controlled by the village *panchayats* and are considered village 'commons' which cannot be used for individual resource needs; 13 are within Reserved Forest lands, which, however, are perceived to belong to the deity, despite of the fact that they are government controlled; 11 are privately owned by local villagers who may or may not be the *pujari* who is the only person permitted to perform rituals at the shrine; 1 is controlled by a temple trust in which a large temple has been built with funds received from people who visit the site and from the sale of timber by cutting the trees from the grove.

The current concept that there are complete bans on resource extraction has not been validated for this area, except for a few groves. In several groves, people are allowed to collect fallen dry wood, fruit from the forest floor, honey, sap (by tapping *Caryota urens* to make an alcoholic beverage) and other products. In some groves, cattle grazing is permitted. In most groves, however, timber cannot be felled without the express permission of the deity, which is obtained through a ritual process known as *kaul*. This is usually done to get timber to build a superstructure for the shrine. In most groves women are not permitted to enter the grove, while in some cases entry is banned only during the menstrual period. The ban on the entry of village women explains why most of the dry wood is left to decay, even where its collection is permitted, as it is the women who are the fuelwood collectors of the settlement. Men rarely collect fuelwood as it is considered a task for women. This practice could well curtail the use of small timber and fuelwood from the grove. A local medicine man has stated that he is reluctant to collect certain types of plants from the groves as local people may not like this to be done.

The deities, ceremonial rites and local taboos

The deities are associated with feminine forms and are thought to be fearsome and vengeful. The shrines consist of a few undressed rocks placed in small clearings where the undergrowth has been removed. Occasionally they are raised on a platform and in a few there is a small brick structure with a wooden roof. Artefacts such as earthen pots, flags and bells are frequently dedicated to the shrine. The deities are usually non-brahminical or may be associated with certain Hindu deities. The name *'Waghjai'* appears in 7 of the 40 groves, and is said to be a tiger goddess or a goddess who uses a tiger as a steed. Several shrines are dedicated to unique deities whose names are not encountered elsewhere.

The rites performed by the local priest frequently use either flowers from the forest, or grain. A *kaul* ceremony is used for requesting permission to extract resources from the grove, or it can be performed on behalf of an individual who wants the *pujari* to enlist the deities' help to take correct decisions or predict the future. The methods used during the ceremony are expected to yield a yes or no answer, which represents the wish of the deity who must then be obeyed. Most ceremonies would have a 50 percent chance of a positive or a negative answer as this is left purely to chance. The *kaul* ceremony varies from shrine to shrine, and may be used to decide if a tree should be cut to support the roof of the shrine, or to collect money for building a superstructure around the deity, or to allow the passage of a road through the grove. The level of belief is obvious as the reply is adhered to and the wishes of the deity are respected even today.

Most of the local folklore has two basic themes, which relate to the origins of the shrine, or to stories that depict the effects on people who have desecrated the grove by felling a tree or the punishment inflicted for violating the sanctuary of the grove in any form. For example, the story of Balgirbua as told by local people depicts an era during the British period when this semi-mythical holy man was supposed to have lived in Lawarde village. He is said to have had to travel a great distance to pray every day, and as he aged he requested God to help him. God is said to have told him that he would follow him every day and protect him from harm on the condition that he should not try to look behind him to confirm His presence. In a moment of weakness, the devotee turned back to see his God who instantly turned into stone and disappeared underground. Prayers for forgiveness resulted in a vision in which Balgirbua was asked to dig at the site. When villagers dug deep into the earth they struck the stone idol which is supposed to have filled the pit with blood. The idol was then installed in a shrine and is venerated today in a small grove of trees.

Several stories tell of the appearance of a shrine out of nowhere through supernatural events. Others describe the occurrence of illness after someone has cut a tree in the grove. In one grove a huge python is said to have surrounded a clump of bamboo on two occasions when people tried to cut it. When a road was being constructed by labour brought in from outside, they fell violently ill until the deity 'permitted' construction after a *kaul* ceremony. Stories of blindness and sickness when women entered the grove recur frequently in the area.

Conclusion

Most authors have claimed that there are strict taboos governing resource collection from sacred groves. Our experience shows that these restrictions vary from one village to another within the same region, from fairly strict prohibition to controlled sustained use, through the establishment of set principles based on religious customs. There is little to substantiate the statement that local people who support these values have objectives that are aimed at the conservation of species. Conservation is rather a side-effect of customs that dedicate forest resources to their deities. This is maintained out of mixed feelings that can be described as a complex of emotional sentiments ranging from love to veneration and even fear. Most frequently, conservation can be ascribed to a sentiment that states that 'this is our custom'.

The findings of this study indicate that these sacred groves have not been established primarily for species preservation but constitute a highly complex social mechanism with several sentiments that are related to the presence of a deity that is venerated by the local community. Most authors appear to suggest that the preservation of these groves is a result of an innate value among traditional communities to preserve biological diversity for the future. This paper in contrast has shown that conservation is a spin-off resulting from other traditional values and sentiments that are unrelated to a desire for species conservation.

In the past, only a small proportion of a region's resources were sequestered in sacred groves. The surrounding landscape could provide relatively easily for all major resource needs, making the cost of preserving the grove insignificant. The perceived benefit was the protection and good will afforded by the deity in return for not disturbing the sanctum. In recent times, however, the cost of preservation has become relatively higher as the surrounding landscape has been degraded or land-use patterns have changed to include more intensive farming. Most recently, urbanization and tourism development have led to a serious depletion of resources as the phenomenal rise in the price of land provides a great incentive to sell the grove irrespective of the sentiment that was at one time sufficiently strong to preserve it.

The survey of 40 sacred groves has shown that although local people hold the deities in deep reverence, changing land-use patterns in the vicinity of the groves reduces their intactness. Groves owned by individuals have recently been sold to develop five-star tourist complexes. Further degradation is inevitable unless traditional beliefs are strengthened through newer 'real world' values for the conservation of biological diversity through local support strategies. An awareness programme at the local level is perhaps the most vital key to the long-term survival of sacred groves in developing areas.

[Field Research Team: Bharati Vidyapeeth Institute of Environment Education and Research (BVIEER) and Tata Electric Companies. Chaudhuri Sujoy, Elangbam Athoiba Singh, Kumar Shamita, Kadapatti Sanjay, Pokharkar Pramod, Surendran Anita, Mukund and Watve Aparna.]

Peasant cosmovisions and biodiversity: some reflections from South Asia (Pramod Parajuli)

'... *jungle jaminka chadar he*' (... forest is a shawl that covers the land)
(Jhari Ghatwal, a peasant from Madhupur, Santal Pragannas)

Indian ecological historian Ram Chandra Guha once commented sarcastically that, 'the Indian environmental debate is an argument in the cities about what is happening in the countryside' (as in Bhaviskar 1995: 229). What, then, are the ways peasants and ecosystem people – those who have to establish a healthy relationship with their immediate environment – comprehend nature and define human–nature relationships? What is their stake in nurturing and maintaining biodiversity? By biodiversity I mean 'a totality of species, populations, communities and ecosystems, both wild and domesticated, that constitute the life of any one area' (Dasmann 1991: 8-9). Thus it is not merely the sum of the biological ecosystem but also the human modifications of this ecological niche. I look closely at the larger ecosystem and in particular at the specific agro-ecosystems. The central message of this chapter is that biodiversity can be maintained within the farms and at the intersection of farms and forestry. As Netting and Stone (1996: 54) argue, 'even high population densities and land use do not eliminate wild plant species but may instead lead to the incorporation of many of them, by protecting them from competition and through conscious management'.

Peasant views are often portrayed as romantic and/or utopian. In actuality I find quite the opposite to be the case. Proposals put forward by environmental movements of the poor are practical and achievable. The only obstacle such proposals face is the way the mainstream biodiversity and nature conservation programmes separate human interests from the so-called conservation of nature. The idea of 'extractive reserves', proposed by the late Chico Mendes in the Brazilian Amazon, is a case in point. He tried to undo the human–nature dichotomy through a proposal to combine the livelihood of rubber tappers with sustaining the Amazon forest ecosystem. By 'extractive reserves' he meant that the land is under public ownership but the rubber tappers and others who live on the land should have the right to live and work there (Mendes 1989).

I would like to consider this question using an example from the Jharkhand region of east-central India.

The Jharkhand region (*Jhar* meaning 'forest' and *Khand* meaning 'area') existed as an ecological unit long before it emerged as a political concept. Situated at an elevation of 500 to 1,000 metres above sea level, Jharkhand's hilly terrain hosts its own unique technologies of forestry, agriculture and irrigation. Suited to those technologies, it has also developed distinctive labour and political arrangements. The inhabitants of Jharkhand are called by various names such as *adivasi* (the indigenous peoples), *vanajati* (forest dwellers), *girijan* (mountain dwellers), *paharia* (hill dwellers) and *sadans*. Although used as cultural categories, they are rooted in the ecological make-up of the region. The Jharkhand region in India offers a unique milieu to juxtapose the peasant 'style of cognition' with the so-called scientific style of resource management. Here we can see that indigenous peasant knowledge is not only about the knowledge of particular species of plants or animals, or their knowledge of medicines or crop-diversity; it is also about the holistic way in which such traditions of knowledge operate in the context of their place within the entire cosmos.

Although Jharkhandi *adivasis* are settled agriculturalists and wet rice cultivators, everything in their culture still speaks of the forest and their wanderings in it. They weave a protective ring around the forest through ritual, religion, folklore and tradition (Bosu-Mullick 1991). I have heard several versions of the stories about how their ancestors wandered from one forest to another 'feeding like silkworms'. How could *adivasis* who settled down by clearing the forest maintain a healthy relationship with the forest? What emerges is that forests are not pristine wilderness untouched by humans: there is a depth of interdependence between forests and *adivasi* peasants, whose struggle to survive led to a distinct mode of livelihood and corresponding cultural and political arrangements. 'Forest is the shawl that covers the land' is perhaps the most appropriate way of representing the peasant view about the connection between agricultural land and forest. Forests provide materials for homes, ploughs and other agricultural implements. Land peripheral to forests is used as pasture. In many parts of Jharkhand it was customary for tribals to burn sections of the forest so that the ashes could be washed down to the neighbouring fields. This was one of their chief sources of organic fertilizer. Moreover, forests provide many medicinal herbs, derived from as many as 250 tree and plant species.

The beliefs held by the *adivasi* peasants of Jharkhand about land, water and forest are based on soundly utilitarian views, enriched by moral economies of reciprocity, restraint and prudence. By ecological prudence, I refer to a system of thinking and practice that maintains the natural process of renewal, the notion of sacredness, and the compulsion of utility in day-to-day life, worked together as a moral constraint against destructive resource use.

Protection from spirits, and against epidemics of drought, floods and crop failure, is sought by maintaining the delicate balance between society and the natural and the supernatural world. The people collectively maintain social discipline in terms of taboos and fear of retribution from the spirits. For example, both the natural and the supernatural spheres interact while contracting a marriage. A marriage cannot be successful without appeasing the creator, the village guard and the ancestral spirits. All these have to be balanced in order to build a new hut or to move to a different settlement. The deities are believed to cause rain and look after livestock and human life.

The natural sphere provides the material base for human activity, but while being acted upon and altered it also gradually alters human consciousness. The wider the intersecting area between human, natural and supernatural spheres, the healthier and more robust is the ecological cosmovision of a community. For the Jharkhandi *adivasis* there is no causal link of one sphere – human, natural or supernatural – determining the others; rather, they mutually reinforce the system as a whole. Neither are there any distinct external and/or internal factors. It is a dialectical development involving all three components.

How then can we conceive of environmental discourses as perceived by those who have a stake in protecting the environment? What emerges is a peculiar peasant style of cognition, based on the moral economy of nature use. In a poem, Ram Dayal Munda, a poet and political activist from Jharkhand writes:

> I know, new forests will also be 'planted'
> But they can't make a sal forest,
> There will be commercial trees,
> which means, people who have nothing to do with commerce
> have no use for these trees.
> But the purpose behind the creation of us trees goes far beyond commerce.
> The thought behind it is divine:
> the roots, flowers and fruits from us are for all,
> The herbs, the bark and leaves belong to everyone
> The fertility flowing down with the rain from the hill top belongs to all.

The peasant view of the forest is soundly utilitarian, but is enriched by moral economies of reciprocity, restraint and prudence. Such wise ways of using nature and harvesting prudently were instituted through totemic relations. Jharkhandi *adivasi* and non-*adivasi* peasants practice several kinds of prudence. In some cases, peasants have provided total protection to some biological communities or habitat patches: these may include pools along river courses, sacred ponds, sacred mountains, meadows and forests. The *Sarna* grove, and practices of protecting around the mountain top, water sources and sacred religious sites are examples. The other example of prudence is the way communities provide protection to certain selected species. For example in Jharkhand, *sal*, *Mahuwa*, and *bel* trees are considered sacred and are harvested judiciously for specific purposes and in specific seasons. Some trees and animals may be protected in critical life history stages, such as not fishing the entire stock in a pool or protecting a pregnant deer during the hunt. Finally, *adivasis* organize resource harvests under the supervision of a local expert.

There is a qualitative difference between modern systems of forest or biodiversity management and the way forest is embedded into the *adivasi* modes of livelihood. Not surprisingly, in the world view of *adivasi* peasants of Jharkhand, forests are not a 'frontier' to be won; they are, rather, yet another addition to their agrarian 'horizon'. They are simply extensions of farms, where people visit for additional foods and medicinal herbs, and for hunting. Cognitively, forest is both a supernatural and natural sphere for *adivasis*; it is both a context and a condition of survival. The *bongas* (spirits) live in the forest, and they have to be propitiated. Yet you have to hunt, collect tubers, and collect household materials in the forest. For example, Mahuwa flowers constitute at least one-fourth of the nutrition of *adivasi* households. When asked how he felt about harvesting large amounts of Mahuwa flowers in season, Bijju Topno, an Oraon *adivasi* of Palamu district responded, 'These trees are like our adopted kin; we take care of them and they give us our sustenance'. Here, a forest serves not one, but multiple functions. Thus a monocultured forest is not acceptable to *adivasis*. Forests must be diverse in form, as well as in function.

Jharkhandi *adivasis* retain a patch of original forest as *Sarna* where the main deities of the clan are kept. *Sarna* is a pan-Indian institution, especially in the hilly and forested regions. *Sarna* is usually a patch of original forest and is dedicated to the main deity of the community. This concept is supposed to be pre-Vedic in origin (about 3000BC). Gadgil and Vartak (1981) counted over 400 groves in Maharashtra, and about the same in Madhya Pradesh. In the Sarguja district of Madhya Pradesh, every village has a grove of about 20 hectares in extent where both plants and animal life receive absolute protection. These are known as *Sharana* forests, meaning sanctuary (Gadgil and Chandran 1992). Like the *Sarna*, *adivasis* also preserve and protect *sasan* (graveyards), *akharas* (dancing or assembly grounds), water sources, and sometimes mountains and watersheds. Among the Santals, the *jahera*, or the sacred grove, symbolizes the remnants of the original village forest, the sacred grove where most of the communal worship takes place.

Sarna must be distinguished from the notion of parks and sanctuaries of modern conservation programmes. A *Sarna* is not a place to be protected from human use, but is intrinsically related to the continuation and regeneration of body, land and community. It is ironic that successive British colonial Forest Acts since 1865 tried to utilize the widespread evidence of maintaining sacred groves in rural India as a justification for creating 'reserved' and 'protected' forests, and ironically to keep the forest out of the reach of the same local communities (Grove 1995; Sivaramakrishnan 1995a, 1995b). For *adivasis* who are settled cultivators, *Sarna* is not an area that is off limits; it is at the very heart of recognizing the *adivasi* self and regenerating *adivasi* identity. Not surprisingly, protecting *Sarna* groves from the axes of timber contractors or from submersion by mines and/or large dams has become one of the main struggles in Jharkhand today. In this context, the *Sarna* embodies an agrarian environmentalism that is different from both the progressive conservationism that places human needs above nature, and deep ecology which subordinates humans to nature (Guha 1996).

How do human, natural and supernatural spheres dialogue with each other and interact? The relationship seems to be one not of hierarchy and dissonance (as in the Cartesian frame of mind), but of equivalence and conformity. Farms, forests and settlements are not considered distinct from each other in the Jharkhandi landscape, either. They are experienced as a continuum of relationship within the general ecology of the area. In addition to human beings, farms, pastures and forests, there are animals that are kept for milk, manure and ploughing the field. Regrettably, even the (mainstream) environmental movements and discourses have been caught in the web of the Cartesian rationality of the nature/culture divide, which has posed an even deeper anomaly for ecosystem people. As in North America or Europe, in India the mainstream environmentalists share the view of protecting nature by fencing it off from use by ecosystem people. Sharma (1996), a leader of the National Alliance for Tribal Self-Rule in India, shows the dilemma such environmentalism poses for *adivasi* peasants in the Bastar district of Madhya Pradesh:

The environmentalists may fight against commercialization of the forest and will be able to muster international support for the creation of wildlife sanctuaries, national parks, natural/bio-diversity reserves and such like. But the common feature between the environmentalists and market-oriented forestry interests will be exclusion of tribal people. Thus, the forest will become either economic plantations (teak, eucalyptus, pine, and such like) serving the global market, or environmental preserves serving the global environmental need. The tribal will be the loser in either case (Sharma 1996). Medha Patkar, the leader of the Narmada Bachao Andolan, confronts this dilemma in the Narmada valley:

> 'What the tribals are saying today is, don't kick us out of the national parks and sanctuaries because you are the ones who have already destroyed the forest. The little that remains today is because we are living here – non-consuming protectors, users but not ambitious users like you. The whole problem is basically one of life-style. There can be no solution unless the conflict between life-styles, between immediate and future interests, are resolved. Because even when a certain section of the population says the wildlife and forest should be preserved, they have not changed their basic vision of life.'

How can conservationists who talk of wildlife protection on the one hand, and use toilet paper, plastic and air conditioning on the other, be credible? An ecological discourse that the ecosystem people might claim as their own may have been born.

The construction and destruction of 'indigenous' knowledge in India's Joint Forest Management Programme (Nandini Sundar)

In recent years, predominantly tribal areas like Bastar in Central India have seen their demographics change as non-tribal immigrants have flooded the area seeking their fortunes in the trade in timber and non-timber forest produce, in mining and illegal tin smelting, and of course in that vast money-making enterprise called development. Attempts by some local groups to demand an autonomous district council with predominant representation for tribals has prompted indignant responses from these immigrants. The question of who is indigenous and who is not therefore becomes a volatile political question.

There is an increasing wealth of historical and ethnographic evidence that indicates the absence of any distinctly identifiable category called an *indigene*, or *adivasi* (original inhabitant). Yet those certified as 'scheduled tribes' by the government constitute a large percentage of those displaced or destroyed by development. Historically, the construction of 'tribal' in India has been similar to that in Africa, drawing on evolutionary classifications, perceived primitiveness of modes of production, and denigration of any indigenous kinship in favour of an acephalous, kinship bound society. The categories 'tribal', 'primitive', 'savage' or 'wild' are seen as steeped in domination and anxiety, providing both the justification for colonial rule and the means towards it (Spurr 1993; White 1978; Taussig 1987).

In Bastar, while it is possible to distinguish local people, scheduled castes and scheduled tribes, from later immigrants, this is not on the grounds of antiquity or historicity. It is possible to trace tribal migrations to the region that follow multi-caste assemblies and high Hinduism. Subsequent tribal kingdoms were born out of internal stratification and relations with neighbouring populations that sometimes involved plunder and destruction of their environment – so much for traditional indigenous eco-friendly wisdom (Sundar 1997). To use the term 'indigenous' therefore, in the Indian context, is to be complicit in a history of somebody else's making, and to deny the possibilities of common construction and common destruction by 'indigenous' and non-indigenous alike.

There is, however, 'local knowledge' of local situations that people have by virtue of their intimate acquaintance with local landscapes. This is differentiated by class and gender, age and occupation, specialization and degree of expertise. Local knowledge changes as local environments change, and as people become involved in new projects. This is, obviously, an uneven process as memories of past practices continue to hold good for a while, and knowledge is transmitted through cultural practices which continue to take place even when the features of the world they refer to no longer exist in the same way.

The government is currently seeking to valorize indigenous knowledge systems as part of particular schemes; however, at the same time larger political and economic processes are conspiring to marginalize communities that hold this knowledge, both through physical displacement and through denigration of their life-styles and systems of knowledge. An example of an approach that attempts to valorize local knowledge, while perpetuating many of the forces that marginalize communities, is Joint Forest Management (JFM). Joint Forest Management refers to co-operative agreements between village communities and the local Forest Department to protect a particular patch of (state-owned) forest land and share the final harvest. Since the Government of India circular of 1 June 1990, sixteen states have passed resolutions regarding the implementation of JFM, and one estimate suggests that 10,000 to 15,000 village forest protection and management groups are currently protecting over 1.5 million hectares of state forest land. However, each state resolution has different rules regarding the constitution of Forest Protection Committees (FPCs), their legal status, the kind of land they are given to protect, the shares involved, and so on. Apart from this, differences in culture, in ecology and in the organizational structure of the Forest Department have introduced variations in actual practice.

Common features of Joint Forest Management include a focus on micro-planning, including an emphasis on planting villagers' choice of species, wherever natural regeneration is not enough; and in some states the provision of alternative development inputs to wean people away from dependence on the forest or at least to build up trust in the Forest Department. In theory, villagers are asked what their problems are and what is the best way of addressing them. This is perhaps the first time that local knowledge and action is being *formally* tapped within forest policy, though in practice much of the work of forest departments has depended on the intimate knowledge and expertise of villagers regarding the forests in their vicinity. For instance, although 'scientific forestry' has traditionally managed forests for timber or 'major' forest products at the expense of a diversity of 'minor' forest products, in practice forest departments have always depended for much of their revenue on the trade in minor forest products. And this has always been crucially dependent on the knowledge of villagers regarding the plants available, appropriate gathering practices etc. Other peasant practices too, such as grazing, lopping and firing, which were initially denigrated, were eventually utilized by forest departments towards positive silvicultural advantage. (Guha 1989; Rangarajan 1996; Sundar 1997; Jeffery and Sundar 1995).

Underlying the concept of Joint Forest Management is a certain notion of villagers' needs and patterns of use. A critical part of the Orissa JFM resolution is the preparation of a micro-plan, defined in the following terms. 'It would be based on a diagnostic study of the specific problem of forest regeneration of the locality, and the specific cost-effective solution for the same that may emerge from within the community. The views of all sections of the community, particularly the womenfolk, should be elicited in the PRA ['Participatory Rural Appraisal'] exercise for preparing the micro-plan.'

In practice, however, both usage and needs are incorporated into the JFM programme through a process of negotiation between competing agendas. On the one hand there are limits to the number of FPCs that can be set up, as they are circumscribed by targets, funding and the human resources available to the Forest Department. There are also restrictions on the kind of land that can be given over to village communities to protect – in most states this is restricted to degraded land with less than 40 percent crown density. While micro-plans are drawn up, working plans which lay out in detail the working of the Forest Department over a twenty-year period continue to exercise pre-eminence. Crucially, in at least three of the four sites studied by the JFM project, the major threats to forest land came from commercial interests or large-scale development projects, for instance tourism and bauxite mining in Borra in the Paderu Division of Visakhapatnam; proposed power plants in Sambalpur; and paper mills near Kevdi village in Gujarat. Within all this, the space reserved for the exercise of local knowledge or initiative is extremely limited. In some domains, however, some premium is given to villagers' views; for example in the selection of forest areas to protect and the choice of species to plant in those areas.

In some cases the choice of the particular plot is left to the villagers to decide, based on their usage patterns and their knowledge of their immediate environment, while in other cases, it is the Forest Department that does the apportioning. While the latter is sometimes done on the basis of the availability of degraded land, regardless of which village it belongs to, there are also conceptions of custom that both villagers and foresters subscribe to, such as the idea that forest land within the revenue boundary of a village should be given to it for protection. The manner in which existing use is recognized and addressed by schemes like JFM depends on the relative strength of either of these imperatives at any one point of time. On the other hand, contemporary notions of village boundaries are often determined by settlements made by earlier forest departments, and the 'customary rights' that villagers fight to retain are often rights conferred by earlier administrations. Given the complex interplay of 'customary' boundaries, actual usage and the definition of forest as state property – all of which are invoked at different times – the concept that site selection is a function of 'local' knowledge is highly problematic.

Unlike the selection of forest areas, which may or may not be left to villagers, the selection of species for plantation in the protected areas has been specifically set aside for them. As Sivaramakrishnan puts it, highlighting the manner in which participation is compartmentalized: 'Foresters are intent on preserving their control (in the final instance) through silviculture, a knowledge which through manuals and working plans is claimed as their exclusive preserve. Given the participatory framework, this leaves the awkward question of what should be conceded to the domain of local knowledge, where under the rules of JFM villagers are skilled practitioners. The answer, jointly provided by environmentalists and development specialists, is the knowledge of NTFP [non-timber forest products] collection and processing.' (Sivaramakrishnan 1996).

The emphasis on NTFP in Joint Forest Management arises for several reasons. In most states, villagers generally have rights (or concessions) to collect NTFPs, but no rights to timber. Offering villagers an increase in NTFPs as a part of JFM agreements does not normally require any change in the legal rules or existing balance of power. Secondly, timber products will take a long time (40 years, in the case of teak) before they come to maturity. Thirdly, the degraded lands that are given for JFM may never be capable of producing good marketable timber, whereas even the most degraded patch is probably capable of giving some NTFPs – including fuelwood, grasses, and so on (Jeffery and Sundar 1995). In practice again, targets and availability of species in the Forest Department nursery often determine what gets planted. This could mean that NTFPs are planted but it could also mean that timber species such as eucalyptus get planted, since that may be all the local Forest Department has. Where NGOs are involved they may be better placed to cater to villagers' expressed demands, but there too, the process by which demands are expressed is not a straightforward one. For instance, in the plantation projects of the Aga Khan Rural Support Program (AKRSP) in Gujarat, tree selection is done by village-level extension volunteers through PRA techniques like species ranking. This is carried out with two different groups of men and women and the two are then compared and a consensus formed. However, the very idea of ranking is something that is project oriented and associated with the delivery of benefits, and not necessarily connected with the wide range of uses that villagers make of different plants (Mosse 1995). In one village in Bastar, the women identified some 51 species for which they had various uses. Of these, however, only a few would be nursery species, and even fewer would be provided within a plantation programme.

In several villages in Andhra Pradesh which were practising Joint Forest Management, we found that villagers did not ask for trees that had traditionally grown in their forest areas or even those species from which they collected NTFPs for sale. When asked why, a common reply was that these would regenerate naturally with protection, so it made sense to ask for new species, preferably timber species or commercially valuable species like coffee or jaffra, a red dye plant. In one village, the villagers asked for teak although previously they had had mainly rosewood in their forests, because they had learnt from neighbouring villages and NGOs that teak was an important timber species. Thus one finds villagers in diverse ecological settings asking for the same species, regardless of the fact that it may or may not be suitable to the soil and the climate of the area. And this is often despite the fact that they are aware and knowledgeable about the suitability of particular species to particular soil types. In addition, while villagers may recognize that regeneration might be sufficient to afforest degraded land without additional plantation, they are reluctant to turn down the possibility of getting free seeds from the Forest Department. Forest staff, on the other hand, sometimes complain that they have to persuade the villagers of the benefits of natural regeneration and of traditional NTFP species rather than giving in to their demands for the plantation of exotics like cashew. This is in many ways an interesting reversal of roles in a country where the Forest Department has long been castigated for large-scale plantations of alien species like eucalyptus (Andhra Pradesh 1997).

In other words, villagers mould their demands according to what they feel the project can deliver (Mosse 1995), and the opportunity this gives them, or at least some among them, for entering into the commercial sphere as against continuing with their old subsistence economy. The simple assumption that many environmentalists make – that villagers, if asked, will predominantly choose species that yield NTFPs – ignores the multiplicity of interests that different groups in a village can have. As Madhu Sarin (1996) has made clear, the choice of certain species over others, or the choice of certain silvicultural methods over others, is not just a question of local knowledge or local choice, but a gender and class question. For instance, in West Bengal, Orissa and Bihar, a common source of income for women is the collection, manufacture and sale of sal (*Shorea robusta*) leaf plates. As sal forests regenerate and increase in height to the level that foresters consider good timber, the availability of leaves goes down (Sarin 1996). Commercially valuable timber species are often associated with male elites versus fruit and fodder bearing trees associated with women and lower classes. This has been documented in the Chipko case by Shiva (1988). However, in many cases, especially for poorer women, 'subsistence' included the sale of NTFPs or firewood through headloading, and unless some acknowledgement is made of this, environmental action meant to help women or poorer sections can often end up harming them (Sarin 1996).

Over the past two years in the village of Kilagada in Paderu, a hilly and tribal tract of Visakapatnam district in the state of Andhra Pradesh, successive conversations with villagers about the objectives of JFM have yielded different answers. Initially, fresh from the 'motivation speech' delivered by the forest staff, villagers cited ecological reasons, such as increased rainfall, for their involvement in Joint Forest Management. Another reason cited was the fact that the Forest Department constructed a village meeting room and provided various other goods in order to improve their relations with the villagers. In subsequent meetings, villagers cited the increase in NTFPs as a factor for engaging in protection, and finally came down to coffee as the major reason why they had taken to JFM. In Paderu, the Forest Development Corporation and the Integrated Tribal Development Authority which together constitute the major initiators of 'development' in the area, have flooded it with silver oak, which acts as good shade for coffee. Coffee plantations are an important source of employment. Thus, even where the soil and climate are not suitable for silver oak, villagers often end up asking for it to be planted in their protection areas (Gonduru field notes, Paderu). In general, one might argue though that this request for coffee comes from richer villagers who are not so dependent on the forests for the collection and sale of NTFPs such as *adda* leaves. Apart from providing a valuable lesson in the perils associated with oral history and the practice of ethnography in general (Clifford and Marcus 1986), what this illustrates is the manner in which local knowledge is a deeply problematic entity – differentiated by class and gender and under constant refashioning by the passage of time and the interaction with non-local forces.

The basic point in all this is that even where villagers do exercise initiative, it is under terms dictated by the overall framework of targets and activities prescribed by government rules, which in some sense distorts their agency. As Mosse (1995) said in a different context: 'If projects end up ventriloquizing villagers' needs it is not only, or primarily, because artful and risk averse villagers ask for what they think they will get. It is also because various development agencies are able to project their own various institutional needs onto rural communities'.

Conclusion

Essentially, indigenous or local knowledge is not a frozen, inert and timeless entity but something dependent on the material conditions of those whose knowledge it is, the changing environment in which they find themselves, and the uses to which it is put. Its expression through the prism of certain programmes has more to do with the aims and structures of the programme than with any reservoir of local knowledge. The parallels between this process and the codification of indigenous laws and traditions in the colonial period are also striking – the selection of texts to translate, the choice of religious scholars, all determined what kind of corpus would be institutionalized as 'Hindu' or 'Muslim' law and tradition, which all classes and castes were expected to follow (Cohn 1990). Similarly, the kind of local knowledge that is mobilized within 'participatory' programmes for development action is presumed to be a kind of knowledge that is common to all groups within a village, regardless of their differing interests. In reality, local knowledge – while drawing on a shared environment – constructs that environment in different ways.

Acknowledgements

I am grateful to Roger Jeffery and Siddharth Varadarajan for their inputs and editing. Some of the material for this paper has been collected through the Edinburgh University Joint Forest Management research project in India, funded by the ESRC and titled, 'Organising Sustainability: NGOs and Joint Forest Management in India'. The research for this project is being carried out in Orissa, Andhra Pradesh, Madhya Pradesh and Gujarat, in collaboration with the ICFRE, Dehradun. I am grateful to all my colleagues in the EU project for sharing their research and ideas with me.

Of dancing bears and sacred trees: some aspects of Turkish attitudes toward nature, and their possible consequences for biological diversity (Sargun A. Tont)

The importance of metaphors in the way we view the world and how it empowers our responses to the variety of issues confronting our existence has been well stated by Muir (1994) when he wrote, 'Whether helping us conceive of ourselves as part of an interconnected web, or as subject to the explosive force of the population bomb, or taking a ride on spaceship earth, metaphors can profoundly influence images of the environment and mould the ways that we respond to perceived threats to the earth'. Keeping in mind that metaphors are only thinly camouflaged perceptions, we can enlarge the scope of Muir's hypothesis by adding that not only metaphors, but also the way we value and draw aesthetic gratification from the environment, have a direct bearing on the well-being of the ecosystem that we inhabit.

As in most other societies Turkic peoples' attitudes toward nature are so highly complex and multiform that it is impossible to follow all the major variations. Keeping in mind that religious beliefs are useful indicators of a society's attitude toward nature, and since Turkic people have simultaneously or successively practised several universal religions, including Christianity and Buddhism, before the majority were won over to Islam (Roux 1987), one can appreciate the difficulties involved in finding persistent thought patterns and beliefs held in common by significant portions of the population. Furthermore, conversion to Islam has not resulted in a definite and easily identified set of beliefs toward nature since different sects and creeds within the faith display different and often contradictory attitudes toward nature and its inhabitants.

The Anatolian peninsula supports a rich flora and fauna: there are 9,000 species of plants, 3,000 of which are endemic, and nearly 10,000 animal and insects species. Varying climatic conditions and topography (ranging from sea level to 5,000 metres), a multitude of lakes, and the fact that three-quarters of the country is surrounded by sea, account for this rich biodiversity (Kence 1990). It is not surprising, then, that Turkish proverbs, folk sayings and literature are full of references to animals and plants. Unlike the Buddhist belief, however, where man is considered only a part of the ecosystem on equal footing with the rest of the animal world, the perceived division between human and animal world is very sharp and well defined.

Considering that 26 percent of Turkey is covered by forest (Isyk 1990) it is not surprising that the tree plays a central role in Turkish culture (Box 9.8). Turkish proverbs abound with references to trees expressed in lively metaphors. 'When the axe fell, the tree said "its handle is made from me"'; 'The tree is beautiful only when it has all its leaves'; 'When the tree is full of fruits, it lowers its head' (knowledgeable people are modest); 'The best time to bend a branch is when the tree is young'. The roots of this love affair go back to ancient times and are expressed in several cosmogonies. Turkic tribes living in Central Asia believed that nine races of man were kept under the nine branches of a sacred tree. In a Hun legend, Oğuz Han and three of his brothers, and in Uygur legend Bugu Han and his four brothers, were born in the hollow of a tree. (Banarly 1971). After conversion to Islam, love for trees grew even stronger because the prophet Mohammed compared a good Muslim to a palm tree, and declared that tree planting would be an acceptable substitute for alms (Topaloglu 1994).

In one of the oldest collections of Turkish tales, which make up the *Book of Dede Korkut*, the unknown poet agonizes over his failure to find a more exalted name for his beloved plant:

Tree, tree, do not be embarrassed because I call you by that
[after all] The doors of Mekka and Medina are made of wood
The staff of Moses is also made of wood
The bridges spanning over big rivers are also made from you
The ships which roam the black seas are also of wood.

(Banarly 1971; translated by S. A. Tont)

In Yunus Emre, perhaps the best-known Turkish poet-mystic, the tree reaches sacred proportions:

Its stem is gold
Its leaves are silver
As its branches expand
They invoke the name of the Lord.

(Banarly 1971; translated by S. A. Tont)

Box 9.8: Turkish Nature Poems

My Tree

Orhan Veli Kanik; translated by Murat Nemet-Nejat

In our neighbourhood
If there were other trees
I would not love you so.
But if you knew
How to play hopscotch with me
I would love you much more.

My beautiful tree.
When you die
I hope we'll have moved
To another neighbourhood.

I Became a Tree

Melih Cevdet Anday; translated by Talat Sait Halman

I was going under a tree
It happened in a flash
I fell apart from myself
And became a poppy flower
Bending in the sun,
Tortoise shell, house of wedding
Delirious talk, bevy of names.
I turned into the petal that drags
The wind like a blind God,
I became the century.
A tiny moment like a bug.
I was going under a tree
I became a tree
That propels itself
And saw someone stuck in the ground.

One of the earliest Turkish poems about trees was written by the fifteenth-century folk poet Ismail Safevi:

I am a pine tree
I look lovely
I moved to the valley from the mountains
I am a gift (of nature) to mankind
I am the bridges over the river.
(Bezirci 1993; translated by S. A. Tont)

Some traditions widespread throughout Anatolia have a very positive impact on the state of woodlands. A tree is planted after a child is born, when a son is drafted into military service, after a wedding, and also as a memorial to the deceased. Even a disinterested observer would be impressed by the large number of people, ranging from school children to the heads of major corporations, who donate time and money to reforestation campaigns. But intimate and healthy cultural ties with our environment are only one prerequisite for maintaining a healthy ecosystem. Wars and famine and the vagaries of climate may play havoc with ecosystems no matter how close and powerful our cultural ties with the environment are. After all, as an anonymous Turkish poet once said, 'If a man is hungry enough, he will eat his own hunger'.

Gods of the forest – myth and ritual in community forestry (William Burch Jr.)

'... Human beings from all cultures and all times have looked deeply into the forest for larger meaning, yet, in the end, what we seek to understand is ourselves. Consequently, we interpret the structure and functioning of forested ecosystems in ways that make sense to our specialized vision of the world. Our forest practices and responses to trees are effigies of our predominant social values. They mirror our class system, our attitudes toward social order and personal liberty, the relations between young and aged, men and women, the sacred and profane. And trees and forests are not always viewed positively.

Britons in the sixteenth and seventeenth centuries saw forests as synonymous with wildness and danger, as the word 'savage' was derived from the Latin word *silva*, a wood. North America at that time was a wild forest, 'dreadful', gloomy, uncouth – full of wild beasts and wildness. The forest was home for animals not men. Indeed, the first law of community forestry is that any human society is unable to appreciate the value of woodlands until a substantial proportion of them have been converted to other uses. There is, therefore, a direct correlation between rates of woodland decline and rising rates of concern.

The crucial point is that trees and forests are viewed in a variety of ways that often differ from the perceptions of the professional forester. For a community, a forest takes on sacred, mythical or mystical qualities as well as being a source of natural resource benefits. The forests and its gods may be seen as a source of evil spirits, or as the spirit of life that ensures fertility and joy with predictable consequences as to how the forest will be managed. The whole thrust of modern forestry has been to de-mystify the forest – to make it a rational part of the production process, yet that very rationality has required a strong underpinning of myth and the irrational. For one thing, all professional foresters are trapped in a vision of forest scarcity that emerged in fourteenth-century Germany. For better or worse, the forestry training programmes of North America, Switzerland, Britain and most countries in South Asia and the Pacific, have been substantially influenced by German systems and practices of silviculture. By understanding and appreciating the nature, type and uses of myth in the human community, we may gain a greater insight into our own profession, and the myths upon which our actions are based.

In North American forestry we have myths of heroic origins – heroes battling great evil – and we have myths to order our doubt and ambiguity about our practices (Burch 1971). Firstly, our profession believes that it emerged at a crucial time, when nature was being destroyed by vested interests and ignorant farmers. Timber famines were dark clouds on the horizon that promised decaying villages, economic depression and the destruction of family life. Foresters stepped in, just in time, to stop the rape and pillage of nature and to stop the fires. Heroes like Bernard Fernow, Gifford Pinchot, Teddy Roosevelt and the larger-than-life 'Forest Ranger', grappled with evil forces and defeated them. Yet, timber famines and other threats will keep emerging unless society continues to encourage the dedication and ever-vigilant attention of foresters. This view was in great contrast to the folk myth in North America of Paul Bunyan and 'Babe the Blue Ox', who were praised because they massively cut the great American forest to open land for farms, villages and homes. Forest conversion, not forest protection, was the heroic deed most praised in such myths.

In point of fact, it was not the profession of forestry, but significant changes in the US political economy that 'saved' the forest. People were moving from farm to city, energy sources moved from wood to coal, and primary building material from wood to iron and concrete. The economy moved from primary production to manufacturing. In short, forestry became successful at the very time when it was becoming a minor factor in the US economy. Each of the various predicted timber famines in the US have failed to occur.

The second set of US forest myths has included terms such as 'greatest good for the greatest number in the long run', 'multiple use' and 'sustained yield'. These help to protect foresters from those who might wonder what they do. Such myths serve as ways of adjusting to political challenges by other users and groups that want to take over some of the functions of forestry. Happily, the terms are vague enough not to be subject to empirical test. All organized, persisting human groups require myth to sustain their unity – including professional foresters.

[Adapted from 'Gods of the Forest – Myth, Ritual and Television in Community Forestry', a Seminar Banquet Talk for the Regional Community Forestry Training Center, Asia-Pacific Seminar, Bangkok, Thailand, December 1987. William Burch, Jr., School of Forestry and Environmental Studies, Yale University, New Haven, CT 06511.]

The oak tree of Errol, Scotland (Alexander Porteous)

In addition to the famous and well-known oak trees of England, there are many other oaks, as well as other trees, which have a certain fame but which it is unnecessary to mention here. We must not, however, omit the famous Oak of Errol in Scotland, which has long since disappeared. It was alleged that the fate of the family of the Hays of Errol, in the Carse of Gowrie, was bound up in a mistletoe plant that grew on this tree. One of the descendants of that family, writing in 1822, recorded the belief as follows:

'Among the low county families the badges are now almost generally forgotten; but it appears by an ancient manuscript, and the tradition of a few old people in Perthshire, that the badge of the Hays was the Mistletoe. There was formerly in the neighbourhood of Errol, and not far form the Falcon stone, a vast oak of an unknown age, upon which grew a profusion of the plant: many charms and legends were considered to be connected with the tree, and the duration of the family of Hay was said to be united with its existence. It was believed that a sprig of Mistletoe cut by a Hay on Allhallowmas eve, with a new dirk, and after surrounding the tree three times sunwise, and pronouncing a certain spell, was a sure charm against all glamour or witchery, and an infallible guard in the day of battle. A spray, gathered in the same manner, was placed in the cradle of infants, and thought to defend them from being changed for elfbairns by the Fairies. Finally, it was affirmed that, when the root of the oak had perished, 'the grass should grow in the hearth of Errol, and a raven should sit in the falcon's nest'. The two most unlucky deeds that could be done by one of the name of Hay was to kill a white falcon, and to cut down a limb from the Oak of Errol.'

Thomas the Phymer is credited with having uttered the above prophecy thus:

While the mistletoe bats on Errol's aik,
And that aik stands fast.
The Hays shall flourish, and their good grey hawk
Shall nocht flinch before the blast.
But when the root of the aik decays
And the mistletoe dwines on its withered breast,
The grass shall grow on Errol's hearthstane,
And the corbie roup [croak] in the falcon's nest.

[From: Alexander Porteous (1928). *Forest Folklore*. George Allen & Unwin Ltd., London. New edition published in 1996 under the title *The Lore of the Forest: Myths and Legends*, by The Guernsey Press Co. Ltd., Guernsey.]

Box 9.9: 'On Entering the Forest', 'Thoughts on Manitoba Monoculture' and 'Winds of Change'

On Entering the Forest

Dr Freda Rajotte

Entering into the shadows of the forest floor, Rolando stood still, every sense alert to the symphony of sound, movement, light, colour and perfume. Finally, with awe, he said, 'You are entering now into the womb of creation! Here evolution is taking place all around you. On every slope, in every stream, along each valley and on each hill some unique life form is emerging to fit a specialized niche. It will form a new link in the chain of interdependent life forms that constitute the marvel of the Ecuadorian rainforest – the uniqueness of every slope and every stretch of river.'

Thoughts on Manitoba Monoculture

Dr Freda Rajotte

Before me stretches the monoculture world
designed by the techno-scientific monoculture of the mind,
which asks only how to increase productivity,
how to grow more grain.
Now the forests are cleared, the marshlands drained,
the rivers dammed, soils chemically 'enhanced'.

Gone is the lark from the bright sky singing,
the dappled trout in the shady pool,
purple crocus, bramble, fern and toadstool,
frogs' morning chorus, the bindweed clinging.

And the Devil came to him and said,
'Order these stones to turn into bread'.
But Jesus answered, 'The scripture says,
"Man cannot live on bread alone,
but needs every word that God speaks"'.

Box 9.9 (continued)

Winds of Change
(for the Isle of Man)

Roger Pullin (Words spoken over the Lament and Lullaby by Roger Pullin)

Winds of change
Shape the Trees
Shape the rocks, the shoreline
Winds of change
Shape all who
Take the Island to heart.

Sweeping through the hills and glens
And far across the world.
Winds of change
Shape the way
And the way home.

Winds of change
Blowing strong
Rock the island cradle
Winds of change, make *us* strong
Through the times apart

Travelling on lonely roads
And under different stars
Winds of change
Shape the way
And the way home.

Change
On the outside
But deep within the same
Change
On the outside
But deep within the same
Just like the calm
After a storm at sea
And every year the seasons
The seasons of flowers
And frosts
And the Island strong and fair
Through winds of change
The Island strong and fair
Through winds of change
Change
On the outside
But deep within the same.

CHAPTER 10

AQUATIC AND MARINE BIODIVERSITY

Paul Chambers

CONTENTS

TEXT BOXES

FIGURES AND TABLES

Aquatic and Marine Biodiversity

Paul Chambers

Introduction (Paul Chambers)

Rivers, lakes, seas and oceans and the plant and animal life associated with them are important to every culture on the planet and form an explicit or implicit part of the religious beliefs and cultural heritage of almost all human cultures. They are focal points of myth, folklore, art, cultural and religious imagery, diet, trade, and leisure and recreation.

From the spiritual value of a sacred spring to the daily catch of fish from a tropical reef, many peoples have an ancient and intimate relationship with the water around them and the life it contains. For many, the plants and animals that the sea or rivers provide are valued as sources of food, medicine or income. For an artisanal fisherman, for example, the sea represents a source of food for the family or of goods to trade or sell at the market. Clark (this chapter) describes the essentially utilitarian views of nature in Western Samoa: 'Survival requires pragmatism not aestheticism... Nature exists to feed, to house, then to satisfy cultural obligations'.

Andersson and Ngazi (1995) conducted a detailed survey of the perceptions and values of marine resources by the inhabitants of Mafia Island, Tanzania's first marine National Park. They found that overwhelmingly the most important values given were either direct 'consumption values' – the provision of food or building materials – or 'production values' – the provision of goods that could be sold or exchanged. Spiritual or cultural values may be harder to define or measure but they are surely no less important in determining the priorities for conservation efforts. Cultural and utilitarian values are often closely interlinked so that those resources with great utilitarian value – such as a spring or a particularly important plant or animal species – may be given a religious or ritual significance that ensures their safeguarding.

Among the Hopi of the South-western United States (Whiteley and Masayesva, this chapter), water is the key to their survival in a desert environment and their relationship to water has been enshrined at the core of their culture and beliefs. 'How Hopis get and use water is a major part of their identity, religious beliefs, ritual practices, and their daily engagements and concerns... It is hard to imagine anything more sacred –as substance or symbol– than water in Hopi religious thought and practice.' For the Tukanoan fishermen of the Brazilian Amazon, the integrity of the river-margin spawning grounds of the fish they depend on are crucial to the long-term maintenance of their fisheries. Chernela (this chapter) describes how these very sites are frequently deemed sacred and therefore protected. In the Japanese coastal towns where whaling has been a way of life for centuries, whales have come to acquire great cultural, social and religious importance as well as providing an important source of food. A taboo on hunting whales with calves serves to conserve whale stocks.

The essays in this chapter serve to remind us that the value of the natural environment cannot simply be measured with a check-list of plant and animal species and their abundance and distribution. Roué and Nakashima's study of the effects of a hydro-power project on the Chisasibi Cree of Quebec powerfully illustrates the consequences of ignoring the cultural aspects of biodiversity. In this instance the unique features of the site that once made it so important to the Cree as both a fishing spot and a meeting place were lost, although threatened fish populations were preserved .

Indigenous peoples can, and indeed must, play a central role if marine and aquatic resources are to be effectively conserved. Local communities often have a detailed knowledge of their environment gained by observation over many generations. They may also implement traditional resource management frameworks such as rules of tenure and exclusion which have successfully served to sustain or even augment biotic diversity over many centuries (McNeely 1992; Toledo 1991). 'Biosphere people' (Gadgil 1993a) are the best-motivated guardians of their marine and aquatic environments because they are materially dependent on their long-term preservation.

Local environmental knowledge

Many studies have documented the extensive environmental knowledge of traditional coastal (including river and lake shore) societies. This knowledge is most often practically-oriented and directed towards means of locating and harvesting resources. In this chapter, for example, Raygorodetsky describes the detailed knowledge of the Gwich'in of North-western Canada about the fish they catch, including the best sites and times to fish and the many ways to prepare, preserve and utilize every part of the fish. Ruddle (1994) identifies four main characteristics of local coastal knowledge common to many parts of the world: it is based on long-term, empirical, local observation; it is specifically adapted to local conditions, embraces local variation and is often extremely detailed; it is practical and behaviour-oriented, focusing on important resource types and species; and it is structured such that it is compatible with Western biological and ecological concepts through a clear awareness of ecological links and notions of resource conservation. It is often a dynamic system capable of incorporating an awareness of ecological perturbations or other changes, and of merging this awareness with an indigenous core of knowledge. Toledo (1991) writes about the P'urhépecha surrounding Lake Pátzcuaro in central México: '...[a] salient feature of P'urhépecha indigenous knowledge is its essentially utilitarian character. Nearly all environmental knowledge is linked with one or more practical activities. Thus, ecological, morphological and behavioural knowledge about mammals, aquatic birds and fishes serves as the intellectual means to realize the operational steps of terrestrial hunting and aquatic fishing.'

Kalland (this chapter) describes the detailed knowledge of Japanese whalers: 'In order to hunt whales successfully it is necessary not only to acquire skills in navigation, shooting and handling of meat but also to acquire detailed knowledge of the migratory, mating and feeding behaviour of various species of whales... The whalers have learnt to take account of such natural phenomena as tides, currents, winds, wave patterns, water temperature and colouring. They are also keen observers of fish and bird behaviour. But the knowledge of the whalers goes beyond this. They need an understanding of how whales are part of a larger ecological system in which other maritime organisms as well as humans are also parts. The whalers' knowledge of this ecosystem is based on centuries of accumulated experience and is intimately linked to religious beliefs and practices.'

The knowledge held by traditional communities can provide the human intellectual 'gene pool' with the raw material for adaptation to local environments (McNeely 1992). It should be a key component in conservation and resource-management plans for marine and aquatic diversity (Johannes 1978, 1982; Ruddle 1994). Gadgil (1993b) writes: 'It is vital ... that the value of the knowledge–practice–belief complex of indigenous peoples relating to conservation of biodiversity is fully recognized if ecosystems and biodiversity are to be managed sustainably'. This environmental knowledge is held within the great variety of indigenous cultures in all parts of the world. It is unique and irreplaceable, but is disappearing along with many of those cultures as they are weakened or destroyed. This loss of cultures, or of the traditional knowledge within them, is as serious a problem as the loss of plant or animal species. Crucial knowledge about how different environments might be used to provide benefits on a sustainable basis may be lost for ever (McNeely 1992).

Local resource management systems

Traditional common property systems of fisheries management occur throughout the world (Ruddle 1994). These frequently ensure equitable access and include management measures leading to sustainable resource usage. Management strategies include tenure systems with limited entry, as well as restrictions on fishing gear, seasons and particular species (Johannes 1978, 1982). Johannes (1978), writing about Oceania, states that many of these practices are related to religious or superstitious beliefs and that they serve to conserve fish stocks although it is not possible to judge whether they were designed to do so. Steiner (this chapter) describes the Hawaiian *kapu* system of traditional prohibitions which serve to protect and maintain important natural resources. The 'tragedy of the commons' (Hardin 1968) can be largely avoided by these traditional patterns of resource management (Johannes 1978; Feeny *et al.* 1990). By exercising control over their own members, communities enforce a set of regulations over the resources that serve as a basis for sustainable satisfaction of their own material needs. The richness and accuracy of indigenous ecological knowledge serves as the intellectual means to achieve this end.

It would be misleading to state that indigenous peoples always have a conservationist ethic or that traditional practices always lead to the conservation of resources. Particularly in areas where population densities are low and resource limitation infrequent, wasteful or destructive practices sometimes persist: 'All this should not be taken to mean that Pacific islanders enjoyed a perfect relationship with nature and that all their actions were governed by environmental wisdom and restraint. ...In short, environmentally destructive practices coexisted, as in most societies, with efforts to conserve natural resources' (Johannes 1978). For example, Clark (this chapter) asserts that Western Samoans have a primarily utilitarian view of their environment, contrary to common pre-conceptions about indigenous peoples. The lack of conservationist practices or a reverence for nature may explain the high levels of environmental damage in the Samoan islands. Most subsistence peoples, however, do live sustainably from the resources available to them: 'In contrast to the specialized mode of biological resource use in modern economies, where comparatively few resources are perceived and recognized as having production value, under the indigenous view all of the components of natural landscapes are directly or indirectly useful or usable resources. This aspect alone tends to favour a conservationist attitude towards the environment. Moreover, it seems clear that the multiple use strategy of indigenous communities is an effective mechanism for preserving and even increasing biological diversity by increasing habitat heterogeneity' (Toledo 1991).

The preservation of indigenous peoples and their traditional cultures must therefore have a central part in the conservation of the marine and aquatic biodiversity that they depend on and in turn sustain. 'Perhaps the most difficult problem faced today in the effort to accomplish biological conservation in the Third World is the inability of educated people such as scholars, conservationists and policy-makers to recognize and comprehend the enormous ecological value of the modes of subsistence of traditional cultures that inhabit rural areas. Most recent evidence suggests that the indigenous peoples play a key role in the conservation of biological diversity' (Toledo 1991).

Locally-based resource rights are of great importance for biodiversity conservation. Whenever people with a limited resource base have exclusive control over their own environment they will tend to maintain it to their best material advantage. Given the wide range of environmental services that people need, this generally means managing the environment sustainably and so maintaining a high level of biotic diversity. In practice however, adequate resource tenure is rare. Few indigenous peoples have the legal authority to protect their own resource rights. Without this they are powerless to prevent outside exploitation, even if their own traditional rules of reef or sea tenure, for example, are still applied amongst themselves. In the resulting 'open access' conditions the only rational behaviour is to join in and exploit the dwindling resource base in an unsustainable way. Recognizing locally-based property rights, on the other hand, gives both an incentive and a means for the involvement of local peoples in effective biodiversity conservation.

Conflicts over resource values

Most 'Western' societies have become increasingly alienated from the natural world, both culturally, and as a result of urbanization, also geographically. In the homogenous 'monoculture' that results, little remains of the traditional knowledge of, and attitudes towards, the environment. The differing values accorded to natural resources between such 'modern' cultures and more traditional communities are evident in many of the contributions within this chapter. These differences have inevitably led to conflicts over the use or protection of natural resources. This is never clearer than over attitudes to water. In the first contribution in this chapter, Claus Biegert vividly conveys the contrasting values of water in different cultures; from reverence and respect in some to thoughtless exploitation in others. Whiteley and Masayesva continue this theme, describing how the activities of a coal-mining company have disrupted the precious springs and rivers upon which the Hopi of Arizona depend.

Where these conflicts have occurred, it is rarely 'modern' culture that compromises and accommodates the different values and practices of traditional peoples. In the case of the Chisasibi Cree in Quebec, for example, a large-scale hydro project destroyed traditional fishing grounds and displaced an Indian community. Despite their protests, the material and cultural consequences for the Cree were deemed secondary to the demands for energy of the state.

Most instances of conflict over resource values arise, as in the two examples above, when indigenous people desire to protect their local resources in the face of demands from an industrialized society. However, this is not always the case. Two contributions in this chapter explore recent conflicts generated between traditional resource users – in this instance Japanese and Icelandic whalers or fishermen – and outsiders wishing to protect these same resources – marine mammals – for aesthetic or moralistic reasons. In coastal areas of Japan, whaling is a traditional way of life and has a cultural importance beyond the simple provision of food. Kalland discusses the strong social repercussions of the Western-imposed moratorium on these whaling communities and the resentments that this creates. This is particularly strong as the whale species hunted by the Japanese are no longer endangered. 'That whales should not be killed for ethical reasons is furthermore incomprehensible to most Japanese, who rather would question the ethics of killing domesticated animals for food. To an increasing number of Japanese the moratorium is therefore interpreted as an assault on their culture...'. Einarsson similarly describes the effects of the moratorium on Icelandic whalers and fishermen who feel that outsiders with little experience of, or participation in, their environment are depriving them of their traditional livelihood. Both these examples raise the question: who is biodiversity conservation for? Without a clear answer to this question we will always be faced with irreconcilable conflicts whenever the environmental values of local people do not match those of a wider group – be it a multinational company, the state or the global conservation community.

The contributions in this chapter show that there is a strong link between local peoples – their culture and language, their ecological knowledge, and their traditional management systems – and the biodiversity of their marine and aquatic environments. One lesson we need to learn is that the cultures and ways of life of traditional peoples are as deserving of protection as the biotic diversity that they value and depend on. Local environmental knowledge and traditional resource management practices could be a vital asset as we seek to live sustainably on our planet. All too often, however, local peoples are seen as a potential obstruction or, at worst, as active opponents of biodiversity protection and conservation. To be sustainable and lastingly effective, conservation practice must integrate both cultural and biological conservation; the 'people factor' cannot be simply omitted from considerations of biodiversity. Rarely are these two aims exclusive and, from this chapter, it is apparent that they are inseparably linked.

Sacred waters (Claus Biegert)

'I was in Cleveland in the late sixties and got talking with a Non-Indian about American history', Vine Deloria Jr. remembers. The white businessman said that he was really sorry about what had happened to the Indians, but that there was good reason for it. The continent had to be developed and he felt that Indians had stood in the way and thus had to be removed. 'After all', he remarked, 'what did the Indians do with the land when they had it?'

Deloria didn't understand him at first, but then he read the warnings along the Cuyahoga River and realized that this stream running through Cleveland is inflammable. So many combustible pollutants are dumped into the river that the inhabitants have to take special precautions during the summer to avoid accidentally setting it on fire. Deloria concludes, 'After reviewing the argument of my Non-Indian friend I decided that he was probably correct. Whites had made better use of the land. How many Indians could have thought of creating an inflammable river?'

With this episode Deloria, a Lakota from Standing Rock, opens his provocative publication *We talk, You listen*. As a professor of law and an author of best-selling books, he is troubled by the fact that the members of dominant societies are unable to maintain a spiritual relationship with the natural world. Oren Lyons, Faithkeeper of the Onondaga Longhouse Council of Chiefs, likes to say, 'In the absence of the sacred, anything goes!' Ethical barriers to prevent destruction are simply removed. 'One of the future wars', Lyons predicts, 'will be over clean water.'

Indigenous cultures around the world maintain a special relationship to water: water is sacred. Western civilization could be portrayed by a single sign: *Water Not Potable !* The late Philip Deere from Oklahoma, medicine man of the Muskogee Nation, called rivers and streams the veins of the world. Blood poisoning can be fatal for humans and polluted rivers should be seen the same way, he often said. But Western medicine which has moved from healing to research, is ignoring the patient earth. This separation is not existent in indigenous cultures that consider humankind's well-being and the well-being of Mother Earth as one. From an indigenous view, to destroy ancient aquifers and dump nuclear waste into the oceans is like suicide.

The Maori of New Zealand waged a legal battle on behalf of their rivers. They won, defeating public plans to build a sewage treatment plant along the banks of their sacred Kaituna River. From a Maori point of view, it is unimaginable that effluents and treated wastes should be channelled into a natural body of water. One of the Kaituna's defenders is Eva Waitiki: 'We told the judges about our river, its power, its uniqueness. We told them that the Kaituna is like a brother to us. We even spoke of the *Taniwha*, the protecting water spirits. We explained to them that the sicknesses of the soul which we call *Mate Maori*, can only be healed by a healthy river. And that flushing treated water into the Kaituna would be to rob it of its healing nature and to take away the hopes of those who are sick.'

As a child, Eva Waitiki was brought to the Kaituna to be cured. 'When I was seven years old', she remembers, 'I stepped on a shard of glass and badly cut my foot. The White doctor said that I would be lame for life. Then the elders brought me to the Kaituna. They submerged me entirely, the clear water surging over me. They repeated this every day and after two weeks my foot was completely healed.'

In the highlands of Arizona, the Hopi of Hotevilla on Third Mesa fought a bitter battle against the implementation of a public water system. 'Only when we can visit the spring', say the traditional Hopi, 'can we accord water the respect it is due and use it honourably.' According to the elders, children who only see the water that comes from the faucet – water that has been piped over a course of miles into their homes – can have no real understanding of water's essence. The elders did not win in this controversy: in October 1996 a sewage line was installed on Third Mesa. For the *Kikmonqwis* (spiritual village leaders) and priests it is like an umbilical cord being broken. The *Kachinas*, the messengers to the spirit world, follow a certain pattern between the San Francisco Peaks and the Hopi Mesas. Now, the elders say, their energy lines are disrupted.

In the industrial world, rivers are put to work. The Cree people of James Bay in Northern Quebec are threatened by a gigantic hydro project which, in the late seventies, flooded vast areas of their hunting grounds. For the Cree, the land and the rivers are like a garden which is taken care of and harvested by the people who, temporarily, are the custodians of the land. Now the La Grande River, called Chisasibi (Big River) by the Cree, is drowned by water. While the shores were an important sub-Arctic ecosystem and the stream a rich source of fish, the new shores are just separations between water and land without a specific shoreline vegetation – and the reservoir is a source of fatal sickness since all the fish are heavily contaminated by mercury.

Walter Hughboy, chief of the coastal village Wemindji, remembers flying over the reservoir. 'When I saw the dry land for the first time, which once was covered by water, and when I saw the reservoir, saw the tips of the drowned woods, I was hurt as never before in my life. I don't know if a white person can understand that: something good inside you dies, you lose your innocence. It hits you in your own identity, it hits you exactly where you know who you are and where you belong to. Something like this hits a human being in its essence which is governed by cosmic laws. We Cree have a spiritual relationship with the world around us. Something is taken away from us Cree, something which has its place deep inside you. After the flood you are not the same person any more.'

The relationship with the natural world we have lost cannot simply be brought back at a workshop. Much looking is needed before we can find an honest way to restore it. Meanwhile, help from all sides is needed. A few legal scholars, much in the sense of the traditional Maori, have risen up to defend nature and to grant it a voice. One pioneer of the environmental rights movement is the American Legal Professor Christopher D. Stone. In his provocative book, *Should Trees Have Standing?* he poses the question, 'Why should a mountain or a lake or a forest or an animal species not have the right to be represented by lawyers in a court of law, when a firm – clearly no living entity – is granted such right?' Juristically, land, flora and fauna are yet treated as possessions. Stone demands that not only should such entities have the right to act as plaintiffs, but that nature should be accorded a fundamental standing unmitigated by human values, that natural entities should be accorded a set of intrinsic and inalienable rights regardless of utility or beauty.

It is time that a river that burns will be able to sue.

Paavahu and *Paanaqso'a*: the wellsprings of life and the slurry of death (Peter Whiteley and Vernon Masayesva)

> '[T]his is... one of the most arid countries in the world, and we need that water. That is why we do *Kachina* dances in the summer, just to get a drop of rain. And to us, this water is worth more than gold, or the money. Maybe we cannot stop the mining of the coal, but we sure would like to stop the use of water.' (Dennis Tewa, Munqapi village, at a public hearing on the renewal of the Black Mesa–Kayenta mine permit, Kykotsmovi, 9 August 1989).

The Hopi Indians of north-eastern Arizona are an epitome of human endurance: they are farmers without water. According to their genesis narrative, the Hopi emerged from a layer under the earth into this, the fourth, world by climbing up inside a reed. Upon arrival, they met a deity, Maasaw, who presented them with a philosophy of life, based on three elements: maize seeds, a planting stick and a gourd full of water. *Qa'ö*, maize, was the soul of the Hopi people, representing their very identity. *Sooya*, the planting stick, represented the simple technology they should depend on: there was an explicit warning against over-dependency on technology, which had taken on a life of its own in the third world below, producing destruction through materialism, greed and egotism. *Wikoro*, the gourd filled with water, represented the environment – the land and all its life-forms, the sign of the Creator's blessing if the Hopis would uphold Maasaw's covenant and live right. Maasaw told them life in this place would be arduous and daunting, but through resolute perseverance and industry, they would live long and be spiritually rich.

Persistent occupation of the Hopi mesas for more than a millennium is both remarkable and paradoxical. Unlike the other Pueblos, with no streams or rivers to support agriculture, the Hopi subsistence economy's dependence on maize, beans and squash must seek its water elsewhere. How Hopis get and use water is a major part of their identity, religious beliefs, ritual practices, and their daily engagements and concerns. Much of the complex Hopi religious system is devoted in one way or another to securing necessary blessings of water – in the form of rainfall, snow, spring replenishment, etc. – to sustain living beings, whether humans, animals or plants.

The phrase 'Hopi environmentalism' is practically a redundancy. So much of Hopi culture and thought, both religious and secular, revolves around an attention to balance and harmony in the forces of nature, that environmental ethics are in many ways critical to the very meaning of the word 'Hopi' and its oft-heard opposite *qahopi* ('un-Hopi' – 'badly behaved'). Hopi society is organized into clans, the majority of which are named after, and have specific associations with, natural species and elements, like Bear, Sun, Spider, Parrot, Badger, Corn, Butterfly, Greasewood, Tobacco, Cloud, indicating the utter centrality of environmental forms and ecological relationships in Hopi thought. Myriad usages of natural species and agents in Hopi religious ritual express the depth and detail of this ecological awareness and concern. Ritual performances typically embody and encapsulate key vital principles of the natural world. Even a casual observation of a *Hemis* Kachina spirit at *Niman* (the Home Dance), to just take one case, reveals a being festooned with spruce branches, wild wheat, clouds, butterflies, tadpoles, seashells, and so on. The bringing together of these natural symbols is in many instances designed to both evoke and celebrate the life-giving force of water in the world.

It is hard to imagine anything more sacred – as substance or symbol – than water in Hopi religious thought and practice. To be sure, some elements may appear more prominent: corn, the staff of life, which is ubiquitous in Hopi religious imagery; rattlesnakes from the spectacular Snake Dance; or masked performances by *Kachina* spirits. But intrinsic to these, and underlying much other symbolism in the panoply of Hopi ritual, is the concern with water. Springs (*Paavahu*), water and rain are focal themes in ritual costumes, *kiva* iconography, mythological narratives, personal names, and many songs, which call the cloud chiefs from the various directions to bear their fructifying essence back into the cycle of human, animal and vegetal life. That essence – as clouds, rain and other water forms – manifests the spirits of the dead. When people die, in part they become clouds; songs call to the clouds as ascendant relatives. Arriving clouds are returning ancestors, their rain both communion with and blessing of the living. The waters of the earth (where *Kachina* spirits live) are, then, transubstantiated human life.

In general, springs and ground-water serve as homes for the deity Paalölöqangw, Plumed Water-Snake, who is a powerful patron of the water sources of the earth and the heavens. Paalölöqangw is appealed to in the Snake and Flute ceremonies, and portrayed in religious puppetry during winter night dances. Springs and their immediate surroundings are places of particular religious worship in some instances, like the Flute ceremony, or during *Powamuya* (the Bean Dance) and *Niman* (the Home Dance). The Flute ceremony is specifically devoted to the consecration and regeneration of major springs, and the Lenmongwi, head of the Flute society, in an archetypal gesture, dives to the bottom of a particularly sacred spring to plant prayer-sticks for Paalölöqangw.

Since time immemorial, Hopis have offered blessings of cornmeal and prayers at springs, during specific visits for the purpose or simply while passing through the landscape (say, during herding, hunting or treks to distant cornfields). When blessing a spring, typically a man also scoops a handful of water and splashes it back towards his village or fields as a way to encourage the water to transfer some of its power to where humans most need it. Springs attract the rain and snow to themselves, and thus serve as powerful foci of value in Hopi thought. Indeed, this is why they are sacred places: if much of Hopi religious thought celebrates life, then springs are self-evident indexes of the dynamic process that produces and sustains life. At the winter solstice ceremonies feathered prayer-sticks are placed over major springs around every Hopi village as both protection and supplication.

Springs themselves, like maize in fields, were originally 'planted' in the earth by deities or gifted individuals. There was even a special instrument, a *paa'u'uypi* ('spring planter') known to the elect and used for this purpose. (A spring by Munqapi, for example, is said to have been planted in this way by a man named Kwaavaho – for whom the spring is named – in the late nineteenth century). Pilgrimages to reconsecrate and draw in regenerative power from especially significant springs at distant points are common in the religious calendar. Villages may be named for springs, like the mother village, Songoopavi, 'sand-grass spring place'. Some clans have exceptional responsibilities to springs, like Patkingyam, the Water clan, and some springs are sacred to specific clans or religious societies at the different villages. Clan migration routes from former villages are often retraced – both literally, in pilgrimages, and figuratively, in narratives and songs – at certain times of the year. In many instances, clan associations with springs at their ruins or along the route are mentioned as locations of important historical events. So the Water clan has a series of historical points, marked by springs, along its migration route from the south – like Isva, Coyote springs, south of Dilkon. Similarly, Kiisiwu, Shady Springs, for the Badger and Butterfly clan, Lengyanovi for the Flute clan, Hoonawpa for the Bear clan, and Leenangwva for the Spider clan, are all memorialized in clan tradition and visited in pilgrimage. In this sense, then, the living springs embody Hopi history: they are cultural landmarks, inscribed with significance, and commemorative reminders of the continuing legitimacy of clan rights and interests in specific areas.

In short, springs are key in Hopi social life, cultural values and the conceptualization of the landscape – all of which form the grounds of deeper religious thought and action. Hopis smoke for rain, dance for it, sing for it, and offer many other forms of prayer for it. In the cycle of life, rain-water and snow-melt nourish the plants which feed animals and human beings. So prayers for rain are not abstract; they call the clouds to replenish the waters of the earth so that all life-forms will benefit and 'be happy'. Here, then, is an environment populated not by Western science's instinct-driven organisms without spirit or consciousness, but intentional, spiritual entities which are part and parcel of the same moral system that encompasses human beings. Hopis have, so to speak, both a moral ecology and an ecological morality. As one man put it, 'We pray for rain so that all the animals, birds, insects and other life-forms will have enough to drink too'. The prolific complexity of Hopi ritual attends to springs specifically and in general, as sources of blessing and vehicles of prayer.

Of coal-mines and slurries

The springs, however, are drying up, and with them the essential force of Hopi religious life and culture. Flows have been progressively declining over the last three decades. Numerous springs and seeps have ceased to produce enough water to sustain crops planted below them. The Moenkopi (meaning 'continuously-flowing water place') Wash does not 'continuously flow' any more, and the only major Hopi farming area that depends on irrigation water is in serious jeopardy. In recent years, it has been down to a trickle by late May: not long ago Moenkopi children plunged into swimming holes long into the summer. Even the trickle that does come is supplied only by two upstream tributaries: from the main stream itself, much of the water is channelled into impoundment ponds by Peabody Coal Company.

Peabody's Black Mesa–Kayenta Mine is the only mine in the country that transports its coal by slurry. The strip-mined coal is crushed, mixed with drinking-quality water and then flushed by pipeline to the Mojave Generating Station in Laughlin, Nevada. The cities of Las Vegas and Phoenix – electric oases in the desert – buy some of the power, but most of it goes to the electric toothbrushes, garage-door openers, outsize TV sets and other necessities of life in southern California. Most of the slurry water comes directly from the 'Navajo' or N-aquifer 1,000–3,000 feet within the rock formations of Black Mesa. Peabody uses *c.* 3,700 acre-feet (about 1.2 billion gallons) of water per year for the slurry – ten times as much as the annual water consumption of the entire Hopi community (*c.* 9,000 people).

The pumping, Peabody has claimed, has no effect on the Hopi springs. Those springs, it maintains, are not fed by the N-aquifer but by the overlying 'Dakota' or D-aquifer, and by snow-melt. Hopis do not believe Peabody's position. But an escalating series of letters from Hopi individuals and officials, both traditional leaders and Tribal Council chairs, petitions signed by several hundred Hopis, protests in public hearings, dissenting interpretations by independent geologists and repeated refusal by the Tribal Council to sanction the Department of the Interior's renewal of the mining lease, have all fallen on deaf ears. Flat rebuttals to Hopi protests continue to be retailed by Peabody and its President: 'Changes in the flows from their springs may be the result of drought conditions in the region, and perhaps from the increased pumpage from Hopi community wells located near these springs.... Peabody Western's pumping from wells that are 2,500–3,000 feet deep does not affect these springs.' (*Los Angeles Times* 4–30–1994).

Six months prior to this statement, top US Geological Survey hydrologists concluded that Peabody's ongoing analysis of water impacts was based on a wholly inadequate model. Among other shortcomings: '[T]he model is not sufficient to answer the concerns of the Hopi regarding adverse local, short-term impacts on wetlands, riparian wildlife habitat, and spring flow at individual springs' (Nichols 10–28–1993).

Recent figures (USGS 1995) suggest that declines in water-level of area wells (ranging from 30 feet to 97 feet between 1965 and 1993) are up to two-thirds caused by the mine's pumping. Peabody's claim that throughout the 35-year life of the mine it would use one-tenth of one percent of N-aquifer water, which would naturally recharge itself, is seriously questioned by a USGS recharge study in 1995, which charted a recharge rate 85 percent less than Peabody's estimate. (It has been suggested that Peabody has tried to suppress public release of these discrepant figures since – if verified – the company would be contractually obligated, according to the terms of the lease, to post a bond for aquifer restoration).

It seems evident too that depletion of the N-aquifer has had serious impacts on the D-aquifer, and on the springs themselves: the Moenkopi Wash is directly affected since it is supplied by N-aquifer seepage, and since Peabody impounds surface water at a rate surpassing 1,800 acre feet per year that would otherwise directly supply this Wash. USGS computer simulations predict total drying of some major Hopi wells beginning in the year 2011.

Several alternatives to the slurrying of aquifer water have been proposed, and progress has been made on one: another pipeline from Lake Powell which would provide domestic water for Hopis and Navajos, and industrial uses for Peabody. But Peabody, ever mindful of the bottom line, is evidently using delaying tactics, suspending negotiations and playing off the tribes against each other.

Meanwhile, the Hopis are deeply anxious about all spring declines, for both obvious reasons and deeper metaphysical ones. Hopi moral philosophy, following a covenant entered into with the deity Maasaw upon emergence into the present world, charges people to take care of the earth and all its resources; indeed this is a significant measure of whether one is worthy of the name 'Hopi'. If Hopis break the covenant, cataclysm of cosmic proportions threatens. During the early 1980s when I began ethnographic research, Tsakwani'yma, an older Spider clan man, would sometimes talk about prophecies he had heard from his uncle, Lomayestiwa. He returned to one repeatedly: a time would come when Paalölöqangw, the water serpent deity, would turn over and lash his tail deep within the waters of the earth, and all land-life would tumble back down to the bottom of the ocean. 'Can you interpret it?' he would challenge. 'It means earthquake. But it's also symbolic of the life we are leading today: *koyaanisqatsi*, a life of chaos.' Then in 1987 and 1988, shortly after he passed on, there were two earthquakes on Black Mesa (a rarity), which the Arizona Earthquake Information Center connects to the removal of massive quantities of coal and water. The perception of some elders that this is the result of having their souls sold out from under them – literally, in the link between groundwater and spirits of the dead – causes profound sadness and a sense of intractable religious desecration.

In addition to long-term Hopi interests, regional economic and demographic patterns make the continued pumping of more than a billion gallons of potable water every year for a coal slurry incredibly short-sighted. The coming century will undoubtedly see ever more serious problems of water supply for the rapidly growing conurbations in the West. In this light, Hopi religious concerns with springs become metaphorical of larger issues of global development and natural resource management. But while typically attuned to such universal implications, Hopis in the immediate term are concerned with basic physical, cultural and spiritual survival. If the springs are to be saved, and along with them continued Hopi cultural and religious existence, Peabody's relentless drive toward short-term profits, at the expense of stakeholder concerns, needs a dramatic make-over in line with trends toward local–global balance pursued by more progressive multinationals. In the meantime, the pumps siphon the essence of life from the water-roots of Black Mesa, and the Hopi springs are drying up.

The discourse of ecological correctness: of dam builders 'rescuing' biodiversity for the Cree (Marie Roué and Douglas Nakashima)

The sub-Arctic territory of the Cree Indians of Chisasibi (Quebec, Canada), traditionally a society of hunters, fishers and trappers, has been transformed since the 1970s by a series of hydroelectric megaprojects. This article considers the 'rescue', by the developer, of two fish species threatened by dam construction. Hydro-Quebec presents this action as tangible proof that the utility is 'imbued with a new kind of ethics' and that it 'subscribes to the concept of sustainable development' (Hydro-Quebec 1990). While biodiversity may indeed have been conserved, for principal users of the resource – the Cree – the fish are lost. This paradoxical outcome raises the question 'for whom do we conserve biodiversity?'

Appreciating the social and cultural significance of biodiversity means moving beyond a check-list mentality, the mere quantification of species presence or absence, to apprehend the relationship between people and 'nature'. In hunting or fishing societies, for example, the encounter with animals is laden with symbolic, social and ecological significance. The perpetuation of these fragile moments depends upon the transmission of specialized knowledge and know-how from one generation to the next, as well as the persistence of select places where the encounters can occur. For the Cree of Chisasibi, Upichuun – the First Rapids – was such a place, one

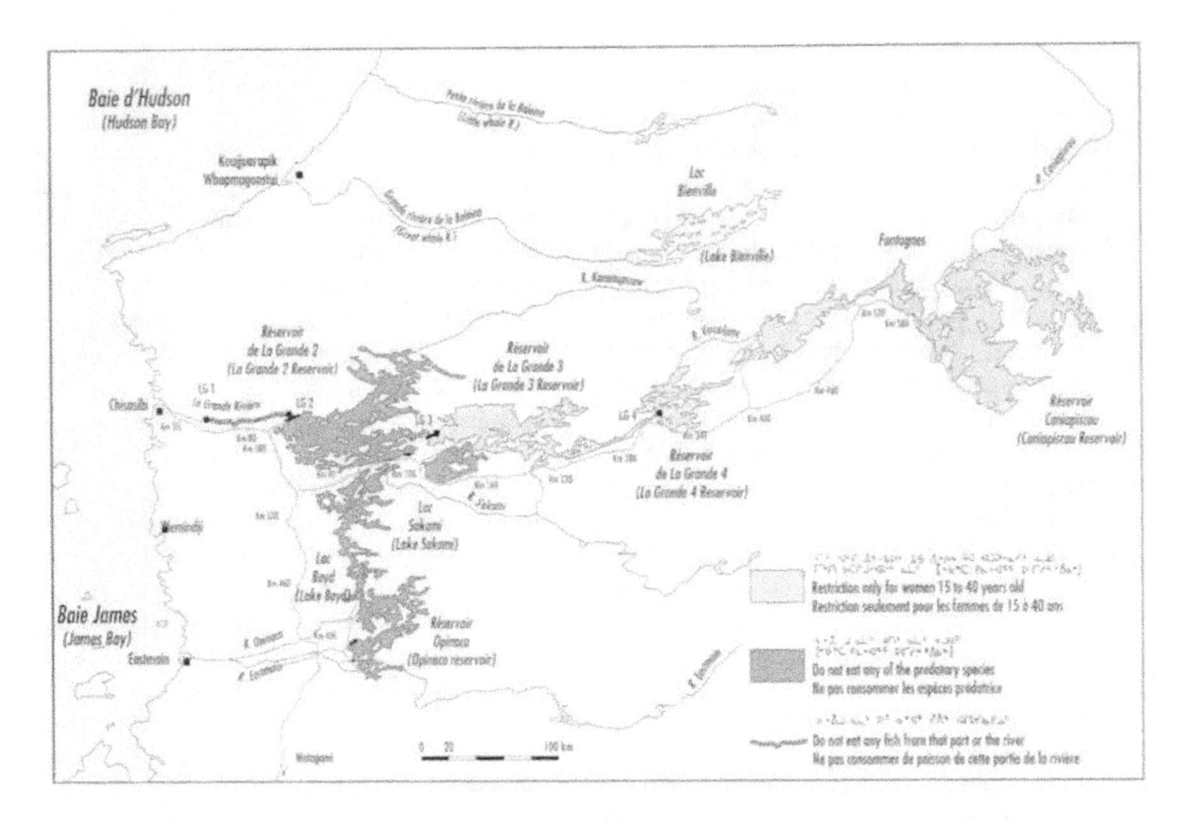

Figure 10.1: Map of eastern James Bay, Quebec, showing the village of Chisasibi near the mouth of La Grande River and the series of reservoirs created by the La Grande Hydroelectric Project. This map is part of a pamphlet prepared by the Cree Board of Health, the Cree Regional Authority and Hydro-Quebec, which instructs the Cree not to eat any fish of any kind caught in the river between the LG1 (First Rapids) and LG2 dams because of high concentrations of mercury in their flesh.

of exceptional cultural importance. It symbolized the separation and unity of 'Coaster' and 'Inlander' Cree, a place of sharing and exchange. Here, amidst the bounty of fish returning fat from the sea and ripening berries thick on the land, the two groups reunited each year to renew social ties, to fish in tandem, to join forces in work and pleasure, and to celebrate in this manner a partnership of long date. Construction of the La Grand 1 powerhouse over the First Rapids obliterated this privileged site of encounter between these complementary groups and the fish that they came together to harvest. Without a place for the encounter, the encounter could no longer take place.

The question 'for whom is biodiversity conserved?' raises the issue of scale. When – under the impetus of the Convention on Biological Diversity – threatened genomes, species or ecosystems are inventoried and actions for conservation are advocated at the planetary level, the concern is for biodiversity as a *global* inheritance. For those who rely directly upon these resources, however, conservation of 'their' animals and plants is a *local* concern, delimited by territories whose culturally-defined boundaries are distinct both from those of the scientist and those of the State.

In the James Bay case, issues of scale and the definition of region or territory are crucial. According to Hydro-Quebec, the State hydroelectric company, the land area flooded by Phase I of the La Grande Hydroelectric Complex is no more than 3.4 percent (Chartrand 1992; Figure 10.1). Through this statistical abstraction, parcels of land are judged to be equivalent, brushing aside the possibility that flooded lands (riparian habitats, valley bottoms, rapids, etc.) may be of exceptional ecological or cultural significance. And fundamental to the issue of scale is the question, '3.4 percent of what?' For the State, the unit of reference is the recently-created administrative zone of 350,000 square kilometres, two-thirds the size of France. This 'James Bay Territory' represents for them 'an immense territory for development', awaiting the progressive exploitation of water, forest, mineral and other natural resources. For the traditional inhabitants of the land, however, the unit of reference has quite another set of boundaries. It is the local hunting band, or more recently the village, which determines territory and identity for the Cree. Cree territories are many times smaller than the James Bay Territory of the State, and for the Chisasibi band in particular the percentage of land that has vanished beneath the La Grande reservoirs is an order of magnitude greater than the figure set forth by Hydro-Quebec.

Upichuun - The First Rapids: place of production and of reproduction

The First Rapids, or Upichuun in the Cree language, located some 35km upstream from the mouth of the La Grande River and the former Cree village of Fort George, was of great importance for fishing from late July to early September. During this summer period between the brief goose-hunting seasons of spring and autumn, fish was and is still today the source of food *par excellence*. But fish and fishing are not by any means important only in summer. They play a critical role throughout the annual Cree subsistence cycle (Weinstein 1976). In sub-Arctic ecosystems, populations of terrestrial animals are subject to drastic, cyclical changes in abundance. Populations of Arctic hare (*Lepus americanus*), for example, pass from maxima to minima with a periodicity of 7 to 10 years. During peak years, Arctic hare may account for up to 25 percent of the Cree subsistence harvest, whereas at the bottom of the cycle the take may be almost nil. Similarly, for caribou (*Rangifer tarandus*), the Northern Quebec population which had greatly decreased by the turn of the century and all but disappeared by 1910 (cf. Nakashima and Roué 1995), began its comeback around 1960 and today numbers more than 600,000 head. Amidst this variability and uncertainty, fish species, along with waterfowl, are stable resources which the Cree can rely upon from one year to the next. But while waterfowl are strictly seasonal, fish are accessible the year around. Many stories recount how people have been saved from famine because they managed to find their way to a reliable fishing spot.

Of all of the fishing places known to the Fort George Cree, the First Rapids of the La Grande River was by far the most productive. The two main species of fish were members of the whitefish family: cisco (*Coregonus artedii*) being the most important, followed by whitefish (*C. clupeaformis*). These anadromous populations returned to the First Rapids from their sojourn in the sea at somewhat differing dates from one year to the next. Nevertheless, the moment of their arrival was made known to the Cree by the flowering of the fireweed, *Epilobium angustifolium*, along the banks of the river close to the fishing site. This use of an 'ecological indicator' has been documented in many cultural traditions. In contrast with the fixed solar calendar, natural phenomena which co-occur in time in response to common environmental factors provide a flexible marker sensitive to year-to-year variation. From long observation the Cree know that when the fireweed comes into flower, it is time to prepare their nets.

Once at the First Rapids, the Cree established camps and fished day in and day out (see Figure 10.2). All accounts attest to the incredible abundance of fish. There was no need for any other food. Some families preferred to paddle back and forth between the rapids and the village on Fort George Island. They did not have to stay long before their canoes were filled with fish. As soon as they had enough, they set off back to the village. Even the residential school at Fort George relied on a stock of fish from the First Rapids to provide for its young Cree boarders: 'We went there by canoe to catch fish. The minister had asked us for some fish, dried fish for the children in the residential school to eat. That is what he told us. We stayed there for one week. We caught so many fish. It seems we could barely keep up with cleaning all the fish'. (Unless indicated otherwise, all citations are words of the Chisasibi Cree from D. Nakashima and M. Roué, 1993-94, Unpublished Field Notes, Chisasibi).

Fish were also conserved by smoking or drying. All parts of the fish including the head and the entrails were used, the latter being boiled before drying. These stocks of dried fish would be carried by the 'Coasters' along the shore to their goose hunting camps in September, or provide for the 'Inlanders' during their long treks back into the interior. Fish scales and any other remaining parts were fed to dogs brought to the First Rapids to be fattened up for autumn and winter when they would be an invaluable source of traction.

A place that both divides and unites, the First Rapids – Upichuun – was of unique significance to the Chisasibi Cree. Along the La Grande River, the principal travel route for the Cree between sea and interior, the First Rapids represented the dividing place between coast and inland. On the way downriver, once Upichuun was passed, no other barriers stood between the people and the sea. On the way inland, Upichuun was the first of a long and arduous series of portages.

Coasters and Inlanders: segmentation and complementarity

Like many other Cree communities along the James Bay coast, the Chisasibi band is divided into complementary groups: Coasters and Inlanders. Turned towards the resources of the sea and the coastal region, the Coasters were hunters of seals and beluga whales, as well as geese, ducks, loons and other migratory birds which each spring and autumn migrated in great flocks along the shores of James and Hudson Bays. They were also more closely associated with the Hudson's Bay Company, as they

Figure 10.2: Series of photographs taken by James Bobbish (a Coaster and at that time Chief of the Band) illustrating the unique seine-fishing technique employed by the Cree at First Rapids.

Awaiting the signal to deploy the seine. When a large wave surges into the small bay, the two men on the river bank (foreground) rapidly pull the seine net across its mouth. The group of men at the tip of the bedrock arm projecting into the river are responsible for carefully feeding the net into the water.

Beginning a sweep of the bay. The two men in the foreground, holding one end of the net, walk along the bank to join the other men out on the arm of bedrock.

Surrounded by onlookers, including numerous children, the men haul the seine net up the smooth face of the bedrock.

As the net is pulled out of the water, the children rush to help gather up the struggling fish.

constituted the principal source of labour for the trading post on Fort George Island. The Inlanders, on the other hand, would only descend the rivers from the interior in late spring, to join the Coasters along the shores of James Bay for the brief summer period.

The following account by an Inlander from Fort Rupert in southern James Bay reveals the nature of the annual spring encounter between the two groups. Inlanders would offer the meat and fat of caribou and receive in exchange the meat and fat of geese. Inlander songs about running rapids would be answered by the Coasters' songs about flights of geese.

> '... the Inlanders, they kill caribou, and dry up the meat and save all the grease. And the Coasters kill the geese, and dry them, and also save the grease ... And then they take the meat and the grease to the Coasters, usually the oldest man in the group. And then the Coasters go to the oldest Inland man, and give him the dried goose meat, and also the grease. And then they build a big wigwam. And when they are finished all of the people join in the feast... The old people start dancing and using their drum. And the Inlanders start singing about shooting down the rapids, and the Coasters start singing about the geese... The Inlanders and the Coasters share their food at this time. And they usually do it in the spring when all of the people from inland come in for supplies.' (cited in Preston 1981)

In this manner, these two groups within the band organize the sharing of biodiversity from two ecological systems. Through an exchange of food and skins, accompanied by songs and stories, they make the most of the ecological, social and cultural diversity forged by the ways-of-life of those in the interior and those on the coast. This dualistic world view which is their shared construction, asserts their differences while reinforcing the cohesion of the band. At Upichuun, everyone camped together, exchanging food and stories and sharing in a fishery which was extraordinarily productive. Now that the rapids have been destroyed, the memories of these joyful periods of reunion are tinged with sadness and regret. A Coaster recalls his father's strong ties to the inland people: 'The part that really hurt me, when I saw (Upichuun after the damming), is that both my parents loved it there. This is why it hurt... Because it was also all those memories that were in that place. I thought of so many people. ... I remembered how happy my father was to be there. He loved the inland people so much. When the inland people came back, he would be gone all day... to the homes of the inland people.'

The Inlanders did not stay long on the coast, but soon headed back to their territories in the interior. They relied on the First Rapids fishery to provide them with a stock of dried fish for the long and hard journey upriver, paddling and portaging. They fished for themselves, but also received fish from those already camped there as fish were part of the traditional cycle of gifts and exchanges. 'People used to get food to bring along when they went inland right here at (Upichuun). Sometimes they had enough food to bring all the way to Caniapiscau (the headwaters of the La Grande River system). This fish was given to them by the ones who hunted here.'

Men's roles - women's roles

The process of production dominated by men is complemented by the preparation and distribution of the fish in which women play the major role. Through this division of techniques and practices, the social relations that create and recreate Cree society are perpetuated. When the Cree reflect back upon the days passed at Upichuun, they recall how the air was filled with excitement and laughter. The abundance of food allowed everyone to share in these carefree moments. While the men fished, the women cleaned the catch, scaled the fish and then hung them up to dry or to be smoked. 'People were walking back and forth to their tents all day, the men bringing their fish back to the tent so the women could clean them and smoke them. And when they finished one batch, there were more.'

Fishing at the First Rapids also coincided with another important event, the ripening of blueberries, *Vaccinium uliginosum* and *V. angustifolium*. When a break could be taken from cleaning fish, the women set off into the nearby bush to enjoy a peaceful moment together, seeking out their favourite berry-picking spots. 'I think about this so often. There was a hilly area on the north side. I used to go there all the time to pick berries. The women always picked berries there. They picked blueberries. They ate these along with the fish.' Plump fish and ripe berries were brought together to make a traditional dish which continues to be highly prized today at feast time. The fish was boiled, the flesh separated from the bones and then shredded and mixed in with the berries.

Biodiversity conserved, but the fish nevertheless are lost

From an engineering viewpoint, the natural narrowing and precipitous drop in height of the river at the First Rapids made this location an ideal site for a power station. For this reason, Hydro-Quebec chose it as the site of the La Grande 1 (LG1) dam. In a bid to save the First Rapids fishery, the Cree managed during a first round of negotiations to have the LG1 dam relocated some 35km upriver. Despite this first reprieve, subsequent negotiations reinstated LG1 at the First Rapids site. Knowing that the outflow of the La Grande River was to be more than doubled by the hydroelectric development, the Cree were informed that the island on which their village was located could be eroded away. In exchange for giving up the Upichuun site, they received Hydro-Quebec's commitment to construct a new village (to be called Chisasibi) on the adjacent mainland, where today virtually all of the people of Fort George now live.

Construction of the La Grande hydroelectric project posed a number of challenging environmental problems. Of serious implication for the fish and the fishery of the First Rapids was the filling of the immense LG2 reservoir, situated immediately upstream. The dilemma was how to proceed with reservoir filling which would require a complete or partial cut-off of the river for about one year. A total cut-off at LG2 would allow saltwater driven by tidal currents and uninhibited by a countering freshwater flow, to penetrate all the way to the First Rapids. This exceptional penetration would eliminate a freshwater stretch below the Rapids. For the estuarine populations of cisco and whitefish, this freshwater pocket is critical as it is their only shelter from the excessively low temperatures that occur in saltwater in winter. If eliminated for even one winter, the cisco and whitefish of the La Grande estuary would be condemned to certain extinction.

How could Hydro-Quebec go about the filling of the LG2 reservoir without endangering the fish populations of such importance to the Cree? Additional research solicited by the Société d'Energie de la Baie James (SEBJ – a branch of Hydro-Quebec) revealed a possible solution. It appeared that if a layer of ice were allowed to form on the river surface before cutting off the flow, the presence of the ice layer would dampen the tidal effect and halt the intrusion of saltwater several kilometres short of the First Rapids. This would allow the persistence of a freshwater pocket and ensure the over-winter survival of the fish. Accordingly, by modifying the project schedule, the development corporation successfully controlled saltwater intrusion and rescued the fish (Berkes 1988; Roy 1982). For this phase of the project, SEBJ had taken remedial measures such that no loss of biodiversity had occurred.

This 'rescue' offers insights into the limitations of technocratic representations of human–nature relationships. It reveals the incapacity of a self-proclaimed leader in sustainable development, and beyond that the incapacity of our society as a whole, to comprehend the linkages between a people and their natural environment. The paradox is striking. On the one hand, the developers boast of their environmental consciousness and technical skill which have allowed them to preserve (at least until proof to the contrary) the anadromous fish of vital importance to the First Rapids fishery of the Cree (SEBJ 1993). Yet at the same time, they proceed with the construction of the LG1 dam which obliterates these same rapids and destroys along with them the sites where fishing takes place. Without a place to fish, the disappearance of the fishery is assured.

In short, biodiversity has been conserved, but the fish have been lost! For even though the populations may still exist, they are inaccessible to the Cree. Of the four fishing sites that existed along the river's banks, three have been blasted away during construction. The fourth and most important can no longer be used. This small bay which the Cree fished with their seines is choked with boulders thrown there by dynamite blasts. Although these rocks were removed as part of remedial measures to reduce project impacts, the specific current patterns which made fishing feasible at that site no longer exist. Furthermore, the river below the dam is subject to severe fluctuations in water level. During a visit to the site in July 1994, the flat bedrock arm from which fishing seines were deployed was completely awash with waves.

As if this were not sufficient to seal the fate of the First Rapids fishery, the spectre of methyl-mercury contamination and Minimata disease now also hangs over the river. Already in the 1970s, the appearance of elevated organic mercury levels in the piscivorous or fish-eating fish of the reservoirs was considered an unexpected impact of the project (Weinstein and Penn 1987; Roué and Nakashima 1994). But Hydro-Quebec biologists were even more astounded to record even higher mercury concentrations below the LG2 dam in all fish types, including non-piscivorous fish. By feeding on the minced remains of large predatory fish such as pike and lake trout that had been drawn into the turbines, non-predatory fish had been transformed into super-predators, bio-magnifying mercury as would organisms at the top of the food chain (Verdon *et al.* 1992). As a result, fishing of all species has been completely prohibited along major stretches of the river.

So biodiversity may indeed have been conserved, but for whom? Without the sites where they can fish, with fish contaminated with mercury, this conservation of biodiversity is meaningless to the Cree.

Nothing can replace the abundance offered by Upichuun. In an attempt to compensate, the Cree are intensifying fishing along the James Bay coast, but these sites do not come close to providing such large numbers of fish with so little effort. Men have even been salaried to fish for those no longer able to fulfil their needs. This economic arrangement whereby fish become a commodity, however, is undermining networks of sharing and exchange. Beyond the value of fish as food, the First Rapids was also a place to camp together and to socialize, a place to transmit to the young a specialized know-how. As the women point out, the fish they consume nowadays are no longer the live fish caught by their own hands and eaten fresh from the water. By erasing practices, knowledge and the bases for sharing and co-operation, the social cohesion of the group is called into question. By troubling the ties between hunter and prey, it is the symbolic relationship with nature that is placed in doubt. The task of enriching the notion of biodiversity with that of cultural diversity still remains before us.

Today, the people of Chisasibi live with many bitter memories, aggravated by the towering silhouette of the LG1 dam which has silenced forever the First Rapids.

[Marie Roué and Douglas Nakashima, CNRS/MNHN, Paris. Research supported by the Programme Environnement, CNRS (France) and the Grand Council of the Crees of Quebec].

Gwich'in attitudes to fish (Gleb Raygorodetsky)

The Gwich'in inhabit Arctic and sub-Arctic regions from the Mackenzie River Valley in Canada's Northwest Territories (NWT) in the east, through the Yukon, and into Alaska, USA, in the west. The Gwich'in Nation still depend on the environment for their livelihood as their ancestors have done for centuries. This relationship has been kept alive through the continuous flow of their traditional knowledge from generation to generation. With changing life-styles in the last half century, however, opportunities for the traditional oral and on-the-land ways of teaching and learning began to disappear. The Gwich'in Environmental Knowledge Project (GEKP), an initiative of the Gwich'in Renewable Resource Board to document local traditional ecological knowledge, began in the summer of 1995 as a result.

The GEKP concentrated on several species of fish, birds and mammals important for Gwich'in subsistence, including: black bear, grizzly bear, beaver, caribou, marten, moose, muskrat, Dall's sheep, wolf, geese, black and mallard ducks, ptarmigan, swan, charr, coney, herring, loche and whitefish. The following two sections extracted from the GEKP report *Gwich'in Words about the Land* (Raygorodetsky 1997) detail Gwich'in knowledge of and attitudes towards two important fish species: charr and whitefish.

Dhik'ii – Dolly Varden Charr (*Salvelinus malma*)

Charr is considered a delicacy because it is very rich and tasty, and it is often hard to come by. People showed the same respect for charr that was given to whitefish and other fish and game. People did not waste charr and they shared their catch with others. They pulled their nets out for a time if they could not process all the charr. Because fishermen shared their charr with big families and old people who could not fish, everybody had something put away for the winter. If one family got far too many charr, several women from different families came together to clean it. Afterwards they shared the fish, so everyone had the same amount. In the 1940s, some people did not have proper nets, and the fishermen with good nets shared their catch. Today it is different – the old way of sharing is dying, and money always comes first. In the old days, people never joked about charr, or any other fish or animal, because their survival depended on having a good relationship with them.

Charr, like any other fish, follow the shortest route along the river – from eddy to eddy. In small rivers, most eddies occur at points where the river makes a turn, and this is where people set their nets. On large rivers, for example the Mackenzie River, eddies develop anywhere along the shore. Charr travel in schools so people catch many at one time. Sometimes the net is loaded to the bottom, and two people have to check it. People fish for charr when the water in the river is low. When the water is high, they have to pull their nets out because of the dirt and debris that comes down the river, and because there are no charr to be caught during this time. During the charr run, nets must be checked at least twice a day, in the morning and evening; otherwise, charr become water-logged and the meat becomes soft and spoils. If there are many charr, the net must be checked three times a day. In the old days, some people set three-inch-mesh herring nets, and caught many young charr. Today, people use four- and four-and-a-half-inch nets so they catch only large charr, allowing the young ones to escape and grow. When the charr season is over, a good fisherman cleans and repairs his net for the next season.

In the old days, people made nets out of willow bark and set fish traps to catch charr. They also swept deep pools along the river with their nets. People also caught charr in the clear waters of the upper Rat River with spears made out of antlers. The antlers were boiled to soften them, and then shaped into two hooks. The hooks, and a central bone spearhead, were attached to the end of a long pole. Today, people can catch charr on store-bought lures.

Over the last several years, some people wanted to catch lots of charr, and so they set many nets, almost blocking the river. Elders said people bother *dhik'ii* too much today. They should leave charr alone for several years, and let them come back and grow in numbers.

Women usually clean and cut up charr the same way as whitefish. Men help if they have time in between looking after the nets. Elders taught boys and girls how to clean and net fish, so when the kids grew up, men knew how to do some of the women's work and women knew how to do some of the men's work.

Charr must be smoked or dried soon after it is caught, otherwise the meat becomes too soft to be used for anything. After a charr is gutted, it is split in half along the back, leaving the halves attached at the tail. Then, a cut on one side is made across the skin at the base of the tail. This way the fish does not slide from the drying poles. *Dhik'ii* is too fat to dry well. After being smoked in a smoke-house for some time, it becomes dry on the surface. *Dhik'ii*, as any other fish, dry best when it is warm, and a little overcast, with a light wind blowing. If it is sunny and hot, the meat simply cooks in the sun and spoils. Charr is good to eat when it is freshly smoked, but if stored in this way for long, it turns rancid because its meat is very oily. Today, many people smoke charr improperly, and the charr comes out too oily and not dry enough. To stop charr from spoiling after smoking it, most people take it to town and put it in a freezer in a plastic bag. In the old days, charr were stored in underground pits or ice-houses for the winter.

The first freshly-caught charr is usually cooked on the fire or boiled right away. Charr liver is cooked with charr eggs and then mashed with cranberries. Charr eggs are also dried. In the old days, when people came back after fishing for charr at the Rat River, they had a feast, where everybody ate fried, boiled and smoked charr. Guts are usually not used for food, although some Elders like to eat them cooked. The guts may also be cooked to render oil from them, which is added to the dog food. This oil is also a very good medicine. For example, people use it as ear drops when they have hearing problems.

Luk zheii – Whitefish (*Coregonus nasus*)

For a long time, whitefish has been important to people because it is a good source of food. People think it is the best fish, because it has the best taste and relatively few bones. People who were raised on *Luk zheii*, say that they cannot live without it. In the past, people made lots of dry fish for their dogs in the winter. Today, whitefish are not as important because very few people still live out on the land and catch fish for themselves or their dogs. Some people, however, still catch whitefish to dry for themselves or for sale.

Long ago, people made fish traps for catching whitefish at the beginning of the fishing season. Usually, they drove posts into the bottom of an eddy, to form a fence. On the upstream side, a long, tubular, chute-like basket was fastened between two posts. Except for an opening into the basket, the posts forming the fence were close together to prevent fish from passing through. The whitefish swimming upstream rest in the eddies, and then enter the trap, ending up in the basket chute. People lifted the basket out of the water and pushed the fish with scoops on to a little platform.

Long before cotton nets were introduced, people made fish nets from willow bark. In the spring, when the willows are full of sap, people peeled the bark off and made long strands out of it, as thin as sinew, and soaked them in water. Then, they used the strands to make nets. These strands had to be kept wet constantly otherwise they broke easily. Later, people made nets using three or four spools of store-bought cotton thread. After the fishing season was over, they washed the nets and dried them for next year.

People caught whitefish during the summer to make dry fish, and in the fall they would freeze freshly caught fish. In the summer, the best place to set nets for whitefish is in the eddies. Big eddies are better for setting nets because they hold more fish. Eddies in front of creeks are also good for setting nets. In the winter, nets are set under the ice anywhere on the river.

A good fisherman works every day, checking his nets regularly. If he cannot check them at least twice a day, he should remove the nets so the fish do not spoil. Fishermen should check their nets three times a day when there are lots of whitefish. During heavy rains and high water, people do not catch whitefish because the fish do not travel in these conditions. People set five- to five-and-a-half-inch mesh nets for whitefish. When the fisherman finds a good eddy in which to set his net, he drives a big wooden pole into the bank or close to the shore. The top and bottom lines of the net are tied to this pole before the fisherman paddles out from the shore, remaining within the eddy, and sets the net from the nose of the boat. Sinkers and buoys are attached along the length of the net, and finally it is checked for tangles and straightened out.

In the fall, when the ice reaches two to three inches thick, nets are set under the ice for whitefish migrating to the coast. Most people take them out by December, or whenever the ice gets too thick, but fish can be caught all winter. To set a net under the ice, people cut a line of holes, eight to ten feet apart, with an axe or a chisel. The net is then pulled under the ice so it stretches between the end holes. If the timing is right, a fisherman may get 500 whitefish from under the ice in one day. Sometimes, the hole in the ice must be enlarged to allow the net to be pulled through with fish in it. When the ice is five to six inches thick, people usually stop setting nets because it becomes hard to draw the nets under the ice.

Long ago, there used to be a fish camp almost every five miles along the Mackenzie River. People still could not catch all the whitefish and reduce their numbers. Today, only a few active fishing camps remain in the delta. Only a commercial fishery would cause the whitefish to disappear in this country. If local people prevent commercial fishing on the rivers, whitefish will always be plentiful.

As with other animals, people should respect whitefish by catching only what they need, and by sharing their catch with others, not joking about whitefish, making use of all parts, and cleaning the fish promptly so there is no waste. Although in the old days there were no laws on how much fish people could catch, they knew how much they needed for their families, relatives and Elders, and caught only what they needed. Some people said when a fisherman catches too many whitefish, or too much of any animal, something bad will happen: for example, the fisherman or somebody in his family will die.

People said that long ago a person always shared his or her catch. When people got lots of whitefish, they froze it and put all the large whitefish aside for Christmas or Easter, at which time they gave each other a fish as a sign of their respect for one another. Some people say that only few people still follow the tradition of sharing, but instead sell their catch.

People use whitefish for food, trap bait and dog food. Dried and frozen whitefish are also a source of income. In the past, people fished more in the summer, when they could dry the fish for themselves and their dogs. When people travelled with dogs to the mountains, it was easier to carry dried whitefish than frozen. Since people do not use dogs any more, they prefer to fish in the fall and freeze the fish for the winter.

Whitefish can be fried, roasted and boiled. Many people think lake whitefish taste better than river whitefish. The flesh of whitefish caught in the spring is very soft, and therefore less desirable. Sometimes, people eat whitefish heads. They collect and wash several of them, and then boil them for food. Whitefish stomach, also called a fish pipe, or *its'agoghoo*, is cut open, rinsed, and then roasted on a stick over a fire. People also fry it in a frying pan with fish eggs. Whitefish stomachs are also used as bait when 'jiggling' for loche. Liver and hearts are also eaten. People collect blood in a bowl from freshly caught whitefish while it is being gutted. This blood, and hearts, may be mixed into a broth made from a boiled whitefish, making good fish soup. Dry fish can be pounded to make pemmican in the same way as is done with caribou meat.

Frozen whitefish eggs are called 'Indian ice-cream'. In the fall, people open up freshly caught female whitefish, and pour the fish eggs into a pan and freeze them. Later, people break them up and eat them frozen, just like ice-cream. To keep whitefish eggs fresh throughout the winter, a fish is left unopened and a small stick is put into the fish's anus. This keeps the eggs from drying up and going rancid. The fish eggs of lake whitefish are creamy and taste better than those of the river whitefish. Whitefish eggs can also be roasted or dried. Some people make a soup out of fish eggs.

Oil made from dried whitefish is good eye medicine. If a person has sore eyes, a drop is put in each eye. Fish oil is also used in tanning skins. It is smeared on the surface of a dried skin and left for several days to make it softer.

Seasonal wetlands (Fleur Ng'weno)

In Africa, seasonal wetlands cover a larger area than permanent fresh water during the rainy season. They play a vital role in the collection, storage, purification and discharge of fresh water. Seasonal wetlands provide people with water, food, building and weaving materials and ceremonial grounds. They serve as breeding and feeding grounds for fish, reptiles, amphibians, invertebrates and birds, including migratory waterfowl. They are especially important in arid and semi-arid areas, where there is little permanent fresh water.

Seasonal wetlands include flooded grassland, seasonal marshes, lakes and springs, temporary pools, flooded rock slabs and seeps. Many are invisible during the dry season. Animals and plants disappear as the water dries. They survive the dry season as eggs, seeds or buried under the mud. All that remains is rock or soil, and sometimes dry plant stalks.

Seasonal wetlands are among the most threatened habitats because they appear dry for much of the year. They are thus converted to agriculture; not reserved during land demarcation; and ignored in road construction and other development activities. Plants and animals in seasonal wetlands live life in the fast lane. They grow rapidly and in great abundance for a short period. Microscopic plants and animals fill the water. They include phytoplankton such as algae and diatoms, and zooplankton, crustaceans, worms, planaria, insects and molluscs. These serve as food for mammals, birds, reptiles, amphibians and fish.

Many fish disperse from the lakes into shallow streams and flooded areas to breed. Seasonal wetlands provide feeding and resting grounds for migratory waterfowl and breeding grounds for waterbirds. Seasonal wetlands provide people with water, food such as fish, crustaceans, molluscs and water plant roots and seeds, grazing for livestock, especially in the dry season, materials for thatching, mats, baskets and other woven products, and ceremonial grounds for religious and cultural ceremonies.

In many parts of Africa, rain falls only occasionally, but it falls with great force. Seasonal wetlands collect the floodwaters, preventing excessive runoff and destructive floods. As the water slowly sinks through mud or porous rock, it is purified. Water is thus stored in seasonal wetlands, and filtered into underground aquifers. During the dry season, the pure stored water is discharged into the environment.

The following story by Beatrice Maloba (Box 10.1) brings out the cultural value of wetlands.

Box 10.1: The Flooding of River Sio

Beatrice B. Maloba

The River Sio empties into Lake Victoria in the Samia district in Western Kenya. The river in spate causes the lake to flood and the fish mentioned are actually lake fish, but only get swept out onto the land when the river floods. There are many causes for reduced annual flooding on the River Sio and in Kenya in general. They include a changing rainfall pattern, clearing of forest and wetlands for agriculture, soil erosion from farmland, cultivating close to river banks and construction along the lake shore. These factors reduce beneficial annual flooding but cause destructive floods every few years.

'A long time ago, the people of Sianja village in Samia looked forward to the River Sio's floods with joy. With joy, because the floods made fish readily available by spilling them into the Sianja swamp. River Sio usually flooded in April and May. It burst its banks just a few kilometres before spilling into Lake Victoria. As soon as the floodwaters subsided, village folk, both fishermen and non-fishermen, rushed to the swamp. There they simply picked up the fish of their choice. They left the unwanted fish to waste away, or for the birds to eat.'

'The village people preferred to collect tilapia, called *engeke*, and lungfish, called *obuyoko* and *emumi*. Tilapia and lungfish were a delicacy, as well as being easy to preserve for future use. Sianja people ate *obuyoko* and *emumi* almost throughout the year. As soon as the people returned home, with special baskets full of fish, the fresh *engeke* was eaten. *Engeke* were cleaned, the scales scraped off, and the stomach cut open. The eggs were used as food, and the rest of the insides were thrown away. The fish was then cut into pieces and cooked in a pot. About six medium-sized tilapia fish could fit into a large pot. It was a big feast for a family.

'After eating there was a bit of rest. Then tasks were divided among the people. The women in the homestead fetched water for cleaning the rest of the fish. In preparation for smoking, the young girls went to collect firewood. The young men split logs for making fires for smoking the roasting fish. Fish was an important food for the Samia people. In Sianja village there were experts in smoking and sun-drying techniques. In each homestead there was a special spot for preparing fish. Tilapia were prepared for smoking in much the same way as for eating fresh. Small lungfish were cleaned and the inner organs removed. Then they were strung on a special stick. Lungfish are long and narrow. The fish's tail was pierced, and the stick passed through it. Then the fish was folded over, and the stick passed through its open mouth. A stick would hold a row of five to eight small lungfish. Depending on the catch, some families would prepare a basket full of sticks, holding a lot of small *obuyoko*. *Emumi*, the large lungfish, were cleaned in the same way. If there were any eggs, they were removed very

Box 10.1 (continued)

carefully and kept separately. Eggs were the most delicious. The lungfish were cut into three segments which were smoked to prevent the fish from rotting. In addition, the smoking process softened some bone, particularly the head bones. Since they were soft, they were chewed when eating smoked fish. The skin of smoked fish acquired a special tasty flavour. Children liked having a bite of the skin of cooked smoked fish to eat. Fish was also sun-dried. The fish was cut on one side and opened out flat. Eggs and internal organs were removed. The fish was flattened, and put in the sun to dry. Sometimes salt was added as a preservative. When a large *emumi* was opened out, its width was more than half a metre! Once dry, it would keep for a long time. Whenever the owner wanted to eat fish a small piece was cut off. It was soaked in water for a while and then cooked to make a meal.'

'Sianja people liked to catch tilapia and lungfish when the river flooded, because at that time the female fish had many eggs. The fresh fish eggs were shaped into mounds, about 15-20cm wide and weighing nearly a kilogram. Such a fish egg cake is called *esiche* (*ebiche* in plural). *Ebiche* were dried in the sun each day preventing the eggs from going bad. Once completely dry, *ebiche* were kept in a storage pot. They were checked now and then to make sure they did not grow mouldy. *Ebiche* in a house was a handy protein, for just a small portion was rich in food value. It was soaked for a while, and washed to remove any dust. Then it was boiled on a slow fire, in order to become soft and for the flavour to form. It was eaten with *obusuma*, a starchy food made from millet.'

'When River Sio flooded, other activities stopped, and all attention was directed towards the fish.'

Competition between ancestors and Chinese traders in the Aru islands, Indonesia (Manon Osseweijer)

The Aru islands are located in the extreme southeast of Maluku province, near Irian Jaya in Indonesia. The archipelago consists of six main islands – Kola, Wokam, Kobroor, Maikoor, Koba and Trangan – which are separated by deep channels, and many smaller islands along the east coast, surrounded by vast coral reefs and sandy seagrass beds. The Aru islands are thinly populated with a population of almost 60,000 in 1994 (Djohani 1996) spread over 8,000 square kilometres with the majority living in one of 125 small villages. Beltubur is one such village on the southernmost island, Trangan.

The Aruese are sedentary coastal foragers, mainly involved in deer and pig hunting, gardening, fishing and gathering in the mangroves, on the beach and at sea. These activities are partly dependent on the two climatic seasons, the monsoons: in general, the east monsoon (May–November) allows people to collect shellfish, crabs and sea cucumbers on the beach in front of the village and on the tidal flats; to do some subsistence fishing; to work in their gardens; and to hunt deer and wild pig in the savannah and forest. During the west monsoon (December–April) people mainly focus on marine resources and are involved in skin-diving for pearl oysters, collecting sea cucumbers on the tidal flats and in deeper areas, and very small-scale commercial fishing (sharks and grouper species). Every village has specific family groups, called *kalaimon* (clan) and *golan* (lineage), which own particular areas of land and sea (including flats and beaches) although marine resources are commonly utilized without asking permission.

Beside the local Aruese there are other, non-local, people active in this region: Buginese and Butonese (from Sulawesi province) fishermen involved in large-scale shark fisheries; Sumatran–Chinese or Taiwanese fishing for *ikan merah* (red snappers); Benjina-based shrimp trawling fishermen; live-fish traders from Hong Kong; Chinese owners of pearl oyster farms; and Chinese traders–shopkeepers in the villages. Marine resources, as well as other products like sago (*Metroxylon*), copra and sometimes meat and vegetables, are all sold to these shopkeepers, either in the villages or in Meror, a small trade settlement. Trade in marine products has long been in the hands of outsiders – first the Buginese, Makassarese and 'Malays', and later, since the second half of the nineteenth century, the Chinese (Van Eijbergen 1866). Aruese villagers collect marine products to sell to Chinese shopkeepers in exchange for cash or to barter for commodities like consumable goods. Traders or shopkeepers allow villagers to take goods from the shop before their foraging trips and then bring back marine produce to settle the debt.

The Chinese shopkeeper can also give credit to buy an engine for a locally made boat (*belang*) or to finance weddings, *adat* ceremonies or feasts such as *Lebaran* (the end of *Ramadan*, the Moslem fasting month). Most of the time people have ongoing debts to the shopkeepers.

For some within the Beltubur community, ancestors are important and people understand the environment as reciprocating; in return for appropriate conduct, nature will provide food and resources. Others leave their fate in the Chinese shopkeepers' hands.

Traditionally, every marine species has an ancestral keeper, and people who want to extract something out of the sea have to adhere to specific behavioural rules. In daily practice, tobacco offerings are made to the sea before pearl-diving or collecting sea cucumbers. Mama Ida explained to me how to behave on the tidal flats and the sandy sea grass beds when collecting crabs and sea cucumbers or when pearl-diving or fishing at sea: 'You see, the sea is like a backyard. Just as people in the village have a yard in front of or behind their house, so do our ancestors of the sea. Just like a mother in the village does not like children making too much noise in the yard and eventually sends them away, so do our ancestors of the sea. When we are making too much noise while foraging at sea, the ancestors will send us away with signs such as big waves, strong winds and, in general, a small harvest. You see, to make a living we are searching the sea, the backyard of our ancestors. Therefore we have to adhere to rules of behaviour by collecting quietly, not shouting, not laughing, not talking about sexual topics, not being greedy, and so on. As long as you respect these rules, you will have good harvests and do not have to be afraid of dangers like getting hurt by sting rays, getting overwhelmed by sudden bad weather, or coming home with no harvest.'

Respecting these rules is the only way to prevent the decrease or disappearance of natural resources. When the ancestors are disappointed about the behaviour of people at sea and in the village, they can decide to hide the resources. For example, in 1996, when Beltubur and neighbouring village Karey were fighting a law suit over sea and beach rights in the region, sea cucumbers were less frequently encountered by villagers – a punishment of the ancestors because of the discord. Certain animals are ancestral or of human origin and cannot be caught or eaten by certain families. These restrictions directly define and regulate foraging behaviour. People from Mangar, for example, are represented by ancestors who had skin like the garotong fish (Serranids – species of grouper) and came from the rainbow. As a result they are prohibited from catching this fish. Ignoring this rule will bring death and disease in the family. Accidents at sea can be ascribed to misbehaviour with regard to these ancestral animals.

In 1991/92 an unknown virus attacked local oysters causing the collapse of the pearl industry. With the loss of their main source of income, the Aruese have been stimulated to collect more species of edible sea cucumbers (*trepang*). Men, women and children are now involved in *trepang* gathering during the west and east monsoons (Osseweijer 1997).

Interestingly, Chinese shopkeepers are given a role in Aruese knowledge of the environment. Some villagers, when asked about the threatening depletion of sea cucumbers from over-exploitation, seemed undisturbed: 'When the sea cucumbers are gone, 'The Chinese' will tell us what to collect next. Until now it has always been this way'. Even people who think their fate is primarily in the shopkeepers' hands cannot follow the Chinese without considering certain customary rules. In villages around Beltubur, for example, there is a prohibition on catching, killing or eating hammerhead sharks (*Sphyrnidae*) and certain grey-spotted sharks (*Orectolobidae*). However, villagers now sell the sharks' fins, saying, 'As long as the carcass is thrown overboard, not taken with you ashore nor eaten, it is quite all right'.

This trend of adapting ancestral rules to present market demands will probably go on. The Chinese traders will instigate the collection of 'new' resources, but ancestors will always play their part in the Aruese image of the natural environment. Whenever things go wrong it will be because of the ancestors, who are the keepers of natural resources and responsible for biodiversity. As explained by a fisherman, 'When sea cucumbers get depleted, it is because we have collected too much, because we were led by greed and therefore the ancestors will take the sea cucumbers away from us'. Still, at present, Aruese people do not seem to be worried about the future of marine resources, and the Chinese shopkeepers continue to gain.

The loss of cultural diversity and marine resource sustainability: the impact in Hawai'i (William Wallace Mokahi Steiner)

The marine ecosystem has always played a focal part in Hawaiian culture. Hawaiians had to rely on many different aspects of their culture in order to traverse ocean realms and open new island systems to colonization. The principles of building ocean-going vessels, sails and ropes, storage systems for food and fibre, etc. would have to be well established in order to undertake the long journeys over water, sometimes lasting for months, completely out of sight of land. Likewise, the knowledge gained of the biology and behaviour of marine life, navigation and astronomy had to withstand the test of time in order for Polynesia to come into existence and survive. Ancestral Hawaiians believed that all life other than human life springs from the gods since it is out of the control of humans. Thus other life-forms and even inanimate forms were viewed as alive with spirit force. There were four main male gods, and two main female deities. Kane was the creator, the father of everything, whose dwelling place was the sun. Lono was the gentle giver of peace and healing, whose abodes were the clouds and plants. Kanaba was lord of the sea and all it held. Ku was the source of power in all its manifestations, and dealt with war. With him came his consort Hina, whose feminine attributes complemented the masculine Ku. And special to Hawaii was Pele; blazing, impulsive, hot-tempered, unpredictable, yet who could take forms as maker of mountains and islands, devourer of men who did not heed her power, and grandmother, Kuku Wahine, who could receive the bodies of men and care for them in gentleness in her soil.

Of interest is the connection the Hawaiians made between marine and terrestrial resources. Dudley (1990) presents this as recognition of a 'principle of dualism' – that things of the cosmos presented as pairs of opposites. Beckwith (1977) discusses the extent to which this occurred, and points out that this enabled Hawaiians to arrive at an organized conceptual form with paired opposites depending on each other to complete the whole. Her examples run the gamut from night versus day, male versus female, land versus water, and *po* (darkness or nothingness) versus *ao* (light or existence). This pairing occurs often in the chants passed down to today. The pairing is often hidden or implicit, allowing chants to carry almost double the information content they would otherwise carry, and suggesting one reason why their careful formulation had to be passed conservatively between generations. One result of this pairing process is that for most marine organisms there was a terrestrial counterpart, the opposite inherent on land versus that found in the marine. In the over-2,000 stanza origin chant, the *kumulipo*, it is significant that the land is seen as rising out of the opposite, that is, the 'fathomless depths of the ocean' (Beckwith 1951). This sets up the ensuing birth of marine and botanical life by stanzas 15 through 43. But in stanzas 35 through 43 the pairings of sea life forms with land forms has begun. These inevitably involve plant pairings, some extremely practical such as in Hawaiian native medicine where if one takes a dose of a land-based medicinal herb, the first food taken afterwards is that paired with it which grows in the sea (Dudley 1990).

This practical basis of dualism was important because one could invoke the terrestrial form or use it in dances and chants and thus allow access to the marine form while on land. Implicit in this is the idea that Polynesian societies recognized the marine world as equally important as the terrestrial. This idea is reinforced in the sharing of the natural world by gods deemed to have counterparts in the opposing realms such as the god Ku-ula-kai (god of abundance in the sea, thought by some to be named after the man who invented the adze for building canoes) and his brother Ku-ula-uka, considered to be the sacred god of cultivators (Beckwith 1940).

Realization of the importance of the marine world is a conceptual breakthrough in the development of the Polynesian culture. Here, the sea is no longer seen as a boundary beyond which humankind can not venture. It becomes instead a herald of opportunity; a provider as source of raw materials for food, and a medium for carrying humans over far horizons to new worlds. Such a breakthrough has to contribute to technological advancement similar to the way the development of the rifle leads to the ability to conceptualize fighting wars at greater and greater distances rather than hand-to-hand. It becomes a new way of adding to the knowledge base of the society and thus contributes to the growth of culture.

The *kapu* system as a response to natural laws

It is likely that the *kapu* system of prohibitions arose for several reasons, including a need to control the common people, a need to provide a warning about attitudes or behaviours considered dangerous to health, and a need to control access to the economic bases of the culture for one reason or another. Placing the harvest of natural resources under *kapu* could be made for any of these reasons. However, the extent of *kapus* placed on fishing and the harvesting of marine life suggests the latter two reasons probably played an inordinate role in making the decision.

The old Hawaiian calendar was a lunar calendar based on a 29 1/2 day cycle with each day of the cycle named. Two seasons were recognized, a dry season and a wet season. Summer, or *Kau*, began in May when the Pleiades set as the sun rose, while winter, or *Ho'oilo*, began in October when the weather turned cool and wet (Cunningham 1994). Knowing the month and day of the old calendar enabled households to observe monthly rituals. Planting, fishing, harvesting, *kapa* (cloth) making and prayer were all observed on a strict schedule. Other days were simply 'off limits', or *kapu*: upon fear of death you did not fish those days at all. Some months were also *kapu* for particular species of fish, and these corresponded to exactly those months when the species was spawning.

Still other *kapus* applied because of the gods. A fishermen never wore red because the god Ku in his form as Ku'ula (god of fishing) found it offensive. A fisherman never took a *kino lau* (representative) of another god out to sea while fishing either; thus bananas as the *kino lau* of Kanaba (god of the ocean) was not taken out in the canoe while fishing. On shore, women would be forbidden to eat *kino lau* of Kanaba such as sea tortoise, porpoise, whale and spotted sting ray lest the ocean take their husbands or sons.

Examination of these *kapus* in some cases yields interesting insights into the culture. Five of the six household *kapus*, for example, would ease the mind of the fisherman and help him focus on the task at hand. That concerning eating the fisherman's bait concerned fishing directly: probably a day would be spent before the actual fishing trip gathering the bait which for offshore fishing would be inshore bait-fish. Red, though it could not be worn, could be used and was effective in making squid or octopus lures; thus the colour would have more than just passing interest in harvesting the resource.

But the most interesting *kapus* are those that relate to harvesting directly. Only eight of the days in any month would be considered good for fishing except during the *Makahiki* (the winter months celebration for the primary god Lono and the period when the Chiefs collected the tribute owed them as a form of taxation) when fishing in general might be avoided, and during the spring months of April and May when fishing was avoided for specific species of fish. In any month, four *kapu* periods apply, sacred in turn to Ku (1st, 2nd and 3rd days), Lono (12th and 13th days), Kanaba (23rd and 24th days) and Kane (27th and 28th days). Thus, harvest restrictions applied which not only acted to protect the resource, but also enabled the resource to spawn, reproduce and replenish itself to provide a continual food source for the people.

Learning the breeding seasons of certain species would have required two things: keen observation on the part of the native populace, and the recognition of its importance. Kirch (1982) and Storrs and James (1991) suggest that the early Hawaiians, after first discovering and settling the islands, had a great impact on the ecology and even geomorphology of the islands. These perturbations include the extinction of large flightless birds of the Family Anatidae, including four large Moa-like birds, six large geese, and at least one duck, presumably by hunting (although this point is not proven as introduced dogs and rats, and fire, may have had a larger impact). The culture was in a relatively stable density upon its discovery, with a population estimated at around 1 million at the high end and 200,000 at the low end (Stannard 1989). (It should be pointed out that Stannard considers it likely that about 800,000 would have been right given the 20:1 loss experienced by other native tribes due to disease introduced by Western cultural contact). If Kirch and Storrs and James are right, it is likely that the Hawaiians, who had developed advanced methods in agriculture, aquaculture and marine fisheries, would have learned what the impact of their tools and techniques could be on the local marine resource. This could, in fact, have given rise to the *kapu* system as a means of protecting the resource and not overfishing it.

Problems in Paradise

Pre-contact Hawaiian culture was highly structured and was governed by strict religious customs, customs that helped relate the Hawaiian to his natural and political environment. Many of the customs related to marine resource harvesting were taught by a class of teachers known as *kahuna*, in schools specially designed to teach the skills, knowledge and techniques and pass them on to future generations. The *kahuna* also related their special knowledge to the gods and to natural elemental forces the gods controlled. These same gods were derived from ancestors who had gone before them and achieved their godlike status from discoveries that enabled the Hawaiian to better survive over time. In a sense, the Hawaiians were derived from a world which they themselves had invented.

Several things happened in the 30 years between 1790 and 1820 which changed the religion, system of *kapus*, and political structure under which the Hawaiians had been living for the previous 500 years, and under which they had achieved some balance between population density, cultural stability and resource use. All are associated with the discovery by and continued contact with Western cultures.

First was the conquering of the islands by the great war-chief, Kamehameha of the Big Island of Hawaii. By bringing all the islands under one rule, Kamehameha set up the political structure that was to allow (a) continued contact (usually on his terms) with Western culture which led to the undermining of traditional ways of living and allowed a new behavioural pattern associated with desire and greed to supplant that of communal sharing; (b) increased exposure to Western diseases such as venereal diseases and cholera which decimated Hawaiians and resulted in possibly a 75 percent loss of the population in this period alone (Gutmanis 1995; see also the cogent arguments of Stannard 1989), and (c) a growing demand on labour and certain components of their natural resources which undermined their use by the native populations for sustenance effort.

The second factor followed close at hand. The Hawaiian populace noticed that the breaking of strict *kapus*, such as that of men and women not eating together, did not harm the foreigners who came to the islands. In addition, these foreigners seemed so much more sophisticated and certainly more technologically advanced. Thus, a basic and growing disbelief in the old *kapu* system began to evolve in the population at large.

Finally, into the heart of this growing dissatisfaction with their plight and their gods, came the missionaries with promises of Utopia and safety. That the Western god seemed to combine the best of the major Hawaiian gods helped the conversion to the new Western religions No longer was it necessary to call to a dozen different gods and *'aumakua* to provide guidance and safety on a daily basis throughout a dozen different rituals common to everyday life.

Today, Hawaii struggles with the loss of the culture and the *kapu* system and what it meant for Hawaiian marine resources. There is no replacement in terms of the religious context in which Hawaiian resource use and conservation took place. The laws that have been enacted to protect resources have been weak or too late at best. Although some headway has recently been made in protecting available and still extant resources, problems still abound – for example, the continuing siltation and pollution of reef flats and the over-fishing of algae-eating fish is leading to reefs becoming overgrown with algae.

Documentation of the decline of Hawaiian fisheries for food fish has just begun. Although in some cases, such as *'ama'ama* or mullet, the decline in fish has been reduced by augmentation with fish from hatcheries, not enough is known about augmenting fisheries for other species although *moi*, an inshore fish, is being considered next for stock releases. In the cases of *'ama'ama* and *moi*, the state hopes aquaculture will maintain these fisheries.

Loss of world

Are there ways to prevent the continued loss of resources and biodiversity we are now experiencing in the marine area? An examination of the differences in management practices between pre-contact and post-contact (modern) Hawaii is summarized in Table 10.1 and gives some insight into what actions may be needed.

Clearly, there are three aspects of Table 10.1 that define the difference between the past and the present in terms of management practice. These include differences in the use made of biological knowledge, the harshness of the restrictions, and the order of precedence in terms of who could draw on the resource. The lessons seem easy. In the first case, we need to establish as much biological knowledge about the resource as we can. Then we need to apply it, and indeed make all harvesting completely dependent on it. Only in this way can we assure that the resource will survive.

The second case is a lesson in itself: we need to make the punishment of infraction of the laws protecting our resources resolute. That is, if they are dealt with harshly it is likely that infractions will decline. This accomplishes two things. It demonstrates to would-be lawbreakers such as high sea drift-net users that violate United Nations resolutions and international treaty agreements that we mean business. Second, it could break the economic means by which lawbreakers can operate and get away with it. Examples include not only the confiscation of boats and equipment but also, as in the new drug trafficking laws, the requirement that lawbreakers give up homes and all property that could have been bought and/or paid for by illegal gains.

TABLE 10.1: PRE-CONTACT AND POST-CONTACT MANAGEMENT PRACTICES IN HAWAI'I.

MANAGEMENT PRACTICE	PRE-CONTACT	POST-CONTACT
Restrictions on harvest	*kapu* days dependent on spawning season or religious restriction	seasonal, not dependent on spawning but some size restrictions
Market	restricted to local population	not necessarily restricted, world-wide in some cases
Sanction to breaking rules	death	monetary fine, licence restriction
Tie to knowledge base	exclusive	occasional
Systematic collection of knowledge resources about the resource	not known	occasional, dependent on monetary (derived)

With respect to precedence, it should be a requirement that in any fishing grounds the needs of local populations take precedent over those of the world at large. Once local demand is satisfied, then additional fish or other natural resources could be sold on broader markets. Or the fishery, if such, could be closed for the remainder of the season to allow stocks to build up. This gives the control of fisheries and other natural resources back to the largest possible number of those who directly benefit by them and takes that control out of the hands of multinational corporations who see only the bottom line and profit.

What we are experiencing is not only the loss of biological diversity, but also the loss of a world upon which humanity in general has evolved. The loss of world experienced by the Hawaiian culture was in part due to the loss of the knowledge base of the culture itself. But today we differ in that we are fully aware of what we are doing, and of the possible consequences of such actions. We differ in another sense as well. There is some evidence, given the *kapu* system which the Hawaiians used, that they were willing to protect their resources on pain of death. We have no such taboos in place, and the economic and legal sanctions we bring to bear have obviously, based on the post-contact experience in the Hawaiian islands, not worked. Today, the marine resources continue to decline: world, our world, continues to be lost. Somewhat worrisome is this idea: with the loss of cultural diversity, do we also lose the many different ways in which world can be perceived, invented, lived in? It the answer is yes, then we have lost more than we realize.

Western Samoan views of the environment (Clark Peteru)

Popular images of tropical islands as paradises are quite misleading. For the residents it is not a life of ease. Few adults spend much time swimming or basking in the sun. For the majority, it is a constant struggle with nature. *'Gagau le vao'* (literally, 'break the bush') is the common encouragement to the young men in a family to clear the bush or forest for cultivation of crops. If one does not plant, one starves. Survival requires pragmatism not aestheticism, and for that reason Samoans have a utilitarian view of nature. Nature exists to feed, to house, then to satisfy cultural obligations. Usually it is only at the end of the day that one can relax and enjoy one's surroundings and the coolness of the approaching evening. To my knowledge the environment is neither revered nor protected for its own sake. The government, along with NGOs, has recently intervened to protect the environment, but the reason is primarily economic rather than aesthetic or cultural. Conservationists now invoke taboos, citing them as traditional conservation devices, but many traditional taboos didn't have a deliberate conservation goal. Samoan beliefs do not reveal a tradition of nature worship.

One of the great heroes of Samoan folklore, Pili, is credited with having made the first fishing net and to have introduced taro planting to Samoa. When he died, the country's principal island was divided amongst his children who were left with three emblems: the planting stick (signifying a planter's role), a club and spear (signifying the role of a warrior), and a staff and fly whisk (the emblems of an orator chief). A fourth child received the office of mediator. Pre-Christian Samoans believed in many gods. The significant ones were known as *'aitu'*. Sixty-five *aitu* have been described but there were probably more than double that number. They took the form of lizards, bats, cuttlefish, stones, leaves, lightning and other objects. The main *aitu* were war gods or general village gods. People's relationship to them was generally one of dread and appeasement. Significantly, none relates to 'mother earth' or represents nature. Even proverbs (of which over 500 have been recorded), invariably quoted by orator chiefs at ceremonies, have practically nothing to say about the virtues of an unspoilt environment.

There is, however, a preoccupation with rank and with food which is aptly illustrated by a proverb that compares Samoa to a fish taken from the sea. Like other living creatures, the parts of a fish, even before it is killed, have already been earmarked for distribution to chiefs and dignitaries in the village. So it is with Samoa: its political divisions have been determined. Taking the proverb further, every man, woman and child within a village has a rank of which he or she is aware: pastor, chief, chief's wife, untitled man, unmarried woman, child, etc.

The reference to food is not to belittle the culture: the observation has been made that food is probably the most culturally valuable product of Oceanic economic systems and that its classification by Polynesians provides an analytically useful and culturally meaningful point of entry into their value systems. This is certainly true of Samoa where on formal occasions rules determining the apportionment, delivery and serving of food must be followed closely.

In the past, people may have been aware of the effect of their activities on the environment and may even have consciously attempted to control or eliminate such activities. But today there are few examples of self-control. The reefs and lagoons have been described as among the most degraded in the Pacific. In 1993 it was estimated that at prevailing rates of depletion, all remaining merchantable timber would be gone in six to seven years. The same report attributed 20 percent of forest clearing to logging, with the remaining 80 percent a result of agriculture and other activities. These impacts are caused by overfishing and expansion of plantations into forested areas.

Perhaps this is not a surprising result given the utilitarian view of nature. Yet in a subsistence economy there is little incentive to work for a surplus because it would have to be shared. So why the environmental crisis?

I attribute resource depletion to several new factors – increasing individualism; a shift from subsistence farming to cash cropping reflecting an increasing need for money; an increase in population, and improved harvesting technology – and one old factor – cultural obligations. Family, church, work, village and other obligations are onerous, requiring contributions of food, money and traditional goods. Chiefs need to contribute according to their status and prefer saving face to saving resources. In all ceremonies, more is better. Yet even if one is aware, for example, of diminishing fish catches from the reef, there is no incentive to conserve because what one fisherman leaves behind in order to breed, another fisherman will most certainly pick up in order to eat, thus accelerating the spiral of resource depletion.

Although the country is becoming increasingly cash dependent, I would not describe the trend as irreversible. Not yet, anyway. Conservation projects have commenced, resource laws are to be drafted, and awareness programmes have been under way for several years now. Hopefully the trends can be reversed. Despite the turmoil, people still survive from their own plot of land rather than the supermarket shelf.

Customary Maori fisheries (Nga Kai O Te Moana, Ministry of Maori Development, 1993)

For Maori, a kinship exists between all elements of the natural world, of which people are one. This Maori world-view has traditionally been brought to the management of natural resources, including fisheries. It is reflected in protocol, rituals and regulations. Maori used fisheries to uphold customary obligations within, and between, *whanau* (extended family), *hapu* (sub-tribe) and *iwi* (tribe). Because the fulfilment of these duties depended on safeguarding their fisheries, Maori developed a system of practical rules, checks and balances to manage this important resource. These practices reflected a philosophy that extended to all the natural world. They formed part of a holistic system of environmental management by which Maori ensured the sustainability of each natural resource.

For Maori, waterways have always had special significance. Waters and their resources are a source of spiritual life:

> 'All water begins as a sacred gift from the deity to sustain life. Waste water is defiled water which must be purified by returning it through the cleansing qualities of the earth. Here fresh water is also the life-giving gift of the Gods and is also used to bless and to heal. Separate water streams are used for cooking, drinking and cleaning. Waste water is purified by return to the earth, ritualistic purification or, with the exception of water containing animal waste, by mixing with large quantities of pure water.'

All waters have their own *mauri* (the physical life force). These waters, and their *mauri*, mix naturally when rivers flow into the sea. They mix unnaturally when, for example, waste waters are discharged into clean waters. Scientists might consider the harm caused by waste-water discharge to result from water temperature changes or contaminants carried in the waste water. Maori have traditionally seen these harmful effects as due to the unnatural mixing of *mauri* from different waters. These are simply different cultural perspectives on the same issue.

The traditional Maori view is that fisheries, like all elements of the natural world, originate from the gods, and are thus imbued with *mana atua* (the prestige and power of the gods). Like all living things they are possessed of *mauri*, they are *mahinga kai* (places of customary food gathering) and because of their origins and utility, they are *taonga* (valued resources). The rules and practices by which Maori managed their waters and fisheries reflected the significance of this view. Conservation has always been important to Maori. Customary Maori fishing practices included measures intended to maintain the habitat, preserve fish stocks, and regulate fisheries use. This knowledge has been retained and these measures are still practised today.

Fisheries and fishing grounds, like other aquatic resources, were not common property available to all. A complex set of rights existed which determined who could harvest certain fish species, fish in particular areas, or at what times harvesting could occur. 'Marine tenure' was no different to Maori than land tenure. Fisheries were seen as *whanau*, *hapu* or *iwi* property. In most cases, ownership rights rested with those in occupation of the adjacent coastal lands. The fisheries themselves were clearly defined areas with known rights of access. The use of physical formal markers was not common, rather the knowledge of boundaries was handed down through generations. The boundaries were minutely known, and included mountains, hills, rocks, trees, streams and rivers. From these geographical features fishing areas were delineated. Knowledge of particular fishing grounds was closely guarded by *whanau* and community and great care was taken to pass on this knowledge. All fishing grounds, banks and rocks had special names which often feature in story, song and proverb. Landmarks were sometimes named after both the species found in the fishing grounds and the season or month in which they could be fished. In that way, marks told the Maori fisher not only where to look, but also when to look and what to look for.

Although fisheries were community-owned, they were subject to traditional forms of authority. This was usually administered by the *Rangatira* (*hapu* or *iwi* head) – for it was their responsibility to ensure the sustainability of the resource. It was much the same with other indigenous peoples throughout the Pacific. *Kaitiaki* (individuals responsible for governing a particular resource) used a system of rules and prohibitions based on spiritual concepts. *Tapu* (a prohibition or restriction) and *rahui* (a temporary *tapu*) found everyday use in the regulating of fisheries. *Rahui* were used, for example, to retire grounds that were in danger of being over-fished. The consequences of breaking these rules ranged from supernatural punishment to the practice of *muru* (ritual punishment). The use of *tapu* and *rahui* was guided by ethics of custodianship and conservation. Where divine retribution failed to deter potential transgressors, more down-to-earth measures like *muru* prevailed.

The importance of *mana* (prestige and social reputation) within Maori society is paramount. Historians have described the life-style of pre-European Maori as involving a constant pursuit of *mana*. For Maori, the *mana* of a people is demonstrated when fulfilling the obligations of *manaakitanga* – a duty that compels a conservation ethic. To *manaaki* someone is to show respect for, and hospitality to, that person. The *mana* of the Maori is based in part on this ability to contribute and share. Naturally Maori developed rules to protect the resources which contributed to *mana*. If their fisheries were thriving, users could express suitable hospitality and enhance their *mana*.

For Maori, the inability to demonstrate appropriate hospitality to guests, visitors or relations means being diminished in their eyes. Potential loss of *mana* remains a compelling factor in the careful and controlled management of fisheries by Maori.

Indigenous knowledge and Amazonian blackwaters of hunger (Janet M. Chernela)

From the point of view of the Eastern Tukanoan speakers of the Brazilian north-west Amazon, the forested river margin is part of the aquatic, not the terrestrial, realm. Accordingly, the Tukano preserve the river margin as a grazing area for fish, rather than deforesting the river margins for agriculture. Indeed, the river and adjacent areas are demarcated into ethnobiological zones thought to be protected by spirits, linking interests to rights, privileges and prohibitions.

From the point of view of the outsider, the decision to maintain the forested river margin as a fish reserve makes good management sense for three reasons: first, the forested river margins are essential to the regeneration of fish stocks as they provide both nutrients and spawning conditions; second, by attracting fish they create favourable conditions for fish harvesting; and third, they are unfavourable locations for agriculture since the floodplain soils are poor in nutrients, and no advantage is derived from removing the riparian forest for farming. In sum, more protein can be derived per unit labour from fishing in the flooded forests than from utilizing such areas for agriculture (Chernela 1982, 1994).

In the Uaupés River, a tributary 1,200km north of the Amazon proper, which is home to about 10,000 Eastern Tukanoan speakers like the Arapaço, fish are less abundant than in the main channel of the Amazon, and the size of the individual fish is smaller (Meggers 1971). Moreover, the region is known for its nutrient-poor rivers, referred to by local inhabitants and researchers alike as Starvation Rivers (*Rios de Fome*) (Meggers 1971; Junk and Furch 1985). It is therefore surprising that the Eastern Tukanoans rely more heavily on fish than does any other group of Amazonian Amerindians (Carneiro 1970; Chernela 1989; Goldman 1963; Jackson 1983; Hugh-Jones 1979; Moran 1991).

The Uaupés River flows 850km south-easterly across the eroded, quartzitic sandstone uplands of Colombia to enter the Rio Negro at São Gabriel da Cachoeira in Brazil. Rising from the tertiary rock base of this catchment area, the waters of the Uaupés are nutrient-deficient, likened by some scientists to distilled water (Chernela 1982; Santos *et al.* 1984). Most of the nutrient sources in this river system are derived not from internal primary production, as they are in temperate rivers, but from sources external to the river (Knoppel 1970; Goulding 1980). The single most important contributor of nutrients is the surrounding forest. While debris from the flanks falls year-round into the river, flooding dramatically augments the terrestrial input into the aquatic system, as the rising river encompasses the riparian forest. The water levels of the Uaupés fluctuate significantly with seasonal rain regimes, averaging seven metres in height per year (Chernela 1989, 1993). Lowlands may be submerged for periods of up to eight months, distributed in two annual peak high-water seasons when waters flow up to 20km into the adjacent forests. The spreading waters carry fish into the forest, where they disperse to feed on the abundant foods only then available.

Tukanoan fishing

The Arapaço Tukanoans are fisher-horticulturists, with fish providing the principal source of protein and garden products the principal source of carbohydrates. Apart from a few small mammals and birds, fish provides almost all the animal protein in the Tukanoan diet. As fishermen dependent on these river systems, the Arapaço are acutely aware of the relationship between the life-cycles of fishes and the local environment, particularly the role played by the adjacent forest in providing nutrient sources that maintain vital fisheries.

In direct opposition to the expectations of many scientists (see, for example, Roosevelt 1980), most fish caught by Tukanoan fishermen derive from flooded forests. Since Amazonian ecologists, economists and anthropologists have associated high waters with reduced fishing yields, it is of special interest that data gathered over one year in the Uaupés demonstrate the dramatically high percentages of fish captured in seasonally flooded forests as opposed to other habitats. Data collected over one year in the Uaupés River showed that a total of 860kg of fish were captured in the flooded forest, compared to 270kg of fish captured in the next most productive habitat, the rapids (Chernela 1993). Thus, approximately three times as much fish was obtained from flooded forests than from any other fishing location. Furthermore, these findings reveal that the average weight per fish obtained in flooded forests is 363 grams, compared to 92 grams per fish for non-flood seasonal uplands, lending further evidence to the strong correspondence of feeding season with water level (Chernela 1993). It is of some interest, therefore, that while the Tukano regard the standing forests as a significant food supply, they prohibit fishing these areas at certain times of year.

Unlike Western thought, the Tukanoan system of classifying living beings subsumes human and animal beings within a single system. The system supposes a nested hierarchy of inclusiveness based upon principles of patrilineal descent where the reference point at each level of inclusiveness is a putative named ancestor who emblemizes or stands for his descendants. At each level of inclusiveness a different pivotal ancestor becomes relevant, and thus a different calculus of membership is applied. Starting at the lowest level of inclusiveness, an individual is a member of a local descent group known as 'the children of X', where X is the founding ancestor of the local branch of a patriclan. All of the settlements of a single patriclan are known to be the children of the founding ancestor of that clan, and all of the clans of a single language group consider themselves to be descended from a single focal ancestor. At a still higher level of inclusiveness, ancestors of several language groups recognize common origin from a single ancestral anaconda-fish. At each level, the name and recognition of a focal, totemic, ancestor unites persons into a putative brotherhood. The size of the brotherhood, defined by common ancestry, expands or contracts in direct relation to the temporal distance to that ancestor.

Topography is socially mapped so that each localized patriclan has a designated location in space, based on principles of historic precedence. For humans as well as animals, the proper 'sitting-and-breathing-place' (*duhinia*) is that where the ancestors 'sat', for their souls are recycled in alternate generations. The absolute dominion of each group in its 'sitting-and-breathing-place' (*duhinia*) must be respected. (Families may be out of place, but this simply reinforces the model of things as they should be and may not always be.) A breach in this rule calls for retaliation. Space is thus socialized through a universal principle of precedence, based in the historicity of space. Lithographs and stone formations in the river indicate ancestral events and the 'houses' of spirit guardians. The river as such is an encapsulation of history, a reading of positions and origins, a codex in which are inscribed the signs of history. The specificity of placement is significant, and disputes over ancestral rights can lead to violence.

Context-dependent and spatially-situated, descent is anchored in space. This is what the totemic guardian figure stands for. The challenge is to maintain intact the socio-spatial configuration of difference based upon patrifiliation as the legitimizing function that links spatial demarcation to past events.

At the basis of a Tukanoan theory of efficacy is the assumption that spatial designations, the result of historic processes, influence future events. A pan-Tukanoan founding myth recounts the placement of clans along the river from a sacred anaconda-fish-canoe that constructed the proper 'seating' (*duhinia*), or settlement, of each local descent group. In this myth, the ancestral anaconda-fish swam upriver from a primordial Water Door until arriving at the headwaters of the Uaupés river. There the great anaconda-fish ancestor turned, with its head downriver and its tail upriver. From the segmentations of its body emerged the first ancestors of each of the localized patriclans of the Uaupés river basin. (Today, local descent groups are situated at high points along the river edge, with front paths leading to the port and rear paths to the gardens.) The birth order of clans from the body of the ancestor is regarded as an order of status fixing the relations between groups and the connections between social units and space. The myth narrates the foundation of a socio-topographical order, a 'territorialization' and emplacement of peoples, animals and spirits.

At the basis of a Tukanoan economy and ecology of resource exchange is the notion of an eternal gauge writ into nature. The sacred 'houses' of fish totems are always located within flood zones. A clan guardian, descended from and exchanged for the 'first ancestor', guards its progeny and avenges those who prey on its members. The rule of retaliation or tit-for-tat prevails: the guardian spirit will take one human offspring for every one of his offspring taken (cf. Reichel-Dolmatoff 1976; Smith 1981). Eating and being eaten are regarded as a single mode of exchange, a retaliatory dialogue between families who trade in life and death. Principles of reciprocity link predator and prey, as killing is a means of rectifying an imbalance, inherited and requiring correction. Only equity is capable of producing and maintaining peace.

There are several lessons for the outsider. The first is that what appears to be the area of food procurement may be greatly reduced by restrictions relating to a concept of the sacred. The second concerns implications for management of fish populations and sustainable harvesting. The consequences of such proscriptions could serve to conserve fish stocks in the long run, if the areas under restriction were critical to the maintenance of the fish populations. This would be the case, for example, if the restricted areas corresponded to spawning sites. Exploring this hypothesis through observation and interview revealed that in the Arapaço case, all zones subject to flooding and known to be spawning sites are heavily restricted in the high-water season, a form of intermittent prohibition and exploitation known as 'pulse fishing'. Fish samples obtained at various sites along the Uaupés revealed that the protected areas corresponded to the spawning grounds of the *Leporinus* species of fish known in *lingua geral* as *aracú*. These findings show a correspondence between restricted areas and spawning grounds. By providing refugia for the reproduction of fish populations, these areas, off-limits to human predation during significant portions of the year, could contribute to the growth and preservation of fish populations.

The floodplain, primary target of fishing effort and related fishing yields, is the same habitat that is selectively regulated during seasons of greatest potential productivity. In the absence of prohibitions, fishing yields might be increased considerably. Instead, a strategy of 'pulse fishing' protects sites during periods critical to fish growth, and may reduce pressures due to localized and specialized fishing.

The agricultural alternative

The Tukanoan practice of preserving river margins for fisheries maintenance is in conflict with the widespread trend among developers to utilize this area for agriculture. It is therefore relevant to examine the agricultural alternative.

The Tukano practise a form of rotational polyculture in which a variety of plant types are grown simultaneously. While other crops, including peach palm (*Bactris gasipaes*) and pineapple (*Ananas comosus*) are grown along paths, in gardens, or alongside houses, manioc (*Manihot esculenta*)occupies 91 percent of all lands in cultivation, furnishing from 85 to 95 percent of daily caloric consumption.

Soils generated by the geologically ancient granitic and gneissic Guiana Shield, across which the Uaupés River flows, are among the least fertile soils in the world. The two predominant soils of the region, white sand spodosols, and red and yellow psamments, are acidic, infertile soils, whose high leaching potential results in aluminium toxicity as well as in deficiencies in important micronutrients, including phosphorus, potassium, calcium, magnesium, sulphur and zinc, all of which are necessary for reasonable agricultural yields. In evaluating these soils, tropical agronomists (Sanchez 1982) conclude that '... the agricultural potential of these soils is very limited and their erodability high'. Pedro Sanchez, known for his optimism regarding the possibilities for agriculture in many tropical soils considered unfit by opponents, recommends against the clearing of either psamments or spodosols in the humid tropics (Sanchez 1982).

Generally, Tukanoan farmers maintain three gardens: one recently-planted, a second of one or two years in age, and a third of two or three years in age. Since soil nutrients are principally derived from the forest cover, and are quickly eroded with clearing, the consequences for agriculture are dramatic. Heavy rainfall leaches the already depauperate soils, declining fertility further and reducing agricultural yields with each succeeding crop. Accordingly, utilization periods are short while fallow periods are long. Tukanoans use the same plot for only two consecutive plantings, a period never exceeding four years. Yields from the second planting are substantially lower and further plantings in the same location are regarded as futile. After the second crop, the plot is abandoned and, except when there is a shortage of land, left to lie fallow for a period upwards of twenty years.

In the depauperate soils of the Uaupés basin, the slow-growing manioc tuber is harvested after eighteen months. Since the crop is susceptible to waterlogging, there is no advantage to utilizing flood zones for manioc production. Nevertheless, national and international funding in agricultural research and development has been dedicated to identifying fast-growing subspecies of manioc capable of reaching maturation between wet seasons. A programme of dedicating the floodplain to agriculture, rather than to fish and fishing, is based upon assumptions gained in observations of fertile whitewater floodplains. The extension of these models to zones other than the principal east-west Amazon River itself reveals a fundamental misunderstanding of the different floodplain ecologies of the basin.

Common representations of the river margin as a fertile plain are based on models derived from the ecologies of whitewater river systems. These models cannot be applied to blackwater systems. Unlike blackwater rivers, whitewater rivers carry suspended sediments across great distances, washing new nutrients onto the river flanks with each flood.

The floodplains of whitewater and blackwater rivers contrast dramatically. In blackwater rivers, nutrient source materials are characteristically poor and the waters are not laden with silts. Unlike the rich, silt-laden whitewaters, flooding does not improve the adjacent soils in blackwater rivers and does not increase their potential agricultural productivity. Rather, the pattern of fertility renewal is the reverse: it is the forest that restores and rejuvenates the river system.

Conclusions

The current status of Amazon fisheries is a growing concern as the rate of exploitation increases and as developers and colonists apply pressure to convert gallery forests to cropland. Fish has been a principal food source for several million residents of the Amazon basin. Moreover, fish is one of Brazil's principal exports. A sizeable portion of these fish are freshwater. Closely tied to the forest are not only rich terrestrial resources but also the aquatic life of vast tracts of water courses.

The Eastern Tukanoan Indians of the Upper Rio Negro basin in the north-west portion of the Brazilian Amazon rainforest have managed the delicate ecology of the floodplain over generations, structuring subsistence activities around the physical and limnochemical limitations of infertile blackwaters. By reserving the natural vegetation of the riparian forest for fisheries maintenance rather than deforesting the margin for agriculture they demonstrate the potential of the forested floodplain as an important source of animal protein. Experimentation shows no comparable potential for agricultural production on blackwater soils. For the Tukanoans, forest maintenance and restricted fish harvesting transforms a potentially infertile system into a relatively productive one.

The Tukanoan view of the environment described here contrasts with Western belief models. For the Tukano, boundaries are respected and reciprocity is the mode of boundary-crossing. Breaching the rules creates imbalance and retaliation.

According to Tukanoan ideology, humans, fish and other living forms are finite and precarious, locked into a carefully monitored relationship of exchange and balance. Nature is socialized into ethnoecological zones pertaining to specific categories (families) of living beings. The logic is based on a schemata underlying the relations of efficacy that include the principles boundedness, precedence and reciprocity. Boundaries are to be respected, according to this ideology. A belief that breaching these rules creates an imbalance that in turn carries dangerous consequences serves to separate realms and reduce intrusion into what might be considered, from the point of view of the outsider–conservationist, refugia.

When such ideologies are lost, the natural resources protected by them are likely to disappear as well. The next step may well be rapid environmental deterioration, unless the underlying lessons learned can be put into practice.

A Japanese view on whales and whaling (Arne Kalland)

In 1982 the International Whaling Commission (IWC) declared a total moratorium on all commercial whaling, beginning with the 1985/86 season. Consequently, Japan sent her last commercial fleet to the Antarctic in 1986, and in 1987 the last Japanese land stations for large whales were closed. In 1988 the ban was extended to include coastal whaling for minke whales. At present, Japan is only conducting scientific whaling of minke whales in the Antarctic and North Pacific, and nine small coastal vessels of less than 48 tons each are licensed to catch a total of about one hundred Baird's beaked whales and pilot whales, which are two species not under the IWC's jurisdiction.

When the IWC imposed a moratorium on commercial whaling, the majority argued that the moratorium, which was to be reconsidered not later than in 1990, was necessary until better population estimates and a new management regime (the Revised Management Procedure, RMP) were available. IWC's own Scientific Committee, however, did not share the opinion of the Commission and found a blanket moratorium on all species unnecessary. Of the species targeted by the Japanese fleets, only the sperm whale was defined as 'endangered' by the US Endangered Species Act of 1973. The other species – the minke, pilot, Bryde's and Baird's beaked whales – were among the 57 species of cetacean that towards the end of the 1980s were at or near their original level of abundance (Aron 1988). The Antarctic stocks of minke whales are particularly abundant (Gulland 1988). Since the moratorium went into effect in 1987, the Scientific Committee has concluded that several whale stocks can be exploited sustainably, and the RMP has been completed. Yet, the moratorium has not been reconsidered as called for in the 1982 decision.

The Japanese have difficulties understanding the logic of the moratorium, which prohibits them from catching the abundant minke whales while permitting the Alaskan Eskimos to catch the endangered bowhead whales. The IWC rationale for this different treatment is that Japanese whaling is 'commercial' whereas Eskimo whaling is classified as 'aboriginal subsistence whaling' (ASW) or whaling for purposes of local aboriginal consumption carried out by or on behalf of aboriginal, indigenous or native peoples who share strong community, familial, social and cultural ties related to the continuing traditional dependence on whaling and the use of whales (IWC 1981). However, nowhere does IWC define what it means by 'aboriginal', 'subsistence' or 'commercial'. As recently as March 1997, an IWC workshop discussing the 'commercial elements' of Japanese coastal whaling refused to define these important concepts.

Japan has since 1986 argued that minke whaling along the coasts (so-called small-type coastal whaling, STCW) shares many characteristics with ASW, not covered by the moratorium. Due to the small size of the vessels, the boats usually leave the harbours early in the morning and return after dark the same day. Most of the boat owners have their own small flensing stations where highly skilled master-flensers do the initial partitioning of the carcass. Less skilled flensers, often women and retired whalers, do the final cuttings. In several of the communities the whales are sold at auction by the local fishing associations. The owners of these small whaling companies, many of the crews, and most of the flensers and the distributors, live in these communities, where much of the meat is consumed.

According to the Japanese position, people in four STCW communities do 'share strong community, familial, social and cultural ties related to the continuing traditional dependence on whaling and the use of whales'. Between 1986 and 1994 Japan presented to the IWC 33 papers documenting this claim, most of them written by social scientists from eight different countries; and several books have been published on the subject (e.g. Takahashi 1988; Akimichi *et al.* 1988; Kalland and Moeran 1992). Japan has even designed an 'Action Plan', placing the distribution of the meat under firm control of management councils appointed from government offices and community organizations. The main purpose is to reduce the commercial elements and prevent any of the whale products from reaching the markets in the expectation that the IWC would grant Japan an interim quota of 50 minke whales for the nine small STCW boats, 'to make possible the continuation of the cultural, social, religious, dietary and historical heritage of the four traditional coastal whaling communities' and thereby 'to alleviate the profound social, cultural and dietary distress suffered' in these communities (GoJ 1995). It is not surprising that the Japanese have argued along two main lines: the cultural significance of whaling and the social, economic and cultural impacts of the whaling moratorium. In the following I will limit myself to discussion of the former. (For discussions of the impacts, see for example GoJ 1989, 1997; Kalland and Moeran 1992; Freeman 1997).

The cultural significance of whaling

Anthropologists usually define culture as shared knowledge, values and norms that are transmitted (usually with some modifications) from one generation to the next through processes of socialization. In addition to knowledge related to the technologies used to catch whales, the Japanese whaling culture comprises a detailed understanding of an ecosystem in which both men and whales have their roles to play; rituals and belief systems that surround whaling activities; an elaborate cuisine based on whale products, and a large number of other activities that relate to the production, distribution and consumption of whales. Within this complex, attention has particularly been given to the importance of local cuisine, to non-commercial exchange, and to rituals associated with the hunts.

Japan is a country poor in natural resources, and the Japanese early in their history had to learn to make the most of the scarce resources they had. This applies also to the utilization of whales. Whale oil, which had long been used for lighting, became in the Tokugawa period (1600–1868) an important insecticide used in the rice fields and was thus an important factor in averting serious famines (Kalland 1995a). The bones and some of the entrails were used as fertilizer, and there developed an important market for such products. Baleen, whale tooth, jaw bones and sinews became raw materials for various handicraft products. But most important, whale meat – including blubber, skin, cartilage, fluke, intestines and genitals – has for centuries been used as food. During the first post-war years, whale meat accounted for 47 percent of the animal protein intake by Japanese, many of whom are convinced that the whale saved them from a major famine. Some of their attachment to whales and whale meat possibly stems from this belief.

The importance of whale meat as food has given rise to a rich culinary tradition. The quality of the meat is finely graded, and the various parts are regarded as suitable for different dishes. Whale meat recipes have been included in Japanese cookery books since at least 1489 when it was mentioned as a superior food. In 1832 a special whale cookery book, *Geiniku chōmihō*, was published in Hirado, north of Nagasaki, and this divided the whale into seventy named parts, each described in terms of taste and method of cooking. According to this source, roasted red meat (*akami*) might 'taste better than geese and ducks' and *unagi* (the outer side of the upper gums near the baleen) is tender and has a 'noble' taste, whereas the trachea (*nodowami*) is 'given to servants in the countryside' and the duodenum (*akawata*) is eaten by the poor. What part of the whale people ate thus signalled their social position in the community and therefore carried important symbolic significance. In 1989, the only wholesaler in Arikawa, not far from Hirado, still dealt in 60 items of whale meat, although he had run out of stocks of twelve of these (Kalland 1989).

Regional food preferences have emerged as a result of the history of whaling in particular communities. Such preferences exist both in terms of species eaten and in method of cooking. In Arikawa the most cherished meat in the past was that of right whale, but because this meat is no longer available, salted blubber of fin whale has become a new favourite. Dried, salted dolphin meat is also regarded as a delicacy, while on Iki Island in the same prefecture dolphin meat is regarded as non-food (*tabemono ja nai*). In Taiji, not far from Osaka, people have developed a special liking for pilot whale, which is often eaten raw as *sashimi*. In Wadaura outside Tokyo, a local speciality is dried, marinated slices of Baird's beaked whale (*tare*), while raw red meat (*sashimi*) of the minke whale is preferred in Auykawa and Abashiri in the north. A special New Year's dish in Abashiri is soup boiled from salted blubber. The meat of sperm whale is the preferred food in some areas of north-eastern Japan.

Few things are as symbolically laden as food, and local cuisine is one of the strongest markers of social identity in Japan. The various ways of preparing meat have become important social markers, and whalers from different communities never seem to grow tired of discussing local whale cuisine. Beside being a staple – even now some people in these communities will try to have a small piece of whale meat daily 'just to get a taste of it' – whale meat is an indispensable part of all types of community gatherings and celebrations. It is extensively served at weddings, funerals, memorials for the ancestors, as well as to celebrate the building of a new house, a child's first day at school, and so on. (See Braund *et al.* 1990 for a complete list of such occasions in Ayukawa.) It is often the typical food for New Year, and in Arikawa about a fifth of the annual sale of whale products occurs during that season. The other peak in consumption occurs in August, which is the month when Bon (All Soul's Day) is celebrated (Kalland 1989). Manderson and Akatsu classify whale meat in Ayukawa as 'super food' due to its double role in being 'highly valued culturally and as a staple' (1993). Moreover, food often figures in the 'special products' of localities; and whale meat serves that purpose in whaling communities. People will travel to whaling communities in order to eat the 'special product', and whale meat is thus a important tourist attraction for these communities. No wonder than that the question of food culture has been one of the main arguments used by the Japanese Government, and a number of the papers presented to the IWC since 1986 focus on this theme (e.g. Braund *et al.* 1990; GoJ 1991a, 1991b, 1992; Ashkenazi 1992).

One of the arguments for permitting Alaskan Eskimos to hunt whale was the non-commercial character of the distribution and consumption of whale products in their society. It is therefore not surprising that many of the Japanese reports focus on the distribution of whale products, and particularly on the non-commercial channels of distribution. The importance of gift-giving in Japan is well known and whale meat makes excellent gifts. There are a number of occasions when whale meat is given away. Although such gifts are known in all kinds of whaling communities (in Arikawa a gunner used to bring 100–120 kg of salted blubber home from a season in the Antarctic), they have been of special significance in the communities engaged in small-type coastal whaling. A large portion of a minke whale caught by an STCW boat never reaches the commercial market, but enters an elaborate system of exchange. The whalers are partly paid in kind, and when a whale is brought ashore for flensing, people will come to lend a hand in return for some meat. Fishermen are given a chunk of meat if the whale has been caught as a consequence of information received from them. The distribution of meat becomes particularly extensive in connection with the first catch of the season (*hatsuryō*) and the launching of new boats. One of the boat-owners in Ayukawa, for example, usually received about 300 bottles of *sake* plus whisky, beer and Coca Cola in connection with the commencement of the minke whaling season in May. Additional bottles were given to crew members and the boat. These bottles were given under the assumption that they would be reciprocated by gifts of whale meat (Kalland and Moeran 1992; see also Akimichi *et al.* 1988; Manderson and Hardacre 1989). This particular owner estimated that between 10 and 20 percent of the total amount of minke meat is distributed in this way.

Most of those who gave *sake* received whale meat on several occasions, relatives more often than friends and neighbours, and these again more often than business associates. Much of the meat received as gifts was further sub-divided and given away to relatives, neighbours and acquaintances. According to a survey in Ayukawa and the neighbouring village of Kugunari, about two-thirds of the households received gifts of whale meat during the minke whaling season (GoJ 1991b). Meat is not given only to relatives and friends, but community institutions also receive a share. In the old days the villages being hosts to whaling operations received meat to compensate for inconveniences this may have caused fishermen and others who did not benefit directly from whaling. More recently, STCW owners in Ayukawa and other communities have frequently given meat to community centres, old people's clubs, schools, the fire brigade and to temples and shrines (Akimichi *et al.* 1988).

Through this extensive gift exchange, in which whale meat is the essential commodity, social connections have been kept 'warm'. Although gifts of whale meat to relatives and friends living outside whaling communities also occur (and serve as an important means through which such social relations are revitalized) the intensity and the particular characteristics of gift exchange within these communities set them firmly apart from villages in their vicinity, thus strengthening the community identity further.

Whaling also contributes to the local identity in other ways than through local cuisine and gift exchange. All the communities that have been engaged in whaling have a set of myths, legends and other stories about whales and whaling. These stories have through the centuries contributed to a rich cultural heritage. They provide a picture – although at times quite distorted, perhaps – of how people lived in these communities in the past and at the same time provide models of behaviour for young whalers today. Many stories tell about legendary harpooners and gunners, and such legends 'provide local residents with an appropriate cultural or folk hero with which to identify' (Akimichi *et al.* 1988: 75). New legends are always in the making.

This common cultural heritage is expressed and reinforced in festivals, songs and dances, through which the present is linked to the past. Rituals give each community its distinct character: the set of Shintō deities is unique to each community, and the festivals are different as well. But they are all variations on common themes based on a conception of the whale as a creature with an immortal soul, and a world-view stressing the interdependence of supernatural, human and animal worlds. According to the prevalent world-view in Japan, all animals are endowed with souls. There is a common belief that people, animals, plants and even inorganic objects have 'souls', or some inert power. In Buddhist doctrine one takes a holistic view and talks about 'Buddha–nature' in all things, while in Shintō one talks about *kami*, a supernatural power that resides in anything and which gives a person a feeling of awe. There is no sharp line, as in much of Judaeo–Christian thinking, between people and the rest. This has, of course, far-reaching implications for Japanese attitudes but has not induced the Japanese to take a 'no-touch' approach to nature, as is the case with much of the recent ecological movement in the West. It is recognized that it is the nature of things that one organism feeds upon another, creating relations of indebtedness in the process (Kalland 1995b). A whale that has been killed is regarded as having given itself up to mankind so that we can live, and in return the whalers become indebted to the whale. The whales are also gifts from nature, which itself is infused by Shintō deities (*kami*). Thus, whaling activities become intimately bound up with religious beliefs, and as a gift the prey has to be utilized to the fullest. To do otherwise would be an insult to the animal. To repay the whales for sacrificing their lives, the whalers have to take care of their souls; otherwise the whale souls can turn into 'hungry ghosts' that might cause illness, accidents and other misfortunes.

It has, therefore, been the practice in many whaling communities to treat the souls of whales in a manner similar to that in which they treat the souls of deceased human beings. The whales have been given posthumous names (*kaimyo*) which have been inscribed on wooden memorial tablets (*ihai*) and included in temples' death registers (*kakocho*). Tombs and memorial stones can be found in at least 48 places, from Hokkaido in the north to Kyushu in the south, and annually at least 25 festivals (*matsuri*) and memorial rites (*kuyō*) are held in honour of whales (Akimichi *et al.* 1988). A tomb at Koganji temple (Yamaguchi prefecture) has been designated a national historical monument. Built in 1962, it marks the burial of 75 foetuses found in whales caught before 1868. The temple, which is dedicated to whales, is moreover the stage for elaborate memorial ceremonies with Buddhist priests reciting *sutras* from 28 April to 2 May every year in order to help the souls of deceased whales to be reborn in a higher existence. Such services have a number of implications, and people may have different interpretations of the rites. The temple priest may perform the memorial service in the belief that the whale will reach enlightenment and thus be released from rebirth into this world and enter Paradise as 'Buddha' (*hotoke*), whereas some people may believe that the whale will be reborn as a human being or as another whale to be hunted. Finally, memorial services are held to ensure that the whalers, and the gunners in particular, are forgiven (*tsukunaru*) for the sin involved in taking life. Memorial services therefore carry special meanings to the gunners, and they frequently go directly to the temple upon returning home in order to conduct memorial services for the whales they have killed.

This rite, then, also serves to secure safe voyages in the future by preventing whale souls from becoming malevolent. Successes as well as failures are explained in relation to the divine. Accidents may be caused by failure to repay the whale's sacrifice through rituals, or by breaking taboos. There are many stories about the malevolent spirits of whales, and some of them are known throughout Japan. The most famous is the disaster that struck Taiji in 1878 when whalers attacked a right whale with calf. One hundred and eleven whalers lost their lives in the following gale. A similar story is told from Ukushima where 72 whalers perished in 1715 after attacking a blue whale with calf, allegedly on their way to worship at a famous temple (Kalland and Moeran 1992). These accidents, which have become part of the communities' cultural heritage, and thus help in giving them their peculiar identities, have reinforced the validity of the taboo, a taboo which, incidentally, has great value in the conservation of the stocks.

A number of lesser ceremonies and rituals are performed in order to repay whales for their personal sacrifices and thus to prevent accidents. The whaling companies used to gather the whalers and their wives for a joint ceremony before the commencement of the season, as well as after, and in some places the wives would go on pilgrimages to local shrines. Daily religious observations are conducted in front of the family Shintō altar (*kamidana*), praying for the husband's safety and for good catches. Similar rituals are performed on the boats, where a piece of the whale's tail may be offered to the Shintō altar. The many rituals tie whalers to each other, to their families, to the whales, giving the local residents both a feeling of the common heritage and meaning to their lives.

Discussion

In order to hunt whales successfully it is necessary not only to acquire skills in navigation, shooting and the handling of meat but also to acquire detailed knowledge about the migratory, mating and feeding behaviour of various species of whales (Takahashi *et al.* 1989; Kalland and Moeran 1992). The whalers have learnt to take account of such natural phenomena as tides, currents, winds, wave patterns, water temperature and colouring. They are also keen observers of fish and bird behaviour. But the knowledge of the whalers goes beyond this. They need an understanding of how whales are part of a larger ecological system in which other maritime organisms as well as humans are also parts. The whalers' knowledge of this ecosystem is based on centuries of accumulated experience and is intimately linked to religious beliefs and practices.

Whaling in Japan, then, is more than just earning a living. It is a way of life. Whaling activities are firmly embedded in the local culture which encompasses religious beliefs and rituals, food habits, special ways to socialize children, and so on. In many ways this culture is common to all communities that have based their economy on whaling. There are, nevertheless, distinctive features unique to each of them. The distinct character of STCW is the very close connection between whaling-related activities and the communities from which the boats operate; the localized distribution network where a large part of the whale meat is used for gift exchange; and the relatively egalitarian work organization found both on the STCW boats and on the flensing stations. In these qualities, Japanese community-based whaling resembles the whaling conducted by the Alaskan Eskimos. With the current moratorium on whaling this culture is being eroded. A culinary tradition is being lost and rituals are disappearing. In Ayukawa, in particular, solidarity is further undermined by the collapse of the intricate gift exchange system. This loss of social identification, coupled with losses in community identity and solidarity, has caused stress-related problems among some whalers and their families.

Social and cultural changes are inevitable, and the whalers and their families might have been able to cope with the situation if they had been convinced that the moratorium was necessary in order to save endangered species. But the species hunted by the Japanese STCW boats are not endangered. That whales should not be killed for ethical reasons is furthermore incomprehensible to most Japanese, who rather would question the ethics of killing domesticated animals for food. To an increasing number of Japanese the moratorium is therefore interpreted as an assault on their culture, and groups have been established to protect 'the Japanese food culture'. In the announcement of one such group it is stated that its purpose is to argue 'strongly against the ethnocentric Western coercion to change Japanese eating habits' (GoJ 1989). The best guarantee for biodiversity might in fact be the maintenance of cultural diversity.

Global preservation and the creation of a pest: Marine mammals and local response to a moratorium in arctic Iceland (Níels Einarsson)

To Icelandic artisanal fishermen, the prime examples of alienated Westerners are whale and seal conservationists, who they see as dangerous extremists. They feel that such people should not be taken seriously as they have no understanding of how nature works or how to survive in it.

While whale watching has become a major industry and an ever more popular pastime, it has gained little ground in Iceland in spite of some potential (see Linquist and Tryggvadóttir 1990). Unlike whale watchers, for fishermen whales are an integral part of everyday reality. In their everyday work artisanal fishermen encounter marine animals in a range of situations. Being out at sea in a small boat often means that they come into close contact with animals under conditions that the fishermen have not chosen. In contrast, most urbanized Westerners leave their cities not for their livelihood, but for recreation.

Whereas the visiting urbanite may seek out wildlife to look at and admire, fishermen are primarily engaged in hunting (the Icelandic word translates into English as both 'to hunt' and 'to fish') and they find little reason for coming close to animals for other purposes. Sometimes contact with animals can indeed be too close for comfort, as in the case of breaching whales. In the village where I worked, a fairly big boat was nearly capsized at night when a whale threw itself against its stern. Had it been a smaller open boat, fishermen say, there would have been little hope for survival as it would have been crushed. In Arctic waters a man can survive for only a few minutes. The fishermen did not know what kind of creature 'attacked' the boat involved but considered it likely to have been a minke whale, the 'jumper' (or *'stökkull'*). When breaching, minkes jump into the air and hit the sea making sounds like cannons being fired.

Icelandic fishermen are not incapable of appreciating the beauty of nature. In fact, one of the reasons small-scale fishermen often give for their choice of occupation is that it gives them the opportunity to experience the beauty of the fjords and the coastline where they fish, and to do so in total tranquillity and at their own pace. It is seen as one of the positive attributes of being a small-scale fisherman. To use nature and animals does not mean to deny that it can at the same time be of aesthetic value although many environmentalists see these as mutually exclusive. They may never have seen a fisherman take a large and fat cod and with a smile of appreciation describe it as particularly beautiful. The word used about a catch of big fish of high monetary value is indeed 'beautiful fish' (*fallegur fiskur*).

Fishermen see whales as predators eating millions of tonnes of fish that would otherwise be available to them, while the fishermen are facing decreasing quotas which pose a serious threat to their livelihood. Focusing on minke whale predation on Norwegian cod and herring fisheries, Flaaten and Stollery (1994) conclude that, 'the cost of predation is large, [with] a 10 percent increase in the 1989 minke whale stock causing estimated damage of over seventeen million $US in value; 5.2 and 12.4 percent respectively of the total gross profits of the Norwegian cod and herring fisheries'. Assuming that the diet of the minkes in Icelandic coastal waters does not differ significantly from Norway then whales in Icelandic waters consume 4.6 to 4.8 million tonnes of fish per year. The Icelandic economy is recovering from its lowest economic slump since World War

II, and from drastic reductions in fish quotas (see Matthiasson 1994). Many small-scale fishermen have been forced to give up fishing since quotas have not sufficed to keep their businesses going, yet scientists at the International Whaling Commission maintain that, given restricted and regulated quotas, the minke could be hunted sustainably. Fishermen claim that there have never been as many whales as today and that numbers are out of control, threatening what they think of as ecological balance in the sea. All this is transforming whales into vermin for Icelandic fishermen.

The effectiveness of protests of various environmental groups concerns Icelanders. Many worry that arguments about animal welfare might be stretched to include all fishing. Fishermen think of their occupation as a natural way of life. They maintain that, as hunters, they are an integral part of the ecosystem and as their methods of fishing are small in scale, they pose no threat to fish stocks or to the environment. They like to juxtapose what they see as a natural way of living with the lives of 'city dwellers', who they claim are becoming ever more removed from nature – living in crime-ridden, inhumane surroundings and buying all their products, including nicely packaged meat, from supermarkets. Fishermen point to the inconsistencies in the treatment of animals by their critics who may even call themselves 'animal lovers' while at the same time tolerating the mass suffering of factory-raised animals: animals that ordinary people never know and never see.

The stigmatizing view fishermen have of whale conservationists is understandable in terms of the organizations and individuals they see as representing the movement. Paul Watson, the leader and founder of Sea Shepherd, a marine mammal conservation group, is one of the militant 'eco-warriors' (a term endorsed by the members) engaged in whale conservation. He firmly believes that in the future people will be able to communicate with whales. After partially succeeding in scuttling a minke whaling boat in Norway in 1992, he wrote to a Norwegian newspaper: 'Do you want your children to hear about what you did to the whales from the very same beings that you have helped to slaughter? It is a certainty that the whales will talk about you in the same vein as Jews now talk of Nazis. For in the eyes of whalekind, there is little difference between the behaviour of the monsters of the Reich and the monsters behind the harpoon' (quoted in *Nordlys*, p.7). Such messages and actions only serve to reinforce the image of environmentalists as potentially dangerous extremists who are not to be taken seriously.

The present moratorium on whaling is leading to the perception of whales as pests. Icelandic fishermen do not share the common feeling of what has been called 'sentimental anthropomorphism' to the extent observed in urban Western culture (see Jasper and Nelkin 1992). They do not believe that whales are especially intelligent or otherwise endowed and that they should therefore be spared from hunting, and they draw the moral circle narrowly around humans. Outside that circle come animals with very different moral rules attached to them. This is seen as self-evident and a part of a life-style that involves fishermen in direct confrontation with animals. These differences in the perception of whales are obviously not the intention of whale conservationists, but may become the actual consequences of total protection. Icelandic small-scale fishermen have an intimate, practically-based experience with the resources from which they gain their livelihood. Their attitudes towards sea animals are shaped by this fact. They are another example of how people construct their moral universe to fit their convenience. Fishermen maintain that they have the right to hunt or otherwise manage animals in the ecosystem in which they participate. It is a matter of principle. To question that principle is to question what fishermen feel is their self-evident right to survive. By referring to the fact that they work in close contact with the natural environment they exploit, they make claims on a truer version of reality and therefore a licence to speak of nature 'as it really is'.

Box 10.2: Theoretically

William Wallace Mokahi Steiner

I have been thinking about this for a long time,
just as centuries of color have, wondering at
why it is the pale ones exploded out of their circles
    engulfing all before them with their own culture,
    their own unique way of reading the world

as if they owned it; there it was, apple for the picking; they
swept all before their wave, not knowing or caring
or sometimes even noticing anyone else was there, that
    world had adjusted to us, that the different songs in the
    different voices promoted the same thing, a kind of

peaceful coexistence with nature if not with each other.
Oh I know about the theories; there is one about the role
of a pale God imaged in their own shape to reduce their
    fear of such things, make him (always him) more
    palatable and recognizable and in his great

parochialism give them the right to seize the world
by its very throat. To give them strength, they said.
And there is the one about survival of the fittest, words
    taken from a context which recognized that rules applied
    and one of those rules meant that the survivors did not have

a thinking brain and reacted only in need which raises the question
what is the need of this tiresome and pale race? And how about
that other theory, the one that was used to subtract the Indian,
    the Hawaiian, the Aborigine from the spirit of some nurturing
    soil or sea, you know, the one which said we did not know

what to do with what we had. Did not know how to be efficient killers
and miners and traders enough so as to control, to own, to build
empires with the resources at our feet; they would teach us how
    even if we did not want their language, their plans, their seed,
    even if we did not need these new thoughts on how to be a human.

And in the thinking about these things I have a new theory, a kind of
pathogenic idea, one which builds on the very best of their own knowledge
and is the only explanation I can see. It goes like this: suppose there
    is a virus that recognizes only certain gene sequences in the
    archetype of the pale ones; suppose it knows how to destroy the spirit of

Moral Conscience.....?

CHAPTER 11

ETHICAL, MORAL AND RELIGIOUS CONCERNS

Jeff Golliher

CONTENTS

TEXT BOXES

Ethical, Moral and Religious Concerns

Jeff Golliher

Introduction (Jeff Golliher)

Environmental philosophy and ecotheology as values criticism

This chapter examines the principal philosophical, cosmological and ethical ideas that inform secular as well as religious views about biodiversity and the unravelling web of life. Special attention is also given to the ecologically-rooted beliefs of the major world religions. Although definitions of religion are especially ambiguous, they provide an appropriate point of departure. Geertz's (1965) classic anthropological definition includes the following: 'a system of symbols whose function is to establish pervasive moods and motivations and to formulate conceptions of a general order of existence with an aura of factuality that those moods and motivations seem realistic'. The function of religion is central to Geertz's definition, while its focus on the 'general order of existence' characterizes its cosmological scope. This view of religion is important because one's appreciation of cosmology is broadly equivalent to our respect for and knowledge of ecological systems (Reichel-Dolmatoff 1976; Metzner 1994).

Most scholars interpret the time frame of 10,000BP as a great cosmological divide and a threshold of significant historical transition in human culture. A number of complex, interrelated cultural changes occurred about this time, for example intensive cultivation based on the domestication of plants and animals, the rise of patriarchies, class stratification, and the origin of cities, to name a few; and with them, a clear emphasis on world rejection emerged on the human scene (Ruether, this chapter). This cultural divide has a bearing on popular assumptions about the meaning and purpose of religion. Traditional cosmologies are sometimes interpreted, especially through Western eyes, as inherently 'spiritual' rather than 'religious', because they effortlessly integrate intimate connections with nature and practical knowledge of ecosystems with everyday patterns of living. Anthropological studies suggest that the term 'religion' as it is often understood in modern industrial societies can be misleading when it is used to describe traditional, indigenous or pre-modern cultures (Getui, this chapter). Moreover, Jensen (this chapter), who notes that indigenous cosmologies cannot be described in a single, monolithic manner, dispels the 'myth of primitive ecological wisdom' on which many modern religious reconstructions of history are based.

The fragmentation of modern industrial societies into separate realms (e.g. political, economic, religious) is a recurring pattern in the historical process of cultural evolution. To maintain unity, modern cultures coalesce around values shared by different institutions in a reciprocal and mutually reinforcing way. Max Weber's (1904) classic sociological account of the relationship between the Protestant work ethic and the rise of entrepreneurial capitalism illustrates this pattern. Thus, 'the religion' of modern societies includes much more than the official beliefs held by religious institutions; it draws upon the values, moods and motivations of the entire interlocking political, economic and religious complex. From an ecofeminist perspective (Ruether, this chapter), this process has been aptly described as the 'uprootedness of development' (Shiva, this chapter). The distinction between institution and experience may be, in itself, a product of the separation of spirit from matter, culture from nature, and ethics from science.

An opposition between spirituality and religion implicitly shapes the nature of ongoing debates about environmental ethics and the meaning of ecology. It underlies the great significance attributed to shamanistic experience among spiritual seekers in the West, but it can also lead to warnings from indigenous peoples about the potentially disingenuous nature of 'white shamanism' (Rose 1992). Western explorations in shamanistic traditions may assist ecologically and spiritually uprooted people to regain connections with the sacred through nature. Yet without a genuine commitment to work on behalf of indigenous peoples in terms of basic human rights, these spiritual explorations amount to another example of Western cultural exploitation.

Individuals in any society are expected to be morally responsible in the context of their cultural norms. A larger perspective, however, is required to account for the mental, spiritual and ecological crisis of modern times. In this regard, Hillman (1988) makes an enlightening distinction between *universe* and *cosmos*. In the mechanistic Cartesian world-view, 'universe' connotes desacralized, empty space filled with 'discrete islands of phenomena requiring complex explanatory connections'. In an ecology-based world-view, 'cosmos' refers to a 'body of life'. In the former, the universe becomes significant when people attribute meaning to it; in the latter, the cosmos is meaningful in its own existence.

It would be misleading to assume that modern industrial culture has been entirely devoid of a cosmology since the onset of the Enlightenment or the rise of Christianity. A Western cosmology exists, but it has been shaped by certain anthropocentric, mechanistic and dualistic assumptions underlying the ideal of unlimited progress. These themes have figured prominently in the politico-economic systems of capitalism and communism, and both have fuelled the systematic genocide of indigenous and traditional peoples and the plunder of much of the natural world – rationalized by the assumption that nature is either dead or simply a warehouse of resources to be consumed for human benefit. Orr (1992) summarized the strangely mirror-image relationship between capitalism and communism in this way: 'Communism has all but collapsed because it could not produce enough; capitalism is failing because it produces too much and shares too little ... Neither system is sustainable in either human or ecological terms'.

Any discussion of ecology and cosmology rests upon difficult questions surrounding four interrelated ethical positions: cultural relativism, moral relativism, moral pluralism and ethical monism. These are controversial and highly abstract subjects, yet they encapsulate assumptions about the nature of unity and diversity in human affairs and continuously shape the conduct of international, intercultural and interreligious relations. Cultural relativism recognizes that different cultures embody different ways of perceiving the nature of right and wrong (Herskovits 1964). The philosophical position of moral relativism, which is more extreme, holds that there is no 'objective' (i.e. rational) standard for making moral judgements (Singer 1979). The complex implications of moral and cultural relativism have a huge bearing on efforts to meet the challenges of the environmental crisis. For example, one would reasonably have expected the inter-faith religious leaders who gathered in Assisi (Jensen, this chapter; see also Appendix 2) to form a unified moral response around environmental issues. Their common interest in reversing ongoing extinctions, for example, would contradict the position of moral relativity, at least among the world's religions. (Appendix 2 provides a summary of statements of the religious groups that participated in the Assisi Conference.) Yet, the moral responses of the Assisi participants might differ or come into conflict when local decisions concerning energy usage, the rights of non-human species, or the environmental impact of economic development projects are taken into account.

Examples like this enter into the debate between the philosophical positions of *ethical monism*, which seeks to discover or derive a set of master principles, a meta-ethic, to guide the global community as a whole, and *moral pluralism* (Stone 1988), which challenges the assumption that a single global ethic can be reconciled with the complex moral and ecological dilemmas of everyday living. These positions not only reflect the complex and politically charged relationship between unity and diversity in religious and ecological terms, but also call attention to the historical legacy of Western and Christian colonialism. Callicott (Box 11.4), for example, argues that the formulation of a global ethic which does not compromise the integrity of local cultures and ecosystems is a necessary and worthy goal; while Jensen (this chapter), who has the evidence of history on his side, adopts a more cautious perspective. He points to the possibility that one or more of the world religions might, even unintentionally, attempt to seize the present ecological crisis as a self-serving opportunity for 'green' proselytizing and/or institutional expansion under the guise of protecting diversity.

Nevertheless, religious and philosophical contributions to the environmental movement aim to raise our holistic awareness of the fact that a global crisis actually exists by deepening our understanding of ecological interdependence. David Quammen (1996) puts it this way: 'the web of life is unravelling'. These arresting words force us to consider both the virtually unimaginable scale of current extinctions and the possibility that the human capacity for creative imagination will diminish as biological diversity is lost. People may disagree about the rate of ongoing extinctions or its causes, but the subject of the debate itself has an eerie, surreal quality that deeply disturbs the human psyche. Disturbances of this kind may be a distinctively spiritual and ecologically-grounded response to an unprecedented event in the cosmos. Eco-psychologists (e.g. Roszak 1992) claim that these disturbances originate, ultimately, not within human consciousness but from the web of life of which human consciousness is a part. The fact that 'disturbance' has also emerged as an organizing principle in the scientific study of ecosystems appears to support this psychological possibility (Capra, this chapter).

Disturbance is no longer regarded as a departure from the normal functioning of ecosystems or communities, whose 'health' was previously defined by the criteria of balance and harmony. Instead, theorists see ecosystems in a state of constant flux

(Pickett and White 1985). From a psychological perspective, the implications of this shift are profound. For example, should psychological disturbance in response to extinctions and ecosystem destruction be interpreted as a form of emotional imbalance or mental aberration? To pose this question does not mean that the terms 'ecosystem health' and 'mental health' should be discarded, but it does suggest that such disturbing responses in the human psyche are signs of hope: an awakening of our forgotten, but instinctual interconnections with nature that give birth to new forms of artistic imagination.

Box 11.1: Art and Environment, or Look Here

Margot McLean

When speaking about art and the environment one must ask a few questions: What constitutes environment? Is it to be perceived mainly as habitat? If so, what habitat? How does the public connect with 'environmental issues', and how does the scientific study of ecological processes bear on the mind and hand of the artist? Does the artist follow trends or advance societal consciousness? What are the spiritual connections, and can a work of art challenge our habitual perceptions of how we look at the world – how we 'see'?

As a visual artist, I would like to take up one of these issues; namely, the challenge to our habits of seeing. How we 'look' at what's there, how we see the world, is directly related to how we regard it, and therefore what we do with it. The more unconscious we are of our surroundings, the less we have to take them into account and the more likely we are to neglect or abuse them. This is one purpose of the artist: to make our perceptions more conscious.

Another purpose is to change perceptual habits, which is not a simple task in a place like Manhattan where the pace is moving faster and faster and information overwhelms our perceptual apparatus. Here, in New York, we are a culture that 'looks' just long enough to identify and classify. It's easy, but do we observe what is actually here? A tree, a mouse, a building – the world becomes generic. Moreover, doesn't this inability to see diversity in the things around us add to our acceptance of a monoculture and the loss of biodiversity? What about that particular tree that has been growing in that particular place for the last 200 years, holding that specific ground together with its roots which spread halfway across that town. Do we even have the capacity to respond to a particular tree rather than to trees in general? I believe we do; however, there is a risk in seeing. The risk is that some unknown emotion, some unknown feeling may be exposed, uncovered. To look carefully is to be surprised.

Artists have always tried to show things in a different way: to break habitual images and make fresh ones. Artists are the culture's image-makers. In our increasingly conservative climate, politically, art has taken a marginal role in the shaping of philosophies and psychologies, and therefore has had little influence on our thinking about the environment. This political climate is self-serving and human-centred. My notion of environmental art is *other than human-centred*. Perhaps it is 'place' centred, allowing specific faces of the world – particular animals and plants – to speak to human consciousness through the work.

For many years, my work focused on the more political aspect of our ecological crisis – which I consider as my finger-pointing work. But I moved away from the more in-your-face kind of preaching art because I realized I had failed to allow the spirit of what was being painted to enter into the piece. It remained a political idea. A good idea, but not what I was trying to accomplish. It felt as if I were worshipping at an altar that was far too strict, closing out other voices and emotions. I wanted to create a prayer to those things other-than-human – a space where the viewer could enter and not be preached at, but be left alone to look. Whatever it was that invoked in me to paint, that is what I wanted to reverberate out to those who looked.

The first step in restoring the value of the 'other-than-human' world is to see the image of things as they are – before scientific knowledge analyses it, before its place in the human scheme of things is assigned, before our judgements about beauty and ugliness, good and bad are formed. The image will reveal meaning.

There is an objectivity in the eye perceiving the image of an animal, a street, a chair – and I think this objectivity is like that of scientists and therefore of equal importance to understanding the world. This creative eye allows things out there to show themselves beyond one's own personal opinions and feelings. David Ignatow once wrote: 'I should be content to look at a mountain for what it is and not as a comment on my life'.

Grounded in the study of ecosystems rather than ecopsychology or ecotheology, the 'biophilia hypothesis' (Wilson 1984; Kellert and Wilson 1993) affirms a profoundly deep bond between humans and the natural world, a connecting tissue that reminds us of our extended family and evolutionary origins. Wilson (1993) suggests these four areas for a biodiversity ethic: (1) biodiversity is the Creation; (2) other species are our kin; (3) the biodiversity of a country is part of its national heritage; and (4) biodiversity is the frontier of the future. They focus our attention clearly on the hopeful possibility that workable alternatives to our present course can be found with the sound application of wisdom and ingenuity. Sagan and Margulis (1993) suggest that present biodiversity losses may be 'balanced by an increase in technological diversity'. This arresting statement offers an alternative to the pessimistic choice between perpetuating our dependence on a few unsustainable forms of technology or pursuing their complete elimination.

Debates about the ecology of disturbance and hope form part of a larger critical discussion of values and a related commentary on the political significance of environmentalism. The writing of Thomas Berry (1988) demonstrates that the greater purpose of religion and philosophy today is to undertake this critical task, and he is joined by a large number of environmental philosophers. Some of these commentaries represent the creative analysis of environmental philosophers conducting their reflective work in good faith. Others, which constitute manipulative attacks on environmentalism, intentionally confuse the issues at hand.

Having said that, the challenge of ecology to the Western emphasis on individualism sometimes results in the charge of 'environmental fascism' (Ferry 1992), specifically in reaction to religious and liberal-democratic movements to extend legal rights to animals. The assumptions that form the basis of this critique are these: (1) an elevation of the moral standing of non-human beings unjustly diminishes humanity's place in the cosmos; (2) the extension of legal rights to nature cannot in itself be understood as a sign that any given culture has become more humane or ecologically benign. To counter charges of this kind, Irene Diamond (1994) points out that if genocidal efforts to eradicate indigenous peoples by the sterilization of women had happened to a species of tree or animal, 'the outcry that would have arisen would have been difficult to ignore, yet because it happened to Indian women little is even known of what was happening'. Moreover, Callicott (1994a) notes that environmental ethics are not a substitute for the humanistic tradition of moral philosophy, but an addition to it.

Fox (1990), who would agree with Diamond, challenges the charge of environmental fascism in his observation that ecosystems place limits on the behaviour of all creatures. Democracies based on ecological principles would encourage their members to pursue their diverse interests, but 'no one is *above* the law' – which is precisely the moral failure of fascist regimes. In addition to this quasi-legal principle, according to Fox, an ecologically-rooted democracy would promote the ideal that 'no one is *above* the ecology'. Ecofeminists make a similar claim with reference to cultural, economic and political systems built on a domination ethic (Ruether, this chapter). For example, efforts to replace the earth's biosphere with a human-created technosphere amount to a form of institutional violence. They not only epitomize the values and goals of patriarchal domination, but also expose its false premise that natural processes can be overcome by technological means.

Many of these complex issues rest on the distinction between our images of nature as opposed to nature itself, and on the question of whether any cultural knowledge of nature can ever be truly objective. Unlike 'post-modern deconstructionism', the question here goes well beyond the recognition that all forms of cultural knowledge are 'interpretations' of reality; it also addresses the ideological impact of these interpretations on the environmental movement and society as a whole (Soulé and Lease 1995). For example, much of the environmental movement internationally is located in and shaped by media-saturated urban subcultures which have little contact with or knowledge of 'natural' ecosystems. Thus, urban subcultures might elevate the significance of cultural meaning to a point where the objective status of the natural world is replaced by myths about it (Soulé 1995). Likewise, it has been wisely argued that distorted images of nature and indigenous ecological knowledge (Jensen, this chapter) might enter into the thinking of the environmental movement (or of its opposition) and this could undermine efforts to protect biodiversity – although it is unclear how this might actually happen. Rapid advances in the technology of 'virtual reality' would complete this nihilistic movement by replacing the web of life with an electronic web of images, symbols and texts (Shepard 1995). Hillman (this chapter) takes up a portion of this complicated matter by addressing the meaning of 'environment' in a way that deconstructs and recreates the richness of human experience in urban settings.

Environmental philosophers generally identify three approaches to the academic field of ethics (e.g. Callicott 1994a): (1) an *anthropocentric* approach in

Box 11.2: We Must Try to Hear Others

Finn Lynge

Humanity is torn between diverse identities on the one level and strong currents towards greater unity on the other, and we are only at the beginning of finding out what is 'good' or 'bad' for humankind as a whole. (Robert Muller 1984)

These prophetic words have been highlighted by remarkable developments and initiatives on the international scene, as policy-makers try to safeguard not only biological diversity, but cultural diversity as well. The 1980s saw the upsurge of basic concerns about the ethical dimensions of conservation and the proper assessment of indigenous peoples in the use and management of nature. Ethical issues seemed to spring primarily from a growing public concern for animal welfare, while questions of defining the proper place for indigenous peoples in the consultation and decision-making processes of nature management appeared from human rights-oriented NGOs.

Obviously, the two concerns ought not be in conflict with one another since both deal with rights and duties on the part of humans. But in practice, the two have often clashed. The anti-seal-skin campaigns, for example, have reoccurred since the 1970s depending upon public sentiment for the endearing pups. These campaigns have resulted in the collapse of the international market for seal skins, followed in turn by the collapse of the traditional Inuit economy. Seal skins are the only traditional product of any consequence that the Inuit of the Arctic are able to market internationally, and Inuit do not take pups. The Inuit seal hunt is certainly encompassed by the two international human rights conventions of 1966, which guarantee that under no circumstances may a people be deprived of their capacity to live by their own natural resources. However, when it comes to the seal-skin issue, practically no government has ever taken occasion of these conventions to assume an appropriate leadership role in the shaping of public opinion.

In Tanzania, Western conservationists have been operative in motivating the government to drive the Maasai off their traditional lands in order to create the – admittedly wonderful – natural reserves of Serengeti and Ngorongoro. There has never been any proper consultation with the aboriginal peoples of these lands, the Maasai or others, not to mention efforts towards creating mechanisms for their participation in the decision-making process. Decidedly, in the Serengeti case, there has been, and still is, a breach of the human rights of the local populations.

Two more examples need to be mentioned because of the attention they command in public opinion: the campaigns against the fur trade, which are threatening the livelihood of remaining Indian cultures of North America, and the campaigns against the trade in elephant ivory, which are contested by States where elephants abound. In both cases, governments are faced with problems that cannot be resolved simply by pointing to the principle of sustainability. There is no problem with the abundance of the animal populations in question. They can readily be harvested in a sustainable manner. The problem lies elsewhere. As it is stated in *Caring for the Earth* (IUCN/UNEP/WWF 1991): 'Perhaps there is no issue over which human rights and animals have collided with such emotional force. Such conflicts reveal radically different cultural interpretations of the ethic of living sustainably. Ethical principles need to be developed to resolve these dilemmas.' Genuine global co-operation over the principles underlying good nature management depends upon 'the recognition of the interdependent wholeness of humanity'.

We have a right to be listened to, and the fact that some countries have the economy to put up satellites and dominate media and communications does not give them the moral right to impose their own conservation rationale on others. That is no easy programme to set up. In the debate for and against consumptive use of nature, polarization has become virtually extreme, and in the rare cases where dialogue is attempted, the counterparts pose rather like two groups of people trying to shout something to one another, each from their side of the Grand Canyon. You can see the others moving their arms and legs, but you cannot hear what they are saying.

Yet, there is no way around it. We must try and hear the others. In matters of conservation and nature management there must be room for cultural diversity. Otherwise things will simply not function. Animal rights as an ethical issue also entails respect for human rights, for the human being is also an animal, and therefore the rights of all animals intertwine. The human rights of the 300 million indigenous

Box 11.2 (continued)

peoples of the world are also an ethical issue. They are entitled to co-decision in matters that touch upon the environments that are theirs.

Nowadays, people actually believe that they have risen above nature, including the act of killing animals. The fact, of course, is that every person who wears a leather belt or leather shoes has an animal on their conscience. It is easy enough to convince people of that. The matter, however, becomes more difficult to deal with when the spotlight is shifted to synthetic materials. Superficially, a nylon jacket is proof that man is in a position to rise above the Stone Age and the necessity to kill an animal in order to protect himself from the elements. The person who wears a fake fur has no animal on his conscience and does not clothe himself at the expense of nature. But is this really true?

No! The overwhelming threat against many plant and animal species – biodiversity – comes from two sides: an ever more imposing invasion of wild animals' habitats by industry and housing, and the long-distance pollution of the pristine lands that remain. All scientific evidence shows that the invisible pollution of the air, water and food chains around the world is a threat of a magnitude and gravity never seen before. Air- and water-borne pollution with dioxin, heavy metals, persistent organic chloride compounds such as PCB and other toxins are not registered in the Arctic where there is no production of any such substances. The result is a decrease in fertility among marine mammals, birds and fish of all sorts.

Where does all this pollution come from? Among the biggest offenders is the synthetic material industry. Very few industrial processes bring with them such obnoxious chemical pollutants as does the production of fake fur, synthetic turf, thick plastics, nylon materials and the like. When one buys a fake fur, no animal has been directly killed for the fulfilling of your wishes. On the other hand, one contributes to an invisible process that is far more dangerous for animals than the activities of hunters.

The worst part of it is that consumers are unaware. We have all seen pictures of top models being hired to protest the use of animals furs. But who has ever heard of protests with nude people bearing signs that read, 'I'd rather be naked than wear a fake fur'? Perhaps it is true that some of us resort to such acts!

The real problem lies deeper. Many view the killing of a few living, beautiful animals as far more repulsive than the more theoretical notion of some species becoming less reproductive – even though the latter, unlike the former, could threaten the survival of the species as a whole. It is a matter of the public having come to consider the killing of animals as by itself unethical. When modern consumers find animal furs repulsive, it is not only due to their wish to avoid animal suffering. Many people simply find the very thought of where the skin comes from extremely unpleasant. They suffer from the anxiety of getting close to nature.

I personally have several times had the pleasure of taking European big-city tourists to the open-door market in small Greenlandic towns where it is possible to pick out your own lunch from among the freshly-shot game. This is where hunters sell their wares. There, one finds eider and auk, halibut and seal, small cetaceans, caribou and musk ox, all just brought in from nature's big storeroom – and whole animals mind you, not neatly cut and plastic-wrapped as in a delicatessen. Here, you can point at a cavity between bloody intestines and say: 'I would like a slice of that liver, please'. It is then delivered to you in a bag that you have brought along for the occasion.

The reaction from the European friends is always clear: This meat market is simply disgusting! It is too close to nature for comfort.

which human welfare and concerns are regarded as the primary measure of value; (2) a *biocentric* approach which considers criteria, e.g. conation or sentience, for establishing the moral standing of individual non-human life-forms; and (3) an *ecocentric* approach which is informed by the ecology of whole communities and their interdependent relationships. Ecocentrism challenges the anthropocentric traditions of Islam, Judaism and Christianity (Appendix 2); however, recent ecotheological work on biodiversity from a biblical perspective also contradicts any justification of human efforts to dominate our fellow creatures (Nash, Hessel, this chapter). Representatives of Eastern traditions (Appendix 2; Halifax and Peale, this chapter), on the other hand, find the ecocentric approach to be generally compatible with their spiritual teachings. Although some environmental ethicists (e.g. Attfield 1991;

Norton 1995) and biologists (Wilson 1993) suggest that the creation of a sustainable society and new world-view may not require a non-anthropocentric model, 'deep ecologists' (Naess 1989; Devall 1988; Fox 1990; Sessions 1995) seek to dislodge the presumption of human uniqueness and self-importance from the modern industrial world-view. In this regard, the term 'environment' can be misleading because it creates a rigid, if not false, distinction between the organism and nature (Hillman, this chapter). Deep ecologists replace the distinction between 'fact' and 'value' (*is* and *ought*) of 'injunctive' philosophy with an 'invitational' approach for the individual to realize his/her identity as part of 'the environment' (Abram 1996 and this chapter). Thus, the *hubris* in asking people 'to take responsibility' for the environment is replaced by an invitation to realize the depth of existing ecological relationships.

Ecofeminism traces the impact of unequal gender and power relations throughout whole cultural systems, and it asserts that anthropocentricism, in theory, amounts to androcentrism (male-centred values and institutions) in practice (e.g. Ruether 1992 and this chapter; Adams 1993; Mies and Shiva 1993; Plaskow and Chirst 1989; Primavesi 1991). The domination of women and nature operate as twin dimensions of the same colonial process. The commonplace use of the term 'nature' as distinct from 'spirit' indicates the oppressive quality that patterns of language can assume as they are shaped by an ethic of domination. Ecofeminist ethics seek to reshape androcentric, dualistic concepts of reality based on the identification of women with the realm of nature (particularly non-human nature) and men with the realm of spirit. Dualistic conceptions such as these have assigned women an inferior status in cosmological terms and resulted in the 'feminization of poverty', which correlates to a high degree with the destruction of ecosystems.

Social ecologists (e.g. Bookchin 1980, 1981) also hold the position that society and ecology are inseparable. They severely criticize efforts to 'integrate' humankind with nature through a biocentric world-view, which Bookchin (1992) criticizes as a form of 'mystical ecology' representing escapist tendencies originally attributed to organized religion by Marxist theory. This implies that politico-economic theory and practice play little part in the creation of environmentally sustainable societies. Instead, social ecologists argue that because human beings are already part of the evolutionary process, strategies for achieving a sustainable society must be based on the transformation of dominant institutions by a dialectical process of ecological consciousness and practice.

Sagoff (1996 and this chapter) bluntly asks: 'Why does society care about the extinction of species?' The same person may respond by giving complex, multi-layered answers that reflect deep conflicts about the meaning and value of nature. For example, the marketplace assigns values to biodiversity based on cost–benefit indicators and consumer preferences, while 'consumers' may have conflicting opinions about nature both as economically useful and as sacred. Conflict in the way people in modern societies make moral judgements about nature is accounted for by the scholarly distinction between *use values* or *instrumental values*, on the one hand, and *intrinsic values*, on the other (Norton, this chapter; Sagoff, this chapter; Callicott 1994b). Based on the difference between usefulness to humankind and attributes other than human usefulness, this distinction is more subtle than one might think. The rationale for biodiversity conservation might be based on instrumental values in the form of economic usefulness (e.g. food or medicine) or because of nature's religious usefulness. Instrumental values, then, reflect a broad anthropocentric pattern of thinking about nature solely with respect to human uses (economic, religious, aesthetic, etc.), rather than because nature is valuable in itself.

Another question addresses the source of nature's intrinsic value: Are intrinsic values attributed (i.e. subjectively) by humans or do they exist independently of human presence (i.e. objectively)? Rolston (1988) labels the view that nature's intrinsic value is attributed by humans as 'anthropogenic' not because it is based on an anthropocentric perspective but because it fails to recognize that people can appreciate values that are 'objectively' given. Two additional questions are of concern to the field of environmental ethics. The first concerns the moral status of instrumental values: Are instrumental values, on which economic behaviour is based, an expression of moral thinking or are they value free? Some market economists maintain that decision-making in the marketplace is 'value free' because it rests on 'factual' or scientific principles. Others maintain that these instrumental values express moral thinking, regardless of the environmental impact of free market economies. The second question concerns the intrinsic value of individuals, classes of individuals (e.g. species) or whole ecosystems: What are the criteria for determining which species qualify as having intrinsic value? This concern follows long-standing efforts within religions and elsewhere to protect the rights of animals (Regenstein 1991). *Sentience* (awareness) and *conation* (the capacity to strive for certain ends) have been suggested as criteria on which legal rights can be extended from humans to other organisms (Goodpaster 1978; Nash, this chapter). Debate continues as to whether those criteria apply to the moral status of whole ecosystems, and the questions themselves imply that a high degree of ethical development in modern society will be required to reduce biodiversity losses.

Deep ecologists (e.g. Fox 1990) play down the distinction between intrinsic and instrumental value as they seek forms of self-identification with nature that collapse the subject/object dichotomy. From a spiritual standpoint, this implies some form of pre-existing cosmological unity which the individual can experience first-hand. For that reason, deep ecology has been compatible with Buddhist-inspired explorations in 'practical ecology' (Halifax and Peale, this chapter), ecopsychology (Abram, this chapter) and artistic reflections on the interplay between extinctions and the creative process in humans (McLean, Box 11.1).

Box 11.3: Ten Theses for Biological Diversity and Ethical Development

Denis Goulet

1. Ethical, or authentic, development requires biological diversity.
2. Ethical development also requires cultural diversity.
3. Ethical development requires plural modes of rationality for two reasons:

   to destroy the monopoly of legitimacy appropriated by scientific and technological rationality, and

   to integrate technical, political, and ethical rationalities in decision-making in a circular pattern of mutual interaction.
4. Ethical development requires plural models of development. There is no single and necessary path to development predicated on energy-intensive, environmentally wasteful, culturally destructive and psychologically alienating models of progress.
5. Ethical development requires a non-reductionist approach to economics. As Schumacher (1973) insists in *Small Is Beautiful*, 'We must conduct economics as if people mattered'.
6. Ethical development requires pluralistic and non-reductionist approaches in technology. Technology is not an absolute value for its own sake which has a mandate to run roughshod over all consideration. As Ellul (1980) urges, we must demythologize technology.
7. Ethical development requires an approach to human beings that is not exclusively instrumental. Human beings are useful to other human beings and, to some degree, are used as aids in satisfying needs. But human beings have their ultimate worth independently of their instrumental value. Indeed, if one universal value exists in human life, it is that humans are precious for their own sake and on their own terms, independently of their utility to others.
8. The biosphere must be kept diverse both as an instrumental value to render ethical development possible and as a value *per se*. Like cultural diversity, biological diversity is a value for its own sake, although it is neither a transcendental nor an absolute value. It is, nevertheless, an end value: it has value not merely as a means or as an instrumentality serving human purposes.
9. The question, 'Is it possible to have piety toward nature without accountability to nature's creator and to a supreme judge of human affairs?' cannot be answered definitively and absolutely. One recalls, however, that all great religions have preached stewardship of the cosmos and responsibility for nature's integrity and survival based on ultimate human accountability to nature's creator or providential conductor.
10. If ethical development is the only adequate support system for biological diversity, reciprocally, biological diversity is the only support system for ethical development.

Source: Goulet, D. 1993. Biological Diversity and Ethical Development. In: Hamilton, L.S. (ed.), *Ethics, Religion, and Biodiversity: Relations between conservation and cultural values*. The White Horse Press, Cambridge, pp. 37–38.

The earth and sustainable communities

In the last fifty years, a systems view of life has received increasing attention as a subject of scientific study. The distinguishing feature of the systems approach is its integrative view of the process of life and of the organization of living systems based on complex networks. Gaia theory, which asserts that the earth is a complex living system, is perhaps the best-known and most controversial example of this line of ecological inquiry (Lovelock 1979). Having rich and diverse intellectual roots (from general systems theory, gestalt psychology, communication theory, cognitive science and the study of complexity, to quantum physics, chemistry and biology), the systems view of life lies at the scientific forefront of the ecological challenge to the mechanistic worldview. The theoretical foundation of this field of study was established largely by Maturana and Varela's (1980, 1987) seminal work on *autopoiesis* as the organization of living systems, i.e. living systems continuously renew their own structure and organizational activity. The richly contextualized networks of living systems make it difficult to argue that any particular part of the whole has more intrinsic value than another. On the other hand, Fox (1990) observes that autopoiesis qualifies as a classical criterion of intrinsic value because living systems cannot be considered as a means to an end; rather, living systems operate as ends in themselves.

Recognizing that ecological systems are 'sustainable communities of plants, animals and microorganisms', Capra (this chapter) infers several characteristics of living systems which might apply as principles of sustainability for human communities. Diversity, which is one of them, indicates resilience and the capacity to adapt to changing conditions. Sustainable communities are those that maintain and renew complex networks of interdependent relationships through which information and resources flow to nurture the whole. The economic, social and political implications of this principle are profound, especially in connection with the 'feminization of poverty' (i.e. the globalized pattern of women's work becoming increasingly unpaid), the growing disparity between the rich and poor, and the power of the media to create uniformity around the values of consumerism – all at the expense of cultural diversity.

The idea that sustainable communities are critical to biodiversity conservation has long historical roots which can be traced to Aldo Leopold's (1949) 'land ethic' and the complementary notion of ecological 'citizenship', i.e. 'all individuals are members of a community of interdependent parts'. Leopold's land ethic represents nothing new to most indigenous peoples, but it has had a clear impact on 'place-based ethics' (Norton, this chapter), the ethos of grass-roots movements and sustainable communities, and discussions about the content and purposes of a global ethic (Callicott, Box 11.4). Although the land ethic is anthropocentric, it can also be seen as ecocentric because the exercise of citizenship occurs within the encompassing biotic community. Much of the sustainable community movement has directed its attention to the political and economic destruction of small farmers as a result of the 'green revolution' (Berry 1977; Jackson *et. al.* 1984), globalized agribusiness, the ideology of inevitability that perpetuates the idea that small, relatively independent communities will disappear. Increasingly, the sustainable community movement attempts to maintain the ecological viability of small towns and farms whose agricultural existence was first threatened during the sixteenth century when the enclosure movement destroyed the European 'commons' (Rifkin 1991).

The land ethic and sustainable community movement strongly caution against the belief that global strategies will be an effective means for solving environmental problems. Wendell Berry's (1989) words on this reflect the common-sense wisdom of the land ethic: humankind is simply 'not smart enough or conscious enough or alert enough to work responsibly on a gigantic scale'. Following Berry, Orr (1992) amplifies the ethical problems of inappropriate scale by noting three of its primary consequences: (1) preoccupation with quantity lessens quality; (2) with increasing scale, the separation between costs and benefits increases, making it easier to forsake ethical responsibilities; (3) increasing scale encourages the accumulation of political power based on domination. In keeping with a more humble sense of proportion, the ethic of *right scale* and the complementary ethic of *sufficiency* occupy a prominent position in the rationale for sustainable communities by establishing them in local ecological knowledge.

The sustainable community movement rests on a single principle of environmental ethics: the way people relate to the earth is reflected in how they relate to each other. This is particularly evident in connection with agricultural systems and farming practices. The technological and economic pattern of replacing polycultural systems with the cultural and economic homogeneity of monoculture not only destroys biodiversity but also uproots local communities and destroys the livelihoods and cultures of millions of people (Shiva, this chapter). Sustainable communities challenge the downsizing and uprooting process of economic globalization. The expansion of corporate control over agricultural systems, most recently through the development of genetically engineered foods and the patenting of seeds and other life-forms, has extended the enclosure movement on an unprecedented scale (Shiva 1995a, 1995b, 1997). This social, economic and political assault on ecosystems further diminishes the prospects for participatory democracy and local sovereignty by extending corporate control into the molecular basis of life itself.

Has a global ethic emerged?

The impetus to formulate codes of global ethics in this century originated in the experience and destruction of the two world wars, the Holocaust, and the nuclear destruction of Hiroshima and Nagasaki. This previously unimaginable scale of death brought with it the realization that the persistent threat of world domination must be countered by a new global institution for managing human affairs peacefully. Following the League of Nations, the United Nations was established to accomplish this goal, and in 1948 the General Assembly adopted the 'Universal Declaration of Human Rights'. Evidence of further destruction was realized shortly thereafter when the environmental crisis captured the attention of visionary scientists like Rene Dubos, whose work advanced the field of ecology as an academic discipline and introduced its principles to international policy-makers. Subsequently, the UN has turned increasing amounts of time and energy to articulating practical measures for meeting the global environmental crisis and to forming an international consensus around a global environmental ethic. Much of this effort came to fruition at the 1992 Earth Summit through the passage of Agenda 21, the Rio Declaration, and the Convention on Biological Diversity (CBD).

The formulation of a practical and morally just global environmental ethic relates to issues of cultural relativism and moral pluralism, and it implies that these can be reconciled with some form of cross-cultural unity. Among environmental ethicists, Callicott (Box 11.4) has argued that a global ethic might be justified, if it is based on an array of diverse cultural interpretations rooted in local ecosystems. The secular, scientific framework of such an ethic was evident at the 1993 World Parliament of Religions where the *Declaration Toward a Global Ethic* was passed. This event was energized by the World Wildlife Fund initiated Assisi Declarations, the tireless commitment of organizers like Martin Palmer (1988; Box 11.6 in Jensen, this chapter) and theologians like Hans Kung (1991). Gatherings such as these demonstrate that a cosmological foundation for environmental ethics must be anchored in empirical knowledge with the understanding that a global ethic cannot be achieved at the expense of diversity.

Roszak (1972) posed essentially the same question in connection with the emerging field of ecology: Will it become 'the last of the old sciences or the first of the new?' He had in mind the emergence of a new intellectual paradigm to replace the dualistic one underlying our geographically expanding materialistic world-view. Twenty-five subsequent years of ecosystem destruction and biodiversity loss not only give his question greater urgency, but also suggest an expansion of its scope: Will the religious, economic and political institutions of today become the last of the old, or will they be first among the new? With questions like these in mind, Todd and Todd (1984) introduced the term 'sacred ecology' to call attention to the unknowable 'metapattern which connects' all life (Bateson 1979). Some might consider the meaning of 'sacred ecology' to be redundant or contradictory, perhaps an expression of romanticized scientism or heresy; nevertheless, it draws together the empirical study of natural systems as a sacred art with the human acts of creative imagination as embodiments of an ecological process.

On the level of international policy-making, the emergence of a new paradigm encapsulated by a global ethic centres on the terms *sustainability* and *sustainable development*, both based on the assumption that development must be understood in terms of present economic, social and political requirements in combination with an environmental ethic that preserves the ecological inheritance of future generations. Thus, sustainable development rests primarily within the ethical domain of economic justice, which has been the source of intense debates between (and sometimes among) nations of the global North and South (South Commission 1990) and within the religious community world-wide (Hallman 1994). This debate turns on the twin assumptions that sustainable development will further the South's technological and economic dependence on the North, while perpetuating a relatively high level of economic prosperity among Northern States and multinational corporations who will continue to exploit natural resources in the South for their own benefit.

Some environmental philosophers dismiss the ethic of sustainable development as little more than a popular cliche; others (e.g. Attfield 1991) acknowledge its faults, but still support its pragmatic outline, ethical framework, and the apparently high degree of political consensus surrounding it. Furthermore, these debates have formed the framework of an ongoing decision-making process and opened the

Box 11.4: Towards a Global Environmental Ethic

J. Baird Callicott

With the current and more ominous global dimension of the twentieth century's environmental crisis now at the forefront of attention, environmental philosophy must strive to facilitate the emergence of a global environmental consciousness that spans national and cultural boundaries. In part, this requires a more sophisticated cross-cultural comparison of traditional and contemporary concepts of the nature of nature, human nature, and the relationship between people and nature, than has so far characterized discussion. I am convinced that the intellectual foundations of the industrial epoch in world history are an aberration, and agree with Fritjof Capra that a new paradigm is emerging that will sooner or later replace the obsolete mechanical world-view and its associated values and technological esprit.

What I envision for the twenty-first century is the emergence of an international environmental ethic based on the theory of evolution, ecology and the new physics, and expressed in the cognitive lingua franca of contemporary science. Complementing such an international, scientifically grounded and expressed environmental ethic – global in scope as well as focus – I also envision the revival of a multiplicity of traditional cultural environmental ethics that resonate with it and that help to articulate it. Thus we may have one world-view and one associated environmental ethic corresponding to the contemporary reality that we inhabit one planet, that we are one species, and that our deepening environmental crisis is common and global. And we may also have a plurality of revived and renewed world-views and associated environmental ethics corresponding to the historical reality that we are many peoples inhabiting many diverse bioregions apprehended through many and diverse cultural lenses. But this one and these many are not at odds. Quite the contrary; they may be regarded as a single but general and abstract metaphysical and moral philosophy expressible in many conceptual modes. Each of the many world-views and associated environmental ethics may crystallize the international ecological environmental ethic in the vernacular of a particular and local cultural tradition. The environmental ethic based on contemporary international science and those implicit in the many indigenous and traditional cultures can thus be fused to form a unified but multifaceted global environmental ethic. Let us by all means think globally and act locally. But let us also think locally as well as globally and try to tune our global and local thinking as the several notes of a single, yet common, chord.

[Excerpted from 'Towards a Global Environmental Ethic', In: Tucker, M. and Grim, J. (eds), *Worldviews and Ecology: Religion, philosophy, and the environment*. Orbis, Maryknoll, 1994.]

possibility of an alternative world-view (Norgaard 1988). Such a process has been guided by the Earth Charter, Agenda 21, and the Convention on Biological Diversity (Norton 1997b). Some points of reference for this process are:

- Sustainable development may be feasible, in a limited way, for economically advantaged nations with the technological and/or financial resources to accomplish it; but it perpetuates the South's economic dependence on dominant Northern economic institutions.
- Sustainable development perpetuates unjust political, social and economic structures and cannot be justified on the basis of spiritual traditions.
- 'Sustainable development' is a contradictory, anthropocentric concept based on the mistaken assumption that the earth's value is primarily instrumental and utilitarian.
- The current practice of sustainable development is based on the assumption that the conservation of nature is dependent on the generation of economic capital, i.e. 'the marketplace will save the earth'.
- 'Future generations' is too remote a time frame and too imprecise as a moral basis ('intergenerational justice') for effective policy-making.

Representatives of the religious community have joined with environmental organizations in recognizing that biodiversity conservation, if it is to be successful, must be addressed within the context of sustainable development (WRI/IUCN/UNEP 1992). However, the assumption that economic growth must be the primary vehicle for achieving sustainable ways of living continues to be a source of debate (Lele 1991). A crucial point of contention is this: Who receives the benefits of economic growth and who pays its costs? Biodiversity conservation depends on appropriate financial incentives, but in practice money alone may not help economically disadvantaged nations (Perrings 1995), where biodiversity losses depend on the creation of sustainable societies based on the principles of ecojustice (IUCN/UNEP/WWF 1991). Dieter Hessel (1996; this chapter) has outlined four eco-justice principles constituting a global ethic: (1) respect for and solidarity with people and other creatures; (2) ecological sustainability as it applies to habitual ways of living; (3) sufficiency, equitable and just patterns of consumption; (4) genuinely participatory decision-making as to resource management and economic development policies and projects.

Disappointing results of the 1997 Rio+5 Conference in New York may indicate that further consensus may be difficult to reach. Non-governmental organizations argue that the integrated goals of Agenda 21 have been undermined in large measure by the following factors: (1) accelerated implementation of a neo-liberal version of economic globalization; (2) enormous environmental resources being regarded as 'externalities' (O'Neill and Holland, this chapter); (3) continuing unsustainable patterns of production and consumption, primarily in the North; (4) economic development projects and global trade policies (GATT, NAFTA, and more recently the Multilateral Agreement on Investment) that ignore local values and deny community participation and rights; and (5) the North's refusal to provide financial assistance and new technologies to the South.

Efforts to implement strategies for sustainable development through agreements made by nations are currently ineffective. The more likely catalysts for change are non-governmental organizations and local communities working with governments and corporations (Viederman, Meffe and Carrol 1994). The Canadian Council of Churches has recently suggested that 'sustainable communities' rather than 'sustainable development' would be a more appropriate organizing principle for a global ethic (Commission on Justice and Peace 1997).

Conclusions: biodiversity and the sacred

Oren Lyons (this chapter), Faithkeeper of the Onondaga Nation, explains that 'biodiversity' really means 'all our relations', and in doing so he addresses two enlightening questions: Who are 'all our relations'? And why should we care about them? These questions have a deeply spiritual and cosmological significance that illumines the meaning of ongoing extinctions as a consequence of human action. These questions are directed to every human being, and they work as a test of wisdom, revealing the depth of our character by the nature of our response.

For the present, most individual and collective responses to the biodiversity crisis are mixed – and this is better than no response. Informed responses are often shaped by the sheer brutality of genocide and ecocide, both being manifestations of religious and political persecution. These responses are usually combined with an ironic blend of hope in and suspicion of the promise of modern science. It is possible, of course, that a mutually supportive relationship among ecofeminism, science and religion may gradually overturn the dualism of matter and spirit – and the domination ethic of people against nature – in favour of a richly diverse living cosmos. And yet this possible outcome is matched by the danger of creating theocratic political regimes. Despite this, dialogue and collective action among indigenous peoples, scientists, economists and theologians are hopeful. Motivated as much by awe before the web of life as by commitments to particular theories or beliefs, people increasingly find common ground in their shared concern and love for the earth.

Can modern institutions gain the wisdom to reverse ongoing extinctions and environmental decline? Perhaps so, and yet it would be dishonest to offer anything but a realistic view. Beginning on a smaller scale, the sustainable community movement bears a certain resemblance to some historic monastic orders in their creativity, commitment to daily spiritual practices, and ecological rootedness, but they alone cannot bear the responsibility of transforming the forces behind modern industrial culture. Similarly, cost-effective strategies for the preservation of species and ecosystems may impact on the global marketplace, but the sheer magnitude of biodiversity loss questions the power of consumerism and modern institutional governance. Perhaps restructuring the anthropocentric, utilitarian institutions of religion, economics and government may reverse biodiversity loss. However, economic globalization proceeds at a rapid pace, claiming more and more of the web of life as ecosystems are degraded into exploitable, consumable 'resources'. To further this process, the telecommunications industry places increasing claims on the consciousness of people. A mutually reinforcing arrangement between the values of the marketplace and the power of the global media – which may qualify as 'religion' according to Geertz's definition – may provide many short-term human benefits, but it is economically unjust and it perpetuates an unsustainable way of living.

In that context, the goal of the world religions, especially those in the West, must be to examine and redefine their genuine purpose, taking into account habitual institutional objectives and then comparing them, for example, with the spirit and content of the Assisi Declarations. This is a radical goal. It represents the courage to rediscover religion's foundation in the web of life, rather than a willingness to prolong the captivity of religion by narrow political and economic interests. It is especially radical at a time when so many institutions eagerly exploit in the name of 'progress', while gaining the support of consumers through 'earth-friendly' public relations strategies. If the world religions choose to follow this familiar route, then they will both forsake the earth and commit a horrible betrayal of the sacred.

Box 11.5: 'All That Is – Is a Web of Being'

William N. Ellis and Margaret M. Ellis

We **belong** to the Webs-of-being – to Earth – to Gaia.

Belonging is the protovalue from which all other values derive.

We **belong** to the physiosphere, to the biosphere, to the ideosphere.

We **belong** to Gaia.

As the aborigines said it, 'We are the ownees of the land, not the owners of the land'.

As Chief Seattle said it, 'We can not own the land, we are part of the land'.

We **belong** to and are inseparable from our culture – from one another – from Gaia.

We are interdependent with all that is.

Belonging is scientific fact; and belonging is more than scientific fact.

Belonging is not merely 'being a member of', but is being subject to – being in partnership with – being responsible for.

We **belong** to – are responsible for – the webs-of-being – the universe – Earth – Gaia.

Belonging-to-Gaia means recognizing that we are enmeshed in the webs-of-being – that our well-being is dependent on the well-being of Gaia – the well-being of one another.

If we destroy Gaia, we destroy ourselves.

Belonging implies 'cooperation' – working with what is – with Gaia – the webs-of-being.

Belonging implies 'community'. In our face-to-face relationships with people we form community – we belong to community.

Belonging implies 'responsibility'. We are responsible for Gaia. We are responsible for one another.

Belonging implies 'Love'.

We can not separate love (agape) from the fact that we **belong** to Gaia.

We love because we must love to preserve Gaia – to preserve ourselves – to preserve the webs-of-being.

Cultures built on values other than **belonging** are doomed to self-destruct.

A culture built on 'domination of the earth, and all the animals therein' is doomed to disappear.

A culture based on 'self-interest' is doomed to disintegrate.

A culture based on 'survival of the fittest' will not survive.

A culture built on 'competition' will destroy itself.

To be stable and sustainable a culture must be based on cooperation, community, responsibility, love, honesty, caregiving, and the other values which are implied by and intertwined with one another and with **belonging**.

We can no more separate ourselves from **belonging** – from Gaia – and remain a viable culture than an oxygen atom can separate itself from hydrogen atoms and retain the qualities of water.

Because the genuine purpose of religion has never been to divide, its urgent purpose today must be to make peace in the present war against the web of life. To reclaim this mediating purpose may be equivalent to the recovery of an ancient cosmological vision rather than the discovery of a new one. Either way, the web is unravelling and the call is one of loyalty to the vibrating presence that is life itself.

All my relations: perspectives from indigenous peoples (Oren Lyons)

The Lakota end all of their prayers with, 'all my relations'. This means more than their families or extended families. It includes *all* life upon this earth. It is the recognition, respect and love for the interconnected 'web of life' that Chief Seattle spoke of. It is instruction to the human community of our relationship to the earth. We call the earth 'mother' to emphasize this relationship. It is the recognition that this mysterious power of the life-force springs from the seed. This is the great regenerative law of life upon this earth.

That is why water is the first law of life; all life needs water to nurture and grow. The laws of the natural world dictate these immutable realities. This is what indigenous peoples understand. By definition, the term 'indigenous' means 'belonging (to)'; as a native; inherent, innate; and for us it means from the land or from the sea. People who live in one place for a long time know who they are. We understand the long-term rhythms of the earth. We have intimate contact with the life that surrounds us because we are dependent upon that life. We personify the natural forces of nature to remind us how we relate and to teach our children respect for these forces. This underscores our connection and our responsibilities as part of these forces to maintain balance and harmony. With indigenous peoples respect *is* a law: without it there is little chance for harmony or community.

With respect comes harmony, justice, law and community: hence, 'all our relations'. Respect for the laws of regeneration ensures endless cycles of life. This is a gift of creation, one that demands absolute fealty. The laws of nature are absolute. There are no lawyers, evidentiary hearings, juries, judges or *habeas corpus*. There is only the law. Violation of these laws demands retribution and we suffer in exact ratio to our transgressions. There is no mercy in the natural law. Human cries go unheeded as do all the cries in the natural world. It is the law of life and death. We are dependent upon the earth our Mother and the nourishment she gives to all life. The creator laid down the laws of regeneration that insures endless cycles of life. The earth does not need us: we need her. She functions under these universal laws that are immense beyond our comprehension. We, the *Haudenosaunee*, were instructed to bind ourselves to these universal laws to ensure the survival of our people. We know that other indigenous nations and peoples understand these laws and instructions.

This knowledge and understanding gives our leaders and peoples great vision. This then is what we can offer humanity – the basis for making long-term decisions. We are responsible to life in the future. We are responsible to coming generations of life that includes our children. This was made specific to us by the great peacemaker who raised our nations into a great democratic confederacy he called the *Haudenosaunee* (the Six Nations Iroquois Confederacy). One of his most important instructions to the leaders he raised was this:

> 'When you sit and council for the welfare of the people, think not of yourself, your family, nor even your generation. But think of those coming whose faces are looking up from the earth and make your decisions on their behalf, even unto the seventh generation coming. This will insure peace for yourself and insure that the seventh generation will enjoy the same things you enjoy today.'

This is good advice for any leader because it ensures long-term decision-making. There is evidence that the peacemaker's words continue today not only in our councils but even to world leaders. At the opening session of the 1992 Earth Summit in Rio, Secretary-General Maurice Strong made direct references to this instruction for the seventh generation and credited the *Haudenosaunee* for its origins.

In 1994, President Clinton invited American Indian leaders to the White House for a discussion on relationships between Indian nations and the United States. He closed his address with the seventh generation instructions from the *Haudenosaunee*. So the instructions given to us by the Great Peacemaker a thousand or more years ago lives on today. Is this not relevant? Is this not proof that indigenous peoples contribute to contemporary life? We know it is, and that we continue to contribute to today's societies in the area of new ideas, democracy and the environment.

It was evident from our first encounters with Christopher Columbus and the white man that our basic philosophies on life were different. This caused conflict which continues today. Why? Because in these past five hundred years, the ethnocide and genocide practised against our peoples were and still are not able to change our basic beliefs or destroy our cultures. We continue to exist and challenge the systems of religions and philosophies of governance of our white brothers from Europe.

Biodiversity is another term for life; it is an all-encompassing term that reflects the technological societies we live in today. It is a scientific term that fills another category in a technological world. We say, 'all our relations'. Both terms talk about the same things. But our term reflects association and love. And that is the basic difference between indigenous peoples and Western societies.

Chief Matchewan of the Algonquins of Barriere Lake said it quite simply: 'Maybe we don't know what government scientists know, but they don't know what we know'. He was referring to the mismanagement of their lands (including La Verendrye Wildlife Reserve) by Quebec officials of Canada.

A contrast to the close relationships of native peoples to land and life, the Christian philosophy of the white man deserves to be quoted in full here. It clearly emphasizes the ego-driven hegemony of Christian doctrine and because of who says it, cutting trees down becomes an ethic of American society. John Adams, one of the founding fathers of the United States, said this as he outlined his family saga for Benjamin Rush (July 19, 1812):

> 'The first Adams in America was named Henry, a congregational dissenter from the Church of England. Persecuted by the intolerant spirit of Archbishop Laud. He came with eight children. This Henry and his son Joseph, my Great-grandfather, and his grandson Joseph, my Grandfather, whom I knew although he died in 1739, and John my father who died in 1761, all buried in the congregational churchyard in Quincy, half a mile from my house. Each of the five generations in this country has beared numerous children, multiplied like the sands of the sea shore or the stars in the Milky Way, to such a degree that I know not who there is in America to whom I am not related. My family I believe has cut down more trees in America than any other name.' (Drinnon 1980)

Another Adams, Henry, author of *The Great History of the United States* (1889–91) and great-grandson of John Adams, wrote of white America in 1880: 'From Lake Erie to Florida, in long unbroken line, pioneers were at work, cutting into forests with the energy of so many beavers, with no more express moral purpose than the beavers they drive away' (Drinnon 1980).

I think that Henry Adams does a disservice to beavers by equating them with moral or immoral human beings. But certainly there is a difference between quiet beaver dams and the industry of timber companies. Does this attitude continue today? Consider a recent report by *USA Today*.

> 'Spotted Owls: The Fish and Wildlife Service today is expected to propose less strict logging restrictions in an effort to save timber jobs in the north-west, home to the rare spotted owl. The revised plan would restrict logging on 6.9 million acres instead of 11.6 million acres proposed in April. And a Federal hearing in Portland Oregon today involved a US Cabinet level panel dubbed 'The God Squad' to allow logging on 4,570 acres of federal forest along the Oregon coast.'

How does this relate to biodiversity? Indigenous peoples understand the forest as communities; communities within communities. For instance, certain plants grow around certain shrubs. Certain shrubs grow around certain trees. This small community may nurture specific insects that will only flourish within such a community. In turn, certain communities attract certain insects and these insects attract certain birds and other wildlife. Thus you have ecological balance that serves a circle of life. Clear-cutting destroys that community and in doing so affects human health by destroying a source of medicinal support; it destroys habitat for insects, birds and wildlife in a sudden calamity. This action challenges the laws of regeneration.

Planting one kind of fast-growing tree in its place with an eye toward future cutting does not replace the communities of different trees (the 'biodiversity') that once were indigenous to that place. The lesson is clear: economic motivation regarding what Western society calls 'natural resources' operates against natural law.

We now come full circle on the issue of perspectives: the indigenous perspective inherent in the phrase 'all my relations', and the Western perspective inherent in the scientific term 'biodiversity'. Both terms refer to the same reality of life on earth. But the two perspectives on how this life is viewed are as far apart today as it they were when Christopher Columbus made his landfall in the Western Hemisphere.

In my opinion, the indigenous perspective encompasses the long-term view of community and life, and respects the laws of regeneration. The scientific term is more interested in commodification as the Christian ethic sees all non-human life as under the 'dominion' of the human beings and more specifically, Christian human beings. Therefore, life becomes 'resources'.

GATT and NAFTA are international trade agreements that among other things specify natural resources as another component of discussion and access. The world market is a commodification of all resources, which includes technologies, human labour and natural resources. There's that word again. 'All my relations' gets lost in this clamour of globalization or the 'free' market. We, the indigenous peoples, have come to understand that nothing is 'free' in the Western world. Someone or something pays for resources. We see the current practices of some of the international corporations plundering these 'resources' at the expense of future generations and life on earth as we know it.

We have had to learn English, and the meanings of the language, often at the expense of our own language. We have had to learn definitions of democracy, capitalism, communism, socialism and Christian doctrine. One very important lesson we learned here in America was that capitalism is *not* democracy. Also, that our fundamental ethic of sharing was against this principle of 'private property' and that individual rights superseded this right of the whole, be it community or future generations. We are alarmed at the lack of balance in today's global markets. Gro Harlem Brundtland spoke to these issues in her report to the United Nations entitled *Our Common Future* (WCED 1987). 'Tribal and indigenous peoples will need special attention as the forces of economic development disrupt their traditional lifestyles ... lifestyles that can offer modern societies many lessons in the management of resources.'

There's that word again – resources. Until humankind can truly understand our relationship to the earth and all of its amazing, beautiful and diversified life we will destroy ourselves. This battle between sustainable life-styles and world commerce is in this and the next generations' hands. The economic forces of corporations are now larger than any of the world nations. The technology of communications 'cyberspace' is the next area of development and will further exacerbate the disparity between the rich and the poor. Knowledge is power and the instant communication of commerce and development will place more power in fewer hands.

The indigenous voice is being heard by some of the international community. The NGO community that has developed around the United Nations is gaining strength and credibility and there are many voices now being raised in unison to meet the environmental crisis we face today. Environmental newsletters often report on one the world's most influential concerns – the insurance industry. It is interesting to note that these voices speak loudly in opposition to the unfettered world market principles of capitalism embodied by this industry. They are paying more and more compensation for hurricane, flood and earthquake damage. That, we say, is a result of the overuse of water, global warming, the imbalance of the human population, and related pressures on the resources of Mother Earth.

The warnings are all about us if we will only listen. These warnings come in many forms. Recently, at a meeting of the Onondaga Council of Chiefs, called to address the antisocial behaviour of a young teenage boy of our community, we all spoke to him to enlighten and give positive direction to his life but it was the eldest member of our council, Louis Farmer, who spoke what I considered the most profound message. Chief Farmer is partially blind and walks very slowly. He stood directly in front of the boy and said:

> 'There are fewer and fewer birds singing in the morning. Each day I listen to their songs, I know their songs, I know them. Now I notice there are fewer and fewer songs. The birds are sad. They are sad because of the way we are acting. They are sad at how we treat one another and they are sad because of our conduct. They are going away and we suffer because of this. We will miss them and our life will be less because of this. When they are gone we will have only ourselves to blame. So I urge you [talking to the boy] to try hard to do better because at some time we will depend on you to take care of us.'

Dah Nayto (Now I am finished)

Joagquishonh (Oren Lyons)
Faithkeeper, Turtle Clan
Onondaga Council of Chiefs
Haudenosaunee

Homeless in the 'global village' (Vandana Shiva)

Global market integration and the creation of the 'level playing field' for transnational capital, creates conditions of homelessness in real and imaginary ways. The transnational corporation executive who finds a home in every Holiday Inn and Hilton, is homeless in terms of the deeper cultural sense of rootedness. But the culturally-rooted tribal is made physically homeless by being uprooted from the soil of her/his ancestors. Two classes of the homeless seem to be emerging in this 'global village'. One group is mobile on a world scale, with no country as its home but the whole world as its property; the other has lost even the mobility within rootedness, and lives in refugee camps, resettlement colonies and reserves. The cumulative displacement caused by colonialism, development and the global market-place has made homelessness a cultural characteristic of the late twentieth century.

Dams, mines, energy plants, military bases – these are the temples of the new religion called 'development', a religion that provides the rationale for the modernizing state, its bureaucracies and technocracies. What is sacrificed at the altar of this religion is nature's life and peoples' life. The sacraments of development are made of the ruins and desecration of other sacreds, especially sacred soils. They are based on the dismantling of society and community, on the uprooting of people and cultures. Since soil is the sacred mother, the womb of life in nature and society, its inviolability has been the organizing principle for societies that 'development' has declared backward and primitive. But these people are our contemporaries. They differ from us not in belonging to a bygone age but in having a different concept of what is sacred and of what must be preserved. The sacred is the bond that connects the part to the whole. The sanctity of the soil must be sustained, limits must be set on human action. From the point of view of the managers of development, the high priests of the new religion, sacred bonds with the soil are impediments and hindrances to be shifted and sacrificed. Because people who hold the soil as sacred will not voluntarily allow themselves to be uprooted, 'development' requires a police state and terrorist tactics to wrench them away from their homes and homelands, and consign them as ecological and cultural refugees into the wasteland of industrial society. Bullets, as well as bulldozers, are often necessary to execute the development project.

In India, the magnitude of this sacrifice is only now becoming evident. Victims of progress have, of course, experienced their own uprooting and have resisted it. But both the victims and the State have perceived each sacrifice as a small one for the larger 'national interest'. Over 40 years of planned development, the planned destruction of nature and society no longer appears negligible; and the larger 'national interest' turns out to be embodied in an elite minority without roots. Fifteen million people have been uprooted from their homelands in India during the past four development decades (Fernandes and Enakshi 1989). They, and their links with the soil, have been sacrificed to accommodate mines, dams, factories and wildlife parks.

One word echoes and reverberates in the songs and slogans of Indian people struggling against 'development': *mati* – soil. For these people soil is not simply a resource, it provides the very essence of their being. For large segments of Indian society the soil is still a sacred mother. 'Development' has meant the ecological and cultural rupture of bonds with nature, and within society; it has meant the transformation of organic communities into groups of uprooted and alienated individuals searching for abstract identities. What today are called ecology movements in the South are actually movements for rootedness, movements to resist uprooting before it begins. And what are generally perceived as ethnic struggles are also, in their own way, movements of uprooted people seeking social and cultural rootedness. These are the struggles of people, taking place in the ruins wrought by development, to regain a sense of selfhood and control over their destinies.

Wherever development projects are introduced, they tear apart the soil and sever the bonds between people and the soil. *'Mati Devata, Dharam Devata'* – The soil is our Goddess; it is our religion'. These are the words of *adivasi* women of the 'Save Gandmardhan' movement (Bahuguna 1986), as they embraced the earth while being dragged away by the police from the blockade sites in the Gandmardhan hills in Orissa. Dhanmati, a 70-year-old woman of the movement had said, 'We will sacrifice our lives, but not Gandmardhan. We want to save this hill which gives us all we need'.

The forests of Gandmardhan are a source of rich plant diversity and water resources. They feed 22 perennial streams which in turn feed major rivers such as the Mahanadi. According to Indian mythology, Gandmardhan is the sacred hill where Hanuman gathered medicinal herbs to save Laxman's life in the epic *Ramayana*. The saviour has now to be destroyed for 'development'. It has to be desecrated by the Bharat Aluminium Company (BALCO) to mine for bauxite. BALCO had come to Gandmardhan after destroying the sanctity and ecology of another important mountain, Amarkantak, the source of the rivers Narmada, Sone and Mahanadi. The destruction of Amarkantak was a high cost to pay for reserves which, in any case, turned out to be much smaller than originally estimated. To feed its 100,000 tonne aluminium plant at Korba in Madhya Pradesh, BALCO has now moved to Orissa to begin the rape of the Gandmardhan hills. Since 1985 the tribals of the region have obstructed the work of the company and refused to be tempted by its offers of employment. Even police help has failed to stop the determined protest.

For communities who derive their sustenance from the soil it is not merely a physical property situated in Cartesian space; for them, the soil is the source of all meaning. As an Australian Aborigine said, 'My land is my backbone. My land is my foundation'. Soil and society, the earth and its people, are intimately interconnected. In tribal and peasant societies, cultural and religious identity derive from the soil, which is perceived not as a mere 'factor of production' but as the very soul of society. Soil has embodied the ecological and spiritual home for most cultures. It is the womb not only for the reproduction of biological life but also of cultural and spiritual life; it epitomizes all the sources of sustenance and is 'home' in the deepest sense.

The Hill Maris tribe in Bastar see *bhum* , or soil, as their home. *Shringar Bhum* is the universe of plants, animals, trees and human beings. It is the cultural spiritual space which constitutes memory, myths, stories and songs that make up the daily life of the community. *Jagha Bhum* is the name for the concrete location of social activities in a village. Savyasaachi reports a village elder as saying: 'The sun, the moon, the air, the trees, are signs of my continuity. Social life will continue as long as these continue to live. I was born as a part of the *bhum*. I will die when this *bhum* dies ... I was born with all others in this *bhum*; I go with them. He who has created us all will give us food. If there is so much variety and abundance in *bhum*, there is no reason for me to worry about food and continuity' (Savyasaachi, forthcoming).

The soil is thus the condition for the regeneration of nature's and society's life. The renewal of society therefore involves preserving the soil's integrity; it involves treating the soil as sacred.

Desacralization of the soil takes place through changes in the meaning of space. Sacred space, the universe of all meaning and living, the ecological source of all sustenance, is transformed into a mere site, a location in Cartesian space. When that site is identified for a development project, it is destroyed as a spiritual and ecological home. There is a story that elders tell to their children in central India to illustrate that the life of the tribe is deeply and intimately linked to the life of the soil and the forest.

> The forest was ablaze. Pushed by the wind, the flames began to close in on a beautiful tree on which sat a bird. An old man, escaping the fire himself, saw the bird and said to it, 'Little bird, why don't you fly away? Have you forgotten you have wings?' And the bird answered, 'Old man, do you see this empty nest above? There is where I was born. And this small nest from which you hear the chirping is where I am bringing up my small child. I feed him with nectar from the flowers of this tree and I live by eating its ripe fruit. And do you see the dropping below on the forest floor? Many seedlings will emerge from them and thus do I help to spread greenery, as my parents before me did, as my children after me will. My life is linked to this tree. If it dies I will surely die with it. No, I have not forgotten my wings.' (Rane 1987)

The fact that people did not move from their ancestral homelands, that they continued to reproduce life in nature and society in sustainable ways, was not seen as the conservation of the earth and of the soil ethic. Instead, it was seen as evidence of stagnation, of an inability to move on – to 'progress'. The stimulation to move on and progress was provided by the development project, and the uprooting and destruction it involved was sanitized under the neo-Cartesian category of 'displacement'.

Peter Berger (1981) has described development as the 'spreading condition of homelessness'. The creation of homelessness takes place both through the ecological destruction of the 'home' and the cultural and spiritual uprooting of peoples from their homes. The word 'ecology' was derived from *oikos*, the household, and ecological destruction in its essence is the destruction of the *bhum* as the spiritual and ecological household. By allocating a Cartesian category to space in substitution for the sacred category it becomes possible for development technocrats and agencies to expand their activities into the management of 'Involuntary Resettlement in Development Projects'. An irreversible process of genocide and ecocide is neutralized by the terms 'displacement' and 'resettlement'. It becomes possible for agencies such as the World Bank to speak of reconciling the 'positive' long-term 'national' interests served by development projects and the 'negative' impacts of displacement borne by 'local' communities through resettlement and rehabilitation projects.

Colonialism and capitalism transformed land and soil from being a source of life and a commons from which people draw sustenance, into private property to be bought and sold and conquered. Development continued colonialism's unfinished task. It transformed man from the role of guest to that of predator. In a sacred space, one can only be a guest, one cannot own it. This attitude to the soil and earth as a sacral home, not private property, is characteristic of most Third World societies.

In the indigenous world-views in Africa, the world in its entirety appears as consisting of a single issue. Man cannot exercise domination over it by virtue of his spirit. What is more, this world is sacralized, and man must be prudent in the use he makes of it. Man must act in this world as a guest and not as an exploiting proprietor.

When the rhythms and patterns of the universe are displaced, the commons are displaced by private property. In most indigenous communities, the entire tribe is the trustee of the land it occupies, and the community or tribe includes not only the currently living members but also the ancestors and future generations. The absence of private property rights and of a territorial concept of space make for easy dispossession of indigenous communities' land.

In defining a sacred space, soil does not define cartographic space on a map, or a territorial unit. As Benedict Anderson (1989) has shown, the creation of territorial space in large areas of the world was an instrument of colonization. Tracing the shift from cultural space to territorial space in Thailand, he shows how, between 1900 and 1915, the traditional words *brung* and *muang* largely disappeared because 'sovereignty' was imaged in terms of sacred sites and discrete population centres. In their place came *parthet*, 'country' which imaged it in the invisible terms of bounded territorial space. Sovereignty thus shifted from the soil and soil-linked communities to sovereignty of the nation state. Laws of nature and their universality were replaced by the laws of a police state which dispossessed peoples of their original homelands to clear the way for the logic of the world market.

In this way organic communities give way to slum dwellers or urban and industrial jungles. Development builds new 'temples' by robbing nature and society of their integrity, and their soul. Development has converted soil from sacred mother into disposable object – to be ravaged for minerals that lie below, or drowned beneath gigantic reservoirs. The soil's children, too, have been made disposable: mines and dams leave behind wastelands and uprooted people. The desacralization of the soil as sacred space was an essential part of colonialism then and of development now. As Rifkin (Rifkin and Peelas 1983) has so aptly stated, 'Desacralization serves as a kind of psychic ritual by which human beings deaden their prey, preparing it for consumption'.

The irony involved in the desacralization of space and the uprooting of local communities is that the secular categories of space, as used in development, transform the original inhabitants into strangers while intruders take over their homes as private property. A political redefinition of people and society is taking place with shifts in the meaning of space. New sources of power and control are being created in relationship to nature and to society. As relationships between nature and society and between different communities are changed and replaced by abstract and rigid boundaries between nature and people and between peoples, power and meaning shift from roots in the soil to links with the nation state and with global capital. These one-dimensional, homogenizing concepts of power create new dualities and new exclusions.

The new borders, evidently, are created for the people who belong to that land. There are no borders for those who come in to colonize and destroy the land. In the words of financial consultant Kenichi Ohmae:

> 'On a political map, the boundaries between countries are as clear as ever. But on a competitive map, a map showing the real flows of financial activity, these boundaries have largely disappeared... Borderless economy... offers enormous opportunities to those who can criss-cross the boundaries in search of better profits. We are finally living in a world where money, securities, services, options, futures, information and patents, software and hardware, companies and know-how, assets and memberships, paintings and brands, are all traded without national sentiments across traditional borders.' (Ohmae 1990)

Spiritual beliefs and cultural perceptions of some Kenyan communities (Mary Getui)

Kenya is a multi-religious society with the following religions represented: Christianity, Islam, African indigenous religion; pockets of Jews, Hindus and Bahai; and recently there has been an upsurge of New Age movements originating mainly from the East. Each one of these religions has its own spirituality which includes views on nature and its conservation.

Kibieho (in Ogutu 1992) has noted that African indigenous religion did not exist as a separate institution but was interlaced in the fabric of every institution and element of life. On the same topic, Bahemuka (1982) emphasized that, 'There was no division between what was religious and what was not religious. Whether a person prayed, ate, danced, sang, cultivated the land or walked, those were religious acts for they were performed by a religious being. It is the level of religious participation that motivated and bound people together'.

These two authors are echoing Mbiti's (1969) famous words that Africans are notoriously religious:

'Religion permeates into all the departments of life so fully that it is not easy or possible to isolate it'. One aspect of importance that may require underlining is that in the African context, harmony and the well-being of one and all are stressed through kinship systems which extend to cover animals, plants and non-living objects (Mbiti 1969). Hence, it can be argued that African views on nature and conservation have a clearly religious significance.

We now turn to an examination of religious and cultural views on certain parts of nature; namely, animals and plants. One community that attaches a lot of importance to animals, especially cattle, are the Maasai. Kipury (1983) notes that livestock provide the Maasai with almost everything, ranging from food requirements to domestic utensils and clothing. The social and economic role that livestock plays in Maasai life is of paramount significance. Livestock is procured and exchanged in kinship relations as well as for bridewealth. The exchange of gifts of livestock between friends and kinsmen is the basis of social relationships. Cattle are also used in the payment of civil transgressions as well as for capital crimes such as homicide. Mbiti (1969) adds that the Maasai firmly believe that since God gave them cattle at the very beginning, nobody else has the right to own cattle. As such, it is their duty to raid cattle from neighbouring peoples without feeling that they are committing theft or robbery. In other communities, domesticated animals, cattle, sheep and goats are also used for sacrificial and other religious purposes, but a Maasai will neither sell nor slaughter his animal without good religious cause.

With regard to wild animals, Kipury (1983) goes on to say that traditionally the Maasai were not hunters and abhorred the consumption of wild animals. The lion hunt was the only form of hunting permitted, mainly to eliminate predators when they posed a danger to livestock, and as a sporting activity. They also attribute a material significance to wild animals through the exchange of wild animal products, e.g. buffalo-hide shields, wildebeest and giraffe tails used by elders as fly-whisks, and ostrich feathers made into warrior head-dresses with their neighbours the Dorobo. While the Maasai appreciate these items, they despise the Dorobo hunters who they see as having an abnormal desire to kill domestic as well as wild animals, instead of leaving them to exist on the basis of their own authentic value.

Two religious features that contribute to the conservation of nature in Africa are taboos and totemic beliefs. Among the Kalenjin, several taboos are observed. The dietary laws are significant in ways that might be interpreted as an economic measure: it is taboo to eat meat and drink milk or blood on the same day. The explanation given for this law is that if one consumes both, the cow or goat which gave the milk will dry up. In reality, the theory surrounding this law arose from the recognition that the human body needs both milk and meat, but neither may be plentiful. The idea was to spread it out – use one today and the other tomorrow (Fish and Fish 1995). Indeed, this is a conservation measure.

Totemic beliefs, which abound in Africa, are based on the special relationship a clan may have with a particular animal or plant, which is not to be cut or killed – even for food. Abagusii of western Kenya have animal totems according to clans:

Mobasi	Zebra	*(Enchage)*
Monchari	Hippopotamus	*(Eugubo)*
Mogirango	Leopard	*(Engo)*
Mogetutu	Elephant	*(Enchogu)*

It is said that Mobasi, son of Mogusii (the forefather of Abagusii), owned many cattle which the neighbouring Maasai envied. The latter organized a raid and made off with all the farmer's cattle. With the help of others, Mobasi set off in pursuit of the animals. A fierce battle followed, and many lives were lost on both sides. The Mobasi stealthily escaped with some cattle, and this time it was the Maasai who gave chase. They caught up with Mobasi's son, also called Mobasi, who had been left behind by the rest of the party because he had been weakened by illness. The younger Mobasi realized the danger he was in and prayed to *Engoro* (God) for protection. As if in answer to his prayer, a multitude of zebra and wildebeest burst across the open plains. Mobasi used the length of his rope to loop himself to a fleeing zebra *(Enchage)*. There are several reasons why the zebra saved Mobasi – he talked to it quietly and reassuringly.

It is also said that if a wild animal cannot smell the blood of animals in a human being it will treat the person as a friend. Mobasi had never eaten meat in his life. The zebra took Mobasi to a safe territory where the two stayed together in confidence for several days. Mobasi then made his way home, riding his *enchage*. Proudly Mobasi introduced *enchage*, his friend, to his family. In grateful thanks for saving his life Mobasi swore to protect the zebra and his descendants forever more (O'Keragori 1995). The same concept of a good relationship between the people and animals prevails in the other three examples of totems.

In addition to cattle and other animals, vegetation also plays an extremely important role among the Maasai. For example, the size, shape, sturdiness and long life of the *Orereti* tree epitomize the idea of life; hence, the tree is sung about and invoked in prayers and blessings. From a very early age, Maasai children are expected to distinguish between various plants and grasses and their uses. Grass is used in to build houses, for grasping in the hand as a sign of peace, and for blessing during rituals. Trees and shrubs provide traditional medicines and herbs, some of which have a special ritual value. Every ailment has a traditional treatment, if not a cure, obtained from the leaves, the bark or the fruit of plants.

Thus, many African indigenous religious beliefs and the cultural perceptions of Kenyan communities are geared towards the conservation of nature. These are aspects of traditional cultures that should be identified and applied in the contemporary situation where nature is exploited, violated and abused. Yet, some of these religious beliefs and cultural perceptions may also be counterproductive to the conservation of the environment. For example, the Maasai attachment towards cattle, which can justify rustling and the keeping of large, low quality herds, can be damaging to the environment. These beliefs and practices require special attention from an ecological standpoint so that the practice may become economically and environmentally viable.

Ecofeminism: domination, healing and world-views (Rosemary Radford Ruether)

Ecofeminism asks how women and nature have been linked in the patriarchal domination that has characterized Western and other societies, and also how the liberation of women and the healing of nature might be interconnected. This is a very complex inquiry that demands a multi-cultural and cross-class/race perspective. The relation of the domination of women and the domination of nature needs to be looked at both culturally and socio-economically.

Anthropological studies have suggested that the identification of women with nature and males with culture is both ancient and wide-spread. This cultural pattern itself expresses a monopolizing of the definition of culture by males. The very word 'nature' in this formula is part of the problem, because it defines nature as a reality below and separated from 'man', rather than one nexus in which humanity itself is inseparably embedded. It is, in fact, human beings who cannot live apart from the rest of nature as our life-sustaining context, while the community of plants and animals both can, and for billions of years did, exist without humans. The concept of humans outside of nature is a cultural reversal of natural reality.

This reversal seems to have emerged gradually, as early gardening changed to plough agriculture, redefining land as male private property, and urbanization, class hierarchy and slavery over the period from 10,000 to 3,000 BC. By the time the patriarchal law codes of the Babylonians, Hebrews and Greeks were formulated between 3000 and 500 BC, a system of ownership had been codified in which women, slaves, animals and land were all seen as types of property and instruments of labour, owned and controlled by male heads of families as a ruling class.

As we look at the mythologies of the ancient Near Eastern, Hebrew, Greek and early Christian cultures, one can see a shifting symbolization of women and nature as spheres to be conquered, ruled over and, finally, repudiated altogether. In the Babylonian creation story, which goes back to the third millennium BC, Marduk, the warrior champion of the gods of the city states, is seen as creating the cosmos by conquering the Mother Goddess Tiamat, pictured as a monstrous female animal. Marduk kills her, treads her body underfoot and then splits her in half, using one half to fashion the starry firmament of the skies, and the other half the earth below. The elemental mother is literally turned into the matter out of which the cosmos is fashioned (the words 'mother' and matter have the same etymological root). She can be used as matter only by being killed; that is, by destroying her as 'wild', autonomous life, and making her life-giving body into 'stuff' possessed and controlled by the architect of a male-defined cosmos.

The view of nature found in Hebrew scripture has several cultural layers, but the overall tendency is to see the natural world, together with human society, as something created, shaped and controlled by God, a God imaged after the patriarchal ruling class. The patriarchal male is entrusted with being the steward and caretaker of nature, but under God, who remains its ultimate creator and Lord. This also means that nature remains partly an uncontrollable realm that can confront human society in destructive droughts and storms. These experiences of nature that transcend human control, bringing destruction to human work, are seen as divine judgement against human sin and unfaithfulness to God.

The image of God as single, male and transcendent, prior to nature, also shifts the symbolic relation of male consciousness to material life. Marduk was a young male god, who was produced out of a process of theogony and cosmogony. He conquers and shapes the cosmos out of the body of an older Goddess that existed prior to himself, within which he himself stands. The Hebrew God exists above and prior to the cosmos, shaping it out of a chaos that is under his control.

When we turn to Greek philosophical myth, the link between mother and matter is made explicit. Plato, in his creation myth, the *Timaeus*, speaks of primal, unformed matter as the receptacle and 'nurse'. He imagines a disembodied male mind as divine architect or *Demiurgos*, shaping this matter into the cosmos by fashioning it after the intellectual blueprint of the Eternal Ideas. These Eternal Ideas exist in an immaterial, transcendent world of Mind, separate from and above the material stuff that he is fashioning into the visible cosmos.

The World Soul is also created by the *Demiurgos*, by mixing together dynamics the Same and the Other. This world soul is infused into the body of the cosmos in order to make it move in harmonic motion. The remnants of this world soul are divided into bits to create the souls of humans. These souls are first placed in the stars, so that human souls will gain knowledge of the eternal ideas. Then the souls are sown in the bodies of humans on earth. The task of the soul is to govern the unruly passions that arise from the body.

If the soul succeeds in this task, it will return at death to its native star and there live a life of leisured contemplation. If not, the soul will be reincarnated into the body of a woman or an animal. It will then have to work its way back into the form of an (elite) male and finally escape from bodily reincarnation altogether, to return to its original disincarnate form in the starry realm above. Plato takes for granted an ontological hierarchy of being: the immaterial intellectual world over material cosmos, and, within this ontological hierarchy, the descending hierarchy of male, female and animal.

In the Greco-Roman era, a sense of pessimism about the possibility of blessing and well-being within the bodily, historical world deepened in Eastern Mediterranean culture, expressing itself in apocalypticism and gnosticism. In apocalypticism God is seen as intervening in history to destroy the present sinful and finite world of human society and nature and to create a new heaven and earth, freed from both sin and death. In gnosticism, mystical philosophies chart the path to salvation by way of withdrawal of the soul from the body and its passions and its return to an immaterial realm outside of and above the visible cosmos.

Early Christianity was shaped by both the Hebraic and Greek traditions, including their alienated forms in apocalypticism and gnosticism. Second-century Christianity struggled against gnosticism, reaffirming the Hebraic view of nature and body as God's good creation. The second-century Christian theologian Irenaeus sought to combat Gnostic anticosmism and to synthesize apocalypticism and Hebraic creationalism. He imaged the whole cosmos as a bodying forth of the Word and Spirit of God, the sacramental embodiment of the invisible God. Sin arises through a human denial of this relation to God. But salvific grace, dispensed progressively through the Hebrew and Christian revelations, allows humanity to heal its relation to God. The cosmos then grows into blessed and immortalized manifestation of the divine Word and Spirit which is its ground of being.

However, Greek and Latin Christianity, increasingly influenced by neo-Platonism, found this materialism distasteful. They deeply imbibed the Platonic eschatology of the escape of the soul from the body and its return to a transcendent world outside the earth. The earth and the body must be left behind in order to ascend to another, heavenly world of disembodied life. Even though the Hebrew idea of resurrection of the body was retained, increasingly this notion was envisioned as a vehicle of immortal light for the soul, not the material body in all its distasteful physical processes, which they saw as the very essence of sin as mortal corruptibility. Eternal life was for the disembodied male soul freed from the material underpinnings in the mortal bodily life, represented by woman and nature.

Classical Christianity was deeply ambivalent about its view of nature. One side of Patristic and Medieval thought retained something of Irenaeus' sacramental cosmos, which becomes the icon of God through feeding on the redemptive power of Christ in the sacraments of bread and wine. But the redeemed cosmos as resurrected body, united with God, is possible only by freeing the body of its sexuality and mortality. The dark side of Medieval thought saw nature as possessed by demonic powers that draw us down to sin and death through sexual temptation. Women, particularly old crones still perversely retaining their sexual appetites, are the vehicles of the demonic power of nature. They are the witches who sell their souls to the Devil in a satanic parody of the Christian sacraments.

The Calvinist Reformation and the Scientific Revolution in England in the late sixteenth and seventeenth centuries represent key turning points in the Western concept of nature. In these two movements the Medieval struggle between the sacramental and the demonic views of nature was recast. Calvinism dismembered the Medieval sacramental sense of nature. For Calvinism nature was totally depraved. There was no residue of divine presence in it that could sustain a natural knowledge or relation to God. Saving knowledge of God descends from on high, beyond nature, in the revealed Word available only in Scripture, as preached by the Reformers. Calvinism dismantled the sacramental world of

Medieval Christianity, but it maintained and re-enforced its demonic universe. The fallen world, especially physical nature and other human groups outside the control of the Calvinist church, lay in the grip of the Devil. In the Calvinist church, women were still the gateway of evil. If women were completely obedient to their fathers, husbands, ministers and magistrates, they might be redeemed as 'goodwives'. But in any independence of women lurks heresy and witchcraft.

The Scientific Revolution at first moved in a different direction, exorcising the demonic powers from nature in order to reclaim it as an icon of divine reason manifest in natural law. But, in the seventeenth and eighteenth centuries, the more animist natural science that unified material and spiritual lost out to a strict dualism of transcendent intellect and dead matter. Nature was secularized. It was no longer the scene of a struggle between Christ and the Devil. Both divine and demonic spirits were driven out of it. In Cartesian dualism and Newtonian physics it becomes matter in motion, dead stuff moving obediently according to mathematical laws knowable to a new male elite of scientists. With no life or soul of its own, nature could be safely expropriated by this male elite and infinitely reconstructed to augment their wealth and power.

This Western scientific industrial revolution has been built on injustice from its beginnings. It has generated enormous affluence and power for a global elite, but one based on exploitation of the land and labour of the many for the benefit of the few, with its high consumption of energy and waste. It cannot be expanded to include the poor without destroying the basis of life of the planet itself. We are literally destroying the air, water and soil upon which human life and planetary life depends. The critical question for ecofeminism is: can this global system be changed and is it useful to focus particularly on the woman–nature connection in our struggle against it?

Most male ecologists focus on new forms of 'technological fixes' that will prevent environmental disaster without really challenging the present system of wealth and power. Any focus on women is seen by such men as trivial at best. Some feminists also look at ecofeminism with suspicion because it seems to ratify patriarchal thought patterns in which women are identified with nature and hence with the non-rational and somatic side of the human–nature split. If women are to be liberated we must affirm women's full humanity. This means affirming women's equal capacity for rationality and agency, not by continuing to identify women with non-human nature.

Ecofeminism needs to evaluate this challenge. Patriarchal culture does not merely denigrate women and nature as inferior or evil, but it also idealizes both women and nature. We have in our cultures two images of woman as nature in complementary tension: woman as evil nature, associated with sin, sex and death, and woman as bountiful, ever-nurturing nature, nature as paradise. Ecofeminists, while repudiating the first stereotype of woman and nature, often uncritically take over the second stereotype. We need to question not only the negative images of women and nature as inferior, as the source of sex, sin and death, but also nature as paradise, as ever-bountiful mother.

In order to shift from a patriarchal world-view that justifies the exploitation of women, other subjugated humans and nature, to one that would support an ecofeminist perspective, I suggest eight interconnected transformations of our world-view:

1. A shift from a conception of God as holding all sovereign power outside of and ruling over nature; to a conception of God who is under and around all things, sustaining and renewing nature and humans together as one creational biotic community.

2. A shift from a mechanistic view of the physical world as composed of inert 'dead' matter pushed and pulled from outside; to a view of the world as an organic living whole, manifesting energy, spirit, agency and creativity.

3. A shift from an ethic that non-human entities on the earth, such as animals, plants, minerals, water, air and soil have only utilitarian use value for humans for industrial development, production, consumption and profit; to a view of all things as having intrinsic value to be respected and celebrated for their own being.

4. A shift from a psychology that splits mind from body, mind from physical nature, setting mind as the superior reality that is to rule over body, the bodies of dominated people and the bodies of non-human nature; to a holistic psychology that recognizes ourselves as psychospiritual–physical wholes in interrelation with the rest of nature as also psychospiritual–physical wholes who are to mutually interdepend in one community of life.

5. A shift from a view that patriarchal dominance is the order of 'nature', for creation and society, the necessary way to keep right order in all relations; to a recognition that patriarchal dominance is the root of distorted relations, and that we must shift to gender equality, equity and mutual interrelations between men and women in all aspects of life.

6. A shift from the concept of one superior culture (white Western Christian) to be imposed on all other peoples to 'save' and 'civilize' them; to a respect for the diversity of human cultures in dialogue and mutual learning, overcoming racist hierarchy and

defending particularly the bioregional indigenous cultures which are on the verge of extinction.

7. A shift from an economy of maximization of profits that treats nature as 'material resources' to be used, and as the depository of waste without accountability; to an economy of sustainable life that will renew nature from generation to generation.

8. A shift from a politics of survival of the fittest that allocates resources and power to the most powerful; to a political community based on participatory democracy, community-based decision-making and representation of the welfare of the whole bio-region in making decisions.

Two conceptions of biodiversity value (John O'Neill and Alan Holland)

Discussions of biodiversity illustrate with clarity a particular approach to environmental values: what might be called the itemizing approach. It is an approach that dominates much ecological and economic thinking about the environment. A list of goods is offered that correspond to different valued features of our environment, such that increasing value is a question of maximizing one's score on different items on the list, or at least of meeting some 'satisfying' score on each. We have something like a score card, with valued kinds of objects and properties, valued goods, a score for the significance of each, and we attempt to maintain and where possible increase the total score, the total amount of value. The approach involves a form of consequentialism: we assess which action is best in a given context solely by its consequences, by the total amount of value it produces.

On one level this approach to biodiversity might seem surprising. Diversity is not itself a discrete item in the world, but appears rather to be a property of the relations between several items. It refers specifically to the existence of significant differences between items in the world. Biodiversity refers to the existence of actual and potential differences between biological entities. It is standard to distinguish between levels of differences: genetic diversity, species diversity, ecosystem diversity and habitat diversity. It is also possible to distinguish within the standard scientific approaches between kinds of differences: numerical or the number of species; dimensional – in the degree of separation, or distinction, along a dimension such as genotype; material – differences in the substance(s) and structural properties of which things are composed; relational – differences in the kinds of interactions between organisms; causal and historical differences in the way in which things have come into existence, e.g. salmon and lungfish are phylogenetically similar, but have very different evolutionary histories and origins. Finally, biodiversity is often used as a concept that refers to the potentials of environments rather than their state at any point of time: to maintain biodiversity is to maintain the capacity of a system to diversify rather than to maintain the actual diversity at a given point of time. Given these features of biodiversity, it is an unlikely candidate for an itemizing approach to environmental choice. However, when it is made operational, it is invariably approached in terms of itemization. Consider in the context of environmental policy-making the following definition of biodiversity formulated by the United States Government Office of Technology Assessment:

> 'Biodiversity refers to the variety and variability among living organisms and the ecological complexes in which they occur. Diversity can be defined as the number of different items and their relative frequency. For biodiversity these items are organized at many levels, ranging from complete ecosystems to the chemical structures that are the molecular basis of heredity. Thus, the term encompasses different eco-systems, species, genes, and their relative abundance.' (US Office of Technology Assessment 1987)

While the definition captures many of the standard observations about biodiversity, it tends to an approach that focuses on the number and relative frequency of different items: ecosystems, species and genes. As it is operationalized, one is offered an itemization of species and habitats and the injunction to maintain or enhance the numbers on the list: hence, red lists of endangered species or lists of threatened habitat types.

This itemizing approach also lends itself to economic valuations of the significance of biodiversity loss – a fact that may help to explain why it has found favour. Environmental valuation requires defined commodities. As Vatn and Bromley (1994) note: 'A precise valuation demands a precisely demarcated object. The essence of commodities is that conceptual and definitional boundaries can be drawn around them and property rights can then be attached – or imagined'. Thus by operationalizing the concept of biodiversity in terms of a list, one can arrive at a surrogate that can be commodified. Like the supermarket list, one can ask how much individuals are willing to pay for each item in order to arrive at a measure of the full economic significance of losses. A trolley full of itemized goods does offer the possibility of putting an economic valuation into negotiations: 'While we cannot say that similar kinds of expressed values will arise for protection of biodiversity in other countries, even a benchmark figure of, say, $10 per person for the rich countries of Europe and North America would produce a fund of $4 billion' (Pearce and Moran 1994).

What is wrong with this itemizing approach? Before attempting to answer that question we should make clear at the outset that we are not objecting to itemization *per se*. There is nothing wrong with producing red lists of endangered species or lists of threatened habitat types to be maintained. They can serve an important function, not least as ways of indicating the significant losses in biological variety. However, they tell at best an incomplete story about environmental values: indeed, their problem is that they omit the story of environmental value entirely. The consequence is a failure to capture significant environmental losses and inadequate policy responses as a result.

Sustainability and natural capital

'Sustainability', like 'biodiversity', has become one of the key phrases of the politics of the environment and everyone is in favour of them. They often appear together – biodiversity goals form a central component in the pursuit of 'sustainability', while in the economic literature sustainability is construed in terms of 'capital'. It requires each generation to leave its successor a stock of capital assets no less than it receives. In other words, it requires that capital – explained also as capital wealth or productive potential – should be constant, or at any rate not decline, over time. A distinction is standardly drawn between natural and man-made capital: man-made capital includes not just physical items such as machines, roads and buildings, but also 'human capital' such as knowledge, skills and capabilities; natural capital includes organic and inorganic resources construed in the widest possible sense to cover not just physical items but also genetic information, biodiversity, ecosystemic functions and waste assimilation capacity. The distinction between the two forms of capital is taken to generate two possible versions of the sustainability requirement, each with variations: (i) that overall capital – the total comprising both natural and man-made capital – should not decline, or (ii) that natural capital in particular should not decline (Pearce *et. al.* 1989).

This distinction generates a debate between 'weak' and 'strong' sustainability. The debate turns on the degree to which 'natural capital' and 'man-made capital' can be substituted for each other. As the distinction is set up, proponents of 'weak' sustainability affirm that natural capital and man-made capital are indefinitely or even infinitely substitutable (Jacobs 1995; cf. Daly 1995). Proponents of 'strong' sustainability, on the other hand, hold that because there are limits to the extent to which natural capital can be replaced or substituted by human-made capital, sustainability requires that we maintain the level of natural capital, or at any rate that we maintain natural capital at or above the level that is judged to be 'critical'. Sustainability is taken to involve the protection of 'critical natural capital' – that biodiversity which cannot be 'readily replaced' and a set of 'constant natural assets' that do have possible substitutes through re-creation and translocation. The terms in this context do have a strained meaning, which is acknowledged.

What are the criteria of adequate 'substitutability' or 'replaceability'? Capital, both natural and man-made, is perceived as a bundle of assets. Insofar as it concerns biodiversity, the concept of 'capital' fits well with the itemizing approach. We have a list of valued items, habitat types, woodlands, heathlands, lowland grasslands, peat-lands and species assemblages. We maintain our natural capital if, for any loss of these, we can recreate another with the same assemblages. The promise of the approach is its flexibility. If a road potentially runs through some rare habitat type, say a meadow land, or an airport runway is to occupy woodland that contains some rare orchid, we can allow the development to take place provided we can recreate or translocate the habitat. The issue becomes one of the technical feasibility of replacement: on this turns the distinction between 'critical natural capital' (CNC) and 'constant natural assets' (CNA).

Time, history and biodiversity

Time and history are not just a technical constraint on the realization of certain results, a certain spread of valued items – species and habitats. The problem here concerns our understanding of what is the value in 'biodiversity', and what is the source of concern in the loss of biodiversity. If it were simply a case of maintaining and enhancing biological diversity understood as lengthening a list of species, habitats and genetic material as such, then we are confined only by the limits of our technical creativity. We can 'create' habitats, say fenlands, for species and the tourist industry. Indeed, if it is just a list a varieties we are after, the creation of Jurassic Park and genetic engineering both would 'enhance' biological diversity in the sense of increasing the total of species and genetic variety we have at our disposal. However, while that might make a contribution to the entertainment industry and, the undoubted risks aside, has some potential instrumental value for human health, it rather misses what is at stake in worries about biodiversity loss.

Time and history do not enter into problems of biodiversity policy solely as technical constraints on the possibility of recreating certain landscapes with certain physical properties; for example, particular habitats are valued precisely because they embody a certain history and processes. The history and processes of their creation matter, not just the physical attributes they display (Goodin 1992; O'Neill 1993, 1997). The temporal 'technical constraints' of the UK biodiversity plan are in fact a source of the very value of habitats that could not in principle be overcome. We value an ancient woodland by virtue of the history of human and natural processes that together went into making it: it embodies the work of human generations and the chance colonization of species and has value because of the processes that made it what it is. No reproduction could have the same value, because its history is wrong. In deliberations about environmental value, history and process matters, and constrains our decisions as to what kind of future is appropriate. We value forests, lakes, mountains, wetlands and other habitats specifically for the particular history they embody. Geological features have histories with no human component, while landscapes often involve the interplay of human use and natural processes. Most nature conservation problems are concerned with flora and fauna that flourish in particular sites that are the result of a specific history of human pastoral and agricultural activity, not with sites that existed prior to human intervention. The past is evident also in the conservation of the embodiments of the work of past generations that are a part of the landscapes of the old world: stone walls, terraces, old irrigation systems and so on. And, at the local level, the past matters in the value we put upon place (Clifford and King 1993). The value of a specific location is often a consequence of the way that the life of a community is embodied within it. Historical ties of community have a material dimension in both the human and natural landscapes within which a community dwells.

The natural world, landscapes humanized by pastoral and agricultural environments, the built environment – all take their value from the specific histories they contain. We do not enter into or live within 'natural capital'. Our lived worlds are rarely natural and are not capital. We live in places – habitats if one likes – that are rich with past histories, the narratives of lives and communities from which our own lives take significance. We need to take the narrative and temporal dimensions seriously. The way not to get hold of these biodiversity problems is to attempt to itemize the 'values' of the various items which feature in the situation and pursue the policy of 'maximizing value'.

Place and nature cannot be faked. What matters is the story of the place. This renders many 'ordinary' places that are technically 'easily reproducible' places less open to substitution. Take ponds, which are a feature of landscapes that matter to many on an everyday level: dew ponds, village ponds, local ponds are relatively easy to recreate at the level of physical features such as species variety. However, the reproductions would simply not be the same places with the same meaning for a community. It is *this* pond that was used by people long ago to water their livestock, where for generations we in this local community have picnicked, fed ducks, looked for frogs and newts – *this* pond that we want to preserve. Another pond built last year could never do as a substitute simply because its history is wrong. We want to preserve an ancient meadowland, not a modern reproduction of an ancient meadowland, not because it is difficult to reproduce but simply because it wouldn't be an ancient meadowland.

While the approach we recommend in some ways adds constraints, the appeal to history also relaxes inappropriate constraints on change. One unsatisfactory feature of a great deal of biodiversity management is the way it attempts to freeze history at a certain point. This point has special relevance for 'biodiversity management' in the context of 'wilderness' or 'nature' preservation in national parks of the 'new' worlds. Nature parks are created and legitimized through an ahistorical wilderness model of nature conservation which puts considerable emphasis on the values of wilderness, understood as nature untouched by humans, and of 'ecological integrity', understood as the integrity of ecological systems free from human interference. This wilderness model has developed historically from an image of nature as an unspoilt wilderness that contrasts with the domesticated environments of Europe (Anderson and Grove 1987; R. Grove 1995; MacKenzie 1988). However, the nature or wilderness to be preserved represents landscapes at a particular point in their history prior to which it is deemed a wilderness. The point is that of European settlement. The parks of South Africa, for example, were set up to protect the pristine landscape 'just as the Voortrekkers saw it' (Reitz, quoted in Carruthers 1989). As the influential report of the Leopold Committee, *Wildlife Management in the National Parks in the U.S.A.*, put it:

> 'As a primary goal we would recommend that the biotic associations within each park be maintained, or where necessary recreated, as

nearly as possible in the condition that prevailed when the area was first visited by the white man. A national park should represent a vignette of primitive America.' (Cited in Runte 1987)

The appeal to wilderness becomes a way of avoiding coming to terms with a troubling historical dimension to environmental evaluation. In the context of conquest and colonization the wilderness model of nature has a legitimating role – to render invisible the people whose home one has appropriated. Consider the fate of some of the Maasai who have been excluded from national parks across Kenya and Tanzania (Monbiot 1994) – for example the Maasai suffering from malnutrition and disease on scrubland bordering the Mkomazi Game Reserve from which they were forcibly evicted in 1988. Attempts to evict indigenous populations from the Kalahari reveal the influence of the same wilderness model: 'Under Botswana land use plans, all national parks have to be free of human and domestic animals' (*The Times*, 5 April 1996). Nor is the policy of enforced eviction confined to Africa. Similar stories are to be found in Asia where the same alliance of local elites and international conservation bodies has led to similar pressures to evict indigenous populations from their traditional lands. In India, the development of wildlife parks has led to a series of conflicts with indigenous populations. Thus, in the Nagarhole National Park there are moves from the Karnataka Forest Department to remove 6,000 tribal people from their forests on the grounds that they compete with tigers for game. The move is supported by international conservation bodies: hence the remark of one of the experts for the Wildlife Conservation Society – 'relocating tribal or traditional people who live in these protected area is the single most important step towards conservation' (cited in Guha 1997). The wilderness model fails to acknowledge the ways parks are not wildernesses but homes for their native inhabitants, and indeed the degree to which the landscapes and ecologies of the 'wildernesses' were themselves the result of human pastoral and agricultural activity.

It is surprising that biodiversity has become identified with an atemporal itemizing approach to environmental values. Even at the biological level, history has a role: salmon and lungfish are phylogenetically similar, but have very different evolutionary histories and origins. The point should be extended to the appreciation of the richness of natural and human histories embodied in the habitats and worlds into which we enter. To render them all managed in a uniform manner for the sake of 'biodiversity' would be to lose one source of richness. Moreover, it would unlikely realize its result. Biodiversity is used as a concept that refers to the *potentials* of environments, not just their state at any point in time: to maintain biodiversity is to maintain the capacity of a system to diversify, rather than the actual diversity at any point of time. For that we need to focus on the processes of change, and the maintenance of plurality in process, not at the static maintenance of systems at some artificially frozen point in time. The belief that there is some defined set of ideal management systems for biodiversity that can be globally exported and imposed on local populations is one that is likely to be inimical to the maintenance of biodiversity.

Kantian and utilitarian approaches to the value of biodiversity (Mark Sagoff)

Two frameworks for decision-making

The *utilitarian approach* to environmental protection, by comparing the benefits and costs of policies, appeals to common sense. It also appeals to the choices people make or would make in trading freely with each other. A perfectly competitive market – that is, one in which property rights to all assets are fully defined, owned and enforced, and in which the costs of bargaining are low – will allocate these assets to those who are willing to pay the most for them, and in that sense those who benefit most from their use. As a general rule, such an 'efficient' market establishes the economic value of resources, including plants and animals, by setting prices in response to supply and demand. The economic value of the resource, in other words, is its 'exchange' or 'marginal' value – the amount it fetches in a market in competition with other goods and services.

The Kantian *deontological approach* to valuation, in contrast, relies on deliberative processes associated with representative democracy, through which society enacts rules that reflect its identity and establish its aspirations. Policy-makers then determine, first, how to characterize a problem, and second, how to respond to it in a way consistent with these politically determined rules, goals and values. In this democratic process, society takes economic factors into account, of course, since to will a particular outcome one must also will the means to achieve it. But a policy, in serving legislated purposes, may be justified primarily on ethical rather than on economic grounds – however clever society may be in using economic analysis to achieve the policy most effectively and at the lowest cost.

These two frameworks for decision-making differ in the way they conceive the decision-maker and in the kinds of choices to which they are applied. In the utilitarian perspective, the decision-maker is the individual; he or she starts with – indeed, is conceived in terms of – an ordered set of preferences. These preferences are assumed to reflect judgements the individual makes about what will benefit her or him. If preferences are always connected in this way with individual welfare, a competitive market, because it maximizes the satisfaction preferences, weighted by willingness to pay, across all members of society, will maximize social utility. If one assumes that the goal of environmental policy is to maximize welfare or utility, then governmental intervention is unnecessary when markets function efficiently. Accordingly, policy analysts conclude that the 'assertion of market failure is probably the most important argument for governmental intervention' in environmental matters where issues of social equity are secondary (Cowen 1992).

Many economists argue that markets fail to 'price' endangered species correctly because they are 'public goods'. (How species are 'public goods' may not be clear, since the habitats of many endangered species may be privately owned land, and the plants and animals themselves, like all plants and animals, can generally be captured, bought and sold.) To 'correct' this market failure, these economists have developed scientific methods for measuring the value of species, primarily by asking how much people are willing to pay to preserve them. If people state a high willingness to pay (WTP), economists infer these people must believe that the existence of the species in some way benefits them. This is because economists equate WTP with expected welfare.

In fact, people generally state a high WTP to protect species because they think extinction is morally wrong rather than because of any consequences they believe extinction may have for them (Kempton *et al.* 1995). This kind of WTP, as we shall see, may have nothing to do with expected welfare or utility. Accordingly, it may have no relation to economic value. People who believe on moral principle that extinction is wrong – for example, out of a religious view about Creation – may be willing to pay a 'fair share' to rescue a species. To act on such a moral principle, however, is not to satisfy a preference related to one's own welfare or well-being. Principles and the demands they make on personal behaviour are the appropriate objects of legislation – hence the Endangered Species Act. Preferences and the expected utility they reflect are appropriately left to markets.

Yet, the market prices of rare or endangered species, such as Furbish's lousewort, even if negligible, reflect their economic value, which is to say, these prices measure, as accurately as possible, the contribution of plants and animals to social welfare. If the point of public policy is to maximize welfare, there is then no need for governments to protect endangered species, since markets generally function efficiently in allocating those resources. For example, a shopping mall developer may bid more than a drug or seed company for the land a rare plant inhabits. If the landowner sells to the developer, the result represents a normal market outcome, not a market failure. Human beings simply benefit more – they realize a greater welfare gain – from the construction of the mall than from the protection of the plant. The developer has a more profitable use of the land than anyone else, including drug, seed or other companies.

The Kantian view recognizes that markets generally succeed as well as is feasible in providing consumer goods and services. Markets generally succeed, in other words, in maximizing the welfare or well-being of individuals within resource constraints. Accordingly, if the purpose of environmental policy were to maximize the welfare of individuals across society, there might be little need for governmental action, especially with respect to endangered species. Markets as they are serve this purpose as well as possible.

The Kantian approach, however, denies that maximizing social welfare is the goal of environmental policy. Kantians insist that public policy should respond to human values, of course, but these include moral principles not just personal preferences. A person who supports the Endangered Species Act, on this view, may do so because he or she thinks the principle it legislates morally binds our nation as a matter of what is objectively right. Of course, the utilitarian also has a view of what is intrinsically or objectively right, namely the satisfaction of preferences weighed by WTP and taken simply as they come. From the premise that only welfare, so defined, possesses intrinsic value, the utilitarian concludes that no one should or perhaps could pursue any other goal for its own sake. This rules out on a priori grounds the possibility that people could value species protection as a goal justified in itself rather than as a means to further their welfare or well-being.

Both the Kantian and the utilitarian agree that public policy should be based entirely on the values of the members of society. Human beings assign all the values. From the premise that only human values count in public policy, the utilitarian too hastily concludes that only human welfare counts in public policy. The utilitarian thus assumes that people can or should value intrinsically only their own well-being. Since any other morality is ruled out, all value is economic value, commensurable as a means to utility.

The Kantian position, in contrast, distinguishes between economic and ethical value. Economic value has to do with what people want to buy to benefit themselves. Moral value, according to the Kantian, involves a judgement a person makes that something is good or bad, right or wrong, from an objective or universal point of view. Thus, a person who believes that Furbish's lousewort will never benefit her or him may, as a matter of principle or moral obligation, insist nevertheless that society should go to great lengths to keep that species from extinction. Through political process, citizens convene upon values, goals and characteristics they aspire to achieve as a community. These may have little to do with the kinds of preferences these individuals typically reveal in markets.

A Kantian theory differs from the utilitarian approach, moreover, in recognizing that markets, however competitive or efficient, may reach morally obnoxious outcomes. When citizens find these outcomes ethically abhorrent – prostitution, child labour and the sale of narcotics are examples – they legislate against them. Laws against prostitution are justified not because the sale of sex is inefficient but because it is (deemed to be) immoral. Similarly, society legislates against extinction on moral not economic grounds. No 'just so' story about market failure or public goods has to be told. Society may be justified in preserving species, the deontologist argues, whether or not the benefits exceed the costs.

This non-utilitarian approach to decision-making applies paradigmatically to objects which, given their symbolic, historic and spiritual significance, are valued more because of their meaning than because of their use. Kant (1959 [1785]) draws this distinction as follows: 'That which is related to general human inclination and needs has a market price... But that which... can be an end in itself does not have mere relative worth, i.e. a price, but an intrinsic worth, i.e. a dignity'.

Both the utilitarian and the Kantian approaches provide frameworks for rational choices – one by emphasizing consequences for preferences, the other by appealing to principles and procedures appropriate to the identity of the decision-maker in the given situation. The first approach asks the economic question: 'What will maximize the well-being of the individual over the long term?' The second approach asks the political and ethical question: 'What do we stand for as a society?' Which conception of the common will or the public interest is correct? Which rules should we follow with respect to problems such as pollution, the extinction of species, or wilderness preservation, given our history, culture and sense of shared identity?

In a recent book, James G. March (1994) provides a theoretical understanding of these two alternative approaches to decision-making. When decision-makers adopt the economic approach, March explains, they choose among given alternatives 'by evaluating their consequences in terms of prior preferences'. When they adopt the principle-based or Kantian framework, decision-makers, 'pursue a logic of appropriateness, fulfilling identities or roles by recognizing situations and following rules that match appropriate behaviour to the situations they encounter'. As members of a society determine who they are and what they stand for as a community and, accordingly, the appropriate rules by which they are to regulate economic activity, the duties of deliberation take precedence over the algorithms of aggregation. In this framework for decision-making, according to March, the reasoning process, 'is one of establishing identities and matching rules to recognized situations'.

Is human well-being the only obligation?

The question before us, then, may be this: Are we to protect species because of the sacredness of nature or the efficiency of markets? The answer depends on who we think we are – what we stand for as a community and a nation. Economic theory tells us one thing, but cultural history tells us another.

Two reasons for protecting endangered species have been identified: (1) because we need, use or like them; and (2) because we ought to do so. It is the fundamental thesis of mainstream economics that the second reason can be reduced to the first. From the mainstream perspective, the nation ought to protect species only to the extent that to do so promotes human welfare. This framework assumes that all values are instrumental values except welfare, which is the one goal that this framework suggests is worth pursuing in itself or for its own sake.

Should one accept the proposition that human welfare or well-being is the only object that has intrinsic value and therefore grounds for obligation? Oddly enough, when one considers what mainstream economists mean by 'welfare' – that is, the satisfaction of preferences taken as they come – one sees that it fails to provide a basis for any moral obligation. Indeed, 'welfare' or 'well-being' as these terms appear in mainstream environmental economics have no connection with anything, such as happiness or contentment, that people outside economic science would think of as a good thing.

When environmental economists speak of 'welfare' or 'well-being', what do they have in mind? They do not mean what utilitarian philosophers like Bentham and Mill meant by these terms, namely, pleasure or happiness. Rather, contemporary economists usually define 'welfare' in terms of the satisfaction of preferences taken as they come, bounded by indifference, and weighed by the amount people are willing to pay to satisfy them. The question arises, then, of what 'welfare' defined in this way – that is, in terms of preference-satisfaction – has to do with welfare or well-being as Bentham, Mill and others understand it, i.e. contentment or happiness. The answer is, nothing. Preference-satisfaction, which mainstream economic theory seeks to maximize, correlates with neither happiness nor with well-being as these terms are understood in the philosophical tradition and in ordinary discourse. Study after study has shown that no covariance holds between preference-satisfaction and any conception of welfare or well-being not simply defined in terms of it. The empirical evidence is conclusive in showing that the satisfaction of preferences, after basic needs are met, does not make us better off, in spite of the contrary assumption of mainstream economics.

If we take the Kantian point of view, in contrast, we have no difficulty in understanding that individuals respect species as good in themselves and that people perceive in these creatures properties that are admirable and therefore worth protecting for their own sake. We have no difficulty understanding that citizens may legislate moral goals other than the one mainstream economists endorse, i.e. the satisfaction of preferences taken as they come weighed by willingness to pay. The laws in which we reflect public morality within constitutional constraints are not to be taken as themselves data for economic analysis, as if they expressed preferences related to welfare. Rather, environmental law simply rejects the approach of environmental economics – it repudiates contemporary utilitarianism in favour of a different conception of public morality and collective choice.

Selling pigeons in the temple: the blasphemy of market metaphors in an ecosystem (Timothy C. Weiskel)

And Jesus entered the temple of God and drove out all who sold and bought in the temple, and he overturned the tables of the money-changers and the seats of those who sold pigeons ... (Matthew 21:12)

There is a growing sense of frustration and disgust bordering on holy outrage among scientists and citizens around the world as we witness the evident collapse of political leadership on matters of grave environmental concern. The situation is both deep and pervasive and it intrudes at the level of our language of analysis and our metaphors of perception.

Over the last several decades, much of our public discourse about the environment has been reduced to a series of catchy clichés and mantras suggesting that nature is an economy, the ecosystem is a marketplace, and that our relationship to it can and should be calculated in terms of cost–benefit analysis. With our money capital we are exhorted to invest in the environment as part of a prudent portfolio from which we are encouraged to think that, with proper management, we can obtain a continuous flow of goods for our infinite gratification.

In this context, the value of biodiversity is extolled for its future market potential as a source of food, materials for our creature comforts, the drugs which may well some day hold the cure to cancer and other illnesses of civilization. As one might expect in an increasingly service-driven economy, we are being made more aware of the services that nature provides for us which would be very expensive if their true costs were to be calculated. In short, the magnificent richness and manifest wonder of biological diversity in creation has been abusively flattened and crudely crammed into the impoverished concepts of 'market commodities' and 'ecosystem services.'

One of the most sophisticated and comprehensive attempts to assess the market value of ecosystem 'goods and services' was recently published by Robert Costanza (1997) and his colleagues, in *Nature*. The article, entitled 'The Value of the World's Ecosystem Services and Natural Capital', has received world-wide attention through an Internet discussion group devoted to it. The thirteen economists and ecologists who compiled the analysis on the basis of reviewing many previous studies acknowledged the problems inherent in assigning market values to nature. As they put it, '... although ecosystem valuation is certainly difficult and fraught with uncertainties, one choice we do not have is whether or not to do it'.

Clearly such a pre-emptive statement generates more disputes than it settles. It may well be true that economists feel they have no choice but to undertake a market valuation study in their role as professional economists. But what does such an exercise by one group of specialists imply about other forms of valuation? What about the valuations of political leaders, scientists, citizens or those who strive valiantly to protect biodiversity for its own sake? Must they – must we – all sign on unquestioningly to the economists' sophisticated commodification of the ecosystem? The implication of the article is, 'if you are rational, of course you must'! [But see article in *Nature* 395, 1 October 1988, 'Audacious bid to value the planet whips up a storm'.]

What is so revealing and disturbing about the logic throughout the article is the blithe assumption that because economists can make their valuations 'explicit', these valuations should somehow be acknowledged by society at large as the pre-eminent basis for determining public policy. The term 'explicit' in this context appears to mean the ability to provide numerical estimates in the form of monetary values. All other forms of valuation by implication are not considered explicit. Of course, all choice is predicated on valuation. But beneath this truism, the unstated implication is that all forms of explicit valuation must necessarily take the form of monetary valuation. Otherwise they can be dismissed because they lack clarity and can lead to moral disputes.

The difficulty arises in that there is a superficial element of truth in talking about aspects of nature with market metaphors. Beans and barley, pork-bellies and pigeons, are regularly sold for cash in markets around the world. The power of these metaphors resides ultimately in the fact that they require no thought. They work instead on a subliminal level, and they therefore succeed in controlling our behaviour by mobilizing primordial beliefs and automaton calculations rather than rational thought and considered judgement.

Social scientists and students of language have provided insightful analyses of just how this subtle process operates. Metaphors are far more important than simple figures of speech (Lakoff and Johnson 1980) because they entail and provoke a whole series of reflexive reactions, associated images and unconscious mental processes. Thought is short-circuited by symbol, and groups of symbols are linked to one another in wide networks of implicit images which channel, direct and sometimes preclude thought. In effect, metaphors come in packages, and the power they exercise in generating thought derives not from their truth claims but rather from their 'extensibility' – that is, their power to spawn an internally coherent, imaginary vocabulary to mirror certain aspects of the parallel external world. The substantive connection between any given metaphor and the reality it purports to illumine may be very tenuous indeed. Indeed, given the selective and partial nature of any metaphor, associative imagery may obscure rather than clarify important aspects of reality. Such is the case when economic metaphors are used to describe ecological realities. Chomsky (1989) calls this 'thinkable thought'.

Economic metaphors function to define the range of so-called 'responsible' public thought. For instance, environmental goals which are widely acknowledged to be desirable and good but which are thought to be 'expensive' are often characterized as 'unrealistic'. In this manner, a certain conception of 'economic reality' is swiftly invoked to label specific kinds of thoughts or proposals about the environment as out of bounds ('unrealistic', 'unreasonable', 'too idealistic', etc.) for public discussion. Whether or not the particular notion of 'economic reality' involved corresponds to anything more than an extended fiction is never questioned because market metaphors have been accepted as the governing framework for responsible discourse.

Perhaps the most insidious feature of the pervasive use of economic metaphors in our environmental thinking takes the form of the subconscious question, 'Can we afford a sustainable environment?' Strictly speaking, of course, this is a meaningless question because all questions of worth are predicated upon the prior and continuing existence of a valuing agent. Thus, it would be logically and physically impossible to 'afford' anything over time if the environment of the valuing agent ceases to exist. The evident absurdity of the question, however, is not confronted within the mind-set of the market mentality because the question itself is never overtly posed. It remains implicit. It lurks in the background, conditioning every decision we make and suggesting on a subliminal level that a viable ecosystem is not a 'luxury' no longer available to the mass of humankind.

Society cannot solve a collective problem simply by multiplying private solutions that try to opt out of that problem. All this accomplishes is to generate the collective problem on yet a bigger scale. Nevertheless, from within the internally coherent fantasy world of market metaphors there is no vantage point from which individuals can see that the private solution is a delusion. Indeed, as consumers we are urged to think that the market alternatives are the only solutions available. Alternate identities that we hold as parents, members of faith communities, volunteers, community members and global citizens, etc. are simply wiped away as we are taught that the only legitimate goal of public policy is to protect the 'rights of the consumer'. Even the much-vaunted defence of the taxpayer on the part of politicians who say they will reduce taxes is publicly justified in order to provide citizens with more money to spend as consumers in the open market. The role of citizen–consumer is so thoroughly conflated in public discourse that we are made to feel that 'buying American' or simply buying anything at all is a patriotic act to keep the economy growing.

Market commentators encourage us to think that self-interest, greed and unbridled competition are an expression of the natural order of things – the only order that has ever existed or could ever possibly exist in the 'real world'. We live in a world, they assure us, characterized by 'the survival of the fittest', by which they seem to mean 'survival of the fattest'. While overt acts of graft and corruption grab the headlines from time to time, there is a much more subtle and pervasive corruption of perspective which has also taken hold of all public leadership. By giving unquestioned priority to market metaphors our leaders have narrowed their vision to such an extent that they know the cost of everything and the value of nothing.

Public leadership needs now to define, declare and defend the public good in terms that transcend private self-interest. There are no doubt connections between public good and private gain, but to justify the former exclusively in terms of the latter is a fundamental mistake in moral reasoning. Without public leadership that can understand this fundamental difference and learn to defend the public good in its own right, industrial civilization will become irretrievably consumed in a scramble for private profit and personal advantage in a dismal world of diminishing resources.

The Secretary General of the United Nations, Kofi Annan, expressed this fear with a riveting sense of urgency in his opening remarks at the Rio+5 conference in New York: 'Failure to act now could damage our planet irreversibly, unleashing a spiral of increased hunger, deprivation, disease and squalor. Ultimately, we could face the destabilizing effects of conflict over vital natural resources ... We must not fail' (IPS 1997).

In past epochs, individual religious and spiritual figures emerged to warn society of this kind of impending doom. Prophets of old inveighed against gluttonous consumption based on inequity and iniquity, and they warned societies of the physical consequences of failing to mend their ways. Perhaps more importantly, they served to remind societies of the natural order of the created world and the proper place of humankind in it. Amos, Jesus of Nazareth and Mohammed of Medina all arose in the ancient Near East with strikingly parallel messages in this regard. Jews, Christians and Muslims to this day retain scriptural traditions that remind them that the earth does not ultimately belong to humans, nor will their mistreatment of the earth or their fellow creatures go unpunished.

In these religious traditions, arrogant, self-centred behaviour with regard to the created order is thought to be morally wrong, however expedient or profitable it may prove to be for individuals in the short run. We are not fully informed by the preserved text, but one suspects that selling pigeons in the temple prompted a sense of moral outrage on the part of Jesus of Nazareth not because the prices were a bit too high but because such activity involved a fundamental confusion of sacred space with the market place.

Nature and culture in the valuation of biodiversity (Bryan G. Norton)

What is the value of biodiversity? The question is deceiving in its simplicity, and it is wise to avoid assuming, prior to careful analysis, the form that its answer will take. Note first that the definite pronoun, 'the', already suggests there is a single answer – that however many ways different people use, enjoy, worship and respect wild living things, there must be a unitary or at least synoptic answer to the question of biodiversity values. This suggestion of unitary value has gained credibility in many circles from a fear of relativism – the view that every valuation of wild life is so conditioned by local situations that judgements of nature's value add up to nothing more than the subjective and irreconcilable feelings of many different persons in many different cultures.

Most discussions of how to evaluate nature in general terms have been based on one or the other of two theories of the value of nature – 'economism' and 'deep ecology'. The former holds that elements of nature have instrumental value only, and should be valued like any other commodities; the latter, that elements of nature have 'inherent' or 'intrinsic' value, and that these elements are therefore deserving of preservation for their own sake.

These two value theories tends to polarize discussions on biodiversity. Environmentalists from developed countries tend to espouse intrinsic values in nature, advocating the preservation of natural areas intact and free of economic usage, while peoples from developing countries tend to value nature for its uses and developmental potential. In international policy forums it is common for spokespersons for the developing world to complain – with considerable justification – that First World countries have already exploited and converted their forests and are now asking the Developing Countries to forgo forest-based development and attendant increases in their standard of living. The tension regarding why and how to value nature therefore has practical effects, making it more difficult to forge North/South and other coalitions to protect biodiversity.

It is important to recognize that these opposed value theories rest on a cluster of highly vulnerable assumptions (Norton 1997b). Both economists and deep ecologists accept a sharp dichotomy between values that are 'inherent' and those that are instrumental; further, both groups proceed to use this sharp dichotomy to separate nature into beings/objects that have, and those that lack, 'moral considerability'. Economists and deep ecologists, then, agree that there must be some special status for those beings that have non-instrumental value – they simply disagree regarding which objects in nature actually have this special status. For economists, the special status is co-extensive with the human race; for the deep ecologists, moral considerability is co-extensive with a much larger subset of nature's components. Either way the sharp distinction between 'instrumental' and 'inherent' values ensures that questions of environmental value are posed in all-or-nothing terms. For the economist, 'Should we protect this river?' becomes, 'Does this river have net positive economic value (for humans) or not?' For the deep ecologist, it becomes, 'Does this river have inherent value?'

Both economists and deep ecologists think there is only one kind of ultimate value; they differ only in how widely they define that value. While economists and deep ecologists sometimes appear to exhaust the possibilities available in the literature for understanding environmental values, and while these proponents espouse positions that seem to be in polar opposition to each other, the important thing to realize is that both positions rest on a complex of highly vulnerable assumptions. First, both positions assume there is a sharp distinction between 'instrumental' and 'inherent' values. Further, both assume that the best way to achieve a unified theory of environmental values is to identify some characteristics shared by entities in the environment that are valued. In fact, both these assumptions are unjustifiable (Norton 1997) as I have argued elsewhere. Furthermore, it seems unlikely to me that either of these all-or-nothing, monolithic theories of value will prove rich enough to guide difficult, real-world choices regarding what should be saved, where conservationists should concentrate their efforts, and how they should set priorities. But should that not be precisely the role of a theory of environmental value in the conservation policy process?

New beginnings

An alternative approach is needed to environmental value theory and to environmental evaluation – avoiding the constraining assumptions of valuation in terms of commitment to a single value principle. Callicott (1989) notes that 'value' is 'a verb, not a noun', and emphasizes the experiential aspect of environmental valuation as the acts of individuals. But he attempts to use these subjective judgements as a basis for the classification of objects into categories of instrumental and intrinsically valued items such as species and ecosystems (Norton 1995). I support a more radical position – that we need a theory of environmental valuation based on the premise that a theory for classifying objects as instrumentally or inherently valuable is unnecessary. Once we adopt the value-as-verb idea and also reject the sharp dichotomy between intrinsic and instrumental valuings, the task of classification will be simpler. The starting point should be individuals, who live and act within their cultures and have positive and negative experiences in their interactions with the processes of nature.

The goal must be to achieve some systematization and clarification, and eventually to provide some means of forging common policy from the diversity of individual valuational acts. In contrast to the usual approaches to environmental valuation discussed above, one must begin by recognizing the diversity and cultural specificity of acts of valuing, respect these diverse starting points, and then work toward a specification of common factors in these valuations. Rather than attempting to fit all locally experienced values into pre-conceived general theories, a more general consensus about nature's values must be developed. In this way it can be seen that environmental values are a product of countless local dialectics between experiencing members of local cultures and the ecological communities in which those cultures are embedded. These local dialectics are driven by a common striving of all people to adapt, to choose appropriate activities and institutions, and to live meaningful, fulfilling lives within their particular contexts. Each local dialectic is dynamic and changing, subject to new information and evidence, both from within the culture and – increasingly in the modern world – from other cultures, but it is *place-based*.

Values are a part of a constant and ongoing process by which people, acting in cultural contexts, develop new and more appropriate strategies and new adaptations. The place-based approach entails respect and gives a sort of 'internal validity' to natural values espoused by a culture at a particular time. These values are also clarified, modified and improved over time through new sources of information and through experimentation with new choices and new strategies. These values may converge, and in the interest of international co-operation we hope they do. But in this process there is no way to conclude, in advance, that such a consensual outcome is unnecessary or pre-destined, and certainly there is no way to specify the nature of that consensus prior to the process of consensus-building.

Theoretically, this approach holds that it is impossible even to value or save natural objects and processes independently of the local interaction of nature and culture, and it steers us away from attempts to identify what is of ultimate value in nature. That determination can only emerge within an ongoing, iterative and re-iterative process, as individuals – acting within the cultures that give their actions meaning – evolve in response to new situations and information.

Valuing biodiversity in place

The idea of place owes much to the Darwinian idea of evolution through natural selection. Darwinian selection necessarily depends upon variation, and upon differential success of differing actors under diverse local conditions – particular individuals with particular characteristics either survive, or fail to survive, in many local interactions with varied environments. All biological evolution is in this sense local. Success in biological evolution, at the organism level, is survival and reproduction of one's genes in subsequent generations. Success on the cultural level is the survival of a viable community in a place.

The place-based idea can also be examined from a historical perspective by drawing insight from the relatively new subdiscipline of ecological history. The formative book from this perspective is Cronon's *Changes in the Land* (1983) in which he challenged on historical grounds the 'wilderness' ideal where natural systems are considered to be most healthy and most diverse in the absence of human disturbance. Cronon showed that the New England landscape had been intentionally managed by the Native American (Indian) tribes of New England for millennia prior to the arrival of Europeans, and that any ideal of a 'pristine' wilderness is a mental and social construct. Cronon argues that, 'The replacement of Indians by predominantly European populations in New England was as much an ecological as a cultural revolution, and the human side of that revolution cannot be fully understood until it is embedded in the ecological one'. Besides forcing us to re-think the nature of 'pristine' reserves as a management ideal in most situations, Cronon's work points toward two further important consequences for biodiversity policy. First, he supports the importance of dynamic models tracing system processes. Second, he shows how social values and institutions shape, and are shaped by, the interaction of human populations with their local environments; and he also shows how external demands, such as the need to produce goods tradable in world markets to obtain money to buy the various necessities, have indirect – but huge – impacts on the landscape.

Another way of examining the idea of place and its function in valuing biodiversity is by focusing on an important aspect of the work of Aldo Leopold (1949) who recognized that the tendency of individuals to value things from a local and personal perspective is susceptible to dangerous bias if it is not balanced by a longer and larger perspective. In arguing that we must learn to think like a mountain, he realized that we naturally focus on dynamics that are relevant on the human, perceptual scale – the scale on which individual humans act economically, seeking food and shelter. The need to integrate these various levels is, in my view, the central idea of Leopold (1949) who recognized that, in addition to perceiving and valuing the world in economic time, we must also think in the multi-generational ('ecological') time of mountains and wetlands.

Learning a new way to perceive is a cultural task and it is an essential element in developing a successful public policy, which can only be accomplished in democratic countries through dialogue among scientists, politicians and the public. The policy dialogue must originate from many local viewpoints, and flow upward through higher levels of generality, integrating individual, economic and environmental concerns on all levels. Creativity in natural systems occurs when an organism or population responds in a new way to the natural dynamic in which it lives. New adaptations are given meaning and significance by virtue of their response to constraints implicit in larger, slower-moving systems. It is this dynamic that creates environmental values.

If there is rapid alteration of the larger context in which those behaviours evolved and gained meaning, today's genetic and cultural adaptations become irrelevant more quickly than new ones can evolve. Leopold illustrated this point by noting that, in the American South-west, the Pre-Columbian Pueblo cultures survived and reproduced themselves for many generations in the fragile, arid ecosystems of the area; but the arrival of European settlers and their grazing culture led to drastic declines in the quality of the environment in just one or two generations, because agricultural methods adapted to Europe were applied in an arid ecosystem. Leopold saw survival in conjunction with a slowly evolving environment as the 'objective' test of whether a culture has found 'the truth'. Note that, on this approach to objectivity, the truth lies in adaptation to local conditions, but that local conditions must be understood as occurring within a longer and larger context.

Thus, nature and culture are intimately and inextricably intertwined. The tendency to value biodiversity by placing elements of nature in culturally independent categories gives way to an acceptance that valuation emerges from culturally shaped actions – attempts to enjoy and adapt to local situations – at a local level. Similarly, the diversity of a landscape – and the associated microhabitats that are available to support diverse organisms and species – is in the modern world a function of the complex dynamic of persons and cultures adapting to changing surroundings. If cultures become more homogenous, we can expect landscapes to become more homogenous as a direct and indirect result. Conversely, if we hope to save the biodiversity of the earth, we must also save the special dialectics that have emerged in those local places. This is not to suggest that we should attempt to halt cultural interaction and the development of more cosmopolitan cultures, which would be futile, but it does mean that the world economy and the world-wide media culture must be considered no more than an 'overlay' on local cultures.

The search for locally based, and culturally diverse, values in nature need not rule out an associated search for generality and universal values in nature. But these should emerge 'from the bottom up' and cannot be imposed from above, based on grand dichotomies and universal principles of value. Advocates of categorical valuation procedures that have dominated economics and environmental ethics are not, therefore, mistaken in their goals – to identify some general principles of environmental valuation – but wrong in that they seek these values by a top-down, a priori method, and in that they attempt to enforce universal values downward from above.

Recognizing that peoples and cultures of the world exist at very different stages of technological and economic 'development', and coupling this recognition with the idea that all environmental values emerge from local dialectics between culture and nature, it is possible to characterize a common source of the values in nature, which I refer to as the 'creativity' of nature (Norton 1997). 'Creativity' includes both the productivity of stocks such as biomass – fruits and lumber, for example – that can be used for food and shelter; and also the ability of natural systems to diversify across time and create new forms and new interactions through ecological and evolutionary forces. The creativity of nature, in this sense, is the sum total of the outputs of ecological and evolutionary processes.

Peoples, in their diversity and in their local adaptations as well as in their individual, day-to-day choices, can only make use of, find meaning in, and enjoy nature and its processes and products if they have a range of options available to them. From the standpoint of a person living within a culture and within a particular natural habitat, who uses or enjoys a biological 'resource', nature is valuable (whatever their uses or enjoyments might be) because it makes available to them a range of choices. Cultural practices in the use of resources simply represent choices that have become habituated. From the viewpoint of nature, on the other hand, these options and choices simply reflect the ways in which local natural systems create and sustain products and experiences. A depauperate landscape offers fewer, and often less culturally meaningful, options. And so it is here, at the nexus of individual choice, cultural practice and nature's productive and creative capacities that we should search for the general and universal aspects of nature's values to humanity.

On bioresponsibility (James A. Nash)

One of the most important questions on the frontier of ethics – and certainly the most perplexing as well as the most neglected one – concerns the moral status of non-human species and their incarnations. The basic question is: do 'otherkind' have moral claims on humans – and, of course, only on humans, as the only moral agents on the planet – for moral consideration of their interests?

In my view, the most fruitful ethical approach to this question is to redefine responsible human relationships with the rest of the planet's beleaguered biota, and to ground these human responsibilities not only, weakly, in human utility (for example, economic arguments) or even generosity (for example, kindness), but also, strongly, in concepts of distributive justice in consideration of the vital interests of otherkind. I call this approach 'bioresponsibility'. It is a basic moral response to the fact of biodiversity. It recognizes that, in justice, our responsibilities arise in response to other's rights; our duties are defined in reference to other's dues.

Bioresponsibility raises a host of questions that are mind-numbing in their novelty, if not their complexity. The main questions seem to be the following: Who or what belongs within the covenant of moral relationships, having justifiable claims on humanity? Does the covenant cover whales, voles, owls, trout, frogs, beetles, nematodes, redwoods, lilies of the field, fungi, bacteria, plankton, rocks and rills, hills and rivers, etc.? Where should we draw the line, if at all, on moral consideration? Do all species and their members have equal moral standing, as biotic egalitarianism argues, or is some graded valuation, proportionate to moral significance, necessary to give due consideration to morally relevant differences among species? Do abiotic elements (such as water and minerals) have a moral status in themselves? In view of the predatory and exploitative character of the biosphere, in which humans must destroy other life-forms and their habitats in order to satisfy our basic needs and to exercise our peculiar powers of cultural creativity, how can we balance the intrinsic and instrumental values of non-human life?

My purpose here is to offer a basic and brief review of biotic rights as one important part – not the whole – of adequate ecological ethics. By 'rights', I mean simply specified standards of just relationships. Biotic rights are the moral claims against humans, and only humans, for the imperative conditions of well-being for other species and their members. Several clarifications about biotic rights are necessary to note at this point.

First, biotic rights are not moral absolutes, contrary to one of the most common misinterpretations of 'rights'. Indeed, the concept of biotic rights is absurd when interpreted absolutistically: that would preclude human survival. Instead, rights are prima facie or presumptive moral claims, which means they can be overridden for a 'just cause' such as self-defence against a pathogen or an agricultural pest, the satisfaction of basic human needs, or even the culling of an alien species to protect an ecosystem. Humans, of course, are unavoidably predatory consumers and self defenders in trophic relationships. But what the concept of biotic rights demands of us in this context is a moral justification for any harm that we do to otherkind, and a limitation of such harm only to the extent that it can be so justified.

Second, biotic rights are *not* equal rights with humans. Indeed, 'biotic egalitarianism' – for example putting mosquitoes, molluscs and mammals, including humans, on the same moral plane, as deep ecologists do – is, in my view, inherently unjust, because it ignores morally relevant differences. We ought to show respect for other species in proportion to their value-experiencing and value-creating capacities. This allows for graded valuations among species and, indeed, a moral preference for humans over all other species in conflict situations, because of our rational and culture-creating capacities. Biotic rights deny the exclusivity of human values and rights, but they do not diminish the fundamental importance of human values and rights.

Third, in focusing on biotic rights I am not suggesting an alternative to ecosystemic concerns. For a fully adequate ecological ethic we need a basis for respecting both life-forms (individuals, populations and species) and collective connections, that is, the 'ecological common good' or diverse and whole ecosystems in a healthy ecosphere. The individualistic and holistic poles are not alternatives; they are complementary sides of a comprehensive ethic. In fact, the idea of biotic rights finds its fullest practicality in the quest of the ecological common good.

Finally, biotic rights are not narrowly individualistic. In fact, in most cases, we cannot respect organisms, such as microbes, individualistically. It is totally impractical, if not absurd, to try. Nevertheless, we can best respect the biotic rights of these and other organisms *collectively or corporately*, by protecting ecosystems from our profligate consumption, production, pollution and reproduction, as well as from our imprudent interventions.

Biotic rights, of course, need a moral justification. Rights cannot be assigned arbitrarily; they need a reasonable basis, some moral status that warrants appropriate entitlements from human beings. The usual criteria for recognizing moral claimants are sentience (above all), rationality, moral agency, self-consciousness, linguistic ability, etc. All of these seem to be sufficient conditions for the recognition of moral claims, but they appear to be anthropomorphically biased and non-ecological. They recognize moral status on the basis of human characteristics and values. Among animal rights advocates, for example, this status covers highly evolved animals, such as mammals, but it leaves the rest of the biota – the vast majority of animals, as well as all plants and species of other taxonomic 'Kingdoms' – with the instrumental status of 'things'.

In contrast, the one necessary criterion for the recognition of biotic rights that I find compelling is *conation* – that is, a *striving* to be and to do, characterized by aims *or* drives, goals *or* urges, purposes *or* impulses, whether conscious or non-conscious, sentient or non-sentient. At this point, organisms can be described as having 'vital interests' – that is, needs or goals – *for their own sakes*. These conative interests provide a necessary and minimally efficient status for at least elementary moral claims against humans.

Conation as the basis of moral rights covers all organisms – fauna, flora, etc. – but it also excludes abiotic elements, such as minerals and gases, because they cannot have interests for themselves. Nevertheless, in excluding abiotic elements from rights-coverage, humans have no licence for abuse. All life-forms are dependent on the abiotic elements in the biosphere. Non-living elements must be treated with care as instrumental or systemic values, because they are the resources and habitations of all creatures. We have indirect duties to ecosystems because we have direct duties to the host of creatures that partially constitute, and are interdependent in, these ecosystems.

Grounded in conation, biotic rights apply to both individuals and species, because individuals and species seem to be constitutive of one another – inseparable and interdependent. Even if a species as a genetic lifeline from the past for the future is not conative in itself (which I doubt), that claim would not contradict a theory of biotic rights grounded in conation. The reason is that a species as a genetic lifeline is not only the aggregation of conation in present populations but also the carrier of conation for all future generations. So, we can argue that recognizing the rights of species is essentially the same as recognizing the rights of future generations. Species can be said to have at least *anticipatory rights* in the sense that we can reasonably expect that they will exist and have vital interests, unless we deprive them of that potential. Therefore, the human community has anticipatory obligations to preserve otherkind's conation for the future.

On these foundations, I propose the following bill of biotic rights as some of the prima facie just claims for the vital interests of all non-human species and their members against the human community. (Additional rights may apply to some highly evolved organisms, such as primates, as they do to humans, if their essential needs so require.)

1. The right to participate in the natural dynamics of existence

This right allows otherkinds to work out their own interactions and adaptations without human protections or interventions, unless one can make an ethical case for these. It includes the right to flourish by being free in natural settings, without unjustified domestication or captivity. It doesn't treat predators, like felines or canines, as moral offenders. The moral role of humans in wildlands is to protect wildlife from injustices, of which humans are the only perpetrators.

2. The right to healthy and whole habitats

These systems of existence, despite ambiguities, provide species and their members with the only possibilities of realizing the good for their kind and performing their special functions for the ecological common good. In fact, the genetic identity of all creatures has been shaped by environmental adaptations, and the interests of species deteriorate with dehabitation. The preservation of healthy habitats, therefore, is the most effective means of promoting the good of otherkind. The ecological common good is a basic right in itself and the precondition of all other biotic rights – or even human rights. At this point of habitat preservation, the concern for individuals and species most clearly intersects with holistic concerns.

3. The right to reproduce their own kind without humanly-induced chemical, radioactive, hybridized or bioengineered aberrations

The powers of genetic reproduction are a fundamental feature of an organism's 'reason for being' and the prerequisite of species' preservation. Thus, this right is respect for genetic integrity, evolutionary legacies, and ecological relationships. It entails, for example, a variety of moral constraints and public regulations on genetic engineering. It means also that if we are proposing to 'artificialize' a species, we need a significant moral justification to do so that goes beyond simplistic versions of utilitarianism.

4. The right to fulfil their evolutionary potential with freedom from human-induced extinctions

Extinctions are a natural part of the evolutionary process, but human-induced extinctions are unjust. They deprive individuals of one of their conative drives and species of future generations. They eliminate new evolutionary radiations. Respect for biodiversity, therefore, entails perpetually sustaining viable populations of non-human species in healthy habitats until the end of their evolutionary time.

5. The right to freedom from human cruelty, flagrant abuse or profligate use

This right implies that kindness is not only a matter of benevolence but also of justice. Equally, it implies the ecological virtue of frugality, caring, constrained usage and, therefore, minimal harm to otherkind – in human production and consumption. Frugality is the earth-affirming instrument of distributive justice, to ensure 'enough' scarce goods on this planet for all species (unless we can provide a powerful justification to the contrary in particular instances, such as destroying a human pathogen).

6. The right to reparations or restitution through managerial interventions to restore a semblance of natural conditions disrupted by human abuse

Under optimum conditions of wildness, it seems best generally to adopt a *laissez-faire* strategy and let 'nature take its course' without the dubious benefit of human managers. Previous human disruptions, however, mean that interventions – for example, pollution controls, captive breeding programmes, and the culling of aliens – are often necessary to enable a return to an approximation of original ecosystemic interactions.

7. The right to a 'fair share' of the goods necessary for the flourishing of species and their populations

Of course, a 'fair share' is a very vague criterion, and defining it is extremely difficult, especially when some destruction of otherkind is necessary.

Nevertheless, it is a criterion that the human community must struggle to define in order to stifle human imperialism over the rest of nature – evident particularly in human over-development and overpopulation. One conclusion, however, seems clear: biotic justice imposes obligations on the human community to limit our economic production and our sexual reproduction to prevent the excessive destruction of wildlife and wildlands. If other species are ends or good for themselves, then our economic and population policies need to pursue what Herman Daly calls the 'biocentric optimum' in contrast to the 'anthropocentric optimum' which presently prevails as the norm (Daly 1993).

Both humans and non-humans are wronged when problems of excessive population and production are 'resolved' by the further sacrifice of non-human species and their habitats. We already use far more than any reasonably defined 'fair share' of this world's goods. These human dilemmas are best resolved not by the tacky tactic of pitting the poor against endangered species and habitats, but rather by confronting directly the prime sources of both poverty and ecological degradation: over-consumption by economic elites, human overpopulation, and economic maldistribution.

The debate about the moral status of otherkind is also finally about ourselves. What is the place of humans in the scheme of things? What are the rights and powers of humankind in relation to the rest of the biophysical world? What are the moral obligations that arise from our being parts and products of nature? These are not biological or ecological questions in themselves: they are metaphysical and ethical questions that cannot be ignored.

Contrary to some critics, a commitment to biotic values does not mean indifference to the human project. The purpose is not to substitute biotic values for anthropic ones, but rather to supplement the latter with the former and to weave them together coherently for the enhancement of both.

Biodiversity, faith and ethics (Dieter T. Hessel)

Globalized economic enterprise generates a pattern that destroys otherkind, threatens humankind, and even diminishes God. As Wallace (1996) puts it:

> 'God is so internally related to the universe that the specter
> of ecocide raises the risk of deicide; to wreak environmental
> havoc on the earth is to run the risk that we will do irreparable,
> even fatal harm to the Mystery we call God.'

Religious views and practices, viewed instrumentally as resources to meet the environmental challenge, are typically evaluated by how well they foster basic human attitudes of wonder, thanksgiving, humility, respect for persons, reverence for other life, and a sense of belonging to planet earth in the cosmos. Environmentalists judge the authenticity of religion by its cultivation of human spirituality (or spirited humanity) that cares deeply for the earth community. But the converse is also importantly true: richly biodiverse places – nature's living temples – are essential (though not in themselves sufficient) for healthy religions and cultures. The human spirit is always shaped in an ecological context, even if organized religion in that place fails to meet its special obligation to care for the earth.

So what ought humans to do? The answer embodied by the ecumenical movement evolving over the last three decades is to foster theological reconstruction and ethical engagement oriented to 'eco-justice' with a rich understanding of 'just, participatory and sustainable societies', and the imperatives of 'justice, peace and integrity of creation'. In this spiritually-grounded moral perspective, all beings on earth make up one household *(oikos)*, which benefits from an economy *(oikonomia)* that takes ecological and social stewardship *(oikonomos)* seriously. Humans everywhere must learn to revalue the natural world, and to welcome diverse cultures as well as the myriad species, while working for a just and sustainable community. 'Eco-justice' provides a dynamic framework for thought and action to foster ecological integrity while struggling for social and economic justice. Eco-justice occurs through constructive human responses that serve environmental health and social equity together – for the sake of human well-being with otherkind. The basic norms of an eco-justice ethic are:

- **solidarity** with other people and creatures – companions, victims and allies – in earth community, reflective of deep respect for creation;
- ecological **sustainability** – (environmentally fitting) habits of living and working that enable life to flourish while using appropriate technology on a human scale;
- **sufficiency** as a standard of organized sharing, which requires basic floors and definite ceilings of equitable or 'fair' consumption;
- socially just **participation** in decisions about how to obtain sustenance and to manage community life for the good in common and good of the commons.

These norms illumine an overarching imperative: humans should pursue – in reinforcing ways – what is both ecologically fitting and socially just.

Solidarity comprehends the full dimensions of earth community and of inter-human obligation. Sustainability gives high visibility to ecological integrity and wise behaviour throughout the resource-use cycle. The third and fourth norms express the distributive and participatory dimensions of basic social justice.

The observance of each ethical norm reinforces and qualifies the others. All four are core values that become corrective criteria to guide personal practice, social analysis and policy advocacy. They provide, of course, only the starting point for hard thought and tough choices, and the four norms sometimes come into conflict. For example, democratic participation is often (mis)used to privilege the role of humans at the expense of the norm of ecological sustainability. As Bakken (1995) puts it, 'The full range of social justice values that belong to the eco-justice paradigm includes liberty, community and dignity, as well as equality or equity. Depending on the situation, each of these [social justice] values can reinforce or be at odds with caring for the biosphere. For instance, environmental protections often restrict the liberty of some people to do certain things, but they can also enlarge liberty by giving persons and communities the freedom to engage their natural environments in more satisfying and responsible ways. With regard to equity, the costs and benefits of environmental protection are often unjustly distributed – but, as the 'environmental justice' movement reminds us, so are the costs and benefits of toxic pollution and environmental degradation'.

The point to be emphasized is that a rounded understanding of sustainable society and community life encompasses all four interrelated eco-justice norms as essential to a healthy future. These norms express moral consensus about what the earth community requires; the norms also have plural expression that is respectful of cultural–biotic diversity and alert to situational need.

The authentic human vocation is to respect created reality, to love near and distant neighbours in the spirited community of being, and to meet common needs on a basis of ecological integrity combined with social equity.

Interbeing: precepts and practices of an applied ecology (Joan Halifax and Marty Peale)

Advocacy on behalf of the earth's ecosystems is often conducted as if we were at war. For this reason, the Fourteen Precepts that Thich Nhat Hanh drafted for the Order of Interbeing (the Tiep Hien Order in the Lam Te (Lin Chi) and Lieu Quan Schools of Zen) in the wake of the Vietnam War are sound guidelines for an environmental activist. The Order is dedicated to alleviating suffering, fostering peace, raising awareness of non-duality, and raising awareness of the interpenetration of all beings. Thich Nhat Hanh's teachings reveal the depth of the suffering he witnessed and felt throughout the war that had raged for a quarter century, leaving 1,500,000 countrymen dead and forests defoliated across his tropical homeland – and they reveal the depth of his compassion. The precepts translate an awareness of 'no separate self' into daily life. They also translate contemplative and engaged practice into daily life as 'no separate act'.

What we see, then, informs our understanding of ourselves as intricately and intimately related in our world. What we do, then, actually weaves each of us more richly into the world as one who listens, one who responds personally and skilfully to changing conditions, and one who moves other forms of Being as s/he is moved. These are key aspects of an applied ecology.

The ramifications of dualism

Environmentalism, by definition, sets up a division between a human being and the rest of the world. 'Environment' means 'that which surrounds'. Although the word presents the world as an integrated context within which each being lives, it is problematic in that it splits organism from surrounding. While we undeniably perceive a distinction, the notion that any two things or two actions are separate is dangerously misleading.

Perhaps, because the word 'environmental' sets up a duality of organism and surroundings, it should come as no surprise that the language of environmental activism has grown divisive and combative. Marty was engaged in environmental work in Alaska and Washington, D.C. for over 15 years. She recounts:

> 'We spoke all the time of 'losing this battle' against logging on Kuiu Island or 'winning that fight' against dumping mining effluent into Berners Bay. We congratulated ourselves for 'outnumbering the opposition' at public hearings. We knew how many congressmen or legislators were 'in our camp' and how many we needed to 'win over to our side' on any given vote. Our strategic thinkers knew where 'the weakest link' was in 'the opposition's' case or coached us to 'go for the jugular'. We disparaged the oil company executives as we listened to their rationale for drilling in the coastal plain of the Arctic National Wildlife Refuge. And we

felt absolutely justified in our vilification of the Fish and Game Commissioner's proposal to hunt whole wolf packs with AK-47s from helicopters. We were at war, and the 'veterans' had their 'war stories'.

In 1986, Marty was working for the Northern Alaska Environmental Center in Fairbanks. The organization was suing the Alaska Department of Environmental Conservation for failure to enforce existing water quality regulations: at issue was chronic sediment and arsenic discharge from placer-mining operations upstream from an Athabascan village. The Environmental Center was not suing miners. It was not advocating enactment of more stringent regulations. It was simply insisting that the State enforce existing regulations. The Northern Center won the lawsuit.

The following day, a woman came into the front room of the little house in which the Center operated. She had two children in tow. They came in and stood there between the shelves of tee-shirts and note cards and the old metal desk, and she said, 'OK then, you feed them'. They were the children of a miner whose work was suspended because the Northern Center had won that lawsuit against the State of Alaska. The woman who brought them so that the environmentalists could see them was their mother. Marty remembers:

'I had to stop and sit down. I knew I could bake bread for them every week – and I knew that wasn't what she meant. They needed more than my generosity; they needed my understanding, and they needed a way to support themselves. I began to see that there was no value in winning the environmental battles, lawsuit-by-lawsuit and bill-by-bill, if hunger, bitterness and hatred flowed in the wake of 'justice'. I began to have a sense of my own insensitivity to other people, and to see the despair and anger that had grown in the absence of personal contact. I began to see that the environmental community was losing ground even as we were winning in the courts.'

Shifting the emphasis to relationships

'Ecology' is 'the science of relationships between organisms and their environments'. Literally, the word means 'the study of home', and the field of study includes humans along with all other components of ecosystems.

Unlike 'environment', the definition of ecology emphasizes relationships. For this reason perhaps, the words 'ecology' and 'ecological' are used, in not quite a strict sense, by many people who wish to invoke patient attention to the details and dynamics of existence without invoking the dichotomy and the resulting emotional charge associated with 'environment' and 'environmental'. While the definition of ecology places an emphasis on relationships between things, it falls short of acknowledging the in-flowings that blur the distinctions by which we rather arbitrarily distinguish things. Thich Nhat Hanh has coined the word 'interbeing' to address this dynamic of interpenetration.

What do we mean when we speak of 'an applied ecology'?

'To apply' means 'to make use of'. In the sciences, 'applied' refers to research that is need-driven, research designed to generate results that will be applied to a current problem. When we refer to 'applied ecology', we invoke an investigation of beings, communities, landscapes, energies and systems that is need-driven. We call for an investigation characterized by attention to detail and relationships. We have a clear intention to apply our observations in our day-to-day lives. And we reaffirm that such an investigation will inform our understanding of ourselves as intricately and intimately related to our world.

There are two bodies of practice in an applied ecology. One is to heighten awareness. The other is to engage – to literally reincorporate – the insights of interconnectedness, as we move in the fabric of Being as a whole.

When activism takes the form of raising awareness about the relationships between our individual, incremental decisions and the signs of strain in the world, then our actions can be constructive. When activism includes listening, it remains rooted in humility. When activism is conducted with respect for the perspectives, the needs and the fears of those who do not see the world the way we do, it can be compassionate. When activism includes a reassessment of the way we live our own lives, it can be community-building.

Let us consider, then, what shift can occur when we apply the Precepts of the Order of Interbeing to environmental activism. The First Precept states,

> 1. Do not be idolatrous about or bound to any doctrine, theory or ideology, even Buddhist ones. Buddhist systems of thought are guiding means; they are not absolute truth.

This suggests that, even if we have studied the ecology of the boreal forest, we must not assume that we know what is best for the forest. It suggests that neither an environmentalist nor an industrial logger knows absolute truth. It reminds us that suffering is based in beliefs that are profoundly misguided. Our world-view and the knowledge informing it is often an obstacle to our deeper understanding. The First Precept encourages us to develop a stance of absolute tolerance and complete openness and to expand the boundaries of our understanding. It invites us into a vital process of inquiry, reminding us that narrowness produces conflict and suffering, and that our true mandate is freedom from suffering for all beings.

> 2. Do not think that the knowledge you presently possess is changeless, absolute truth. Avoid being narrow-minded and bound to present views. Learn and practice non-attachment from views in order to be open to receive others' viewpoints. Truth is found in life and not merely in conceptual knowledge. Be ready to learn throughout your entire life and to observe reality in yourself and in the world at all times.

The Second Precept reminds us that knowledge is relative, and we are shown the door of compassion and interbeing as a way of deepening our understanding and opening ourselves to the viewpoints of other beings. This suggests that we should not enter into a discussion with hopes of changing the minds of people who do not share our views unless we too are open to the possibility of changing our minds in light of what they may have to tell us. It points to the possibility that what we know now might not to be true ten years from now or a hundred years from now. It invites us to be open-minded and aware of the inevitability of change and the interconnectedness of phenomena.

> 3. Do not force others, including children, by any means whatsoever, to adopt your views, whether by authority, threat, money, propaganda or even education. However, through compassionate dialogue, help others renounce fanaticism and narrowness.

This Precept points to freedom, freedom of thought and mind, and a call for us to respect the viewpoints of others. It also calls us to enter into 'compassionate dialogue', which means truly understanding the viewpoints of others. This suggests that personal discussions based in compassion are the most effective means of informing and engaging any individual in change. Neither arguments for economic security nor new legislation are conducive to broadening understanding and building community.

> 4. Do not avoid contact with suffering or close your eyes before suffering. Do not lose awareness of the existence of suffering in the life of the world. Find ways to be with those who are suffering by all means, including personal contact and visits, images, sound. By such means, awaken yourself and others to the reality of suffering in the world.

This Precept indicates that each of us would do well to become more aware of the suffering in our own lives and in the world in which we live, including those who do not agree with our way of thinking and living. It suggests that environmental advocates should look deeply into the lives of those who will be affected by their efforts and engage in constructive initiatives to address hunger, fear, economic insecurity and other forms of suffering that lead to environmental damage. The placer-miner's wife could not have more clearly demonstrated the capacity for personal contact to transform the nature of conflict. It also calls us to look for the causes of suffering and to work with the internal and social forces that give rise to suffering. This Precept is about stepping out of our protected corners and working for the well-being of others. It embodies the true spirit of the *Mahayana*, which is to help ourselves and others be free from ill-being. If we look deeply, we see that when one being suffers, all beings suffer.

> 5. Do not accumulate wealth while millions are hungry. Do not take as the aim of your life fame, profit, wealth or sensual pleasure. Live simply and share time, energy and material resources with those who are in need.

This Precept suggests that environmental activists live their own lives in a way that does not tax the earth or themselves. It invites us to examine the inevitable ramifications of our patterns of consumption as well as the stressful and driven lives that many of us lead. In the *Sutra on the Eight Realizations of the Great Beings*, it says, 'The human mind is always searching for possessions and never feeling fulfilled. This causes impure actions ever to increase'. A 'pure action' is the sharing of time, energy and material resources with those who need them. This lays a foundation for voluntary simplicity and deep ecology, a life that sees and confirms the interconnectedness of beings and actions, and their causal relationships.

> 6. Do not maintain anger or hatred. Learn to penetrate and transform them when they are still seeds in your consciousness. As soon as they arise, turn your attention to your breath in order to see and understand the nature of

> your anger and hatred and the nature of the persons who have caused your anger and hatred.

The Sixth Precept tells us to recognize that the seeds of anger are within us, whether or not we are angry. The practice of mindfulness has the potential for transforming these seeds before they sprout. If anger does arise, we can turn our awareness toward the anger in order to understand its roots. In the environmental movement, there is often a great deal of anger expressed. Environmentalists have been hung in effigy by loggers. Loggers have been vilified by environmentalists. Anger does not solve problems. It creates them. We need to be aware of blaming a person for his or her position or perspective. That person has been shaped by his/her experience as surely as we have been shaped by our own. We are asked to understand the roots of our anger and the roots of the anger expressed by those who confront us. This understanding can give rise to compassion.

> 7. Do not lose yourself in dispersion and in your surroundings. Practice mindful breathing to come back to what is happening in the present moment. Be in touch with what is wondrous, refreshing and healing both inside and round you. Plant seeds of joy, peace and understanding in yourself in order to facilitate the work of transformation in the depths of your consciousness.

This Precept suggests that an environmental activist step out of the business, the urgency and the assumptions that direct his/her activity. Being in the present moment, looking deeply into reality, helps life become real and of value. It invites the practitioner to a more wholesome and kind way of living, strengthening the mind and heart. Thich Nhat Hanh has said that this is the heart of the Precepts, and points to our struggle with forgetfulness and lack of understanding.

> 8. Do not utter words that can create discord and cause the community to break. Make every effort to reconcile all conflicts, however small.

This suggests that advocacy that divides a community will cause more suffering than any suffering it aims to alleviate. Reconciliation is built of small initiatives; peace is built through personal relationships. It suggests that I begin it.

> 9. Do not say untruthful things for the sake of personal interest or to impress people. Do not utter words that cause division or hatred. Do not spread news that you do not know to be certain. Do not criticize or condemn things that you are not sure of. Always speak truthfully and constructively. Have the courage to speak out about situations of injustice, even when doing so may threaten your own safety.

This suggests that antagonistic, divisive language is not an effective tool for increasing understanding about our relationship to the earth; impatience rules us when we reason that our ends justify our means. We are each responsible for increasing awareness in our communities through honest, compassionate, responsible and constructive dialogue. Raising awareness about the injustice which our over-consumption visits upon other species may call for all of us to simplify our lives and relinquish illusions of security. This Precept also points to the value of silence.

> 10. Do not use the Buddhist community for personal gain or transform your community into a political party. A religious community, however, should take a clear stand against oppression and injustice, and should strive to change the situation without engaging in partisan conflicts.

Thich Nhat Hanh has said, 'although religious communities are not political powers, they can use their influence to change society. Speaking out is the first step, proposing and supporting appropriate measures for change is the next. Most important is to transcend all partisan conflicts. The voice of caring and understanding must be distinct from the voice of ambition'. In many ways, for many activists, environmentalism is a religion. Our lives are dedicated to an entity, a whole realm of being, which is far greater than any one of us. We strive to live by a code of ethics. There are places we hold sacred, and we go to them for renewal. But like certain religious organizations, environmental groups can have more than a touch of partisan fanaticism. This Precept suggests that, while we work against injustice, we do so without generalizing, without a platform that does not take into account the well-being of our adversaries, and without an aim to govern others.

> 11. Do not live with a vocation that is harmful to humans and nature. Do not invest in companies that deprive others of their chance to life. Select a vocation that helps realize your ideal of compassion.

This Precept urges us to 'walk our talk'. It directs us to live and work in a way that is not damaging to people or the natural world, even as we oppose injustice. This Precept is about Right Livelihood, which points to practising a vocation that is based in compassion, and engaging in that vocation in a mindful way. Again, we are asked to look at our patterns of consumption. Although we might be a dedicated environmental advocate, we could be supporting others in harmful vocations through our style of life and work.

> 12. Do not kill. Do not let others kill. Find whatever means possible to protect life and to prevent war.

At the 1992 Retreat for Environmentalists, the Twelfth Precept, 'Do not kill', was interpreted as referring to all life. When it is understood to refer to all beings, he explained, several stages of working with the precept unfold. The first step is to 'take the precept', to affirm that one is determined to live by it. The second is to realize that one cannot always live up to this guideline, because the beets we eat, the organisms in the dry wood we burn, and the tiny creatures in the soil upon which we walk die in support of our lives. This realization leads the practitioner straight back to the First Precept which advised us not to be bound to any doctrine, even Buddhist ones. There is no black and white; there is at best a very personal exploration of the landscape of Being. The third stage, Thây said, is to confer with members of our community about the difficulties we face and the insights we have, as we succeed and fail to abide by the guideline not to kill. It is this last practice, he taught, that builds community.

This precept is extremely important for environmental activists, because it reminds us that we cannot afford to be righteous. We, too, are responsible for taking life, consciously or otherwise. We cannot point the finger at offenders as if we were not also implicated in harming. The humility that this engenders, and the honest communication that Thây requests of us, are antidotes to our tendency to alienate ourselves and polarize our communities.

> 13. Possess nothing that should belong to others. Respect the property of others, but prevent others from enriching themselves from human suffering or the suffering of other species on earth.

This Precept means, 'Do not steal'. Like the one before it, we violate this guideline, daily, as we wear leather, eat honey and eggs, build with lumber from the mixed hardwood forest, and clear a corner of a field for a garden plot. Everything that supports our life comes from another creature or another creature's habitat. Even the sunlight we intercept with a solar panel would have otherwise provided energy for a plant or warmed a thrush's nest. This precept sheds light on the depth of suffering created by overconsumption and environmental degradation – and the significance of voluntary simplicity. The standard of living we maintain in 'the First World' is supported by people and ecosystems around the world who grow our coffee and sell their rainforests for a song. It also points to social and economic injustice. How we define and legitimize theft must be called into question by any individual who strives to live by this Precept. And daily practice heightens appreciation of the lives that contribute their energy to our own.

> 14. Do not mistreat your body. Learn to handle it with respect. Do not look on your body as only an instrument. Preserve vital energies (sexual, breath, spirit) for the realization of the Way. Sexual expression should not happen without love and commitment. In sexual relationships, be aware of future suffering that may be caused. To preserve the happiness of others, respect the rights and commitments of others. Be fully aware of the responsibility of bringing new lives into the world. Meditate on the world into which you are bringing new beings.

This Precept affirms that we must conduct our lives and our work in a way that does not lead to the collapse of systems that support life, including our own bodies. Living in a way that leads to mental or physical illness it disrespectful of the gift of life and is harmful to our communities. This Precept also suggests that our lives serve to support our spiritual growth. It suggests that our physical, manual efforts are not our only means of working in the world. It suggests that we not become preoccupied with our bodies or other material aspects of the world, but keep a bearing on the Way, on love and commitment, and on the capacity and opportunity for new beings to flourish in the world and bodies they inherit. It also reminds us that overpopulation is causing immeasurable suffering to beings everywhere. We are admonished to consider the world our children will inherit.

The Precept speaks to activists who exhaust themselves as if their well-being were not as essential for the earth as that of the earth is for them. It speaks also to the activist who might otherwise fall into a habit of assuming that the state of a temperate rainforest or the Bering Sea cod fishery were more important than the calling for her to develop compassion. Marty says that it occurred to her after more than a decade of advocacy that perhaps all the ranks of Douglas fir and wolves, fin whales and condors have stood up and will continue to stand up among humans, one after another, risking their lives to crossfire, to see if we have yet learned to put down our weapons, to see if we have yet learned to engage with compassion.

Psychospiritual effects of biodiversity loss in Celtic culture and its contemporary geopoetic restoration (Alastair McIntosh)

In 1743 the last wild wolf in Scotland was shot by a hunter named McQueen on the territory of my tribal clan, Mackintosh, in the upper reaches of the Findhorn River (Harting 1972). Three years later, and in the same region, the last battle to be fought on mainland British soil, Culloden, put an end to the old culture of the Scottish Highlands in an act of internal colonial conquest by the consolidating British state. This marked the onset of the 'Highland Clearances', whereby some half a million people were forced off their land to make way for sheep ranching and blood sports. Today, Scotland retains a feudal system in which nearly two-thirds of its private land is owned by only about 1,000 people – one fiftieth of one percent of the population.

It was doubtless coincidence that a significant element of ecocide, the wolf's local extinction, preceded cultural genocide. But in terms of psychohistory, it is possible that the events were connected. Gaelic traditions of (Celtic) Scotland and Ireland suggest that we must understand the psychospiritual impact of historical processes of degradation in our relationship to nature in order to create an authentic human ecology. This lies at the core of 'sustainable development' and, from a Celtic cultural perspective, it is in part a spiritual process. Spirituality is about interconnection, thus the loss of the wolf, or any other depletion of biodiversity, can be seen as a loss of an aspect of our extended selves. Such a principle of deep ecology is therefore germane to indigenous Celtic spirituality.

Celtic 'bards' or poets were in touch with the equivalent of our song-lines and dreamtime. A growing body of evidence, much as yet unpublished, points to their shamanic role and technique, including things like the use of *tigh n' alluis* or Irish sweat lodges for *dercad* meditation, leading to a state of *sitchain* or mystical peace (Ellis 1995). However, from the repressive 1609 Statutes of Iona onwards, the bards' role in maintaining cultural and ecological processes was repressed or marginalized. As in so many colonized traditional societies around the world, poetic power, by which the deep Spirit was expressed through socio-political structures, was replaced by the power of money and money as power.

Celtic nature poetry

The earliest recorded Celtic nature poetry reveals an acute sensitivity to the wild which is often intermingled with Christian devotional material. It illustrates Celtic Christianity's most distinctive feature – God and nature. The highest flowering of early Celtic nature poetry, from the eighth to the twelfth century, reveals a creation-centred theology that anticipated, by at least a millennium, modern ecotheology and deep ecology.

Consider the representative passages I have linked together here, mostly from Seamus Heaney's translation of *Sweeney Astray*. Heaney draws on a relatively late, seventeenth century Irish manuscript, but one that had probably taken shape in the ninth century, which starts with the Battle of Moira of 637CE. In it, Suibhne or Sweeney, a seventh-century Irish or possibly Scottish 'king, saint or holy fool', is sent 'mad' in battle by a cleric's curse. This was a shamanic madness or *geilt*. It projects him on a journey where he falls in love with nature and becomes a poet of deep ecology. He reconciles easily such 'paganism' with Christ, but frequently challenges the clerical efforts then being made to build an institutional church. Transformed by the 'curse and miracle' of his affliction into a bird, Suibhne flies around Ireland and the west of Scotland, where he contemplates for six weeks in St. Donan's Cave on the Hebridean Isle of Eigg. And he proclaims:

> 'I perched for rest and imagined cuckoos calling across water, the Bann cuckoos, calling sweeter than church bells that whinge and grind ... From lonely cliff tops, the stag bells and makes the whole glen shake and re-echo. I am ravished. Unearthly sweetness shakes my breast. O Christ, the loving and the sinless, hear my prayer, attend, O Christ, and let nothing separate us. Blend me forever in your sweetness ... I prefer the badgers in their sett to the tally-ho of the morning hunt; I prefer the re-echoing belling of a stag among the peaks to that arrogant horn ... Though you think sweet, yonder in your church, the gentle talk of your students, sweeter I think the splendid talking the wolves make in Glenn Bolcain. Though you like the fat and meat which are eaten in the drinking halls, I like better to eat a head of clean water-cress in a place without sorrow ...' (Heaney 1984: 19, 20, 43; Jackson 1971: 255)

Typical of actual historical accounts that reveal the importance of nature is a Gaelic song, translated by Michael Newton (personal correspondence) as *'O Green Morvern of the Hills'*. It was written by Donald Mackinnon, 'Domhnull Ruadh', a West Highland bard, between 1845 and 1860. MacKinnon laments both the ecological and the social changes heralded after a wealthy Londoner, Octavius Smith, bought the 'property' in 1845 and cleared the indigenous people away to other lands on the tall ships to create space for ecologically devastating sheep ranching.

O green Morvern of the hills
You look full of despair and sorrow
The situation has become very desperate
That you might turn completely into a wasteland

The reason for my sadness
Is to be gazing at your hills
Down beside the Sound of Mull
Toward the ships of tall masts

It is your non-native lords
Who left the natives dispossessed
And who let the people who don't belong to you
Dwell in their place ...

They called you 'land of the woods'
And there was a time when you deserved that
But today your woods have been denuded
By the people of the pale-faces

A modern example, that emphasizes the need for healing such alienation or 'anomie', comes from Clydebank, an industrial town down the river from Glasgow. Here my friend, the contemporary Gaelic poet who was born there, Duncan MacLaren (personal communication), dreams of his people's Hebridean origins and yearns for rustic frugality rather than the multiple deprivation of urban decay where unemployment, during the 1980s, exceeded thirty percent:

Bruach Chluaidh. Bidh Bruadar air uair agam 's tu nad eilean air bhog eadar Barraidh 's ceann an Neimh ... Clydebank – I sometimes dream that you are an island afloat between Barra and the end of Heaven and that the only speech on the tongues of your people is the language of the Hebrides and the mists would put a poultice on your stinking houses and it wouldn't be vomit on the street but bog-cotton and your rusty river would be a dark-green sea. And, in the faces of your people, the wrinkles of their misery would only be the lash of wind and waves and your grinding poverty would somehow be diminished *... agus thigeadh lughdachadh air do bhochdainn chraidh.*

Cauterization of the heart?

The evidence I have touched upon from the bardic record suggests that the human heart became cauterized by historical vicissitudes – broken and sealed off from its cause of suffering. Could it be that, at an unconscious level, most of us still carry such echoes of that painful past? Could it be that we accumulate the effects of bygone extinctions, degradations and colonizations, and that these inhibit our ability to act; bind us in our apparent powerlessness to resolve the major issues of our times? Is there a parallel here with other forms of 'intergenerational trauma' such as sexual abuse, addiction and violence that can be handed down from generation to generation within families?

Robert Burns (1759–1796), the most acclaimed bard of Scotland, lived in the immediate aftermath of Culloden. In his two-verse poem, *Strathallan's Lament* (1767), he brilliantly illustrates the psychology of despair that lies behind cultural genocide. I have alluded to how ecocide depletes culture. Here we see the counterpoint – repression of culture setting in place the preconditions of ecocide. Burns stands himself in the shoes of the fifth Viscount Strathallan whose Highland father was killed by the forces of the British state at Culloden (1746). He portrays this battle as having left behind an emotionally vacant modern world 'without a friend'; one in which the ravishes of neither nature nor friendship (the 'busy haunts of base mankind') could be appreciated. The ability to feel had indeed been cauterized; the very ability to perceive reality, altered. As the thatched houses of peasant farmer and fisher crofters were set ablaze by the 'butcher' Cumberland's vanquishing troops – who had been ordered to give 'Bonnie Prince Charlie's' retreating soldiers of the 1745 Jacobite uprising 'no quarter' – the soul of a people resisting the unacceptable heartlessness of the forces of modernity was dealt a near-mortal blow:

Thickest night, surround my dwelling!
Howling tempests, o'er me rave!
Turbid torrents wintry-swelling,
Roaring by my lonely cave!
Crystal streamlets gently flowing,
Busy haunts of base mankind,
Western breezes softly blowing,
Suit not my distracted mind.

In the cause of Right engaged,
Wrongs injurious to redress,
Honour's war we strongly waged,
But the heavens deny'd success.

Ruin's wheel has driven o'er us;
Not a hope that dare attend,
The wide world is all before us,
But a world without a friend

(Mackay 1993: 287)

Towards a cultural psychotherapy

I consider that the renewal of community and the cultural spirit comes about in considerable degree through re-connection with the deep poetics of place – that is, with the totems and their expression of underlying psychospiritual dynamics. This is broadly what the Scots poet Kenneth White (1992: 174) calls, 'Poetry, geography – and a higher unity: *geopoetics*'. Such a mythopoetic underpinning rehabilitates, after four centuries, the repressed bardic tradition. But can what amounts to a psychotherapy be effective as a change agent?

Conscientization, as Freire (1972) would call it, partly through a geopoetic approach, was used over a seven-year period in helping to bring about land reform on the Isle of Eigg in the Scottish Hebridean islands (McIntosh 1997). A community of some sixty-five people were assisted, by me, many others, and not least themselves, to waken up to their history, grasp a vision of what it could be like if seven generations of oppressive landlordism were put behind them, and develop the unity and media know-how that enabled them to repel the landlord's 1994 attempt to evict 12 percent of the population for no apparent good reason, except, perhaps, that they were starting to speak out and take responsibility for their own emancipation. The process finally resulted in the island coming into community ownership on 12 June 1997. Partners with the community in the process were both the local authority, Highland Council, and the Scottish Wildlife Trust – a unique unity of social and conservation interests.

Part of the community's empowerment process involved erecting two standing stones – one of which had last stood some 5,000 years previously when the ancient forests were at their maximum post-glacial extent. Another part entailed an only-half-joking identification with legends about Amazonian 'big women' on Eigg by some of the women activists who played lead roles. A further element was the organizing of the first traditional *feis* (music festival) to be held in modern times. This restored a respect for indigenous arts and knowledge, as communicated by some of the elderly residents who had thought it would die with them. For some, the two pairs of eagles resident on the island, and the oystercatcher associated with St. Bride (Bridgit, the Goddess Brigh), and the little flowers through which 'God can be seen', played privately empowering roles.

I have used a similar geopoetic process in my work opposing the proposed Isle of Harris 'super-quarry'. This would turn the biggest mountain in the south of this National Scenic Area in the Outer Hebrides into road aggregate. Initially, the community were 90 percent in favour of the quarry because of promised jobs. It was clear that conventional arguments alone were not sufficiently going to shift local opinion. Working in close liaison with other objectors, I arranged for the Rev. Prof. Donald MacLeod of the Free Church College in Edinburgh and Canadian Mi'Kmaq Warrior Chief and sacred peace pipe carrier, Sulian Stone Eagle Herney, to give evidence at the inquiry with me on theological grounds. This approach reached to the taproot of vernacular values in what is a deeply Calvinist culture. The relevance of Stone Eagle was that, although not a Christian, his people were also alienated from their tribal lands – ironically, in large measure, by cleared Highland settlers. They are now subject to a super-quarry threat at their sacred mountain, *Kluscap* (Kelly's Mountain). In effect, the theological testimony drew out a Judeo-Christian ecology which received extensive local, national, and international media coverage.

At the conclusion of the government public inquiry on Harris (which has yet to report its findings), the island's chosen representative, John MacAulay, summed up the cultural, economic and environmental importance of the mountain. He said, 'It is not a 'holy mountain' but is certainly worthy of reverence for its place in Creation'. ('Let the people of Harris decide on their own future', *West Highland Free Press*, 9 June 1995, 5).

Reverence towards nature and one another is the keystone of 'sustainable development'. As if to corroborate, the Irish-born Scottish-based theologian Fr. Noel O'Donoghue makes use of words almost identical to those chosen in Mr MacAulay's full statement to describe what he calls 'the mountain behind' such mountains: He writes in the context of discussing Kathleen Raine's poetry in his book which uses as its title an expression of hers, *The Mountain Behind the Mountain*:

> 'The mountain behind the mountain is not the perfect or ideal mountain in some Platonic sense ... [I]t is neither an ideal nor a mythical mountain, nor is it exactly a holy or sacred mountain made sacred by theophany or transfiguration. No, it is a very ordinary, very physical, very material mountain, a place of sheep and kine (cattle), of peat, and of streams that

one might fish in or bathe in on a summer's day. It is an elemental mountain, of earth and air and water and fire, of sun and moon and wind and rain. What makes it special for me and for the people from which I come is that it is a place of Presence and a place of presences. Only those who can perceive this in its ordinariness can encounter the mountain behind the mountain.' (O'Donoghue 1993, 30–31).

Thus the ancient Celtic world-view has a place even in modern political debate. It suggests that to open ourselves to sustainable life-ways, we must cut through the conspiracy of silence that marginalizes the cultural and the spiritual in modernist and postmodernist nihilism. Only then can we hope to know the wolf that remains alive in our souls; the mountain behind the mountain; the ecological self underlying the urbanized self. Only then can we glimpse the further reaches of reality, which is the sole font of values often perceived but dimly.

The ecology of animism (David Abram)

Late one evening, I stepped out of my little hut in the rice paddies of eastern Bali and found myself falling through space. Over my head the black sky was rippling with stars, densely clustered in some regions, almost blocking out the darkness between them, and more loosely scattered in other areas, pulsing and beckoning to each other. Behind them all streamed the great river of light, with its several tributaries. But the Milky Way churned beneath me as well, for my hut was set in the middle of a large patchwork of rice paddies, separated from each other by narrow two-foot high dikes, and these paddies were all filled with water. The surface of these pools, by day, reflected perfectly the blue sky – a reflection broken only by the thin, bright green tips of new rice. But by night, the stars themselves glimmered from the surface of the paddies, and the river of light whirled through the darkness underfoot as well as above; there seemed no ground in front of my feet, only the abyss of star-studded space falling away forever.

I was no longer simply beneath the night sky, but also above it. The immediate impression was of weightlessness. I might perhaps have been able to reorient myself, to regain some sense of ground and gravity, were it not for a fact that confounded my senses entirely: between the galaxies below and the constellations above drifted countless fireflies, their lights flickering like the stars, some drifting up to join the clusters of stars overhead, others, like graceful meteors, slipping down from above to join the constellations underfoot, and all these paths of light upward and downward were mirrored, as well, in the still surface of the paddies. I felt myself at times falling through space, at other moments floating and drifting. I simply could not dispel the profound vertigo and giddiness; the paths of the fireflies, and their reflections in the water's surface, held me in a sustained trance. Even after I crawled back to my hut and shut the door on this whirling world, I felt that now the little room in which I lay was itself floating free of the earth.

Fireflies! It was in Indonesia, you see, that I was first introduced to the world of insects, and there that I first learned of the great influence that insects, such diminutive entities, could have upon the human senses. I had travelled to Southeast Asia in order to study magic; more precisely, to study the relationship between magic and medicine, first among the traditional sorcerers, or *dukuns*, of the Indonesian archipelago, and later among the *dzankris*, the traditional shamans of Nepal. One aspect of my research was somewhat unique: I was journeying through rural villages not outwardly as an anthropologist or academic researcher, but as a magician in my own right, in hopes of gaining a more direct access to the local sorcerers. I had been a professional sleight-of-hand magician for five years back in the United States, helping to put myself through college by performing in clubs and restaurants throughout New England. I had, as well, taken a year off from my studies in the psychology of perception to travel as a street magician through Europe, and toward the end of that journey had spent some months in London, England, exploring the use of sleight-of-hand magic in psychotherapy, as a means of engendering communication with distressed individuals largely unapproachable by clinical healers. The success of this work suggested to me that sleight-of-hand might lend itself well to the curative arts, and I became, for the first time, interested in the relation, largely forgotten in the West, between folk-medicine and magic.

It is thus that, two years later, I embarked upon my sojourn as a magician in rural Asia. There, my sleight-of-hand skills proved invaluable as a means of stirring the curiosity of the local shamans. For magicians, whether modern entertainers or indigenous, tribal sorcerers, have in common the fact that they work with the malleable texture of perception. When the local sorcerers gleaned that I had at least some rudimentary skill in altering the common field of perception, I was invited into their homes, asked to share secrets with them, and eventually encouraged, even urged, to participate in various rituals and ceremonies.

But the focus of my research gradually shifted from questions regarding the application of magical techniques in medicine and ritual curing, toward a deeper pondering of the relation between traditional magic and the animate natural world. This broader concern seemed to hold the keys to the earlier questions. For none of the several island sorcerers that I came to know in Indonesia, nor any of the *dzankris* with whom I lived in Nepal, considered their work as ritual healers to be their major role or function within their communities. Most of them, to be sure, were the primary healers or 'doctors' for the villages in their vicinity, and they were often spoken of as such by the inhabitants of those villages. But the villagers also sometimes spoke of them, in low voices and in very private conversations, as witches (or *lejaks* in Bali), as dark magicians who at night might well be practising their healing spells backwards, (or while turning to the left instead of to the right) in order to afflict people with the very diseases that they would later work to cure by day. Such suspicions seemed fairly common in Indonesia, and often were harboured with regard to the most effective and powerful healers, those who were most renowned for their skill in driving out illness. For it was assumed that a magician, in order to expel malevolent influences, must have a strong understanding of those influences and demons: even, in some areas, a close rapport with such powers. I myself never consciously saw any of those magicians or shamans with whom I became acquainted engage in magic for harmful purposes, nor any convincing evidence that they had ever done so. (Few of the magicians that I came to know even accepted money in return for their services, although they did accept gifts in the way of food, blankets and the like.) Yet I was struck by the fact that none of them ever did or said anything to counter such disturbing rumours and speculations, which circulated quietly through the regions where they lived. Slowly I came to recognize that it was through the agency of such rumours that the sorcerers were able to maintain a basic level of privacy. If the villagers did not entertain certain fears about the local sorcerer then they would likely come to obtain her or his magical help for every little malady and disturbance, and since a more potent practitioner must provide services for several large villages, he would be swamped from morning to night with requests for ritual aid.

This privacy, in turn, left the magician free to attend to what he acknowledged as his primary craft and function. A clue to this function may be found in the circumstance that such magicians rarely dwell at the heart of their village; rather their dwellings are commonly at the spatial periphery of the community or, more often, out beyond the edges of the village – amid the rice fields, or in a forest, or a wild cluster of boulders. I could easily attribute this to the just-mentioned need for privacy, yet for the magician in a traditional culture it seems as well to serve another purpose, providing a spatial expression of his or her symbolic position with regard to the community. For the magician's intelligence is not encompassed within the society; its place is at the edge of the community, mediating between the human community and the larger community of beings upon which the village depends for its nourishment and sustenance. This larger community includes, along with the humans, the multiple non-human entities that constitute the local landscape, from the myriad plants and the diverse animals, birds, mammals, fish, reptiles and insects that inhabit or migrate through the region, to the particular winds and weather patterns that inform the local geography, as well as the various landforms, forests, rivers, caves and mountains that lend their specific character to the surrounding earth.

The traditional or tribal magician, I came to discern, acts as an intermediary between the human community and the larger ecological field, ensuring that there is an appropriate flow of nourishment, not just from the landscape to the human inhabitants, but also from the human community back to the local earth. By his constant rituals, trances, ecstasies and 'journeys' he ensures that the relation between human society and the larger society of beings is balanced and reciprocal, and that the village never takes more from the living land than it returns to it, not just materially but also with prayers, propitiations and praise.

The scale of a harvest or the size of a hunt are always negotiated between the tribal community and the natural world that it inhabits. To some extent every adult in the community is engaged in this process of listening and attuning to the other presences that surround and influence daily life. But the shaman or sorcerer is the exemplary voyager in the intermediate realm between the human and the more-than-human worlds; the primary strategist and negotiator in any dealings with the Others.

And it is only as a result of her continual engagement with the animate powers that dwell beyond the human community that the traditional magician is able to alleviate many individual illnesses that arise within that community. The sorcerer derives her ability to cure ailments from her more continuous practice of 'healing' or balancing the community's relation with the surrounding land. Disease, in many such cultures, is conceptualized as a kind of systemic imbalance within the sick person, or more vividly as the intrusion of a demonic or malevolent presence into his body. There are, at times, malevolent influences within the village or tribe itself that disrupt the health and emotional well-being of susceptible individuals within the community. Yet such destructive influences within the human community are commonly traceable to a disequilibrium between that community and the larger field of forces in which it is embedded. Only those persons who, by their everyday practice, are involved in monitoring and maintaining the relations between the human village and the larger animate environment are able to appropriately diagnose, treat and ultimately relieve personal ailments and illnesses arising within the village. Any healer who was not simultaneously attending to the intertwined relation between the human community and the larger, more-than-human field, would likely dispel an illness from one person only to have the same problem arise (perhaps in a new guise) somewhere else in the community. Hence, the traditional magician or medicine-person functions primarily as an intermediary between human and non-human worlds, and only secondarily as a healer. Without a continually adjusted awareness of the relative balance or imbalance between the human group and its non-human environment, along with the skills necessary to modulate that primary relation, any 'healer' is worthless; indeed, not a healer at all. The medicine-person's primary allegiance, then, is not to the human community, but to the earthly web of relations in which that community is embedded. It is from this that her or his power to alleviate human illness derives, and this sets the local magician apart from other persons.

The most sophisticated definition of 'magic' that now circulates through the American counter-culture is 'the ability or power to alter one's consciousness at will'. Oddly, there is no mention made of any reason for altering one's consciousness. Yet in tribal cultures that which we call 'magic' takes all of its meaning from the fact that humans, in an indigenous and oral context, experience their own consciousness as simply one form of awareness among many others. The traditional magician cultivates an ability to shift out of his or her common state of consciousness precisely in order to make contact with the other organic forms of sensitivity and awareness with which human existence is entwined. Only by temporarily shedding the accepted perceptual logic of his culture can the shaman hope to enter into relations with other species on their own terms; only by altering the common organization of his senses will he be able to enter into a rapport with the multiple non-human sensibilities that animate the local landscape. It is this, we might say, that defines a shaman: the ability to readily slip out of the perceptual boundaries that demarcate his or her particular culture – boundaries reinforced by social customs, taboos, and most importantly the common speech or language – in order to make contact with, and learn from, the other powers in the land. His/her magic is precisely this heightened receptivity to the meaningful solicitations – songs, cries, gestures – of the larger, more-than-human field.

Magic, then, in its most primordial sense, is the experience of existing in a world made up of multiple intelligences, the intuition that every form one perceives – from the swallow swooping overhead to the fly on a blade of grass, and indeed the blade of grass itself – is an experiencing form, an entity with is own predilections and sensations, albeit sensations that are very different from our own.

In the course of living with traditional magicians and shamans I found my own sensory experience beginning to shift and transform; I became increasingly susceptible to the solicitations of non-human things. In the course of struggling to decipher the magicians' odd gestures or to fathom their constant spoken references to powers unseen and unheard, I began to see and to hear in a manner I never had before. When a magician spoke of a power or 'presence' lingering in the corner of his house, I learned to notice the ray of sunlight that was then pouring through a chink in the roof, illuminating a column of drifting dust, and to realize that that column of light was indeed a power, influencing the air currents by its warmth, and indeed influencing the whole mood of the room: although I had not consciously seen it before, it had already been structuring my experience. My ears began to attend, in a new way, to the songs of birds – no longer just a melodic background to human speech, but meaningful speech in its own right, responding to and commenting on events in the surrounding earth. I became a student of subtle differences: the way a breeze may flutter a single leaf on a whole tree, leaving the others silent and unmoved (had not that leaf, then, been brushed by a magic?); or the way the intensity of the sun's heat expresses itself in the precise rhythm of the crickets. Walking along the dirt paths, I learned to slow my pace in order to feel the difference between one nearby hill and the next, or to taste the presence of a particular field at a certain time of day when, as I had been told by a local *dukun*, the place had a special power and proffered unique gifts. It was a power communicated to my senses by the way the shadows of the trees fell at that hour, and by smells that only then lingered in the tops of the grasses without being wafted away by the wind, and by other elements I could only isolate after many days of stopping and listening.

And gradually, then, other animals began to intercept me in my wanderings, as if some quality in my posture or the rhythm of my breathing had disarmed their wariness; I would find myself face to face with monkeys, and with large lizards that did not slither away when I spoke, but leaned forward in apparent curiosity. In rural Java I often noticed monkeys accompanying me in the branches overhead, and ravens walked toward me on the road, croaking. While at Pangandaran, a nature preserve on a peninsula jutting out from the south coast of Java ('a place of many spirits', I was told darkly in the nearby fishing village), I stepped out from a clutch of trees and found myself looking into the face of one of the rare and beautiful bison that exist only on that island. Our eyes locked. When it snorted, I snorted back; when it shifted its shoulders, I shifted my stance; when I tossed my head, it tossed its head in reply. I found myself caught in a non-verbal conversation with this Other, a gestural duet with which my conscious awareness had very little to do. It was as if my body in its actions was suddenly being motivated by a wisdom older than my thinking mind, as though it was held and moved by a logos – deeper than words – spoken by the Other's body, the trees, the wind and the stony ground on which we stood.

Our inability to discern the shaman's allegiance to the more-than-human natural world has led to a curious circumstance in the 'developed world' today, where many persons in search of spiritual understanding are enrolling for workshops in 'shamanic' methods of personal discovery and revelation. Psychotherapists and some physicians have begun to specialize in 'shamanic healing techniques'. 'Shamanism' has thus come to connote an alternative form of therapy; the emphasis, among these practitioners, is on personal insight and curing.

These are noble aims, to be sure, yet they are secondary to, and entirely derivative from, the primary role of the indigenous shaman, a role that cannot be fulfilled without long and sustained exposure to wild nature, to its patterns and vicissitudes. Mimicking the indigenous shaman's curative methods without his or her intimate knowledge of the wider natural community cannot, if I am correct, do anything more than trade certain symptoms for others, or shift the locus of dis-ease from place to place within the human community. For the source of stress lies in the relation between the human community and the natural landscape in which it is embedded.

Contemporary civilization, of course, with its massive scale and rapidly globalizing economy, can hardly be seen in relation to any particular landscape or ecosystem; the more-than-human ecology with which it is directly engaged is the biosphere itself. Sadly, our civilization's relation to the earthly biosphere can in no way be considered a reciprocal or balanced one: with thousands of acres of non-regenerating forest disappearing every hour, and hundreds of our fellow species becoming extinct each month as a result of our civilization's excesses, we can hardly be surprised by the amount of epidemic illness in Western culture, from increasingly severe immune dysfunctions and cancers, to widespread psychological distress, depression, and ever more frequent suicides, to the accelerating number of household killings and mass murders committed for no apparent reason by otherwise coherent individuals.

From an animistic perspective, the clearest source of all this distress, both physical and psychological, lies in the aforementioned violence needlessly perpetrated by our civilization on the ecology of the planet; only by alleviating the latter will we be able to heal the former. While this may sound at first like a simple statement of faith, it makes eminent and obvious sense as soon as we acknowledge our thorough dependence upon the countless other organisms with whom we have evolved. Caught up in a mass of abstractions, our attention hypnotized by a host of human-made technologies that only reflect us back to ourselves, it is all too easy for us to forget our carnal inherence in a more-than-human matrix of sensations and sensibilities. Our bodies have formed themselves in delicate reciprocity with the manifold textures, sounds and shapes of an animate earth – our eyes have evolved in subtle interaction with other eyes, as our ears are attuned by their very structure to the howling of wolves and the honking of geese. To shut ourselves off from these other voices, to continue by our life-styles to condemn these other sensibilities to the oblivion of extinction, is to rob our own senses of their integrity, and to rob our minds of their coherence. We are human only in contact, and conviviality, with what is not human. Only in contact with what is other do we begin to heal ourselves.

Where is the environment? (James Hillman)

According to dictionary definitions, the word 'environment' means 'surrounding'. It comes from a French word, *viver*, meaning 'turns around, rings and circles'. All kinds of things turn around in rings and circles, and the moment we think of that, we think about ourselves in the middle of that circle. So the narrow sense of environment is the immediate world of things that are mainly natural – things close at hand. The middle sense of environment is a habitat, an ecosystem, on which anything depends; it's interactive; we're involved with it: things that turn on each other or evolve together, coevolution, turning around on each other. The widest sense of environment we find in psychology; for example, with Jung's sense of the psyche or the collective unconscious or Freud's id. Theodore Roszak made a beautiful point of saying that the id of Freud and the collective unconscious are actually the physical environment. It's out there. Or you can say it is the Gaia world of Lovelock and Margulis or what David Abram calls the *ruach elohim*, the breath of God, the whole air, aquaspheric, spirit, breath world. That is the wider sense of what encircles or is around.

But where does it stop? Where does the environment stop? Are there degrees of the environment? Is anything around environment? Is anything in this room, is everything in this room, the environment? And where does it stop physically, geographically? My breakfast bananas from Ecuador, where do they stop? Are they the environment when I'm sitting with them in the kitchen or I'm eating them? Or does that include Ecuador where they're grown? For instance, the sea sounds outside your bedroom, is that part of the environment? Where do they stop? And is the inorganic also part of the environment? Because we tend to think mainly of the organic world as environment.

Then, another major question: How do all these parts of the environment form a whole? We might call everything in this room the environment but is it only because it's in this room? What forms the whole of an environment? It could be that it's a matter of a foreground and background, that some things are in the foreground and other things in the background, and they are less environmental because they're in the background. So we take them as less significant in the environment. But then, what puts things in the foreground and some things in the background? In our world today, we take certain things as very important in the foreground – the air or noise pollution or the light pollution. For other people or other parts of the world or other creatures of the world, these may not be in the foreground at all.

What forms the whole environment? Is it a composition? Is there some kind of formal law that makes for an environment? This would bear on artwork. Is there a composition that goes into making an environment so that we can speak of 'an environment'. Must there be a relationship with it? Is it an I-thou relation? Is that important for environment? Or is there environment without any relation: We're just there in it. Or it's just there.

Now the *gestalt* philosophers would say that there has to be some sort of formal quality that makes for a whole. It would have to be an expressive form or an aesthetic frame. Or there would have to be organic mutuality: The insects depend on the plants and the plants depend on the soil and the soil requires a certain amount of moisture and the insects live in the soil and the roots of the plant so there is an organic mutuality. That is an environment. So you can't disrupt any piece of that. But again, what cuts some things out, or do you include everything in? These are also aesthetic questions because certainly when you're working on a piece of art, you're cutting some things out and not including everything that is there.

Or, could there be an inherent tension that requires certain things to stay together and other things to drop out – an inherent tension that holds a drop of water? There are barriers all the way through the world, not just the natural, but there are things that close in on themselves and keep other things out. And that becomes an environment. There is a skin to it, so to speak.

Other questions come up: the relation between parts of the environment. Whatever you're using as the environment, your kitchen or the garden or Gaia, the planet, whatever world you're using – are the parts related internally or externally? This is a philosophical question. If the parts are related internally to each other, they are necessary to each other and they are co-present to each other. You can't have one without the other. If they are externally related, then you have to raise the question of how does this relate to that. So you have to create some kind of force or some kind of modality, some bridge, a third component, that connects these two things because those are external relationships.

An internal relationship would be like Yin and Yang. They are necessary to each other: they are co-present or coterminate or required. That may be what is essential for an environment: the parts are internally related, one to another. So I think 'environment' can be expressed not just biologically; it can be expressed philosophically as an internal relationship. And, if an environment is internally related, its parts could be related not just logically but also invisibly, and that is what I'm particularly interested in.

There are problems posed by this word 'environment'. The power of the word, ringed around or encircled, tends to put a 'me' in the middle. And therefore, because of the way we tend to think, I can do what I want and not notice too much how I affect the other components of the environment. That's one problem posed by the word. Another is that the word 'environment', ringed around, evokes the archetypal image of the enclosed garden, a *hortus inclusus*, the Catholic image some may remember. Mary was considered an enclosed garden. You see a lot of medieval images of an enclosed garden with a unicorn in it or Mary sitting in a walled garden. Or Eden, paradise, the garden surrounded by a river, the four rivers. This is the idea that the environment is in some way a sanctuary. That archetypal image of the enclosed garden motivates a lot of green work by environmental, green people. The restoration of the perfect world, the Garden. When the environment becomes nature and nature becomes touched by that archetypal fantasy of a walled perfect closed space, then nature becomes an idealized place.

We return now to the idea that the items that make up an environment are related or belong together or fit invisibly. What is the logic of the invisible? I'm trying to get away from the idea that an environment is just simply what's there, what's hanging around and happens to be there and that anything is the environment. For there's some kind of internal necessity that this word is covering. This word is a mask covering something much more than simply the bushes and trees that we call the 'environment' or what environmentalists call 'the environment'.

I want to get at what this word 'environment' is doing. What is its power? There may be formal laws that hold. Something invisible is determining the visible. The word 'fitting' is one of those words Plato was concerned with – appropriate, fitting. Now perhaps this invisible holding of the environment is more biological or bioregional, i.e. a mutual organic necessity that forms a bioregion. Yellowstone Park, for example, doesn't stop with the geographic borders of Yellowstone Park as marked on the map, but it's the whole bioregion around it. So when wolves or bison are protected in Yellowstone and they go over the border of the park, they're shot because they're outside of the park, but they're not outside of the bioregion.

Or, is what holds it together a simple location? That is, all these things are in the same perceptual field. They are all in this room so the simple fact that they're all here is the law that holds them together. Simple location. They're here, present in this place. It seems to me that that's not enough of an idea, just simple location. Because it's meaningless that these phenomena all happen to be in this room at the same time, accidental, a mere collection of stuff. Unless you're mystical and say God put them all here at this place and that's it.

Maybe what holds them all together is a common symbolic meaning under a dominant theme such as the environment of a church where all the things belong because they are held together by the church. Or your kitchen, where they're all held together by the theme of the kitchen – the utensils, the pots, the pans – all the things that belong to the environment.

Or, perhaps it is 'time' that gives them an internal relationship. Then, the invisible glue is time, simultaneity. They are all there at the same time. It's just chance but they're all there at the same time and that makes this environment or that environment.

These are not biological ways of thinking. I am attempting to get the word 'environment' freed from where it is, trapped in biology and then trapped in nature – in our fantasies of nature. So that we can think more aesthetically. I want to break down the green approach to the environment and get back to the city and all the other kinds of environments that are so crucial.

So what is the invisible logic or force that holds together the discrete items in a surrounding. Let's take a look at the word 'invisible'. The word has this little prefix 'in', which means two very different things. It means on the one hand, 'not'. It's not visible – in the same way as inarticulate or inept or incapable or inconclusive or insensitive mean not any of those things. The other meaning of 'in' as a prefix has to do with direction, motion, or situation: into, innate, inherent, interior, inside, going in. This leads to a lot of confusion. The first meaning of invisible is what is not presented to the senses. At the same time, in our minds, it also means what is therefore inside or interior. If what's not visible, is not present to the senses, then you go in to find it. The inside is not allowed to be presented. But maybe it is!

There are many things that are not visible. We live under the power of invisibles – all sorts of ideas, gods, laws, principles, rules. Much of what dominates us is present and invisible. Life is governed by invisibles. And I'm suggesting now, to make it a little shorter here, that 'environment' is one of these invisibles that has come to replace the very idea of the invisible. It is the presenting of something else in the midst of the visible. I am suggesting that the power of the word 'environment' is a substitute for something even more powerful: the invisible. That is why we can't quite define what it is. Environment is not just whatever is around because what is really always around is something we are unable to seize, the invisible, which many cultures recognize all the time, do dances for, perform rituals for, propitiation and smudges, paying careful attention, because of this invisible that sustains and supports, or curses if it's not treated right. Yes, this idea of invisibility invites a more religious and aesthetic, even pagan, approach.

In our modern materialistic culture today, we're using the environment and the foggy notions of it to replace the invisible. So we need to sophisticate our thinking; we need to make more specific what environment is. For instance, I'm curious to know why our civilized minds insist that environment equals 'nature'. Why do we feel when we talk about the environment that we're talking about natural things – rain, rivers, leaves, wood, birds. And yet we live in cities! By not realizing that the environment is not merely nature, but is something invisible besides, we have something that needs to be teased out or defined in some other way. We say the gods are in nature, not in the city – the devil is in the city. The only fantasies we can have then about what to do about cities is to make them more like the woods or nature and bring in the parks, green belts, waterwalls, and so on – as if that were the way to return goodness to cities. I think it is important to sophisticate our notion of environment beyond identifying it with nature. Therefore this invisibility idea seems to be very important.

I'm working very hard here at getting rid of the split between city and nature. I'm trying to understand environment *as a copresencing of the invisible and the visible*. And I would prefer to drop the word environment and use the word copresence. But that isn't satisfactory either. I think what we go to nature for is this invisible. And we haven't had enough teaching in how to discover the invisible in the city. We have such condemnation of the city, such hatred of the city and disgust about the city, that we let the city go and don't pick up enough of the urban invisible, the radiance of what Plotinus calls 'divine enhancement' in the city.

Reconnecting with the web of life: deep ecology, ethics and ecological literacy (Fritjof Capra)

As our century draws to a close, our great challenge is to create sustainable communities – social, cultural and physical environments in which we can satisfy our needs and aspirations without diminishing the chances of future generations (Brown 1981). A sustainable community is designed in such a way that its economy, physical structures and technologies do not interfere with nature's inherent potential to sustain life.

One of the main reasons why we do not live sustainably today is that industrial society is dominated by a mechanistic view of the world, which has led us to treat the natural environment as if it consisted of separate parts, to be exploited by different interest groups. Moreover, we have extended this fragmented view to human society, dividing it into different nations, races, religious and political groups (Capra 1982). The belief that all these fragments – in ourselves, in our environment, and in our society – are really separate has alienated us from nature and from our fellow human beings, and thus has diminished us. To regain our full humanity, we have to regain our experience of connectedness with the natural world, with the entire 'web of life', as suggested in the celebrated speech attributed to Chief Seattle:

> This we know.
> The Earth does not belong to us;
> we belong to the Earth.
> This we know.
> All things are connected
> like the blood which unites one family.
>
> Whatever befalls the Earth
> befalls the sons and daughters of the Earth.
> We did not weave the web of life;
> we are merely a strand in it.
> Whatever we do to the web,
> we do to ourselves
>
> (Gifford and Cook 1992).

Deep ecology and ethics

Naess's distinction between 'shallow' and 'deep' ecology is now widely accepted as a very useful terminology for referring to a major division within contemporary environmental thought (Naess 1973). Shallow ecology is anthropocentric. It views humans as above or outside nature, as the source of all value, and ascribes only instrumental or use value to nature. Deep ecology does not separate humans from the natural environment, nor does it separate anything else from it. It does not see the world as a collection of isolated objects but rather as a network of phenomena that are fundamentally interconnected and interdependent. Deep ecology recognizes the intrinsic values of all living beings and views humans as just one particular strand in the web of life. It recognizes that we are all embedded in, and ultimately dependent upon, the cyclical processes of nature (Devall and Sessions 1985).

The question of values is crucial to deep ecology: it is, in fact, its central defining characteristic. Deep ecology is grounded in 'ecocentric' values. It is a world-view that acknowledges the inherent value of non-human life and sees all human beings as members of *oikos* , the Earth Household, a community bound together in a network of interdependencies. When this deep ecological perception becomes part of daily awareness, a radically new system of ethics emerges (Fox 1990).

Within the context of deep ecology, the view that values are inherent in all of living nature is based on the experience that nature and the self are one. This expansion of the self all the way to the identification with nature is the grounding of deep ecology, as Arne Naess clearly recognized (Fox 1990). What this implies, as Warwick Fox has pointed out, is that the connection between an ecological perception of the world and corresponding behaviour is not a logical connection but a psychological connection. Logic does not lead us from the fact that we are an integral part of the web of life to certain norms of how we should live. However, if we have deep ecological awareness, or experience, of being part of the web of life, then we will (as opposed to should) be inclined to care for all of living nature. Indeed, we can scarcely refrain from responding in this way.

Ecological literacy

Reconnecting with the web of life also means that we can learn valuable lessons from nature about how to build sustainable communities, because nature's ecosystems are sustainable communities of plants, animals and micro-organisms. To understand these lessons, we need to learn the basic principles of ecology. We need to become, as it were, ecologically literate (Orr 1992). Being ecologically literate, or 'ecoliterate', means understanding the principles of organization of ecological communities (i.e. ecosystems) and using those principles for creating sustainable human communities. We need to revitalize our communities – including our educational communities, business communities and political communities – so that the principles of ecology become manifest in them as principles of education, management and politics (Capra and Pauli 1995).

To understand the principles of ecology, we need a new way of seeing the world and a new way of thinking – thinking in terms of relationships, connectedness and context. This new way of thinking is known as systems thinking. It emerged during the first half of the century in several disciplines, in which scientists explored living systems – living organisms, ecosystems and social systems – and recognized that all these living systems are integrated wholes whose properties cannot be reduced to those of smaller parts (Capra 1996).

Systems thinking has been raised to a new level during the past twenty years with the development of a new science of complexity, including a whole new mathematical language and a new set of concepts to describe the complexity of living systems (Capra 1996). The emerging new theory of living systems is the theoretical foundation of ecological literacy. Instead of seeing the universe as a machine composed of elementary building blocks, scientists have discovered that the material world, ultimately, is a network of inseparable patterns of relationships; that the planet as a whole is a living, self-regulating system (Lovelock 1991). The view of the human body as a machine and of the mind as a separate entity is being replaced by one that sees not only the brain, but also the immune system, the bodily tissues, and even each cell, as a living, cognitive system (Capra 1996). Evolution is no longer seen as a competitive struggle for existence, but rather as a co-operative dance in which creativity and the constant emergence of novelty are the driving forces (Margulis and Sagan 1986).

The new theory of living systems provides a conceptual framework for the link between ecological communities and human communities. Both are living systems that exhibit the same basic principles of organization. They are 'autopoietic' networks that are organizationally closed, but open to the flows of energy and resources (Maturana and Varela 1980). They maintain themselves far from equilibrium and are capable of spontaneously producing new forms of order at critical points of instability (Prigogine and Stengers 1980). Their structures are determined by their histories. They are intelligent because of the cognitive dimensions inherent in the processes of life (Maturana and Varela 1980).

Of course, there are many differences between ecosystems and human communities. There is no self-awareness in ecosystems, no language, no consciousness, and no culture; and therefore no justice, nor democracy; but also no greed, nor dishonesty. We cannot learn anything about those human values and shortcomings from ecosystems. But what we can learn and must learn from them is how to live sustainably. During more than three billion years of evolution, the planet's ecosystems have organized themselves in subtle and complex ways so as to maximize their ecological sustainability. This wisdom of nature is the essence of ecoliteracy.

The new systemic understanding of life allows us to formulate a set of principles of organization that may be identified as the basic principles of ecology, and to use them as guidelines to build sustainable human communities (Capra 1996). The first of those principles is interdependence. All members of an ecological community are interconnected in a vast and intricate network of relationships, the web of life. They derive their essential properties, and in fact their very existence, from their relationships to other things. Interdependence – the mutual dependence of all life processes on one another – is the nature of all ecological relationships. The behaviour of every living member of the ecosystem depends on the behaviour of many others. The success of the whole community depends on the success of its individual members, while the success of each member depends on the success of the community as a whole.

Understanding ecological interdependence means understanding relationships. It requires the shifts of perception that are characteristic of systems thinking – from the parts to the whole, from objects to relationships, from contents to patterns. A sustainable human community is aware of the multiple relationships among its members. Nourishing the community means nourishing those relationships. The fact that the basic pattern of life is a network pattern means that the relationships among the members of an ecological community are non-linear, involving multiple feedback loops. Linear chains of cause and effect exist very rarely in ecosystems. Thus a disturbance will not be limited to a single effect but is likely to spread out in ever-widening patterns. It may even be amplified by interdependent feedback loops, which may completely obscure the original source of the disturbance.

The cyclical nature of ecological processes is another important principle of ecology. The ecosystem's feedback loops are the pathways along which nutrients are continually recycled. Being open systems, all organisms in an ecosystem produce wastes, but what is waste for one species is food for another, so that the ecosystem as a whole remains without waste. Communities of organisms have evolved in this way over billions of years, continually using and recycling the same molecules of minerals, water and air.

The lesson for human communities here is obvious. A major clash between economics and ecology derives from the fact that nature is cyclical, whereas our industrial systems are linear. Our businesses take resources, transform them into products plus waste, and sell the products to consumers, who discard more waste when they have consumed the products. Sustainable patterns of production and consumption need to be cyclical, imitating the cyclical processes in nature. To achieve such cyclical patterns, we need to fundamentally redesign our businesses and our economies (Hawken 1993).

Ecosystems differ from individual organisms in that they are largely (but not completely) closed systems with respect to the flow of matter, while being open with respect to the flow of energy. The primary source for that flow of energy is the sun. Solar energy, transformed into chemical energy by the photosynthesis of green plants, drives most ecological cycles.

The implications for maintaining sustainable human communities are again obvious. Solar energy in its many forms – sunlight for solar heating and photo-voltaic electricity, wind and hydropower, biomass, etc. – is the only kind of energy that is renewable, economically efficient, and environmentally benign. By disregarding this ecological fact, our political and corporate leaders again and again endanger the health and well-being of millions around the world. The 1991 war in the Persian Gulf, for example, which killed hundreds of thousands, impoverished millions, and caused unprecedented environmental disasters, had its roots to a large extent in misguided energy policies.

Partnership is an essential characteristic of sustainable communities. The cyclical exchanges of energy and resources in an ecosystem are sustained by pervasive co-operation. Indeed, since the creation of the first nucleated cells over two billion years ago, life on earth has proceeded through ever more intricate arrangements of co-operation and coevolution. Partnership – the tendency to associate, establish links, live inside one another, and co-operate – is one of the hallmarks of life. In the words of Lynn Margulis and Dorion Sagan (1986), 'Life did not take over the globe by combat, but by networking'.

In human communities, partnership means democracy and personal empowerment, because each member of the community plays an important role. Combining the principle of partnership with the dynamic of change and development, we may also use the term 'coevolution' metaphorically in human communities. As a partnership proceeds, each partner better understands the needs of the other. In a true, committed partnership both partners learn and change – they coevolve. Here again we notice the basic tension between the challenge of ecological sustainability and the way in which our present societies are structured, between economics and ecology. Economics emphasizes competition, expansion and domination; ecology emphasizes co-operation, conservation and partnership.

The principles of ecology mentioned so far – interdependence, the cyclical flow of resources, co-operation and partnership – are all different aspects of the same pattern of organization. This is how ecosystems organize themselves to maximize sustainability. Once we have understood this pattern, we can ask more detailed questions. For example, what is the resilience of these ecological communities? How do they react to outside disturbances? These questions lead us to two further principles of ecology – flexibility and diversity – that enable ecosystems to survive disturbances and adapt to changing conditions.

The flexibility of an ecosystem is a consequence of its multiple feedback loops, which tend to bring the system back into balance whenever there is a deviation from the norm, due to changing environmental conditions. For example, if an unusually warm summer results in increased growth of algae in a lake, some species of fish feeding on these algae may flourish and breed more, so that their numbers increase and they begin to deplete the algae. Once their major source of food is reduced, the fish will begin to die out. As the fish population drops, the algae will recover and expand again. In this way, the original disturbance generates a fluctuation around a feedback loop, which eventually brings the fish/algae system back into balance.

Disturbances of that kind happen all the time, because things in the environment change all the time, and thus the net effect is continual fluctuation. All the variables we can observe in an ecosystem – population densities, availability of nutrients, weather patterns, etc. – always fluctuate. This is how ecosystems maintain themselves in a flexible state, ready to adapt to changing conditions. The web of life is a flexible, ever-fluctuating network. The more variables are kept fluctuating, the more dynamic is the system; the greater is its flexibility; the greater its ability to adapt to changing conditions.

All ecological fluctuations take place between tolerance limits. There is always the danger that the whole system will collapse when a fluctuation goes beyond those limits and the system can no longer compensate for it. The same is true of human communities. Lack of flexibility manifests itself as stress. In particular, stress will occur when one or more variables of the system are pushed to their extreme values, which induces increased rigidity throughout the system. Temporary stress is an essential aspect of life, but prolonged stress is harmful and destructive to the system. These considerations lead to the important realization that managing a social system – a company, a city, or an economy – means finding the optimal values for the system's variables. If one tries to maximize any single variable instead of optimizing it, this will invariably lead to the destruction of the system as a whole.

The principle of flexibility also suggests a corresponding strategy of conflict resolution. In every community there will invariably be contradictions and conflicts, which cannot be resolved in favour of one or the other side. For example, the community will need stability and change, order and freedom, tradition and innovation. Rather than by rigid decisions, these unavoidable conflicts are much better resolved by establishing a dynamic balance. Ecological literacy includes the knowledge that both sides of a conflict can be important, depending on the context, and that the contradictions within a community are signs of its diversity and vitality, and thus contribute to the system's viability.

In ecosystems, the role of diversity is closely connected with the system's network structure. A diverse ecosystem will also be resilient, because it contains many species with overlapping ecological functions that can partially replace one another. When a particular species is destroyed by a severe disturbance so that a link in the network is broken, a diverse community will be able to survive and reorganize itself, because other links in the network can at least partially fulfil the function of the destroyed species. In other words, the more complex the network is, the more complex its pattern of interconnections, the more resilient it will be.

In ecosystems, the complexity of the network is a consequence of its biodiversity, and thus a diverse ecological community is a resilient community. In human communities, ethnic and cultural diversity may play the same role. Diversity means many different relationships, many different approaches to the same problem. A diverse community is a resilient community, capable of adapting to changing situations.

However, diversity is a strategic advantage only if there is a truly vibrant community, sustained by a web of relationships. If the community is fragmented into isolated groups and individuals, diversity can easily become a source of prejudice and friction. But if the community is aware of the interdependence of all its members, diversity will enrich all the relationships and thus enrich the community as a whole, as well as each individual member. In such a community information and ideas flow freely through the entire network, and the diversity of interpretations and learning styles – even the diversity of mistakes – will enrich the entire community.

These, then, are some of the basic principles of ecology – interdependence, recycling, partnership, flexibility, diversity and, as a consequence of all those, sustainability. As our century comes to a close and we go toward the beginning of a new millennium, the survival of humanity will depend on our ability to reconnect with the web of life, to understand the principles of ecology and live accordingly.

Forming 'the alliance of religions and conservation' (Tim Jensen)

'One day in 1953, two men stood on the summit of Mt. Everest: Sir Edmund Hillary, a Western scientist, and Sherpa Tenzing, a Himalayan Buddhist. Separated as they were by culture and beliefs, they had together scaled the highest mountain in the world and had, for the first time in history, reached its summit. What they did speaks volumes for the

Box 11.6: Of Pandas and Religion: The Road to Assisi

Martin Palmer, Secretary General, The Alliance of Religions and Conservation

It took a certain kind of courage in the mid-1980s to do what the World Wide Fund for Nature (then known as World Wildlife Fund) International did. In an era still dominated by a belief in the power of statistics to shock people into environmental concern and action, WWF saw another way forward. Ironically, in an era that saw highly emotive and often misguided use of apocalyptic imagery, Garden of Eden visions and quasi-religious language being used indiscriminately by environmental organizations trying to awaken public concern, WWF went directly to the religious worlds and asked for help.

By the mid-1980s it was clear to many in conservation that no matter how well-intentioned the environmental movement might be, it was failing to deliver the message to the vast majority of people. More importantly, it was failing to help or make people change their life-styles in order to help save the world. As David Bellamy, the high-profile British conservationist said at the time, he had been fighting environmental issues for twenty years and they had all got worse, not better!

It was in this spirit – a spirit of recognizing the need to involve new partners and to seek new motivations for change – that WWF International, at the suggestion of its President HRH The Prince Philip, invited the International Consultancy on Religion, Education and Culture (ICOREC) to bring together five of the world's major faiths in Assisi, Italy, to see what could be done to draw the faiths into active ecological work. It was a unique gathering in that the conservation/environment world had never before sat down with the religions in this way. Indeed, never had so many religions come together at this level on the topic of the environment. As Prince Philip said:

'We came to Assisi to find vision and hope. Vision to discover a new and caring relationship with the rest of our living world, and hope that the destruction of nature can be stopped before all is wasted and lost. I believe that today, in this famous shrine of the patron saint of ecology, a new and powerful alliance has been forged between the forces of religion and the forces of conservation. I am convinced that secular conservation has learnt to see the problems of the natural world from a different perspective and, I hope and believe, that the spiritual leaders have learnt that the natural world of Creation cannot be saved without their active involvement. Neither can ever be quite the same again.' (Assisi Liturgy 1986)

All the faiths expressed grave reservations about the direction and implicit values of conventional conservation that gave prominence to economics as the main driving force and rationale for conservation work. Many faith communities found the mythology of free enterprise capitalism dismissive of any attempt to work from more altruistic, compassionate or even self-sacrificing models of how to deal with crises. All too often, faiths were told that what they did on their own lands or within their own faith communities was fine, but could never work in the 'real world'. Some faith groups expressed alarm at the anthropocentric nature of the environmental argument; others at the reliance upon facts over ethos. The coming together of the faiths presaged an enormous questioning of the assumptions of conservation.

But in return, the faiths had to face the criticism of the environmental groups, that whilst each faith had fine teachings, it had done very little about them for centuries. This was one reason why it was so hard in 1985–1986 to find religious leaders and organizations who could reflect upon their long neglected teachings in the light of real engagement with environmental issues.

The Assisi Declarations reflect this and provide a framework for pragmatic programmes of religiously led or inspired projects. Within just a couple of years, the Orthodox Church particularly under the personal and theological guidance of His All-Holiness the Ecumenical Patriarch Dimitrios, had produced its own environmental statement. The Thai Forest monks, Sri Lankan Buddhist activists, and the newly emerging Network of Engaged Buddhists all developed their own articulations based upon experience. By 1995, the United Nations could comment that the Network on Conservation and Religion had reached untold millions who could not be reached by any other network. It is estimated that by 1995, some 120,000 religion-based communities had engaged in some way or another with environmental issues.

real differences between them and their cultures. Edmund Hillary stuck a Union Jack, the flag of Great Britain, in the snow and claimed to have 'conquered' Mt. Everest. Sherpa Tenzing sank to his knees and asked forgiveness of the gods of the mountain for having disturbed them.' (Edwards and Palmer 1997)

This story illustrates the central part of the spirit, philosophy and reasoning behind an alliance between religions and conservation which led to a call to religious leaders by the World Wildlife Fund (WWF) to meet with representatives from environmentalist groups in Assisi in Italy in 1986. Their goal was to explore how the world's religions could help in the struggle to save the natural world (see Palmer, Box 11.6). A diverse, variegated and colourful train of eco-pilgrims made its way towards the birthplace of St. Francis, the Christian saint, who in 1979 had become the patron saint for many ecologists. Representatives from the five major religions – Buddhism, Christianity, Hinduism, Islam and Judaism – mixed with secular grass-root conservationists in an environmental rally and pilgrimage to discuss the following:

- how the environmental crisis is a mental and ethical crisis due, in part, to powerful, predominantly Western and Christian world-views that encourage materialistic, dualistic, anthropocentric and utilitarian concepts of nature;
- that environmental organizations and politicians are victims of the same economic and technological thinking that provoked the crisis;
- that alternative world-views and ethics must be respected to counter current dominant thinking; and,
- that the world's religions constitute enormous human and spiritual potentials.

The gathering opened by dramatizing the destruction of indigenous peoples and Third World environments: formal apologies were made by the participants to a Maori warrior obstructing the entrance to the Basilica of St. Francis as they proceeded into the basilica. After elaborate, specially designed ceremonies, including thanksgiving for creation, repentance, celebration of the religions' visions of the future and dedication to implement the visions, the Assisi Declarations were produced. The Declarations stated the past neglect of the religions to environmental issues, their present good-will, and the benign potentials of their religious traditions for the future. 'The New Alliance', as it was most fittingly called by then, was a fact. The cross-cultural aspects, the theme of the diversity of visions, and the theme of 'unity in diversity' were integral to the project, and were also stressed in the introductory speech by the Minister General of the Franciscans: 'We are convinced of the inestimable value of our respective traditions and of what they can offer to re-establish ecological harmony; but at the same time we are humble enough to learn from each other. The very richness of our diversity lends strength to our shared concern and responsibility for Planet Earth.

Over the next few years, four other religions joined (the Baha'is in 1987, the Sikhs in 1989, the Jains in 1991 and the Taoists in 1995). The Assisi event was followed up by conferences and interfaith meetings in many places throughout the world, including Canterbury, Washington, D.C. and Copenhagen. In 1990 the WWF and the religions published statements on biodiversity, and in 1997 thousands of religiously based environmental projects are running world-wide – some assisted by WWF, some springing directly from local sources. In 1994 the United Nations Environment Programme (UNEP) awarded Prince Philip one of the 'UNEP 500' awards for outstanding contributions to conservation, stating that the initiative had 'helped reach untold millions world-wide with a conservation message through religious channels'. In 1995 two more meetings took place: one in Ohito in Japan and the other, called 'The World Summit of Religion and Conservation', in England. Representatives from the nine religions came up with a joint revised and revitalized commitment and with revised editions of their former declarations. ARC was created as a fund-raising structure, new programmes were designed and old projects evaluated and continued. According to the latest newsletter issued by the ARC, the work has now (1997) inspired the World Bank to initiate consultations with representatives from the religions and the consultants of the WWF.

Appendix 2.2 and Appendix 2.3 provide summaries of the Assisi Declarations and the subsequent statements by the world's major faiths.

Common key-concepts of the religions

Environmental ethics are hard to separate from general ethics, just as a way of seeing the natural world cannot be separated from a general world-view. Likewise, the way of perceiving the relation of humans and nature cannot be separated from the relation between humans and the divine. The Declarations, consequently, refer to virtually all the central ideas in the entire cosmological and ethical systems of the religions, and this is the secret behind their motivating power.

To confer an idea of the relevant notions and to demonstrate the common grounds on which the diverse religions meet, a list of cross-cultural and cross-religious notions of explicit environmental relevance is rendered below. The list is based on an analysis of the WWF Declarations and the classical scriptures they refer to, but other source material has also been included. The notions are presented according to an arrangement of the religions in 'families'. A list showing the similarities across the two 'families' of the five religions normally designated 'world religions' is added.

Buddhism and Hinduism

(a) ***Mysticism.*** In both religions the perception of the relationship between humans and nature is deeply influenced by the mystic traditions referring to experiences and ideas of totality, oneness and unity.

(b) ***Pantheism.*** In both traditions we find the notion that the 'ultimate reality', 'truth' or the divine pervades everything. In Hinduism this finds its classical expression in the formula *atman = brahman* (the eternal innermost being of humankind, *atman*, and the innermost universal being, *brahman*, is one and the same). In Buddhism a similar notion is expressed in concepts of the 'emptiness' of everything or in the concepts of an all-pervading 'Buddha–nature'.

(c) ***Ahimsa.*** In Hinduism as well as in Buddhism, the ideal of absolute and total, active nonviolence (*ahimsa*) is central to ethics in general and to environmental ethics. Humankind and nature are, in the last analysis and final liberating insight, made of the same 'soul', the same 'emptiness' or the same 'elements'. By way of *karma* and *samsara* and due to the central (especially Buddhist) notion of the interrelatedness of all things, everybody and everything must be treated with care and compassion.

(d) ***Karma-samsara* and *reincarnation.*** The idea of eternal cycles of rebirths, on the universal as well as the individual level, and the idea that the thoughts and actions of humans determine the mode and way of future lives, is – in spite of all differences concerning the notion of soul (*atman* and *anatman*) – of the uttermost importance in both religions and plays a central role in connection with their environmental ethics.

(e) ***The world as a cosmos.*** The world is conceived of as a well-organized totality, a cosmos. It is part of humankind's place and religious duty in this cosmos, and incorporated in the scheme for attaining final liberation, to help in upholding the cosmic order.

(f) ***Morality can be 'read' in nature.*** In both religions we find the notion that the reason why nature is having a bad time is due to the fact that humans do not stick to *dharma*, the universal and moral order and to the teachings of the Buddha. Social justice and order within the city walls mean order and fruitfulness outside.

(g) ***Humans are not the owners of the world.*** Though this notion is met primarily in theistic versions of Hinduism, we find it also in Buddhism. Humankind's greed and ignorance, and the development of differences between those who have and those who have not, lead to the present world of suffering if not balanced by the utmost aim of final liberation. The blind attachment to material wealth is a hindrance to the final insight and liberation and contributes to the environmental crisis.

(h) ***Ignorance is a sin.*** Ignorance is a sin leading to wrong and detrimental conceptions and actions. It is often stated that the environmental crisis is due to humankind's ignorance and false conceptions of the true relationship between the (falsely perceived) 'ego-self', the 'Buddha–nature' and the environment.

Christianity, Islam and Judaism

(a) ***The concept of God.*** In spite of the differences (cf. the Christian idea of The Holy Trinity and the Incarnation), the three religions share the notion of the one and only, transcendent, eternal God and Creator. The basic notion of the world as God's creation is immensely important for all further discussions of an environmental ethic, and all the traditions deal with the 'problem' of the created world as different from, but given by, God. In all three religions, traditions of the mystics convey notions and experiences of the relationship between humans, nature and God bridging the gap between the transcendent God, humans and the created world, notions that at times get close to pantheism or to pan-en-theism.

(b) ***The notion of the creation.*** In accordance with the notion of the one God as the Creator, they all share notions of the created world as a cosmos, an integrated whole. The discourse on the 'Integrity of Creation' is common to all three.

(c) ***Man's place in the created world.*** Man is something special and certainly differs from the rest of creation by means of his freedom and capacity for rational thinking, and the 'fruits of the earth' certainly are there to be used by humans. In contrast to previous dominant interpretations, especially of the Biblical texts, modern environmentally oriented Christians, Muslims and Jews are eager to stress the idea of humans not as tyrants but as sentient, responsible and rational stewards and caretakers.

(d) ***Destroying the world is a sin.*** On the background of these notions it becomes evident that destruction of God-given nature is a religious and moral fault, a sin. It links up with the notion of the original fall from the mythical, paradisaical closeness to God as a consequence of human arrogance and rebellion. Humans wanted to rise above their allotted place to become god-like creators.

(e) ***Morality can be read in nature.*** As in Hinduism and Buddhism we find the idea that the morality of humans can be 'read' in the environment. Peace and social justice show in fertile fields and women. Unjust leadership, social injustice and arrogance show themselves in wastelands, barren women, plagues – and environmental crises.

The similarities within the families of religions have to do with the historical developments and past processes of syncretism. The similarities across the families, however, must be explained otherwise. To this end and for the present purpose, it suffices to mention a few of the most relevant of such explanations:

- In all the religions, humans relate to nature and the environment *vis-a-vis* a third factor (the divine, the truth, ultimate reality). This is the most common specifying criterion in most external definitions of religions as cultural interpretative systems *and* the decisive factor in the internal self-understanding of the religions, also in the light of the environmental crisis. No matter the empirical specifics nor the epistemological and ontological difficulties connected with it: it is the intricate relations of 'nature' and 'the supernatural' that matter and make the religions alternative or supplementary systems of interpretation, ethics and action.
- In all the religions, popular ideas and practices co-exist with the theological systems. The popular ideas and practices are often closely knit to the historical influences of primitive cultures and religions and to believers whose religious aims are more closely related to an economy based on their direct relations to the near-environment.
- In all the religions, traditions of mysticism influence the discourse on the divine, humans and nature. These traditions, no matter how 'otherworldly' they may be, often contribute to bridging the opposites of the dualistic schemes otherwise separating the transcendent divine from the immanent human and nature. The ascetic ideals and practices furthermore function as models for a non-materialistic, non-consumerist attitude to life.

Religiously based projects

Hundreds of projects (e.g. reforestation projects; ecologically sound adjustment of clerical administrations; creation of educational bodies and environmental prizes; organic farming, etc.) differ from secular ones only because they are initiated and run by members of a religious community or take place on lands belonging to religious communities. These are of immense value in helping to safeguard biological and cultural diversity.

Special inter-religious events, like inter-religious services in central and famous churches, have been celebrated; and even special, cross-religious rituals have been designed to honour unity and diversity. Diversity has been displayed and celebrated in several festivals and rituals within each religion. Festivals and rituals linked to the seasons and symbols referring to the natural as well as the supernatural have been used to introduce a collective awareness and responsibility for the environment. There are plenty of examples: only a few can be mentioned:

- The traditional Christian harvest festivals have been redesigned and reinvented in order to make participants remember the 'gift of God', the obligation to share the fruits of the earth with the poor, and the neglected responsibility toward the creation of God.
- The Christian Mass (Eucharist, Communion) has been celebrated and reinterpreted to stress that the doctrines of Incarnation and final Resurrection relate to more than the human being; namely, the sanctification of the whole of creation and the resurrection of the cosmos. Conversely, to destroy the Integrity of Creation becomes equivalent to committing a sin for which you have to repent before you can participate in communion.
- The Jewish Sabbath has been celebrated to underscore its cosmic and re-creative implications. Other Jewish festivals (e.g. *Shavuot* and *Sukkkot*) have been used to strengthen the bonds between humans, nature and God. *Tu Bishvat*, the Festival of Trees, has been celebrated in connection with active reforestation programmes and the planting of trees in the deserts of the Holy Land, in Brooklyn, and elsewhere.
- In India, Hindus celebrating the god Krishna during the pilgrimage to his sacred forests in Vrindavan have combined this with a tree-planting project. Participants entering the sacred precinct are handed trees to plant and are given information on the deplorable status of the birthplace of the god.

Several of the conservation programmes linked to land considered sacred reveal that sometimes it is the religious concepts and practices that have contributed to the crisis, and this is acknowledged by the religious groups participating. Examples include the Vrindavan project where millions of pilgrims have for centuries celebrated their god at the expense of the forest, and reformed practices in connection with the grossly polluted Ganges. The restoration and conservation of sacred mountains, especially the Taoist and Buddhist Sacred Mountains of China, is a project supported by ARC. A recent off-shoot of this kind of religious environmentalism is the large-scale British project called the 'Sacred Land Project'. Its aim is to rediscover, reinvigorate and replant old pilgrimage routes and other sacred spaces, and it links up with many other projects of turning Christian churchyards and Muslim cemeteries into ecologically sound spaces. These projects and devices are all species of a common cross-cultural and cross-religious 'Divine Command Ethics' and a globalization of the concept of nature and the environment, although the exact means of communication and understanding are bound to local traditions. Biodiversity, then, has become a word in a common environmental language, but the message and the means are based on the diverse cultural and religious symbolic languages.

The 'greening' of world religions

Academic studies of religion and environmentalism (and public and popular discourses on the matter) have focused mainly on environmentalist interpretations and applications of so-called primitive religions, New Age environmentalist interpretations of these, and the Eastern religions. The involvement of groups from within the so-called world religions, however, so far has attracted few academics, and studies on this aspect of religion, environmentalism and biodiversity are rare and mostly quite recent.

Before I move on to a survey and critical assessment of some of the ways in which the world religions are involved in environmentalism and adding to biological as well as spiritual diversity, a few words are needed about the criticism of the environmentalist interpretations and applications of so-called primitive religions and Eastern religions (especially Buddhism and Taoism), including the so-called New Age environmentalist interpretations and applications.

The majority of critics have focused on what is called the 'myth of primitive ecological wisdom' and the kind of 'primitivism' used by and practised by environmentalists as well as by scholars. The critics voice a wide range of scholarly, ethical–political and environmentalist concerns. Sometimes the academic and analytical criticism is primarily trying to 'get down to facts', and frequently the scholars are able to demonstrate that the 'myth of primitive ecological wisdom' is, indeed, more akin to the world of religious myths than it is to the world of realities. No matter how intimate a relation to what we call 'nature', the 'nature-religions' and 'nature-people' also demonstrate the general differences between ideals and practices, and their relationship to nature cannot be reduced to the wishes of the twentieth-century urbanized environmentalist. Sometimes the academic and analytic criticism is combining the academic interest in facts versus myth with ideological and ethical (humanistic and Enlightenment ideas and ideals) and sometimes it is combined with environmentalist concerns as to how the environmental crisis may actually most rationally and efficiently be handled.

Adopting this last-mentioned perspective on the matter I shall say only this much on the 'myth of the ecological wisdom of primitive cultures and religions' and 'primitivism': primitive religions and cultures, often conceived of as constituting one single and the earliest form of religion, have constantly functioned as the positive or negative counterpart to Western civilization and life. In the period of environmentalism they have predominately functioned as positive, sometimes even paradisaical, models for an ecologically sound world-view and society. The period of environmentalism coincides with a period of New Age thinking and with many a first people's fight for cultural and political independence and rights to traditional land. Consequently, it is no wonder that we find an amalgam of their and Western concepts of nature and sound ecological thinking in environmentalist organizations, among first peoples themselves and among many a theologian within the world religions.

That the popular notions of 'nature-religions', the 'nature-worship' and the 'ecological wisdom' of the primitive societies are loaded with problems and further questions is, thanks to scholars of religion and anthropologists, a well-established fact, and I see no reason to elaborate much upon this. Judging from the history of religions, it actually does seem that there is no limit to the phenomena, animate and inanimate, of nature, which have sometimes somewhere been conceived of as sacred and allotted a place in mythologies and ritualized worship. A journey through the discipline consequently takes us through discussions on totemism, animism and dynamism, through the corridors of Astral- and naturmythologische Schulen, through theories of Korn-damonen, sky gods, sacred trees, waters, earth – and sacred cows, of course. Yet, after all these years, the need for empirical studies in the various relations between the 'natural' and the 'supernatural' and for theoretical and methodological reconsiderations of essential analytical tools and concepts like 'religion', 'nature', 'the supernatural', 'environment', 'cosmos' and 'chaos' has only become more pressing.

A combination of the academic interest in religion with an interest in conservation calls for more empirical studies evaluating the positive and negative results of the past and present diffusion and application of the 'myth of the primitive ecological wisdom'. Studies are needed using the traditional methodologically agnostic approach – permitting the scholar of religion to bracket the questions of the truth of the myth while at the same time taking account of the truth it transmits to believers and users. More historical and critical analyses of the possible neo-imperialist and neo-colonialist implications, the political use and abuse of the myth, in the West and elsewhere, are also needed. Studies are also needed to estimate the degree to which the diffusion and use of the myth actually has had a benign or malign effect on environmentalism, biodiversity and cultural diversity.

Moving to the related topic of the more general religious dimensions in contemporary environmentalist thinking and acting, I think that religious studies as well as studies trying to pave the way for practical environmentalism, could benefit from intensified work in this area. To give a hint of the possible directions and possibilities of such work, I shall use an example. My 'American friend', who naturally uses computer-steered ventilation of the compost, tells about one of his 'peak-experiences':

> *... a month or two ago, as I was standing on top of a 2.000 cubic yard of compost, one side of which was loaded with tiny little red master composters (worms), I realised how humbling it is ... that to get the best compost ... the best finished product for growing plants ... 162 pounds for me, 5 years of college, 1265 years old (born 731), all the reading I had done, able to turn over 1.000 CY of compost a day, all this ... and I am dependent on ... worms ... to make the best compost. Very humbling .. for me.*

The expressions and postulated experiences can be interpreted as parallels to traditional religious symbolism of the Axis Mundi, the centre of the world, and to the claims of mystics of their integration in divine cosmic cycles; and I believe there is plenty of such unused source material waiting for the scholar of religion in the world of environmentalism. More work in this field could also contribute to a better understanding of the relation between (the combination of) modern scientific and modern religious attitudes to the environment and to more empirically grounded discussion about the practical implications of this relationship: How, where, when and why does the religious dimension become a hindrance to rational environmentalism, and how, where, when and why does it further it?

Turning, finally, to questions about the environmentally benign potentials of the greening of the world religions, as manifested for instance in the ARC, I find it important to stress that it certainly has some benign and powerful potentials. The greening of the world religions is good for studies in religion; good for the persons and organizations who have of their own free will decided to participate; good and useful because it contributes to our general understanding of several important questions about the relation between world-views, environmentally benign or malign, and the actual state of the environment; and good and useful because it adds to our understanding of more or less effective means of remedy.

I find the reasoning behind the WWF project fairly sound and balanced. The participation of the world religions in environmental projects does, in the light of the dimensions of the crisis and from a pragmatic point of view, add a substantial and much-needed amount of manpower to global and local projects. That representatives from various cultures, or if you like, various interpretative communities (the various religions, the communities of natural scientists, economists and environmentalists) co-operate can lead to mutual inspiration and a balancing of opposite views and various means. To return to Mt. Everest: since Mt. Everest in the time after it was conquered evidently was not protected (at least not very well if judged by the present state of affairs) neither by spirits or other Buddhist or materialistic notions, one may of course say that this only proves the general discrepancy between world-views and practice, be they religious or not, and that the reinterpretative efforts of the religions are not only somewhat pathetic but also futile. Admitting the relevance and relative truth of such arguments does not, however, exclude the possibility that religiously based concepts of nature and a religiously based environmental ethics may be of some help, nor that it is a bad idea to combine religious and non-religious ideas and means. The co-operation, so to say, of Edmund Hillary and Sherpa Tenzing may, in spite of one's personal opinions about modern or even New Age eclecticism, work better than no co-operation.

The cross-cultural, cross-religious and inter-religious flows of ideas and alternatives lead to new amalgams and new concepts and practices. Rational analyses and religious views may very well constitute that unity in diversity, and that combination of local and global, that is needed. If biological diversity is good for the environment, maybe conceptual and practical, cultural and religious diversity and interaction is good too.

Naturally, our American friend on top of the mountain of compost may, like some Buddhists, Hindus or Christians, entertain some religious ideas and sentiments, e.g. about the god-given 'naturally natural', which is a hindrance to rational and ecologically sound solutions. He or they may insist on concepts of purity running contrary to scientifically proven methods and means. On the background of my analyses of the factual environmental projects within the religions, I must however say that it is not my impression that the leading figures believe, so to speak, that faith alone can move mountains. Notions, not only of the former failure of the religions to realize their environmentally benign attitudes and views, but also of the fact that not all believers are good or practising believers, are frequently expressed in the material. In general, the greening of the religions cannot be interpreted as an irrational return to some 'supernatural' belief in 'faith healing'. It is, in my opinion, a rather rational and balanced hope that faithfulness to certain religious beliefs combined with pragmatic, rational and scientifically aided thinking and acting, may be conducive to more environmentally benign thinking and acting. The religions and the adherers certainly will not be able to make mountains move. They may, however, contribute to efforts to stop mountains from turning into heaps of garbage.

Criticism pointing to the fact that none of the world religions hosted concepts of 'nature' and 'environment' and 'biodiversity' in the period where they and their classical scriptures were born, is a criticism that misses a very important fact about religion: religion and religions are dynamic, historic and cultural variables, interpretative systems creating and recreating meaning. The present reinterpretations done by the religions are to be expected and quite natural. Mountains may not move, but religions do, and as stated earlier, the whole idea of the 'Alliance of Religion and Conservation' is future-oriented towards the mobilization of the spiritual and human potentials within and across the world religions.

Today the religions have re-entered the scene where global ethical, and therefore political, issues are discussed and negotiated. I think the environment may profit from this, but it is, naturally, also important to ask if the religiously based environmental ethics will continue to respect the democratic and open-ended process of pluralistic negotiations? Will the religious world-views add to the celebrated diversity and plurality, or will they – in the long run – represent a threat to pluralism, democracy and a rational, secular discourse? So far the ARC seems to balance the religious and secular interests and one can only hope that the co-operation continues to be more of a co-operation than a competition of values and a struggle for the absolute right and power to decide and define what counts as really real, really good, really natural. Cultural diversity and biological diversity can only be upheld if there is also respect for religious diversity – and for world-views and views of nature that are not religious.

Box 11.7: 'Communing Before Supermarkets', and 'Bringing in the Sheaves'

Carter Revard

'Communing Before Supermarkets'

It's probably because we were always trying
to have enough money to eat
that I can taste and smell the truckloads
of summer that came by and sometimes
turned jouncing up the long
dirt lane from U.S. Sixty to our house –
they saw kids swarming out in the yard,
white house with a green roof and a big white
two-story garage, haybarn and cowbarn,
nothing around but meadow, no crops, no
rows of corn or hills of watermelons, a lot of hungry kids
that would be wanting what they were taking round
from their truck farms or orchards –
elephantine loads of melons, sometimes the light
green long ones, the striped ones, the dark
green cannonballs, incredible abundance,
or old swaying trucks loaded with bushels of peaches,
apples and apricots, with grapes and pears that I
remember. I wonder, now, where they came from –
over in Sand Creek Valley by the little town
in the Osage Hills, the hamlet really,
they called it Okesa where we
drove once: there we saw a hillside full
of orchards, berry bushes, the sandy bottomland shaggy
with watermelon vines where the great green melons rounded
heavy and warm on the loam –
it struck me staring from the car, how strange
that dirt does turn into their sweet crisp red flesh
and juice in the mouth, that those long vines
can draw the dark earth up and make it melons, and I said
to myself, how does the seed know to make
a watermelon and not an apricot? Then we had brought
our dimes and pennies for a summer's day, we took

Box 11.7 (continued)

the silver and copper and we turned
them into two huge melons that the blond boy went casually
out into the field and pulled, just those we wanted,
he took our thirty cents and we –
I think we drove away back down to Sandy Creek and in
the pebbly shallow ford we drove out in the water and killed
the engine and we took
the melons from the trunk and in the shallow ripples splashed
each other and the car, we washed the car, the melons
we took them out onto the bank and sat
on a blanket spread across the grass and stuck
a great long butcher knife into the first green melon and it split,
it was so ripe it cracked almost before the knife
could cut it open, the red heart
looked sugar-frosted, dewy with juice and the pieces
broke to our fingers better than to knives,
in the mouth crisp and melting fragrant, spicy nearly,
as pieces of rind were scattered and the ants reporting mountains of
manna climbed and swarmed and buried themselves in our leavings
as we stripped to shorts and underthings and waded down into
the deeper colder pool below the ford where the springs
welled slowly out from under the bouldery bank
at the bend, and swimming I thought,
now the melon is turning into me, and my sisters and brothers,
my mother and father and uncles and aunts and into the
ants feasting there on the melon rinds,
and into the grass and trees growing there, and into the dirt –
and Sand Creek is turning, this day is turning to night, so now
when we go home I'll remember and it will be turned
into words, and maybe sometime
it would all grow again a long way off, a long way into
the future, and that's what a few pennies and dimes can do
if you have them, a few seeds, a little rain where creekwaters rise,
and the whole world turns into food for all
the different beings in their times.

"Communing Before Supermarkets" from **An Eagle Nation**, by Carter C. Revard. Reprinted by permission of the University of Arizona Press.

Box 11.7 (continued)

'Bringing in the Sheaves'

The '49 dawn set me high on a roaring yellow tractor,
slipping the clutch or gunning the twenty-foot combine
to spurt that red-gold wheat into Ceres's mechanical womb.
I'd set her on course and roll for a straight two miles
before turning left, and it got monotonous as hell –
at first all the roar and dust and the jiggling stems collapsing
to whisk up that scything platform and be stripped of their seed,
then even the boiling from under of rats and rabbits scrambling
to hide again in their shrinking island of tawny grain
as the hawks hung waiting for their harvest of torn fur and blood.
So I'd play little god with sunflowers drooping their yellow heads –
would see a clump coming, and spin the wheel right, left, right, straight,
so the shuddering combine swiveled on its balljoint hitch
first right, then left, that great chatter of blades would go swinging
so the tip barely brushed those flowers and left their clump standing
like a small green nipple out from the golden breastline, and next time past
reversing wheel-spins cut free a sinuous lozenge left for the bumblebees,
its butter-and-black-velvet tops limp-nodding over wilted leaves.
But sunflowers weren't enough, I left on the slick stubble islets
of blue-flowered chicory, scarlet poppies, and just for the hell of it cockleburrs:
"From now on, kid, you run that sumbitch straight!' the farmer said.
You know, out on that high prairie I bet the goldfinches,
bobwhites and pheasants still are feasting in that farmer's fields
from the flower-seeds I left out, summer, fall and winter harvests
that make the bread I eat taste better by not
being ground up with it, then or now.

CHAPTER 12

RIGHTS, RESOURCES AND RESPONSES

Graham Dutfield

CONTENTS

TEXT BOXES

Rights, Resources and Responses

Graham Dutfield

Introduction (Graham Dutfield)

One of the great ironies for traditional communities is that while interest in their ecological knowledge and resource management practices is increasing, human cultural diversity is eroding at an accelerating rate as the world steadily becomes more biologically and culturally uniform. One of the main reasons for this situation is that traditional communities have so often been the victims, rather than the beneficiaries, of development, nature conservation and scientific and commercial research.

Until recently the conventional view among many governments, development agencies and conservation organizations was that traditional communities were part of the problem (sometimes even *the* problem), rather than part of the solution (Adams 1990; Colchester 1997; Pimbert and Pretty 1997). Consequently, it was sometimes deemed necessary to impose alien land tenure systems, to expel traditional peoples from ancestral lands to create protected areas, and to deny the right of communities to change and adapt on their own terms. Even now, the rights of traditional peoples in conservation and development projects are sometimes ignored or given low priority. Roch (this chapter) shows that traditional communities in Indonesia are still threatened with expulsions in the name of nature conservation. According to Born (this chapter), in Brazil national conservation laws and policies infringe on the rights of traditional communities to use and dispose of their resources as they see fit, while new intellectual property rights laws fail to provide for the protection of their knowledge.

Although outsiders have collected knowledge and biological resources from traditional peoples for centuries, 'bioprospecting' (the search for and collection of biological material and traditional knowledge for commercial ends, with particular reference to the pharmaceutical, biotechnological and agricultural industries) has intensified in recent years (Juma 1989; Gray 1991; Reid *et al.*. 1993; Chadwick and Marsh 1994; Joyce 1994; RAFI 1994a, 1994b; Posey 1995; Posey and Dutfield 1996; Balick *et al.*. 1996). Traditional communities to a large extent welcome interest in their knowledge as long as this means greater respect for their rights as holders of this knowledge. But they condemn the 'mining' of their knowledge by commercial concerns and scientists who do not respect their rights, ensure that benefits flow back to their communities, or help to stem the erosion of their cultures. Intellectual property rights laws are seen as a serious threat because they appear to *encourage* and *legitimize* 'biopiracy' (the alienation and unauthorized commercial exploitation of their knowledge and biological resources). Biopiracy and the patenting of living material derived from traditional peoples' knowledge and resources – and even from their own bodies – are not only exploitative in the economic sense but also violate the spiritual values of many traditional peoples (Mead 1994).

There are similarities between seizing territories and displacing their traditional inhabitants, purportedly 'for the good of the biosphere', and taking traditional knowledge in the public domain and patenting 'inventions' based upon this knowledge 'for the benefit of humankind'. In each case, territories, ecosystems, plant varieties (whether domesticated or not) and traditional knowledge are treated as if they are *res nullius* (the property of nobody) before their 'discovery' by explorers, scientists, governments, corporations and conservation organizations.

During the colonial period, sparsely populated 'wildernesses' were regarded as being to all legal intents and purposes vacant prior to colonization. Settler societies, such as those in Australia, built up legal systems based upon the *terra nullius* (the land of nobody) doctrine. Even today, traditional forest communities in some countries (e.g. Latin America) can more easily acquire legal title to their lands if they 'improve' them by removing the trees so that they are no longer 'virgin forests'.

As with land, open access is the rule for traditional knowledge and resources, whereas enclosure is the rule as soon as these are proved to have economic value. Patent systems assign the monopoly rights either to the first to invent (in the United States) or the first to file (elsewhere), not to the providers of the biological raw material or lead information, who may be forbidden to use the invention without the permission of the patent owner. Again, the fact that traditional communities have their own property and resource allocation systems essential to maintaining the link between a people and its environment is ignored (see Monbiot, Vira, this chapter) or not understood (see Wiener, this chapter). Wiener's contribution makes the point that understanding is often hampered by ethnocentric assumptions that ignore the social and ethical dimensions of traditional property systems.

There is a tendency to assume that the property regimes prevailing in Westernized and traditional societies are polar opposites with individualism, privatization and enclosure the norm in the former, and communal property systems in the latter societies, in which resources and information are shared freely among community members. In fact such an assumption is simplistic. Nevertheless, as Monbiot explains (this chapter) the enclosure of the commons is a global phenomenon that has resulted in the loss of local control over lands, resources and knowledge systems. Ironically, this process has often been stimulated by policies that were intended to benefit the local people. Time after time, though, the result is deepening poverty, decreased self-sufficiency and further environmental degradation.

Bioprospecting or biopiracy?

Recent advances in biotechnology have increased the ability of scientists to investigate organisms at the molecular and genetic levels and to find ways to commercialize products developed from such investigations. This is recognized by the increasing number of companies involved in bioprospecting. Several recent economic analyses (e.g. Aylward and Barbier 1992; Pearce 1993) have shown that enormous profits are possible from bioprospecting, and that such commercial potential can be a powerful conservation incentive. Whether or not this optimism is justified, there has been a resurgence of natural product-based research, especially by the pharmaceutical industry (Reid *et al.*, 1993).

Bioprospecting is hardly new (see Juma 1989; Joyce 1994), but a difference with historical 'gene hunting' is that nowadays much bioprospecting can be carried out without even visiting the places from where resources were originally collected. Botanic gardens and agricultural research institutions, such as the member institutions of the Consultative Group on International Agricultural Research (CGIAR), hold large *ex situ* collections of plant genetic resources. Many scientific institutions throughout the world hold substantial collections of microbial genetic resources (culture collections), animal genetic resources (Heywood 1995), and even human genetic resources. Neither does traditional knowledge have to be acquired directly from traditional communities. A great deal can be gleaned from literature searches carried out from a computer terminal located almost anywhere in the world.

This situation is highly beneficial for scientific research institutions and corporations. Producers of seeds and agrochemicals have benefited substantially from free access to germplasm and knowledge from biodiversity-rich areas. The market value of the seed germplasm utilizing traditional landraces is estimated by Rural Advancement Foundation International (RAFI) at $50 billion per year in the US alone (1994a). Consequently, traditional communities have been providing subsidies to a modern agricultural system that barely recognizes their contributions. In 1985, the market value of plant-based medicines sold in developed countries was estimated at a total of $43 billion (Principe 1989). Similar situations exist with timber and non-timber forest products, as well as other natural product markets such as personal care, foods, industrial oils, essences, pesticides and preservatives.

Corporations involved in bioprospecting will normally seek intellectual property right protection for their products so as to maximize returns from their investments in research and development. Fortunately for these corporations, the recent popularity of bioprospecting has coincided with a period in which developing countries are under tremendous pressure to adopt intellectual property rights (IPR) regimes matching the strong levels of protection permitted in North America, Europe and Japan. Inter-governmental negotiations on the standardization of IPRs, which took place during the Uruguay Round of the General Agreement on Tariffs and Trade (GATT), culminated in an international agreement commonly known as the 1994 GATT Final Act. The Act, which established the World Trade Organization (WTO) and has been signed by over 120 governments, includes the Agreement on Trade-Related Aspects of Intellectual Property Rights (TRIPs). The main effect of TRIPs is to accelerate the world-wide adoption of IPR regimes similar to those of the developed countries, which provide greatly enhanced protection for biotechnology products and processes. The utilitarian justification for IPRs is that such legal systems benefit not only the rights holders, but also society as a whole (see Hettinger 1989). However, certain features of IPRs and the new international intellectual property order being overseen by the WTO appear to create barriers to equitable benefit sharing, and may even be threatening to biodiversity.

Patent-holders, who must be legal persons (i.e. individuals or legally recognized entities such as corporations), have the right to exclude others from selling products based upon their inventions. Only inventions that can be dated and attributed to an individual or small group of people can be patented. In traditional societies, the sources of traditional knowledge may be attributable to individuals, kinship- or gender-based groups (see Reichel, Chapter 3), or to single communities. In theory such knowledge may be patentable. However, a great deal of traditional knowledge is not traceable to a specific community or geographical area and is ineligible for patent protection. Whether widely known or not, once traditional knowledge is recorded and publicly disseminated, its use and application is beyond the control of the original knowledge providers. If a researcher investigates a piece of published traditional knowledge and then improves upon it in a practical way, the result may well become a patentable 'invention'.

In theory, the rights of communities to continue using a resource in accordance with traditional practices cannot be affected by a patent. Derivatives of neem (*Azadirachta indica*) seeds, which have been used since time immemorial by Indian farmers as a pesticide, have been patented by companies and research institutions *for the same use* as the farmers (Axt *et al.*, 1993; Shiva and Holla-Bhar 1993; Natarajan, Chapter 6). Nevertheless, these farmers are not precluded from using neem seeds to protect their crops. However, the recent patenting of turmeric powder in the United States reflects a dangerous trend, since the 'invention' described in the patent is identical to a useful application of turmeric that has been known to millions of people in India for centuries. Consequently, people from India living in the United States who continued using turmeric for wound healing as their ancestors did technically became patent infringers. Although this patent was revoked following an expensive and time-consuming challenge from the Council of Scientific and Industrial Research of India, it is probable that many other equally dubious patents are awarded in the United States and elsewhere.

Another serious concern is the possibility of a negative link between enhanced intellectual property rights over biological material, and the state of the world's biological diversity. It is often suggested that there may be linkages between the availability of IPR protection for plant varieties and the replacement of complex, diverse agro-ecosystems with monocultures of single 'improved' varieties. The genetic uniformity of large monocultures leads to vulnerability to crop diseases that can devastate large areas. Increasingly, the new varieties have been genetically engineered to be resistant to a herbicide being marketed by the same company (Bell 1996; Kloppenburg 1988). Both the herbicide and the seed for which it is 'designed' are likely to be IPR-protected. Excessive use of the herbicide is likely to result in other plant varieties and species growing nearby being killed (Bell 1996).

Shankar (this chapter) emphasizes the need to consider the cultural and political aspects of cross-cultural transfers of knowledge and resources through bioprospecting, and argues that Western conceptions of IPRs reinforce the unbalanced and inequitable nature of these transfers. Indeed, very few, if any, corporations involved in bioprospecting are willing to consider sharing their intellectual property rights with traditional community knowledge or resource providers. Unless such companies are prepared to provide generous compensation for the communities whose knowledge and resources they depend on, traditional communities are unlikely to be favourably disposed towards bioprospecting. In fact, indigenous peoples have initiated an international campaign for a moratorium on bioprospecting on their lands. The genesis of this movement was the 1993 Mataatua Declaration on Cultural and Intellectual Property Rights of Indigenous Peoples (see Appendix 1.5) which was adopted by indigenous representatives from the Pacific, Asia and the Americas. This call for a moratorium was reiterated at the United Nations Development Programme's Consultation on Indigenous Peoples' Knowledge and Intellectual Property Rights, held in 1995 in Suva, Fiji (in PCRC 1995), and also by representatives of indigenous peoples' organizations attending the Third Meeting of the Conference of the Parties to the Convention on Biological Diversity.

The Convention on Biological Diversity and Intellectual Property Rights

The main international agreement dealing with biological diversity is the Convention on Biological Diversity (CBD), which came into force in 1993 and has now been ratified by over 170 countries. The CBD has three main objectives (Article 1): (i) the conservation of biological diversity, (ii) the sustainable use of its components, and (iii) the fair and equitable sharing of the benefits arising out of the utilization of genetic resources.

Article 15, which recognizes that 'the sovereign rights of States over their natural resources, [and] their authority to determine access to genetic resources, rests with the national governments and is subject to national legislation', has aroused controversy. Indigenous peoples have argued that this assertion of sovereignty by States implicitly infringes upon their right to self-determination, and conflicts with international human rights law, which upholds the right of all peoples to 'freely dispose of their natural wealth and resources' (Articles 1.2 of both the International Covenant on Economic, Social and Cultural Rights and the International Covenant on

Civil and Political Rights). On the other hand, a commonly-expressed view in many Third World countries is that Article 15 is favourable for traditional communities, since national sovereignty rights over genetic resources are a pre-condition for an effective community rights system (e.g. Nijar 1996; Gene Campaign/Forum for Biotechnology and Food Security 1997). There are two reasons for such an opinion: firstly, only governments can enact national legislation; secondly, traditional communities would otherwise be left vulnerable to the influence of transnational corporations.

The CBD has some potentially far-reaching concessions for indigenous peoples and local communities. With respect to the 'knowledge, innovations and practices of indigenous and local communities embodying traditional lifestyles', Article 8(j) requires governments to: 'respect, preserve and maintain knowledge, innovations and practices of indigenous and local communities embodying traditional lifestyles relevant for the conservation and sustainable use of biological diversity, and promote the wider application with the approval and involvement of the holders of such knowledge, innovations and practices and encourage the equitable sharing of the benefits arising from the utilization of such knowledge, innovations and practices'. The wording is somewhat unclear, but the word 'holders' implies acceptance by the contracting parties that these communities have legal entitlements over their knowledge, innovations and practices, just as companies have over their inventions (Costa e Silva 1995). This interpretation is reinforced by Article 18(4), which affirms the need for contracting parties to 'encourage and develop methods of co-operation for the development and use of technologies, *including traditional and indigenous technologies...*'. Since it is agreed that indigenous and traditional technologies have a role to play in biodiversity conservation, there is no justification for denying that such technologies have a lower status than other technologies relevant to the Convention; nor should they be any less entitled to legal protection (Posey 1996).

The question is whether, given the realities explained in the previous section, intellectual property rights can ever protect the rights of the holders of traditional knowledge, innovations, practices and technologies. In fact, several imaginative proposals have been made either to modify the patent system, or to use trade secret law (Box 12.1).

Although these proposals are interesting, the lack of economic self-sufficiency of many traditional communities, the unequal power relations between them and the corporate world, and the high cost of litigation, would make it very difficult for them to protect their IPRs (Posey and Dutfield 1996). In any case, current IPRs cannot accommodate the complexity of many non-Western proprietary systems. These systems are sometimes assumed to be collective or communally-based, but in fact any assumption that there exists a generic form of non-Western, indigenous collective property rights ignores the intricacies and sheer diversity of indigenous proprietary systems (see Vira, Wiener, this chapter). According to the North American indigenous peoples' organization, the Four Directions Council (1996), '[i]ndigenous peoples possess their own locally-specific systems of jurisprudence with respect to the classification of different types of knowledge, proper procedures for acquiring and sharing knowledge, and the rights and responsibilities which attach to possessing knowledge, all of which are embedded uniquely in each culture and its language'. Rather than attempting to devise uniform IPR guidelines for protection of traditional knowledge, the Four Directions Council urges governments to 'agree that traditional knowledge must be acquired and used in conformity with the customary laws of the peoples concerned'.

The Indigenous Peoples Biodiversity Network (IPBN 1996) believes that it would be mistaken to construe Article 8 (j) only in terms of its support for the intellectual property rights of traditional communities. For Article 8(j) to have constructive value to indigenous peoples, governments must go much further than this, to: (i) accept the right to self-determination of indigenous peoples; (ii) observe indigenous peoples' human rights and fundamental freedoms; (iii) facilitate capacity-building of indigenous and local communities; (iv) promote the validation of indigenous knowledge; (v) acknowledge that indigenous knowledge is an intellectual property in the broadest sense but is inadequately protected by existing IPR regimes; and (vi) develop national legislation to protect indigenous knowledge including the establishment of *sui generis* systems.

Some traditional groups consider Article 8(j) to be a powerful statement of support for the right of communities to control access to their territories, resources and knowledge. Stephenson (this chapter) explains how the Loita Maasai of Kenya are using the Article as the basis for their claim that their sacred forest be guaranteed legal protection. However, Article 10(c) may constitute a more powerful assertion of rights. It requires parties to, 'protect and encourage customary use of biological resources in accordance with traditional cultural practices that are compatible with conservation or sustainable use requirements'. This is an acknowledgement that accommodating customary laws and practices

Box 12.1: Using IPRs to Protect Traditional Biodiversity-Related Knowledge: Some Recent Proposals

1. Certificates of origin have been proposed by a Peruvian non-governmental organization, Sociedad Peruana de Derecho Ambiental (Peruvian Society for Environmental Law), in order to make patent law more compatible with provisions in the CBD on national sovereignty, prior informed consent, and the rights of indigenous peoples and local communities (Tobin and Ruiz 1996; Tobin 1997). Administrative requirements for filing patent applications based on use of genetic resources and/or traditional knowledge should require inclusion of: (i) a sworn statement as to the genetic resources and associated knowledge, innovations and practices of indigenous peoples and local communities utilized, directly or indirectly, in the research and development of the subject matter of the IPR application; and (ii) evidence of prior informed consent of the country of origin and/or indigenous or local community, as appropriate.

International standardization of these conditions would be achieved through an international certification system. Accordingly, countries providing resources and/or traditional knowledge would issue certificates indicating that all obligations to the source country and the relevant indigenous people or local community had been fulfilled, such as prior informed consent, equitable benefit sharing, and perhaps other conditions imposing limitations on the use of the genetic material or knowledge. Patent applications would then need to include these certificates without which they would automatically be rejected. The system would not affect indigenous communities' right to veto access to and use of their knowledge or resources.

2. Innovation registers in conjunction with a low-cost patent system have been proposed by Anil Gupta of the Society for Research and Initiatives for Sustainable Technologies and Institutions (SRISTI), who emphasizes the adaptive and creative nature of so-called 'traditional' systems of knowledge and resource management (Gupta, this chapter). Rejecting arguments that all traditional knowledge should be treated as common property, Gupta asserts that some individuals within communities possess and generate considerably more knowledge than their neighbours do, and strive to conserve biodiversity while encouraging young people to learn and further develop traditional knowledge. Therefore, entitlements are not equal for all community members. Gupta (1996a) believes it is possible to work within existing intellectual property rights regimes by establishing a global register of innovation along the lines of SRISTI's local innovations database, with each entry bar coded to prevent data from being misappropriated. The register would be made available to all national patent offices to ensure that patents claims duplicating innovations contained in the register are rejected. Gupta suggests various reward mechanisms for grassroots innovators. One of these is the petty patent, a low-cost type of patent that is currently available in several countries (Gupta 1996b).

3. Transforming traditional knowledge into trade secrets is the title of an InterAmerican Development Bank-supported project based in Ecuador, the aim of which is to enable indigenous peoples to benefit from bio-prospecting though effective IPR protection of their knowledge (Vogel 1997). Knowledge from communities wishing to participate in the project will be catalogued and deposited in a restricted access database. Each community will have its own file in the database. Checks will be made to see whether each entry is not already in the public domain and whether other communities have the same knowledge. If communities with the same knowledge were to compete rather than collaborate, there would be a price war that would benefit only the corporate end-users. To overcome this danger, the project envisages the creation of a cartel comprising those communities bearing the same trade secret. The trade secret can then be negotiated in a Material Transfer Agreement (MTA) with the benefits shared between the government and the cartel members.

relating to genetic resources use and environmental management within national laws can enhance biodiversity conservation.

A provision that is frequently overlooked in the context of community resource rights is in Article 11 (Incentive Measures), according to which parties are expected to, 'adopt economically and socially sound measures that act as incentives for the conservation and sustainable use of components of biological diversity'. Equitable benefit sharing should be seen as a way of enabling governments to fulfil this article, but incentives should not *only* be in terms of economic gains. It may be that granting greater social justice for traditional communities would be the most effective incentive measure. Vogel (this chapter) considers means of achieving the fair and equitable sharing of *economic* benefits. In contrast, McAfee (this chapter) is highly critical of economistic approaches to biodiversity conservation, which she believes dominate policy-making. Similarly, Scholz and Chapela argue that perceiving the conservation and sustainable use of biodiversity in narrow economistic terms fails to reflect the conceptual difficulties surrounding the word 'biodiversity', and ignores the reality that cultural, spiritual and other non-economic values of biodiversity frequently guide economic considerations.

The CBD, the World Trade Organization and the FAO International Undertaking on Plant Genetic Resources

To review implementation of the CBD, the Conference of the Parties (i.e. all those countries that have ratified the CBD) meets at regular intervals. At the Third Meeting of the Conference of the Parties to the CBD (COP-3), two of the agenda items were, 'implementation of Article 8(j)' and 'intellectual property rights'. At the opening plenary, representatives of traditional peoples demanded that: (i) within an open, transparent and democratic process, guidelines be established to help national governments draw up legislation to implement Article 8(j); (ii) a process of dialogue be established with indigenous peoples to develop alternatives to the existing IPR system for the protection of indigenous knowledge systems; (iii) in the meantime, the claims of non-indigenous peoples to intellectual property rights over the processes and products associated with indigenous peoples' knowledge and genetic resources be prohibited; (iv) a moratorium be established on bioprospecting and ethnobotanical collections within indigenous peoples' territories until adequate protection mechanisms for indigenous knowledge are established; and (v) an open-ended inter-sessional working group on indigenous peoples and biodiversity, with the full participation of indigenous peoples' representatives, be established.

COP-3 agreed on the need to 'develop national legislation and corresponding strategies for the implementation of Article 8(j) in consultation with representatives of their indigenous and local communities' (UNEP 1997b [Decision III/14]). The Parties also agreed to establish an 'inter-sessional process' to advance implementation of Article 8(j) and related provisions with a view to producing a report for consideration at COP-4, in May 1998, in Bratislava, Slovakia. [In the event, Parties attending COP-4 agreed to establish 'an *ad hoc* inter-sessional working group', part of whose mandate was to provide advice on the application and development of legal and other appropriate forms of protection for the knowledge, innovations and practices of indigenous and local communities. For details of COP-4 decisions see the Website of the CBD Secretariat: http://www.biodiv.org] As part of this process, the CBD Secretariat was requested to arrange a Workshop on Traditional Knowledge and Biodiversity. The Workshop took place in Madrid, Spain in November 1997, and was attended by representatives of 148 indigenous and local community organizations. According to Burgiel, Prather and Schmidt (1997), the Workshop 'provided a unique opportunity for indigenous peoples and governments to engage in a dialogue on equal terms and under relatively open and flexible circumstances, and laid the groundwork for meaningful dialogue and future collaboration between Parties and indigenous and local communities'. Hopefully, this inter-sessional process will culminate in an effective and influential permanent inter-sessional working group or subsidiary body to provide urgently needed advice to governments on implementation of Article 8(j).

The decision on intellectual property rights (Decision III/17) calls, *inter alia*, for dissemination of case studies on the relationships between IPR and the knowledge, innovations and practices of indigenous and local communities. It is suggested that these case studies consider matters such as (i) the role and potential of *existing* IPR systems in enabling 'interested parties', including indigenous and local communities, to determine access and equitable benefit sharing, and (ii) the development of IPR, such as *sui generis* systems.

The World Trade Organization

One of the challenges to the development of creative alternatives to conventional IPRs is the influence of the World Trade Organization and its highest authority and supreme decision-making body, the Ministerial Conference. The Ministerial Declaration from the first meeting of the Ministerial Conference in Singapore in December 1996 stated that 'each Member should carefully review all its existing or proposed legislation, programmes and measures to ensure their full compatibility with the WTO obligations' (Paragraph 12). This statement should be considered in the light of Article 16(5) of the CBD which recognizes that 'patents and other intellectual property rights may have an influence on this Convention' and requires States to co-operate 'subject to national legislation and international law in order to ensure that such rights are supportive of and do not run counter to its [i.e. the CBD's] objectives'. Since the WTO's Committee on Trade and Environment has not yet been able to confirm that the TRIPs provisions are supportive of and do not in any way run counter to the CBD's objectives – and has consequently recommended further investigation of the matter (WTO 1996) – it was very presumptuous of the ministers in Singapore to issue such a statement, especially since most members of the WTO are also contracting parties to the CBD.

It seems that for the time being at least, the international community of States gives priority to the WTO as the main forum for consideration of all trade-related issues, including intellectual property rights. Nevertheless, *sui generis* systems to protect the knowledge of traditional communities are not incompatible with the TRIPs Agreement (see Correa, this chapter; Dutfield 1997). In fact, Article 27.3(b) of TRIPs specifically allows for an 'effective' *sui generis* system for plant variety protection as an alternative to patents. This is an opportunity for national governments to introduce a community-based rather than individual-based IPR system that is consistent with the CBD's objectives and its provisions on indigenous peoples and local communities, and also with FAO Farmers' Rights.

The Food and Agriculture Organization of the United Nations

In 1983, the Food and Agricultural Organization of the United Nations (FAO) established an intergovernmental Commission on Plant Genetic Resources (CPGR) and adopted a non-binding International Undertaking on Plant Genetic Resources (IUPGR) to 'ensure safe conservation and promote the availability and sustainable utilization of plant genetic resources for present and future generations, by providing a flexible framework for sharing the benefits and burdens' (UNEP 1994). The IUPGR included the concept of 'Farmers' Rights' as an attempt to acknowledge 'the contribution farmers have made to the conservation and development of plant genetic resources, which constitute the basis of plant production throughout the world' (ibid). Resolution 5/89 defined Farmers' Rights as: 'rights arising from the past, present and future contributions of farmers in conserving, improving and making available plant genetic resources particularly those in the centres of origin/diversity. Those rights are vested in the international community, as trustees for present and future generations of farmers, and supporting the continuation of their contributions as well as the attainment of overall purposes of the International Undertaking [on Plant Genetic Resources]'.

In 1993, the CPGR Resolution 93/1 called for a revision of the IUPGR to harmonize the latter with the CBD. To this effect, the Commission, now called the Commission on Genetic Resources for Food and Agriculture (CGRFA), is working in close co-operation with the CBD to revise the International Undertaking accordingly. If governments so decide, the revision could be converted into a legally-binding instrument or Protocol to the CBD (Glowka *et al.* 1994). A legally-binding IUPGR would ensure that Farmers' Rights finally have legal recognition. However, to harmonize with the CBD, the concept of Farmers' Rights must be redefined, and probably renamed (Posey 1996). The first step is to widen the scope of 'farmers' beyond its conventional meaning to recognize that: (a) plant genetic resources conserved by farmers include not only field crops but also non-timber forest products, and medicinal and herbal plants [the CBD has in effect done this]; and (b) indigenous peoples and other traditional groups and local communities, such as fisher folk, hunters, pastoralists, nomads and gatherers, are also included in any benefit sharing and protection system.

Integrating rights for community control

Indigenous peoples, local communities and their supporters are responding to international agreements like the CBD, TRIPs and the IUPGR by: (i) actively opposing trends in intellectual property and international trade law; (ii) advocating equitable benefit-sharing from biotechnological research through use of contracts and covenants, and the development of ethical guidelines and codes of practice; or (iii) by using emerging international environmental and human rights law as part of a campaign aimed at empowering traditional communities (Sutherland 1997a).

Many indigenous peoples and local communities around the world have their own effective systems of conservation, access regulation and (non-monetary) benefit-sharing. A policy of non-interference in these systems is a practical way to conserve biodiversity, and also a way to respect the rights of the people concerned. Accommodation of customary law in national legal systems is therefore important, but this alone does not guarantee local control of territory, knowledge and resources. In Papua New Guinea, for example, where local customary law is part of the so-called 'Underlying Law' of the country, and most of the land is owned by local communities, local communities are still dominated and exploited by logging and mining companies (Donigi 1994; Nadarajah, this chapter).

Craig and Nava (1995) assert that a broader-based 'integrated rights approach', bringing together a series of rights that are gaining acceptance in international environmental and human rights law, may be the most effective approach to securing local control over biodiversity and traditional knowledge, innovations and practices. A more holistic conception of traditional communities' rights is needed because: 'indigenous sustainable development is an integrated cultural concept that is part of everyday customary laws and practices. It cannot be separated from other indigenous rights such as human rights, rights to land and resources and their management, intellectual and cultural property rights, and the right to self-government, which are being asserted in international law and policy' (Craig and Nava 1995; see also Shelton 1995).

Traditional Resource Rights (TRRs) is a concept founded on such a broad-based integrated rights approach (see Posey 1996; Posey and Dutfield 1996). TRRs are more of a process than a product; not so much a system as a framework of principles that can serve, not as a single rigid 'solution' to which the Four Directions Council is opposed, but as the basis for fairer laws and agreements that support the rights of traditional communities. According to Posey (1996), TRRs emerge from four processes: (1) identifying 'bundles of rights' expressed in existing moral and ethical principles; (2) recognizing rapidly evolving 'soft law' influenced by customary practice and legally non-binding agreements, declarations, and covenants; (3) harmonizing existing legally-binding international agreements; and, (4) 'equitizing' to provide marginalized indigenous, traditional, and local communities with favourable conditions to influence all levels and aspects of policy planning and implementation.

By acknowledging communities' rights to control access to traditional resources and territories, TRR-guided negotiations and legal processes can offer opportunities and mechanisms for new partnerships based on increased respect for traditional communities, and guide governments in effecting their international obligations on trade, environment and development, and their human rights responsibilities. TRRs bring together principles or 'bundles of rights' that are widely recognized by international agreements, thereby integrating environmental conservation with intellectual property rights and human rights. These include: (i) the right to self-determination; (ii) territorial and land rights; (iii) human rights (individual and collective); (iv) the right to development; (v) accommodation of customary law and practice; and (vi) cultural rights.

The emerging international law of sustainable development is rapidly evolving and increasingly embraces individual and collective human rights principles such as the right to development, the right to a healthy environment, equity, access to information, and public participation (DPCSD 1996). International law, therefore, already provides a considerable measure of support for those pursuing rights-based approaches to securing local control over biodiversity and traditional knowledge, innovations and practices. However, there is much more to be done. As Craig and Nava (1995) assert: '[t]he challenge for international law and policy is to learn from the unique experience and culture of indigenous peoples in relation to sustainable development and to develop international environmental law in a way that respects established and emerging international rights and standards applicable to indigenous peoples. International recognition is an important moral and legal force in achieving national and local implementation of indigenous sustainable development'.

The right to self-determination

A distinction between indigenous peoples and other local communities is that the former claim that the right to self-determination – a well-established principle of international law – applies to them as much as it does to the nation states that emerged from the post-World War II decolonization era. The principle of self-determination is set forth in the Charter of the United Nations, and in Article 1 of the International Covenant on Economic, Social and Cultural Rights (ICESCR) and the International Covenant on Civil and Political Rights (ICCPR), according to which: 'all peoples have the right to self-determination. By virtue of that right they freely determine their political status and freely pursue their economic, social and cultural development'.

Full self-determination would allow indigenous people to (see Posey and Dutfield 1996): (i) be self-governing; (ii) enact legislation; (iii) control access to the territory and the resources existing within territorial boundaries; and (iv) enter into legally-binding international treaties.

Perhaps the broadest sovereignty rights for indigenous peoples are exercised by the people of Greenland under the 1979 Home Rule Act passed by the Danish parliament (Nuttall 1994; Petersen 1994). The Inuit of Nunavut in northern Canada will enjoy similar rights from 1999. In the United States, native tribes recognized by the federal government have sufficient sovereignty rights to allow tribal courts to adjudicate violations of customary law committed by both Indians and non-Indians. The right to self-determination includes recognition and full implementation of historical treaties made between indigenous peoples and settler states such as the United States, Canada, Australia and New Zealand, where the 1840 Treaty of Waitangi is still in force and the basis for an ongoing claim by the Maori that their cultural and intellectual property rights be respected (see Solomon, this chapter).

Territorial and land rights

The right of ownership and control over traditional lands and resources should be considered a fundamental right. A number of international texts and agreements provide support for this, including Article 14 of the International Labour Organization's Convention 169 on Indigenous and Tribal Peoples (1989), which states: '[t]he rights of ownership and possession of the peoples concerned over the lands which they traditionally occupy shall be recognized. In addition, measures shall be taken in appropriate cases to safeguard the right of the peoples concerned to use lands not exclusively occupied by them, but to which they have traditionally had access for their subsistence and traditional activities. Particular attention shall be paid to the situation of nomadic peoples and shifting cultivators in this respect'.

The UN Draft Declaration on the Rights of Indigenous Peoples (Appendix 1.18) further states the following in Part VI, Paragraphs 25–26: '[I]ndigenous peoples have the right to maintain and strengthen their distinctive spiritual and material relationship with the lands, territories, waters and coastal seas and other resources which they have traditionally occupied or used, and to uphold their responsibilities to future generations in this regard'. And: '[I]ndigenous peoples have the right to own, develop, control and use the lands and territories, including the total environment of the lands, air, waters, coastal seas, sea-ice, flora and fauna and other resources which they have traditionally owned or otherwise occupied or used. This includes the right to the full recognition of their laws, traditions and customs, land-tenure systems and institutions for the development and management of resources, and the right to effective measures by States to prevent any interference with, alienation of, or encroachment upon, these rights'.

Nadarajah's contribution on Papua New Guinea (this chapter) serves as a timely warning that even if communities enjoy territorial security as recognized customary landowners, this may not be enough to protect them from exploitation. Many development project designers there are either unwilling or unable to make land owners equal partners in their projects, and the communities are in a weak position to negotiate agreements with companies that favour their long-term interests. Nevertheless, recognition of land rights is of vital importance to communities wishing to safeguard their resources.

Human rights

The extent to which Western notions of human rights are compatible with more collectively oriented societies has evoked considerable debate. Whether or not the individualistic orientation of Western human rights is problematic in the context of traditional societies, human rights as they exist in international law are not entirely individualistic. The right to self-determination is a collective right, and development rights and cultural rights have strong collective aspects.

According to Barsh (1995), 'indigenous peoples generally think in terms of the freedom of individuals to be what they were created to be, rather than being free from certain kinds of state encroachments. Along with this highly individualized notion of 'rights' is a sense of unique personal responsibilities to kin, clan and nation. Each individual's 'rights', then, consists of the freedom to *exercise responsibilities towards others*, as she or he understands them, without interference.' In short, indigenous societies often consider each member as having *individual* rights and *collective* responsibilities that are linked inextricably.

According to Article 8 paragraph 2 of the International Labour Organization's Convention Concerning Indigenous and Tribal Peoples in Independent Countries (ILO 169): '[t]hese [i.e. indigenous and tribal] peoples shall have the right to retain their own customs and institutions, where these are not incompatible with fundamental rights defined by the national legal system and with internationally recognized human rights'. Therefore, recognizing the collective rights of peoples does not diminish the entitlement of members of such peoples to the individual rights guaranteed by international law and the laws of that country.

The right to development

The principle of the right to development is enshrined in international law in the two main human rights treaties, the International Covenants on Civil and Political Rights, and on Economic, Social and Cultural Rights. Article 1(2) of both these documents calls for the recognition of development rights *for all peoples*, who 'may, for their own ends, freely dispose of their natural wealth and resources without prejudice to any obligations arising out of international economic co-operation, based upon the principle of mutual benefit, and international law. In no case may a people be deprived of its own means of subsistence'.

The right to development for traditional communities includes, (a) the right of access to resources on their territories, and (b) the right to seek development on their own terms. It is a very important principle because government agencies and NGOs concerned with conservation sometimes deny communities the right to exploit and to commercialize local resources. The need for traditional communities to secure rights to their subsistence base, enforced by the state and backed by international law, is a precondition for fostering and maintaining traditional systems of ecosystem management and *in situ* biodiversity conservation. According to the UN Declaration on the Right to Development (adopted by the General Assembly in 1986): 'The right to development is an inalienable human right by virtue of which every human person and all peoples are entitled to participate in, and enjoy, economic, social, cultural and political development' (Article 1).

ILO Convention 169 states: [t]he peoples concerned shall have the right to decide their own priorities for the process of development as it affects their lives, beliefs, institutions and spiritual well-being and the lands they occupy or otherwise use, and to exercise control, to the extent possible, over their own economic, social and cultural development. In addition, they shall participate in the formulation, implementation and evaluation of plans and programmes for national and regional development which may affect them directly' (Article 7[1]). Also, '[t]he rights of the peoples concerned to the natural resources pertaining to their lands shall be specially safeguarded. These rights include the right of these peoples to participate in the use, management and conservation of these resources' (Article 15[1]). Agenda 21 also contains powerful assertions of development rights with specific reference to indigenous peoples, pointing out that such rights are necessary for sustainable development (see Posey 1996).

Some conservationists are sceptical that recognition of the right to development encourages sustainable community-based practices. But experience shows that when traditional communities enjoy territorial security and freedom to make their own decisions, they tend to exercise their right to development in ways that provide long-term environmental benefits. Conservationists and developers must be prepared to share or even relinquish decision-making powers to traditional communities, while seeking to understand local priorities and criteria for sustainable development.

Accommodation of customary law and practice

Communally shared concepts and communally owned property are fundamental aspects of many traditional societies. Traditional proprietary systems are often highly complex, and varied, but a common characteristic is collective responsibility for land and territory. Individuals and families may hold lands, resources or knowledge for their own use, but ownership is often subject to customary law and practice and based on the collective consent of the community. According to the Coordinating Body of Indigenous Organizations of the Amazon Basin (COICA 1994): '[f]or members of indigenous peoples, knowledge and determination of the use of resources are collective and intergenerational. No indigenous population, whether of individuals or communities, nor the government, can sell or transfer ownership of resources which are the property of the people and which each generation has an obligation to safeguard for the next'.

Traditional communities sometimes have their own juridical institutions, and often find that their own laws are not recognized by States, forcing them to conform to laws with which they are unfamiliar, and which may be inappropriate or conflict with their own laws (see Morris, this chapter). Recognition of customary law is vital not only for the cultural integrity of indigenous peoples but also for the conservation of biodiversity. According to Alejandro Argumedo of the Indigenous Peoples Biodiversity Network (personal communication, 1994), 'the legal systems of our nations have grown from, and continue to be inextricably bound up in, systems of management and conservation of biodiversity. Therefore these laws, far from being anthropological curiosities, should guide all work related to the conservation of biodiversity'.

Cultural and heritage rights

The knowledge, innovations and practices of indigenous peoples and local communities are manifestations of their cultures. Protecting a peoples' culture means maintaining those conditions that allow a culture to thrive and develop further. This is why framing rights in terms of 'cultural rights' or 'heritage rights' may be useful in some situations. A United Nations study on the protection of the cultural and intellectual property of indigenous peoples (Daes 1993) adopted the term 'heritage' to refer to: 'everything that belongs to the distinct identity of a people and which is theirs to share, if they wish, with other peoples. It includes all of those things which international law regards as the creative production of human thought and craftsmanship, such as songs, stories, scientific knowledge and artworks. It also includes inheritances from the past and from nature, such as human remains, the natural features of the landscape, and naturally-occurring species of plants and animals with which a people has long been connected'. Therefore, protecting a peoples' cultural heritage involves *inter alia* maintaining the link between a people and 'natural features of the landscape and naturally-occurring species of plants and animals'.

Although the term 'cultural rights' does not exist *per se* in international law, Prott (1988) identifies a bundle of cultural rights that are supported by international hard or soft law (i.e. non-legally binding) instruments, almost all of which are also covered by the UN Draft Declaration on the Rights of Indigenous Peoples. Prott notes that most of these rights are collective peoples' rights.

Article 27 of the International Covenant on Civil and Political Rights links cultural rights to the rights of minorities: '[i]n those States in which ethnic, religious or linguistic minorities exist, persons belonging to such minorities shall not be denied the right, in community with the other members of their group, to enjoy their own culture, to profess and practice their own religion, or to use their own language'. Kingsbury (1992) warns that this is a limited provision and that many of the leading cases have been brought by individual members of minorities seeking protection from the policies of the minority community itself (1992). Suagee (1994: 196 [fn. 17]) mentions a case in which a Native Canadian was forbidden to live on her band's reserve because of her marriage to a non-Indian. She filed a successful complaint with the UN Human Rights Committee on the basis that her rights as set out in Article 27 of the ICCPR had been violated. Nevertheless, the article has been interpreted in ways that support claims brought by indigenous people against unjust land and other policies of their governments (ibid; Sutherland 1997b). In one case, a decision of the Human Rights Committee implied that the right of an individual Saami to engage in reindeer husbandry as a member of a Saami community was protected by Article 27 (ibid). In another case brought by a group of indigenous people from Canada whose way of life and culture were jeopardized by imposed territorial restrictions, the persistence of 'historical inequities' was violated (Article 27). This implies, according to Kingsbury, that: '[t]he right of members of a group to enjoy their own culture may be violated where they are not allocated the land and control of resource development necessary to pursue economic activities of central importance to their culture, such as hunting and trapping'.

It is evident that laws and policies to protect traditional knowledge, innovations and practices are urgently needed. Such laws and policies should be based on an integrated rights approach guided by the principles outlined in the introduction. Maintaining traditional communities' cultural and spiritual values relating to biodiversity is in large part a question of social justice. Recognition of self-determination for indigenous peoples and local autonomy for non-indigenous local communities are the fundamental conditions for maintaining 'the inextricable link' between biological and cultural diversity. Traditional communities justifiably demand recognition of their land and territorial rights, including the right to refuse access and to veto projects, plans and biological resource collections on these territories. These communities also demand greater accommodation within national laws for their customary laws and practices. Without respect for these rights, traditional communities will continue to be exploited and marginalized. Their cultures will be eroded irreversibly and their unique and diverse systems of knowledge will be lost forever.

The tragedy of enclosure (George Monbiot)

During the long dry seasons in the far north-west of Kenya, the people of the Turkwel River keep themselves alive by feeding their goats on the pods of the acacia trees growing on its banks. Every clump of trees is controlled by a committee of elders, who decide who should be allowed to use them and for how long. Anyone coming into the area who wants to feed his goats on the pods has to negotiate with the elders. Depending on the size of the pod crop, they will allow him in or tell him to move on. If anyone overexploits the pods, or tries to browse his animals without negotiating with the elders first, he will be driven off with sticks: if he does it repeatedly he may be killed. The acacia woods are a common: a resource owned by many families. Like all the commons of the Turkana people, they are controlled with fierce determination.

In the 1960s and 1970s the Turkana were battered by a combination of drought and raiding by enemy tribes armed with automatic weapons. Many people came close to starvation, and the Kenya Government, the United Nations Development Programme and the UN's Food and Agriculture Organization decided that something had to be done to help them. The authorities knew nothing of how the Turkana regulated access to their commons. What they saw, in the acacia forests, and the grass and scrublands of the savannahs, was a succession of unrelated people moving in, taking as much as they wanted, then moving out again. If the Turkana tried to explain how it worked, their concepts were lost in translation. It looked like a free-for-all, and the experts blamed the lack of regulation for the disappearance of the vegetation. This was, in fact, caused not by people but by drought.

They decided that the only way to stop the people from over-using their resources was to settle them down, get rid of most of their animals and encourage them to farm. On the banks of the Turkwel River they started a series of irrigation schemes, where ex-nomads could own a patch of land and grow grain. They spent US$60,000 per hectare in setting up the scheme.

People flocked in; not, on the whole, to farm, but to trade, to find paid labour or to seek protection from their enemies. With the first drought the irrigation scheme collapsed. The immigrants reverted to the only certain means of keeping themselves alive in the savannahs: herding animals. They spread along the banks, into the acacia woods. Overwhelmed by their numbers, the elders could do nothing to keep them away from their trees. If they threatened to kill anyone for taking pods without permission, they were reported to the police. The pods and the surrounding grazing were swiftly exhausted and people started to starve. The commons had become a free-for-all. The authorities had achieved exactly what they set out to prevent.

The overriding of commoners' rights has been taking place, often with similarly disastrous consequences, for centuries, all around the world. But in the last three decades it has greatly accelerated. The impetus for much of this change came from a paper whose title has become a catch-phrase among developers.

In *The Tragedy of the Commons* (Hardin 1968), Garrett Hardin argued that common property will always be destroyed because the gain that individuals make by over-exploiting it will outweigh the loss they suffer as a result of its over-exploitation. He used the example of a herdsman keeping his cattle on a common pasture. With every cow the man added to his herds he would gain more than he lost: he would be one cow richer, while the community as a whole would bear the cost of the extra cow. He suggested that the way to prevent this tragedy was to privatize or nationalize common land.

The article had an enormous impact. It neatly encapsulated a prevailing trend of thought, and appeared to provide some of the answers to the growing problem of how to prevent starvation. For authorities such as the World Bank and Western governments it provided a rational basis for the widespread privatization of land. In Africa, among newly-independent governments looking for dramatic change, it encouraged the massive transfer of land from tribal peoples to the State or to individuals. In Africa, Asia, Europe and the Americas, developers hurried to remove land from commoners and give it to people they felt could manage it better. The commoners were encouraged to work for those people as waged labour or to move to the towns, where, in the developing world, they could become the workforce for the impending industrial revolutions.

But the article had one critical flaw. Hardin had assumed that individuals can be as selfish as they like in a commons, because there is no one to stop them. In reality, traditional commons are closely regulated by the people who live there. There are two elements to common property: common and property. A common is the property of a particular community which, like the Turkana of the Turkwel River, decides who is allowed to use it and to what extent they are allowed to exploit it.

Hardin's thesis works only where there is no ownership. The oceans, for example, possessed by no one and poorly regulated, are over-fished and polluted, as every user tries to get as much out of them as possible, and the costs of their exploitation are borne by the world as a whole. But these are not commons but free-for-alls. In a true commons, everyone watches everyone else, for they know that anyone over-exploiting a resource is exploiting them.

The effects of dismantling the commons to prevent Hardin's presumed tragedy of over-exploitation from running its course can scarcely be overstated. In Brazil, for example, peasant communities are being pushed off their land to make way for agro-industry. Land that supported thousands of people becomes the exclusive property of one family or corporation. Mechanization means that hardly any permanent labour is needed. Some of the dispossessed go to the cities where, instead of an industrial revolution, they find unemployment and destitution. Others go to the forests, where they may try to move into the commons belonging to the Indians, defying their regulations by cutting and burning the forests.

No group has suffered more than the people singled out by Hardin's paper: the traditional herders of animals, or pastoralists. In Kenya, the Maasai have been cajoled into privatizing their commons: in some parts every family now owns a small ranch. Not only is this destroying Maasai society, as tight communities are artificially divided into nuclear families, but it has also undercut the very basis of their survival.

In the varied and changeable savannahs, the only way a herder can survive is by moving. Traditionally the Maasai followed the rain across their lands, leaving an area before its resources were exhausted and returning only when it had recovered. Now, confined to a single plot, they have no alternative but to graze it until drought or over-use brings the vegetation to an end. When their herds die, entrepreneurs move in, buy up their lands for a song and either plough them for wheat and barley, exhausting the soil within a few years, or use them as collateral for securing business loans.

In Somalia, Siad Barre's government nationalized the commons, nullifying the laws devised by Somali communities to protect their grazing lands from people of other clans. When charcoal burners moved in to cut their trees, the local people found that there was nothing they could do to stop them. The free-for-all with which the commons regime was replaced was one of the reasons for the murderous chaos that overtook the country.

These changes in the ownership of land lie at the heart of our environmental crisis. Traditional rural communities use their commons to supply most of their needs: food, fuel, fabrics, medicine and housing. To keep themselves alive they have to maintain a diversity of habitats: woods, grazing lands, fields, ponds, marshes and scrub. Within these habitats they need to protect a wide range of species: different types of grazing, a mixture of crops, trees for fruit, fibres, medicine or building. The land is all they possess, so they have to look after it well. But when the commons are privatized they pass into the hands of people whose priority is to make money. The most efficient means of making it is to select the most profitable product and concentrate on producing that.

As land changes hands, so does power. When communities own the land they make the laws, and develop them to suit their own needs. Everyone is responsible for ensuring that everyone else follows these laws. As landlords take over, it is their law that prevails, whether or not it leads to the protection of local resources. The language in which the old laws were expressed gives way to the language of outsiders. With it go many of the concepts and cautionary tales encouraging people to protect their environment. Translated into the dominant language they appear irrational and archaic. As they disappear, so does much that makes our contact with the countryside meaningful: it becomes a series of unrelated resources, rather than an ecosystem of which we, economically, culturally and spiritually, are a part. For human beings, as for the biosphere, the tragedy of the commons is not the tragedy of their existence but the tragedy of their disappearance.

Property rights: clarifying the concepts (Bhaskar Vira)

According to Bromley, there are few concepts in economics that are more confused than those of property, rights, and property rights. He attempts to provide some clarity with the following definition (Bromley 1991: 2): 'property is a benefit (or income) stream, and a property right is a claim to a benefit stream that the state will agree to protect through the assignment of duty to others who may covet, or somehow interfere with, the benefit stream. ... Property is not an object but rather is a social relation that defines the property holder with respect to something of value (the benefit stream) against all others'.

Once society makes a distinction between property and mere physical possession, it has effectively described property as a social relation, or a *right*. Property rights are defined here as socially legitimized claims to a benefit stream. This is a communitarian definition, in contrast to the Lockean tradition of 'natural' rights theories, which justify private property and ownership without reference to any form of social contract. The social nature of property becomes clear from the distinction between property and *access*, where the latter is not supported by protection of claims by the collectivity. To be defined as property, a claim to the use or benefit of something must be enforceable and honoured by society, either by custom, convention or law. While access may be pre-social, property is a social relation.

Property rights, then, refer not to relations between agents and things, but sanctioned and enforceable behavioural relations among agents 'that arise because of the existence of things and pertain to their use' (Furubotn and Pejovich 1972: 1139). For Bromley (1989: 206), 'the core of property is not the physical objects but the rights, the expectations, the duties and the obligations that must exist in any society before property can exist'. Hallowell (1943: 130) stresses that the concept of property extends beyond the exercise of rights and duties with respect to objects of value by members of society, since 'it also embraces the specific social sanctions which reinforce the behaviour that makes the institution a going concern'.

Claims other than those made with respect to a benefit stream may receive social recognition, but these constitute *rights* and not *property rights*. It would be conceptually mistaken and, arguably, morally repugnant, to conflate the two. In most declarations of rights, including the Universal Declaration of Human Rights, the right to property is one among many. Defenders of civil liberties would be loath to subsume their claims for freedom of speech, equality before the law, freedom of movement, and the like, under claims for property rights. At the very least, it can be argued that failing to distinguish *rights* from *property rights* implies that the qualifying prefix 'property' is meaningless. More precisely, the notion of rights is a more encompassing one that includes, among others, the right to property.

For Becker, property rights are the rights of ownership, and he argues that 'to have a property right in a thing is to have a bundle of rights that defines a form of ownership' (Becker 1980: 189-90). Becker modifies Honoré's (1961) analysis of the concept of full, liberal ownership to provide an overview of his notion of property rights. Becker's list of the bundle of rights that constitute the concept of ownership are: (i) right to *possess*; (ii) right to *use*, i.e. the owner's personal use and enjoyment of the benefit stream; (iii) right to *manage*; i.e. the authority to decide how and by whom something may be used, including decisions about whether others are allowed physical access; (iv) right to the *income* where the owner foregoes personal use and allows others access; (v) right to *consume or destroy*, i.e. the right to annihilate the benefit stream; (vi) right to *modify*; (vii) right to *alienate*, i.e. to transfer the benefit stream and to abandon ownership; (viii) right to *transmit*, i.e. to bequeath his/her interest in a benefit stream; (ix) right to *security*, i.e. immunity of the owner from arbitrary appropriation of the benefit stream.

There are four qualifications to the exercise of these rights: (i) *absence of term*, i.e. the length of ownership rights is not determinate; (ii) *prohibition of harmful use*, i.e. the exercise of ownership rights must not cause harm to oneself or to others; (iii) *liability to execution*; i.e. the benefit stream may be taken away to repay debts that have been incurred by the owner; and (iv) *residuary rules*, which refers to situations where ownership rights have lapsed.

Becker describes full ownership as the existence of all the elements, but suggests that apart from the right to security, 'any of the remaining eight rights ... can stand as variety of legal ownership when it is supplemented by some version of the right to security' (Becker 1980: 192). The qualifications clarify that even full, liberal ownership (or 'unattenuated' property rights) does not imply that owners have no restrictions upon their activity. The most explicit restrictions are those that are imposed because of the prohibition of harmful use and the liability to execution. It is unlikely that there are empirical circumstances where ownership has ever been absolute, and most societies impose some form of restrictions upon the exercise of ownership. There is little evidence that supports the contention of the *universality* of property rights, if this universality goes beyond the descriptive and makes claims about the *functional* uniformity of the social relations that can be so described.

An important implication of this description of the elements of property is that it applies in a variety of cultural settings. Munzer (1990: 25–7) suggests that the classification is useful in examining social settings that differ from the established Western legal tradition without referring to them as representing some form of primitive communism. This is especially important when considering the first assumptions that followed from the colonial encounter in many parts of the world, where the settlers faced established systems of customary law that were radically different from those with which they were familiar. This led the colonialists to assume that they were faced with primitive societies that had *no* system of law, ownership or property. The reaction was to 'civilize' the colonies by imposing a system of Western law, which was completely alien to the indigenous communities. This was the reason for extremely inequitable processes of settlement and control of land and resources, rejecting established community systems of control. The present framework provides a useful way to interpret the concepts of property and ownership that exist in Western law as well as in customary law, and to identify differences between these legal traditions without assuming the necessary superiority of any particular system. As attempts are made to integrate traditional law into codified national legal systems, such a common framework of analysis is particularly useful to facilitate reasoned debate about culturally embedded attitudes to ownership and property.

In some everyday usage, property rights are taken as synonymous with individual, private property rights. The current debate over the potential commercial exploitation of traditional ecological knowledge of indigenous peoples is illustrative. Posey (1994) argues that the conventional instruments of Intellectual Property Rights (IPRs) are inappropriate for the protection of indigenous peoples' rights, and proposes an alternative concept of 'Traditional Resource Rights' or TRR. Significantly, TRR does *not* refer to *property*, because of the common association with *individual* rights to alienate and exclude, which is particularly inappropriate because traditional knowledge systems are often *community*-based and knowledge may be handed down through the generations. Similarly, Nijar (1994: 17) suggests the term 'Community Intellectual Rights' because IPR 'connotes commoditization and ownership in private hands primarily for commercial exchange'. Thus, the right to exclude, if it exists, can be viewed only as a *group* right, which rests with the community.

However, members of the community may have the *secure* (individual) right to *use* their traditional knowledge, and this can be viewed as giving rise to a benefit stream. In this limited sense, Becker's descriptive framework would allow the relationship to be interpreted as a property right. The merit of describing this as a form of property right, but not as the individual right to alienate and exclude, is to focus on the specific features of the system that distinguish it from the Western legal tradition. This protects against the assertions of bio-prospectors who claim that the non-existence of private property rights in the Western legal sense implies that there are *no* property rights, and who argue that knowledge is freely accessible until protected by conventional IPR instruments. Customary law does protect access to traditional knowledge, and some aspects of this protection can be analysed as property rights, although these are not individual rights to exclude and alienate. Other dimensions of traditional knowledge are part of the cultural identity of indigenous groups, and cannot be analysed as generating streams of benefits; these include myths, stories, rituals, sacred sites, and other non-material items of knowledge which are an intrinsic part of the society, but are not reducible to property rights.

Macpherson (1973) traces the rise of the liberal conception of property to a specific historical process, the development of the capitalist market economy. According to him, 'property as exclusive, alienable, 'absolute', individual, or corporate rights in things was required by the full market society because and in so far as the market was expected to do the whole work of allocation of natural resources and capital and labour among possible uses' (Macpherson 1973: 133). Thus, an outcome of the rise of the capitalist economy was a reduction of the conception of property to *private* property, viewed as an individual (or corporate) right to exclude others from the benefit or use of something. Demsetz (1988: 24) defines 'full' private ownership as permitting exclusivity and alienability by private decision. However, there are circumstances where the definition of exclusive private rights, even if it is desirable, may be physically impossible or extremely costly. As Dasgupta elaborates, with renewable natural resources such as 'forests, coastal waters, threshing grounds, grazing lands, village ponds and tanks, rivulets and aquifers ... private property rights ... are often difficult to define. Even when they are definable, they are difficult to enforce' (Dasgupta 1993: 146). This is extremely important for the present discussion of biodiversity, since the physical characteristics of natural resources suggest that exclusive individual ownership is not to be expected. However, this does *not* imply that no form of property rights can be defined with respect to these resources.

While the second element emphasized by Demsetz, alienability, is an important element of many property systems, this should not be taken to imply that there are no reasons to believe that alienability should be restricted. Reeve (1986: 154) argues, for instance, that if an owner's descendants have a 'strong right to inherit property, then the present owner may not be entitled to dispose of it in his lifetime'. This raises an interesting issue. If future generations have a 'strong right' to inherit biodiversity, *and* if allowing alienability increases the exploitation of biodiversity, there may be an argument against alienability. In fact, if such resources are viewed as 'natural capital', then there would be reason to restrict not only the right to alienate, but also the right to consume or destroy, and, arguably, the right to modify the natural environment, since many ecologists believe that there is imperfect substitutability between natural and man-made capital.

Bromley (1978) applies Calabresi and Melamed's (1972) legal analysis of 'inalienable' entitlements in the context of environmental policy, and argues that the existence of significant third-party effects may be one reason to suggest the adoption of an inalienability rule. Although both parties to a voluntary trade would be better off after the exchange, there may be other interested groups who prefer the pre-trade allocation, and are unrepresented in the transaction. In general, the right to alienate may be restricted because of reasons of *morality*, *paternalism* or *self-paternalism* (Calabresi and Melamed 1972). For instance, Goodin suggests that pollution permits and environmental taxes can be viewed as 'selling indulgences' for activities that degrade nature. For Goodin (1993: 581), such indulgences belong to the category of 'things that ought not to be bought and sold', and this is a strong moral reason to reject the market as an allocation mechanism when referring to the natural environment and biodiversity. Paternalism may be used to protect groups from unequal exchange in markets that are not competitive, or are inequitable. Those who argue that certain indigenous populations should be protected by inalienable rights to their ancestral resources are using such a justification for their claim. Populations close to the level of subsistence may be tempted to part with their resources if they are offered very large sums of money, and while this would be a voluntary relinquishment of resources, they may be exchanged at a price that does not reflect their true value. Here, inalienability serves the long-term interests of groups who have insufficient material resources to resist the short-term temptations of selling their property for immediate returns, and who participate in the market on unequal terms.

More generally, arguments in favour of alienability appear to rest on the assumed superiority of decentralized allocation decisions over those that would be made under some alternative social process. Clearly this cannot be *assumed* to hold for all resources and in all societies. As Reeve (1986: 24–5) argues, 'many aspects of what persons find important about property are ... not connected with exchangeability. ... That something has value, and that it is exchangeable, are two distinct requirements'. Becker (1980: 207–9) refers to Polanyi's classification of economic systems as predominantly social, political or market, and argues that alienability is required only for market economies. The point here is stronger, since it suggests that even in a predominantly market economy, the individual right to alienate may not be desirable in certain circumstances, such as when referring to biodiversity.

Can biodiversity be 'owned'?

An interesting debate in the property rights literature revolves around *what* may be owned in a system of property rights. Concern sometimes arises because of conventional morality, which may require that certain types of transactions are kept free from commercial influences. For instance, Titmuss (1970) argues that there are ethical reasons to believe that blood should be donated, and not sold through the market. Munzer (1994: 284) argues a case against property rights in body parts which is based on the Kantian premise that all human beings have dignity, and 'if an entity has dignity, then treating that entity or some part of it as a commodity is morally objectionable if the treatment offends dignity'. Similarly, although slave ownership and child labour are considered morally abhorrent today, this has not been true for all human society. These examples suggest that there are strictly *moral* reasons for preventing the commoditization of certain benefit streams, even if it is economically efficient to do so. It has been suggested that there are similar moral reasons not to allow rights of exclusion and alienation over the environment (Sagoff 1988; Vatn and Bromley 1995). All cultures do not conceive of the human–environment relationship as one of *ownership*, and it is ethically unacceptable for them to reduce the environment to the status of a commodity that can be traded and exchanged.

A case can be made against defining property rights over biodiversity by pointing out that not all aspects of such resources are necessarily reducible to *benefit streams*. Pearce and Turner (1990: 129–37) identify three potential sources of human value for natural resources and biodiversity, namely use value, option value and existence value. Now, both use values (this includes aesthetic values) and option values can be understood to contribute to a stream of benefits that flow from the existence of biodiversity. However, existence value is more difficult to classify as a benefit stream. Pearce and Turner (op cit.: 131) define existence values as 'values expressed by individuals such that those values are unrelated to use of the environment, or future use by the valuer or the valuer on behalf of some future person'. Similarly, Aldred (1994: 394) adopts a definition of existence value as a residual that captures all values that are not reducible to a benefit stream. Thus, if there is an existence value attributed to biodiversity, then this cannot be analysed as a benefit stream, and hence cannot be the subject of a property right as it has been defined here. Thus, if one of the reasons for humans to value biodiversity is existence value, then biodiversity cannot be owned.

Property, agency, time, culture, spirit (John Wiener)

Despite the challenges and profound resistance to traditional values and ways of life, the North remains inhabited by thousands devoted to keeping their cultures alive (Bosworth 1995; Caulfield 1992; Huntington 1992; Langdon 1986; Feit 1991). The ethnographies consistently report such values, and it is useful to compare those to places with more diversity of species, although not necessarily more complex challenges to be met (Cooch *et al.* 1987). Consideration of ideas about property, agency, and time and substitution show there are genuine differences between the West and non-Western socio-cultural situations. The inseparability of the social, the cultural and the political is shown in current trends in scholarship. Then, the cultural and spiritual claims can be more effectively understood. The point is this: working from our own microeconomics, economic anthropology, and cultural/political ecology, as well as the available behavioural and physical evidence, we have no theoretical or generic reason for wholesale rejection of important claims about very different relationships experienced between persons in other cultures, and other persons and their environments.

We must not overlook the stakes involved: if we believe that our own preferences are 'human nature', for all humans and all times, then our questions (and therefore answers) about how to value biodiversity, the future, and human others, are directly affected. If we accept real differences, our questions (and answers) will change. The host of gender, class, non-Western and other 'post-modern' voices and issues entering current discourse achieve their importance and conviction from this perspective. The conventional use of 'we' and 'our' often implies that 'society' is somehow singular, and that there is either some generic or some 'scientific' consensus. However, my use of 'our' and 'we' is strictly a short-hand reference to the current political consensus, and the mainstream positions in academic disciplines and economics, which are very well presented in the *Global Biodiversity Assessment* (Heywood 1995), to which this essay replies. By contrasting my position with a 'stylized' version of mainstream economics, I present this argument rather briefly, but the goal is to provide the overview rather than the details. I also want to note that local ecological knowledge is enormously valuable for our purposes (DeWalt 1994; Booth and Kessler 1996; Field 1996; Heywood 1995 [Section 12]), but those understandings have value beyond their potential as yet another resource for the world economy (Hornborg 1996).

There are two critical limits on our own understanding. First, our own disciplines and traditions are dynamic and incomplete; we tend to lose the long perspective. Because of faith in our methods, we often presume that our current best word is the last word. Second, we are limited in our ability to think abstractly outside of our own native language (Whorf 1956), and must therefore be careful about translation problems. The fundamental insight is that one can only understand in one's own terms, literally and metaphorically. Language is the perceptual and conceptual mediator between reality and thought; put differently, conscious thought is carried on both in and with language. Our linguistic blind-spots do not completely overlap with those of others, fortunately, and this alone is reason enough for conservation of cultural diversity (Cohen 1995). And it means that we must be especially careful of the values, cultural and spiritual, that are rendered obscure or invisible to us. It is easy to reject what we do not understand, but in the light of our own theories of culture and human relationships with the world, we cannot accept glib naiveté about false objectivity.

The purpose of this essay is to support the more explicit claims made in the rest of this volume (and elsewhere) about who people can be, and what they can value in the most meaningful ways. The magnificent heritage of biodiversity that existed worldwide until very recently, and the behaviour of remaining traditional peoples, are powerful evidence of the values humans put on their environments (Kellert and Wilson 1993), just as the mistakes and catastrophes are evidence of fallibility or worse.

The social nature of property is discussed elsewhere in this chapter (Vira). But it is interesting to note in this context that North American societies expressed concepts of themselves and nature as within a mutually supportive and constitutive relationship, including humans and the biosphere (e.g. Nelson 1986, 1983, 1969; Brody 1987, 1982; and Feit 1994, 1991, 1982). Perhaps because the premises are genuinely different from our own, other views were generally and conveniently (for Europeans) denied, declaring that resources were 'unoccupied' or 'unclaimed' and free for the taking (Bodley 1982, 1988; Burger 1987). The allocations of property are profoundly political and strategic, which is to say, cultural. The full set of social conventions that are relevant in any given case must be empirically determined, and those that evolved historically and politically in other societies would almost certainly have little similarity to our own (Leacock and Lee 1982; Ingold *et al.* 1988; Johnson and Earle 1987; Pryor 1977). In fact, a wide range of thinking in development studies (Arndt 1987), economic anthropology, and other disciplines concerns the problems of confrontation and mixing of genuinely different social orders (Champagne 1989; White 1983) including property systems in the North (Usher 1995; Usher and Bankes 1986; Chance 1990; Brody 1987).

The point that rights are accompanied by corollary obligations is particularly important (Schmid 1995, 1987; Bromley 1989, 1991). Euro-Americans generally deny rights in each other, except in limited senses involving children, safety and contracts. This was a huge change 'from status to contract' (Maine 1861). In contrast, the essential feature of the non-market economies, and the non-market parts of mixed economies such as those in the North, is the claims, obligations and situations created by status (Callaway 1995; Wenzel 1995; Langdon 1995; Burch and Ellanna 1994; Lee and DeVore 1968). In an Inupiat, Kutchin or Koyukon village, a person is a harvester or processor, a sister or brother, a child or parent, a neighbour or partner, in good health or bad, old or young; these are the basics of identity, not as limiters but as beginnings (e.g. Nelson 1969, 1983, 1986; Chance 1990; Damas 1973; Langdon 1986). Security and risk management were basic to these arrangements, and the well-being of the one was intimately related to the well-being of the group.

Individual and collective agency, and the village as firm

Individualized rights to resources were not the norm in most societies in most times (Johnson and Earle 1987; Pryor 1977; Lee and Devore 1968; Leacock and Lee 1982). The individual was not the only economic actor, and not a resource-managing agent; the agent, like the current business firm, was the group. Groups held territories, or resources, perhaps moving around with the resources (Albers and Kay 1987; Albers 1993; Shinkwin 1984; Burch and Ellanna 1994). The benefits of localized experience and ecological knowledge from territorial or resource specialization are fairly obvious. It is a small step to appreciate that the people involved understood this, and also understood that where there were limits from the environment or the neighbours, management had better be conservative (in the true sense of conservative).

A great deal of effort in economics has been devoted to explaining the firm (Coase 1937; McManus 1975; Demsetz 1997). Why would individuals submit to exploitation in the economic sense of producing value greater than they receive in exchange for their labour, and why not cheat as much as possible? However, the real puzzle is explaining individuals; for millions of years we lived in groups. The benefits of familiarity and stability in extended interaction turn out to be substantial, and the costs of institutions needed to support frequent one-time market exchanges are far greater (Polanyi 1944; Dalton 1969, 1968, 1961; Platteau 1994a, 1994b; Ortiz 1983; Bennett and Bowen 1988). Groups work. Note also that complete co-operation is not required to sustain the benefits of effective group operation (Runge 1986, 1985; Schelling 1978); working groups provide a set of incentives from which there is no internal incentive to defect.

The theory of common property regimes is now firmly established (McCay and Acheson 1987; Berkes 1989; Berkes *et al.* 1989; Bromley 1992). Private property (*res propria*), state property (*res publica*), and 'open access' (*res nullius*) or 'non-property' conditions had long been recognized. The empirical demonstration of the fourth category, however, has been overwhelming; it is 'common property' in the current sense, not the 'commons' referred to in 'the tragedy of the commons', which was open access; see Monbiot (above) and Aguilera-Klink (1994). Common property (*res communes*) is held by a defined and limited user group, neither individual nor governmental, of interest beyond economics to political science, development and resource management (Ostrom 1990; Ostrom, Feeny and Picht 1988).

The local group, perhaps a village in the North, acts as a firm and achieves the benefits of sole ownership. These benefits include substantially increased resilience in both human and harvested populations, through harvesting at the level of maximum economic yield (Wiener 1995; Townsend and Wilson 1987). By resilience we refer to the capacity of the system to preserve all of its elements despite perturbations, as opposed to stability which is the ability to return to a prior state despite perturbations (Holling 1986). Systems with high stability may be fragile, in having low resilience, and vice versa. Maximum economic yield is the highest possible return on effort; in harvesting wildlife, this will almost always be significantly (in terms of resilience and economics) lower than the maximum sustainable yield, which is the biological estimate of the maximum harvest removable on a steady basis. Harvest at the lower maximum economic yield means that restoration of the prey species is relatively easier than harvest at the higher level, providing bigger margins for error, and it also means that if additional harvest is needed it is easier for the humans than additional harvest starting from the higher level; and finally, additional harvest above the economic level may still be within the easy replacement capacity of the prey, but by definition will not be if it is added to the maximum sustainable yield.

The village acting as a firm can thus achieve excellent results, in economic terms, without ideological or economic individualism, and without individualized and expensive establishment or renegotiation of property rights (that is, at the lowest transactions costs). These arrangements are a fair characterization of the operating rules and practices reported in the ethnographies, and the moral values that are apparently operating. Notice also that there are no incentives in such a regime to harvest in excess, and no benefits from accumulation in the present, compared to the future, where individuals have only partial claim to their own harvests, and storage 'on the hoof' is far more productive than in other forms. In fact, one of the most bitterly misunderstood points of difference between Northern traditional groups such as the Dene and Euro-Americans has been clearly explained by Asch (1989, 1982), who showed that local animals in Dene practice and understanding are far more like domesticated stock than 'wildlife' in the European sense (see also Feit 1973; Brody 1982). The Dene own their land and its resources, in their terms, and they reject the Canadian idea that wildlife is subject to no private claims until legally 'taken'. The intrusion of cash (an exchange medium with a very long 'shelf-life') and individualized trading opportunities have powerful effects on these situations, but agreeing that they can be (and often have been) altered (in whole or in part) does not require agreeing that they were maladaptive or undesirable.

As the *Global Biodiversity Assessment* illustrates, the divergence between social versus individual costs and benefits is a fundamental problem. But where the agent is the group, this divergence may be far less troublesome. Individual preferences and interests are far more overlapping in such a situation. The group agent is socially produced and reproduced, like language and other aspects of culture. Long-term interests are held by the firm-like group, and personal and shorter-term interests are held by individuals (often as delegates of the group; Bromley 1992). The firm-like group agent is not a super-organic entity, but simply a social structure with supporting ideas and traditions that provide operating as well as constitutional rules. It is the central institution in a group's cultural adaptation to the environment.

Humans in the environment

The security (and confinement) of the group, and the group as the basic risk management institution, must surely provide a profoundly different sense of identification and psychological background of solidarity than we experience. Some of the very first inquiries of post-Enlightenment social sciences included alienation, isolation and 'anomie', and considered the relations of individual and social setting (Bottomore and Nisbet 1978; Giddens and Turner 1987; Heilbroner 1992; Roll 1992; Harris 1968; Lasch 1991; Nelson 1991; Postman 1985).

Secondly, there may be an identification of the self as a part of nature, a part belonging in nature, and a participant in something, rather than as parasite or predator, or a supra-natural being with preordained dominion and no moral relations with nature. The ethical implications are significant, because they involve claims about human nature, the choice or inevitability of environmental destruction or transformation, and the qualities of institutional arrangements needed. Classic works in ethnography describe cultural institutions that are enormously difficult to translate, but which provide glimpses into different perceptions of human nature and the human place in the world, such as Briggs' work on Inuit socialization of young children (1982), and the varying explanations of band territoriality and relationships to land from Mbuti (Lee 1982), Aborigines (Hamilton 1982), and so on (e.g. Descola and Palsson 1996). In the Arctic and Sub-Arctic, a wealth of study related to the territorial aspects of the fur trade compared to pre-Columbian times has shown the level of interest in whether 'they' were really different (e.g. Leacock and Lee 1982; Feit 1991).

In distorted images of inhuman angels miraculously stabilized in their presumably static environments, or incompetent devils struggling to over-harvest and heedlessly destroy, limited only by their inability to use more destructive technologies, 'others' are portrayed in extremes for strategic reasons. Wenzel (1995) discussed the use of images of savage fur slaughters. Brosius (1997) considered abuse of images of idyllic sacred life among the unspoiled rainforests. These issues are emotionally loaded and manipulated because of their implications for our own choices.

Time, substitution and capital mobility

The village as firm and economic agent may have, in relative isolation, a neutral or zero time preference, meaning that a moose in the future (or a good run of salmon) does not have lower value than a moose or a good run in the present. This is very important. A neutral or negative time preference expresses long-term values, and makes decisions different from the short-term preference for the present.

Conventional economic thought played two serious tricks. First, we started from methodological individualism and stopped there, ending up in effect with ontological, epistemological and ideological individualism; the tool became the end rather than the means. The realities of common property user groups, 'discovered' all over the world in the last decade, did not suddenly appear – they were just very hard to see when thinking as an economist. Second, the positive time preference of the individual was accepted as an inevitable and 'natural' fact of economics. There is substantial debate in the literature about the appropriate rate of discount, but effectively none about a zero or negative rate of discount.

To understand the other human cultures we must empathize with systems in which the economic agent, the group itself, functions in relatively static conditions, in two important ways. First, there is no evidence suggesting a universal value that more humans are always better than fewer in a given place. Second, there is no evidence that every group had a positive time preference; on the contrary, as this volume demonstrates, we are finally becoming aware of the sophistication of local ecological knowledges and experience, and there is no showing of universal resource destruction. There is substantive meaning in ideas such as 'management for the seventh generation ahead' (see Lyons in Little Bear *et al.* 1984).

Another fundamental idea is the presumption of 'mobility of capital': that not only can we measure all values against each other, or commensurate them, in order to make effective choices by accurate comparisons, we can also presume that those exchanges can be realized! This is not true, and there is no evidence to show that other peoples believed this (see also Vatn and Bromley 1995). We have testimony (e.g. in this volume) from all over the world to the contrary, about long-term values, and values in and for parts of the environment as they are. Only after the penetration of the market economy is there serious commitment to universal currencies, although the cases are various as one would expect (Plattner 1989). Exchange and specialization do not require mobility of capital. Perhaps refusing to accept mobility helps to avoid irreversibility problems (see Pearce and Turner 1990; Bromley 1995), but this is not yet formally demonstrated.

Whatever the motives, the very long histories of many indigenous peoples world-wide show both extinctions now and then, and also very effective management for conservation as well as human use. As Denevan (1992) concluded, the Americas were anything but 'virginal' in 1492, but nevertheless there was enormous richness of wildlife that had not been destroyed, despite millennia of use of technologies such as fire and fish-traps (Gray 1993; Berkes 1989; Driver 1969; Thomas 1956; Meyer and Turner 1994). When we hear claims such as 'the land is our bank', as the North American Natives said over and over in the Alaska Native Review Commission (1984; Berger 1985, 1988) and the Berger Inquiry in Canada (Berger 1977), for example, we might assume strategic behaviour (lying) and dismiss the claims, but this is not science; it is what psychologists call 'projection'. For those unable to accept testimony, the empirical evidence is also substantial, from the diversity and abundance of wildlife that greeted newcomers to the Americas (Gray 1993), to evidence of continued occupation of places throughout the North for literally millennia (Schaaf 1995; Langdon 1995; Burch and Ellanna 1994, for example).

Agreeing that there were genuinely other values in operation does not require accepting a romantic fantasy (see Shinkwin 1984 and Fienup-Riordan 1994, for example, on Inupiat and Yupik territories and warfare); in the end it simply requires accepting human capacity to live by values different from those the worst examples display.

Culture and spirituality - cultural, social, political, analytical?

Wallerstein (1990) has urged 'unthinking social sciences', beginning with dissolving distinctions from the nineteenth century between the social (including 'civil society'), the economic, and the political. Many of the conceptual difficulties facing human scientists stem from the analytical categories employed. Originally strategic, as with the microeconomic model, the conceptual tools have taken on accidental ontological status. One example, which is useful for present purposes, comes from the '30 years' war' in economic anthropology, recently summarized by Plattner (1989). It matters because the real point was whether everyone is rational or not.

The argument was between the 'substantivists', following Karl Polanyi and his successor George Dalton, whose claims, in short, were that only in the uniquely distinguished market economies could one reasonably separate and treat the market in formal economic terms; but in all other societies, including those using (generalized) reciprocity, such as the traditional societies described above, and those using redistribution, there was no sensible separation between the social and economic. The 'substantive' workings of society included provision of material needs as part of other relationships. Only in the market economies is the 'material provisioning of society' a set of activities with rules supposedly uniquely economic. In reciprocity systems, the key social structure is kinship, and in redistributive systems the key social structures are political (Johnson and Earle 1987; Pryor 1977). Stridently opposed to this were the 'formalists' who insisted that all production and distribution must be understood in terms of current formal economic theory, as the universal logic of choice.

After acrimonious debate, attention finally dwindled when it was sufficiently clear that formal models of choice based on rationality were indeed applicable to a vast array of behaviours when those were sufficiently well examined in their context, and that indeed the social structure and context establish the range of acceptable choices. The correct answer was, 'yes, both', but it was harder to see before the economists had got to their current understandings of property as construct, and the anthropologists had to escape their own heritage of understanding by analogy to their own social structures. And the underlying issue resolved was that rational explanations could be found with adequate understanding of the human context; this was essentially a translation problem.

Parallel to this rejoining, two other major trends should be appreciated as evidence of the inseparability of economy and culture. First, human and cultural ecology, as formerly separate traditions, are converging very quickly. Human ecology (Young 1983) began with biologically-based questions and explorations and eventually included other humans and the social structure as powerful determinants in the effective environment (Moran 1984, 1990). Cultural ecology, in contrast, was concerned less with populations, calories and the biological model, and more with relations between the environment and the cultural adaptation to it (Steward 1955; Netting 1986); however, the method itself led to increasing convergence and inclusiveness. Effective explanation or at least appreciation of a cultural outcome and its physical manifestations cannot omit parts of the determinant forces, whether human or not. The resulting empirically-oriented fusion, with roots in development (e.g. Blaikie and Brookfield 1987) as well as anthropology (Moran 1990) and geography (Nietschmann 1979; Grossman 1993, 1977), is often called political ecology, in recognition of the powerful external human influences acting on a given local group or place. Whether starting with biology or with culture, almost no human outcome is explicable without the complex suite of contextual factors; prematurely dividing these factors into sets with presumed lack of connections is unhelpful.

Second, there are currently two profound transitions in social theory itself, including on the one hand 'the return of grand theory' (Skinner 1985) as willingness to work in terms of social structure *per se*, and to bring such large visions to bear on empirical cases, in increasingly rich interdisciplinary efforts (e.g. Castells 1983, and the florescence of sociologically and politically informed urban studies, for example; see Giddens and Turner 1987). And on the other hand, there has been widely-recognized progress in the development of theory that reconciles the dual nature of social structure as both medium (constraints and opportunities) in which lives take place, and as outcome (produced and reproduced by people living in compliance and deviation, material and intellectual change), following Giddens (1984) and others (e.g. Bourdieu 1977a).

Finally, in terms of our current scholarship there is a groundswell of recognition for what Norgaard (best developed in 1994) and others have called the 'co-evolution' of humans and their environments. Similar ideas have included 'the ecological transition' (Bennett 1976, 1993), or 'the appropriation of nature' (Ingold *et al.*. 1988). The common conviction is that humans have modified and influenced their environments for millennia, as well as themselves (Marsh 1864; Thomas 1956; Turner *et al.*. 1990; Meyer and Turner 1994).

Given the inseparability of culture, economy and policy, what are the policy implications? The substantial increase in interest in civil society and social capital in development respond to questions of how people organize themselves for group action (Putnam 1993; Platteau 1994a; Balland and Platteau 1996). In terms of biodiversity and its cultural and spiritual values, the question is how people's lives are organized and affected by those values. The cultures of the Northerners, according to them, include their activity as well as their art and music, and they say it is their lands and knowledge. They say their bank is the land. They say their footprints are everywhere, and we begin to see that they are. And they say their identities are rooted in the land and life; the ethnographies are loaded with testimony of a different relation to the world (see the works of Nelson 1969, 1983, 1986, 1991; Brody 1982, 1987; Chance 1990; Peterson and Johnson 1995; Langdon 1986; The Transcripts of the Alaska Native Review Commission; and Berger 1977, 1985, 1988 for countless examples). We cannot deny this, on theoretical grounds.

Biodiversity - cornucopia of knowledge (Astrid Scholz and Ignacio Chapela)

The current debate about the need for biodiversity preservation, as laid out in the *Convention on Biological Diversity* (CBD) and UNEP's *Global Biodiversity Assessment*, is fraught with conceptual difficulties surrounding the notion of biodiversity. The term itself was coined in the 1980s by a group of ecologists and biologists with the intention of getting it onto the global conservation agenda (Takacs 1996). Furthermore, the subsequent division of this vast and all-encompassing term into the three realms of genetic, species and ecological diversity (Heywood 1995) is as artificial as it is convenient. It underlies the dichotomization of biodiversity into economic and non-economic uses, into exchange and 'other' values which have shaped the conservation debate.

'Biodiversity' as a natural resource – infinite if used sustainably – allows for the apparently unequivocal and easily operationalizable application of existing economic valuation methods. These have concentrated mostly on species as the valuation unit of choice (Chichilnisky 1996; Simpson *et al.* 1996) whose value is a function of useful genetic information contained therein. However, biodiversity is different from other natural resources in a number of important ways. First and foremost, biodiversity involves uncertainty at an unprecedented scale, making standard decision-making algorithms obsolete. Not only do we not know the absolute number of what is out there, but we also lack knowledge about the distribution of biodiversity and how it is dynamically connected at its various stages of organization. Furthermore, at the largest scale of diversity recognized in the *Global Biodiversity Assessment* (Heywood 1995), that of the biome, organisms interact in complex, spatially and temporally heterogeneous ways that defy economic models of natural resources. A classic example of the intractable problems faced at the interface between ecology and economics are low biological growth rates. Purely due to their slow growth rates do some species such as whales become 'unfeasible' if compared to the discount rates on other forms of capital, as the existence of pure time preferences prescribes.

Furthermore, species cannot be assumed to serve as substitutes for one another with regard to their useful chemical 'functions'. Trade-offs can only be established *a posteriori*, based on a successful 'hit' and the analytical tools this enables (Bergstrom *et al.*. 1995). Similarly, species cannot be assumed to be independent with regard to their medicinal value. Polasky and Solow (1995) illustrate how the discovery of taxol, a powerful anti-cancer agent in the bark of *Taxus brevifolia* led to a rush to screen other species in the genus *Taxus* and subsequent discoveries. Finally, the difficulties with fitting biodiversity into the standard economic valuation models exemplify that economic value only constitutes a small part of its total value. Gowdy's (1997) survey of economic and other values of biodiversity to humans and ecosystems emphasizes that we cannot hope to find a single measure of the value of biodiversity. Hence the singling-out of economic values as somehow most important and counterbalancing all 'other' values is in itself a value judgement. Once this is made explicit, and ethical judgements underlying the valuation methodology used are recognized (Funtowicz and Ravetz 1994), the dominance of the efficiency principle in environmental policy decisions can be put into context and other values made come to bear.

The kind of information relevant for pharmaceutical research goes beyond the incremental gain of quantitative knowledge about a lead. Rather, the various principles used in the bioprospecting and discovery stages of drug design involve qualitative 'meanings' of biodiversity, progressively coming to bear on the different stages of drug design and referring to varying levels of biological and other information about the compound. The many small decisions characterizing the design process over its lengthy period are generally structured by the ordering principles of systematics. Through making sense of complex information at a number of biologically relevant levels, systematics orders the drug design process. This information increasingly includes geographic, ecological, cultural and other information not commonly associated with the 'use' value of biodiversity. However, not only is there an explicit role for these realms of knowledge, but they also provide the avenue for diverse values coming to bear. It is entirely conceivable that aesthetic preferences play a role in deciding where to send a bioprospecting expedition. And even in the lab, various values come to bear at the increasingly more complex steps of the drug design process. In this sense, information is more than the byte-sized information contained in genes. It gives meaning, in the original sense of the word *informare* (Chapela 1996), to the medicinally interesting compound. Yet meanings are embedded in many personally and culturally idiosyncratic values. Consider, for example, the Western work ethic in making Fleming go to work despite a flu, which ultimately led to the discovery of *Penicillin*.

To understand the pervasive role of 'other' values in the commercial exploitation of biodiversity, it helps to reconsider the question of what is the relevant 'piece' of biodiversity. As the practical implications of drug design reveal, a mixture of chemical features, value-adding gene, interspecific interaction, geographic details and cultural information defines a composite object that is potentially economically interesting. The principles at work in the pharmaceutical industry illustrate that even the apparently unequivocally commercial 'mining' of biodiversity involves values other than purely economic ones. In fact, under the uncertainty inherent in biodiversity, decision-making at a number of drug design steps needs to rely more or less heavily on values other than economic cost-benefit considerations. Beginning with the decision about what and where to prospect, values usually alien to economics such as cultural and aesthetic ones enter into the drug design process. Furthermore, considering the principles of systematics employed to put relative weights on lead compounds, these can be extended to encompass relevant information, which in turn defines the pharmacophore. Extrapolating from the pharmaceutical industry, since the uncertainty documented above pervades other and all areas of biodiversity protection, it is high time to recognize the important role of 'other' values.

Selling nature to save it: biodiversity and the global economic paradigm (Kathy McAfee)

The global environmental–economic paradigm today offers itself as a powerful, pan-planetary metric for valuing and prioritizing natural resources and services and for managing their international exchange. It tries to encompass environmental issues in a neoclassical economic framework, imputing commodity-like characteristics to all elements of nature. According to the paradigm, it is only when local natural resources are brought into the world-wide circuit of commodity relations that their values can be realized, and the 'benefits of biodiversity' can be calculated and shared. But by valuing nature in relation to international markets – denominating diversity in dollars, euros or yen – it reinforces the claims of global elites, those with the greatest purchasing power, to the greatest share of the earth's biomass and all it contains. The global economic paradigm thus justifies environmental injustice on a world scale.

Other contributors to this volume eloquently attest to the extent of what the world will lose if 'environmental investments' are based solely on the dominant scientific paradigm – or any single knowledge system – elevated to the status of universal supremacy. Yet, at the very moment when many modern scientists are noticing the contributions of alternative 'local' or 'pre-industrial' systems of knowledge about nature, the global economic paradigm appears poised to subsume them all, 'Western' and 'non-Western' alike.

The rise to prominence of the economic paradigm reflects the environmental aspects of 'globalization', the increasingly transnational interconnectedness of communications, trade and other economic activities that affect or depend upon the natural world, and the consequent need for 'global' environmental policies and institutions. Its bold ambition also reflects the triumph of a particular economic system, capitalism, with its astonishing ability to translate all things – produced goods and the hours it takes to produce them, food and the land it grows on, etc. – into a single, universal, common denominator: money.

At the heart of the global economic paradigm is the ideal of the world as a vast marketplace, in which all human–nature interactions, as well as all social interactions, can best be understood as market-type exchanges, the cumulative effect of which is the most efficient possible distribution and use of all goods, services, information and natural resources. The paradigm puts forward a conceptualization of environmental problems as amenable to market logic and to management by market-oriented international institutions and principles, especially free trade and internationally tradable property rights, including intellectual property rights (IPRs) to components of nature and knowledge about nature (Vogel 1994; cf. Swanson 1995).

Diversity as a commodity

On this basis, the global economic paradigm offers the promise of ways to enable nature to pay its own way in an increasingly marketized and integrated world economy. It encourages schemes by which biodiversity would earn its own right to survival, such as conservation projects financed by the sale of ecological services and the rights to use them (including permits to pollute), the marketing of access to sites for ecotourism and research, and the export of intellectual property rights to genetic information and cultural knowledge about nature. This approach has obvious appeal to policy-makers charged with devising ways to combine fiscal austerity with environmental gains. Many ecologists and biologists, however reluctantly, are coming to believe that the attaching of monetary prices to everything from molecules to mountainscapes may help achieve conservation goals or garner research funds.

A central tenet of the global economic paradigm is that ecological problems can be conceptualized separately from social, political and cultural conflicts, and managed without addressing international or local inequalities. Thus, the paradigm provides a rationalization for the continued pursuit of environmental goals without changes in existing political structures, and without reversal of the long-term continuing net transfer of financial and material resources from the global 'South' to the 'North', and from rural to urban areas everywhere. In this way, the global economic paradigm helps to make environmentalism safe for the world-wide expansion of capitalism, and for the interests of the relatively small number of multinational firms that dominate the international economy.

In this context, many NGOs and scholars are convinced that it is in the best interests of beleaguered indigenous communities to act quickly to claim individual or collective property rights to their forests, fields and traditional knowledge, and then bargain for the best deals they can make, either directly with commercial bioprospectors seeking genetic information on behalf of transnational pharmaceutical and agrochemical firms, or through NGOs or parastatal intermediaries such as Costa Rica's INBIO. Some local communities have entered such arrangements. Others have opposed them, in some cases on the grounds that privatization and sale of cultural knowledge and living nature would cause unacceptable damage to their communities' social cohesion or spiritual well-being.

The issue is not whether 'traditional people' should or should not enter the market. There are few such communities that are not already involved in some sort of market production and exchange. That they ought to receive the highest negotiable return for what they sell, be it maize or bush medicines – should they choose to sell – is the only morally defensible position. It is up to local communities to make those choices, and it is the obligation of others to help ensure that they can make those choices with full information and without political and economic coercion.

Rather, the issues posed by the intellectual property approach to saving diversity are: (i) can the tactic of 'selling nature to save it', i.e. using the proceeds of bioprospecting deals or other natural resource and service exports to finance ecological and cultural conservation, be the basis of a strategy for saving biological diversity? And (ii) can the commercialization of genetic resources provide compensation that can, in any sense, be construed as 'fair and equitable'? The answer to both questions is 'no', for reasons rooted in the structural inequalities stressed at the beginning of this article.

First, there can be few, if any, winners in an international competition among would-be exporters of biological diversity. Those who would sell their biodiversity in the global marketplace, whether would-be 'developing' countries or local communities struggling for survival, are almost certain to emerge even poorer, both in terms of biodiversity and in terms of relative economic strength, for much the same reason that no country in history has ever advanced up the international economic ladder by exporting primary commodities on 'free market' terms.

In the emerging international 'genetic resources' market, prices are primarily determined by the short-term interests of pharmaceutical and agrochemical corporations, based on those companies' estimates of the potential profits to be made from them. In addition, the same or similar biological source materials are often found in more than one place, making it easier for bioprospectors to play off one source country against another. Even if some diversity-rich countries and communities gain the capacity to develop more specialized (screened and/or refined) natural product samples and extracts, and in this way add some market value to their living heritage, they will remain at a great competitive disadvantage *vis-à-vis* foreign bio-buyers.

Further, biotechnology advances enable corporations to alter the genetic make-up of natural samples, creating products that are – at least technically – 'innovations' and therefore immune to legal patent challenges or demands by the original suppliers of genetic information for a share of the profits from those products. New techniques for screening and then imitating, synthesizing or growing natural pharmaceuticals enable companies to obtain the materials they want more quickly and cheaply. As this potential develops, it will become even easier for bioprospectors to mine ecosystems for items that interest them and then move on without making large investments or long-term commitments to source countries or communities.

There are, therefore, good reasons to expect that bioprospecting and the market for genetic 'green gold' will go through a cycle of commodity boom, market saturation, and then bust – just as happened in the cases of indigo, rubber, sugar, and so many other once-touted tropical miracle crops – leaving the exporting countries poorer and their ecosystems degraded. Genetic resources would thus become another instance of the oft-tried – and always failed – strategy of export-dependent development, in which priorities are determined by outsiders rather than by the needs of local people.

Secondly, contrary to the premise of the global economic paradigm, there is no way to set up a universal metric for comparing and exchanging the 'real values' of nature among different groups of people from different cultures, and with vastly different degrees of political and economic power. Nor is there any way to place a price – even one that includes aesthetic, option, existence and even spiritual values – on any element of biological diversity torn out of its social and ecological context. The trait shared by 'local and indigenous communities' is that their economically productive activities take place at least partially outside of the international market economy, and, at least to some degree, their world-views and cultural practices – including the ways they interact with their natural environments – reflect this fact. This, after all, is why their 'traditional life-styles' are of such great interest to defenders of biological diversity more broadly.

'Globalized' prices cannot adequately reflect the values of biological diversity to people who live in direct interdependence with that diversity. The various categories of meanings and 'values' of nature – subsistence values, exchange values on local and regional markets, and symbolic values – are specific to each eco-social system, at least until they are swallowed by transnational capitalism and supplanted by commodity relations. This is particularly true with regard to the non-use values of ecosystems and resources. Citizens of the global 'North', as well as many 'traditional' peoples, place high values on the aesthetic or spiritual aspects of nature, but those values cannot be captured, much less compared, in monetary terms. It is simply not meaningful to weigh the amount that a professional earning US$50,000 a year is 'willing to pay' for the continued existence of a tropical ecosystem against the 'willingness to accept compensation' for the loss of her ancestral homeland of a resident of that same ecosystem who has little or no cash income, and a vastly different world-view.

The economic paradigm and the Convention on Biological Diversity

The influence of the global economic paradigm can be seen in the text and subsequent decisions of the Convention on Biological Diversity (CBD), the world's most ambitious effort to stem the destruction of biological and related cultural diversity. By conceptualizing nature as an internationally tradable commodity (along with goods, services, labour and information) the global economic paradigm suggests methods for putting price tags on the vast and various 'benefits of biodiversity'. This appears to some eco-diplomats to suggest a basis for addressing the diplomatically thorny issue of how the 'benefits' of the 'utilization of genetic resources' are to be 'equitably' shared, as called for in the Convention.

Three interrelated notions – so-called genetic resources, biodiversity benefits and intellectual property rights – have a pivotal position in the CBD, and together reflect the influence of the global economic paradigm. The concept of genetic resources in itself represents a discursive conquest by paradigm's short-sighted instrumentalism. It reduces biological diversity to its purported essence as a commodity, presumably separable from its complex relationships with other 'units' of nature, and valuable only to the extent that it is consumed. The notion of biodiversity 'benefit-sharing', explicitly linked in the CBD text to genetic resources and promised to Southern countries and communities as a reward for their willingness to enforce the IPRs of foreign commercial interests, steers emphasis toward those aspects of nature that can be removed from their local context, 'developed' by means of private industrial technology, and sold for a profit on international markets.

If, as the global economic paradigm suggests, such market-based valuation and exchange provides a method for the just and efficient world-wide allocation of 'biodiversity benefits', then cultural differences, economic inequalities – North–South, urban–rural, landed–landless – and disputes about the authority of states over indigenous and local communities, all become irrelevant to the task of international environmental management through the CBD and other multilateral institutions. Indeed, much of the attractiveness of the global economic paradigm to international policy-makers lies in the fact it does not draw attention to the specific agents and beneficiaries of environmentally destructive policies and practices, but instead provides a language and a set of concepts for blaming biodiversity destruction on abstractions such as 'market failures' and 'policy failures'.

However, the CBD's conservation objectives, and especially the goals of Article 8 and other CBD commitments to 'indigenous and traditional' knowledges and practices, and to *in situ* conservation, resist being subsumed within the environmental economic paradigm. The paradigm has been designed to produce 'globally' applicable criteria for mapping the world and ordering and ranking its contents. It is entirely unable to take account of the infinitely variable, site-specific nature of biological diversity or of the human-linked dimensions of that diversity.

Defending diversity: On whose terms?

The global economic paradigm fosters the illusion that we can 'green the planet' while continuing to grow along demonstrably unsustainable economic trajectories. It offers a rationale for the notion that biological diversity can be 'saved' without fundamental changes in present distributions of political power. It purports to provide an objective metric for estimating the values of all components of nature worldwide, but actually offers values determined by the powers and desires of international elites.

The global economic paradigm pins the fate of diversity on the outcome of competition among economically powerful bidders in the global marketplace, who may at best have a temporary interest in the conservation of one or a few elements of diversity excised from their eco-social context. By promoting the commoditization of nature, the global economic paradigm helps to legitimize and speed the extension of homogenizing market relations into diverse and complex eco-social systems, with material and cultural consequences that do more to diminish than to conserve biodiversity.

To the extent that economic and political circumstance permit, many indigenous and local communities will continue to maintain their diversity-based cultural and livelihood practices (cf. Dove 1996). But no culture remains frozen in space and time. As they continue to evolve and interact with the world beyond the 'local', many will seek access to wider markets and to 'Western' knowledge and technology. Whether these processes necessarily lead to major diversity losses depends upon the terms on which they take place: those of local peoples, or those of the 'globalized' market.

The greater the political autonomy, access to information, and economic self-reliance of local and indigenous communities, the more likely it is that they will be able to set their own terms, and on this basis adopt, adapt and invent mixes of old and new cultural forms and technologies that work with nature instead of at nature's expense. A multiplicity of site-specific, information-intensive technologies that rely more on inputs of intelligence than of agrochemicals and large machines can foster the conservation and sustainable use of biological diversity. They can also do more to increase food production than can the one-package-fits-all approach that characterizes green revolution technology, including current versions based on genetically engineered 'super seeds'. The latter, unsustainable, approach will prevail – until it collapses – if the international market determines how resources are used. Only the full involvement of a plurality of perspectives and centres of power – not unified 'global' criteria for valuing and managing biodiversity – can make the more sustainable alternative possible.

If diversity's chance of survival depends on calculations of the prices at which its various component resources might be sold to ecotourists; timber, oil and mining companies; researchers, or pharmaceutical firms, then diversity is doomed. Strategies to save biodiversity must derive from the economic and cultural importance of natural resources to the people who gather, farm, fish, worship, and otherwise live in direct interdependence with those resources. It is they who have the most direct and longest-term stake in their conservation.

The Convention on Biological Diversity and equitable benefit-sharing: an economic analysis (Joseph Henry Vogel)

Almost as soon as 'privatization' entered the lexicon of policy-making in Latin America, it became a dirty word – more associated with the rigged sale of state-owned assets and the enrichment of corrupt officials than with its economic meaning, i.e. the conversion of social costs and social benefits into private costs and private benefits. However, only in economic theory is privatization so symmetrical. When external benefits exist, the powerful will attempt to internalize the externality; when external costs exist, the powerful will resist internalization.

In the case of genetic resources, Northern industry has been able to privatize the benefits of biotechnologies that derive from these resources, while socializing the costs of access. Before the CBD, genetic resources were free *de jure* under the doctrine of 'the common heritage of mankind'. Now, after the CBD, genetic resources are no longer free *de jure*; they are free *de facto* through the elimination of economic rents in competitive Material Transfer Agreements (MTAs). Some of these MTAs have been reported as low as 0.2 percent (one fifth of one percent).

If patents, copyrights and trademarks are accepted as legitimate instruments to enable the emergence of a market for information goods, then Southern countries should exercise oligopoly rights over genetic resources to enable the emergence of a market for habitats. Countries that supply biological resources should set up a cartel to fix a royalty rate and distribute economic rents according to their ability to have provided the patented biochemical. Whereas TRIPs is the legal vehicle to achieve monopolies through patents, copyrights and trademarks, the CBD can become the legal vehicle to achieve an oligopoly over biological diversity through the establishment of a Special Protocol.

To accomplish cartelization, the CBD could endorse a two-tier reward structure for bioprospecting. At the higher level is the value-added to genetic information through taxonomy and quality control; at the lower level is the economic rent for the ability to supply the raw material. Vogel (1994, 1995, 1997) has suggested a royalty of 15 percent on net sales of biotechnologies with 2 percent going to the country of contact for the value added to the genetic information in taxonomy and preparation of extracts and the other 13 percent to be divided among all countries that could have supplied the same piece of biological diversity in proportion to the existence of that genetic information in the country. Besides providing real incentives to conserve biological diversity, this seemingly high royalty rate may also succeed in persuading the CBD parties that they must give up some of their hard-won, albeit illusory, sovereignty over biological diversity and subject themselves to the rigour of a cartel.

Just as a cartel is needed to prevent countries from competing in a price war for the provision of biological diversity, so too is a cartel needed to protect the knowledge of traditional communities. Such a cartel would need to overcome three basic problems: (i) that much of the traditional knowledge associated with genetic resources is already in the public domain and beyond legal claim; (ii) that much traditional knowledge is diffused among communities and ethnic groups and competition will drive the price of access down to the marginal cost of being interviewed; and (iii) that according to the CBD, the State is sovereign over the genetic resources while the communities can only withhold approval from accessing knowledge associated with those genetic resources. This means that the State can collect randomly without the consent and participation of the communities but the communities cannot perform ethno-bioprospecting without the consent and participation of the State.

Rather than interpreting the right of each community to withhold approval as a right to engage in ethno-bioprospecting, the CBD should specify that it is the right to join in a cartel over ethno-bioprospecting and receive a portion of the economic rents with other communities that also share the same knowledge. The institutional details of achieving such a cartel are highly complex and would include the establishment of regional databases, the filtration of inputted traditional knowledge against the contents of NAPRALERT (the ethnopharmacological database held at the University of Illinois at Chicago, USA) in order to determine what is and is not in the public domain, and the securement of Prior Informed Consent from traditional communities (see Vogel, in press).

How much should the State (or more accurately, the Cartel of States) receive for being sovereign over the genetic resource? And how much should the communities (or more accurately, the Cartel of Communities) share for their associated knowledge? A 50–50 split between the State and the Communities will probably be acceptable to both parties. A similar problem to that of sharing royalties between the State and the communities is the sharing of benefits within any given community. The easiest solution for sharing benefits within any given community would be a disbursement of money among all the families of that community. But traditional knowledge is seldom evenly distributed within a community; usually it is concentrated in the shaman. Although a *pro rata* division of benefits would not contradict the CBD, such a division would not leave the shaman with very much incentive to participate in ethno-bioprospecting; indeed, he or she may even become resentful that others within the same community are benefiting equally. A solution to the problem is to give the shaman a large voice in the ranking of community projects to be funded through the community share of bioprospecting economic rents.

In sum, the potential exists within the CBD to achieve an equitable and efficient framework for bioprospecting. No country or community can do it alone as competition will assure that MTAs do not reflect the opportunity costs of the conservation of biodiversity and associated knowledge. A cartel is needed and is justified in the same economic reasoning embodied in monopoly intellectual property rights.

The importance of the Convention on Biological Diversity to the Loita Maasai of Kenya (David J. Stephenson, Jr.)

Since the Convention on Biological Diversity was adopted at the 1992 Earth Summit, the ripples from this monumental event have spread to every corner of the world, including a pristine forest in south-eastern Kenya inhabited by the Loita Maasai. These people are promoting this legal instrument in a desperate plea to save their society and their indigenous ecosystem from extinction (Loita Naimina Enkiyio Conservation Trust Company 1994: 1,7 [hereinafter Loita 1994]).

The Loita Maasai comprise approximately 22,000 people who maintain a semi-nomadic life, steeped in Maasai age group, clan and religious traditions, and rooted in a pastoral economy. They live in scattered settlements, or *bomas*, on hillsides that surround the Loita Forest, located approximately 320 kilometres south-west of Nairobi, Kenya (Ololtisatti n.d.: 11). This dense indigenous forest can be approached only on foot, and is considered sacred by the Loita Maasai. Not only is it home for their traditional sacred rituals, but it is also the source of traditional medicinal plants and herbs, and a vital component of the ecosystem upon which their pastoral livelihood depends. It provides essential water, trees, green leaves and grass (Loita 1994: 1). The Loita Maasai imbue this sacred forest with an elaborate mythology. The springs and streams that emerge from the forest symbolize enduring hope, and a Loita Maasai legend tells about a child who became lost forever in the forest while herding her father's cattle (Loita 1994: 1). Hence, it is known as 'The Forest of the Lost Child' (Loita 1994).

This sacred forest, however, is in grave danger of being taken away from the Loita Maasai who worship, preserve and protect it, by the local Narok County Council, which wants to convert the forest into an extension of the largest game preserve in Kenya, the Masai Mara, and thereby develop the forest for mass tourism. The Loita Maasai believe that if this conversion occurs they will be unable to continue to access the forest either for their sacred ceremonies or for critical water resources during the dry season (Loita 1994: 1). They foresee the inevitable consequences of this transformation of the forest as the destruction of their society.

The Loita Maasai have responded in several ways to the concerted efforts of the federal and local governments to expropriate the sacred Loita Forest. Loita Maasai have authored pamphlets and articles documenting their plight (Loita 1994; Ololtisatti n.d.). They have encouraged others to spread word of their plight. They have formed networks with other disenfranchised pastoralists in Kenya, and they have sought the assistance of the international community (Loita 1994). Their most focused and significant effort, however, has been to seek to obtain legal recognition of the Loita Enkiyio Conservation Trust Company (hereinafter 'the Trust') (Loita 1994: 7; Ololtisatti n.d.: 12).

The purpose of this Trust is to preserve the sacred forest for the benefit of its guardian, the Loita Maasai (Loita 1994: 7). The Trust is controlled by ten Loita Maasai elders, and its membership is defined as, 'all the bona fide residents of the administrative Loita location' (Loita 1994: 7). The Trust has filed a lawsuit (*Loita Conservation Trust vs. Narok County Council*, High Court Misc. Civil Application No. 361 (1994)) seeking a declaratory judgement that it, and not the Narok County Council, is the true owner of the sacred Loita Forest and the entity that should safeguard its fate (Loita 1994: 7; Ololtisatti n.d.: 16).

In this lawsuit seeking formal legal entitlement to the Loita Forest, the Trust has invoked the following language from Article 8(j) of the CBD, which requires each signatory, 'subject to its national legislation', to 'respect, preserve and maintain knowledge, innovations and practices of indigenous peoples and local communities embodying traditional life-styles relevant for the conservation and sustainable use of biological diversity and promote their wider application with the approval and involvement of the holders of such knowledge, innovations and practices and encourage the equitable sharing of the benefits arising from the utilization of such knowledge, innovations and practices'.

The Loita Maasai have not accidentally, idly, or casually invoked this language from the CBD. They have done it with a profound, sophisticated awareness of its critical relationship to their desperate struggle for survival. They specifically state, 'In our view, implementing this obligation is a key to the success of the Convention' (Loita 1994: 7). They further state: 'We recognize that we are just one of many indigenous peoples and local communities throughout the world whose traditions and cultures have ensured the survival of their forests and the rich biodiversity within them. Yet so many of these communities are now threatened with extinction along with the forests and lands they have protected for generations. Our own traditions and practices are unique to the Loita and Naimina Enkiyio, but the threat we now face is symptomatic of a global sickness. The Convention on Biological Diversity recognizes this and provides the words – the building blocks – to begin healing the sickness. It represents our hope for the future.'

They also underscore that President Moi of Kenya was one of the first signatories of the CBD (Loita 1994: 7). The success of their lawsuit, however, besides its dependence on the vagaries of Kenyan politics, is also dependent on how the Kenyan courts reconcile CBD Article 8(j) with Kenya's national legislation, because the Article, by its own terms, is subject to such national legislation. A Loita Maasai lawyer, Ole Kamuaro Olotisatti (n.d.) has critically dissected the relevant national legislation of Kenya in this context. Specifically, the critical national legislation is the Kenyan Constitution, which defines Trust Lands and establishes their administrative parameters; and the Trust Lands Act of 1939 (Olotisatti n.d).

The Trust Lands Act is derived from the Crown Lands Ordinance of 1902, which effectively expropriated land from indigenous, native Kenyans in favour of colonialists. This expropriation was accomplished through administrative local county councils, such as the Narok County Council, whose decisions were considered final and not subject to court appeal. The Act expressly contradicts section 115(2) of the Kenyan Constitution, which states: '[e]ach County Council shall hold the Trust Land vested in it for the benefit of the persons ordinarily resident on that land and shall give effect to such rights, interests and benefits in respect to the land as may, under African customary law for the time being in force and applicable thereto, be vested in any tribe, group, family or individual' (Olotisatti n.d.).

The classification and registration of land in Kenya, however, was done without regard to the title to land that was vested to indigenous Kenyans under their customary law (Olotisatti n.d.).

The Trust Lands Act itself also contains internal contradictions, in some places recognizing land rights vested under customary law (Section 69) and elsewhere (Section 59) or even within the same subsection (also in Section 69) 'checking' those rights against 'provisions of any law for the time being in force', thus making traditional land tenure and the rights of the people indigenous to local land subject to the vagaries of national politics (Olotisatti n.d.: 14). Thus, local county councils, such as the Narok County Council, with the support of national government, can simply, and relatively arbitrarily, invoke the Trust Lands Act to justify the usurpation of local lands from the people native to those lands for the expedient ends of the local councils and the federal government – and arguably in violation of the Kenyan Constitution. The net result has been the 'systematic degradation' of finely balanced indigenous ecosystems that have been protected, and even worshipped, by generations of people indigenous to those ecosystems, in favour of the short-term commercial exploitation of the natural resources within those ecosystems by non-natives, and – in the case of many traditional Maasai lands – for the benefit of tourists, who may pass through those ecosystem homelands of indigenous peoples only once in their lives, for only a few days, and whose tourist dollars are mostly grabbed by a few elite, mostly non-indigenous, politicians and power brokers.

The litigation brought by the Loita Maasai is an effort by them to force the Kenyan Government to face up both to the contradictions between its Constitution and its land laws with respect to indigenous peoples' rights and to the systematic degradation of indigenous ecosystems that is fostered by the political economy supporting the actions of local county councils. This latter focus has been considerably enhanced by invoking Article 8(j) of the CBD, which places the sacred Maasai Forest ecosystem in the context of threatened indigenous homelands throughout the world and which places the Loita Maasai struggle for survival in the context of emerging human rights laws that transcend national boundaries and that are truly global and international in scope.

This legal effort is admittedly a challenge (Loita 1994: 7). However, in a recent ruling in this ongoing litigation (*Loita Conservation Trust v. Narok County Council*, High Court Civil Case No. 1679 (1994)), the right of the Loita Maasai to institute their action against the Narok County Council was upheld, with the Honourable Justice G. S. Pall noting, 'The law should be amended to allow for applicants with deserving cases, such as the Loita, to be heard in judicial review'. As the Kenyan courts begin to give a voice to the 'deserving' struggle of the Loita Maasai to protect their sacred forest, the voices of similarly situated indigenous peoples throughout the world are given new strength, and the power of emerging international human rights law, such as the CBD, is confirmed for all of us for all time.

Beyond TRIPs: protecting communities' knowledge (Carlos M. Correa)

The need to develop some form of protection of communities' knowledge has gained growing recognition in the last ten years [1988–1998]. Approaches and proposals range from the creation of new types of intellectual property rights (IPRs) to the simple option of legally excluding all forms of appropriation. However, little has been achieved and only a few countries have begun to address the complex conceptual and operational problems involved in the recognition of communities' rights over their knowledge. The exceptions are a small number of countries that require communities' consent to provide access to genetic resources found in their territories, such as the Philippines and the Andean Pact countries.

The Philippines Executive Order No. 247, and the Andean Pact's Common System on Access to Genetic Resources empower the communities to participate in the process of admission of access requests, but do not create any type of rights in the knowledge or materials under the communities' control. There is no conflict, hence, with existing IPRs, or with the TRIPs Agreement. The Philippines Executive Order provides for the granting of compulsory licences in case the knowledge or materials obtained are subsequently patented. Such licences are explicitly permitted under the TRIPs Agreement, Article 31 (Correa 1994a).

The obligation to disclose, in a patent application, the origin of the germplasm used and improved may be the basis of a right to compensation for the benefit of the concerned communities. This is, in fact, one of the options open to national legislation to deal with this issue (Correa 1994b). But some proposals to deal with communities' rights have gone further than a right to compensation, and imply the creation of a new, *sui generis*, form of protection. Such is the case of a draft Bill by the Government of Thailand, which would recognize rights to traditional healers and medicinal natural resources. The draft is based on the concept of 'collective rights' which is gaining growing support among non-governmental organizations and researchers working on community issues. The Thai proposal includes the registration of traditional medicines and some form of benefit-sharing in cases where medical or scientific researchers make commercial use of the protected knowledge. This proposal encountered an early and unexpected challenge by the United States Government, which in April 1997 officially requested the Thai Government to explain the relationship of the proposed rights to the granting of patent and plant variety protection in Thailand. In the communication to the Thai Government, the US Government suggested that to the extent that only Thai nationals will be able to take advantage of the registration process under the proposal, the latter could violate the WTO's national treatment provisions. The main concern of the US Government relates to whether the proposed legislation would: (i) restrict access to traditional knowledge; (ii) imply costs for US researchers and companies; and (iii) jeopardize existing pharmaceutical patents based on compounds or knowledge derived from traditional medicine. The paradox is that the United States has pioneered the extension of IPRs to a whole range of new areas, paying little attention to the interests of the users of the newly protected matter or to the interests of other parties.

A proposal like the Thai draft legislation, to create a new modality of protection in favour of indigenous/local communities, does not violate any international convention, including the TRIPs Agreement. Several arguments support this assertion. First, as mentioned before, communities' rights have been recognized by the CBD, an internationally binding instrument. National laws can implement them in accordance with the applicable legal systems and practices. Second, at least some aspects of communities' knowledge may be protected as 'works of folklore', in accordance with the 'Model Provisions for National Laws for the Protection of Expressions of Folklore against Illicit Exploitation and other Prejudicial Actions' developed jointly by UNESCO and the World Intellectual Property Organization (WIPO). This type of protection has been implemented in several countries, and so far no questions about its legitimacy have been reported. Third, the TRIPs Agreement sets forth the *minimum* standards of protection in most, but not all, areas today recognized as components of the field of 'intellectual property'. For instance, the TRIPs Agreement does not prevent any WTO Member from protecting *utility models* (petty patents), which are not included in TRIPs. Fourth, TRIPs does not disallow the creation of new types of IPRs or the provision of protection more extensive than existing IPRs (Article 1). Fifth, national treatment only applies to the IPRs explicitly covered by the Agreement, with the exceptions allowed by the applicable international conventions. Rights not covered by the TRIPs Agreement are not subject to the Agreement's principles and rules. In sum, any WTO Member may provide protection beyond the TRIPs Agreements standards, and is fully empowered to create new titles of IPRs, or new forms of protection, to the extent that this does not diminish or neutralize the protection to be granted in the areas covered by the Agreement.

The development of a new regime to protect medicinal plants or other expressions of indigenous/traditional knowledge, faces significant difficulties relating to the nature, scope and effects of the rights to be conferred, the determination of beneficiaries, and the enforcement of rights both domestically and internationally. The design of new forms of protection for communities' knowledge is *not* incompatible with the TRIPs Agreement. Moreover, initiatives to create new rights or expand those granted in existing areas may be seen as an outcome of the paradigm that developed countries wish to universalize through the TRIPs Agreement, i.e. that valuable knowledge should not be misappropriated or used without the consent of those who created it.

There are, then, sufficient grounds and justification to establish a *misappropriation regime* with respect to communities' knowledge, not necessarily based on the concept of 'property' but on a right to condemn illegal modes of acceding and using such knowledge. There is nothing in the international rules in force that would prevent a country from implementing a new legal regime based on that concept.

Cultural and political dimensions of bioprospecting (Darshan Shankar)

Bioprospecting initiatives are controversial because of a general failure to acknowledge the source of materials and information and the frequent absence of equitable benefit-sharing agreements. Practical problems arise when seeking to identify the 'rightful' owners of indigenous knowledge in attempts to reward the holders of indigenous knowledge and owners of the resources. Most often, traditional knowledge of a particular resource is widely distributed beyond a community. This can be seen in the case of the plant *Phyllanthus niruri*, whose use for treatment of infectious hepatitis is known throughout southern India. The Fox Chase Research Center in the USA has filed a patent claim on a hepatitis drug developed from this plant. The practical problem in such cases is how one shares benefits with a widely distributed owner constituency. One possible solution is to place the benefits in a common community biodiversity fund. This proposal raises further issues, such as who would control the biodiversity fund and how it would be applied equitably.

But a central issue missed in these debates is that of serious cultural erosion taking place while one culture is prospecting on the intellectual and biological resources of another. In effect, science and its carriers (a Western knowledge system) are prospecting on a traditional knowledge system of non-Western origin. One culture is considered advanced and the other viewed as essentially backward. One culture is assumed to be the creator of new and superior knowledge and products; the other, the donor of raw material and imperfect, crude and unrefined knowledge which is only good enough to provide 'leads'. Economic and political power lies within the prospecting culture, whereas the donor culture belongs to a society that is politically and economically weak. Given the domination of the prospecting culture, intellectual property rights are also only defined in terms of the parameters of one cultural tradition. The parameters, categories and concepts of diverse ethnic knowledge systems cannot be applied to claim IPRs under the rules of the currently expanding global market.

When bioprospecting is viewed as a cross-cultural transaction, the question that arises is: can any financial 'compensation' and reward to the donor ethnic culture for permitting itself to be prospected upon, thus demeaning its own integrity, identity and the value of its heritage, make up for the erosion and loss of its own culture? Whereas mutually respectful exchanges across cultures are to be welcomed, the political, sociological and epistemological foundations for such cross-cultural dialogue have not yet been established, and bioprospecting represents the typical example of a one-sided transaction.

Southern countries must act politically to change the terms of cross-cultural discourse, especially in the context of 'globalization'. Globalization should: (i) involve multi-cultural exchange of diverse cultural goods and services; (ii) promote substantial economic investments in the diverse social cultures of the world, so that they can retain their integrity and creativity; and (iii) promote a modern world order where cultural diversity can flourish and global unity is founded not on a 'uniformity' of economic, political, social and technological forms, but on a sharing of diversity.

Globalization based on a sharing of diversity would reduce the size of the world market and encourage local markets and even non-market (nature–culture) relations to re-manifest themselves amongst the ecosystem peoples of the world. Cultural diversity is as essential for human cultural evolution as genetic diversity is for biological evolution and therefore for the long-term survival of human societies.

Managing environments sustainably through understanding and assimilating local ecological knowledge: the case of the Honey Bee (Anil K. Gupta)

There is widespread concern that natural resources are rapidly being eroded. However, activities and investments to deal with the problem are often focused only on resources. The knowledge that people have accumulated for conserving these resources over many generations has not been given adequate attention. The Honey Bee Network was started about six years ago, primarily to arrest the erosion of ecological and technological knowledge and also to document and disseminate the contemporary innovations produced by people for sustainable natural resource management. The network has two basic values: (i) to collect knowledge from people so that they do not complain, just as flowers do not complain when the honey bee collects their pollen; and (ii) to connect farmer to farmer in local languages, just as the honey bee connects flower to flower through pollination. The Honey Bee Network has now extended to 75 countries and is today one of the world's largest networks of indigenous innovators.

SRISTI (Society for Research and Initiatives for Sustainable Technologies and Institutions) was set up to strengthen the Honey Bee Network and pursue research, action and advocacy around the issues of knowledge and resource rights of people. Several kinds of activity have been organized among young students, adults, scholars, public administrators, grassroots functionaries and farmers, which demonstrate that the spirit of competition to produce excellence is compatible with the spirit of co-operation. These include:

(a) Biodiversity contests

The idea is to encourage young schoolchildren to collect knowledge about plant biodiversity along with its uses. In the presence of a jury comprising local teachers, voluntary workers, herbalists, etc., each child brings the list of plants that they know about, with or without uses, in addition to samples of the plants that they can identify. They are evaluated on the basis of five parameters: (a) number of plants listed; (b) number of plants brought; (c) awareness about habitat; (d) familiarity with uses, and (e) style of presentation. Those who excel are given prizes, as well as certificates of honour. Many children ask their parents or their grandparents for help – and in that process knowledge transfer takes place. Children do not compete only to win: often the spirit of participation dominates the spirit of competition. We have noted that ecological knowledge and academic excellence were not necessarily correlated (Shukla, Chand and Gupta 1995), and that children from backward and scheduled castes knew twice as much about plant diversity as children from higher castes, possibly due to their greater dependence on natural resources for their survival.

(b) Indigenous innovation documentation

SRISTI documents indigenous innovations by farmers using local biodiversity for developing non-chemical sustainable technologies for agriculture, livestock, fisheries, agriculture and food processing, etc. SRISTI now has one of the largest databases in the world on farmers' innovations, with names and addresses of the innovators/communicators of the ideas where these are drawn from traditional knowledge systems. These innovations have been collected from several different parts of the world, but mostly from India. More than 2,000 villages have been surveyed with the help of undergraduate students on summer vacation. In addition to documenting innovations through the students, SRISTI organizes state-wide competitions among grassroots functionaries as well as farmers. The process generates humility and respect for indigenous innovators.

(c) Innovators as researchers

Several artisans and innovators who have developed innovative uses of local biodiversity and other materials were requested to scout around for other people of their kind. This process has generated a very participatory way of learning from local innovations. Often, innovators discovered in this way would otherwise have remained unknown.

(d) Institutions for conserving, regenerating and diversifying biodiversity

The Honey Bee Network has been building databases on (i) indigenous technological innovations; (ii) institutional innovations, and (iii) literature on indigenous ecological knowledge systems. The institutional innovations are no less important than the technological innovations. In fact, one could argue that without understanding the institutional context of technological innovations, it is impossible to speculate on the scope of sustainability.

(e) Linking formal and informal science to generate incentives for conservation

Peoples' knowledge systems about natural resources are not only multi-functional but also multi-dimensional. *Tulasi* (see Natarajan, this volume) is revered as a sacred plant but it is also found useful for various purposes in human, livestock and agricultural systems. Sometimes, the scientific mind refuses to take note of the sacred and insists on dealing only with the secular aspect of knowledge systems. Just like the double-helix structure of DNA, the secular and sacred are intertwined. The sacred space provides identity to nature in which the ecocentric view can override the anthropocentric view. The secular knowledge system generally gives priority to human preferences in dealing with nature.

The principles of sustainability and bio-ethics that underlie the Honey Bee approach are:

(a) Sustainability of spirit is the key

Even with technologies that can help us to use resources sustainably, this does not mean that appropriate institutions will emerge if the spirit is absent.

(b) Sustainability requires acknowledgement of the rights of the 'others', i.e. all sentient beings, including unborn human and non-human life

In most societies and cultures, strands of philosophy are found which justify the rights of 'perfect strangers' like the unborn and other living forms which constitute biodiversity. It is necessary for us to understand the process through which such a consciousness is ingrained in the day to day use of resources and observance of boundaries.

(c) Sustainability through creative culture bounds indigenous institutions of common property resources management

Most of the sustainable arrangements for natural resource management require group action through some kind of common property resource institutions. While many of the available frameworks for analysing such institutional arrangements have emphasized either game theory or utilitarian perspectives, it is necessary to attach at least as much importance to the process of rule-making as to rules *per se*. Further, there is an admixture, a 'double-helical intertwining', of explicit and implicit, secular and sacred, and 'this' and the 'other' worldly consciousness in these indigenous institutions.

(d) Sustainability through multi-functional institutions of restraint, reciprocity and respect generating collective responsibility for nature

There is a custom that people in Bhutan go together to the forest to collect shingle wood on a particular day. There are several implications: (i) if somebody falls down a steep slope, there are people around to help in an emergency; (ii) everybody monitors everybody else's collection of wood; (iii) since collection of wood has to be done keeping in mind the age, health and condition of the tree, corrective restraint helps in maintaining those conditions; (iv) some people are either too old, too handicapped, too weak, or their requirements are greater than they can manage on their own: groups can carry the extra burden; (v) some sites may suffer damage due to rain, landslides or other phenomena: since such sites are observed together, mobilization of the collective will for corrective action is more likely; and (vi) in addition to the utilitarian dimensions mentioned above, the group action is its own reward when there is music, fun and laughter around.

Thus, an emphasis on only the economic part of a resource would not provide sufficient information or insights for building institutions that can help in managing resources sustainably. Development is possible only through creative institutions that constrain individual choices to some extent and yet provide scope for entrepreneurship.

(e) Sustainability through blending of holistic and reductionist perspectives for regenerating resources

We need both the reductionist and the holistic perspective. Any theory-building process requires drawing a boundary that renders the phenomena being studied as partial. On the other hand we need a holistic view so that interconnections of different parts of nature can be seen. Sustainability requires balancing the see-saw of these two ends of the same spectrum.

(f) Bio-ethics for sustainability

The sustainability of a resource use requires development and demonstration of an ethic that guides decisions regarding current versus future consumption of resources. The conception of nature and the relationship between human and non-human, animate and inanimate, born and unborn etc., are defined if not determined by this ethic. The bio-ethics can raise the following choices: (i) do I draw a natural resource at a rate that the resource can renew itself within a short cycle? (ii) do I draw as much as I can as long as it is available? (iii) do I draw less than what can be used so as not to impair the ability of the resource to renew itself? (iv) do I draw only as much of a resource as I need whilst ensuring that the genuine needs of others are also met, and that the resource is renewed before it drains down to its critical limits? (v) do I draw as much as possible, hoard it if feasible, and then market it at a very high price to ensure some kind of rationing of its use? (vi) do I develop an institution which through its inefficiency (or coercion, or both) generates a constraint on the maximum sustainable yield? These vectors of human choices confront every decision-maker involved in resource restoration. To what extent these choices actually influence the design of organizations is a matter to be pursued further.

One reason why many externally induced interventions fail is because the local knowledge system is often discounted, and even if considered is seen only in a utilitarian perspective (Gupta 1980, 1981, 1987, 1989; Richards 1985; Verma and Singh 1969). This realization dawned on development planners some time ago, but the mechanisms chosen to build upon local knowledge are often worse than the problem. Various short-cut methods popularly called rapid rural appraisal (RRA) are invoked to get a handle on the local situation. These methods have been critiqued on ethical as well as efficiency grounds. Organizations of creative people whether in the form of networks or informal co-operatives or just loose associations can generate a very different pressure on society for sustainable development. The spirit of excellence, critical peer group appraisal, competitiveness and entrepreneurship, so vital for self-reliant development, can emerge only in the networks of local 'experts', innovators and experimenters. It is true that every farmer or artisan does experiment, but not every one is equally creative and not in the same resource-related fields.

The organizational principles that guide collective action in different regions have some common elements but also many uncommon dimensions. The institution-building process involves simultaneous intervention in eight dimensions of organizational change: (i) leadership, (ii) stake building, (iii) value reinforcement, (iv) clarifying norms and rule-making processes, (v) capacity building, (vi) innovation and creativity, (vii) self -renewal, and (vii) networking. The theory of institution-building (IB) has to be significantly remodelled for historical reasons. The IB processes were evolved to increase the capacity of Third World organizations to receive funds/aid and use it efficiently and effectively. The problem was defined from the external perspective and resolved or sought to be resolved accordingly. Such a perspective provided only limited insights for strengthening the capacities of organizations that have emerged autonomously at the local level.

In conclusion, sustainable management of natural resources requires widening of decision-making choices and extending the time-frame. The utilitarian logic by itself is unlikely to provide the long-time framework necessary for the purpose. Similarly, decision-making options cannot be widened without bringing in the tools and techniques that are available in modern science. Thus, while choice can be widened by modern science when blended with informal science, granting the rights of future generations and the non-human sentient beings can extend the time frame.

Traditional rights to land in Indonesia: a high potential for conflicts (Delphine Roch)

With the huge population growth and the accompanying scarcity of land, the issues of land utilization and access to land have assumed a high potential for conflict in Indonesia during recent years. Local communities can hardly be expected to conserve the environment when their rights over the lands and natural resources they depend on are not respected. No matter how long they have occupied the land, these people live under the constant threat of being arbitrarily displaced, under 'legal' eviction. The main source of the problem is the grant of concession rights in areas that belong to the local peoples under their customary (*adat*) laws. As a result of State forest policy and the inability of the local communities to obtain secure, registered title to traditional lands, concessionaires gain unlimited access to these lands for agriculture, mining, logging or road construction. The Basic Forestry Law of 1967 places forests and other natural resources under the 'primary legal jurisdiction' of the Ministry of Forestry. However, support for the recognition of *adat* law within State forests comes from Governmental Regulation Number 21/1970 which requires forest concession holders to meet with representatives of communities to determine the 'nature and implementation' of the traditional rights. In reality, though, such meetings are never held (Zerner 1990). Huge short-term profits have thus been made by the economic elites at the expense of local communities, whose access to their natural riches are restricted for the duration of the concession contract. Furthermore, illegal logging or land purchases make things even worse, and even community leaders are sometimes involved in these criminal schemes. So far, support from legislation and levels of law enforcement are inadequate. Consequently, many encroachments on protected areas are neither discouraged nor prevented.

In addition, the assumption that local use is incompatible with nature conservation is used to 'justify' the eviction of current residents from areas designated for protection or conservation. Forest protection and biological resource regulations, and the Ecosystem Conservation Act 5/1990, have placed strict provisions on the use and occupancy of these areas, 'without any consideration given to the fact that indigenous forest dwellers live within some of the reserves' (Moniaga 1993).

The case of Bugis farmers and fishers in East Kalimantan's Kutai National Park is a good illustration of the official view of settlers within protected forest areas. Most of the Buginese immigrated from South Sulawesi to the eastern coastal zone of the reserve in the mid-1920s, so they were already well established within the unique lowland rainforests when the park became a 'protected area' in 1936. Government regulations do not provide for settlements within national parks, and a plan to evict and relocate all settlers in Kutai Park is now under study. The strategy is 'to close all options for expansion, initiate gentle intimidation and provide settlement incentives elsewhere with compensation but with a penalty for latecomers' (Ministry of Forestry 1991). So far, the settlers have not been consulted, nor informed in a significant way, and even those who escape eviction may have to contend with mass tourism. Indigenous peoples often have to sacrifice their traditional property rights without adequate or just compensation for the development of luxurious tourism resorts. Conversion of forests, transmigration and mining constitute other major threats that need to be addressed for effective protection of the environment.

Adat land use rights and national land laws exist side by side, giving rise to conflicts that can lead to violence. However, international law provides a basis for the recognition and protection of community-based tenure systems, at least insofar as 'indigenous' peoples are concerned (see above). In Indonesia, there are also laws mandating recognition of community tenurial systems, but most *adat* land tenure experts feel they have not yet been effectively enforced. Moreover, the national land registration system does not accommodate communal rights.

Dispossession of indigenous communities is increasing, and most of the traditional systems of control of forest resources are probably condemned to disappear. The *adat* is being gradually abandoned by the present generation, and the forest of today is no longer mythic or mystical, but the domain of the forest administrators and often of others who come to benefit from the communities' ancestral resources (Michon *et al.*. 1995).

The deforestation issue has become particularly difficult to address as it results as well from the actions of traditional communities (illegal logging, encroachments on conservation areas, or forest fires caused by burning brush to clear fallow agricultural lands). In addition, the fact that *adat* rights to land are often abused, providing developers with easy legal access to natural resources, means that Indonesia is experiencing rapid deforestation. Therefore, the important contribution of the wood industry to the national economy can no longer mask the negative effects of forest management on ecological or social plans. Many studies demonstrate that the confrontation between the State, the market, the indigenous societies and the private companies results in increasing over-exploitation of forest resources. Natural riches can most effectively be conserved under the control of the peoples who depend on these for their own needs. Between an institutional framework limiting their access to natural resource, and an economic reality of intensifying use of their resources, the forest-dwellers have nevertheless often created systems that promote sustainable and environmentally sound development. Successful examples of community-based tenure have been under-estimated or misunderstood. For effective conservation, it is vital that the two levels of law (State and *adat*) be integrated so that the former does not negate the latter.

The critical importance of people's participation in Papua New Guinea's development plans (Tahereh Nadarajah)

Traditional systems of land tenure in Melanesia are by no means uniform. In Papua New Guinea (PNG), most land is clan owned, yet systems vary from place to place, some even allowing individual members of the clan to own land. However, the concept of 'ownership' often does not exist. Instead, traditional inhabitants exercise 'custodianship' over the land (Faracklas 1993). The people belong just as much to the land as the land does to them. This relationship is not normally expressed in terms of an alienable (commodifiable) possession, but rather in the context of 'familial' or even corporal association. In fact, the terms that Melanesians traditionally use are 'children', 'siblings' or 'parents' of the ground, rather than 'landowners'. In this system, buying and selling land is not envisioned. What has transformed this concept into a more commodity-based idea? Two factors may be involved: the most immediate one is outside economic pressure; the second is rural-urban migration which will gradually (and some believe, inevitably) assist land alienation, because the bond with the land is no longer an essential element to living and working in towns. Because the people are not using the land, they will probably be prepared to sell it. It will be interesting to see how clan solidarity will withstand this test, or adopt new measures as a result.

One of the most important elements of meaningful participation is learning (Arbab, cited by Momen n.d.). Arbab argues that, 'this is not the same as saying that training is a necessary component of every development project'. Too many global programmes have decided that the poor simply need to be trained to carry out specific productive tasks or simple services. However, a little reflection shows that for the vast majority of human beings, training to carry out orders without accompanying spiritual and intellectual development only reinforces the present division between the modern and traditional sectors. The mark of development of an individual is, undoubtedly, the attainment of true understanding. Burkey (1993) has also described participation as, 'not the mere mobilization of labour forces, or coming together to hear about predetermined plans'. Participation is 'an essential part of human growth' (ibid.).

The role of knowledge in facilitating participation

Papua New Guineans have to decide on the nature of the development process, as well as the role of each sector. According to Vick (1989), 'true development is the mobilization of society to transform itself'. There are two basic steps to achieving this: awakening a desire for social change, and creating the confidence that it can be achieved. Fostering self-reliance is much more difficult than spending US$100,000 on a two-year, expatriate-led development project.

Achieving sustainable development requires the fostering of local initiatives and action. There is a need for greater faith in local people's analytical capabilities. The applicability and success of technical assistance should make a shift from 'transferring technology' to 'enhancing local capabilities' (Chambers 1993). The personal, professional and institutional challenge is to change attitudes. Chambers states that, 'we have been holding the stick for too long'. Instead, it is necessary to listen more to the different views of foresters and farmers.

There are many development projects, in Papua New Guinea and other countries, that have not worked because the people did not act for themselves to make the project successful. Avei (1994), who investigated the development pattern of the Boera Community in Central Province, has identified those projects that were successful and were based on traditional knowledge and technologies of proven efficacy for communities over many generations, and those that did not stand the test of time. An analysis of the two types of project makes it abundantly clear that the success stories were those projects that had originated from the people according to their knowledge and needs, and thus the will to carry them out was there. In contrast, the unsuccessful projects were initiated and imposed by outside development agencies.

Such considerations are often overlooked by development agencies, but if left to the people themselves they would plan their needs in a way to suit their social patterns of life. For Boera village, Avei (1994) perceives the problem to be a lack of proper leadership at the community level, because the decision-making process, 'has been left wide open to the younger levels of the community, without proper grooming as in the times past'. Some government officials in the recent working group for the development of a National Sustainable Development Strategy for Papua New Guinea argued that the proper decision-making structure, the clan, exists, but the government has not been able to use those mechanisms effectively to implement their policies.

The valuation of modern technical knowledge and indigenous knowledge should be the focus of attention. The ideal solution is to find the complementary role of each system. Relating knowledge, power and the environment is like relating the modern system of knowledge to a fox that knows many things (Banuri and Marglin 1993). Like the fox, this system has many strengths. However, it has one 'fatal' flaw; that in its quest to master nature, it is slowly but surely destroying the basis of life upon this planet. The success of this system of thinking has come at a considerable cost. The so-called modern societies view indigenous knowledge as 'inferior and regressive'. Traditional systems evolved to ensure survival of human and other life-forms on this planet. Traditional knowledge is 'embedded in the social, cultural and moral milieu of any community, whereas the modern system of knowledge distinguishes these different dimensions'.

The attainment of any objective is conditional upon knowledge, volition and action. By expressing the knowledge and will of the community, their activities gain momentum (Baha'i Writings, cited by Vick 1989).

When thinking about the three requirements for achieving any objective: knowledge, volition and action, the communities need to reach their own vision of growth, arrive at unity of thought, and devise plans of action which they are capable of carrying out themselves (Baha'i International Teaching Centre, pers. comm. 1994). The implications of this principle for Papua New Guinea's biodiversity projects are to include local initiatives right from the start. The aid agencies need to be sensitive to existing capacities and provide the back-up support in accessing the information and technology. The current funding trend of aid agencies was described by one of the participants at the 1993 Waigani Seminar as 'a boomerang' and in need of a complete shift. Instead of spending all of the aid money on infrastructure and salaries for 'expatriate experts', it should be used to identify and develop local capacities.

Conclusion

Although the Papua New Guineans are the customary owners of the land, they have had very little chance of participating in the plans that affect their resources. This is due to the unavailability of knowledge that would equip them for meaningful participation in the nation's development plans. The other reason is the lack of a mechanism that can make landowners equal partners in resource development projects. There is a need for a change of attitude towards the resource owners on the part of the other partners in development. There is a strongly held perception both by academics and by development agencies that Papua New Guinea landowners are willing partners in the destruction of their resources. This may be so, but the system has failed because of wrong attitudes and assumptions – primarily the assumption that conservation is a matter of going to landowners with a ready-made conservation package and expecting them to implement it. If the conservation project still holds the perception of US-vs.-THEM (forest industry), of course this will oil the already existing polarization of the community on this issue. It simply has not worked. The volition for the conservation of the forest must come from them. They own the land, but they must also own the concept, and the goals of the conservation projects. The project design should be borne out of consultation with the people.

Perceptions of natural resource use rights and intellectual property: the case of Jureia (Gemima Born)

Traditional communities inhabit the ecological station of Juréia-Itatins, Brazil. These people, who have lived there since long before the ecological station's establishment, are descendants of white, black and indigenous frontiersmen (*caboclo*) known as 'hillbillies' (*caipira*), beach (*caiçara*), riverside (*ribeirinho*) and hill dwellers (*capuava*). According to environmental legislation dealing with protected areas, the government should own the lands, but the majority of the lands are in fact privately owned. In spite of restrictions imposed by such legislation, the relationship of the communities to the natural environment is still fairly close and harmonious. The accumulated ethnobotanical knowledge demonstrates various categories of utilization based on the material to be used, i.e. timber for lumber and construction; fibres for cording; dyes, and so on. These plant species are harvested under certain conditions and processed in different ways. In all, researchers from a Brazilian NGO, Vitae Civilis, have identified over 1,000 different applications for about 400 different species falling into 21 useful categories. Regarding the ethnopharmacological knowledge, a whole range of illnesses are treated with plants or other natural ingredients, prepared in different ways and administered in specified doses, for a determined period and with certain precautions.

The communities maintain their traditional knowledge about the management of natural resources in spite of the increasing risk of attrition due to legal restrictions, the imbalance created by the new situation, and the exodus of the population, especially the young. Consequently, Vitae Civilis was faced with a dilemma: how to guarantee a better standard of living for the communities, including their participation in the management of the natural resources they know so well and utilize, and how to make use of their knowledge for the benefit of humanity, guaranteeing IPRs to the communities and compensation – either financial or in the form of other benefits.

In order to identify opinions, projections and alternatives, a qualitative study was developed together with some key participants of Juréia involved with the subject. Government, NGOs, academic/research and private (media) representatives, and members of the local communities, expressed their opinions, projections and alternatives for action regarding the right to use natural resources and IPRs. The representatives of the different stakeholder groups all agreed that it is necessary and important to protect nature against people's predatory actions. The communities opposed this, as they do not believe that conservation measures are necessary. Almost all the inhabitants of the communities complained of losses and damages to their livelihood caused by the creation of the ecological station, and refused to accept that they themselves cause any negative impact on the environment. The government sector and the conservation NGOs agreed that rights to use natural resources must be linked to rules for environmental protection. The communities claim they have the absolute right to use the resources, because: (i) they were already there before the creation of the ecological station, and have always utilized the resources; and (ii) this is the way they should live, utilizing the resources, for this is what they were taught, and this is how their ancestors lived.

Most of the inhabitants of the ecological station were unaware of the importance, and the possibility of utilization, of their knowledge, although some felt they were in possession of knowledge, or 'heritage', of great importance. Medicinal learning was considered to be the most significant knowledge, and something that should be used by urban society as long as compensation measures are in place. They also pointed out that: (i) the existing legal apparatus is inadequate for the protection of traditional knowledge; and (ii) that identifying the precise origin of the knowledge is difficult. The communities were more concerned with their practical welfare than with an abstract concept of 'intellectual property'. The stakeholders presented many doubts and were generally incapable of making suggestions for solving the issue due to general ignorance and a failure to recognize the immediacy and importance of the subject. This led to the general conclusion that there is a need to delve further into the issue, and involve the communities further in this general discussion.

Considerations and proposals

Traditional communities play an important role in the survival of the human species due to the collective transgenerational contribution of their knowledge about conservation and use of natural resources. Traditional knowledge also assumes an economic value in capitalist societies. Because of the possibility of the knowledge producing wealth, it becomes a legally justifiable asset, independently of the existence of specific laws, and so is the *legacy of its owner*. The economic exploitation of this asset by third parties, without some form of benefit to its owner, is a form of illicit or unjust gain. Consequently, mere acknowledgement of the value of traditional peoples and their learning for the conservation of biodiversity is not sufficient. Sustainable development depends not only on the conservation of biodiversity, but also on respect, appreciation and harmonious articulation of cultural and ethnic diversity.

Legal means of redress for the unauthorized use of traditional knowledge by third parties are necessary. Otherwise, these peoples will disappear due to their poverty and marginalization. There is still no Brazilian law to implement Article 8(j) of the CBD, but the Federal Bill of the New Indigenous Statute (PL 2057/91) grants access and use of traditional indigenous knowledge only with the community's prior written agreement. Unfortunately, the 'New Indigenous Statute' applies only to indigenous peoples, and does not include other forest inhabitants (riverside and beach dwellers, hillbillies, frontiersmen, etc.).

Senate Bill 306/95, which deals with the control of access to Brazil's genetic resources, contains a chapter dedicated to the protection of traditional knowledge. It affirms that local communities have the right to benefit collectively from their traditions and knowledge and to be compensated adequately for conserving biological and genetic resources.

Legal, anthropological, ethnological and political studies are necessary to identify formal options that will safeguard traditional knowledge and regulate its use by society. A new law should stipulate minimum guarantees in contractual relationships to prevent unfair exploitation of traditional communities, and be the subject of broad discussion among traditional communities and other interested parties. Such an instrument will guarantee esteem to the culture of these communities by returning their respect, dignity and pride, and will allow them to reproduce and further develop their knowledge at their own pace. Otherwise, the culture and knowledge of traditional communities cannot survive.

Maori cultural and intellectual property claim: Wai 262 (Maui Solomon)

The Waitangi Tribunal in Aotearoa/New Zealand was established in 1975 as a Commission of Inquiry to investigate claims by Maori people that they have been prejudicially affected by actions or omissions of the Crown (government) that are contrary to the Treaty of Waitangi signed in 1840 between the Maori Chiefs and tribes of New Zealand and Her Majesty Queen Victoria. The Treaty of Waitangi enabled the Westminster (British) style of government and the Common Law to be established in New Zealand in return for Maori being guaranteed the 'full, exclusive and undisturbed possession' of their lands, forests, fisheries and all other *taonga* (treasured possessions) for so long as they wished to retain them. For 150 years, New Zealand ignored its obligations to the Maori people under the Treaty. But Maori people never gave up hope of having the Treaty honoured and eventually, in 1975, the Waitangi Tribunal was established. Ten years later in 1985 the

Tribunal was given retrospective powers to investigate claims back to 6 February 1840 (the date the Treaty was signed). Since then, the Tribunal has heard a wide array of claims involving fisheries, lands, language and broadcasting, minerals and forestry to name but a few.

Maori, like many other cultures, possess an indigenous body of knowledge which describes the origins of the universe and the place of humans within the universe. Two fundamental aspects of this relationship are *whakapapa* (genealogy) and the personification of natural phenomenon. Complex genealogical relationships explain both the time before and the time after the origin of the universe, including the creation of life. The Maori scholar, Peter Buck (1949), described the relationships in these terms:

> 'Most versions consist of recitations of creation events arranged in genealogical order. Some *wananga* (schools of learning) begin with a description of *Te Kore* (the realm of 'chaos' or nothingness; of 'potential being'). In this realm dwelt Io, the supreme being from whose *iho* ('essence') the subsequent voids were conceived. Thus from *Te Kore* arose *Te Po* (the night realm), and from thence the twilight dawn, the *Te Ao Marama* (the full light of day). Io then created a single being or ancestor from whence came Rangi and Papa (who after separation, became known as *Ranginui e tu nei*, the male principle or 'Sky Father' and *Papatuanuku*, the female principle or 'Earth Mother').
>
> From these two primal parents arose many offspring, all supernatural beings, each responsible for, or guardians of, particular natural phenomena. Tane was the most important... Personified as Tane Mahuta (god of the standing forest) he engaged in numerous procreation events with supernatural female deities. For example, from Hinewaoriki came the *kahikatea* and *matai* trees, and from Mumuhunga the *totara* tree; in all a total of eight wives produced nine species of large trees. With Punga he produced the insects and other small creatures of the forest, while from Parauri came the *tui*. Further cohabitations produced all other birds indigenous to Aotearoa.
>
> Tangaroa was the god of the sea and all sea creatures. All fishes are descended from one of his grandchildren (Ikatere), and reptiles from another (Tutewehiwehi).
>
> Tawhirimatea was god ancestor of the winds and all other meteorological aspects, while Tumatauenga had authority over warfare and human affairs. He is also the progenitor of human beings in some *whakapapa* (genealogy). Rongomatane, god of agriculture, was responsible for all cultivated foods, especially the *kumara*, taro, *hue* or gourd and the *uhi* or yam. To this function was added that of god of peace.'

In 1991, a number of representatives of different tribes throughout New Zealand filed the 'Wai 262' claim seeking recognition, restoration and protection of Maori cultural and intellectual heritage rights (*taonga tuku iho*) including those in relation to the ownership, protection and use of native flora and fauna. The claim is a wide-ranging one and applies also to the use of Maori designs, images, icons and traditional material such as carvings, mythology and traditional patterns. The claim is for *tino rangatiratanga* (Maori authority/control) over all of these things. It seeks recognition that Maori have a unique relationship and world-view of these cultural treasures which are invariably alien to – and cannot be provided for within – the Western regime of intellectual property rights.

In relation to native flora and fauna, the claimants have concerns about the genetic manipulation of native species both past and present. Maori people regard themselves as guardians of their natural world and advocates on behalf of the creator gods such as Tane Mahuta – god of the forests, and Tangaroa – god of the sea. As indigenous peoples they have a holistic relationship with the environment and everything within it. The advent of genetic manipulation of bio-material therefore raises serious moral and ethical issues. A recent example in New Zealand is a proposal to inject the DNA of an extinct giant flightless bird, the Moa, into chickens. Another was the approval by the Minister for the Environment to approve applications for field testing of genetically modified canola containing the 'Roundup Ready' gene for herbicide resistance for the 1996/97 growing season. The potential for patenting the genes of indigenous peoples is also of particular concern to Maori. Tampering with the genetic material of plants and peoples has unknown risks that could threaten the very fabric of human society.

The Wai 262 claim seeks (among other things) to raise awareness within New Zealand of the different cultural perceptions that Maori as indigenous people have of their natural environment and the manner in which that relationship is expressed in modern-day terms. The IPR system, which focuses on individual property rights, must be reassessed in light of the growing need to devise a system that recognizes and protects communal indigenous knowledge systems and beliefs. Cultural diversity is increasingly seen as being vital to the preservation of biological diversity and should be given more than just lip service.

The claim also challenges the underlying ethos of the GATT-TRIPs Agreements that have been incorporated into national legislation. The claim challenges the underlying principle of universalization and commoditization of indigenous knowledge and the myth of equitable sharing of resources. Indigenous people are, without exception, at the bottom of the socio-economic hierarchy in most countries. This has resulted from colonization and the consequent alienation of the traditional peoples from their lands and culture. Maori are no exception in this regard. Globalization of markets and increased competition assumes a 'level playing field' mentality. Maori people, as with indigenous peoples elsewhere, do not have access to the substantial resource base and equity investment required to take advantage of this so-called level playing field. Indeed the process of colonization has largely stripped indigenous peoples of their resources and thus the means to generate wealth. Many Maori regard the world-wide efforts to record, analyse and capture indigenous peoples' knowledge of their environment and traditions as potentially another form of colonization. Maori are not opposed to development and advancement. In fact their own culture fully embraced and was highly adaptive to new technology. But Maori, like indigenous peoples elsewhere, seek a greater degree of control over the decision-making processes that will determine the direction of future developments in this field.

A full law (Christine Morris)

Any conflict that arises between peoples stems from a lack of respect for each other's law. This lack of respect emanates from fear of the unknown rather than one wanting dominance over 'the other'. Similarly, in the case of the biodiversity argument in Australia, at the core of the dispute are differences in law. These core differences are best explored through the concept of custodianship, for it is custodianship over the maintenance of the world's biodiversity that is everyone's concern, whether they wish to nurture it or to exploit it. Before discussing this difference however I will tell you of my place in indigenous Australia; indigenous Australians being the custodians of the oldest continuous culture in the world.

I am a descendant of the Kombumerri/Munaljahlai clans of the east coast of Australia, who are among the 200 or more clans on the Australian continent. I take the responsibility of maintaining the world's oldest continuous culture very seriously. This responsibility was made very pertinent to me when it became the duty of my grandmother's people to fight the local university to regain and re-bury the remains of 200 of my ancestors. This activity was seen as the Dead calling on the Kombumerri to become involved more actively in the overall maintenance of indigenous culture in Australia. By the 1990s many of the descendants were in positions important to cultural maintenance, language preservation being the most prominent. In recent times, these initiatives are moving more towards law.

From a biodiversity or land perspective my clans are situated in and around the Gold Coast, Australia's number one tourist destination. The Gold Coast has been popular with Australians ever since *Dugais* (non-indigenous Australians) came to my grandmother's country. At first they sought the cedar, but then the area became a seaside resort for the landed gentry. Because of its status as a holiday resort, the *Dugais* have in general tried to maintain the natural beauty of the landscape. This has allowed the continuation of the annual migration, to the area, of hundreds of bird species from around the world. I am fortunate enough to live near a mangrove colony known as the Coombabah lakes system, and also one of Australia's best examples of heath growth. The biggest danger to our natural species, however, is canal development. The Kombumerri clan has been very active in trying to arrest this development, but the core differences about how the custodial relationship to land is put into practice prevent us from carrying out our custodial duties to protect these important species breeding areas. Therefore the biodiversity debate is part of my daily reality, and the energy associated with it draws me into the argument.

The two laws

Firstly I will set out what I see as a diverging focus of the two laws, which leads to tension and acts of domination.

On the one hand there is Indigenous Law, which I term 'a full law', i.e. a law that is applicable to both the seen reality and the unseen reality. The unseen reality is as valid and impacts on the indigenous world-view as much as does the seen. This law, which is always prefaced, 'The Law' when used by the indigenous, stems from a belief that humans must reciprocate with every aspect of life on earth and the spiritual realm. The penalty for a breach of this law is called 'payback'. This colloquial term goes to the heart of the guiding principal of reciprocity, which in turn shapes the custodial ethic.

On the other hand I would suggest that the Australian Common Law is a 'half law', as it applies only to the seen reality. This is not meant to be a derogatory statement but rather a reflection of what Australians in general wish their law to cover. This law stems from a monotheist belief system that has as its central dogma the supremacy of the human species as the custodian of the earth. The guiding principal of law in Australia is to protect the individual and his goods so that they may be an effective contributor to the economic system. Any breach of this law attracts a penalty of financial retribution or a restriction of the person's liberty.

I feel that these two laws actually could coexist in our Land if proper protocols were adhered to. Protocols I have found lead one into formalities that give both parties time to appraise the situation without commitment. By having this space, the two groups are allowed time to come to terms with their own fears about control of the relationship. Also, it allows peoples who cannot conceive of another people's perception of reality (in the Australian case, the influence of the spirit world) to digest and come to terms with the fact that nobody is asking them to believe in a spirit realm; they are just being asked to have a mature attitude to someone else's reality. After all, is this not what human rights are about?

The clans that brought me up have taught me that conciliatory and inclusive behaviour leads to a more aware society. This inclusivity includes respecting the law of 'the other'. However, in the case of Australian Common Law it is impossible for the indigenous people to carrying out their custodial duties if they are subject to a half law. And more importantly, it puts them in breach of their own custodial duties and Law. Therefore it is the inadequacy of the half law that is at the core of the tension in the biodiversity debate, not the fact that it is the law of the invaders.

The ego and the killing fields

Ms Mary Graham, one of the clan elders of my grandmother's people, constantly addresses in her cross-cultural lectures the issue of the ego and its sense of being a discrete entity in the world, and the damage it has done to the earth in its narcissistic pursuit of its own productivity and creativity. It is not that these pursuits are wrong; it is just that they have been allowed to run rampant, with no apparent boundaries. She further adds that indigenous societies adhered strongly to formalities and ritual in relation to caring for the land, one of the intentions being to give the ego a place to express itself in a more orderly manner.

Let me add here that formalities are the training ground for the maturing of the ego, not the killing fields of the individual's potential. Reverential behaviour is there to remind the ego that you are who you are because: (i) the dead who have gone before you still influence you from the grave (as have my ancestors); (ii) you as a child spirit-being chose the clan into which you wished to be born and nurtured; (iii) that clan passed down to you a genetic heritage of concepts and experiences as well as your physiological features; (iv) that clan gives you your intellectual property through its stories, patterns, dances and songs; and (v) you are a custodian of a particular tract of land and in the future will become a custodian for other lands (through marriage, adoption or other circumstances).

But most importantly – and this is a point that is reiterated in all forums by indigenous Australians – Land is the most important reference point in one's world-view. Land, therefore, like the Law, is always prefaced. Land is not valued directly for its utility value but rather because it is our teacher – it teaches us to be human, it gives us our place, and it fulfils the human longing to be needed for something meaningful.

Within the indigenous world in which I move, the Law provides rules and regulations that allow me to be part of a clan. The Law does not protect my chattels or even my individual being, but rather protects a social system in which the individual feels safe and nurtured. It is a system that gives me a lot of responsibility and very few rights.

Due to the all-pervasiveness of the Law one never has a sense of being alone, because a spirit, an animal or a plant is with you. So there is no opportunity to develop a world-view that would allow you to take, destroy or move another aspect of the earth without some contrary thought as to whether this is correct or not. I am not saying it stops the person but it is no surprise to the person if the Law asks for payback.

I myself have observed the elders throughout Australia for many years and have learnt that when one is subject to a law that dictates one's every waking moment and movement, one does not become a burdened ass but rather a refined and dignified human being who carries no malice for the oppressors nor expectations on the young. Rather, one sees a world constructed around a divine comedy in which humans are hapless actors believing in their own self-importance. I have observed these elders take great delight in watching younger people come to terms with what the young think is the yolk of the laws, responsibilities and atrocities of the past.

These old people patiently enter the real-life dramas of the young and wait in expectation for the day the young realize the Law is not their enemy or yolk which does not protect their person or personal possessions, but the dreaming track that leads them to a world full of wonderment, challenges, opportunities and most importantly humour with which to bring out the best in themselves, so that they can be role models for the next generation, the next carers of the fertility of the Land.

Acknowledgements

My bibliography is made up of oral citations over many years of learning, but particularly in the past seven years in which intense experiential learning has occurred. Prior to the last seven years it was the elders of the Kombumerri and Munaljahlai people. In the last seven years the most influential have been the following personal communications: Ms Mary Graham – Kombumerri clan elder; Ms Lillia Watson – birth spirit country Gungalu elder; Mr David Mownjarlai – Custodian of the Wandjinas; Mr Galarrwuy Yunupingu – Chairman of the Northern Land Council, through personal communications, oratories and songs (Yothu Yindi Band).

Box 12.2: 'Choice, No Choice?'

Janis Alcorn

Dreams of life, green forests,
health and laughter among leaves and water,
conjuring visions of diversity,
interacting, companions of a union
not capitalized.

Visions in the heads
of those of
us who
work with
old holy men in buffer zones, tigers in swamps,
butterflies trapped on falling trees,
children old at 7 who still laugh in the morning, we are
born in a time when there is choice, either
act or deny, speak, or be silent and prepare to pay.

Just as the saddhu said,
he left his waterfall in the forest when it became a park and
was drained of the villages that gave it life,
he could have stayed,
deemed to be at one with nature, but
he went with the ousted ones
into the buffer zone
to buffer the ousted ones
against the police and the beat guards and the bureaucrats
who eat money, like the tigers stalking prey in the park.
Just as the saddhu said,
tigers, leaves, and water cannot live alone.
Choices.
It's all about the tigers.

Box 12.2 (continued)

It's about leaves crushed fragrant
into a wound
Roots steeped with bark and five kinds of flowers
Seeds ground into paste tied under leaves
found only near clear flowing springs where water drawn at
first light
is needed for the offering and the words that must bring
strength to the medicine stolen at midnight from the park
where jeep drivers bring
tourists who do not hear the songs of the healing, only the
jingle of coins and clicking of cameras
when a tiger is sighted
before retiring to afternoon tea.
It's all about the tourists.

Like someone idly ripping
small tears in the fabric of life,
deaf to the sound of the ripping,
unawares that
cost will follow,
tourists finance separation
so the mothers of feverish children
must sneak across park borders under darkness
to find medicine under threat of
thrashing should they disturb tourists' pleasure,
tigerlovers' coffee table picture books, pocketbooks.
And the tigers get killed.

Us,
we chose to act
to keep company,
to bear witness to the wound,
to the shreds of cloth,
to the water flowing alone
no longer amplifying the sound of
laughing children bathing in the late afternoon,
to the silenced songs,
and the need for healing.

Only by choosing to keep company,
to know that only together
compañeros en ese momento,
can any one,
tiger, tourist, mother, saddhu, weaver, child
enjoy
healing waters running clear in
forests green,
fabric whole,
choices.

CONCLUSION

MAINTAINING THE MOSAIC

Darrell Addison Posey

Maintaining the Mosaic

Darrell Addison Posey

THIS VOLUME COMPRISES A RICH MOSAIC OF DOCUMENTAtion by people and peoples around the world who deplore the loss of cultural and biological diversity. The wide range of contributors shows that those concerned with globalization of markets and economic values, linked to devastation of local communities and the environments upon which they depend, are not just isolated purists, cultural conservatives and idealistic environmentalists. The volume has provided depositions from scientists, scholars, political and spiritual leaders, indigenous elders, traditional farmers, shamans and curers, poets, artists, songwriters, journalists and others. Together they speak for the vast majority of global citizens who are worried about the world we are bequeathing to future generations of *Homo sapiens* and other life-forms.

Recently a Special Session of the United Nations General Assembly was convened to evaluate progress five years after the Earth Summit. Most observers found that the tally for Nation/State action was poor to mediocre. But some of the parallel processes that have accompanied the global environmental movement are finding considerable successes. Indigenous peoples, for example, have become significant players in debates on sustainability, trade, environment and human rights. Traditional farmers have become well-organized and their demands increasingly heard through debates in the CBD, FAO and elsewhere. Environmental NGOs now administer considerable funds for, and influence, policies on sustainable development and conservation. Professional societies and businesses are increasingly concerned with ethics and environmental justice. And citizens' groups have mobilized to impact significantly on how biosafety and patents-on-life issues are viewed by politicians and governments.

But there is much to be done. Indigenous peoples, traditional farmers and local communities are further and further marginalized from economic benefits and political power. Traditional Ecological Knowledge is still not recognized for its worth, or taken seriously by the scientific, development and political communities. Languages continue to be lost at unprecedented rates – and the knowledge and resources that are described in those languages are likewise eroded, only to be replaced by 'monocultures of the mind'. Resources are limited for investigation of the wider use and application of what the CBD identifies as 'indigenous and traditional technologies'. International laws that guarantee protection of traditional knowledge and genetic resources are inadequate to non-existent, making most bioprospecting, food and agricultural development, trade and alternative product initiatives predatory on local communities, rather than supportive of them.

Equity has yet to play the dominant role it must play in the orientation of global initiatives to use and conserve biodiversity. Human rights have yet to become the guiding factor they must be for equitable trade and commercial exchanges. And diversity is still seen as an enemy to be vanquished by globalization, patents and trade monopolies.

This volume has shown that the best way to conserve the diversity of cultures and nature is through the empowerment of the people and peoples whose local knowledges and experiences form the foundations that conserve much of the earth's remaining biological and ecological diversity. It would be nice to think that adept anthropologists and ethnoecologists (aided by advances in information technology) might one day be able to adequately describe TEK systems. But scientists themselves admit that they will never get more than a inkling of the whole, intricate webs of symbols, values, practices and information that have evolved in unique systems for each society. The only way to employ all the force and sophistication of local communities is to allow them to develop and design their own systems for change, conservation, land and resource use. The volume has shown that this can be done best through communities in equitable relationships (true partnerships, if you will) with scientific and technical advisors – and that it works best when the 'scientific experts' are in the role of advisors, not commanders.

We are faced with some serious multi-disciplinary and multi-cultural dilemmas. Some of these are methodological, as well as philosophical and political. How can indigenous concepts be used as 'criteria and indicators' in the development of baseline studies of biodiversity? And how can these become central, for example, to environmental impact assessments, monitoring activities and national biodiversity surveys? How can spiritual and cultural values be incorporated into planning and policy decisions? Can any of this be assigned monetary value?

If not, how can other value systems be respected and weighed?

These questions do not depend on political will alone to implement change, but also require considerable intellectual work to develop integrated methodologies to guide the practical tasks for such studies. Likewise the legal basis for protection of indigenous, traditional and local community rights is far from complete, given that the peoples involved have rarely been seriously consulted on what the basic principles for codes of conduct, and standards of practice, and new, appropriate laws might be. The latter, it must be said, depends more upon changes in political and economic policies than on methodological difficulties.

This conclusion really ends with the fundamental question: Where do we begin?

The answers come in the diversity of voices heard in this volume, and scores of others that cannot be heard at all because they speak languages unknown to those in power; are non-literate or find it inappropriate to write contributions; are too distant from political and economic centres to state their concerns; have no access to the media; are imprisoned for their beliefs, or whose ideas fall on deaf ears because they have cultural and spiritual values that differ from our own.

The most feasible and practical solutions are often the most simple. Some of the major steps forward simply require a recognition of scientific fact: that many indigenous, traditional and local groups *already* employ principles of sustainability and strategies for successful biodiversity conservation; and that by learning from them conservation and sustainable development projects, programmes and policies can be improved to the benefit of all.

So, first and foremost is the recognition of the basic rights of indigenous and traditional peoples to their own cultures, customs, languages, lands, territories and resources. Implementation of International Labor Organisation Convention 169, ratification by the UN General Assembly of the Draft Declaration of Rights of Indigenous Peoples, recognition of 'Farmers' Rights', support for the proposed United Nations Permanent Forum on Indigenous Peoples, and development of Traditional Resource Rights, are simple but decisive steps forward.

Other basic but essential steps would be to redirect financial and political support from government economic and sustainability plans to support the programmes and projects conceived and implemented by indigenous and local communities themselves. This volume has given examples of how such projects work and how extremely successful they can be. But some of the most flourishing initiatives may be hidden from our view – and almost none get the support and recognition they deserve.

Global discourse would lead us to believe that conservation of biological diversity is of the highest priority. If that were indeed so, then it is clear that highest priority should be given to the protection of those remaining cultures and societies that are struggling to preserve the precious biodiversity that remains in their care.

Likewise, it would seem that to prioritize support for restoration of areas already damaged or destroyed by unsuccessful development and exploitation would be a way to steer economic activities away from the more fragile areas inhabited by indigenous and traditional peoples and direct them toward the impoverished communities that suffer most from degradation of their traditional resource bases. Traditional Ecological Knowledge could be employed as the basis for such restoration and renovation projects, providing an opportunity to use and apply the 'traditional technologies' – while in the hands and under the control of the people who perfected them – as called for in the Convention on Biological Diversity and Agenda 21.

It is increasingly evident that the 'minority' and disenfranchised peoples of the earth are the ones who speak for all humanity. They speak clearly and with dedication and conviction, because they know their lives and immediate futures – as well as the well-being of future generations – depend upon the environments in which they live and the biodiversity upon which they depend.

Technological society seems to have lost this basic reality. We must, therefore, allow the 'Voices of the Earth' to become our intellectual and spiritual guides, so that we too can relearn what they know and practice: that the future of humanity depends upon the maintenance of the mosaic of biological, cultural, linguistic and spiritual diversity.

APPENDICES

APPENDIX 1

DECLARATIONS OF THE INDIGENOUS PEOPLES' ORGANIZATIONS

IN CHRONOLOGICAL ORDER:

1.1: Declaration of Principles of the World Council of Indigenous Peoples

This declaration was ratified by the IV General Assembly of the World Council of Indigenous Peoples.

1. All human rights of Indigenous Peoples must be respected. No form of discrimination against Indigenous Peoples shall be allowed.

2. All Indigenous Peoples have the right to self-determination. By virtue of this right they can freely determine their political, economic, social, religious and cultural development in agreement with the principles stated in this declaration.

3. Every nation-state within which Indigenous Peoples live shall recognize the population, territory and institutions belonging to said peoples.

4. The culture of Indigenous Peoples is part of mankind's cultural patrimony.

5. The customs and usages of the Indigenous Peoples must be respected by the nation-states and recognized as a legitimate source of rights.

6. Indigenous Peoples have the right to determine which person(s) or group(s) is (are) included in its population.

7. Indigenous Peoples have the right to determine the form, structure and jurisdiction of their own institutions.

8. The institutions of Indigenous Peoples, like those of a nation-state, must conform to internationally recognized human rights, both individual and collective.

9. Indigenous Peoples and their individual members have the right to participate in the political life of the nation-state in which they are located.

10. Indigenous Peoples have inalienable rights over their traditional lands and over the use of their natural resources which have been usurped, or taken away without the free and knowledgeable consent of Indian peoples, shall be restored to them.

11. The rights of the Indigenous Peoples to their lands includes: the soil, the sub-soil, coastal territorial waters in the interior, and coastal economic zones, all within the limits specified by international legislation.

12. All Indigenous Peoples have the right to freely use their natural wealth and resources in order to satisfy their needs and in agreement with principles 10 and 11 above.

13. No action or process shall be implemented which directly and/or indirectly would result in the destruction of land, air, water, glaciers, animal life, environment, or natural resources, without the free and well informed consent of the affected Indigenous Peoples.

14. Indigenous Peoples will reassume original rights over their material culture, including archaeological zones, artefacts, designs and other artistic expressions.

15. All Indigenous Peoples have the rights to be educated in their own language and to establish their own educational institutions. Indigenous Peoples' languages shall be respected by nation-states in all dealings between them on the basis of equality and non-discrimination.

16. All treaties reached through agreement between Indigenous Peoples and representatives of the nation-states will have total validity before national and international law.

17. Indigenous Peoples have the rights by virtue of their traditions, to freely travel across international boundaries, to conduct traditional activities and maintain family links.

18. Indigenous Peoples and their designated authorities have the right to be consulted and to authorize the implementation of technological and scientific research conducted within their territories and the right to be informed about the results of such activities.

The aforementioned principles constitute the minimal rights to which Indigenous Peoples are entitled, and must be complemented by all nation-states.

1.2: Charter of the Indigenous–Tribal Peoples of the Tropical Forests

Article 1: We, the Indigenous/Tribal Peoples of the tropical forests, present this charter as a response to hundreds of years of continual encroachment and colonization of our territories and the undermining of our lives, livelihoods and cultures caused by the destruction of the forests that our survival depends on.

Article 2: We declare that we are the original peoples, the rightful owners and the cultures that defend the tropical forests of the world.

Article 3: Our territories and forests are to us more than an economic resource. For us, they are life itself and have an integral and spiritual value for our communities. They are fundamental to our social, cultural, spiritual, economic and political survival as distinct peoples.

Article 4: The unity of people and territory is vital and must be recognized.

Article 5: All policies towards the forests must be based on a respect for cultural diversity, for a promotion of indigenous models of living, and an understanding that our peoples have developed ways of life closely attuned to our environment.

Therefore, we declare the following principles, goals and demands:

Respect for our rights

Article 6: Respect for our human, political, social, economic and cultural rights; respect for our right to self-determination and to pursue our own ways of life.

Article 7: Respect for our autonomous forms of self-government, as differentiated political systems at the community, regional and other levels. This includes our right to control all economic activities in our territories.

Article 8: Respect for our customary laws and that they be incorporated in national and international law.

Article 9: Where the peoples so demand, nation states must comply with the different treaties, agreements, covenants, awards and other forms of legal recognition that have been signed with us Indigenous Peoples in the past, both in the colonial period and since independence, regarding our rights.

Article 10: An end to violence, slavery, debt peonage and land grabbing; the disbanding of all private armies and militias and their replacement by the rule of law and social justice; the means to use the law in our own defence, including the training of our people in the law.

Article 11: The approval and application of the Draft Declaration on the Rights of Indigenous Peoples, which must affirm and guarantee our right to self-determination, being developed by the United Nations, and the setting up of an effective international mechanism and tribunal to protect us against the violation of our rights and guarantee the application of the principles set out in this charter.

Article 12: There can be no rational or sustainable development of the forests and of our peoples until our fundamental rights as peoples are respected.

Territory

Article 13: Secure control of our territories, by which we mean a whole living system of continuous and vital connection between man and nature; expressed as our right to the unity and continuity of our ancestral domains; including the parts that have been usurped, those being reclaimed and those that we use; the soil, subsoil, air and water required for our self-reliance, cultural development and future generations.

Article 14: The recognition, definition and demarcation of our territories in accordance with our local and customary systems of ownership and use.

Article 15: The form of land tenure will be decided by the people themselves, and the territory should be held communally, unless the people decide otherwise.

Article 16: The right to the exclusive use and ownership of the territories which we occupy. Such territories should be inalienable, not subject to distraint and unnegotiable.

Article 17: The right to demarcate our territories ourselves and that these areas be officially recognized and documented.

Article 18: Legalize the ownership of lands used by non-Indigenous Peoples who live within and on the forests' margins in the areas that are available once title has been guaranteed to the Indigenous Peoples.

Article 19: Land reforms and changes in land tenure to secure the livelihoods of those who live outside the forests and indigenous territories, because we recognize that landlessness outside the forests puts heavy pressure on our territories and forests.

Decision-making

Article 20: Control of our territories and the resources that we depend on: all development in our areas should only go ahead with the free and informed consent of the indigenous people involved or affected.

Article 21: Recognition of the legal personality of our representative institutions and organizations, that defend our rights, and through them the right to collectively negotiate our future.

Article 22: The right to our own forms of social organization; the right to elect and revoke the authorities and government functionaries who oversee the territorial areas within our jurisdiction.

Development policy

Article 23: The right to be informed, consulted and above all to participate in the making of decisions on legislation or policies: and in the formulation, implementation or evaluation of any development project, be it at local, national or international level, whether private or of the state, that may affect our futures directly or indirectly.

Article 24: All major development initiatives should be preceded by social, cultural and environmental impact assessments, after consultation with local communities and Indigenous Peoples. All such studies and projects should be open to public scrutiny and debate, especially by the Indigenous Peoples affected.

Article 25: National or international agencies considering funding development projects which may affect us must set up tripartite commissions – including the funding agency, government representatives and our own communities as represented through our representative organizations to carry through the planning implementation, monitoring and evaluation of the projects.

Article 26: The cancellation of all mining concessions in our territories imposed without the consent of our representative organizations. Mining policies must prioritize, and be carried out under our control, to guarantee rational management and a balance with the environment. In the case of the extraction of strategic minerals (oil and radioactive minerals) in our territories, we must participate in making decisions during planning and implementation.

Article 27: An end to imposed development schemes and fiscal incentives or subsidies that threaten the integrity of our forests.

Article 28: A halt to all imposed programmes aimed at resettling our peoples away from their homelands.

Article 29: A redirection of the development process away from large-scale projects towards the promotion of small-scale initiatives controlled by our peoples. The priority for such initiatives is to secure our control over our territories and resources on which our survival depends. Such projects should be the cornerstone of all future development in the forests.

Article 30: The problems caused in our territories by international criminal syndicates trafficking in products from plants such as poppy and coca must be confronted by effective policies which involve our peoples in decision-making.

Article 31: Promotion of the health systems of the Indigenous Peoples, including the revalidation of traditional medicine, and the promotion of programmes of modern medicine and primary health care. Such programmes should allow us to have control over them, providing suitable training to allow us to manage them ourselves.

Article 32: Establishment of systems of bilingual and intercultural education. These must revalidate our beliefs, religious traditions, customs and knowledge; allowing our control over these programmes by the provision of suitable training, in accordance with our cultures, in order to achieve technical and scientific advances for our peoples, in tune with our own cosmo-visions, and as a contribution to the world community.

Article 33: Promotion of alternative financial policies that permit us to develop our community economies and develop mechanisms to establish fair prices for the products of our forests.

Article 34: Our policy of development is based, first, on guaranteeing our self-sufficiency and material welfare, as well as that of our neighbours; a full social and cultural development based on the values of equity, justice, solidarity and reciprocity, and a balance with nature. Thereafter, the generation of a surplus for the market must come from a rational and creative use of natural resources, developing our own traditional technologies and selecting appropriate new ones.

Forest policy

Article 35: Halt all new logging concessions and suspend existing ones that affect our territories. The destruction of forests must be considered a crime against humanity and a halt must be made to the various antisocial consequences, such as roads across indigenous cultivations, cemeteries and hunting zones; the destruction of areas used for medicinal plants and crafts; the erosion and compression of soil; the pollution of our environment; the corruption and enclave economy generated by the industry; the increase of invasions and settlement in our territories.

Article 36: Logging concessions on lands adjacent to our territories, or which have an impact on our environment, must comply with operating conditions – ecological, social, of labour, transport, health and others – laid down by the Indigenous Peoples, who should participate in ensuring that these are complied with. Commercial timber extraction should be prohibited in strategic and seriously degraded forests.

Article 37: The protection of existing natural forests should take priority over reforestation.

Article 38: Reforestation programmes should be prioritized on degraded lands, giving priority to the regeneration of native forests, including the recovery of all the functions of tropical forests, and not being restricted only to timber values.

Article 39: Reforestation programmes on our territories should be developed under the control of our communities. Species should be selected by us in accordance with our needs.

Biodiversity and conservation

Article 40: Programmes related to biodiversity must respect the collective rights of our peoples to cultural and intellectual property, genetic resources, gene banks, biotechnology and knowledge of biological diversity. This should include our participation in the management of any such project in our territories, as well as control of any benefits that derive from them.

Article 41: Conservation programmes must respect our rights to the use and ownership of the territories we depend on. No programmes to conserve biodiversity should be promoted on our territories without our free and informed consent as expressed through our representative organizations.

Article 42: The best guarantee of the conservation of biodiversity is that those who promote it should uphold our rights to the use, administration, management and control of our territories. We assert that guardianship of the different ecosystems should be entrusted to us, the Indigenous Peoples, given that we have inhabited them for thousands of years and our very survival depends on them.

Article 43: Environmental policies and legislation should recognize indigenous territories as effective 'protected areas', and give priority to their legal establishment as indigenous territories

Intellectual property

Article 44: Since we highly value our traditional technologies and believe that our biotechnologies can make important contributions to humanity, including 'developed' countries, we demand guaranteed rights to our intellectual property, and control over the development and manipulation of this knowledge.

Research

Article 45: All investigations in our territories should be carried out with our consent and under joint control and guidance according to mutual agreement; including the provision for training, publication and support for indigenous institutions necessary to achieve such control.

Institutions

Article 46: The international community, particularly the United Nations, must recognize us Indigenous Peoples as peoples, as distinct from other organized social movements, non-governmental organizations and independent sectors, and respect our right to participate directly and on the basis of equality, as Indigenous Peoples, in all fora, mechanisms, processes and funding bodies so as to promote and safeguard the future of the tropical forests.

Education

Article 47: The development of programmes to educate the general public about our rights as Indigenous Peoples and about the principles, goals and demands in this charter. For this we call on the international community for the necessary recognition and support.

Article 48: We Indigenous Peoples will use this charter as a basis for promoting our own local strategies for action.

Penang, Malaysia , 15 February 1992

1.3: The Kari-Oca Declaration

Preamble

The World Conference of Indigenous Peoples on Territory, Environment and Development (25-30 May 1992).

The Indigenous Peoples of the Americas, Asia, Africa, Australia, Europe and the Pacific, united in one voice at Kari-Oca Villages express our collective gratitude to the Indigenous Peoples of Brazil. Inspired by this historical meeting, we celebrate the spiritual unity of the Indigenous Peoples with the land and ourselves. We continue building and formulating our united commitment to save our Mother the Earth. We, the Indigenous Peoples, endorse the following declaration as our collective responsibility to carry our indigenous minds and voices into the future.

Declaration

We the Indigenous Peoples, walk to the future in the footprints of our ancestors.

From the smallest to the largest living being, from the four directions, from the air, the land and the mountains, the creator has placed us, the Indigenous Peoples, upon our Mother the Earth.

The footprints of our ancestors are permanently etched upon the land of our peoples.

We the Indigenous Peoples, maintain our inherent rights to self-determination.

We have always had the right to decide our own forms of government, to use our own laws to raise and educate our children, to our own cultural identity without interference.

We continue to maintain our rights as peoples despite centuries of deprivation, assimilation and genocide.

We maintain our inalienable rights to our lands and territories, to all our resources – above and below – and to our waters. We assert our ongoing responsibility to pass these on to the future generations.

We cannot be removed from our lands. We, the Indigenous Peoples, are connected by the circle of life to our land and environments.

We the Indigenous Peoples, walk to the future in the footprints of our ancestors.

Signed at Kari-Oca, Brazil, on the 30th day of May 1992

1.4: Indigenous Peoples' Earth Charter

Human rights and international law

1. We demand the right to life.
2. International law must deal with the collective human rights of Indigenous Peoples.
3. There are many international instruments which deal with the rights of individuals but there are no declarations to recognize collective human rights. Therefore, we urge governments to support the United Nations Working Group on Indigenous Peoples' (UNWGIP) Universal Declaration of Indigenous Rights, which is presently in draft form.
4. There exist many examples of genocide against Indigenous Peoples. Therefore, the convention against genocide must be changed to include the genocide of Indigenous Peoples.
5. The United Nations should be able to send Indigenous Peoples' representatives, in a peace keeping capacity, into indigenous territories where conflicts arise. This would be done at the request and consent of the Indigenous Peoples concerned.
6. The concept of *Terra Nullius* must be eliminated from international law usage. Many state governments have used internal domestic laws to deny us ownership of our own lands. These illegal acts should be condemned by the World.
7. Where small numbers of Indigenous Peoples are residing within state boundaries, so-called democratic countries have denied Indigenous Peoples the right of consent about their future, using the notion of majority rule to decide the future of Indigenous Peoples. Indigenous Peoples' right of consent to projects in their areas must be recognized.
8. We must promote the term 'Indigenous Peoples' at all fora. The use of the term 'Indigenous Peoples' must be without qualifications.
9. We urge governments to ratify International Labour Organization (ILO) Convention 169 to guarantee an international legal instrument for Indigenous Peoples. (Group 2 only.)
10. Indigenous Peoples' distinct and separate rights within their own territories must be recognized.
11. We assert our rights to free passage through state imposed political boundaries dividing our traditional territories. Adequate mechanisms must be established to secure this right.
12. The colonial systems have tried to dominate and assimilate our peoples. However, our peoples remain distinct despite these pressures.
13. Our indigenous governments and legal systems must be recognized by the United Nations, State governments and International legal instruments.
14. Our right to self-determination must be recognized.
15. We must be free from population transfer.
16. We maintain our right to our traditional way of life.
17. We maintain our right to our spiritual way of life.
18. We maintain the right to be free from pressures from multinational (transnational) corporations upon our lives and lands. All multinational (transnational) corporations which are encroaching upon indigenous lands should be reported to the United Nations Transnational Office.
19. We must be free from racism.
20. We maintain the right to decide the direction of our communities.
21. The United Nations should have a special procedure to deal with issues arising from violations of indigenous treaties.
22. Treaties signed between Indigenous Peoples and non-Indigenous Peoples must be accepted as treaties under international law.
23. The United Nations must exercise the right to impose sanctions against governments that violate the rights of Indigenous Peoples.
24. We urge the United Nations to include the issue of Indigenous Peoples in the agenda of the World Conference of Human Rights to be held in 1993. The work done so far by The United Nations Inter-American Commission of Human Rights and the Inter-American Institute of Human Rights should be taken into consideration.
25. Indigenous Peoples should have the right to their own knowledge, language, and culturally appropriate education, including bicultural and bilingual education. Through recognizing both formal and informal ways, the participation of family and community is guaranteed.

26. Our health rights must include the recognition and respect of traditional knowledge held by indigenous healers. This knowledge, including our traditional medicines and their preventive and spiritual healing power, must be recognized and protected against exploitation.
27. The World Court must extend its powers to include complaints by Indigenous Peoples.
28. There must be a monitoring system from this conference to oversee the return of delegates to their territories. The delegates should be free to attend and participate in International Indigenous Conferences.
29. Indigenous Women's Rights must be respected. Women must be included in all local, national, regional and international organizations.
30. The above mentioned historical rights of Indigenous Peoples must be guaranteed in national legislation.

* Please note, for the purposes of the Declaration and this statement any use of the term 'Indigenous Peoples' also includes Tribal Peoples.

Land and territories

31. Indigenous Peoples were placed upon our Mother, the earth by the Creator. We belong to the land. We cannot be separated from our lands and territories.
32. Our territories are living totalities in permanent vital relation between human beings and nature. Their possession produce the development of our culture. Our territorial property should be inalienable, unceasable and not denied title. Legal, economic and technical back up are needed to guarantee this.
33. Indigenous Peoples' inalienable rights to land and resources confirm that we have always had ownership and stewardship over our traditional territories. We demand that these be respected.
34. We assert our rights to demarcate our traditional territories. The definition of territory includes space (air), land, and sea. We must promote a traditional analysis of traditional land rights in all our territories.
35. Where indigenous territories have been degraded, resources must be made available to restore them. The recuperation of those affected territories is the duty of the respective jurisdiction in all nation states which can not be delayed. Within this process of recuperation the compensation for the historical ecological debt must be taken into account. Nation states must revise in depth their agrarian, mining and forestry policies.
36. Indigenous Peoples reject the assertion of non-indigenous laws onto our lands. States cannot unilaterally extend their jurisdiction over our lands and territories. The concept of **Terra Nullius** should be forever erased from the law books of states.
37. We, as Indigenous Peoples, must never alienate our lands. We must always maintain control over the land for future generations.
38. If a non indigenous government, individual or corporation wants to use our lands, then there must be a formal agreement which sets out the terms and conditions. Indigenous Peoples maintain the right to be compensated for the use of their lands and resources.
39. Traditional indigenous territorial boundaries, including the waters, must be respected.
40. There must be some control placed upon environmental groups who are lobbying to protect our territories and the species within those territories. In many instances, environmental groups are more concerned about animals than human beings. We call for Indigenous Peoples to determine guidelines prior to allowing environmental groups into their territories.
41. Parks must not be created at the expense of Indigenous Peoples. There is no way to separate Indigenous Peoples from their lands.
42. Indigenous Peoples must not be removed from their lands in order to make it available to settlers or other forms of economic activity on their lands.
43. In many instances, the numbers of Indigenous Peoples have been decreasing due to encroachment by non-Indigenous Peoples.
44. Indigenous Peoples should encourage their peoples to cultivate their own traditional forms of products rather than to use imported exotic crops which do not benefit local peoples.
45. Toxic wastes must not be deposited in our areas. Indigenous Peoples must realize that chemicals, pesticides and hazardous wastes do not benefit the peoples.
46. Traditional areas must be protected against present and future forms of environmental degradation.
47. There must be a cessation of all uses of nuclear material.
48. Mining of products for nuclear production must cease.
49. Indigenous lands must not be used for the testing or dumping of nuclear products.
50. Population transfer policies by state governments in our territories are causing hardship. Traditional lands are lost and traditional livelihoods are being destroyed.

51. Our lands are being used by state governments to obtain funds from the World Bank, the International Monetary Fund, the Asian Pacific Development Bank and other institutions which have led to a loss of our lands and territories.
52. In many countries our lands are being used for military purposes. This is an unacceptable use of the lands.
53. The colonizer governments have changed the names of our traditional and sacred areas. Our children learn these foreign names and start to lose their identity. In addition, the changing of the name of a place diminishes respect for the spirits which reside in those areas.
54. Our forests are not being used for their intended purposes. The forests are being used to make money.
55. Traditional activities, such as making pottery, are being destroyed by the importation of industrial goods. This impoverishes the local peoples.

Biodiversity and conservation

56. The Vital Circles are in a continuous interrelation in such a way that the change of one of its elements affects the whole.
57. Climatic changes affect Indigenous Peoples and all humanity. In addition, ecological systems and their rhythms are affected which contributes to the deterioration of our quality of life and increases our dependency.
58. The forests are being destroyed in the name of development and economic gains without considering the destruction of ecological balance. These activities do not benefit human beings, animals, birds and fish. The logging concessions and incentives to the timber, cattle and mining industries affecting the ecosystems and the natural resources should be cancelled.
59. We value the efforts of protection of the Biodiversity but we reject to be included as part of an inert diversity which pretend to be maintained for scientific and folkloric purposes.
60. The Indigenous Peoples' strategies should be kept in a reference framework for the formulation and application of national policies on environment and biodiversity.

Development strategies

61. Indigenous Peoples must consent to all projects in our territories. Prior to consent being obtained the peoples must be fully and entirely involved in any decisions. They must be given all the information about the project and its effects. Failure to do so should be considered a crime against the Indigenous Peoples. The person or persons who violate this should be tried in a world tribunal within the control of Indigenous Peoples set for such a purpose. This could be similar to the trials held after World War II.
62. We have the right to our own development strategies based on our cultural practices and with a transparent, efficient and viable management and with economical and ecological viability.
63. Our development and life strategies are obstructed by the interests of governments and big companies and by neoliberal policies. Our strategies have, as a fundamental condition, the existence of International relationships based on justice, equity and solidarity between human beings and nations.
64. Any development strategy should prioritize the elimination of poverty, the climatic guarantee, the sustainable manageability of natural resources, the continuity of democratic societies and the respect of cultural differences.
65. The global environmental facility should assign at best 20% for Indigenous Peoples' strategies and programs of environmental emergency, improvement of life quality, protection of natural resources and rehabilitation of ecosystems. This proposal in the case of South America and the Caribbean should be concrete in the Indigenous development fund as a pilot experience in order to be extended to the Indigenous Peoples of other regions and continents.
66. The concept of development has meant the destruction of our lands. We reject the current definition of development as being useful to our peoples. Our cultures are not static and we keep our identity through a permanent recreation of our life conditions; but all of this is obstructed in the name of so-called developments.
67. Recognizing Indigenous Peoples' harmonious relationship with Nature, indigenous sustainable development strategies and cultural values must be respected as distinct and vital sources of knowledge.
68. Indigenous Peoples have been here since the time before time began. We have come directly from the Creator. We have lived and kept the Earth as it was on the First Day. Peoples who do not belong to the land must go out from the lands because those things (so called 'Development' on the land) are against the laws of the Creator.
69. (a) In order for Indigenous Peoples to assume control, management and administration of their resources and territories, development

projects must be based on the principles of self-determination and self-management.

(b) Indigenous Peoples must be self-reliant.

70. If we are going to grow crops, we must feed the peoples. It is not appropriate that the lands be used to grow crops which do not benefit the local peoples.

    (a) Regarding indigenous policies, State Government must cease attempts of assimilation and integration.

    (b) Indigenous Peoples must consent to all projects in their territories. Prior to consent being obtained, the peoples must be fully and entirely involved in any decisions. They must be given all the information about the project and its effects. Failure to do so should be considered a crime against Indigenous Peoples. The person or persons responsible should be tried before a World Tribunal, with a balance of Indigenous Peoples set up for such a purpose. This could be similar to the Trials held after the second World War.

71. We must never use the term 'land claims'. It is the non-Indigenous Peoples which do not have any land. All the land is our land. It is non-Indigenous Peoples who are making claims to our lands. We are not making claims to our lands.

72. There should be a monitoring body within the United Nations to monitor all the land disputes around the World prior to development.

73. There should be a United Nations Conference on the topic of 'Indigenous Lands and Development'.

74. Non-Indigenous Peoples have come to our lands for the purpose of exploiting these lands and resources to benefit themselves, and to the impoverishment of our peoples. Indigenous Peoples are victims of development. In many cases Indigenous Peoples are exterminated in the name of a development program. There are numerous examples of such occurrences.

75. Development that occurs on indigenous lands, without the consent of Indigenous Peoples, must be stopped.

76. Development which is occurring on indigenous lands is usually decided without local consultation by those who are unfamiliar with local conditions and needs.

77. The Eurocentric notion of ownership is destroying our peoples. We must return to our own view of the world, of the land and of development. The issue cannot be separated from Indigenous Peoples' rights.

78. There are many different types of so-called development: road construction, communication facilities such as electricity, telephones. These allow developers easier access to the areas, but the effects of such industrialization destroy the lands.

79. There is a world wide move to remove Indigenous Peoples from their lands and place them in villages. The relocation from the traditional territories is done to facilitate development.

80. It is not appropriate for governments or agencies to move into our territories and to tell our peoples what is needed.

81. In many instances, the state governments have created artificial entities such as 'district councils' in the name of the state government in order to deceive the international community. These artificial entities are then consulted about development in the area. The state government, then, claims that Indigenous Peoples were consulted about the project. These lies must be exposed to the international community.

82. There must be an effective network to disseminate material and information between Indigenous Peoples. This is necessary in order to keep informed about the problems of other Indigenous Peoples.

83. Indigenous Peoples should form and direct their own environmental network.

Culture, science and intellectual property

84. We feel the Earth as if we are within our mother. When the Earth is sick and polluted, human health is impossible. To heal ourselves, we must heal the Planet and to heal the Planet, we must heal ourselves.

85. We must begin to heal from the grassroots level and work towards the international level.

86. The destruction of the culture has always been considered an internal, domestic problem within national states. The United Nations must set up a tribunal to review the cultural destruction of the Indigenous Peoples.

87. We need to have foreign observers come into our indigenous territories to oversee national state elections to prevent corruption.

88. The human remains and artefacts of Indigenous Peoples must be returned to their original peoples.

89. Our sacred and ceremonial sites should be protected and considered as the patrimony of Indigenous Peoples and humanity. The establishment of a set of legal and operational instruments at both national and international levels would guarantee this.

90. The use of existing indigenous languages is our right. These languages must be protected.

91. States that have outlawed indigenous languages and their alphabets should be censored by the United Nations.
92. We must not allow tourism to be used to diminish our culture. Tourists come into the communities and view the people as if Indigenous Peoples were part of a zoo. Indigenous Peoples have the right to allow or to disallow tourism within their areas.
93. Indigenous Peoples must have the necessary resources and control over their own education systems.
94. Elders must be recognized and respected as teachers of the young people.
95. Indigenous wisdom must be recognized and encouraged.
96. The traditional knowledge of herbs and plants must be protected and passed onto future generations.
97. Traditions cannot be separated from land, territory or science.
98. Traditional knowledge has enabled Indigenous Peoples to survive.
99. The usurping of traditional medicines and knowledge from Indigenous Peoples should be considered a crime against peoples.
100. Material culture is being used by the non-indigenous to gain access to our lands and resources, thus destroying our cultures.
101. Most of the media at this conference were only interested in the pictures which will be sold for profit. This is another case of exploitation of Indigenous Peoples. This does not advance the cause of Indigenous Peoples.
102. As creators and carriers of civilizations which have given and continue to share knowledge, experience and values with humanity, we require that our right to intellectual and cultural properties be guaranteed and that the mechanism for each implementation be in favour of our peoples and studied in depth and implemented. This respect must include the right over genetic resources, gene banks, biotechnology and knowledge of biodiversity programs.
103. We should list the suspect museums and institutions that have misused our cultural and intellectual properties.
104. The protection, norms and mechanisms of artistic and artisan creation of our peoples must be established and implemented in order to avoid plunder, plagiarism, undue exposure and use.
105. When Indigenous Peoples leave their communities, they should make every effort to return to the community.
106. In many instances, our songs, dances and ceremonies have been viewed as the only aspects of our lives. In some instances, we have been asked to change a ceremony or a song to suit the occasion. This is racism.
107. At local, national and international levels, governments must commit funds to new and existing resources for education and training for Indigenous Peoples, to achieve their sustainable development, to contribute and to participate in sustainable and equitable development at all levels. Particular attention should be given to indigenous women, children and youth.
108. All kinds of folkloric discrimination must be stopped and forbidden.
109. The United Nations should promote research into indigenous knowledge and develop a network of indigenous sciences.

1.5: The Mataatua Declaration on the Cultural and Intellectual Property Rights of Indigenous Peoples

June 1993

In recognition that 1993 is the United Nations International Year for the World's Indigenous Peoples:

The Nine Tribes of Mataatua in the Bay of Plenty Region of Aotearoa New Zealand convened the First International Conference on the Cultural and Intellectual Property Rights of Indigenous Peoples (12–18 June 1993, Whakatane).

Over 150 Delegates from fourteen countries attended, including indigenous representatives from Ainu (Japan), Australia, Cook Islands, Fiji, India, Panama, Peru, Philippines, Surinam, USA and Aotearoa.

The Conference met over six days to consider a range of significant issues, including; the value of indigenous knowledge, biodiversity and biotechnology, customary environmental management, arts, music, language and other physical and spiritual cultural forms. On the final day, the following Declaration was passed by the Plenary.

Preamble

Recognizing that 1993 is the United Nations International Year for the World's Indigenous Peoples;

Reaffirming the undertaking of United Nations Member States to:

'Adopt or strengthen appropriate policies and/or legal instruments that will protect indigenous intellectual and cultural property and the right to preserve customary and administrative systems and practices.' United Nations Conference on Environmental Development; UNCED Agenda 21 (26.4b);

Noting the Working Principles that emerged from the United Nations Technical Conference on Indigenous Peoples and the Environment in Santiago, Chile 18–22 May 1992 (E/CN.4/Sub.2/1992/31); and

Endorsing the recommendations on Culture and Science from the World Conference of Indigenous Peoples on Territory, Environment and Development, Kari-Oca, Brazil, 25-30 May 1992;

We

Declare that Indigenous Peoples of the world have the right to self determination: and in exercising that right must be recognized as the exclusive owners of their cultural and intellectual property;

Acknowledge that Indigenous Peoples have a commonality of experiences relating to the exploitation of their cultural and intellectual property;

Affirm that the knowledge of the Indigenous Peoples of the world is of benefit to all humanity;

Recognize that Indigenous Peoples are capable of managing their traditional knowledge themselves, but are willing to offer it to all humanity provided their fundamental rights to define and control this knowledge are protected by the international community;

Insist that the first beneficiaries of indigenous knowledge (cultural and intellectual property rights) must be the direct indigenous descendants of such knowledge;

Declare that all forms of discrimination and exploitation of Indigenous Peoples, indigenous knowledge and indigenous cultural and intellectual property rights must cease.

1. Recommendations to indigenous peoples

In the development of policies and practices, Indigenous Peoples should:

1.1 Define for themselves their own intellectual and cultural property.

1.2 Note that existing protection mechanisms are insufficient for the protection of Indigenous Peoples' Intellectual and Cultural Property Rights.

1.3 Develop a code of ethics which external users must observe when recording (visual, audio, written) their traditional and customary knowledge.

1.4 Prioritize the establishment of indigenous education, research and training centres to promote their knowledge of customary environmental and cultural practices.

1.5 Re-acquire traditional indigenous lands for the purpose of promoting customary agricultural production.

1.6 Develop and maintain their traditional practices and sanctions for the protection, preservation and revitalization of their traditional intellectual and cultural properties.

1.7 Assess existing legislation with respect to the protection of antiquities.

1.8 Establish an appropriate body with appropriate mechanisms to:
   (a) preserve and monitor the commercialism or otherwise of indigenous cultural properties in the public domain;
   (b) generally advise and encourage Indigenous Peoples to take steps to protect their cultural heritage;
   (c) allow a mandatory consultative process with respect to any new legislation affecting Indigenous Peoples' cultural and intellectual property rights.

1.9 Establish international indigenous information centres and networks.

1.10 Convene a Second International Conference (Hui) on the Cultural and Intellectual Property Rights of Indigenous Peoples to be hosted by the Coordinating Body for the Indigenous Peoples Organizations of the Amazon Basin (COICA).

2. Recommendations to states, national and international agencies

In the development of policies and practices, States, National and International Agencies must:

2.1 Recognize that Indigenous Peoples are the guardians of their customary knowledge and have the right to protect and control dissemination of that knowledge.

2.2 Recognize that Indigenous Peoples also have the right to create new knowledge based on cultural traditions.

2.3 Note that existing protection mechanisms are insufficient for the protection of Indigenous Peoples' Cultural and Intellectual Property Rights.

2.4 Accept that the cultural and intellectual property rights of Indigenous Peoples are vested with those who created them.

2.5 Develop in full co-operation with Indigenous Peoples an additional cultural and intellectual property rights regime incorporating the following:
   - collective (as well as individual) ownership and origin

- retroactive coverage of historical as well as contemporary works
- protection against debasement of culturally significant items
- co-operative rather than competitive framework
- first beneficiaries to be the direct descendants of the traditional guardians of that knowledge
- multi-generational coverage span

Biodiversity and Customary Environmental Management

2.6 Indigenous flora and fauna is inextricably bound to the territories of indigenous communities and any property right claims must recognize their traditional guardianship.

2.7 Commercialization of any traditional plants and medicines of Indigenous Peoples must be managed by the Indigenous Peoples who have inherited such knowledge.

2.8 A moratorium on any further commercialization of indigenous medicinal plants and human genetic materials must be declared until indigenous communities have developed appropriate protection mechanisms.

2.9 Companies, institutions both governmental and private must not undertake experiments or commercialization of any biogenetic resources without the consent of the appropriate Indigenous Peoples.

2.10 Prioritize settlement of any outstanding land and natural resources claims of Indigenous Peoples for the purpose of promoting customary, agricultural and marine production.

2.11 Ensure current scientific environmental research is strengthened by increasing the involvement of indigenous communities and of customary environmental knowledge.

Cultural objects

2.12 All human remains and burial objects of Indigenous Peoples held by museums and other institutions must be returned to their traditional areas in a culturally appropriate manner.

2.13 Museums and other institutions must provide, to the country and Indigenous Peoples concerned, an inventory of any indigenous cultural objects still held in their possession.

2.14 Indigenous cultural objects held in museums and other institutions must be offered back to their traditional owners.

3. Recommendations to the United Nations

In respect for the rights of Indigenous Peoples, the United Nations should:

3.1 Ensure the process of participation of Indigenous Peoples in United Nations fora is strengthened so their views are fairly represented.

3.2 Incorporate the Mataatua Declaration in its entirety in the United Nations Study on Cultural and Intellectual Property of Indigenous Peoples.

3.3 Monitor and take action against any States whose persistent policies and activities damage the cultural and intellectual property rights of Indigenous Peoples.

3.4 Ensure that Indigenous Peoples actively contribute to the way in which indigenous cultures are incorporated into the 1995 United Nations International Year of Culture.

3.5 Call for an immediate halt to the ongoing Human Genome Diversity Project (HUGO) until its moral, ethical, socio-economic, physical and political implications have been thoroughly discussed, understood and approved by Indigenous Peoples.

4. Conclusion

The United Nations, International and National Agencies and States must provide additional funding to indigenous communities in order to implement these recommendations.

1.6: Recommendations from the conference 'Voices of the Earth: Indigenous Peoples, New Partners and the Right to Self-determination in Practice'

Amsterdam, Netherlands, 10-11 November 1993

Preamble

We, the Indigenous Peoples assembled at the Congress 'Voices of the Earth: Indigenous Peoples, New Partners, the Right to Self-determination in Practice' hereby declare the results of our deliberations as an important contribution and milestone in our struggle for promotion, protection and recognition of our inherent rights.

We, the indigenous participants consider the outcome of our meeting as a continuation of ALL indigenous conferences during this important United Nations Year of the World's Indigenous Peoples.

We, the Indigenous Peoples express our deep gratitude to the moral and political support of those who have contributed to this Congress.

As we continue to walk to the future in the footprints of our ancestors, we spoke in Amsterdam on November 10 and 11, 1993.

Recommendations

Political rights

1. The right of Indigenous Peoples to self-determination as stated in the Preamble of the Kari-Oca Declaration and Indigenous Peoples' Earth Charter and in Article 3 of the Draft UN Declaration on the Rights of Indigenous Peoples must be fully recognized.
2. Indigenous Peoples are clearly to be distinguished from minorities. Therefore the protection of their rights cannot be adequately considered under Article 27 of the Covenant on Civil and Political Rights.
3. Procedures should be developed for Indigenous Peoples to bring conflicts with national government concerning political self-determination and other questions before an independent international body such as the International Court of Justice. The European Community, the Dutch Government and all other governments should take the initiative to work toward the establishment of those procedures.
4. Indigenous Peoples should be provided with legal and technical assistance, at their request, to effectively defend their rights.
5. The European Community, the Dutch Government and all other governments should fully support the UN Draft Declaration on the Rights of Indigenous Peoples (UN doc.E/CN.4/Sub.2/1993/29) that will be for adoption by the UN Working Group on Indigenous Populations at its 1994 session.
6. The European Community, the Dutch Government and all other governments should work toward facilitating open access and full participation for Indigenous Peoples in the entire process of debate concerning the adoption of the UN Declaration and in all other forums discussing indigenous issues.
7. The European Community, the Dutch Government and all other governments should support the designation of an International Decade of Indigenous Peoples by the UN General Assembly. This Decade should start in 1995 with a preparatory year in 1994.
8. The European Community, the Dutch Government and all other governments should take the initiative for the implementation of the recommendation of the Vienna World Conference on Human Rights that a permanent forum be established in the United Nations for the rights of Indigenous Peoples, in co-operation with the representatives of Indigenous Peoples.
9. The European Community should also recognize the full right to self-determination of the Indigenous Peoples presently living on European Community territory (New Caledonia, French Polynesia and French Guyana).

Economic rights

The effective enjoyment of the economic rights of Indigenous Peoples depends on a recognition of their right to self-determination.

Territories:

1. Indigenous Peoples' rights to their territories, meaning full ownership of their lands and natural resources above and below the earth and waters, must be recognized.

Control:

2. Indigenous Peoples' rights to control the use of resources in their territories must be fully recognized.

Trade offs:

3. These rights are non negotiable and cannot be traded off in the name of development of the nation state or other sectors. However, Indigenous Peoples may choose to promote the use of their resources in ways that benefit others: they need to be assured that they enter such discussions from a position of power.

Private sector:

4. i) The private sector must assume responsibility for its activities. A wider notion of profit should be a condition of investment practice, giving emphasis to the quality of life, not just the quantity of money.
   (ii) NGOs monitoring transnational corporations should focus more on Indigenous Peoples and share information widely with them.
   (iii) In developing Codes of Conduct, companies must engage in dialogue with Indigenous Peoples and create mechanisms that allow public scrutiny of their adherence to these codes.
   (iv) An organization parallel to the International Centre for the Settlement of Investment Disputes must be established to resolve

conflicts between Transnational Corporations and Indigenous Peoples.

Role of the State:

5. States should provide adequate assistance to Indigenous Peoples to enable them to develop their own economic base and power. Control over this process must be vested with the Indigenous Peoples concerned to avoid the creation of dependency.

Environment:

6. Bearing in mind the two major international human rights covenants of December 1966, according to which, Part 1, Article 1 in both covenants, no peoples may under any circumstances be deprived of its own means of subsistence,

   Conscious that the 1992 Rio Summit recognized the valuable role of Indigenous Peoples in maintaining a sustainable use of natural resources, and underlined in Principle 22, the pressing need for Indigenous Peoples' active participation in environmental management,

   Acknowledging the Brundtland Commission report's recommendation of 1987 about the empowerment of vulnerable groups,

   Aware that the World Conservation Strategy of 1991, 'Caring for the Earth', supports a special role for Indigenous Peoples in global efforts for a sound environment,

   Mindful that the World Conservation Union (IUCN) on its 18th General Assembly unanimously adopted two resolutions supporting the Indigenous Peoples' cause, including their right to use nature's resources wisely,

   Conscious of the Biodiversity Convention and ILO Convention 169, both of which lend support to Indigenous Peoples and their role in sustainable development,

   Pointing to the fact that as a general rule, ecosystems that appear as the most sound, are also those which are under indigenous control,

   Now, therefore, the 'Voices of the Earth' Congress, assembled in Amsterdam, calls on governments;

   (i) to heed the concerns of Indigenous Peoples world-wide,

   (ii) to give effect in their respective national policies to the above cited international instruments to which they have given their assent,

   (iii) to properly protect the market access for Indigenous Peoples' products derived from a sustainable and wise use of nature, and

   (iv) to give financial support to the UN's decade for Indigenous Peoples.

International legislation:

7. States should recognize the Declaration on the Rights of Indigenous Peoples as presently drafted. It was suggested that an ombudsman be nominated to oversee the adherence of States to this Declaration. An independent tribunal might also review adherence to the Declaration.

Demilitarization:

8. There should be a demilitarization of indigenous territories. In this respect it is the special responsibility of the Dutch Government to immediately stop the low-level flying activities of the Royal Dutch Air Force above the territories of the Innu people in Canada. Compulsory military service for indigenous people must be abolished.

Dutch Government's responsibilities:

9. In addition to observing the above recommendations, the Dutch Government is urged to press for an enhanced allocation to Indigenous Peoples of the resources of the UN agencies and other multilateral bodies.

Cultural, scientific and intellectual property

1. All relevant agencies and programmes of the Dutch Government, European Community and the United Nations (e.g. World Bank, WIPO, UPOV UNCTAD, UNEP, UNDP, Human Rights Center, ILO, GATT etc.) should develop a common policy, based on dialogue with and consent of Indigenous Peoples, on how protection of and compensation for indigenous intellectual, scientific, and cultural property can be established and effected.
2. A Council on Indigenous Intellectual, Cultural and Scientific Property Rights, composed of indigenous people, should be established, funded and given special international status in order to:

   (a) develop educational materials on intellectual, cultural and scientific property rights;

   (b) develop mechanisms for protection and compensation;

   (c) advise indigenous and traditional communities on legal and political actions;

   (d) monitor unethical activities by individuals, institutions and governments that are misusing intellectual, scientific and cultural property;

(e) develop mechanisms for enforcement of rules, regulations and laws for protection and compensation, including legal advice and counsel; and

(f) establish a network to exchange information about successful and unsuccessful attempts by local communities to secure their rights.

3. Governmental and non-governmental organizations, as well as scientific and professional groups, should develop Codes of Ethics and Conduct regarding respect for Indigenous Peoples and their intellectual, cultural and scientific property. Funding agencies should require that effective measures for protection and compensation for intellectual, cultural and scientific property be an integral part of all projects and such measures be a requirement for funding.

4. Rights of Indigenous Peoples to their traditional properties supersede the rights of anyone, including the rights of museums to possess these properties. No international or national agencies may infringe on the right of Indigenous Peoples to refuse to share their intellectual, cultural and scientific properties.
   Museums all over the world should co-operate fully with Indigenous Peoples to re-identify their cultural heritage and recognize their right to repossess it.

5. All governments, international institutions, non-governmental organizations and Indigenous Peoples are called upon to establish the 'University of the Earth' which shall incorporate the values and the knowledge of both indigenous and non-Indigenous Peoples. This University need not have a specific location but would take the form of a Global Network of journalists, farmers, foresters, engineers, shaman, hunters, scientists, artists and others who will exchange information through journals, television, films, videos, conferences and other forms of mass-media. The mission of this 'University of the Earth' will be to enhance all peoples' respect for and knowledge of the Earth. The European Community and the Dutch Government are called upon to strengthen Indigenous Peoples newspapers and other forms of information dissemination.

Right to self-development

1. Effective enjoyment of Indigenous Peoples' right to self-development depends on the recognition of the right of Indigenous Peoples to self-determination.
2. International institutions and funding agencies should adopt [sic.] their requirements, structures and policies to the cultures, needs and aspirations of Indigenous Peoples.
3. Indigenous Peoples must have full control over the planning, implementation, monitoring, evaluation and follow-up of projects affecting them.
4. Indigenous Peoples' knowledge and culture should be fully taken into consideration before entering into development relations with Indigenous Peoples.
5. Results of studies, carried out with the full participation of Indigenous Peoples, concerning the impacts of development projects on Indigenous Peoples should be carefully taken into account before implementing a proposed project.
6. The European Community, the Dutch Government and all other governments should respect the Indigenous Peoples' social and political organizations, and assist them to give these institutions an impulse by institution building for the sake of sustainable, 'grassroots' development.
7. A code of conduct for international institutions such as the World Bank, the IMF, the EC Development Fund and the UNDP, must be established in collaboration with Indigenous Peoples to ensure that funding for development activities does not infringe on the territorial and environmental integrity of Indigenous Peoples.
8. The European community, the Dutch Government and all other governments should take into consideration the actual situation of Indigenous Peoples in developed countries. Indigenous Peoples in developed countries should not be overlooked or discriminated against by funding institutions because they may be in circumstances similar to those in developing countries.
9. The European Community, the Dutch Government and all other governments, international institutions and funding agencies should take into consideration the specific interests of indigenous women and children in the planning and implementation of development projects.

1.7: Julayinbul Statement on Indigenous Intellectual Property Rights

On 27 November 1993, at Jingarrba, North-Eastern coastal region of the continent of Australia, it was agreed that:

Indigenous Peoples and Nations share a unique spiritual and cultural relationship with Mother Earth which recognizes the inter-dependence of the total environment and is governed by the natural laws which determine our perceptions of intellectual property.

Inherent in these laws and integral to that relationship is the right of Indigenous Peoples and Nations to reaffirm their right to define for themselves their own intellectual property, acknowledging their own self-determination and the uniqueness of their own particular heritage.

Within the context of this statement, Indigenous Peoples and Nations also declare that we are capable of managing our intellectual property ourselves, but are willing to share it with all humanity provided that our fundamental rights to define and control this property are recognized by the international community.

Aboriginal Common Law and English/Australian Common Law are parallel and equal systems of law.

Aboriginal intellectual property, within Aboriginal Common Law, is an inherent inalienable right which cannot be terminated, extinguished, or taken.

Any use of the intellectual property of Aboriginal Nations and Peoples may only be done in accordance with Aboriginal Common Law, and any unauthorized use is strictly prohibited.

Just as Aboriginal Common Law has never sought to unilaterally extinguish English/ Australian Common Law, so we expect English/Australian Common Law to reciprocate.

We, the delegates assembled at this conference urge Indigenous Peoples and Nations to develop processes and strategies acceptable to them to facilitate the practical application of the above principles and to ensure the dialogue and negotiation which are envisaged by the principles.

We also call on governments to review legislation and non-statutory policies which currently impinge upon or do not recognize indigenous intellectual property rights. Where policies, legislation and international conventions currently recognize these rights, we require that they be implemented.

1.8: Declaration of the international meeting around the First World Gathering of Elders and Wise Persons of Diverse Indigenous Traditions

Amautic Janajpacha Community, Cochabamba, Bolivia, 1-11 August 1994

Invocation

Nourished by the luminous rays of Tata Inti, sheltered by the purity of the Pachamama, illuminated by the wisdom of our beloved elders, inspired by the joyful enthusiasm of our younger brothers and sisters, we have come from diverse indigenous traditions and countries of the world, to this meeting with Mother Nature in our hearts. With the presence and permission of the Great Spirit, we open our most intimate and beautiful voices so that with the help of our brother the Wind, this message of brotherhood and solidarity may be propagated to the four cardinal directions. With the conviction that this will flourish vigorously in every humble and open heart, which is building the Path toward a human being interwoven and One with Nature.

Those signatories here below, united around the 'First World Gathering of Elders and Wise Persons of Diverse Indigenous Traditions' taking place in the 'Amautic Janajpacha Community' in Cochabamba, Bolivia, between the 1st and 11th August 1994, for the purpose of sharing experiences and teachings under the inspiration of the wise and divine Mother Nature, which is enshrined in the sacred shamanic wisdom of the ancient peoples, and under the spiritual guidance of the Inca-Tiwanakota tradition, with the aim of analysing the grave problems which confront humanity and the Earth, as well as looking for solutions based in love, we declare the following:

1. Evoking the words of the North American Indian Chief Seattle, who said: 'This we know, the Earth does not belong to man; man belongs to the Earth. We are all interwoven like the blood that unites a family. All is connected... everything that happens to the Earth happens to the children of the Earth', we want to reaffirm that human beings are only the custodians of the riches and natural resources which Mother Nature has nobly and generously provided for the fulfilment of our basic needs. Under no circumstances can we have a patent to destructive exploitation of nature; this would be irreverent and suicidal.
2. For this reason we reject all attempts to privatize for monopolies of scientific, commercial or industrial use of the riches and primary natural resources that Mother Nature has made available for the use of all humankind. We reject in particular, as abhorrent and absolutely at odds with morals and ethics, the patenting of living organisms to which the industry of genetic engineering wishes to lead us, with their scientific pretensions. Even less can they pretend to monopolize life. Life is a dynamic process which depends upon the divine breath, unable to be captured, therefore for the pretensions of ownership of a mistaken and irresponsible ego or human greed.
3. We emphatically reject the attempts to instil in Indigenous Peoples, values of property rights or the exclusive use of common goods or knowl-

edge, contributed by the natural heritage, the result of the joint efforts of humans and Nature, as an expression of a universal wisdom. In our custody and use of the resources of Mother Nature, we consider it to be of primary importance, above all, to think in terms of duties, rather than rights: duties of caring, duties of not harming, duties of sharing with fellow human beings in need; duties ultimately of reverence to the Creator and His designs, so that love and sharing may reign in society, in fulfilment of our superior evolution as human beings, and not egotism and competition in a race toward destruction.

4. To the attempts to privatize common goods or knowledge of our Mother Earth, we respond with a call for the popularization of the reverent use of these goods and knowledge, for the benefit of all humankind. In this context we call for the propagation throughout the world of this indigenous knowledge concerning the use of prodigious plants, both nutritional and medicinal, and that their cultivation be propagated to any parcel of land available, in every home. This action will be the best way to respond to the attempts of merchants to patent or expropriate Nature.
5. They can never patent or expropriate the sacred indigenous cultures which have inspired or guided our action. They can only take isolated elements of the same for their pretensions of commercialists, inventive appropriation, for example, the 'active ingredients' or chemical substances of our medicinal plants, which will never have the effectiveness of the plants utilized as a whole, and in harmony with the natural and cultural surroundings, as the indigenous cultures use them. Moreover, this fragmental use of medicinal plants may cause damage to users or consumers.
6. Definitely, the decadent abuses of human greed and irresponsible destruction which we are faced with, are part of the collapse of an order, which has brought untold damage to Mother Nature and to the physical, psychic and spiritual integrity of human beings. We trust that the regeneration of humankind and of the planet, which is now being manifested, will ultimately prevail, so that the forces of Light, Love and Humility will overcome the shadows, hatred and irreverence. This is what has been envisioned by the wise indigenous elders, and in them we take refuge, with hope and determination. We are resolved to do that which we must for the healing of humanity and of the planet.
7. We reject the merchants and irresponsible distortionists of the sacred and authentic wisdom of our indigenous ancestors. We respect the cultural diversity in which such wisdom has been manifested in different parts of the world., since we know that these manifestations come from a common source. Let us be instruments of a convergent effort, destined to add to and not subtract from the aspirations of the peoples.

   The efforts of all, above our human limitations and short-sightedness, is imperative. With humility, reverence, and foremost with self-responsibility and personal example in daily life, we can succeed.

   May there be many other gatherings in the world such as the one that has taken place here in Cochabamba, so that indigenous wisdom can propagate to the four winds and fulfil its contribution to the salvation of humankind and the planet, envisioned by our venerable elders under the inspiration of God and Mother Nature.

Movimiento Pachamama Universal
Comunidad Janajpacha

1.9: Statement from the COICA/UNDP Regional Meeting on Intellectual Property Rights and Biodiversity

Santa Cruz de la Sierra, Bolivia, 28–30 September 1994

I. Basic points of agreement

1. Emphasis is placed on the significance of the use of intellectual property systems as a new formula for regulating North-South economic relations in pursuit of colonialist interests.
2. For Indigenous Peoples, the intellectual property system means legitimation of the misappropriation of our peoples' knowledge and resources for commercial purposes.
3. All aspects of the issue of intellectual property (determination of access to national resources, control of the knowledge or cultural heritage of peoples, control of the use of their resources and regulation of the terms of exploitation) are aspects of self-determination. For Indigenous Peoples, accordingly, the ultimate decision on this issue is dependent on self-determination. Positions adopted under a trusteeship regime will be of a short-term nature.
4. Biodiversity and a people's knowledge are concepts inherent in the idea of indigenous territo-

riality. Issues relating to access to resources have to be viewed from this standpoint.

5. Integral indigenous territoriality, its recognition (or restoration) and its reconstitution are prerequisites for enabling the creative and inventive genius of each Indigenous People to flourish and for it to be meaningful to speak of protecting such peoples. The protection, reconstitution and development of indigenous knowledge systems call for additional commitments to the effort to have them reappraised by the outside world.
6. Biodiversity and the culture and intellectual property of a people are concepts that mean indigenous territoriality. Issues relating to access to resources and others have to be viewed from this standpoint.
7. For members of Indigenous Peoples, knowledge and determination of the use of resources are collective and intergenerational. No indigenous population, whether of individuals or communities, nor the Government, can sell or transfer ownership of resources which are the property of the people and which each generation has an obligation to safeguard for the next.
8. Prevailing intellectual property systems reflect a conception and practice that is:
    - colonialist, in that the instruments of the developed countries are imposed in order to appropriate the resources of Indigenous Peoples;
    - racist, in that it belittles and minimizes the value of our knowledge systems;
    - usurpatory, in that it is essentially a practice of theft.
9. Adjusting indigenous systems to the prevailing intellectual property systems (as a world-wide concept and practice) changes the indigenous regulatory systems themselves.
10. Patents and other intellectual property rights to forms of life are unacceptable to Indigenous Peoples.
11. It is important to prevent conflicts that may arise between communities from the transformation of intellectual property into a means of dividing indigenous unity.
12. There are some formulas that could be used to enhance the value of our products (brand names, appellations of origin), but on the understanding that these are only marketing possibilities, not entailing monopolies of the product or of collective knowledge. There are also some proposals for modifying prevailing intellectual property systems, such as the use of certificates of origin to prevent use of our resources without our prior consent.
13. The prevailing intellectual property systems must be prevented from robbing us, through monopoly rights, of resources and knowledge in order to enrich themselves and build up power opposed to our own.
14. Work must be conducted on the design of a protection and recognition system which is in accordance with the defence of our own conception, and mechanisms must be developed in the short and medium term which will prevent appropriation of our resources and knowledge.
15. A system of protection and recognition of our resources and knowledge must be designed which is in conformity with our world view and contains formulas that, in the short and medium term, will prevent appropriation of our resources by the countries of the North and others.
16. There must be appropriate mechanisms for maintaining and ensuring rights of Indigenous Peoples to deny indiscriminate access to the resources of our communities or peoples and making it possible to contest patents or other exclusive rights to what is essentially indigenous.
17. There is a need to maintain the possibility of denying access to indigenous resources and contesting patents or other exclusive rights to what is essentially indigenous.
18. Discussions regarding intellectual property should take place without distracting from priorities such as the struggle for the right to territories and self-determination, bearing in mind that the indigenous population and the land form an indivisible unity.

II. Short-term recommendations

1. Identify, analyse and systematically evaluate from the standpoint of the indigenous world view different components of the formal intellectual property systems, including mechanisms, instruments and forums, among which we have:

Intellectual property mechanisms

- Patents
- Trademarks
- Authors' rights
- Rights of developers of new plant varieties
- Commercial secrets
- Industrial designs
- Labels of origin

Intellectual property instruments

- The Agreement on Trade-Related Intellectual Property Rights (TRIPS) of the Gen-

eral Agreement on Tariffs and Trade (GATT)
- The Convention on Biological Diversity, with special emphasis on the following aspects: environmental impact assessments, subsidiary scientific body, technological council, monitoring, national studies and protocols, as well as on rights of farmers and *ex situ* control of germplasm, which are not covered under the Convention.

Intellectual property forums

Define mechanisms for consultation and exchange of information between the indigenous organizational universe and international forums such as:
- The Treaty for Amazonian Co-operation
- The Andean Pact
- The General Agreement on Tariffs and Trade
- The European Patent Convention
- The United Nations Commission on Sustainable Development
- The Union for the Protection of New Varieties of Plants
- The World Intellectual Property Organization (WIPO)
- The International Labour Organization (ILO)
- The United Nations Commission of Human Rights

2. Evaluate the possibilities offered by the international instruments embodying cultural, political, environmental and other rights that could be incorporated into a sui generis legal framework for the protection of indigenous resources and knowledge.
3. Define the content of consultation with such forums.
4. Define the feasibility of using some mechanisms of the prevailing intellectual property systems in relation to:
   - Protection of biological/genetic resources
   - Marketing of resources
5. Study the feasibility of alternative systems and mechanisms for protecting indigenous interests in their resources and knowledge.
   Sui generis systems for protection of intellectual property:
   - Inventor's certificates
   - Model legislation on folklore
   - New deposit standards for material entering germplasm banks
   - Commissioner for intellectual property rights
   - Tribunals
   - Bilateral and multilateral contracts or conventions
   - Material Transfer Agreements
   - Biological prospecting
   - Defensive publication
   - Certificates of origin
6. Seek to make alternative systems operational within the short term, by establishing a minimal regulatory framework (for example bilateral contracts).
7. Systematically study, or expand studies already conducted of, the dynamics of Indigenous Peoples, with emphasis on:
   - Basis of sustainability (territories, culture, economy)
   - Use of knowledge and resources (collective ownership systems, community use of resources)
   - Community, national, regional and international organizational bases

   That will make it possible to create mechanisms within and outside Indigenous Peoples capable of assigning the same value to indigenous knowledge, arts and crafts as to western science.
8. Establish regional and local indigenous advisory bodies on intellectual property and biodiversity with functions involving legal advice, monitoring, production and dissemination of information and production of materials.
9. Identify national intellectual property organizations, especially in areas of biodiversity
10. Identify and draw up a timetable of forums for discussion and exchange of information on intellectual property and/or biodiversity. Seek support for sending indigenous delegates to participate in such forums. An effort will be made to obtain information with a view to the eventual establishment of an Information, Training and Dissemination Centre on Indigenous Property and Ethical Guidelines on contract negotiation and model contracts.

III. Medium-term strategies

1. Plan, programme, establish timetables and seek financing for the establishment of an indigenous programme for the collective use and protection of biological resources and knowledge. This programme will be developed in phases in conformity with areas of geographical coverage.
2. Plan, draw up timetables for and hold seminars and workshops at the community, national and regional levels on biodiversity and prevailing intellectual property systems and alternatives.

3. Establish a standing consultative mechanism to link community workers and indigenous leaders, as well as an information network.
4. Train indigenous leaders in aspects of intellectual property and biodiversity.
5. Draw up a Legal Protocol of Indigenous Law on the use and community knowledge of biological resources.
6. Develop a strategy for dissemination of this Legal Protocol at the national and international levels.

1.10: UNDP Consultation on the Protection and Conservation of Indigenous Knowledge

Sabah, East Malaysia, 24–27 February 1995

Basic points of agreement on the issues faced by the Indigenous Peoples of Asia

From the deliberations it is clear that self-determination is most important to the Indigenous People. The definition of self-determination is different in different countries, ranging from land rights, autonomy, self rule without secession, autonomy under a federal system, to independence. Indigenous Peoples' struggle and right to self-determination are being threatened by repressive governments (e.g. Burma); development policies and projects such as large dams (e.g. North Thailand, Sarawak in East Malaysia); unjust land laws (e.g. Hill Tribes of Thailand, Malaysia, Vietnam); genocide (e.g. Chittagong Hill tribes, Bangladesh); religion and the dominant culture.

Land, in particular native customary or ancestral land, is significant to Indigenous Peoples because it is the source of their livelihood and the base of their indigenous knowledge, spiritual and cultural traditions.

The Indigenous Peoples' struggle for self-determination is a very strong counter-force to the intellectual property rights system vis-a-vis indigenous knowledge, wisdom and culture. Therefore, the struggle for self-determination cannot be separated from the campaign against intellectual property rights systems, particularly their applications on life forms and indigenous knowledge.

Specific points raised on Indigenous Knowledge and Intellectual Property Rights (IPR)

For the Indigenous Peoples of Asia, the intellectual property rights system is not only a very new concept but it is also very Western. However, it is recognized that the threats posed by the intellectual property rights systems are as grave as the other problems faced by the Indigenous Peoples at present. When, in the past, Indigenous Peoples' right to land has been eroded through the imposition of exploitative laws imposed by outsiders; with intellectual property rights, alien laws will also be devised to exploit the indigenous knowledge and resources of the Indigenous Peoples.

The prevailing intellectual property rights system is seen as a new form of colonization and a tactic by the industrialized countries of the North to confuse and to divert the struggle of Indigenous Peoples from their rights to land and resources on, above and under it.

The intellectual property rights system and the (mis)appropriation of indigenous knowledge without the prior knowledge and consent of Indigenous Peoples evoke feelings of anger, of being cheated, and of helplessness in knowing nothing about intellectual property rights and indigenous knowledge piracy. This is akin to robbing Indigenous Peoples of their resources and knowledge through monopoly rights.

Indigenous Peoples are not benefiting from the intellectual property rights system. Indigenous knowledge and resources are being eroded, exploited and/or appropriated by outsiders in the likes of transnational corporations (TNCs), institutions, researchers and scientists who are after the profits and benefits gained through monopoly control.

The technological method of piracy is too sophisticated for Indigenous Peoples to understand, especially when indigenous communities are unaware of how the system operates and who are behind it.

For Indigenous Peoples, life is a common property which cannot be owned, commercialized and monopolized by individuals. Based on this world view, Indigenous Peoples find it difficult to relate intellectual property rights issues to their daily lives. Accordingly, the patenting of any life forms and processes is unacceptable to Indigenous Peoples.

The intellectual property rights system is in favour of the industrialized countries of the North who have the resources to claim patent and copyright, resulting in the continuous exploitation and appropriation of genetic resources, indigenous knowledge and culture of the Indigenous Peoples for commercial purposes. The intellectual property rights system totally ignores the contribution of Indigenous Peoples and the peoples of the South in the conservation and protection of genetic resources through millennia.

The intellectual property rights system totally ignores the close inter-relationship between Indigenous Peoples, their knowledge, genetic resources and their environment. The proponents of intellectual property rights are only concerned with the benefits that they will gain from the commercial exploitation of these resources.

The Indigenous Peoples of Asia strongly condemn the patenting and commercialization of their cell lines or body parts, as being promoted by the scientist and institutions behind the Human Genome Diversity Project (HGDP).

Plan of actions proposed by the Asian consultation workshop

The Consultation recognizes that the struggle for self-determination is closely connected to retaining rights over ancestral lands and the entire way of life of the Indigenous Peoples. The threats that Indigenous Peoples have been facing in this regard are very clear, and they have their own plans of action to address these concerns.

The Consultation also recognizes that indigenous knowledge is closely linked to land which can be taken away from Indigenous Peoples. Thus, the need to protect and conserve indigenous knowledge is just as important as the struggle for self-determination.

In a broad sense, therefore, the Indigenous Peoples of Asia have one common aspiration – to reclaim their right to self-determination and to their indigenous knowledge. The question of sovereignty is traditionally understood as land but now it also encompasses indigenous knowledge since the two are very closely linked.

Towards this end, the Consultation has suggested the following actions and strategies:

A. Plan of action at the local level

Noting the different experiences, prevailing realities in the political environment and varied situations that the Indigenous Peoples of Asia currently find themselves in, the methods for achieving their aspirations may again differ, or be in different stages of expression at the local or national level. In such circumstances, it was generally felt that the general plan of action be disseminated to Indigenous Peoples organizations for them to implement them in their own ways, based on their specific realities.

However, it became clear during the Consultation that there is a need to emphasize the following aspects in the activities related to indigenous knowledge at the local level:

- Strengthen the Indigenous Peoples' organizations and communities to be able to collectively address local concerns related to indigenous knowledge and intellectual property rights.
- Continue the Indigenous Peoples' struggle for self-determination since this can be a strong counter force against the threats posed by intellectual property rights systems on indigenous knowledge and genetic resources.
- Raise the awareness of Indigenous Peoples' organizations and communities on the global trends and developments in intellectual property rights systems, especially as they apply to life forms and indigenous knowledge.

B. General plan of action

Immediate/short-term strategies

- Issue a statement to the European Parliament calling for the rejection of the patenting of life forms in the European Union, in time for its voting on the issue on 1 March 1995.
- Disseminate information pertaining to the Asian Consultation Workshop to the local mass media for publication and wider mass awareness.
- Organize follow-up workshops at the community level to raise awareness of local farmers and Indigenous Peoples on the prevailing intellectual property systems.
- Organize local or national conferences on customary laws to explore indigenous mechanisms and systems of effectively protecting and conserving indigenous knowledge.
- Plan regional meetings for follow-up discussion and exchange of information on indigenous self-determination and related issues such as indigenous knowledge, intellectual property rights systems and the patenting of life forms. At the outset, the Alliance of Taiwan Aborigines (ATA) has expressed its plan to initiate a regional meeting on these issues in Taiwan in 1996. The ATA will look for funding sources and will welcome financial support from the United Nations Development Programme (UNDP).

Medium-term strategies

- Intensify advocacy and campaign works against intellectual property systems and the Human Genome Diversity Project (HGDP) at national and international levels.
- Provide updates on the HGDP and patenting, to be disseminated to Indigenous Peoples, indigenous organizations and non-governmental organizations sympathetic to the cause of Indigenous Peoples. The Rural Advancement Foundation International (RAFI) has been requested to collaborate with local and Asia-based regional organizations to produce and disseminate materials in popular forms, written in the

local languages and based on the local context. The Southeast Asia Regional Institute for Community Education (SEARICE) will also distribute their monographs on the impact of global developments on the Indigenous Peoples, and will assist in information dissemination.

- Develop capacity of the Asian Indigenous Peoples Pact (AIPP), a forum for Indigenous Peoples' movements in Asia. In this respect, national Indigenous Peoples' organizations will contribute human and material resources, as well as identify members for short to medium-term internship programmes.
- AIPP to co-ordinate and monitor activities and developments related to the plans formulated for the region.
- Build alliances and network with groups within Asia and outside, such as the AIPP, RAFI, SEARICE and the Indigenous Peoples' Biodiversity Network (IPBN).
- Indigenous Peoples to design their own educational curriculum that will help promote their culture and indigenous knowledge. Such an educational curriculum will instil a deep awareness and pride among Indigenous Peoples, especially children, on the importance of their indigenous knowledge, culture and resources.

1.11: Declaración de Jujuy

Declaración del Consejo Indio de Sudamérica
Aprobada en su III Sesión del Consejo Directivo CISA ampliada
Realizada en San Salvador de Jujuy, Argentina, 27 de marzo de 1995

Amparados en que la proclamación del Decenio Internacional de los Pueblos Indígenas, se invocó a los Estados tomar medidas para la realización plena de los Derechos de los Pueblos Indígenas del mundo.

Teniendo en cuenta que, en todos los países de Sudamérica han preexistido y existen Pueblos originarios indios, que participaron en los procesos de independencia, en la perspectiva de descolonizar el continente y retomar su propio rumbo histórico.

Habiendo constatado, que transcurrido mas de ciento ochenta años, en una situación de independencia para los países inmersos en nuestros territorios ancestrales, sin que estos hayan logrado consolidar sus proyectos de Estado-Nación que garanticen para nuestros Pueblos el ejercicio de sus Derechos Fundamentales.

Preocupados por la situación de extrema pobreza por la que atraviesan nuestros Pueblos, sin que de nuestra parte podamos tomar medidas de gestión propia para superarla, por estar privados de nuestros derechos fundamentales como pueblos que, en otros continentes han sido restituidos, en virtud de la Declaración de Banjul en 1981.

Teniendo presente, que las Naciones Unidas, a través del Grupo de Trabajo sobre Pueblos Indígenas, ha proyectado aprobar la Declaración de la ONU proclamando los Derechos de los Pueblos Indígenas del Mundo, en un texto elevado a la Asamblea.

Considerando que, en particular la situación de nuestros pueblos se ha visto afectada por una serie de conflictos recientes, en materia de demarcación fronteriza, tráfico de armas, defraudación a fondos de los Estados, guerras, tráfico de drogas, terrorismo, desplazamientos forzados, exagerado centralismo, hacinamiento urbano, mafias electorales; en los que, indios de nuestros pueblos están siendo objeto de nuevas formas de exterminio y esclavitud.

Consternados porque tales hechos recientes, ponen en peligro la unidad nacional y la integridad territorial en los países de Sudamérica. Y estando nuestros pueblos, actualmente sometidos a sus respectivas jurisdicciones, somos los que mas sufrimos las consecuencias de tales males, por estar privados del ejercicio de nuestros derechos fundamentales como pueblos, para reaccionar en legítima defensa.

Convencidos de que ha llegado el momento en que las Repúblicas de Sudamérica, en un acto de justicia histórica, y reconociendo el aporte de los pueblos indios en el proceso de descolonización; permitan que éstos asuman por si mismos la gestión de su propio destino, reivindicando el ejercicio de sus Derechos fundamentales de que fueron privados hace quinientos dos años.

Por tanto, declaramos:

PRIMERO. Asumir para los Pueblos Indios de Sudamérica, los principios proclamados en la "Declaración de Banjul de 1981", como derechos fundamentales de todos los pueblos, invocando el respaldo de la opinión pública internacional.

SEGUNDO. Iniciar la reorganización de nuestro continente basados en principios realistas, equitativos y científicos conducentes a reivindicar una vida mas digna para nuestros pueblos y sus integrantes.

TERCERO. Respetar y hacer respetar la indisoluble unidad de nuestro continente, integrado ancestralmente por los Andes y la Foresta Amazónica, bajo un modelo indianista de gestión de territorio, de medio ambiente y la sociedad.

CUARTO. Somos pueblos pre existentes a la colonia, actualmente privados de nuestro sagrado derecho a la libre determinación política, y con una propuesta originaria que seguiremos manteniendo hasta lograr que se restablezca la harmonía en nuestro continente.

San Salvador de Jujuy (Argentina) 27 de Marzo de 1995.

Firman las siguientes organizaciones participantes:

Comunidad de extensión y desarrollo de cultivo andinos CEDCA (PERU), Asociación Indígena Urbana Pacha Aru (CHILE), Instituto Quechua Jujuy Manta (ARGENTINA), Centro Paraguayo en Jujuy (PARAGUAY), Comisión Interamericana de Jurista Indígenas (ARGENTINA), Movimiento indio Tupac Katari MITKA (BOLIVIA), Asociación de Ganaderos artesanos mineros aborígenes de la puna AGAYMAPU, Cruzada Nativista indoamericana, Comunidad Kolla los Ayrampos (QOLLA SUYO), Organización Willkanina (NACION QUECHUAYMARA), Sacha Sacha Wawau (CHACO ARGENTINA).

Derechos fundamentales de los Pueblos

1. Todos los Pueblos tienen Derecho a la existencia
2. Todos los Pueblos son Iguales
3. Todos los Pueblos tendrán los mismos Derechos
4. Todos los Pueblos gozarán del mismo respeto
5. Ningún Pueblo tiene Derecho a oprimir a otro Pueblo
6. Todos los Pueblos tienen Derecho a la autodeterminación de su Status Político.
7. Todos los Pueblos tienen Derecho a perseguir su desarrollo económico y social de acuerdo a las políticas que libremente hayan elegido.
8. Todos los Pueblos tienen Derecho a la asistencia solidaria para lograr su liberación.
9. Todos los Pueblos tienen Derecho de disponer libremente a sus riquezas y recursos naturales.
10. Todos los Pueblos tiene Derecho a recuperar su propiedad, cuando haya sido expoliada.
11. Todos los Pueblos tienen Derecho al intercambio equitativo por sus recursos naturales.
12. Todos los Pueblos tienen Derecho a beneficiarse de las ventajas derivadas de los Recursos Nacionales.
13. Todos los Pueblos tienen Derecho a protegerse de los Monopolios Extranjeros.
14. Todos los Pueblos tienen Derecho a su desarrollo económico, social y cultural y a que se respete su identidad.
15. Todos los Pueblos tienen Derecho al disfrute igualitario del patrimonio común de la humanidad.
16. Todos los Pueblos tienen Derecho a que se les asegure el ejercicio del Derecho al Desarrollo.
17. Todos los Pueblos tienen Derecho a la paz y seguridad nacional e internacionales.
18. Todos los Pueblos tienen Derecho a participar en la paz, solidaridad y relaciones amistosas entre pueblos.
19. Todos los Pueblos tienen Derecho a un satisfactorio ambiente favorable para su desarrollo.
20. Todos los Pueblos tienen Derecho a asumir la gestión y la administración de sus Recursos Naturales..

San Salvador de Jujuy (Argentina) 27 de Marzo de 1995

1.12: Final statement from the UNDP consultation on Indigenous Peoples' Knowledge and Intellectual Property Rights

Suva, April 1995

Preamble

We the participants at the Regional Consultation on Indigenous Peoples' Knowledge and Intellectual Property Rights held in April 1995 in Suva, Fiji, from independent countries and from non-autonomous colonized territories hereby:

Recognize that the Pacific Region holds a significant proportion of the world's indigenous cultures, languages and biological diversity,

Support the initiatives of the Mataatua Declaration (1993), the Kari-Oca Declaration (1992), Julayabinul Statement (1993), and the South American and Asian Consultation Meetings,

Declare the right of Indigenous Peoples of the Pacific to self-governance and independence and ownership of our lands, territories and resources as the basis for the preservation of Indigenous Peoples' knowledge,

Recognize that Indigenous Peoples of the Pacific exist as unique and distinct peoples irrespective of their political status,

Acknowledge that the most effective means to fulfil our responsibilities to our descendants is through the customary transmission and enhancement of our knowledge,

Reaffirm that imperialism is perpetuated through intellectual property rights systems, science and modern technology to control and exploit the lands, territories and resources of Indigenous Peoples,

Declare that Indigenous Peoples are willing to share our knowledge with humanity provided we determine when, where and how it is used. At present the international system does not recognize or respect our past, present and potential contributions,

Assert our inherent right to define who we are. We do not approve of any other definition,

Condemn attempts to undervalue Indigenous Peoples traditional science and knowledge,

Condemn those who use our biological diversity for commercial and other purposes without our full knowledge and consent,

Propose and seek support for the following plan of action:

1. Initiate the establishment of a treaty declaring the Pacific Region to be a life forms patent-free zone.
   - 1.1 Include in the treaty protocols governing bioprospecting, human genetic research, *in situ* conservation by Indigenous Peoples, *ex situ* collections and relevant international instruments.
   - 1.2 Issue a statement announcing the treaty and seeking endorsement by the South Pacific Forum and other appropriate regional and international fora.
   - 1.3 Urge Pacific governments to sign and implement the treaty.
   - 1.4 Implement an educational awareness strategy about the treaty's objectives.
2. Call for a moratorium on bioprospecting in the Pacific and urge Indigenous Peoples not to co-operate in bioprospecting activities until appropriate protection mechanisms are in place.
   - 2.1 Bioprospecting as a term needs to be clearly defined to exclude Indigenous Peoples' customary harvesting practices.
   - 2.2 Assert that *in situ* conservation by Indigenous Peoples is the best method to conserve and protect biological diversity and indigenous knowledge, and encourage its implementation by indigenous communities and all relevant bodies.
   - 2.3 Encourage Indigenous Peoples to maintain and expand our knowledge of local biological resources.
3. Commit ourselves to raising public awareness of the dangers of expropriation of indigenous knowledge and resources.
   - 3.1 Encourage chiefs, elders and community leaders to play a leadership role in the protection of Indigenous Peoples' knowledge and resources.
4. Recognize the urgent need to identify the extent of expropriation that has already occurred and is continuing in the Pacific.
   - 4.1 Seek repatriation of Indigenous Peoples' resources already held in external collections, and seek compensation and royalties from commercial developments resulting from these resources.
5. Urge governments who have not signed the General Agreement on Tariffs and Trade (GATT) to refuse to do so, and encourage those governments who have already signed to protest against any provisions which facilitate the expropriation of Indigenous Peoples' knowledge and resources and the patenting of life forms.
   - 5.1 Incorporate the concerns of Indigenous Peoples to protect their knowledge and resources into legislation by including 'Prior Informed Consent or No Informed Consent' (PICNIC) procedures and exclude the patenting of life forms.
6. Encourage the South Pacific Forum to amend its rules of procedure to enable accreditation of Indigenous Peoples and NGOs as observers to future Forum officials meetings.
7. Strengthen indigenous networks. Encourage the United Nations Development Programme (UNDP) and regional donors to continue to support discussions on Indigenous Peoples' knowledge and intellectual property rights.
8. Strengthen the capacities of Indigenous Peoples to maintain their oral traditions, and encourage initiatives by Indigenous Peoples to record their knowledge in a permanent form according to their customary access procedures.
9. Urge universities, churches, governments, non-governmental organizations, and other institutions to reconsider their roles in the expropriation of Indigenous Peoples' knowledge and resources and to assist in their return to their rightful owners.
10. Call on the governments and corporate bodies responsible for the destruction of Pacific biodiversity to stop their destructive practices and to compensate the affected communities and rehabilitate the affected environment.
    - 10.1 Call on France to stop definitively its nuclear testing in the Pacific and repair the damaged biodiversity.

1.13: Charter of 'Farmers Rights'

Issued at the Convention of Farmers, 30th January 1997. Bangalore, Karnataka, India.

We, the farmers of India, including the landless, the rural artisans and the tribals, have gathered in

Bangalore to resist the attempts, by the forces of neo-globalization, at legislation of the expropriation of their knowledge and resources and to save themselves and the country from these external assaults. We also want to alert all the governments of the world that monopolization of resources and knowledge by a handful of these forces of globalization threatens the future of humanity.

Farmers' Rights have a deep historic character, have existed since man developed agriculture to serve his needs, have remained vital for the conservation of biodiversity, and we have been renewing them with our constant generation of new resources and their improvement. We have been guarding genetic resources and supporting the evolution of species. We are the inheritors of these efforts and knowledge of generations which have created this biological wealth and on the basis of this status of ours, we re-proclaim and reiterate this charter of Farmers' Rights.

1. Farmers' rights include the right over resources and associated knowledge, united indivisibly, and mean accepting traditional knowledge, respect for cultures, and recognition that these are the basis of the creation of knowledge.
2. The right to control, the right to decide the future of genetic resources, the right to define the legal framework of property rights of these resources.
3. Farmers' Rights are of an eminently collective nature and for this reason should be recognized in a different framework than that of private property.
4. These rights should have a national application, and the Union Government should promote legislation to this effect, respecting the sovereignty of our country, to establish local laws based on these principles, rejecting intellectual property rights and the patenting of any form of life or of knowledge.
5. Rights to the means to conserve biodiversity and arrive at food security, such as territorial rights, right to land, right to water, to air.
6. The right to participate in the definition, elaboration, and execution of policies and programmes linked to genetic resources, at the national, regional and international levels.
7. The right to appropriate technology as well as participation, design and management of research programmes.
8. The right to define the control and handling of benefits derived from the use, conservation, and management of resources.
9. The right to use, choose, store and freely exchange genetic resources.
10. The right to develop models of sustainable agriculture that protect biodiversity and to influence the policies that support it.

Food security is now one of the great worries of humanity. Food security is only possible if it supports agricultural biodiversity, whose conservation and sustainable use we farmers have realized through generations of practising Farmers' Rights. All that remains is to recognize them.

We pledge ourselves to protect and perpetuate these rights.

1.14: Statement from Indigenous Peoples participating at the Fourth Session of the Commission on Sustainable Development (CSD-4)

18 April – 3 May 1996, New York

Mr Chairman, distinguished delegates and colleagues,

In recognition of this, the International Decade of the World's Indigenous Peoples, we the indigenous representatives participating at CSD-4, would like to direct your attention to General Assembly resolution A/Res/50/157 which emphasizes:

1. the important role of international co-operation in promoting the goals and activities of the Decade and the rights, the well being and sustainable development of Indigenous Peoples;
2. enabling Indigenous Peoples to achieve greater responsibility for their own affairs and an effective voice in decisions on matters that affect them;
3. that an official observance of the Decade be undertaken at all international conferences related to the aims and themes of the Decade;
4. that implementation of Chapter 26 of Agenda 21, adopted by the United Nations Conference on Environment and Development, and the relevant provisions of the Convention on Biological Diversity and other relevant future high level conferences.

Based upon these precedents, and to ensure the continuance of a co-ordinated follow-up to the UN Conference on Environment and Development, we continue to be concerned with:

1. the limited participation of Indigenous Peoples within the CSD-4 sessions and other regional forums due to the fact that:
   a) few indigenous communities and organizations are aware of the CSD and its mandate;
   b) the crucial obstacle for indigenous participation remains the lack of financial re-

sources. Few indigenous communities and organizations can afford to participate without external financial contributions;

2. dominant sector societies' lack of understanding of Indigenous Peoples' values and our special relationship with the Earth, and that we have developed traditional technologies and subsistence systems which are as relevant today as they have been for thousands of years. This demonstrates our deep spiritual connection to our ancestral homelands which is essential to sustainability;
3. the results of globalization and the implementation by various countries of liberalized policies and programmes which are intended to benefit developed countries and further marginalize Indigenous Peoples;
4. biotechnology projects under the guise of increased food production, leading to greater profits for big business in industrialized countries and further endangering indigenous societies;
5. the majority policies of state governments, multilateral development banks and TNCs are not sensitive to the rights, issues and concerns of Indigenous Peoples;
6. the lack of effective consultation of Indigenous Peoples by relevant UN agencies involved in development projects which affect them.

Recommendations

We reiterate the recommendations of the Indigenous Peoples' Preparatory Committee (IPCC) for the CSD, submitted 2 January 1996:

Indigenous Peoples recommend the creation of new policies for environmental and resource management. In this regard, the following principles should be considered.

1. That the international community recognize, protect and support the rights of Indigenous Peoples, including their cultural, territorial and human rights to own, control and manage their lands and resources and also Indigenous Peoples' right to sustainable development. Specifically, the CSD should push for the speedy adoption of the UN Draft Declaration on the Rights of Indigenous Peoples by the General Assembly as a concrete indication of its commitment to ensuring the implementation of sustainable development for Indigenous Peoples.
2. That Indigenous Peoples must participate fully in land, resource and development decision-making and policy development at international, regional, national and local levels, including UN processes such as the CSD and Convention on Biological Diversity.
3. The support and implementation of long term sustainable development, based on Indigenous Peoples' values, knowledge and technologies, to ensure full and intergenerational equity of resource allocations.
4. That Indigenous Peoples' cultural diversity, traditional values and ways of life be protected through Agenda 21 and the institutions and policies which will and do fulfil its objectives.
5. It must be recognized that the survival and development of Indigenous Peoples is directly connected to biological diversity and, as such, ensuring the protection of this diversity must be a priority for the international community.
6. That a new partnership be created between Indigenous Peoples and the international community based on mutual respect, reciprocity and a harmonious, accessible and equitable process.
7. Finally, we invite governments and UN agencies to support Indigenous Peoples in implementing Agenda 21, to recognize and support our rights to land, resources and development.

1.15: Declaration of the Consultation Meeting 'Indigenous Peoples, Mother Earth and Spirituality'

May 31st 1996, San Jose, Costa Rica

We, the original Peoples of Abya Yala (the Americas), solemnly declare that the Creator placed us here and that we carry the truth of this land in our hearts. The knowledge of Mother Earth flows through our veins. We are children of the Creator and Mother Earth who were instructed to live in balance and harmony with the sky, Mother Earth, plants, trees, rocks, minerals, winged ones, water beings, and the four legged.

We reaffirm our commitment to live under the natural laws of the Creator that govern Creation and maintain balance between all forms of life. We are thankful to the will of the Creator and to the sustenance of Mother Earth that we have survived.

We call upon the world community to legally recognize our collective rights to a homeland / territory; demand that Sacred Sites be respected and protected and that unrestricted access be granted to conduct ceremonies. As equal members of the human family, we exercise the right to worship, according to our cosmo-vision; therefore acknowledging the strength of our spiritual values and cultural traditions.

We demand from the international community, the reaffirmation and achievement of the Rio Agreements, in particular those that relate to Indigenous Peoples. Also, we call for the establishment and ratification of International Conventions for the protection and defence of our original Spirituality and intellectual cultural property rights.

We recognize and respect that the natural forces of the universe are restoring global balance, fulfilling their sacred duty to perpetuate life. We understand from the heart that humanity must strive to achieve inner balance, by learning to walk harmoniously upon Mother Earth.

Our children have a right to a healthy, sustainable future, free of environmental degradation. In the spirit of our ancestors, we call upon all the inhabitants of Mother Earth to sustain themselves in a respectful manner, to ensure the survival of future generations.

Having created a Spiritual Consultative Council on this 1st day of June, 1996, the above is ratified by spiritual leaders and representatives of the Bri-Bri, Cree, Garifuna, Keetowah, Ketchua, Kolla, Kuna, Macuna, Mapuche, Maya, Nahuat and Shuar Nations.

Indigenous Peoples, Mother Earth and Spirituality
Earth Council – Indigenous Peoples' Programme

1.16: The Leticia Declaration and proposals for action from the international meeting of indigenous and other forest-dependent peoples on the management, conservation and sustainable development of all types of forests

Leticia, Colombia, 13 December 1996.

A Contribution to the Intergovernmental Panel on Forests.

Co-sponsored by the Governments of Colombia and Denmark. Organized by the International Alliance of Indigenous-Tribal Peoples of the Tropical Forests, the Co-ordinating Body of Indigenous Organizations in the Amazon Basin (COICA) and the Organization of Indigenous Peoples of the Colombian Amazon-region (OPIAC).

All peoples are descendants of the forest. When the forest dies we die.

We are given responsibility to maintain balance within the natural world.

When any part is destroyed, all balance is cast into chaos. When the last tree is gone, and the last river is dead, then people will learn that we cannot eat gold and silver. To nurture the land is our obligation to our ancestors, who passed this to us for future generations.

Preamble

We, participants of the International Meeting of Indigenous Peoples and Other Forest-Dependent Peoples on the Management, Conservation and Sustainable Development of All Types of Forests:

Recognising that Indigenous Peoples enjoy the right of self-determination and by virtue of that right they freely determine their political status and freely pursue their economic, social and cultural development;

Recognising that forests are the homes of many Indigenous Peoples and other forest-dependent peoples and are fundamental to their survival as distinct peoples, forming the basis for their livelihoods, cultures and spirituality;

Recognising that nearly all forests are inhabited;

Recognising that the maintenance of cultural diversity and the welfare of Indigenous Peoples and forest-dependent peoples is a fundamental aspect of sustainable forest management;

Recognising that secure and long-term guarantees to lands and territories are essential to sustainable forest management;

Recognising that human rights, sustainable forest management and peace are interdependent and indivisible;

Recognising, therefore, that the interests of Indigenous Peoples and other forest-dependent peoples should have priority in any decisions about forests.

Therefore this meeting calls upon Governments, Intergovernmental Organizations, Non-Governmental Organizations and Major Groups involved in implementing Agenda 21 to support the following principles, conclusions and proposals for action.

General principles

That the rights, welfare, viewpoints and interests of Indigenous Peoples and other forest-dependent peoples should be central to all decision-making about forests at local, national, regional and international levels.

That their rights to their lands, territories, forests and other natural resources should be recognised, secured, respected and protected.

That they should have full control over the management, use and conservation of these resources.

That the representative institutions of Indigenous Peoples and other forest-dependent peoples should be fully recognized and respected.

That Indigenous Peoples and other forest-dependent peoples need strong and autonomous organizations and support for the consolidation of indigenous systems and cultural institutions.

That new mechanisms should be established to ensure the equal participation of Indigenous Peoples and other forest-dependent peoples in decision-making on forests at all levels.

That Indigenous Peoples and other forest peoples constitute an important cross-cutting theme in the forest agenda, affecting many other issues.

That United Nations bodies, when dealing with indigenous issues, should not narrowly work on sectoral themes but continue to engage in dialogue to enhance mutually enriching discussions, and to strengthen co-ordination.

That there should be wholehearted support for the current Draft United Nations Declaration on the Rights of Indigenous Peoples being discussed by the United Nations Commission on Human Rights, and its relevance for the goals of sustainable development and sustainable forest management.

That these principles shall be applied in the management, conservation and sustainable development of all types of forests.

National Forest and Land Use Programmes

National Forest and Land Use Programmes should promote broad participation and be decentralized to ensure a wide involvement in implementation.

Indigenous Peoples and other forest-dependent peoples must play an integral part in national forest and land use planning.

Rights to and respect for Indigenous Peoples and other forest-dependent peoples' lands and territories and tenure systems must be guaranteed in forms appropriate to them.

Within the framework of national and regional planning, decisions on land use must be devolved to the local level so that Indigenous Peoples and other forest-dependent peoples can assert effective customary systems of sustainable forest management.

Indigenous Peoples and other forest-dependent peoples' rights to rotational farming, hunting, gathering, fishing, grazing and other land use should be respected and secured within national land use and forest programmes.

All forest and land use planning activities that affect Indigenous Peoples and other forest-dependent peoples must fully respect their customary systems of dispute settlement.

National land use and forest programmes should uphold international standards on the rights of Indigenous Peoples, with due consideration for the wishes of the peoples concerned, including appropriate national legislation and implementation.

National governments are urged to respect indigenous communities and cultures across national frontiers and not continue with practices that divide Indigenous Peoples.

Proposals for Action

1. The IPF should acknowledge the importance of indigenous land use and resource rights, as well as customary law and indigenous legal systems, as inherently related to the forest issue.
2. To establish participatory mechanisms for national forest and land use planning which include the following elements:
   - Equal access and full participation by Indigenous Peoples and other forest-dependent peoples in national land use planning and forest programmes, at all stages of planning, implementation, monitoring, assessment and evaluation.
   - Initiatives, programmes and projects must take into account customary land use and tenure systems of indigenous and other forest-dependent peoples, including rotational farming, hunting, gathering, fishing, grazing and others.
   - No activities must take place on Indigenous Peoples' territories without their full and informed consent through their representative institutions, including the power of veto.
   - No activities for resource utilization or conservation, including the establishment of protected areas, must be initiated on lands of Indigenous Peoples and other forest-dependent peoples without security and full respect for their territorial rights.
   - Indigenous Peoples and other forest-dependent peoples must be included in the decision making process at all levels in all areas that affect them, including policy decisions of international development agencies, multilateral development banks and all trans-national corporations.
   - Any benefits from territories of Indigenous Peoples and other forest-dependent peoples must primarily be for their own local use and in accordance with principles of benefit-sharing established by them.

Underlying Causes of Deforestation and Forest Degradation:

Deforestation and forest degradation is exacerbated by a lack of understanding of the holistic world views and ways of life of Indigenous Peoples and other forest-dependent peoples.

The underlying causes of deforestation and forest degradation which urgently need to be addressed include the following:

- The failure of governments and other institutions to recognise and respect the rights of Indigenous Peoples and other forest-dependent peoples to their territorial lands, forests and other resources.
- The increasing problem of landlessness among impoverished peasants who are denied access to land outside forest areas due to inequitable land ownership patterns, and who also have no alternate economic opportunities;
- Government policies and those of private sector industry are geared to exploit forest and mineral resources to the fullest extent for purely economic gain. These policies are often incompatible with other existing forest conservation policies. Such policies include substituting forests with industrial tree plantation for the pulp industry; oil and gas exploration by trans-national corporations; uncontrolled mining operations, and establishing nuclear waste storage sites on indigenous territories.

Proposals for Action

1. Genuine participatory mechanisms need to be developed which allow Indigenous Peoples and other forest-dependent peoples a decisive voice in evaluations of deforestation processes and the evolution of appropriate policy responses.
2. Governments and other institutions are urged to pursue measures which can reduce pressure from industrial societies on forests by, inter alia, reducing consumption, reusing, recycling and substituting forest products as appropriate.

Traditional Forest-Related Knowledge:

Traditional forest-related knowledge is essential to sustainable forest management practices.

Traditional forest-related knowledge is intimately bound up with Indigenous Peoples and other forest-dependent peoples' ownership and control of their lands and territories and their continued management, use and conservation of all types of forests.

The contributions of women in the development, promotion and protection of indigenous knowledge must be acknowledged and supported.

The ownership of Indigenous Peoples and other forest-dependent peoples of their forest related traditional knowledge and contemporary innovations should be recognised and secured.

Traditional knowledge must remain alive, cultures must continue to develop, and indigenous contemporary knowledge and technologies must also be respected.

Use of this knowledge should not be made without the prior informed consent of the Peoples concerned;

Indigenous Peoples oppose the patenting of life forms, sacred plants and the Human Genome Diversity Project and the imposition of private intellectual property rights on collective indigenous knowledge and resources.

Proposals for action

1. New legislative frameworks and sui generis systems that recognise and effectively protect the cultural heritage and traditional forest related knowledge of Indigenous Peoples and other forest-dependent peoples must be established. These must be based on customary law and governance structures.
2. To note and support the actions and recommendations agreed by the Conference of Parties to the Convention on Biological Diversity during its third meeting relating to the implementation of Article 8j and other related areas.
3. To establish in the Amazon region a world university of Indigenous Peoples with the support of the international community. This University shall function under the direct administration of the Indigenous Peoples with the purpose to study, promote and protect indigenous cultures, cosmo-vision and traditional knowledge. It shall be open to all peoples and cultures.
4. IPF4 should support the Principles and Guidelines for the Protection of the Heritage of Indigenous Peoples contained in the Final Report of the Special Rapporteur, Mrs. Erica- Irene Daes (E/CN.4/Sub.2/1995/26) covering definitions, transmission of heritage, recovery and restitution of heritage, national programmes and legislation, researchers and scholarly institutions, business and industry, artists, writers and performers, public information and education and international organizations.
5. To undertake a series of expert meetings under the auspices of the CSD on sustainable development and territorial management, and the related issues of partnerships, agreements and legal frameworks for cultural heritage protection.
6. National governments and the international community are urged to support education programmes which promote the sustainable forest management practices of Indigenous Peoples and other forest-dependent peoples.
7. To implement broad technical capacity-building programmes on sustainable land use and forest management, giving special attention to settlers.

8. To support information exchange programmes between Indigenous Peoples and local communities on issues of intellectual and cultural rights, knowledge, innovations and practices.
9. To develop association agreements on sustainable forest management between Indigenous Peoples, other forest-dependent peoples and groups living in the forest.
10. To establish an information centre and dissemination system to enable Indigenous Peoples and other forest-dependent peoples to access all relevant information. Efforts should be made to make this information freely and widely available in languages and forms accessible to Indigenous Peoples.

Financial Assistance and Technology Transfer

Proposals for Action

1. Financial assistance and mechanisms must be publicly accountable, transparent and unconditional. The IPF must also create possibilities for Indigenous Peoples and other forest-dependent peoples to design their own financial mechanisms.
2. Indigenous Peoples and forest-dependent peoples should directly access and receive technical assistance and support upon their own request. When such assistance is proposed by government agencies and other bodies it should only proceed with the approval of the Indigenous Peoples and forest-dependent peoples.
3. Donor criteria should go beyond project financing, and also support the capacity- and institution-building of Indigenous Peoples and other forest-dependent peoples.
4. To set up an independent fund to enable Indigenous Peoples and other forest-dependent peoples to enjoy full participation in the international forest policy debate. The funding mechanism must ensure independent, fair and equitable participation from all regions and all types of forests. Indigenous Peoples must be part of the administration of this fund.

Criteria and Indicators for Sustainable Forest Management

All criteria and indicators for sustainable forest management and certification principles must secure the spiritual, cultural, social and material well-being of Indigenous Peoples and other forest-dependent peoples.

Proposals for Action

1. Indigenous Peoples and other forest-dependent peoples should be supported to carry out their own inventory of forest resources, and to define locally appropriate criteria and indicators for sustainable forest management.

International Instruments and Mechanisms

All future international instruments and mechanisms dealing with forests should involve Indigenous Peoples and other forest-dependent peoples in all stages of decision-making as equal partners.

The key areas that need to be addressed are participation and representation at the international level, and access to relevant information and funding. In order to effectively participate in decision-making processes, governments and other institutions must provide necessary support to help strengthen organizations of Indigenous Peoples and other forest-dependent peoples.

Proposals for Action

1. Indigenous Peoples and other forest dependent peoples must continue to have improved representation at all upcoming UN forums in the immediate future, including IPF4, CSD 5 and the Special Session of the UN General Assembly in June 1997. All these meetings should register our participation as 'Indigenous Peoples', a recognized Major Group in the implementation of Agenda 21.
2. Full support should be given to establishing a UN Permanent Forum for Indigenous Peoples during the UN Decade for Indigenous Peoples which will draw together human rights, environment and development issues.
3. Agenda 21 and the Forest Principles should be interpreted to harmonize with the UN Draft Declaration on the Rights of Indigenous Peoples, with particular emphasis on the use of the term 'peoples' in its documentation.

Action commitments

The participants committed themselves to undertake concrete plans of action including the following.

- Identification of concrete participatory mechanisms for Indigenous Peoples and other forest-dependent peoples to ensure full and equal participation in any post IPF forest processes and also the Convention on Biological Diversity.
- Creation of a global, regional and local network for the flow of appropriate information and proposals for action.
- Plan and seek support for a programme of action to facilitate Indigenous Peoples and forest-dependent peoples.

- Identification of locally appropriate criteria and indicators for sustainable forest management.
- Preparation of proposals for the inclusion of traditional forest-related knowledge by Indigenous Peoples and other forest-dependent peoples into tertiary education and professional training programmes within all nations.
- Prepare proposals at the national level for the education of all peoples, especially the young people about forest destruction and degradation and the role of Indigenous Peoples and forest-dependent peoples in managing and protecting our forest resources.
- Prepare proposals for regional meetings between Indigenous Peoples and other forest-dependent peoples and UN-agencies and other bodies to explore the implications of international instruments and agreements at the national and local level. National Indigenous Peoples' organizations should consider forming alliances with universities to create national forums for research, study and action.

Conclusions

The meeting requests the IPF to incorporate these principles and proposals for action in the final IPF Report during its fourth session.

The meeting also requested that the Leticia Declaration and Proposals for Action also be considered by the UN Commission on Sustainable Development during the 5th Session, the General Assembly Special Session, and all other international forest policy deliberations.

The meeting requested the broadest dissemination of the Leticia Declaration and Proposals for Action to all relevant bodies.

1.17: International Workshop on Indigenous Peoples and Development

Ollantaytambo, Qosqo, Peru, 21–26 April 1997.

We, the participants of the International Workshop on Indigenous Peoples and Development held at Ollantaytambo, Qosqo, Peru, from 21 to 26 April 1997:

Considering the importance of cultural diversity and of Indigenous People's values and philosophies for a new paradigm for sustainable development;

Considering the importance of Indigenous Peoples' knowledge and practices for the maintenance and conservation of biological and cultural diversity for future generations;

Considering the International Decade of the World's Indigenous Peoples, Draft Declaration on the Rights of Indigenous Peoples, Chapter 26 of Agenda 21 of the Rio Declaration, ILO Convention 169, the Convention on Biological Diversity, the Convention on Desertification, the Copenhagen Social Summit Declaration, Article 30 of the Convention on the Rights of the Child and other relevant international agreements and covenants that recognize the crucial role of Indigenous Peoples in Conservation and development;

Considering that multilateral development banks, international development agencies, national and international non-governmental development organizations have programmes and projects targeting Indigenous Peoples that are not widely known;

Considering the demographic importance of Indigenous Peoples in the different countries of the world;

Considering that the governments of the world do not usually speak and act according to the needs, interests and aspiration of Indigenous Peoples;

Have agreed that:

On values

1. Our vision of development is based on indigenous philosophy and values. Our collective vision includes a philosophy that reflects each unique indigenous culture that incorporates all components of life for the enrichment of peoples in their collective realization of life tasks that ensures the reciprocity and balance between peoples and their environment.
2. Our philosophy and values on indigenous development exist within a paradigm based on our knowledge of the cosmos. This knowledge becomes the foundation of our territoriality, our identity, our cultures and practices, our language, our spirituality, our indigenous systems and institutions, and our technologies which enables us to respond to problems and to protect, preserve and continue our survival as peoples and communities of the land.
3. Our inherent spirituality is the driving force of our action. Reciprocity --to give in order to receive – solidarity, the respect for all living forms, and the coexistence with Mother Earth, are values that are always among us.
4. Our culture is an accumulated experience and proven ways of doing things. It is the foundation upon and around which we shall build our development and other desired changes which also ensures gender justice as well as equality for all.
5. As Indigenous Peoples, we have an inherent responsibility to utilize our available resources

in a sustainable manner that will ensure a balance and harmony for us and our future generations. Our existence is dependent upon this holistic practice and approach.

6. Indigenous education is the basis in the formation of our own knowledge, the learning of our environment and society based on our traditional practices, uses, and our philosophical, historical, technological and artistic heritage.

On methods and technologies

7. Indigenous knowledge, and the methods and technologies that derive from it, should be conserved, protected, promoted and improved for the benefit of indigenous communities.
8. Methods and technologies should strengthen, consolidate and rehabilitate indigenous values for the communities who have lost or are losing their indigenous identities.
9. Methods and technologies should be integrated, encompassing all the indigenous visions and knowledge, taking into consideration: the cultural dynamic of Indigenous Peoples, the total collective well-being, the promotion of equality, and the sustainable management and use of resources for present and future generations.
10. The adoption and application of methods and technologies should be based on indigenous people's social structures and organizations, guaranteeing the equal participation of both women and men.
11. The adoption and application of methods and technologies should be the result of a collective decision-making process, taking into account the advice from knowledge keepers and practitioners according to the needs and interest of the local community.
12. Training in methods and technologies should include strategies that raise awareness and strengthen self-esteem in our indigenous identity and our cultural and organizational practices.

On organizational follow-up

13. We shall create an international network having the following objectives:
    a) To defend the interests of Indigenous Peoples at the national, regional and international levels.
    b) To research, monitor and disseminate information among Indigenous Peoples on the use and development of Indigenous People's resources, including intellectual property rights.
    c) To exchange experience from different peoples and regions to find common strategies for development.
    d) To develop and defend a common position at international fora.
14. An interim committee has been created to elaborate on practical mechanisms and formalization of the network. The interim international coordination of the network will be facilitated by the Asociacion Andes, Qosqo, Peru. The regional focal points are: AFRICA: International Centre for Environmental and Forest Studies (Cameroon); ASIA: Asia Indigenous Peoples Pact; CENTRAL AMERICA: Federation Indega Tawahka (Honduras); NORTH AMERICA: En'owkin Centre (Canada); PACIFIC: Te Kawau Maro (Aotearoa); SOUTH AMERICA: Andes (Peru)

Ollantaytambo, Qosqo, Peru, 26 April 1997.

For further information, please contact:

Longgena Ginting
PLASMA
Jl. Pertahanan No. 1 - Kompleks Yeschar
Samarinda 75119 - Kalimantan Timur
Tel/Fax: +62-541-35753
E-mail: danum@indo.net.id

1.18: UN draft declaration on the rights of Indigenous Peoples

As agreed upon by members of the Working Group on Indigenous Populations at its 11th session, 1993.

Affirming that Indigenous Peoples are equal in dignity and rights to all other peoples, while recognizing the right of all peoples to be different, to consider themselves different, and to be respected as such;

Affirming also that all peoples contribute to the diversity and richness of civilizations and cultures, which constitute the common heritage of humankind;

Affirming further that all doctrines, policies and practices based on or advocating superiority of peoples or individuals on the basis of national origin, racial, religious, ethnic or cultural differences are racist, scientifically false, legally invalid, morally condemnable and socially unjust;

Reaffirming also that Indigenous Peoples, in the exercise of their rights, should be free from discrimination of any kind;

Concerned that Indigenous Peoples have been deprived of their human rights and fundamental freedoms, resulting, inter alia, in their colonization and dispossession of their lands, territories and resources, thus preventing them from exercising, in particular, their right to development in accordance with their own needs and interests;

Recognizing the urgent need to respect and promote the inherent rights and characteristics of Indigenous Peoples, especially their rights to their lands, territories and resources, which derive from their political, economic and social structures, and from their cultures, spiritual traditions, histories and philosophies;

Welcoming the fact that Indigenous Peoples are organizing themselves for political, economic, social and cultural enhancement and in order to bring an end to all forms of discrimination and oppression wherever they occur;

Convinced that control by Indigenous Peoples over developments affecting them and their lands, territories and resources will enable them to maintain and strengthen their institutions, cultures and traditions, and to promote their development in accordance with their aspirations and needs;

Recognizing also that respect for indigenous knowledge, cultures and traditional practices contributes to sustainable and equitable development and proper management of the environment;

Emphasizing the need for demilitarization of the lands and territories of Indigenous Peoples, which will contribute to peace, economic and social progress and development, understanding, and friendly relations among the nations and peoples of the world;

Recognizing in particular the right of indigenous families and communities to retain shared responsibility for the upbringing, training, education and well-being of their children;

Recognizing also that Indigenous Peoples have the right freely to determine their relationships with States in a spirit of coexistence, mutual benefit and full respect;

Considering that treaties, agreements and other arrangements between States and Indigenous Peoples are properly matters of international concern and responsibility;

Acknowledging that the Charter of the United Nations, the International Covenant on Economic, Social and Cultural Rights and the International Covenant on Civil and Political Rights affirm the fundamental importance of the right of self-determination of all peoples, by virtue of which they freely determine their political status and freely pursue their economic, social and cultural development;

Bearing in mind that nothing in this Declaration may be used to deny any peoples their right of self-determination;

Encouraging States to comply with and effectively implement all international instruments, in particular those related to human rights, as they apply to Indigenous Peoples, in consultation and cooperation with the peoples concerned;

Emphasizing that the United Nations has an important and continuing role to play in promoting and protecting the rights of Indigenous Peoples; and

Believing that this Declaration is a further important step forward for the recognition, promotion and protection of the rights and freedoms of Indigenous Peoples and in the development of relevant activities of the United Nations system in this field,

Solemnly proclaims the following United Nations Declaration on the Rights of Indigenous Peoples:

Articles

Part I

1. Indigenous Peoples have the right to the full and effective enjoyment of all human rights and fundamental freedoms recognized in the Charter of the United Nations, the Universal Declaration of Human Rights and international human rights law.
2. Indigenous individuals and peoples are free and equal to other individuals and peoples in dignity and rights, and have the right to be free from any kind of adverse discrimination, in particular that based on their indigenous origin or identity.
3. Indigenous Peoples have the right of self-determination. By virtue of that right they freely determine their political status and freely pursue their economic, social and cultural development.
4. Indigenous Peoples have the right to maintain and strengthen their distinct political, economic, social and cultural characteristics, as well as their legal systems, while retaining their rights to participate fully, if they so choose, in the political, economic, social and cultural life of the State.
5. Every indigenous individual has the right to a nationality.

Part II

6. Indigenous Peoples have the collective right to live in freedom, peace and security as distinct peoples and to full guarantees against genocide or any other act of violence, including the removal of indigenous children from their families and communities under any pretext. In addition, they have the individual rights to life, physical and mental integrity, liberty and security of person.

7. Indigenous Peoples have the collective and individual right not to be subjected to ethnocide and cultural genocide, including prevention of and redress for:
   (a) any action which has the aim or effect of depriving them of their integrity as distinct peoples, or of their cultural values or ethnic identities;
   (b) any action which has the aim or effect of dispossessing them of their lands, territories or resources;
   (c) any form of population transfer which has the aim or effect of violating or undermining any of their rights;
   (d) any form of assimilation or integration by other cultures or ways of life imposed on them by legislative, administrative or other measures;
   (e) any form of propaganda directed against them.
8. Indigenous Peoples have the collective and individual right to maintain and develop their distinctive identities and characteristics, including the right to identify themselves as indigenous and to be recognized as such.
9. Indigenous Peoples and individuals have the right to belong to an indigenous community or nation, in accordance with the traditions and customs of the community or nation concerned. No disadvantage of any kind may arise from the exercise of such a right.
10. Indigenous Peoples shall not be forcibly removed from their lands or territories. No relocation shall take place without the free and informed consent of the Indigenous Peoples concerned and after agreement on just and fair compensation and, where possible, with the option of return.
11. Indigenous Peoples have the right to special protection and security in periods of armed conflict. States shall observe international standards, in particular the Fourth Geneva Convention of 1949, for the protection of civilian populations in circumstances of emergency and armed conflict, and shall not:
   (a) recruit indigenous individuals against their will into the armed forces and, in particular, for use against other Indigenous Peoples;
   (b) recruit indigenous children into the armed forces under any circumstances;
   (c) force indigenous individuals to abandon their lands, territories or means of subsistence, or relocate them in special centres for military purposes;
   (d) force indigenous individuals to work for military purposes under any discriminatory purposes.

Part III

12. Indigenous Peoples have the right to practice and revitalize their cultural traditions and customs. This includes the right to maintain, protect and develop the past, present and future manifestations of their cultures, such as archaeological and historical sites, artefacts, designs, ceremonies, technologies and visual and performing arts and literature, as well as the right to the restitution of cultural, intellectual, religious and spiritual property taken without their free and informed consent or in violation of their laws, traditions and customs.
13. Indigenous Peoples have the right to manifest, practice, develop and teach their spiritual and religious traditions, customs and ceremonies; the right to maintain, protect, and have access in privacy to their religious and cultural sites; the right to the use and control of ceremonial objects; and the right to the repatriation of human remains.
    States shall take effective measures, in conjunction with the Indigenous Peoples concerned, to ensure that indigenous sacred places, including burial sites, be preserved, respected and protected.
14. Indigenous Peoples have the right to revitalize, use, develop and transmit to future generations their histories, languages, oral traditions, philosophies, writing systems and literatures, and to designate and retain their own names for communities, places and persons.
    States shall take effective measures, whenever any right of Indigenous Peoples may be threatened, to ensure this right is protected and also to ensure that they can understand and be understood in political, legal and administrative proceedings, where necessary through the provision of interpretation or by any other appropriate means.

Part IV

15. Indigenous children have the right to all levels and forms of education of the State. All Indigenous Peoples also have this right and the right to establish and control their educational systems and institutions providing education in their own languages, in a manner appropriate to their cultural methods of teaching and learning.
    Indigenous children living outside their communities have the right to be provided with access to education in their own culture and language.

States shall take effective measures to provide appropriate resources for these purposes.

16. Indigenous Peoples have the right to have the dignity and diversity of their cultures, traditions, histories and aspirations appropriately reflected in all forms of education and public information.

    States shall take effective measure, in consultation with the Indigenous Peoples concerned, to eliminate prejudice and discrimination and to promote tolerance, understanding and good relations among Indigenous Peoples and all segments of society.

17. Indigenous Peoples have the right to establish their own media in their own languages. They also have the right to equal access to all forms of non-indigenous media.

    States shall take effective measures to ensure that State-owned media duly reflect indigenous cultural diversity.

18. Indigenous Peoples have the right to enjoy fully all rights established under international labour law and national labour legislation.

    Indigenous Peoples have the right not to be subjected to any discriminatory conditions of labour, employment or salary.

Part V

19. Indigenous Peoples have the right to participate fully, if they so choose, at all levels of decision-making in matters which may affect their rights, lives and destinies through representatives chosen by themselves in accordance with their own procedures, as well as to maintain and develop their own indigenous decision-making institutions.

20. Indigenous Peoples have the right to participate fully, if they so choose, through procedures determined by them, in devising legislative or administrative measures that may affect them.

    States shall obtain the free and informed consent of the peoples concerned before adopting and implementing such measures.

21. Indigenous Peoples have the right to maintain and develop their political, economic and social systems, to be secure in the enjoyment of their own means of subsistence and development, and to engage freely in all their traditional and other economic activities.

    Indigenous Peoples who have been deprived of their means of subsistence and development are entitled to just and fair compensation.

22. Indigenous Peoples have the right to special measures for the immediate, effective and continuing improvement of their economic and social conditions, including in the areas of employment, vocational training and retraining, housing, sanitation, health and social security.

    Particular attention shall be paid to the rights and special needs of indigenous elders, women, youth, children and disabled persons.

23. Indigenous Peoples have the right to determine and develop priorities and strategies for exercising their right to development. In particular, Indigenous Peoples have the right to determine and develop all health, housing and other economic and social programmes affecting them and, as far as possible, to administer such programmes through their own institutions.

24. Indigenous Peoples have the right to their traditional medicines and health practices, including the right to the protection of vital medicinal plants, animals and minerals.

    They also have the right to access, without discrimination, to all medical institutions, health services and medical care.

Part VI

25. Indigenous Peoples have the right to maintain and strengthen their distinctive spiritual and material relationships with the lands, territories, waters and coastal seas and other resources which they have traditionally owned or otherwise occupied or used, and to uphold their responsibilities to future generations in this regard.

26. Indigenous Peoples have the right to own, develop, control and use the lands and territories, including the total environment of the lands, air, waters, coastal seas, sea-ice, flora and fauna and other resources which they have traditionally owned or otherwise occupied or used. This includes the right to the full recognition of their laws, traditions and customs, land-tenure systems and institutions for the development and management of resources, and the right to effective measures by States to prevent any interference with, alienation of, or encroachment upon these rights.

27. Indigenous Peoples have the right to the restitution of the lands, territories and resources which they have traditionally owned or otherwise occupied or used; and which have been confiscated, occupied, used or damaged without their free and informed consent. Where this is not possible, they have the right to just and fair compensation. Unless otherwise freely agreed upon by the peoples concerned, compensation shall take the form of lands, territories and resources equal in quality, size and legal status.

28. Indigenous Peoples have the right to the conservation, restoration and protection of the total environment and the productive capacity of their lands, territories and resources, as well as to assistance for this purpose from States and through international co-operation. Military activities shall not take place in the lands and territories of Indigenous Peoples, unless otherwise freely agreed upon by the peoples concerned.
    States shall take effective measures to ensure that no storage of hazardous materials shall take place in the lands and territories of Indigenous Peoples.
    States shall also take effective measures to ensure, as needed, that programmes for monitoring, maintaining and restoring the health of Indigenous Peoples, as developed and implemented by the peoples affected by such materials, are duly implemented.
29. Indigenous Peoples are entitled to the recognition of the full ownership, control and protection of their cultural and intellectual property. They have the right to special measures to control, develop and protect their sciences, technologies and cultural manifestations, including human and other genetic resources, seeds, medicines, knowledge of the properties of fauna and flora, oral traditions, literatures, designs and visual and performing arts.
30. Indigenous Peoples have the right to determine and develop priorities and strategies for the development or use of their lands, territories and other resources, including the right to require that States obtain their free and informed consent prior to the approval of any project affecting their lands, territories and other resources, particularly in connection with the development, utilization or exploitation of mineral, water or other resources. Pursuant to agreement with the Indigenous Peoples concerned, just and fair compensation shall be provided for any such activities and measures taken to mitigate adverse environmental, economic, social, cultural or spiritual impact.

Part VII

31. Indigenous Peoples, as a specific form of exercising their right to self-determination, have the right to autonomy or self-government in matters relating to their internal and local affairs, including culture, religion, education, information, media, health, housing, employment, social welfare, economic activities, land and resources management, environment and entry by non-members, as well as ways and means of financing these autonomous functions.
32. Indigenous Peoples have the collective right to determine their own citizenship in accordance with their customs and traditions. Indigenous citizenship does not impair the right of indigenous individuals to obtain citizenship of the States in which they live.
    Indigenous Peoples have the right to determine the structures and to select the membership of their institutions in accordance with their own procedures.
33. Indigenous Peoples have the right to promote, develop and maintain their institutional structures and their distinctive juridical customs, traditions, procedures and practices, in accordance with internationally recognized human rights standards.
34. Indigenous Peoples have the collective right to determine the responsibilities of individuals to their communities.
35. Indigenous Peoples, in particular those divided by international borders, have the right to maintain and develop contacts, relations and co-operation, including activities for spiritual, cultural, political, economic and social purposes, with other peoples across borders.
    States shall take effective measures to ensure the exercise and implementation of this right.
36. Indigenous Peoples have the right to the recognition, observance and enforcement of treaties, agreements and other constructive arrangements concluded with States or their successors, according to their original spirit and intent, and to have States honour and respect such treaties, agreements and other constructive arrangements. Conflicts and disputes which cannot otherwise be settled should be submitted to competent international bodies agreed to by all parties concerned.

Part VIII

37. States shall take effective and appropriate measures, in consultation with the Indigenous Peoples concerned, to give full effect to the provisions of this Declaration. The rights recognized herein shall be adopted and included in national legislation in such a manner that Indigenous Peoples can avail themselves of such rights in practice.
38. Indigenous Peoples have the right to have access to adequate financial and technical assistance, from States and through international co-operation, to pursue freely their political, economic, social, cultural and spiritual devel-

opment and for the enjoyment of the rights and freedoms recognized in this Declaration.

39. Indigenous Peoples have the right to have access to and prompt decision through mutually acceptable and fair procedures for the resolution of conflicts and disputes with States, as well as to effective remedies for all infringements of their individual and collective rights. Such a decision shall take into consideration the customs, traditions, rules and legal systems of the Indigenous Peoples concerned.
40. The organs and specialized agencies of the United Nations system and other intergovernmental organizations shall contribute to the full realization of the provisions of this Declaration through the mobilization, inter alia, of financial co-operation and technical assistance. Ways and means of ensuring participation of Indigenous Peoples on issues affecting them shall be established.
41. The United Nations shall take the necessary steps to ensure the implementation of this Declaration including the creation of a body at the highest level with special competence in this field and with the direct participation of Indigenous Peoples. All United Nations bodies shall promote respect for and full application of the provisions of this Declaration.

Part IX

42. The rights recognized herein constitute the minimum standards for the survival, dignity and well-being of the Indigenous Peoples of the world.
43. All the rights and freedoms recognized herein are equally guaranteed to male and female indigenous individuals.
44. Nothing in this Declaration may be construed as diminishing or extinguishing existing or future rights Indigenous Peoples may have or acquire.
45. Nothing in this Declaration may be interpreted as implying for any State, group or person any right to engage in any activity or to perform any act contrary to the Charter of the United Nations.

1.19: Principles and Guidelines for the Protection of the Heritage of Indigenous Peoples

Elaborated by the Special Rapporteur of the Subcommission Mrs. Erica-Irene Daes (E/CN.4/Sub.2/1995/26 — 21 June 1995)

Principles

1. The effective protection of the heritage of the Indigenous Peoples of the world benefits all humanity. Cultural diversity is essential to the adaptability and creativity of the human species as a whole.
2. To be effective, the protection of Indigenous Peoples' heritage should be based broadly on the principle of self-determination, which includes the right and the duty of Indigenous Peoples to develop their own cultures and knowledge systems, and forms of social organization.
3. Indigenous Peoples should be recognized as the primary guardians and interpreters of their cultures, arts and sciences, whether created in the past, or developed by them in the future.
4. International recognition and respect for the Indigenous Peoples' own customs, rules and practices for the transmission of their heritage to future generations is essential to these peoples' enjoyment of human rights and human dignity.
5. Indigenous Peoples' ownership and custody of their heritage must continue to be collective, permanent and inalienable, as prescribed by the customs, rules and practices of each people.
6. The discovery, use and teaching of Indigenous Peoples' knowledge, arts and cultures is inextricably connected with the traditional lands and territories of each people. Control over traditional territories and resources is essential to the continued transmission of Indigenous Peoples' heritage to future generations, and its full protection.
7. To protect their heritage, Indigenous Peoples must control their own means of cultural transmission and education. This includes their right to the continued use and, wherever necessary, the restoration of their own languages and orthographies.
8. To protect their heritage, Indigenous Peoples must also exercise control over all research conducted within their territories, or which uses their people as subjects of study.
9. The free and informed consent of the traditional owners should be an essential precondition of any agreements which may be made for the recording, study, use or display of Indigenous Peoples' heritage.
10. Any agreements which may be for the recording, study, use or display of Indigenous Peoples' heritage must be revocable, and ensure that the peoples concerned continue to be the primary beneficiaries of commercial application.

Guidelines

11. The heritage of Indigenous Peoples is comprised of all objects, sites and knowledge the nature or use of which has been transmitted

from generation to generation, and which is regarded as pertaining to a particular people or its territory. The heritage of an indigenous people also includes objects, knowledge and literary or artistic works which may be created in the future based upon its heritage.

12. The heritage of Indigenous Peoples includes all moveable cultural property as defined by the relevant conventions of UNESCO; all kinds of literary and artistic works such as music, dance, song, ceremonies, symbols and designs, narratives and poetry; all kinds of scientific, agricultural, technical and ecological knowledge, including cultigens, medicines and the rational use of flora and fauna; human remains; immovable cultural property such as sacred sites, sites of historical significance, and burials; and documentation of Indigenous Peoples' heritage on film, photographs, videotape or audio tape.
13. Every element of an Indigenous Peoples' heritage has traditional owners, which may be the whole people, a particular family or clan, an association or society, or individuals who have been specially taught or initiated to be its custodians. The traditional owners of heritage must be determined in accordance with Indigenous Peoples' own customs, laws and practices.

Transmission of heritage

14. Indigenous Peoples' heritage should continue to be learned by the means customarily employed by its traditional owners for teaching, and each Indigenous Peoples' rules and practices for the transmission of heritage and sharing of its use should be incorporated in the national legal system.
15. In the event of a dispute over the custody or use of any element of an Indigenous Peoples' heritage, judicial and administrative bodies should be guided by the advice of indigenous elders who are recognized by the indigenous communities or peoples concerned as having specific knowledge of traditional laws.
16. Governments, international organizations and private institutions should support the development of educational, research, and training centres which are controlled by indigenous communities, and strengthen these communities' capacity to document, protect, teach and apply all aspects of their heritage.
17. Governments, international organizations and private institutions should support the development of regional and global networks for the exchange of information and experience among Indigenous Peoples in the fields of science, culture, education and the arts, including support for systems of electronic information and mass communication.
18. Governments, with international co-operation, should provide the necessary financial resources and institutional support to ensure that every indigenous child has the opportunity to achieve full fluency and literacy in his/her own language, as well as an official language.

Recovery and restitution of heritage

19. Governments, with the assistance of competent international organizations, should assist Indigenous Peoples and communities in recovering control and possession of their moveable cultural property and other heritage.
20. In co-operation with Indigenous Peoples, UNESCO should establish a programme to mediate the recovery of moveable cultural property from across international borders, at the request of the traditional owners of the property concerned.
21. Human remains and associated funeral objects must be returned to their descendants and territories in a culturally appropriate manner, as determined by the Indigenous Peoples concerned. Documentation may be retained, displayed or otherwise used only in such form and manner as may be agreed upon with the peoples concerned.
22. Moveable cultural property should be returned wherever possible to its traditional owners, particularly if shown to be of significant cultural, religious or historical value to them. Moveable cultural property should only be retained by universities, museums, private institutions or individuals in accordance with the terms of a recorded agreement with the traditional owners for the sharing of the custody and interpretation of the property.
23. Under no circumstances should objects or any other elements of an Indigenous Peoples' heritage be publicly displayed, except in a manner deemed appropriate by the peoples concerned.
24. In the case of objects or other elements of heritage which were removed or recorded in the past, the traditional owners of which can no longer be identified precisely, the traditional owners are presumed to be the entire people associated with the territory from which these objects were removed or recordings were made.

National programmes and legislation

25. National laws should guarantee that Indigenous Peoples can obtain prompt, effective and afford-

able judicial or administrative action in their own languages to prevent, punish and obtain full restitution and just compensation for the acquisition, documentation or use of their heritage without proper authorization of the traditional owners.

26. National laws should deny to any person or corporation the right to obtain patent, copyright or other legal protection for any element of Indigenous Peoples' heritage without adequate documentation of the free and informed consent of the traditional owners to an arrangement for the sharing of ownership, control, use and benefits.
27. National laws should ensure the labelling and correct attribution of Indigenous Peoples' artistic, literary and cultural works whenever they are offered for public display or sale. Attribution should be in the form of a trademark or an appellation of origin, authorized by the peoples or communities concerned.
28. National laws for the protection of Indigenous Peoples' heritage should be adopted following consultations with the peoples concerned, in particular the traditional owners and teachers of religious, sacred and spiritual knowledge, and wherever possible, should have the informed consent of the peoples concerned.
29. National laws should ensure that the use of traditional languages in education, arts and the mass media is respected and, to the extent possible, promoted and strengthened.
30. Governments should provide indigenous communities with financial and institutional support for the control of local education, through community-managed programmes, and with use of traditional pedagogy and languages.
31. Governments should take immediate steps, in co-operation with the indigenous peoples concerned, to identify sacred and ceremonial sites, including burials, healing places, and traditional places of teaching, and to protect them from unauthorized entry or use.

Researchers and scholarly institutions

32. All researchers and scholarly institutions should take immediate steps to provide Indigenous Peoples and communities with comprehensive inventories of the cultural property, and documentation of Indigenous Peoples' heritage, which they may have in their custody.
33. Researchers and scholarly institutions should return all elements of Indigenous Peoples' heritage to the traditional owners upon demand, or obtain formal agreements with the traditional owners for the shared custody, use and interpretation of their heritage.
34. Researchers and scholarly institutions should decline any offers for the donation or sale of elements of Indigenous Peoples' heritage, without first contracting the peoples or communities directly concerned and ascertaining the wishes of the traditional owners.
35. Researchers and scholarly institutions must refrain from engaging in any study of previously undescribed species or cultivated varieties of plants, animals or microbes, or naturally occurring pharmaceuticals, without first obtaining satisfactory documentation that the specimens were acquired with the consent of the traditional owners.
36. Researchers must not publish information obtained from Indigenous Peoples or the result of research conducted on flora, fauna, microbes or materials discovered through the assistance of Indigenous Peoples, without identifying the traditional owners and obtaining their consent to publication.
37. Researchers should agree to an immediate moratorium on the Human Genome Diversity Project. Further research on the specific genotypes of Indigenous Peoples should be suspended unless and until broadly and publicly supported by Indigenous Peoples to the satisfaction of United Nations human rights bodies.
38. Researchers and scholarly institutions should make every possible effort to increase Indigenous Peoples' access to all forms of medial, scientific and technical education, and participation in all research activities which may affect them or be of benefit to them.
39. Professional associations of scientists, engineers and scholars, in collaboration with Indigenous Peoples, should sponsor seminars and disseminate publications to promote ethical conduct in conformity with these guidelines and discipline members who act in contravention.

Business and Industry

40. In dealings with Indigenous Peoples, business and industry should respect the same guidelines as researchers and scholarly institutions.
41. Business and industry should agree to an immediate moratorium on making contracts with Indigenous Peoples for the rights to discover, record and use previously undescribed species or cultivated varieties of plants, animals or microbes, or naturally occurring pharmaceuticals. No further contracts should be negotiated until Indigenous Peoples and communities themselves

are capable of supervising and collaborating in the research process.

42. Business and industry should refrain from offering incentives to any individuals to claim traditional rights of ownership or leadership within an indigenous community, in violation of their trust within the community and the laws of the Indigenous Peoples concerned.
43. Business and industry should refrain from employing scientists or scholars to acquire and record traditional knowledge or other heritage of Indigenous Peoples in violation of these guidelines.
44. Business and industry should contribute financially to the development of educational and research institutions controlled by Indigenous Peoples and communities.
45. All forms of tourism based on Indigenous Peoples' heritage must be restricted to activities which have the approval of the peoples and communities concerned, and which are conducted under their supervision and control.

Artists, writers and performers

46. Artists, writers and performers should refrain from incorporating elements derived from indigenous heritage into their works without the informed consent of the traditional owners.
47. Artists, writers and performers should support the full artistic and cultural development of Indigenous Peoples, and encourage public support for the development of greater recognition of indigenous artists, writers and performers.
48. Artists, writers and performers should contribute, through their individual works and professional organizations, to the greater public understanding and respect for the indigenous heritage associated with the country in which they live.

Public information and education

49. The mass media in all countries should take effective measures to promote understanding of and respect for Indigenous Peoples' heritage, in particular through special broadcasts and public-service programmes prepared in collaboration with Indigenous Peoples.
50. Journalists should respect the privacy of Indigenous Peoples, in particular concerning traditional religious, cultural and ceremonial activities, and refrain from exploiting or sensationalizing Indigenous Peoples' heritage.
51. Journalists should actively assist Indigenous Peoples in exposing any activities, public or private, which destroy or degrade Indigenous Peoples' heritage.
52. Educators should ensure that school curricula and textbooks teach understanding and respect for Indigenous Peoples' heritage and history and recognize the contribution of Indigenous Peoples to creativity and cultural diversity.

International organizations

53. The Secretary-General should ensure that the task of co-ordinating international co-operation in this field is entrusted to appropriate organs and specialized agencies of the United Nations, with adequate means of implementation.
54. In co-operation with Indigenous Peoples, the United Nations should bring these principles and guidelines to the attention of all Member States through, inter alia, international, regional and national seminars and publications, with a view to promoting the strengthening of national legislation and international conventions in this field.
55. The United Nations should publish a comprehensive annual report, based upon information from all available sources, including Indigenous Peoples themselves, on the problems experienced and solutions adopted in the protection of Indigenous Peoples' heritage in all countries.
56. Indigenous Peoples and their representative organizations should enjoy direct access to all intergovernmental negotiations in the field of intellectual property rights, to share their views on the measures needed to protect their heritage through international law.
57. In collaboration with Indigenous Peoples and Governments concerned, the United Nations should develop a confidential list of sacred and ceremonial sites that require special measures for their protection and conservation, and provide financial and technical assistance to Indigenous Peoples for these purposes.
58. In collaboration with Indigenous Peoples and Governments concerned, the United Nations should establish a trust fund with a mandate to act as a global agent for the recovery of compensation for the unconsented or inappropriate use of Indigenous Peoples' heritage, and to assist Indigenous Peoples in developing the institutional capacity to defend their own heritage.
59. United Nations operational agencies, as well as the international financial institutions and regional and bilateral development assistance programmes, should give priority to providing financial and technical support to indigenous communities for capacity-building and exchanges of experience focused on local control of research and education.

60. The United Nations should consider the possibility of drafting a convention to establish international jurisdiction for the recovery of Indigenous Peoples' heritage across national frontiers, before the end of the International Decade of the World's Indigenous People.

1.20: Indigenous Peoples and Knowledge of the Forest

Contribution submitted by the International Alliance of the Indigenous-Tribal Peoples of the Tropical Forests to the Intergovernmental Panel on Forests (IPF/II), Geneva, 1996.

We indigenous peoples have lived in the tropical forests of the world since time immemorial. We are the first owners of the forest and throughout history have nurtured its biodiversity through skills and practices based on wide experience, a broad variety of knowledge, and a holistic understanding of our environment. Our forests provide us with life, they shape our identity as peoples and frame the boundaries of our territories. The fate of the forests is intimately bound up with our survival and requires the recognition of our indigenous rights.

Our tropical forests, homelands and territories are the areas of the world with the highest cultural diversity. Rainforests of the Amazon, Central Africa, Asia and Melanesia contain over half the total global spectrum of indigenous peoples and at the same time contain some of the highest species biodiversity in the world. We and the forests have survived because throughout the centuries they have provided us with the means of life and in return we have managed them sustainably according to our own scientific knowledge and customary practices.

We do not see humans and natural species as distinct elements for classification and utilization; we are part of a broader system binding us to the forest through social, cultural, political, economic and ecological ties – all expressed through our indigenous spirituality. This holistic view binds indigenous peoples closely with any discussion about forests. We are the front line, defending the forests, and need the support, recognition and respect of the international community for this task.

Whereas our distinct cultures, territories and production methods make us unique guardians of the forest, yet, throughout the world, we are oppressed, our resources are plundered, and we face devastation. Deforestation is a major concern to us because as our forests are destroyed, our capacity to survive withers and we die. The only way to ensure that indigenous peoples and forests survive is to recognize our rights as indigenous peoples living in our own territories, respecting our distinct cultures, political institutions and customary legal systems, and allowing us the means to carry out our own sustainable self-development. These rights can be summarized in the concept of self-determination, which is the process whereby we control our own lives and destinies. Only from a position of respect can we negotiate genuine partnerships with the governments, other forest dwellers, and other institutions in the states where we live, for the sustainable use and development of forests.

It is our experience that a central underlying cause of deforestation is the systematic denial by states, national and foreign companies and international agencies of our rights. Forest policies and national development strategies that deny the customary rights of forest inhabitants simultaneously deprive forests of protection and facilitate unregulated access to forest resources by outsiders. The reason for this is that the forests are a part of us; their life is our life.

In spite of international statements in favour of participation, we indigenous peoples have been sidelined and marginalized in the drafting of Agenda 21 and the Forest Principles and in the discussions at the Commission on Sustainable Development and the Inter-governmental Panel on Forests. No decision relating to indigenous peoples should be taken without our full and free participation at all levels and in all discussions relating to our concerns.

The only area of the UN system where participation has been satisfactory has been the thirteen-year process of discussion on indigenous rights at the UN Working Group on Indigenous Populations (WGIP). The conclusion of this participatory deliberation has been that the recognition of rights is fundamental to all questions involving indigenous peoples. In this way full participation within a process goes hand in hand with the recognition of our rights.

The UN system has gone some way towards recognizing the rights of indigenous peoples. Of particular importance is the draft Declaration on the Rights of Indigenous Peoples which was approved by the Sub-Commission on the Prevention of Discrimination and the Protection of Minorities in 1994. This document arose out of the participatory process of the WGIP and has been built on a general understanding of our rights. This began with the unsatisfactory integrationist ILO Convention 107 and the more recent Convention 169. The UN Draft Declaration is currently the clearest statement of indigenous rights in the UN system and should be seen as a starting point for all subsequent discussions to avoid duplication of a successful thirteen-year process.

Chapter 26 of Agenda 21 which was drafted at the Rio Conference in 1992, acknowledges in its first paragraph that indigenous peoples have 'developed over many generations a holistic traditional scientific knowledge of their lands, natural resources and environment. Indigenous people and their communities shall enjoy the full measure of human rights and fundamental freedoms without hindrance or discrimination'. Agenda 21 and the Forest Principles recognize indigenous rights to lands, to intellectual and cultural property, and to current customary practices. They advocate empowerment of indigenous peoples and promote participation. The Convention on Biological Diversity, also signed at Rio in 1992, also recognizes rights to intellectual and cultural property. However these documents also contain several features that give us cause for concern.

1. State sovereignty

Similar to the CBD, Agenda 21 and the Forest Principles place much emphasis on the rights of States to exploit 'their own' resources. For example, Article 2a of the Forest Principles says 'States have the sovereign and inalienable right to utilize, manage and develop their forests in accordance with their development needs'. We underline that a monolithic notion of State sovereignty undermines the rights of indigenous peoples who live in those forests.

When the State wants to 'develop' an indigenous-owned forest territory without the consent of the people living there, we find ourselves faced with the plundering of our resources. National forests and land use plans are particularly problematic. Too often these are imposed upon us from above in the name of state sovereignty. We indigenous peoples have our own forms of sovereignty based on our indigenous institutions and our legal systems of customary law. These must be respected by national and international bodies wishing to work with indigenous peoples.

2. The question of peoples

Throughout Agenda 21 and the Forest Principles, reference is consistently made to 'indigenous people'. This is not in harmony with the rest of the UN system where the ILO Convention 169 concerns Indigenous and Tribal Peoples, and the draft UN Declaration is on the 'Rights of Indigenous Peoples'. The term 'people' only considers the individual aspect of our being and denies our whole collective identity as peoples. If we are only treated as individuals, it is impossible for us collectively to participate, make contracts or come to agreement with governments as urged by Agenda 21 and the Forest Principles.

3. Territories

The concept of territory does not appear in Agenda 21 or the Forest Principles. This is problematic because the use of the terms 'environment', 'lands' and 'resources' only cover limited aspects of our relationship with the great variety of ecosystems within which we live. We indigenous peoples conceptualize our worlds holistically, and the most suitable term which embraces the economic, socio-cultural, legal, spiritual and political aspects of our relationship with the environment is the term 'territory'. Without the recognition of our territories, we cannot defend the holistic approach to forests emphasized in Agenda 21 and the Forest Principles.

4. Objects of study and development

We do not want to be objects of development, but the subjects of self-development. However, in Chapter 11 of Agenda 21, our knowledge and practices are seen as the 'objects' of non-indigenous research. Not only does this have the effect of commodifying our cultural heritage, but it artificially takes our knowledge out of context and misrepresents the information and ourselves. If we are no more than objects of study, we will continue to be plagued by intrusive investigators and economic interests which aim to utilize our knowledge for personal and commercial gain.

Agenda 21 and the Forest Principles continuously refer to 'participation'. This is frequently linked with the notion of 'partnership'. However, these terms are very ambiguous and, unless they are made clear, make us feel portrayed as passive individuals who have been granted the right to take part in outside initiatives on our lands and territories.

Traditional forest-related knowledge

As an introduction to this subject, we would like to draw attention to the specific needs of indigenous peoples who live in forest areas. All peoples who live in forests have rights; however, those of us who are indigenous have a clearly defined set of rights which have been under discussion in international fora for over twenty years. Two useful documents in this respect are by the UN Special Rapporteur, Erica-Irene Daes. The first (Daes 1994) is a study on the 'Protection of the Cultural and Intellectual Property of Indigenous Peoples' which lays out in a broad and succinct manner our main concerns. The principles and guidelines arising from this study were updated in the final report (Daes 1995).

We consider that any discussions of our knowledge, its uses and benefits, have to be placed within the framework established in these principles and guidelines. They place our rights at the forefront of the discussion – in particular our right to self-determination, our identity as peoples, the inalienable, collective and permanent ownership and custody of our heritage, our territorial rights, and our control and consent over transmission, research, use or commercial application.

However, this does not mean that we are the only forest-dwellers. In the same way that the forest is made up of a variety of species with their different needs which work together within a range of ecosystems, we feel that the needs and desires of different forest peoples – indigenous, peasant farmers, and others, should be outlined and supported in a complementary fashion according to their respective circumstances. In this way forest protection can be strengthened and utilized in a sustainable manner.

The nature of indigenous knowledge

Indigenous Peoples are often depicted as having one 'traditional knowledge'. But this is not the case; we have 'knowledges', diverse systems based on our own scientific principles, built up through thousands of years of empirical observation and experiment. Indigenous knowledge is culturally embedded in its local context.

Unfortunately, when our knowledge is discussed by outsiders, it invariably becomes incorporated into an alien classification system which denies our diversity and is then treated derogatively. For example, our sustainable systems of rotational agriculture have worked in harmony with forests for millennia. Yet now, when colonists misuse these sophisticated techniques, we are accused of destroying our own forests through 'slash and burn agriculture'.

Knowledge for indigenous peoples is grounded in particular territories and is thus extremely diverse. The collective relationship to a territory is what binds us together as distinct peoples. Thus knowledge, territory and identity as peoples are interrelated. Knowledge cannot, therefore, be separated from the human and natural environment.

We indigenous peoples do not learn and develop our understanding of the world through abstract prescriptions but through practical experience. The rules for use and the information exist, but they are the results of a multiplicity of activities and long-term observations which are largely tacit, embodying a multitude of skills and practicalities. Far from being rigid, these are constantly up-dated and changing. All this change takes place within a framework grounded in our indigenous institutions and customary legal systems which express continuity within a culturally appropriate framework.

Any reference to knowledge has to refer to the practitioners of that knowledge. Our concern is that we do not want to be categorized in such a way that we are accused of inauthenticity and consequently of no longer being indigenous when we modify our customs and practices to suit changing circumstances. The impression is of a largely static corpus of information and rules passing down the generations with minor adjustments, constituting the notion of 'traditional'. We are uncomfortable about this approach and the use of the word 'tradition' to relate to our distinct cultures because we easily become portrayed as fossilized survivals from the past. This does not reflect the reality of our lives as living vibrant peoples.

Knowledge is passed on according to our own indigenous educational principles which are rooted in social activities and practical experience passed from our ancestors to our descendants. This temporal transfer of collective experience is not institutionalized as in national educational methods, but is rooted in our activities, languages and oral heritage.

Complementary to this is the receiving of knowledge from revelations from the spirit world. We have used shamanic practices for thousands of years to draw us into personal and collective relationships with the world of spirituality. They provide a perfect example of how methods of indigenous knowledge are quite distinct from those of Western scientific methodology.

Our sharing of knowledge binds us together and forms our identity as indigenous peoples, constituting our distinct and unique cultures. For this reason, our insistence on being identified as peoples and not as people or communities is essential for understanding our knowledge.

Knowledge also includes the creative, innovatory and experimental approach to the world, bound up with our own technologies. This knowledge not only arises from ourselves, but also from trading relationships with other Indigenous and non-indigenous peoples. Providing we control it, trade and technology can be transformed into culturally appropriate change.

We reject the non-indigenous classificatory abstract notions of knowledge used by those who wish to take our knowledge out of context for commodification and exploitation. Knowledge, such as species recognition, understanding of forest resources or practical use, is based on an understanding of how the different aspects of knowledge interrelate. This is what we mean by holistic.

Knowledge is bound up with the notion of territoriality; knowledge is practical and flexible and should not be defined rigidly as changeless tradition;

knowledge is collective and a shared aspect of our cultural identity as distinct peoples; knowledge and tradition have to be defined by indigenous peoples ourselves; and knowledge in its long-term and holistic world-view is an important aspect of our fundamental right to self-determination.

Indigenous knowledge and property rights

The conclusions and declarations of the regional meetings between indigenous peoples under the auspices of the United Nations Development Programme in Latin America, Asia and the Pacific take indigenous rights as the starting point for looking at indigenous knowledge and property rights. Each of the documents assert that the protection of indigenous peoples' knowledge is based on the right to self-determination and territoriality; it is collective and pertains to indigenous peoples as peoples.

The International Alliance strongly rejects the patenting of life-forms and are outraged to learn that applications for patents have been made on our medicinal plants such as *Banisteriopsis*. The revision of plant variety rights could be extended to our knowledge systems, but we strongly oppose the limited notion of Farmers' Rights held by the FAO. Under the FAO's system, benefits do not accrue to local farmers and the whole system has been ineffective.

Indigenous knowledge must be recognized as being inherent in a collectivity, rather than the property of individuals, and consequently information cannot be used without the consent of the people concerned. Group ownership should therefore be recognized in law and access to indigenous knowledge should only take place by agreement between the parties involved. However, there can be no partnership until our rights are recognized: our collective rights as peoples to our territories; our free access and control over our resources; and our prior and informed consent before entering into any partnership.

Participation and partnership

There are three areas where participation and partnership can arise relating to our knowledge. The first is management of protected areas. Throughout the world we indigenous peoples find our territories being declared protected areas unilaterally. We consider this to be a form of expropriation because inevitably decisions over these areas are made without our full participation and frequently this involves curtailment of our customary economic practices and, even worse, forcible relocation from our ancestral territories. The creation of protected areas without our consent can be as threatening as large-scale development activities such as dams, mines or oil prospecting. We consider that management in partnership has to recognize our rights to our territories.

Another area of partnership is research on our territories. We are pleased to collaborate with researchers with whom we have agreed to work; however, this does not mean that we become 'stakeholders' on our territories. To research in a collaborative way does not mean that we are sharing our lands and resources with outside interests. Throughout history, sharing rights has been the first step in the colonization our peoples.

A third area of partnership consists of defining the practical use of an area and takes the form of zoning. Dividing territories into different areas of use is something which we indigenous peoples have been doing for centuries. However, rather than operate in terms of large-scale zoning techniques, we organize our territories in a multitude of different ways reflecting the nature of the lands and resources. Indigenous patterns of resource management encountered throughout the world can be highly sophisticated. In an indigenous context it is counter-productive to impose externally designed patterns of zoning. Indigenous management and organizational practices must be used as the basis for any management partnership.

We want a form of partnership and participation that is based on the principles of control over our own resources and which acknowledges that no activity takes place in our territories without our free and informed consent.

Biodiversity prospecting: the commodification of knowledge

Indigenous knowledge can help biodiversity prospectors create new goods and services that might be patented and sold. The International Alliance agrees with the statement made at the Pacific Indigenous Peoples' UNDP consultation which calls for a 'moratorium on bio-prospecting in the Pacific and urges indigenous peoples not to co-operate in bio-prospecting activities until appropriate mechanisms are in place'. As the statement says, we indigenous peoples are willing to share our knowledge with humanity provided we determine when, where and how it is used. At present the international system does not recognize or respect our past, present and potential contribution.'

We are not interested in the commercial significance for multinational companies but in the welfare of humanity. We have the right to determine how we should deal with companies and we consider that our consent is a prerequisite for any access or commercial contract. This consent can only be forthcoming if our rights are fully recognized. The forest resources are a community resource and we are the peoples who can provide access to the resources on our territories. No single individual has the right to grant anyone access: it has to be done collectively because of our inalienable rights to our ancestral territories.

Three topics relate to the commercial use of our resources:

a) Trade

We are not opposed to trading. Markets have been the mainstay of our economic survival for centuries, and our trading systems have provided our economies with a basis of sustainability. We want to retain a diversity of markets and not become dependent on one monolithic set of market forces. Our goal is a fair trading system which recognizes our rights to resources and products, in particular sustainable harvesting. However, we do not want to see our resources stolen by outsiders in order to improve profits. Any trading of our resources must take place under our own control. This principle must be the basis for discussions on the relationship between trade and the environment.

Benefit-sharing is clearly a desirable goal, but this should not operate under the assumption that profit distribution to indigenous peoples is a return for automatic access to our resources. Furthermore, we object strongly to attempts by countries and governments to convert benefit-sharing, which is ours by right, into development aid, which we solicit. The sharing of benefits has to be negotiated fully, fairly and transparently. We are tired of multinationals entering our territories with the connivance of governments eager to sell off our resources and share the spoils between them. We are not prepared to pick up a few crumbs of profit to justify this despoliation and plundering of our territories.

Any form of trade, whether bio-prospecting or the trade of non-wood products, can easily draw indigenous peoples into a situation of dependency. Clearly in many parts of the world we need access to the market economy, but this must take place on our terms or else we will speedily become integrated into a system we cannot control. Our resources, our sole sustenance, will disappear rapidly. Instead, we consider that prior to thinking of contracts with bio-prospecting companies or multinational companies, the following economic strategy is appropriate.

i) Concentrate on strengthening and empowering indigenous economies so that, regardless of the fluctuations of the markets with which we trade, we can enter trading relationships from positions of self-sufficiency, not of poverty and dependency.
ii) That trading starts from our own local and national economies, so that the benefits arising from the trade go to strengthening our position within our immediate social and ecological environment, and not to extracting our resources for the profits of outsiders.

b) Control of companies

The converse side of trading is the relationship between those who buy our resources and those who sell them. In order to ensure fair trading, companies should be controlled. The Alliance is not convinced that development contracts have yet been proven successful. Even though the Alliance appreciates that some small companies have striven to ameliorate the harm caused by exploitative trading patterns, these do not really solve the problem of how profits should be shared among indigenous peoples, nor to whom the knowledge belongs. Although there are innovative approaches to benefit-sharing by small companies, these are distorted through trusteeship arrangements which do not correspond to indigenous peoples' demands for direct ownership and control of our knowledge and its uses.

Criteria and indicators of sustainable forestry and processes of certification should ensure that companies respect the environment and deal openly and fairly with us. However, unless these measures are particularly stringent, their effect will be worse than useless. The discussion on scientific assessment of forests, criteria and indicators for sustainable forestry, and the complementary process of certification, should not become simply a discussion on maximizing logging within certain limits. All definitions of 'sustainability' must include a recognition of the capacity and requirements of local peoples to survive in a sustainable manner. This means that criteria, indicators and certification must include the social and material well-being of local indigenous peoples.

We are very concerned that certification and the use of criteria and indicators can be misused to promote the extraction of our resources. We therefore insist that we can only accept them if our indigenous rights are clearly accepted as criteria and indicators of sustainability and that certification is based on recognition of our rights and welfare.

Finally we would like to make a plea that the international community does something to stop the uncontrolled entry of multinational corporations into our communities: mining, logging and oil exploration are serious environmental and social problems. Indeed, at the moment these are causing more harm than bio-prospecting. They constitute one of the greatest threats to indigenous knowledge because they stand to destroy the forests in which we live and consequently threaten our very existence.

c) Intellectual Property Rights and the GATT

We are extremely concerned that the Trade-Related Aspects of Intellectual Property Rights (TRIPs) section of the GATT treaty is going to open up our resources to yet more despoliation. We are threatened by the world trade system because it facilitates companies and prospectors to expropriate our knowledge and resources while opening up the possibility of patenting life forms.

While we are aware that Article 27 ('Patentable Subject Matter') allows for an 'effective *sui generis* system' for plant variety protection, this refers to the right of States to define their own systems of protection. Whereas in the best cases this might allow for community-based systems of resource protection for indigenous peoples, this requires the good will of our governments, most of whom are reluctant to recognize even our basic rights. Relating to this question, the Alliance endorses the conclusion of the Pacific Regional meeting which urges 'those Pacific governments who have not signed the General Agreement on Tariffs and Trade (GATT) to refuse to do so, and encourage those governments who have already signed to protest against any provisions which facilitate the expropriation of indigenous peoples' knowledge and resources and the patenting of life forms'.

We consider that we already have indigenous *sui generis* systems within our own customary law and indigenous legal systems. We urge strongly that our own indigenous forms of property protection are recognized and respected. This is essentially the basis on which our control over our knowledge can be maintained and our consent can be obtained for its use.

Sharing good ideas

This refers to non-commercial information which could be of widespread benefit if shared. We agree that our knowledge could be useful if shared with concerned groups and institutions but insist that we should be acknowledged for our contribution to any collaborative initiatives. Our concern is that the sharing is not between equal parties. Too often, we find ourselves in the invidious position that we are either treated as ignorant and unable to understand our environment, or else our knowledge is recognized and coveted. Only rarely do we receive the respect due to our information and are able to enter into genuinely collaborative arrangements with non-indigenous researchers. Any form of computerization of our knowledge must be recognized as pertaining to our cultural heritage and means must be set aside so that information referring to us is held in our territories.

Several indigenous peoples of the Amazon have agreed in their assemblies that all researchers in their territories have to receive the prior informed consent of any community they work in and the written consent of the local indigenous organization. Copies of data have to be deposited in the organization's archive and all work has to be completely transparent. Whereas monitoring this in practice is difficult, it is important that researchers are responsible enough to treat these requests seriously.

Our concern is even greater when researchers carry out assessments and evaluations of our resources. Whereas we have been told that it is in our interests to see the value of our territories, we cannot see how our resources can be assessed and evaluated out of context by outside interests. Our own values are not based on money and costs, but a priceless combination of all the facets of forests – economic, political, cultural, social and spiritual. The subsequent misrepresentation of our knowledge arising from external assessments and other research presents a problem because we are rarely in a position to answer back and defend our integrity.

The principles behind indigenous sharing of knowledge should reflect the UNESCO Declaration of the Principles of International Cultural Co-operation (1966) where the free exchange of cultural knowledge is linked to 'respect' and 'reciprocity' among cultures. Furthermore, our right to privacy should be protected, as in Article 17 of the International Covenant on Civil and Political Rights; and, as in Article 19 of the same Covenant, we also expect protection from representation of indigenous peoples that is injurious to our 'rights or reputations'. This means transparency and accuracy in sharing our knowledge.

We feel that the primary interest of governments and international institutions is in what information, insights and knowledge they can get out of us; yet when we ask for what we desperately need – recognition of our rights as indigenous peoples – we almost inevitably encounter a lack of interest, or even open hostility. Recognition of our basic rights is a part of this whole process; without our rights clearly respected, there will be no forests, no indigenous peoples, and nothing for science or broader humanity to learn.

Obstacles to further progress

Obstacles to further progress arise from the lack of implementation of indigenous peoples' rights, as recognized in the draft Declaration of the Rights of Indigenous Peoples, and a lack of co-operation within the United Nations. We would suggest that each UN body provides the starting point for open negotiations with indigenous peoples in the transparent manner that the UN Working Group on Indigenous Populations has done over the last decade.

The rights initiatives from the Commission on Human Rights should flow through the other bodies of the UN. Similarly, conclusions on sustainable development and insights into Agenda 21 can flow to the CBD and human rights bodies. The CBD can look at questions of access, use and benefit-sharing; however, these should not be exclusive but mutually enriching discussions. The effect would be a holistic approach to indigenous concerns which reflects our own experiences and practices. Chapter 26 of Agenda 21 proposes to 'appoint a special focal point within each international organization' for indigenous peoples. An indigenous office in these institutions would be extremely useful in bringing together forest questions with those of indigenous peoples. However, of even more importance is the proposal currently under discussion in the UN for the formation of a Permanent Forum on Indigenous Peoples which will deal not only with human rights questions, but also with development and environment questions. This would have the capacity of drawing together the different strands of the UN's work with indigenous peoples and provide a forum where indigenous representatives can come and present their opinions and perspectives on forest questions.

Another obstacle to implementation is financial. Currently the Global Environmental Facility (GEF) seems to be the main candidate for implementing the financial aspects of international environmental negotiations. As it stands, the GEF is dominated by the World Bank which, since the 1970s, has been carrying out projects directly inimical to the needs and desires of indigenous peoples. Furthermore, the GEF can only fund 'the incremental costs of global benefits' (above national benefits). It is thus unsuitable for funding projects with indigenous peoples whose interests are local and national. To insist on the GEF's approach could be highly problematic for indigenous peoples who are striving for a constructive and beneficial relationship with governments, not an abrogation of responsibility. Any future financial mechanism has to be accountable, transparent, and without conditions on incremental costs.

Conclusions and recommendations

We consider that our participation as indigenous peoples in the UN environmental negotiations has been totally deficient. The result is that we are not considered as peoples with territories who should be subjects of our own development. This unsatisfactory situation leaves us without voice and ignored.

Any discussion on peoples and forests has to address the totality of our rights: our identity as peoples, our territoriality, our cultural heritage, our customary law and our political institutions which are framed by our fundamental right to self-determination. The document which best reflects our rights is the draft UN Declaration on the Rights of Indigenous Peoples.

Indigenous peoples' knowledge about forests is bound up with our whole lives, our world views and our identity as indigenous peoples acting and reflecting on our territories. Knowledge is a living thing, which changes and adapts to our own circumstances. It is embedded in practical activities and cannot be incorporated into universal notions such as 'traditional'. Tradition arises from our distinct identities and cannot be defined by outsiders. Indigenous knowledge is a part of our culture and heritage.

Partnership and participation can only take place between equals and in conditions where our fundamental rights remain intact. We are right-holders, not stakeholders. No activity should take place on our territories without our free and informed consent: we insist that we have the right to control our own resources.

Our consent is a prerequisite for any access agreement or commercial contract. Whereas we support benefit-sharing, this must be based on principles of fair trade, with a priority focused on empowering our diverse local economies and markets.

Criteria and indicators for sustainable development or certification too easily become a means for accepting the exploitation of our resources. We insist that these should ensure indigenous peoples' rights and welfare.

The threat of GATT/TRIPs to our cultural heritage and forest-related knowledge is serious. Even the possibility of *sui generis* approaches to intellectual property rights under Article 27 of GATT/TRIPs could end up reflecting state priorities and not provide the protections which we need.

In the work of the environmental negotiations, too much emphasis is given to what governments and international institutions can get out of us. All knowledge that is shared must be based on principles of reciprocity, consent, transparency, accuracy, respect and sensitivity.

Without recognition of our territorial rights, our own indigenous sovereign institutions and our customary legal systems, sustainable forest management will be impossible, the process of colonization will continue, and States and multinational corporations will have free access to take our resources without our consent. The result will be increased deforestation and our disappearance as peoples. The recognition of our rights is a prerequisite for any solution to this problem. In this way a notion of pluriculturality can emerge within a framework of co-existing sovereignties.

This has to be the starting point for constructive agreements between indigenous peoples and governments in dealing with forests. States must recognize indigenous peoples' political institutions so that we have the means to express our approval and prior informed consent before any extraneous activities take place on our lands and territories.

APPENDIX 2

FAITH STATEMENTS ON RELIGION AND ECOLOGY

2.1 General introduction (Joanne O'Brien, Martin Palmer and James Wilsdon, ICOREC)

In 1985 WWF UK published its first RE (Religion and Environment) book dedicated to exploring the relationships between religious belief and ecology. The book, *Worlds of Difference*, not only brought environmental studies into RE in schools, and religious studies into ecology, but also launched a world-wide programme of work with the major religions. As a direct result of *Worlds of Difference* WWF International decided to make its 25th anniversary celebrations a religious event. Five major faiths were invited to meet in Assisi in Italy, and from this sprang the work, which WWF world-wide has sponsored and assisted, of environmental programmes and projects arising from faith communities.

In 1995 the United Nations honoured this unique work with an award, saying that through the religion and ecology project, the environmental message had reached "untold millions who could not have been reached by other methods".

Today, over 120,000 religiously based environmental projects exist around the world. But work with RE and with schools is still a priority for WWF and for its religious advisers, ICOREC (International Consultancy on Religion, Education and Culture). In recent years, new developments have taken place. The original five religions have grown to nine and the depth of reflection by the world's faiths has also grown.

The original Assisi Declarations have since been overtaken by new developments within each religion. The four faiths who joined after Assisi (Baha'is, Sikhs, Jains and Taoists) have also published their initial Statements and these are included here.

Section 2.2 of this Appendix starts with the five momentous Assisi Declarations, and adds the declarations of the Baha'is (1987), Sikhs (1989) and Jains (1991).

Section 2.3 brings in the new Statements of the major religions and adds a new one, Taoism. These are now considered the most authoritative statements by each Faith.

2.2: The Assisi Declarations

Introduction

In the late summer of 1986, an historic event occurred in the Umbrian hill town of Assisi. For the first time, representatives of the great faiths of the world came together in pilgrimage in order to hear what religion had to offer to conservation and what conservation had to share with religion.

Pilgrims, religious leaders, musicians and dancers marched into Assisi alongside scientists and conservationists, their divisions overcome by their shared commitment to the natural world. As they entered the town, the marchers were met by a glorious cacophony of sound and colour as singers and performers from many different parts of the world joined in the celebrations.

The following day, 800 people gathered in the Basilica of St Francis of Assisi to participate in an inspiring ceremony, at which representatives of five great religions, the WWF family, the pilgrims and many other guests affirmed their commitment to safeguarding the integrity of creation.

The central message of the ceremony was that ecology needs the deep truths which lie within the major faiths of the world if people are to recognize their responsibility for and with nature. As Father Serrini, Minister General of the Franciscans said in his introduction, 'We are convinced of the inestimable value of our respective traditions and of what they can offer to re-establish ecological harmony; but, at the same time, we are humble enough to desire to learn from each other. The very richness of our diversity lends strength to our shared concern and responsibility for our Planet Earth'.

The final part of the Assisi celebrations, which shifted the focus from the present to the future, was the issuing of the Declarations. These are the authoritative statements on ecology and nature from the five major faiths – Buddhism, Christianity, Hinduism, Islam and Judaism – sections of which are reprinted below.

The members of each faith then went back and reflected on their own teachings and on the events they had witnessed at Assisi. Many sought to put what they had learned into practice through practical conservation projects and programmes of environmental education. A vast number of such projects are now operating throughout the world, ensuring that the ecological message reaches millions of people.

Soon after Assisi, three more faiths – Baha'i, Jainism and Sikhism – produced their own Declarations to accompany those of the other religions, and these are included below. The overall significance of what had taken place at Assisi was summed up by HRH The Duke of Edinburgh when he announced during the ceremony that, 'a new and powerful alliance has been forged between the forces of religion and the forces of conservation'.

Buddhism (1986)

There is a natural relationship between a cause and its resulting consequences in the physical world. In the life of the sentient beings too, including animals, there is a similar relationship of positive causes bringing about happiness, while undertakings generated through ignorance and negative attitude bring about suffering and misery. And this positive human attitude is, in the final analysis, rooted in genuine and unselfish compassion and loving kindness that seeks to bring about light and happiness for all sentient beings. Hence Buddhism is a religion of love, understanding and compassion, and committed towards the ideal of non-violence. As such, it also attaches great importance to wild life and the protection of the environment on which every being in this world depends for survival.

We regard our survival as an undeniable right. As co-inhabitants of this planet, other species too have this right of survival. And since human beings as well as other non-human sentient beings depend upon the environment as the ultimate source of life and well-being, let us share the conviction that the conservation of the environment, the restoration of the imbalance caused by our negligence in the past, be implemented with courage and determination.

Christianity (1986)

God declared everything to be good; indeed, very good. He created nothing unnecessarily and has omitted nothing that is necessary. Thus, even in the mutual opposition of the various elements of the universe, there exists a divinely willed harmony because creatures have received their mode of existence by the will of their Creator, whose purpose is that through their interdependence they should bring to perfection the beauty of the universe. It is the very nature of things considered in itself, without regard to humanity's convenience or inconvenience, that gives glory to the Creator.

Humanity's dominion cannot be understood as licence to abuse, spoil, squander or destroy what God has made to manifest His glory. That dominion cannot be anything other than a stewardship in symbiosis with all creatures. On the one hand, humanity's position verges on a viceregal partnership with God; on the other, his self-mastery in symbiosis with creation must manifest the Lord's exclusive and absolute dominion over everything, over humanity and its stewardship. At the risk of destroying itself, humanity may not reduce to chaos or disorder, or, worse still, destroy God's bountiful treasures.

Every human act of irresponsibility towards creatures is an abomination. According to its gravity, it is an offence against that divine wisdom which sustains and gives purpose to the interdependent harmony of the universe.

Hinduism (1986)

Hinduism believes in the all-encompassing sovereignty of the divine, manifesting itself in a graded scale of evolution. The human race, though at the top of the evolutionary pyramid at present, is not seen as something apart from the earth and its multitudinous life-forms.

The Hindu viewpoint on nature is permeated by a reverence for life, and an awareness that the great forces of nature – the earth, the sky, the air, the water and fire – as well as various orders of life including plants and trees, forests and animals, are all bound to each other within the great rhythms of nature. The divine is not exterior to creation, but expresses itself through natural phenomena. In the Mudaka Upanishad the divine is described as follows:

'Fire is his head, his eyes are the moon and sun; the regions of space are his ears, his voice the revealed Veda; the wind is his breath, his heart is the entire universe; the earth is his footstool, truly he is the inner soul of all.'

The natural environment has received the close attention of the ancient Hindu scriptures. Forests and groves were considered sacred, and flowering trees received special reverence. Just as various animals were associated with gods and goddesses, different trees and plants were also associated in the Hindu pantheon. The Mahabharata says that, 'even if there is only one tree full of flowers and fruits in a village, that place becomes worthy of worship and respect'.

Islam (1986)

The essence of Islamic teaching is that the entire universe is God's creation. Allah makes the waters flow upon the earth, upholds the heaven, makes the rain fall and keeps the boundaries between day and night. The whole of the rich and wonderful universe belongs to God, its maker. It is God who created the plants and the animals in their pairs and gave them the means to multiply. Then God created humanity – a very special creation because humanity alone was created with reason and the power to think, and even the means to turn against the Creator. Humanity has the potential to acquire a status higher than that of the angels or sink lower than the lowliest of the beasts.

For the Muslim, humanity's role on earth is that of a 'khalifa', vice-regent or trustee of God. We are God's stewards and agents on Earth. We are not masters of this Earth, it does not belong to us to do what we wish. It belongs to God and He has entrusted us with its safekeeping. Our function as vice-regents, 'khalifa' of God, is only to oversee the trust. The 'khalifa' is answerable for his/her actions, for the way in which he/she uses or abuses the trust of God.

Allah is Unity; and His Unity is also reflected in the unity of humanity, and the unity of humanity and nature. His trustees are responsible for maintaining the unity of His creation, the integrity of the Earth, its flora and fauna, its wildlife and natural environment. Unity cannot be had by discord, by setting one need against another or letting one end predominate over another; it is maintained by balance and harmony.

Judaism (1986)

When God created the world, so the Bible tells us, He made order out of primal chaos. The sun, the moon, and the stars; plants, animals, and ultimately humanity, were each created with a rightful and necessary place in the universe. They were not to encroach on each other. 'Even the divine teaching, the Torah, which was revealed from on high, was given in a set measure' (Vayikra Rabbah 15:22) and even these holy words may not extend beyond their assigned limit.

The highest form of obedience to God's commandments is to do them not in mere acceptance but in the nature of union with Him. In such a joyous encounter between man and God, the very rightness of the world is affirmed. The encounter of God and man in nature is thus conceived in Judaism as a seamless web with man as the leader and custodian of the natural world.

There is a tension at the centre of the Biblical tradition, embedded in the very story of creation itself, over the question of power and stewardship. The world was created because God willed it, but why did He will it? Judaism has maintained, in all of its versions, that this world is the arena that God created for man, half beast and half angel, to prove that he could behave as a moral being. The Bible did not fail to demand even of God Himself that He be bound, as much as humanity, by the law of morality. Thus Abraham stood before God, after He announced that He was about to destroy the wicked city of Sodom, and Abraham demanded of God Himself that He produce moral justification for this act: 'Shall not the judge of all the earth do justice?' Comparably, man was given dominion over nature, but he was commanded to behave towards the rest of creation with justice and compassion. Humanity lives, always, in tension between his/her power and the limits set by conscience.

The following three statements, from the Baha'is, Sikhs and Jains, were issued in the years following Assisi.

Baha'i (1987)

'Nature in its essence is the embodiment of My Name, the Maker, the Creator. Its manifestations are diversified by varying causes, and in this diversity there are signs for men of discernment. Nature is God's Will and is its expression in and through the contingent world. It is a dispensation of Providence ordained by the Ordainer, the All-Wise.' (Baha'i writings.)

With those words, Baha'u'llah, Prophet-founder of the Baha'i faith, outlines the essential relationship between humanity and the environment: that the grandeur and diversity of the natural world are purposeful reflections of the majesty and bounty of God. For Baha'is, there follows an implicit understanding that nature is to be respected and protected, a divine trust for which we are answerable.

As the most recent of God's revelations, however, the Baha'i teachings have a special relevance to present-day circumstances when the whole of nature is threatened by man-made perils ranging from the wholesale destruction of the world's rainforests to the final nightmare of nuclear annihilation.

A century ago, Baha'u'llah proclaimed that humanity has entered a new age. Promised by all the religious Messengers of the past, this new epoch will ultimately bring peace and enlightenment for humanity. To reach that point, however, humankind must first recognize its fundamental unity – as well as the unity of God and of religion. Until there is a general recognition of this wholeness and interdependence, humanity's problems will only worsen.

Sikhism (1989)

Since the beginning of the Sikh religion in the late fifteenth century, the faith has been built upon the message of the 'oneness of Creation'. Sikhism believes the universe was created by an almighty God. He himself is the creator and the master of all forms in the universe, responsible for all modes of nature and all elements in the world.

Sikhism firmly believes God to be the source of the birth, life and death of all beings. God is the omniscient, the basic cause of the creation and the personal God of them all.

From the Divine command occurs the creation and the dissolution of the universe. (p117 Guru Granth Sahib)

As their creator, the natural beauty which exists and can be found in all livings things whether animals, birds, fish, belongs to Him, and He alone is their master and without His *Hukum* (order) nothing exists, changes or develops.

Having brought the world into being, God sustains, nourishes and protects it. Nothing is overlooked. Even creatures in rocks and stones are well provided for. Birds who fly thousands of miles away leaving their young ones behind know that they would be sustained and taught to fend for themselves by God (Guru Arjan, in Rehras). The creatures of nature lead their lives under God's command and with God's grace.

Jainism (1991)

The Jain ecological philosophy is virtually synonymous with the principle of *ahimsa* (non-violence) which runs through the Jain tradition like a golden thread.

Ahimsa is a principle that Jains teach and practise not only towards human beings but towards all nature. It is an unequivocal teaching that is at once ancient and contemporary.

There is nothing so small and subtle as the atom nor any element so vast as space. Similarly, there is no human quality more subtle than non-violence and no virtue of spirit greater than reverence for life.

The teaching of *ahimsa* refers not only to physical acts of violence but also to violence in the hearts and minds of human beings, their lack of concern and compassion for their fellow humans and for the natural world. Ancient Jain texts explain that violence (*himsa*) is not defined by actual harm, for this may be unintentional. It is the intention to harm, the absence of compassion, that makes an action violent. Without violent thought there could be no violent actions.

Jain cosmology recognizes the fundamental natural phenomenon of symbiosis or mutual dependence. All aspects of nature belong together and are bound in a physical as well as a metaphysical relationship. Life is viewed as a gift of togetherness, accommodation and assistance in a universe teeming with interdependent constituents.

2.3 Religious statements

We are grateful to the World Wildlife Fund, United Kingdom (WWF-UK) and the Alliance of Religions and Conservation (ARC) for allowing us to use the following papers.

Bahá'í and the environment

> 'Nature in its essence is the embodiment of My Name, the Maker, the Creator. Its manifestations are diversified by varying causes, and in this diversity there are signs for men of discernment. Nature is God's Will and is its expression in and through the contingent world. It is a dispensation of Providence ordained by the Ordainer, the All-Wise.' (Bahá'í writings.)

With those words, Bahá'u'lláh, Prophet-founder of the Bahá'í faith, outlines the essential relationship between man and the environment: that the grandeur and diversity of the natural world are purposeful reflections of the majesty and bounty of God. For Bahá'ís, there follows an implicit understanding that nature is to be respected and protected, as a divine trust for which we are answerable.

Such a theme, of course, is not unique to the Bahá'í faith. All the world's major religions make this fundamental connection between the Creator and His creation. How could it be otherwise? All the major independent religions are based on revelations from one God – a God who has successively sent His Messengers to earth so that humankind might become educated about His ways and will. Such is the essence of Bahá'í belief.

As the most recent of God's revelations, however, the Bahá'í teachings have a special relevance to present-day circumstances when the whole of nature is threatened by man-made perils ranging from the wholesale destruction of the world's rainforests to the final nightmare of nuclear annihilation.

A century ago, Bahá'u'lláh proclaimed that humanity has entered a new age. Promised by all the religious Messengers of the past, this new epoch will ultimately bring peace and enlightenment for humanity. To reach that point, however, humankind must first recognize its fundamental unity, as well as the unity of God and of religion. Until there is a general recognition of this wholeness and interdependence, humanity's problems will only worsen.

'The well-being of mankind, its peace and security, are unattainable unless and until its unity is firmly established', Bahá'u'lláh wrote. 'The earth is but one country, and mankind its citizens.'

The major issues facing the environmental movement today hinge on this point. The problems of ocean pollution, the extinction of species, acid rain and deforestation, not to mention the ultimate scourge of nuclear war, respect no boundaries. All require a transnational approach.

While all religious traditions point to the kind of co-operation and harmony that will indeed be necessary to curb these threats, the religious writings of the Bahá'í faith also contain an explicit prescription for the kind of new world political order that offers the only long-term solution to such problems.

'That which the Lord hath ordained as the sovereign remedy and mightiest instrument for the healing of the world is the union of all its people in one universal Cause', Bahá'u'lláh wrote.

Built around the idea of a world commonwealth of nations, with an international parliament and executive to carry out its will, such a new political order must also, according to the Bahá'í teachings, be based on the principle of economic justice, equality between the races, equal rights for women and men, and universal education.

All these points bear squarely on any attempt to protect the world's environment. The issue of economic justice is an example. In many regions of the world, the assault on rainforests and endangered species comes as the poor, legitimately seeking a fair share of the world's wealth, fell trees to create fields. They are unaware that, over the long term and as members of a world community which they know little about, they may be irretrievably damaging rather than improving their children's chances for a better life. Any attempt to protect nature must, therefore, also address the fundamental inequities between the world's rich and poor.

Likewise, the uplifting of women to full equality with men can help the environmental cause by bringing a new spirit of feminine values into decision-making about natural resources. The scriptures of the Bahá'í faith note that, 'Man has dominated over woman by reason of his more forceful and aggressive qualities both of body and mind. But the balance is already shifting; force is losing its dominance, and mental alertness, intuition and the spiritual qualities of love and service, in which woman is strong, are gaining ascendancy. Hence the new age will be an age less masculine and more permeated with feminine ideals'.

Education, especially an education that emphasizes Bahá'í principles of human interdependence, is another prerequisite to the building of a global conservation consciousness. The earth's theology of unity and interdependence relates specifically to environmental issues. Again, to quote Bahá'í sacred writings:

> 'By nature is meant those inherent properties and necessary relations derived from the realities of things. And these realities of things, though in the utmost diversity, are yet intimately connected one with the other. Liken the world of existence to the temple of man. All the organs of the human body assist one another, therefore life continues. Likewise among the parts of existence there is a wonderful connection and interchange of forces which is the only cause of life of the world and the continuation of these countless phenomena.'

The very fact that such principles should come with the authority of religion and not merely from human sources is yet another piece of the overall solution to our environmental troubles. The impulse behind the Assisi declarations on nature is testimony to this idea.

There is perhaps no more powerful impetus for social change than religion. Bahá'u'lláh said, 'Religion is the greatest of all means for the establishment of order in the world and for the peaceful contentment of all that dwell therein.' In attempting to build a new ecological ethic, the teachings of all religious traditions can play a role in helping to inspire their followers.

Bahá'u'lláh, for example, clearly addresses the need to protect animals. 'Look not upon the creatures of God except with the eye of kindliness and of mercy, for Our loving providence hath pervaded all created things, and Our grace encompassed the earth and the heavens.'

'He Himself expressed a keen love and appreciation for nature, furthering the connection between the environment and the spiritual world, in Bahá'í theology. 'The country is the world of the soul, the city is the world of bodies', Bahá'u'lláh said.

This dichotomy between spirituality and materialism is a key to understanding the plight of humankind today. The major threats to our world environment, such as the threat of nuclear annihilation, are manifestations of a world-encompassing sickness of the human spirit, a sickness that is marked by an overemphasis on material things and a self-centredness that inhibits our ability to work together as a global community. The Bahá'í faith seeks above all else to revitalize the human spirit and break down the barriers that limit fruitful and harmonious co-operation among men and women, whatever their national, racial or religious background. For Bahá'í the goal of existence is to carry forward an ever-advancing civilization. Such a civilization can only be built on an earth that can sustain itself. The Bahá'í commitment to the environment is fundamental to our faith.

Buddhism and the environment

All Buddhist teachings and practice come under the heading of *Dharma* which means Truth and the path to Truth. The word *Dharma* also means 'phenomena' and in this way we can consider everything to be within the sphere of the teachings. All outer and inner phenomena, the mind and its surrounding environment, are understood to be inseparable and interdependent.

In his own lifetime, the Buddha came to understand that the notion that one exists as an isolated entity is an illusion. All things are interrelated, all interconnected, and do not have autonomous existence. Buddha said, 'This is because that is; this is not because that is not; this is born because that is born; this dies because that dies'. The health of the whole is inseparably linked with the health of the parts, and the health of the parts is inseparably linked with the whole. Everything in life arises through causes and conditions.

Many Buddhist monks such as His Holiness the Dalai Lama, Venerable Thich Nhat Hanh, Venerable Kim Teng, and Venerable Phra Phrachak emphasize the natural relationship between deep ecology and Buddhism. According to the Vietnamese monk Venerable Thich Nhat Hanh:

> 'Buddhists believe that the reality of the interconnectedness of human beings, society and nature will reveal itself more and more to us as we gradually recover – as we gradually cease to be possessed by anxiety, fear and the dispersion of the mind. Among the three – human beings, society and nature – it is us who begin to effect change. But in order to effect change we must recover ourselves, one must be whole. Since this requires the kind of environment favourable to one's healing, one must seek the kind of life-style that is free from the destruction of one's humanness. Efforts to change the environment and to change oneself are both necessary. But we know how difficult it is to change the environment if individuals themselves are not in a state of equilibrium.'

In order to protect the environment we must protect ourselves. We protect ourselves by opposing selfishness with generosity, ignorance with wisdom, and hatred with loving kindness. Selflessness, mindfulness, compassion and wisdom are the essence of Buddhism. We train in Buddhist meditation which enables us to be aware of the effects of our actions, including those destructive to our environment. Mindfulness and clear comprehension are at the heart of Buddhist meditation. Peace is realized when we are mindful of each and every step.

In the words of Maha Ghosananda:

> 'When we respect the environment, then nature will be good to us. When our hearts are good, then the sky will be good to us. The trees are like our mother and father; they feed us, nourish us, and provide us with everything; the fruit, leaves, the branches, the trunk. They give us food and satisfy many of our needs. So we spread the *dharma* (truth) of protecting ourselves and protecting our environment, which is the *dharma* of the Buddha.
>
> When we accept that we are part of a great human family – that every being has the nature of Buddha – then we will sit, talk, make peace. I pray that this realization will spread throughout our troubled world and bring humankind and the earth to its fullest flowering. I pray that all of us will realize peace in this lifetime and save all beings from suffering.
>
> The suffering of the world has been deep. From this suffering comes compassion. Great compassion makes a peaceful heart. A peaceful heart makes a peaceful person. A peaceful person makes a peaceful family. A peaceful family makes a peaceful community. A peaceful community makes a peaceful nation. A peaceful nation makes a peaceful world. May all beings live in happiness and peace.'

Buddhism as an ecological religion or a religious ecology

The relationship between Buddhist ideals and the natural world can be explored within three contexts: nature as teacher; nature as a spiritual force, and nature as a way of life.

Nature as teacher

> 'Like the Buddha, we too should look around us and be observant, because everything in the world is ready to teach us. With even a little intuitive wisdom we will be able to see clearly through the ways of the world. We will come to understand that everything in the world is a teacher. Trees and vines, for example, can all reveal the true nature of reality. With wisdom there is no need to question anyone, no need to study. We can learn from Nature enough to be enlightened, because everything follows the way of Truth. It does not diverge from Truth.' (Ajahn Chah, Forest Sangha Neauldtff)

Buddha taught that respect for life and the natural world is essential. By living simply one can be in harmony with other creatures and learn to appreciate the interconnectedness of all that lives. This simplicity of life involves developing openness to our environment and relating to the world with awareness and responsive perception. It enables us to enjoy without possessing, and mutually benefit each other without manipulation.

However, the Buddha was no romantic idealist. He also saw and realized that all phenomena, including the natural world, was a pit of suffering. He saw creatures struggling for survival in a precarious world. He saw death and fear, the strong preying on the weak, and the devastation of thousands of beings as one lonely figure ploughed the earth to reap the harvest. He also saw impermanence. As Ajahn Chah has written:

> '... take trees for example... first they come into being, then they grow and mature, constantly changing, until they finally die as every tree must. In the same way, people and animals are born, grow and change during their lifetimes until they eventually die. The multitudinous changes which occur during this transition from birth to death show the Way of Dharma. That is to say, all things are impermanent, having decay and dissolution as their natural condition.'
> (*Buddha-Nature*)

Nature is not independent and unchanging and neither are we. Change is the very essence of nature. In the words of Stephen Batchelor, 'We each believe we are a solid and lasting self rather than a short-term bundle of thoughts, feelings and impulses.' (*The Sands of the Ganges*).

We do not exist independently, separate from everything else – all things in the universe come into existence, 'arise' as a result of particular conditions. It is surely a mistake to see fulfilment in terms of external or personal development alone.

Buddha taught us to live simply, to cherish tranquillity, to appreciate the natural cycle of life. In this universe of energies, everything affects everything else. Nature is an ecosystem in which trees affect climate, the soil and the animals, just as the climate affects the trees, the soil, the animals and so on. The ocean, the sky, the air, are all inter-related, and inter-dependent – water is life and air is life. A result of Buddhist practice is that one does not feel that one's existence is so much more important than anyone else's. The importance of the individual, and emphasis on self is, in the West, a dominant outlook which is moving to the East as 'development' and consumerism spreads. Instead of looking at things as a seamless undivided whole we tend to categorize and compartmentalize. Instead of seeing nature as our great teacher we waste and do not replenish, and we forget that Buddha learned 'Wisdom from Nature'.

Once we treat nature as our friend, to cherish it, then we can see the need to change from the attitude of dominating nature to an attitude of working with nature. We are an intrinsic part of all existence rather than seeing ourselves as in control of it.

Nature as a spiritual force

For Shantiveda in eighth-century India, dwelling in nature was obviously preferable to living in a monastery or town:

> When shall I come to dwell in forests
> Amongst the deer, the birds and the trees,
> That say nothing unpleasant
> And are delightful to associate with.
> (*A Guide to the Bodhisattva's Way of Life*)

Patrul Rinpoche, one of the greatest Tibetan Buddhist teachers of the nineteenth century writes:

> Base your mind on the Dharma,
> Base your Dharma on a humble life,
> Base your humble life on the thought of death,
> Base your death on a lonely cave.
> (*The Words of My Perfect Teacher*)

The Buddha taught that the balance of nature is achieved by the functions of the forest. Survival of the forest is vital to the survival of natural harmony, balance, morality and environment.

Buddhist teachers and masters have constantly reminded us of the importance of living in tune with nature, to respect all life, to make time for meditation practice, to live simply and use nature as a spiritual force. Buddha stressed the four boundless qualities: loving kindness, compassion, sympathetic joy (delight in the well-being of others) and equanimity (impartiality).

Venerable Asabho has spoken of the value of living in retreat in Hammer Woods, Chithurst. The forest has its own rhythms and after a few days the metabolism and sleeping patterns adjust and the senses begin to sharpen to this new and unfamiliar setting. Ear and nose play a more important role when not having any comforts of life – gas, electricity, artificial light and the like. Living in the fast and furious pace of the twentieth century, our true nature is often dulled by the massive sensory impact unavoidable in modern urbanized living. Living close to nature is a very healing experience – to have few activities, few distractions. Learning to trust yourself and being more of a friend than a judge one develops a lightness of being, a light confidence. One realizes the truth of the notion of impermanence – the sound of animals, the texture of trees, the subtle changes in the forest and land, the subtle changes in your own mind. [Time spent in] retreats, or simply living in the forest with nature:

> 'helps people get back to earth, to calm you down – just living with the unhurried rhythms of nature. With Nature, everything – birth, growth, degeneration and decay is just as it is, and in that holistic sense everything is all right. Touching lightly is the right touch, the natural touch in which blame, praise, crises, retreats, progress, delays are just as it is and so all right.'
> (Talks given at Chithurst Buddhist Monastery)

Living in this way we can appreciate the fragility of all we love, the fickleness of security. Retreat and solitude complement our religious practice and give the opportunity of deepening, refining and strengthening the mind. By being mindful about the daily routine one pays attention to the flow of life – and sees nature as a positive, joyful, spiritual force.

Nature as a way of life

The Buddha commended frugality as a virtue in its own light. Skilful living avoids waste and we should try to recycle as much as we can. Buddhism advocates a simple, gentle, non-aggressive attitude towards nature – reverence for all forms of nature must be cultivated.

Buddha used examples from nature to teach. In his stories the plant and animal world are created as part of our inheritance, even as part of ourselves (as Krishnamurti said, 'We are the world, the world is us.'). By starting to look at ourselves and the lives we are living we may come to appreciate that the real solution to the environmental crisis begins with us. Craving and greed only bring unhappiness – simplicity, moderation and the middle way bring liberation and hence equanimity and happiness. Our demands for material possessions can never be satisfied – we will always need to acquire more: there is not enough in the universe to truly satisfy us and give us complete satisfaction and contentment, and no government can fulfil all our desires for security.

Buddhism, however, takes us away from the ethos of the individual and its bondage to materialism and consumerism. When we try to conquer greed and desire we can start to have inner peace and be at peace with those around us. The teaching of the Buddha, the reflections on *Dharma*, relate to life as it actually is. To be mindful – receptive, open, sensitive and not fixed to any one thing but able to fix on things according to what is needed in that time and at that place.

By developing the right actions of not killing, stealing or misconduct in sexual desires perhaps we can begin to live with nature, without breaking it or injuring the rhythm of life. In our livelihoods by seeking work that does not harm other beings, refraining from trading in weapons, in breathing things, meat, alcohol and poisons we can feel more at one with nature.

Our minds can be so full, so hyperactive, we never allow ourselves a chance to slow down to be aware of our thoughts, feelings and emotions, to live fully in the present moment. We need to live as the Buddha taught us to live, in peace and harmony with nature; but this must start with ourselves. If we are going to save this planet we need to seek a new ecological order, to look at the life we lead and then work together for the benefit of all; unless we work together no solution can be found. By moving away from self-centredness, to sharing wealth more, being more responsible for ourselves, and agreeing to live more simply, we can help decrease much of the suffering in the world. As the Indian philosopher Nagarjuna said, '... things derive their being and nature by mutual dependence and are nothing in themselves'.

> Breathing in, I know I'm breathing in,
> Breathing out, I know
> as the in-breath grows deep,
> the out-breath grows slow.
> Breathing in makes me calm.
> Breathing out makes me ease.
> With the in-breath, I smile.
> With the out-breath, I release.
> Breathing in, there is only the present moment
> Breathing out is a wonderful moment.
> (From a poem by Thich Nhat Hanh in *Buddhism and Ecology*)

Christianity and the environment

Christianity teaches that all of creation is the loving action of God, who not only willed the creation but also continues to care for all aspects of existence. As Jesus says in the Gospel of Luke, 'Are not five sparrows sold for two pennies? Yet not one of them is forgotten by God. Indeed the very hairs of your head are all numbered.' (Chapter 12, verses 6-7). Yet sadly, many Christians have been more interested in the last part of what Jesus said. 'Don't be afraid, you are worth more than many sparrows.'

There exists within Christianity a tension between God's creative, loving powers and humanity's capacity and tendency to rebel against God. Christianity, drawing upon the Biblical imagery of Genesis I and 2 and Genesis 9, is unambiguous about the special role of humanity within creation. But this special role has sometimes been interpreted as giving free rein to mastership. As the World Council of Churches said in the document from the Granvollen, Norway, meeting of 1988:

> The drive to have 'mastery' over creation has resulted in the senseless exploitation of natural resources, the alienation of the land from people and the destruction of indigenous cultures.... Creation came into being by the will and love of the Triune God, and as such it possesses an inner cohesion and goodness. Though human eyes may not always discern it, every creature and the whole creation in chorus bear witness to the glorious union and harmony with which creation in endowed. And when our human eyes are opened and our tongues unloosed, we too learn to praise and participate in the life, love, power and freedom that is God's continuing gift and grace.'

In differing ways, the main churches have sought to either revise or re-examine their theology and as a result their practice in the light of the environmental crisis. For example, Pope Paul VI in his Apostolic Letter, *Octogesima Adventeins*, also comments in a similar manner:

> 'by an ill-considered exploitation of nature, he [humanity] risks destroying it and becoming, in his turn, the victim of this degradation,... the flight from the land, industrial growth, continual demographic expansion and the attraction of urban centres bring about concentrations of population difficult to imagine.'

In his New Year message, 1990, His Holiness the Pope stated:

> 'Christians, in particular, realize that their responsibility within creation and their duty towards nature and the Creator are an essential part of their faith.'

In Orthodoxy this is brought out even more strongly, especially in the document produced by the Ecumenical Patriarchate, *Orthodoxy and the Ecological Crisis* (1990). The Orthodox Church teaches that humanity, both individually and collectively, ought to perceive the natural order as a sign and sacrament of God. This is obviously not what happens today. Rather humanity perceives the natural order as an object of exploitation. There is no one who is not guilty of disrespecting nature, for to respect nature is to recognize that all creatures and objects have a unique place in God's creation. When we become sensitive to God's world around us, we grow more conscious also of God's world within us. Beginning to see nature as a work of God, we begin to see our own place as human beings within nature. The true appreciation of any object is to discover the extraordinary in the ordinary.

The Orthodox Church teaches that it is the destiny of humanity to restore the proper relationship between God and the world as it was in Eden. Through repentance, two landscapes – the one human, the other natural – can become the objects of a caring and creative effort. But repentance must be accompanied by soundly-focused initiatives which manifest the ethos of Orthodox Christian faith.

The World Council of Churches, predominantly Protestant, but also with full Orthodox participation, issued the following when they called their member churches together in 1990 to consider the issues of Justice, Peace and the integrity of Creation:

> **Affirmation VII**
>
> 'We affirm the creation as beloved of God.
>
> We affirm that the world, as God's handiwork, has its own inherent integrity; that land, waters, air, forests, mountains and all creatures, including humanity, are 'good' in God's sight. The integrity of creation has a social aspect which we recognize as peace with justice, and an ecological aspect which we recognize in the self-renewing, sustainable character of natural ecosystems.
>
> We will resist the claim that anything in creation is merely a resource for human exploitation. We will resist species extinction for human benefit; consumerism, and harmful mass production; pollution of land, air and waters; all human activities which are now leading to probable rapid climate change; and the policies and plans which contribute to the disintegration of creation.
>
> Therefore we commit ourselves to be members of both the living community of creation in which we are but one species, and members of the covenant community of Christ; to be full co-workers with God, with moral responsibility to respect the rights of future generations; and to conserve and work for the integrity of creation both for its inherent value to God and in order that justice may be achieved and sustained.'

Implicit in these affirmations is the belief that it has been human selfishness, greed, foolishness or even perversity that has wrought destruction and death upon so much of the planet. This is also central to Christian understanding. As far as we can tell, human beings are the only species capable of rebelling against what God has revealed as the way in which we should live. This rebellion takes many forms, but one of these is the abuse of the rest of creation. Christians are called to recognize their need to be liberated from those forces within themselves and within society which militate against a loving and just relationship one with another and between humans and the rest of creation. The need to repent for what has been done and to hope that change can really transform the situation are two sides of the same coin. The one without the other becomes defeatist or romantic – neither of which is ultimately of much use to the rest of the world.

The Orthodox Churches pursue this in their own line of theology and reflection concerning creation, and expressed their commitment in the document, *Orthodoxy and the Ecological Crisis* (1990).

> 'We must attempt to return to a proper relationship with the Creator and the creation. This may well mean that just as a shepherd will in times of greatest hazard lay down his life for his flock, so human beings may need to forego part of their wants and needs in order that the survival of the natural world can be assured. This is a new situation – a new challenge. It calls for humanity to bear some of the pain of creation as well as to enjoy and celebrate it. It calls first and foremost for repentance – but of an order not previously understood by many.' (Pages 10-11)

The hope comes from a model of our relationship with nature which turns the power we so often use for destruction into a sacrificial or servant power – here using the image of the priest at the Eucharist:

> 'Just as the priest at the Eucharist offers the fullness of creation and receives it back as the blessing of Grace in the form of the consecrated bread and wine, to share with others, so we must be the channel through which God's grace and deliverance is shared with all creation. The human being is simply yet gloriously the means for the expression of creation in its fullness and the coming of God's deliverance for all creation.' (Page 8)

For Christians, the very act of creation and the love of God in Christ for all creation, stands as a constant reminder that, while we humans are special, we are also just a part of God's story of creation. To quote again from the World Council of Churches, from the report of the 1991 General Assembly on the theme Come Holy Spirit – Renew the Whole Creation:

> 'The divine presence of the Spirit in creation binds us human beings together with all created life. We are accountable before God in and to the community of life, an accountability which has been imagined in various ways: as servants, stewards and trustees, as tillers and keepers, as priests of creation, as nurturers, as co-creators. This requires attitudes of compassion and humility, respect and reverence.'

For some Christians, the way forward lies in a rediscovery of distinctive teachings, life-cycles and insights contained within their tradition. For others, it requires a radical rethinking of what it means to be Christian. For yet others, there is still a struggle to reconcile centuries of human-centred Christian teaching with the truths which the environmentalists are telling us about the state of the world we are responsible for creating. For all of them, the core remains the belief in the Creator God who so loved the world that he sent his only begotten Son, that whoever believes in him should have eternal life (John 3:16). In the past, we can now see, this promise of life eternal has often been interpreted by the churches as meaning only human life. The challenge to all Christians is to discover anew the truth that God's love and liberation is for all creation, not just humanity; to realize that we should have been stewards, priests, co-creators with God for the rest of creation but have actually often been the ones responsible for its destruction; and to seek new ways of living and being Christians that will restore that balance and give the hope of life to so much of the endangered planet.

Hinduism and the environment

Part one: Basic hindu environmental teachings
Sustaining the balance (Swami Vibudhesha Teertha)
These days it looks as if human beings have forgotten that a particular natural condition on Earth enabled life to come into existence and evolve to the human level. Humanity is disturbing this natural condition on which his existence, along with the existence of all other forms of life, depends. This is like the action of a wood-cutter cutting a tree at the trunk, on the branch on which he is sitting. According to Hindu religion, *'dharanath dharma ucyate'* – that which sustains all species of life and helps to maintain harmonious relationship, among them is

dharma – and that which disturbs such ecology is *adharma*.

Hindu religion wants its followers to live a simple life. It does not allow people to go on increasing their material wants. People are meant to learn to enjoy spiritual happiness, so that to derive a sense of satisfaction and fulfilment they need not run after material pleasures and disturb nature's checks and balances. They have to milk a cow and enjoy; not cut at the udder of the cow with greed to enjoy what is not available in the natural course. Do not use anything belonging to nature, such as oil, coal or forest, at a greater rate than you can replenish it. For example, do not destroy birds, fish, earthworms and even bacteria which play vital ecological roles: once they are annihilated you cannot re-create them. Thus only can you avoid becoming bankrupt and the life cycle can continue for a long, long time.

'Conserve ecology or perish' is the message of the *Bhagavad Gita* – a dialogue between Sri Krishna and Arjuna which is a clear and precise Life Science. It is narrated in the third chapter of this great work that a life without contribution towards the preservation of ecology is a life of sin and a life without specific purpose or use. The ecological cycle is explained in verses 3.14-16:

> 'All living bodies subsist on food grains, which are produced from rains. Rains are produced from performance of *yajna* (sacrifice), and *yajna* is born of prescribed duties. Regulated activities are prescribed in the Vedas, and the Vedas are directly manifested from the Supreme Personality of Godhead. Consequently the all-pervading Transcendence is eternally situated in acts of sacrifice. My dear Arjuna, one who does not follow in human life the cycle of sacrifice thus established by the Vedas certainly lives a life full of sin. Living only for the satisfaction of the senses, such a person lives in vain.'

Life is sustained by different kinds of food; rainfall produces food; timely movement of clouds brings rains; to get the clouds moving on time, *yajna* – religious sacrifice – helps; *yajna* is performed through rituals; those actions which produce rituals belong only to God; God is revealed by the Vedas; the Vedas are preserved by the human mind; and the human mind is nourished by food. This is the cycle which helps the existence of all forms of life on this globe. One who does not contribute to the maintenance of this cycle is considered as a destroyer of all life here. When the Lord desired to create life, He created the Sun, Moon and Earth, and through them a congenial atmosphere for life to come into being. Therefore the Sun, Moon, Earth, Stars and all objects in the universe jointly, not individually, create the atmosphere for the creation, sustenance or destruction of everything in the universe. The Earth is the only daughter of the Sun to produce children. The Moon is essential for the creation of the right atmosphere for those children to exist and evolve. This we say because of the influence of the Moon in the high and low tides of our rivers and oceans. This is narrated also in the *Bhagavad* Gita: 'I become the Moon and thereby supply the juice of life to all vegetables'. 15.13)

We cannot refute this influence of the Moon on life. It is proved by the movement of all liquid on this globe depending on the movement of the Moon. Therefore ecology in totality must be preserved: just a part of it would not suffice.

Hinduism is a religion which is very near to nature. It asks its followers to see God in every object in the Universe. Worship of God in air, water, fire, Sun, Moon, Stars and Earth is especially recommended. Earth is worshipped as the spouse of God. Hence very dear and near to God. All lives on Earth are considered as children of God and Earth.

Sri Krishna in the *Bhagavad Gita* says,

> 'I am pervading the Universe. All objects in the Universe rest on me as pearls on the thread of a garland.'

The *Upanishads* narrate that after creating the Universe, the creator entered into each and every object to help them maintain their inter-relationship. The *Upanishad* says '*tat sristva ta devanu pravisat*': after creating the universe He entered into every object created. Therefore to contribute towards the maintenance of this inter-relationship becomes worship of God. Hindus believe that there is soul in all plants and animals. One has to do penance even for killing plants and animals for food. This daily penance is called '*visva deva*'. *Visva deva* is nothing but an offering of prepared food to the Creator, asking His pardon.

The Hindu religion gives great importance to protecting cattle. At every Hindu house there is a cow and it is worshipped. The cow is a great friend of humans. It nourishes us through its milk and provides manure to grow our food. This it does without any extra demand – it lives on the fodder got while growing our food. Advanced countries have started to realize the harmful effects of consuming food grown with chemical manure. When we use chemical manure, the top soil loses its fertility. This generation has no right to use up all the fertility of the soil and leave behind an unproductive land for future generations.

There is no life that is inferior. All lives enjoy the same importance in the Universe and all play their fixed roles. They are to function together and no link in the chain is to be lost. If some link is lost, the whole ecological balance would be disturbed. All kinds of life – insects, birds and animals – contribute towards the maintenance of ecological balance, but what is man's contribution towards this? He is an intelligent animal, therefore his contribution should be the biggest. But we find the absence of his contribution. On the other hand he is nullifying the benefits of the contributions made by other species of life. He is disturbing the balance because of his greed for material enjoyment and his craze for power. He does not allow earthworms and bacteria to maintain the fertility of the soil by using chemical manures and insecticides which kill them. He destroys plants and forests indiscriminately and comes in the way of plants providing oxygen essential for the very existence of life. By destroying plants and forests he becomes an agent for increasing the deadly carbon dioxide in the atmosphere. He pollutes the air by burning oil for all sorts of machines. He produces unhealthy sounds through his various machineries and instruments which cause sound pollution. By building towns and cities on the banks of rivers he pollutes all water in rivers. The Hindu religion holds all rivers as holy; polluting them is a big sin. It encourages the planting of trees like tulasi, neem, peepal and the like which are rich in medicinal properties.

Rishis gave the Lotus sprung out of the navel of Vishnu place for Brahma, the creator, and to the sustainer Vishnu, they gave the heart as His abode. The destroyer, Shiva, is given control of the brain. By doing this they wanted us to know that the language of the heart only can sustain us – when we start speaking through the language of the mind our destruction becomes inevitable. Therefore, a thinking animal has to be very careful while it uses its menial abilities: these are to be applied only with spiritual background. Mind is to act as our friend and not as our enemy. It is to function under our control – we should not succumb to its control. '*Mana eva manusvanam karanam bandha moksayoh*' – 'for man, mind is the cause of bondage and mind is the cause of liberation' (*Amrita Bindhu Upanishad* 2).

There should be a purpose for the creation of man. What it might be! Man could be the sustainer of inter-relationships among numerous life species on Earth. He is one who can see God, and all objects, as the controller and sustainer of ecological balance. All other animals play their roles without knowing what they are doing, but man does everything with full consciousness. God created man's mind to see His own reflection as in a mirror. Man's mind can meditate on God and know him more and more. When he develops consciousness of the presence of God and His continuous showering of blessings on the universe, man develops deep love for Him. To enjoy this nectar of love, God created man. Only man has time–space conception. Therefore, he only can see God, pervading time–space, conserving the ecological balance which is the greatest boon bestowed on the universe by God. Though he can not contribute towards the conservation in the same way as other animals do, man can help all lives and other objects in the universe to play their roles effectively by persuading God through prayers of love to grant them the required energy and directions.

'*Yavat bhumandalam datte samrigavana karnanam, tavat tisthati medhinyam santatih putra pautriki*': 'so long as the Earth preserves her forests and wildlife, man's progeny will continue to exist'. This is the Hindu approach towards the conservation of ecology.

Sacrifice and protection (Dr Sheshagiri Rao)

Sacrifice

> 'The Creator, in the beginning, created men together with sacrifice, and said, 'By this you shall multiply. Let this be your cow of plenty and give you the milk of your desires. With sacrifice you will nourish the gods, and the gods will nourish you. Thus you will obtain the Highest Good.' (*Bhagavad Gita* 3, 10-11)

Sacrifice does not just mean ritual worship – it means an act which protects life. Personal health depends on eyes, ears and other sense organs working together in harmony; human prosperity and happiness depend on a well-ordered society and nature; the universe is sustained by the cosmic powers such as the sun and moon working together in unison. Sacrifice reinvigorates the powers that sustain the world by securing cosmic stability and social order. It activates the positive forces of the universe, and protects the Earth from degeneration.

Non-violence

God's creation is sacred. Humanity does not have the right to destroy what it cannot create. Humans have to realize the interconnectedness of living entities and emphasize the idea of moral responsibility to oneself, one's society and the world as a whole.

In our cosmic journey, we are involved in countless cycles of births and deaths. Life progresses into higher forms or regresses into lower forms of life based upon our good or bad *karma*. Kinship exists between all forms of life. Reincarnation warns us against treating lower forms of life with cruelty.

Cow protection

Man has evolved from lower forms of life. He is, therefore, related to the whole of creation. The principle of cow protection symbolizes human responsibility to the sub-human world. It also indicates reverence for all forms of life. The cow serves humans throughout its life, and even after death. The milk of the cow runs in our blood. Its contributions to the welfare of the family and the community are countless. Hindus pray daily for the welfare of cows. When the cows are cared for, the world at all levels will find happiness and peace.

Earth as mother

Hindus revere the Earth as mother. She feeds, shelters and clothes us. Without her we cannot survive. If we as children do not take care of her we diminish her ability to take care of us. Unfortunately the Earth herself is now being undermined by our scientific and industrial achievements.

Breaking the family (Shrivatsa Goswami)

> 'Let there be peace in the heavens, the Earth, the atmosphere, the water, the herbs, the vegetation, among the divine beings and in Brahman, the absolute reality. Let everything be at peace and in peace. Only then will we find peace.'

According to Hindu philosophy, the goal of human life is the realization of the state of peace. *Dharma*, loosely translated as religion, is the source by which peace can be fully realized. This peace is not the stillness of death; it is a dynamic harmony among all the diverse facets of life. Humanity, as part of the natural world, can contribute through *dharma* to this natural harmony.

The natural harmony which should exist in the play of energies between humanity and the natural world is now disrupted by the weakest player in the game – humanity. Although it is the totality of this game that provides our nourishment, through ignorance of our own natural limits we destroy this source of nourishment.

This awareness of ecological play or playful ecology is inseparable from awareness of the need for friendship and play as the real basis for human relationship. The family within which these relationships are nourished is not limited to its human members. Just as the human child has to be nourished by Mother Nature, and the human spirit has to be embraced and loved by beautiful nature, so the human being who has grown old or sick has to be supported by caring nature. If humans distress the mother, rape the beauty and beat the caring nurse, what happens? The relationship collapses, and the family is broken.

The Sanskrit for family is *parivara*, and environment is *paryavarana*. If we think of the environment as our home and all of its members as our family it is clear that the key to conserving nature is devotion, love – giving and serving. Nature, *prakiiti*, as the feminine can give and serve. But the role of humanity, *purusha*, is then to protect. Nowadays *purusha*, humanity, is interested not in protecting but in exploiting, so *prakriti*, nature, has to defend herself. This is why we see nature in her furious manifestation – in drought, floods or hurricanes. If we rape the mother's womb she has convulsions, and we blame her for devastating earthquakes. If we denude her of her lush hair and beautiful skin, she punishes us by withholding food and water.

As it is through ignorance that we destroy our relationships in the family and within the environment, that ignorance becomes the root cause of our suffering. The best way to get rid of this ignorance is to unlearn what is wrong. This unlearning is shaped not only in the school but in the family and community, and it has to begin with the very young.

Traditional Hindu education covers all facets of life – economic, political, cultural and above all religious. Whether we speak of Krishna, of Chaitanya, or of Gandhi, we see that they drew no clear division between the economic or political and the religious or cultural facets of life. The body and mind are in the service of the heart. In the same way politics and economics are rooted in and guided by religion and culture, and ultimately by spiritual experience.

Part two: The present-day context

Pollution of mind and planet

Mental pollutions are the underlying cause of these physical and environmental ravages. Mental pollutions come from lust, greed and anger which lead to the desire for profit, power and aggrandisement. They result in disharmony, conflict, sickness and degradation of nature. Truthfulness, humility, unity of humanity, reverence for life, and care for the environment are all expressions of the moral and spiritual laws of life which form part of the very structure of the universe. When these give way to mental pollutions, negative forces are released.

Now, human short-sightedness and selfishness have landed us in an ecological crisis of immense proportion. Environmental resources are being depleted rapidly on a global scale. Industries are polluting the atmosphere, releasing chemical wastes into water, using up non-renewable energy sources without a thought for the future and causing acid rain. They pollute and deplete natural resources. Chemical fertilizers, pesticides and herbicides pollute the vegetation and the soil, harm human health and cause genetic damage. The disappearing ozone layer will bring cancer, blindness and death to our children.

Role of religious leadership

Materialistic and selfish outlook has afflicted the Indian mind. Religious values have gone down in the scale of priorities. Now, religious leaders have to take the initiative and provide, once again, moral guidance. Drawing upon the wisdom of saints and sayings of scriptures, they should strengthen the movement towards conservation and address the environmental problems urgently.

Hindu tradition has no clerical hierarchy and ecclesiastical organization. There is no head of the Hindu community who can speak for or give dictates to all the members of the community. However, the words of spiritual leaders do carry a lot of weight. Traditional religious institutions still exercise great influence over customs and practices. They have considerable authority and command dedication and devotion. They can help enlist the participation of people at grass-roots level in ecological programmes, especially in local and regional projects

Traditional health care

The value of herbs and plants to provide remedies to diseases both known and unknown through the Ayurvedic system of preventative health-care is not sufficiently understood or encouraged. Broad-leaved trees absorb moisture and encourage rain. Plants are a necessary link in the web of life. Organic farming is good for health and for the soil. Non-violence and vegetarianism are embedded in the Hindu ideal of life. A vegetarian diet feeds more people and is less wasteful of the Earth's resources.

Traditional ecology

Religious leaders can and must impress upon their followers the importance of looking after the sacred elements of nature, of keeping rivers clean, of protecting forests and mountains. Rivers such as the Ganga, Yamuna, Godavari and Kauveri and mountains such as the Himalayas, Vindhva and Malava are sacred, because they give and sustain life. The same goes for oceans, trees and rocks.

There is a need to tap the wisdom of traditional practices as well as the best of modern technology. Many traditional environmental practices are still relevant today:

- Dig wells
- Excavate water tanks
- Plant trees: whenever a body is cremated, plant a tree
- Don't cut green trees
- Create parks and flower gardens (don't pick flowers at night)
- Don't disturb water at night (allow pollution to settle)
- Don't pollute (defecate on) river banks
- Adopt a simple, non-violent life-style
- Reduce consumption and harm to the environment
- Recycle or reuse

At a time like this, Sri Krishna's childhood among the cowherds is a source of inspiration. He is concerned with preserving the life-sustaining waters of the Yamuna, the life-promoting hill of Govardhan and the life supporting animals of Vrindavan. The Supreme Divinity plays with simple cowherds and endears himself not just to humans but to the whole natural world.

The *Isa Upanishad* says, 'The whole universe is pervaded by God'. Nature has spiritual significance. It has implications for the survival of humanity as well. Protection of nature is protection of self.

Islam and the environment

Man's most primordial concepts of religion relate to the environment. Man's history on planet Earth is, in a geological scale, very short indeed. Planet Earth itself is a mere 3,800 million years old. Man only appeared one or maybe two million years ago. Most of the physical patterns of planet Earth were probably in place – broadly speaking – by the time Man evolved. Apart from what he first saw, he also probably witnessed some spectacular changes himself. He must, at the very least, have gone through one Ice Age and seen some graphic volcanic eruptions – assuming he was able to avoid their consequences. The environment, therefore, very probably induced the first thoughts of a Super-Being – a God, if you like – whose manifestations lay in Man's immediate surroundings.

Man's environment also provided another dimension in his relationship with Nature. To survive in a given environment, Man had to adjust what he takes from that environment to what can give him sustainable yields on (at the very least) an annual basis. In effect this meant that early Man had to learn to conserve at an early age. Being largely dependent on what was available rather than from what he could cultivate, he entered into a partnership with the environment. To take more than the regenerative capacity of the environment could lead to serious subsequent exhaustion – quite rightly seen as harsh retribution from an angry God. The converse situation of exploitation with moderation led to sustained yields which were (again, quite rightly) taken as having pleased God.

This relationship between conservation and religion is thus not only a natural one but also probably as old as the proverbial hills. But when we quickly open most of the pages of Man's history on planet Earth and come to the last three hundred years or so, we find the advent of the Industrial Revolution. The Industrial Revolution made possible the production of large quantities of goods in a very short time. That meant that raw materials in ever-increasing quantities had to be found to feed the hungry mills ready to convert them into finished or semi-finished goods. The consequences of the Industrial Revolution were many – economic, social and environmental. The material achievements of the human race in the past one hundred or so years have overshadowed the contributions made by all past civilizations. The Industrial Revolution that took place in Europe in the eighteenth and nineteenth centuries exacted a high social and environmental cost. Now these costs are even higher and more universal, being manifestly so in the great urban centres of the world. The paradox of 'progress' today is the easily perceived correlation between complex consumer societies and the degeneration of the human being. Or as John Seymour puts it:

'We see men now wherever we look, so blinded by arrogance and the worship of man as God that they are doing things no one but the insane would do. Men maddened by the belief that they are both omniscient and omnipotent, that they are indeed, God.'

The Industrial Revolution also proclaimed a new revival of another God – Mammon. Mammon regrettably has no respect for environmental integrity – nor do his followers. The last 250 years have seen a growing decimation of ever more pristine areas of Nature to feed the insatiable industrial cuckoo and its resultant consumerism. Forests – particularly tropical forests – have been systematically hewn down, the seas ransacked, the lands made totally dependent on a host of inorganic fertilizers and pesticides for food production, wastes galore have filled the seas, the rivers and the lakes not to mention the land-fills.

We must also take note that the 'unmatched material progress' of this century we are often fond of talking about has only been possible for the few: that is the population of the Northern Hemisphere and a small minority among the peoples of the South. This is usually translated as less than 25 per cent of the world's population consuming over 75 per cent of the world's resources. This rate of consumption by a minority of the human species has caused unparalleled climatic changes, eco-system disintegration and species extinction, and as a report by WWF (World Wide Fund For Nature) observes: 'loss of biodiversity world-wide, and the combination of global warming with other human pressures, will present the greatest challenge in conservation for decades to come'.

This would lead us to conclude that there is a profound and inherent contradiction in the efforts made by the 'North' to keep ahead of the rest as consumers, and the push by the remaining 75 per cent of the world's population to catch up. Given this scenario, if just Eastern Europe or Russia or India or China managed to raise their standards of living by just a few percentage points, then the consequences of putting this extra load on the earth's eco-system, which is already under severe strain, would be catastrophic.

This is the background against which the followers of the relatively ancient, environmentally conscious (indeed environmentally concerned) God have gathered to re-examine and to restate their own commitment to environmental integrity from their own individual religion's standpoint. We for our part will look at the underpinnings of conservation in Islam.

Islam and conservation

There are several Islamic principles which, when taken individually, seem to have little bearing on conservation. Together, however, they add up to a clear concept of the Islamic view on conservation. We shall now annotate these principles briefly.

Tawheed

The first Islamic principle which relates to conservation is that of the Oneness of Allah, or *Tawheed*. This principle is absolutely fundamental to Islam. Every Muslim must believe in this Oneness of Allah. It is said by some *Ulamaa* that some two-thirds of Prophet Muhammad's (SAW) early preaching – and indeed of the Qur'an itself – were and are dedicated purely to endorsing this very Oneness of Allah. One indivisible God means to a Muslim that there is no separate deity for each of the many attributes that to Muslims belong to the One Universal God who is also God of the Universe.

Tawheed is the monotheistic principle of Islam and it begins by declaring that 'there is no God but God' (the second half of this declaration asserts that 'Muhammad is His Messenger'). We are for the present concerned with the first part which affirms that there is nothing other than the Absolute, the Eternal, All Powerful Creator. This is the bedrock statement of the Oneness of the Creator from which stems everything else.

> 'It is the primordial testimony of the unity of all creation and the interlocking grid of the natural order of which man is intrinsically a part'. (*Islam, Ecology and the World Order*)

> God says in the Qur'an:

> 112.001. Say: He is Allah the One and Only.
> 112.002. Allah the Eternal Absolute.
> 112.003. He begetteth not nor is He begotten.
> 11 2.004. And there is none like unto Him.

God is real, not an abstract idea or concept; He is One, the Everlasting Refuge for all creation.

Man's relation to God

The emphasis on *Tawheed* is significant unto itself, but it is even more relevant to the present discussion by virtue of defining a Muslim's relationship to Allah. The Omniscience and Omnipotence of Allah means, by definition, that a Muslim's relationship to Allah is total. To Him – and to Him only – should Man refer for all his needs: physical, mental and spiritual. Indeed, Allah would not have it any other way. As He says in the Qur'an:

> 004.048. Allah forgiveth not that partners should be set up with him; but He forgiveth anything else to whom He pleaseth; to set up partners with Allah is to devise a sin most heinous indeed. But Allah is not only the One Indivisible God. He is also the Universal God as well as the Lord of the Universe.

> 001.002. Praise be to Allah, Lord of the Worlds.

And again:

> 006.071. Say: 'Allah's guidance is the (only) guidance and we have been directed to submit ourselves to the Lord of the worlds'.

> 006.072. To establish regular prayers and to fear Allah; for it is to Him that we shall be gathered together.

> 006.073. It is He Who created the heavens and the earth in true (proportions): the day He saith 'Be' Behold! it is. His Word is the truth. His will be the dominion he day the trumpet will he blown, He knoweth the Unseen as well as that which is open. For He is the Wise well acquainted (with all things).

To Allah belongs the earth and the heavens

Yet another principle which underpins Islamic commitment to the conservation of nature and natural resources is the principle of divine ownership of all that exists on earth and in the heavens – animate and inanimate.

There are countless verses in the Holy Qur'an which state this. A few are given below. In the celebrated Ayatul Kursiyy:

> 1002.255. Allah! there is no Allah but He the living the Self-subsisting Eternal. No slumber can seize him nor sleep. His are all things in the heavens and on earth. Who is there can intercede in His presence except as He permitteth? He knoweth what (appeareth to his creatures as) before or after or behind them. Nor shall they compass aught of his knowledge except as He willeth. His throne doth extend over the heavens and the earth and He feeleth no fatigue in guarding and preserving them. For He is the Most High, the Supreme (in glory)."

And again:

> 004.171. To Him belong all things in the heavens and on earth. And enough is Allah as a Disposer of affairs.

> 006.013. To Him belongeth all that dwelleth (or lurketh) in the night and the day. For He is the One Who heareth and knoweth all things.

> 020.006. To Him belongs what is in the heavens and on earth and all between them and all beneath the soil.

> 021.019. To Him belong all (creatures) in the heavens and on earth: even those who are in His (very) Presence are not too proud to serve Him nor are they (ever) weary (of His service).

But we are reminded that all things animate and inanimate, in their own ways, submit themselves to the Glory of Allah. There are many verses in the Qur'an about this:

030.026. To Him belongs every being that is in the heavens and on earth: all are devoutly obedient to Him.

And again:

062.001. Whatever is in the heavens and on earth doth declare the Praises and Glory of Allah the Sovereign the Holy One the Exalted in Might the Wise.

Thus Allah, the One Indivisible God, the Universal God and the Lord of the Universe, is the Owner also of all that is in the universe, including Man. After all, we are reminded to say constantly:

002.155. Be sure We shall test you with something of fear and hunger some loss in goods or lives or the fruits (of your toil) but give glad tidings to those who patiently persevere.

002.156. Who say when afflicted with calamity, 'To Allah we belong and to Him is our return'.

The above set of principles – all taken from Islam's ultimate authority, the Holy Qu'ran – set the perspectives of the relationship of Man to God and of God to the environment in its totality. A second set of principles that the Holy Qur'an enunciates prescribe Man's relationship to the environment after, of course, Man has accepted the preceding principles.

Man and the Khalifa

The most important of this second set of principles is that which defines man's role and his responsibilities in the natural order that Allah provided. The appointment of Man as a Khalifa, or guardian, is the sacred duty God has given to the human race. The appointment of Man to this elevated position gives rise to the one occasion when the angels actually questioned Allah's decision, as seen in the following verses.

002.030. Behold the Lord said to the angels, 'I will create a vice-regent on earth'. They said, 'Wilt thou place therein one who will make mischief therein and shed blood? Whilst we do celebrate Thy praises and glorify Thy holy (name)?' He said, 'I know what ye know not'.

002.031. And He taught Adam the nature of all things; then He placed them before the angels and said, 'Tell Me the nature of these if ye are right'.

002.032. They said, 'Glory to Thee of knowledge: we have none save that Thou hast taught us: in truth it is Thou who art perfect in knowledge and wisdom'.

002.033. He said, 'O Adam! tell them their natures'. When he had told them Allah said, 'Did I not tell you that I know the secrets of heaven and earth and I know what ye reveal and what ye conceal?'

002.034. And behold We said to the angels, 'Bow down to Adam'; and they bowed down. Not so Iblis: he refused and was haughty: he was of those who rejected Faith'.

Clearly Allah preferred the unprogrammed free will of Man to the pre-programmed goodness of angels!

And again:

006.165. It is He who hath made you (His) agents inheritors of the earth: He hath raised you in ranks some above others: that he may try you in the gifts He hath given you: for thy Lord is quick in punishment: yet He is indeed Oft-Forgiving Most Merciful."

The exercise of the vice-regency is defined in the Qur'an by another set of principles between which Man's privileges as well as his responsibilities are clearly defined. We shall deal with these briefly in the following paragraphs.

Mizaan

One of the most important attributes conferred on Man is the faculty of reasoning. This, above all, might well be the deciding fact or in his appointment as God's vice-regent on Earth. The relevant verses reminding Man of this faculty are below:

055.001. (Allah) Most Gracious!
055.002. It is He Who has taught the Qur'an.
055.003. He has created man.
055.004. He has taught him speech (and intelligence).
055.005. The sun and the moon follow courses (exactly) computed.
055.006. And the herbs and the trees – both (alike) bow in adoration.
055.007. And the Firmament has He raised high and He has set up the
balance (of justice).
055.008. In order that ye may not transgress (due) balance.

055.009. So establish weight with justice and fall not short in the balance.
055.010. It is He Who has spread out the earth for (His) creatures:
055.011. Therein is fruit and date-palms producing spathes (enclosing dates).
055.012. Also corn with (its) leaves and stalk for fodder and sweet-smelting plants.
055.013. Then which of the favours of your Lord will ye deny?

Man was not created to function exclusively on instinct. The 'explanation' was taught to him because he had the capacity to reason and understand. There is order and purpose in the whole pattern of creation. The Sun and Moon following stable orbits make this possible. The whole universe is in submission to the Creator – the stars that enable us to steer courses and the trees that give us sustenance, shelter and other uses. The world functions only because creation follows a pre-ordained pattern. Man then has a responsibility by virtue of being able to reason, to behave justly, 'to transgress not in the balance'. We owe this to ourselves as much as for the rest of creation.

Justice

The capacity to reason and to balance intellectual judgement would in itself be insufficient without the additional moral commitment to justice. And this is what the Qur'an prescribes for Muslims.

004.135. 0 ye who believe! stand out firmly for justice as witnesses to Allah even as against yourselves or your parents or your kin and whether it be (against) rich or poor: for Allah can best protect both. Follow not the lusts (of your hearts) lest ye swerve and if ye distort justice or decline to do justice verily Allah is well-acquainted with all that ye do.

And again:

004.085. Whoever recommends and helps a good cause becomes a partner therein: and whoever recommends and helps an evil cause shares in its burden: and Allah hath power over all things.

004.058. Allah doth command you to render back your trusts to those to whom they are due; and when ye judge between man and man that ye judge with justice: verily how excellent is the teaching which He giveth you for Allah is He who heareth and seeth all things.

And again:

005.009. 0 ye who believe! stand out firmly for Allah as witnesses to fair dealing and let not the hatred of others to you make you swerve to wrong and depart from justice. Be just: that is next to Piety: and fear Allah for Allah is well-acquainted with all that ye do.

005.045. (They are fond of) listening to falsehood, of devouring anything forbidden. If they do come to thee, either judge between them or decline to interfere. If thou decline they cannot hurt thee in the least. If thou judge, judge in equity between them; for Allah loveth those who judge in equity.

Use but do not abuse

Several times in the Qur'an, Man is invited to make use of the nourishing goods that Allah has placed on earth for him, but abuse – particularly through extravagance and excess – is strictly forbidden. Sometimes this is stated in one breath, so to speak. Sometimes they are stated separately. But the message is the same as the following verse indicates:

007.031. 0 children of Adam!... eat and drink: but waste not by excess for Allah loveth not the wasters.

There are as many invitations to partake of Nature as provided for Man and for other creatures of the Earth as there are for the avoidance of wasteful extravagance. Time and again, Allah reminds us that He loveth not wasters.

006.14I. It is He who produceth gardens with trellises ... and dates and tilth with produce of all kinds, and olives and pomegranates similar (in kind) and different (in variety): eat of their fruit in their season but render the dues that are proper on the day that the harvest is gathered. But waste not by excess: for Allah loveth not the wasters.

Fitra

The last principle which we bring forth in our examination of the Qur'anic underpinnings of conservation is that of *Fitra*. *Fitra* can be taken as perhaps the most direct injunction by Allah to Man to conserve the environment and not to change the balance of His creation. This is specifically contained in the verse below:

030.030. So set thou thy face steadily and truly to the Faith: (Establish) Allah's handiwork according to the pattern on which He has made mankind: no change (let there be) in the work (wrought) by Allah: that is the standard Religion: but most among mankind understand not.

Thus, Islam teaches that humanity is an integral part of the environment, it is part of the creation of Almighty God. We remain deeply locked into the natural domain despite the fact that there is talk of bringing the environment to the people as though we are independent of it.

The power given to man by God is seen in Islam to be used ... not only towards God and other men and women, but also towards the rest of creation.

Seyyed Hossein Nasr says:

> The Divine Law (*al shariah*) is explicit in extending the religious duties of man to the natural order and the environment.' (*The Need for a Sacred Science*)

The conclusion

As we indicated at the beginning, there are a number of Qur'anic principles which, taken separately, do not have an obvious connection with conservation. But taken in their totality, they state in clear terms that Allah, the One True God, is the Universal God and the Creator of the Universe and indeed, the Owner of the Universe. To Him belong all the animate and inanimate objects, all of whom should or do submit themselves to Him. Allah, in His Wisdom, appointed Man, the creature that He has conferred with the faculty of Reason and with Free will, to be His vice-regent on Earth. And while Allah has invited Man to partake of the fruits of the Earth for his rightful nourishment and enjoyment, He has also directed Man not to waste that which Allah has provided for him – for He loves not wasters. Furthermore, Allah has also ordered Man to administer his responsibilities with justice. Above all, Man should conserve the balance of Allah's creation on Earth. By virtue of his intelligence, Man (when he believes in the One Universal Allah, the Creator of the Universe), is the only creation of Allah to be entrusted with the responsibility of maintaining planet Earth in the overall balanced ecology that Man found. If biologists believe that Man is the greatest agent of ecological change on the surface of the Earth, is it not Man who, drawn from the brink, will – for his own good – abandon Mammon and listen to the prescriptions of God on the conservation of his environment and the environment of all the creatures on Earth? The Islamic answer to this question is decisively in the affirmative.

Jainism and the environment

History of Jainism

Jainism is one of the oldest living religions. The term *jain* means 'the follower of the *Jinas*' (spiritual victors), human teachers who attained omniscience. These teachers are also called *Tirthankaras* (fordmakers), those who help others escape the cycle of birth and death.

The twenty-fourth *Tirthankara*, called Mahavira, was born in 599 BC. At the age of thirty, he left home on a spiritual quest. After twelve years of trials and austerities, he attained omniscience. Eleven men became his *ganadha* or chief disciples. At seventy-two Mahavira died and attained nirvana, that blissful state beyond life and death. Mahavira was not the founder of a new religion. He consolidated the faith by drawing together the teachings of the previous Tirthankarma, particularly those of his immediate predecessor, Parsva, who lived about 250 years earlier at Varnasi.

Initially, the followers of Jainism lived throughout the Ganges Valley in India. Around 250 BC, most Jains migrated to the city of Mathura on the Yamuna River. Later, many travelled west to Rajasthan and Gujarat and south to Maharashtra and Karnataka where Jainism rapidly grew in popularity. The Jain population throughout the world is less than ten million, of which about one hundred thousand have settled overseas in North America, United Kingdom, Kenya, Belgium, Singapore, Hong Kong and Japan.

Jain practices

Jains believe that to attain the higher stages of personal development, lay people must adhere to the three jewels (*ratna-traya*) namely enlightened world view, true knowledge, and conduct based on enlightened world view and true knowledge. They must endeavour to fulfil the *anuvratas* (small vows). These vows are:

Ahimsa (non-violence)

This is the fundamental vow and runs through the Jain tradition like a golden thread. It involves avoidance of violence in any form through word or deed, not only to human beings but to all nature. It means reverence for life in every form, including plants and animals. Jains practice the principle of compassion for all living beings (*Jiva-daya*) at every step in daily life. Jains are vegetarians.

Satya (truthfulness)

Tirthankara Mahavira said, *Sachham Bhagwam* (Truth is God).

Asteya (not stealing)
This is the principle of not taking what belongs to another. It means avoidance of greed and exploitation.

Brahmacharya (chastity)
This means practising restraint and avoiding sexual promiscuity.

Aparigraha (non-materialism)
For lay Jains, this means limiting their acquisition of material goods and contributing one's wealth and time to humanitarian charities and philanthropic causes.

Jain beliefs

Anekantavada (non-one-sidedness)
This philosophy states that no single perspective on any issue contains the whole truth. It emphasizes the concept of universal interdependence and specifically recommends that one should take into account the viewpoints of other species, other communities and nations, and other human beings.

Loka (the universe)
Space is infinite but only a finite portion is occupied by what is known as the universe. Everything within the universe, whether sentient (*jiva*) or insentient (*ajiva*) is eternal, although the forms that a thing may take are transient. Jains preach and practice the principle of the duty of every human being to promote universal well being (*sarvamangalya*).

Jiva (soul)
All living beings have an individual soul (*jiva*) which occupies the body, a conglomerate of atoms. At the time of death, the soul leaves the body and immediately takes birth in another. Attaining nirvana and thereby terminating this cycle of birth and death is the goal of Jain practice.

Ajiva (non-soul)
Ajiva is everything in the universe that is insentient, including matter, the media of motion and rest, time and space.

Karma
Is understood as a form of subtle matter which adheres to the soul as a result of individual actions of body, speech and mind. This accumulated *karma* is the cause of the soul's bondage in the cycle of birth and death.

Moksha or niverna (eternal liberation through enlightenment)
The ultimate aim of life is to emancipate the soul from the cycle of birth and death. This is done by exhausting all bound *hatuas* and preventing further accumulation. To achieve *moksha*, it is necessary to have enlightened world view, knowledge and conduct

Jainism is fundamentally a religion of ecology and has turned ecology into a religion. It has enabled Jains to create an environment-friendly value system and code of conduct. Because of the insistence on rationality in the Jain tradition, Jains are always ready and willing to look positively and with enthusiasm upon environmental causes. In India and abroad, they are in the forefront of bringing greater awareness and putting into practice their cardinal principles on ecology. Their programmes have been modest and mostly self-funded through volunteers.

Judaism and the environment

The Environment – Jewish Perspectives (Professor Nabum Rakover)

> 'Consider the work of God; for who can make that straight, which man has made crooked? (Eccles. 7:13)
>
> 'When God created Adam, he showed him all the trees of the Garden of Eden and said to him: "see my works, how lovely they are, how fine they are. All I have created, I created for you. Take care not to corrupt and destroy my universe, for if you destroy it, no one will come after you to put it right". (Eccles. Rabbah 7)

The present paper is concerned with the vast and complex problem of protecting our natural environment from pollution and destruction, so that we can live in God's world while enjoying its beauty and deriving from it the maximum physical and spiritual benefit.

In Jewish sources, the rationale for man's obligation to protect nature may he found in the biblical expression, 'For the earth is Mine' (Lev. 23:23). The Bible informs us that the earth is not subject to man's absolute ownership, but is rather given to man 'to use and protect' (Gen. 2.15).

From biblical sources which refer to man's 'dominion' over nature, it would appear as though man was granted unlimited control of his world, as we find in Genesis 1:26.

> 'And God said, "Let us make man in our image, after our likeness; and let them have dominion over the fish of the sea, and over the fowl of the air, and over the cattle and over all the earth,

and over every creeping thing that creeps upon the earth".

And again in Genesis 1:28

> 'And God blessed them [Adam and Eve] "Be fruitful, and multiply, and fill the earth and subdue it; and have dominion over the fish of the sea, and over the fowl of the air, and over every living thing that creeps upon the earth".

Rav Kook (R. Avraham Yitzhak hakohen Kook, 1865–1935, first chief rabbi of Palestine) has an insightful understanding of the idea (Kook 1961):

> 'There can be no doubt to any enlightened or thoughtful person, that the 'dominion' mentioned in the Bible in the phrase, 'and have dominion over the fish of the sea, and over the fowl of the air, and over every living thing that creeps upon the earth', is not the dominion of a tyrant who deals harshly with his people and servants in order to achieve his own personal desires and whims. It would be unthinkable to legislate so repugnant a subjugation and have it forever engraved upon the world of God, who is good to all and whose mercy extends to all. He has created, as is written, "the earth is founded upon mercy" (Ps. 89.3).

Three things which grant man tranquillity

The Sages of the Talmud gave expression to the environment's effect on man's spirit in their statement (*Berakhot* 57b).

> 'Three things restore a man's consciousness: [beautiful] sounds, sights, and smells. Three things enlarge a man's spirit: a beautiful dwelling, a beautiful wife, and beautiful clothes.'

Life in the city

The Sages of the Talmud also noted that the environment undergoes more damage in large cities than in small towns. In explaining a law of the Mishnahi that a spouse may not compel his mate to move from a village to a large city, the Talmud cites the reasoning of R Yosi Ben Hanina (*Ketubot* 110b). 'Life is more difficult in the city', Rashi (Rashi, ad IGC., s.v. *Yeshivat*) explains,

> '... because so many live there, and they crowd their houses together, and there is no air, whereas in villages there are gardens and orchards close to the homes, and the air is good.'

One who threw stones into the public domain

We learn of the obligation of the individual to protect the public domain from a story brought in the Tosefta (*Tosefta Baba Kama* 2:1 0).

> It happened that a certain person was removing stones from his ground onto public ground when a pious man found him and said, 'Fool, why do you remove stones from ground which is not yours to ground which is yours?'. The man laughed at him. Sometime later he was compelled to sell his field, and when he was walking on that public ground, he stumbled over the stones he had thrown there. He then said, 'How well did that pious man say to me, "Why do you remove stones from ground which is not yours to ground which is yours?"'

In other words, the terms 'public domain' and 'private domain' are not necessarily identical to the concepts 'mine' and 'not mine'. What was once my private domain might one day not be mine, while the public domain will always remain my domain.

Protection of the environment and the love of man

In addition to the rules governing man's relations with his fellow man, which are based upon the biblical imperative, 'Love your neighbour as yourself' (Lev.19:18), norms were established for man's treatment of plants, animals and even the inanimate elements of nature.

When approaching the subject of environmental protection, we must be careful to maintain the proper balance between protection of the environment and protection of man. The proper balance in this context is certainly not one of equality between man and nature. The relationship between man and nature is one of ownership – albeit limited. In our enthusiasm for protecting the environment, we must not forget man's interests or his role in the scheme of creation. Love of nature may not take precedence over love of man. We must avoid at all costs the error of those who were known as lovers of animals yet perpetrated the worst crimes imaginable against their fellow men.

The proper balance must also be maintained between individual interests and the interest of the public. Sometimes an individual's act may harm the community, as when a person builds a factory that pollutes the environment with industrial waste. Sometimes, however, it is the community that is interested in a factory although it constitutes a serious infringement upon an individual's ability to enjoy his own home and surroundings.

When discussing the quality of the environment,

we must remember that the environment also comprises the persons living in it – individuals and community. Protection of the environment, by itself, cannot solve conflicts of interest, though it can extend the range of factors considered when seeking solutions to problems. Solutions must, in the final analysis, be based upon economic, social and moral considerations.

In our survey, we examine Jewish sources that relate to our topic. We shall mention the limitations imposed on acts that do harm to nature, one's neighbours and society at large.

A number of the subjects we investigate are rooted in the laws governing relations among neighbours and the laws of torts. These laws are numerous and complex, and a comprehensive discussion of them all is far beyond the scope of the present survey. We shall, however, attempt to cover briefly a number of the guiding principles in these areas. And even if we do not find solutions for all the problems raised, we hope that we can at least refine the questions and pose challenges for further analysis of the issues.

Protecting nature

Man and his environment

> 'I recall the early days, from 1905 onward, when it was granted me by the grace of the blessed Lord to go up to the Holy Land, and I came to Jaffa. There I first went to visit our great master R Abraham Isaac Kook (of blessed memory), who received me with good cheer, as it was his hallowed custom to receive everyone. We chatted together on themes of Torah study. After the afternoon service, he went out, as was his custom, to stroll a bit in the fields and gather his thoughts; and I went along. On the way, I plucked some branch or flower. Our great master was taken aback; and then he told me gently, "Believe me. In all my days I have taken care never to pluck a blade of grass or flower needlessly, when it had the ability to grow or blossom. You know the teaching of the Sages that there is not a single blade of grass below, here on earth, which does not have a heavenly force telling it 'Grow'. Every sprout and leaf of grass says something, conveys some meaning. Every stone whispers some inner, hidden message in the silence. Every creation utters its song (in praise of the Creator)." Those words, spoken from a pure and holy heart, engraved themselves deeply on my heart. From that time on, I began to feel a strong sense of compassion for everything.' (R Aryeh Levine, in Simcha Raz 1976)

Rav Kook's attitude towards each individual plant and to creation in general is based upon a comprehensive philosophical approach to man's relationship with nature. This position was well articulated by the noted mystic R Moshe Cordovero (1522-1570: the leading kabbahst of Safed in the period preceding R. Yitzhak Luria.) in his work *Tomern Devorah.*

> 'One's mercy must extend to all the oppressed. One must not embarrass or destroy them, for the higher wisdom is spread over all that was created: inanimate, vegetable, animal and human. For this reason were we warned against desecrating foodstuffs, and in the same way, one must not desecrate anything, for all was created by His wisdom – nor should one uproot a plant, unless there is a need, or kill an animal unless there is a need.'

The sabbatical year

The idea of conservation may be found in the biblical institution of the sabbatical year (Lev. 25:1-5).

> And the Lord spoke unto Moses on Mount Sinai, saying, 'Speak unto the children of Israel, and say unto them: When you come into the land which I give you, then shall the land keep a Sabbath unto the Lord. Six years shall you sow your field, and six years shall you prune your vineyard, and gather in the produce thereof. But in the seventh year shall be a Sabbath of solemn rest for the Land, a Sabbath unto the Lord; you shall neither sow your field nor prune your vineyard. That which grows of itself of your harvest you shall not reap, and the grapes of your undressed vine you shall not gather; it shall be a year of solemn rest for the land.' (See also Lev. 25:6.7 and Exod. 23:10-11.)

R. Moshe ben Maimon Maimonides (1135-1204: the most distinguished Jewish authority of the Middle Ages), in his *Guide for the Perplexed* suggests a reason for the sabbatical year:

> With regard to all the commandments that we have enumerated in Laws concerning the Sabbatical year and the jubilee, some of them are meant to lead to pity and help for all men – as the text has it: 'That the poor of the people may eat; and what they leave, the beasts of the field shall eat...' (Exod. 23.-11) – and are meant to make the earth more fertile and stronger through letting it fallow.

In other words, one of the goals of ceasing all agricultural activity is to improve and strengthen the land.

Another reason for the sabbatical year which emphasizes man's relationship with his environment is suggested by the author of *Sefer haHinnukh* in his explanation of the obligation to declare all produce ownerless, such that anyone may enter any field and take from its produce during the sabbatical year. (*Sefer haHinnukh* commandment 84 (ed. Chavel, commandment 69). Cf. *Sefer haHinnukh* on the Jubilee Year, commandment 330 (ed. Chavel, commandment 326): 'God wished to teach His people that all belongs to Him, that ultimately all things are turned to the one to whom God wished to give them at the outset. For the world is His, as is written (Ex. 19:5), "... for all the earth is Mine"'.)

To the reasons for the Sabbatical Year, Rav Kook adds restoration of the proper balance among man, God, and nature. In the Sabbatical year, according to Rav Kook,

> '... man returns to the freshness of his nature, to the point where there is no need to heal his illnesses, most of which result from destruction of the balance of life as it departs ever further from the purity of spiritual and material nature" (Introduction to *Shabbat haAretz*, pp. 8-9).

To establish in our hearts and make a strong impression on our thoughts that the world was created as a new entity out of nothing, 'that in six days God made the heaven and the earth' (Exod. 20:11);

> '... and on the seventh day, when He created nothing, he decreed rest for Himself. And, therefore, the Holy One commanded [us] to declare all produce of the earth ownerless during this (sabbatical year in addition to cessation of agricultural work, so that man will remember that the earth which yields its produce for him each year, does not do so on its own strength or of itself, but rather there is one who is Master over the land and its owners, and when He wishes, He commands that the produce be ownerless.'

It is worth noting that the institution of the sabbatical year is practised into modern times within observant circles; the last Sabbatical observed was in 1993-94 (corresponding to the Hebrew calendar year 5754.)

Altering creation
In addition to refraining from over-exploitation of the earth's resources, we must also be mindful of preserving the natural balance of creation. This is the approach taken by R Avraham ibn Ezra (*c.* 1089–1164: biblical commentator, poet, philosopher and physician who lived in Spain) in his explanation of the biblical prohibition against mixing species. In Leviticus (19:19) we find:

> 'You shall keep my statutes. You shall not let your cattle gender with a diverse kind; you shall not sow your field with two kinds of seed; neither shall there come upon you a garment of two kinds of stuff mingled together.'

One aspect of preventing changes in the creation finds expression in the effort to avoid causing the extinction of any animal. The presumption that everything that was created was created for some purpose denies us the possibility of eliminating from the world any species. So writes Nahmanides concerning the prohibition of mixing species (Lev. 19:19):

> 'The reason for the prohibition against mixing species is that God created all the species of the earth and gave them the power to reproduce so that their species could exist forever, for as long as He wishes the world to exist, and He created for each one the capacity to reproduce its own species and not change it, forever. And this is the reason for sexual reproduction among animals, to maintain the species; so too among humans, it is for the purpose of being fruitful and multiplying.'

Wasteful destruction
a) The prohibition of wasteful destruction
An additional expression of man's obligation to preserve his natural environment may be found in the commandment against wasteful destruction, *bal tash'hit*. In general, the commandment prohibits the destruction of anything from which humans may benefit. It applies to the destruction of animals, plants, and even inanimate objects (see *Encyclopedia Talmudit*, s.v. *Bal tash'hit;* Nahum Rakover. *A Bibliography of Jewish Law - Otzar haMishpat*, s.v. *Bal tash'hit*, vol. I, p. 285; vol. II, p. 278).

Instructive remarks are found in *Sefer haHinnukh*'s discussion of the prohibition of cutting down fruit-bearing trees. The discussion opens with a discourse on the scope of the commandments (Commandment 529 (ed. Chavel, commandment 530)):

> 'We have been prevented from cutting down trees when we lay siege to a city in order to press and bring pain to the hearts of its resi-

> dents, as is said: ".. you shall not destroy the trees thereof... and you shall not cut them down" (Deut. 20:19). Included in this prohibition is destruction of every type, such as burning or tearing a garment, or breaking a vessel for no reason.'

The author of *Sefer HaHinnukh* then goes on to explain the reason for the prohibition:

> It is known that this commandment is meant to teach us to love the good and the useful and cling to them, and in this way goodness will cling to us, and we will avoid all that is bad and decadent. And this is the way of the pious: they love peace and rejoice in the good fortune of others, and bring everyone near to the Torah, and do not waste even a mustard seed, and they are pained by all destruction and waste that they see. And they save anything they can from destruction with all their might. But the wicked are different. They are the allies of those who destroy, they rejoice in destruction of the world and they destroy themselves: 'with the kind of measure a man measures, so shall he be measured'. (*Mishnah Sotah* 1:7).

The source of the prohibition of wasteful destruction is the biblical prohibition of cutting down fruit-bearing trees, which will be discussed below. The prohibition of wasteful destruction, however, is more comprehensive than the prohibition of destroying fruit-bearing trees, and it extends to anything that has use. In other words, the prohibition includes the destruction of man-made objects, and is not restricted to the preservation of nature.

b) Cutting down fruit-bearing trees

In the book of Deuteronomy (20:19), among the laws concerning the waging of war, we find:

> 'When you shall besiege a city a long time, in making war against it to take it, you shall not destroy the trees thereof by wielding an axe against them; for you may eat of them, but you shall not cut them down; for is the tree of the field man, that it should be besieged by you?'

The Bible thus warns that even in time of war, it is forbidden to destroy fruit-bearing trees.

The author of *haKetav vehaKabbalah* explains the prohibition (Deut. 20:19)

> 'It is not proper to use some created thing for a purpose diametrically opposed to the purpose for which it was created, as has been stated (Exodus 20:22.) "for if you lift up your sword to it, you have profaned it" – the altar was created to prolong the life of man, and iron was created to shorten the life of man; thus it is not fitting that something which shortens man's life be used upon that which lengthens it. So too a tree, which was created to make fruit to nourish men and animals, should have nothing done to it that destroys man.'

c) For man is a tree of the field

The relationship of God, man, and nature is depicted in the biblical expression, 'For man is a tree of the field'. Various interpretations have been given to this relationship: even plants are subject to divine Providence; both man and the tree are God's creatures. Sifrei asserts, 'This shows that man's living comes from trees'.

The Sages also compared the death of the tree to the departure of man's soul from his body:

> 'There are five sounds that go from one end of the world to the other, though they are inaudible. When people cut down the wood of a tree that yields fruit, its cry goes from one end of the world to the other, and the sound is inaudible. When the soul departs from the body, the cry goes forth from one end of the world to the other, and the sound is inaudible.'

On the basis of this passage, R Menahem Recanad (an early kabbalist active in Italy at the end of the thirteenth century and the beginning of the fourteenth century) comments that, 'when man wreaks destruction in the material world, destruction is wreaked in the metaphysical world as well, and that this is what was meant by "For man is a tree of the field"'.

Polluting the environment

Smoking

Smoking constitutes a serious environmental pollutant and danger to health. Public awareness of this problem has led to legislation against smoking in public places. (In 1983, the Israeli Knesset enacted the Restriction in Public Places Law, which was supplemented in 1994 by an executive order signed by the minister of health (*Kovetz Takkanot*, 21 July 1994, pp. 1197-98). The executive order confines smoking in the work place to specially designated areas where there are no non-smokers, where there is adequate ventilation, and where smoking does not cause a nuisance to other parts of the work place.)

Jewish legal authorities have considered whether it is prohibited to smoke in places where the smoke might bother others. One authority who absolutely prohibited smoking in public places was R Moshe Feinstein (1895-1986: considered the spiritual leader of American Orthodoxy and American Jewry's leading authority on Jewish law in recent years). It was his opinion that even if smoking was irritating only to those who are hypersensitive, it would nevertheless be prohibited in public places. The precedent for this is the talmudic case of R Yosef, who was hypersensitive to noise. If it is possible to restrain particular actions on the basis of hypersensitivity, R Feinstein reasons, certainly it is possible to do so where there is pain or injury. Thus, where smoking is harmful to others, it is certainly prohibited.

Beauty

On seeing creatures that are beautiful or exceptionally well formed or goodly trees, one says, 'Blessed are You, O Lord our God, King of the universe, who has such as these in His world'. If one goes out into the fields or gardens during the month of Nisan (i.e. the Spring) and sees the trees budding and the flowers in bloom, one says, 'Blessed are You, O Lord our God, king of the universe, who has made Your world lacking in nought and created therein beautiful creatures and goodly trees for the benefit of mankind'(Maimonides, M.T., *Berakhol* 10:13).

Aesthetic beauty appears in Jewish sources not only as a value worthy of fostering in the life of the individual and the community, but also as the basis for a variety of legal obligations. The obligations derive from biblical regulations and from rabbinical legislation.

In the *Pentateuch* (Num. 35:2-5), we find instructions regarding city planning, which required designation of open spaces free of all obstruction. Rashi describes the purpose of the open strip as being 'for the beautification of the city, that it have air' (Rashi, *Sotah* 27b, s.v. *Migrash*. Cf. Rashi's comments on Num. 35:2).

Later rabbinical legislation expanded the applicability of this rule to include cities other than those mentioned in the Bible (Maimonides, M.T., *Shemitah veYovel* 13:1-2,4-5).

Sikhism and the environment

> 'Creating the world, God has made it a place to practice spirituality.' (*Guru Granth Sahib*, page 1035.)

The Sikh scripture declares that the purpose of human beings is to achieve a blissful state and be in harmony with the earth and all creation. It seems, however, that humans have drifted away from that ideal. For the earth is today saturated with problems. It is agonizing over the fate of its inhabitants and their future! It is in peril as never before. Its lakes and rivers are being choked, killing its marine life. Its forests are being denuded. A smoky haze envelops the cities of the world. Human beings are exploiting human beings.

There is a sense of crisis in all parts of the world, in various countries and amongst various peoples. The demands of national economic growth and individual needs and desires are depleting the natural resources of the earth. There is serious concern that the earth may no longer be a sustainable biosystem. The major crises facing the earth – the social justice crisis and the environmental crisis – together are heading the earth towards a disastrous situation. The social injustice crisis is that of humanity's confrontation with itself and the environmental crisis is caused by humanity's confrontation with nature. The social justice crisis is that poverty, hunger, disease, exploitation and injustice are widespread. There are economic wars over resources and markets. The rights of the poor and the marginal are violated. Women constituting half the world's population have their rights abused. The environmental crisis caused by humanity's exploitation of nature is leading to the depletion of renewable resources, destruction of forests, over-use of land for agriculture and habitation. Today pollution is contaminating air, land and water. Smoke from industries, homes and vehicles is in the air. Industrial waste and consumer trash is affecting streams and rivers, ponds and lakes. Much of the waste is a product of modern technology; it is not biodegradable, not re-usable and its long term consequences are unknown. The viability of many animal and plant species, and possibly that of the human species itself, is at stake.

This crisis cries out for an immediate and urgent solution. The crisis requires a going back to the basic question of the purpose of human beings in this universe and an understanding of ourselves and God's creation. We are called to the vision of Guru Nanak which is a World Society comprising God-conscious human beings who have realized God. To these spiritual beings the earth and the universe are sacred; all life is unity, and their mission is the spiritualization of all.

Guru Nanak laid the foundation of Sikhism in the late fifteenth century. His writings, those of other human Gurus who succeeded him, and other spiritual leaders, are included in the scripture – *Guru Granth Sahib*. Guru Granth has been the Guru and Divine Master of the Sikhs since 1708, when Guru Gobind Singh declared that there would be no more human Gurus. Guru Nanak and his successors during their lifetime worked towards creating an ideal society that has as its basis spiritual awareness and ethical integrity. The name 'Sikh' means disciple or learner of the Truth. Guru Nanak in his philosophy states that the reality that humans create around themselves is a reflection of their inner state. The current instability of the natural system of the earth – the external environment of human beings, is only a reflection of the instability and pain within humans. The increasing barrenness of the earth's terrain is a reflection of the emptiness within humans.

The solution to problems manifest in our world is in prayer and accepting God's *hukam*. It is difficult to translate certain Sikh concepts accurately. *Hukam* is one such concept – it may be best described as a combination of God's will, order, or system. With an attitude of humility, and surrender to the Divine Spirit, conscientious human beings can seek to redress the current crises of the environment and of social justice. In the Sikh Way this is done through the guidance of the Guru, who is the Divine Master and messenger of God. A Sikh theologian, Kapur Singh, explains that Sikhism has three postulates implicit in its teachings:

- that there is no essential duality between spirit and matter;
- that humans have the capacity to consciously participate in the process of spiritual progression, and
- that the highest goal of spiritual progression is harmony with God, while remaining earth-conscious, so that the world itself may be transformed to a spiritual plane of existence.

Unity of spirit and matter and the interconnectedness of all creation

The Sikh view is that spirit and matter are not antagonistic. Guru Nanak declared that the spirit was the only reality and matter is only a form of spirit. Spirit takes on many forms and names under various conditions.

> 'When I saw truly, I knew that all was primeval. Nanak, the subtle (spirit) and the gross (material) are, in fact, identical.' (*Guru Granth Sahib*, page 281)

That which is inside a person, the same is outside; nothing else exists.

> 'By divine prompting look upon all existence as one and undifferentiated; the same light penetrates all existence.' (*Guru Granth Sahib*, page 599)

The chasm between the material and the spiritual is in the minds of humans only. It is a limitation of the human condition that spirit and matter appear as duality, and their unity is not self-evident.

The material universe is God's creation. Its origin was in God and its end is in God; and it operates within God's *hukam*. Guru Nanak declares that God alone knows the reasons for and the moment of earth's creation. The origin of the universe is unknowable. The act of creation itself, the creation of the primeval atom, was instantaneous, caused by the Will of God.

Further description of the universe and its creation in Sikh scriptures are remarkably similar to recent scientific speculation about the universe and its origin. One of the basic hymns in the Sikh scripture, which may be called the 'Hymn of the Genesis" describes the indeterminate void before the existence of this universe. Guru Nanak speaks of innumerable galaxies, of a limitless universe, the boundaries of which are beyond human ability to comprehend. God alone knows the extent of creation.

God created the universe and the world, for reasons best known to Him. And being the results of God's actions all parts of the universe are holy. God is an all-pervasive being manifest through various elements of creation.

Having created this universe and the world, God directs them. All actions take place within God's *hukam*. God alone knows how and why. God, however, not only directs this vast and massive theatre, but also watches over with care and kindness – the benign, supportive parent!

> Men, trees, pilgrimage places, banks of sacred streams, clouds, fields.
> Islands, spheres, universes, continents, solar systems.
> The sources of creation, egg-born, womb-born, earth-born,
> sweat-born, oceans, mountains and sentient beings.
> He, the Lord, knows their condition, O Nanak.
> Nanak, having created beings, the lord takes care of them all.
> The creator who created the world, He takes thought of it as well.
> (*Guru Granth Sahib*, page 466)

The world, like all creation, is a manifestation of God. Every creature in this world, every plant, every form is a manifestation of the Creator. Each is part of God and God is within each element of creation. God is the cause of all and He is the primary connection between all existence.

The Creator created himself.
And created all creation in which he is manifest.
You Yourself the bumblebee, flower, fruit and the tree.
You Yourself the water, desert, ocean and the pond.
You Yourself are the big fish, tortoise and the Cause of causes.
Your form can not be known.
(*Guru Granth Sahib*, page 10 1 6)

In the world God has created he has also provided each species and humans with means of support and nurturing.

In Sikh beliefs, a concern for the environment is part of an integrated approach to life and nature. As all creation has the same origin and end, humans must have consciousness of their need to guide themselves through life with love, compassion and justice. Becoming one and being in harmony with God implies that humans endeavour to live in harmony with all of God's creation.

The second postulate is that humans, practising a highly disciplined life, while remaining active in the world, are capable of further spiritual progression. It is important that Sikhs retain the primacy of spirit over matter, while it is desirable that they do not deny matter or material existence. It is not required that humans renounce the world. They must maintain their life in the world and uphold all responsibilities in the world. Humans should be renouncers of plenty and maintain a simple life. Further spiritual progress fundamentally starts with an individual conquering himself/herself with the guidance of the Guru. The emphasis is on mastery over the self and the discovery of the self; not mastery over nature, external forms and beings. Sikhism teaches against a life of conspicuous, wasteful consumption. The Guru recommends a judicious utilization of material and cultural resources available to humans.

Then why get attached to what you will leave behind.
Having wealth, you indulge in pleasures bout,
From that, tell me, who will bail you out?
All your houses, horses, elephants and luxurious cars,
They are just pomp and show, all totally false.
(*Guru Granth Sahib*)

The Gurus taught humans to be aware of and respect the dignity in all life, whether human or not. Such a respect for life can only be fostered where one can first recognize the Divine spark within oneself, see it in others, cherish it, nurture and fulfil it.

This little shrine of the human body
This great opportunity of life!
The object is to meet the Beloved, thy Master
(*Guru Granth Sahib*)

Spiritual discipline

Humans have the capability to further their spiritual progression through conscious choice and it is important to identify the method by which they might do so. The method suggested by Guru Nanak is one of spiritual discipline, meditation and prayer, and sharing. Sikhism emphasizes mastering five negative forces: lust, anger, worldly or materialistic attachment, conceit and greed. These together constitute what Sikhs term *Haumai* -'I am-ness'. Mastering *haumai* is achieved by developing five positive forces: compassion, humility, contemplation, contentment and service, without expecting any material or spiritual reward. The guiding principles are love and forgiveness. Every decision in life has to be based on rationality and a personal code of ethics. Guru Nanak's philosophy of values inspires the individual to transcend his/her existence through this spiritual discipline. Sikh religion preaches strong family involvement. A person pursuing this spiritual discipline must also work to create an atmosphere for other members of the family to progress spiritually.

The ideal Sikh – someone who has an intense desire to do good

The third postulate is that the true end of human beings is in their emergence as God-conscious beings, who remain aware of the earth and operate in the mundane material world, with the object of transforming and spiritualizing it into a higher plane of existence. In this spiritual state, individuals are motivated by an intense desire to do good, transforming their surroundings.

Through a life based on the method prescribed by the Gurus, humans may achieve a higher spiritual state. Such truly emancipated, vibrant and enlightened spirits (*aivan-mukta, brahma-gyani*) become the true benefactors of humanity and the world around them. Such an individual would not exploit another human or sentient being, as each is a manifestation of the eternal and the supreme. In this God-conscious state they see God in all and everything.

I perceive Thy form in all life and light;
I perceive Thy power in all spheres and sight.
(*Guru Granth Sahib*, page 464)

Spiritualization is a liberation from material compulsions and attractions. It means an awareness of the Cosmic Order and striving towards the execution of Divine Will. So, the spiritualized human is creative and constructive. Therefore a Sikh life is a life of harmony with other individuals and with other beings and other forms. For an enlightened individual the world has only one purpose – to practice spirituality. That is the ultimate objective of all humans.

Such a person is involved in human problems and society and has to prove his or her effectiveness there. Such a person lives with a mission – and works for the emancipation of all. A true Sikh is for individual human rights, the environment, and justice for all.

> 'The God-conscious persons is animated with an intense desire to do good in this world.' (*Guru Granth Sahib*, page 273)

Practising the philosophy

Integrated approach: care of the environment without social justice is not possible

Environmental concerns may be viewed as part of the broader issue of human development and social justice. Many environmental problems, particularly the exploitation of environmental resources in developing nations, are due to the poverty of large parts of the population. Therefore an integrated approach is necessary.

Sikhism opposes the idea that the human race's struggle is against nature and that its supremacy lies in the notion of 'harnessing' nature. The objective is harmony with the eternal – God – which implies a life of harmony with all existence. Aspiring for a life of harmony, therefore, also implies a life of supporting individual rights, environmentalism, a life that works against injustice towards anybody and anything.

The tenth Guru founded the Order of the Khalsa in 1699, consisting of those who practise the spiritual discipline of Sikhism and are committed to ensure the preservation and prevalence of a World Society. Over the last three centuries the members of Khalsa Order have stood up for the rights of the oppressed and the disenfranchised even at the cost of their own lives. The Khalsa vision of the World Society is:

> Henceforth such is the Will of God:
> No man shall coerce another;
> No person shall exploit another.
> Each individual has the inalienable birth-right to seek and pursue happiness and self-fulfilment.
> Love and persuasion is the only law of social coherence.
> (*Guru Granth Sahib*, page 74)

The Khalsa have opposed any force that has threatened the freedom and dignity of human beings. In the eighteenth century it was the oppressive rulers of northern India, and invaders from Afghanistan; in the nineteenth and twentieth centuries they have struggled against the oppression of European colonists and Indian governments. For the Khalsa justice requires the participation and inclusion of all in obtaining and enjoying the fruits of God's creation. Justice achieved through co-operative effort is desirable. The ideal for the Khalsa is to strive for justice for all, not merely for themselves.

The Institution of Sangat, Pangat and Langar

The Sikh Gurus, through their lives, provided role models for the Sikhs. They all actively worked to stress the equality of all humans and challenged the rigid social stratification of the caste system in India. The very existence of the Sikh religion is based on challenging (1) inequality in society, and (2) the exploitation of the poor and the marginal by the religious and political establishment.

Sikh Gurus provided many examples of standing by their principles and confronting exploitation and oppression. They stood by the 'low' and the 'poor', for, as Guru Nanak said:

> There are the lowest men among the low-castes. Nanak, I shall go with them. What have I got to do with the great?
> God's eye of mercy falls on those who take care of the lowly.
> (*Guru Granth Sahib*, page 15)

Sikh Gurus challenged the status quo and came into conflict with the entrenched elite – political, social, religious and economic. The Gurus were most sympathetic to the down-trodden of society, the untouchables and those of lower caste. They vehemently opposed the division of society on the basis of caste, which had been and is still significantly present. They identified themselves with the poor in full measure and were critical of those responsible for their misery. In the course of their travels they preferred to live in the homes of those who made an honest living to the homes of the rich who thrived on exploitation.

Two Sikh Gurus were martyred by the regimes of their period for challenging the contemporary authorities. One, Guru Tegh Bahadur, was martyred when he stood up for the religious freedom of the

Hindu inhabitants of Kashmir who were being forced to accept Islam by the rulers.

Sikh Gurus also moulded traditional life-styles to exemplify a more equitable society. They created many institutions that form the basis of Sikh society and are based on the equality of all. The Sikh Gurus invited people of all castes and creeds to meditate together. That would be called *Sangat*. Either before or after the meditation, people were asked to sit and eat together irrespective of their social background to create a sense of equality. That process would be called *Parigat*. Sikh Gurus started a tradition of free distribution of food to the rich or poor through the Sikh meeting areas. That would be called *Langar*. These three ideas were in contrast to the Indian society which had separate temples or water wells for social outcasts. These changes by the Sikh Gurus created a lot of opposition from the religious establishment. These changes are still much alive in Sikh practices today. Through the creation of the Khalsa, the Gurus established a system that would protect and maintain a free and just order.

Equality of women

Women and their rights have been ignored for too long. Any approach to solve problems of social justice and the environment must be sensitive to women's concerns, and must include women as equals.

Often piece-meal solutions to environmental problems focus on limiting population growth and family planning programmes. Most family planning measures end up abusing women's rights and should be rejected on those grounds alone. Meanwhile they spread mistrust of family planning among women.

Guru Nan took steps to implement these principles. Guru Nanak denounced the idea that spirituality was only for men, and not for women. The first Sikh Guru in his preaching and writings made direct statements emphasizing that women were no less than men:

> After the death of one's wife, one seeks another, and through her social bonds are cemented.
> Why should we condemn women who give birth to leaders and rulers?
> Everyone is born of a woman and a woman alone.
> Nobody is born otherwise.
> God alone is an exception to this rule.
> (*Guru Granth Sahib*, page 473)

Guru Amardas strongly opposed the custom of *Sati* in the sixteenth century and also advocated widow marriages. *Sati* was the Indian practice whereby widows burned themselves with their husband's corpse at cremation. Guru Amar Das appointed and ordained a large number of women preachers, and at least one bishop – Mathura Devi – four hundred years ago. The Sikh Gurus also raised their voices against the *purdah* or veil. Guru Amardas did not even allow the Queen of Haripur to come into the religious assembly wearing a veil.

The immediate effect of these reforms was that women gained an equal status with men. Those who lived as grovelling slaves of society became fired with a new hope and courage to lift themselves with the belief that they were not helpless creatures but responsible beings endowed with a will of their own with which they could do much to mould the destiny of society.

Women came forward as the defenders of their honour and dignity. They also became the rocks that stood against tyrants. Without the burden of unnecessary and unreasonable customs, Sikh women became the temporal and spiritual supporters of men, often acting as the 'conscience of men'. Sikh women proved themselves the equals of men in service, devotion, sacrifice and bravery.

Since the late nineteenth century, Sikh community organizations have made efforts at expanding educational opportunities for all. Individual Sikhs, men and women, in various cities and towns, took the initiative to start and operate women's colleges and schools. Women's education was part of the drive to improve education among the Sikhs, initiated by Sikh organizations in the 1920s. In towns and villages in the Punjab, and in cities with significant Sikh populations, there exist schools and colleges operated by Sikh organizations.

Community-based sharing of resources

Traditional modes of farming, and traditional modes of life in Northern India, have been dependent upon limited resources. As there exist circumstances where there are large numbers of people dependent upon relatively unexploited resources, the traditional life-style ensures use of the least resources, considerable re-use, and recycling of resources. In a culture based on organic resources and materials, recycling is an intrinsic and natural part of the resource cycle. There are strong traditions of sharing resources.

There have been traditional practices that maintained lands and forests as community property within proximity of human habitation. For instance, in traditional rural India and Punjab, two of the most important centres of human activity have been the Sikh *gurudwara*, and a source of water – pond, tank, pool or running water. Both of these sites were surrounded by community land, not owned by anyone and not used for agriculture. This was where there were trees and plants – groves or small forests. They provided shade and shelter, and were a source of firewood within easy reach of habitation.

The Gurus established towns and cities, each created around a religious centre. The focus was on a life-style based on sharing: a life-style that promoted equity among people, and optimum utilization of resources. Even today, even rural Punjab families share resources with their neighbours. This is particularly evident on large family occasions such as weddings, when the entire village may play host to guests and share living space, beds, etc.

Most *gurudwaras* in India were specifically designed to have a water tank, or were near running waters (rivers or pools) which were always considered a community resource. For instance, Ainritsar grew up around the Harimandir (ordinarily referred to as the Golden Temple) and the Arnrit Samovar (The pool of nectar – the water). The cities and towns that grew around *gurudwaras* were ideally centred on a spiritual life-style based on sharing.

Since the time of the Gurus, Sikh *gurudwaras* have included institutionalized practices that emphasize sharing of resources. *Gurudwaras*, in addition to being places for congregation for prayer and meditation, are (1) a place to stay for travellers and others; (2) a community kitchen – *langar*; (3) a place for dispensing medication and medical care; (4) a place to impart education to the young. *Gurudwaras* have always been places of shelter for travellers and visitors. Most major *gurudwaras* have rooms where visitors may stay. In addition, Sikh *gurudwaras* stock extra beds, pots and pans, etc. At weddings and other family events, the *gurudwaras* are a source for borrowing sheets, beds, pots and pans.

There has always been great emphasis on avoiding waste. Traditionally the community kitchen served food on plates made from leaves and cups made from clay. Today they tend to use steel plates and utensils that are re-used. The kitchens have always been stocked by ordinary people – farmers, traders, others in the community – on a voluntary basis.

Sikhism against smoking

It is now a known fact that smoking is both a primary and secondary health hazard. In addition to the environment, it has seriously deleterious effects on the person who smokes, to the bystander who breathes the second-hand smoke, and to the unborn foetus of the female smoker. Though this only been scientifically verified in the last half-century, Guru Gobind Singh, the last living Guru of the Sikhs, listed the use of tobacco as one of the four major acts forbidden to initiated adherents of the Sikh region. Though tobacco was introduced into India only in the mid-1600s, he had the wisdom to specifically interdict it in 1699. From its very beginning, Sikhism had forbidden the use of any intoxicants or mind-altering substances for any purpose, except medicinal.

Conclusion

The ideal for Sikhism is a society based upon mutual respect and co-operation and providing an optimal atmosphere for individuals to grow spiritually. Sikhism regards a co-operative society as the only truly religious society, as the Sikh view of life and society is grounded in the worth of every individual as a microcosm of God. Therefore, an individual must never be imposed upon, coerced, manipulated or engineered:

> If thou wouldst seek God, demolish and distort not the heart of any individual.
> (*Guru Granth Sahib*, page 1384)

All life is interconnected. A human body consists of many, parts, every one of them has a distinct name, location and function and all of them are dependent upon each other. In the same way, all the constituents of this universe and this earth are dependent upon each other. Decisions in one country or continent cannot be ignored by others. Choices in one place have measurable consequences for the rest of the world. It is part of the same system.

Life, for its very existence and nurturing, depends upon a bounteous nature. A human being needs to derive sustenance from the earth; not to deplete, exhaust, pollute, burn or destroy it. Sikhs believe that an awareness of that sacred relationship between humans and the environment is necessary for the health of our planet, and for our survival. A new 'environmental ethic' dedicated to conservation and wise use of the resources provided by a bountiful nature can only arise from an honest understanding and dedicated application of our old, tried and true spiritual heritage.

Taoism and the environment

Part one: General statement on taoism

Taoism emerged on the basis of what are known as the One Hundred Schools of Thought during the period 770-221 BC. Starting with the formal setting up of Taoist organizations in the East Han period (25-220), it has a history of nearly 2,000 years. Taoism has been one of the main components of Chinese traditional culture, and it has exerted great influence on the Chinese people's way of thinking, working and acting. It is no exaggeration to say that in every Chinese person's consciousness and subconsciousness, the factors of Taoism exist to a greater or lesser degree.

Because of its deep cultural roots and its great social impact, Taoism is now one of the five recognized religions in China; namely Buddhism, Christianity, Catholicism, Islam and Taoism. Even more, the influence of Taoism has already transcended the Chinese-speaking world, and has attracted international attention.

According to our statistics, more than 1,000 Taoist temples have now opened to the public (this number does not include those in Taiwan, Hong Kong and Macao) and about 10,000 Taoists live in such communities. There are about 100 Taoist associations all over China, affiliated to the China Taoist Association. Several colleges have also been established to train Taoists, and many books and periodicals on the study and teaching of Taoism have been published. All Taoists work hard in order that Taoism should develop and flourish. They take an active part in mobilizing the masses, carrying forward the best in Taoist tradition and working for the benefit of human society.

Like every major world religion, Taoism has its own outlook on the universe, human life, ideals of virtue and ultimate purpose. Due to its distinctive cultural and historical background, it has its own striking characteristics. It can be briefly summarised in the following two precepts.

Give respect to the Tao above everything else

Tao simply means 'the way'. Taoism considers that *Tao* is the origin of everything, and *Tao* is the ultimate aim of all Taoists. This is the most fundamental tenet of Taoism. *Tao* is the way of Heaven, Earth and Humanity. The *Tao* took form in the being of the Grandmother Goddess. She came to Earth to enlighten humanity. She taught the people to let everything grow according to its own course without any interference. This is called the way of no-action, no-selfishness (*wu-wei*), and this principle is an important rule for Taoists. It teaches them to be very plain and modest, and not to struggle with others for personal gain in their maternal life. This kind of virtue is the ideal spiritual kingdom for which the followers of Taoism long.

Give great value to life

Taoism pursues immortality. It regards life as the most valuable thing. Master Chang Daoling (c 2nd century AD) said that life is another expression of *Tao*, and the study of *Tao* includes the study of how to extend one's life. With this principle in mind, many Taoists have undertaken considerable exploration in this regard. They believe that life is not controlled by Heaven, but by human beings themselves. People can prolong life through meditation and exercise. The exercises include both the moral and the physical sides. People should train their will, discard selfishness and the pursuit of fame, do good deeds, and seek to become a model of virtue (*te*).

Taoism considers that the enhancement of virtue is the precondition and the first aim of practising the *Tao*. The achievement of immortality is a reward from the gods for practising worthy acts. With a high moral sense and with systematic exercise in accordance with the Taoist method and philosophy of life, people can keep sufficient life essence and energy in their bodies all their lives. The Taoist exercise of achieving immortality has proved very effective in practice. It can keep people younger and in good health. But there is one point which cannot be neglected: a peaceful and harmonious natural environment is a very important external condition.

Part two: Taoist ideas about nature

With the deepening world environmental crisis, more and more people have come to realize that the problem of the environment is not only brought about by modern industry and technology, but it also has a deep connection with people's world outlook, with their sense of value, and with the way they structure knowledge. Some people's ways of thinking have, in certain ways, unbalanced the harmonious relationship between human beings and nature, and over-stressed the power and influence of the human will. People think that nature can be rapaciously exploited.

This philosophy is the ideological root of the current serious environmental and ecological crisis. On the one hand, it brings about high productivity; on the other hand, it brings about an exaggerated sense of one's own importance. Confronted with the destruction of the earth, we have to conduct a thorough self-examination on this way of thinking.

We believe that Taoism has teachings which can be used to counteract the shortcomings of currently prevailing values. Taoism looks upon humanity as the most intelligent and creative entity in the universe (which is seen as encompassing humanity, Heaven, Earth within the *Tao*).

There are four main principles which should guide the relationship between humanity and nature.

1. In the *Tao te Ching*, the basic classic of Taoism, there is this verse: 'Humanity follows the Earth, the Earth follows Heaven, Heaven follows the *Tao*, and the *Tao* follows what is natural'. This means that the whole of humanity should attach great importance to the earth and should obey its rule of movement. The earth has to respect the changes of

Heaven, and Heaven must abide by the *Tao*. And the *Tao* follows the natural course of development of everything. So we can see that what human beings can do with nature is to help everything grow according to its own way. We should cultivate in people's minds the way of no-action in relation to nature, and let nature be itself.

2. In Taoism, everything is composed of two opposite forces known as *Yin* and *Yang*. *Yin* represents the female, the cold, the soft and so forth: *Yang* represents the male, the hot, the hard, and so on. The two forces are in constant struggle within everything. When they reach harmony, the energy of life is created. From this we can see how important harmony is to nature. Someone who understands this point will see and act intelligently. Otherwise, people will probably violate the law of nature and destroy the harmony of nature.

There are generally two kinds of attitude towards the treatment of nature, as is said in another classic of Taoism, *Bao Pu Zi* (written in the 4th century). One attitude is to make full use of nature, the other is to observe and follow nature's way. Those who have only a superficial understanding of the relationship between humanity and nature will recklessly exploit nature. Those who have a deep understanding of the relationship will treat nature well and learn from it. For example, some Taoists have studied the way of the crane and the turtle, and have imitated their methods of exercise to build up their own constitutions. It is obvious that, in the long run, the excessive use of nature will bring about disaster, even the extinction of humanity.

3. People should take into full consideration the limits of nature's sustaining power, so that when they pursue their own development they have a correct standard of success. If anything runs counter to the harmony and balance of nature, even if it is of great immediate interest and profit, people should restrain themselves from doing it, so as to prevent nature's punishment. Furthermore, insatiable human desire will lead to the over-exploitation of natural resources. So people should remember that to be too successful is to be on the path to defeat.

4. Taoism has a unique sense of value in that it judges affluence by the number of different species. If all things in the universe grow well, then a society is a community of affluence. If not, this kingdom is on the decline. This view encourages both government and people to take good care of nature. This thought is a very special contribution by Taoism to the conservation of nature.

To sum up, many Taoist ideas still have positive significance for the present world. We sincerely hope that the thoughts of all religions which are conducive to the human being will be promoted, and will be used to help humanity build harmonious relationships between people and nature. In this way eternal peace and development can be maintained in the world.

PARTICIPANTS IN THE PEER REVIEW PROCESS

ADIMIHARDJA, Kusnaka
Professor
Padjadjaran University
JI, Cigadung Raya Tengah 3A
Bandung 4019
Indonesia
Nationality: Indonesian
Tel: +62 22 251 0018
Fax: +62 22 250 8592

BRAMWELL, Martyn J.
Consultant Production Editor
Anagram Editorial Service
26 Wherwell Road
Guildford GU2 5JR
UK
Nationality: British
Tel: +44 1483 533497
Fax: +44 1483 306848
Email: mjb@anagram.u-net.com

CARIÑO, Joji
Alliance of the Indigenous–Tribal Peoples of the Tropical Forest
14 Rudolf Place
Miles Street
London SW8 1RP
UK
Nationality: Philipino
Tel: +44 171 587 3737
Fax: +44 171 793 8686
Email: morbeb@gn.apc.org

CLARK, Norman G.
Professor of Environmental Studies
Wolfson Centre
University of Strathclyde
Glasgow G4 0NW
UK
Nationality: British
Tel: +44 141 548 4079
Fax: +44 141 552 5498
Email: n.g.clark@strat.ac.uk

DANIELS, R. J.
Principal Scientific Officer
M. S. Swaminathan Research Foundation
3rd Cross Street
Taramani Institutional Area
Taramani
Madras 600113
India
Nationality: Indian
Tel: +91 44 2351229 / 698
Fax: +91 44 2351319
Email: mdsaaa51@giasmd01.vsnl.net.in

GAMBOA, Maria L.
Rainforest Rescue
San Rafael S/m (front of A.M.Y.)
P. O. Box 17-12-105
Quito
Ecuador
Nationality: Ecuadorian
Tel: +593 2 343 725
Fax: +593 2 449 772

GOLLIHER, Rev'd Canon Jeffrey Mark
Canon, The Cathedral of St. John the Divine, New York;
Director, The Dubos Center for Spiritual Ecology
150 West End Avenue #30M
New York 10023
New York
USA
Nationality: American
Tel: +1 212 316 7573
Fax: +1 212 932 7348

GONZALES, Dr Tirso
Native American Studies Dept.
Hart Hall
University of California
Davis, CA 95616
USA
Tel: +1 916 752 2915
Fax: +1 916 752 9097
Email: tagonzalez@ucdavis.edu
tirso_2000@yahoo.com

HEYWOOD, Vernon H.
Professor Emeritus
School of Plant Sciences
University of Reading
Whiteknights
P. O. Box 221
Reading RG6 2AS
UK
Nationality: British
Tel: +44 118 931 6640
Fax: +44 118 931 6640
Email: v.h.heywood@reading.ac.uk

KABUYE, Christine
C/o National Museums of Kenya
P. O. Box 40658
Nairobi
Kenya
Nationality: Ugandan
Tel: +254 2 743 513
Fax: +254 2 741 424

MADSEN, Hanne-Rie
Fund Management Officer
Biodiversity Unit
UNEP
P. O. Box 30552
Nairobi
Kenya
Nationality: Danish
Tel: +254 2 623 878
Fax: +254 2 624 296
Email: madsenh@unep.org

MANUKA-SULLIVAN, Margaret
Director
Maori Programme
P. O. Box 5023
Wellington
New Zealand
Nationality: New Zealander
Tel: +64 4 472 0997
Fax: +64 4 473 0890

MASINDE, Isabella
Programme Officer
Biodiversity Unit
UNEP
P. O. Box 30552
Nairobi
Kenya
Nationality: Kenyan
Tel: +254 2 623 264
Fax: +254 2 623 926
Email: masindei@unep.org

McNEELY, Jeffrey A.
Chief Scientist
IUCN
Rue Mauverney 28
1196 Gland
Switzerland
Nationality: American
Fax: +41 22 999 0025

MOLES, Jerry
Co-Director
Neosynthesis Research Center
20093 Eney Drive
Topanga
CA 90290
USA
Nationality: American
Tel: +1 310 455 9377
Fax: +1 310 455 7098
Email: jmoles@igc.apc.org

MUGABE, John
Executive Director
African Centre for Technology Studies (ACTS)
P. O. Box 45917
Nairobi
Kenya
Nationality: Kenyan
Tel: +254 2 522 984 / 521 450
Fax: +254 2 522 987 / 521 001
Email: acts@cgiar.org

MWANUNDU, Sheila
Senior Environmentalist
Environment & Sustainable Development Unit
01 BP 1387
Abidjan
Côte d'Ivoire
Nationality: Kenyan
Tel: +225 20 44 40
Fax: +225 20 50 33
Email: oesuadb@africaonline.co.ci

PADOCH, Christine
Curator
New York Botanical Garden
Bronx
New York 10458
New York
USA
Nationality: American
Tel: +1 718 817 8975
Fax: +1 718 220 1029
Email: cpadoch@nybg.org

PERRINGS, Charles
Professor and Head of Department
Department of Environmental Economics
and Environmental Management
Heslington
York YO1 5DD
UK
Nationality: British
Tel: +44 1904 432 997
Fax: +44 1904 432 998
Email: cap8@york.ac.uk

POSEY, Darrell A.
Professor, Oxford Centre for the Environment, Ethics and Society
Mansfield College
Oxford OX1 3TF
UK
Nationality: American
Tel: +44 1865 284 665
Fax: +44 1865 284 665
Email: posey@mansfield.ox.ac.uk

SENANAYAKE, Ranil
41 1/1 Gregorys Road
Colombo 7
Sri Lanka
Nationality: Sri Lankan
Tel: +941 693 613
Email: 100232.3435@compuserve.com

SLIKKERVEER, L Jan
Professor
Leiden Ethnosystems and Development Programme (LEAD)
Institute of Cultural and Social Studies
Leiden University
P. O. Box 9555
2300 RB Leiden
Netherlands
Nationality: Dutch
Tel: +31 71 527 3469 / 3472
Fax: +31 71 527 3619
Email: slikkerveer@rulfsw.fsw.LeidenUniv.nl

TAVERA, Carmen
Programme Officer
Biodiversity Unit
UNEP
P. O. Box 30552
Nairobi
Kenya
Nationality: Colombian
Tel: +254 2 624 182
Fax: +254 2 624 268
Email: carmen.tavera@unep.org

VAN DER HAMMEN, Maria Clara
Tropenbos-Colombia
Avenida 22#39-28 (Piso 2)
Santafe de Bogota, DC
Colombia
Nationality: Colombian
Fax: +571 288 0128
Email: ftropenb@colomsat.net.co

APPENDIX 4

THE CONTRIBUTORS

Chapter 1

MORRIS, Christine
Research Fellow, Australian Key Centre for Cultural and Media Policy
Arts Faculty
Humanities Bld.
Griffith University
Nathan 4111
Australia
Tel: +61 387 57338
Fax: +61 387 55511
Email: christine.morris@hum.gu.edu.au

POSEY, Darrell A.
Professor, Oxford Centre for the Environment, Ethics & Society
Mansfield College
University of Oxford
OX1 3TF
UK
Tel: +44 1865 282904
Fax: +44 1865 282904
Email:
darrell.posey@mansfield.ox.ac.uk
and
Departamento de Ciências Biológicas
Universidade Federal do Maranhão
Campus-Bacanga
65.000 São Luís
Maranhão
Brazil

Chapter 2

ANDRIANARIVO, Jonah
Lemur Dolls
1333 Arlington Bld
Davis, CA 95616
USA
Email: zaizabe@cwnet.com

MAFFI, Dr Luisa
Northwestern University
Department of Psychology
102 Swift Hall
2029 Sheridan Road
Evanston, Il 60208-2710
USA
Tel: +1 847 467 6513
Fax: +1 847 491 7859
Email: maffi@nwu.edu

SKUTNABB-KANGAS, Dr Tove
Roskilde University, 3.2.4.
Dept of Languages and Culture
P.O. Box 260
4000 Roskilde
Denmark
Phone: +45 46 742740
Fax: +45 46 743061
Email: tovesk@babel.ruc.dk

Chapter 3

BARSH, Russel Lawrence
Associate Professor
Department of Native American Studies
University of Lethbridge
Lethbridge
Alberta T1K 3M4
Canada
Email: barsh@hg.uleth.ca

BENNETT, Dr David H.
Executive Director
The Australian Academy of the Humanities
P. O. Box 93
Canberra ACT 2601
Australia
Email: kestrels@dynamite.com.au

CARIÑO, Ms Joji
Alliance of the Indigenous–Tribal Peoples of the Tropical Forests
14 Rudolf Place
Miles Street
London SW8 1RP
England
UK
Tel: +44 171 587 3737
Fax: +44 171 793 8686
Email: morbeb@gn.apc.org

CHAPESKIE, Andrew J.
The Taiga Institue for Land, Culture and Economy
300-120 Second St. S.
Kenora, ON
P9N 1E9
Canada
Tel: +1 807 468 9607
Fax: +1 807 468 4893
Email: taiga-institute@voyageur.ca

GRAY, Dr Andrew
IWGIA
Fiolstraede 10
1171 Copenhagen K
Denmark
Tel: 00 45 33124724
Fax: 00 45 33147749
Email: iwgia@iwgia.org
and
Forest Peoples Programme
1c Fosseway Business Centre
Stratford Road
Moreton-in-Marsh
GL56 9NQ
UK
Tel: +44 1608 652893
Fax: +44 1608 652878
Email: wrm@gn.apc.org

McGREGOR, Davianna Pomaika'i
Associate Professor of Ethnic Studies
University of Hawai'i – Manoa
1859 East–West Road
Honolulu
Hawai'i 9822
Email: davianna@hawaii.edu

MEAD, Aroha Te Pareake
Manager, External Relations Unit
Treaty Compliance Group
Ministry of Maori Development
Aotearoa New Zealand
Tel: +64 4 479 7781
Fax: +64 4 494 7106
Email: aroham@nzonline.ac.nz

PERSOON, Gerard
Leiden Ethnosystems and Development Programme (LEAD)
Institute of Cultural and Social Studies
Leiden University
P. O. Box 9555
2300 RB Leiden
Netherlands
Tel: +31 71 527 3469 / 3472
Fax: +31 71 527 3619
Email: persoon@rulcml.leidenuniv.nl

PETERSON, Richard B.
Institute for Environmental Studies
University of Wisconsin-Madison
Room 70 Science Hall
550 N. Park St.
Madison, WI 53706
USA
Email: rbpeters@students.wisc.edu

REICHEL D., Dr Elizabeth
Corporación Propuesta Ambiental - Por Una Cultura Nueva
Santafé de Bogotá
Colombia
Email: ereichel@openway.com.co

RODRIGUEZ-MANASSE, Jaime
Avenida 33, #14-22 (Teusaquillo)
Santafé de Bogotá
Colombia

SHEPARD Jr., Glenn
Department of Anthropology
University of California at Berkeley
USA
Email: GShepardJr@aol.com

SUZUKI, David
Suzuki Foundation
2477 Point Gray Road
Vancouver, BC
Canada V6K 1A1

URIBE MARIN, Mónica del Pilar
Apartado Aéreo 4662
Santafé de Bogotá
Colombia
Tel/Fax: +57 1 251 1751
Email: uribemar@colnodo.apc.org

WALKER, Dr Marilyn
Department of Anthropology
Mount Allison University
Sackville, New Brunswick
Canada E4L 1A7
Tel: +1 506 364 2287
Fax: +1 506 364 2625
Email: mwalker@mta.ca

WHITT, Dr Laurie Anne
Michigan Technology University
1400 Townsend Drive
Houton, MI 49931-1295
USA
Email: lawhitt@mtu.edu

WINTHROP, Dr Robert H.
Cultural Solutions
PO Box 401, Ashland, OR 97520
USA
Tel: +1 541 482 8004
Fax: +1 541 552 0825
Email: rhwinth@mind.net

Chapter 4

MASINDE, Isabella
Programme Officer
Biodiversity Unit
UNEP
P. O. Box 30552
Nairobi
Kenya
Tel: +254 2 623 264
Fax: +254 2 623 926
Email: isabella.masinde@unep.org

MOLES, Jerry
Co-Director
Neosynthesis Research Center
20093 Eney Drive
Topanga
CA 90290
USA
Tel: +1 310 455 9377
Fax: +1 310 455 7098
Email: jmoles@igc.apc.org

SENANAYAKE, Ranil
41 1/1 Gregorys Road
Colombo 7
Sri Lanka
Tel: +941 693 613
Email:
100232.3435@compuserve.com

TAVERA, Carmen
Programme Officer
Biodiversity Unit
UNEP
P.O. Box 30552
Nairobi
Kenya
Tel: +254 2 624 182
Fax: +254 2 624 268
Email: carmen.tavera@unep.org

Chapter 5

ADIMIHARDJA, Kusnaka
Professor
Padjadjaran University
Jl. Cigadung Raya Tengah 3A
Bandung 4019
Indonesia
Tel: +62 22 251 0018
Fax: +62 22 250 8592

AGRAWAL, Arun
Department of Political Science
Yale University
124 Prospect Street
P.O. Box 208301
New Haven CT 06520
USA
Tel: +1 203 432 5011
Fax: +1 2-3 432 6196
Email: arun.agrawal@yale.edu

ALCORN, Dr Janis B.
Biodiversity Support Program, WWF
24th Street NW
Washington, DC 20037
USA
Tel: +1 202 293 4800
Fax: +1 202 293 9211
Email: janis.alcorn@wwfus.org

BALASUBRAMANYAN, Dr A. V.
Director, CIKS
No. 2, 25th East Street
Tiruvanmiyur
Madras-600 041
India
Tel: +91 44 401 5909
Fax: +91 44 491 6316

BROOKFIELD, Harold
Department of Anthropology
Division of Society and Environment
The Australian National University
Canberra, ACT 0200
Australia
Fax: +61 2 6249 4896
Email: hbrook@coombs.anu.edu.au

CHAMBI, Nestor
Asociación Chuyma de Apoyo Rural "Chuyma Aru"
Puno
Peru

CLARK, John F. M.
School of History
Rutherford College
University of Kent at Canterbury
Canterbury, Kent CT2 7NX
UK
Tel: +44 122 776 4000
Email: j.f.m.clark@ukc.ac.uk

DAVIS, Dr Stephen D.
Centre for Economic Botany
Royal Botanic Gardens
Kew, Richmond
Surrey TW9 3AE
UK
Email: s.davis@rbgkew.org.uk

ELLEN, Dr R.F.
Department of Anthropology
Eliot College
University of Kent at Canterbury
Canterbury
Kent CT2 7NS
UK
Tel: +44 1227 764000 extn 3421
Fax: +44 1227 827289
Email: R.F.Ellen@ukc.ac.uk

FAIRHEAD, Dr James
Department of Anthropology and Sociology
School of Oriental and African Studies
University of London
Thornhaugh Street
Russell Square
London WC1H 0XG
UK
Tel: +44 171 323 6328
Fax: +44 171 436 3844
Email: jf18@soas.ac.uk

FOLLER, Dr Maj-Lis
Grundtvig Institutet
Vasaparken
Box 100
SE-405 30 Göteborg
Tel: +46 31 773 1310
Email: grundtvig@adm.gu.se

GARI, Josep-Antoni
School of Geography
University of Oxford
Mansfield Road
Oxford OX1 3TB
UK
Email: josep.gari@geography.ac.uk

GIRARDET, Herbert
93 Cambridge Gardens
London W10 6JE
UK
Email: herbie@easynet.co.uk

GONZALES, Dr Tirso
Native American Studies Dept.
Hart Hall
University of California
Davis, CA 95616
USA
Tel: +1 916 752 2915
Fax: +1 916 752 9097
Email: tagonzalez@ucdavis.edu
tirso_2000@yahoo.com

HARRIS, Holly
Department of Anthropology
Eliot College
University of Kent at Canterbury
Canterbury
Kent CT2 7NS
UK

HAY-EDIE, Terence
Department of Anthropology
University of Cambridge
Cambridge
UK
Email: tph22@cus.cam.ac.uk

HILL, Rosemary
School of Tropical Environment Studies and Geography
and
Co-operative Research Centre for Tropical Rainforest Ecology and Management
James Cook University
Cairns Campus
P. O. Box 6811
Cairns 4870
Australia
Tel: +61 7 40 421196
Fax: +61 7 40 421284
Email: Rosemary.Hill@jcu.edu.au

KOTHARI, Ashish
Kalpavriksh Environmental Action Group
Apartment 5, Shree Dutta Krupa
908 Deccan Gymkhana
Pune 411004, Maharashtra
India
Tel/fax: +91 212 354239
Email: ashish@nda.vsnl.net.in

LEACH, Dr Melissa
Institute of Development Studies
University of Sussex
Falmer, Brighton
East Sussex BN1 9RE
UK
Tel: +44 1273 678685 / 606261
Fax: +44 1273 621202 / 691647
Email: m.leach@ids.ac.uk

MACHACA, Marcela
Asociación Bartolom Aripaylla
Quispillacta
Ayacucho
Peru

MAHALE, Prabha and SOREE, Hay
Centre for Indigenous Knowledge Systems
No.2, 25th East Street
Tiruvanmiyur
Madras 600 041
India

McNEELY, Jeffrey A.
Chief Biodiversity Officer
IUCN-The World Conservation Union
rue Mauverney 28
CH-1196 Gland
Switzerland
Tel: +4122 999 0285
Fax: +4122 999 0025
Email: jam@hq.iucn.org

PIEROTTI, Dr Raymond
Department of Ecology and Evolutionary Biology
University of Kansas
Lawrence
Kansas 66045–2106
USA
Tel: +1 785 864 4326
Fax: +1 785 864 5321
Email: pierotti@falcon.cc.ukans.edu

PIMBERT, Dr Michel P.
Chemin en Purian 3
Prangins, CH-1197
Switzerland
Tel: +41 22 362 6389
Fax: +41 22 361 6349
Email: pimbert@bluewin.ch

PRENDERGAST, Hew D. V.
Centre for Economic Botany
Royal Botanic Gardens
Kew, Richmond
Surrey TW9 3AE
UK
Email:
H.Prendergast@lion.rbgkew.org.uk

PRETTY, Jules N.
Director, Centre for Environment and Society
Department of Biological Sciences
University of Essex
John Tabor Laboratories
Wivenhoe Park
Colchester, Essex C04 3SQ
UK
Email: pretj@essex.ac.uk

SLIKKERVEER, Dr L. Jan
Leiden Ethnosystems & Development Programme (LEAD)
Institute of Cultural & Social Studies,
University of Leiden,
P.O. Box 9555,
2300 RB Leiden
Netherlands
Tel: +31 715 273 469/273 472
Fax: +31 715 273 619
Email:
slikkerveer@rulfsw.fsw.LeidenUniv.nl

SMYTH, Dermot M.
Honorary Research Fellow
Department of Tropical Environmental Studies and Geography
James Cook University
P. O. Box 1202
Atherton
Queensland 4883
Australia
Tel: +61 7 40915408
Fax: +61 7 40915408
Email: smyth.bahrdt@bigpond.com

WAY, Michael
Seed Conservation Section
Royal Botanic Gardens, Kew
Wakehurst Place
Ardingly
West Sussex RH19 6TN
UK

WILDCAT, Daniel R.
Department of Natural and Social Sciences
Haskell Indian Nations University
Lawrence
Kansas 66044
USA
Tel: +1 785 749 8428
Fax: +1 785 832 6613

Chapter 6

ARVIGO, Rosita
Ix Chel Tropical Research Foundation
San Ignacio Cayo
Belize
Central America

BAHUCHET, Serge
Museum of Natural History
Paris
France

BALICK, Dr Michael J.
Institute of Economic Botany
New York Botanical Garden
Bronx
NY 10458-5126
USA
Email: mbalick@nybg.org

BARBIRA-FREEDMAN, Dr Francoise
Department of Social Anthropology
Free School Lane
Cambridge, CB2 3RF
UK
Tel: +44 1223 334599
Fax: +44 1223 335993
Email: fb205@cam.ac.uk

BERLIN, Dr Brent
University of Georgia
Department of Anthropology
Baldwin Hall
Athens
Georgia 30602-1619
USA
Email: obberlin@uga.cc.uga.edu

BERLIN, Dr Elois Ann
University of Georgia
Department of Anthropology
Baldwin Hall
Athens
Georgia 30602-1619
USA
Email: eaberlin@uga.cc.uga.edu

BODEKER, Dr Gerard
GIFTS of Health
Health Services Research Unit
Institute of Health Sciences
University of Oxford
Oxford
England, UK
Tel: +44 1865 226 880
Fax: +44 1865 226 711
Email: gerry.bodeker@green.ox.ac.uk

GOODWILL, Jean
Fort Qu'Appelle
Saskatchewan
Canada SOG 1S0
Tel: +1 306 332 6261
Fax: +1 306 332 5985
Email: k.goodwill@sk.sympatico.ca

HARRISON, Kathleen
PO Box 807
Occidental CA 95465
USA
Tel: +1 707 874 1531
Email: kat@wco.com

HUGH-JONES, Dr Stephen
Department of Social Anthropology
Free School Lane
Cambridge, CB2 3RF
UK
Tel: +44 1223 334599
Fax: +44 1223 335993
Email: sh116@cam.ac.uk

IWU, Dr Maurice
Director
Bioresources Development and
Conservation Programme
11303 Amherst Avenue, Suite 2
Silver Spring, Maryland 20902
USA
Tel: +1 301 962 6201
Fax: +1301 962 6205
Email: iwum@igc.org

NATARAJAN, Bhanumathi
School of Pharmacy
University of Oslo
Divn C
Post Box 1068
0316 Blindern
Oslo
Norway
Fax: +47 22 854402
Email:
bhanumathi.natarajan@farmasi.uio.no

SHANKAR, Mr Darshan
Foundation for the Revitalisation of
Local Health Traditions
50 MSH Layout, 2nd Stage, 3rd Main
Anandnagar
Bangalore 560024
India
Tel: +91 80 336909
Fax: +91 80 334167
Email: darshan@frlht.ernet.in

SHROPSHIRE, Gregory
Ix Chel Tropical Research Foundation
San Ignacio Cayo
Belize
Central America

THOMAS, Jacqueline M. C.
Associate Professor
Department of Geography
McGill University
805 Sherbrooke W
Montreal, PQ
Canada H3A 2K6
Tel: +1 514 398 4219
Fax: +1 514 398 7437
Email:
meredith@felix.geog.McGill.ca

Chapter 7

ALTIERI, Miguel A.
Associate Professor
Department of Environmental Science,
Policy and Management
University of California at Berkeley
California 94720
USA
Tel: +1 510 642 9802
Fax: +1 510 642 7428
Email:
agroeco3@nature.berkeley.edu

BROKENSHA, David
Tanrhocal House
86 Newland
Sherborne
Dorset DT9 3DT
Email:
101655.3717@compuserve.com

CHRISTIE, Ms Jean
Rural Advancement Foundation
International (RAFI)
110 Osborne Street, Suite 202
Winnipeg
Manitoba R3L 1Y5
Canada
Tel: +1 204 453 5259
Fax: +1 204 452 7456
Email: rafican@web.net

ENRIS, Ideng
PAN Indonesia
Jl. Persada Raya No.1
Menteng Dalam
Jakarta 12870
Indonesia

HOND, Kuli
Halalinga Village
c/o Tiliba Mission
P.O. Box 221
Mendi
Southern Highlands Province
Papua New Guinea

KABUYE, Dr Christine H. S.
National Museums of Kenya
P.O. Box 40658
Nairobi
Kenya
Tel: +254 2 742 161/4 or 742 131/
4
Fax: +254 2 741 424

KLINE, Elsie
8940 County Road 235
Fredericksburg, OH 44627
USA

LAWRENCE, Kathy
c/o Just Food
625 Broadway, Suite 9C
New York, New York 10012
USA

MANN, Peter
c/o World Hunger Year
505 Eighth Avenue, 21st Floor
New York, New York 10018-6582
USA
Email: pmann@isc.org

MOONEY, Mr Pat
Rural Advancement Foundation
International (RAFI)
110 Osborne Street, Suite 202
Winnipeg
Manitoba R3L 1Y5
Canada
Tel: +1 204 453 5259
Fax: +1 204 452 7456
Email: rafican@web.net

MOORE, Richard H.
Department of Anthropology
The Ohio State University
124 W. 17th Ave.
Columbus, OH 43210
USA
Tel: +1 614 292 0230
Email: moore.11@osu.edu

PELIA, Jim
See entry for HOND this Chapter

PLENDERLEITH, Kristina
Working Group on Traditional Resource
Rights
Oxford Centre for the Environment,
Ethics and Society
Mansfield College
University of Oxford
Oxford OX1 3TF
UK
Tel: +44 1865 284665
Fax: +44 1865 284665
Email:
106550.2460@compuserve.com

RIBEIRO DE OLIVEIRA, Rogerio
GEHECO/UFRJ e DIVEA/FEEMA
Estrada da Vista Chinesa, 741. 20531-410
Rio de Janeiro, RJ
Brazil
Email: machline@geo.puc-rio.br

RICHARDS, Dr Paul
Department of Anthropology
University College London
UK
and
Technology & Agrarian Change Group
Wageningen Agricultural University
Netherlands
Email: paul.richards@tao.tct.wau.nl

SILLITOE, Paul
Department of Anthropology
University of Durham
Durham
UK
Email: paul.sillitoe@durham.ac.uk

STINNER, Deborah H.
Department of Entomology
Ohio Agricultural Research and Development Center
The Ohio State University
1680 Madison Ave.
Wooster, Ohio 44691-4096
USA
Tel: +1 330 263 3724
Fax: +1 330 263 3686
Email: stinner.2@osu.edu

THRUPP, Dr Lori Ann
World Resources Institute
10 G Street, NE, 8th floor
Washington, D.C. 20006
USA
Email: ann@wri.org

TJAHJADI, Riza V.
National Co-ordinator
Pesticide Action Network (PAN)
Indonesia
Jl. Persada Raya No. 1 Menteng Dalam
Jakarta 12870
Indonesia
Tel: +61 21 8296545
Fax: +61 21 8296545
Email: biotani@rad.net.id

TOFFOLI, Daniel di Giorgi
Mestrando em Geografia - PPGG/UFRJ
Rua Viuva Lacerda, 128/402 HUMAITA
22261-050
Rio de Janeiro, RJ
Brazil
Tel: +55 21 538 0926/286 7656
Fax: +55 21 286 8394
Email: toffoli@trip.com.br

UNGUTIP, Wabis
See entry for HOND this Chapter

WABIS, Ben
See entry for HOND this Chapter

Chapter 8

BERNBAUM, Dr Edwin
Research Associate, University of California at Berkeley
Senior Fellow, The Mountain Institute
home address:
1846 Capistrano Ave.
Berkeley, CA 94707
USA
Tel: +1 510 527 1229
Fax: +1 510 527 1290
Email:
bernbaum@socrates.berkeley.edu

HALBERTSMA, Tjalling
Co-Director, Sacred Mountains of China Project
Alliance of Religions and Conservation
9A Didsbury Park
Manchester M20 5LH
UK
Tel: +44 161 434 0828
Fax: +44 161 434 8374
Email: icorec@icorec.nwnet.co.uk.

MARTIN, Evelyn
Fellow, Center for Respect of Life and Environment
Home address:
3718 Chanel Rd.
Annandale, VA 22003
USA
Tel: +1 703 354 5224, 703 354 0077
Fax: +1 703 354 1331

PALMER, Martin
Secretary General of the Alliance of Religions and Conservation (ARC)
9A Didsbury Park
Manchester M20 5LH
UK
Tel: +44 161 434 0828
Fax: +44 161 434 8374
Email: icorec@icorec.nwnet.co.uk.

SHENGJI, Dr Pei
Head, Mountain Natural Resources Division
ICIMOD
GPO Box 3226
Kathmandu
Nepal
Fax: +977 1 52 4509
Email: icimod@mos.com.np

PUROHIT, A. N.
Director
High Altitude Plant Physiology Research Centre,
H.N.B. Garhwal University
Srinagar Garhwal - 246174, U.P.
India
Tel: +91 1388 2172 (w)
Fax: +91 1388 2061 and 1388 2072
Email: happrc@nde.vsnl.net.in

SCHAAF, Thomas
Program Specialist, UNESCO
Division of Ecological Sciences
MAB Programme
1, rue Miollis
F-75732 Paris Cedex 15
France
Tel: +33 1 45 68 40 65
Fax: +33 1 45 68 58 04
Email: t.schaaf@unesco.org

SNOW, Chief John
Chief of the Wesley Band of Stoney Indians
PO Box 30
Morely, Alberta
TOL 1NO
Canada

STEVENS, Stanley, F.
Adjunct Associate Professor
Department of Geosciences
Box 35820
University of Massachusetts
Amherst, MA 01003-5820
USA
Tel: +1 413 545 0768 (office)
Fax: +1 413 545 1200
Email: sstevens@geo.umass.edu

Chapter 9

BHARUCHA, E.
Director, Bharati Vidyapeeth Institute of Environment Education and Research
Member, Tata Electric Companies, Afforestation Program
22, Saken, Valentina Society
North Main Road
Koregaon Park, Pune - 411001
India
Email: c/o ashok@gagan.dcc.dartnet.com

BURCH Jr., William
Email:
102771.3456@compuserve.com

DOSSOU-GLEHOUENOU, Bernadette
Assistant - Faculty of Agronomic Science
Universit Nationale du Bnin
BP 526 - Cotonou-Benin
Tel: +229 30 30 84

DOVE, Dr Michael R.
Professor of Social Ecology
School of Forestry & Environmental Studies
Yale University
205 Prospect Street
New Haven, Connecticut 06511-2189
USA
Tel: +1 203 432 3463
Fax: +1 203 432 3817
Email: michael.dove@yale.edu

FALCONER, Julia
World Bank
South Asia Rural Development Department
1818 H Street, NW
Washington, D.C.
USA
Tel: +1 202 458 1300
Email: jfalconer@worldbank.org
and
Collaborative Forest Management Unit
Planning Branch
P. O. Box 1457
Forestry Department
Kumasi
Ghana

KABUYE, Dr Christine H. S.
See Chapter 7 entry

LAIRD, Sarah A.
P.O. Box 4004
Westport, MA 02790 USA
Tel: +1 508 636 8017
Fax: +1 508 636 8017
Email: sarahlaird@aol.com

MEREDITH, T. C.
Department of Geography
McGill University
805 Sherbrook W
Montreal, P.Q.
Canada H3A 2K6
Email:
meredith@felix.Geog.McGill.CA

PARAJULI, Dr Pramod
117 Euclid Terrace
Syracuse
New York 13210
New York
USA
Tel: +1 315 426 8858
Email: pparajul@maxwell.syr.edu]

POSEY, Dr Darrell A.
See Chapter 1 entry

PULLIN, Roger
Email: r.pullin@cgiar.org

RICHARDS, Dr Paul
See Chapter 7 entry

RIVAL, Dr Laura
Department of Anthropology
Eliot College
University of Kent at Canterbury
Canterbury
Kent CT2 7NS
UK
Email: l.m.rival@ukc.ac.uk

SHANLEY, Patricia
Durrell Institute for Conservation and the Environment
University of Kent
Canterbury
UK
Email: trishanely@aol.com

SHENGJI, Dr Pei
Head, Mountain Natural Resources Division
ICIMOD
GPO Box 3226
Kathmandu
Nepal
Fax: +977 1 52 4509
Email: icimod@mos.com.np

SMITH, Wynet
Lands Manager
Kitikmeot Inuit Association
Box 315
Kugluktuk, NT
XOE OEO
Canada
Tel: +1 867 982 3310
Fax: +1 867 982 3311
Email: wynetsm@polarnet.ca

SUNDAR, Nandini
Institute of Economic Growth
University Enclave
Delhi - 110007
India
Fax: 91 11 7257410
Email:
Nandini.Sundar@jfm.Sprintrpg.ems.vsnl.net.in
Email: nsundar@ieg.ernet.in

TONT, Sargun A.
Department of Biology
Middle East Technical University
06531, Ankara
Turkey
Tel: +90 312 2105165
Fax: +90 312 210 1289
Email: tont@rorqual.cc.metu.edu.tr

YOUNG, Kenny
Lazy Moon Farm
Upper Brailes, Oxon
UK
Email: kennyyoung@compuserve.com

ZOUNDJIHEKPON, Jeane
WWF - 08 BP 1776
Abidjan 08
Côte d'Ivoire
Tel: +225 44 87 86
Fax: +225 44 87 74
Email: wwfwarpo@AfricaOnline.Co.Ci

Chapter 10

BIEGERT, Claus
Weissenburgstrasse 40
D-81667 Munich
Germany
Email:
101515.3370@compuserve.com

CHAMBERS, Dr Paul
Department of the Environment, Transport and the Regions
Eland House
Stag Place
London SW1E 5DU
UK
Tel: +44 171 890 3000
Fax: +44 171 890 6290
Email: paulc@matahari.u-net.com

CHERNELA, Dr Janet M.
Associate Professor
Dept. of Sociology and Anthropology
Florida International University
North Miami, FL 33181-3600
USA
Tel: +1 305 919 5861
Fax: +1 305 919 5964
Email: chernela@fiu.edu

EINARSSON, Niels
Director
Stefansson Arctic Institute
Sólborg
600 Akureyri
Iceland
Tel: +354 463 0581
Fax: +354 463 0589
Email: ne@unak.is

KALLAND, Dr Arne
Department and Museum of Anthropology
PO Box 1091 Blindern
N-0317 Oslo
Norway
Fax: +47 22 85 45 02
Tel: +47 22 85 52 29
Email: arne.kalland@sum.uio.no

NAKASHIMA, Douglas
Programme Specialist
Environment and Development in Coastal Regions and in Small Islands (CSI)
UNESCO
1, rue Miollis,
75732 Paris Cedex 15
France
Tel: +33 1 4568 3993
Fax: +33 1 4568 5808
Email: d.nakashima@unesco.org

NG'WENO, Fleur
Chair, Bird Committee of Nature Kenya
the Eats Africa Natural History Society
P.O. Box 42271
Nairobi
Kenya
Tel: +254 2 724 166
Fax: +254 2 724 349
Email:
Fleur_Ngweno@africaonline.co.ke

OSSEWEIJER, Manon
Centre of Environmental Science
Leiden University
P.O. Box 9518
2300 RA Leiden
The Netherlands

PETERU, Clark
Email: peteru@samoa.net

RAYGORODETSKY, Gleb
WCS (Asia Program) & Columbia University (CERC)
P.O. Box 250552
New York, NY 10025
USA
Tel: +1 212 854 8186
Fax: +1 212 854 8188
Email: cherylc@earthlink.net

ROUE, Marie
Director, UMR 8575 du CNRS
APSONAT Appropriation et Socialisation de la Nature
Laboratoire d'Ethnobiologie
Museum National d'Histoire Naturelle
57, rue Cuvier
75231 Paris Cedex 05
France
Tel: +33 1 40 79 36 68 or 82
Fax: +33 1 40 79 36 69
Email: roue@mnhn.fr

STEINER, William Wallace Mokahi
Pacific Island Ecosystems Research Center
Biological Resources Division
USGS
406 Gilmore
3050 Maile Way
Honolulu
HI 96822
USA
Email: steiner@hawaii.edu

WHITELEY, Peter
Department of Anthropology
Sarah Lawrence College
Bronxville, NY 10708
USA
Email: pwhitele@mail.slc.edu

Chapter 11

CALLICOT, Dr J. Baird
Professor
Philosophy and Religious Studies
University of North Texas
Tel: +1 940 565 2266

GETUI, Mary
Department of Religious Studies
Kenyatta University
Nairobi
Kenya

GOLLIHER, The Rev'd Canon Jeff
The Cathedral of St. John the Divine, New York; Associate for the Environment, The Office of the Anglican Observer at the United Nations
150 West End Avenue #30M
New York 10023
New York
USA
Tel: +1 212 316 7573
Fax: +1 212 932 7348
Email: golliher@pop.interport.net

GOULET, Dr Denis
O'Neil Professor in Education for Justice
University of Notre Dame
USA
Tel: +1 219 631 5250

HALIFAX, Joan
c/o UPAYA
1404 Cerro Gordo
Santa Fe, New Mexico 87501
USA

HESSEL, Dieter T.
1 Astor Ct.
Princeton, New Jersey 08540
USA

HILLMAN, James
P.O. Box 216
Thompson, Connecticut 06277-0216
USA

HOLLAND, Alan
Senior Lecturer
Philosophy Department
lancaster University
Lancaster
UK
Tel: +44 1524 592446
Fax: +44 1524 592503
Email: a.holland@lancaster.ac.uk

JENSEN, Dr Tim
Centre for Religious Studies
University of Odense
Odense
Denmark
Email: t.jensen@filos.ou.dk

LYONS, Chief Oren
Onondaga Nation
Box 200
Nedrow, New York 13120
USA

McINTOSH, Alastair
Trustee, Isle of Eigg Trust
Craigencalt Farm
Kinghorn Loch
Fife KY3 9YG
Scotland
Tel: +44 1592 891829
Email: alastair@gn.apc.org

McLEAN, Margot
55 N. Moore Street #4F
New York, N. Y. 10013
USA
Tel: +1 212 431 9141
Fax: +1 212 431 5673
Email: mmclean@interport.net

NASH, James
39 Sunset Drive
Burlington, Massachusetts 01803
USA

NORTON, Dr Bryan G.
School of Public Policy
Georgia Institute of Technology
Atlanta, GA 30135
USA
Email: bryan.norton@pubpolicy.gatech.edu

O'NEILL, John
Professor of Philosophy
Philosophy Department
Lancaster University
Lancaster
UK
Tel: +44 1524 592490
Fax: +44 1524 592503
Email: j.oneill@lancaster.ac.uk

PALMER, Martin
See Chapter 8 entry

PEALE, Marty
1028 Don Diego
Santa Fe, New Mexico 87501
USA

RADFORD RUETHER, Rosemary
Garrett - Evangelical Theological Seminary
2121 Sheridan Road
Evanston, Illinois 60201
USA

REVARD, Carter
6638 Pershing Ave.
St. Louis, Missouri 63130
USA
Email: ccrevard@artsci.wustl.edu

SAGOFF, Mark
Senior Research Scholar
Institute for Philosophy and Public Policy
University of Maryland
College Park
MD 20742
USA
Email: MSAGOFF@puafmail.umd.edu

SHIVA, Dr Vandana
Research Foundation for Science, Technology & Ecology
A-60, 2nd floor, Hauz Khas,
New Delhi 110016
India
Tel: +91 11 665 003
Fax: +91 11 685 6795
Email: vandana@twn.unv.ernet.in

WEISKEL, Dr Timothy C.
Director, Harvard Seminar on Environmental Values
45 Francis Avenue
Cambridge, MA 02138
USA
Tel: +1 617 496 5208
Fax: +1 617 496 3668
Email: tweiskel@div.harvard. edu

Chapter 12

ALCORN, Dr Janis B.
Biodiversity Support Programme, WWF
24th Street NW,
Washington, DC 20037
USA
Tel: +1 202 293 4800
Fax: +1 202 293 9211
Email: janis.alcorn@wwfus.org

BORN, Gemima C. Cabral
Vitae Civilis - Institute for Development, Environment and Peace
Caixa Postal 11.260 - CEP 05422-970
So Paulo, SP
Brazil
Tel: +55 11 8696528
Fax: +55 11 8692941
Email: vcivilis@ax.apc.org

CHAPELA, Ignacio
Department of Environmental Science, Policy and Management
University of California at Berkeley
334 Hilgard Hall
Berkeley CA 94720
USA
Email: ichapela@nature.berkeley.edu

CORREA, Carlos
Professor of the Economics of Science and Technology
Universidad de Buenos Aires
Buenos Aires
Argentina
Email: quies@wamani.apc.org

DUTFIELD, Mr Graham
Working Group on Traditional Resource Rights
Oxford Centre for the Environment, Ethics and Society
Mansfield College
Oxford University
Oxford OX1 3TF
UK
Tel: +44 1865 282904
Fax: +44 1865 282904
Email:
wgtrr.ocees@mansfield.ox.ac.uk

GUPTA, Professor Anil K.
Society for Research & Initiatives for Sustainable Technologies and Institutions (SRISTI)
Indian Institute of Management
Ahmedabad, 380 015
India
Tel: +91 272 407241
Fax: +91 272 427896
Email: anilg@iimahd.ernet.in

LYNGE, Finn
Consultant in Greenland Affairs
Asiatisk Plads 2
DK-1448 Copenhagen K
Denmark
Tel: +45 3392 0441
Fax: +45 3392 1585

MCAFEE, Kathy
Department of Geography
University of California at Berkeley
501 McCone Hall
Berkeley CA 94707
USA
Email: kmcafee@uclink.berkeley.edu

MONBIOT, Mr George
82 Percy Street
Oxford OX4 3AD
UK
Tel: +44 1865 724 360
Email: g.monbiot@zetnet.co.uk

MORRIS, Ms Christine
Research Fellow
Australian Key Centre for Culture and Media Policy
Faculty of Humanities,
Griffith University
Australia
Email:
christine.morris@hum.gu.edu.au

NADARAJAH-DJAFARI, Ms Tahereh
Ulaanbaatar International College
Ulaanbaatar
Mongolia
Email: nadarajah@magicnet.mn

ROCH, Delphine
Project Consultant,
UNESCO Office Jakarta
Indonesia
Tel: +62 21 3141308 ext. 805
Fax: +62 21 3150382
Email: uhst1@unesco.org

SCHOLZ, Astrid
Energy and Resources Group
310 Barrows Hall
University of California at Berkeley
Berkeley CA 94720
USA
tel: +1 510 642 6886
Fax: +1 510 642 1085
Email: ajscholz@socrates.berkeley.edu

SHANKAR, Mr Darshan
Director
Foundation for Revitalisation of Local Health Traditions
No.50, MSH Layout,
Bangalore
India
Tel: +91 80 336909
Fax: +91 80 334167
Email: darshan@frlht.ernet.in

SOLOMON, Mr Maui
Molesworth Chambers
34 Molesworth Street
P.O. Box 3458
Wellington
Aotearoa New Zealand
Tel: +64 4 472 6744
Fax: +64 4 499 6172
Email: moriori@nzonline.ac.nz

STEPHENSON, Jr., Dr David J.
Senior Associate
Wagenlander and Associates
Denver, Colorado
USA
Tel: +1 303 329 6090
Email: DavidS23@aol.com

VIRA, Dr Bhaskar
Department of Geography
University of Cambridge
Cambridge
UK
Email: bv101@cam.ac.uk

VOGEL, Dr Joseph H.
Facultad Latinoamericana de Ciencias Sociales (FLACSO-Ecuador)
Eloy Alfaro 266 y Berlin, 102
Quito
Ecuador
Email: henvogel@uio.satnet.net

WIENER, Dr John D.
Program on Environment and Behavior
Institute of Behavioral Science
University of Colorado
Boulder, CO 80309-0468
USA
Tel: +1 303 492 6746
Fax: +1 303 492 1231
Email: wienerj@spot.Colorado.edu

Appendices

RAO, Dr P. K.
Center for Development Research
302 Nassau Street #5
Princeton
NJ 09542
USA
Tel/Fax: +1 609 252 0795
Email: pinninti@aol.com

Editorial Consultant

HEYWOOD, Prof. Vernon H.
President, IUBS / ICMAP
School of Plant Sciences
University of Reading
Whiteknights
P. O. Box 221
Reading RG6 6AS
UK
Tel: +44 118 931 6640
Fax: +44 118 931 6640
Email: v.h.heywood@reading.ac.uk

Production Editor

BRAMWELL, Martyn J.
Anagram Editorial Service
26 Wherwell Road
Guildford
Surrey GU2 5JR
UK
Tel: +44 1483 533497
Fax: +44 1483 306848
Email: mjb@anagram.u-net.com

Colour Plates

CHERNELA, Janet
See Chapter 10 entry

GONZALES, Tirso
See Chapter 5 entry

McLEAN, Margot
See Chapter 11 entry

MOORE, Richard
See Chapter 7 entry

NAKASHIMA, Douglas
See Chapter 10 entry

NATARANJAN, Bhanumathi
See Chapter 6 entry

NEWING, Helen
DICE, Dept. of Anthropology,
University of Kent at Canterbury
Canterbury Kent CT2 7NZ
UK
Tel: +44 122 776 4000

PRENDERGAST, Hew
See Chapter 5 entry

RODRÍGUEZ-MANASSE, Jaime
See Chapter 3 entry

TOAQUIZA, Alfredo
TOAQUIZA, Gustavo
TOAQUIZA, Julio
c/o Jean G. Colvin
(Representative of the Tigua Artist Cooperative)
Director
University Research Expeditions Programme (UREP)
University of California
Davis, CA 95616
USA
Tel: +1 530 752 0692
Fax: +1 530 752 0681
Email: jgcolvin@ucdavis.edu

TOFFOLI, Daniel
See Chapter 7 entry

APPENDIX 5

ABBREVIATIONS AND ACRONYMS

AICA	Arizona-Idaho Conservation Act
AIPP	Asian Indigenous Peoples Pact
AIRFA	American Indian Religious Freedom Act
AKRSP	Aga Khan Rural Support Program
ARC	Alliance of Religions and Conservation
ASW	Aboriginal Subsistence Whaling
ATA	Alliance of Taiwan Aborigines
BALCO	Bharat Aluminium Company
BBA	*Beej Bachao Andolan*, or 'Save the Seed Movement'
CBD	Convention on Biological Diversity
CBRs	Community Biodiversity Registers
CCR	Community-Controlled Research
CGIAR	Consultative Group on International Agricultural Research
CGRFA	Commission on Genetic Resources for Food and Agriculture
CIKS	Centre for Indian Knowledge Systems
CIPL	Permanent International Committee of Linguists
CIPSH	International Council for Philosophy and Humanistic Studies
CNA	Constant Natural Assets
CNC	Critical Natural Capital
CNPPA	Commission on National Parks and Protected Areas (IUCN)
COICA	Coordinating Body of Indigenous Organisations of the Amazon Basin
CONAIE	Confederación de Nacionalidades Indígenas del Ecuador
COP	Conference of the Parties
CPGR	Commission on Plant Genetic Resources (FAO)
CR	Collaborative Research
CSD	Commission on Sustainable Development
CTA	China Taoist Association
DDRIP	Draft Declaration on the Rights of Indigenous Peoples
ECOSOC	Economic and Social Council (UN)
ELCI	Environment Liaison Centre International
FAO	Food and Agriculture Organization of the UN
FENAMAD	Federación Nativa del Madre de Dios
FERC	Federal Energy Regulatory Commission
FES	Field of Ethnological Study
FPCs	Forest Protection Committees
GATT	General Agreement on Tariffs and Trade
GBA	Global Biodiversity Assessment
GBKS	Gender-Based Knowledge Systems
GBPHED	G.B. Plant Institute of Himalayan Environment and Development
GEF	Global Environment Facility
GEKP	Gwich'in Environmental Knowledge Project

GIFTS	Global Initiative for Traditional Systems
GIS	Geographic Information Systems
GP	General Practice
GPA	Global Plan of Action
HP	Historical Perspective
HYVs	High Yielding Varieties
HGDP	Human Genome Diversity Project
IARI	International Agricultural Research Institutes
IB	Institutional Building
ICCPR	International Covenant on Civil and Political Rights
ICESCR	International Covenant on Economic, Social and Cultural Rights
ICHEL	International Clearinghouse for Endangered Languages
ICOREC	International Consultancy on Religion, Education and Culture
ICDP	Integrated Conservation and Development Project
IEK	Indigenous Ecological Knowledge
IFF	Intergovernmental Forum on Forests
IHS	Indian Health Service
IK	Indigenous Knowledge
IKS	Indigenous Knowledge Systems
ILO	International Labour Organization
INDAKS	Indigenous Agricultural Knowledge Systems
IPBN	Indigenous Peoples Biodiversity Network
IPED	International Panel of Experts on Desertification
IPF	Intergovernmental Panel on Forests
IPRs	Intellectual Property Rights
IRMS	Indigenous Resource Management Systems
ITFF	Inter-Agency Task Force on Forests
IUCN	The World Conservation Union
IUPGR	International Undertaking on Plant Genetic Resources
IWC	International Whaling Commission
IWGIA	International Working Group for Indigenous Affairs
JFM	Joint Forest Management
KIPOC	Korongoro Integrated Peoples Oriented to Conservation
LCSs	Local Community Systems
LG	La Grande
LHRs	Linguistic Human Rights
LK	Local Knowledge
MTA	Material Transfer Agreement
NAFTA	North American Free Trade Agreement
NDRs	Non-Domesticated Resources
NGOs	Non-Governmental Organizations
NHS	National Health Service
NTFPs	Non-Timber Forest Products
NWT	North-West Territories
OPIAC	Organization of Indigenous Peoples of the Colombian Amazon Region
OSCE	Organization for Security and Co-operation in Europe

PHPA	*Perlindungan Hutan dan Pelestarian Alam* the Directorate General of Forest Protection and Natural Preserves
PIC/NIC	Prior Informed Consent/No Informed Consent
PLEC	People, Land Management and Environmental Change
PNG	Papua New Guinea
POPs	Persistent Organic Pollutants
PR	*Parana*
PRA	Participatory Rural Appraisal
PV	Participants View
RAFI	Rural Advancement Foundation International
RCIADIC	Royal Commission into Aboriginal Deaths in Custody
RD	Research and Development
RE	Religion and Environment
RMP	Revised Management Procedure
RRA	Rapid Rural Appraisal
SEARICE	Southeast Asia Regional Institute for Community Education
SBSTTA	Subsidiary Body on Scientific, Technical and Technological Advice
SEBJ	*Societe d'Energie de la Baie James*
SEPASAL	Survey of Economic Plants of Arid and Semi-Arid
Lands	
SP	Sao Paulo
SRISTI	Society for Research and Initiatives for Sustainable Technologies & Institutions
STCW	Small-type Coastal Whaling
TEK	Traditional Ecological Knowledge
TK	Traditional Knowledge
TNCs	Transnational Corporations
TOLN	Taxonomically-Organized Local Names
TOT	Transfer of Technology
TRIPs	Trade-Related Aspects of Intellectual Property Rights
TRRs	Traditional Resource Rights
UDHR	Universal Declaration of Human Rights
U-KTNP	Uluru-Kata Tjuta National Park
UN	United Nations
UNCED	United Nations Conference on Environment and Development
UNCTAD	United Nations Commission on Trade and Development
UNDP	United Nations Development Programme
UNEP	United Nations Environment Programme
UNESCO	United Nations Education and Scientific Organization
UNU	United Nations University
UNWGIP	United Nations Working Group on Indigenous Peoples
USGS	United States Geological Survey
WGTRR	Working Group on Traditional Resource Rights
WHC	World Heritage Convention
WHO	World Health Organization
WIPO	World Intellectual Property Organization
WRI	World Resources Institute
WTO	World Trade Organization
WTP	Willingness to Pay
WTWHA	Wet Tropics of Queensland World Heritage Area
WWF	World Wide Fund for Nature

BIBLIOGRAPHY

ABA (1993) Asociación Bartolomé Aripaylla. Diversificación de Germoplasma Agrícola en Quispillacta. Ao Agricola 1991-1992 (manuscript), Asociación Bartolomé Aripaylla de Ayacucho.

Abbadie, L., Lepage, M. and le Roux, X. (1992) Soil fauna at the forest-savanna boundary: the role of termite mounds in nutrient cycling. In: Furley, P. A., Proctor, J and Ratter, J. A. (eds) *Nature and dynamics of forest-savanna boundaries.* Chapman & Hall, London. pp. 473-484.

Abbiw, D. K. (1990) *Useful Plants of Ghana: West African Uses of Wild and Cultivated Plants.* Intermediate Technology Publications, London.

Abdussalam, A. S. (1998) Human language rights: An Islamic perspective. In: Benson, P., Grundy, P. and Skutnabb-Kangas, T. (eds), Language Rights. Special volume. *Language Sciences* 20(1): 55-62.

Abram, D. (1996) *The Spell of the Sensuous: Perception and language in a more-than-human world.* Pantheon, New York.

Abramowitz, J. and Nichols, R. (1993) 'Women and Agrobiodiversity' *SID Journal on Development* .

Abu-Zeid, M. O. (1973) Continuous cropping in areas of shifting cultivation in Southern Sudan. *Tropical Agriculture* 50:285-90.

Adams, C. (ed.) (1993) *Ecofeminism and the Sacred.* Continuum, New York.

Adams, W. M. (1990) *Green Development: Environment and Sustainability in the Third World.* Routledge, London.

Adams, W. M. (1996) Capture and Disengagement: Indigenous Irrigation and Development in Sub-Saharan Africa. In: Adams, W. M. and Slikkerveer, L. J. (eds) *Indigenous Knowledge and Change in African Agriculture*, Studies in Technology and Social Change No. 26, TSC Programme, Iowa State University, Ames, Iowa, USA.

Adams, W. M. and Slikkerveer, L. J. (eds) (1996) *Indigenous Knowledge and Change in African Agriculture*, Studies in Technology and Social Change No. 26, TSC Programme, Iowa State University, Ames, Iowa, USA.

Adams, W. M., Potanski, T. and Sutton, J. E. G. (1994) Indigenous farmer-managed irrigation in Sonjo, Tanzania. *Geographical Journal* 160: 17-32.

Adeola, M. O. (1992) Importance of wild animals and their parts in the culture, religious festivals and traditional medicine of Nigeria. *Environmental Conservation* 19, 125-134.

Adimihardja, K. (1992) *Kasepuhan: Yang Tumbuh Di Atas Yang Luruh. Pengelolaan Lingkungan Secara Tradisional Di Kawasan Gunung Halimun, Jawa Barat,* PT. Tarsito, Bandung.

Adimihardja, K. (1992) The Traditional Agricultural Rituals and Practices of the *Kasepuhan* Community of West Java. In: Fox, J. J. (ed) *The Heritage of Traditional Agriculture Among the Western Ausronesians*. The Australian National University, Canberra.

Adimihardja, K. (1993) Kasepuhan: Ecological Influences on Traditional Agriculture and Social Organization in Mt. Halimun of West Java, Indonesia. Paper for the International Scientific Congress on Human Evolution in its Ecological Context, 26 June – 1 July 1993, Leiden.

Adimihardja, K. (1995) Indigenous Agricultural Knowledge in Mount Halimun Area, West Java. INDAKS Project Report, INRIK, Bandung.

Adimihardja, K. and Iskandar, J. (1993) *Kasepuhan: Ecological Influences on Traditional Agriculture and Social Organization in Mt Halimun Area of West Java, Indonesia.* Paper for the International Scientific Congress on Human Evolution in its Ecological Context, 26 June – 1 July 1993, Leiden.

Agrawal, A. (1995) Dismantling the divide between indigenous and scientific knowledge. *Development and Change* 26:413-439.

Aguilera-Klink, F. (1994) Some notes on the misuse of classic writings in economics on the subject of common property. *Ecological Economics* 9: 221-228.

Ahern, J. and Boughton, J. (1994) Wildflower Meadows as Suitable Landscapes. In: Platt, R. H., Rowntree, R. A. and Muick, P. C. (eds) *The Ecological City: Preserving and Restoring Urban Biodiversity*, pp. 172-187. University of Massachusetts Press, Amherst.

Akerele, O., Heywood, V. and Synge, H. (eds) (1991)*Conservation of Medicinal Plants*. Cambridge University Press, Cambridge.

Akimichi, T., Asquith, P. J., Befu, H., Bestor, T. D., Braund, S. R., Freeman, M. M. R., Hardacre, H., Iwasaki, M., Kalland, A., Manderson, L., Moeran, B. D. and Takahashi, J. (1988) *Small type coastal whaling in Japan*. Occasional Paper Number 27. Boreal Institute for Northern Studies, Edmonton, Alberta.

Akinnaso, F. N. (1994) Linguistic unification and language rights. *Applied Linguistics* 15(2): 139-168.

Aksoy, M. (1971) *Atasozleri ve deyimler sozlugu*. Turk Dil Kurumu Yayynlary.

Alaska Native Review Commission (1984) Transcripts of Hearings (36 Volumes). Inuit Circumpolar Conference, Anchorage, Ak.

Albers, P. (1993) 'Symbiosis, merger, and war – contrasting forms of intertribal relationship among historic Plains Indians'. In: Moore, J. H. (ed.) *The Political Economy of North American Indians*. 94-132. University of Oklahoma Press, Norman, OK.

Albers, P. and Kay, J. (1987) 'Sharing the land: a study in American Indian territoriality'. In: Ross, T. E. and Moore, T. G. (eds) *A Cultural Geography of North American Indians*. Westview Press, Boulder.

Alcorn J. (1989) 'Making Use of Traditional Farmers Knowledge'. In: *Common Futures, Proceedings of an International Forum on Sustainable Development.* Toronto: Pollution Probe.

Alcorn, J. (1990) 'Indigenous Agroforestry Strategies Meeting Farmers' Needs'. In: Anderson, A. (ed.). *Alternatives to Deforestation: Steps Toward Sustainable Use of the Amazon Rain Forest.* Columbia University Press, New York.

Alcorn, J. (1993) Indigenous peoples and conservation. *Conservation Biology* 7(2):424-426.

Alcorn, J. B. (1981) 'Huastec noncrop resource management: implications for prehistoric rainforest management' *Human Ecology* 9: 395-417.

Alcorn, J. B. (1994) Noble savage or noble state? Northern myths and southern realities in biodiversity conservation. *Etnoecologica* 2: 7-19.

Alcorn, J. B. (1994) Noble Savages or Noble State?: Northern Myths and Southern Realities in Biodiversity Conservation. In: Toledo, V. M. E. (ed) *Ethnoecologica*, Vol. II No.3 April.

Alderete, W. and Guevara, G. (1996) South and Central America Regional Workshop on Traditional Health Systems: GIFTS of Health. Conclusions and Recommendations, *J. Altern and Complem. Med.*, 2,3. 398-401. Mary Ann Liebert Publishers Inc., Larchmont, NY.

Aldred, J. (1994) Existence Value, Welfare and Altruism. *Environmental Values*, 3: 381-402.

Alfonso, N. and McAllister, D. E. (1994) *Biodiversity and the Great Whale Hydroelectric Project.* Great Whale Environmental Assessment, Background Paper No. 11. Great Whale Public Review Support Office, Montreal.

Allan, D. J. and Flecker, A. S. (1993) Biodiversity Conservation in Running Waters: Identifying the Major Factors that Threaten Destruction of Riverine Species and Ecosystems, *Bioscience* 43(1): 32-43.

Allan, W. (1965) *The African Husbandman* . Oliver and Boyd, Edinburgh.

Altieri, M. A. (1991) Traditional Farming in Latin America. *The Ecologist* 21(2):93-6.

Altieri, M. A. (1994) *Biodiversity and pest management in agroecosystems.* Haworth Press, New York.

Altieri, M. A. (1995) *Agroecology: the science of sustainable agriculture* Westview Press, Boulder.

Altieri, M. A. (1996) *Enfoque agroecologico para el desarrollo de sistemas de produccion sostenibles en los Andes.* Centro de Investigacion, Educacion y Desarrollo, Lima.

Altieri, M. A. and Hecht, S. B. (1991) *Agroecology And Small Farm Development..* CRC Press, Boca Raton

Altieri, M. A., Anderson, M. K. and Merrick, L. C. (1987) Peasant agriculture and the conservation of crop and wild plant resources. *Conservation Biology* 1: 49-58.

Altieri, Miguel (1987) *Agroecology: The Scientific Basis of Alternative Agriculture.* Westview Press, Boulder.

Alvares, M. (c. 1615/1990) Ethiopia Minor and a geographical account of the Province of Sierra Leone. Transcrption from an unpublished manuscript by Avelino Teixeira da Mota and Luis de Matos. Translation by P. E. H. Hair. ms. Department of History, University of Liverpool. 1990.

Amanor, K. S. (1993) 'Farmer experimentation and changing fallow ecology in the Krobo district of Ghana'. In: de Boef, W. *et al.* (eds) *Cultivating Knowledge: Genetic Diversity, Farmer Experimentation and Crop Research.* Intermediate Technology Publications, London.

Amend, S. and Amend, T. (eds) (1992) *Spacios sin Habitatas?: Parques Nacionales de America del Sur aAeas (Habitats Without Inhabitants?: National Parks of South America).* IUCN, Gland, Switzerland, 497 pp.

Anderson, A. (ed.) (1990) *Alternatives to Deforestation: Steps Toward Sustainable Use of the Amazon Rain Forest.* Columbia University Press, New York.

Anderson, A. B. and Posey, D. A. (1985) 'Manejo de cerrado pelos Indios Kayapó. *Boletim do Museu Paraense Emilio Goeldi, Botânica* 2: 77-98.

Anderson, A. B. and Posey, D. A. (1987) Management of a tropical scrub savanna by the Gorotire Kayapó of Brazil. *Advances in Economic Botany* 7, 159-173.

Anderson, A., May, P. H. and Balick, M. J. (1991) *The Subsidy from Nature; Palm Forests, Peasantry, and Devleopment on an Amazonian Frontier.* Columbia University Press, New York.

Anderson, B. (1989) *Nationalism.* Paper presented at WIDER Seminar on Systems of Knowledge as Systems of Power. Karachi.

Anderson, D. and Grove, R. (1987) 'The scramble for Eden: past, present, and future in African conservation'. In: Anderson, D. and Grove, R. (eds), *Conservation in Africa.* Cambridge University Press, Cambridge.

Anderson, E. N. (1996) *Ecologies of the Heart: Emotion, Belief and the Environment.* Oxford University Press, Oxford.

Anderson, J. C. (1984) The Political and Economic Basis of Kuku-Yalanji Social History, St Lucia. MA thesis, Department of Anthropology and Sociology, University of Queensland.

Andersson, J. E. C. and Ngazi, Z. (1995) Marine resource use and the establishment of a marine park: Mafia Island, Tanzania. *Ambio* 24: 475-481.

Andrianarivo, J. A. (1987) Standardization of Malagasy names for plant and animal species. Paper presented at the Academie Nationale Malagasy (July 1987). Antananarivo, Madagascar.

Annamalai, E. (1986) 'A typology of language movements and their relation to language planning'. In: Annamalai, E., Jernudd, B. and Rubin, J. (eds), *Language planning: Proceedings of an Institute.* 6-17. Central Institute of Indian Languages and East-West Center, Mysore and Honolulu.

Anon. (1996a) Inspired by nature: Taught by a titmouse. *Blackboard Bulletin*, March, p.19.

Anon. (1996b) The Puzzle Page: The nature walk. *Blackboard Bulletin*, October, p.23.

Ansre, G. (1979) Four rationalisations for maintaining European languages in education in Africa. *African Languages /Langues Africaines* 5(2): 10-17.

Antommarchi, Dr F. (1825) *The Last Days of the Emperor Napoleon*, 2 vols. Henry Colburn, London, pp. 315-316. Napoleon substituted the honey-bee for the fleur-de-lis as the Bourbon family emblem. See Charles L. Hogue, 'Cultural Entomology', *Annual Review of Entomology*, 32 (1987).

Apffel Marglin, F. and Mishra, P. C. (1993) Sacred Groves: Regenerating the Body, the Land, the Community. In: Sachs, W. (ed) *Global Ecology: A New Arena of Political Conflict*, Zed Books, London.

Appadurai, A. (ed.) (1988) *The social life of things: Commodities in cultural perspective*. Cambridge University Press, Cambridge.

Applebaum, H. (ed.)(1987) 'Perspectives in Cultural Anthropology'. Albany, State University of New York Press.

Araujo, H., Brack-Egg, A. and Grillo, E. (1989) *Ecologia, agricultura y autonomia campesina en los Andes*. Fundación Alemana para el Desarrollo Internacional, Feldafing-Lima-Hohenheim.

Arias, J. (1991) *El Mundo Numinoso de los Mayas*. 2nd ed. Serie Antropología. Tuxtla Gutierrez, Chiapas: Talleres Gráficos del Estado.

Aris, M. (1990) Man and Nature in Buddhist Himalayas. In: Rustomji, N. K. and Ramble, C. *Himalayan Environment and Culture*. Indian Institute of Advanced Study in association with Indus Publishing Company, New Delhi, pp. 85-101.

Arndt, H. W. (1987) *Economic Development: The History of an Idea*. University of Chicago, Chicago.

Arnold, J. E. M. and Stewart, W. C. (1991) Common Property Resource Management in India. *Tropical Forest Papers* No 24.Oxford Forestry Institute, Oxford.

Arnold, M. (1991) *The Long Term Global Demand for and Supply of Wood*. Forestry Commission Occasional Paper, 36.

Aron, W. (1988) The commons revisited: thoughts on marine mammal management. *Coastal Management* 16(2):99-110.

Aryal, M. (1993) *Diverted wealth: the trade in Himalayan herbs*. HIMAL, Kathmandu, Jan/Feb p.10.

Asch, M. I. (1982) Dene self-determination and the study of hunter-gatherers in the modern world. In: Leacock, E. and Lee, R. (eds) (1982) *Politics and History in Band Societies*. 347-372. Cambridge University Press, Cambridge.

Asch, M. I. (1989) Wildlife: defining the animals the Dene hunt and the settlement of aboriginal rights claims. *Canadian Public Policy* 15(2): 205-219.

Ashkenazi, M. (1992) Summary of whale meat as a component of the changing Japanese diet in Hokkaido. IWC Document IWC/44/SEST2, published in *Papers on Japanese small-type coastal whaling submitted by the Government of Japan to the International Whaling Commission, 1986-1996*, 212-219. The Government of Japan, Tokyo.

Asociación Bartolomé Aripaylla. (1993) Diversificacion de Germoplasma Agrìcola en Quispillacta. Año Agrìcola (1991-1992) Asociación Bartolomé Aripaylla De Ayacucho (Manuscript).

Assisi Litugy. WWF, Gland, Switzerland.

Atran, S. (1990) *Cognitive Foundations of Natural History: Towards an Anthropology of Science*, Cambridge University Press, Cambridge.

Atran, S. (1993) Itza Maya tropical agro-forestry. *Current Anthropology* 34: 633-700.

Atran, S. (1996) The commons breakdown. Paper presented at the working conference 'Endangered Languages, Endangered Knowledge, Endangered Environments', Berkeley, California, October 25-27.

Atran, S. and Medin, D. L. (1997) 'Knowledge and action: Cultural models of nature and resource management in Mesoamerica'. In: Bazerman, M. H., Messick, D. M., Tenbrunsel, A. E. and Wade-Benzoni, K. A. (eds). *Environment, Ethics, and Behavior*. 171-208. New Lexinton Press, San Francisco.

Atran, Scott. (1990) *Cognitive Foundations of Natural History: Towards an Anthropology of Science*. Cambridge University Press, Cambridge.

Attfield, R. (1991) *The Ethics of Environmental Concern*. The University of Georgia Press, Athens.

Attfield, R. and Dell, K. (eds) (1989) *Values, Conflict and the Environment*.. Ian Ramsey Centre, Oxford, and Centre for Applied Ethics, Cardiff.

Aubert, V. (1959) Chance in social affairs. *Inquiry* 2:1-24.

Aumeeruddy, Yildiz and Jet Bakels. (1994) Management of a Sacred Forest in Kerinci Valley, Central Sumatra: An Example of Conservation of Biological Diversity and its Cultural Basis. *Journal d'agriculture traditionnelle et de botanique appliquée*, nouvelle série, 36(2): 39-65.

Ausubel, K. (1994) *Seeds of Change: The living treasure*. Harper, San Francisco.

Avei, D. (1994) Participation for Community Re-education: The Use of Community Story Framework Analysis, unpublished. Waigani: The National Research Institute.

Axt, J. R., Corn, M. L., Lee, M. and Ackerman, D. M. (1993) *Biotechnology, Indigenous Peoples and Intellectual Property Rights*. Congressional Research Service. The Library of Congress, Washington D.C.

Aylward, B. A. and Barbier, E. B. (1992)*What is Biodiversity Worth to a Developing Country? Capturing the Pharmaceutical Value of Species Information*. London Environmental Economics Centre, Paper DP 92-05.

Ba, A. H. (1981) The living tradition. In: J. Ki-Zerbo (ed.) *General History of Africa 1: Methodology and African Prehistory*. London: Heinemann, London California and UNESCO.

Baer, G. (1984) *Die Religion der Matsigenka, Ost Peru*. Wepf & Co.AG Verlag, Basel.

Baer, G. (1992) The one who intoxicates himself with tobacco. In: *Portals of Power: Shamanism in South America*. (eds) Matteson-Langdon, J. and Baer, G. University of New Mexico Press, Albuquerque.

Bahai International Community (1995) The Prosperity of Humankind, a statement prepared by the community's Office of Public Information, presented to the World Summit for Social Development in Copenhagen, March (1995)

Bahemuka, J. (1982) *Our Religious Heritage*. Thomas Nelson and Sons, Edinburgh.

Bahuchet, S. (1985) *Les Pygmées Aka et la forêt centrafricaine*. SELAF, Paris.

Bahuchet, S. (1993) 'History of the inhabitants of the Central African rainforest: Perspectives from comparative linguistics'. In: Hladik C. M., Hladik, A., Linars, O. F., Pagezy, H., Semple, A. and Hadley, M. *Tropical Forests, People and Food: Biocultural Interactions and Applications to Development*. Man and the Biosphere Series, Vol. 13. UNESCO and Parthenon Press, Paris.

Bahuguna, S. (1986) In: *Chipko News*. Mimeo, Navjeevan Ashram, Silyara.

Bailey, F. G. (1961) Tribe and Caste in India. In: *Contributions to Indian Sociology* 5(1).

Bailey, F. G. (1963) *Politics and Social Change*. University of California Press, Berkeley.

Bailey, R. C., Head, G., Jenike, M., Owen, B., Rechtman, R. and Zechente, E. (1989) Hunting and Gathering in Tropical Rain Forest: Is it Possible? *American Anthropologist* 91:59-82.

Bakken, P. (1995) The eco-justice movement in Christian theology: patterns and processes. *Theology and Public Policy* 7,14-19.

Bakken, P., Engel, J. and Engel, J. (1995) *Ecology, Justice, and Christian Faith: A critical guide to the literature*. Greenwood Press, Westport, Conn.

Balée, W. (1988) Indigenous adaptation to Amazonian palm forests. *Principes* 32(2):47-54.

Balée, W. (1989a) 'Cultura na vegetação da Amazônia', *Boletim do Museu Paraense Emilio Goeldi* 6: 95-110 (Coleção Eduardo Galvão).

Balée, W. (1989b) 'The culture of Amazonian forests'. In: Posey, D. A. and Balée, W. (eds) Resource Management in Amazonia: Indigenous and Folk Strategies. *Advances in Economic Botany* 7: 1-21.

Balée, W. (1992) People of the Fallow: A Historical Ecology of Foraging in Lowland South America. In: Redford, K. H.and Padoch, C. (eds) *Conservation of Neotropical Forests: Working from Traditional Use*. Columbia University Press, pp. 35-57.

Balée, W. (1994) *Footprints of the Forest: Ka'apor Ethnobotany-The Historical Ecology of Plant Utilization by an Amazonian People*. Columbia University Press, New York.

Balée, W. (1995) Historical Ecology of Amazonia. In: Sponsel, L. E. (ed.) *Indigenous Peoples and the Future of Amazonia*. University of Arizona Press, Tucson, pp. 97-110.

Balée, W. (ed) (1997) *Principles of Historical Ecology*. Columbia University Press, New York.

Balée, W. and Gély, A. (1989) Managed forest succession in Amazonia: the Ka'apor case. In: Posey, D. A. and Balée, W. (eds) Resource Management in Amazonia: Indigenous and Folk Strategies. *Advances in Economic Botany* 7: 129-148.

Balick, M. and Cox, P. (1996) *Plants, People, and Culture: The Science of Ethnobotany*. Scientific American Library, NY.

Balick, M. J. and Mendelssohn, R. (1992) Assessing the economic value of traditional medicines from tropical rain forests. *Conservation Biology* 6(1):128-130.

Balick, M. J. Elisabetsky, E. and Laird, S. A. (eds) (1996) *Medicinal Resources of the Tropical Forest: Biodiversity and its Importance to Human Health*.. Columbia University Press, New York.

Balkanu (1997) Draft Statement of Principles, Cape York Development Corporation Pty Ltd., Cairns.

Balland, J. and Platteau, J. (1996) *Halting Degradation of Natural Resources: Is there a role for rural communities?* UN Food and Agriculture Organization, and Clarendon Press, Oxford.

Ballón Aguirre, E., Cerrón-Palomino, R.and Chambi Apaza, E. (1992) *Vocabulario Razonado de la Actividad Agraria Andina, Terminología Agraria Quechua*. Centro de Estudios Regionales Andinos, Bartolomé de Las Casas, Cusco.

Bama Wabu (1996) *Reasonable Expectations or Grand Delusions?* Cape York Land Council, Cairns.

Bamgbose, A. (1991) *Language and the Nation. The Language Question in Sub-Saharan Africa*. Edinburgh University Press, Edinburgh.

Banarly, Nihat Sami. (1971) *Resimli Turk Edebiyaty Tarihi*. 28, 402. Milli Egitim Basymevi.

Banuri, T. and Marglin, F. A. (eds) (1993) *Who Will Save the Forests? Knowledge, Power, and Environmental Destruction*. The United Nations University and World Institute for Development Economic Research (UNU/WIDER). Zed Books, London and New Jersey.

Barbira Freedman, F. C. (1997) 'Jungle pharmacy: Comoditisation of indigenous knowledge in Amazonia'. Working paper, Department of Social Anthropology, University of Cambridge.

Barbira Freedman, F. (in press) 'El caduceo del bien y de la maldad en el vegetalismo'. Revista Takiwasi, Tarapoto, Peru.

Barker, D., Oguntoyinbo, J. S. and Richards, P. (1977) The Utility of the Nigerian Peasant Farmer's Knowledge in the Monitoring of Agricultural Resources. *General Report Series* No. 4, Monitoring and Assessment Research Centre, Chelsea College, London.

Bärmark, J. (1988) Om Kunskapens Kulturberoende ('On The Cultural Contextualization of Knowledge'), *Vest*, 5-6:32-42.

Bärmark, J. (1993) Tibetan Buddhist Medicine and Anthropology of Knowledge. In Hultberg, J. (ed) *New Genres in Science Studies*, Report No 179, Department of Theory of Science, University of Goteborg, Goteborg, pp. 23-34.

Barnett, T. (1977) *The Gezira Scheme: an Illusion of Development*. Frank Cass, London.

Barr, J. (1972) Man and nature: The ecological controversy and the Old Testament. *Bulletin of the John Rylands Library* 55, 9-32.

Barreiro, J. (1992) The Search for Lesson. *Akwekon* IX(2):21.

Barsh, R. L. (1995) Indigenous peoples and the idea of individual human rights. *Native Studies Review* 10(2), pp.35-55.

Basso, K. H. (1996) *Wisdom Sits in Places: Landscape and Language among the Western Apache*. University of New Mexico Press, Albuquerque.

Bastien, J. W. (1978) *Mountain of the Condor: Metaphor and Ritual in an Andean Ayllu*. West. Pub. Co., St. Paul.

Bastien, J. W. (1986) The Human Mountain. In: Tobias, M. (ed.), *Mountain People*. 45-57. University of Oklahoma Press, Norman and London.

Bates, Crispin. (1995) 'Race, Caste and Tribe in Central India: Early Origins of Indian Anthopometry'. In: Peter Robb (ed.) *The Concept of Race in India*. Oxford University Press, Delhi.

Bateson, G. (1972) *Steps to An Ecology of Mind*. Chandler Publishing Company, New York.

Bateson, G. (1972) *Steps to an Ecology of Mind*. Ballantine Books, New York.

Bateson, G. (1979) *Mind and Nature: A necessary unity*. Dutton, New York.

Batibo, H. (1996) Patterns of language shift and maintenance in Botswana: The critical dilemma. Paper presented at the working conference 'Endangered Languages, Endangered Knowledge, Endangered Environments', Berkeley, California, October 25-27.

Batibo, H. (1998) The endangered languages of Africa: A case study from Botswana. In: Maffi, L. (ed.) Language, Knowledge, and the Environment: The Interdependence of Biological and Cultural Diversity. Submitted to Smithsonian Institution Press.

Battaglia, Debbora. (1995) 'Problematizing the Self: A Thematic Introduction' In: *Rhetorics of Self-Making*. University of California Press, Berkeley.

Baudry, J. (1989) Interactions between agricultural and ecological systems at the landscape level. *Agricultural Ecosystems and Environment*, 27:119-130.

Baum, B. (1996) 'Saving the Soul of the Salmon'. *Ashland Daily Tidings*, November 5, 1996:7.

Baxter, P. T. W. (1989) with Richard Hogg (eds) *Property, Poverty and People: Changing Rights in Property and Problems of Pastoral Development*. Manchester, Dept. of Social Anthropology (includes comprehensive bibliography).

Baxter, P. T. W. (ed.) (1991) *When the Grass is Gone: Development Interventions in African Arid Lands.*. Scandinavian Institute of African Studies, Uppsala, Sweden.

Beal, G. M., Dissanayake, W. and Konoshima, S (1986) *Knowledge Generation, Exchange and Utilisation*. Westview Press, Boulder.

Bebbington, A. J. (1993) Modernisation from Below: An Alternative Indigenous Development?, *Economic Geography* 16: 274-292.

Becker, L. C. (1980) 'The Moral Basis of Property Rights'. In: Pennock, J. R. and Chapman, J. W. (eds) *Property*. New York University Press, New York.

Beckerman, W. (1994) Sustainable development: is it a useful concept? *Environmental Values* 3: 191-209.

Beckham, S. *et al.* (1988) *Prehistory and History of the Columbia River Gorge National Scenic Area, Oregon and Washington.*. Heritage Research Associates, Eugene, Oregon.

Beckwith, M. W. (1940) *Hawaiian Mythology*. Yale University Press, New Haven. pp. 571.

Beckwith, M. W. (1951) *The Kumuilpo: a Hawaiian Creation Chant*. University of Hawaii Press, Honolulu: pp. 257.

Beckwith, M. W. (1977) *Hawaiian Mythology*. University of Hawaii Press.

Beekman, E. M. (ed. and transl.) (1981) *The Poison Tree: Selected Writings of Rumphius on the Natural History of the Indies*. University of Massachusetts Press, Amherst.

Beets, W. C. (1990) *Raising and Sustaining Productivity of Smallholders Farming Systems in the Tropics*. AgBe Publishing, Holland.

Begue, L. (1937) Contribution - l'étude de la végétation forestière de la haute Côte d'Ivoire. *Publ. Com. Hist. Sc. Afr.* Occ. Fr. Ser. B. 4.

Begue, L. (1937) Contribution a l'etude de la vegetation forestière de la haute Côte d'Ivoire. *Publ. Com. Et. Hist. Sc. Afr. Oxx. Fr.* Ser. B. 4.

Bell, J. (1996) 'Genetic engineering and biotechnology in industry'. In: Baumann, M., Bell, J., Koechlin, F. and Pimbert, M. (eds) *The Life Industry: Biodiversity, People and Profits.*. Intermediate Technology Publications, London. 31-52.

Bellah, R. (1964) Religious evolution. *Amer. Soc. Rev.* 29: 358-374.

Bellman, B. (1984) *The language of secrecy: symbols and metaphors in Poro ritual*. Rutgers University Press, New Brunswick.

Belshaw, D. (1980) 'Taking Indigenous Knowledge Seriously: The Case of Inter-Cropping Techniques in East Africa'. In: Brokensha, D. *et al.* (1980):195-202 (op cit).

Bennagen, P. L. (1996) 'Consulting the Spirits, working with nature, sharing with others: an overview of indigenous resource management'. In: Bennagen and Lucas-Fernan (eds) *Consulting the Spirits, Working with Nature, Sharing with Others: indigenous resource management in the Philippines*. Sentro Para sa Ganap na Pamayanan.

Bennett, B. (1992) Plants and people of the Amazonian rainforests: the role of ethnobotany in sustainable development. *BioScience* 42(8).

Bennett, D. H. (1996) 'Native title and intellectual property' In: *Land, Rights, Laws: Issues of Native Title*, (ed.) P. Burke. Issues paper No 10. Native Title Research Unit, Australian Institute of Aboriginal and Torres Strait Islander Studies, Canberra.

Bennett, F. J., Mugalula-Mukibi, A. A., Lutwama, J. S. W. and Nansubuga, G. (1965) An inventory of Kiganda Foods. *The Uganda Journal* 29, 1 45-53.

Bennett, J. W. (1976) *The Ecological Transition*. Pergamon Press, New York.

Bennett, J. W. (1993) *Human Ecology as Human Behavior - Essays in Environmental and Development Anthropology*. Transaction Publishers, New Brunswick, N.J.

Bennett, J. W. and Bowen, J. R. (eds) (1988) *Production and Autonomy*. University Press of America, Lanham, Md.

Berg, H. van den (1991) *La Terra no da Asì no Màs. Los Ritos Agrìcolas en la Religión de los Aymara-cristianos*, CEDLA, Latin American Studies 51, Amsterdam.

Berger, P. *et al.* (1981) *The Homeless Mind*. Pelican Books, London.

Berger, T. R. (1977) *Northern Frontier, Northern Homeland*. The Report of the Mackenzie Valley Pipeline Inquiry: Volume One. Ministry of Supply and Services Canada, Ottawa.

Berger, T. R. (1985) *Village Journey*. Hill and Wang, New York.

Berger, T. R. (1988) Conflict in Alaska. *Natural Resources Journal* 28: 37-62.

Bergstrom, J. D., Dufresne, C., Bills, G. F., Nallinomstead, M. and Byrne, K. (1995) 'Discoveries, biosynthesis and mechanisms of action of the zaragozic acids: Potent inhibitors of squalene synthase' pp. 607-39. In: Ornstron, L. N. (ed.) *Annual Review of Microbiology* vol. 49 Annual review Inc.: Palo Alto.

Berkes, F. (1988) 'The intrinsic difficulty of predicting impacts: lessons from the James Bay Hydro Project'. *Environ. Impact Assess. Rev.* 8: 201-220.

Berkes, F. (1989) (ed.) *Common Property Resources, Ecology and Community-Based Sustainable Development*. Belhaven Press, London.

Berkes, F. (1993) Traditional Ecological Knowledge in Perspective. In: Inglis, J. T. (ed) *Traditional Ecological Knowledge: Concepts and Cases*, International Development Research Centre, Canada.

Berkes, F., Feeny, D., McCay, B. J. and Acheson, J. M. (1989) The benefits of the commons. *Nature* v.340 (13 July 1989): 91-93.

Berkes, F., George, P. and Preston, R. J. (1991) Co-management: the evolution in theory and practice of the joint administration of living resources. *Alternatives* 18(2):12-39.

Berlin, B. (1973) The relation of folk systematics to biological classification and nomenclature. *Annual Review of Ecology and Systematics*. 4:259-271.

Berlin, B. (1992) *Ethnobiological Classification: Principles of Categorization of Plants and Animals in Traditional Societies*. Princeton University Press, Princeton, NJ.

Berlin, B. (1999). 'One Maya Indian's view of the plant world: How a folk botanical system can be both natural and comprehensive'. In: Medin, D. L. and Atran, S. (eds) *Folkbiology*. MIT Press, Cambridge, Mass.

Berlin, B. and Berlin, E. A. (1996) *Medical Ethnobiology of the Highland Maya of Chiapas, Mexico: The Gastrointestinal diseases*. Princeton University Press, Princeton, NJ.

Berlin, B., Breedlove, D. E. and Raven, P. H. (1974) *Principles of Tzeltal Plant Classification: An Introduction to the Botanical Ethnography of a Mayan Speaking Community in Highland Chiapas*. Academic Press, New York.

Berlin, L. M., Guthrie, T., Weider, A.*et al.* (1955) 'Studies in human cerebral function: The effects of mescaline and lysergic acid on cerebral processes pertinent to creativity'. *Journal of Nervous and Mental Disease* 122:487-491.

Berman, H. (1993) 'The Development of International Recognition of the Rights of Indigenous Peoples' In: '..*Never drink from the same cup*' (eds) Veber, H., Dahl, J., Wilson, F. and Waehle, E. Document 74. IWGIA and the Centre for Development Research, Copenhagen.

Bernard, F. (1972) *East of Mount Kenya: Meru Agriculture in Transition*. Weltforum, Munich.

Bernard, R. (1992) Preserving language diversity. *Human Organization* 51(1): 82-89.

Bernbaum, E. (1990) *Sacred Mountains of the World*. Sierra Club Books, San Francisco.

Bernbaum, E. (1996) Sacred Mountains: Implications for Protected Area Management. *Parks* 6(1): 41-48.

Berndt, R. M. (1970) 'Traditional Morality as Expressed Through the Medium of an Australian Aboriginal Religion'. In: *Australian Aboriginal Anthropology: modern studies in the social anthropology of the Australian Aborigines*.(ed.) R.M. Berndt. Nedlands, University of Western Australia Press for the Australian Institute of Aboriginal Studies: 216-247.

Berndt, R. M. (1974) *Australian Aboriginal Religion*.. E.J. Brill, Leiden.

Berry, B. J. L. (1990) Urbanization. In: Turner, B. L., Clark, W. C., Kates, R. W., Richards, J. F., Mathews, J. T. and Meyer, W. B. (eds) *The Earth as Transformed by Human Action*. Cambridge University Press, Cambridge, pp. 103-121.

Berry, T. (1988) *The Dream of the Earth*. Sierra Club Books, San Francisco.

Berry, W. (1977) *The Unsettling of America: Culture and agriculture*. Avon Books, New York.

Berry, W. (1987) *Home Economics*. North Point Press, San Francisco.

Berry, W. (1989) The futility of global thinking. *Harper's Magazine*, September, 16-22.

Berry, W. (1993) Decolonizing rural america. *Audubon Magazine*, March/April, 100-105.

Beversluis, J. (proj. ed.) (1995) *A Sourcebook for Earth's Community of Religions*. CoNexus, Grand Rapids, and Global Education Associates, New York.

Bezirci, Asym. (1993) *Turk Halk Siiri*. 24. Say Yayynlary.

Bhagavatam, Shrimad (1988) TWN and CAP - Third World Network and Consumers' Association of Penang. 'Health'. In: *Modern Science in Crisis: A Third World Response*. Third World Network, Penang, Malaysia.

Bhardwaj, S. M. (1973) *Hindu Places of Pilgrimage in India: A Study in Cultural Geography*. University of California Press, Berkeley.

Bierhorst, J. (1994) *The Way of the Earth: Native America and the Environment*. William Morrow & Co., New York.

Binford, M. W. and Buchenau, M. J. (1993) Riparian Greenways and Water Resources. In: *Ecology of Greenways*, University of Minnesota Press, Minneapolis/ London, pp. 69-104.

Biodiversity Unit (1993) 'Biodiversity and its value', *Biodiversity Series*, Paper No. 1. Biodiversity Unit, Department of the Environment, Sport and Territories, Canberra.

Blacker, C. (1975) *The Catalpa Bow: A Study of Shamanistic Practices in Japan*. Allen and Unwin, London.

Blaikie, P. and Brookfield, H. (1987) *Land Degradation and Society*. Methuen, London.

Bledsoe, C. (1984) The political use of Sande ideology and symbolism. *American Ethnologist* 455-472.

Bobaljik, J. D. and Pensalfini, R. (1996) Introduction. In: Bobaljik, J. D., Pensalfini, R. and Storto, L. (eds) *Papers on Language Endangerment and the Maintenance of Linguistic Diversity*. MIT Working Papers in Linguistics vol. 28. 1-24. MIT, Department of Linguistics, Cambridge, MA.

Bocco, G. and Toledo, T. M. (1997) Integrating peasant knowledge and geographic informations systems: A spatial approach to sustainable agriculture. *Indigenous Knowledge Development Monitor* 5(2): 10-13.

Bodeker, G. (1996) Global Health Traditions. In: Micozzi, M. (ed.) *Fundamentals of Alternative and Complementary Medicine*. Churchill-Livingstone, New York.

Bodeker, G. (ed.) (1996) The GIFTS of Health Reports. *J. Altern. and Complem. Med.*, 2,3, 397-4-5 and 435-47.

Bodley, J. (1976) *Anthropology and Contemporary Human Problems*. Benjamin Cummings Publishing, Menlo Park.

Bodley, J. H. (1982) *Victims of Progress*. Mayfield Publishing Co., Palo Alto, CA.

Bodley, J. H. (1988) (ed.) Tribal Peoples and Development Issues – A global overview. Mayfield Publishing, Mountain View, CA.

Bodley, J. H. (1990) *Victims of Progress*. Mayfield Publishing Co., Mountain View, CA.

Boff, L. (1995) *Ecology and Liberation: A new paradigm*. Orbis Books, Maryknoll.

Bookchin, M. (1980) *Towards an Ecological Society*. Black Rose Books, Montreal.

Bookchin, M. (1981) *The Ecology of Freedom*. Cheshire Books, Palo Alto, California.

Bookchin, M. (1990) *Remaking society: Pathways to a green future*. Boston, South End Press.

Bookchin, M. (1992) A philosophical naturalism. *Society and Nature* 2(1): 60-88.

Boom, B. M. (1989) Use of plant resources by the Chacobo. In: Posey, D. A. and Balée, W. (eds) Resource Management in Amazonia: Indigenous and Folk Strategies. *Advances in Economic Botany* 7 8-96.

Booth, A. L. and Kessler, W. B. (1996) Understanding linkages of people, natural resources, and ecosystem health. In: Ewert, A. W. (ed.), *Natural Resource Management: The human dimension*. 231-248. Westview Press, Boulder, CO.

Borrini-Feyerabend, G. (1996) *Collaborative Management of Protected Areas: Tailoring the Approach to the Context*, IUCN, Gland.

Boster, J. (1996) Human Cognition as a Product and Agent of Evolution. In: Ellen, R. and Fukui, K (eds) *Redefining Nature: Ecology, Culture and Domestication*, Berg, Oxford, pp. 269-89.

Boster, J. S. (1984) 'Classification, cultivation, and selection of Aguaruna cultivars of *Manihot esculenta* (Euphorbiaceae)'. *Advances in Economic Botany* 1:34-47.

BOSTID (1986) *Proceedings of the Conference on Common Property Resource Management*, 21-26 April 1985, Annapolis, Maryland/National Academy Press, Washington, DC.

Bosu Mullick, Sanjay (ed.) (1991) *Cultural Chotanagpur: Unity in Diversity*. Uppal Publishing House, New Delhi.

Bosworth, R. (1995) Biology, politics, and culture in the management of subsistence hunting and fishing: an Alaskan case history. In: Peterson, D. L. and Johnson, D. R. (eds), *Human Ecology and Climate Change — People and Resources in the Far North*. 245-260. Taylor and Francis, Washington, D.C.

Bottomore, T. and Nisbet, R. (1978) *A History of Sociological Analysis*. Basic Books, New York.

Bourdieu, P. (1977a) *Outline of a Theory of Practise*. Cambridge University Press, Cambridge.

Bourdieu, P. (1977b) The economics of linguistic exchange. *Social Science Information* 16(6): 645-668.

Bouwkamp, J. C. (ed.) (1985) *Sweet potato products: a natural resource for the tropics*. CRC Press, Boca Raton, Florida. 10-33.

Bowler, P. J. (1989) 'Holding Your Head Up High: Degeneration and Orthogenesis in Theories of Human Evolution'. In: James R. Moore (ed) *History, Humanity and Evolution*. Cambridge University Press, Cambridge.

Boxberger, D. and Robbins, L. (1994) *An Archival and Oral History Inventory of the White Salmon and Klickitat Rivers*. Prepared for the Columbia River Gorge National Scenic Area. Western Washington University, Bellingham, Wash.

Boyer, M. F. (1996) *Tree-Talk, Memory, Myths and Timeless Customs*. Thames and Hudson, London.

Braccho, F. (1995) Plants, Food and Civilization: The lessons of indigenous Americans. *J. Altern. and Complem. M ed.* 1,2, 125-130.

Branson, J. and Miller, D. (1995) Sign language and the discursive construction of power over the deaf through education. In: Corson, D. (ed.) *Discourse and Power in Educational Settings*. Hampton Press, Creskill, New Jersey.

Braund, S. R., Takahashi, J. and Freeman, M. M. R. (1990) Quantification of local need for minke whale meat for the Ayukawa-based minke whale fishery. IWC Document TC/42/SEST8, published in *Papers on Japanese small-type coastal whaling submitted by the Government of Japan to the International Whaling Commission, 1986-1996*, 175-190. The Government of Japan, Tokyo.

Bremam, Jan (1974) *Patronage and Exploitation: Changing Social Relations in South Gujarat*. Oxford University Press, Delhi.

Brenner, N. (1996) Monarch butterflies, my summer hobby. *Family Life*, Sept., 29.

Briggs, J. L. (1982) Living dangerously: the contradictory foundations of value in Canadian Inuit society. In: Leacock, E. and Lee, R. (eds), *Politics and History in Band Societies*. 109-132. Cambridge University Press, Cambridge.

Brody, H. (1982) *Maps and Dreams*. Pantheon Books, New York.

Brody, H. (1987) *Living Arctic: Hunters of the Canadian North*. Douglas and McIntyre, Vancouver.

Brokensha, D., Warren, D. M. and Werner, O. (eds) (1980) *Indigenous Knowledge Systems and Development*, University Press of America, Lanham, 460 pp.

Bromley, D. W. (1978) 'Property Rules, Liability Rules, and Environmental Economics', *Journal of Economic Issues*, 12, 43-60.

Bromley, D. W. (1989) *Economic Interests and Institutions: the conceptual foundations of public policy*. Blackwell, Oxford.

Bromley, D. W. (1991) *Environment and Economy: property rights and public policy*. Blackwell, Oxford.

Bromley, D. W. (ed.) (1992) Making the Commons Work: Theory, Practise and Policy. Institute for Contemporary Studies, San Francisco.

Bromley, D. W. (ed.) (1995) *The Handbook of Environmental Economics*. Blackwell Publishers, Oxford.

Bromley, D. W. and Cernea, M. M. (1989) The Management of Common Property Natural Resources, Discussion Paper No.57, The World Bank, Washington, D.C.

Bronowski, J. (1981) *The Ascent of Man*, Futura Publications, London, 287 pp.

Brookfield, H. and Padoch, C. (1994) Appreciating Agrodiversity: A Look at the Dynamism and Diversity of Indigenous Farming Practices. *Environment* 36 (5), pp. 6-11; 37-45.

Brosius, J. P. (1996) 'Analyses and Interventions: Anthropological Engagements with Environmentalism'. Presented to *Human Dimensions of Environmental Change: Anthropology Engages the Issues*. AAA, San Francisco.

Brosius, J. P. (1997) Endangered forest, endangered people: environmentalists' representations of indigenous knowledge. *Human Ecology* 25(1): 47-71.

Brosse, Jacques. (1989) *Mythologie des Arbres*. Payot, Paris.

Browder, J. O. (1986) *Logging the Rainforest: A Political Economy of Timber Extraction and Unequal Exchange in the Brazilian Amazon*. University of Pennsylvania Dissertation.

Browder, J. O. (1992) The limits of extractivism; tropical forest strategies beyond extractive reserves. *Bioscience* 42(3):174-182.

Brown, A. H. D. (1978) 'Isozymes, Plant Population Genetic Structure, and Genetic Conservation'. *Theoretical Applied Genetics* 52: 145-57, cited in Cleveland *et al.* (1994).

Brown, C. H. (1984) *Language and Living Things: Uniformities in Folk Classification and Naming*. Rutgers University Press, New Brunswick, N.J.

Brown, L. (1981) *Building a Sustainable Society*. Norton, New York.

Brown, L. R. and Jacobson, J. (1987) Assessing the Future of Urbanization. In: Brown, L. R. *et al.* (eds) *State of the World*, W. W. Norton, New York, pp. 38-56.

Brown, M.F. (1978) 'From hero's bones: Three Aguaruna hallucinogens and their uses'. In: *The Nature and Status of Ethnobotany. Ann Arbor: Anthropology Museum*, (ed.) R.I. Ford. University of Michigan (Anthropological Papers; Vol. 67), 119-36.

Brown, N. and Quiblier, P. (eds) (1994) *Ethics and Agenda 21: Moral implications of a global consensus*. United Nations Publications, New York.

Bruggermann, J. (1997) National Parks and Protected Area Management in Costa Rica and Germany: A Comparative Analysis. In: Ghimire, K. B. and Pimbert, M. P. (eds) *Social Change and Conservation*, UNRISD and Earthscan, London.

Brundtland, G. H. (1987) *Our Common Future*. World Commission on Environment and Development. Oxford University Press, Oxford, 383 pp.

Brush, S. B. (1989) Rethinking Crop Genetic Resource Conservation. *Conservation Biology*, 3 (1), pp. 19-29.

Brush, S. B. and Stabinsky, D. (eds). (1996) *Valuing Local Knowledge: Indigenous People and Intellectual Property Rights*.. Island Press, Washington, D.C.

Brush, S. B. *et al.* (1981) Dynamics of Andean potato agriculture. *Economic Botany* 35: 70-88.

Buck, P. (1949) *The Coming of the Maori*. Whitcombe and Tombs, Wellington.

Budiansky, S. (1995) *Nature's Keepers: The New Science of Nature Management*, Weidenfeld and Nicolson, London.

Buffington, A. F. (1939) Pennsylvania German: Its relation to other German dialects. *American Speech*, December, p.276-86.

Buller, G. (1983) Comanche and Coyote: The Culture Maker. In: Swann, B. (ed) *Smoothing the Ground*, University of California Press, Berkeley, pp. 245-254.

Bulmer, R. N. H. (1974) Folk biology in the New Guinea Highlands. *Social Science Information* 13: 9-28.

Burch, E. S. Jr. and Ellanna, L. J. (1994) (eds) *Key Issues in Hunter-Gatherer Research*. Berg Publishers, Oxford.

Burch, W. R. (1971) *Daydreams and Nightmares*. Harper and Row, New York.

Burch, W. R. (1983) Summary: Toward Social Forestry in the Tropics. In: Mergen, F. (ed.) *Tropical Forests: Utilization and Conservation*. Proceedings of an International Symposium held at Yale University, New Haven, CT.

Burch, W. R. (1987) *Gods of the Forest – Myth, Ritual and Television in Community Forestry*. Seminar Banquet Talk for the Regional Community Forestry Training Center - Asia–Pacific Seminar, Bangkok, Thailand.

Burger, J. (1987) *Report from the Frontier: the State of the World's Indigenous Peoples*. Zed Books, London.

Burgiel, S., Prather, T. and Schmidt, K. (1997) Summary of the Workshop on Traditional Knowledge and Biological Diversity: 24-28 November (1997) *Earth Negotiations Bulletin* 9(75). International Institute for Sustainable Development, Winnipeg.

Burke, P. (1995) *The Skills of Native Title Practice*, Australian Institute of Aboriginal and Torres Strait Islander Studies, Canberra.

Burkey, S. (1993) *People First Guide to Self-Reliance, Participatory Rural Development*. Zed Books Ltd., London.

Burkill, I. H. (1935) *A Dictionary of Economic Products of the Malay Peninsula*, Crown Agents for the Colonies, London (2 volumes).

Burley, J. (1994) Forests and biodiversity: the researcher's perspective. Prepared for IVA International Symposium on Forest Products and Future Society Development and Values, Stockholm.

Burnham, P. (1993) *The Cultural Context of Rain Forest Conservation in Cameroon*. Paper presented at the 36th annual meeting of the African Studies Association, Boston, Mass. December 1993.

Burnham, P., Richards, P., Rowlands, M.and Sharpe, B. (1995) *The Cultural Context of Rain Forest Conservation in West Africa.* Final report to the Global Environmental Change programme. Economic and Social Research Council, UK.

Byler, M. (1996) A quiz on birds' nests. *Young Companion*, Aug., p.19.

Byler, M. (1997) Winter birds quiz. *Young Companion*, Feb., p.17.

Caillat, C. and Kumar, R. (1981) *The Jain Cosmology*. Harmony Books, New York.

Calabresi, G. and Douglas Melamed, A. 1972 'Property Rules, Liability Rules and Inalienability: one view of the cathedral', *Harvard Law Review*, 85, 1085-128.

Caldecott, J. (1988) *Hunting and Wildlife Management in Sarawak.* The IUCN Tropical Forest Programme, Cambridge, UK.

Callaway, D. (1995) Resource use in rural Alaskan communities. In: Peterson, D. L. and Johnson, D. R. (eds), *Human Ecology and Climate Change — People and Resources in the Far North*. 155-168. Taylor and Francis, Washington, D.C.

Callenbach, E. *et.al.* (1993) *EcoManagement*. Berrett-Koehler, Berkeley.

Callicott, J. (1989) *In Defense of the Land Ethic: Essays in environmental philosophy.* State University of New York Press, Albany.

Callicott, J. (1991) Genesis and John Muir. In: Robb, C. and Casebolt, C. (eds), *Covenant for a New Creation: Ethics, religion, and public policy*. Orbis, Maryknoll, New York.

Callicott, J. (1991) The wilderness idea revisited: the sustainable development alternative. *Environmental Professional* 13: 235-247.

Callicott, J. (1994a) Conservation values and ethics. In: Meffe, G. and Carroll, R. (eds), *Principles of Conservation Biology*. Sinauer Associates, Inc., Sunderland, Mass.

Callicott, J. (1994b) Toward a global environmental ethic. In: Tucker, M. and Grim, J. *Worldviews and Ecology: Religion, philosophy, and the environment.* Orbis Books, Maryknoll.

Calme-Griaule, G. (1965) *Ethnologie et langage: la parole chez les Dogon.* Gallimard, Paris.

Calvet, L.-J. (1974) *Linguistique et Colonialisme: Petit Traité de Glottophagie.* Payot, Paris.

Cannell, R. (1997) Raft Above the Rainforest, *Glaxo Wellcome World*, 5: 23-25.

Capotorti, F. (1979) *Study of the Rights of Persons Belonging to Ethnic, Religious and Linguistic Minorities.* United Nations, New York.

Capra, F. (1982) *The Turning Point*. Simon and Schuster, New York.

Capra, F. (1996) *The Web of Life: A new scientific understanding of living systems*. Doubleday, New York.

Capra, F.and Pauli, G. (eds) (1995) *Steering Business toward Sustainability*. United Nations University Press, Tokyo.

Caraka Samhita (1994) Trans. Sharma, R. K. and Dash, B. Chowkhamba Sanskrit Series Office, Varanasi, India. 3rd edition.

Carneiro, R. L. (1970) The transition from hunting to horticulture in the Amazon Basin. *Eighth International Congress of Anthropological and Ethnological Sciences* 3:244-8.

Carrithers, M., (1992) *Why humans have Cultures: Explaining Anthropology and Social Diversity.* Oxford University Press, Oxford.

Carruthers, J. (1989) Creating a national park. 1910-1926. *Jour. of S. Afr. Studies* 2, 208.

Carter, J. (1996) Recent approaches to participatory forest resource assessment. *Rural Development Forestry Study Guide 2*, Overseas Development Institute, London.

Carter, Paul. (1987) *The Road to Botany Bay: An Exploration of Landscape and History.* University of Chicago Press, Chicago.

Castells, M. (1983) *The City and the Grassroots*. University of California Press, Berkeley.

Castillo Caballero, D. (1989) *Mito y sociedad en los Bari.* Salamanca, Amarú Ediciones.

Castro, A. P. (1995) *Facing Kirinyaga: A Social History of Forest Commons in Southern Mount Kenya*. Intermediate Technology Publications, London.

Caulfield, R. A. (1992) Alaska's subsistence management regimes. *Polar Record* 28(164): 23-32.

CBD Clearing House (1997) Secretariat to the Convention on Biological Diversity. *Forest Biological Diversity*. Clearing House Mechanism, Montreal, Canada.

CBD Secretariat (1996) *The Impact of Intellectual Property Rights Systems on the Conservation and Sustainable Use of Biological Diversity and on the Equitable Sharing of Benefits from its use*. UNEP/CBD/COP/3/22.

Centre for Aboriginal and Torres Strait Islander Participation (1996) Guidelines on Research Ethics Regarding Aboriginal and Torres Strait Islander Cultural, Social, Intellectual and Spiritual Property, R&D, James Cook University of North Queensland, Townsville.

Cernea, M. M. (1993) Culture and Organisation: The Social Sustainability of Induced Development, *Sustainable Development*, 1(2) 18-29.

CGIAR (1994) *Partners in Selection*. Washington, D.C: Consultative Group on International Agricultural Research.

Chadwick, D. J. and Marsh, J. (ed.). (1994) *Ethnobotany and the Search for New Drugs*. John Wiley and Sons, Chichester.

Chakrabarti (1981) *Around the Plough.*

Chalk, F. and Jonassohn, K. (1990) *The History and Sociology of Genocide: Analyses and Case Studies.*Concordia University Press, Montreal.

Chambers, R. (1983) *Rural Development: Putting The Last First*. Longman, London and New York.

Chambers, R. (1993) 'Local People Know'. In: *People's Management of Natural Resources and the Environment: Voices from a Workshop*. Stockholm: Lundberg, J.and Schlebrugge von K. (eds).Swedish Academy for Research Cooperation with Developing Countries (SAREC). Stockholm.

Chambers, R. (1993) *Challenging the Professions: Frontiers for Rural Development*, Intermediate Technology Publications, London.

Chambers, R. and Guijt, I. (1995) PRA-five years later. Where are we now? *Forests, Trees and People Newsletter* 26/27:4-14.

Chambers, R. and Richards, P. (1995) Preface. In: Warren, D. M., Slikkerveer, L. J. and Brokensha, D. (eds) *The Cultural Dimension of Development: Indigenous Knowledge Systems*, Intermediate Technology Publications, London.

Chambers, R., Pacey, A. and Thrupp, L. A. (1989) *Farmer First: Farmer Innovation and Agricultural Research*, Intermediate Technology Publications, London.

Chambers, Robert (1994) The Origins and Practice of Participatory Rural Development. In: *World Development*, 22(7): 953-969.

Chambi, N. and Chambi, W. (1995) *Ayllu y Papas: Cosmovisión, Religiosidad y Agricultura en Conima, Puno*, Asociación Chuyma de Apoyo Rural 'Chuyma Aru'.

Chambi, N., Chambi, W., Quiso, V., Cutipa, S., Apaza, J. and Gordillo, V. (1996) La Crianza de las Yoqchi as en los Distritos de Conima, Tilali e Ilave, Puno. In: PRATEC (ed) *La Cultura Andina de la Biodiversidad*, Proyecto Andino de Tecnologias Campesinas, Lima, pp. 43-84.

Champagne, D. (1989) *American Indian Societies – Strategies and conditions of political and cultural survival*. Cultural Survival, Cambridge, MA.

Chance, N. (1990) *The Inupiat and Arctic Alaska: An ethnography of development*. Holt, Rinehart and Winston, Fort Worth, TX.

Chandler, T. and Fox, G. (1974) *3000 Years of Urban Growth*, The Academic Press, New York.

Chandrakanth, M. G. and Romm, J. (1991) Sacred Forests, Secular Forest Policies and People's Actions. *Natural Resources Journal*, 31 (4): 741-756.

Chang, G. C. C. (trans.) (1970)*The Hundred Thousand Songs of Milarepa*. Harper Colophon, New York.

Chapela, I. (1996) 'The informatic value of biodiversity: towards a market for bioprospecting'. Unpublished: UC Berkeley.

Chapeskie, A. J. (1990) Indigenous law, state law and the management of natural resources: wild rice and the Wabigoon Lake Ojibway Nation. *Law & Anthropology (Internationales Jahrbuch fur Rechtsanthropologie)* 5: 129-166.

Chapeskie, A. J. (1995) *Land, Landscape, Culturescape: Aboriginal Relationships to Land and the Co-Management of Natural Resources* (Paper prepared for the Royal Commission on Aboriginal Peoples) (Available on CD-ROM as part of: *For Seven Generations: An Information Legacy of the Royal Commission on Aboriginal Peoples*. Libraxus Inc., Ottawa, 1997).

Chartrand, N. (1992) Ouverture du colloque In: Chartrand, N. and Thérien, N. (eds). *Les enseignements de la phase 1 du Complexe La Grande*. pp. 6-9. Universitéde Sherbrooke and Hydro-Québec, Sherbrooke.

Chavannes, E. (1910) *Le T'ai Chan: Essai de monographie d'un culte Chinois*. Ernest Leroux, Paris.

Chavez, J., Gomez, S. and Malagamba, M. (1989) *Propuesta de agricultura organica para la sierra*. IDEASCONYCET, Lima.

Chernela, J. (1993) *The Wanano Indians of the Brazilian Amazon: A Sense of Space*. University of Texas Press, Austin.

Chernela, J. M. (1982) An indigenous system of forest and fish management in the Uaupés Basin of Brazil. *Cultural Survival Quarterly* 6:17-18.

Chernela, J. M. (1989) Managing Rivers of Hunger, the Importance of the Blackwater River Margin. In: Balee, W. and Posey, D. (eds), *Resource Management in Amazonia. Indigenous and Folk Strategies. Advances in Economic Botany*, New York Botanical Garden 7:238-248.

Chernela, J. M. (1993) *The Wanano Indians of the Brazilian Amazon. A Sense of Space*. University of Texas Press, Austin.

Chernela, Janet (1994) 'Tukanoan Know-how: The Importance of the Forested River Margin to Neotropical Fishing Populations'. *National Geographic Research & Exploration* 10 (4) 440-457.

Chichilnisky, G. (1996) 'The Economic Value of the Earth's Resources' *Trends in Ecology and Evolution* 11 135-40.

Chichlo, Boris (1987) Tunguz religion. In: Eliade, Mircea. (ed.) *Encyclopaedia of Religion*. 13:83-86

Chittenden, A. (1992) Aboriginal Involvement in National Parks: Aboriginal Rangers' Perspectives of Wujal Wujal Community. In: Birckhead, J., De Lacy, T. and Smith, L. (eds) *Aboriginal Involvement in Parks and Protected Areas*. Aboriginal Studies Press, Canberra, p. 35.

Chokechaijaroenporn, O. N., Bunyapraphatsara, N. and Kongchuensin, S. (1994) Mosquito repellent activities of *Ocimum* volatile oils. *Phytomedicine* 1,135-

Chomsky, N. (1966) *Cartesian Linguistics*, New York, Harper and Row.

Chomsky, N. (1989) Necessary illusions: Thought control in democratic societies, Boston, South End Press, p. 48.

Chopa, R. N., Nayar, S. L. and Chopra, I. C. (1956)*Glossary of Indian medicinal plants*. Council of Scientific and Industrial Research. New Delhi.

Christie, J. and Mooney, P. (1994) *Enclosures of the Mind - Intellectual Monopolies*. Rural Advancement Foundation International. Ottawa. Electronic format on RAFI's website at http://www.rafi.ca

Christner, S. (1996a) Along Nature's Trails: The innumerable hosts. *Family Life*, June, pp.10-11.

Christner, S. (1996b) Along Nature's Trails: The innumerable hosts. *Family Life*, July, pp.9-11.

Christner, S. (1996c) Along Nature's Trails: The way of the turtle. *Family Life*, Oct., pp.10-12.

Ciba Foundation (1994) *Ethnobotany and the Search for New Drugs*. John Wiley and Sons, New York.

Cicourel, A. V. (1964) *Mesure and Measurement in Sociology*, McGraw Hill, New York.

CIDE/ACTS (1990) *Participatory Rural Appraisal Handbook*. Washington, D.C.

Cipriani, L. (1966) *The Andaman Islanders*, Wiedenfeld and Nicolson, London.

Clad, J. (1984) Conservation and indigenous peoples: a study of convergent interests. *Cultural Survival Quarterly* 8: 68-73.

Clarfield, G. (1996) *Articulating indigenous indicators: An NGO Guide for community-driven project evaluation, based on a case study among the Ariaal of Kenya's arid rangelands.* ELCI, Nairobi.

Clark, D. and Williamson, R. (eds) (1996) *Self-Determination: International Perspectives*. The Macmillan Press, London.

Clark, J. F. M. (1997) '"A little people, but exceedingly wise?" Taming the Ant and the Savage in Nineteenth-Century England'. *La Lettre de la Maison Française d'Oxford*, no. 7 (Trinity Term, 1997), pp. 65-83, esp. 72-75.

Clarke, G. E. (1995) 'Development in Nepal: Mana from Heaven', In: *Festschrift pour Alexander W. MacDonald*, Samten G Karmay and Philip Sagant (eds), Laboratoire d'ethnologie et de sociologie comparative, Université de Paris-X, Nanterre.

Clarkson, L., Morrissette, V. and Regallet, G. (1992) *Our Responsibility to the Seventh Generation: Indigenous Peoples and Sustainable Development.* International Institute for Sustainable Development, Winnipeg.

Clastres, Pierre. (1972) La Societé Contre l' Etat. Editions de Minuit, Paris.

Clawson, D. L. (1985) Harvest security and intraspecific diversity in traditional tropical agriculture. *Economic Botany* 39: 56-57.

Clay, J. (1996) *Generating Income and Conserving Resources: Twenty Lessons from the Field.*

Clay, J. and Clement, C. R. (1993) *Selected Species and Strategies to Enhance Income Generation from Amazonian Forests.* FAO Forestry Paper (draft), Rome, Italy.

Clay, J. W. (1988) Indigenous peoples and tropical forests. *Cultural Survival.* Cambridge, MA.

Clay, J. W. (1993) Looking back to go forward: Predicting and preventing human rights violations. In: Miller, M. S. (ed.), *State of the Peoples: A Global Human Rights Report on Societies in Danger*. 64-71. Beacon Press, Boston.

Cleary, D. M. (1992) Overcoming socio-economic and political constraints to 'wise forest management': lessons from the Amazon. In: Miller, F. and Adam, K. L. (eds) *Wise Management of Tropical Forests.* Proceedings of the Oxford Forestry Institute Conference on Tropical Forests.

Clement, C. R. (1989) A center of crop genetic diversity in western Amazonia: a new hypothesis of indigenous fruit crop distribution. *Bioscience* 39:624-630.

Cleveland, D. *et al.* (1994) 'Do Folk Crop Varieties Have a Role in Sustainable Agriculture?' *Bioscience* 44(11):740-51.

Clifford, James (1988) *The Predicament of Culture.* Cambridge University Press, Cambridge.

Clifford, James and Marcus, George. (1986) *Writing Culture.* The University of California Press, Berkeley.

Clifford, S. and King, A. (eds) (1993) *Local Distinctiveness: Place, particularity and identity*. Common Ground, London.

Coase, R. H. (1937) The nature of the firm. *Economica* : 386-405.

Cobarrubias, J. (1983) Ethical issues in status planning. In: Cobarrubias, J. and Fishman, J. A. (eds) *Progress in Language Planning: International Perspectives*. 41-85. Mouton, Berlin.

Cody, M. L. (1986) Diversity, Rarity and Conservation in Mediterranean-Climate Regions. In: Soulé, M. E. (ed) *Conservation Biology: The Science of Scarcity and Diversity.* Sinauer Assoc. Inc., Massachusetts, pp. 122-152.

Cohen, F. (1986) *Treaties on Trial: The Continuing Controversy over Northwest Indian Fishing Rights.* University of Washington Press, Seattle.

Cohen, S. J. (1995) An interdisciplinary assessment of climate change on northern ecosystems: the Mackenzie Basin Impact Study. In: Peterson, D. L. and Johnson, D.R. (eds), *Human Ecology and Climate Change — People and Resources in the Far North*. 301-316. Taylor and Francis, Washington, D.C.

Cohn, Bernard. (1990) *An Anthropologist Among the Historians and Other Essays.* Oxford University Press, Delhi.

COICA (1994) Coordinadora de las Organizaciones Indigenas de la Cuenca Amazonica (COICA). Papers from a Regional Meeting sponsored by COICA and UNDP on Intellectual Property Rights and Biodiversity, 28-30 September 1994, Santa Cruz de la Sierra, Bolivia.

Colchester, M. (1989) 'Indian Development in Amazonia: Risks and Strategies,' *The Ecologist*, 19 (6) 249-254.

Colchester, M. (1994) *Salvaging Nature: Indigenous Peoples, Protected Areas and Biodiversity Conservation*, UNRISD-WRM-WWF, UNRISD Discussion Paper No.55, Geneva.

Colchester, M. (1997) Salvaging nature: indigenous peoples and protected areas. In: Ghimire, K. B. and Pimbert, M. P. *Social Change and Conservation*. Earthscan, London, and UNRISD, 97-130.

Commission on Justice and Peace (1997) *Toward Sustainable Community: Five years since the earth summit*. Taskforce on the Churches and Corporate Responsibility, Toronto.

Conan Doyle, A. (1981) 'His Last Bow' In: *The Penguin Complete Sherlock Holmes.* Penguin Books, London. p. 978.

Conklin, B. A. and Graham, L. R. (1995) The Shifting Middle Ground: Amazonian Indians and Eco-Politics *American Anthropologist* 97:4: 695-710.

Conklin, H. (1957) Hanunoo Agriculture: A Report on an Integral System of Shifting Cultivation in The Philippines. Forestry Development Paper 12, FAO, Rome.

Conklin, H. C. (1954) The Relation of Hanunóo Culture to the Plant World. Unpublished Ph.D. dissertation thesis, Yale University.

Conklin, Harold C. (1954) An Ethnoecological Approach to Shifting Agriculture. *Transactions of the New York Academy of Sciences* Ser. 2, 17(2):XXX.

Convention on Biological Diversity CBD (1992), UNCED, Rio de Janeiro.

Conway, G. R. (1985) Agroecosystem analysis. *Agricultural Administration* 20:31-55.

Cooch, F. G., Gunn, A. and Stirling, I. (1987) Faunal processes. In: Nelson, J. G., Needham, R. and Norton, L. (eds), *Arctic Heritage: Proceedings of a Symposium August 24-28*, Banff, Alberta, Canada. 95-111. Association of Canadian Universities for Northern Studies, Ottawa.

Cordeiro, A. (1993) 'Rediscovering local varieties of maize: challenging seed policy in Brazil'. In: de Boef *et al.*, *Cultivating Knowledge: genetic diversity, farmer experimentation and crop research.*

Corlett, R. (1992) Conserving the Natural Flora and Fauna in Singapore. In: Chua Beng Huat and Edwards, N. (eds) *Public Space: Design, Use and Management.* Singapore University Press, Singapore, pp. 128-137.

Correa, Carlos (1994a) The GATT Agreement on Trade-Related Aspects of Intellectual Property Rights: New Standars for Patent Protection, *European Intellectual Property Review*, vol.16, Issue 8.

Correa, Carlos (1994b)*Sovereign and Property Rights over Plant Genetic Resources*, FAO, Commission on Plant Genetic Resources, First Extraordinary Session, Rome.

Correll, E. (1925) *Das scheirzerische Taufermennonitentum.* Tubingen: Verlag von J.C.B. Morh.X pp.110-121.

Corson, D. (1992) Bilingual education policy and social justice. *Journal of Education Policy* 7(1): 45-69.

Costa e Silva, E. (1995) The protection of intellectual property for local and indigenous communities. *European Intellectual Property Review*, 11, 546-549.

Costanza, R. (ed.) (1991) *Ecological Economics: The science and management of sustainability.* Columbia University Press, New York.

Costanza, R. *et al.* (1997) The value of the world's ecosystem services and natural capital. *Nature*, 15 May.

Costanza, R., Norton, B. and Haskell, B. (1992) *Ecosystem Health: New goals for environmental management.* Island Press, Washington, D.C.

Cotton, C. M. (1996) *Ethnobotany: Principles and Applications.* Wiley, Chichester.

Coulibaly, S. (1978) *Le pays Sénoufo.* Nouvelles Editions Africaines. Abidjan-Dakar: 245p.

Courlander, H. (1996) *A Treasury of Afro-American Folklore.* Marlowe and Company, New York.

Covarrubias, M. (1946) *Island of Bali.* Alfred A. Knopf, New York.

Cowen, T. (1992) *Public Goods and Market Failures.* Transaction, New Brunswick, New Jersey.

Cox, P. A. and Elmqvist, T. Ecocolonialism and indigenous knowledge systems: village controlled rainforest preserves in Samoa. *Forum Essays.*

Craig, C. G. (1992) Miss Nora, rescuer of the Rama language: A story of power and empowerment. In: Hall, K., Bucholtz, M. and Moonwomon, B., *Locating Power.* Proceedings of the 2nd Women and Language Conference, April 4-5, (1992) Berkeley Women and Language Group, University of California, Berkeley.

Craig, D. and Nava, D. P. (1995) Indigenous peoples' rights and environmental law. In: UNEP's *New Way Forward: Environmental Law and Sustainable Development.* UNEP, Nairobi, pp. 115-146.

Croll, E. and Parkin, D. (eds) (1992) *Bush Base: Forest Farm - Culture, Environment and Development*, Routledge, London.

Crone, G. R. (1937) *The voyages of Cadamosto.* Hakluyt, London.

Cronk, S. (1981) Gelassenheit: the rites of the redemptive process in Old Order Amish and Old Order Mennonite communities. *The Mennonite Quarterly Review* 55:5-44.

Cronon, W. (1983) *Changes in the Land.* Hill and Wang, New York.

Cronon, William. (ed.) (1995) *Uncommon ground: Toward reinventing nature.* W. W. Norton, New York.

Cros, M. (1990) *Anthropologie du sang en Afrique.* Harmattan, Paris.

Cros, M. (1990) *Anthropologie du sang en Afrique.* l'Harmattan, Paris.

CSD (1995) Secretariat to the Commission on Sustainable Development. *Intergovernmental Panel on Forests: its mandate and how it works.* United Nations, New York.

Cummins, J. (1989) *Empowering Minority Students.* California Association for Bilingual Education, Sacramento.

Cummins, J. (1996) *Negotiating Identities: Education for Empowerment in a Diverse Society.* California Association for Bilingual Education, Ontario, California.

Cunningham, A. B. (1989) Indigenous plant use: balancing human needs and resources. In: Huntley, B. (ed.) *Conserving Biological Diversity in Southern Africa.* Oxford University Press, Oxford.

Cunningham, A. B. (1990) 'Man and medicines: the exploitation and conservation of traditional Zulu medicinal plants'. *Mitteilungen aus dem Institut für allgemeine Botanik, Hamburg*, 23:979-990. Proceedings of the 1988 AETFAT Congess, Hamburg.

Cunningham, A. B. (1991) Development of a conservation policy on commercially exploited medicinal plants: a case study from southern Africa. In: Akerele, O., Heywood, V. and Synge, H. (eds) *The Conservation of Medicinal Plants.* Cambridge University Press, Cambridge, U.K.

Cunningham, A. B. (1992) *People, Park and plant Use: Research and Recommendations for multiple-Use Zones and Development Alternatives Around Bwindi-Impenetrable National Park, Uganda.* Report prepared for CARE-Internation, Kampala, Uganda.

Cunningham, A. B. (1993) *Ethics, Ethnobiological Research and Biodiversity: Guidelines for Equitable Partnerships in New Natural Products Development.* WWF-International, Gland.

Cunningham, A. B. (1993) *Ethics, Ethnobiological Research, and Biodiversity.* WWF/UNESCO/Kew People and Plants Initiative, WWF International, Gland, Switzerland.

Cunningham, A. B. (1993a) 'African Medicinal Plants: Setting Priorities at the Interface between Conservation and Primary Health Care'. *People and Plants Working Paper 1*, UNESCO, Paris.

Cunningham, A. B. and Milton, S. J. (1987) Effects of basket-weaving on Mokola Palm and dye plants in Northwestern Botswana. *Economic Botany* 41, 386-402.

Cunningham, S. (1994) *Hawaiian Religion & Magic.* Llewellyn Publications, St. Paul. pp.231.

d'Ohsson, M. (1788-1824) *Tableau General de l'Empire Othoman*. 4:307.

D'Souza, M. (1993) *Tribal Medicine*. Society for promotion of wastelands development. New Delhi.

Daes, E. (1994) *Study on the Protection of the Cultural and Intellectual Property of Indigenous Peoples.* United Nations Sub-commission on Prevention of Discrimination and Protection of Minorities, Forty-Sixth Session, E/CN.4/Sub.2/1994/31.

Daes, E. (1995) *Protection of the Heritage of Indigenous Peoples:* Final Report of the Special Rapporteur, Mrs. Erica-Irene Daes, in conformity with Sub-Commission resolution 1993/44 and decision 1994/105 of the Commission on Human Rights, Sub-Commission on Prevention of Discrimination and Protection of Minorities, Forty-Seventh Session, E/CN.4/Sub.2/1995/26.

Daes, E.-I. (1993) *Study on the Protection of the Cultural and Intellectual Property of Indigenous Peoples.* United Nations Sub-commission on Prevention of Discrimination and Protection of Minorities, Forty-Fifth Session, E/CN.4/Sub.2/1993/28.

Daes, E.-I. (1995) 'Redressing the balance: The struggle to be heard'. Paper presented at the Global Cultural Diversity Conference, Sydney, 26-28 April (1995)

Dahaban, Z., Nordin, M. and Bennett, E. L. (1992) *Immediate effects on wildlife is selective loggin in a hill dipterocarp forest in Sarawak: mammals.* Department of Zoology, University Kebangsaan, Malaysia and Wildlife Conservation Society.

Dalton, G. (1961) Economic theory and primitive society. *American Anthropologist* 63: 1-25.

Dalton, G. (1968) Economics, economic development and economic anthropopology. *Journal of Economic Issues* 2: 173-186.

Dalton, G. (1969) Theoretical issues in economic anthropology (with comment and reply). *Current Anthropology* 10(1): 63-102.

Daly, H. (1993) Elements of environmental macroeconomics. In: *Sustainable Growth: A contradiction in terms?* Report of the Visser't Hooft Memorial Consultation, Visser't Hooft Endowment Fund for Leadership Development, Geneva.

Daly, H. (1995) On Wilfred Beckermann's critique of sustainable development. *Environmental Values* 4,1,49-55.

Daly, H. and Cobb, J. (1989) *For the Common Good: Redirecting the economy toward community, the environment, and a sustainable future.* Beacon Press, Boston.

Dalziel, J. M. (1937) *The Useful Plants of West Tropical Africa* 113.

Damas, D. (1973) Environment, history and central Eskimo society. In: Cox, B. (ed.) *Cultural Ecology*. 269-300. McClelland and Stewart, Toronto.

Dange, A. S. (1990) *Encyclopedia of puranic beliefs and practices*. Navrang, New Delhi.

Dankelman, I. (1993) Women, children and environment: implications for sustainable development. In: Steady, F. C. (ed.), *Women and Children First*. Schenkman Brooks, Rochester, Vermont.

Daphney, H. and Royee, F. (1992) 'Kowanyama and Malanbarra Views on Nature Conservation Management.' In: *Aboriginal Involvement In Parks and Protected Areas* (eds) Birckhead, J., De Lacy, T and Smith, L. Aboriginal Studies Press, Canberra:39-44.

Darwin, C. (1859) *The Origin of Species by Means of Natural Selection.* Studio Editions Ltd., London.

Dasgupta, P. (1991) The environment as a commodity. In: Blasi, P. and Zamagni, S. (eds). *Man-Environment and Development: Towards a global approach.* 149-180. Nova Spes International Foundation Press, Rome.

Dasgupta, P. (1993) *An Inquiry into Well-Being and Destitution.* Clarendon Press, Oxford.

Dasmann, F. L. F. (1984) *Environmental Conservation*, 5th Edition, John Wiley, New York.

Davidson, D. W. and McKey, D. (1993) 'Ant-plant symbioses: Stalking the Chuyachaqui.' *Trends in Ecology and Evolution* 8: 326-332.

Dawkins, R. (1989) *The Selfish Gene*. Oxford University Press, Oxford.

Dawkins, Richard. (1995) *River Out of Eden*. Basic Books, New York.

Daxl, R. and Swezey, S. (1983) *Breaking the Circle of Poison: The IPM Revolution in Nicaragua.* Institute of Food and Development Policy, San Francisco.

De Avila, A. (1996) The biogeography of Mesoamerican textiles. Paper presented at the 19th Annual Conference of the Society of Ethnobiology, Santa Barbara Museum of Natural History, March 27-29, 1996.

De Beer, J. H. and McDermott, M. J. (1989) *The Economic Value of Non-Timber Forest Products in Southeast Asia with Emphasis on Indonesia, Malaysia and Thailand.* Netherlands Committee for IUCN, Amsterdam.

De Boef, W., Amanor, K. and Wellard, K. (1993) *Cultivating Knowledge: Genetic Diversity, Farmer Experimentation And Crop Research*. Intermediate Technology Publications, London.

De Josselin de Jong, P. E. (1980) The Concept of the Field of Ethnological Study. In: Fox, J. (ed) *The Flow of Life.* Harvard University Press, Cambridge, Ma., USA.

De Lamartine, A. (1897) *Voyagen en Orient*. 2:259.

De Schlippe, P. (1956) *Shifting Cultivation in Africa: The Zande System of Agriculture.* Routledge and Kegan Paul. London.

Deihl, C. (1985) Wildlife and the Masai. *Cultural Survival*, 9(1).

de Jong, W. D. and Mendelssohn, R. (1992) *Managing the Non-Timber Forest Products of Southeast Asia.* Paper prepared for the World Bank, Southeast Asia Division.

Delacour, J. (1947) *Birds of Malysia*. MacMillan, New York.

Delamonica, P. (1997) *Florística e estrutura de floresta atlântica secundária - Reserva Biológica Estadual da Praia do Sul, Ilha Grande, RJ.* M.Sc. Thesis, Universidade de São Paulo. 113 p.

Déléage, J.-P. (1992) *Histoire de l'Écologie*. Une science de l'homme et de la nature. Editions La découverte: 330p.

Deloria Jr., V. (1990) Knowing and Understanding: Traditional Education in the Modern World. *Winds of Change* 5(1): 12-14.

Deloria Jr., V. (1992) The Spatial Problem of History. In: *God is Red*. North American Press, Golden, CO, pp. 114-134.

Deloria Jr., V. (1995a) The Bering Straight and Narrow. *Winds of Change*, 10(1):68-75.

Deloria Jr., V. (1995b) *Red Earth, White Lies*. Harper and Row, New York.

Deloria, V. (1992) 'Out of Chaos' in *Becoming Part of It*. (eds) Dooling, D. and Jordan-Smith, P. Harper, San Francisco.

Demsetz, H. (1988) *Ownership, Control and the Firm: the organisation of economic activity, Volume I*. Basil Blackwell, Oxford.

Demsetz, H. S. (1997) The firm in economic theory: a quiet revolution. *American Economic Review* 87(2): 426-429.

Denevan, D. W. (1995) Prehistoric agricultural methods as models for sustainability. *Advance Plant Pathology* 11: 21-43.

Denevan, W. M. (1992) The pristine myth: the landscape of the Americas in 1492. *Annals of the Association of American Geographers* 82(3): 369-385.

Denevan, W. M. and Padoch, C. (eds) (1988) Swidden-fallow agroforestry in the Peruvian Amazon. *Advances in Economic Botany* 5.

Denslow, J. S. and Padoch, C. (1988) *People of the Tropical Rain Forest*, University of California Press, Berkeley.

Department for Policy Coordination and Sustainable Development (DPCSD) (1996) Report of the Expert Group Meeting on Identification of Principles of International Law for Sustainable Development. Geneva, Switzerland, 26-28 September (1995). Prepared for the Division for Sustainable Development for the Commission on Sustainable Development, Fourth Session, 18 April-3 May, New York. DPCSD, New York.

Department for Policy Coordination and Sustainable Development (DPCSD) (1997) *Interagency Partnership on Forests: Implementation of IPF Proposals for Action by the ITFF*, June (1997) United Nations, New York.

Descola, P. (1992) 'Societies of nature and the Nature of Society'. In: *Conceptualizing Society* Kuper, A.(ed). Routledge, London:107-126.

Descola, P. and Palsson, G. (eds) (1996) *Nature and Society — Anthropological Perspectives*. Routledge, London.

Descola, Philippe. (1986) *La Nature Domestiqué: Symbolisme et praxis dans l'ecologie des Achuar*. Editions de la Maison des Sciences de l'Homme, Paris.

Deshpande, N. A. (1990) *The Padma-Purana, Part V*. Motilas Banarsidass Publishers Pvt. Ltd. Delhi.

Deshpande, N. A. (1991) *The Padma-Purana, Part VIII*. Motilas Banarsidass Publishers Pvt. Ltd. Delhi.

Deshpande, N. A. (1992) *The Padma-Purana, Part X*. Motilas Banarsidass Publishers Pvt. Ltd. Delhi.

Desjarlais, R. R. (1992) *Body and Emotion*. Philadelphia.

Deslandes, J. C., Fortin, R., Verdon, R., Roy, D. and Belzile, L. (1992) Evolution de la communauté de poissons du réservoir de la Grande 2 à la suite de la mise en eau. In: Chartrand, N. and Thérien, N. (eds) *Les enseignements de la phase 1 du Complexe La Grande*. pp. 106-120. Université de Sherbrooke and Hydro-Québec, Sherbrooke.

Desmond, A. and Moore, J. (1991) *Darwin*, Michael Joseph, London.

Devall, B. (1988) *Simple in Means, Rich in Ends: Practicing deep ecology*. Gibbs Smith, Salt Lake City.

Devall, B. and Sessions, G. (1985) *Deep Ecology: Living as if nature mattered*. Gibbs M. Smith Publisher, Salt Lake City, Utah.

DeWalt, B. R. (1994) Using indigenous knowledge to improve agriculture and natural resource management. *Human Organization* 53(2): 123-131.

Dey, B. B. and Choudhuri, M. A. (1984) Essential oil of Ocimum sanctum and its antimicrobial activity. *Indian Perfumer*. 28, 82-87.

Diamond, I. (1994) *Fertile Ground: Women, earth, and the limits of control*. Beacon Press, Boston.

Diamond, J. (1987) The Environmentalist Myth, *Nature* 324:9-20.

Diamond, J. (1990) Bach, God and the jungle. *Natural History* 12 (90):22-27.

Diamond, J. (1993) Speaking with a single tongue. *Discover* 14(2): 78-85.

Diamond, J. (1997) *Guns, Germs and Steel: A Short History of everybody for the Last 13,000 Years*. Jonathan Cape, London.

Diamond, J. M. (1986) Introductions, Extinctions, Exterminations and Invasions. In: Case, T. J. and Diamond, J. M. (eds) *Community Ecology*, Harper and Row, New York, pp. 65-79.

Diamond, J. M. (1986) Rapid Evolution of Urban Birds, *Nature* 3 24:107-104.

Diegues, A. C. (1994) *O Mito da Natureza Intocada*. NUPAUB, Universidade de São Paulo. 163p.

Dieterlen, G. (1987) Les témoignage des dogon. In: Solange de Ganay, A., J-P Lebeuf and Dominique Zahan (eds.) *Hommages ≠ Marcel Gruiaule*. Hermann, Paris.

Dixon, R. M. W. (1976) Tribes, Languages and Other Boundaries in North East Queensland. In: Peterson, N. (ed) *Tribes and Boundaries in Australia*, Australian Institute of Aboriginal Studies, Canberra, pp. 207-234.

Djohani, R. (1996) *An Evaluation of Conservation and Development Opportunities for the Aru islands. Fact finding mission Aru Tenggara Marine Reserve*. Jakarta. Maluku Conservation and Natural Resources Project (MACONAR). The World Bank.

Dobkin de Rios, M. (1972) Visionary Vine: Psychedelic Healing in the Peruvian Amazon. Chandler Publications for Health Sciences, San Francisco.

Dobkin de Rios, M. and Winkelman, M. (1989) 'Shamanism and altered states of consciousness: An introduction'. In: *Shamanism and Altered States of Consciousness* (eds) Dobkin de Rios, M. and Winkelman, M. *Journal of Psychoactive Drugs* 21(1).

Donigi, P. (1994) *Indigenous or Aboriginal Rights to Property: A Papua New Guinea Perspective*. International Books, Utrecht.

Dorian, N. (1995) Sharing expertise and experience in support of small languages. *International Journal of the Sociology of Language* 114: 129-137.

Dorian, N. (1997) Lexical loss among the final speakers of an obsolescent language: A formerly-fluent speaker and a semi-speaker compared. *News and Views from Terralingua: Partnerships for Linguistic and Biological Diversity* (electronic newsletter) 1(4): n.p.

Dossou, B. (1992) 'Problématique et politique du bois-énergie au Bénin'. Thèse de Doctorat (PhD) de Philosophie. Université de Laval, Canada: 414p.

Doubleday, N. C. (1993) Finding Common Grounds: Natural Law and Collective Wisdom. In: Inglis, J. T. (ed.) *Traditional Ecological Knowledge: Concepts and Cases*. International Development Research Centre, Canada.

Douglas, I. (1990) Sediment Transfer and Siltation. In: Turner, B .L., Clark, W. C., Kates, R. W., Richards, J. F., Mathews, J. T., Meyer, W. B. (eds) *The Earth as Transformed by Human Action*. Cambridge University Press, Cambridge, pp. 215-234.

Douglas, W. O. (1951) *Of Men and Mountains*. Victor Gollancz, London.

Dounias, E. (1993) 'Dynamique et gestion du système de production a dominante agricole des Mvae du sud Cameroun forestier'. Doctoral thesis, University of Montpellier.

Dove, M. R. (1983) Theories of Swidden Agriculture and the Political Economy of Ignorance. *Agroforestry Systems* 1 (3), 85-99.

Dove, M. R. (1993) Uncertainty, humility and adaptation to the tropical forest: The agricultural augury of the Kantu. *Ethnology* 40 (2):145-167.

Dove, M. R. (1996) Center, periphery, and biodiversity: a paradox of governance and developmental challenge. In: Brush, S. B. and Stabinsky, D. (eds) *Valuing local knowledge*. Island Press, Washington.

Drijver, C. A., Van Wetten, J. C. J. and De Groot, W. T. (1995) Working with nature: local fishery management on the Logone floodplain in Chad and Cameroon. In: Van Den Breemer, J. M. P., Drijver, C. A. and Verma, L. B. (eds) *Local resource management in Africa*. 30-45. Wiley, Chichester.

Drinnon, A. (1980) *Facing West: The metaphysics of indian busting and empire building*. University of Minnesota Press, Minneapolis.

Driver, H. E. (1969) *Indians of North America*, 2d (ed.) , University of Chicago Press, Chicago.

Drouin, J.-M. (1992) L'image des sociétés d'insectes en France à l'époque de la Revolution. *Revue de Synthèse*, 4 pp. 333-45.

Druviete, I. (1994) Language policy in the Baltic States: a Latvian case. In: *Valodas politika Baltijas Valstīs/ Language Policy in the Baltic States*. 151–160. Krājumu sagatavojis, Latvijas Republikas Valsts valodas centrs, Riga.

Dubos, R. (1972) *A God Within*. Charles Scribner's Sons, New York.

Dubos, R. (1981) *Celebrations of Life*. McGraw-Hill Book Company, New York.

Dudley, M. K. (1990) *Man, Gods, and Nature*. Na Kane O Ka Malo Press, Honolulu. pp.151.

Dullforce, W. (1990) EC suggests draft text of law on intellectual property. *Financial Times*, Mar. 7. from The Life Industry, Biodiversity, people and profits. (ed.) by Miges Bauman *et al.* Intermediate Technology Publications, London.

Dumont, Louis (1996) Process versus product in Kantu'augury: A traditional knowledge system's solution to the problem of knowing. In: Fukui, K. and Elolen, R.F. (eds), *Redefining Nature: Ecology, culture, domestication*. 557-596, Berg Publishers, Oxford.

Dumont, Louis. (1962) Reply to F. G. Bailey's 'Caste and Tribe in India'. In: *Contributions to Indian Sociology 6*.

Dundas, P. (1992) *The Jains*. Routledge, London and New York.

Durning, A. T. (1992) *Guardians of the Land: Indigenous Peoples and the Health of the Earth*. Worldwatch Paper 112. Worldwatch, Washington, D.C.

Durning, A. T. (1993) Supporting indigenous peoples. In: *State of the World 1993: A Worldwatch Institute Report on Progress Toward a Sustainable Society*. 80-100. Norton and Co., New York.

Durrenberger, E. and Palsson, G. (1986) Finding Fish: The Tactics of Icelandic Skippers. *American Ethnologist* 13(2), 213-229.

Dutfield, G. (1997) *Can the TRIPS Agreement Protect Biological and Cultural Diversity?* Biopolicy International Series No. 19, African Centre for Technology Studies, Nairobi, Kenya.

Dwivedi, O. P. (1994) *Environmental Ethics*. Sanchar Publishing House, New Delhi.

Earls, J. (1989) *Planificacion agricola Andina*. COFIDE, Lima.

Earth Prayers (1991) Harper, San Francisco.

Ecologist (1992) 'Whose Common Future', Vol.22, No.4 (July/Aug.) p.182

Edwards, J. (1985) *Language, Society and Identity*. Oxford University Press, Oxford.

Edwards, J. (1994) *Multilingualism*. Routledge, London/New York.

Edwards, J. and Palmer M. (eds) (1997) *Holy Ground, The Guide to Faith and Ecology*. Bilkington Press.

Eggeling, W. J. (1951) *The Indigenous Trees of the Uganda Protectorate* 255.

Ehrenfeld, D. (1988) Why put a value on biodiversity? In: Wilson, E. and Peter, F. (eds), *Biodiversity*. National Academy Press, Washington, D.C, pp. 212-216.

Ehrenfeld, D. (1993) *Beginning Again: People and nature in the new millenium*. Oxford University Press, New York.

Eide, A. (1995) Economic, social and cultural rights as human rights. In: Eide, A., Krause, C. and Rosas, A. (eds). *Economic, Social and Cultural Rights. A Textbook*. 21-40. Martinus Nijhoff Publishers, Dordrecht, Boston and London.

Eijbergen, H. C. van (1866) Verslag eener reis naar de Aroe-en Key eilanden. *Tijdschrift voor Indische Taal-, Land- en Volkenkunde*. 15(1): 220-361. Lange and Co, Batavia.

Einarsen, J. (1995) *The Sacred Mountains of Asia*, pp. 125-30. Shambala, Boston and London. [Originally published as *Kyoto Journal* No. 25, 1993.]

Eisenberg, D., Kessler, R. D., Foster, C., Norlock, F. E., Calkins, D. R. and Delbanco, T. L. (1993) Unconventional Medicine in the United States. *New England Journal of Medicine*, 328, 246-252.

Ekka, F. (1984) Status of minority languages in the schools of India. *International Education Journal* 1(1): 1-19.

Elanga (1995) Elanga, Fulani Aloni. Interviewed by author, 1 August 1995, Ebale, Zaire. *Tape recording.*

Elder, J. and Wong, H. D. (1994) *Family of Earth and Sky: Indigenous Tales of Nature from Around the World.* Beacon Press, Boston.

Eliot, G. (1871) *Middlemarch*. Penguin Books.

Elisabetsky, E. (1991) Sociopolitical, economic and ethical issues in medicinal plant research. *Journal of Ethnopharmacology* 32.

Elizabetsky, E. and Shanley, P. (1994) Ethnopharmacology in the Brazilian Amazon. *Pharmacology and Therapeutics* 64: 201-214.

Elkana, Y. (1981) A Programmatic Attempt at an Anthropology of Knowledge. In: Mendelsohn, R. and Elkana, Y. (eds) *Sciences and Cultures*, Reidel, pp. 1-76.

Ellen, R. F. (1979) Introductory Essay. In: Ellen, R. F. and Reason, D. A. (eds) *Classifications in their Social Context.* Academic Press, London.

Ellen, R. F. (1986) What Black Elk Left Unsaid: On the Illusory Images of Green Primitivism. *Anthropology Today* 2(6): 8-12.

Ellen, R. F. (1988) 'Foraging, starch extraction and the sedentary lifestyle in the lowland rainforest of central Seram.' In: *History, evolution and social change in hunting and gathering societies.* (eds) Woodburn, J., Ingold, T. and Riches, D. Berg, London:117-34.

Ellen, R. F. (1993) 'Rhetoric practice and incentive in the face of the changing times: a case study of Nuaulu attitudes to conservation and deforestation'. In: *Environmentalism: the view from anthropology*. (ed.) Milton, K. Routledge, London: 126-43.

Ellen, R. F. (in press) 'Forest knowledge, forest transformation: political contingency, historical ecology and the renegotiation of nature in central Seram'. *Transforming the Indonesian uplands*. (ed.) T. Li. (Studies in Environmental Anthropology, 2). Harwood, London.

Ellis, A. (1965) The Ewe-speaking peoples of the slave coast of West Africa. Benin Press, Chicago.

Ellis, A. B. (1965) *The Ewe speaking peoples of the slave coast of West Africa: their religion, manners, customs, laws, languages, etc.* Benin Press, Chicago.

Ellis, P. B. (1995) *Celtic Women: Women in Celtic Society and Literature*. Constable, London.

Ellul, J. (1980) *The Technological System*. Continuum, New York.

English Nature (1993) *Position Statement on Sustainable Development*. English Nature, Peterborough.

Erickson, C. L. and Chandler, K. L. (1989) 'Raised fields and sustainable agriculture in the lake Titicaca basin of Peru'. In: Browder, J. O. (ed.), *Fragile Lands of Latin America*. 230243. Westview Press, Boulder.

Escobar, A. (1995), *Encountering Development; The Making and the Unmaking of the Third World.* Princeton University Press, Princeton, New Jersey.

Escobar, A. (1996a) Constructing Nature: Elements for a Poststructural Political Ecology. In: Peet, R. and Watts, M. (eds) *Liberation Ecologies: Environment, Development, Social Movements.* Routledge, London and New York.

Escobar, A. (1996b) Conclusion: Towards a Theory of Liberation Ecology. In: Peet, R. and Watts, M. (eds) *Liberation Ecologies: Environment, Development, Social Movements.* Routledge, London and New York.

Etkin, N. (1997) Antimalarial plants used by Hausa in northern Nigeria. *Tropical Doctor* 27 (suppl.1), 12-16.

Evans-Pritchard, E. E. (1937) *Witchcraft, Oracles and Magic among the Azande*. Oxford.

Evans-Pritchard, E. E. (1937) *Witchcraft, Oracles and Magic among the Azande.* The Clarendon Press, Oxford.

Evers, H. D. (ed) (1988) *Teori Masyarakat. Proses Peradaban Dalam Sistem Dunia Modern.* Yayasan Obor Indonesia, Jakarta.

Evers, L. and Molina, F. S. (1987) *Maso Bwikam/ Yaqui Deer Songs: A Native American Poetry*. Sun Track and University of Arizona Press, Tucson.

Evliya Celebi. *19 Seyehatname*. 1:585-587. Zuhuri Danysman.

Eyzaguirre, P. and Myer, L. (1997) 'European Research on Human Aspects of Plant Genetic Resources'. *Newsletter for Europe*, 10:April:7. IPGRI Rome.

Fabrega, H. Jr. (1970) 'Dynamics of Medical Practice in a Folk Community'. *Millbank Memorial Fund Quarterly* 48(4): 391-412.

Fabrega, H. Jr. and Silver, D. B. (1970) 'Some Social and Psychological Properties of Zinacanteco Shamans'. *Behavioral Science* 15: 471-486.

Fabrega, H. Jr. and Silver, D. B. (1973) *Illness and Shamanistic Curing in Zinacantán.* Stanford University Press, Stanford, CA.

Fabrega, H. Jr., Metzger, D. and Williams, G. (1970) 'Psychiatric Implications of Health and Illness in a Maya Indian Group: A Preliminary Statement'. *Social Science and Medicine* 3: 609-626.

Fairclough, N. (1989) *Language and Power*. Longman, Harlow.

Fairhead, J. (1992) *Indigenous Technical Knowledge and Natural Resources Management in Sub-Saharan Africa.* Natural Resources Institute, Chatham.

Fairhead, J. and Leach, M. (1994) Declarations of Difference. In: Scoones, I. and Thompson, J. (eds) *Beyond Farmer First.* Intermediate Technology Publications, London.

Fairhead, J. and Leach, M. (1994) Représentations culturelles africaines et gestion de l'environnement. *Politique africaine*, 53: 11-24.

Fairhead, J. and Leach, M. (1996) Misreading the African Landscape: Society and ecology in a forest-savanna mosaic. Cambridge University Press, Cambridge.

Fairhead, J. and Leach, M. (1996) *Misreading the African Landscape: society and ecology ina forest-savanna mosaic.* CUP, Cambridge.

Falconer, J. (1990) *The Major Significance of 'Minor' Forest Products; The Local Use and Value of Forests in the West African Humid Forest Zone.* FAO Community Forestry Note No. 6, Rome.

Falconer, J. (1992) A study of the non-timber forest products of Ghana's forest zone. In: *The Rainforest Harvest: Sustainable Strategies for Saving the Tropical Forests?* Friends of the Earth, London.

Faracklas, N. (1993) From structural adjustment to land mobilization to expropriation. Is Melanesia the World Bank/International Monitory Fund's latest victim? (unpublished).

Feeny, D., Berkes, F., McCay, B. J. and Acheson, J. M. (1990) The tragedy of the commons: twenty-two years later. *Human Ecology* 18: 1-19.

Feit, H. A. (1973) The ethno-ecology of the Waswanipi Cree: or how hunters can manage their resources. In: Cox, B. (ed.), *Cultural Ecology*. 125-156. McClelland and Stewart, Toronto.

Feit, H. A. (1982) The future of hunters within nation-states: anthropology and the James Bay Cree. In: Leacock, E. and Lee, R. (eds), *Politics and History in Band Societies.* 373-412. Cambridge University Press, Cambridge.

Feit, H. A. (1991) The construction of Algonquian hunting territories: private property as moral lesson, policy advocacy, and ethnographic error. In: Stocking, G. W. Jr. (ed.), *Colonial Situations: Essays in the Contextualization of Ethnographic Knowledge*. University of Wisconsin Press, Madison, WI.

Feit, H. A. (1994) The enduring pursuit: land, time and social relationships in anthropological models of hunter-gatherers and in Sub-Arctic hunters' images. In: Burch, E. S. Jr. and Ellanna, L. J. (eds), *Key Issues in Hunter-Gatherer Research.* Berg Publishers, Oxford.

Feld, S. (1996) Waterfalls of Song: An Acoustemology of Place Resounding in Bosavi, Papua New Guinea. In: Feld, S. and Basso, K. (eds) *Senses of Place*. School of American Research Press, New Mexico.

Fernandes, W. and Enakshi, G. (1989) *Development, Displacement and Rehabilitation*. Indian Social Institute, p. 80.

Ferry, L. (1992) *The New Ecological Order.* University of Chicago Press, Chicago.

Fesl, E. M. D. (1993) *Conned! A Koorie Perspective.* University of Queensland Press, St Lucian, Queensland.

Fettes, M. (1998) Language planning and education. In: Wodak, R. and Corson, D. (eds) Language Policy and Political Issues in Education. *Encyclopedia of Language and Education,* Vol. 1. Kluwer Academic, Boston.

Field, A. J. (1984) Microeconomics, Norms, and Rationality. *Economic Development and Cultural Change,* Vol. 32: 683-711.

Field, D. R. (1996) Social science: a lesson in legitimacy, power and politics in land management agencies. In: Ewert, A. W. (ed.) *Natural Resource Management: The human dimension*. 249-256. Westview Press, Boulder, CO.

Fienup-Riordan, A. (1994) Eskimo war and peace. In: Fitzhugh, W. W. and Chaussonnet, V. (eds), *Anthropology of the North Pacific Rim.* Smithsonian Institution, Washington, D.C.

Finger, A. and Ghimire, K. (1997) Local Development and Parks in France. In: Ghimire, K. B. and Pimbert, M. P. (eds) *Social Change and Conservation,* UNRISD and Earthscan, London.

Fish, C. and Fish Gerald W. (1995) *The Kalenjin Heritage.* Africa Gospel Church, Kericho.

Fisher, S. (1994) *Women Play Key Role in Food Security*. USC Canada, Ottawa.

Fishman, J. A. (1982) Whorfianism of the third kind: Ethnolinguistic diversity as a worldwide societal asset. *Language in Society* 11: 1-14.

Fishman, J. A. (1989) *Language and Ethnicity in Minority Sociolinguistic Perspective*. Multilingual Matters, Clevedon, UK.

Fishman, J. A. (1991) *Reversing Language Shift. Theoretical and Empirical Foundations of Assistance to Threatened Languages*. Multilingual Matters, Clevedon, UK.

Fishman, J. A. (1996) What do you lose when you lose your language? In: Cantoni, G. (ed.), *Stabilizing Indigenous Languages*. 80-91. Center for Excellence in Education, Northern Arizona University, Flagstaff, AZ.

Flaaten, O. and Stollery, K. (1994) The economic effects of biological predation. Theory and application to the case of the north-east Atlantic minke whale's (*Balaenoptera acutorostrata*) consumption of fish. (Manuscript)

Flannery, T. F. (1996) *The Future Eaters: An Ecological History of the Australasian Lands and People*. Secker and Warburg, London.

Florey, M. J. and Wolff, X. Y. (1998) Incantations and herbal medicines: Alune ethnomedical knowledge in a context of change. *Journal of Ethnobiology*.

Floyd, C. N., D'Souza, E. J. and LeFroy, R. D. B. (1987) *Composting and crop production on volcanic ash soils in the Southern Highlands of Papua New Guinea.* Department of Primary Industry Technical Report 87/6. Port Moresby.

Follér, M-L. (1990) *Environmental Changes and Human Health: A Study among the Shipibo-Conibo in Eastern Peru.* Goteborg University, Goteborg.

Follér, M-L. (1993) Health, Healing and Illness among the Maya Population in Yucatán, Mexico. In: Hultberg, J. (ed) *New Genres in Science Studies: Papers from the 1992 Society for Social Studies of Science / European Association for the Study of Science and Technology Conference.* University of Göteborg, Goteborg. Report No 179, pp. 39-53.

Follér, M-L. (1997) 'Protecting Nature in Amazonia. Local Knowledge as a Counterpoint to Globalization'. In: Finn Arler and Ingeborg Svennevig (eds) *Cross-Cultural Protection of Nature and the Environment.* Odense University Press, pp. 134-147.

Follér, M-L. and Garret (1996) Modernization, Health and Local Knowledge: The Case of the Cholera Epidemic among the Shipibo-Conibo in Eastern Peru. In: Follér, M-L. and Hansson, L. O. (eds) *Human Ecology and Health. Adaptation to a Changing World.* University of Göteborg, pp. 135-166.

Forster, R. R. (1973) *Planning for Man and Nature in National Parks.* IUCN Publications, New Series No 26, Switzerland, Morges.

Fortmann, L. and Bruce, J. W. (eds) (1988) *Whose Trees? Proprietary Dimensions of Forestry.* Westview Press, Boulder.

Foster, G. M. and Anderson, B. G. (1978) *Medical Anthropology.* John Wiley & Sons, New York.

Foucault, M. (1970) *The Order of Things: An Archaeology of the Human Sciences.* English Translation. Tavistock, London.

Foucault, M. (1980) *Power/Knowledge: Selected Interviews and Other Writings, 1972-1977*, edited by Colin Gordon. Pantheon Books, New York.

Four Directions Council. (1996) *Forests, Indigenous Peoples and Biodiversity: Contribution of the Four Directions Council.* Draft paper submitted to the Secretariat of the Convention on Biological Diversity.

Fourmile, H. (1996) Protecting Indigenous Property Rights in Biodiversity. *Current Affairs Bulletin* February/March: 36-41.

Fourmile, H., Schnierer, S. and Smith, A. (eds) (1995) *An Identification of Problems and Potential for Future Rainforest Aboriginal Cultural Survival and Self-determination in the Wet Tropics.* Centre for Aboriginal and Torres Strait Islander Participation Research and Development, James Cook University of North Queensland, Townville.

Fowler, C. and Mooney, P. (1990) *Shattering: Food, Politics, and the Loss of Genetic Diversity.* University of Arizona Press, Tucson: p.104.

Fox, J. J. (1980) *The Flow of Life.* Harvard University Press.

Fox, R. L. (1981) 'Soils with variable charge: agronomic and fertility aspects'. In: Theng, B. K. G. (ed.) *Soils with variable charge.* New Zealand Society of Soil Science, Lower Hutt: 195-244.

Fox, W. (1990) *Toward a Transpersonal Ecology: Developing new foundations for environmentalism.* Shambala, Boston.

Francis, C. A. (1986) *Multiple cropping systems.* MacMillan, New York.

Freeman, J. D. (1960) Iban augury. In: Smythies, B.E. (ed.), *The Birds of Borneo.* 73-98, Oliver and Boyd, Edinburgh.

Freeman, M. M. R. (1997) A review of documents on small-type whaling, submitted to the international whaling commission by the Government of Japan, 1986-95. IWC Workshop on Japanese Community-based Whaling, IWC Document TC/M97/CBW75.

Freire, Paulo. (1972) *Pedagogy of the Oppressed.* Penguin, Harmondsworth.

Frickel, P. (1959) 'Agricultura dos índios Mundurukú'. *Boletim do Museu Paraense Emilio Goeldi,* n.s. No. 4.

Funtowizcz, S. and Ravetz, J. (1994) 'The Value of a Songbird: Ecological Economics as a Post-Normal Science' *Ecological Economics* 10:3 197-208

Furubotn, E. G. and Pejovich, S. (1972) 'Property Rights and Economic Theory: a survey of recent literature'. *Journal of Economic Literature,* 10, 1137-62.

Gadgil, M. (1985) Cultural Evolution of Ecological Prudence. *Landscape Planning* 12: 285-300.

Gadgil, M. (1987a) Diversity: Cultural and Biological. In: *TREE* Vol.2, No. 12, December. Elsevier Publications, Cambridge.

Gadgil, M. (1987b) Culture, Perceptions and Attitudes to the Environment. *Technical Report* No. 30, Centre for Ecological Sciences, Indian Institute of Science, Bangalore.

Gadgil, M. and Iyer, P. (1989) 'On the Diversification of Common Property Resource Use by the Indian Society' in Berkes, F. (ed), *Common Property Resources,* Belhaven Press, London.

Gadgil, M. (1993a) Biodiversity and India's Degraded Lands. *Ambio* 22: 167-172.

Gadgil, M. (1993b) Indigenous Knowledge for Biodiversity Conservation. *Ambio* 22: 151-156.

Gadgil, M. and Berkes, F. (1991) 'Traditional Resource Management Systems'. *Resource Management Optimization* 8:3-4:127-141.

Gadgil, M. and Chandran, S. (1992) Sacred Groves. *Indian International Quarterly* 19: 1-2: 183-188.

Gadgil, M. and Guha, R. (1992) *This Fissured Land: An Ecological History of India.* Oxford University Press, New Delhi.

Gadgil, M. and Iyer, P. (1989) On the Diversification of Common Property Resource Use by the Indian Society. In: Berkes, F. (ed) *Common Property Resources.* Belhaven Press, London.

Gadgil, M. and Thapar, R. (1990) Human ecology in India: some historical perspectives. *Interdisciplinary Science Reviews* 15: 209-223.

Gadgil, M. and Vartak, V. D. (1976) Sacred Groves of India: A Plea for Continued Conservation. *Journal of the Bombay Natural History Society* Vol. 72 No. 2.

Gadgil, M. and Vartak, V.D. (1981) Sacred groves of Maharashtra: an inventory. In: *Glimpses of Indian Ethnobotany,* pp.279-294.

Gadgil, M., Berkes, F. and Folke, C. (1993) Indigenous knowledge for biodiversity conservation. *Ambio* 22:151-156.

Galarrwuy Yunupingu, J. (1995) Quoted from Australian Catholic Social Justice Council's 'Recognition: The Way Forward', in *Native Title Report: January - June (1994)*. Aboriginal and Torres Strait Islander Social Justice Commissioner, Canberra, Australian Government Publishing Service.

Galtung, J. (1995) *Peace by Peaceful Means. Peace and Conflict, Development and Civilization.* International Peace Research Institute, Oslo, and Sage, London/ Thousand Oaks, New Delhi.

Garfinkel, H. (1964) *Studies in Ehnomethodology.* Englewood Cliffs, New York.

Geertz, C. (1965) Religion as a cultural system. In: Banton, M. (ed.), *Anthropological Approaches to the Study of Religion.* Tavistock Publications, London.

Geertz, C. (1973) *Interpretation of Cultures.* Basic Books, New York.

Geertz, C. (1983) *Local Knowledges.* Basic Books, New York.

Geertz, C. (1993) *Local Knowledge: Further Essays in Interpretive Anthropology.* Fontana, London.

Gellner, E. (1983) *Nations and Nationalism*. Blackwell, Oxford, UK and Cambridge, USA.

Gemmell, R. P. (1980) The Origin and Botanical Importance of Industrial Habitats. In: Bornkamm, R., Lee, J. A. and Seaward, M. R. D. (eds), *Urban Ecology*. Blackwell Scientific Publications, London, pp. 33-39.

Gene Campaign and Forum for Biotechnology and Food Security (1997) Draft Act Providing for the Establishment of Sovereign Rights over Biological Resources.

Gentry, A. H. (ed.) (1990) *Four Neotropical Forests*. Yale University Press, New Haven.

Gentry, A. H. and Vasquez, R. (1988) Where have all the *Ceibas* Gone? A Case History of Mismanagement of a Tropical Forest Resource. *Forest Ecology and Management* 23: 73-76.

Ghimire, K. (1994) Parks and people: livelihood issues in national parks management in Thailand and Madagascar. *Development and Change*, vol. 25, January, pp. 195-229.

Ghimire, K.B. and Pimbert, M. P. (1997) *Social Change and Conservation: Environmental Politics and Impacts of National Parks and Protected Areas*. Earthscan and UNRISD, London.

Ghotage, N. and Ramdas, S. (1997) Rural Communities as Protectors of Biological Diversity of Livestock Resources in India. Paper presented at the Second Congress on Traditional Sciences and Technologies, Madras, January 1996.

Giddens, A. (1984) *The Constitution of Society*. Polity Press, Cambridge, and University of California Press, Berkeley.

Giddens, A. (1987) *Sociology and Modern Social Theory*. Stanford University Press, in association with Polity Press and Blackwell, Oxford.

Giddens, A. and Turner, J. H. (eds) (1987) *Social Theory Today*. Stanford University Press, Stanford, CA.

Gifford, E. and Cook, R. (eds) (1992) *How Can One Sell the Air?* The Book Publishing Company, Summertown, Tennessee.

Gilbert, R., Stevenson, D., Girardet, H. and Stren, R. (1996) *Making Cities Work*. Earthscan, London.

Gill, S. D. and Sullivan, I. F. (1992) *Dictionary of Native American Mythology*. ABC-CLIO, Santa Barbara, CA.

Gillespie, J. and Shepherd, P. (1995) *Establishing for Identifying Critical Natural Capital in the Terrestrial Environment*.. English Nature, Peterborough.

Girardet, H. (1992 and 1996) *The Gaia Atlas of Cities*. Gaia Books, London.

Gliwitcz, J., Goszczynski, J. and Luniak, M. (1994) Characteristic Features of Animal Populations under Synurbanization – The Case of the Blackbird and of the Striped Field Mouse. In: G.M. Barker, M. Luniak, P. Trojan and H. Zimny (eds) Proceedings of the Second European Meeting of the International Network for Urban Ecology. *Memorabilia Zoologica* 49, Warsaw, pp. 237-244.

Glowka, L. and Burhenne-Guilmin, F. in collaboration with Synge, H., McNeely, J. A. and Gundling, L. (1994) *A Guide to the Convention on Biological Diversity*. Environmental Policy and Law Paper No. 30. Gland: IUCN.

Godoy, R. and Lubowski, R. (1992) Guidelines for the Economic Valuation of Nontimber Tropical-Forest Products. *Current Antrhopology* 33(4):4323-430.

GoJ (Government of Japan) (1989) Report to the working group on socio-economic implications of a zero catch limit. IWC Document IWC/41/21, published in *Papers on Japanese small-type coastal whaling submitted by the Government of Japan to the International Whaling Commission, 1986-1996*, 17-69. The Government of Japan, Tokyo.

GoJ (Government of Japan) (1991a) The cultural significance of everyday food use. IWC Document TC/43/SEST1, published in *Papers on Japanese small-type coastal whaling submitted by the Government of Japan to the International Whaling Commission, 1986-1996*, 195-201. The Government of Japan, Tokyo.

GoJ (Government of Japan) (1991b) Age difference in food preference with regard to whale meat. Report of a questionnaire survey in Oshika township. IWC Document TC/43/SEST4, published in *Papers on Japanese small-type coastal whaling submitted by the Government of Japan to the International Whaling Commission, 1986-1996*, 209-212. The Government of Japan, Tokyo.

GoJ (Government of Japan) (1992) The importance of everyday food use. IWC document IWC/44/SEST4, published in *Papers on Japanese small-type coastal whaling submitted by the Government of Japan to the International Whaling Commission, 1986-1996*, 223-228. The Government of Japan, Tokyo.

GoJ (Government of Japan) (1995) 'Action Plan for Japanese community-based whaling (CBW)'. IWC document IWC/47/SEST 1. Published in *Papers on Japanese small-type coastal whaling submitted by the Government of Japan to the International Whaling Commission 1986-1996*, 285-288. Government of Japan, Tokyo, 1997.

GoJ (Government of Japan) (1997) *Papers on Japanese small-type coastal whaling submitted by the Government of Japan to the International Whaling Commission, 1986-(1996)* The Government of Japan, Tokyo.

Golan, C. (1993) Agricultural risk management through diversity: field scattering in Cuyo Cuyo, Peru. *Culture and Agriculture* 45: 8-13.

Goldman, I. (1963) The Cubeo. Indians of the Northwest Amazon. *Illinois Studies in Anthropology Number 2*, University of Illinois Press.

Gomes, E. H. (1911) *Seventeen Years Among the Sea Dyaks of Borneo: A record of intimate association with the natives of the Borneana jungles*. Seeley and Co., London.

Gomez-Pompa, A. (1991) Learning from traditional ecological knowledge: insights from Mayan silviculture. In: Gomez-Pompa, A. Whitmore, T. C., and Hadley, M. (eds) *Rain Forest Regeneration and Management*. UNESCO and The Parthenon Publishing Group Limited, Paris.

Gomez-Pompa, A. and Burley, F. W. (1991) The management of natural tropical forests. In: Gomez-Pompa, A., Whitmore, T. C. and Hadley, M. (eds) *Rain Forest Regeneration and Management*. UNESCO and The Parthenon Publishing Group Limited, Paris.

Gomez-Pompa, A. and Kaus, A. (1990) Traditional Management of Tropical Forests in Mexico. In: Anderson, A. (ed.) *Alternatives to Deforestation: Steps Toward Sustainable Use of the Amazon Rain Forest.* Columbia University Press, New York.

Gomez-Pompa, A. and Kaus, A. (1992) Taming the Wilderness Myth. *Bioscience*, 42 (4): 271-279.

Gonzales, T. (1987) The Political Economy of Agricultural Research and Education in Peru, (1902-1980). MA thesis, Departments of Sociology and Rural Sociology, University of Wisconsin, Madison.

Gonzales, T. (1996) Political Ecology of Peasantry, the Seed, and NGOs in Latin America: A Study of Mexico and Peru, (1940-1995). PhD thesis, Department of Sociology, University of Wisconsin, Madison.

Goodenough, W. (1956) Componential Analysis and the Study of Meaning. *Language* 32.2:195-216.

Goodin, R. (1992) *Green Political Theory*. Polity Press, Cambridge.

Goodin, R. (1993) 'Selling Environmental Indulgences'. *Kyklos*, 47, 573-96.

Goodland, R. (1990) Environmental sustainability in economic development - with emphasis on Amazonia. In: Goodland, R. (ed.), *Race to Save the Tropics: Ecology and economics for a sustainable future.* 171-189. Island Press, Washington, D.C.

Goodpaster, K. (1978) On being morally considerable. *J. Philos.* 75: 308-325.

Gore, A. (1992) *Earth in the Balance: Ecology and the Human Spirit.* Penguin Books, New York.

Gose, P. (1994) *Deathly Waters and Hungry Mountains: Agrarian Ritual and Class Formation in an Andean Town.* University of Toronto Press.

Goudie, A. (1993) *The Human Impact on the Natural Environment.* (4th Edition). Blackwell, Oxford.

Goulding, Michael (1980) *The Fishes and The Forest: Explorations in Amazonian Natural History.* University of California Press, Berkeley, California.

Goulet, D. (1993) Biological diversity and ethical development. In: Hamilton, L. (ed.), *Ethics, Religion, and Biodiversity.* The White Horse Press, Cambridge.

Gowdy, J. M. (1997) The Value of Biodiversity: Markets, Society and Ecosystems. *Land Economics* 73 (1): 25-41.

GRAIN (1995) *Framework for a full articulation of farmers' rights.* Discussion paper, Genetic Resources Action International, Barcelona.

GRAIN. (1994) Biodiversity in Agriculture: Some Policy Issues. *IFOAM Ecology and Farming.* January: 14.

GRAIN. (1997) Biodiverse Farming Produces More. *Seedling* 14:3:6-14 October (1997).

Granberg-Michaelson, W. (1992) *Redeeming the Creation, The Rio Earth Summit challenge to the churches.* WCC Publications, Geneva.

Granoff, P. (1995) Jain stories inspiring renunciation. In: Lopez, D. (ed.), *Religions of India in Practice.* Princeton University Press, Princeton.

Grant, B. R. and Grant, P. R. (1989) *Evolutionary Dynamics of a Natural Population.* University of Chicago Press, Chicago.

Grapard, A. G. (1982) Flying Mountains and Walkers of Emptiness: Toward a Definition of Sacred Space in Japanese Religions.*History of Religions*, 21(3):195-221.

Gray, A. (1991) The Impact of Biodiversity Conservation on Indigenous Peoples. In: Shiva, V., Anderson, P., Schuking, H. Gray, A., Lohmann, L. and Cooper, D. (eds), *Biodiversity: Social and Ecological Perspectives .* Zed Books Ltd./World Rainforest Movement: London, UK and New Jersey, USA / Penang, Malaysia.

Gray, A. (1995) 'The Indigenous Movement in Asia', In: *Indigenous Peoples of Asia.* (eds) Barnes, R. H., Gray, A. and Kingsbury, B. Association for Asian Studies, Inc., Michigan:35-58.

Gray, A. (1996) *Mythology, Spirituality and History.* Berghahn Books, Oxford.

Gray, A. (1997a) *The Last Shaman: Change in an Amazonian Community.* Berghahn Books, Oxford.

Gray, A. (1997b) *Indigenous Rights, Development and Self-determination in an Amazonian Community.* Berghahn Books, Oxford.

Gray, F. (1995) Starving children. *The New Yorker*, p. 51.

Gray, G. G. (1993) *Wildlife and People: The Human Dimensions of Wildlife Ecology.* University of Illinois Press, Urbana, IL.

Gray, R. F. (1963) *The Sonjo of Tanganika: an anthropological study of an irrigation-based society.* Oxford University Press, New York.

Griaule, M. (1965) Conversations with Ogotemelli.

Griaule, M. and Dieterlen, G. (1991) *Le renard Pale. tome 1. Le mythe cosmogonique.* Paris: Travaux et Memoires de l'Institut d'Ethnologies 72 (2nd edition, 1991).

Grigg, D. B. (1974) *The Agricultural Systems Of The World.* Cambridge University Press, London.

Grillo, E. (1991) La Cosmovisión Andina de Siempre y la Cosmologla Occidental Moderna, *Serie: Documentos de Estudio* N-21, Abril, PRATEC, Lima, Peru.

Grillo, E. (1993) Afirmación Cultural: Digestión del Imperialismo en los Andes (manuscript).

Grillo, E. (1994) Cultural Affirmation: Digestion of Imperialism in the Andes (manuscript).

Grimes, A., Loomis, S., Jahnige, P., Burnham, M., Onthank, K., Alarcon, R., Palacios, W., Ceron, C., Neill, D., Balick, M., Bennett, B. and Mendelssohn, R. (1993) *Valuing the Rain Forest: The Economic Value of Non-Timber Forest Products in Ecuador.* Yale School of Forestry and Environmental Studies and The New York Botanical Garden.

Grimes, B. (ed.) (1992) *Ethnologue: Languages of the World.* 12th ed. Summer Institute of Linguistics, Dallas.

Grimes, B. (ed.) (1996) *Ethnologue: Languages of the World.* 13th ed. Available on the World Wide Web @ http://www.sil.org/ethnologue/. Summer Institute of Linguistics, Dallas.

Grinde, D. A. and Johansen, B. E. (1995) *Ecocide of Native America.* Clear Light, Santa Fe.

Groombridge, B. (ed.) (1992) *Global Biodiversity: Status of the Earth's Living Resources.* World Conservation Monitoring Centre, Chapman and Hall, London.

Grossman, L. S. (1977) Man-environment relationships in anthropology and geography. *Annals of the Association of American Geographers*: 67 (1): 126-144.

Grossman, L. S. (1993) The political ecology of banana exports and local food production in St. Vincent, Eastern Caribbean. *Annals of the Association of American Geographers* 83(2): 347-367.

Grove, R. (1995) *Green Imperialism: Colonial Expansion, Tropical Island Edens and the Origins of Environmentalism, 1600-1860*. Cambridge University Press, Cambridge.

Grove, R. (1996) Indigenous Knowledge and the Significance of South-west India for Portuguese and Dutch Constructions of Tropical Nature. *Modern Asian Studies* 30(1) 121-143.

Guer. (1846). *Moeurs et Usages des Turcs*. 369.

Guertin, G. (1990) 'Incidences sur l'environnement du projet d'aménagement hydro-électrique de la Baie James'. In: *The impact of large water projects on the environment – Proceedings of an International Symposium convened by Unesco and UNEP.* UNESCO, Paris.

Guha, R. (1996) Two phases of American environmentalism: a critical history. In: Apffel-Marglin and S. Marglin (eds). *Decolonizing Knowledge*. 110-141.

Guha, R. (1997) The Authoritarian biologist and the arrogance of anti-humanism. *The Ecologist* 27, 1.

Guha, Ramachandra. (1989) *The Unquiet Woods. Ecological Change and Peasant Resistance in the Himalaya.* Oxford University Press, Delhi.

Guidelines (1992) *Guidelines – Environmental Impact Statement for the Proposed Great Whale River Hydroelectric Project, 1992.* Great Whale Public Review Support Office, Montreal.

Guinko, S. (1984) *Contribution a l'etude de la vegetation et de la flore du Burkina Faso, IV - La vegetation des Termitieres 'Cathedrales'.* Notes et documents Burkinabe 15, 4, 1-11

Guinko, S. (1984) Contribution a l'étude de la végétation et de la flore du Burkina Faso, IV ñ La végétation des Termitières Cathedralesí. *Notes et Documents Burkinabe* 15, 4, 1-11.

Guiteras-Holmes, C. (1961) *Perils of the Soul.* The Free Press of Glencoe, New York.

Gulati, B. and Sinha, G. K. (1989) Studies on some important species of Ocimum. *Proceedings 11th International Congress of Essential Oils, Fragrances, Flavours*. 4,197-206.

Gulland, J. (1988) The end of whaling? *New Scientist* October 29: 42-47.

Gupta, A. (1992) *Building Upon People's Ecological Knowledge: Framework for Studying Culturally Embedded CPR institutions*. Indian Institute of Management, Centre for Management in Agriculture, Ahmedabad.

Gupta, A. K. (1980) *Communicating with Farmers - Cases in Agricultural Communication and Institutional Support Measures,* IIPA, New Delhi, (1980)

Gupta, A. K. (1981) A Note on Internal Resource Management in Arid Regions: Small Farmers Credit Constraints: A Paradigm. *Agricultural Systems (UK),* Vol.7 (4) 157-161

Gupta, A. K. (1987) Role of Women in Risk Adjustment in Drought Prone Regions with Special Reference to Credit Problems. October 1987, IIM Working Paper No. 704.

Gupta, A. K. (1989) *Managing Ecological Diversity, Simultaneity, Complexity and Change: An Ecological Perspective*. W.P.No. 825. IIM Ahmedabad. P 115, 1989, Third survey on Public Administration, Indian Council of Social Science Research, New Delhi.

Gupta, A. K. (1996a) *Getting Creative Individuals and Communities their Due: Framework for Operationalizing Articles 8(j) and Article 10(c)*. Submission to the Secretariat for the Convention on Biological Diversity.

Gupta, A. K. (1996b) Technologies, institutions and incentives for conservation of biodiversity in non-OECD countries: assessing needs for technical cooperation. Presented at the OECD *Conference on Biodiversity Conservation Incentive Measures*. Cairns, March 1996.

Gupta, M. S. (1971) *Plant myths and traditions in India.* E. J Brill, Leiden.

Gurugita. (1991) SYDA Foundation, South Fallsburg, New York.

Gutmanis, J. (1995) *Kahuna La 'au Lapa'au.* Island Heritage Books, Aiea, Hawaii. pp.144.

Gwich'in Words about The Land. 250 pp. Gwich'in Geographics Ltd., Inuvik.

Gyekye, Kwame. (1987) *An essay on African philosophical thought: The Akan conceptual theme*. Cambridge University Press, Cambridge.

Haarmann, H. (1986) *Language in Ethnicity: A View of Basic Ecological Relations*. Mouton de Gruyter, Berlin/New York/Amsterdam.

Habermas, J. (1987) *The Philosophical Discourses of Modernity: Twelve Lectures.* Polity/Blackwell, Cambridge.

Hackel, J. D. (1993) Rural Change and Nature Conservation in Africa: A Case Study from Swaziland. *Human Ecology,* 21:295-312.

Hails, C. J. (1992) Improving the Quality of Life in Singapore by Creating and Conserving Wildlife Habitats. In: Chua Beng Huat and Edwards, N. (eds) *Public Space: Design, Use and Management.* Singapore University Press, Singapore, pp. 138-154.

Haimendorf, F.von (1985) *Tribal Populations and Cultures of the Indian Sub-Continent.* E.J. Brill, Leiden/Köln.

Hale, K. (1992) On endangered languages and the safeguarding of diversity. *Language* 68(1): 1-3.

Hall, R. L. (1986) Alcohol treatment in American Indian populations: An indigenous treatment modality compared with traditional approaches. *Annals of the New York Academy of Science*, 472, 168-178.

Hallman, D. (1994) (ed.) *Ecotheology: Voices from south and north*. World Council of Churches, Geneva, and Orbis Books, Maryknoll.

Hallowell, A. I. (1943) The Nature and Function of Property as a Social Institution. *Journal of Legal and Political Sociology,* 1, 115-38.

Hambly, W. D. (1931) *Serpent worship in Africa.* Field Museum of Natural History, Chicago, Publication 289 Anthropological Series vol 21, 1.

Hamel, R. E. (1994a) Linguistic rights for Amerindian peoples in Latin America. In: Skutnabb-Kangas, T. and Phillipson, R. (eds), *Linguistic Human Rights. Overcoming Linguistic Discrimination*. 289-303. Mouton de Gruyter, Berlin and New York.

Hamel, R. E. (1994b) Indigenous education in Latin America: Policies and legal frameworks. In: Skutnabb-Kangas, T. and Phillipson, R. (eds), *Linguistic Human Rights. Overcoming Linguistic Discrimination*. 271-287. Mouton de Gruyter, Berlin and New York.

Hames, R. (1991) Wildlife conservation in tribal societies. In: Oldfield, M. L. and Alcorn, J. B. (eds) *Biodiversity: Culture, Conservation, and Ecodevelopment*. Westview Press, Boulder, San Francisco and Oxford, pp.172-199.

Hamilton, A. (1982) Descended from father, belonging to country: rights to land in the Australian Western Desert. In: Leacock, E. and Lee, R. (eds). *Politics and History in Band Societies*. 85-108. Cambridge University Press, Cambridge.

Hamilton, L. (ed.) (1993) *Ethics, Religion, and Biodiversity: Relations between conservation and cultural value*. The White Horse Press, Cambridge.

Hammond, P. (ed.) (1985) *The Sacred in a Secular Age: Toward revision in the scientific study of religion*. University of California Press, Berkeley.

Hanefi (1993) 'Urai kerei di pulau Siberut Mentawai'. In: *Seni Pertunjukan Indonesia*, pp. 139-157. Grasindo, Jakarta.

Hanes, R. (1995) *American Indian Interests in the Northern Intermontane of Western North America*. Draft 2/95. Interior Columbia Basin Ecosystem Management Project Office, Walla Walla, Washington.

Harborne, B. J. and Baxter, H. (eds) (1993) *Phytochemical dictionary*. Taylor and Francis, London, Washington, D.C.'

Hardesty, D. L. (1977) *Ecological Anthropology*. Wiley & Sons, New York.

Hardin, G. (1968) The Tragedy of the Commons. *Science*, 162: 1243-44.

Harman, R. C. (1974) *Cambios Midicos y Sociales en una Comunidad Maya-Tzeltal*. Mixico, D. F. Instituto Nacional Indigenista.

Harmon, D. (1995) The status of the world's languages as reported in *Ethnologue*. *Southwest Journal of Linguistics* 14: 1-33.

Harmon, D. (1996a) Losing species, losing languages: Connections between biological and linguistic diversity. *Southwest Journal of Linguistics*.

Harmon, D. (1996b) The converging extinction crisis: Defining terms and understanding trends in the loss of biological and cultural diversity. Paper presented at the colloquium 'Losing Species, Languages, and Stories: Linking Cultural and Environmental Change in the Binational Southwest', Arizona-Sonora Desert Museum, Tucson, AZ, April 1-3, 1996.

Harrington, G. N. and Sanderson, K. D. (1994) Recent Contraction of Wet Sclerophyll Forest in the Wet Tropics of Queensland due to Invasion by Rainforest. *Pacific Conservation Biology* 1:319-327.

Harris, H. (1996) A Critical Analysis of the Concept of 'Indigenous Knowledge', within current Development Discourse. MA thesis, University of Kent, Canterbury.

Harris, M. (1968) *The Rise of Anthropological Theory*. Harper and Row, New York.

Harris, S. (1990) *Two-way Aboriginal Schooling. Education and Cultural Survival*. Aboriginal Studies Press, Canberra.

Harris, W. V. (1971) (2nd), *Termites: their recognition and control*. Longman, London.

Harrison, P. (1990) *The Greening of Africa: Breaking Through the Battle for Land and Food*. IIED/Earthscan, Paladin Crafton Books, London.

Harrison, R. P. (1992) *Forests: The Shadow of Civilization*. The University of Chicago Press, Chicago.

Harrison, T. (1960) Men and birds in Borneo. In: Smythies, B. E. (ed.) *The Birds of Borneo. 20-61*, Oliver and Boyd, Edinburgh.

Harting, J. E. (1972) *British Animals Extinct Within Historic Times with Some Account of British Wild White Cattle*. Paul Minet, Chicheley.

Hartshorn, G. (1990) Natural Forest Management by the Yanesha Forestry Cooperative. In: Anderson, A. (ed.) (1990) *Alternatives to Deforestation: Steps Toward Sustainable Use of the Amazon Rain Forest*. Columbia University Press, New York.

Harwood, R. R. (1979) *Small Farm Development*. Westview Press, Boulder.

Hassanpour, A. (1992) *Nationalism and Language in Kurdistan (1918-1985)*. The Edwin Mellen Press, New York.

Haugen, Einar (1972) *The Ecology of Language*. Stanford University Press, Stanford.

Hauser, P. (1978) *l'Action des Termites en Milieu de Savane Sèche*. Cahiers O. R. S.T.O.M. sér. Sciences Humaines 15, 1, 35-49.

Haverkort, B. (1995) Agricultural Development with a Focus on Local Resources: ILEIA's View on Local Knowledge. In: Warren, D. M., Slikkerveer, L. J. and Brokensha, D. (eds) (1995) *The Cultural Dimension of Development: Indigenous Knowledge Systems*. Intermediate Technology Publications, London, pp. 454-457.

Haverkort, B. and Millar, D. (1994) Constructing Diversity: The Active Role of Rural People in Maintaining and Enhancing Biodiversity. *Ethnoecologica* 2(3): 51-64.

Hawken, P. (1993) *The Ecology of Commerce*. Harper Collins, New York.

Hawksworth, D. L. (1990) The Long-term Effects of Air Pollution on Lichen Communities in Europe and North America. In: Woodwell, G. M. (ed.) *The Earth in Transition*. Cambridge University Press, Cambridge, pp. 45-64.

Hays, T. E. (1982) Utilitarian/adaptationist explanations of folk biological classification: Some cautionary notes. *Journal of Ethnobiology* 2: 89-94.

Head, L. (1992) 'Australian Aborigines and a Changing Environment – Views of the Past and Implications for the Future.' In: *Aboriginal Involvement In Parks and Protected Areas*, (eds) Birckhead, J. ,Lacy, T. De and Smith, L. Aboriginal Studies Press, Canberra, 47-56.

Heaney, S. (1984) *Sweeney Astray*. Faber and Faber, London.

Hecht, S. and Cockburn, A. (1989) *The Fate of the Forest; Developers, Destroyers and Defenders of the Amazon.* Verso Press, London.

Hecht, S. and Posey, D.A. (1989) 'Preliminary Results on Soil Management Techniques of the Kayapó Indians' *Advances in Economic Botany* 7:174-188.

Hecht, S., Thrupp, L. A. and Browder, J. (1996) 'Diversity and Dynamics of Shifting Cultivation: Myths, Realities, and Human Dimensions.' Draft paper. World Resources Institute, Washington, D.C.

Heilbroner, R. L. (1992) *The Worldly Philosophers*. Simon and Schuster, New York.

Heine-Geldern, R. (1956) *Conceptions of State and Kingship in Southeast Asia*. Data Papers 13, Ithaca, NY, Southeast Asia Program, Dept. of Far Eastern Studies, Cornell University.

Heinrich, B. (1988) Food Sharing in the Raven, *Corvus Corax*. In: Slobodchikoff, C. N. (ed.) *The Ecology of Social Behavior*. Academic Press, New York, pp. 285-311.

Hemming, J. (1978) (1995) *Red Gold: The Conquest of the Brazilian Indians.* Papermac, London.

Henkel, H. and Stirrat, R. L. Undated paper. 'Participation as Spiritual Duty; the Religious Roots of the New Development Orthodoxy', presented at The University of Edinburgh, 50 years of Anthropology celebrations, (1996)

Henthorn, W. E. (1971) *A History of Korea*. Free Press, New York.

Herman, E. and Chomsky, N. (1988) *Manufacturing Consent: the political economy of the mass media*. Pantheon Books, New York.

Herskovits, M. (1964) *Cultural Relativism: Perspectives in cultural pluralism.* Random House, New York.

Hessel, D. (1992) *After Nature's Revolt: Eco-justice and theology.* Fortress Press, Minneapolis.

Hessel, D. (1996) Ecumenical ethics for earth community. *Theology and Public Policy* 8: 1, 2.

Hessel, D. (1996) *Theology for Earth Community: A field guide.* Orbis Books, Maryknoll, NY.

Hettinger, E. C. (1989) Justifying Intellectual Property. *Philosophy and Public Affairs* 18, 31-52.

Heywood, V. H. (executive ed.) (1995) *Global Biodiversity Assessment.* United Nations Environment Programme and Cambridge University Press, Cambridge.

Hildebrand, Martin von, and Pachón, J. (1987)*Introducción a Colombia Amerindia*. Instituto Colombiano de Antropología, Bogotá.

Hildebrand, Martin von, and Reichel, E. 'Indigenas del Miriti-paraná' (eds) F. Correa

Hildebrand, Martin von. (1987) 'Datos etnográficos sobre la Astronomia de los indigenas Tanimuka' J. Arias and E. Reichel, Editors Etnoastronomias Americanas, Universidad Nacional de Colombia, Bogotá.

Hill, J. H. (1995) The loss of structural differentiation in obsolescent languages. Paper presented at the symposium 'Endangered Languages', Annual Meeting of the American Association for the Advancement of Science, Atlanta, GA, 18 February 1995.

Hill, J. H. (1996) 'Dimensions of attrition in language death'. Paper presented at the working conference 'Endangered Languages, Endangered Knowledge, Endangered Environments', Berkeley, California, October 25-27, 1996.

Hill, J. H. (1997) The meaning of linguistic diversity: Knowable or unknowable? *Anthropology Newsletter* 38(1): 9-10.

Hill, J. H. (1998) Dimensions of attrition in language death. In: Maffi, L. (ed.) Language, Knowledge, and the Environment: The Interdependence of Biological and Cultural Diversity. Submitted to Smithsonian Institution Press.

Hill, S. Y., Krattiger, F. A., Lesser, L. W., McNeely, J. A., Miller, K. R. and Senanayake, R. (eds) (1994) *Widening Perspectives on Biodiversity.* IUCN, Gland, Switzerland and International Academy of the Environment, Geneva, Switzerland.

Hillman, J. (1988) Cosmology for Soul: From universe to cosmos. In: *Cosmos Life Religion: Beyond humanism.* Tenri University Press, Tenri.

Hinton, L. (1994) *Flutes of Fire: Essays on California Indian Languages*. Heyday Books, Berkeley, CA.

Hinton, L. (1994/1995) Preserving the future: A progress report on the Master-Apprentice Language Learning Program. *News from Native California* 8(3): 14-20.

Hinton, L. (1996) Language revitalization and environmental philosophy. Paper presented at the working conference 'Endangered Languages, Endangered Knowledge, Endangered Environments', Berkeley, California, October 25-27, 1996.

Hladik, C. M., Hladik, A., Linares, O. F., Pagezy, H., Semple, A. and Hadley, M. (1993) *Tropical Forests, People & Food.*

HM Government (1990) *This Common Inheritance*. Cm.1200 HMSO, London.

HM Government (1994) *Biodiversity Action Plan*. Cm. 2428 HMSO, London.

Hobart, M. (1993) Introduction. In: Hobart, M. (ed.) *The Growth of Ignorance? An Anthropological Critique of Development.* Routledge, London, pp. 1-30.

Hobsbawm, E. J. (1991) *Nations and Nationalism Since 1780: Programme, Myth, Reality.* Cambridge University Press, Cambridge.

Hole, R. (1989) *Pulpits, Politics and Public Order in England 1760-1832.* Cambridge University Press, Cambridge.

Holland, A. and O'Neill, J. (forthcoming) The ecological integrity of nature over time: some problems. *Global Bioethics*.

Holland, A. and Rawles, K. (1994) *The Ethics of Conservation*. Report presented to the Countryside Council of Wales. Thingmount Series, No. 1, Lancaster University: Department of Philosophy.

Holland, W. R. (1963) *Medicina Maya en Los Altos de Chiapas.* Instituto Nacional Indigenista, Mexico City.

Holland, W. R. and Tharp, R. G. (1964) 'Highland Maya Psychotherapy'. *American Anthropologist* 66: 41-59.

Holling, C. S. (1986) The resilience of terrestrial ecosystems: local surprise and global change. In: Clark, W. C. and Munn, R. E. (eds), *Sustainable Development of the Biosphere*. International Institute for Applied Systems Analysis, and Cambridge University Press, Cambridge.

Holling, C. S., Taylor, P. and Thompson, M. (1991) From Newton's sleep to Blake's fourfold vision: Why the climax community and the rational bureaucracy are not the ends of the ecological and social-cultural roads. *Annals of the Earth* 9(3) 19-21.

Honey Bee, c/o Professor Anil Gupta, Indian Institute of Management, Vastrapur, Ahmedabad 380 015, India. Email: honeybee@iimahd.ernet.in

Honoré, A. M. (1961) 'Ownership'. In: Guest, A. G. (ed.) *Oxford Essays in Jurisprudence*, Oxford University Press, Oxford.

Hornborg, A. (1996) Ecology as semiotics: outlines of a contextualist paradigm for human ecology. In: Descola, P. and Palsson, G. (eds), *Nature and Society — Anthropological perspectives*. Routledge, London.

Horsfall, N. and Fuary, M. (1988) The Cultural Heritage Values of Aboriginal Archaeological Sites and Associated Themes in and Adjacent to the Area Nominated for World Heritage Listing in the Wet Tropics Rainforest Region of North East Queensland. Unpublished Report to the State of Queensland, 12th December 1988, Permit No. 31. EIS.84.

Horsfall, N. and Hall, J. (1990) People and the Rainforest: An Archaeological Perspective. In: Webb, L. J. and Kikkawa, J. (eds) *Australian Tropical Rainforests*. CSIRO, Melbourne, pp. 33-39.

Hose, C. and McDougall, W. (1902) The relations between men and animals in Sarawak. *Journal of the Anthopological Institute* 31: 173-213.

Hostetler, J. A. (1993) *Amish Society*. The John Hopkins University Press, Baltimore, MD, 4th ed., 435pp.

Houfe, S. (1978) *The Dictionary of British Book Illustrators and Caricaturists 1800-1914*. Baron Publishing, Woodbridge, Suffolk.

Houis, M. (1958) Conte Kiniagui: la fille orgueilleuse. *Notes Africaines* 1958, 112-114

Hoy, D. C. (1986) Power, Repression, Progress: Foucault, Lukes and the Frankfurt School. In: Hoy, D. C. (ed.) *Foucault: A Critical Reader.* Basil Blackwell, Oxford, pp. 123-44.

Hugh-Jones, C. (1979) *From the Milk River. Spatial and Temporal Processes in Northwest Amazonia*. Cambridge University Press, Cambridge.

Hughes, C. C. (1968) Ethnomedicine. In: *International Encyclopedia of the Social Sciences*. New York, Macmillan, pp. 87-93.

Hunn, E. (1980) 'Sahaptin Fish Classification' *Northwest Anthropological Research Notes* 4 (1):1-19.

Hunn, E. (1982) Mobility as a Factor Limiting Resource Use in the Columbia Plateau of North America. In: *Resource Managers: North American and Australian Hunter-Gatherers*. (eds) Williams, N. and Hunn, E. Westview Press, Boulder.

Hunn, E. (1993) What is Traditional Ecological Knowledge? In: Williams, N. and Baines, G. (eds) *Traditional Ecological Knowledge: Wisdom for Sustainable Development*. Centre for Resource and Environmental Studies, ANU, Canberra, pp. 13-15.

Hunn, E. and Selam, J. (1990) *Nch'i-Wana, "The Big River": Mid-Columbia Indians and Their Land*. University of Washington Press, Seattle.

Hunn, E. S. (1977) *Tzeltal Folk Zoology: The Classification of Discontinuities in Nature*. Academic Press, New York.

Hunn, E. S. (1982) The utilitarian factor in folk biological classification. *American Anthropologist* 84: 830-847.

Hunn, E. S. (1996) Columbia Plateau place names: What can they teach us? *Journal of Linguistic Anthropology* 6(1): 3-26.

Hunn, E. S. (1999). Size as limiting the recognition of biodiversity in folk biological classifications: One of four factors governing the cultural recognition of biological taxa. In: Medin, D. L. and Atran, S. (eds), *Folkbiology*. MIT Press, Cambridge, Mass.

Hunt, A. (1992) 'Foucault's Expulsion of Law: Towards a Retrieval', *Law and Society Inquiry*. 17, No. 1: 1-62.

Huntington, G. E. (1956) 'Dove at the Window: a study of an Old Order Amish community in Ohio'. Ph.D. dissertation, Yale University, 1212 pp.

Huntington, H. P. (1992) *Wildlife Management and Subsistence Hunting in Alaska*. University of Washington Press, Seattle.

Huntington, S. P. (1996) *The Clash of Civilisations And The Remaking Of World Order*. Simon and Schuster, New York.

Hussein, Mian. (1994) 'Regional Focus News - Bangladesh'. *Ecology and Farming: Global Monitor, IFOAM*. January: 20.

Huxley, L. (ed) (1900) *Life and Letters of Thomas Henry Huxley*, 2 vols. Macmilland and Co., London, pp 267-268.

Huxley, T. H. Collection, Imperial College, London, vol 10, ff 220-21. For a similar review of 'insect socialism' see Florence Fenwick Miller, 'Insect Communists', *The National Review*, 15 (May 1890) 392-403.

Hvalkof, S. (1989) 'The Nature of Development: Native and Settler Views in Gran Pajonal, Peruvian Amazon'. *Folk* (31):125-150.

Hydro-Quebec (1990) Hydro-Québec and the Environment - Proposed Hydro-Québec Development Plan 1990-1992 - Horizon 1999. Hydro-Québec, Montréal.

Ibo, J. G. (1992) L'expérience historique de protection de la nature en milieu traditionnel africain: le cas de la société Abron-Koulango. *Compte-rendu du séminaire sur l'aménagement intégré des forêts denses humides et des zones agricoles périphériques*. Wooren Séries Tropenbos 1:180-189.

Ibo, J. (1996) Bibliographie sur les méthodes traditionnelles de gestion de la biodiversité en Côte d'Ivoire. Document interne du WWF WARPO: 16p - Toffin, G. (1987) De la nature au surnaturel. *Etudes rurales*, n° 107-108: 9-26.

IFOAM. (1994) 'Biodiversity: Crop Resources at Risk in Africa'. *Ecology and Farming – Global Monitor.* January: 5.

Ignatieva, M. E. (1994) Investigation of the Flora of St. Petersburg's Green Areas. In: Barker, G. M., Luniak, M., Trojan, P. and Zimny, H. (eds) Proceedings of the Second European Meeting of the International Network for Urban Ecology. Warsaw, *Memorabilia Zoologica* 49:139-142.

II(ED.) (1995) *Hidden Harvests Project Overview*. Sustainable Agriculture Programme. London.

IIED (1995) *Whose Eden? An Overview of Community Approaches to Wildlife Management.* London.

IIPA (1989) *Management of National Parks and Sanctuaries in India: A Status Report.* Indian Institute of Public Administration, New Delhi, 297 pp.

IIPA (1993) Conservation of Biological Diversity in India (draft).

Illich, I. (1981) Taught mother tongue and vernacular tongue. In: Pattanayak, D. P. *Multilingualism and Mother-Tongue Education.* 1-46. Oxford University Press, Delhi.

ILO (International Labour Organization) (1989) ILO Convention 169.

Indigenous Knowledge and Development Monitor (1995-1996), CIRAN/NUFFIC, The Hague.

Indigenous Peoples' Biodiversity Network (1996) Indigenous peoples, indigenous knowledge and innovations and the Convention on Biological Diversity. Conference of the Parties to the Convention on Biological Diversity, 3rd Meeting, Buenos Aires, 4–15 November, 1996 (IPBN/CBD/COP3/Nov. 5, 1996).

Indrawan, M. and Wirakusumah , S. (1995) Jakarta Urban Forest as Bird Habitat: A Conservation View. *Tiger Paper* 22(1):29-32.

Inglis, T. (ed.) (1993) *Traditional Ecological Knowledge: Concepts and Cases*, IDRC, Ottawa, 142 pp.

Ingold, T. (1993) 'Preface to the Paperback Edition', in *What is an Animal?*, op. cit. (3), pp. xviii-xxiv, on xx and xxiv.

Ingold, T., Riches, D. and Woodburn, J. (eds) (1988) *Hunters and Gatherers 1: History, Evolution and Social Change*; and *Hunters and Gatherers 2: Property, Power and Ideology*. Berg Publishers, Oxford.

Innis, D. (1997) *Intercropping and the Scientific Basis of Traditional Agriculture*. Intermediate Technology Publications, London.

International Alliance of the Indigenous–Tribal Peoples of the Tropical Forests (1992) *Charter of the Indigenous–Tribal Peoples of the Tropical Forests.* IAI-TPTF, Penang.

International Alliance of the Indigenous–Tribal Peoples of the Tropical Forests (1996) *Indigenous Peoples, Forests and Biodiversity.* IWGIA, Copenhagen.

International Labour Organization (ILO) (1989) *Convention Concerning Indigenous and Tribal Peoples in Independent Countries.* International Labour Office, Geneva.

International Labour Organization Convention 169 (1989) Preamble and Art. 13-1).

International NGO Consultation on the Mountain Agenda, Lima, Peru, April 1995. *Summary Report and Recommendations to the United Nations Commission on Sustainable Development* (copies available from The Mountain Institute, Franklin, WV).

IPR (no date) *Agriculture and Trade Policy*. No. 6 and 7, 1315 5th Forest Suits, Minneapolis, Minnesota.

IPS (Inter-Press Service) (1997) *Environment - U.N.: Leaders find little progress*. 24 June, 20:14 p(ed.)

Iroko, A. F. (1982) Le rule des termitières dans l'histoire des peuples de la République Populaire du Bénin des origines * nos jours. *Bulletin de l'I. F. A. N.* 44 (B) 1-2, 1982, 50-75

Iroko, A. F. (1994) *Une histoire des hommes et des moustiques en Afrique: Côte des Esclaves* (XVIe-XIX siècles) Harmattan, Paris.

Iroko, A. F. (1994) *Une histoire des hommes et des moustiques en Afrique: Côte des Esclaves (XVIe-XIX siècles.* Harmattan, Paris.

Irvine, D. (1989) Succession management and resource distribution in an Amazonian rain forest. In: Posey, D. A. and Balee, W. (eds) *Resource management in Amazonia: indigenous and folk strategies. Advances in Economic Botany* 7:223-237.

Isyk, K. (1990) *Ormanlar ve milli parklar.* In: Kence, Aykut (ed.) *Turkiye'nin Biyolojik Zenginlikleri.* 95. Turkiye Cevre sorunlary Vakfy Yayyny.

ITFF (1997) Department for Policy Coordination and Sustainable Development (DPCSD) *Interagency Partnership on Forests: Implementation of IPF Proposals for Action by the ITFF,* June (1997) United Nations, New York.

IUCN (1990) *United Nations List of National Parks and Protected Areas*. IUCN, Gland, Switzerland. 275 pp.

IUCN (1994) *Guidelines for Protected Area Management Categories*. IUCN, WCMC, Gland, Switzerland and Cambridge, UK, 261 pp.

IUCN (1994) *Report of the Global Biodiversity Forum.* IUCN, Gland.

IUCN Inter-Commission Task Force on Indigenous Peoples (1997) *Indigenous peoples and sustainability: Cases and actions*.International Books, Utrecht.

IUCN/UNEP/WWF (1980) *World Conservation Strategy*. IUCN, Gland, Switzerland.

IUCN/UNEP/WWF (1991) *Caring for the Earth: A Strategy for Sustainable Living.* Gland, Switzerland.

IWC (International Whaling Commission) (1981) Report on the ad hoc technical committee working group on development of management principles and guidelines for subsistence catches of whales by indigenous (aboriginal) peoples. IWC/33/14.

IWGIA (International Work Group for Indigenous Affairs) (1995) 'Consultation on Indigenous Peoples' Knowledge and Intellectual Property Rights', *Indigenous Affairs*. 4:26.

Jackson, J. (1983) *The Fish People*. Cambridge Studies in Social Anthropology. Cambridge University Press, Cambridge.

Jackson, J. E. (1984) The Impact of the Small-Scale Societies. *Studies in Comparative International Development* 19:2:3-32.

Jackson, K. H.(1971) *A Celtic Miscellany*. Penguin, Harmondsworth.

Jackson, W. (1994) *Becoming Native to this Place.* University of Kentucky Press, Lexington, KY.

Jackson, W., Berry, W. and Colman, B. (eds) (1984) *Meeting the Expectations of the Land: Essays in sustainable agriculture and stewardship*. North Point Press, San Francisco.

Jacobi, H. (trans.) (1895) Acaranga sutra. In: *Jaina Sutras*. Clarendon Press, Oxford.

Jacobs, M. (1995) Sustainable development, capital substitution and economic humility: a response to Beckerman. *Environmental Values* 4: 57-68.

Jaimes, M. (ed.) (1992) *The State of Native America: Genocide, colonization, and resistance*. South End Press, Boston.

Jain, S. K. (1991) *Dictionary of Indian folk medicine and ethnobotany*. Deep Publications. New Delhi.

Jain, S. K. (1992) 'Folk Beliefs in Conservation of Nature and Culture'. Paper presented at the International Congress of Ethnobiology, Mexico.

Jaini, P. (1979) *The Jaina Path of Purification*. University of California Press, Berkeley.

Jardhari, V. and Kothari, A. (1996) Conserving Agricultural Biodiversity: The Case of Tehri Garhwal and Implications for National Policy. In: Sperling, L. and Loevinsohn, M. (eds) *Using Diversity: Enhancing and Maintaining Genetic Resources On-Farm*. International Development Research Centre, New Delhi.

Jasper, J. M. and Nelkin, D. (1992) *The animal rights crusade: The growth of a moral protest*. The Free Press, New York.

Jeanrenaud, S. J. (1998) 'Can the leopard change its spots? Exploring people-oriented conservation in WWF'. Unpublished PhD thesis, University of East Anglia.

Jeffery and Sundar, with Tharakan, Mishra and Peter. (1995) 'A Move from Major to Minor: Competing Discourses of Non Timber Forest Products in India'. Paper presented to the SSRC Conference on Environmental Discourses and Human Welfare in South and Southeast Asia.

Jensen, E. (1974) *The Iban and Their Religion*. Oxford Monographs on Social Anthropology. Oxford University Press, Englewood Cliffs, N.J.

Jensen, T. (1997) Familiar and Unfamiliar Challenges to the Study and Teaching of Religions. In: Holm, N. G. (ed.) *The Familiar and the Unfamiliar in the World Religions: Challenges for Religious Education Today*. Åbo: The Academy of Åbo Press, 199-223.

Jiggins, J. (1989) An Examination of the Impact of Colonialism in Establishing Negative Values and Attitudes towards Indigenous Agricultural Knowledge. In: Warren, D. M., Slikkerveer, L. J. and Titilola, S. O. (eds) *Indigenous Knowledge Systems: Implications for Agriculture and International Development*. Studies in Technology and Social Change No. 11 Ames, Iowa State University, pp. 68-7

Jodha, N. S. (1990) *Rural Common Property Resources: Contributions and Crisis*. ICIMOD, Kathmandu.

Johannes, R. E. (1978) Traditional marine conservation methods in Oceania and their demise. *Annual Review of Ecology and Systematics* 9: 349-364.

Johannes, R. E. (1982) Traditional conservation methods and protected marine areas in Oceania. *Ambio* 11: 258-261.

Johannes, R. E. (1987) Primitive Myth. *Nature* 325: 474.

Johannes, R. E. (ed.) (1989) *Traditional Ecological Knowledge*. IUCN, The World Conservation Union, Cambridge.

Johannes, R. E. and Hatcher, B.G. (1986) Shallow tropical marine environments. In: Michael E. Soulé (ed.) *Conservation Biology: The Science of Scarcity and Diversity*, 371-382. Sinauer Associates, Sunderland, MA.

Johannes, R. E. and Ruddle, K. (1993) Human interactions in tropical and marine areas: lessons from traditional resource use. In: Price, A. and Humphreys, S. (eds) *Applications of the Biosphere Reserve Concept to Coastal Marine Areas*. 19-25. IUCN, Gland.

John, A. D. (1992) Species conservation in managed tropical forests. In: Whitmore, T. C., and Sayer, J. A. (eds) *Tropical Deforestation and Species Extinction*. Chapman and Hall, London.

Johns, T., Mhoro, E. B. and Sanaya, P. (1996) Food Plants and Maticants of the Batemi of Ngorongoro District, Tanzania. *Economic Botany* 50: 115-121.

Johns, T., Mhoro, E. B., Sanaya, P. and Kimanani, E. K. (1994) Quantitative appraisal of the herbal remedies of the Batemi of Ngorongoro District, Tanzania. *Economic Botany* 48: 90-95.

Johnson, A. W. and Earle, T. (1987) *The Evolution of Human Societies: From Foraging Group to Agrarian State*. Stanford University Press, Stanford, CA.

Johnson, M. (1992) *Lore: Capturing Traditional Environmental Knowledge*. Dene Cultural Institute/IDRC, Hay River.

Jones, S. (1978) Tribal Underdevelopment in India. In: *Development and Change 9*.

Joplin, J. and Girardet, H. (1996) *Creating a Sustainable London*. Sustainable London Trust, London.

Joralemon, D. (1983) 'The Symbolism and Physiology of Ritual Healing in a Peruvian Coastal Village' [Ph.D. dissertation]. Dept. of Anthropology, University of Michigan, Ann Arbor.

Jordan, D. (1988) Rights and claims of indigenous people. Education and the reclaiming of identity: The case of the Canadian natives, the Sami and Australian Aborigines. In: Skutnabb-Kangas, T. and Cummins, J. (eds), *Minority Education: From Shame to Struggle*. 189-222. Multilingual Matters, Clevedon, UK.

Joyce, C. (1994) *Earthly Goods: Medicine-Hunting in the Rainforest*. Little, Brown and Co., Boston, New York, Toronto and London.

Juma, C. (1989) *The Gene Hunters: Biotechnology and the Scramble for Seeds*. African Centre for Technology Studies Research Series 1, Zed Books Ltd., London.

Juma, C. (1989) *The Gene Hunters: Biotechnology and the Scramble for Seeds*. Princeton University Press, Princeton, NJ.

Juma, Calestous (1997) *Statement by the Executive Secretary of the Secretariat of the Convention on Biological Diversity, to the second Committee of the United Nations General Assembly*, 6 November (1997)

Junk, W. J. and Furch, K. (1985) The physical and chemical properties of Amazonian waters. In: Prance, G. T. and Lovejoy T. E. (eds), *Key environments. Amazonia*. Pergamon Press.

Kakonge, J. O. (1995) 'Traditional African values and their use in implementing Agenda 21'. *Indigenous Knowledge and Develoment Monitor*. 3:2:19-22.

Kalantzis, M. (1995) Coming to grips with the implications of multiculturalism. Paper presented at the Global Cultural Diversity Conference, Sydney, 26-28 April, 1995.

Kalland, A. (1989) Arikawa and the impact of a declining whaling industry. *NIAS-Report* , Nordic Institute of Asian Studies, Copenhagen. 1:94-138.

Kalland, A. (1994a) Indigenous – local knowledge: prospects and limitations. In: Hansen, B. V. (ed.) *Arctic Environment: Report on the Seminar on Integration of Indigenous Peoples' Knowledge, Reykjavik, September 20-23, 1994*. 150-167. Ministry for the Environment (Iceland), Ministry of the Environment (Denmark) and The Home Rule of Greenland (Denmark Office), Reykjavik and Copenhagen.

Kalland, A. (1994b) Aboriginal subsistence whaling: a concept in the service of imperialism. In: High North Alliance *11 Essays on Whales and Man* (2nd Edition). High North Alliance, Reine Lofoten.

Kalland, A. (1995a) *Fishing villages in Tokugawa Japan*. Curzon Press, London / University of Hawaii Press, Honolulu.

Kalland, A. (1995b) Culture in Japanese nature. In: Bruun, O. and Kalland, A. (eds), *Asian perceptions of nature: A critical approach*. Curzon Press, London.

Kalland, A. and Moeran, B. (1992) *Japanese whaling. End of an era?* Curzon Press, London.

Kamanā, K. and Wilson, W. H. (1996) Hawaiian language programs. In: Cantoni, G. (ed.), *Stabilizing Indigenous Languages*. 153-156. Center for Excellence in Education, Northern Arizona University, Flagstaff, AZ.

Kant, I. (1959) *Foundations of the Metaphysics of Morals*. Woldd, R. (ed.), Beck, L. (trans.), Bobbs-Merrill, Indianapolis, Indiana.

Kāretu, T. (1994) Maori language rights in New Zealand. In: Skutnabb-Kangas, T. and Phillipson, R. (eds) *Linguistic Human Rights. Overcoming Linguistic Discrimination*. 209-218. Mouton de Gruyter, Berlin and New York.

Kaur, J. (1985) *Himalayan Pilgrimages and the New Tourism*. Himalayan Books, New Delhi.

Kay, P. (1966) Ethnography and Theory of Culture. *Bucknell Review* 14: 106-113.

Kellert, S. T., and Wilson, E. O. (eds) (1993) *The Biophilia Hypothesis*. Island Press, Washington D.C.

Kellogg, E. L. (ed.) (1995) *The Rain Forests of Home: An Atlas of People and Place. Part 1: Natural Forests and Native Languages of the Coastal Temperate Rain Forest*. Ecotrust/Pacific GIS/Conservation International, Portland, OR/Washington, D.C.

Kemf, E. (ed.) (1993) *Indigenous Peoples and Protected Areas - The Law of Mother Earth*. Earthscan Publications Ltd., London.

Kemf, E. (ed.) (1993) *The Law of the Mother: Protecting Indigenous Peoples in Protected Areas*. Sierra Club Books, San Francisco, in association with WWF and the IUCN.

Kempton, W., Boster, J. and Hartley, J. (1995) *Environmental values in American culture*. MIT Press, Cambridge, Massachusetts.

Kence, Aykut (ed.) (1990) *Turkiye'nin Biyolojik Zenginlikleri*. Turkiye Cevre Sorunlary Vakfy Yayyny.

Kendall, L. (1985) *Shamans, Housewives, and Other Restless Spirits: Women in Korean Ritual Life*. University of Hawaii Press, Honolulu.

Kessel, J. van, and Condori Cruz, D. (1992) *Criar la Vida: Trabajo y Tecnología en el Mundo Andino*. Vivarium: Santiago, Chile.

Khubchandani, L. M. (1991) *Language, Culture and Nation-Building. Challenges of Modernisation*. Indian Institute of Advanced Study, in association with Manohar Publications, Shimla and New Delhi.

Khubchandani, L. M. (1994) 'Minority' cultures and their communication rights. In: Skutnabb-Kangas, T. and Phillipson, T. (eds), *Linguistic Human Rights. Overcoming Linguistic Discrimination*. 305-315. Mouton de Gruyter, Berlin and New York.

Kimber, A. J. (1970) 'Some cultivation techniques affecting yield response in sweet potato'. In: Plucknett, D.K. (ed.) *Proceedings of the Second International Symposium on Tropical Root and Tuber Crops*. University of Hawaii, Honolulu. 32-36.

Kimber, A. J. (1971) 'Cultivation practices with sweet potato'. *Harvest*. 1:1: 31-33.

King, V. (1977) Unity, formalism and structure: Comments on Iban augury and related problems. *Bijdragen* 133(1): 63-87.

Kingsbury, B. (1992) Claims by non-state groups in international law. *Cornell International Law Journal* 25(3) 481-513.

Kipury, N. (1983) *Oral Literature of the Maasai*. East African Educational Publishers Ltd., Nairobi.

Kirby, W. (1825) 'Introductory Address', *Zoological Journal*, 2, 1-8, on 2 and 5.

Kirch, P. V.(1982) The impact of the prehistoric Polynesians on the Hawaiian ecosystem. *Pacific Science* 36:1-14.

Kirdar, U. (1997) (ed) *Cities Fit For People*. United Nations Publications, New York.

Kirschenmann, F. (1997) Expanding the vision of sustainable agriculture. In: Madden, J. and Chaplowe, S. (eds), *For All Generations: Making world agriculture more sustainable*. A World Sustainable Agriculture Publication.

Kiss, A. (ed.) (1990) Living with Wildlife: Wildlife Resource Management with Local Participation in Africa. *Technical Paper* No.130, World Bank, Washington, D.C.

Kleinman, A.. (1988) *The Illness Narratives*. Basic Books, New York.

Kline, D. (1990) *Great Possessions: An Amish Farmer's Journal*. North Point Press, San Francisco.

Kline, D. (1994) On Nature: Fall mushrooms can be tasty (or deadly). *Daily Record Newspaper*, Wooster, OH, Oct. 8, (1994)

Kline, D. (1995a) 'A theology for living'. Paper presented at Creation Summit. Mennonite Environmental Task Force, WI, USA, February 19, 1995.

Kline, D. (1995b) On Nature: Hayfield is a community unto itself. *Daily Record Newspaper*, Wooster, OH, June 1995.

Kline, D. (1995c) On Nature: Memories of rare butterflies past. *Daily Record Newspaper*, Wooster, OH, Aug. 6, 1995.

Kline, D. (1996) *Birds of Holmes and Wayne County*. 10543 CR 329, Shreve, Ohio 44676.

Kline, D. (1996) On Nature: Monarch returns, seeking milkweed. *Daily Record Newspaper*, Wooster, OH, June 8.

Kline, D. (1997) *Scratching the Woodchuck. Nature on an Amish Farm*. The University of Georgia Press, Athens, G.A. 205pp.

Kloppenburg, J. Jr (1991) No Hunting: Biodiversity, indigenous rights and scientific poaching. *Cultural Survival Quarterly* (Summer). 14-18.

Kloppenburg, J. R. Jr (1988) *First the Seed: The Political Economy of Plant Biotechnology*. Cambridge University Press, Cambridge.

Klotz, S. (1990) Species/Area and Species/Inhabitants Relations in European Cities. In: Sukopp, H., Mejny, S. and Kowarik, I. (eds) *Urban Ecology: Plants and Plant Communities in Urban Environments*. SPB Academic Publishing, The Hague, pp. 45-74; 99-103.

Knight, C. G. (1974) *Ecology and Change: Rural Modernization in an African Community*. Academic Press, New York.

Knight, D. (1981) *Ordering the World: A History of Classifying Man*. Andre Deutsch, London.

Knobloch, K. and Herrmann-Wolf, B. (1985) Biology and essential oil of Ocimum species. *Topics in Flavour Research*. April, 277-279.

Knoppel, H. A. (1970) 'Food of Central Amazonian Fishes: Contribution to the nutrient-ecology of Amazonian rain forest streams'. *Amazoniana* 2 (3): 257-352.

Knudtson, P. and Suzuki, D. (1992) *Wisdom of the Elders*. Stoddart Publishing Co. Limited, Toronto.

Koch, E. (1997) Ecotourism and Rural Reconstruction in South Africa: Reality or Rhetoric? In: Ghimire, K. B and Pimbert, M. P. (eds) *Social Change and Conservation*. UNRISD and Earthscan, London.

Kocherry, T. (1994) The Conservation of Fishing Communities and Fisheries Resources in India. In: Shiva, V. (ed) *Biodiversity Conservation: Whose Resources? Whose Knowledge?* INTACH, New Delhi.

Kohen, J. (1995) *Aboriginal Environmental Impacts*. University of New South Wales Press, Sydney.

Konstant, T. L., Sullivan, S. and Cunningham, A. B. (1995) The effects of utilization by people and livestock on *Hyphaene petersania* (Arecaceae) basketry resources in the palm savanna of North-central namibia. *Economic Botany* 49, 345-356.

Kook, R. (1961) 'Hazon hatzimhonut vehashalom, Afikim baNegev II'. In: *Lahai Ro'i*. Fried, Y. and Riger, A. (eds). Jerusalem.

Korombara, N. C. (1995) *Los ancianos cuentan/Saaiimayí iibaeedariin*. Maracaibo.

Kothari, A. (1997) *Understanding Biodiversity: Life, Equity, and Sustainability*. Orient Longman, Tracts for the Times, New Delhi.

Kothari, A. and Kothari, M. (1995) *Sacrificing Our Future: The New Economic Policy and the Environment*. CUTS, Calcutta.

Kothari, A., Pande, P., Singh, S. and Dilnavaz, R. (1989) *Management of National Parks and Sanctuaries in India*. Status Report, Indian Institute of Public Administration, New Delhi.

Kothari, A., Singh, N. and Suri, S. (1995) Conservation in India: A New Direction. *Economic and Political Weekly*, Vol. XXX, No. 43, 28 October.

Kowarik, I. (1990) Some Responses of Flora and Vegetation to Urbanization in Central Europe. In: Lal, B., Vats, S. K., Singh, R. D. and Gupta, A. K. (1994) Plants Used As Ethnomedicine by Gaddis in Kangra and Chamba, Abstract in Ethnobiology. In: *Human Welfare*, Fourth International Congress Of Ethnobiology, 17-21 November 1994, Abstracts, National Botanical Research Institute, Lucknow.

Krauss, M. (1992) The world's languages in crisis. *Language* 68(1): 4-10.

Kropotkin, P. (1914) *Mutual Aid: A Factor of Evolution*. rept. The Penguin Press, Allen Lane, London.

Kuhns, L. (1989) 'Farming technology and its ominous pose to the Amishman's society'. *Small Farming Journal*. Spring 13 (2): 20-22.

Kulkami, D. K., Kulkami M. B. and Kumbhojkar, M. S. (1996) Quantification of Tree Resources Utilized for Medicinal Purposes by Mahadeo Kolis of Western Maharashtra, India. In: Jain, S. K. (ed.) *Ethnobiology in Human Welfare*, pp.429-436. Deep Publications, New Delhi.

Kull, A. C. (1996) The evolution of conservation efforts in Madagascar. *International Environmental Affairs* 8(1): 50-86.

Kung, H. (1991) *Global Responsibility: In search of a new world ethic*. Crossroad, New York.

Kuo Hsi (1935) *An Essay on Landscape Painting*. John Murray, London.

Laguna, F. D. (1972) *Under Mount Saint Elias: The History and Culture of the Yakutat Tlingit*. Smithsonian Institution Press, Washington, D.C.

Laird, S. A. (1995) The Natural Management of Tropical Forests for Timber and Non-Timber Products. *OFI Occasional Papers* 49. Oxford Forestry Institute, Department of Plant Sciences, University of Oxford.

Laird, S. A. and Sunderland, T. C. H. (1996) 'The Over-Lapping Uses of 'Medicinal' Species in Southwest Cameroon: Implications for Forest Management'. Paper presented at the Society of Economic Botany 37th annual meeting, London.

Lakoff, G. and Johnson, M. (1980) *Metaphors We Live By*. University of Chicago Press, Chicago.

Lal, R. (1975) 'Role of mulching techniques in tropical soil and water management'. *International Institute of Tropical Agriculture Bulletin* 1.

Lal, R. (1987) *Tropical ecology and physical edaphology*. Wiley, Chichester.

Lamb, D. (1991) Combining traditional and commercial uses of rain forests. *Nature and Resources* 27(2):3-11.

Lane & Lane (1981) Lane & Lane Associates, with Douglas Nash. *The White Salmon River Indian Fisheries and Condit Dam*. Prepared for the Bureau of Indian Affairs, Portland, Oregon.

Lane, M. B. and Rickson, R. E. (1997) Resource Development and Resource Dependency of Indigenous Communities: Australia's Jawoyn Aborigines and Mining at Coronation Hill. *Society and Natural Resources* 10:121-142.

Langdon, S. (1995) Increments, ranges, and thresholds: human population responses to climate change in northern Alaska. In: Peterson, D. L. and Johnson, D. R. (eds) *Human Ecology and Climate Change — People and Resources in the Far North*. 139-154. Taylor and Francis, Washington, D.C.

Langdon, S. J. (ed.) (1986) *Contemporary Alaskan Native Economies*. University Press of America, Lanham, Md.

Lange, D. (1996) In: Newsletter of the Medicinal Plant Study Group of the World Conservation Union (IUCN).

Langton, M. (1996) Art, wilderness and *terra nullius*. In: *Ecopolitics IX: Perspectives on Indigenous Peoples' Management of Environmental Resources*. Conference papers and resolutions. Northern Territory University, Darwin, 1-3 September, 1995.

Lanning, E. C. (1959) Bark Cloth Hammers. *The Uganda Journal*. 23, 1:79-82.

Laperre, P. E. (1971) A study of soils and the occurrence of termitaria and their role as an element in photointerpretation for soil survey purposes in a region in the Zambesi Delta, Mocambique. ms. Library of Natural Resources Institute.

Lasch, C. (1991) *The True and Only Heaven — Progress and its Critics*. W. W. Norton, New York.

Latreille, P.-A. (1798) *Essai sur l'histoire des fourmis de la France*. rept. Cite des Science et de l'Industrie / Champion-Slatkine, Paris, 1989.

Lawrence, R. D. (1976) *In Praise of Wolves*. Bantam Publishing, New York.

Leacock, E. and Lee, R. (eds) (1982) *Politics and History in Band Societies*. Cambridge University Press, Cambridge.

Leakey, L. S. B. (1934) *Kenya: Contrasts and Problems*. MacMillan, London.

Leakey, R. E and Slikkerveer, L. J. (eds) (1991) Origins and Development of Indigenous Agricultural Knowledge Systems in Kenya, East Africa. *Studies in Technology and Social Change*, No. 19, Iowa State University, Ames, Iowa, USA.

Leakey, R. E. and Slikkerveer, L. J. (1993). *Man-Ape Ape-Man: The Quest for Human's Place in Nature and Dubois' 'Missing Link'*. Kenya Wildlife Service Foundation, Leiden, 182 pp.

Lébré Okou, H. (1995) Forêt et équilibre des sociétés traditionnelles. *Actes du premier Forum international sur la forêt*. SODEFOR, Abidjan: 67-69.

Lee, K. E. (1990) 'The Diversity of Soil Organisms'. In: Hawksworth, D. K. (ed.). *The Biodiversity of Microorganisms and Invertebrates: Its Role in Sustainable Agriculture*. CAB International, London.

Lee, Kai N. (1993) *Compass and Gyroscope: Integrating Science and Politics for the Environment*. Island Press, Washington, D.C.

Lee, R. (1982) Risk, reciprocity and social influences on !Kung San economics. In: Leacock, E. and Lee, R. (eds). *Politics and History in Band Societies*. 37-60. Cambridge University Press, Cambridge.

Lee, R. and DeVore, I. (eds) (1968) *Man the Hunter*. Aldine, Chicago, IL.

Légré Okou, H. (1995) 'Forêt et équilibré des sociétés traditionnelles', *Actes du Premier Forum International Sur la Forêt*. Abidjan, SODEFOR.

Leidy, R. A. and Fiedler, P. L. (1985) Human Disturbance and Patterns of Fish Species Diversity in the San Francisco Bay Drainage, California. *Biological Conservation* 33:247-267.

Lele, S. (1991) Sustainable development: A critical review. *World Development* 19, 607-621.

Lemmonier, P. (1993) The Eel and the Ankave-Anga of Papua New Guinea: Material and Symbolic Aspects of Trapping. In: Hladik, C. M., Hladik, A., Linares, O. F., Pagezey, H., Semple, A. and Hadley, M. (eds) Tropical Forests, People and Food: Biocultural Interactions and Applications to Development. *Man and Biosphere*, Vol 13, UNESCO, Paris.

Leng, A. S. (1982) Maintaining fertility by putting compost into sweet potato mounds. *Harvest* 8:2: 83-84.

Leopold, A. (1923) Some fundamental of conservation in the southwest. *Environmental Ethics* 1, 131-141.

Leopold, A. (1948) Thinking Like a Mountain. In: *A Sand County Almanac*. Oxford University Press, New York, pp. 129-133.

Leopold, A. (1949) *A Sand County Almanac and Sketches Here and There*. Oxford University Press, New York.

Lepage, M., Abbadie, L. and Zaidi, Z. (1989) Significance of Hypogeous Nests of Macrotermitinae in a Guinea Savanna Ecosystem, Ivory Coast (Isoptera). *Sociobiology* 15, 2, 1989, p. 267.

Leskien, D. and Flitner, M. (1997) Intellectual Property Rights and Plant Genetic Resources: Options for a *Sui Generis* System. *Issues in Genetic Resources* No. 6. International Plant Genetic Resources Institute (IPGRI), Rome.

Leslie C.(1985) *Paths to Asian Medical Knowledge*, UC Berkeley Press, Berkeley.

Leslie, A. J. (1987) The economic feasibility of natural management of tropical forests. In: Mergen, F. and Vincent, J. R. (eds) *Natural Management of Tropical Moist Forests; Silvicultural and Management Prospects of Sustained Utilization*. Yale University, School of Forestry and Environmental Studies, New Haven, CT.

Levi-Strauss, Claude (1962) *La Pensée Sauvage*. Plon, Paris.

Levi-Strauss, Claude (1984) *La Potiere Jalouse*. Plon, Paris.

Lewis, H. T. (1989) A Parable of Fire: Hunter-Gatherers in Canada and Australia. In: Johannes, R. E. (ed.) *Traditional Ecological Knowledge: A Collection of Essays*. IUCN, Gland, Switzerland.

Lewis, H. T. (1993) Traditional ecological knowledge – Some definitions. In: Baines, G. and Williams, N. (eds), *Traditional Ecological Knowledge: Wisdom for Sustainable Development*. 8-11. Centre for Resource and Environmental Studies, Australian National University, Canberra.

Lewis, H. T. (1994) Management Fires vs. Corrective Fires in Northern Australia: An Analogue for Environmental Change. *Chemosphere* 29 (5):949-963.

Lindquist, O. and Tryggvadotttir, M. (1990) 'Whale watching in Iceland: A feasibility study'. Manuscript.

List, F. (1885) *The National System of Political Economy*. London.

Little Bear, L., Boldt, M. and Long, J. A. (eds) (1984) *Pathways to Self-Determination: Canadian Indians and the Canadian State*. University of Buffalo Press, Toronto.

Livingston, J. A. (1996) *Rogue Primate: An Exploration of Human Domestication*. Key-Porter Books.

Lizarralde and Beckerman (1982) Historia contemporánea de los Barí. *Antropológia* 58: 3-52.

Lobry de Bruyn, L. A. and Conacher, A. J. (1990) The role of termites and ants in soil modification: a review. *Australian Journal of Soil Research* 28, pp. 55-93.

Loita Naimana Enkiyio Conservation Trust (1994) Forest of the Lost Child: A Maasai Conservation Success Threatened by Greed. Kenya, Narok, 8 pp.

Loneragan, O. W. (1990) *Historical review of Sandalwood (Santalum spicatum) research in Western Australia.* Research Bulletin No. 4, Western Australia Department of Conservation and Land Management.

Lopez, B. H. (1978) *Of Wolves and Men*. Charles Scribner's Sons, New York.

Lord, N. (1996) Native tongues. *Sierra* 81(6): 46-69.

Loucks, O. L. (1994) Sustainability in Urban Ecosystems: Beyond an Object of Study. In: Platt, R. H., Rowntree, R. A. and Muick, P. C. (eds) *The Ecological City: Preserving and Restoring Urban Biodiversity.* University of Massachusetts Press, Amherst, pp 49-64.

Loucou, J.-N. (1984) *Histoire de la Côte d'Ivoire. La formation des peuples*. CEDA, Abidjan: 208p.

Lovelock, J. (1979) *Gaia, A new look at life on earth.* Oxford University Press, Oxford.

Lovelock, J. (1991) *Healing Gaia*. Harmony Books, New York.

Lubbock, J. (1856) 'On the Objects of a Collection of Insects', *Entomologist's Annual* for 1856, 115-21.

Lucas, P. H. C. (1993) History and Rationale for Mountain Parks as Exemplified by Four Mountain Areas of Aotearoa (New Zealand). In: Hamilton, S. *et al.* (eds). *Parks, Peaks, and People*. 24-28. East-West Center, Program on Environment, Honolulu.

Ludwig, D., Hilborn, R. and Walters, C. (1993) Uncertainty, resource exploitation and conservation: Lessons from history. *Science* 260: 17-36.

Lumbreras, L. G. (1992) Cultura, Tecnologia y Modelos Alternativos de Desarrollo. *Comercio Exterior*, 42, 3 (Marzo):199-205.

Luna, L. E. (1986) Vegetalismo: Shamanism among the Mestizo population of the Peruvian Amazon. Stockholm, Almquist & Weiskell International.

Luniak, M. (1994) The Development of Bird Communities in New Housing Estates in Warsaw. In: Barker, G. M., Luniak, M., Trojan, P. and Zimny, H. (eds). Proceedings of the Second European Meeting of the International Network for Urban Ecology, Warsaw. *Memorabilia Zoologica* 49:257-267.

Lydon, B. (1996) Fast food hunger. *WHY Magazine*.

Mabey, R. (1980) *The Common Ground*. Dent, London.

MacDonald, D. W. and Newdick, M. T. (1980) The Distribution and Ecology of Foxes in Urban Areas. In: Bornkamm, R., Lee, J. and Seaward, M. R. D. (eds) *Urban Ecology.* Blackwell Scientific Publications, London, pp. 123-135.

Machaca, M. (1992) *El Agua y los Quispillactinos*, Documento Presentado en el Curso de Formacion en Agricultura Andina-PRATEC, Asociacion Bartolome Aripaylla, Ayacucho (manuscript).

Machaca, M. (1993) *Actividades de Crianza de Semillas en la Comunidad Campesina de Quispillacta, Ayacucho,* Acompaoadas por la Asociacion Bartolome Aripaylla, Campaoa (1992-93). In: PRATEC (ed) *Afirmacion Cultural Andina*, Lima, Peru.

Machaca, M. (1996) La Crianza de la Biodiversidad y la Cultura Andina, in: PRATEC (ed) *La Cultura Andina de la Biodiversidad*., Lima, Peru, pp. 101-122.

Machaca, M. and Machaca, M. (1994) *Crianza Andina de la Chacra en Quispillacta, Semillas ñ Plagas y Enfermedades,* Asociacion Bartolome, Ayacucho, Aripaylla.

Mackay, J. A. (ed.) (1993) *Robert Burns: The complete poetical works*. Alloway Publishing, Darvel.

MacKenzie, J. (1988) *The Empire of Nature: Hunting, conservation and British imperialism*. Manchester University Press, Manchester.

MacLennan, A. H., Wilson, D. W. and Taylor, A. W. (1996) Prevalence and cost of alternative medicine in Australia. *Lancet*, 347, 569-572.

Macpherson, C. B. (1973) *Democratic Theory: essays in retrieval*, Clarendon Press, Oxford.

Maffi, L. (1996) Position paper for the working conference 'Endangered Languages, Endangered Knowledge, Endangered Environments', Berkeley, California, 25-27 October, 1996. Terralingua Position Paper #1, published electronically.

Maffi, L. (1996a) Position paper for the working conference 'Endangered Languages, Endangered Knowledge, Endangered Environments', Berkeley, California, October 25-27, 1996. Terralingua Position Paper #1, published electronically.

Maffi, L. (1996b) Domesticated land, warm and cold: Linguistic and historical evidence on Tzeltal Maya ethnoecology. In: Blount, B. G. and Gragson, T. S. *Ethnoecology: Knowledge, Resources and Rights*. Under consideration by Iowa University Press.

Maffi, L. (1997) Language, knowledge and the environment: Threats to the world's biocultural diversity. *Anthropology Newsletter* 38(2): 11.

Maffi, L. (1999) Domesticated land, warm and cold: Linguistic and historical evidence on Tzeltal Maya ethnoecology. In: Blount, B. G. and Gragson, T. S., *Ethnoecology: Knowledge, Resources and Rights.* Georgia University Press, Athens.

Magga, O.-H. (1994) The Sámi Language Act. In: Skutnabb-Kangas, T. and Phillipson, T. (eds) *Linguistic Human Rights. Overcoming Linguistic Discrimination*. 219-233. Mouton de Gruyter, Berlin and New York.

Maheswari, J. K., Singh, K. K. and Saha, S. *The ethnobotany of the tharus of kheri district, Uttar Pradesh*. Economic Botany Information Service. National Botanical Research Institute. Lucknow.

Maine, H.S. (1861 and 1873) *Ancient Law: Its Connection with the Early History of Society and its Relation to Modern Ideas*. Holt, New York.

Malaka, S. S. Omo (1972) Some measures applied in the control of termites in parts of Nigeria. *Nigerian Entomological Magazine* 2, 137-141.

Malhotra, K. C., Gadgil, M. and Khomne, S. B. (1978) Social Stratification and Caste Ranking among the Nandiwallas of Maharashtra. *Proceedings of the Seminar on Nomads in India*, Mysore.

Malhotra, K. C., Khomne, S. B. and Gadgil, M. (1983) 'On the role of hunting in the nutrition and economy of certain nomadic populations of Maharashtra.' *Man in India* 63.

Malhotra, K. C., Poffenberger, M., Bhattacharya, A. and Dev, D. (1991) Rapid appraisal methodology trials in Southwest Bengal: assessing natural forest regeneration patterns and non-wood forest product harvesting practices. *Forests, Trees and People Newsletter* 15-16.

Malik, K. (1996) One man talking. Utne Reader July-August 1996: 35-37.

Manderson, L. and Akatsu, H. (1993) Whale meat in the diet of Ayukawa villagers. *Ecology of Food and Nutrition* 30: 207-220.

Manderson, L. and Hardacre, H. (1989) Small-type coastal whaling in Ayukawa. Draft report of research. IWC Document IWC/41/SE3, published in *Papers on Japanese small-type coastal whaling submitted by the Government of Japan to the International Whaling Commission, 1986-1996*, 85-103. The Government of Japan, Tokyo

Mandlebaum, David. (1984) *Society in India*. Popular Prakashan, Bombay.

Mando, A., van Driel, W. F. and N. Prosper Zombré. Le role des termites dans la restauration des sols ferrugineux tropicaux encroutes au Sahel. Contribution au 1ere Colloque International de l'AOCASS: Gestion Durable des Sols et de l'Environnement en Afrique Tropicale, Ouagadougou, 6-10 Décembre 1993.

Mann, R. D. 'Time Running Out: The Urgent Need for Tree Planting in Africa'. *The Ecologist* 20(2): 48-53.

Manriquez, L. F. (1996) Silent no more: California Indians reclaim their culture – and they invite you to listen. Paper presented at the working conference 'Endangered Languages, Endangered Knowledge, Endangered Environments'. Berkeley, California, October 25-27, 1996.

Manriquez, L. F. (1998) Silent no more: California Indians reclaim their culture – and they invite you to listen. In: Maffi, L. (ed.), *Language, Knowledge, and the Environment: The Interdependence of Biological and Cultural Diversity*. Submitted to Smithsonian Institution Press.

Maori Congress (1992) *Maori Congress Position Paper to the UN Conference on Environment & Development*. Wellington.

March, J. (1994) *A Primer in Decision Making*. The Free Press, New York.

Margulis, L. and Sagan, D. (1986) *Microcosmos*. Summit, New York.

Margulis, L. and Sagan, D. (1995) *What Is Life?* Simon and Schuster, New York.

Markham, A. (1994) *A Brief History of Pollution*. Earthscan Publications Ltd., London.

Markham, Sir Clements (1913) Published in English as *'Colloquies on the simples and drugs of India'*. Garcia da Orta, trans. Henry Southern, London.

Marks, S. A. (1994) Managerial Ecology and Lineage Husbandry: Environmental Dilemmas in Zambia's Luanga Valley. In: Hufford, M. (ed.) *Conserving Culture*. University of Illinois Press, Urbana (Illinois).

Marsh, G. P. (1864) *Man and Nature, or Physical Geography as Modified by Human Action*. Belknap Press of Harvard, Cambridge, MA.

Marshall, J. III. (1995) *On Behalf of the Wolf and the First Peoples*. Red Crane Books, Santa Fe, New Mexico.

Marshall-Thomas, E. (1994) *The Tribe of Tiger: Cats and their Culture*. Simon and Schuster, New York.

Marten, G. G. (1986) *Traditional Agriculture in Southeast Asia*. In: Marten, G. G. (ed.) Westview Press, Boulder, CO.

Martin, C. (1978) *Keepers of the Game*. University of California Press, Berkeley.

Martin, E. (1993) The Last Mountain, *American Forests*, March/April: 44-54.

Martin, G. (1991) *The Rainforests of West Africa: ecology, threats conservation*. Birkhauser Verlag, Basel.

Martin, G. J. (1995) *Ethnobotany: A Methods Manual*. Chapman and Hall, London.

Martin, P. S. (1984) Prehistoric overkill: the global model. In: Martin, P. S. and Klein, R. G. (eds) *Quaternary Extinctions: A Prehistoric Revolution*. 354-403. University of Arizona Press, Tucson.

Martin, P. S. and Klein, R. G. (eds) (1984) *Quaternary Extinctions: A Prehistoric Revolution*. University of Arizona Press, Tucson.

Martineau, H. (1859) 'Female Industry'. *The Edinburgh Review*, 109 (April 1859) pp. 293-336.

Martinez, D. (1994) Traditional Environmental Knowledge Connects Land and Culture. *Winds of Change* 9(4):89-94.

Martini, A. M. Z., Rosa, N.de A., Uhl, C. (1992) *A First Attempt to Predict Amazonian Tree Species Threatened by Logging Pressure*. Instituto do Homen e Meio Ambiente da Amazonian (IMAZON), Belem, Brazil.

Mateene, K. (1980) Failure in the obligatory use of European languages in Africa and the advantages of a policy of linguistic independence. In: Mateene, K. and Kalema, J. (eds), *Reconsideration of African Linguistic Policies*. 9-41. OAU/BIL Publication 3. OAU Interafrican Bureau of Languages, Kampala.

Mateene, K. (1985) Colonial languages as compulsory means of domination, and indigenous languages as necessary factors of national liberation and development. In: Mateene, K., Kalema, J. and Chomba, B. (eds). *Linguistic Liberation and Unity of Africa*. 60-69. OAU/BIL Publication 6. OAU Bureau of Languages, Kampala.

Mathias-Mundy, E. (ed) (1992) Indigenous Knowledge and Sustainable Development. *Proceedings of the International Symposium of the Regional Program for the Promotion of Indigenous Knowledge in Asia*. IIRR, Silang, Cavite, The Philippines.

Matos, M. J. (1991) Los Pueblos Indios de America. *Pensamiento Iberoamericano* 19:181-200.

Matowanyika, J. Z. Z. (1997) Resource management and the Shona people in rural Zimbabwe. In: Posey, D. A. and Dutfield, G. *Indigenous Peoples and Sustainability: Cases and Actions*. 257-266. IUCN and International Books, Utrecht.

Matthews, W. (1897) *Navajo Legends*, Boston, American Folklore Society, 5.

Matthíasson, P. (1994) En sivilisert l'sning! Aftenposten 9th July.

Mattson, R. (1996) quoted in Heiberg, R. 'The Inner gardener' *Museletter*, 64.

Maturana, H. and Varela, F. (1980) *Autopoiesis and Cognition*. D. Reidel, Dordrecht, Holland.

Maturana, H. and Varela, F. (1987) *The Tree of Knowledge*. Shambhala, Boston.

Mauss, M. (1970) *The Gift: Forms and functions of exchange in archaic societies*. Cohen and West, London.

May, R. J. (1984) *Kaikai aniani: a guide to bush foods, markets and culinary arts of Papua New Guinea*. R. Brown and Associates, Bathurst.

Mayrhofer, M. (1953) *Kurzgefaßtes etymologisches Wörterbuch des Altindischen* (A concise etymological sanskrit dictionary). Heidelberg.

Mazur, R. E. and Titilola, S. O. (1992) Social and Economic Dimensions of Local Knowledge Systems in African Sustainable Agriculture. *Sociologia Ruralis*, 32 (2-3): 264-286.

Mbiti, J. (1969) *African Religions and Philosophy*. Heinemann, London.

Mbiti, J. (1975) *Introduction to African Religion*. Heinemann, London.

McCay, B. J. and Acheson, J. M. (eds) (1987) *The Question of the Commons*. University of Arizona Press, Tucson, Az.

McGean, B. and Poffenberger, M. (1996) *Village Voices and Forest Choices: Joint Forest Management in India*. Oxford University Press, New Delhi.

McIntosh, A. (1997) Colonised Land, Colonised Mind, *Resurgence*, No. 184, 28-30.

McIntyre, R. (1995) *War Against the Wolf: America's Campaign to Exterminate the Wolf*. Voyageur Press, Stillwater, Minnesota.

McIvor, C. (1997) Management of Wildlife, Tourism and Local Communities in Zimbabwe. In: Ghimire, K. B. and Pimbert, M. P. (eds) *Social Change and Conservation*. UNRISD and Earthscan, London.

McKenna, T. K. (1992) Food of the Gods: The search for the original tree of knowledge, a radical history of plants, drugs and evolution. London, Rider.

McLuhan, T. C. (1971) *Touch the Earth: Native American Teachings*. Promontory Press, New York.

McManus, J. C. (1975) The costs of alternative economic organizations. *Canadian Journal of Economics* 8: 334-350.

McNeely, J. (1988) *The Economics of Biological Diversity: Developing and using economic incentives to conserve biological resources*. IUCN, Gland, Switzerland.

McNeely, J. (1992) Nature and culture: conservation needs them both. *Nature & Resources* 28:37-43.

McNeely, J. A. (1988) *Economics and Biological Diversity: Developing and Using Economic Incentives to Conserve Biological Resources*. IUCN, Gland, Switzerland.

McNeely, J. A. (1993) Economic Incentives for Conserving Biodiversity: Lessons for Africa. *Ambio* 22(2-3): 144-150.

McNeely, J. A. (1993) *Parks for Life: Report of the IVth World Congress on National Parks and Protected Areas*. IUCN, Gland, Switzerland.

McNeely, J. A. (1993) People and Protected Areas: Partners in Prosperity'. In: Kemf, E.*The Law of the Mother: Protecting Indigenous Peoples in Protected Areas*. Sierra Club Books, San Francisco.

McNeely, J. A. (1994) Lessons from the Past: Forests and Biodiversity. *Biodiversity and Conservation* 3: 3-20.

McNeely, J. A. *et al.* (1990) *Conserving the World's Biological Diversity*. IUCN/WRI/CI/WWF-US/The World Bank, Gland, Switzerland/Washington, D.C.

McPherson, E. G. (1994) Cooling Urban Heat Islands with Sustainable Landscapes. In: Platt, R.H., Rowntree, R. A.and Muick, P. C.(eds) *The Ecological City: Preserving and Restoring Urban Biodiversity*. University of Massachusetts Press, Amherst, pp. 151-171.

Mead, A. (1994) 'Indigenous rights to Land and Biological Resources'. Paper presented to the Conference on Biodiversity: Impacts on Government, Business and the Economy, Auckland.

Mead, A. (1995) 'Letter to Darryl Macer', 22 September. native-l@gnosys.svle.ma.us.

Mead, A. (1996) 'Genealogy, Sacredness and the Commodities Market' In: *Cultural Survival Quarterly*.

Mead, A. T. P. (1994) *Misappropriation of Indigenous Knowledge: The Next Wave of Colonisation*. Otago Bioethics Report, 3(1), 4-7.

Meggers, B. (1977) *Amazônia: A Ilusão de um Paraíso*. (ed.) Civilização Brasileira, Rio de Janeiro. 234p.

Meggers, B. J. (1971) *Amazonia: Man and culture in a counterfeit paradise*. Aldine Press.

Meggers, B. J. (1996) *Amazonia: man and culture in a counterfeit paradise*. Revised edition. Smithsonian Institution Press, Washington and London.

Mehta, G. (1995) A Culture of Trees. In *Resurgence* 168: 31-35.

Meilleur, B.A. (1996) 'Selling Hawaiian Crop Cultivars'. In Brush and Stabinsky (1996)*Valuing Local Knowledge: Indigenous People and Intellectual Property Rights*.. Island Press, Washington, D.C.

Melnyk, M. (1995) 'The contribution of forest foods to the livelihoods of the Piaroa Amerindians of southern Venezuela'. *Indigenous Knowledge and Development Monitor*, 3:2:26.

Mendiratta, P. K., Dewan, V., Bhattacharyya, S. K., Gupta, V. S., Maiti,. P. C.and Sen, P. (1988) 'Effect of Ocimum sanctum L. on humoral immune responses'. *Indian Journal of Medical Research*. 87, April, 384-386.

Meninick, J. and Winthrop, R. H. (1995) *Talking with Anglos: Cultural Barriers to Communicating an Indian Perspective on Environmental Impacts*. Society for Applied Anthropology, Albuquerque, New Mexico.

Merchant, C. (1980) *The Death of Nature: Women, ecology, and the scientific revolution*. Harper and Row, New York.

Mercier, P. (1968) *Tradition, changement, histoire. Les 'Somba' du Dahomey septentrional*. Editions Anthropos, Paris.

Mergen, F. and Vincent, J. R. (eds) (1987) *Natural Management of Tropical Moist Forests; Silvicultural and Management Prospects of Sustained Utilization*. Yale University, School of Forestry and Environmental Studies, New Haven, CT.

Messerli, B. and Ives, J. (eds), (1997) *Mountains of the World: A Global Priority*. Parthenon, New York and London.

Metcalf, P. (1976) Birds and deities in Borneo. *Bijdragen* 132(1): 96-123.

Metzger, D. and Williams, G. (1963) 'Tenejapa Medicine: The Curer'. *SW Journal of Anthropology* 19: 216-234.

Metzner, R. (1994) The emerging cosmological worldview. In: Tucker, M. and Grim, J. (eds), *Worldviews and Ecology: Religion, philosophy, and the environment*. Orbis Books, Maryknoll.

Meyer, W. B. and Turner, B. L. II (eds) (1994) *Changes in Land Use and Land Cover: A Global Perspective*. Cambridge University Press, Cambridge.

Meyers, T. J. (1983) 'Amish origins and persistence: the case of agricultural innovation'. Paper presented at the Rural Sociological Society, Lexington, Kentucky, August 17-20, 1983.

Michon, G. and de Foresta, H. (1990) 'Complex Agroforestry Systems and the Conservation of Biological Diversity'. In: *Harmony with Nature. Proceedings of International Conference on Tropical Biodiversity*. SEAMEO-BIOTROP, Kuala Lumpur, Malaysia.

Michon, G., de Foresta , H. and Levang, P. (1995) Stratégies agroforestières paysannes et développement durable : les agroforêts à *damar* de Sumatra, *Nature-Sciences-Sociétés*, 3 (3).

Middleton, J. (1994) Effects of Urbanization on Biodiversity in Canada. In: Biodiversity Science Assessment Team, *Biodiversity in Canada: A Science Assessment*. Environment Canada, Ottawa, pp. 116-120.

Midgley, M. (1992) Science as Salvation. *Routledge, London.*

Mielke, H. W and Mielke, P. W. (1982) Termite mounds and chitemene agriculture: a statistical analysis of their association in southwestern Tanzania. *Journal of Biogeography* 9, 499-504.

Mies, M. and Shiva, V. (1993) *Ecofeminism*. Zed Books. London.

Millar, D. (1996) 'The Role of Women's Cosmovision in Indigenous Experimentation'. In COMPAS, *Agri-Culture and Cosmovision*.

Miller, A. G. and Morris, M. (1988) *Plants for Dhofar. The Southern Region of Oman. Traditional economic and medicinal uses*. The Office of the Adviser for Conservation of the Environment, Diwan of Royal Court, Sultanate of Oman.

Miller, A. N. (1996) 'Song of the Spider'. *Family Life*. May, p.26.

Miller, F. and Adam, K. L. (eds) (1992) 'Wise Management of Tropical Forests'. *Proceedings of the Oxford Forestry Institute Conference on Tropical Forests*.

Milliken, W. (1997) Malaria and antimalarial plants in Roraima, Brazil. *Tropical Doctor*, 27 (suppl.1), 20-25.

Milliken, W., Miller, R. P., Pollard, S. R. and Wandelli, E. V. (1992) *Ethnobotany of the Waimiri Atroari Indians of Brazil*. Royal Botanic Gardens, Kew.

Millimouno, D. (1991) Portée philosophique des mythes en milieu traditionnel Kissi (Centre d'application Kissidougou). Mémoire de Diplôme de Fin d'Etudes. Université de Kankan, République de Guinea.

Milton, K. (1996) Cultural and ecological diversity of forest-based Amazonian societies in Brazil. Paper presented at the working conference 'Endangered Languages, Endangered Knowledge, Endangered Environments', Berkeley, California, October 25-27, 1996.

Ministry of Forestry, Directorate General of Forest Protection and Nature Conservation (1991) Kutai National Park, Kalimantan Timur, Indonesia - Developmental plan 1992-1996, operating plan 1992, framework for management - September

Minta, S. C., Minta, K. A. and Lott, D. F. (1992) Hunting Associations Between Badgers and Coyotes. *Journal of Mammalogy* 73:814-820.

Mishkin, B. (1940) Cosmological Ideas among the Indians of the Southern Andes, *Journal of American Folklore*, 53(210):225-41.

Mishra, P. K. (1980) Boundary Maintainence among Cholanaickan: The Caveman of Kerala. In: *Man in India*, Vol.60.

Mitra, A. and Pahl, S. (1994) The Spirit of the Sanctuary. In: *Down to Earth* 31:21-36.

Mitra, A. and Pal, S. (1994) The Spirit of the Sanctuary. *Down To Earth*, Vol 2, No 17, 31 January 1994.

Mittermeier, R. A. *et al.* (1994) *Lemurs of Madagascar*. Conservation International, Washington, D.C.

Mohanty, A. K. (1994a) *Bilingualism in a Multilingual Society, Psycho-social and Pedagogical Implications*. Central Institute of Indian Languages, Mysore.

Mohanty, A. K. (1994b) Bilingualism in a multilingual society: Implications for cultural integration and education. Keynote address, 23rd International Congress of Applied Psychology, July 17-22 1994, Madrid, Spain.

Molina. F. S. (1996) 'Wa huya ania ama vutti yo'oriwa – the wilderness world is respected greatly: The Yoeme (Yaqui) truth from the Yoeme communities of Arizona and Sonora, Mexico'. Paper presented at the working conference 'Endangered Languages, Endangered Knowledge, Endangered Environments', Berkeley, California, October 25-27, 1996.

Molina. F. S. (1998) 'Wa huya ania ama vutti yo'oriwa – the wilderness world is respected greatly: The Yoeme (Yaqui) truth from the Yoeme communities of Arizona and Sonora, Mexico'. In: Maffi, L. (ed.) *Language, Knowledge, and the Environment: The Interdependence of Biological and Cultural Diversity*. Submitted to Smithsonian Institution Press.

Molyneaux, B. L. (1995) The Sacred Earth – Spirits of the Landscape, Ancient Alignments and Sacred Sites, Creation and Fertility. In: Vitebsky, P. (Series Consultant) *Living Wisdom*. Macmillan in association with Duncan Baird Publishers, London.

Momen, M. (n.d.) *Bahai'i Focus on Development*. Bahai'i Publishing Trust, Leicestershire.

Monbiot, G. (1994) *No Man's Land*. Macmillan, London.

Mondjannagni, A. C. (1975) Vie rurale et rapports ville-campagne dans le Bas-Dahomey. Thèse pour le doctorat d'Etat es Lettres. Paris, 2 tomes. 720 pages.

Moniaga, S. (1993) Toward community-based forestry and recognition of *adat* property rights in the outer islands of Indonesia, *East-West Centre Program on Environment*, Paper No.16.

Monier-Williams, M. (1899) *A Sanskrit-English dictionary* (Reprint 1964). Oxford.

Montecinos, C. (1996) *Sui Generis* - a Dead End Alley. *Seedling*, 13:4:19-28. GRAIN.

Moore, D. (1996) 'The systematic audio-video documentation of indigenous languages in Brazil'. Paper presented at the working conference 'Endangered Languages, Endangered Knowledge, Endangered Environments', Berkeley, California, October 25-27, 1996.

Moore, D. (1998) A tape documentation project for native Brazilian languages. In: Maffi, L. (ed.) *Language, Knowledge, and the Environment: The Interdependence of Biological and Cultural Diversity*. Submitted to Smithsonian Institution Press.

Moore, Richard H. (1996) 'Sustainability and the Amish: Chasing butterflies?' *Culture and Agriculture*. 53:24-25.

Moran, E. (1991) Human Adaptive Strategies in Amazonian Blackwater Ecosystems. *American Anthropologist* 93: 361-382.

Moran, E. (1996) 'Nurturing the Forest Among Native Amazonians'. In: Roy Ellen and Katsuyoshi Fuki (eds) *Redefining Nature: Ecology, Culture, and Domestication*. Berg: Washington, D.C.

Moran, E. F. (1990) 'Levels of analysis and analytical level shifting: examples from Amazonian ecosystem research'. In: Moran, E.F. (ed.), *The Ecosystem Approach in Anthropology: From concept to practise*. 279-309. University of Michigan Press, Ann Arbor, MI.

Moran, E. F. (ed.) (1984) *The Ecosystem Concept in Anthropology*. Westview Press, Boulder, CO.

Moran, E. F. (ed.) (1990) *The Ecosystem Approach in Anthropology: From concept to practise*. University of Michigan Press, Ann Arbor, MI.

Morawetz, W., Henzl, M. and Wallnfer, B. (1992) 'Tree killing by herbicide producing ants for the establishment of pure Tococa occidentalis populations in the Peruvian Amazon.' *Biodiversity and Conservation* 1: 19-33.

Morris, G. T. (1995) 12th Session of UN Working Group on Indigenous Peoples. The Declaration Passes and the US Assumes a New Role. *Fourth World Bulletin. Issues in Indigenous Law and Politics*. University of Colorado at Denver. 4(1-2): 1ff.

Morrison, J. (1997) Protected Areas, Conservationists and Aboriginal Interests in Canada. In: Ghimire, K. B. and Pimbert, M. P. (eds) *Social Change and Conservation*. UNRISD and Earthscan, London.

Mosse, David (1994) Authority, Gender and Knowledge: Theoretical Reflections on the Practice of Participatory Rural Appraisal. *Development and Change* 25: 497-526.

Mosse, David (1995) People's Knowledge in Project Planning: The Limits and Social Conditions of Participation in Planning Agricultural Development. *Odi Network Paper 58*. London.

Motte, E. (1979) 'Thérapeutique chez les Pygmées Aka de Mongoumba'. In: Bahuchet, S. (ed.) *Pygmées d'Afrique Centrale*. Pp. 77-108. SELAF, Paris.

Motte, E. (1980) 'A propos des thérapeutes Pygmées Aka de la région de la Lobaye (Centrafrique)'. Journal d'Agriculture traditionalle et de botanique appliquée, 27: 113-132.

Motte-Florac, Bahuchet S., and Thomas, J. M. C. (1993) 'The role of food in the therapeutics of the Central African Republic'. In: Hladik, C. M., Hladik, A., Linars, O. F., Pagezy, H., Semple, A. and Hadley, M.*Tropical Forests, People and Food: Biocultural Interactions and Applications to Development*. Man and the Biosphere Series, Vol. 13. UNESCO and Parthenon Press, Paris,

Mountford, C. P. (1965) *Ayers Rock: Its People, Their Beliefs and Their Art*. East-West Center Press, Honolulu.

Mshegeni, K. E., Nkunya, M. H. H., Fupi, V., Mahunnah, R. L. A. and Mshiu, E. N. (1991) *Proceedings of an International Conference of Experts from Developing Countries on Traditional Medicinal Plants*. Dar Es Salaam University Press.

Muffet, T. (1658) *The Theater of Insects*, appended to Edward Topsell, *History of Foure-footed Beasts and Serpents*, revised and enlarged by J. R[owland]. E. Cotes, London, p. 1075.

Mühlhäusler, P. (1990) 'Reducing' Pacific languages to writings. In: Joseph, J. E. and Taylor, T. J. (eds) *Ideologies of Language*. 189-205. Routledge, London.

Mühlhäusler, P. (1995a) The interdependence of linguistic and biological diversity. In: Myers, D. (ed.) *The Politics of Multiculturalism in the Asia/Pacific*. 154-161. Northern Territory University Press, Darwin, Australia.

Mühlhäusler, P. (1995b) The importance of studying small languages. *The Digest of Australian Languages and Literacy Issues* 13, May 1995.

Mühlhäusler, P. (1996) *Linguistic Ecology: Language Change and Linguistic Imperialism in the Pacific Rim*. Routledge, London.

Muir, J. (1901) *Our National Parks*. Houghton Mifflin, Boston and New York.

Muir, J. (1916) *A Thousand-Mile Walk to the Gulf*. Bade, W. (ed.), Houghton Mifflin, New York.

Muir, J. (1987) (1911 reprint) *My First Summer in the Sierra*. Penguin, New York.

Muir, J. (1988) (1914 reprint) *The Yosemite*. Sierra Club Books, San Francisco.

Muir, Star (1994) 'The Web and the Spaceship: Metaphors of the Environment'. *Et Cetera*, 51:145.

Munthali, S. M. (1993) Traditional and Modern Wildlife Conservation in Malawi: The Need for an Integrated Approach. *Oryx* 27:185-187.

Munzer, S. R. (1990) *A Theory of Property.* Cambridge University Press, Cambridge.

Munzer, S. R. (1994) 'An Uneasy Case against Property Rights in Body Parts'. In: Ellen Frankel Paul *et al.* (eds) *Property Rights.* Cambridge University Press, Cambridge.

Mussolini, G. (1980) *Ensaios de Antropologia Indígena e Caiçara.* (ed.) Paz e Terra, Rio de Janeiro. 289p.

Myers, F. R. (1986) *Pintubi Country, Pintubi self: sentiment, place, and politics among Western Desert Aborigines.* Australian Institute of Aboriginal Studies, Canberra.

native-l@gnosys.svle.ma.us.

Nabhan, G. P. (1996) Discussion paper for the colloquium 'Losing Species, Languages, and Stories: Linking Cultural and Environmental Change in the Binational Southwest', Arizona-Sonora Desert Museum, Tucson, AZ, April 1-3, 1996.

Nabhan, G. P. and Carr, J. L. (eds) (1994) *Ironwood: an ecological and cultural keystone of the Sonoran desert.* Conservation International, Wahington, D.C.

Nabhan, G. P. and St. Antoine, S. (1993) The loss of floral and faunal story: The extinction of experience. In: Kellert, S. R.and Wilson, E. O. *The Biophilia Hypothesis.* 229-250. Island Press, Washington, D.C.

Nadkarni, A. K. (1982) *Indian Materia Medica. Vol. 1.* Popular Prakashan, Bombay.

Naess, A. (1973) The shallow and the deep, long-range ecology movement: a summary. *Inquiry* 16.

Naess, A. (1989) *Ecology, Community and Lifestyle: outline of an ecosophy / Arne Naess.* Translated and revised by David Rothenburg. Cambridge University Press, Cambridge.

Naess, A. (1992) Deep ecology and ultimate premises. *Society and Nature* 1(2): 108-119.

Nakashima, D. and Roué, M. (1993) (1994) Unpublished field notes, Chisasibi.

Nakashima, D. and Roué, M. (1995) Allées et venues dans l'espace humain: déclin des populations de caribou et notion de cycle chez les scientifiques et les Inuit du Québec arctique, *Anthropozoologica* 21: 21-30.

Nanda, Bikram (1994) *Contours of Continuity and Change: The story of the Bonda Highlanders.* Sage, New Delhi.

Nanda, N. (1990) 'Who is Destroying the Himalayan Forests?' In: Rusomji, N. K. and Ramble, C. (eds) *Himalayan Environment and Culture.* 48-60. Indian Institute of Advanced Study and Indus Publishing Company, Shimla and New Delhi.

Nash, J. (1991) *Loving Nature: Ecological integrity and christian responsibility.* Abingdon Press, Nashville.

Nash, J. (1993) Biotic rights and human ecological responsibilities. *The Annual of the Society of Christian Ethics,* SCE, Boston.

Nash, R. (1989) *The Rights of Nature: A history of environmental ethics.* University of Wisconsin Press, Madison.

Nasr, S. (1990) *Man and Nature: The spiritual crisis in modern man.* Unwin Hyman Limited, London.

National Research Council (1989) *Alternative Agriculture.* National Academy Press, Washington, D.C.

Neihardt, J. G. (1932) *Black Elk Speaks: Being the Life Story of a Holy Man of the Oglala Sioux.* University of Nebraska Press, Lincoln.

Neihardt, J. G. (1959) *Black Elk Speaks.* Washington Square Press.

Nelson, R. H. (1991) *Reaching for Heaven on Earth.* Rowman and Littlefield, Lanham, Md.

Nelson, R. K. (1969) *Hunters of the Northern Ice.* University of Chicago Press, Chicago, Il.

Nelson, R. K. (1983) *Make Prayers to the Raven: A Koyukon View of the Northern Forest.* University of Chicago Press, Chicago, Il.

Nelson, R. K. (1983) *Make Prayers to the Raven: A Koyukon View of the Northern Forest.* University of Chicago Press, Chicago.

Nelson, R. K. (1986) *Hunters of the Northern Forest,* 2d (ed.) University of Chicago Press, Chicago, Il.

Nepstad, D. C. and Schwartzman, S. (eds) (1992) Non-Timber Products from Tropical Forests; Evaluation of a Conservation and Development Strategy. *Advances in Economic Botany,* 9. The New York Botanical Garden, New York.

Netting, R. M. (1986) *Cultural Ecology,* 2d (ed.) Waveland Press, Prospect Heights, Il.

Netting, Robert McC. (1968) *Hill Farmers of Nigeria: Cultural Ecology of the Kofyar of the Jos* Plateau. University of Washington Press, Seattle.

Ngũgĩ, wa T. (1986) *Decolonising the Mind: The Politics of Language in African Literature.* James Currey, London.

Niamir, M. (1990) Community Forestry: Herders' Decision Making in Natural Resource Management in Arid and Semi-arid Africa. *Community Forestry Note* 4, FAO, Rome.

Nichols, J. (1992) *Linguistic Diversity in Space and Time.* University of Chicago Press, Chicago, Il.

Nichols, W. D. (Western Region Ground Water Specialist, U.S.G.S.). 10-28-(1993) letter to W. M. Alley, Chief, Office of Ground Water, Water Resources Division, U.S.G.S.

Nicholson-Lord, D. (1987) *The Greening of the Cities.* Routledge and Kegan Paul, London/New York.

Nietschmann, B. (1979) *Caribbean Edge: The Coming of Modern Times to Isolated People and Wildlife.* Bobbs-Merrill, New York.

Nietschmann, B. (1985) 'Torres Strait Islander sea resource management and sea rights'. In: Ruddle, K. and Johannes, R. E. (eds) *The Traditional Knowledge and Management of Coastal Systems in Asia and the Pacific.* 124-154. UNESCO, Jakarta.

Nietschmann, B. Q. (1992) The Interdependence of Biological and Cultural Diversity. Occasional Paper #21, Center for World Indigenous Studies, December 1992.

Nijar, G. S. (1994) *Towards a Legal Framework for Protecting Biological Diversity and Community Intellectual Rights - a Third World perspective.* Third World Network, Penang.

Nijar, G. S. (1996) *In Defence of Biodiversity and Indigenous Knowledge: A Conceptual Framework and the Essential Elements of a Rights Regime.* Third World Network, Penang.

Nijar, G. S. (1996) *In Defence of Local Community Knowledge and Biodiversity: A Conceptual Framework and the Essential Elements of a Rights Regime.* Third World Network, Penang.

Norgaard, R. (1988) Sustainable Development: A co-evolutionary view. *Futures* (December).

Norgaard, R. B. (1994) *Development Betrayed.* Routledge, London.

Northern Land Council (1996) *Ecopolitics IX: Perspectives on Indigenous Peoples' Management of Environmental Resources.* NLC, Casuarina.

Norton, B. (1988) The constancy of Leopold's land ethic. *Conservation Biology* 2, 93-102.

Norton, B. (1989) Intergenerational equity and environmental decisions: a model using Rawls' veil of ignorance. *Ecological Economics* 1(2): 137-159.

Norton, B. (1991) *Toward Unity among Environmentalists.* Oxford University Press, New York.

Norton, B. (1992) Sustainability, human welfare and ecosystem health. *Environmental Value* 1, 97-111.

Norton, B. (1995) Why I am not a nonanthropocentrist. *Environmental Ethics* 17, 341-358.

Norton, B. (1997a) *Can There Be a Universal Earth Ethic? Reflections on values ror the proposed earth charter.* Working Paper No. 92. Man and Nature Center, Odense, DK.

Norton, B. (1997b) Environmental values: a place-based theory. *Environmental Ethics* 19, 227-245.

Norton, H. *et al.* (1983) 'The Klickitat Trail of South-Central Washington: A Reconstruction of Seasonally Used Resource Sites'. In: *Prehistoric Places in the Southern Northwest Coast.* (ed.) Greengo, R. Burke Memorial Museum Research Report, no. 4. University of Washington, Seattle.

Noteman, T. (1769) See *'An account of the diseases, natural history and medicine of the East Indies, translated from the Latin of James Bontius, Physician to the Dutch settlement at Batavia, to which are added annotations by a physician'.* T. Noteman, London.

NRC (1989) *Alternative Agriculture.* National Academy Press, Washington, D.C.

Ntiamo-Baidu, Y., Gyiamfi-Fenteng, L. J. and Abbiw, W. (1992) *Management Strategies for Sacred Groves in Ghana.* Report prepared for the World Bank and EPC Ghana.

Numata, M. (1980) Changes in Ecosystems Structure and Function in Tokyo. In: Bornkamm, R., Lee, J. A. and Seaward, M. R. D.(eds) *Urban Ecology.* Blackwell Scientific Publications, London, pp. 139-147.

Nuttall, M. (1994) 'Greenland: emergence of an Inuit homeland'. In: Minority Rights Group (ed.). *Polar peoples: self-determination and development.* Minority Rights Publications, London, UK.

Nvorteva, P. (1971) The Synantrophy of Birds as an Expression of the Ecological Cycle Disorder Caused by Urbanization. *Annales Zoologici Fennici* 8:547-553.

Nye, P. H. and Greenland, D. J. (1960) *The soil under shifting cultivation.* Commonwealth Bureau of Soils Technical Bulletin No.51, Harpenden.

O'Barr, W. M. (1982) *Linguistic Evidence: Language, Power and Strategy in the Courtroom.* Academic Press, New York.

O'Donoghue, N. D. (1993) *The Mountain Behind the Mountaln: Aspects of the celtic tradition.* T. and T. Clark, Edinburgh.

O'Keragori, A. (1995) *Totems of the Kisii.* Jacaranda Designs, Nairobi.

O'Neill, J. (1993) *Ecology, Policy and Politics: Human well-being and the natural world.* Routledge, London.

O'Neill, J. (1997) 'Time, Narrative and Environmental Politics'. In: R. Gottlieb ed. Ecological Community, London, Routledge.

O'Neill, J. (1997) Managing without prices: On the monetary valuation of biodiversity. *Ambio* 26, pp. 546-550.

Odum, E. P. (1962) 'Relationships Between Structure and Function in the Ecosystem'. *Japanese Journal of Ecology.* 12, No.3:108-118.

Odum, E. P. (1997) *Ecology: A Bridge Between Science and Society.* Sinauer Associates, Sunderland, MA. 283 pp.

Ofori, C. S. (1973) 'Decline in fertility status in a tropical ochrosol under continuous cropping'. *Experimental Agriculture* 9: 15-22.

Ogutu, N. (ed.) (1992) *God, Humanity and Mother Earth.* The New Ecumenical Research Association, New York.

Ohio Amish Directory: Holmes County and Vicinity 1996 Edition. Carlisle Printing, Walnut Creek, Ohio.

Ohmae, K. (1990) *The Borderless World.* Collins, London. p. 18.

Oikonomou, D., Oswald, P.and Palmer, M. (1993) *Final Report of the Ecumenical Patriarchate, WWF International and WWF Greece Team on the Ecological Status of the Monasteries of Mount Athos.*

Oldfield, M. L. and Alcorn, J. B. (1991) 'Conservation of traditional agroecosystems'. In: Oldfield, M. L. and Alcorn, J. B. (eds) *Biodiversity: Culture, Conservation and Ecodevelopment,* 37-58. Westview Press, Boulder, San Francisco.

Oliveira, R. R. and Coelho Netto, A. L. (1996) 'O Rastro do Homem na Floresta. A construção da paisagem da Reserva Biológica Estadual da Praia do Sul a partir de intervenções antrópicas'. *Albertoa* 4 (10):109-116.

Oliveira, R. R., Lima, D. F., Delamonica, P., Toffoli, D. D. G. and Silva, R. F. (1994) 'Roça Caiçara: um sistema 'primitivo' auto-sustentável'. *Ciência Hoje* 18 (104):44-51.

Ollikainen, M. (1995) *Vankkurikansan perilliset. Romaanit, Euroopan unohdettu vähemmistö.* Ihmisoikeusliitto r.y.:n julkaisuja 3. Helsinki University Press, Helsinki.

Ololtisatti, O. K. (n.d.) *State and Community Conflicts in Natural Resource Management: Trust Lands Cap. 288 Laws of Kenya (Cap. IX of Kenya Constitution),* World Resources Institute, Washington, D.C.

Omo-Fadaka, Jimoh (1990) 'Communalism: The moral factor in African development'. In: *Ethics of environment and development: Global challenge, international response.* (eds) J.R. and J.G. Engel. Belhaven Press, London. 176-182.

Ontario Ministry of Natural Resources (1987) *Wild Rice Report 1986 (Northwestern Region of Ontario)*. Queen's Printer for Ontario, Toronto.

Onyewuenyi, Innocent (1991) 'Is there an African philosophy?'. In: *African philosophy: The essential readings.* (ed.) Serequeberhan, T. Paragon House, New York. 29-46.

Opoku, K. A. (1978) *West African Traditional Religion.* FEP International Pvt., Lagos.

Orbell, M. (1985) *The Natural World of the Maori.* Collins/ Bateman, Auckland.

Orlove, Benjamin S. (1996) 'Anthropology and the Conservation of Biodiversity'. *Annual Review of Anthropology* 25:329-52.

Orr, D. (1992) *Ecological Literacy: Education and the transition to a postmodern world.* State University of New York Press, Albany.

Orr, D. (1994) *Earth in Mind: On education, environment, and the human prospect.* Island Press, Washington, D.C.

Ortiz, S. (ed.) (1983) *Economic Anthropology*. University Press of America, Lanham, Md.

Osseweijer, M. (1997) 'We wander in our ancestors' yard: sea cucumber gathering in the Aru archipelago, Indonesia'. Paper presented at the third East-West Environmental Linkages Workshop on Indigenous Environmental Knowledge and its Transformations, 8-10 May, University of Kent, Canterbury.

Ostrom, E. (1990) *Governing the Commons*. Cambridge University Press, Cambridge.

Ostrom, E. (1990) *Governing the Commons: The Evolution of Institutions for Collective Action.* Cambridge University Press, New York.

Ostrom, V., Feeny, D. and Picht, H. (eds) (1988) *Rethinking Institutional Analysis and Development..* Institute for Contemporary Studies, San Francisco.

Pacific Concerns Resource Center (1995) *Proceedings of the Indigenous Peoples' Knowledge and Intellectual Property Rights Consultation*, 24-27 April, Suva, Fiji. PCRC, Suva.

Padilla, R. and Benavides, A. H. (eds) (1992) *Critical Perspectives on Bilingual Education Research.* Bilingual Review Press / Editorial Bilingüe, Tempe, Arizona.

Padoch, C. (1992) 'Marketing of non-timber forest products in western Amazonian: general observations and research priorities'. In: Nepstad, D. C. and Schwartzman, S. (eds) *Non-Timber Products from Tropical Forests; Evaluation of a Conservation and Development Strategy.* Advances in Economic Botany, 9. The New York Botanical Garden, New York.

Padoch, C. and de Jong, W. (1992) 'Diversity, variation, and change in Ribereno agriculture'. In: Redford, K. H. and Padoch, C. (eds) *Conservation of Neotropical Forests; Working from Traditional Resource Use.* Columbia University Press, New York.

Paley, W. (1802) *Natural Theology: or, Evidences of the Existence and Attributes of the Deity, Collected from the Appearances of Nature.* R. Faulder, London.

Palmer, M. (1988) *Genesis or Nemesis.* Dryad Press, London.

Palmer, P., Sánchez, J. and Mayorga, G. (eds) (1993) *Taking care of Sibo's gifts: An environmental treatise from Costa Rica's Kekoldi Indigenous Reserve.* Asociación de Desarrollo Integral de la Reserva Indigena Cocles/ KéköLdi,, San José, Costa Rica.

Paques, V. (1953) Bouffons sacrés du cercle de Bououni (Soudan Français). *Journal de la Société des Africanistes*, Paris, 23, 1& 2, 63-110.

Parajuli, P (1997) Discourse on Knowledge, Dialogue and Diversity: Peasant Cosmovision and the Science of Nature Conservation. In: Worldviews, Environment, Culture and Religion. 1: 3: 189-210.

Pareek, S. K., Gupta, Rajendra and Maheswari, M. L. (1982) A eugenol rich sacred basil (Ocimum sanctum L.) - Its domestication and industrial application. *PAFAI Journal.* 4- 13-16.

Parkin, D. (1991) *Sacred Void – Spatial Images of Work and Ritual among the Giriama of Kenya.* Cambridge University Press, Cambridge.

Parren, M. and de Graaf, N. R. (1995) The quest for natural forest management in Ghana, Côte d'Ivoire and Liberia. *Tropenbos Series 13.* Tropenbos, Wapeningen, NL.

Passmore, J. (1974) *Man's Responsibility for Nature: Ecological problems and western traditions.* Scribner's, New York. (2nd edn., 1980).

Pattanayak, D. P. (1981) *Multilingualism and Mother-Tongue Education.* Oxford University Press, Delhi.

Pattanayak, D. P. (1988) Monolingual myopia and the petals of the Indian lotus: Do many languages divide or unite a nation? In: Skutnabb-Kangas, T. and Cummins, J. (eds) *Minority Education: From Shame to Struggle.* 379-389. Multilingual Matters, Clevedon, UK.

Pattanayak, D. P. (1991) Tribal education and tribal languages: a new strategy. In: Pattanayak, D. P. *Language, Education and Culture.* 166-177. Central Institute of Indian Languages, Mysore.

Paul, F. (1994) Spruce Root Basketry of the Alaska Tlingit. Second reprint edition 1991. Friends of the Sheldon Jackson Museum, Sitka, Alaska, USA.

Pawley, A. (1996) Some problems of describing linguistic and ecological knowledge. Paper presented at the working conference 'Endangered Languages, Endangered Knowledge, Endangered Environments', Berkeley, California, October 25-27, 1996.

Pawley, A. (1998) Some problems of describing linguistic and ecological knowledge. In: Maffi, L. (ed.) *Language, Knowledge, and the Environment: The Interdependence of Biological and Cultural Diversity.* Submitted to Smithsonian Institution Press.

PCRC (1995) Pacific Concerns Resource Center. *Proceedings of the Indigenous Peoples' Knowledge and Intellectual Property Rights Consultation*, 24-27 April 1995, Suva, Fiji. PCRC, Suva, Fiji.

Pearce, D. (1993) *Economic Values and the Natural World.* Earthscan, London.

Pearce, D. and Moran, D. (1994) *The Economic Value of Biological Diversity.* Earthscan, London.

Pearce, D. W. and Turner, R. K. (1990) *Economics of Natural Resources and the Environment..* Harvester Wheatsheaf, London.

Pearce, D., Markandya, A. and Barbier, E. (1989) *Blueprint for a Green Economy*. Earthscan, London.

Pearson, N. (1995) World Heritage and Indigenous Peoples. In: Fourmile, H., Schnierer, S. and Smith, A. (eds) *An Identification of Problems and Potential for Future Rainforest Aboriginal Cultural Survival and Self-determination in the Wet Tropics*. Centre for Aboriginal and Torres Strait Islander Participation Research and Development, James Cook University of North Queensland, Townsville.

Peet, R. and Watts, M. (1996) *Liberation Ecologies: Environment, Development, Social Movements*. Routledge, London and New York. p. 52.

Peeters, A. (1979) Nomenclature and Classification in *Rumphius's Herbarium Amboinense*. In: Ellen, R. F. and Reason, D.(eds) *Classifications in their Social Contexts*. Academic Press, London.

Pei Shengji (1993) 'Managing for Biological Diversity Conservation in Temple Yards and Holy Hills: The Traditional Practices of the Xishuangbanna Dai Community, Southwest China'. In: Hamilton, L. S. (ed.) *Ethics, Religion and Biodiversity*. The White Horse Press, Knapwell, Cambridge.

Pelevin, V. (1996) *The Life of Insects* (translated from the Russian by Andrew Bromfield). Harbord Publishing, London.

Peluso, N. L. (1992) *Rich Forests, Poor People; Resource Control and Resistance in Java*. University of California Press, Berkeley.

Penguine Books, New York.

Pereira, W. (1993) *Tending the Earth: Traditional, Sustainable Agriculture in India*. Earthcare Books, Bombay.

Pereira, W. and Gupta, A. K. (1993) A dialogue on indigenous knowledge. *Honey Bee* 4: 6-10.

Pérez, M. Ruiz and Arnold, J. E. M. (eds) (1996) *Current Issues in Non-Timber forest Products Research*. ODA, Center for International Forestry Research.

Perrings, C. (1995) 'Economic values of biodiversity'. In: Heywood, V. H. (ed.) *Global Biodiversity Assessment*.. UNEP, and Cambridge University Press, Cambridge.

Perrings, C. *et al.* (eds) (1995) *Biodiversity Loss: Economic and Ecological Issues*. Cambridge University Press, Cambridge.

Perrot, Cl-H. (1989) Le système de gestion de la pêche en lagune Aby (Côte d'Ivoire) au XIXème siècle. *Cahiers des Sciences Humaines*. Vol. 25, No.1-2: 177-188.

Persoon, G. A. (1994) *Vluchten of Veranderen. Processen van ontwikkeling en verandering bij tribale groepen in Indonesië*. FSW, Leiden.

Peters, C. M. (1994) *Sustainable Harvest of Non-timber Plant Resources in Tropical Moist Forest: An Ecological Primer.* Biodiversity Support Program, Washington, D.C.

Peters, C. M., Gentry, A. H. and Mendelssohn, R. O. (1989) Valuation of an Amazonian rainforest. *Nature* 339:655-656.

Petersen, T. S. (1994) 'The home rule situation in Greenland'. In: van der Vlist, L. (ed.). *Voices of the earth: indigenous peoples, new partners and the right to self-determination in practice*. Netherlands Centre for Indigenous Peoples, Amsterdam, Netherlands. Pp. 113–123.

Peterson, D. L. and Johnson, D. R. (eds) (1995) *Human Ecology and Climate Change – People and Resources in the Far North*.. Taylor and Francis, Washington, D.C.

Phelan, P. L., Mason, J. F. and Stinner, B. R. (1995) *Agriculture, Ecosystems and Environment* 56:1-8.

Phelan, P. L., Norris, K. H. and Mason, J. F. (1996) 'Soil-management history and host preference by *Ostrinia nubilalis*: evidence for plant mineral balance mediating insect-plant interactions'. *Environ. Ent.* 25(6):1329-1336.

Phillips, O. and Gentry, A. H. (1993a) The Useful plants of Tambopata, Peru: I. Statistical hypotheses tests with a new quantitative technique. *Economic Botany* 47(1): 15-32.

Phillips, O., Gentry, A. H., Reynel, C., Wilkin, P. and Galvez-Durand, B. C. (1994) Quantitavie Ethnobotany and Amazonian Conservation. *Conservation Biology* 8(1): 225-48.

Phillipson, R. (1992) *Linguistic Imperialism*. Oxford University Press, Oxford.

Pickett, S. and White, P. (eds) (1985) *The Ecology of Natural Disturbance and Patch Dynamics*. Academic Press, San Diego.

Pierotti, R. (1991) Ravens in Winter. *Condor* 93:788-789.

Pimbert, M. P. (1994) The Need for Another Research Paradigm. *Seedling* 11:2 (July) 20-32.

Pimbert, M. P. (1994a) Editorial, Guest Editor, *Etnoecologica*, II:3 (Abril) 3-5.

Pimbert, M. P. and Pretty, D. (1995) Parks, People and Professionals: Putting 'Participation' into Protected Area Management. UNRISD-IIED-WWF, *UNRISD Discussion Paper* No.57, Geneva, 60 pp.

Pimbert, M. P. and Pretty, J. N. (1997) 'Parks, People and Professionals: Putting 'Participation' into Protected Area Management'. In: Ghimire, K. B. and Pimbert, M. P. *Social Change and Conservation*. Earthscan Publications, London, and UNRISD, 297-330.

Pimbert, M. P. and Toledo, V. (1994) Indigenous People and Biodiversity Conservation: Myth or Reality? Special Issue of *Ethnoecologica*, 2 (3). Mexico, 96 pp.

Pimentel, D. *et al.* (1992) 'Conserving Biological Diversity in Agricultural/Forestry Systems'. *Bioscience*. 42(5):360.

Pinchot, G. (1947) *Breaking New Grouond*. Harcourt, Brace, New York.

Pinendo-Vasquez, M., Zarin, D., Jipp, P. and Chota-Inuma, J. (1990) Use-values of tree species in a communal forest reserve in northeast Peru. *Conservation Biology* 4(4): 405-416.

Pinzon, C. and Garay, G. (1990) 'Por los senderos de la construccion de la verdad y la memoria'. In: *Por las rutas de nuestra America*. Universidad Nacional de Bogota, Bogota, Colombia.

Plaskow, J. and Chirst, C. (eds) (1989) *Weaving the Visions: New patterns in feminist spirituality*. Harper and Row, San Francisco.

Platt, R. H., Rowntree, R. A. and Muick, P. C.(eds) (1994) *The Ecological City: Preserving and Restoring Urban Biodiversity*. University of Massachusetts Press, Amherst.

Platteau, J. P. (1994a) Behind the market stage where real societies exist, 1: the role of public and private order institutions. *Journal of Development Studies* 30(3): 533-578.

Platteau, J. P. (1994b) Behind the market stage where real societies exist, 2. The role of moral norms. *Journal of Development Studies* 30(4): 753-818.

Plattner, S. (ed.) (1989) *Economic Anthropology*. Stanford University Press, Stanford, CA.

Ploeg, J. D. van der (1993) Potatoes and Knowledge. In: Hobart, M. (ed.) *An Anthropological Critique of Development: The Growth of Ignorance*. Routledge, London/New York.

Plotkin, M. and Famolare, L. (eds) (1992) *Sustainable Harvest and Marketing of Rain Forest Products.* Conservation International and Island Press, Washington, D.C.

Plowman, T., Leuchtmann, A., Blaney, C. and Clay, K. (1990) 'Significance of the fungus Balansia cyperi infecting medicinal species of Cyperus (Cyperaceae) from Amazonia'. *Economic Botany* 44(4):452-462.

Plucknett, D. and Horne, M. E. (1992) 'Conservation of Genetic Resources'. *Agriculture, Ecosystems, and the Environment.* 42: 75-92.

Pocaterra, F. (1995) Address to the Global Initiative For Traditional Systems (GIFTS) of Health Regional Conference for South and Central America, Unpublished report, GIFTS of Health, Oxford.

Poerbo, H. (1986) Mencari Pendekatan Pengelolaan Lingkungan Kota Yang Lebih Efektif. *Prisma*, No 5 Tahun XV, pp. 81-92.

Poffenberger, Mark and McGean, Betsy. (1996) *Village Voices, Forest Choices: Joint Forest Management in India.* Oxford University Press, New Delhi.

Polanyi, K. (1944) *The Great Transformation*. Rinehart, New York.

Polasky, S and Solow, A R. (1995) 'On the value of a collection of species'. *Journal of Environmental Economics and Management* 29 298-303

Poole, P. J. (1993) Indigenous Peoples and Biodiversity Protection. In: Davis, S. H. The Social Challenge of Biodiversity Conservation. *Working Paper* No.1, Global Environment Facility, pp. 14-25.

Poole, P. J. (1995) *Indigenous People, Mapping, and Biodiversity Conservation.* Biodiversity Support Program, World Wildlife Fund, Washington, D.C.

Poore, D., Burgess, P., Palmer, J., Rietbergen, S. and Synnott, T. (1989) *No Timber Without Trees; sustainability in the Tropical Forest.* Earthscan Publications Ltd., London.

Porteous, A. (1928) *The Lore of the Forest: Myths and Legends*. Guernsey Press Co. Ltd., Guernsey.

Posey, D. A. (1982) Keepers of the Forest. *New York Botanical Garden Magazine* 6(1): 18-24.

Posey, D. A. (1983) 'Indigenous ecological knowledge and development of the Amazon'. In: Moran, E. F. (ed.) *The Dilemma of Amazonian Development.* Westview Press, Colorado.

Posey, D. A. (1983) Indigenous knowledge and development: an ideological bridge to the future? *Ciência e Cultura* 35: 877-894.

Posey, D. A. (1984) 'Ethnoecology as Applied Anthropology in Amazonian Development'. *Human Organization* 43 (2).

Posey, D. A. (1985) Indigenous Management of Tropical Forest Ecosystems: The Case of the Kayapó Indians of the Brazilian Amazon. *Agfroforestry Systems* 3: pp. 139-154.

Posey, D. A. (1987) Contact before contact: typology of post-Colombian interaction with Northern Kayapó of the Amazon Basin. *Boletim do Museu Paraense Emilio Goeldi, Serie Antropologica* 3: 135-154.

Posey, D. A. (1988) 'The Declaration of Belem'. In: *Proceedings of the First International Congress of Ethnobiology*, (eds) Posey, D. A. and Overal, W. Museu Paraense Goeldi, Belem.

Posey, D. A. (1990) Intellectual Property Rights and Just Compensation for Indigenous Knowledge. *Anthropology Today*, 6(4): 13-16.

Posey, D. A. (1990) *The Application of Ethnobiology in the Conservation of Dwindling Natural Resources: Lost Knowledge or Options for the Survival of the Planet.* Proceedings from the First International Congress of Ethnobiology, Brazil: 47-61.

Posey, D. A. (1992) Interpreting and Applying the 'Reality' of Indigenous Concepts: What is Necessary to Learn from the Natives? In: Redford, K. H. and Padoch, C. (eds) *Conservation of Neotropical Forests: Working from Traditional Use*. Columbia University Press, pp. 21-34.

Posey, D. A. (1993) Indigenous knowledge and development: an ideological bridge to the future. *Ciência e Cultura*. 35(7):877-894, São Paulo.

Posey, D. A. (1993) The Importance of Semi-Domesticated Species in Post-Contact Amazonia: Effects of Kayapo Indians on the Dispersal of Flora and Fauna. In: Hladik, C. M., Hladik, A., Linares, O. F., Pagezey, H., Semple, A. and Hadley, M.(eds) Tropical Forests, People and Food: Biocultural Interactions and Applications to Development. *Man and Biosphere*, Vol 13, UNESCO, Paris, pp. 63-71.

Posey, D. A. (1994) 'Traditional Resource Rights: de facto self-determination for indigenous peoples'. In: Leo van der Vlist (ed.) *Voices of the Earth: indigenous peoples, new partners & the right to self-determination in practice.* International Books, Utrecht.

Posey, D. A. (1995) Indigenous Peoples and Traditional Resource Rights: A Basis for Equitable Relationships? *Proceedings of a Workshop*, Green College Centre for Environmental Policy and Understanding, 28th June 1995, Oxford, UK.

Posey, D. A. (1996) 'Provisions and Mechanisms of the Convention on Biological Diversity for Access to Traditional Technologies and Benefit Sharing for Indigenous and Local Communities Embodying Traditional Lifestyles'. Oxford Centre for the Environment, Ethics and Society (OCEES) Research Paper No. 6.

Posey, D. A. (1996) Commonwealth Forestry Paper, Oxford Forestry Institute, Oxford.

Posey, D. A. (1996) Indigenous knowledge, biodiversity, and international rights: learning about forests from the Kayapo Indians of the Brazilian Amazon. In: Grayson, A. J. (ed.). *The Commonwealth Forestry Review* 76(1): 53-60. The Commonwealth Forestry Association, Oxford.

Posey, D. A. (1996) Indigenous Peoples and Traditional Resource Rights: A Basis for Equitable Relationships? In: Sultan, R., Josif, P., Mackinolty, C. and Mackinolty, J. (eds) *Ecopolitics IX, Perspectives on Indigenous Peoples' Management of Environment Resources.* Conference Papers and Resolutions, Northern Land Council, Casuarina, pp. 43-60.

Posey, D. A. (1996) Protecting Indigenous Peoples' Right to Biodiversity: People, Property, and Bioprospecting. *Environment*, Vol. 38(8): 6-9, 37-45.

Posey, D. A. (1996) *Traditional Resource Rights: International Instruments for Protection and Compensation for Indigenous Peoples and Local Communities*. (with contributions by G. Dutfield, K. Plenderleith, E. da Costa e Silva and A. Argumedo). IUCN. Gland, Switzerland, and Cambridge, UK.

Posey, D. A. (1997) The Kayapó. In: *Indigenous Peoples and Sustainability: Cases and Actions* IUCN Inter-Commission Task Force on Indigenous Peoples. International Books, Utrecht. 240-254.

Posey, D. A. and Balée, W. (eds) (1989) 'Resource Management in Amazonian: Indigenous and Folk Strategies'. *Advances in Economic Botany* 7.

Posey, D. A. and Dutfield, G. (1996) *Beyond Intellectual Property Rights: Toward International Resource Rights for Indigenous Peoples and Local Communities*. IDRC, Ottawa, 303 pp.

Posey, D. A. and Dutfield, G. (1997) *Indigenous Peoples and Sustainability: Cases and Actions.* IUCN and International Books, Utrecht.

Posey, D. A. and Overal, W. (eds) (1990) *Ethnobiology: Implications and Applications: Proceedings of the First International Congress of Ethnobiology. Volume 1 (Theory and Practice; Ethnozoology)*. MPEG/CNPq/MCT, Belém).

Posey, D. A., Frechione, J., Eddins, J., Francelino-Da Silva, L., Myers, D., Case, D. and MacBeath, P. (1994) 'Ethnoecology as Applied Anthropology in Amazonian Development'. *Human Organization* 43(2):95-107.

Postman, N. (1985) *Amusing Ourselves to Death — Public discourse in the age of show business*. Viking Penguin, New York.

Prain, Gordon, Fujisaka, Sam and Warren, D. Michael (eds) (forthcoming). *Biological and Cultural Diversity: The Role of Indigenous Agricultural Experimentation in Development*. Intermediate Technology Publications, London.

Prance, G. T., Balee, W., Boom, B. M. and Carneiro, R. L. (1987) 'Quantitative Ethnobotany and the case for conservation in Amazonia'. *Conservation Biology* 14: 269-310.

PRATEC (1993) *Afirmacin Cultural Andina*, Proyecto Andino de Tecnologĺas Campesinas (ed) Lima, Peru.

PRATEC (1993) *Desarrollo o Descolonización en los Andes?* Lima: Proyecto Andino de Tecnologĺas Campesinas (ed) Lima, Peru.

PRATEC (1995) *Crianza Andina de la Chacra*, Proyecto Andino de Tecnologias Campesinas (ed) Lima, Peru.

PRATEC - PPEA (PNUMA) (1989) *Manejo Campesino de Semillas en los Andes*. PRATEC, (eds) Proyecto Piloto de Ecosistemas Andinos, PPEA, and Proyecto Andino de Tecnologias Campesinas, Lima, Peru.

PRATEC - PPEA (PNUMA) (1991) *Vigorización de la Chacra Andina*, Proyecto Andino de Tecnologias Campesinas (ed) Proyecto Piloto de Ecosistemas Andinos, PPEA, Programa de Naciones Unidas Para el Medio Ambiente, PNUMA, Cajamarca.

Preston, R. J. (1981) 'East Main Cree', pp. 196-207, In: Helm, J. (ed.) *Handbook of North American Indians - Subarctic*, Vol. 6. Smithsonian Institution, Washington, D.C.

Preston, S. R. (1990) 'Investigation of compost/fertiliser interactions in sweet potato grown on volcanic ash soils in the highlands of Papua New Guinea'. *Tropical Agriculture* 67:3: 239-42.

Pretty, J. (1995) *Regenerating Agriculture*. World Resources Institute. Washington, D.C.

Pretty, J. N. (1994) Alternative Systems of Inquiry for Sustainable Agriculture. IDS, University of Sussex, *IDS Bulletin* 25(2): 37-44.

Price, L. W. (1981) *Mountains & Man: A Study of Process and Environment*. University of California Press, Berkeley and Los Angeles.

Price, M. (1996) Saving China's holy mountains. *People & the Planet* 5(1):12-13.

Prigogine, I. and Stengers, I. (1980) *Order out of Chaos*. Bantam, New York.

Primavesi, A. (1991) *From Apocalypse to Genesis: Ecology, feminism, and christianity*. Fortress Press, Minneapolis.

Principe, P. P. (1989) 'Valuing the biodiversity of medicinal plants'. In: Akerele, O., Heywood, V., Synge, H. (eds). *The conservation of medicinal plants*. Cambridge University Press, Cambridge, UK.

Proctor, J. (ed.) (1989) *Mineral Nutrients In Tropical Forest And Savanna Ecosystems*. Blackwell, Oxford.

Prott, L. V. (1988) 'Cultural Rights as Peoples' Rights in International Law'. In: Crawford, J. (ed.) *The Rights of Peoples*. Clarendon Press, Oxford. pp. 93-106.

Pryor, F. L. (1977) *The Origins of the Economy*. Academic Press, New York.

Pujol, R. (1975) Definition d'un Ethnoecosysteme avec Deux Exemples: Etude Ethnozoobotanique des Carderes (Dipsacus) et Interrelations Homme-Animal-Truffe. In: Pujol, R. (ed) *L'homme et Animal: Premier Colloque d'Ethnozoologie*. Institut International d'Ethnosciences, Paris: pp. 91-114.

Purseglove, J. W. (1975) *Tropical Crops Monocotyledons*. 142-143.

Putnam, R. (1993) *Making Democracy Work: Civic Traditions in Modern Italy*. Princeton University Press, Princeton, N.J.

Quammen, D. (1996) *The Song of the Dodo: Island biogeography in an age of extinction*. Simon and Schuster, New York.

Quansah, P., Andrianarisata, S. and Quansah, N. (1996) The applied Ethnobotanical research approach to biodiversity conservation: The Madagsacan example (abs.). The GIFTS of Health Reports, *J. Altern. and Complem. Med.* 2,3, 397-4-5 and 435-447.

Quenum, J. F. (1980) *Milieu naturel et mise en valeur agricôle entre Sakété et Pobé dans le Sud-Est du Bénin (Afrique Occidentale).* Université Louis Pasteur. UER de Geographie-AmÈnagement régional et développement. Strasbourg.

Quinn, D. (1992) *Ishmael: An Adventure of the Mind and Spirit.* Bantam Publishing, New York.

Raber, Ben J. (1996) *The New American Almanac: For the Year of Our Lord 1996 Leap Year* (Twenty Seventh Edition). Raber's Book Store Baltic, Ohio 43804.

Rackham, O. (1996) Corpus Christi College, Cambridge. From a letter to the Editor, *Tree News,* Spring 1996. The Tree Council Magazine, 51 Catherine Place, London.

Raeburn, P. (1995) *The Last Harvest: The Genetic Gamble that Threatens to Destroy American Agriculture.* Simon and Schuster, New York. p. 40.

RAFI (1997) 'Sovereignty or Hegemony? Africa and Security: negotiating from reality'. *RAFI Communique,* May/June.

RAFI (Rural Advancement Foundation International) (1994a) *Conserving indigenous knowledge: integrating two systems of innovation.* An independent study by the Rural Advancement Foundation International. Commissioned by the United Nations Development Programme, New York.

RAFI (Rural Advancement Foundation International) (1994b) *An Overview of Bioprospecting.* Pittsboro, NC.

RAFI – Christie, J. and Mooney, P. (1994) *Enclosures of the Mind - Intellectual Monopolies.* Rural Advancement Foundation International. Ottawa. Electronic format on RAFI's website at http://www.rafi.ca

Rainforest Aboriginal Network (eds) (1993) *Julayinbul Aboriginal Intellectual and Cultural Property: Definitions, Ownership and Strategies for Protection in the Wet Tropics World Heritage Area.* Rainforest Aboriginal Network, Daintree.

Rajasekaran, M., Sudhakaran, C., Pradhan, S. C., Bapna, J. S. and Nair, A. G. R. (1989) Mast cell protective activity of ursolic acid-a triterpene from the leaves of Ocimum sanctum Journal of Drug Development *2, 3,179-182.*

Ramakrishnan, P. S. (1984) Tribal Man in Humid Tropics of the Northeast. *Man in India,* Vol.64.

Ramakrishnan, P. S. (1992) *Shifting Agriculture and Sustainable Development of North-eastern India.* UNESCO-MAB Series, Paris, Parthenon Publishers, Carnforth, UK.

Ramakrishnan, P. S. (1996) Conserving the sacred: from species to landscapes. *Nature & Resources.* 32(1):11-20.

Ramaprasad, V. (1994) *Navdanya: A Grassroot Movement for Conservation of Biodiversity: the Lifeline of Women and Rural Poor.* In: Shiva, V. (ed.) *Biodiversity Conservation: Whose Resources? Whose Knowledge?* INTACH, New Delhi.

Ramble, C. (1995) Gaining Ground: Representations of Territory in Bon and Tibetan Popular Tradition. *Tibet Journal* Vol.20:83-124.

Rambo, A. T. (1981) *The Involvement of Social Scientists in Ecology Research on Tropical Agroecosystems in Southeast Asia.* Environment and Policy Institute, The East West Center, Honolulu.

Ramirez, J. D., Pasta, D. J., Yuen, S. D., Billings, D. K. and Ramey, D. R. (1991) *Longitudinal Study of Structured Immersion Strategy, Early-exit, and Late-exit Bilingual Education Programs for Language Minority Children.* Vols. 1-2. Aguirre International, San Mateo, CA.

Ramsey, J. (1977) *Coyote was going There: Indian Literature of the Oregon Country.* University of Washington Press, Seattle, 295 pp.

Rane, U. (1987) The Zudpi Factor. *Sanctuary* 7, 4.

Rangan, Haripriya (199?) Romancing the Environment: Popular Environmental Action in the Garhwal Himalayas. In: *In Defense of Livelihood: Comparative Studies on Environmental Action.* West Hartford, CT.

Rangarajan, Mahesh (1996) *Fencing the Forest: Conservation and Ecological Change in India's Central Provinces, 1860-1914.* Oxford University Press, Delhi.

Rannut, M. (1994) Beyond linguistic policy: The Soviet Union versus Estonia. In: Skutnabb-Kangas, T. and Phillipson, T. (eds) *Linguistic Human Rights. Overcoming Linguistic Discrimination.* 179-208. Mouton de Gruyter, Berlin and New York.

Rasmussen, L. (1996) *Earth Community, Earth Ethics.* Orbis Books, Maryknoll.

Ravetz, J. R. (1986) 'Usable knowledge, usable ignorance: Incomplete science with policy implications'. In: Clar, W. C. and Munn, R. E. (eds) *Sustainable Development of the Biosphere,* 415-434. Cambridge University Press, Cambridge, for the International Institute for Applied Systems Analysis.

Raychaudhuri, B. (1980) *The Moon and the Net: Study of a Transient Community of Fishermen at Jambudwip.* Anthropological Survey of India, Calcutta.

Raygorodetsky, G. (1997). *Nanh' Kak Geenjit Gwich'in Ginjik:*

Raz, Simcha (1976) *A Tzadik in Our Times,* trans. Charles Wengrov, Jerusalem.

Redclift, M. (1987) *Sustainable Development: Exploring the Contradictions.* Methuen, New York.

Redford, K. H. (1991) The ecologically noble savage. *Cultural Survival Quarterly* 15: 46-48.

Redford, K. H. and Padoch, C. (1992) *Conservation of Neotropical Forests; Working from Traditional Resource Use.* Columbia University Press, New York.

Redford, K. H. and Stearman, A. M. (1993) Forest-dwelling native Amazonians and the conservation of biodiversity: interests in common or in collision? *Conservation Biology* 7(2):248-255.

Rees, W. and M. Wackernagel (1996) *Our Ecological Footprint: Reducing Human Impact on the Earth,* New Society Publishers, Philadelphia PA.

Reeve, A. (1986) *Property.* Macmillan, London.

Regan, T. and Singer, P. (eds) (1976) *Animal Rights and Human Obligations.* Prentice-Hall, Englewood Cliffs, New Jersey.

Regenstein, L. (1991) *Replenish the Earth.* Crossroad, New York.

Reichel, Elizabeth (1976) Manufactura del Budare entre la Tribu Tanimuka, Amazonas. *Revista Colombiana de Antropologia,* Vol. XX, Bogotá.

Reichel, Elizabeth (1987a) 'Etnografia de los Grupos Indigenas Contemporaneos'. In: *Colombia Amazonica*. Universidad Nacional de Colombia and FEN, Bogota.

Reichel, Elizabeth (1987b) 'Etnoastronomía Yukuna-Matapi'. In: *Etnoastronomias Americanas*. Arias, J. and Reichel, E. (eds) Universidad Nacional de Colombia, Bogotá.

Reichel, Elizabeth (1989) 'La Danta y el Delfin: Manejo Ambiental e intercambio entre dueños de maloca y chamanes: El caso Yukuna-Matapi (Amazonas)'. *Revista de Antropologia*, Departamento de Antropologia, Universidad de los Andes, Vol. V no 1-2, Bogotá.

Reichel, Elizabeth (1992) 'La Eco-politica en conceptos indigenas de territorio en la Amazonia colombiana'. In: Sanchez, E. (ed.) *Antropologia Juridica: Normas Formales, Costumbres Legales en Colombia*. Parcomun, Bogota.

Reichel, Elizabeth (1993) 'The Role of Indigenous Women in the Reproduction of Plant Biodiversity in the Colombian Amazon'. Report to the Biodiversity Support Program and WWF, Washington.

Reichel, Elizabeth (1996) 'Diversidad Amerindia, Saberes y Ciencia Indígena'. In: Angarita, C. (ed.) *Derechos, Etnias y Ecología*. Misión de Ciencia, Educación y Desarrollo, Presidencia de Colombia y Colciencias, Bogotá.

Reichel, Elizabeth (1997) 'The Eco-politics of Yukuna and Tanimuka Cosmology (Northwest Amazon, Colombia)'. Ph.D. Dissertation. Cornell University, Ithaca.

Reichel-Dolmatoff, G. (1971) *Amazonian Cosmos: The Sexual and Religious Symbolism of the Tukano Indians*. University of Chicago Press, Chicago, Il.

Reichel-Dolmatoff, G. (1976) Cosmology as ecological analysis: a view from the rainforest. *Man* 11: 307-318.

Reichel-Dolmatoff, G. (1996) *The Forest Within: The World View of the Tukano Amazonian Indians*. Council Oak Books, Tulsa, OK.

Reichel-Dolmatoff, Gerardo (1996) *The Forest Within: The Worldview of the Tukano Amazonian Indians*. Themis Books, London.

Reichhardt, K.L., E. Mellink, G.P. Nabhan, and A. Rea (1994) Habitat Heterogeneity and Biodiversity Associated with Indigenous Agriculture in the Sonoran Desert. *Ethnoecologica*, 2(3): 21-34.

Reid, W. V. and Miller, K. R. (1993) *Keeping Options Alive: The Scientific Basis for Conserving Biodiversity*. World Resources Institute, Washington, D.C.

Reid, W. V., Laird, S. A., Meyer, C. A., Gámez, R., Sittenfeld, A., Janzen, D. H., Gollin, M. A. and Juma, C. (eds) (1993) *Biodiversity Prospecting: Using Genetic Resources for Sustainable Development*. World Resources Institute, Washington, D.C.

Reinhard, J. (1985) Sacred Mountains: An Ethno-archaeological Study of High Andean Ruins. *Mountain Research and Development*, 5(4):299-317.

Reining, C., Heinzman, R. M., Madrid, M. C., Lopez, S. and Solorzano, A. (1992) *Non-Timber Forest Products of the Maya Biosphere reserve, Peten Guatemala*. Conservation Internation, Washington, D.C.

Reinjtes, C., Haverkort, B. and Waters-Bayer, A. (1992) *Farming for the Future*. MacMillan, London.

Religion & Nature Interfaith Ceremony. WWF, Gland, Switzerland.

Rengifo, G. and Regalado, E. (1991) *Vigorización de la chacra Andina*. PRATEC-PPEA, Lima.

Renman, G. and Mörtberg, U. (1994) Avifauna – Relation to Size, Configuration and Habitat Conditions of Green Urban Areas in Stockholm. In: Barker, G. M., Luniak, M., Trojan, P. and Zimny, H. (eds) Proceedings of the Second European Meeting of the International Network for Urban Ecology. *Memorabilia Zoologica* 49, Warsaw, pp. 245-256.

Revilla, J. L. and Paucar, R. S. (1996) 'Sharing the Fruits of Mother Earth'. *COMPAS Newsletter*. March/April (1996) Special Workshop Issue pp.8-9.

Rhoades, R. E. (1991) 'The world's food supply at risk'. *National Geographic* 179(4):74-105. (April 1991 issue).

Richards, A. (1939) *Land, Labour and Diet in Northern Rhodesia: An Economic Study of the Bemba Tribe*. Oxford University Press, London (for the International African Institute).

Richards, A. (1972) Iban augury. *Sarawak Museum Journal* 20: 63-81.

Richards, M. (1993) The potential of non-timber forest products in sustainable natural forest management in Amazonian. *Commonwealth Forestry Review* 72(1): 21-27.

Richards, P. (1985) *Indigenous Agricultural Revolution: Ecology and Food-Crop Farming in West Africa*. Hutchinson, London.

Richards, P. (1992) Saving the rainforest: contested futures in conservation. In: Sallman, S. (ed.) *Contemporary futures*. Routledge, London.

Richards, P. (1992) Saving the Rainforest? Contested Futures in Conservation. In: Wallman, S. (ed.) *Contemporary Futures: Perspectives from Social Anthropology*. Routledge, London. pp. 138-158.

Richards, P. (1993) Cultivation: Knowledge or Performance? In: Hobart, M. (ed.) *An Anthropological Critique of Development: The Growth of Ignorance*. Routledge, London/New York.

Richards, P. (1995) 'Local understandings of primates and evolution: some Mende beliefs concerning chimpanzees' In: Corbey, R. and Theunissen, B. (eds) *Ape, Man, Apeman: changing views since 1600*. Department of Prehistory, Leiden University, Leiden.

Richards, P. (1996) *Fighting for the Rain Forest: War, Youth, and Resources in Sierra Leone*. The International African Institute, London.

Richards, P., Slikkerveer, L. J. and Phillips, A. O. (1989) Indigenous Knowledge Systems for Agriculture and Rural Development: The CIKARD Inaugural Lectures. *Studies in Technology and Social Change* No. 13, Iowa State University, Ames, Iowa, USA, 40 pp.

Richards, P.W. (1996) *The Tropical Rain Forest: and ecological study*. (2nd edition) Cambridge University Press, Cambridge.

Richards, R. J. (1989) *Darwin and the Emergence of Evolutionary Theories of Mind and Behavior*. Univ. of Chicago Press, Chicago and London.

Rifkin, J. (1991) *Biosphere Politics: A cultural odyssey from the middle ages to the new age*. Harper, San Francisco.

Rifkin, J. (1994) Ethnography and Ethnocide: A Case Study of the Yanomami. *Dialectical Anthropology* 19:2-3:295-327.

Rifkin, J. and Peelas N. (1983) *Algeny*. The Viking Press, New York

Rigg, J. (1862) *A Dictionary of the Sunda Language of Java*. Lage, Batavia.

Riley, B. and Brokensha, D. (1988) *The Mbeere in Kenya, Vol. I, Changing Rural Ecology*. University Press of America, Lanham, Maryland, and Institute for Development Anthropology.

Rival, L. (1993) Confronting Petroleum Development in the Ecuadorian Amazon: The Huaorani Human Rights and Environmental Protection, *Anthropology in Action* - Journal of the Association for Social Anthropology in Policy and Practice (BASPP) Issue No.16, Autumn, pp. 14-15.

Rival, Laura (1993) 'The Growth of Family Trees: Understanding Huaorani Perceptions of the Forest'. *Man* 28(4):635-52.

Rival, Laura (ed.) (1998) *The Social Life of Trees: Anthropological Perspectives on Tree Symbolism*. Berg, Oxford.

Rivera, J. V. (1995) 'Andean Peasant Agriculture: Nurturing a Diversity of Life in the Chacra'. In: *Regeneration in the Andes*. Apffel-Marlin, F. and Rivera, J. V. INTERculture XXVIII(1):25.

Robb, C. and Casebolt, C. (eds) (1991) *Covenant for a New Creation: Ethics, religion, and public policy*. Orbis Books, Maryknoll.

Robins, R. H. and Uhlenbeck, E. M. (eds) (1991) *Endangered Languages*. Berg, Oxford.

Rockefeller, S. (1996) Global ethics, international law, and the Earth Charter. *Earth ethics* 7: 1-3.

Rockefeller, S. and Elder, J. (eds) (1992) *Spirit and Nature: Why the environment is a religious issue: an interfaith dialogue*. Beacon Press, Boston.

Rodríguez Manasse, J. (1996) Memoria testimonial Barí. Field Research in Barí Nation: Saimadoyi, Karañakaék (Venezuela), and Karikachaboquirá, Ichirridankairá (Colombia). mss.

Roe, Peter (1982) *The Cosmic Zygote. Cosmology in the Amazon Basin*. Rutgers University Press, New Brunswick, NJ.

Roll, E. (1992) *A History of Economic Thought*. Faber and Faber, London.

Rolston, H. III (1988) *Environmental Ethics: Duties to and values in the natural world*. Temple University Press, Philadelphia.

Roosevelt, A. (ed) (1994) *Amazonian Indians: From Prehistory to the Present – Anthropological Perspectives*. University of Arizona Press, Tucson.

Roosevelt, A. C. (1980) *Parmana. Prehistoric Maize and manioc Subsistence Along the Amazon and Orinoco*. Academic Press, New York.

Rose, B. (1995) *Land management issues: Attitudes and perceptions amongst Aboriginal people of central Australia*. Central Land Council, Alice Springs.

Rose, D. B. (1996) *Nourishing Terrains: Australian Aboriginal views of landscape and wilderness*. Australian Heritage Commission, Canberra.

Rose, D. B. (1996) The Public, the Private, and the Secret across Cultural Difference. In: Finlayson, J. and Jackson-Nakano, A. (eds) *Heritage and Native Title: Anthropological and Legal Perspectives*. Australian Institute of Aboriginal and Torres Strait Islander Studies, Canberra, pp. 113-124.

Rose, W. (1992) The great pretenders: further reflections on white shamanism. In: Jaimes, M. A. (ed.) *The State of Native America: Genocide, colonization, and resistance*. South End Press, Boston.

Rosengren, D. (1987) 'In the Eyes of the Beholder: Leadership and Social Construction of Power and Dominance among the Matsigenka of the Peruvian Amazon'. Göteborgs Etnografiska Museum, Etnologiska Studier 39. Göteborg.

Rössler, M. (1993) Tongariro: first cultural landscape on the World Heritage List. *World Heritage Newsletter* 4:15.

Roszak, T. (1972) *Where the Wasteland Ends*. Anchor, New York.

Roszak, T. (1992) *The Voice of the Earth*. Simon and Schuster, New York.

Roué, M. and Nakashima, D. (1994) 'Pour qui préserver la biodiversité?: A propos du complexe hydroélectrique La Grande et des Indiens Cris de la Baie James'. Special issue on Cultural Diversity, Biological Diversity, JATBA *Revue d'Ethnobiologie* 36: 211-235.

Roux, Jean-Paul (1987) Turkic Religions. In: Eliade, Mircea (ed.) *Encyclopaedia of Religion*. 15:87-94.

Roy, D. (1982) 'Répercussions de la coupure de la Grande Rivière à l'aval de LG2'. *Naturaliste Can. (Rev. Ecol. Syst.)* 109: 883-891.

Royal Commission into Aboriginal Deaths in Custody. (RCIADIC) Australia. (1991) *National Report*, vol.2. Australian Government Publishing Service, Canberra.

Rubagumya, C. (1990) *Language in Education in Africa: A Tanzanian Perspective*. Multilingual Matters, Clevedon, UK.

Ruddle, K. (1994) Local knowledge in the future management of inshore tropical marine resources and environments. *Nature & resources* 30: 28-37.

Ruddle, K. and Johannes, R. E. (eds) (1985) *The Traditional Knowledge and Management of Coastal Systems in Asia and the Pacific*. UNESCO, Jakarta.

Ruether, R. (1992) *Gaia and God: An ecofeminist theology of earth healing*. Harper, San Francisco.

Runge, C. F. (1985) The innovation of rules and the structure of incentives in open access resources. *American Journal of Agricultural Economics* 67: 368-372.

Runge, C. F. (1986) 'Common property and collective action in economic development'. In: Panel on Common Property Resource Management, National Research Council, *Proceedings of the Conference on Common Property Resource Management*. National Academy Press, Washington, D.C.

Runte, A. (1987) *National Parks: The American experience*. 2nd (ed.) University of Nebraska Press, Lincoln.

Rural Advancement Foundation International (1994a) *Conserving Indigenous Knowledge: Integrating Two Systems of Innovation* (Commissioned by the United Nations Development Programme). UNDP, New York.

Rural Advancement Foundation International (1994b) *An Overview of Bioprospecting*. Pittsboro, NC.

Rural Advancement Foundation International (1997a) RAFI Communiqué, issue on The Human Tissue Trade, Jan.-Feb. 1997 (citing: USA: Cell Cultures are Increasingly Important to Manufacturing in the Biotechnology Industry, Businesswire, 28 May 1996).

Rural Advancement Foundation International (1997b) *Enclosures of the mind*. RAFI, Ottawa.

Rusden, M. (1679) *A Further Discovery of Bees. Treating of the Nature, Government, Generation & Preservation of the Bee*. London, pp. A2-3, 1, 4, 25, 23, 28.

Saberwal, V. K. (1996) Pastoral Politics: Gaddi Grazing, Degradation, and Biodiversity Conservation in Himachal Pradesh, India. *Conservation Biology*, Vol. 10, No.3.

Sagan, D. and Margulis, L. (1993) God, gaia, and biophilia. In: Kellert, S. and Wilson, E. (eds) *The Biophilia Hypothesis*. Island Press, Washington, D.C.

Sagoff, M. (1988) *The Economy of the Earth: Philosophy, Law and the Environment*. Cambridge University Press, Cambridge.

Sagoff, M. (1996) On the value of endangered and other species. *Environmental Management* 20(6): 897-911.

Sahai, R. (1993) Animal Genetic Resource Scenario of India. Paper presented at the National Seminar on Animal Genetic Resources and their Conservation, 22-23 April (1993) Karnal, Haryana, National Institute of Animal Genetics, National Bureau of Animal Genetic Resources and Nature Conservator.

Sahlins, M. (1972) *Stone Age Economics*. Tavistock Publications, London.

Sakar, S. (1996) *Medicinal plants and the law*. Centre for Environmental Law, WWF-India, New Delhi, India.

Sale, K. (1991) *Dwellers in the Land: The bioregional vision*. New Society Publishers, Philadelphia.

Salick, J. (1992a) 'Amesha forest use and management: an integration of indigenous use and natural forest management'. In: Redford, K.H. and Padoch, C. (eds) *Conservation of Neotropical Forests; Working from Traditional Resource Use*. Columbia University Press, New York.

Salick, J. (1992b) 'The sustainable management of nontimber rain forest products in the Si-a-Paz Peace Park, Nicaragua'. In: Plotkin, M. and Famolare, L. (eds) *Sustainable Harvest and Marketing of Rain Forest Products*. Island Press, Washington D.C.

Salick, J. and Merrick, L. C. (1990) Use and Maintenance of Genetic Resources: Crops and their wild Relatives. In: Carroll, R.C., Vandermeer, J. H.and Rosset, P. M.(eds) *Agroecology*. McGraw-Hill, New York, pp. 517-544.

Sampath, C. K. (1984) 'Evolution and development of siddha medicine'. In: Pillai, N. K. *Heritage of the tamils-siddha medicine*. International Institute of Tamil Studies, Taramani, Madras.

Sanayal, J. M. (trans.) (1973) Shrimad Bhagavatam, Munshiram Manoharlal Publishers Pvt. Ltd., New Delhi.

Sanchez, J. B. (1989) *Conservacion de recursos naturales, produccion de alimentos y organizacion campesina*. CIED, Lima, Peru.

Sanchez, P. (1982) *Humid tropical ecosystems*. Committee on Selected Biological Problems in the Humid Tropics. Ecological aspects of development in the humid tropics. National Academy Press, Washington, D. C.

Sanchez, P. A. (1976) *Properties and management of soils in the tropics*. Wiley, New York.

Sandin, B. (1980) *Iban Adat and Augury*. Universiti Sains Malaysia, Penang.

Sanguino, D. (1994) *Significado de los mitos Barí a luz de la analogía*. Cúcuta.

Santmire, H. (1985) *The Travail of Nature: The ambiguous ecological promise of Christian theology*. Fortress Press, Philadelphia.

Santos, U., de Menezes, S. R., Bringel, B., Bergamin Filho, H., Ges Ribeiro, M. N. and Bananeira, M. (1984) Rios da Bacia Amazonica I. Afluentes do Rio Negro. *Acta Amazonica*,14: 222-237.

Sarin, M. (1996) *Who Gains? Who Loses? Gender and Equity Concerns in Joint Forest Management*. Mimeo.

Šatava, L. (1992) Problems of national minorities. In: Plichtová, J. (ed.) *Minorities in Politics: Cultural and Languages Rights. The Bratislava Symposium II/(1991) 78*-81. Czechoslovak Committee of the European Cultural Foundation, Bratislava.

Saunders, D. A., Hobbs, R. J.and Margules, C. R. (1991) Biological Consequences of Ecosystem Fragmentation: A Review. *Biological Conservation*. 5(1):18-32.

Savyasaachi (forthcoming) In: Marglin, F. and Banuri, T. (eds) *Dominating Knowledge*. Zed Books, London.

Sayer, J. (1991) *Rain Forest Buffer Zones: Guidelines for Protected Area Managers*. IUCN, Gland, Switzerland.

Sayer, J. A. and Whitmore, T. C. (1991) Tropical moist forests; destruction and species extinction. *Biological Conservation* 55: 199-213.

Schaaf, J. (1995) 'Understanding northern environments and human populations through cooperative research: a case study in Beringia'. In: Peterson, D.L. and Johnson, D.R. (eds) *Human Ecology and Climate Change – People and Resources in the Far North*. 229-244. Taylor and Francis, Washington, D.C.

Schama, S. (1995) *Landscape and Memory*. Alfred A. Knopf, New York.

Schama, S. (1995) *Landscape and Memory*. Harper Collins, London.

Schefold, R. (1973) 'Schlitztrommeln und Trommelsprache in Mentawai'. *Zeitschrift für Ethnologie* vol. 98, 1, S. 36-73.

Schefold, R. (1988) *Lia, das grosse Ritual auf den Mentawai-Inseln*. Dietrich Reimer Verlag, Berling.

Schefold, R. (in press): The two faces of the forest: Visions of the wilderness on Siberut (Mentawai) in a comparative Southeast Asian perspective.

Schelling, T. S. (1978) *Micromotives and Macrobehavior*. W.W. Norton, New York.

Scheuer, J. (1993) Biodiversity: beyond Noah's ark. *Conserv. Biol.* 7: 206-207.

Schillebeeckx, E. (1979) *Jesus: An experiment in christology*. Crossroads, New York.

Schlippe, P.de (1956) *Shifting Cultivation in Africa*. Routledge and Kegan Paul, London.

Schmid, A. A. (1987) *Property, Power and Public Choice*. Praeger, New York.

Schmid, A. A. (1995) 'The environment and property rights issues'. In: Bromley, D.W. (ed.) *The Handbook of Environmental Economics*. 45-60. Blackwell Publishers, Oxford.

Schmid, J. A. (1994) Wetlands in the Urban Landscape of the United States. In: Platt, R. H., Rowntree, R. A. and Muick, P. C. (eds) (1994) *The Ecological City: Preserving and Restoring Urban Biodiversity*. University of Massachusetts Press, Amherst, pp. 106-136.

Schneemann, G. (1992) *Kirekat. Brieven uit Siberut.*. Hadewijch, Antwerpen/Baarn.

Schoonemaker, P. K., von Hagen, B. and Wolf, E. C. (eds) (1997) *The Rain Forests of Home: Profile of a North American Bioregion*. Island Press, Covelo, CA.

Schreiber, W. I. (1962) *Our Amish neighbors*. University of Chicago Press. pp. 186-191.

Schultes, R. E. (1991) Ethnobotanical Conservation and Plant Diversity in the Northwest Amazon. *Diversity* 7: 1&2: 69-72.

Schultes, R. E. (1992) 'Ethnobotany and technology in the northwest Amazon: a partnership'. In: Plotkin, M. and Famolare, L. (eds) *Sustainable Harvest and Marketing of Rain Forest Products*. Island Press, Washington, D.C.

Schumacher, E. F. (1973) *Small is Beautiful: a study of economics as if people mattered*. Bond and Briggs, London.

Science News (1995) Letters to the Editor. *Science News* 147(18): 275.

Scoones, Ian and Thompson, John (eds) (1994) *Beyond Farmer First: Rural people's knowledge, agricultural research and extension practice*. Intermediate Technology Publications, London.

Scott, D. A. and Poole, C. M. (1989) *A Status Overview of Asian Wetlands*. Asian Wetland Bureau, Kuala Lumpur, Malaysia.

Scott, James (1985) *Weapons of the Weak*. Yale University Press, New Haven.

Scudder, T. (1962) *The Ecology of the Gwembe Tonga*. Manchester University Press, Manchester.

SEBJ (1993) *Le milieu de l'eau – Insertion du complexe La Grande dans l'environnement du Moyen Nord*, No. 1. Société d'Energie de la Baie James, Montreal.

Secord, J. (1981) Nature's Fancy: Charles Darwin and the Breeding of Pigeons. *Isis* 72:163-86.

Secord, J. (1985) Darwin and the Breeders: A Social History. In: Kohn, D. (ed.) *The Darwinian Heritage*. Princeton University Press, Princeton, New Jersey, pp. 519-42.

Secretariat to the Commission on Sustainable Development. (1995)*Intergovernmental Panel on Forests: its mandate and how it works*. United Nations, New York.

Secretariat to the Convention on Biological Diversity (1996) *The Biodiversity Agenda: Decisions from the Third Meeting of the Conference of the Parties to the Convention on Biological Diversity*, November 1996, Buenos Aires.

Secretariat to the Convention on Biological Diversity (1997)*Forest Biological Diversity*. Clearing House Mechanism, Montreal, Canada.

Seeds of Survival (1997) 'Eleven Countries represented at USC-Canada/Mali Workshop'. *African Diversity Update, Seeds of Survival*. 14:January 1997, pp.9-11.

Seguy, J. (1973) 'Religion and agricultural success: the vocational life of the French Mennonites from the seventeenth to the nineteenth centuries'. Translated by Michael Shank. *Mennonite Quarterly Review* 47: 182-224.

Sengupta, Nirmal (1986) *The March of an Idea: Evolution and Impact of the Dichotomy Tribe-Mainstream*. MIDS Offprint, Madras.

Sengupta, Nirmal (1988) Reappraising Tribal Movements - I, II, III, IV. In:*Economic and Political Weekly*, May (1988)

Sequoyah v. Tennessee Valley Authority, (1980) 620 F.2d 1159.

Sessions, G. (ed.) (1995) *Deep Ecology for the 21st Century*. Shambhala, Boston.

Shankar, D. (1994) Medicinal Plants and Biodiversity Conservation. In: Shiva, V. (ed.) *Biodiversity Conservation: Whose Resources? Whose Knowledge?*. INTACH, New Delhi.

Shanley, P. (in press) Where are the Fruit Trees? Obstacles and opportunities for marketing forest fruits. *Natural History*, New York.

Shanley, P. and Hohn, I. (1995) *Receitas sem Palavras: Plantas Medicinais da Amazonia*. EMBRAPA/Woods Hole Research Center, Belém, Brazil.

Shanley, P., Luz, L., Galvão, J. and Cymerys, M. (1996) Translating Dry Data for Forest Communitites: Science Offers Incentives for Conservation. *ODI Rural Development Forestry Network Paper* 19e.

Sharma, A. B. (1989) Tulasi. Krishnadas Ayurveda Series. Krishnadas Academy, Varanasi.

Sharma, B. D. (1996) *Globalization: The Tribal Encounter*. Har-Ananda Publications, Delhi.

Sharma, H. Maharishi Ayurveda (1995) In: Micozzi, M. (ed.) *Fundamentals of Alternative and Complementary Medicine*. Churchill-Livingstone, New York.

Sharpe, B. (1996) *Forest people and conservation initiatives: the cultural context of rainforest conservation in West Africa*. GEC/ESRC Rainforest Research Group, Department of Anthropology, University College, London.

Sharpe, B. (1997) 'Forest people and conservation initiatives: the cultural context of rainforest conservation in West Africa'. In: Goldsmith, B. *Rainforests: A Wider Perspective*.

Sheldon, J. W., Balick, M. J. and Laird, S. A. (1997) *Medicinal Plants: Can Utilization and Conservation Co-Exist?* Advances in Economic Botany. The New York Botanical Garden, New York.

Shelton, D. (1995) *Fair Play, Fair Pay: Strengthening Local Livelihood Systems through Compensation for Access to and Use of Traditional Knowledge and Biological Resources*. World Wide Fund for Nature, Gland.

Shengji, P. (1991) Conservation of Biological Diversity in Temple-Yards and Holy Hills by the Dai Ethnic Minorities of China. *Ethnobotany*, 3: 27-35.

Shengji, P. (1993) 'Managing for Biological Diversity Conservation in Temple Yards and Holy Hills: The Traditional Practices of the Xishuangbanna Dai Community, Southwest China'. In: Hamilton, L. S. (ed.) *Ethics, Religion and Biodiversity*. The White Horse Press, Knapwell, Cambridge.

Shepard, P. (1995) 'Virtually hunting reality in a forest of simulacra'. In: Soule, M. and Lease, G. (eds) *Reinventing Nature? Responses to postmodern deconstructionism*. Island Press, Washington, D.C.

Shepherd, P. A. (1994) A Review of Plant Communities of Derelict Land in the City of Nottingham, England, and their Value for Nature Conservation. In: Barker, G. M., Luniak, M., Trojan, P. and Zimny, H. (eds) Proceedings of the Second European Meeting of the International Network for Urban Ecology, Warsaw. *Memorabilia Zoologica* 49:129-134.

Shinkwin, A. (1984) 'Traditional Alaska Native societies'. In: Case, D. S. *Alaska Natives and American Laws*. Ch. 8. University of Alaska Press, Anchorage.

Shiva, V. (1991a) 'The Green Revolution in the Punjab'. *The Ecologist*. 21(2): 57-60.

Shiva, V. (1991b) *The Violence of the Green Revolution: Third world agriculture, ecology and politics*. Third World Network, Penang.

Shiva, V. (1993) *Monocultures of the Mind: Perspectives on Biodiversity and Biotechnology*. Zed Books, London and Atlantic Heights, NJ., Third World Network, Penang.

Shiva, V. (1995a) *Captive Minds, Captive Lives: Ethics, ecology, and patents on life*. Research Foundation for Science, Technology, and Natural Resource Policy, Dehra Dun, India

Shiva, V. (1995b) *Monocultures of the Mind: Perspectives on biodiversity and biotechnology*. Zed Books Ltd., London, and Third World Network, Penang.

Shiva, V. (1996) 'The losers' perspective'. In: Miges Baumann, Janet Bell, Florianne Koechlin and Michel Pimbert (eds).*The Life Industry. Biodiversity, people and profits*. Intermediate Technology Publications, London.

Shiva, V. (1997) *Biopiracy: The plunder of nature and knowledge*. South End Press, Boston, Mass.

Shiva, V. (ed.) (1994) *Biodiversity Conservation: Whose Resources? Whose Knowledge?*. INTACH, New Delhi.

Shiva, V. (n.d.) *Biodioversity, A third world perspective*. Third World Network, Penang.

Shiva, V. and Holla-Bhar, R. 1993 Intellectual piracy and the neem tree. *The Ecologist*, 23, 223-27.

Shiva, Vandana (1988) *Staying Alive: Women, Ecology and Survival in India*. Kali for Women, New Delhi.

Shiva,V., Jafri, A., Bedi, G and Holla-Bhar, R. (1997) *The enclosure and recovery of the commons*. Research Foundation for Science, Technology and Ecology, New Delhi.

Shukla, S., Chand, V. S. and Gupta, A. K. (1995) *Teachers as Transformers*.

Shweder, R. A. (1991) *Thinking Through Cultures. Expeditions in Cultural Psychology*. Harvard University, Cambridge, Mass.

Shweder, R. A. (1991) *Thinking Through Cultures: Expeditions in Cultural Psychology*. Harvard University Press, Cambridge, Massachusetts.

Siebert, S. (1987) 'Land use intensification in tropical uplands: effects on vegetation, soil fertility and erosion'. *Forest Ecology and Management* 21: 37-56.

Sikana, P. (1994) 'Indigenous Soil Characterisation in Northern Zambia'. In: Scoones, I. and Thompson, J. *Beyond Farmer First*. Intermediate Technology Publications, London.

Sillitoe, P. (1983) *Roots of the earth: crops in the highlands of Papua New Guinea*. Manchester University Press, Manchester.

Silver, D. B. (1996a) 'Enfermedad y Curacisn en Zinacantan: Esquema Provisional'. In: Vogt, E. Z. (ed.) *Los Zinacantecos: Un Pueblo Tzotzil de los Altos de Chiapas*. Mixico D. F. Instituto Nacional Indigenista.

Silver, D. B. (1996b) 'Zinacanteco Shamanism'. Unpublished PhD dissertation in anthropology, Harvard University, Cambridge, Mass.

Simons, E. L. (1988) A new species of *Propithecus* (Primates) from Northeast Madagascar. *Folia Primatologica* 50: 143-151.

Simpson, D., Sedjo, R. and Reid, J. W. (1996) 'Valuing biodiversity for use in pharmaceutical research'. *Journal of Political Economy* 104:1 163-85

Sindima, H. (1990) 'Community of life: Ecological theology in African perspective'. In: W. Eakin, C. Birch and J. McDaniel (eds) *Liberating life: Contemporary approaches to ecological theology*. Orbis Books, Maryknoll, NY. 137-147.

Singer, P. (1979) *Practical Ethics*. Cambridge University Press, Camridge.

Singer, P. (1992) *Animal Liberation: A new ethics for our treatment of animals*. Avon Books, New York.

Singh, R. (1997) Experiences of Community-Based Conservation in the Alwar District of Rajasthan, India. Paper presented at the Regional Workshop on Community-Based Conservation, 9-11 February 1997. Indian Institute of Public Administration, New Delhi.

Siskind, J. (1973) (1975) *To Hunt in the Morning*. Oxford University Press, London.

Sit, V. F. S. (ed.) (1988) *Chinese Cities: The Growth of the Metropolis since 1949*. Oxford University Press, Hong Kong.

Sivarajan, V.V. and Balachandran, I. (1994) *Ayurvedic drugs and their plant sources*. International Science Publisher. U.S.A.

Sivaramakrishnan, K. (1995a) 'The Making and Unmaking of Scientific Forestry in Bengal'. Paper presented to the *SSRC Conference on Environmental Discourses and Human Welfare in South and Southeast Asia*.

Sivaramakrishnan, K. (1995b) Colonialism and forestry in India: imagining the past in present politics. *Comparative Study of Society and History* 37 (1): 3-40.

Sivaramakrishnan, K. (1996) 'Joint Forest Management: The Politics of Representation in West Bengal'. Paper presented to the *Workshop on Participation and the Micropolitics of Development Encounters*, Harvard Institute of International Development, Harvard.

Sizer, N. (1992) *Extractive Reserves in Amazonian; the Necessary Transformation of a Tradition.* The Botany School, University of Cambridge.

Skinner, Q. (ed.) (1985) *The Return of Grand Theory in the Human Sciences.* Cambridge University Press, Cambridge.

Skutnabb-Kangas, T. (1984) *Bilingualism or Not: The Education of Minorities.* Multilingual Matters, Clevedon, UK.

Skutnabb-Kangas, T. (1988) Multilingualism and the education of minority children. In: Skutnabb-Kangas, T. and Cummins, J. (eds) *Minority Education. From Shame to Struggle.* 9-44. Multilingual Matters, Clevedon, UK.

Skutnabb-Kangas, T. (1989) Har ursprungsbefolkningar rätten till språk och kultur? (Do indigenous peoples have the right to language and culture?). *Mennesker og Rettigheter, Nordic Journal on Human Rights* 1: 53-57.

Skutnabb-Kangas, T. (1990) *Language, Literacy and Minorities.* The Minority Rights Group, London.

Skutnabb-Kangas, T. (1996a) Promotion of linguistic tolerance and development. In: Legér, S. (ed.) *Vers un Agenda Linguistique: Regard Futuriste sur les Nations Unies / Towards a Language Agenda: Futurist Outlook on the United Nations.* 579-629. Canadian Center of Language Rights, Ottawa.

Skutnabb-Kangas, T. (1996b) Educational language choice: Multilingual diversity or monolingual reductionism? In: Hellinger, M. and Ammon, U. (eds) *Contrastive Sociolinguistics.* 175-204. Mouton de Gruyter, Berlin and New York.

Skutnabb-Kangas, T. (1996c) The colonial legacy in educational language planning in Scandinavia: From migrant labour to a national ethnic minority? *International Journal of the Sociology of Language* 118: 81-106. Dua, H. (ed.), Language Planning and Political Theory.

Skutnabb-Kangas, T. (forthcoming) *Linguistic Genocide in Education: Do Minority Languages Have a Future?* Series Contributions to the Sociology of Language. Mouton de Gruyter, Berlin and New York.

Skutnabb-Kangas, T. and Bucak, S. (1994) Killing a mother tongue: How the Kurds are deprived of linguistic human rights. In: Skutnabb-Kangas, T. and Phillipson, R. (eds) in collaboration with M.Rannut. *Linguistic Human Rights. Overcoming Linguistic Discrimination.* 347-370. Mouton de Gruyter, Berlin and New York.

Skutnabb-Kangas, T. and Cummins, J. (eds) (1988) *Minority Education. From Shame to Struggle.* Multilingual Matters, Clevedon, UK.

Sligh, M. (1996) Suspended between the damages of the green revolution and the promises of the gene revolution. *WHY Magazine*, winter.

Slikkerveer, L. J. (1989) Changing Values and Attitudes of Social and Natural Scientists towards Indigenous Peoples and their Knowledge Systems. In: Warren, D. M., Slikkerveer, L. J.and Titilola, S. O.(eds.) Indigenous Knowledge Systems: Implications for Agriculture and International Development. *Technology and Social Change Monograph Series* No. 11, Iowa State University, Ames, Iowa, USA, pp. 121-134.

Slikkerveer, L. J. (1990) *Plural Medical Systems in The Horn of Africa: The Legacy of 'Sheikh Hippocrates.* Kegan Paul International, London/Boston, 352 pp.

Slikkerveer, L. J. (1991) *New Perspectives on Development Research in The Netherlands.* IMWOO, The Hague, 85pp.

Slikkerveer, L. J. (1992) Bio-Cultural Diversity: Indigenous Knowledge in Agriculture and Rural Development. Paper presented to the African Studies Association of the United Kingdom (ASAUK), University of Stirling, Stirling, Scotland, 21 pp.

Slikkerveer, L. J. (1992) Indigenous Agricultural Knowledge Systems in Kenya: Origins, Development and Prospects. In: Mathias-Mundy, E. (ed.) Indigenous Knowledge and Sustainable Development. *Proceedings of the International Symposium of the Regional Program for the Promotion of Indigenous Knowledge in Asia,* IIRR, Silang, Cavite, The Philippines.

Slikkerveer, L. J. (1996a) Indigenous Agricultural Knowledge Systems for Sustainable Agriculture in Developing Countries. (INDAKS), *Progress Report* No. 6, Leiden/Brusssels.

Slikkerveer, L. J. (1996b) Indigenous Agricultural Knowledge Systems in Kenya: Towards Conservation of Bio-Cultural Diversity in East Africa. In: Adams, W. M. andSlikkerveer, L. J. (eds) *Indigenous Knowledge and Change in African Agriculture.* Studies in Technology and Social Change No. 26, TSC Programme, Iowa State University, Ames, Iowa, USA.

Smith, D. S. and Helmund, P. C. (eds) (1993) *Ecology of Greenways.* University of Minnesota Press, Minneapolis.

Smith, E. A. (1996) 'Cultural, linguistic, and biological diversity: Do they coevolve?' Paper presented at the working conference 'Endangered Languages, Endangered Knowledge, Endangered Environments', Berkeley, California, October 25-27, 1996.

Smith, E. A. (1998) On the coevolution of cultural, linguistic, and biological diversity. In: Maffi, L. (ed.) *Language, Knowledge, and the Environment: The Interdependence of Biological and Cultural Diversity.* Submitted to Smithsonian Institution Press.

Smith, H. C. (1957) *The story of the Mennonites.* 4th ed. Mennonite Publication Office, Newton, Kansas. 846 pp.

Smith, N. (1981) *Man, Fishes, and the Amazon.* Columbia University Press, New York.

Smith, N. J. H., Williams, J. T., Plucknett, D. L. and Talbot, J. H. P. (1992) *Tropical Forests and Their Crops.* Cornell University Press, Ithaca, NY.

Smith, Wynet, Meredith, T. C. and Johns, T. (1996) Use and conservation of woody vegetation by the Batemi of Ngorongoro District, Tanzania. *Economic Botany* 50: 290-299.

Smith, Wynet, Meredith, T. C. and Johns, T. (submitted) 'Assessing rapid means for Biodiversity and Conservation'.

Smolicz, J. J. (1979) *Culture and Education in a Plural Society.* Curriculum Development Centre, Canberra.

Smolicz, J. J. (1986) National policy on languages. *Australian Journal of Education* 30(1): 45-65.

Smolicz, J. J. (1990) Language and economy in their cultural envelope. *Vox: The Journal of the Australian Advisory Council on Languages and Multicultural Education* 4: 65-79.

Smythies, B. E. (ed.) (1960) *The Birds of Borneo.* Oliver and Boyd, Edinburgh.

Snow, Chief John (1977) *These Mountains are our Sacred Places: The Story of the Stoney People*. Samuel Stevens, Toronto and Sarasota.

Sodero, P. (1996) 'Biodiversidade e dinâmica evolutiva em rozas de caboclos amazónicos'. Paper presented at the *Simposium Diversidade Biol—gica e Cultural da Amazónia em um Mundo em Transformacao*. Museu Paraense Emilio Goeldi, Belem, 23-27 September (1996)

Soedjatmoko (1986) *Dimensi Manusia Dalam Pembangunan.*, LP3ES, Jakarta.

Sofowora, A. (1993) *Medicinal Plants and Traditional Medicine in Africa*. Spectrum Books Ltd., Lagos.

Sokpon, N., Agbo, V. and Sodegla, H. (1993) *Approches locales de gestion des forêts sacrées au BénIn: Etude de cas dans le département du Mono*. Les Cahiers du C.R.A., Université Paris I.

Sorensen, A. P. Jr. (1972) Multilingualism in the northwest Amazon. In: Pride, J. B. and Holmes, J. (eds) *Sociolinguistics*. 78-93. Penguin Books, Harmondsworth.

Soulé, M. (1995) The social siege of nature. In: Soule, M. and Lease, G. (eds) *Reinventing Nature?: Responses to postmodern deconstructionism*. Island Press, Washington, D.C.

Soulé, M. and Lease, G. (eds) (1995) *Reinventing Nature?: Responses to postmodern deconstructionism*. Island Press, Washington, D.C.

Soulé, M. and Wilcox, B. (eds) (1980) *Conservation Biology: An evolutionary-ecological perspective*. Sinauer Associates, Sunderland, MA.

Soulé, M. E. and Simberloff, D. (1986) What do Genetics and Ecology Tell Us About the Design of Nature Reserves? *Biological Conservation*. 35:19-40.

South Commission (1990) *The Challenge to the South: The report of the South Commission*. Oxford University Press, New York.

Southall, Aidan (1970) The Illusion of the Tribes. *Journal of Asian and African Studies*.

Spear, T. T. (1978) *The Kaya Complex: A History of the Mijikenda Peoples of the Kenya Coast to 1900*. Kenya Literature Bureau, Nairobi.

Sponsel, L. E. (1995) 'Relationships Among the World System, Indigenous Peoples, and Ecological Anthropology in the Endangered Amazon'. In: Sponsel, L. E. (ed.) *Indigenous Peoples and the Future of Amazonia*. University of Arizona Press, Tucson. 263-293.

Sponsel, L. E., Headland, T. N. and Bailey, R. C. (eds) (1996) *Tropical Deforestation: The Human Dimension*. Columbia University Press, New York.

Spore (1995) Termites: the good, the bad and the ugly. *Spore* 1995, 64, 4.

Spore (anonymous) (1995) Termites: the good, the bad and the ugly. *Spore* 64, 4.

Spurr, David (1993) *The Rhetoric of Empire*. Duke University Press, Durham.

Stairs, A. (1988) Beyond cultural inclusion: An Inuit example of indigenous education development. In: Skutnabb-Kangas, T. and Cummins, J. (eds) *Minority Education. From Shame to Struggle*. 308-327. Multilingual Matters, Clevedon, UK.

Standing Bear, L. (1993) *Land of the Spotted Eagle*. University of Nebraska Press, Lincoln, Nebraska. (Reprint of Standing Bear (1933), Houghton Mifflin, Boston.)

Stannard, D. E. (1989) *Before the horror: the population of Hawaii on the eve of Western contact*. SSRI Publications, University of Hawaii Press, Honolulu. p. 149.

Stannard, D. E. (1992) *American Holocaust: The Conquest of the New World*. University of Hawaii, Honolulu. Oxford University Press, New York.

Stanner, W. E. H. (1976) 'Some Aspects Of Aboriginal Religion'. *Colloquium*, 9:19-35.

Stanner, W. E. H. (1979) *White Man Got No Dreaming: essays 1938-(1973)* Australian National University Press, Canberra.

Stanton, J. P. (1994) Common Perceptions and Misconceptions of QDEH Fire Management Programs. In: McDonald, K. and Batt, D. (eds) *Fire Management on Conservation Reserves in Tropical Australia*. Queensland Department of Environment and Heritage, Brisbane.

Stark, N. and Jordan, C. F. (1978) 'Nutrient retention by root mass of an Amazonian rain forest'. *Ecology*. 59:437-439.

State of the Environment Advisory Council (SoE) (1996) *Australia State of the Environment (1996)* CSIRO Publishing, Collingwood, Vic.

Stavenhagen, R. (1990) *The Ethnic Question*. United Nations University Press, Tokyo.

Stavenhagen, R. (1990) *The Ethnic Question: Conflicts, Development, and Human Rights*. United Nations University Press, Tokyo.

Stavenhagen, R. (1995) Indigenous Peoples: Emerging Actors in Latin America – Ethnic Conflict and Governance in Comparative Perspective. *Working Paper Series* No. 215, Latin American Program, Woodrow Wilson International Center for Scholars. Washington, D.C.

Stavenhagen, R. (1996) *Ethnic Conflicts and the Nation-State*. MacMillan Press/St. Martin's Press, London/New York.

Stearman, A. M. (1992) Neotropical hunters and their neighbors: effects of non-indigenous settlement patterns on three native Bolivian societies. In: Redford, K. H. and Padoch, C. (eds) *Conservation of Neotropical Forests: Building on Traditional Resource Use*. 108-128. Columbia University Press, New York.

Stephenson, R. O. (1982) Nunamiut Eskimos, Wildlife Biologists and Wolves. In: Harrington, F. H.and Pacquet, P. C. (eds) *Wolves of the World*. Noyes Publ., Park Ridge, New Jersey, pp. 434-439.

Stevens, S. F. (1993) *Claiming the High Ground: Sherpas, Subsistence, and Environmental Change in the Highest Himalaya*. University of California Press, Berkeley and Los Angeles.

Stevens, S. F. (1996) *Claiming the High Ground – Sherpas, Subsistence and Environmental Change in the Highest Himalaya*. Montilal Banarsidass, Delhi.

Steward, J. H. (1955) *Theory of Culture Change: The Method of Multilinear Evolution*. University of Illinois Press, Urbana, IL.

Stewart, S. (1984) *On Longing: Narratives of the Miniature, the Gigantic, the Souvenir, the Collection*. rept. Duke University Press, Durham, 1993.

Stinner, D. H., Glick, I. and Stinner, B. R. (1992) 'Forage legumes and cultural sustainability: lessons from history'. *Agriculture, Ecosystems and Environment* 40: 233-248.

Stinner, D. H., Paoletti, M. G. and Stinner, B. R. (1989) 'Amish agriculture and implications for sustainable agriculture'. *Agriculture Ecosystems and Environment* 27:77-90.

Stinner, D. H., Stinner, B. R. and Martsolf, E. (1997) 'Biodiversity as an organizing principle in agroecosystem management: case studies of Holistic Resourse Management practitioners in the USA'. *Agriculture, Ecosystems and Environment*, (in press).

Stockdale, Mary and Corbett, Jonathan (1997) *Participatory Mapping and Inventory in Two Villages in Indonesia*. Draft submitted to Lund, G. (ed.) IUFRO guidelines for designing multiple resource inventories.

Stocking, George (ed.) (1983) *Observers Observed: Essays on Ethnographic Fieldwork*. University of Wisconsin Press, Madison.

Stocking, George (ed.) (1984) *Functionalism Historicized: Essays on British Social Antrhopoloty*. University of Wisconsin Press, Madison.

Stocking, George (ed.) (1991) *Colonial Situations: Essays on the Contextualisation of Ethnographic Knowledge*. University of Wisconsin Press, Madison.

Stoffle, R. and Evans, M. (1990) 'Holistic Conservation and Cultural Triage: American Indian Perspectives on Cultural Resources'. *Human Organization* 49(2):91-99.

Stone, C. (1974) *Should Trees Have Standing?* William Kaufman, Los Altos, California.

Stone, C. D. (1988) *Earth and other Ethics: The case for moral pluralism*. Harper and Row, New York.

Stone, R. D. (1991) *Wildlands and Human Needs: Reports from the Field*. World Wildlife Fund, Washington, D.C.

Storrs, L. O.and James, H. F. (1991) Descriptions of thirty-two new species of birds from the HawaIlan Islands: Part 1. Non-Passiformes. *Ornithological Monographs* 45: 7-17, 81-88.

Strathern, Marilyn (1996) Potential Property. Intellectual rights and property in persons. *Social Anthropology* 4(1): 17-32.

Strathern, Marilyn (ed) (1995)*Shifting Contexts: Transformations in Anthropological Knowledge*. Routledge and Kegan Paul, London.

Strehlow, T. G. H. (1950) *An Australian Viewpoint*. Hawthorn Press, Adelaide.

Stross, B. (1973) Acquisition of botanical terminology by Tzeltal children. In: Edmonson, M. S. (ed.) *Meaning in Mayan Languages*. 107-141. Mouton, The Hague.

Suagee, D. B. (1994) 'Human Rights and Cultural Heritage: Developments in the United Nations Working Group on Indigenous Populations'. In: Greaves, T. (ed.) *Intellectual Property Rights for Indigenous Peoples: A Sourcebook*. SfAA, Oklahoma City. Pp. 191-208.

Sukopp, H. and Werner, P. (1982) Nature in Cities. *Nature and Environment Series* No. 28, Council of Europe, Strassbourg.

Sultan, R., Craig, D. and Ross, H. (1997) Aboriginal joint management of Australian national parks: Uluru-kata Tjuta. In: Posey, D. A. and Dutfield, G. *Indigenous Peoples and Sustainability: Cases and Actions*. 326-338, IUCN and International Books, Utrecht.

Sundar, Nandini (1997) *Subalterns and Sovereigns: An Anthropological History of Bastar, 1854-(1996)* Oxford University Press, New Delhi.

Supalla, T. (1993) *Report on the Status of Sign Language*. World Federation of the Deaf, Helsinki.

Survival International (1996) *Mobil threatens uncontacted Indians in the Peruvian Amazon* Urgent Action Bulletin. June.

Sutherland, J. (1997a) 'Global politics, genetic resources and traditional resource rights: 1996 and beyond'. In: Elliott, L. (ed.) *Ecopolitics X Conference Proceedings*. Australian National University, Canberra.

Sutherland, J. (1997b) (with assistance from D. Craig and D. Posey) Emerging new legal standards for comprehensive rights. *Environmental Policy and Law* 27(1), pp. 13-30.

Suttles, W. and Ames, K. (1997) Pre-European history. In: Schoonemaker, P. K., von Hagen, B. and Wolf, E. C. (eds) *The Rain Forests of Home: Profile of a North American Bioregion*. 255-274. Island Press, Covelo, CA.

Suzuki, D. and Knudtson, P. (1992) *Wisdom of the Elders: Sacred Native Stories of Nature*. Bantam Books, New York.

Swanson, T. (1995) 'The appropriation of evolutions' values: an institutional analysis of intellectual property regimes and biodiversity conservation'. In: *Intellectual property regimes and biodiversity conservation*. Cambridge University Press, Cambridge.

Swanson, T. (1996) The Reliance of Northern Economies on Southern Biodiversity, Biodiversity as Information. *Ecological Economics*, 17: 1-4.

Sweeney, D. (1993) 'Fishing, Hunting and Gathering Rights of Aboriginal Peoples in Australia'. *University of New South Wales Law Journal* 16(1):97-160.

Swezey, S., Daxl, R. and Murray, D. (1986) 'Nicaragua's revolution in pesticide policy'. *Environment* 28 (1).

Tabor, J. A. and Hutchinson, C. F. (1994) Using indigenous knowledge, remote sensing and GIS for sustainable development. *Indigenous Knowledge Development Monitor* 2(1): 2-6.

Tafoya, T. (1982) 'Coyote's Eyes: Native Cognition Styles'. *Journal of American Indian Education*, February: 21-33.

Takacs D. (1996) *The Idea of Biodiversity: Philosophies of Paradise* The Johns Hopkins University Press, Baltimore.

Takahashi, J. (1988) *Women's tales of whaling. Life stories of 11 Japanese women who live with whaling*. Japan Whaling Association, Tokyo.

Takahashi, J., Kalland, A., Moeran, B. and Bestor, T.C. (1989) Japanese whaling culture: continuities and diversities. *Maritime Anthropological Studies* 2: 105-133.

Takoya, T., (1982) 'Coyote's Eyes: Native Cognition Styles'. *Journal of American Indian Education*. February.

Tamang, S. (1995) Asian Regional Consultation/Workshop on the Protection and Conservation of Indigenous Knowledge in Sabah, East Malaysia, February 1995.

Tapia, M. E. and Rosas, A. (1993) 'Seed fairs in the Andes: a strategy for local conservation of plant genetic resources'. In: de Boef, W. *et al.* (eds) *Cultivating Knowledge: Genetic Diversity, Farmer Experimentation and Crop Research*. Intermediate Technology Publications, London.

Taswell, R.,1(986) Geothermal Development in Hawaii Threatens Religion and Environment. *Cultural Survival Quarterly*, 10(1):54-56.

Taussig, M. (1987) *Shamanism, Colonialism, and the Wild Man: A Study in Terror and Healing*. University of Chicago Press, Chicago.

Taussig, Michael (1993) *Mimesis and Alterity*. Routledge, New York.

Taylor, P. (1992) *Respect for Nature*. Princeton University Press, New Jersey.

Taylor, Paul (1986) *Respect for nature*. Princeton University Press.

ten Kate, K. and Laird, S. (1997) *Placing Access and Benefit Sharing within the Commercial Context: Private Sector Practices and Perspectives*. World Resources Institute, Washington, D.C.

The Manyoshu: The Nippon Gakujutsu Shinkokai Translation of One Thousand Poems (1969) Columbia University Press, New York.

The New York Times (1997) June 27, A11.

Theriault, M. K. (1992) *Moose to Moccasins: The Story of Ka Kita Wa Pa No Kwe*. Natural Heritage/Natural History Inc.,Toronto.

Thieberger, N. (1990) Language maintenance: Why bother? *Multilingua* 9(4): 333-358.

Third World Network and Consumers' Association of Penang (1988) Health. In: *Modern Science in Crisis, A third world response*. Third World Network, Penang, Malaysia.

Thomas, J. M. C. (1987) 'Des goûts et dégoûts chez les Akas, Ngbaka et autres (Centrafrique)'. In: Koechlin, B., Sigaut, F., Thomas, J. M. C. and Toffin, G. (eds) *De la voûte céleste au terroir, du jardin au foyer.* Pp. 489-504. EHESS, Paris.

Thomas, J. M. C. (1991) 'La société Aka'. In: Thomas, J. M. C. and Bahuchet, S. *Encyclopédie des Pygmées Aka. Techniques et langage des chasseurs-cueilleurs de la forêt centrafricaine*. Vol. 1, Fas. 3. SELAF, Paris.

Thomas, J. M. C. and Bahuchet, S. (1981-1992) *Encyclopédie des Pygmées Aka. Techniques et langage des chasseurs-cueilleurs de la forêt centrafricaine*. Vol. 1, Fas. 3. SELAF, Paris.

Thomas, K. (1983) *Man and the Natural World: Changing Attitudes in England 1500-1800*. Penguin Books, Harmondsworth.

Thomas, W. L. (ed.) (1956) *Man's Role in Changing the Face of the Earth*. University of Chicago Press, Chicago.

Thompson, A. D. F. (1934) Bark Cloth Making in Buganda. *The Uganda Journal* 1,1 17-21.

Thornton, R. (1992) 'The Rhetoric of Ethnographic Holism'. In: *Rereading Cultural Anthropology*. (ed.) Marcus, G. Duke University Press, Durham. 15-33.

Thrupp, L. (1997) *Linking Biodiversity and Agriculture: Challenges and opportunities for sustainable food security*. The World Resources Institute.

Thrupp, L. A. (1984) 'Women, Wood, and Work'. In: 'Kenya and Beyond', *Unasylva. FAO Journal of Forestry*. December.

Thrupp, L. A. (1996) *New Partnerships For Sustainable Agriculture*. World Resources Institute, Washington, D.C.

Thrupp, L. A. (1997) *Linking Biodiversity and Agriculture: Challenges and Opportunities for Sustainable Food Security*. World Resources Institute: Issues and Ideas, March 1997, Washington, D.C.

Thrupp, L. A., Cabarle, B. and Zazueta, A. (1994) *Participatory Methods in Planning and Political Processes: Linking the Grassroots and Policies for Sustainable Development*..

Tickoo, M. L. (1994) Kashmiri, a majority-minority language: An exploratory essay. In: Skutnabb-Kangas, T. and Phillipson, R. (eds) *Linguistic Human Rights. Overcoming Linguistic Discrimination*. 317-333. Mouton de Gruyter, Berlin and New York.

Tilley, C. (1994) *A Phenomenology of Landscape – Places, Paths and Monuments.* Explorations in Anthropology Series. Berg Publishers, Oxford/prodence, USA.

Tillman, D., Wedline, D. and Knops, J. (1996) 'Productivity and Sustainability Influenced by Biodiversity in Grassland Ecosystems'. *Nature*. 379(22):718-720.

Titmuss, R. M. (1970) *The Gift Relationship*. Penguin, London.

Tobias, M. (ed.) (1986) *Mountain People*. University of Oklahoma Press, Norman and London.

Tobias, M. C. and Drasdo, H. (eds) (1979) *The Mountain Spirit*. The Overlook Press, Woodstock, NY.

Tobin, B. (1997) 'Certificates of Origin: A Role for IPR Regimes in Securing Prior Informed Consent'. In: Mugabe, J., Barber, C. V., Henne, G., Glowka, L. and La Vina, A. (eds) *Access to Genetic Resources: Strategies for Sharing Benefits*. ACTS Press, Nairobi. pp. 329-340.

Tobin, B. and Ruiz, M. (1996) 'Access to genetic resources, prior informed consent, and conservation of biological diversity: the need for action by recipient nations'. Presented at the ERM *Stakeholder Workshop on Implementation of Articles 15 and 16 of the Convention on Biological Diversity by the European Union*. London, February 1996.

Todd, N. and Todd, J. (1984) *Bioshelters, Ocean Arks, and City Farming*. Sierra Club Books, San Francisco.

Toffoli, D. D. G. (1996) *Roça Caiçara: uma abordagem etnoecolgica de um sistema agrícola de herança indígena.* Monografia de Bacharelado em Geografia e Meio Ambiente - Pontifweia Universidade Católica/PUC-Rio. 76p.

Toledo, V. M. (1991) 'Patzcuaro's Lesson: Nature, Production and Culture in an Indigenous Region of Mexico'. In: Oldfield, M. L. and Alcorn, J. B. (eds), *Biodiversity: Culture, Conservation and Ecodevelopment.* pp. 147-171. Westview Press, Boulder.

Toledo, V. M. (1992) What is ethnoecology? Origins, scope and implications of a rising discipline. *Etnoecológica* 1(1): 5-21.

Toledo, V. M., Carabias, J., Mapes, C. and Toledo, C. (1985) *Ecologia y Autosuficiencia alimentaria.* Siglo Veintiuno Editores, Mexico City.

Topaloglu, Bekir (1994) 'Islamda Agac.' In: *Turkiye Diyanet Vakfy Islam Ansiklopedisi.* 1: 457-459. Turkiye Diyanet Vakfy.

Tothill, J. D. (ed.) (1948) *Agriculture in the Sudan.* Oxford University Press.

Townsend, R. and Wilson, J. A. (1987) 'An economic view of the tragedy of the commons'. In: McCay, B. J. and Acheson, J. M. (eds) *The Question of the Commons.* 311-326. University of Arizona Press, Tucson.

Trojan, P. (1994) The Shaping of the Diversity of Invertebrate Species in the Urban Green Spaces of Warsaw. In: Barker, G. M., Luniak, M., Trojan, P. and Zimny, H.(eds), Polska Akademia Mauk Muzeum i Institut Zoologii, Warszawa. *Memorabilia Zoologica* 49:167-173.

Tu, W. (1993) 'Toward the possibility of a global community'. In: Hamilton, L. S. (ed.) *Ethics, Religion, and Biodiversity.* The White Horse Press, Cambridge.

Tucker, M. E and Grim, J. (eds) (1994) *Worldviews and Ecology: Religion, philosophy, and the environment.* Orbis Books, Maryknoll.

Turnbull, C. (1961) (1988) *The Forest People.* Triad Paladin Grafton Books, London.

Turnbull, C. (1961) (1988) *The Forest People.* Triad Paladin Grafton Books, London.

Turner, B. L. II, Clark, W. C., Kates, R. W., Richards, J. F., Mathews, J. T. and Meyer, W. B. (eds) (1990) *The Earth as Transformed by Human Action; Global and regional changes in the biosphere over the past 300 years.* Cambridge University Press, Cambridge.

Turner, I. M., Tan, H. T. W., Wee, Y. C., Ibrahim, A. B., Chiew, P. T. and Corlett, R. T. (1994) A Study of Plant Species Extinction in Singapore: Lessons for the Conservation of Tropical Biodiversity. *Conservation Biology* 8(3):705-712.

Twahirva, A. (1994) *Politiques et Pratiques Linguistiques en Afrique: Rapport d'un Travail Réalisé pour l'UNESCO.* UNESCO, Division of Arts and Cultural Life, Paris.

TWN and CAP (1988) - Third World Network and Consumers' Association of Penang. 'Health'. In: *Modern Science in Crisis: A Third World Response.* Third World Network, Penang, Malaysia.

Tyler, S. A. (ed.) (1969) *Cognitive Anthropology.* Holt, Rinehart and Winston, New York.

Tynell, J. (1997) Case study of Ecuador. In: Fleischer Michaelsen, S., Munive, J., Pihlajamäki, M., Søndergaard, K., Tynell, J. and Vogeliues Wiener, C.*Towards Bilingualism*. Project report. 105-127. Roskilde University, International Cultural Studies, Roskilde.

U.S. Geological Survey (1995) Results of Ground-Water, Surface-Water, and Water-Quality Monitoring, Black Mesa Area, Northeastern Arizona 1992-93. *Water Resources Investigations Report* 95-4156. U.S.G.S., Tucson.

Übelacker, M. (1984) *Time Ball: A Story of the Yakima People and the Land.* The Yakima Nation [Toppenish, Wash.]

Uchiyamada, Yasushi. (in press). 'The Grove is Our Temple'. In: Laura Rival (ed.) *The Social Life of Trees: Anthropological Perspectives on Tree Symbolism.* Berg, Oxford.

Uhl, C. and Vieira, I. C. G. (1989) Ecological Impacts of Selective Logging in the Brazilian Amazon: A case Study from the Paragominas Region of the State of Para. *Biotropica* 21(2):98-106.

Uibopuu, V. (1988) *Finnougrierna och deras spraak.* Lund: Studentlitteratur.

UN (1986) *Study of the Problem of Discrimination Against Indigenous Populations.* E/CN.4/Sub.2/1986/7 and Add.1-4.

UN (1995) *World Urbanization Prospects: The 1994 Revision.* Annex Tales, United Nations, New York.

UN ECOSOC (1986) *Study of the Problem of Discrimination Against Indigenous Populations.* United Nations Economic and Social Council, E/CN.4/Sub.2/1986/7 and Add. 1-4.

UNCED (1992) *Agenda 21*, Preamble and Chapters 26-1, Rio de Janeiro, UN, New York.

UNDP (1992) *Benefits of Diversity: An Incentive Toward Sustainable Agriculture.* UN Development Programme, New York.

UNDP (1994) *Conserving Indigenous Knowledge: Integrating two systems of innovation.* An independent study by the Rural Advancement Foundation International.

UNDP (1995) Agroecology: Creating the Synergisms for Sustainable Agriculture. New York: United Nations, p.7. Citing Francis, C. A. (ed.) (1986) *Multiple Cropping Systems.* MacMillan, New York.

UNDP (1996) *Urban Agriculture – New York, The Habitat Agenda.* UN, New York.

UNEP (1992) Convention on Biological Diversity. UNEP, Nairobi.

UNEP (1992) *The State of the Environment 1972-1992: Saving Our Planet.* UNEP, Nairobi.

UNEP (1997) *The Biodiversity Agenda*, Decisions from the Third Meeting of the Conference of the Parties to the Convention on Biological Diversity, Buenos Aires, 1996.

UNEP (1997a) *Compilation of National Contributions on Agricultural Biological Diversity.* UNEP/CBD/SBSTTA/3/Inf.9.

UNEP (1997b) *Recommendations for a Core Set of Indicators of Biological Diversity.* UNEP/CBD/SBSTTA/3/9.

UNEP (1997c) *Convention on Biological Diversity: Workshop on Traditional Knowledge and Biological Diversity.* 24-28 November 1997, Madrid, Spain. UNEP/CBD/TKBD/1/Rev-1.

UNEP-CBD (1994) *Convention on Biological Diversity.* Text and Annexes, UNEP Interim Secretariat for the Convention on Biological Diversity, Geneva, Switzerland.

UNESCO (1993) *Amendment to the Draft Programme and Budget for 1994-1995 (27 C/5), Item 5 of the Provisional Agenda* (27 C/DR.321). UNESCO, Paris.

UNESCO (1994) *Operational Guidelines for the Implementation of the World Heritage Convention.* UNESCO, Paris.

UNESCO (1996) *Sacred Sites - Cultural Integrity, Biological Diversity*. Programme proposal, November 1996, Paris.

UNESCO/WHC/2/Revised/1995 UNESCO (1996) *Revised Operational Guidelines for the Implementation of the World Heritage Convention*. Intergovernmental Committee for the Protection of the World Cultural and Natural Heritage (Item 38 WHC/2/Revised).

Union of Concerned Scientists (1992) *Scientist's Warning to Humanity*. UCS, Cambridge.

United Nations Development Programme (1996) *Urban Agriculture: Food, Jobs, and Sustainable Cities*. Publication Series for Habitat II.

United Nations Environment Programme (1994) Progress Report on Resolution 3 of the Nairobi Final Act: *Ex Situ* Collections and Farmers' Rights. 2nd Session of the Intergovernmental Committee on the Convention on Biological Diversity, 20 June - 1 July, Nairobi.

United Nations Environment Programme (1997) The Biodiversity Agenda. Decisions from the Third Meeting of the Conference of the Parties to the Convention on Biological Diversity, Buenos Aires, November (1996) UNEP, Geneva.

UNO (United Nations Organization) (1992a) Convention on Biological Diversity. UNCED, Rio de Janeiro.

UNO (United Nations Organization) (1992b) Rio Declaration on Environment and Development. UNCED, Rio de Janeiro.

UNO (United Nations Organization) (1992c) Agenda 21. ENCED, Rio de Janeiro.

US Office of Technology Assessment (1987) *Technologies to Maintain Biological Diversity*. Government Printing Office, Washington, D.C.

USGS (1995) Results of Ground-Water, Surface-Water, and Water-Quality Monitoring, Black Mesa Area, Northeastern Arizona, 1992-93. *Water Resources Investigations Report 95-4156*. United States Geological Survey, Tucson.

Usher, P. J. (1995) 'Co-management of natural resources: some aspects of the Canadian experience'. In: Peterson, D. L. and Johnson, D. R. (eds) *Human Ecology and Climate Change — People and Resources in the Far North*. 197-206. Taylor and Francis, Washington, D.C.

Usher, P. J. and Bankes, N. D. (1986) *Property: the Basis of Inuit Hunting Rights*. Inuit Committee on National Issues, Ottawa.

Valladolid, J. (1986) *Cultivos Andinos: importancia y posibilidades de su recuperacion y desarrollo*. Mimeo, Ayacucho.

Valladolid, J. (1992) Las Plantas en la Cultura Andina y en Occidente Moderno. *Serie Documentos de* Estudio. No. 23, Marzo; Proyecto Andino de Tecnologias Campesinas, PRATEC, Lima, Peru.

Valladolid, J. (1995) Andean Peasant Agriculture: Nurturing a Diversity of Life in the Chacra. *INTERculture*, XXVIII:1 (Winter no. 126):18-56.

Van Buitenen, J. A. B. (trans. and ed.) (1975) *The Mahabharata*, 3 vols. University of Chicago Press, Chicago, Il.

Van Gelder, Sarah (1993) 'Remembering our purpose: An interview with Malidoma Some'. *In Context* (34): 30-34.

Van, C. and Lee, D. (1996) 'Prospects for Self-Determination of Indigenous Peoples in Latin America: Questions of Law and Practice'. *Global Governance* 2(1):43-64.

Vansina, J. (1990) *Paths in the Rainforest: towards a history of political tradition in equatorial Africa*. James Currey, London.

Vansina, J. (1990) *Paths in the rainforests: Towards a history of political tradition in Equatorial Africa*. University of Wisconsin Press, Madison.

Vansina, J. (1992) Habitat, economy, and society in the Central African rain forest. *Berg Occasional Papers in Anthropology* 1: na.

Varese, S. (1996) Parroquialismo y Globalización: Las Etnicidades Indigenas ante el Tercer Milenio. In: Varese, S. (Co-ord.) *Pueblos Indios, Soberanìa y Globalismo*. Ediciones Abya Ayala, Quito, Ecuador.

Vartak, V. D and Upadhye, A. (nd) *Floristic studies on genus Ocimum L. from western Mahastra and Goa*. Krishna Tulas-A Monograph. University of Poona.

Vartak, V. D. (1996) Sacred Groves for In situ Conservation. In: Jain S. K. (ed.) *Ethnobiology in Human Welfare*, pp300-302. Deep Publications, New Delhi.

Vartak, V. D. and Gadgil, M. (1981) Studies on sacred groves along the Western Ghats from Maharashtra and Goa. Role of Beliefs and Folklore. In: *Glimpses of Indian Ethnobotany*, pp.272-278.

Vasquez, R. and Gentry, A. H. (1989) Use and misuse of forest-harvested fruits in the Iquitos area. *Conservation Biology* 3(4): 350-361.

Vatn, A. and Bromley, D. (1994) Choices without prices without apologies. *Journal of Environmental Economics and Management* 26: 129-148.

Vatn, A. and Bromley, D. W. (1995) Choices without prices without apologies. In: Bromley, D. W. (ed.) *The Handbook of Environmental Economics*. 3-25. Blackwell Publishers, Oxford.

Vavilov, N. I. (1949) *Chronica Botanica: An International Collection of Studies in the Method and History of Biology and Agriculture*. Vol. 13. Waltham, MA.

Vavilov, N. I. (1951) *The Origin, Variation, Immunity and Breeding of Cultivated Plants*. Roland Press, New York.

Vayda, A. P. (1980) Interaction Between People and Forest in East Kalimantan. *Impact of Science on Society*. Vol. 30:179-190.

Vayda, A. P. and Rappaport, R. A. (1968) Ecology, Culture, and Non-Culture. In: Clifton, J. A. (ed.) *Essays in the Scope and Methods of the Science of Man*. Houghton Mifflin, Boston.

Venkatesananda, S. (trans.) (1984)*The Supreme Yoga: Yoga Vasishta*. The Chiltern Yoga Trust (Aust.), P.O. Box 2, South Fremantle 6162, Western Australia.

Vennum, T. Jr. (1986) *Wild Rice and the Ojibway People*. Minnesota Historical Society Press, St. Paul.

Verdon, R., Schetagne, R., Demers, C., Brouard, D. and Lalumière, R. (1992) 'Evolution de la concentration en mercure des poissons du complexe La Grande', pp. 66-78, In: Chartrand, N. and Thérien, N. (eds) *Les enseignements de la phase 1 du Complexe La Grande*. Univeristé de Sherbrooke and Hydro-Québec, Sherbrooke.

Verma, M. R. and Singh, Y. P. (1969) A Plea for studies in Traditional Animal Husbandry. *Allahabad Farmer*, Vol. XL III (2), pp. 93-98.

Vick, H. (1989) *Social and Economic Development.* George Ronal, Oxford.

Vickers, W. T. (1983) Tropical Forest Mimicry in Swiddens: A Reassessment of Geertz's Model with Amazonian Data. *Human Ecology*, 11:35-46.

Vidyarthi, L. P. (1963) *The Maler: A Study in Nature-Spirit-Man Complex*, Sanchar Publishing House, Patna.

Viederman, S., Meffe, G. and Carroll, C. R. (1994) 'The role of institutions and policymaking in conservation'. In: Meffe, G. and Carroll, C. (ed.) *Principles of Conservation Biology.* Sinauer Associates Ltd., Sunderland, Mass.

Vijayalakshmi, K. (1994) Conserving People's Agricultural Knowledge. In: Shiva, V. (ed.) *Biodiversity Conservation: Whose Resources? Whose Knowledge?* INTACH, New Delhi.

Vijayalaksmi, K. and Shyam Sunder, K. M. (1993) *Vrkshayurveda: An Introduction to Indian Plant Science.* Monograph No. 9. Lok Swasthya Parampara Samvardhan Samithi, Madras.

Vijayalaksmi, K. and Shyam Sunder, K. M. (1994)*Pest Control and Disease Management in Vrkshayurveda*. Lok Swasthya Paramapara Samvardhan Samithi, 2, 25 East Street, Thiruvanmiyur, Madras 600 041.

Villajero, Alejandro (1952) *Asi Es la Selva.* Samarti and Ciz., Lima.

Vision Statement (1995) IFAD: International Fund for Agricultural Development. Conference on Hunger and Poverty, Brussells.

Vogel, J. H. (1995) A market alternative to the valuation of biodiversity: the example of Ecuador. *Association of Systematics Collection Newsletter* October: 66-70.

Vogel, J. H. (1997) 'The Successful Use of Economic Instruments to Foster Sustainable Use of Biodiversity: Six Case Studies from Latin America and the Caribbean'. White Paper commissioned by the Biodiversity Support Program on behalf of the Inter-American Commission on Biodiversity and Sustainable Development in preparation for the Summit of the Americas on Sustainable Development, Santa Cruz de la Sierra, Bolivia. *Biopolicy Journal*, volume 2, Paper 5 (PY97005), Online Journal. URL · http://www.bdt.org/bioline/py. British Library ISSN# 1363-2450.

Vogel, Joseph Henry (1994) *Genes for Sale: Privatization as a Conservation Policy*. Oxford University Press, New York.

Vogel, Joseph Henry (ed.) (in press). *From Traditional Knowledge to Trade Secrets: Prior Informed Consent and Bio-prospecting*. Inter-American Development Bank-Consejo Nacional de Desarrollo/CARE-SUBIR/EcoCiencia, Quito, Ecuador.

Vogt, E. Z. (1966) (ed.) *Los Zinacantecos.* Mixico D. F. Instituto Nacional Indigenista.

Vogt, E. Z. (1969) *Zinacantan: a Maya Community in the Highlands of Chiapas.* The Belknap Press of Harvard University Press, Cambridge, Mass.

Vogt, E. Z. (1970) *The Zinacantecos of Mexico.* Holt, Rinehart and Winston, New York.

Vogt, E. Z. (1976) *Tortillas for the Gods.* Harvard University Press, Cambridge, Mass. and London.

Voisin, A. (1960) *Better Grassland Sward*. Crosby, Lockwood and Son, London.

Waddell, E. (1972) *The mound builders: agricultural practices, environment, and society in the Central Highlands of New Guinea*. University of Washington Press, Seattle.

Wagler, E. (1996) 'The Puzzle Page: Our feathered friends'. *Blackboard Bulletin*, April, p.23.

Walker, H. J. (1990) The Coastal Zone. In: Turner, B. L., Clark, W. C., Kates, R. W., Richards, J. F., Mathews, J. T. and Meyer, W. B. (eds) *The Earth as Transformed by Human Action.* Cambridge University Press, Cambridge, pp. 271-294.

Wallace, E. and Hoebel, E. A. (1952) *The Comanches: Lords of the South Plains*. University of Oklahoma Press, Norman, Oklahoma.

Wallace, M. (1996) *Fragments of the Spirit*. Continuum, New York.

Wallerstein, I. (1990) *Unthinking Social Science: The Limits of Nineteenth Century Paradigms*. Polity Press, in association with Basil Blackwell, London.

Walsh, M., Shackley, S. and Grove-White, R. (1996) *Fields Apart? What Farmers Think of Nature Conservation in the Yorkshire Dales: A report for English Nature and the Yorkshire Dales National Park Authority*. Center of the Study of Environmental Change, Lancaster.

Warnaen, S. (ed) (1987) *Pandangan Hidup Orang Sunda Seperti Tercermin Dalam Tradisi Lisan dan Sastra Sund,* Proyek Penelitian dan Pengembangan Kebudayaan Sunda, Bandung.

Warren, D. M. (1989) The Impact of Nineteenth Century Social Science in Establishing Negative Values and Attitudes towards Indigenous Knowledge Systems. In: Warren, D. M., Slikkerveer, L. J. and Titilola, S. O. (eds.) Indigenous Knowledge Systems: Implications for Agriculture and International Development. *Technology and Social Change Monograph Series* No. 11, Iowa State University, Ames, Iowa, USA, pp. 154-162.

Warren, D. M. (1995) Indigenous Knowledge, Biodiversity Conservation and Development. In: Bennun, L. A., Aman, R. A. and Crafter, S. A. (eds) *Conservation of Biodiversity in Africa: Local Initiatives and Institutional Roles*. National Museums of Kenya, Nairobi.

Warren, D. M., Slikkerveer, L. J. and Brokensha, D. (eds) (1995) *The Cultural Dimension of Development: Indigenous Knowledge Systems*. Intermediate Technology Publications, London. 582 pp.

Warren, D. M., Slikkerveer, L. J. and Titilola, S. O. (eds) (1989) *Indigenous Knowledge Systems: Implications for Agriculture and International Development.* Studies in Technology and Social Change No. 11, Iowa State University, Ames, Iowa, USA.

Watson, P. (1993) 'An open letter to Norwegians'. In: *Nordlys*, p. 7.

Watt, G. (1989-1996) *Dictionary of the economic products of India.* 6 volumes. Superintendent of Government Printing, Calcutta.

Watts, M. and Peet, R. (1996) 'Towards a theory of liberation ecology'. In: Peet, R. and Watts, M. (eds) *Liberation Ecologies: Environment, Development, Social Movements.* Routledge, London and New York.

WCED (1987) *Our Common Future: Report of the world commission on environment and development.* Oxford University Press, Oxford.

WCMC (World Conservation Monitoring Centre) (1994) *Data Sheet.* Compiled by the WCMC, Cambridge.

Wear, A. (1995) Epistemology and Learned Medicine in Early Modern England. In: Bates, D. (ed.) *Knowledge and the Scholarly Medical Traditions.* Cambridge University Press, Cambridge. pp. 151-73.

Weaver, J. (ed.) (1996) *Defending Mother Earth: Native american perspectives on environmental justice.* Orbis Books, Maryknoll.

Weber, M. (1904) *The Protestant Ethic and the Spirit of Capitalism.*

Wee, Y. C. and Corlett, R. C. (1986) *The City and the Forest.* Singapore University Press, National University of Singapore, Singapore.

Weiner, J. (1993) *The Beak of the Finch.* Vintage Books, Random House, New York.

Weinstein, M. and Penn, A. (1987) *Mercury and the Chisasibi Fishery,* draft 2 - A report prepared for the Cree Regional Authority, Montreal.

Weinstein, M. S. (1976) *What the Land Provides : An examination of the Fort George subsistence economy,* Report of the Fort George Resource Use and Subsistence Economy Study, Grand Council of the Crees, Montreal.

Welch, J. (1986) *Fools Crow.* Penguin Books, London.

Wells, M. and Brandon, K., with Hannah, L. (1992) *People and Parks: Linking Protected Area Management with Local Communities.* World Bank, WWF US and US Agency for International Development, Washington, D.C.

Wenger, J. C. (1956) *The Complete Writings of Menno Simmons c. 1496-1561.* Herald Press, Scottdale, PA. 1092 pp.

Wenzel, G. W. (1995) 'Warming the Arctic: environmentalism and Canadian Inuit'. In: Peterson, D. L. and Johnson, D. R. (eds) *Human Ecology and Climate Change – People and Resources in the Far North.* 169-184. Taylor and Francis, Washington, D.C.

Werber, B. (1996) *Empire of the Ants,* translated from the French by Margaret Rocques. Bantam Press, London.

West, P. C. and Brechin, S. R. (eds) (1991) *Resident People and National Parks: Social Dilemmas and Strategies of International Conservation.* University of Arizona press, Tucson.

Westman, W. E. (1990) Detecting Early Signs of Regional Air Pollution Injury to Coastal Sage Scrub. In: Woodwell, G. M. (ed.) *The Earth in Transition.* Cambridge University Press, Cambridge, pp. 323-346.

WGTRR (1996) Aalbersberg, B. 1996. Medicinal Plant Conservation and Development in Fiji. *Bulletin of the Working Group on Traditional Resource Rights.* 3:13. Working Group on Traditional Resource Rights, Oxford.

Whitaker, R. and Andrews, H. (1994) Report prepared for the First Meeting of IUCN/SSC Specialist Group on Sustainable Use of Wild Species, Gland, Switzerland.

White, Hayden (1978) *Tropics of Discourse.* John Hopkins University Press, Baltimore.

White, K. (1992) Elements of geopoetics. *Edinburgh Review* 88: 163-181.

White, L. Jr. (1967) The historical roots of our ecological crisis. *Science* 155: 1203-1207.

White, N. and Meehan, B. (1993) 'Traditional ecological knowledge: a lens on time'. In: Williams, N. and Baines, G. (eds) *Traditional Ecological Knowledge: wisdom for sustainable development..*Centre for Resource and Environmental Studies, Canberra, ANU: 31-40.

White, R. (1983) *The Roots of Dependency: Subsistence, environment, and social change among the Chotaws, Pawnees, and Navajos.* University of Nebraska Press, Lincoln, NE.

Whiteley, D. (1994) The State of Knowledge of the Invertebrates of Urban Areas in Britain, with Examples Taken from the City of Sheffield. In: Barker, G. M., Luniak, M., Trojan, P. and Zimny, H. (eds) Proceedings of the Second European Meeting of the International Network for Urban Ecology, Warsaw. *Memorabilia Zoologica* 49:207-220.

Whitmore, T. C. and Sayer, J. A. (eds) (1992) *Tropical Deforestation and Species Extinction.* Chapman and Hall, London.

Whitt, L. A., (1995) 'Cultural Imperialism and the Marketing of Native America'. *American Indian and Culture Research Journal.* 19(3):1-31.

Whitten, T. (1980) *The gibbons of Siberut.* Dent, London.

WHO (1978) International Conference on Primary Health Care. Alma Ata, WHO, Geneva, Switzerland.

WHO (1978)*Traditional Medicine.* World Health Organization, Geneva,

WHO (1993)*Guidelines for the Evaluation of Herbal Medicines.* WHO Regional Office, Manila.

Whorf, B. L. (1956) *Language, Thought and Reality.* John Wiley, New York.

Wiener, J. D. (1995) 'Common property resource management and northern protected areas'. In: Peterson, D. L. and Johnson, D. R. (eds) *Human Ecology and Climate Change – People and Resources in the Far North.* 207-218. Taylor and Francis, Washington, D.C.

Wigboldus, J. S. and Slikkerveer, L. J. (1991) Agrohistory and Anthropology in Africa: The Wageningen SADH/HODA Approach Related to the Leiden Ethnosystems Perspective. In: Leakey, R. E. and Slikkerveer, L. J. (eds.) Origins and Development of Agriculture in East Africa: The Ethnosystems Approach to the Study of Early Food Production in Kenya. *Studies in Technology and Social Change* No. 19, Iowa State University, Ames, Iowa, USA.

Wilbert, J. (1987) *Tobacco and Shamanism in South America.* Yale University Press, New Haven.

Wilcove, D. S., McLellan, C. H. and Dobson, A. P. (1986) Habitat Fragmentation in the Temperate Zone. In: Soulé, M. E. (ed.) *Conservation Biology: the Science of Scarcity and Diversity.* Sinauer Assoc. Inc., Massachusetts, pp. 237-256.

Wilken, G. C. (1987) *Good Farmers: Traditional Agricultural Resource Management In Mexico And Guatemala.* University of California Press, Berkeley.

Wilkie, D. S., Sidle, J. G. and Boundzanga, G. C. (1992) Mechanized logging, market hunting, and a bank loan in Congo. *Conservation Biology* 6(4):570-580.

Wilkins, D. (1993) Linguistic evidence in support of a holistic approach to traditional ecological knowledge. In: Williams, N. M. and Baines, G. (eds) *Traditional Ecological Knowledge: Wisdom for Sustainable Development*. 71-93. Centre for Resource and Environmental Studies, Australian National University, Canberra.

Willers, B. (ed.) (1991) *Learning to Listen to the Land*. Island Press, Washington, D.C.

Willett, A. B. J. (1993) Indigenous Knowledge and its Implications for Agricultural Development and Agricultural Extension: A Case Study of the Vedic Tradition in Nepal. PhD thesis, Iowa State University, Ames, Iowa, USA.

Williams, B. and Ortiz Solorio, C. (1981) 'Middle American folk soil taxonomy'. Annals of the Association of American Geographers. *71: 335- 358.*

Williams, F. L. (1936) 'Sowing and Harvesting in Ankole'. *The Uganda Journal*. 3,3:203-210.

Wilson, A. (1993) 'Sacred Forests and the Elders'. In: Kemf, E. *The Law of the Mother: Protecting Indigenous Peoples in Protected Areas*. Sierra Club Books, San Francisco.

Wilson, C. (1996) *The Atlas of Holy Places and Sacred Sites*. Dorling Kindersley, New York.

Wilson, E. (1984) *Biophilia: The human bond with other species*. Harvard University Press, Cambridge, MA.

Wilson, E. (1988) *Biodiversity*. National Academy Press, Washington, D.C.

Wilson, E. (1993) 'Biophilia and the Conservation Ethic'. In: Kellert, S. and Wilson, E. (eds) *The Biophilia Hypothesis*. The Island Press, Washington, D.C.

Wilson, E. (1994) In: Ausubel, K. *Seeds of Change: The living treasure*. Harper, San Francisco, pp. 19-20.

Wilson, E. O. (1984) *Biophilia*. Harvard University Press, Cambridge.

Wilson, E. O. (1992) *The Diversity of Life*. Cambridge: Belknap Press of Harvard University Press.

Wilson, E. O.(ed.) (1988) *Biodiversity*. National Academy Press, Washington.

Winkelman, M. (1986) 'Trance states: A theoretical model and cross cultural analysis'. *Ethos* 14:174-203.

Woenne-Green, S., Johnston, R., Sultan, R. and Wallis, A. (1994) *Competing Interests Aboriginal Participation in National Parks and Conservation Reserves in Australia: A Review*. Australian Conservation Foundation, Melbourne.

Wolf, E. R. (1982) *Europe and the People without History*. University of California Press, Berkeley.

Wolf, Eric (1982) *Europe and the People Without History*. University of California Press, Berkeley.

Wolff, P. and Medin, D. L. (1998) Measuring the evolution and devolution of folkbiological knowledge. In: Maffi, L. (ed.) *Language, Knowledge, and the Environment: The Interdependence of Biological and Cultural Diversity*. Submitted to Smithsonian Institution Press.

Wollock, J. (1997) How linguistic diversity and biodiversity are related. *News and Views from Terralingua: Partnerships for Linguistic and Biological Diversity* (electronic newsletter) 1(3): n.p.

Wood, Denis (1992) *The Power of Maps*. Guilford Press, New York.

Wood, P. M. (1997) Biodiversity as the Source of Biological Resources: A New Look at Biodiversity Values. *Environmental Values*, 6: 251-268.

Woodbury, A. C. (1993) A defense of the proposition, 'When a language dies, a culture dies'. In: Queen, R. and Barrett, R. (eds) *SALSA I: Proceedings of the 1st Annual Symposium about Language and Society*. 101-129. TLF 33.

Woolgar, S. (1988) *Science - the very Idea*. Ellis Horwood, Chichester.

World Bank (1996) 'Integrated Pest Management: Strategy and Options for Promoting Effective Implementation'. Draft document, Washington, D.C.

World Commission on Environment and Development (1987) *Our Common Future*. Oxford University Press, Oxford and New York.

World Resources Institute (1994) *World Resources (1994-1995) A Guide to the Global Environment*. Oxford University Press, Oxford.

World Trade Organization - Committee on Trade and Environment (1996) *Excerpt from the report of the meeting held on 21-22 June 1995: record of the discussion on Item 8 of the Committee on Trade and Environment's work programme*. Geneva: WTO (WT/CTE/M/3+Corr.1).

World Wildlife Fund (1980) *Saving Siberut: a conservation master plan*. WWF, Bogor.

Wouden, F. A. E. van (1935) *Types of Social Structure in Eastern Indonesia*. Translation Series No. 11, KITLV. Martinus Nijhoff, The Hague.

WRI/IUCN/UNEP (1992) *Global Biodiversity Strategy*. World Resources Institute, Washington, D.C.

WTMA (1992) *Wet Tropics Plan: Strategic Direction*. Wet Tropics Management Authority, Cairns.

WTMA (1995) *Draft Wet Tropics Plan. Protection Through Partnerships*. Wet Tropics Management Authority, Cairns.

Wurm, S. A. (1991) Language death and disappearance: Causes and circumstances. In: Robins, R. H. and Uhlenbeck, E.M. (eds) *Endangered Languages*. 1-18. Berg, Oxford.

WWF (1980) *Saving Siberut: a conservation master plan*. WWF, Bogor.

WWF (1986) *Assisi Litugy*. WWF, Gland, Switzerland.

WWF (1986) *Religion & Nature Interfaith Ceremony*. WWF, Gland, Switzerland.

Wynne, G., Avery, M., Campbell, L., Gubbay, S., Hawkswell, S., Juniper, T., King, M., Newbery, P., Smart, J., Steel, C., Stones, T., Stubbs, A., Taylor, J., Tydeman, C. and Wynde, R. (1995) *Biodiversity Challenge* (Second Edition). Royal Society for the Protection of Birds, Sandy.

Xaxa, V. (1992) Religion, Customs, and Environment. *Indian International Center Quarterly* 19:1-2:101-112.

Yampolsky, Ph. (1995) *Music from the forests of Riau and Mentawai*. Music from Indonesia vol. 7. Smithsonian/ Folkways, Washington.

Yarrow, D. (1996) Review of Aboriginal Involvement in the Management of the Wet Tropics World Heritage Area. Unpublished Consultancy Report to the Wet Tropics Management Authority, Cairns.

Yearly, S. (1996) *Sociology, Environmentalism, Globalization: Reinventing the Globe.* Sage, London.

Yen, D. E. (1974) *The sweet potato and Oceania: an essay in ethnobotany*. B.P. Bishop Museum Bulletin No.236. Honolulu.

Yoon, H.-K. (1986) *Maori Mind, Maori Land: Essays on the Cultural Geography of the Maori People from an Outsider's Perspective*. Peter Lang, Berne, Frankfurt am Main, New York, Paris.

Young, G. L. (ed.) (1983) *Origins of Human Ecology*. Hutchinson Ross, Stroudsburg, PA.

Yunupingu (1995) Galarrwuy Yunupingu, J. Quoted from Australian Catholic Social Justice Council's 'Recognition: The Way Forward', in Aboriginal and Torres Strait Islander Social Justice Commissioner, Native Title Report: January - June 1994. Australian Government Publishing Service, Canberra.

Zahan, D (1960) *Sociétés d'initiation Bambara: Le N'Domo, Le Kore.* Mouton, Paris.

Zarin, D. J. (1995) Diversity Measurement Methods for the PLEC Clusters. *PLEC News and Views* 4 (1995): 11-21.

Zent, S. (1996a) 'The quandary of conserving ethnoecological knowledge: A Piaroa example'. In: Blount, B. G. and Gragson, T. S., *Ethnoecology: Knowledge, Resources and Rights*. Under consideration by Iowa University Press.

Zent, S. (1996b) *Acculturation and ethnobotanical knowledge loss among the Piaroa of Venezuela.* Paper presented at the working conference 'Endangered Languages, Endangered Knowledge, Endangered Environments', Berkeley, California, October 25-27, 1996.

Zent, S. (1998) Acculturation and ethnobotanical knowledge loss among the Piaroa of Venezuela: Demonstration of a quantitative method for the empirical study of TEK change. In: Maffi, L. (ed.) *Language, Knowledge, and the Environment: The Interdependence of Biological and Cultural Diversity.* Submitted to Smithsonian Institution Press.

Zent, S. (1999) The quandary of conserving ethnoecological knowledge: A Piaroa example. In: Blount, B. G. and Gragson, T. S. Ethnoecology: Knowledge, Resources and Rights. University of Georgia Press, Athens.

Zepeda, O. and Hill, J. H. (1991) The condition of Native American languages in the United States. In: Robins, R. H. and Uhlenbeck, E. M. (eds) *Endangered Languages*. 135-155. Berg, Oxford.

Zerner, C. (1990) Legal options for the Indonesian forestry sector. *FAO Forestry Studies* UTF/INS/065, April.

Zézé-Béké, P. (1989) 'Les interdits alimentaires chez les Nyabwa de Cote d'Ivoire'. *Journal des Africanistes* 59(1 - 2): 229-235.

Zimmermann, F. (1995) The Scholar, the Wise Man, and Universals: Three Aspects of Ayurvedic Medicine. In: Bates, D. (ed.) *Knowledge and the Scholarly Medical Traditions.* Cambridge University Press, Cambridge, pp. 297-319.

Zweifel, H. (1997) 'Biodiversity and the Appropriation of Women's Knowledge'. *Indigenous Knowledge and Development Monitor*, V:5:1.

INDEX

Please note: index entries referring to figures and tables are indicated by *italic* page numbers; entries referring to text boxes are indicated by **bold** page numbers.

A

C

D

E

F

G

H

I

J

K

L

M

N

O

P

Q

R

S

T

U

V

W

X

Y

Z

www.ingramcontent.com/pod-product-compliance
Lightning Source LLC
Jackson TN
JSHW061914140125
77033JS00049B/610

* 9 7 8 1 8 5 3 3 9 3 9 7 6 *